Homology, Cohomology, and Sheaf Cohomology for Algebraic Topology, Algebraic Geometry, and Differential Geometry

Other World Scientific Titles by the Author

Linear Algebra and Optimization with Applications to Machine Learning
Volume I: Linear Algebra for Computer Vision, Robotics, and Machine Learning
ISBN: 978-981-120-639-9
ISBN: 978-981-120-771-6 (pbk)

Linear Algebra and Optimization with Applications to Machine Learning
Volume II: Fundamentals of Optimization Theory with Applications to Machine Learning
ISBN: 978-981-121-656-5

Homology, Cohomology, and Sheaf Cohomology for Algebraic Topology, Algebraic Geometry, and Differential Geometry

Jean Gallier
Jocelyn Quaintance

University of Pennsylvania, USA

NEW JERSEY • LONDON • SINGAPORE • BEIJING • SHANGHAI • HONG KONG • TAIPEI • CHENNAI • TOKYO

Published by

World Scientific Publishing Co. Pte. Ltd.
5 Toh Tuck Link, Singapore 596224
USA office: 27 Warren Street, Suite 401-402, Hackensack, NJ 07601
UK office: 57 Shelton Street, Covent Garden, London WC2H 9HE

British Library Cataloguing-in-Publication Data
A catalogue record for this book is available from the British Library.

HOMOLOGY, COHOMOLOGY, AND SHEAF COHOMOLOGY FOR ALGEBRAIC TOPOLOGY, ALGEBRAIC GEOMETRY, AND DIFFERENTIAL GEOMETRY

ISBN 978-981-124-502-2 (hardcover)
ISBN 978-981-124-503-9 (ebook for institutions)
ISBN 978-981-124-504-6 (ebook for individuals)

For any available supplementary material, please visit
https://www.worldscientific.com/worldscibooks/10.1142/12495#t=suppl

Desk Editor: Liu Yumeng

Preface

The main topics of this book are cohomology, sheaves, and sheaf cohomology. Why? Mostly because for more than thirty years the senior author has been trying to learn algebraic geometry. To his dismay, he realized that since 1960, under the influence and vision of A. Grothendieck and his collaborators, in particular Serre, the foundations of algebraic geometry were built on sheaves and cohomology. But the invasion of these theories was not restricted to algebraic geometry. Cohomology was already a pillar of algebraic topology but sheaves and sheaf cohomology sneaked in too.

For a novice the situation seems hopeless. Even before one begins to discuss curves or surfaces, one has to spend years learning:

(1) Some algebraic topology (especially homology and cohomology).
(2) Some basic homological algebra (chain complexes, cochain complexes, exact sequences, chain maps, *etc.*). Some commutative algebra (injective and projective modules, injective and projective resolutions).
(3) Some sheaf theory.

This book represents the result of an unfinished journey in attempting to accomplish the above. What we discovered on the way is that algebraic topology is a captivating and beautiful subject. We also believe that it is hard to appreciate sophisticated concepts such as sheaf cohomology without prior exposure to fundamentals of algebraic topology, simplicial homology, singular homology, and CW complexes, in particular.

With the above motivation in mind, this book consists of two parts. The first part consisting of the first seven chapters gives a crash-course on the homological and cohomological aspects of algebraic topology, with a bias in favor of cohomology. Unfortunately homotopy theory is omitted.

Generally we do not provide proofs, with the exception of the homological tools needed later in the second part (such as the "zig-zag lemma"). Instead we try to provide intuitions and motivations, but we still provide rigorous definitions.

We conclude this overview of algebraic topology with a presentation of Poincaré duality, one of the jewels of algebraic topology. We follow Milnor and Stasheff's exposition [Milnor and Stasheff (1974)] using the cap product, occasionally supplemented by Massey [Massey (1991)]. Contrary to the previous chapters we provide almost all proofs.

Hopefully this approach will not be frustrating to the reader. Our advice is to keep a copy of Hatcher [Hatcher (2002)] or Munkres [Munkres (1984)] and Massey [Massey (1991)] at hand. Omitted details will be found in these references. Spanier [Spanier (1989)] may also be helpful for some of the more advanced topics.

The second part is devoted to presheaves, sheaves, Čech cohomology, derived functors, sheaf cohomology, and spectral sequences.

Every book on algebraic geometry that goes beyond the classical material known before 1960 discusses sheaves and cohomology. The classic on the subject is Hartshorne [Hartshorne (1977)]. The joke in certain circles is that most people are so exhausted after reading Chapters II and III that they never get to read the subsequent chapters on curves and surfaces.

It appears that after almost seventy five years it is not easy to find thorough expositions of sheaf cohomology designed for a "general" audience, with the exception of Rotman (second edition) [Rotman (2009)]. Godement was already lamenting about this in the preface of his book [Godement (1958)] published in 1958. He says that ironically, someone with expertise in functional analysis (him) was compelled to give a complete exposition, that is, less incomplete than the other existing expositions of sheaf theory.

Godement writes in French in the Bourbaki style, which means that the exposition is terse, motivations are missing, and examples are few. This is very unfortunate because Godement's book contains some interesting material that is not easily found elsewhere, such as the spectral sequence of a differential sheaf and the spectral sequence of Čech cohomology. We discuss these topics in Chapter 15.

Our own experience is that the process of learning sheaves is facilitated by proceeding in stages. The first stage is to just define presheaves and sheaves and to give several examples. We do this in Chapter 8.

The second stage is to define the Čech cohomology of sheaves. Čech cohomology is combinatorial in nature and quite concrete so one can see how sheaves provide varying coefficients. This is the approach followed by Bott and Tu [Bott and Tu (1986)]. It is even possible without getting too technical to explain why De Rham cohomology is equivalent to Čech cohomology with coefficients in $\mathbb{R}$ by introducing the double complex known as the *Čech–de Rham complex*. This material is discussed in Chapter 9.

The third stage is to explain the sheafification process, making a presheaf into a sheaf, and the approach to sheaves in terms of stalk spaces due to Lazard and Cartan. One would like to define the notion of exact sequence of sheaves, but unfortunately the obvious notion of image of a sheaf is not a sheaf in general, so the sheafification process can't be avoided. The right way to define the image of a sheaf is to define the notion of cokernel map of a sheaf and to define the image as the kernel of the cokernel map. These gymnastics are inspired by the notion of image of a map in an abelian categories, so we proceed with a basic presentation of the notions of categories, additive categories, and abelian categories. This way we can rightly claim that sheaves form an abelian category.

Personally, we find Serre's explanation of the sheafification process to be one of the clearest and we borrowed much from his famous paper FAC [Serre (1955)] (actually, his presentation of Čech cohomology of sheaves is also very precise and clear). The above material is presented in Chapter 10.

Having the machinery of sheaves at our disposal, the next step is to introduce sheaf cohomology. This can be done in two ways:

(1) In terms of resolutions by injectives.
(2) In terms of resolutions by flasque sheaves, a method invented by Godement [Godement (1958)].

In either case it is not possible to escape discussing the concept of resolution. We decided that we might as well go further and present some notions of homological algebra, namely projective and injective resolutions, as well as the notion of derived functor. Given a module A, a resolution is an exact sequence starting with A involving projective and injective modules. Projective and injective modules are modules satisfying certain extension properties. Given an additive functor T and an object A, it is possible to define uniquely some homology groups $L_nT(A)$ induced by projective resolutions of A and independent of such resolutions. It is also possible to define uniquely some cohomology groups $R^nT(A)$ induced by injective resolutions of A and independent of such resolutions; see Chapter 11. As special cases we obtain the Tor modules (associated with the tensor product) and the Ext modules (associated with the Hom functor). The modules Tor and Ext play a crucial role in the universal coefficient theorems; see Chapter 12. Our presentation of the homological algebra given in Chapters 11 and 12 is heavily inspired by Rotman's excellent exposition [Rotman (2009)]. Although Mac Lane's presentation is more concise it is still a very reliable and elegantly written source which also contains historical sections [Mac Lane (1975)].

Having gone that far, we also discuss Grothendieck's notion of δ-functors and universal δ-functors. The significance of this notion is that *the machinery of universal δ-functors can be used to prove that different kinds of cohomology theories yield isomorphic groups.* This technique will be used in Chapter 13 to prove that sheaf cohomology and Čech cohomology are isomorphic for paracompact spaces.

Grothendieck's legendary Tohoku paper [Grothendieck (1957)] is written in French in a very terse style and many proof details are omitted (there are also quite a few typos). We are not aware of any source that gives detailed proofs of the main results about δ-functors (in particular, Proposition 2.2.1 on Page 141 of [Grothendieck (1957)]). Lang [Lang (1993)] gives a fairly complete proof but omits the proof that the construction of the required morphism is unique. We fill in this step using an argument communicated to us by Steve Shatz; see Chapter 11.

Having the machinery of resolutions and derived functors at our disposal we are in the position to discuss sheaf cohomology quite thoroughly in Chapter 13. We show that the definition of sheaf cohomology in terms of derived functors is equivalent to the definition in terms of resolutions by flasque sheaves (due to Godement). We prove the equivalence of sheaf cohomology and Čech cohomology for paracompact spaces. We also discuss soft and fine sheaves, and prove that for a paracompact topological space, singular cohomology, Čech cohomology, Alexander–Spanier cohomology, and sheaf cohomology (for a suitable constant sheaf) are equivalent.

The purpose of Chapter 14 is to present various generalizations of Poincaré duality. These versions of duality involve taking direct limits of direct mapping families of singular cohomology groups which, in general, are not singular cohomology groups. However, such limits are isomorphic to Alexander–Spanier cohomology groups, and thus to Čech cohomology groups. These duality results also require relative versions of homology and cohomology.

The last chapter of our book (Chapter 15) is devoted to spectral sequences. A spectral sequence is a tool of homological algebra whose purpose is to approximate the cohomology (or homology) $H(M)$ of a module M endowed with a family $(F^pM)_{p\in\mathbb{Z}}$ of submodules called a filtration. The module M is also equipped with a linear map $d\colon M \to M$ called differential such that $d \circ d = 0$, so that it makes sense to define

$$H(M) = \mathrm{Ker}\, d/\mathrm{Im}\, d.$$

We say that (M, d) is a differential module. To be more precise, the filtration induces cohomology submodules $H(M)^p$ of $H(M)$, the images of $H(F^pM)$ in $H(M)$, and a spectral sequence is a sequence of modules E_r^p (equipped with a differential d_r^p), for $r \geq 1$, such that E_r^p approximates the "graded piece" $H(M)^p/H(M)^{p+1}$ of $H(M)$.

Actually, to be useful, the machinery of spectral sequences must be generalized to filtered cochain complexes. Technically this implies dealing with objects $E_r^{p,q}$ involving three indices, which makes it quite challenging to follow the exposition.

Many presentations jump immediately to the general case, but it seems pedagogically advantageous to begin with the simpler case of a single filtered differential module. This is the approach followed by Serre in his dissertation [Serre (2003)] (Pages 24–104, *Annals of Mathematics*, 54 (1951), 425–505), Godement [Godement (1958)], and Cartan and Eilenberg [Cartan and Eilenberg (1956)].

Spectral sequences were first introduced by Jean Leray in 1945 and 1946. Paraphrazing Jean Dieudoné [Dieudonné (1989)], Leray's definitions were cryptic and proofs were incomplete. Koszul was the first to give a clear definition of spectral sequences in 1947. Independently, in his dissertation (1946), Lyndon introduced spectral sequences in the context of group extensions.

Detailed expositions of spectral sequences do not seem to have appeared until 1951, in lecture notes by Henri Cartan and in Serre's dissertation [Serre (2003)], which we highly recommend for its clarity (Serre defines homology spectral sequences, but the translation to cohomology is immediate). A concise but very clear description of spectral sequences appears in Dieudonné [Dieudonné (1989)] (Chapter 4, Section 7, Parts D, E, F). More extensive presentations appeared in Cartan and Eilenberg [Cartan and Eilenberg (1956)] and Godement [Godement (1958)] around 1955.

There are several methods for defining spectral sequences, including the following three:

(1) Koszul's original approach as described by Serre [Serre (2003)] and Godement [Godement (1958)]. In our opinion it is the simplest method to understand what is going on.
(2) Cartan and Eilenberg's approach [Cartan and Eilenberg (1956)]. This is a somewhat faster and slicker method than the previous method.
(3) Exact couples of Massey (1952). This somewhat faster method for defining spectral sequences is adopted by Rotman [Rotman (1979, 2009)] and Bott and Tu [Bott and Tu (1986)]. Mac Lane [Mac Lane (1975)], Weibel [Weibel (1994)], and McCleary [McCleary (2001)] also present it and show its equivalence with the first approach. It appears to be favored by algebraic topologists. This approach leads to spectral sequences in a quicker fashion and is more general because exact couples need not arise from a filtration, but our feeling is that it is even more mysterious to a novice than the first two approaches.

We will primarily follow Method (1) and present Method (2) and Method (3) in starred sections (Method (2) in Section 15.15 and Method (3) in Section 15.14). All three methods produce isomorphic sequences, and we will show their equivalence. We will also discuss the spectral sequences induced by double complexes and give as illustrations the spectral sequence of a differential sheaf and the spectral sequence of Čech cohomology. These spectral sequences are discussed in Godement [Godement (1958)].

We hope that the reader who read this book, especially the second part, will be well prepared to tackle Hartshorne [Hartshorne (1977)] or comparable books on algebraic geometry. But we will be even happier if our readers found the topics of algebraic topology and homological algebra presented lovable (as Rotman hopes in his preface), and even beautiful.

In the second part of our book, except for a few exceptions we provide complete proofs. We did so to make this book self-contained, but also because we believe that no deep knowledge of this material can be acquired without working out some proofs. However, our advice is to skip some of the proofs upon first reading, especially if they are long and intricate.

The chapters or sections marked with the symbol ⊛ contain material that is typically more specialized or more advanced, and they can be omitted upon first (or second) reading.

Acknowledgement: We would like to thank Ching-Li Chai, Ron Donagi, Herman Gluck, David Harbater, Alexander Kirillov, Julius Shaneson, Jim Stasheff, and Wolfgang Ziller for their encouragement, advice, and what they taught us. Special thanks to Pascal Adjamagbo and Steve Shatz for reporting typos. Steve Shatz also provided several proofs.

Contents

Chapter 1

Introduction

One of the main problems, if not "the" problem of topology, is to understand when two spaces X and Y are similar or dissimilar. A related problem is to understand the connectivity structure of a space in terms of its holes and "higher-order" holes. Of course, one has to specify what "similar" means. Intuitively, two topological spaces X and Y are similar if there is a "good" bijection $f: X \to Y$ between them. More precisely, "good" means that f is a continuous bijection whose inverse f^{-1} is also continuous; in other words, f is a *homeomorphism*. The notion of homeomorphism captures the notion proposed in the mid 1860s that X can be deformed into Y without tearing or overlapping. The problem then is to describe the equivalence classes of spaces under homeomorphism; it is a *classification problem*.

The classification problem for surfaces was investigated as early as the mid 1860s by Möbius and Jordan. These authors discovered that two (compact) surfaces are equivalent iff they have the same *genus* (the number of holes) and orientability type. Their "proof" could not be rigorous since they did not even have a precise definition of what a 2-manifold is! We have to wait until 1921 for a complete and rigorous proof of the classification theorem for compact surfaces; see Gallier and Xu [Gallier and Xu (2013)] for a historical as well as technical account of this remarkable result.

What if X and Y do not have the nice structure of a surface or if they have higher-order dimension? In the words of Dieudonné, the problem is a "hopeless undertaking;" see Dieudonné's introduction [Dieudonné (1989)].

The reaction to this fundamental difficulty was the creation of algebraic and differential topology, whose major goal is to associate "invariant" objects to various types of spaces, so that homeomorphic spaces have "isomorphic" invariants. If two spaces X and Y happen to have some distinct invariant objects, then for sure they are not homeomorphic.

Poincaré was one of the major pioneers of this approach. At first these invariant objects were integers (Betti numbers and torsion numbers), but it was soon realized that much more information could be extracted from invariant algebraic structures such as groups, ring, and modules.

Three types of invariants can be assigned to a topological space:

(1) Homotopy groups.
(2) Homology groups.
(3) Cohomology groups.

The above are listed in the chronological order of their discovery. It is interesting that the first homotopy group $\pi_1(X)$ of the space X, also called *fundamental group*, was invented by Poincaré (*Analysis Situs*, 1895), but homotopy basically did not evolve until the 1930s. One of the reasons is that the first homotopy group is generally nonabelian, so harder to study.

On the other hand, homology and cohomology groups (or rings, or modules) are abelian, so results about commutative algebraic structures can be leveraged. This is true in particular if the ring R is a PID, where the structure of the finitely generated R-modules is completely determined.

There are different kinds of homology groups. They usually correspond to some geometric intuition about decomposing a space into simple shapes such as triangles, tetrahedra, *etc.* Cohomology is more abstract because it usually deals with functions on a space. However, we will see that it yields more information than homology precisely because certain kinds of operations on functions can be defined (cup and cap products).

As often in mathematics, some machinery that is created to solve a specific problem, here a problem in topology, unexpectedly finds fruitful applications to other parts of mathematics and becomes a major component of the arsenal of mathematical tools, in the present case *homological algebra* and *category theory*. In fact, category theory, invented by Mac Lane and Eilenberg, permeates algebraic topology and is really put to good use, rather than being a fancy attire that dresses up and obscures some simple theory, as often is the case.

In view of the above discussion, it appears that algebraic topology might involve more algebra than topology. This is great if one is quite proficient in algebra, but not so good news for a novice who might be discouraged by the abstract and arcane nature of homological algebra. After all, what do the zig-zag lemma and the five lemma have to do with topology?

Unfortunately, it is true that a firm grasp of the basic concepts and results of homological algebra is essential to really understand what are the homology and the cohomology groups and what are their roles in topology.

One of our goals is to attempt to demystify homological algebra. For those of us fond of puns, keep this simple analogy in mind and all trepidation will (hopefully) fade. Homology groups describe what man does in his home; in French, l'homme au logis. Cohomology groups describe what co-man does in his home; in French, le co-homme au logis, that is, la femme au logis. Obviously this is not politically correct, so cohomology should be renamed. The big question is: what is a better name for cohomology?

In the following sections we give a brief description of the topics that we are going to discuss in this book, and we try to provide motivations for the introduction of the concepts and tools involved. These sections introduce topics in the same order in which they are presented in the book. All historical references are taken from Dieudonné [Dieudonné (1989)]. This is a remarkable account of the history of algebraic and differential topology from 1900 to the 1960s which contains a wealth of information.

1.1 Exact Sequences, Chain Complexes, Homology and Cohomology

There are various kinds of homology groups (simplicial, singular, cellular, *etc.*), but they all arise the same way, namely from a (possibly infinite) sequence called a *chain complex*

$$0 \xleftarrow{d_0} C_0 \xleftarrow{d_1} C_1 \longleftarrow \cdots \xleftarrow{d_{p-1}} C_{p-1} \xleftarrow{d_p} C_p \xleftarrow{d_{p+1}} C_{p+1} \longleftarrow \cdots$$

in which the C_p are vector spaces, or more generally abelian groups (typically freely generated), and the maps $d_p \colon C_p \to C_{p-1}$ are linear maps (homomorphisms of abelian groups) satisfying the condition

$$d_p \circ d_{p+1} = 0 \quad \text{for all } p \geq 0. \qquad (*_1)$$

The elements of C_p are called *p-chains* and the maps d_p are called *boundary operators (or boundary maps)*. The intuition behind Condition $(*_1)$ is that elements of the form $d_p(c) \in C_{p-1}$ with $c \in C_p$ are *boundaries*, and "a boundary has no boundary." For example, in $\mathbb{R}^2$, the points on the boundary of a closed unit disk form the unit circle, and the points on the unit circle have no boundary.

Since $d_p \circ d_{p+1} = 0$, we have $B_p(C) = \text{Im}\, d_{p+1} \subseteq \text{Ker}\, d_p = Z_p(C)$ so the quotient $Z_p(C)/B_p(C) = \text{Ker}\, d_p/\text{Im}\, d_{p+1}$ makes sense. The quotient module

$$H_p(C) = Z_p(C)/B_p(C) = \text{Ker}\, d_p/\text{Im}\, d_{p+1}$$

is the *pth homology module* of the chain complex C. Elements of Z_p are called *p-cycles* and elements of B_p are called *p-boundaries*; see Figure 1.1.

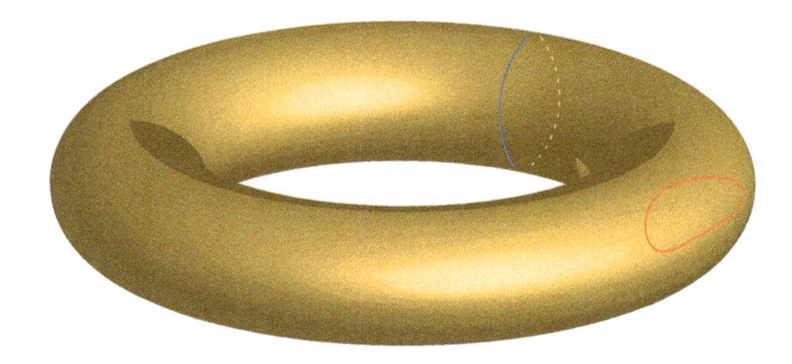

Fig. 1.1 Let X be the surface of the torus. Elements of Z_1 are geometrically represented by curves which are homeomorphic to S^1. Thus both the red and blue curves are 1-cycles. The red curve is also a 1-boundary since it is the boundary of a region in X which is homeomorphic to the closed unit disk.

A condition stronger that Condition $(*_1)$ is that

$$\text{Im}\, d_{p+1} = \text{Ker}\, d_p \quad \text{for all } p \geq 0. \tag{$**_1$}$$

A sequence satisfying Condition $(**_1)$ is called an *exact sequence*. Thus, we can view the homology groups as a measure of the failure of a chain complex to be exact. Surprisingly, exact sequences show up in various areas of mathematics, especially abstract algebra.

In the case of many homology theories, chain complexes are constructed by "nicely" mapping simple geometric objects into a given topological space X. For singular homology the C_p's are the abelian groups $C_p = S_p(X;\mathbb{Z})$ consisting of all (finite) linear combinations of the form $\sum n_i\sigma_i$, where $n_i \in \mathbb{Z}$ and each σ_i, a *singular p-simplex*, is a continuous function $\sigma_i \colon \Delta^p \to X$ from the p-simplex Δ^p to the space X. A 0-simplex is a single point, a 1-simplex is a line segment, a 2-simplex is a triangle, a 3-simplex is a tetrahedron, and a p-simplex is a higher-order generalization of a tetrahedron; see Figure 1.2.

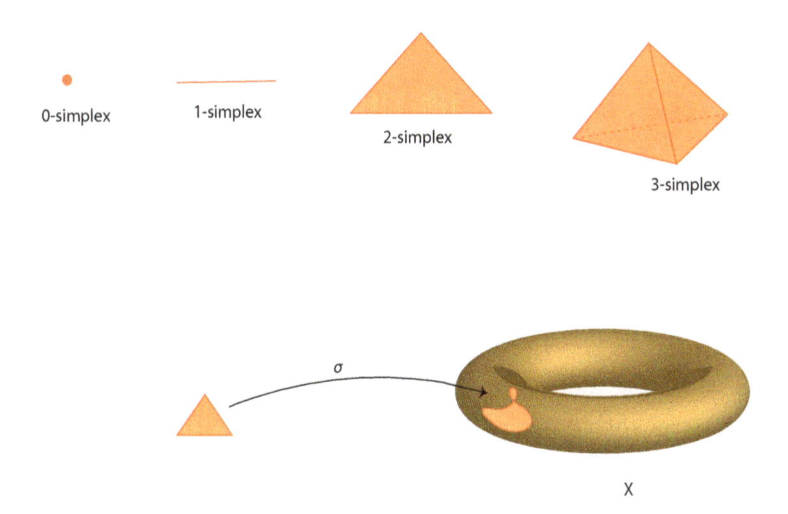

Fig. 1.2 The top row illustrates lower order p-simplicies while the bottom figure illustrates a singular 2-simplex within the 2-dimensional torus.

A p-simplex Δ^p has $p+1$ *faces*, and the ith face is a $(p-1)$-singular simplex $\sigma \circ \phi_i^{p-1} \colon \Delta^{p-1} \to X$ defined in terms of a certain function $\phi_i^{p-1} \colon \Delta^{p-1} \to \Delta^p$; see Section 4.1. In the framework of singular homology, the boundary map d_p is denoted by ∂_p, and for any singular p-simplex σ, $\partial\sigma$ is the singular $(p-1)$-chain given by

$$\partial\sigma = \sigma \circ \phi_0^{p-1} - \sigma \circ \phi_1^{p-1} + \cdots + (-1)^p \sigma \circ \phi_p^{p-1}.$$

A simple calculation confirms that $\partial_p \circ \partial_{p+1} = 0$. Consequently the free abelian groups $S_p(X;\mathbb{Z})$ together with the boundary maps ∂_p form a chain complex denoted $S_*(X;\mathbb{Z})$ called the *simplicial chain complex* of X. Then the quotient module

$$H_p(X;\mathbb{Z}) = H_p(S_*(X;\mathbb{Z})) = \mathrm{Ker}\,\partial_p/\mathrm{Im}\,\partial_{p+1},$$

also denoted $H_p(X)$, is called the *pth homology group* of X. Singular homology is discussed in Chapter 4, especially in Section 4.1.

Historically, singular homology did not come first. According to Dieudonné [Dieudonné (1989)], singular homology emerged around 1925 in the work of Veblen, Alexander and Lefschetz (the "Princeton topologists," as Dieudonné calls them), and was defined rigorously and in complete generality by Eilenberg (1944). The definition of the homology modules $H_p(C)$ in terms of sequences of abelian groups C_p and boundary homomorphisms $d_p \colon C_p \to C_{p-1}$ satisfying the condition $d_p \circ d_{p+1} = 0$ as quotients $\mathrm{Ker}\, d_p / \mathrm{Im}\, d_{p+1}$ seems to have been suggested to H. Hopf by Emmy Noether while Hopf was visiting Göttingen in 1925. Hopf used this definition in 1928, and independently so did Vietoris in 1926, and then Mayer in 1929.

The first occurrence of a chain complex is found in Poincaré's papers of 1900, although he did not use the formalism of modules and homomorphisms as we do now, but matrices instead. Poincaré introduced the homology of *simplicial complexes*, which are combinatorial triangulated objects made up of simplices; see Figure 1.3.

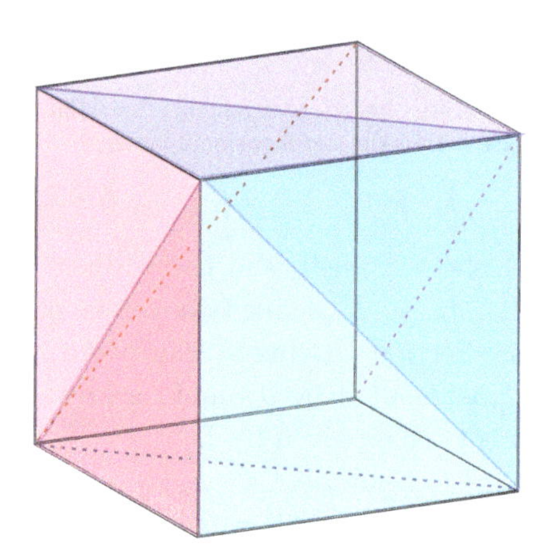

Fig. 1.3 The surface of a cube as a simplicial complex consisting of 12 triangles (2-simplicies), 18 edges (1-simplicies), and 8 vertices (0-simplices).

Given a simplicial complex K, we have free abelian groups $C_p(K)$ consisting of $\mathbb{Z}$-linear combinations of oriented p-simplices, and the boundary maps $\partial_p\colon C_p(K) \to C_{p-1}(K)$ are defined by

$$\partial_p \sigma = \sum_{i=0}^{p} (-1)^i [\alpha_0, \ldots, \widehat{\alpha_i}, \ldots, \alpha_p],$$

for any oriented p-simplex, $\sigma = [\alpha_0, \ldots, \alpha_p]$, where $[\alpha_0, \ldots, \widehat{\alpha_i}, \ldots, \alpha_p]$ denotes the oriented $(p-1)$-simplex obtained by deleting vertex α_i. Then we have a *simplicial chain complex* $(C_p(K), \partial_p)$ denoted $C_*(K)$, and the corresponding homology groups $H_p(C_*(K))$ are denoted $H_p(K)$ and called the *simplicial homology groups* of K. Simplicial homology is discussed in Chapter 5. We discussed singular homology first because it subsumes simplicial homology, as shown in Section 5.2.

A simplicial complex K is a purely combinatorial object, thus it is not a space, but it has a *geometric realization* K_g, which is a (triangulated) topological space. This brings up the following question: if K_1 and K_2 are two simplicial complexes whose geometric realizations $(K_1)_g$ and $(K_2)_g$ are homeomorphic, are the simplicial homology groups $H_p(K_1)$ and $H_p(K_2)$ isomorphic?

Poincaré conjectured that the answer was "yes," and this conjecture was first proved by Alexander. The proof is nontrivial, and we present a version of it in Section 5.2.

The above considerations suggest that it would be useful to understand the relationship between the homology groups of two spaces X and Y related by a continuous map $f\colon X \to Y$. For this, we define mappings between chain complexes called chain maps.

Given two chain complexes C and C', a *chain map* $f\colon C \to C'$ is a family $f = (f_p)_{p\geq 0}$ of homomorphisms $f_p\colon C_p \to C'_p$ such that all the squares of the following diagram commute:

$$\begin{array}{ccccccccccccc}
0 & \xleftarrow{d_0} & C_0 & \xleftarrow{d_1} & C_1 & \longleftarrow & \cdots \xleftarrow{d_{p-1}} & C_{p-1} & \xleftarrow{d_p} & C_p & \xleftarrow{d_{p+1}} & C_{p+1} & \longleftarrow \cdots \\
 & & \downarrow{\scriptstyle f_0} & & \downarrow{\scriptstyle f_1} & & & \downarrow{\scriptstyle f_{p-1}} & & \downarrow{\scriptstyle f_p} & & \downarrow{\scriptstyle f_{p+1}} & \\
0 & \xleftarrow[d'_0]{} & C'_0 & \xleftarrow[d'_1]{} & C'_1 & \longleftarrow & \cdots \xleftarrow[d'_{p-1}]{} & C'_{p-1} & \xleftarrow[d'_p]{} & C'_p & \xleftarrow[d'_{p+1}]{} & C'_{p+1} & \longleftarrow \cdots,
\end{array}$$

that is, $f_p \circ d_{p+1} = d'_{p+1} \circ f_{p+1}$, for all $p \geq 0$.

A chain map $f\colon C \to C'$ induces homomorphisms of homology

$$H_p(f)\colon H_p(C) \to H_p(C')$$

for all $p \geq 0$. Furthermore, given three chain complexes C, C', C'' and two chain maps $f \colon C \to C'$ and $g \colon C' \to C''$, we have

$$H_p(g \circ f) = H_p(g) \circ H_p(f) \quad \text{for all } p \geq 0$$

and

$$H_p(\mathrm{id}_C) = \mathrm{id}_{H_p(C)} \quad \text{for all } p \geq 0.$$

We say that the map $C \mapsto (H_p(C))_{p \geq 0}$ is *functorial* (to be more precise, it is a functor from the category of chain complexes and chain maps to the category of abelian groups and groups homomorphisms).

For example, in singular homology, a continuous function $f \colon X \to Y$ between two topological spaces X and Y induces a chain map $f_\sharp \colon S_*(X;\mathbb{Z}) \to S_*(Y;\mathbb{Z})$ between the two simplicial chain complexes $S_*(X;\mathbb{Z})$ and $S_*(Y;\mathbb{Z})$ associated with X and Y, which in turn yield homology homomorphisms usually denoted $f_{*,p} \colon H_p(X;\mathbb{Z}) \to H_p(Y;\mathbb{Z})$. Thus the map $X \mapsto (H_p(X;\mathbb{Z}))_{p \geq 0}$ is a functor from the category of topological spaces and continuous maps to the category of abelian groups and groups homomorphisms. Functoriality implies that if $f \colon X \to Y$ is a homeomorphism, then the maps $f_{*,p} \colon H_p(X;\mathbb{Z}) \to H_p(Y;\mathbb{Z})$ are *isomorphisms*. Thus, the singular homology groups are topological invariants. This is one of the advantages of singular homology; topological invariance is basically obvious.

This is not the case for simplicial homology where it takes a fair amount of work to prove that if K_1 and K_2 are two simplicial complexes whose geometric realizations $(K_1)_g$ and $(K_2)_g$ are homeomorphic, then the simplicial homology groups $H_p(K_1)$ and $H_p(K_2)$ isomorphic.

One might wonder what happens if we reverse the arrows in a chain complex? Abstractly, this is how cohomology is obtained, although this point of view was not considered until at least 1935.

A *cochain complex* is a sequence

$$0 \xrightarrow{d^{-1}} C^0 \xrightarrow{d^0} C^1 \xrightarrow{d^1} \cdots \ C^p \xrightarrow{d^p} C^{p+1} \xrightarrow{d^{p+1}} C^{p+2} \longrightarrow \cdots,$$

in which the C^p are abelian groups, and the maps $d^p \colon C^p \to C^{p+1}$ are homomorphisms of abelian groups satisfying the condition

$$d^{p+1} \circ d^p = 0 \quad \text{for all } p \geq 0. \tag{$*_2$}$$

The elements of C^p are called *cochains* and the maps d^p are called *coboundary maps*. This time it is not clear how coboundary maps arise naturally.

Since $d^{p+1} \circ d^p = 0$, we have $B^p = \mathrm{Im} d^p \subseteq \mathrm{Ker}\, d^{p+1} = Z^{p+1}$, so the quotient $Z^p/B^p = \mathrm{Ker}\, d^{p+1}/\mathrm{Im}\, d^p$ makes sense and the quotient module

$$H^p(C) = Z^p/B^p = \mathrm{Ker}\, d^{p+1}/\mathrm{Im}\, d^p$$

is the *pth cohomology module* of the cochain complex C. Elements of Z^p are called *p-cocycles* and elements of B^p are called *p-coboundaries*.

There seems to be an unwritten convention that when dealing with homology we use subscripts, and when dealing with cohomology we use with superscripts. Also, the "dual" of any "notion" is the "co-notion."

As in the case of a chain complex, a condition stronger that Condition $(*_2)$ is that

$$\mathrm{Im}\, d^p = \mathrm{Ker}\, d^{p+1} \quad \text{for all } p \geq 0. \tag{$**_2$}$$

A sequence satisfying Condition $(**_2)$ is also called an *exact sequence*. Thus, we can view the cohomology groups as a measure of the failure of a cochain complex to be exact.

Given two cochain complexes C and C', a *(co)chain map* $f\colon C \to C'$ is a family $f = (f^p)_{p\geq 0}$ of homomorphisms $f^p\colon C^p \to C'^p$ such that all the squares of the following diagram commute:

$$\begin{array}{ccccccccccccc}
0 & \xrightarrow{d^{-1}} & C^0 & \xrightarrow{d^0} & C^1 & \xrightarrow{d^1} & \cdots C^{p-1} & \xrightarrow{d^{p-1}} & C^p & \xrightarrow{d^p} & C^{p+1} & \xrightarrow{d^{p+1}} & \cdots \\
 & & \downarrow{\scriptstyle f^0} & & \downarrow{\scriptstyle f^1} & & \downarrow{\scriptstyle f^{p-1}} & & \downarrow{\scriptstyle f^p} & & \downarrow{\scriptstyle f^{p+1}} & & \\
0 & \xrightarrow[d'^{-1}]{} & C'^0 & \xrightarrow[d'^0]{} & C'^1 & \xrightarrow[d'^1]{} & \cdots C'^{p-1} & \xrightarrow[d'^{p-1}]{} & C'^p & \xrightarrow[d'^p]{} & C'^{p+1} & \xrightarrow[d'^{p+1}]{} & \cdots,
\end{array}$$

that is, $f^{p+1} \circ d^p = d'^p \circ f^p$ for all $p \geq 0$. A chain map $f\colon C \to C'$ induces homomorphisms of cohomology

$$H^p(f)\colon H^p(C) \to H^p(C')$$

for all $p \geq 0$. Furthermore, this assignment is functorial (more precisely, it is a functor from the category of cochain complexes and chain maps to the category of abelian groups and their homomorphisms).

At first glance cohomology appears to be very abstract so it is natural to look for explicit examples. A way to obtain a cochain complex is to apply the operator (functor) $\mathrm{Hom}_{\mathbb{Z}}(-, G)$ to a chain complex C, where G is any abelian group. Given a fixed abelian group A, for any abelian group B we denote by $\mathrm{Hom}_{\mathbb{Z}}(B, A)$ the abelian group of all homomorphisms from B to A. Given any two abelian groups B and C, for any homomorphism

$f\colon B \to C$, the homomorphism $\mathrm{Hom}_{\mathbb{Z}}(f, A)\colon \mathrm{Hom}_{\mathbb{Z}}(C, A) \to \mathrm{Hom}_{\mathbb{Z}}(B, A)$ is defined by

$$\mathrm{Hom}_{\mathbb{Z}}(f, A)(\varphi) = \varphi \circ f \quad \text{for all } \varphi \in \mathrm{Hom}_{\mathbb{Z}}(C, A);$$

see the commutative diagram below:

$$\begin{array}{ccc} B & \xrightarrow{f} & C \\ & \searrow^{\mathrm{Hom}_{\mathbb{Z}}(f,A)(\varphi)} & \downarrow \varphi \\ & & A. \end{array}$$

The map $\mathrm{Hom}_{\mathbb{Z}}(f, A)$ is also denoted by $\mathrm{Hom}_{\mathbb{Z}}(f, \mathrm{id}_A)$ or even $\mathrm{Hom}_{\mathbb{Z}}(f, \mathrm{id})$. Observe that the effect of $\mathrm{Hom}_{\mathbb{Z}}(f, \mathrm{id})$ on φ is to precompose φ with f.

If $f\colon B \to C$ and $g\colon C \to D$ are homomorphisms of abelian groups, a simple computation shows that

$$\mathrm{Hom}_R(g \circ f, \mathrm{id}) = \mathrm{Hom}_R(f, \mathrm{id}) \circ \mathrm{Hom}_R(g, \mathrm{id}).$$

Observe that $\mathrm{Hom}_{\mathbb{Z}}(f, \mathrm{id})$ and $\mathrm{Hom}_{\mathbb{Z}}(g, \mathrm{id})$ are composed in the reverse order of the composition of f and g. It is also immediately verified that

$$\mathrm{Hom}_{\mathbb{Z}}(\mathrm{id}_A, \mathrm{id}) = \mathrm{id}_{\mathrm{Hom}_{\mathbb{Z}}(A,G)}.$$

We say that $\mathrm{Hom}_{\mathbb{Z}}(-, \mathrm{id})$ is a *contravariant functor* (from the category of abelian groups and group homomorphisms to itself). Then given a chain complex

$$0 \xleftarrow{d_0} C_0 \xleftarrow{d_1} C_1 \longleftarrow \cdots \xleftarrow{d_{p-1}} C_{p-1} \xleftarrow{d_p} C_p \xleftarrow{d_{p+1}} C_{p+1} \longleftarrow \cdots,$$

we can form the cochain complex

$$0 \xrightarrow{\mathrm{Hom}_{\mathbb{Z}}(d_0,\mathrm{id})} \mathrm{Hom}_{\mathbb{Z}}(C_0, G) \longrightarrow \cdots$$

$$\mathrm{Hom}_{\mathbb{Z}}(C_p, G) \xrightarrow{\mathrm{Hom}_{\mathbb{Z}}(d_{p+1},\mathrm{id})} \mathrm{Hom}_{\mathbb{Z}}(C_{p+1}, G) \longrightarrow \cdots$$

obtained by applying $\mathrm{Hom}_{\mathbb{Z}}(-, G)$, and denoted $\mathrm{Hom}_{\mathbb{Z}}(C, G)$. The coboundary map d^p is given by

$$d^p = \mathrm{Hom}_{\mathbb{Z}}(d_{p+1}, \mathrm{id}),$$

which means that for any $f \in \mathrm{Hom}_{\mathbb{Z}}(C_p, G)$, we have

$$d^p(f) = f \circ d_{p+1}.$$

Thus, for any $(p+1)$-chain $c \in C_{p+1}$ we have

$$(d^p(f))(c) = f(d_{p+1}(c)).$$

We obtain the cohomology groups $H^p(\mathrm{Hom}_{\mathbb{Z}}(C, G))$ associated with the cochain complex $\mathrm{Hom}_{\mathbb{Z}}(C, G)$. The cohomology groups $H^p(\mathrm{Hom}_{\mathbb{Z}}(C, G))$ are also denoted $H^p(C; G)$.

This process was applied to the simplicial chain complex $C_*(K)$ associated with a simplicial complex K by Alexander and Kolmogoroff to obtain the *simplicial cochain complex* $\mathrm{Hom}_{\mathbb{Z}}(C_*(K); G)$ denoted $C^*(K; G)$ and the *simplicial cohomology groups* $H^p(K; G)$ of the simplicial complex K; see Section 5.6. Soon after, this process was applied to the singular chain complex $S_*(X; \mathbb{Z})$ of a space X to obtain the *singular cochain complex* $\mathrm{Hom}_{\mathbb{Z}}(S_*(X; \mathbb{Z}); G)$ denoted $S^*(X; G)$ and the *singular cohomology groups* $H^p(X; G)$ of the space X; see Section 4.8.

Given a continuous map $f \colon X \to Y$, there is an induced chain map $f^\sharp \colon S^*(Y; G) \to S^*(X; G)$ between the singular cochain complexes $S^*(Y; G)$ and $S^*(X; G)$, and thus homomorphisms of cohomology $f^* \colon H^p(Y; G) \to H^p(X; G)$. Observe the reversal: f is a map from X to Y, but f^* maps $H^p(Y; G)$ to $H^p(X; G)$. We say that the map $X \mapsto (H^p(X; G))_{p \geq 0}$ is a contravariant functor from the category of topological spaces and continuous maps to the category of abelian groups and their homomorphisms.

So far our homology groups have coefficients in $\mathbb{Z}$, but the process of forming a cochain complex $\mathrm{Hom}_{\mathbb{Z}}(C, G)$ from a chain complex C allows the use of coefficients in any abelian group G, not just the integers. Actually, it is a trivial step to define chain complexes consisting of R-modules in any commutative ring R with a multiplicative identity element 1, and such complexes yield homology modules $H_p(C; R)$ with coefficients in R. This process immediately applies to the singular homology groups $H_p(X; R)$ and to the simplicial homology groups $H_p(K; R)$. Also, given a chain complex C where the C_p are R-modules, for any R-module G we can form the cochain complex $\mathrm{Hom}_R(C, G)$ and we obtain cohomology modules $H^p(C; G)$ with coefficients in any R-module G; see Section 4.8 and Chapter 12.

We can generalize homology with coefficients in a ring R to modules with coefficients in a R-module G by applying the operation (functor) $- \otimes_R G$ to a chain complex C, where the C_p's are R-modules, to get the chain complex denoted $C \otimes_R G$. The homology groups of this complex are denoted $H_p(C, G)$. We will discuss this construction in Section 4.7 and Chapter 12.

If the ring R is a PID, given a chain complex C where the C_p are R-modules, the homology groups $H_p(C; G)$ of the complex $C \otimes_R G$ are determined by the homology groups $H_{p-1}(C; R)$ and $H_p(C; R)$ *via* a formula called the *Universal Coefficient Theorem for Homology*; see Theorem 12.1. This formula involves a term $\mathrm{Tor}_1^R(H_{n-1}(C); G)$ that corresponds to the fact that the operation $- \otimes_R G$ on linear maps generally does not preserve injectivity ($- \otimes_R G$ is not left-exact). These matters are discussed in Chapter 11.

Similarly, if the ring R is a PID, given a chain complex C where the C_p are R-modules, the cohomology groups $H^p(C; G)$ of the complex $\mathrm{Hom}_R(C, G)$ are determined by the homology groups $H_{p-1}(C; R)$ and $H_p(C; R)$ *via* a formula called the *Universal Coefficient Theorem for Cohomology*; see Theorem 12.6. This formula involves a term $\mathrm{Ext}_R^1(H_{n-1}(C); G)$ that corresponds to the fact that if the linear map f is injective, then $\mathrm{Hom}_R(f, \mathrm{id})$ is not necessarily surjective ($\mathrm{Hom}_R(-, G)$ is not right-exact). These matters are discussed in Chapter 11.

One of the advantages of singular homology (and cohomology) is that it is defined for *all* topological spaces, but one of its disadvantages is that in practice it is very hard to compute. On the other hand, simplicial homology (and cohomology) only applies to triangulable spaces (geometric realizations of simplicial complexes), but in principle it is computable (for finite complexes). One of the practical problems is that the triangulations involved may have a large number of simplices. J.H.C. Whiteahead invented a class of spaces called *CW complexes* that are more general than triangulable spaces and for which the computation of the singular homology groups is often more tractable. Unlike a simplicial complex, a CW complex is obtained by gluing spherical cells as shown in Figure 1.4. CW complexes are discussed in Chapter 6.

There are at least four other ways of defining cohomology groups of a space X by directly forming a cochain complex without first forming a homology chain complex and then dualizing by applying $\mathrm{Hom}_{\mathbb{Z}}(-, G)$:

(1) If X is a smooth manifold, then there is the *de Rham complex* which uses the modules of smooth p-forms $\mathcal{A}^p(X)$ and the exterior derivatives $d^p \colon \mathcal{A}^p(X) \to \mathcal{A}^{p+1}(X)$. The corresponding cohomology groups are the *de Rham cohomology groups* $H^p_{\mathrm{dR}}(X)$. These are actually real vector spaces. de Rham cohomology is discussed in Chapter 3.
(2) If X is any space and $\mathcal{U} = (U_i)_{i \in I}$ is any open cover of X, we can define the *Čech cohomology groups* $\check{H}^p(X, \mathcal{U})$ in a purely combinatorial fashion. Then we can define the notion of *refinement* of a cover and

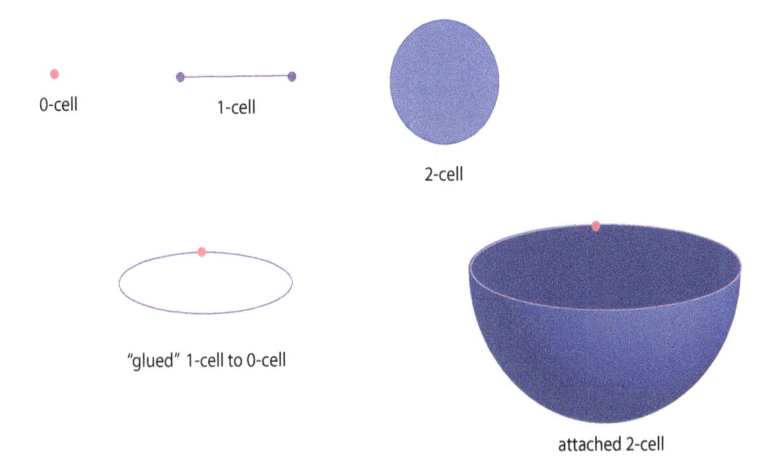

Fig. 1.4 The spherical hemisphere is a CW complex consisting of a point (0-cell), a line segment (1-cell), and a closed unit disk (2-cell).

define the *Čech cohomology groups* $\check{H}^p(X, G)$ with values in an abelian group G using a limiting process known as a *direct limit* (see Section 8.3, Definition 8.10). Čech cohomology is discussed in Chapter 9.

(3) If X is any space, then there is the *Alexander–Spanier* cochain complex which yields the *Alexander–Spanier cohomology groups* $A^p_{\text{A-S}}(X; G)$. Alexander–Spanier cohomology is discussed in Section 13.8 and in Chapter 14.

(4) Sheaf cohomology, based on derived functors and injective resolutions. This is the most general kind of cohomology of a space X, where cohomology groups $H^p(X, \mathcal{F})$ with values in a sheaf $\mathcal{F}$ on the space X are defined. Intuitively, this means that the modules $\mathcal{F}(U)$ of "coefficients" in which these groups take values may vary with the open domain $U \subseteq X$. Sheaf cohomology is discussed in Chapter 13, and the algebraic machinery of derived functors is discussed in Chapter 11.

We will see that for topological manifolds, all these cohomology theories are equivalent; see Chapter 13. For paracompact spaces, Čech cohomology, Alexander–Spanier cohomology, and derived functor cohomology (for constant sheaves) are equivalent (see Chapter 13). In fact, Čech cohomology and Alexander–Spanier cohomology are equivalent for any space; see Chapter 14.

1.2 Relative Homology and Cohomology

In general, computing homology groups is quite difficult so it would be helpful if we had techniques that made this process easier. Relative homology and excision are two such tools that we discuss in this section.

Lefschetz (1928) introduced the *relative homology groups* $H_p(K, L; \mathbb{Z})$, where K is a simplicial complex and L is a subcomplex of K. The same idea immediately applies to singular homology and we can define the relative singular homology groups $H_p(X, A; R)$ where A is a subspace of X. The intuition is that the module of p-chains of a relative chain complex consists of chains of K modulo chains of L. For example, given a space X and a subspace $A \subseteq X$, the *singular chain complex* $S_*(X, A; R)$ *of the pair* (X, A) is the chain complex in which each R-module $S_p(X, A; R)$ is the quotient module

$$S_p(X, A; R) = S_p(X; R)/S_p(A; R).$$

It is easy to see that $S_p(X, A; R)$ is actually a free R-module; see Section 4.3.

Although this is not immediately apparent, the motivation is that the groups $H_p(A; R)$ and $H_p(X, A; R)$ are often "simpler" than the groups $H_p(X; R)$, and there is an exact sequence called the *long exact sequence of relative homology* that can often be used to come up with an inductive argument that allows the determination of $H_p(X; R)$ from $H_p(A; R)$ and $H_p(X, A; R)$. Indeed, we have the following exact sequence as shown in Section 4.3 (see Theorem 4.9):

$$\begin{array}{l} \cdots \longrightarrow H_{p+2}(X, A; R) \xrightarrow{\partial_{*p+2}} \\ H_{p+1}(A; R) \xrightarrow{i_*} H_{p+1}(X; R) \xrightarrow{j_*} H_{p+1}(X, A; R) \xrightarrow{\partial_{*p+1}} \\ H_p(A; R) \xrightarrow{i_*} H_p(X; R) \xrightarrow{j_*} H_p(X, A; R) \xrightarrow{\partial_{*p}} \\ H_{p-1}(A; R) \longrightarrow \cdots \end{array}$$

ending in

$$H_0(A; R) \longrightarrow H_0(X; R) \longrightarrow H_0(X, A; R) \longrightarrow 0.$$

Furthermore, if (X, A) is a "good pair," then there is an isomorphism

$$H_p(X, A; R) \cong H_p(X/A, \{\text{pt}\}; R),$$

where the space X/A, *called a quotient space*, is obtained from X by identifying A with a single point, and where pt stands for any point in X.

The long exact sequence of relative homology is a corollary of one the staples of homology theory, the "zig-zag lemma." The zig-zag lemma says that for any short exact sequence

$$0 \longrightarrow X \xrightarrow{f} Y \xrightarrow{g} Z \longrightarrow 0$$

of chain complexes X, Y, Z there is a long exact sequence of cohomology

$$\begin{array}{l} \cdots \longrightarrow H^{p-1}(Z) \xrightarrow{\delta^{p-1}} \\ H^p(X) \xrightarrow{f^*} H^p(Y) \xrightarrow{g^*} H^p(Z) \xrightarrow{\delta^p} \\ H^{p+1}(X) \xrightarrow{f^*} H^{p+1}(Y) \xrightarrow{g^*} H^{p+1}(Z) \xrightarrow{\delta^{p+1}} \\ H^{p+2}(X) \longrightarrow \cdots \end{array}$$

The zig-zag lemma is fully proven in Section 2.7; see Theorem 2.22. There is also a homology version of this theorem.

Another very important aspect of relative singular homology is that it satisfies the *excision axiom*, another useful tool to compute homology groups. This means that removing a subspace $Z \subseteq A \subseteq X$ which is clearly inside of A, in the sense that Z is contained in the interior of A, does not change the relative homology group $H_p(X, A; R)$. More precisely, there is an isomorphism

$$H_p(X - Z, A - Z; R) \cong H_p(X, A; R);$$

see Section 4.5 (Theorem 4.14). A good illustration of the use of excision and of the long exact sequence of relative homology is the computation of the homology of the sphere S^n; see Section 4.6. Relative singular homology also satisfies another important property: the *homotopy axiom*, which says that if two spaces are homotopy equivalent, then their homology is isomorphic; see Theorem 4.8.

Following the procedure for obtaining cohomology from homology described in Section 1.1, by applying $\mathrm{Hom}_R(-, G)$ to the chain complex $S_*(X, A; R)$ we obtain the cochain complex $S^*(X, A; G) = \mathrm{Hom}_R(S_*(X, A; R), G)$, and thus the *singular relative cohomology groups* $H^p(X, A; G)$; see Section 4.9. In this case we can think of the elements of $S^p(X, A; G)$ as linear maps (with values in G) on singular p-simplices in X that vanish on singular p-simplices in A.

Fortunately, since each $S_p(X, A; R)$ is a free R-module, it can be shown that there is a long exact sequence of relative cohomology (see Theorem 4.36):

$$\begin{aligned}
\cdots &\longrightarrow H^{p-1}(A; G) \xrightarrow{\delta^*_{p-1}} \\
&\longrightarrow H^p(X, A; G) \xrightarrow{(j^\top)^*} H^p(X; G) \xrightarrow{(i^\top)^*} H^p(A; G) \xrightarrow{\delta^*_p} \\
&\longrightarrow H^{p+1}(X, A; G) \xrightarrow{(j^\top)^*} H^{p+1}(X; G) \xrightarrow{(i^\top)^*} H^{p+1}(A; G) \xrightarrow{\delta^*_{p+1}} \\
&\longrightarrow H^{p+2}(X, A; G) \longrightarrow \cdots
\end{aligned}$$

Relative singular cohomology also satisfies the excision axiom and the homotopy axioms (see Section 4.9).

1.3 Duality; Poincaré, Alexander, Lefschetz

Roughly speaking, duality is a kind of symmetry between the homology and the cohomology groups of a space. Historically, duality was formulated only for homology, but it was later found that more general formulations are obtained if both homology and cohomology are considered. We will discuss two duality theorems: Poincaré duality, and Alexander–Lefschetz duality. Original versions of these theorems were stated for homology and applied to special kinds of spaces. It took at least thirty years to obtain the versions that we will discuss.

The result that Poincaré considered as the climax of his work in algebraic topology was a *duality theorem* (even though the notion of duality was not very clear at the time). Since Poincaré was working with finite simplicial complexes, for him duality was a construction which, given a simplicial complex K of dimension n, produced a "dual" complex K^*; see Munkres [Munkres (1984)] (Chapter 8, Section 64). If done the right way, the matrices of the boundary maps $\partial\colon C_p(K) \to C_{p-1}(K)$ are transposes of the matrices of the boundary maps $\partial^*\colon C_{n-p+1}(K^*) \to C_{n-p}(K^*)$. As a consequence, the homology groups $H_p(K)$ and $H_{n-p}(K^*)$ are isomorphic. Note that this type of duality relates homology groups, not homology and cohomology groups as it usually does nowadays, for the good reason that cohomology did not exist until about 1935.

Around 1930 de Rham gave a version of Poincaré duality for smooth orientable, compact manifolds. If M is a smooth, oriented, and compact n-manifold, then there are isomorphisms

$$H^p_{\mathrm{dR}}(M) \cong (H^{n-p}_{\mathrm{dR}}(M))^*,$$

where $(H^{n-p}(M))^*$ is the dual of the vector space $H^{n-p}(M)$. This duality is actually induced by a nondegenerate pairing

$$\langle -, - \rangle\colon H^p_{\mathrm{dR}}(M) \times H^{n-p}_{\mathrm{dR}}(M) \to \mathbb{R}$$

given by integration, namely

$$\langle [\omega], [\eta] \rangle = \int_M \omega \wedge \eta,$$

where ω is a differential p-form and η is a differential $(n-p)$-form. For details, see Chapter 3, Theorem 3.8. The proof uses several tools from the arsenal of homological algebra: the zig-zag lemma (in the form of Mayer–Vietoris sequences), the five lemma, and an induction on finite "good covers."

Around 1935, inspired by Pontrjagin's duality theorem and his introduction of the notion of nondegenerate pairing (see the end of this section), Alexander and Kolmogoroff independently started developing cohomology, and soon after this it was realized that because cohomology primarily deals with functions, it is possible to define various products. Among those, the *cup product* is particularly important because it induces a multiplication operation on what is called the *cohomology algebra* $H^*(X; R)$ of a space X, and the *cap product* yields a stronger version of Poincaré duality.

Recall that $S^*(X; R)$ is the R-module $\bigoplus_{p\geq 0} S^p(X; R)$, where the $S^p(X; R)$ are the singular cochain modules. For all $p, q \geq 0$, it possible to define a function

$$\smile\colon S^p(X; R) \times S^q(X; R) \to S^{p+q}(X; R),$$

called *cup product*. These functions induce a multiplication on $S^*(X; R)$ also called the cup product, which is bilinear, associative, and has an identity element. The cup product satisfies the following equation

$$\delta(c \smile d) = (\delta c) \smile d + (-1)^p c \smile (\delta d),$$

reminiscent of a property of the wedge product. (In the above equation δ is the coboundary map, i.e. $\delta^p \colon S^p(X; R) \to S^{p+1}(X; R)$.) This equation can be used to show that the cup product is a well defined on cohomology classes:

$$\smile \colon H^p(X; R) \times H^q(X; R) \to H^{p+q}(X; R).$$

These operations induce a multiplication operation on $H^*(X; R) = \bigoplus_{p \geq 0} H^p(X; R)$ which is bilinear and associative. Together with the cup product, $H^*(X; R)$ is called the *cohomology ring* of X. For details, see Section 4.10.

The cup product for simplicial cohomology was invented independently by Alexander and Kolmogoroff (in addition to simplicial cohomology) and presented at a conference held in Moscow in 1935. Alexander's original definition was not quite correct and he modified his definition following a suggestion of Čech (1936). This modified version was discovered independently by Whitney (1938), who introduced the notation $\smile$. Eilenberg extended the definition of the cup product to singular cohomology (1944).

The significance of the cohomology ring is that two spaces X and Y may have isomorphic cohomology modules but nonisomorphic cohomology rings. Therefore the cohomology ring is an invariant of a space X that is finer than its cohomology.

Another product related to the cup product is the cap product. The *cap product* combines cohomology and homology classes, it is an operation

$$\frown \colon H^p(X; R) \times H_n(X; R) \to H_{n-p}(X; R);$$

see Section 7.2.

The cap product was introduced by Čech (1936) and independently by Whitney (1938), who introduced the notation $\frown$ and the name *cap product*. Again Eilenberg generalized the cap product to singular homology and cohomology.

The cup product and the cap product are related by the following equation:

$$a(b \frown \sigma) = (a \smile b)(\sigma)$$

for all $a \in S^{n-p}(X; R)$, all $b \in S^p(X; R)$, and all $\sigma \in S_n(X; R)$, or equivalently using the bracket notation for evaluation as

$$\langle a, b \frown \sigma \rangle = \langle a \smile b, \sigma \rangle,$$

which shows that $\frown$ is the adjoint of $\smile$ with respect to the evaluation pairing $\langle -, - \rangle$.

The reason why the cap product is important is that it can be used to state a sharper version of Poincaré duality. But first we need to talk about orientability.

If M is a topological manifold of dimension n, it turns out that for every $x \in M$ the relative (singular) homology groups $H_p(M, M - \{x\}; \mathbb{Z})$ are either (0) if $p \neq n$, or equal to $\mathbb{Z}$ if $p = n$. An *orientation* of M is a choice of a generator $\mu_x \in H_n(M, M - \{x\}; \mathbb{Z}) \cong \mathbb{Z}$ for each $x \in M$ which varies "‘continuously" with x. A manifold that has an orientation is called *orientable*.

Technically, this means that for every $x \in M$, locally on some small open subset U of M containing x there is some homology class $\mu_U \in H_n(M, M - U; \mathbb{Z})$ such that all the chosen $\mu_x \in H_n(M, M - \{x\}; \mathbb{Z})$ for all $x \in U$ are obtained as images of μ_U; see Figure 1.5.

Fig. 1.5 A schematic representation which shows μ_x as the image of μ_U.

If such a μ_U can be found when $U = M$, we call it a *fundamental class* of M and denote it by μ_M; see Section 7.3. Readers familiar with differential geometry may think of the fundamental class as a discrete analog to the notion of volume form. The crucial result is that a compact manifold of dimension n is orientable iff it has a unique fundamental class μ_M; see Theorem 7.7.

The notion of orientability can be generalized to the notion of R-orientability. One of the advantages of this notion is that every manifold is $\mathbb{Z}/2\mathbb{Z}$ orientable. We are now in a position to state the Poincaré duality theorem in terms of the cap product.

If M is compact and orientable, then there is a fundamental class μ_M. In this case (if $0 \leq p \leq n$) we have a map

$$D_M \colon H^p(M;\mathbb{Z}) \to H_{n-p}(M;\mathbb{Z})$$

given by

$$D_M(\omega) = \omega \frown \mu_M.$$

Poincaré duality asserts that the map

$$D_M \colon \omega \mapsto \omega \frown \mu_M$$

is an isomorphism between $H^p(M;\mathbb{Z})$ and $H_{n-p}(M;\mathbb{Z})$; see Theorem 7.16.

Poincaré duality can be generalized to compact R-orientable manifolds for any commutative ring R, and to coefficients in any R-module G. It can also be generalized to noncompact manifolds if we replace cohomology by cohomology with compact support (the modules $H^p_c(X;R)$); see Sections 7.3, 7.4, and 7.5. If $R = \mathbb{Z}/2\mathbb{Z}$, Poincaré duality holds for all manifolds, orientable or not.

Another kind of duality was introduced by Alexander in 1922. Alexander considered a compact proper subset A of the sphere S^n ($n \geq 2$) which is a curvilinear cell complex (A has some type of generalized triangulation). For the first time he defined the homology groups of the open subset $S^n - A$ with coefficients in $\mathbb{Z}/2\mathbb{Z}$ (so that he did not have to bother with signs), and he proved that for $p \leq n-2$ there are isomorphisms

$$H_p(A;\mathbb{Z}/2\mathbb{Z}) \cong H_{n-p-1}(S^n - A;\mathbb{Z}/2\mathbb{Z});$$

see Figure 1.6. Since cohomology did not exist yet, the original version of Alexander duality was stated for homology.

Around 1928 Lefschetz started investigating homology with coefficients in $\mathbb{Z}, \mathbb{Z}/m\mathbb{Z}$, or $\mathbb{Q}$, and defined relative homology. In his book published in

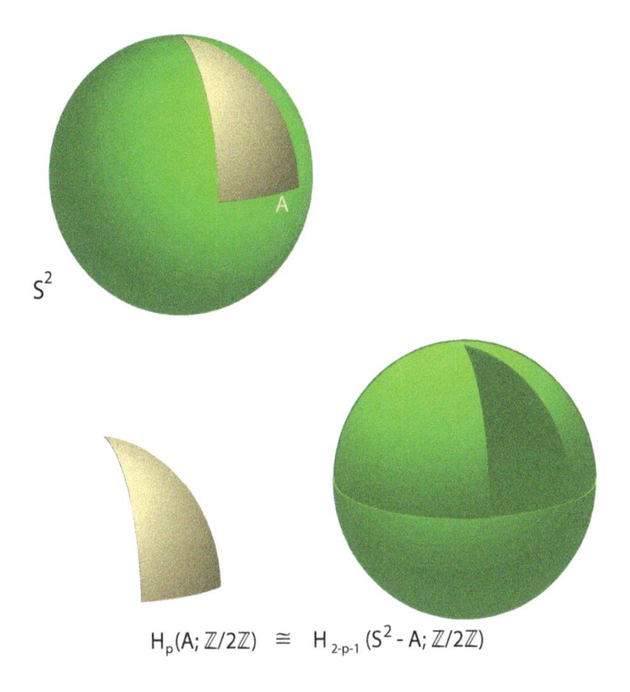

Fig. 1.6 Let A be the golden spherical triangle in S^2. The original version of Alexander duality compares the homology of the golden spherical triangle with the homology of the surface consisting of $S^2 - A$.

1930, using completely different methods from Alexander, Lefschetz proved a version of Alexander's duality in the case where A is a subcomplex of S^n. Soon after he obtained a homological version of what we call the Lefschetz duality theorem in Section 14.5 (Theorem 14.9):

$$H^p(M, L; \mathbb{Z}) \cong H_{n-p}(M - L; \mathbb{Z}),$$

where M, L are complexes and L is a subcomplex of M; see Figure 1.7.

Both Alexander and Lefschetz duality can be generalized to the situation where in Alexander duality A is an arbitrary closed subset of S^n, and in Lefschetz duality L is any compact subset of M and M is orientable, but new kinds of cohomology need to be introduced: *Čech cohomology and Alexander–Spanier cohomology*, which turn out to be equivalent. This is a nontrivial theorem due to Dowker [Dowker (1952)]. Then a duality theorem generalizing both Poincaré duality and Alexander–Lefschetz duality can be proven. These matters are discussed in Chapter 9, Section 13.8, and Chapter 14.

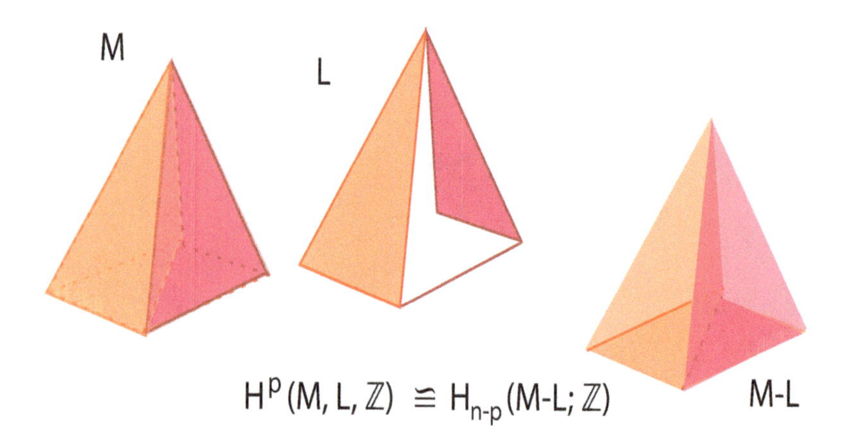

Fig. 1.7 A representation of Lefschetz duality when M is the simplicial complex consisting of two solid tetrahedra while L is the subcomplex consisting of the front left peach face, the back right pink face, and the solid red edge.

Proving the general version of Alexander–Lefschetz duality takes a significant amount of work because it requires defining relative versions of Čech cohomology and Alexander–Spanier cohomology, and to prove their equivalence as well as their equivalence to another definition in terms of direct limits of singular cohomology groups (see Definition 14.13 and Proposition 14.7).

When discussing the notion of duality, we would be remiss if we did not mention the important contributions of Pontrjagin. In a paper published in 1931 Pontrjagin investigates the duality between a closed subset A of $\mathbb{R}^n$ homeomorphic to a simplicial complex and $\mathbb{R}^n - A$. Pontrjagin introduces for the first time the notion of a *nondegenerate pairing* $\varphi\colon U \times V \to G$ between two finitely abelian groups U and V, where G is another abelian group (he uses $G = \mathbb{Z}$ or $G = \mathbb{Z}/m\mathbb{Z}$). This is a bilinear map $\varphi\colon U \times V \to G$ such that if $\varphi(u, v) = 0$ for all $v \in V$, then $u = 0$, and if $\varphi(u, v) = 0$ for all $u \in U$, then $v = 0$. Pontrjagin proves that U and V are isomorphic for his choice of G, and applies the notion of nondegenerate pairing to Poincaré duality and to a version of Alexander duality for certain subsets of $\mathbb{R}^n$. Pontrjagin also introduces the important notion of *direct limit* (see Section 8.3, Definition 8.10) which, among other things, plays a crucial role in the definition of Čech cohomology and in the construction of a sheaf from a presheaf (see Chapter 10).

In another paper published in 1934, Pontrjagin states and proves his famous duality theory between discrete and compact abelian topological groups. In this situation, U is a discrete group, $G = \mathbb{R}/\mathbb{Z} \cong S^1$, and $V = \widehat{U} = \mathrm{Hom}(U, \mathbb{R}/\mathbb{Z})$ (with the topology of simple convergence). Pontrjagin applies his duality theorem to a version of Alexander duality for compact subsets of $\mathbb{R}^n$ and for a version of Čech homology (cohomology had not been defined yet).

One notion that we still need to address, especially since it has appeared numerous times in our aforementioned discussions, is Čech cohomology. We will do so in the next section. It turns out that Čech cohomology accommodates very general types of coefficients, namely *presheaves and sheaves*. In Chapters 8, 9 and 10 we introduce these notions that play a major role in many area of mathematics, especially algebraic geometry and algebraic topology.

One can say that from a historical point of view, all the notions we presented so far are discussed in the landmark book by Eilenberg and Steenrod [Eilenberg and Steenrod (1952)] (1952). This is a beautiful book well worth reading, but it is not for the beginner. The next landmark book is Spanier's [Spanier (1989)] (1966). It is easier to read than Eilenberg and Steenrod but still quite demanding.

The next era of algebraic topology begins with the introduction of the notion of sheaf by Jean Leray around 1946.

1.4 Presheaves, Sheaves, and Čech Cohomology

The machinery of sheaves is applicable to problems designated by the vague notion of "passage from local to global properties." When some mathematical object attached to a topological space X can be "restricted" to any open subset U of X, and that restriction is known for sufficiently small U, what can be said about that "global" object? For example, consider the continuous functions defined over $\mathbb{R}^2$ and their restrictions to open subsets of $\mathbb{R}^2$.

Problems of this type had arisen since the 1880s in complex analysis in several variables and had been studied by Poincaré, Cousin, and later H. Cartan and Oka. Beginning in 1942, Leray considered a similar problem in cohomology. Given a space X, when the cohomology $H^*(U; G) = \bigoplus_{p \geq 0} H^p(U; G)$ is known for sufficiently small U, what can be said about $H^*(X; G) = \bigoplus_{p \geq 0} H^p(X; G)$?

Leray devised some machinery in 1946 that was refined and generalized by H. Cartan, M. Lazard, A. Borel, Koszul, Serre, Godement, and others to yield the notions of presheaves and sheaves.

Given a topological space X and a class $\mathbf{C}$ of structures (a category), say sets, vector spaces, R-modules, groups, commutative rings, *etc.*, a *presheaf on X with values in* $\mathbf{C}$ consists of an assignment of some object $\mathcal{F}(U)$ in $\mathbf{C}$ to every open subset U of X and of a map $\mathcal{F}(i)\colon \mathcal{F}(U) \to \mathcal{F}(V)$ of the class of structures in $\mathbf{C}$ to every inclusion $i\colon V \to U$ of open subsets $V \subseteq U \subseteq X$, such that

$$\begin{aligned}\mathcal{F}(i \circ j) &= \mathcal{F}(j) \circ \mathcal{F}(i)\\ \mathcal{F}(\mathrm{id}_U) &= \mathrm{id}_{\mathcal{F}(U)},\end{aligned}$$

for any two inclusions $i\colon V \to U$ and $j\colon W \to V$, with $W \subseteq V \subseteq U$; see Figure 1.8.

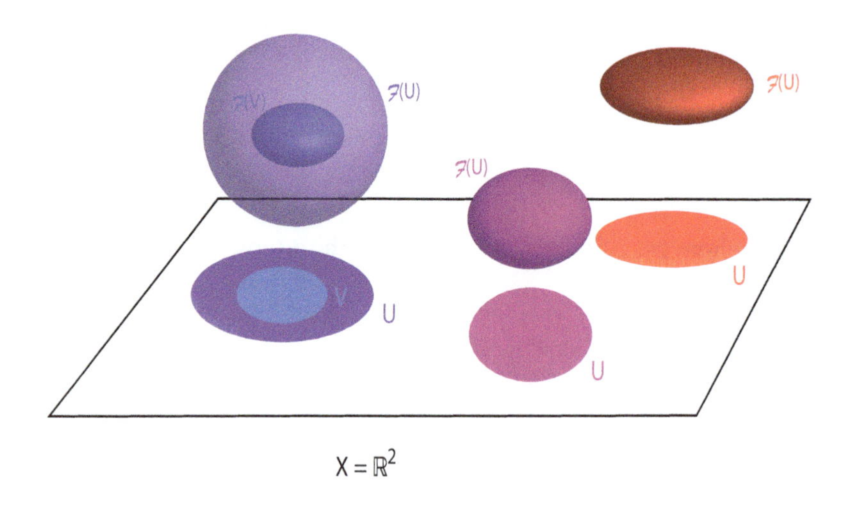

Fig. 1.8 A schematic representation of the presheaf of continuous real valued function on $X = \mathbb{R}^2$. An open set U is a circle in the plane while $\mathcal{F}(U)$ is the "balloon" of functions floating above U.

Note that the order of composition is switched in $\mathcal{F}(i \circ j) = \mathcal{F}(j) \circ \mathcal{F}(i)$.

Intuitively, the map $\mathcal{F}(i)\colon \mathcal{F}(U) \to \mathcal{F}(V)$ is a *restriction map* if we think of $\mathcal{F}(U)$ and $\mathcal{F}(V)$ as sets of functions (which is often the case). For this reason, the map $\mathcal{F}(i)\colon \mathcal{F}(U) \to \mathcal{F}(V)$ is also denoted by $\rho^U_V\colon \mathcal{F}(U) \to \mathcal{F}(V)$, and the first equation of the above definition is expressed by

$$\rho^U_W = \rho^V_W \circ \rho^U_V.$$

Presheaves, as defined above and in Section 8.1, are typically used to keep track of local information assigned to a global object (the space X). It is usually desirable to use consistent local information to recover some global information, but this requires a sharper notion, that of a sheaf.

As stated at the beginning of Section 8.2, the motivation for the extra condition that a sheaf should satisfy is this. Suppose we consider the presheaf of continuous functions on a topological space X. If U is any open subset of X and if $(U_i)_{i\in I}$ is an open cover of U, for any family $(f_i)_{i\in I}$ of continuous functions $f_i\colon U_i \to \mathbb{R}$, if f_i and f_j agree on every overlap $U_i \cap U_j$, then the f_i patch to a unique continuous function $f\colon U \to \mathbb{R}$ whose restriction to U_i is f_i.

Given a topological space X and a class $\mathbf{C}$ of structures (a category), say sets, vector spaces, R-modules, groups, commutative rings, *etc.*, a *sheaf on X with values in* $\mathbf{C}$ is a presheaf $\mathcal{F}$ on X such that for any open subset U of X, for every open cover $(U_i)_{i\in I}$ of U (that is, $U = \bigcup_{i\in I} U_i$ for some open subsets $U_i \subseteq U$ of X), the following conditions hold:

(G) (Gluing condition) For every family $(f_i)_{i\in I}$ with $f_i \in \mathcal{F}(U_i)$, if the f_i are consistent, which means that

$$\rho^{U_i}_{U_i\cap U_j}(f_i) = \rho^{U_j}_{U_i\cap U_j}(f_j) \quad \text{for all } i, j \in I,$$

then there is some $f \in \mathcal{F}(U)$ such that $\rho^U_{U_i}(f) = f_i$ for all $i \in I$; see Figure 1.9.

(M) (Monopresheaf condition) For any two elements $f, g \in \mathcal{F}(U)$, if f and g agree on all the U_i, which means that

$$\rho^U_{U_i}(f) = \rho^U_{U_i}(g) \quad \text{for all } i \in I,$$

then $f = g$.

Many (but not all) objects defined on a manifold are sheaves: the smooth functions $C^\infty(U)$, the smooth differential p-forms $\mathcal{A}^p(U)$, the smooth vector fields $\mathfrak{X}(U)$, where U is any open subset of M.

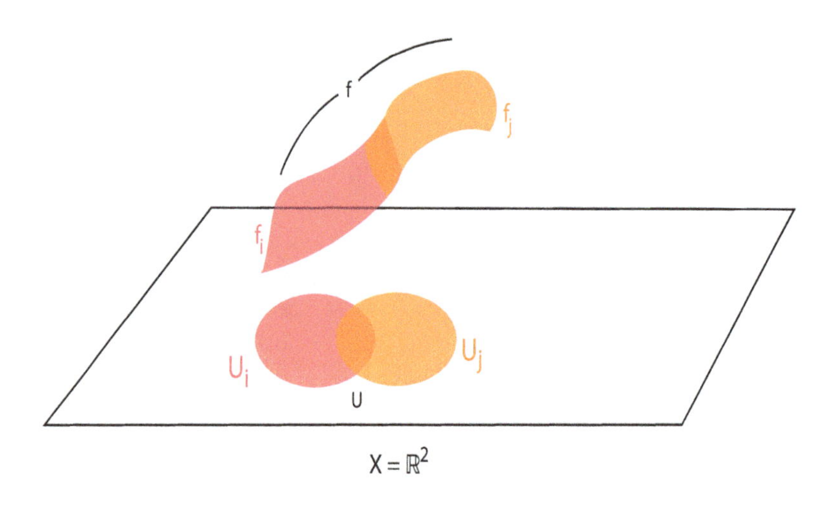

Fig. 1.9 Let $\mathcal{F}$ be the sheaf of continuous real valued functions on $X = \mathbb{R}^2$. Let $U = U_1 \cup U_2$. The graph of the pink function f_1 and the peach function f_2 glue together over $U_1 \cap U_2$ to form the continuous function $f\colon U \to \mathbb{R}^2$.

Given any commutative ring R and a fixed R-module G, the *constant presheaf* G_X is defined such that $G_X(U) = G$ for all nonempty open subsets U of X, and $G_X(\emptyset) = (0)$. The *constant sheaf* $\widetilde{G}_X$ is the sheaf given by $\widetilde{G}_X(U) =$ the set of locally constant functions on U (the functions $f\colon U \to G$ such that for every $x \in U$ there is some open subset V of U containing x such that f is constant on V), and $\widetilde{G}_X(\emptyset) = (0)$; see Figure 1.10.

In general a presheaf is not a sheaf. For example, the constant presheaf is not a sheaf. However, there is a procedure for converting a presheaf to a sheaf. We will return to this process in Section 1.5.

Čech cohomology with values in a presheaf of R-modules involves open covers of the topological space X; see Chapter 9.

Apparently, Čech himself did not introduce Čech cohomology, but he did introduce Čech homology using the notion of open cover (1932). Dowker, Eilenberg, and Steenrod introduced Čech cohomology in the early 1950s.

Given a topological space X, a family $\mathcal{U} = (U_j)_{j \in J}$ is an *open cover* of X if the U_j are open subsets of X and if $X = \bigcup_{j \in J} U_j$. Given any finite sequence $I = (i_0, \ldots, i_p)$ of elements of some index set J (where $p \geq 0$ and the i_j are not necessarily distinct), we let

$$U_I = U_{i_0 \cdots i_p} = U_{i_0} \cap \cdots \cap U_{i_p}.$$

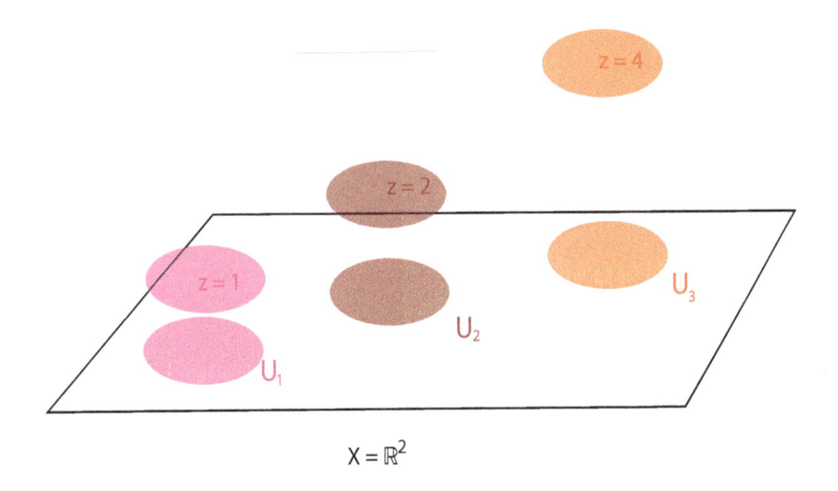

Fig. 1.10 Let $\mathcal{F}$ be the sheaf of continuous real valued functions over $X = \mathbb{R}^2$, and let $U = U_1 \cup U_2 \cup U_3$, a disjoint union. The function f is locally constant over U since it takes a constant value over each U_i, where $1 \leq i \leq 3$.

Note that it may happen that $U_I = \emptyset$. We denote by $U_{i_0 \cdots \widehat{i_j} \cdots i_p}$ the intersection

$$U_{i_0 \cdots \widehat{i_j} \cdots i_p} = U_{i_0} \cap \cdots \cap \widehat{U_{i_j}} \cap \cdots \cap U_{i_p}$$

of the p subsets obtained by omitting U_{i_j} from $U_{i_0 \cdots i_p} = U_{i_0} \cap \cdots \cap U_{i_p}$ (the intersection of the $p+1$ subsets); see Figure 1.11.

Now given a presheaf $\mathcal{F}$ of R-modules, the R-module of *Čech p-cochains* $C^p(\mathcal{U}, \mathcal{F})$ is the set of all functions f with domain J^{p+1} such that $f(i_0, \ldots, i_p) \in \mathcal{F}(U_{i_0 \cdots i_p})$; in other words,

$$C^p(\mathcal{U}, \mathcal{F}) = \prod_{(i_0, \ldots, i_p) \in J^{p+1}} \mathcal{F}(U_{i_0 \cdots i_p}),$$

the set of all J^{p+1}-indexed families $(f_{i_0, \ldots, i_p})_{(i_0, \ldots, i_p) \in J^{p+1}}$ with $f_{i_0, \ldots, i_p} \in \mathcal{F}(U_{i_0 \cdots i_p})$. Observe that the coefficients (the modules $\mathcal{F}(U_{i_0 \cdots i_p})$) can "vary" from open subset to open subset.

We have $p+1$ inclusion maps

$$\delta^p_j \colon U_{i_0 \cdots i_p} \longrightarrow U_{i_0 \cdots \widehat{i_j} \cdots i_p}, \quad 0 \leq j \leq p.$$

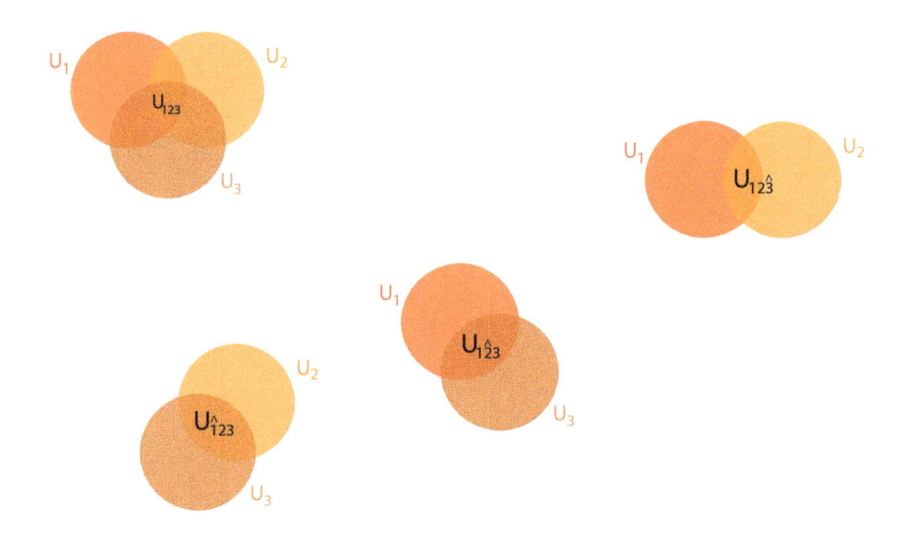

Fig. 1.11 An illustration of the notation $U_{123} = U_1 \cap U_2 \cap U_3$ and the three cases of $U_{i_0\cdots\widehat{i_j}\cdots i_p} = U_{i_0} \cap \cdots \cap \widehat{U_{i_j}} \cap \cdots \cap U_{i_p}$, where $i_0 = 1$ and $i_p = 3$.

Each inclusion map $\delta_j^p\colon U_{i_0\cdots i_p} \longrightarrow U_{i_0\cdots\widehat{i_j}\cdots i_p}$ induces a map

$$\mathcal{F}(\delta_j^p)\colon \mathcal{F}(U_{i_0\cdots\widehat{i_j}\cdots i_p}) \longrightarrow \mathcal{F}(U_{i_0\cdots i_p})$$

which is none other that the restriction map $\rho_{U_{i_0\cdots i_p}}^{U_{i_0\cdots\widehat{i_j}\cdots i_p}}$ which, for the sake of notational simplicity, we also denote by $\rho^j_{i_0\cdots i_p}$.

Given a topological space X, an open cover $\mathcal{U} = (U_j)_{j\in J}$ of X, and a presheaf of R-modules $\mathcal{F}$ on X, the *coboundary maps* $\delta^p_{\mathcal{F}}\colon C^p(\mathcal{U}, \mathcal{F}) \to C^{p+1}(\mathcal{U}, \mathcal{F})$ are given by

$$\delta^p_{\mathcal{F}} = \sum_{j=0}^{p+1} (-1)^j \mathcal{F}(\delta_j^{p+1}), \quad p \geq 0.$$

More explicitly, for any p-cochain $f \in C^p(\mathcal{U}, \mathcal{F})$, for any sequence $(i_0, \ldots, i_{p+1}) \in J^{p+2}$, we have

$$(\delta^p_{\mathcal{F}} f)_{i_0,\ldots,i_{p+1}} = \sum_{j=0}^{p+1} (-1)^j \rho^j_{i_0\cdots i_{p+1}}(f_{i_0,\ldots,\widehat{i_j},\ldots,i_{p+1}}).$$

Unravelling the above definition for $p = 0$ we have

$$(\delta^0_{\mathcal{F}} f)_{i,j} = \rho^0_{ij}(f_j) - \rho^1_{ij}(f_i),$$

and for $p = 1$ we have

$$(\delta^1_{\mathcal{F}} f)_{i,j,k} = \rho^0_{ijk}(f_{j,k}) - \rho^1_{ijk}(f_{i,k}) + \rho^2_{ijk}(f_{i,j}).$$

It is easy to check that $\delta^{p+1}_{\mathcal{F}} \circ \delta^p_{\mathcal{F}} = 0$ for all $p \geq 0$, so we have a chain complex of cohomology

$$0 \xrightarrow{\delta^{-1}_{\mathcal{F}}} C^0(\mathcal{U}, \mathcal{F}) \xrightarrow{\delta^0_{\mathcal{F}}} C^1(\mathcal{U}, \mathcal{F}) \longrightarrow \cdots$$

$$\xrightarrow{\delta^{p-1}_{\mathcal{F}}} C^p(\mathcal{U}, \mathcal{F}) \xrightarrow{\delta^p_{\mathcal{F}}} C^{p+1}(\mathcal{U}, \mathcal{F}) \xrightarrow{\delta^{p+1}_{\mathcal{F}}} \cdots$$

and we can define the Čech cohomology groups as follows.

Given a topological space X, an open cover $\mathcal{U} = (U_j)_{j \in J}$ of X, and a presheaf of R-modules $\mathcal{F}$ on X, the *Čech cohomology groups* $\check{H}^p(\mathcal{U}, \mathcal{F})$ *of the cover* $\mathcal{U}$ *with values in* $\mathcal{F}$ are defined by

$$\check{H}^p(\mathcal{U}, \mathcal{F}) = \mathrm{Ker}\, \delta^p_{\mathcal{F}} / \mathrm{Im}\, \delta^{p-1}_{\mathcal{F}}, \quad p \geq 0.$$

The *classical Čech cohomology groups* $\check{H}^p(\mathcal{U}; G)$ *of the cover* $\mathcal{U}$ *with coefficients in the* R*-module* $G/$ are the groups $\check{H}^p(\mathcal{U}, G_X)$, where G_X is the constant sheaf on X with values in G.

The next step is to define Čech cohomology groups that do not depend on the open cover $\mathcal{U}$. This is achieved by defining a notion of refinement on covers and by taking *direct limits* (see Section 8.3, Definition 8.10). Čech had used such a method in defining his Čech homology groups, by introducing the notion of *inverse limit* (which, curiously, was missed by Pontrjagin whose introduced direct limits!).

Without going into details, given two covers $\mathcal{U} = (U_i)_{i \in I}$ and $\mathcal{V} = (V_j)_{j \in J}$ of a space X, we say that $\mathcal{V}$ *is a refinement of* $\mathcal{U}$, denoted $\mathcal{U} \prec \mathcal{V}$, if there is a function $\tau \colon J \to I$ such that

$$V_j \subseteq U_{\tau(j)} \quad \text{for all } j \in J.$$

Under this notion of refinement, the open covers of X form a directed preorder, and the family $(\check{H}^p(\mathcal{U}, \mathcal{F}))_{\mathcal{U}}$ is what is called a direct mapping family so its direct limit

$$\varinjlim_{\mathcal{U}} \check{H}^p(\mathcal{U}, \mathcal{F})$$

makes sense. We define the *Čech cohomology groups* $\check{H}^p(X, \mathcal{F})$ *with values in* $\mathcal{F}$ by

$$\check{H}^p(X, \mathcal{F}) = \varinjlim_{\mathcal{U}} \check{H}^p(\mathcal{U}, \mathcal{F}).$$

The *classical Čech cohomology groups* $\check{H}^p(X; G)$ *with coefficients in the* R-*module* G/ are the groups $\check{H}^p(X, G_X)$ where G_X is the constant presheaf with value G. All this is presented in Chapter 9.

A natural question to ask is how does the classical Čech cohomology of a space compare with other types of cohomology, in particular singular cohomology. In general Čech cohomology can differ from singular cohomology, but for manifolds it agrees. Classical Čech cohomology also agrees with de Rham cohomology of the constant presheaf $\mathbb{R}_X$. These results are hard to prove; see Chapter 13.

1.5 Sheafification and Stalk Spaces

One of the major goals of this book is to introduce sheaf cohomology. This means we need to develop a deeper understanding of mappings between sheaves. A *map (or morphism)* $\varphi\colon \mathcal{F} \to \mathcal{G}$ of presheaves (or sheaves) $\mathcal{F}$ and $\mathcal{G}$ on X consists of a family of maps $\varphi_U\colon \mathcal{F}(U) \to \mathcal{G}(U)$ of the class of structures in $\mathbf{C}$, for any open subset U of X, such that

$$\varphi_V \circ (\rho_{\mathcal{F}})^U_V = (\rho_{\mathcal{G}})^U_V \circ \varphi_U$$

for every pair of open subsets U, V such that $V \subseteq U \subseteq X$. Equivalently, the following diagrams commute for every pair of open subsets U, V such that $V \subseteq U \subseteq X$

$$\begin{array}{ccc} \mathcal{F}(U) & \xrightarrow{\varphi_U} & \mathcal{G}(U) \\ {\scriptstyle (\rho_{\mathcal{F}})^U_V}\downarrow & & \downarrow{\scriptstyle (\rho_{\mathcal{G}})^U_V} \\ \mathcal{F}(V) & \xrightarrow[\varphi_V]{} & \mathcal{G}(V). \end{array}$$

The notion of kernel $\operatorname{Ker}\varphi$ and image $\operatorname{Im}\varphi$ of a presheaf or sheaf map $\varphi\colon \mathcal{F} \to \mathcal{G}$ is easily defined. The presheaf $\operatorname{Ker}\varphi$ is defined by $(\operatorname{Ker}\varphi)(U) = \operatorname{Ker}\varphi_U$, and the presheaf $\operatorname{Im}\varphi$ is defined by $(\operatorname{Im}\varphi)(U) = \operatorname{Im}\varphi_U$. In the case of presheaves, they are also presheaves, but in the case of sheaves, the kernel $\operatorname{Ker}\varphi$ is indeed a sheaf, but the image $\operatorname{Im}\varphi$ is **not** a sheaf in general.

This failure of the image of a sheaf map to be a sheaf is a problem that causes significant technical complications. In particular, it is not clear what it means for a sheaf map to be surjective, and a "good" definition of the notion of an exact sequence of sheaves is also unclear.

Fortunately, there is a procedure for converting a presheaf $\mathcal{F}$ into a sheaf $\widetilde{\mathcal{F}}$ which is reasonably well-behaved. This procedure is called *sheafification*. There is a sheaf map $\eta\colon \mathcal{F} \to \widetilde{F}$ which is generally not injective.

The *sheafification* process is universal in the sense that given any presheaf $\mathcal{F}$ and any sheaf $\mathcal{G}$, for any presheaf map $\varphi\colon \mathcal{F} \to \mathcal{G}$, there is a unique sheaf map $\widehat{\varphi}\colon \widetilde{\mathcal{F}} \to \mathcal{G}$ such that

$$\varphi = \widehat{\varphi} \circ \eta_{\mathcal{F}}$$

as illustrated by the following commutative diagram

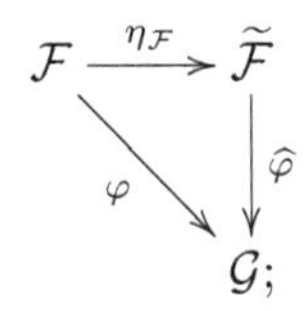

see Theorem 10.12.

The *sheafification* process involves constructing a topological space $S\mathcal{F}$ from the presheaf $\mathcal{F}$ that we call the *stalk space* of $\mathcal{F}$; see Figure 1.12. Godement calls it the *espace étalé*. The stalk space is the disjoint union of sets (modules) $\mathcal{F}_x$ called *stalks*. Each stalk $\mathcal{F}_x$ is the direct limit $\varinjlim(\mathcal{F}(U))_{U\ni x}$ of the family of modules $\mathcal{F}(U)$ for all "small" open sets U containing x (see Definition 10.1).

There is a surjective map $p\colon S\mathcal{F} \to X$ which, under the topology given to $S\mathcal{F}$, is a local homeomorphism, which means that for every $y \in S\mathcal{F}$, there is some open subset V of $S\mathcal{F}$ containing y such that the restriction of p to V is a homeomorphism. The sheaf $\widetilde{\mathcal{F}}$ consists of the continuous sections of p, that is, the continuous functions $s\colon U \to S\mathcal{F}$ such that $p \circ s = \mathrm{id}_U$, for any open subset U of X. This construction is presented in detail in Sections 10.1, 10.2, and 10.4.

The construction of the pair $(S\mathcal{F}, p)$ from a presheaf $\mathcal{F}$ suggests another definition of a sheaf as a pair (E, p), where E is a topological space and $p\colon E \to X$ is a surjective local homeomorphism onto another space X. Such a pair (E, p) is often called a *sheaf space*, but we prefer to call it a *stalk space*. This is the definition that was given by H. Cartan and M. Lazard around 1950. The sheaf ΓE associated with the stalk space (E, p) is defined as follows: for any open subset U or X, the *sections* of ΓE are the continuous sections $s\colon U \to E$, that is, the continuous functions such that $p \circ s = \mathrm{id}$; see Figure 1.13. We can also define a notion of map between two stalk spaces. Stalk spaces are discussed in Section 10.3.

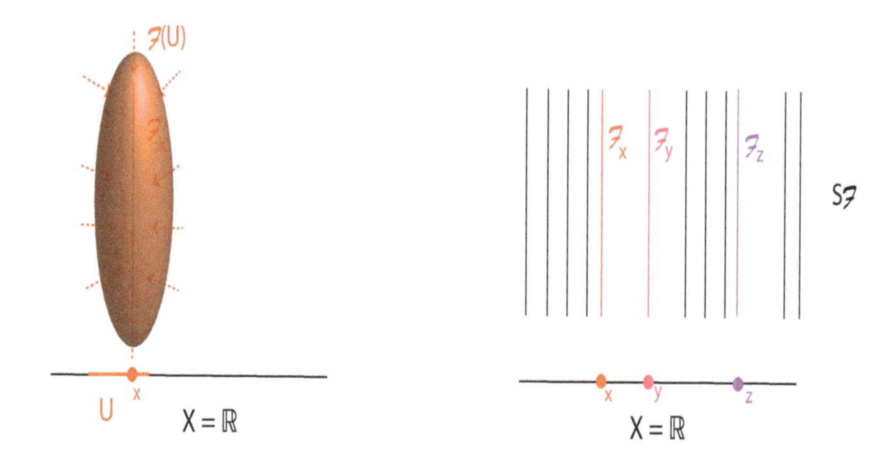

Fig. 1.12 Let $X = \mathbb{R}$ and $\mathcal{F}$ be the sheaf of real valued continuous functions. An element $\mathcal{F}(U)$ is represented by the floating balloon. By "collapsing" the balloon (via the direct limiting process), we form the stalk $\mathcal{F}_x$, which is represented as a vertical line.

As this stage, given a topological space X we have three categories (classes of objects):

(1) The category $\mathbf{Psh}(X)$ of presheaves and their morphisms.
(2) The category $\mathbf{Sh}(X)$ of sheaves and their morphisms.
(3) The category $\mathbf{StalkS}(X)$ of stalk spaces and their morphisms.

There is also a functor

$$S\colon \mathbf{PSh}(X) \to \mathbf{StalkS}(X)$$

from the category $\mathbf{PSh}(X)$ to the category $\mathbf{StalkS}(X)$ given by the construction of a stalk space $S\mathcal{F}$ from a presheaf $\mathcal{F}$, $(S(\mathcal{F}) = S\mathcal{F})$, and a functor

$$\Gamma\colon \mathbf{StalkS}(X) \to \mathbf{Sh}(X)$$

from the category $\mathbf{StalkS}(X)$ to the category $\mathbf{Sh}(X)$, given by the sheaf ΓE of continuous sections of E. Here we are using the term functor in an informal way. A more precise definition is given in Sections 1.7 and 10.10.

Note that every sheaf $\mathcal{F}$ is also a presheaf, and that every map $\varphi\colon \mathcal{F} \to \mathcal{G}$ of sheaves is also a map of presheaves. Therefore, we have an inclusion map

$$i\colon \mathbf{Sh}(X) \to \mathbf{PSh}(X),$$

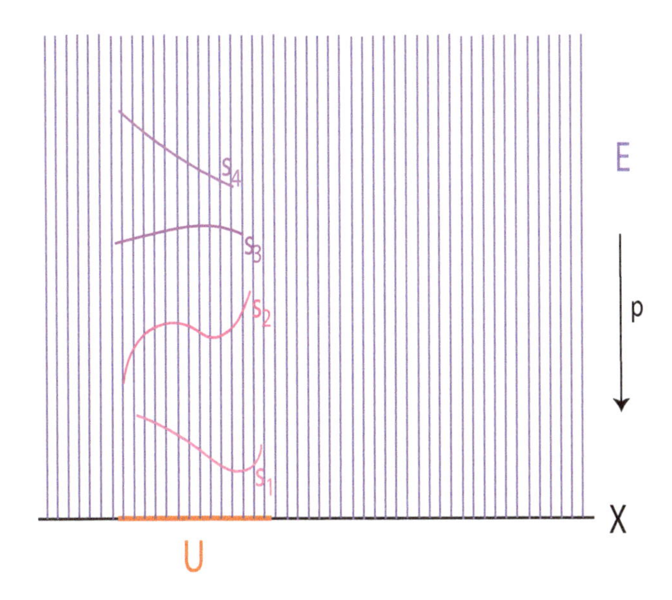

Fig. 1.13 A schematic representation of a stalk space (E, p). We drew four sections over U, where each section is a colored curve such that $p \circ s = \mathrm{id}$.

which is a functor. As a consequence, S restricts to an operation (functor)

$$S\colon \mathbf{Sh}(X) \to \mathbf{StalkS}(X).$$

There is also a map η which maps a presheaf $\mathcal{F}$ to the sheaf $\Gamma S(\mathcal{F}) = \widetilde{\mathcal{F}}$. This map η is a natural isomorphism between the functors id (the identity functor) and ΓS from $\mathbf{Sh}(X)$ to itself. In other words, if we take $\mathcal{F}$, form the stalk space $S\mathcal{F}$, then turn this stalk space into the sheaf of continuous sections $\Gamma S\mathcal{F}$, this new sheaf is *isomorphic* to $\mathcal{F}$.

We can also define a map ϵ which takes a stalk space (E, p) and makes the stalk space $S\Gamma E$. The map ϵ is a natural isomorphism between the functors id (the identity functor) and $S\Gamma$ from $\mathbf{StalkS}(X)$ to itself. In other words, if we take the stalk space (E, p), form the sheaf of continuous sections ΓE, then form the stalk space of ΓE, namely $S\Gamma E$, this new stalk space is *isomorphic* to (E, p).

Then we see that the two operations (functors)

$$S\colon \mathbf{Sh}(X) \to \mathbf{StalkS}(X) \quad \text{and} \quad \Gamma\colon \mathbf{StalkS}(X) \to \mathbf{Sh}(X)$$

are almost mutual inverses, in the sense that there is a natural isomorphism η between ΓS and id and a natural isomorphism ϵ between $S\Gamma$ and id. In such a situation, we say that the classes (categories) $\mathbf{Sh}(X)$ and $\mathbf{StalkS}(X)$ are *equivalent*. The upshot is that it is basically a matter of taste (or convenience) whether we decide to work with sheaves or stalk spaces. In fact, for the aspects of sheaf cohomology that deal with soft and fine sheaves (Sections 13.5 and 13.6), it is best to use the stalk space construction of a sheaf.

We also have the operator (functor)

$$\Gamma S\colon \mathbf{PSh}(X) \to \mathbf{Sh}(X)$$

which "sheafifies" a presheaf $\mathcal{F}$ into the sheaf $\widetilde{\mathcal{F}}$. Theorem 10.12 can be restated as saying that there is an isomorphism

$$\mathrm{Hom}_{\mathbf{PSh}(X)}(\mathcal{F}, i(\mathcal{G})) \cong \mathrm{Hom}_{\mathbf{Sh}(X)}(\widetilde{\mathcal{F}}, \mathcal{G}),$$

between the set (category) of maps between the presheaves $\mathcal{F}$ and $i(\mathcal{G})$ and the set (category) of maps between the sheaves $\widetilde{\mathcal{F}}$ and $\mathcal{G}$. In fact, such an isomorphism is natural, so in categorical terms, i and $\widetilde{\ } = \Gamma S$ are *adjoint functors*.

All this is explained in Sections 10.3 and 10.4.

1.6 Cokernels and Images of Sheaf Maps

We still need to define the image of a sheaf map in such a way that the notion of exact sequence of sheaves makes sense. Recall that if $f\colon A \to B$ is a homomorphism of modules, the *cokernel* $\mathrm{Coker} f$ of f is defined by $B/\mathrm{Im} f$. It is a measure of the surjectivity of f. We also have the projection map $\mathrm{coker}(f)\colon B \to \mathrm{Coker}\, f$, and observe that

$$\mathrm{Im}\, f = \mathrm{Ker}\,\mathrm{coker}(f).$$

The above suggests defining notions of cokernels of presheaf maps and sheaf maps. For a presheaf map $\varphi\colon \mathcal{F} \to \mathcal{G}$ this is easy, and we can define the *presheaf cokernel* $\mathrm{PCoker}(\varphi)$. It comes with a presheaf map $\mathrm{pcoker}(\varphi)\colon \mathcal{G} \to \mathrm{PCoker}(\varphi)$.

If $\mathcal{F}$ and $\mathcal{G}$ are sheaves, we define the *sheaf cokernel* $\mathrm{SCoker}(\varphi)$ as the sheafification of $\mathrm{PCoker}(\varphi)$. It also comes with a presheaf map $\mathrm{scoker}(\varphi)\colon \mathcal{G} \to \mathrm{SCoker}(\varphi)$.

Then it can be shown that if $\varphi\colon \mathcal{F} \to \mathcal{G}$ is a sheaf map, $\mathrm{SCoker}(\varphi) = (0)$ iff the stalk maps $\varphi_x \colon \mathcal{F}_x \to \mathcal{G}_x$ are surjective for all $x \in X$; see Proposition 10.19.

It follows that the "correct" definition for the image $\mathrm{SIm}\,\varphi$ of a sheaf map $\varphi\colon \mathcal{F} \to \mathcal{G}$ is

$$\mathrm{SIm}\,\varphi = \mathrm{Ker}\,\mathrm{scoker}(\varphi).$$

With this definition, a sequence of sheaves

$$\mathcal{F} \xrightarrow{\varphi} \mathcal{G} \xrightarrow{\psi} \mathcal{H}$$

is said to be exact if $\mathrm{SIm}\,\varphi = \mathrm{Ker}\,\psi$. Then it can be shown that

$$\mathcal{F} \xrightarrow{\varphi} \mathcal{G} \xrightarrow{\psi} \mathcal{H}$$

is an exact sequence of sheaves iff the sequence

$$\mathcal{F}_x \xrightarrow{\varphi_x} \mathcal{G}_x \xrightarrow{\psi_x} \mathcal{H}_x$$

is an exact sequence of R-modules (or rings) for all $x \in X$; see Proposition 10.24. This second characterization of exactness (for sheaves) is usually much more convenient than the first condition.

The definitions of cokernels and images of presheaves and sheaves as well as the notion of exact sequences of presheaves and sheaves are discussed in Sections 10.6, 10.7, 10.8, 10.9, and 10.10.

1.7 Injective and Projective Resolutions; Derived Functors

In order to define, even informally, the concept of derived functor, we need to describe what are functors and exact functors.

Suppose we have two types of structures (categories) $\mathbf{C}$ and $\mathbf{D}$ (for concreteness, think of $\mathbf{C}$ as the class of R-modules over some commutative ring R with an identity element 1 and of $\mathbf{D}$ as the class of abelian groups), and we have a transformation T (a functor) which works as follows:

(i) Each object A of $\mathbf{C}$ is mapped to some object $T(A)$ of $\mathbf{D}$.

(ii) Each map $A \xrightarrow{f} B$ between two objects A and B in $\mathbf{C}$ (of example, an R-linear map) is mapped to some map $T(A) \xrightarrow{T(f)} T(B)$ between the objects $T(A)$ and $T(B)$ in $\mathbf{D}$ (for example, a homomorphism of abelian groups) in such a way that the following properties hold:

(a) Given any maps $A \xrightarrow{f} B$ and $B \xrightarrow{g} C$ between objects A, B, C in $\mathbf{C}$ such that the composition $A \xrightarrow{g \circ f} C = A \xrightarrow{f} B \xrightarrow{g} C$ makes sense, the composition $T(A) \xrightarrow{T(f)} T(B) \xrightarrow{T(g)} T(C)$ makes sense in $\mathbf{D}$, and

$$T(g \circ f) = T(g) \circ T(f).$$

(b) If $A \xrightarrow{\mathrm{id}_A} A$ is the identity map of the object A in $\mathbf{C}$, then $T(A) \xrightarrow{T(\mathrm{id}_A)} T(A)$ is the identity map of $T(A)$ in $\mathbf{D}$; that is,

$$T(\mathrm{id}_A) = \mathrm{id}_{T(A)}.$$

Whenever a transformation $T\colon \mathbf{C} \to \mathbf{D}$ satisfies the Properties (i), (ii), (a), (b), we call it a *(covariant) functor* from $\mathbf{C}$ to $\mathbf{D}$.

If $T\colon \mathbf{C} \to \mathbf{D}$ satisfies Properties (i), (b), and if Properties (ii) and (a) are replaced by the Properties (ii') and (a') below

(ii') Each map $A \xrightarrow{f} B$ between two objects A and B in $\mathbf{C}$ is mapped to some map $T(B) \xrightarrow{T(f)} T(A)$ between the objects $T(B)$ and $T(A)$ in $\mathbf{D}$ in such a way that the following properties hold:

(a') Given any maps $A \xrightarrow{f} B$ and $B \xrightarrow{g} C$ between objects A, B, C in $\mathbf{C}$ such that the composition $A \xrightarrow{g \circ f} C = A \xrightarrow{f} B \xrightarrow{g} C$ makes sense, the composition $T(C) \xrightarrow{T(g)} T(B) \xrightarrow{T(f)} T(A)$ makes sense in $\mathbf{D}$, and

$$T(g \circ f) = T(f) \circ T(g),$$

then T is called a *contravariant functor* from $\mathbf{C}$ to $\mathbf{D}$.

An example of a (covariant) functor is the functor $\mathrm{Hom}(A, -)$ (for a fixed R-module A) from R-modules to R-modules which maps a module B to the module $\mathrm{Hom}(A, B)$ and a module homomorphism $f\colon B \to C$ to the module homomorphism $\mathrm{Hom}(A, f)$ from $\mathrm{Hom}(A, B)$ to $\mathrm{Hom}(A, C)$ given by

$$\mathrm{Hom}(A, f)(\varphi) = f \circ \varphi \quad \text{for all } \varphi \in \mathrm{Hom}(A, B).$$

Another example is the functor T from R-modules to R-modules such that $T(A) = A \otimes_R M$ for any R-module A, and $T(f) = f \otimes_R \mathrm{id}_M$ for any R-linear map $f\colon A \to B$.

An example of a contravariant functor is the functor $\mathrm{Hom}(-, A)$ (for a fixed R-module A) from R-modules to R-modules which maps a module B to the module $\mathrm{Hom}(B, A)$ and a module homomorphism $f\colon B \to C$ to the module homomorphism $\mathrm{Hom}(f, A)$ from $\mathrm{Hom}(C, A)$ to $\mathrm{Hom}(B, A)$ given by

$$\mathrm{Hom}(f, A)(\varphi) = \varphi \circ f \quad \text{for all } \varphi \in \mathrm{Hom}(C, A).$$

Categories and functors were introduced by Eilenberg and Mac Lane, first in a paper published in 1942, and then in a more complete paper published in 1945.

Given a type of structures (category) $\mathbf{C}$, let us denote the set of all maps from an object A to an object B by $\mathrm{Hom}_{\mathbf{C}}(A, B)$. For all the types of structures $\mathbf{C}$ that we will dealing with, each set $\mathrm{Hom}_{\mathbf{C}}(A, B)$ has some additional structure; namely it is an abelian group.

Intuitively speaking an *abelian category* is a category in which the notion of kernel and cokernel of a map makes sense. Then we can define the notion of image of a map f as the kernel of the cokernel of f, so the notion of exact sequence makes sense, as we did in Section 1.6. The categories of R-modules and the categories of sheaves (or presheaves) are abelian categories. For more details, see Sections 10.10 and 10.11.

A sequence of R-modules and R-linear maps (more generally objects and maps between objects in an abelian category)

$$0 \longrightarrow A \xrightarrow{\ f\ } B \xrightarrow{\ g\ } C \longrightarrow 0 \tag{$*$}$$

is a *short exact sequence* if

(1) f is injective.
(2) $\mathrm{Im}\ f = \mathrm{Ker}\, g$.
(3) g is surjective.

According to Dieudonné [Dieudonné (1989)], the notion of exact sequence first appeared in a paper of Hurewicz (1941), and then in a paper of Eilenberg and Steenrod and a paper of H. Cartan, both published in 1945. In 1947, Kelly and Pitcher generalized the notion of exact sequence to chain complexes, and apparently introduced the terminology *exact sequence*. In their 1952 treatise [Eilenberg and Steenrod (1952)], Eilenberg and Steenrod took the final step of allowing a chain complex to be indexed by $\mathbb{Z}$ (as we do in Section 2.5).

Given two types of structures (categories) **C** and **D** in each of which the concept of exactness is defined (abelian categories), given an additive functor $T\colon \mathbf{C} \to \mathbf{D}$, by applying T to the short exact sequence $(*)$ we obtain the sequence

$$0 \longrightarrow T(A) \xrightarrow{T(f)} T(B) \xrightarrow{T(g)} T(C) \longrightarrow 0, \qquad (**)$$

which is a chain complex (since $T(g) \circ T(f) = 0$). Then the following question arises:

Is the sequence $(**)$ also exact?

In general, the answer is **no**, but weaker forms of preservation of exactness suggest themselves.

A functor $T\colon \mathbf{C} \to \mathbf{D}$, is said to be *exact* if whenever the sequence

$$0 \longrightarrow A \longrightarrow B \longrightarrow C \longrightarrow 0$$

is exact in **C**, then the sequence

$$0 \longrightarrow T(A) \longrightarrow T(B) \longrightarrow T(C) \longrightarrow 0$$

is exact in **D**; *left exact* if whenever the sequence

$$0 \longrightarrow A \longrightarrow B \longrightarrow C$$

is exact in **C**, then the sequence

$$0 \longrightarrow T(A) \longrightarrow T(B) \longrightarrow T(C)$$

is exact; and *right exact* if whenever the sequence

$$A \longrightarrow B \longrightarrow C \longrightarrow 0$$

is exact in **C**, then the sequence

$$T(A) \longrightarrow T(B) \longrightarrow T(C) \longrightarrow 0$$

is exact.

If $T\colon \mathbf{C} \to \mathbf{D}$ is a contravariant functor, then T is said to be *exact* if whenever the sequence

$$0 \longrightarrow A \longrightarrow B \longrightarrow C \longrightarrow 0$$

is exact in **C**, then the sequence

$$0 \longrightarrow T(C) \longrightarrow T(B) \longrightarrow T(A) \longrightarrow 0$$

is exact in **D**; *left exact* if whenever the sequence

$$A \longrightarrow B \longrightarrow C \longrightarrow 0$$

is exact in $\mathbf{C}$, then the sequence

$$0 \longrightarrow T(C) \longrightarrow T(B) \longrightarrow T(A)$$

is exact; and *right exact* if whenever the sequence

$$0 \longrightarrow A \longrightarrow B \longrightarrow C$$

is exact in $\mathbf{C}$, then the sequence

$$T(C) \longrightarrow T(B) \longrightarrow T(A) \longrightarrow 0$$

is exact.

For example, the functor $\mathrm{Hom}(-, A)$ is (contravariant) *left-exact* but not exact in general (see Section 2.1). Similarly, the functor $\mathrm{Hom}(A, -)$ is *left-exact* but not exact in general (see Section 2.4).

Modules for which the functor $\mathrm{Hom}(A, -)$ is exact play an important role. They are called *projective modules*. Similarly, modules for which the functor $\mathrm{Hom}(-, A)$ is exact are called *injective modules*.

The functor $- \otimes_R M$ is *right-exact* but not exact in general (see Section 2.4). Modules M for which the functor $- \otimes_R M$ is exact are called *flat*.

A good deal of homological algebra has to do with understanding how much a module fails to be projective or injective (or flat).

Injective and projective modules are also characterized by extension properties. As we will see later, these extension characterizations can be used to define injective and projective objects in an abelian category.

(1) A module P is projective iff for any surjective linear map $h\colon A \to B$ and any linear map $f\colon P \to B$, there is some linear map $\widehat{f}\colon P \to A$ lifting $f\colon P \to B$ in the sense that $f = h \circ \widehat{f}$, as in the following commutative diagram:

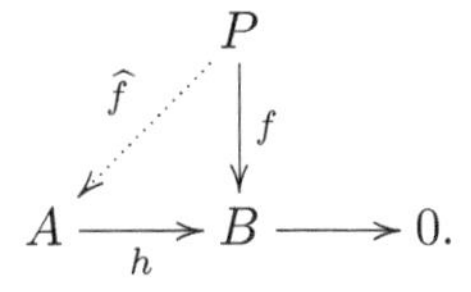

(2) A module I is injective iff for any injective linear map $h\colon A \to B$ and any linear map $f\colon A \to I$, there is some linear map $\widehat{f}\colon B \to I$ extending

$f\colon A \to I$ in the sense that $f = \widehat{f} \circ h$, as in the following commutative diagram:

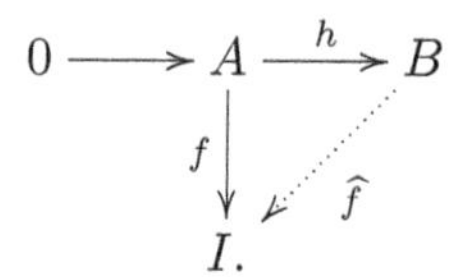

See Section 11.1.

Injective modules were introduced by Baer in 1940 and projective modules by Cartan and Eilenberg in the early 1950s. Every free module is projective. Injective modules are more elusive. If the ring R is a PID, an R-module M is injective iff it is divisible (which means that for every nonzero $\lambda \in R$, the map given by $u \mapsto \lambda u$ for $u \in M$ is surjective).

One of the most useful properties of projective modules is that every module M is the image of some projective (even free) module P, which means that there is a surjective homomorphism $\rho\colon P \to M$. Similarly, every module M can be embedded in an injective module I, which means that there is an injective homomorphism $i\colon M \to I$. This second fact is harder to prove (see Baer's embedding theorem, Theorem 11.6).

The above properties can be used to construct inductively projective and injective resolutions of a module M, a process that turns out to be remarkably useful. Intuitively, projective resolutions measure how much a module deviates from being projective, and injective resolutions measure how much a module deviates from being injective.

Hopf introduced free resolutions in 1945. A few years later Cartan and Eilenberg defined projective and injective resolutions.

Given any R-module A, a *projective resolution* of A is any exact sequence

$$\cdots \longrightarrow P_n \xrightarrow{d_n} P_{n-1} \xrightarrow{d_{n-1}} \cdots \longrightarrow P_1 \xrightarrow{d_1} P_0 \xrightarrow{p_0} A \longrightarrow 0 \tag{$*_1$}$$

in which every P_n is a projective module. The exact sequence

$$\cdots \longrightarrow P_n \xrightarrow{d_n} P_{n-1} \xrightarrow{d_{n-1}} \cdots \longrightarrow P_1 \xrightarrow{d_1} P_0$$

obtained by truncating the projective resolution of A after P_0 is denoted by $\mathbf{P}^A$, and the projective resolution $(*_1)$ is denoted by

$$\mathbf{P}^A \xrightarrow{p_0} A \longrightarrow 0.$$

Given any R-module A, an *injective resolution* of A is any exact sequence

$$0 \longrightarrow A \xrightarrow{i_0} I^0 \xrightarrow{d^0} I^1 \xrightarrow{d^1} \cdots \longrightarrow I^n \xrightarrow{d^n} I^{n+1} \longrightarrow \cdots \qquad (**_1)$$

in which every I^n is an injective module. The exact sequence

$$I^0 \xrightarrow{d^0} I^1 \xrightarrow{d^1} \cdots \longrightarrow I^n \xrightarrow{d^n} I^{n+1} \longrightarrow \cdots$$

obtained by truncating the injective resolution of A before I^0 is denoted by $\mathbf{I}_A$, and the injective resolution $(**_1)$ is denoted by

$$0 \longrightarrow A \xrightarrow{i_0} \mathbf{I}_A.$$

Now suppose that we have a functor $T\colon \mathbf{C} \to \mathbf{D}$, where $\mathbf{C}$ is the category of R-modules and $\mathbf{D}$ is the category of abelian groups. If we apply T to $\mathbf{P}^A$ we obtain the chain complex

$$0 \longleftarrow T(P_0) \xleftarrow{T(d_1)} T(P_1) \xleftarrow{T(d_2)} \cdots \; T(P_{n-1}) \xleftarrow{T(d_n)} T(P_n) \longleftarrow \cdots, \qquad \text{(Lp)}$$

denoted $T(\mathbf{P}^A)$. The above is no longer exact in general but it defines homology groups $H_p(T(\mathbf{P}^A))$.

Similarly, if we apply T to $\mathbf{I}_A$ we obtain the cochain complex

$$0 \longrightarrow T(I^0) \xrightarrow{T(d^0)} T(I^1) \xrightarrow{T(d^1)} \cdots \; T(I^n) \xrightarrow{T(d^n)} T(I^{n+1}) \longrightarrow \cdots, \qquad \text{(Ri)}$$

denoted $T(\mathbf{I}_A)$. The above is no longer exact in general but it defines cohomology groups $H^p(T(\mathbf{I}_A))$.

The reason why projective resolutions are so special is that even though the homology groups $H_p(T(\mathbf{P}^A))$ appear to depend on the projective resolution $\mathbf{P}^A$, in fact they don't; *the groups $H_p(T(\mathbf{P}^A))$ only depend on A and T*. This is proven in Theorem 11.28.

Similarly, the reason why injective resolutions are so special is that even though the cohomology groups $H^p(T(\mathbf{I}_A))$ appear to depend on the injective resolution $\mathbf{I}_A$, in fact they don't; *the groups $H^p(T(\mathbf{I}_A))$ only depend on A and T*. This is proven in Theorem 11.27.

Proving the above facts takes some work; we make use of the *comparison theorems*; see Section 11.2, Theorem 11.17 and Theorem 11.21. In view of the above results, given a functor T as above, Cartan and Eilenberg were led to define the *left derived functors* $L_n T$ of T by

$$L_n T(A) = H_n(T(\mathbf{P}^A)),$$

for any projective resolution $\mathbf{P}^A$ of A, and the *right derived functors* $R^n T$ of T by

$$R^n T(A) = H^n(T(\mathbf{I}_A)),$$

for any injective resolution $\mathbf{I}_A$ of A. The functors $L_n T$ and $R^n T$ can also be defined on maps. If T is right-exact, then $L_0 T$ is isomorphic to T (as a functor), and if T is left-exact, then $R^0 T$ is isomorphic to T (as a functor).

For example, the left derived functors of the right-exact functor $T_B(A) = A \otimes B$ (with B fixed) are the "Tor" functors. We have $\mathrm{Tor}_0^R(A, B) \cong A \otimes B$, and the functor $\mathrm{Tor}_1^R(-, G)$ plays an important role in comparing the homology of a chain complex C and the homology of the complex $C \otimes_R G$; see Chapter 12. Čech introduced the functor $\mathrm{Tor}_1^R(-, G)$ in 1935 in terms of generators and relations. It is only after Whitney defined tensor products of arbitrary $\mathbb{Z}$-modules in 1938 that the definition of Tor was expressed in the intrinsic form that we are now familiar with.

There are also versions of left and right derived functors for contravariant functors. For example, the right derived functors of the contravariant left-exact functor $T_B(A) = \mathrm{Hom}_R(A, B)$ (with B fixed) are the "Ext" functors. We have $\mathrm{Ext}_R^0(A, B) \cong \mathrm{Hom}_R(A, B)$, and the functor $\mathrm{Ext}_R^1(-, G)$ plays an important role in comparing the homology of a chain complex C and the cohomology of the complex $\mathrm{Hom}_R(C, G)$; see Chapter 12. The Ext functors were introduced in the context of algebraic topology by Eilenberg and Mac Lane (1942).

Everything we discussed so far is presented in Cartan and Eilenberg's groundbreaking book, Cartan–Eilenberg [Cartan and Eilenberg (1956)], published in 1956. It is in this book that the name *homological algebra* is introduced. MacLane [Mac Lane (1975)] (1975) and Rotman [Rotman (1979, 2009)] give more "gentle" presentations (see also Weibel [Weibel (1994)] and Eisenbud [Eisenbud (1995)]). A more sophisticated presentation of homological algebra is found in Gelfand and Manin [Gelfand and Manin (2003)].

Derived functors can be defined for functors $T \colon \mathbf{C} \to \mathbf{D}$, where $\mathbf{C}$ or $\mathbf{D}$ is a more general category than the category of R-modules or the category of abelian groups. For example, in sheaf cohomology, the category $\mathbf{C}$ is the category of sheaves of rings. In general, it suffices that **C** and **D** are abelian categories.

We say that **C** *has enough projectives* if every object in **C** is the image of some projective object in **C**, and that **C** *has enough injectives* if every object in **C** can be embedded (injectively) into some injective object in **C**.

There are situations (for example, when dealing with sheaves) where it is useful to know that right derived functors can be computed by resolutions involving objects that are not necessarily injective, but T-acyclic, as defined below.

Given a left-exact functor $T\colon \mathbf{C} \to \mathbf{D}$, an object $J \in \mathbf{C}$ is *T-acyclic* if $R^nT(J) = (0)$ for all $n \geq 1$.

The following proposition shows that right derived functors can be computed using T-acyclic resolutions.

Proposition *Given an additive left-exact functor $T\colon \mathbf{C} \to \mathbf{D}$, for any $A \in \mathbf{C}$ suppose there is an exact sequence*

$$0 \longrightarrow A \xrightarrow{\ \epsilon\ } J^0 \xrightarrow{\ d^0\ } J^1 \xrightarrow{\ d^1\ } J^2 \xrightarrow{\ d^2\ } \cdots \qquad (\dagger)$$

in which every J^n is T-acyclic (a right T-acyclic resolution $\mathbf{J}^A$ formed by truncating ($\dagger$) before J^0). Then for every $n \geq 0$ we have an isomorphism between $R^nT(A)$ and $H^n(T(\mathbf{J}_A))$.

The above proposition is used several times in Chapter 13.

1.8 Universal δ-Functors

The most important property of derived functors is that short exact sequences yield long exact sequences of homology or cohomology. This property was proven by Cartan and Eilenberg, but Grothendieck realized how crucial it was and this led him to the fundamental concept of *universal δ-functor*. Since we will be using right derived functors much more than left derived functors we state the existence of the long exact sequences of cohomology for right derived functors.

Theorem *Assume the abelian category **C** has enough injectives, let $0 \longrightarrow A' \longrightarrow A \longrightarrow A'' \longrightarrow 0$ be an exact sequence in **C**, and let $T\colon \mathbf{C} \to \mathbf{D}$ be a left-exact (additive) functor.*

(1) Then for every $n \geq 0$, there is a map

$$(R^nT)(A'') \xrightarrow{\ \delta^n\ } (R^{n+1}T)(A'),$$

and the sequence

$$
\begin{array}{l}
0 \longrightarrow T(A') \longrightarrow T(A) \longrightarrow T(A'') \xrightarrow{\delta^0} \\
(R^1T)(A') \longrightarrow \cdots \longrightarrow \cdots \longrightarrow \\
(R^nT)(A') \longrightarrow (R^nT)(A) \longrightarrow (R^nT)(A'') \xrightarrow{\delta^n} \\
(R^{n+1}T)(A') \longrightarrow \cdots \longrightarrow \cdots \longrightarrow \cdots
\end{array}
$$

is exact. This property is similar to the property of the zig-zag lemma from Section 1.2.

(2) If $0 \longrightarrow B' \longrightarrow B \longrightarrow B'' \longrightarrow 0$ *is another exact sequence in* $\mathbf{C}$, *and if there is a commutative diagram*

$$
\begin{array}{ccccccccc}
0 & \longrightarrow & A' & \longrightarrow & A & \longrightarrow & A'' & \longrightarrow & 0 \\
 & & \downarrow & & \downarrow & & \downarrow & & \\
0 & \longrightarrow & B' & \longrightarrow & B & \longrightarrow & B'' & \longrightarrow & 0,
\end{array}
$$

then the induced diagram beginning with

$$
\begin{array}{cccccccc}
0 & \longrightarrow & T(A') & \longrightarrow & T(A) & \longrightarrow & T(A'') & \xrightarrow{\delta_A^0} \\
 & & \downarrow & & \downarrow & & \downarrow & \\
0 & \longrightarrow & T(B') & \longrightarrow & T(B) & \longrightarrow & T(B'') & \xrightarrow[\delta_B^0]{}
\end{array}
$$

and continuing with

$$
\begin{array}{ccccccccccc}
\cdots & \longrightarrow & R^nT(A') & \longrightarrow & R^nT(A) & \longrightarrow & R^nT(A'') & \xrightarrow{\delta_A^n} & (R^{n+1}T)(A') & \longrightarrow & \cdots \\
 & & \downarrow & & \downarrow & & \downarrow & & \downarrow & & \\
\cdots & \longrightarrow & R^nT(B') & \longrightarrow & R^nT(B) & \longrightarrow & R^nT(B'') & \xrightarrow[\delta_B^n]{} & (R^{n+1}T)(B') & \longrightarrow & \cdots
\end{array}
$$

is also commutative.

The proof of this result (Theorem 11.31) is fairly involved and makes use of the horseshoe lemma (Theorem 11.25).

The previous theorem suggests the definition of families of functors originally proposed by Cartan and Eilenberg [Cartan and Eilenberg (1956)] and then investigated by Grothendieck in his legendary "Tohoku" paper [Grothendieck (1957)] (1957).

A *δ-functors* consists of a countable family $T = (T^n)_{n\geq 0}$ of functors $T^n\colon \mathbf{C} \to \mathbf{D}$ that satisfy the two conditions of the previous theorem. There is a notion of map, also called morphism, between δ-functors.

Given two δ-functors $S = (S^n)_{n\geq 0}$ and $T = (T^n)_{n\geq 0}$, a *morphism* $\eta\colon S \to T$ between S and T is a family $\eta = (\eta^n)_{n\geq 0}$ of natural transformations $\eta^n\colon S^n \to T^n$ such that a certain diagram commutes; see Definition 11.21.

Grothendieck also introduced the key notion of universal δ-functor; see Grothendieck [Grothendieck (1957)] (Chapter II, Section 2.2).

A δ-functor $T = (T^n)_{n\geq 0}$ is *universal* if for every δ-functor $S = (S^n)_{n\geq 0}$ and every natural transformation $\varphi\colon T^0 \to S^0$, there is a *unique* morphism $\eta\colon T \to S$ such that $\eta^0 = \varphi$; we say that η *lifts* φ.

The reason why universal δ-functors are important is the following kind of uniqueness property that shows that a universal δ-functor *is completely determined by the component* T^0; see Proposition 11.38.

Proposition *Suppose $S = (S^n)_{n\geq 0}$ and $T = (T^n)_{n\geq 0}$ are both universal δ-functors and there is an isomorphism $\varphi\colon S^0 \to T^0$ (a natural transformation φ which is an isomorphism). Then there is a unique isomorphism $\eta\colon S \to T$ lifting φ.*

One might wonder whether (universal) δ-functors exist. Indeed there are plenty of them; see Theorem 11.39.

Theorem *Assume the abelian category $\mathbf{C}$ has enough injectives. For every additive left-exact functor $T\colon \mathbf{C} \to \mathbf{D}$, the family $(R^nT)_{n\geq 0}$ of right derived functors of T is a δ-functor. Furthermore T is isomorphic to R^0T.*

In fact, the δ-functors $(R^nT)_{n\geq 0}$ are universal.

Grothendieck came up with an ingenious sufficient condition for a δ-functor to be universal: the notion of an *erasable* functor. Since Grothendieck's paper is written in French, this notion defined in Section 2.2 (Page 141) of [Grothendieck (1957)] is called *effaçable*, and many books and paper use it. Since the English translation of "effaçable" is "erasable," as advocated by Lang, we will use the English word.

A functor $T\colon \mathbf{C} \to \mathbf{D}$ is *erasable* (or *effaçable*) if for every object $A \in \mathbf{C}$ there is some object M_A and an injection $u\colon A \to M_A$ such that $T(u) = 0$. In particular this will be the case if $T(M_A)$ is the zero object of $\mathbf{D}$. If the category $\mathbf{C}$ has enough injectives, it can be shown that T is erasable iff $T(I) = (0)$ for all injectives I.

Our favorite functors, namely the right derived functors R^nT, are erasable by injectives for all $n \geq 1$. The following result due to Grothendieck is crucial:

Theorem *Assume the abelian category* $\mathbf{C}$ *has enough injectives. Let* $T = (T^n)_{n\geq 0}$ *be a* δ*-functor between two abelian categories* $\mathbf{C}$ *and* $\mathbf{D}$*. If* $T^n(I) = (0)$ *for every injective* I*, for all* $n \geq 1$*, then* T *is a universal* δ*-functor.*

Finally, by combining the previous results, we obtain the most important theorem about universal δ-functors:

Theorem *Assume the abelian category* $\mathbf{C}$ *has enough injectives. For every left-exact functor* $T\colon \mathbf{C} \to \mathbf{D}$*, the right derived functors* $(R^nT)_{n\geq 0}$ *form a universal* δ*-functor such that* T *is isomorphic to* R^0T*. Conversely, every universal* δ*-functor* $T = (T^n)_{n\geq 0}$ *is isomorphic to the right derived* δ*-functor* $(R^nT^0)_{n\geq 0}$*.*

After all, the mysterious universal δ-functors are just the right derived functors of left-exact functors. As an example, the functors $\mathrm{Ext}^n_R(A, -)$ constitute a universal δ-functor (for any fixed R-module A).

The machinery of universal δ*-functors can be used to prove that different kinds of cohomology theories yield isomorphic groups.* If two cohomology theories $(H^n_S(-))_{n\geq 0}$ and $(H^n_T(-))_{n\geq 0}$ defined for objects in a category $\mathbf{C}$ (say, topological spaces) are given by universal δ-functors S and T in the sense that the cohomology groups $H^n_S(A)$ and $H^n_T(A)$ are given by $H^n_S(A) = S^n(A)$ and $H^n_T(A) = T^n(A)$ for all objects $A \in \mathbf{C}$, and if $H^0_S(A)$ and $H^0_T(A)$ are isomorphic, then $H^n_S(A)$ and $H^n_T(A)$ are isomorphic for all $n \geq 0$. This technique will be used in Chapter 13 to prove that sheaf cohomology and Čech cohomology are isomorphic for paracompact spaces.

In Section 1.10 we will further see how the machinery of right derived functors can be used to define sheaf cohomology (where the category $\mathbf{C}$ is the category of sheaves of R-modules, the category $\mathbf{D}$ is the category of abelian groups, and T is the left exact "global section functor").

1.9 Universal Coefficient Theorems

Suppose we have a homology chain complex

$$0 \xleftarrow{d_0} C_0 \xleftarrow{d_1} C_1 \xleftarrow{d_{p-1}} \cdots\; C_{p-1} \xleftarrow{d_p} C_p \xleftarrow{d_{p+1}} C_{p+1} \longleftarrow \cdots,$$

where the C_i are R-modules over some commutative ring R with a multiplicative identity element (recall that $d_i \circ d_{i+1} = 0$ for all $i \geq 0$). Given another R-module G we can form the homology complex

$$0 \xleftarrow{d_0 \otimes \mathrm{id}} C_0 \otimes_R G \xleftarrow{d_1 \otimes \mathrm{id}} C_1 \otimes_R G \longleftarrow \cdots \xleftarrow{d_p \otimes \mathrm{id}} C_p \otimes_R G \longleftarrow \cdots,$$

obtained by tensoring with G, denoted $C \otimes_R G$, and the cohomology complex

$$0 \xrightarrow{\mathrm{Hom}_R(d_0,G)} \mathrm{Hom}_R(C_0, G) \longrightarrow \cdots \longrightarrow \mathrm{Hom}_R(C_p, G) \xrightarrow{\mathrm{Hom}_R(d_{p+1},G)} \mathrm{Hom}_R(C_{p+1}, G) \longrightarrow \cdots$$

obtained by applying $\mathrm{Hom}_R(-, G)$, and denoted $\mathrm{Hom}_R(C, G)$.

The question is: what is the relationship between the homology groups $H_p(C \otimes_R G)$ and the original homology groups $H_p(C)$ in the first case, and what is the relationship between the cohomology groups $H^p(\mathrm{Hom}_R(C, G))$ and the original homology groups $H_p(C)$ in the second case?

The ideal situation would be that

$$H_p(C \otimes_R G) \cong H_p(C) \otimes_R G \text{ and } H^p(\mathrm{Hom}_R(C, G)) \cong \mathrm{Hom}_R(H_p(C), G),$$

but this is generally not the case. If the ring R is nice enough and if the modules C_p are nice enough, then $H_p(C \otimes_R G)$ can be expressed in terms of $H_p(C) \otimes_R G$ and $\mathrm{Tor}_1^R(H_{p-1}(C), G)$, where $\mathrm{Tor}_1^R(-, G)$ is a one of the left-derived functors of $- \otimes_R G$, and $H^p(\mathrm{Hom}_R(C, G))$ can be expressed in terms of $\mathrm{Hom}_R(H_p(C), G))$ and $\mathrm{Ext}_R^1(H_{p-1}(C), G)$, where $\mathrm{Ext}_R^1(-, G)$ is one of the right-derived functors of $\mathrm{Hom}_R(-, G)$; both derived functors are defined in Section 11.2 and further discussed in Example 11.1. These formulae known as universal coefficient theorems are discussed in Chapter 12.

1.10 Sheaf Cohomology

Given a topological space X, we define the *global section functor* $\Gamma(X, -)$ such that for every sheaf of R-modules $\mathcal{F}$,

$$\Gamma(X, \mathcal{F}) = \mathcal{F}(X).$$

This is a functor from the category $\mathbf{Sh}(X)$ of sheaves of R-modules over X to the category of abelian groups.

A sheaf $\mathcal{I}$ is *injective* if for any injective sheaf map $h\colon \mathcal{F} \to \mathcal{G}$ and any sheaf map $f\colon \mathcal{F} \to \mathcal{I}$, there is some sheaf map $\widehat{f}\colon \mathcal{G} \to \mathcal{I}$ extending $f\colon \mathcal{F} \to \mathcal{I}$ in the sense that $f = \widehat{f} \circ h$, as in the following commutative diagram:

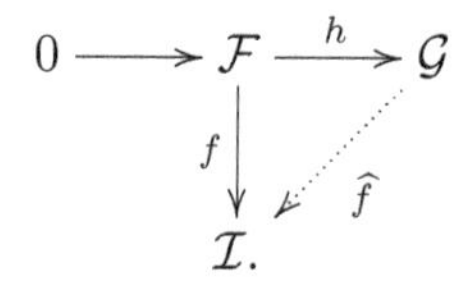

This is the same diagram that we used to define injective modules in Section 1.7, but here, the category involved is the category of sheaves.

A nice feature of the category of sheaves of R-modules is that its has enough injectives.

Proposition *For any sheaf $\mathcal{F}$ of R-modules, there is an injective sheaf $\mathcal{I}$ and an injective sheaf homomorphism $\varphi\colon \mathcal{F} \to \mathcal{I}$.*

As in the case of modules, the fact that the category of sheaves has enough injectives implies that any sheaf has an injective resolution.

On the other hand, the category of sheaves does not have enough projectives. This is the reason why projective resolutions of sheaves are of little interest.

Another good property is that the global section functor is left-exact. Then as in the case of modules in Section 1.7, the cohomology groups induced by the right derived functors $R^p\Gamma(X, -)$ are well defined.

The *cohomology groups* of the sheaf $\mathcal{F}$ (or the *cohomology groups of X with values in $\mathcal{F}$*), denoted by $H^p(X, \mathcal{F})$, are the groups $R^p\Gamma(X, -)(\mathcal{F})$ induced by the right derived functor $R^p\Gamma(X, -)$ (with $p \geq 0$).

To compute the sheaf cohomology groups $H^p(X, \mathcal{F})$, pick *any* resolution of $\mathcal{F}$

$$0 \longrightarrow \mathcal{F} \longrightarrow \mathcal{I}^0 \xrightarrow{d^0} \mathcal{I}^1 \xrightarrow{d^1} \mathcal{I}^2 \xrightarrow{d^2} \cdots$$

by injective sheaves $\mathcal{I}^n$, apply the global section functor $\Gamma(X, -)$ to obtain the complex of R-modules

$$0 \xrightarrow{\delta^{-1}} \mathcal{I}^0(X) \xrightarrow{\delta^0} \mathcal{I}^1(X) \xrightarrow{\delta^1} \mathcal{I}^2(X) \xrightarrow{\delta^2} \cdots,$$

and then

$$H^p(X, \mathcal{F}) = \operatorname{Ker} \delta^p / \operatorname{Im} \delta^{p-1}.$$

By Theorem 11.47 (stated in the previous section) the right derived functors $R^p\Gamma(X, -)$ constitute a universal δ-functor, so all the properties of δ-functors apply.

In principle, computing the cohomology groups $H^p(X, \mathcal{F})$ requires finding injective resolutions of sheaves. However injective sheaves are very big and hard to deal with. Fortunately, there is a class of sheaves known as *flasque* sheaves (due to Godement) which are $\Gamma(X, -)$-acyclic, and every sheaf has a resolution by flasque sheaves. Therefore, by Proposition 11.34 (stated in the previous section), the cohomology groups $H^p(X, \mathcal{F})$ can be computed using flasque resolutions.

Then we compare sheaf cohomology (defined by derived functors) to the other kinds of cohomology defined so far: de Rham, singular, Čech (for the constant sheaf $\widetilde{G}_X$).

If the space X is paracompact, then it turns out that for any sheaf $\mathcal{F}$, the Čech cohomology groups $\check{H}^p(X, \mathcal{F})$ are isomorphic to the cohomology groups $H^p(X, \mathcal{F})$. Furthermore, if $\mathcal{F}$ is a presheaf, then the Čech cohomology groups $\check{H}^p(X, \mathcal{F})$ and $\check{H}^p(X, \widetilde{\mathcal{F}})$ are isomorphic, where $\widetilde{\mathcal{F}}$ is the sheafification of $\mathcal{F}$. Several other results (due to Leray and Henri Cartan) about the relationship between Čech cohomology and sheaf cohomology will be stated.

When X is a topological manifold (thus paracompact), for every R-module G, we will show that the singular cohomology groups $H^p(X; G)$ are isomorphic to the cohomology groups $H^p(X, \widetilde{G}_X)$ of the constant sheaf $\widetilde{G}_X$. Technically, we will need to define *soft* and *fine* sheaves.

We will also define Alexander–Spanier cohomology and prove that it is equivalent to sheaf cohomology (and Čech cohomology) for paracompact spaces and for the constant sheaf $\widetilde{G}_X$.

In summary, for manifolds, singular cohomology, Čech cohomology, Alexander–Spanier cohomology, and sheaf cohomology all agree (for the constant sheaf $\widetilde{G}_X$). For smooth manifolds, we can add de Rham cohomology to the above list of equivalent cohomology theories, for the constant sheaf $\widetilde{\mathbb{R}}_X$. All these results are presented in Chapter 13.

1.11 Alexander and Alexander–Lefschetz Duality

The goal of Chapter 14 is to present various generalizations of Poincaré duality. These versions of duality involve taking direct limits of direct mapping families of singular cohomology groups which, in general, are not singular cohomology groups. However, such limits are isomorphic to Alexander–Spanier cohomology groups, and thus to Čech cohomology groups. These duality results also require relative versions of homology and cohomology.

1.12 Spectral Sequences

A spectral sequence is a tool of homological algebra whose purpose is to approximate the cohomology (or homology) $H(M)$ of a module M endowed with a family $(F^pM)_{p\in\mathbb{Z}}$ of submodules such that $F^{p+1}M \subseteq F^pM$ for all p and

$$M = \bigcup_{p\in\mathbb{Z}} F^pM,$$

called a filtration. The module M is also equipped with a linear map $d\colon M \to M$ called differential such that $d \circ d = 0$, so that it makes sense to define

$$H(M) = \operatorname{Ker} d/\operatorname{Im} d.$$

We say that (M, d) is a differential module. To be more precise, the filtration induces cohomology submodules $H(M)^p$ of $H(M)$, the images of $H(F^pM)$ in $H(M)$, and a spectral sequence is a sequence of modules E^p_r (equipped with a differential d^p_r), for $r \geq 1$, such that E^p_r approximates the "graded piece" $H(M)^p/H(M)^{p+1}$ of $H(M)$.

Actually, to be useful, the machinery of spectral sequences must be generalized to filtered cochain complexes. Technically this implies dealing with objects $E^{p,q}_r$ involving three indices, which makes its quite challenging to follow the exposition.

Many presentations jump immediately to the general case, but it seems pedagogically advantageous to begin with the simpler case of a single filtered differential module. This the approach followed by Serre in his dissertation [Serre (2003)] (Pages 24–104, *Annals of Mathematics*, 54 (1951),

425–505), Godement [Godement (1958)], and Cartan and Eilenberg [Cartan and Eilenberg (1956)]. Spectral sequences are discussed in great detail in Chapter 15.

There are several methods for defining spectral sequences, including the following three:

(1) Koszul's original approach as described by Serre [Serre (2003)] and Godement [Godement (1958)]. In our opinion it is the simplest method to understand what is going on.
(2) Cartan and Eilenberg's approach [Cartan and Eilenberg (1956)]. This is a somewhat faster and slicker method than the previous method.
(3) Exact couples of Massey (1952). This somewhat faster method for defining spectral sequences is adopted by Rotman [Rotman (1979, 2009)] and Bott and Tu [Bott and Tu (1986)]. Mac Lane [Mac Lane (1975)], Weibel [Weibel (1994)], and McCleary [McCleary (2001)] also present it and show its equivalence with the first approach. It appears to be favored by algebraic topologists. This approach leads to spectral sequences in a quicker fashion and is more general because exact couples need not arise from a filtration, but our feeling is that it is even more mysterious to a novice than the first two approaches.

We will primarily follow Method (1) and present Method (2) and Method (3) in starred sections (Method (2) in Section 15.15 and Method (3) in Section 15.14). All three methods produce isomorphic sequences, and we will show their equivalence.

1.13 Suggestions On How to Use This Book

This book basically consists of two parts. The first part covers fairly basic material presented in the first seven chapters. The second part deals with more sophisticated material including sheaves, derived functors, sheaf cohomology, and spectral sequences.

Chapter 3 on de Rham cohomology, Chapter 5 on simplicial homology and cohomology, and Chapter 6 on CW-complexes, are written in such a way that they are pretty much independent of each other and of the rest of book, and thus can be safely skipped. Readers who have never heard about differential forms can skip Chapter 3, although of course they will miss a nice facet of the global picture. Chapter 5 on simplicial homology and cohomology was included mostly for historical sake, and because they have a strong combinatorial and computational flavor. Chapter 6 on CW-complexes was included to show that there are tools for computing homology groups and to compensate for the lack of computational flavor of singular homology. However, CW-complexes can't really be understood without a good knowledge of singular homology.

Our feeling is that singular homology is simpler to define than the other homology theories, and since it is also more general, we decided to choose it as our first presentation of homology.

Our main goal is really to discuss cohomology, but except for de Rham cohomology, we feel that a two step process where we first present singular homology, and then singular cohomology as the result of applying the functor $\mathrm{Hom}(-, G)$, is less abrupt than discussing Čech cohomology (or Alexander–Spanier cohomology) first. If the reader prefers, he/she may to go directly to Chapter 9.

In any case, we highly recommend first reading the first four sections of Chapter 2. Sections 2.7 and 2.2 can be skipped upon first reading. Next, either proceed with Chapter 3, or skip it, but read Chapter 4 entirely.

After this, we recommend reading Chapter 7 on Poincaré duality, since this is one of the jewels of algebraic topology.

Knowledge about manifolds is not necessary to read this book but definitely useful since manifolds form a large class of spaces for which all the main cohomology theories are equivalent. Among the many books that cover manifolds, we suggest (in alphabetic order) Lee [Lee (2006)], Morita [Morita (2001)], Tu [Tu (2008)], and Warner [Warner (1983)]. A detailed presentation, first at a basic level and then at a more advanced level is also provided in Gallier and Quaintance [Gallier and Quaintance (2020a)]. Chapter 3 requires knowledge of differential forms on smooth manifolds. Differential forms are discussed in Tu [Tu (2008)], Morita [Morita (2001)], Madsen and Tornehave [Madsen and Tornehave (1998)], and Bott and Tu [Bott and Tu (1986)]. A detailed exposition, including an extensive review of tensor algebra, is also provided in Gallier and Quaintance [Gallier and Quaintance (2020b)]. A firm grasp of linear algebra and of some commutative algebra, at the level discussed in texts such as Artin [Artin (1991)] and Dummit and Foote [Dummit and Foote (1999)], is required.

The second part, starting with presheaves and sheaves in Chapter 8, relies on more algebra, especially Chapter 11 on derived functors and Chapter 15 on spectral sequences. However, this is some of the most beautiful material, so do not be discouraged if the going is tough. Skip proofs upon first reading and try to plow through as much as possible. Stop to take a break, and go back!

One of our goals is to fully prepare the reader to read books like Hartshorne [Hartshorne (1977)] (Chapter III). Others have expressed the same goal, and we hope to be more successful.

We have borrowed some proofs of Steve Shatz from Shatz and Gallier [Shatz and Gallier (2016)], and many proofs in Chapter 11 are borrowed from Rotman [Rotman (1979, 2009)]. Generally, we relied heavily on Bott and Tu [Bott and Tu (1986)], Bredon [Bredon (1993)], Godement [Godement (1958)], Hatcher [Hatcher (2002)], Milnor and Stasheff [Milnor and Stasheff (1974)], Munkres [Munkres (1984)], Serre [Serre (1955)], Spanier [Spanier (1989)], Tennison [Tennison (1975)], and Warner [Warner (1983)]. These are wonderful books, and we hope that reading our book will prepare the reader to study them. We express our gratitude to these authors, and to all the others that have inspired us (including, of course, Dieudonné).

Since we made the decision not to include all proofs (this would have doubled if not tripled the size of the book!), we tried very hard to provide precise pointers to all omitted proofs. This may be irritating to the expert, but we believe that a reader with less knowledge will appreciate this. The reason for including a proof is that we feel that it presents a type of argument that the reader should be exposed to, but this often subjective and a reflection of our personal taste. When we omitted a proof, we tried to give an idea of what it would be, except when it was a really difficult proof. This should be an incentive for the reader to dig into these references.

Chapter 2

Homology and Cohomology

This chapter is an introduction to the crucial concepts and results of homological algebra needed to understand homology and cohomology in some depth. The two most fundamental concepts of homological algebra are:

(1) exact sequences.
(2) chain complexes.

Exact sequences are special kinds of chain complexes satisfying additional properties and the purpose of cohomology (and homology) is to "measure" the extent to which a chain complex fails to be an exact sequence. Remarkably, when this machinery is applied to topological spaces or manifolds, it yields some valuable topological information about these spaces.

In their simplest form chain complexes and exact sequences are built from vector spaces but a more powerful theory is obtained (at the cost of minor complications) if the vector spaces are replaced by R-modules, where R is a commutative ring with a multiplicative identity element $1 \neq 0$. In particular, if $R = \mathbb{Z}$, then each space is just an abelian group. By a linear map we mean an R-linear map.

In Section 2.1 we introduce exact sequences and prove some of their most basic properties. In this chapter we prove two of their most important properties, namely the "five lemma" and the "zig-zag lemma" for cohomology, or long exact sequence of cohomology.

2.1 Exact Sequences and Short Exact Sequences

We begin with the notion of exact sequence.

Definition 2.1. A $\mathbb{Z}$-indexed sequence of R-modules and R-linear maps between them

$$\cdots \longrightarrow A_{p-1} \xrightarrow{f_{p-1}} A_p \xrightarrow{f_p} A_{p+1} \xrightarrow{f_{p+1}} A_{p+2} \longrightarrow \cdots$$

is *exact* if $\mathrm{Im}\, f_p = \mathrm{Ker}\, f_{p+1}$[1] for all $p \in \mathbb{Z}$. A sequence of R-modules

$$0 \longrightarrow A \xrightarrow{f} B \xrightarrow{g} C \longrightarrow 0$$

is a *short exact sequence* if it is exact at A, B, C, which means that

(1) $\mathrm{Im}\, f = \mathrm{Ker}\, g$.
(2) f is injective.
(3) g is surjective.

Observe that being exact at A_{p+1}, that is $\mathrm{Im}\, f_p = \mathrm{Ker}\, f_{p+1}$, implies that $f_{p+1} \circ f_p = 0$.

Given a short exact sequence

$$0 \longrightarrow A \xrightarrow{f} B \xrightarrow{g} C \longrightarrow 0,$$

since g is surjective, f is injective, and $\mathrm{Im}\, f = \mathrm{Ker}\, g$, by the first isomorphism theorem we have

$$C \cong B/\mathrm{Ker}\, g = B/\mathrm{Im}\, f \cong B/A.$$

Thus a short exact sequence amounts to a module B, a submodule A of B, and the quotient module $C \cong B/A$.

The quotient module $B/\mathrm{Im}\, f$ associated with the R-linear map $f\colon A \to B$ is a kind of "dual" of the submodule $\mathrm{Ker}\, f$ which often comes up when dealing with exact sequences.

Definition 2.2. Given any R-linear map $f\colon A \to B$, the quotient module $B/\mathrm{Im}\, f$ is called the *cokernel* of f and is denoted by $\mathrm{Coker}\, f$.

Observe that $\mathrm{Coker}\, f = B/\mathrm{Im}\, f \cong C = \mathrm{Im}\, g$. Then given an exact sequence

$$\cdots \longrightarrow A_{p-2} \xrightarrow{f_{p-2}} A_{p-1} \xrightarrow{f_{p-1}} A_p \xrightarrow{f_p} A_{p+1} \xrightarrow{f_{p+1}} A_{p+2} \longrightarrow \cdots,$$

we obtain short exact sequences as follows: if we focus on A_p, then there is a surjection $A_p \longrightarrow \mathrm{Im}\, f_p$, and since $\mathrm{Im}\, f_p = \mathrm{Ker}\, f_{p+1}$ this is a surjection

[1]A good mnemonic for this equation is *ikea*; *i* is the first letter in Im, and *k* is the first letter in Ker.

$A_p \longrightarrow \operatorname{Ker} f_{p+1}$, and by the first isomorphism theorem and since $\operatorname{Im} f_{p-1} = \operatorname{Ker} f_p$, we have an isomorphism

$$A_p/\operatorname{Im} f_{p-1} = A_p/\operatorname{Ker} f_p \cong \operatorname{Im} f_p = \operatorname{Ker} f_{p+1}.$$

This means that we have the short exact sequence

$$0 \longrightarrow \operatorname{Im} f_{p-1} \longrightarrow A_p \longrightarrow \operatorname{Ker} f_{p+1} \longrightarrow 0. \qquad (*_{\mathrm{Im}})$$

By a previous remark $\operatorname{Coker} f_{p-2} \cong \operatorname{Im} f_{p-1}$, so we obtain the short exact sequence

$$0 \longrightarrow \operatorname{Coker} f_{p-2} \longrightarrow A_p \longrightarrow \operatorname{Ker} f_{p+1} \longrightarrow 0. \qquad (*_{\mathrm{cok}})$$

Short exact sequences of this kind often come up in proofs (for example, the universal coefficient theorems).

If we are dealing with vector spaces (that is, if R is a field), then a standard result of linear algebra asserts that the isomorphism $A_p/\operatorname{Ker} f_p \cong \operatorname{Im} f_p$ yields the direct sum

$$A_p \cong \operatorname{Ker} f_p \oplus \operatorname{Im} f_p = \operatorname{Im} f_{p-1} \oplus \operatorname{Im} f_p.$$

As a consequence, if A_{p-1} and A_{p+1} are finite-dimensional, then so is A_p.

Some of the fundamental and heavily used results about exact sequences include the "zig-zag lemma" and the "five lemma." We will encounter these lemmas later on. The following (apparently unnamed) result is also used a lot.

Proposition 2.1. *Consider any diagram of R-modules*

$$\begin{array}{ccccc} A & \xrightarrow{f} & B & \xrightarrow{g} & C \\ \Big\downarrow{\scriptstyle \alpha} & & \Big\downarrow{\scriptstyle \beta} & & \Big\downarrow{\scriptstyle \gamma} \\ A' & \xrightarrow[f']{} & B' & \xrightarrow[g']{} & C \end{array}$$

in which the left and right squares commute and α, β, γ are isomorphisms. If the top row is exact, then the bottom row is also exact.

Proof. The commutativity of the left and right squares implies that

$$\gamma \circ g \circ f = g' \circ f' \circ \alpha.$$

Since the top row is exact, $g \circ f = 0$, so $g' \circ f' \circ \alpha = 0$, and since α is an isomorphism, $g' \circ f' = 0$. It follows that $\operatorname{Im} f' \subseteq \operatorname{Ker} g'$.

Conversely assume that $b' \in \operatorname{Ker} g'$. Since β is an isomorphism there is some $b \in B$ such that $\beta(b) = b'$, and since $g'(b') = 0$ we have

$$(g' \circ \beta)(b) = 0.$$

Since the right square commutes, $g' \circ \beta = \gamma \circ g$, so

$$(\gamma \circ g)(b) = 0.$$

Since γ is an isomorphism, $g(b) = 0$. Since the top row is exact, $\operatorname{Im} f = \operatorname{Ker} g$, so there is some $a \in A$ such that $f(a) = b$, which implies that

$$(\beta \circ f)(a) = \beta(b) = b'.$$

Since the left square commutes $\beta \circ f = f' \circ \alpha$, and we deduce that

$$f'(\alpha(a)) = b',$$

which proves that $\operatorname{Ker} g' \subseteq \operatorname{Im} f'$. Therefore, $\operatorname{Im} f' \subseteq \operatorname{Ker} g'$, as claimed. ☐

When the R-module C is free, a short exact sequence

$$0 \longrightarrow A \xrightarrow{f} B \xrightarrow{g} C \longrightarrow 0$$

has some special properties that play a crucial role when we dualize such a sequence.

Definition 2.3. A short exact sequence of R-modules

$$0 \longrightarrow A \xrightarrow{f} B \xrightarrow{g} C \longrightarrow 0$$

is said to *split* (or to be a *short split exact sequence*) if the submodule $f(A)$ is a direct summand in B, which means that B is a direct sum $B = f(A) \oplus D$ for some submodule D of B.

If a short exact sequence as in Definition 2.3 splits, since $\operatorname{Im} f = \operatorname{Ker} g$, f is injective and g is surjective, then the restriction of g to D is a bijection onto C so there is an isomorphism $\theta\colon B \to A \oplus C$ defined so that the restriction of θ to $f(A)$ is equal to f^{-1} and the restriction of θ to C is equal to g.

Proposition 2.2. *Let*

$$0 \longrightarrow A \xrightarrow{f} B \xrightarrow{g} C \longrightarrow 0$$

be a short exact sequence of R-modules. The following properties are equivalent.

(1) The sequence splits.
(2) There is a linear map $p\colon B \to A$ such that $p \circ f = \mathrm{id}_A$.
(3) There is a linear map $j\colon C \to B$ such that $g \circ j = \mathrm{id}_C$.

Symbolically, we have the following diagram of linear maps:

$$0 \longrightarrow A \underset{p}{\overset{f}{\rightleftarrows}} B \underset{j}{\overset{g}{\rightleftarrows}} C \longrightarrow 0.$$

Proof. It is easy to prove that (1) implies (2) and (3). Since $B = f(A) \oplus D$ for some submodule D of B, if $\pi_1\colon A \oplus D \to A$ is the first projection and $f^{-1} \oplus \mathrm{id}_D\colon f(A) \oplus D \to A \oplus D$ be the isomorphism induced by f^{-1}, then let $p = \pi_1 \circ (f^{-1} \oplus \mathrm{id}_D)$. It is clear that $p \circ f = \pi_1 \circ (f^{-1} \oplus \mathrm{id}_D) \circ f = \mathrm{id}_A$. Define $j\colon C \to D$ as the inverse of the restriction of g to D (which is bijective, as we said earlier). Obviously $g \circ j = \mathrm{id}_C$.

If (2) holds, let us prove that

$$B = f(A) \oplus \mathrm{Ker}\, p.$$

For any $b \in B$, we can write $b = f(p(b)) + (b - f(p(b)))$. Obviously $f(p(b)) \in f(A)$, and since $p \circ f = \mathrm{id}_A$ we have

$$p(b - f(p(b))) = p(b) - p(f(p(b))) = p(b) - (p \circ f)(p(b)) = p(b) - p(b) = 0,$$

so $(b - f(p(b))) \in \mathrm{Ker}\, p$, which shows that $B = f(A) + \mathrm{Ker}\, p$. If $b \in f(A) \cap \mathrm{Ker}\, p$, then $b = f(a)$ for some $a \in A$, so $0 = p(b) = p(f(a)) = a$, and thus $b = f(0) = 0$. We conclude that $B = f(A) \oplus \mathrm{Ker}\, p$, as claimed.

If (3) holds, let us prove that

$$B = f(A) \oplus \mathrm{Im}\, j.$$

Since $\mathrm{Im}\, f = \mathrm{Ker}\, g$, this is equivalent to

$$B = \mathrm{Ker}\, g \oplus \mathrm{Im}\, j.$$

For any $b \in B$, we can write $b = (b - j(g(b))) + j(g(b))$. Clearly $j(g(b)) \in \mathrm{Im}\, j$, and since $g \circ j = \mathrm{id}_C$ we have

$$g(b - j(g(b)) = g(b) - g(j(g(b))) = g(b) - (g \circ j)(g(b)) = g(b) - g(b) = 0,$$

so $(b - j(g(b))) \in \mathrm{Ker}\, g$. If $b \in \mathrm{Ker}\, g \cap \mathrm{Im}\, j$, then $b = j(c)$ for some $c \in C$, and so $0 = g(b) = g(j(c)) = c$, thus $b = j(c) = j(0) = 0$. We conclude that $B = \mathrm{Ker}\, g \oplus \mathrm{Im}\, j$. ☐

Corollary 2.3. *Let*

$$0 \longrightarrow A \stackrel{f}{\longrightarrow} B \stackrel{g}{\longrightarrow} C \longrightarrow 0$$

be a short exact sequence of R-modules. If C is free, then the exact sequence splits.

Proof. Pick a basis $(e_i)_{i\in I}$ in C. Define the linear map $j\colon C \to B$ by choosing any vector $b_i \in B$ such that $g(b_i) = e_i$ (since g is surjective, this is possible) and setting $j(e_i) = b_i$. Then

$$(g \circ j)(e_i) = g(b_i) = e_i,$$

so $g \circ j = \mathrm{id}_C$, and by Proposition 2.3 the sequence splits since (3) implies (1). $\square$

The following example is an exact sequence of abelian groups ($\mathbb{Z}$-modules) that does not split

$$0 \longrightarrow m\mathbb{Z} \stackrel{i}{\longrightarrow} \mathbb{Z} \stackrel{\pi}{\longrightarrow} \mathbb{Z}/m\mathbb{Z} \longrightarrow 0,$$

where i is the inclusion map and π is the projection map such that $\pi(n) = n \bmod m$, the residue of n modulo m (with $m \geq 1$). Indeed, any surjective homomorphism p from $\mathbb{Z}$ to $m\mathbb{Z}$ would have to map 1 to m, but then $p \circ i \neq \mathrm{id}$.

Any decent introduction to homological algebra must discuss the "five lemma" (due to Steenrod). Together with the zig-zag lemma discussed in Section 2.7, this one of its most useful results.

2.2 The Five Lemma

As a warm up, let us consider the "short five lemma," from Mac Lane [Mac Lane (1975)] (Chapter I, Section 3, Lemma 3.1).

Proposition 2.4. *(Short Five Lemma) Consider the following diagram (of R-modules) in which the rows are exact and all the squares commute.*

$$\begin{array}{ccccccccc} 0 & \longrightarrow & A & \stackrel{f}{\longrightarrow} & B & \stackrel{g}{\longrightarrow} & C & \longrightarrow & 0 \\ & & \downarrow{\scriptstyle \alpha} & & \downarrow{\scriptstyle \beta} & & \downarrow{\scriptstyle \gamma} & & \\ 0 & \longrightarrow & A' & \xrightarrow[f']{} & B' & \xrightarrow[g']{} & C' & \longrightarrow & 0 \end{array}$$

If α and γ are isomorphisms, then β is also an isomorphism.

Proof. First we prove that β is injective. Assume that $\beta(b) = 0$ for some $b \in B$. Then $g'(\beta(b)) = 0$, and since the right square commutes, $0 = g'(\beta(b)) = \gamma(g(b))$. Since γ is injective (it is an isomorphism), $\gamma(g(b)) = 0$ implies that

$$g(b) = 0.$$

Since the top row is exact and $b \in \operatorname{Ker} g = \operatorname{Im} f$, there is some $a \in A$ such that

$$f(a) = b. \tag{$*_1$}$$

Here is a summary of the situation so far.

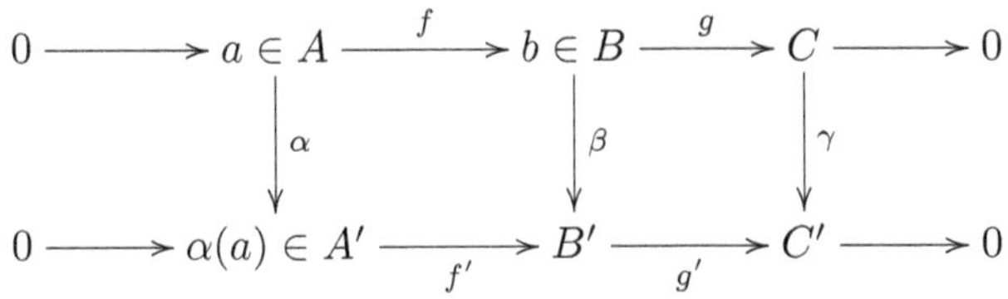

$$\begin{array}{ccccccccc}
0 & \longrightarrow & a \in A & \xrightarrow{f} & b \in B & \xrightarrow{g} & C & \longrightarrow & 0 \\
 & & \downarrow{\scriptstyle\alpha} & & \downarrow{\scriptstyle\beta} & & \downarrow{\scriptstyle\gamma} & & \\
0 & \longrightarrow & \alpha(a) \in A' & \xrightarrow[f']{} & B' & \xrightarrow[g']{} & C' & \longrightarrow & 0
\end{array}$$

Since the left square commutes, using $(*_1)$ we have

$$f'(\alpha(a)) = \beta(f(a)) = \beta(b) = 0.$$

Since the bottom row is exact, f' is injective so $\alpha(a) = 0$, and since α is injective (it is an isomorphism), $a = 0$. But then by $(*_1)$ we have $b = f(a) = 0$, which shows that β is injective.

We now prove that β is surjective. Pick any $b' \in B'$. Since γ is surjective (it is an isomorphism), there is some $c \in C$ such that

$$\gamma(c) = g'(b'). \tag{$*_2$}$$

Since the top row is exact, g is surjective so there is some $b \in B$ such that

$$g(b) = c. \tag{$*_3$}$$

Since the right square commutes, by $(*_2)$ and $(*_3)$ we have

$$g'(\beta(b)) = \gamma(g(b)) = \gamma(c) = g'(b'),$$

which implies $g'(\beta(b) - b') = 0$. Since the bottom row is exact and $\beta(b) - b' \in \operatorname{Ker} g' = \operatorname{Im} f'$ there is some $a \in A'$ such that

$$f'(a') = \beta(b) - b'. \tag{$*_4$}$$

Since α is surjective (it is an isomorphism), there is some $a \in A$ such that

$$\alpha(a) = a'. \tag{$*_5$}$$

Here is a summary of the situation so far.

$$\begin{array}{ccccccccc} 0 & \longrightarrow & a\in A & \xrightarrow{f} & b \in B & \xrightarrow{g} & c\in C & \longrightarrow & 0\\ & & \downarrow{\scriptstyle\alpha} & & \downarrow{\scriptstyle\beta} & & \downarrow{\scriptstyle\gamma} & & \\ 0 & \longrightarrow & a'\in A' & \xrightarrow[f']{} & \beta(b)-b'\in B' & \xrightarrow[g']{} & g'(b')\in C' & \longrightarrow & 0\end{array}$$

Since the left square commutes, using $(*_4)$ and $(*_5)$ we obtain

$$\beta(f(a)) = f'(\alpha(a)) = f'(a') = \beta(b) - b',$$

which implies that $b' = \beta(b - f(a))$, showing that β is surjective. $\square$

Observe that the proof shows that if α and γ are injective, then β is injective, and if α and γ are surjective, then β is surjective.

Proposition 2.5. *(Five Lemma) Consider the following diagram (of R-modules) in which the rows are exact and all the squares commute.*

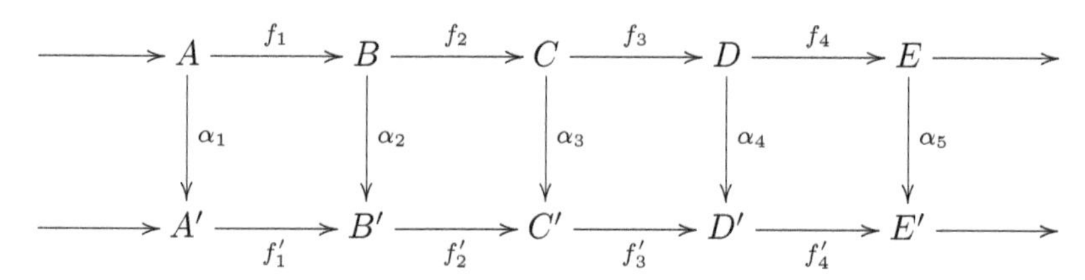

If $\alpha_1, \alpha_2, \alpha_4, \alpha_5$ are isomorphisms, then α_3 is also an isomorphism.

Proof. The proof of Proposition 2.5 can be found in any book on homological algebra, for example Mac Lane [Mac Lane (1975)], Cartan–Eilenberg [Cartan and Eilenberg (1956)], and Rotman [Rotman (1988, 2009)], but the reader may be put off by the fact that half of the proof is left to the reader (at least, Rotman proves the surjectivity part, which is slightly harder, and Mac Lane gives a complete proof of the short five lemma). The five lemma is fully proven in Spanier [Spanier (1989)] and Hatcher [Hatcher (2002)]. Because it is a "fun" proof by diagram-chasing we present the proof in Spanier [Spanier (1989)] (Chapter 4, Section 5, Lemma 11).

First we prove that α_3 is injective. Assume that $\alpha_3(c) = 0$ for some $c \in C$. Then $f_3' \circ \alpha_3(c) = 0$, and by commutativity of the third square, $\alpha_4 \circ f_3(c) = 0$. Since α_4 is injective (it is an isomorphism),

$$f_3(c) = 0.$$

Since the top row is exact and $c \in \operatorname{Ker} f_3 = \operatorname{Im} f_2$, there is some $b \in B$ such that

$$f_2(b) = c.$$

Since the second square commutes,

$$f_2' \circ \alpha_2(b) = \alpha_3 \circ f_2(b) = \alpha_3(c) = 0,$$

and since the bottom is exact and $\alpha_2(b) \in \operatorname{Ker} f_2' = \operatorname{Im} f_1'$, there is some $a' \in A'$ such that

$$f_1'(a') = \alpha_2(b). \tag{$*_1$}$$

Since α_1 is surjective (it is an isomorphism) there is some $a \in A$ such that

$$\alpha_1(a) = a'.$$

Here is a summary of the situation so far.

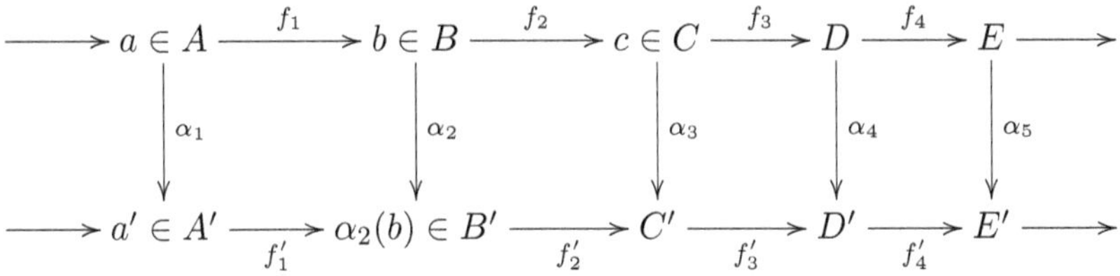

$$\begin{array}{ccccccccccc}
\longrightarrow & a \in A & \xrightarrow{f_1} & b \in B & \xrightarrow{f_2} & c \in C & \xrightarrow{f_3} & D & \xrightarrow{f_4} & E & \longrightarrow \\
& \downarrow{\scriptstyle \alpha_1} & & \downarrow{\scriptstyle \alpha_2} & & \downarrow{\scriptstyle \alpha_3} & & \downarrow{\scriptstyle \alpha_4} & & \downarrow{\scriptstyle \alpha_5} & \\
\longrightarrow & a' \in A' & \xrightarrow[f_1']{} & \alpha_2(b) \in B' & \xrightarrow[f_2']{} & C' & \xrightarrow[f_3']{} & D' & \xrightarrow[f_4']{} & E' & \longrightarrow
\end{array}$$

By the commutativity of the first square and ($*_1$),

$$\alpha_2 \circ f_1(a) = f_1' \circ \alpha_1(a) = f_1'(a') = \alpha_2(b),$$

and since α_2 is injective (it is an isomorphism), $b = f_1(a)$. Since the top row is exact $f_2 \circ f_1 = 0$, so

$$c = f_2(b) = f_2 \circ f_1(a) = 0,$$

proving that α_3 is injective.

Next we prove that α_3 is surjective. Pick $c' \in C'$. Since α_4 is surjective (it is an isomorphism) there is some $d \in D$ such that

$$\alpha_4(d) = f_3'(c'). \tag{$*_2$}$$

Since the bottom row is exact $f_4' \circ f_3' = 0$ and since the fourth square commutes we have

$$0 = f_4' \circ f_3'(c') = f_4' \circ \alpha_4(d) = \alpha_5 \circ f_4(d).$$

Since α_5 is injective (it is an isomorphism),

$$f_4(d) = 0,$$

and since the top row is exact and $d \in \operatorname{Ker} f_4 = \operatorname{Im} f_3$, there is some $c \in C$ such that

$$f_3(c) = d. \tag{$*_3$}$$

Since the third square commutes, using $(*_3)$ and $(*_2)$ we have

$$f_3' \circ \alpha_3(c) = \alpha_4 \circ f_3(c) = \alpha_4(d) = f_3'(c'),$$

so $f_3'(\alpha_3(c) - c') = 0$. Since the bottom row is exact and $\alpha_3(c) - c' \in \operatorname{Ker} f_3' = \operatorname{Im} f_2'$, there is some $b' \in B'$ such that

$$f_2'(b') = \alpha_3(c) - c'. \tag{$*_4$}$$

Since α_2 is surjective (it is an isomorphism) there is some $b \in B$ such that

$$\alpha_2(b) = b'. \tag{$*_5$}$$

Here is a summary of the situation so far.

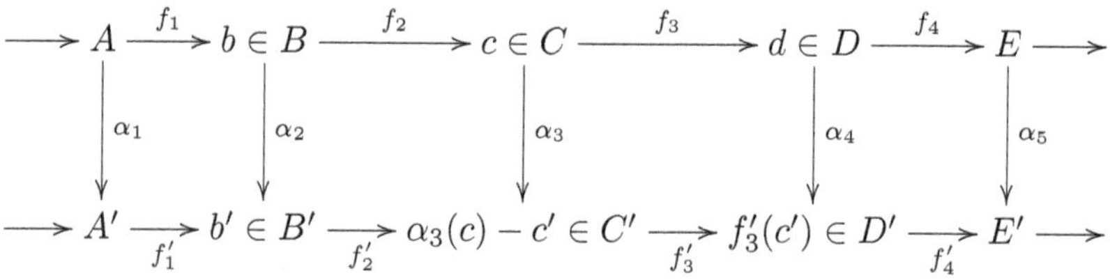

$$\begin{array}{ccccccccccc}
\longrightarrow A & \xrightarrow{f_1} & b \in B & \xrightarrow{f_2} & c \in C & \xrightarrow{f_3} & d \in D & \xrightarrow{f_4} & E \longrightarrow \\
\downarrow \alpha_1 & & \downarrow \alpha_2 & & \downarrow \alpha_3 & & \downarrow \alpha_4 & & \downarrow \alpha_5 \\
\longrightarrow A' & \xrightarrow[f_1']{} & b' \in B' & \xrightarrow[f_2']{} & \alpha_3(c) - c' \in C' & \xrightarrow[f_3']{} & f_3'(c') \in D' & \xrightarrow[f_4']{} & E' \longrightarrow
\end{array}$$

Then using $(*_4)$ and $(*_5)$ and the fact that the second square commutes we have

$$\alpha_3(f_2(b)) = f_2'(\alpha_2(b)) = f_2'(b') = \alpha_3(c) - c',$$

which implies that $c' = \alpha_3(c - f_2(b))$, showing that α_3 is surjective. □

Remark: The hypotheses of the five lemma can be weakened. One can check that the proof goes through if α_2 and α_4 are isomorphisms, α_1 is surjective, and α_5 is injective.

2.3 Duality and Exactness

A common way to define cohomology is to apply duality to homology so we review duality in R-modules to make sure that we are on firm grounds.

Definition 2.4. Given an R-module A, the R-module $\operatorname{Hom}(A, R)$ of all R-linear maps from A to R (also called *R-linear forms*) is called the *dual* of A. Given any two R-modules A and B, for any R-linear map $f\colon A \to B$, the R-linear map $f^\top\colon \mathrm{Hom}(B,R) \to \mathrm{Hom}(A,R)$ defined by

$$f^\top(\varphi) = \varphi \circ f \quad \text{for all } \varphi \in \mathrm{Hom}(B,R)$$

is called the *dual linear map* of f; see the commutative diagram below:

$$\begin{array}{ccc} A & \xrightarrow{f} & B \\ & \searrow^{f^\top(\varphi)} & \downarrow \varphi \\ & & R. \end{array}$$

The dual linear map $f^\top$ is also denoted by $\mathrm{Hom}(f,R)$ (or $\mathrm{Hom}(f,\mathrm{id}_R)$).

If $f\colon A \to B$ and $g\colon B \to C$ are linear maps of R-modules, a simple computation shows that

$$(g \circ f)^\top = f^\top \circ g^\top.$$

Note the reversal in the order of composition of $f^\top$ and $g^\top$. It is also immediately verified that

$$\mathrm{id}_A^\top = \mathrm{id}_{\mathrm{Hom}(A,R)}.$$

Here are some basic properties of the behavior of duality applied to exact sequences.

Proposition 2.6. *Let $g\colon B \to C$ be a linear map between R-modules.*

(a) If g is an isomorphism, then so is $g^\top$.
(b) If g is the zero map, then so is $g^\top$.
(c) If the sequence

$$B \xrightarrow{g} C \longrightarrow 0$$

is exact, then the sequence

$$0 \longrightarrow \mathrm{Hom}(C,R) \xrightarrow{g^\top} \mathrm{Hom}(B,R)$$

is also exact.

Proof. Properties (a) and (b) are immediate and left as an exercise.

Assume that the sequence $B \xrightarrow{g} C \longrightarrow 0$ is exact which means that g is surjective. Let $\psi \in \mathrm{Hom}(C,R)$ and assume that $g^\top(\psi) = 0$, which means that $\psi \circ g = 0$, that is, $\psi(g(b)) = 0$ for all $b \in B$. Since g is surjective, we have $\psi(c) = 0$ for all $c \in C$, that is, $\psi = 0$ and $g^\top$ is injective. $\square$

Proposition 2.7. *If the following sequence of R-modules*

$$A \xrightarrow{f} B \xrightarrow{g} C \longrightarrow 0$$

is exact, then the sequence

$$0 \longrightarrow \mathrm{Hom}(C,R) \xrightarrow{g^\top} \mathrm{Hom}(B,R) \xrightarrow{f^\top} \mathrm{Hom}(A,R)$$

is also exact. Furthermore, if

$$0 \longrightarrow A \xrightarrow{f} B \xrightarrow{g} C \longrightarrow 0$$

is a split short exact sequence, then

$$0 \longrightarrow \mathrm{Hom}(C,R) \xrightarrow{g^\top} \mathrm{Hom}(B,R) \xrightarrow{f^\top} \mathrm{Hom}(A,R) \longrightarrow 0$$

is also a split short exact sequence.

Proof. Since g is surjective, by Proposition 2.6(c), $g^\top$ is injective. Since $\mathrm{Im}\, f = \mathrm{Ker}\, g$, we have $g \circ f = 0$, so $f^\top \circ g^\top = 0$, which shows that $\mathrm{Im}\, g^\top \subseteq \mathrm{Ker}\, f^\top$. Conversely, we prove that if $f^\top(\psi) = 0$ for some $\psi \in \mathrm{Hom}(B,R)$, then $\psi = g^\top(\varphi)$ for some $\varphi \in \mathrm{Hom}(C,R)$.

Since $f^\top(\psi) = \psi \circ f$, if $f^\top(\psi) = 0$, then ψ vanishes on $f(A)$. Thus ψ induces a linear map $\psi' \colon B/f(A) \to R$ such that $\psi = \psi' \circ \pi$ where $\pi \colon B \to B/f(A)$ is the canonical projection. The exactness of the sequence implies that g induces an isomorphism $g' \colon B/f(A) \to C$, and we have the following commutative diagram:

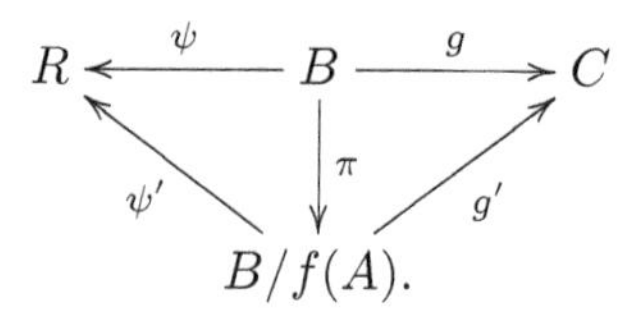

If we let $\varphi = \psi' \circ (g')^{-1}$, then we have a linear form $\varphi \in \mathrm{Hom}(C,R)$, and

$$g^\top(\varphi) = \varphi \circ g = \psi' \circ (g')^{-1} \circ g = \psi,$$

as desired. Therefore, the dual sequence is exact at $\mathrm{Hom}(B,R)$.

If our short exact sequence is split, then by Proposition 2.2 there is a map $p \colon B \to A$ such that $p \circ f = \mathrm{id}_A$, so we get $f^\top \circ p^\top = \mathrm{id}_{\mathrm{Hom}(A,R)}$, which shows that $f^\top$ is surjective, and $p^\top \colon \mathrm{Hom}(A,R) \to \mathrm{Hom}(B,R)$ splits the dual sequence. ☐

If $f \colon A \to B$ is injective, then $f^\top \colon \mathrm{Hom}(B,R) \to \mathrm{Hom}(A,R)$ is not necessarily surjective. For example, we have the following short exact sequence

$$0 \longrightarrow \mathbb{Z} \xrightarrow{\times 2} \mathbb{Z} \xrightarrow{\pi} \mathbb{Z}/2\mathbb{Z} \longrightarrow 0,$$

where $\times 2(n) = 2n$, but the map $(\times 2)^\top$ is not surjective. This is because for any $\varphi \in \mathrm{Hom}(\mathbb{Z},\mathbb{Z})$ we have $(\times 2)^\top(\varphi) = \varphi \circ \times 2$ and this function maps $\mathbb{Z}$ into $2\mathbb{Z}$. Thus the image of $(\times 2)^\top$ is not all of $\mathrm{Hom}(\mathbb{Z},\mathbb{Z})$.

Combining Corollary 2.3 and Proposition 2.7 we get the following result.

Proposition 2.8. *If*

$$0 \longrightarrow A \xrightarrow{f} B \xrightarrow{g} C \longrightarrow 0$$

is a short exact sequence and if C is a free R-module, then

$$0 \longrightarrow \mathrm{Hom}(C,R) \xrightarrow{g^\top} \mathrm{Hom}(B,R) \xrightarrow{f^\top} \mathrm{Hom}(A,R) \longrightarrow 0$$

is a split short exact sequence.

The proposition below will be needed in the proof of the universal coefficient theorem for cohomology (Theorem 12.6).

Let M and G be R-modules, and let $B \subseteq Z \subseteq M$ be some submodules of M. Define B^0 and Z^0 by

$$\begin{aligned} B^0 &= \{\varphi \in \mathrm{Hom}(M,G) \mid \varphi(a) = 0 \quad \text{for all } b \in B\} \\ Z^0 &= \{\varphi \in \mathrm{Hom}(M,G) \mid \varphi(z) = 0 \quad \text{for all } z \in Z\}. \end{aligned}$$

Proposition 2.9. *For any R-modules M, G, and $B \subseteq Z \subseteq M$, if $M = Z \oplus Z'$ for some submodule Z' of M, then we have an isomorphism*

$$\mathrm{Hom}(Z/B,G) \cong B^0/Z^0.$$

Proof. Define a map $\eta \colon B^0 \to \mathrm{Hom}(Z/B,G)$ as follows: for any $\varphi \in B^0$, that is any $\varphi \in \mathrm{Hom}(M,G)$ such that φ vanishes on B, let $\eta(\varphi) \in \mathrm{Hom}(Z/B,G)$ be the linear map defined such that

$$\eta(\varphi)(\alpha) = \varphi(z) \quad \text{for any } z \in \alpha \in Z/B.$$

For any other $z' \in \alpha$ we have $z' = z + b$ for some $b \in B$, and then

$$\varphi(z+b) = \varphi(z) + \varphi(b) = \varphi(z)$$

since φ vanishes on B. Therefore any map $\varphi \in B^0$ is constant on the each equivalence class in Z/B, and $\eta(\varphi)$ is well defined. The map η is surjective because if f is any linear map in $\mathrm{Hom}(Z/B, G)$, we can define the linear map $\varphi_0 \colon Z \to G$ by

$$\varphi_0(z) = f([z]) \quad \text{for all } z \in Z.$$

Since $f \in \mathrm{Hom}(Z/B, G)$, we have $\varphi_0(b) = f([b]) = 0$ for all $b \in B$. Since $M = Z \oplus Z'$, we can extend φ_0 to a linear map $\varphi \colon M \to G$, for example by setting $\varphi \equiv 0$ on Z', and then φ is a map in $\mathrm{Hom}(M, G)$ vanishing on B, and by definition $\eta(\varphi) = f$, since

$$\eta(\varphi)([z]) = \varphi(z) = \varphi_0(z) = f([z]) \quad \text{for all } [z] \in Z/B.$$

Finally, for any $\varphi \in B^0$, since φ is constant on any equivalence class in Z/B, we have $\eta(\varphi) = 0$ iff $\eta(\varphi)([z]) = 0$ for all $[z] \in Z/B$ iff $\varphi(z) = 0$ for all $z \in Z$, iff $\varphi \in Z^0$. Therefore $\mathrm{Ker}\, \eta = Z^0$, and consequently by the first isomorphism theorem,

$$B^0/Z^0 \cong \mathrm{Hom}(Z/B, G),$$

as claimed. ☐

We will also need the next proposition. Let M and G be R-modules, and let B be a submodule of M. As above, let

$$B^0 = \{f \in \mathrm{Hom}(M, G) \mid f|B \equiv 0\},$$

the set of R-linear maps $f \colon M \to G$ that vanish on B.

Proposition 2.10. *Let M and G be R-modules, and let B be a submodule of M. There is an isomorphism*

$$\kappa \colon B^0 \to \mathrm{Hom}(M/B, G)$$

defined by

$$(\kappa(f))([u]) = f(u) \quad \textit{for all } [u] \in M/B.$$

Proof. We need to check that the definition of $\kappa(f)$ does not depend on the representative $u \in M$ chosen in the equivalence class $[u] \in M/B$. Indeed, if $v = u + b$ some $b \in B$, we have

$$f(v) = f(u + b) = f(u) + f(b) = f(u),$$

since $f(b) = 0$ for all $b \in B$. The formula $\kappa(f)([u]) = f(u)$ makes it obvious that $\kappa(f)$ is linear since f is linear. The mapping κ is injective. This is because if $\kappa(f_1) = \kappa(f_2)$, then

$$\kappa(f_1)([u]) = \kappa(f_2)([u])$$

for all $u \in M$, and because $\kappa(f_1)([u]) = f_1(u)$ and $\kappa(f_2)([u]) = f_2(u)$, we get $f_1(u) = f_2(u)$ for all $u \in M$, that is, $f_1 = f_2$. The mapping κ is surjective because given any linear map $\varphi \in \mathrm{Hom}(M/B, G)$, if we define f by

$$f(u) = \varphi([u])$$

for all $u \in M$, then f is linear, vanishes on B, and clearly, $\kappa(f) = \varphi$. Therefore, we have the isomorphism $\kappa\colon B^0 \to \mathrm{Hom}(M/B, G)$, as we claimed. □

Remark: Proposition 2.10 is actually the special case of Proposition 2.9 where $Z = M$, since in this case $Z^0 = M$ and $Z' = (0)$. We feel that it is still instructive to give a direct proof of Proposition 2.10.

If we look carefully at the proofs of Propositions 2.6 through 2.8, we see that they go through with the ring R replaced by any fixed R-module A. This suggests looking at more general versions of Hom.

2.4 The Functors Hom$(-,A)$, Hom$(A,-)$, and $-\otimes A$

In this section we consider several operators T on R-modules that map a module A to another module $T(A)$, and a module homomorphism $f\colon A \to B$ to a module homomorphism $T(f)\colon T(A) \to T(B)$, or to a homomorphism $T(f)\colon T(B) \to T(A)$ (note the reversal). Given any two module homomorphism $f\colon A \to B$ and $g\colon B \to C$, if T does not reverse the direction of maps then $T(g \circ f) = T(g) \circ T(f)$, else $T(g \circ f) = T(f) \circ T(g)$. We also have $T(\mathrm{id}_A) = \mathrm{id}_{T(A)}$ for all A. Such operators are called *functors* (*covariant* in the first case, *contravariant* if it reverses the direction of maps). The reader may want to review Section 1.7 for the notion of a functor.

We begin with the $\mathrm{Hom}_R(-,A)$-functor, which reverses the direction of the maps.

Definition 2.5. Given a fixed R-module A, for any R-module B we denote by $\mathrm{Hom}_R(B,A)$ the R-module of all R-linear maps from B to A. Given any two R-modules B and C, for any R-linear map $f\colon B \to C$, the R-linear map $\mathrm{Hom}_R(f,A)\colon \mathrm{Hom}_R(C,A) \to \mathrm{Hom}_R(B,A)$ is defined by

$$\mathrm{Hom}_R(f,A)(\varphi) = \varphi \circ f \quad \text{for all } \varphi \in \mathrm{Hom}_R(C,A);$$

see the commutative diagram below:

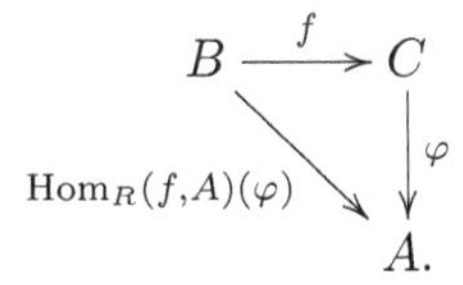

Observe that $\mathrm{Hom}_R(f, A)(\varphi)$ is φ composed with f, that is its result is to *pull back* along f any map φ from C to A to a map from B to A.[2] The map $\mathrm{Hom}_R(f, A)$ is also denoted by $\mathrm{Hom}_R(f, \mathrm{id}_A)$, or for short $\mathrm{Hom}_R(f, \mathrm{id})$. Some authors denote $\mathrm{Hom}_R(f, A)$ by f^*.

If $f \colon B \to C$ and $g \colon C \to D$ are linear maps of R-modules, a simple computation shows that

$$\mathrm{Hom}_R(g \circ f, A) = \mathrm{Hom}_R(f, A) \circ \mathrm{Hom}_R(g, A).$$

Observe that $\mathrm{Hom}_R(f, A)$ and $\mathrm{Hom}_R(g, A)$ are composed in the reverse order of the composition of f and g. It is also immediately verified that

$$\mathrm{Hom}_R(\mathrm{id}_A, A) = \mathrm{id}_{\mathrm{Hom}_R(A,A)}.$$

Thus, $\mathrm{Hom}_R(-, A)$ is a (contravariant) functor. To simplify notation, we usually omit the subscript R in $\mathrm{Hom}_R(-, A)$ unless confusion arises.

Proposition 2.11. *Let A be any fixed R-module and let $g \colon B \to C$ be a linear map between R-modules.*

(a) If g is an isomorphism, then so is $\mathrm{Hom}(g, A)$.
(b) If g is the zero map, then so is $\mathrm{Hom}(g, A)$.
(c) If the sequence

$$B \xrightarrow{g} C \longrightarrow 0$$

is exact, then the sequence

$$0 \longrightarrow \mathrm{Hom}(C, A) \xrightarrow{\mathrm{Hom}(g,A)} \mathrm{Hom}(B, A)$$

is also exact.

The proof of Proposition 2.11 is identical to the proof of Proposition 2.6.

Proposition 2.12. *Let M be any fixed R-module. If the following sequence of R-modules*

$$A \xrightarrow{f} B \xrightarrow{g} C \longrightarrow 0$$

is exact, then the sequence

$$0 \longrightarrow \mathrm{Hom}(C, M) \xrightarrow{\mathrm{Hom}(g,M)} \mathrm{Hom}(B, M) \xrightarrow{\mathrm{Hom}(f,M)} \mathrm{Hom}(A, M)$$

[2]A trick to remember that $\mathrm{Hom}_R(f, A)$ composes φ on the left of f is that f is the leftmost argument in $\mathrm{Hom}_R(f, A)$.

is also exact. Furthermore, if

$$0 \longrightarrow A \xrightarrow{f} B \xrightarrow{g} C \longrightarrow 0$$

is a split short exact sequence, then

$$0 \longrightarrow \mathrm{Hom}(C, M) \xrightarrow{\mathrm{Hom}(g,M)} \mathrm{Hom}(B, M) \xrightarrow{\mathrm{Hom}(f,M)} \mathrm{Hom}(A, M) \longrightarrow 0$$

is also a split short exact sequence.

The proof of Proposition 2.12 is identical to the proof of Proposition 2.7. We say that $\mathrm{Hom}(-, M)$ is a *left-exact* functor.

Remark: It can be shown that the sequence

$$A \xrightarrow{f} B \xrightarrow{g} C \longrightarrow 0$$

is exact iff the sequence

$$0 \longrightarrow \mathrm{Hom}(C, M) \xrightarrow{\mathrm{Hom}(g,M)} \mathrm{Hom}(B, M) \xrightarrow{\mathrm{Hom}(f,M)} \mathrm{Hom}(A, M)$$

is exact for all R-modules M; see Dummit and Foote [Dummit and Foote (1999)] (Chapter 10, Theorem 33).

Proposition 2.13. *Let M be any fixed R-module. If*

$$0 \longrightarrow A \xrightarrow{f} B \xrightarrow{g} C \longrightarrow 0$$

is a short exact sequence and if C is a free R-module, then

$$0 \longrightarrow \mathrm{Hom}(C, M) \xrightarrow{\mathrm{Hom}(g,M)} \mathrm{Hom}(B, M) \xrightarrow{\mathrm{Hom}(f,M)} \mathrm{Hom}(A, M) \longrightarrow 0$$

is a split short exact sequence.

There is also a version of the Hom-functor, $\mathrm{Hom}_R(A, -)$, in which the first slot is held fixed.

Definition 2.6. Given a fixed R-module A, for any R-module B we denote by $\mathrm{Hom}_R(A, B)$ the R-module of all R-linear maps from A to B. Given any two R-modules B and C, for any R-linear map $f\colon B \to C$, the R-linear map $\mathrm{Hom}_R(A, f)\colon \mathrm{Hom}_R(A, B) \to \mathrm{Hom}_R(A, C)$ is defined by

$$\mathrm{Hom}_R(A, f)(\varphi) = f \circ \varphi \quad \text{for all } \varphi \in \mathrm{Hom}_R(A, B);$$

see the commutative diagram below:

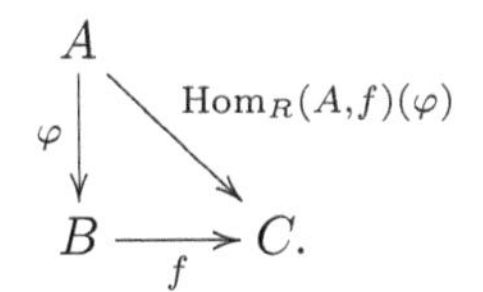

Observe that $\mathrm{Hom}_R(A, f)(\varphi)$ is f composed with φ, that is its result is to *push forward* along f any map φ from A to B to a map from A to C.[3] The map $\mathrm{Hom}_R(A, f)$ is also denoted by $\mathrm{Hom}_R(\mathrm{id}_A, f)$, or for short $\mathrm{Hom}_R(\mathrm{id}, f)$. Some authors denote $\mathrm{Hom}_R(A, f)$ by f_*.

If $f\colon B \to C$ and $g\colon C \to D$ are linear maps of R-modules, a simple computation shows that

$$\mathrm{Hom}_R(A, g \circ f) = \mathrm{Hom}_R(A, g) \circ \mathrm{Hom}_R(A, f).$$

It is also immediately verified that

$$\mathrm{Hom}_R(\mathrm{id}_A, A) = \mathrm{id}_{\mathrm{Hom}_R(A,A)}.$$

Thus, $\mathrm{Hom}_R(A, -)$ is a (covariant) functor.

The $\mathrm{Hom}_R(A, -)$-functor has properties analogous to those of the $\mathrm{Hom}_R(-, A)$-functor, except that sequences are not reversed. Again, to simplify notation, we usually omit the subscript R in $\mathrm{Hom}_R(A, -)$ unless confusion arises.

Proposition 2.14. *Let M be any fixed R-module. If the following sequence of R-modules*

$$0 \longrightarrow A \xrightarrow{f} B \xrightarrow{g} C$$

is exact, then the sequence

$$0 \longrightarrow \mathrm{Hom}(M, A) \xrightarrow{\mathrm{Hom}(M,f)} \mathrm{Hom}(M, B) \xrightarrow{\mathrm{Hom}(M,g)} \mathrm{Hom}(M, C)$$

is also exact. Furthermore, if

$$0 \longrightarrow A \xrightarrow{f} B \xrightarrow{g} C \longrightarrow 0$$

is a split short exact sequence, then

$$0 \longrightarrow \mathrm{Hom}(M, A) \xrightarrow{\mathrm{Hom}(M,f)} \mathrm{Hom}(M, B) \xrightarrow{\mathrm{Hom}(M,g)} \mathrm{Hom}(M, C) \longrightarrow 0$$

is also a split short exact sequence.

The proof of Proposition 2.14 is left as an exercise. We say that $\mathrm{Hom}(M, -)$ is a *left-exact* functor.

[3] A trick to remember that $\mathrm{Hom}_R(A, f)$ composes φ on the right of f is that f is the rightmost argument in $\mathrm{Hom}_R(A, f)$.

If $f\colon A \to B$ is surjective, then $\mathrm{Hom}(C, f)\colon \mathrm{Hom}(C, A) \to \mathrm{Hom}(C, B)$ is not necessarily surjective. For example, we have the following short exact sequence

$$0 \longrightarrow \mathbb{Z} \xrightarrow{\times 2} \mathbb{Z} \xrightarrow{\pi} \mathbb{Z}/2\mathbb{Z} \longrightarrow 0,$$

where $\times 2(n) = 2n$, but if $C = \mathbb{Z}/2\mathbb{Z}$, the map

$$\mathrm{Hom}(\mathbb{Z}/2\mathbb{Z}, \pi)\colon \mathrm{Hom}(\mathbb{Z}/2\mathbb{Z}, \mathbb{Z}) \to \mathrm{Hom}(\mathbb{Z}/2\mathbb{Z}, \mathbb{Z}/2\mathbb{Z})$$

is not surjective. This is because any map $\varphi\colon \mathbb{Z}/2\mathbb{Z} \to \mathbb{Z}$ must map 1 to 0. In $\mathbb{Z}/2\mathbb{Z}$ we have $1 + 1 = 0$, so $\varphi(1 + 1) = \varphi(0) = 0$, but if $\varphi(1) \neq 0$, then $\varphi(1 + 1) = \varphi(1) + \varphi(1) = 2\varphi(1) \neq 0$ in $\mathbb{Z}$, a contradiction. Therefore, $\mathrm{Hom}(\mathbb{Z}/2\mathbb{Z}, \mathbb{Z}) = (0)$, and yet $\mathrm{Hom}(\mathbb{Z}/2\mathbb{Z}, \mathbb{Z}/2\mathbb{Z})$ contains the identity map.

Remark: It can be shown that the sequence

$$0 \longrightarrow A \xrightarrow{f} B \xrightarrow{g} C$$

is exact iff the sequence

$$0 \longrightarrow \mathrm{Hom}(M, A) \xrightarrow{\mathrm{Hom}(M,f)} \mathrm{Hom}(M, B) \xrightarrow{\mathrm{Hom}(M,g)} \mathrm{Hom}(M, C)$$

is exact for all R-modules M. See Dummit and Foote [Dummit and Foote (1999)] (Chapter 10, Theorem 28).

Proposition 2.15. *Let M be any fixed R-module. If*

$$0 \longrightarrow A \xrightarrow{f} B \xrightarrow{g} C \longrightarrow 0$$

is a short exact sequence and if C is a free R-module, then

$$0 \longrightarrow \mathrm{Hom}(M, A) \xrightarrow{\mathrm{Hom}(M,f)} \mathrm{Hom}(M, B) \xrightarrow{\mathrm{Hom}(M,g)} \mathrm{Hom}(M, C) \longrightarrow 0$$

is a split short exact sequence.

A more complete discussion of the functor $\mathrm{Hom}(-, A)$ is found in Munkres [Munkres (1984)] (Chapter 5, §41), and a thorough presentation in Mac Lane [Mac Lane (1975)], Cartan–Eilenberg [Cartan and Eilenberg (1956)], Rotman [Rotman (1979, 2009)], and Weibel [Weibel (1994)].

Another operation on modules that plays a crucial role is the tensor product. Let M be a fixed R-module. For any R-module A, we have the

R-module $A \otimes_R M$, and for any R-linear map $f\colon B \to C$ we have the R-linear map $f \otimes_R \mathrm{id}_M\colon B \otimes_R M \to C \otimes_R M$. To simplify notation, unless confusion arises, we will drop the subscript R on $\otimes_R$.

If $f\colon B \to C$ and $g\colon C \to D$ are linear maps of R-modules, a simple computation shows that

$$(g \otimes \mathrm{id}_M) \circ (f \otimes \mathrm{id}_M) = (g \circ f) \otimes \mathrm{id}_M.$$

It is also immediately verified that

$$\mathrm{id}_M \otimes \mathrm{id}_M = \mathrm{id}_{M \otimes M}.$$

Definition 2.7. For any fixed R-module M, we define $- \otimes M$ as the (covariant) functor that takes any R-module A and produces the R-module $A \otimes M$.

Similarly we have the functor $M \otimes -$ obtained by holding the first slot fixed. This functor has the same properties as $- \otimes M$ so we will not consider it any further.

We would like to understand the behavior of the functor $- \otimes M$ with respect to exact sequences.

A crucial fact is that if $f\colon B \to C$ is injective, then $f \otimes \mathrm{id}_M$ may not be injective. For example, if we let $R = \mathbb{Z}$, then the inclusion map $i\colon \mathbb{Z} \to \mathbb{Q}$ is injective, but if $M = \mathbb{Z}/2\mathbb{Z}$, then

$$\mathbb{Q} \otimes_{\mathbb{Z}} \mathbb{Z}/2\mathbb{Z} = (0),$$

since we can write

$$a \otimes b = (a/2) \otimes (2b) = (a/2) \otimes 0 = 0.$$

Thus, $i \otimes \mathrm{id}_M\colon \mathbb{Z} \otimes \mathbb{Z}/2\mathbb{Z} \to \mathbb{Q} \otimes \mathbb{Z}/2\mathbb{Z} = i \otimes \mathrm{id}_M\colon \mathbb{Z} \otimes \mathbb{Z}/2\mathbb{Z} \to (0)$, which is not injective. Thus, $- \otimes M$ is not left-exact. However, it is right-exact, as we now show.

Proposition 2.16. *Let $f\colon A \to B$ and $f'\colon A' \to B'$ be two R-linear maps. If f and f' are surjective, then*

$$f \otimes f'\colon A \otimes A' \to B \otimes B'$$

is surjective, and its kernel $\mathrm{Ker}\,(f \otimes f')$ *is spanned by all tensors of the form $a \otimes a'$ for which either $a \in \mathrm{Ker}\, f$ or $a' \in \mathrm{Ker}\, f'$.*

Proof. Let H be the submodule of $A \otimes A'$ spanned by all tensors of the form $a \otimes a'$ for which either $a \in \mathrm{Ker}\, f$ or $a' \in \mathrm{Ker}\, f'$. Obviously, $f \otimes f'$ vanishes on H, so it factors through a R-linear map

$$\Phi\colon (A \otimes A')/H \to B \otimes B'$$

as illustrated in the following diagram:

$$\begin{array}{ccc} A \otimes A' & \xrightarrow{\pi} & (A \otimes A')/H \\ & \searrow^{f \otimes f'} & \downarrow \Phi \\ & & B \otimes B'. \end{array}$$

We prove that Φ is an isomorphism by defining an inverse Ψ for Φ. We begin by defining a function

$$\psi\colon B \times B' \to (A \otimes A')/H$$

by setting

$$\psi(b, b') = \overline{a_1 \otimes a_1'}$$

for all $b \in B$ and all $b' \in B'$, where $a_1 \in A$ is any element such that $f(a_1) = b$ and $a_1' \in A'$ is any element such that $f'(a_1') = b'$, which exist since f and f' are surjective. We need to check that ψ does not depend on the choice of $a_1 \in f^{-1}(b)$ and $a_1' \in (f')^{-1}(b')$. If $f(a_2) = b$ and $f'(a_2') = b'$, with $a_2 \in A$ and $a_2' \in A'$, since we can write

$$a_1 \otimes a_1' - a_2 \otimes a_2' = (a_1 - a_2) \otimes a_1' + a_2 \otimes (a_1' - a_2'),$$

and since $f(a_1 - a_2) = f(a_1) - f(a_2) = b - b = 0$, and $f'(a_1' - a_2') = f'(a_1') - f'(a_2') = b' - b' = 0$, we see that $a_1 \otimes a_1' - a_2 \otimes a_2' \in H$, thus

$$\overline{a_1 \otimes a_1'} = \overline{a_2 \otimes a_2'}$$

an ψ is well defined. We check immediately that ψ is R-bilinear, so ψ induces a R-linear map

$$\Psi\colon B \otimes B' \to (A \otimes A')/H.$$

It remains to check that $\Phi \circ \Psi$ and $\Psi \circ \Phi$ are identity maps, which is easily verified on generators. $\square$

We can now show that $- \otimes M$ is right-exact.

Proposition 2.17. *Suppose the sequence*

$$A \xrightarrow{f} B \xrightarrow{g} C \longrightarrow 0$$

is exact. Then the sequence

$$A \otimes M \xrightarrow{f \otimes \mathrm{id}_M} B \otimes M \xrightarrow{g \otimes \mathrm{id}_M} C \otimes M \longrightarrow 0$$

is exact. If f is injective and the first sequence splits, then $f \otimes \mathrm{id}_M$ is injective and the second sequence splits.

Proof. Since the first sequence is exact, g is surjective and Proposition 2.16 implies that $g \otimes \mathrm{id}_M$ is surjective, and that its kernel H is the submodule of $B \otimes M$ spanned by all elements of the form $b \otimes z$ with $b \in \mathrm{Ker}\, g$ and $z \in M$. On the other hand the image D of $f \otimes \mathrm{id}_M$ is the submodule spanned by all elements of the form $f(a) \otimes z$, with $a \in A$ and $z \in M$. Since $\mathrm{Im}\, f = \mathrm{Ker}\, g$, we have $H = D$; that is, $\mathrm{Im}\,(f \otimes \mathrm{id}_M) = \mathrm{Ker}\,(g \otimes \mathrm{id}_M)$.

Suppose that f is injective and the first sequence splits. By Proposition 2.2, let $p\colon B \to A$ be a R-linear map such that $p \circ f = \mathrm{id}_A$. Then

$$(p \otimes \mathrm{id}_M) \circ (f \otimes \mathrm{id}_M) = (p \circ f) \otimes (\mathrm{id}_M \circ \mathrm{id}_M) = \mathrm{id}_A \otimes \mathrm{id}_M = \mathrm{id}_{A \otimes M},$$

so $f \otimes \mathrm{id}_M$ is injective and $p \otimes \mathrm{id}_M$ splits the second sequence. $\square$

Proposition 2.17 says that the functor $- \otimes M$ is *right-exact*. A more complete discussion of the functor $- \otimes M$ is found in Munkres [Munkres (1984)] (Chapter 6, §50), and a thorough presentation in Mac Lane [Mac Lane (1975)], Cartan–Eilenberg [Cartan and Eilenberg (1956)], Rotman [Rotman (1979, 2009)], and Weibel [Weibel (1994)].

2.5 Abstract Cochain Complexes and Their Cohomology

The notion of a cochain complex is obtained from the notion of an exact sequence by relaxing the requirement $\mathrm{Im}\, f_p = \mathrm{Ker}\, f_{p+1}$ to $f_{p+1} \circ f_p = 0$.

Definition 2.8. A *(differential) complex* (or *cochain complex*) is a $\mathbb{Z}$-graded R-module

$$C = \bigoplus_{p \in \mathbb{Z}} C^p,$$

together with a R-linear map

$$d\colon C \to C$$

such that $dC^p \subseteq C^{p+1}$ and $d \circ d = 0$. We denote the restriction of d to C^p by $d^p\colon C^p \to C^{p+1}$. A cochain complex is denoted as a diagram with increasing superscripts and arrows going from left to right as shown below:

$$\cdots \longrightarrow C^{p-1} \xrightarrow{d^{p-1}} C^p \xrightarrow{d^p} C^{p+1} \xrightarrow{d^{p+1}} C^{p+2} \longrightarrow \cdots$$

A cochain complex is *positive* if $C^p = (0)$ for all $p < 0$, *negative* if $C^p = (0)$ for all $p > 0$.

Given a complex (C, d), we define the $\mathbb{Z}$-graded R-modules

$$B^*(C) = \operatorname{Im} d, \qquad Z^*(C) = \operatorname{Ker} d.$$

Since $d \circ d = 0$, we have

$$B^*(C) \subseteq Z^*(C) \subseteq C$$

so the quotient spaces $Z^p(C)/B^p(C)$ make sense and we can define cohomology.

Definition 2.9. Given a differential complex (C, d) of R-modules, we define the *cohomology space* $H^*(C)$ by

$$H^*(C) = \bigoplus_{p \in \mathbb{Z}} H^p(C),$$

where the pth cohomology group (R-module) $H^p(C)$ is the quotient space

$$H^p(C) = (\operatorname{Ker} d \cap C^p)/(\operatorname{Im} d \cap C^p) = \operatorname{Ker} d^p / \operatorname{Im} d^{p-1} = Z^p(C)/B^p(C).$$

Elements of C^p are called *p-cochains* or *cochains*, elements of $Z^p(C)$ are called *p-cocycles* or *cocycles*, and elements of $B^p(C)$ are called *p-coboundaries* or *coboundaries*. Given a cocycle $a \in Z^p(C)$, its *cohomology class* $a + \operatorname{Im} d^{p-1}$ is denoted by $[a]$. A complex C is said to be *acyclic* if its cohomology is trivial, that is $H^p(C) = (0)$ for all p, which means that C is an exact sequence.

We often drop the complex C when writing $Z^p(C)$, $B^p(C)$ of $H^p(C)$.

Typically, when dealing with cohomology we consider positive cochain complexes ($C^p = (0)$ for all $p < 0$):

$$0 \xrightarrow{d^{-1}} C^0 \xrightarrow{d^0} C^1 \xrightarrow{d^1} \cdots \xrightarrow{d^{p-1}} C^p \xrightarrow{d^p} C^{p+1} \xrightarrow{d^{p+1}} C^{p+2} \longrightarrow \cdots$$

We can deal with homology by assuming that we have a negative cochain complex ($C^p = (0)$ for all $p > 0$). In this case, we have a cochain complex of the form

$$\cdots C^{-(p+1)} \xrightarrow{d^{-(p+1)}} C^{-p} \xrightarrow{d^{-p}} C^{-(p-1)} \cdots \longrightarrow C^{-1} \xrightarrow{d^{-1}} C^0 \xrightarrow{d^0} 0.$$

It is customary to use positive indices and to convert the above diagram to the diagram shown below called a *chain complex* in which every negative upper index $-p$ is replaced by the positive lower index p

$$\cdots \longrightarrow C_{p+1} \xrightarrow{d_{p+1}} C_p \xrightarrow{d_p} C_{p-1} \cdots \longrightarrow C_1 \xrightarrow{d_1} C_0 \xrightarrow{d_0} 0.$$

An equivalent diagram is obtained by also reversing the direction of the arrows:

$$0 \xleftarrow{d_0} C_0 \xleftarrow{d_1} C_1 \longleftarrow \cdots\ C_{p-1} \xleftarrow{d_p} C_p \xleftarrow{d_{p+1}} C_{p+1} \longleftarrow \cdots .$$

Which diagram is preferred is a matter of taste.[4]

Definition 2.10. A *chain complex* is a $\mathbb{Z}$-graded R-module

$$C = \bigoplus_{p\in\mathbb{Z}} C_p,$$

together with a R-linear map

$$d\colon C \to C$$

such that $dC_{p+1} \subseteq C_p$ and $d \circ d = 0$. We denote the restriction of d to C_p by $d_p\colon C_p \to C_{p-1}$. A chain complex is denoted as a diagram with increasing subscripts and arrows going from right to left as shown below:

$$\cdots \longleftarrow C_{p-1} \xleftarrow{d_p} C_p \xleftarrow{d_{p+1}} C_{p+1} \xleftarrow{d_{p+2}} C_{p+2} \longleftarrow \cdots .$$

A chain complex is *positive* if $C_p = (0)$ for all $p < 0$, *negative* if $C_p = (0)$ for all $p > 0$.

A cochain complex can be converted to a chain complex, and conversely, by changing C^p to C_{-p} and d^p to d_{-p} and changing the direction of the arrows. The cochain complex

$$\cdots \longrightarrow C^{p-1} \xrightarrow{d^{p-1}} C^p \xrightarrow{d^p} C^{p+1} \xrightarrow{d^{p+1}} C^{p+2} \longrightarrow \cdots$$

becomes the chain complex

$$\cdots \longleftarrow C_{-(p+2)} \xleftarrow{d_{-(p+1)}} C_{-(p+1)} \xleftarrow{d_{-p}} C_{-p} \xleftarrow{d_{-(p-1)}} C_{-(p-1)} \longleftarrow \cdots .$$

Conversely we get a chain complex from a cochain complex by changing C_p to C^{-p} and d_p to d^{-p} and changing the direction of the arrows.

When it is clear from the context, we simply use the term complex, omitting the prefix chain or cochain.

Remark: Given a $\mathbb{Z}$-graded R-module

$$C = \bigoplus_{p\in\mathbb{Z}} C_p,$$

[4]Notice that applying $\mathrm{Hom}(-, R)$ to the second diagram reverses all the arrows so that a complex of cohomology is obtained. For this reason, we have a slight preference for the second diagram.

a R-linear map

$$d\colon C \to C$$

such that $dC_p \subseteq C_{p+r}$ for all $p \in \mathbb{Z}$ for some fixed $r \in \mathbb{Z}$ is said to have *degree* r. The map d is called a *differential* if $d \circ d = 0$. Thus we see that a chain complex is a $\mathbb{Z}$-graded R-module with a differential d of degree -1, and a cochain complex is a $\mathbb{Z}$-graded R-module with a differential d of degree $+1$. Differentials of degree $r \neq -1, 1$ occur in spectral sequences.

Definition 2.11. Given a chain complex (C_p) and the corresponding cochain complex (C^{-p}), we denote the space $H^{-p}(C)$ by $H_p(C)$ and call it the pth *homology space*. More explicitly

$$H_p(C) = \mathrm{Ker}\, d_p / \mathrm{Im}\, d_{p+1},$$

and if we write $Z_p(C) = \mathrm{Ker}\, d_p$ and $B_p(C) = \mathrm{Im}\, d_{p+1}$, we also have

$$H_p(C) = Z_p(C)/B_p(C),$$

elements of C_p are called *chains*, elements of $Z_p(C)$ are called *cycles*, and elements of $B_p(C)$ are called *boundaries*.

Singular homology defined in Section 4.8 is such an example.

Remark: When dealing with cohomology, it is customary to use superscripts for denoting the cochains groups C^p, the cohomology groups $H^p(C)$, the coboundary maps d^p, *etc.*, and to write complexes with the arrows going from left to right so that the superscripts increase. However, when dealing with homology, it is customary to use subscripts for denoting the chains groups C_p, the homology groups $H_p(C)$, the boundary maps d_p, *etc.*, and to write homology complexes with increasing indices and arrows going from right to left (or decreasing indices and arrows going from left to right). In homology the boundary maps $d_p\colon C_p \to C_{p-1}$ are usually denoted by ∂_p, and in cohomology the coboundary maps $d^p\colon C^p \to C^{p+1}$ are usually denoted by δ^p.

Given two cochain complexes (X, d_X) and (Y, d_Y), the complex $X \oplus Y$ consists of the modules $X^p \oplus Y^p$ and of the maps

$$X^p \oplus Y^p \xrightarrow{d_X^p \oplus d_Y^p} X^{p+1} \oplus Y^{p+1}$$

defined such that $(d^p_X \oplus d^p_Y)(x + y) = d^p_X(x) + d^p_Y(y)$, for all $x \in X^p$ and all $y \in Y^p$. It is immediately verified that $(d^{p+1}_X \oplus d^{p+1}_Y) \circ (d^P_X \oplus d^p_Y) = 0$. The following proposition is easy to prove.

Proposition 2.18. *For any two cochain complexes (X, d_X) and (Y, d_Y), we have isomorphisms*

$$H^p(X \oplus Y) \cong H^p(X) \oplus H^p(Y)$$

for all p.

Sketch of proof. It is easy to check that

$$\begin{aligned} \operatorname{Ker} d^p_{X\oplus Y} &\cong \operatorname{Ker} d^p_X \oplus \operatorname{Ker} d^p_Y \\ \operatorname{Im} d^p_{X\oplus Y} &\cong \operatorname{Im} d^p_X \oplus \operatorname{Im} d^p_Y, \end{aligned}$$

from which the results follows. □

2.6 Chain Maps and Chain Homotopies

We know that homomorphisms between R-modules play a very important role in the theory of R-modules. There are two notions of maps between chain complexes that also play an important role in homology and cohomology theory.

Definition 2.12. Given two cochain complexes (C, d_C) and (D, d_D), a *chain map*[5] $f : C \to D$ is a family $f = (f^p)$ of R-linear maps $f^p : C^p \to D^p$ such that

$$d_D \circ f^p = f^{p+1} \circ d_C \quad \text{for all } p \in \mathbb{Z},$$

equivalently all the squares in the following diagram commute.

$$\begin{array}{ccccccccccc} \cdots & \xrightarrow{d_C} & C^{p-1} & \xrightarrow{d_C} & C^p & \xrightarrow{d_C} & C^{p+1} & \xrightarrow{d_C} & C^{p+2} & \xrightarrow{d_C} & \cdots \\ & & \downarrow{\scriptstyle f^{p-1}} & & \downarrow{\scriptstyle f^p} & & \downarrow{\scriptstyle f^{p+1}} & & \downarrow{\scriptstyle f^{p+2}} & & \\ \cdots & \xrightarrow[d_D]{} & D^{p-1} & \xrightarrow[d_D]{} & D^p & \xrightarrow[d_D]{} & D^{p+1} & \xrightarrow[d_D]{} & D^{p+2} & \xrightarrow[d_D]{} & \cdots \end{array}$$

[5]It would be more logical to call a map between cochain complexes a cochain map. Spanier uses the term cochain map but this does not appear to be the usual practice.

A chain map of cochain complexes $f\colon C \to D$ induces a map $f^*\colon H^*(C) \to H^*(D)$ between the cohomology spaces $H^*(C)$ and $H^*(D)$, which means that each map $f^p\colon C^p \to D^p$ induces a homomorphism $(f^p)^*\colon H^p(C) \to H^p(D)$.

Proposition 2.19. *Given a chain map of cochain complexes $f\colon C \to D$, for every $p \in \mathbb{Z}$, the function $(f^p)^*\colon H^p(C) \to H^p(D)$ defined such that*

$$(f^p)^*([a]) = [f^p(a)] \quad \text{for all } a \in Z^p(C)$$

is a homomorphism. Therefore, $f\colon C \to D$ induces a homomorphism $f^*\colon H^*(C) \to H^*(D)$.

Proof. First we show that if $[a]$ is a cohomology class in $H^p(C)$ with $a \in Z^p(C)$ (a is a cocycle), then $f^p(a) \in Z^p(D)$; that is, $f^p(a)$ is a cocycle. Since $a \in Z^p(C)$ we have $d_C(a) = 0$, and since by the commutativity of the squares of the diagram of Definition 2.12

$$d_D \circ f^p = f^{p+1} \circ d_C,$$

we get

$$d_D \circ f^p(a) = f^{p+1} \circ d_C(a) = 0,$$

which shows that $f^p(a) \in Z^p(D)$, that is $f^p(a)$ is a cocycle.

Next we show that $[f^p(a)]$ does not depend on the choice of a in the equivalence class $[a]$. If $[b] = [a]$ with $a, b \in Z^p(C)$, then $a - b \in B^p(C)$, which means that $a - b = d_C(x)$ for some $x \in C^{p-1}$. We have

$$d_D \circ f^{p-1} = f^p \circ d_C,$$

which implies that

$$f^p(a - b) = f^p \circ d_C(x) = d_D \circ f^{p-1}(x),$$

and since f^p is linear we get $f^p(a) - f^p(b) = d_D \circ f^{p-1}(x)$, that is, $f^p(a) - f^p(b) \in \operatorname{Im} d_D$, which means that $[f^p(a)] = [f^p(b)]$. Thus, $(f^p)^*([a]) = [f^p(a)]$ is well defined.

The fact that $(f^p)^*$ is a homomorphism is standard and follows immediately from the definition of $(f^p)^*$. ☐

There are situations, for instance when defining Čech cohomology groups, where we have different maps $f\colon C \to D$ and $g\colon C \to D$ between two (cochain) complexes C and D and yet we would like the induced maps $f^*\colon H^*(C) \to H^*(D)$ and $g^*\colon H^*(C) \to H^*(D)$ to be identical, that is,

$f^* = g^*$. A sufficient condition is the existence of a certain kind of map between C and D called a chain homotopy.

Definition 2.13. Given two chain maps $f: C \to D$ and $g: C \to D$, a *chain homotopy* between f and g is a family $s = (s^p)_{p\in\mathbb{Z}}$ of R-linear maps $s^p: C^p \to D^{p-1}$ such that

$$d_D \circ s^p + s^{p+1} \circ d_C = f^p - g^p \quad \text{for all } p \in \mathbb{Z}.$$

As a diagram, a chain homotopy is given by a family of slanted arrows as below, where we write $h = f - g$:

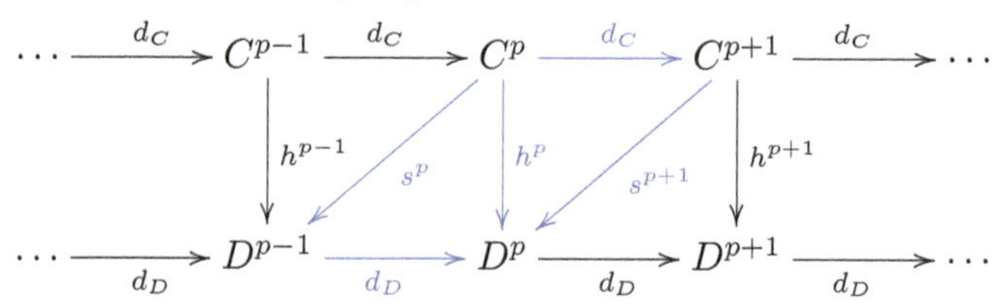

The following proposition clarifies this somewhat mysterious definition.

Proposition 2.20. *Given two chain maps $f: C \to D$ and $g: C \to D$ between two cochain complexes C and D, if s is a chain homotopy between f and g, then $f^* = g^*$.*

Proof. If $[a]$ is a cohomology class in $H^p(C)$, where a is a cocycle in $Z^p(C)$, that is $a \in C^p$ and $d_C(a) = 0$, we have

$$((f^p)^* - (g^p)^*)([a]) = [f^p(a) - g^p(a)] = [d_D \circ s^p(a) + s^{p+1} \circ d_C(a)],$$

and since a is a cocycle $d_C(a) = 0$ so

$$((f^p)^* - (g^p)^*)([a]) = [d_D \circ s^p(a)] = 0,$$

since $d_D \circ s^p(a)$ is a coboundary in $B^p(D)$. □

2.7 The Long Exact Sequence of Cohomology or Zig-Zag Lemma

The following result is the first part of one of the most important results of (co)homology theory.

Proposition 2.21. *Any short exact sequence*

$$0 \longrightarrow X \stackrel{f}{\longrightarrow} Y \stackrel{g}{\longrightarrow} Z \longrightarrow 0$$

of cochain complexes X, Y, Z yields a cohomology sequence

$$H^p(X) \stackrel{f^*}{\longrightarrow} H^p(Y) \stackrel{g^*}{\longrightarrow} H^p(Z)$$

which is exact for every p, which means that $\operatorname{Im} f^* = \operatorname{Ker} g^*$ *for all p.*

Proof. Consider the following diagram where the rows are exact:

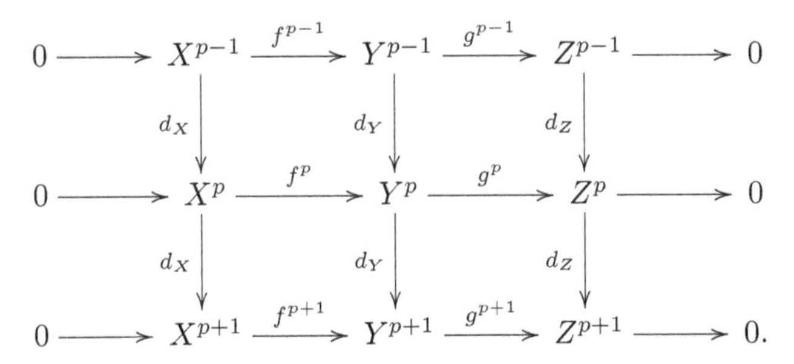

Since we have a short exact sequence, f^p is injective, g^p is surjective, and $\mathrm{Im}\, f^p = \mathrm{Ker}\, g^p$ for all p. Consequently $g^p \circ f^p = 0$, and for every cohomology class $[a] \in H^p(X)$, we have

$$g^* \circ f^*([a]) = g^*([f^p(a)]) = [g^p(f^p(a))] = 0,$$

which implies that $\mathrm{Im}\, f^* \subseteq \mathrm{Ker}\, g^*$. To prove the inclusion in the opposite direction, we need to prove that for every $[b] \in H^p(Y)$ such that $g^*([b]) = 0$ (where $b \in Y^p$ is a cocycle) there is some $[a] \in H^p(X)$ such that $f^*([a]) = [b]$.

If $g^*([b]) = [g^p(b)] = 0$ then $g^p(b)$ must be a coboundary, which means that $g^p(b) = d_Z(c)$ for some $c \in Z^{p-1}$. Since g^{p-1} is surjective, there is some $b_1 \in Y^{p-1}$ such that $c = g^{p-1}(b_1)$. Now g being a chain map the top right square commutes, that is

$$d_Z \circ g^{p-1} = g^p \circ d_Y,$$

so

$$g^p(b) = d_Z(c) = d_Z(g^{p-1}(b_1)) = g^p(d_Y(b_1)),$$

which implies that

$$g^p(b - d_Y(b_1)) = 0.$$

By exactness of the short exact sequence, $\mathrm{Im}\, f^p = \mathrm{Ker}\, g^p$ for all p, and there is some $a \in X^p$ such that

$$f^p(a) = b - d_Y(b_1).$$

If we can show that a is a cocycle, then

$$f^*([a]) = [f^p(a)] = [b - d_Y(b_1)] = [b],$$

proving that $f^*([a]) = [b]$, as desired.

Thus, we need to prove that $d_X(a) = 0$. Since f^{p+1} is injective, it suffices to show that $f^{p+1}(d_X(a)) = 0$. But f is a chain map so the left lower square commutes, that is

$$d_Y \circ f^p = f^{p+1} \circ d_X,$$

and we have

$$f^{p+1}(d_X(a)) = d_Y(f^p(a)) = d_Y(b - d_Y(b_1)) = d_Y(b) - d_Y \circ d_Y(b) = 0$$

since b is a cocycle, so $d_Y(b) = 0$ and $d_Y \circ d_Y = 0$ since Y is a differential complex. □

In general, a short exact sequence

$$0 \longrightarrow X \stackrel{f}{\longrightarrow} Y \stackrel{g}{\longrightarrow} Z \longrightarrow 0$$

of cochain complexes does not yield an exact sequence

$$0 \longrightarrow H^p(X) \stackrel{f^*}{\longrightarrow} H^p(Y) \stackrel{g^*}{\longrightarrow} H^p(Z) \longrightarrow 0$$

for all (or any) p. However, one of the most important results in homological algebra is that a short exact sequence of cochain complexes yields a so-called *long exact sequence* of cohomology groups.

This result is often called the "zig-zag lemma" for cohomology; see Munkres [Munkres (1984)] (Chapter 3, Section 24). The proof involves a lot of "diagram chasing." It is not particularly hard, but a bit tedious and not particularly illuminating. Still, this is a very important result so we provide a complete and detailed proof.

Theorem 2.22. *(Long exact sequence of cohomology or zig-zag lemma for cohomology) For any short exact sequence*

$$0 \longrightarrow X \stackrel{f}{\longrightarrow} Y \stackrel{g}{\longrightarrow} Z \longrightarrow 0$$

of cochain complexes X, Y, Z, there are homomorphisms $\delta^p \colon H^p(Z) \to H^{p+1}(X)$ such that we obtain a long exact sequence of cohomology of the following form:

$$
\begin{array}{l}
\cdots \longrightarrow H^{p-1}(Z) \xrightarrow{\delta^{p-1}} \\
H^p(X) \xrightarrow{f^*} H^p(Y) \xrightarrow{g^*} H^p(Z) \xrightarrow{\delta^{p}} \\
H^{p+1}(X) \xrightarrow{f^*} H^{p+1}(Y) \xrightarrow{g^*} H^{p+1}(Z) \xrightarrow{\delta^{p+1}} \\
H^{p+2}(X) \longrightarrow \cdots
\end{array}
$$

(for all p).

Proof. The main step is the construction of the homomorphisms $\delta^p \colon H^p(Z) \to H^{p+1}(X)$. We suggest that upon first reading the reader looks at the construction of δ^p and then skips the proofs of the various facts that need to be established.

Consider the following diagram where the rows are exact.

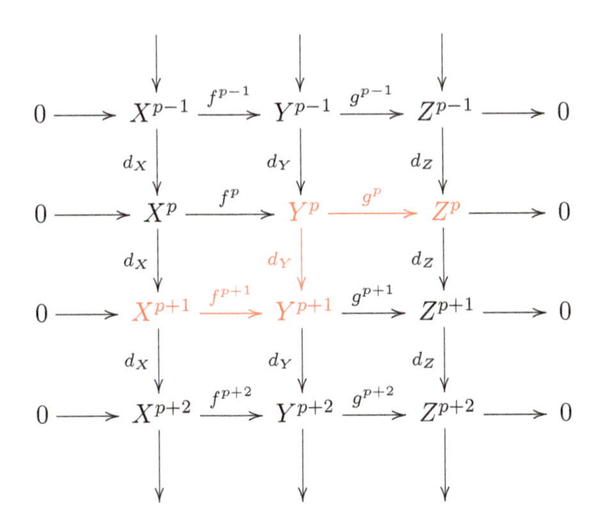

To define $\delta^p([c])$ where $[c] \in H^p(Z)$ is a cohomology class ($c \in Z^p$ is a cocycle, that is $d_Z(c) = 0$), pick any $b \in Y^p$ such that $g^p(b) = c$, push b

down to Y^{p+1} by applying d_Y obtaining $d_Y(b)$, and then pull $d_Y(b)$ back to X^{p+1} by applying $(f^{p+1})^{-1}$, obtaining $a = (f^{p+1})^{-1}(d_Y(b))$. Then set

$$\delta^p([c]) = [a].$$

Schematically, starting with an element $c \in Z^p$, we follow the path from right to left in the diagram below.

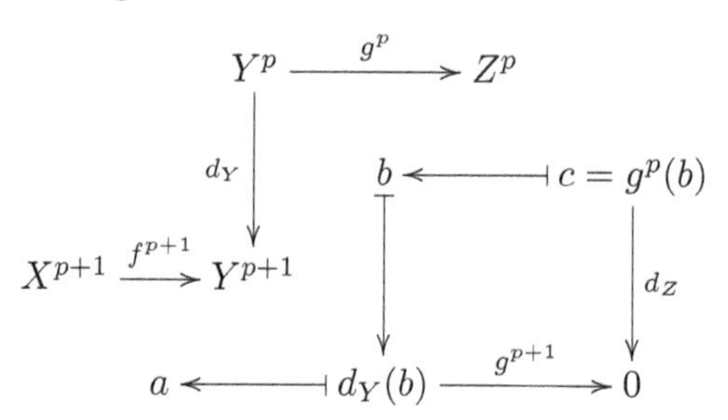

In order to ensure that δ^p is well defined, we must check five facts:

(a) For any $c \in Z^p$ such that $d_Z(c) = 0$ and any $b \in Y^p$, if $g^p(b) = c$, then $d_Y(b) \in \operatorname{Im} f^{p+1}$. This guarantees that $a = (f^{p+1})^{-1}(d_Y(b))$ is well-defined since f^{p+1} is injective.
(b) The element $a \in X^{p+1}$ is a cocycle; more precisely, if $f^{p+1}(a) = d_Y(b)$ for some $b \in Y^p$, then $d_X(a) = 0$.
(c) The cohomology class $[a]$ does not depend on the choice of b in $(g^p)^{-1}(c)$; that is, for all $b_1, b_2 \in Y^p$ and all $a_1, a_2 \in X^{p+1}$, if $g^p(b_1) = g^p(b_2) = c$ and $f^{p+1}(a_1) = d_Y(b_1), f^{p+1}(a_2) = d_Y(b_2)$, then $[a_1] = [a_2]$.
(d) The map δ^p is a linear map.
(e) The cohomology class $[a]$ does not depend on the choice of the cocycle c in the cohomology class $[c]$. Since δ^p is linear, it suffices to show that if c is a coboundary in Z^p, then for any b such that $g^p(b) = c$ and any $a \in X^{p+1}$ such that $f^{p+1}(a) = d_Y(b)$, then $[a] = 0$.

Recall that since f and g are chain maps, the top, middle, and bottom left and right squares commute.

(a) Since $\operatorname{Im} f^{p+1} = \operatorname{Ker} g^{p+1}$, it suffices to show that $g^{p+1}(d_Y(b)) = 0$. However, since the middle right square commutes and $d_Z(c) = 0$ (c is a cocycle),

$$g^{p+1}(d_Y(b)) = d_Z(g^p(b)) = d_Z(c) = 0,$$

as desired.

(b) Since f^{p+2} is injective, $d_X(a) = 0$ iff $f^{p+2} \circ d_X(a) = 0$, and since the lower left square commutes

$$f^{p+2} \circ d_X(a) = d_Y \circ f^{p+1}(a) = d_Y \circ d_Y(b) = 0,$$

so $d_X(a) = 0$, as claimed.

(c) Assume that $g^p(b_1) = g^p(b_2) = c$. Then $g^p(b_1 - b_2) = 0$, and since $\operatorname{Im} f^p = \operatorname{Ker} g^p$, there is some $\widetilde{a} \in X^p$ such that $b_1 - b_2 = f^p(\widetilde{a})$. Using the fact that the middle left square commutes we have

$$\begin{aligned} f^{p+1}(a_1 - a_2) &= f^{p+1}(a_1) - f^{p+1}(a_2) \\ &= d_Y(b_1) - d_Y(b_2) = d_Y(b_1 - b_2) \\ &= d_Y(f^p(\widetilde{a})) = f^{p+1}(d_X(\widetilde{a})), \end{aligned}$$

and the injectivity of f^{p+1} yields $a_1 - a_2 = d_X(\widetilde{a})$, which implies that $[a_1] = [a_2]$.

(d) The fact that δ^p is linear is an immediate consequence of the fact that all the maps involved in its definition are linear.

(e) Let $c \in Z^p$ be a coboundary, which means that $c = d_Z(\widetilde{c})$ for some $\widetilde{c} \in Z^{p-1}$. Since g^{p-1} is surjective, there is some $b_1 \in Y^{p-1}$ such that $g^{p-1}(b_1) = \widetilde{c}$, and since the top right square commutes $d_Z \circ g^{p-1} = g^p \circ d_Y$, and we get

$$c = d_Z(\widetilde{c}) = d_Z(g^{p-1}(b_1)) = g^p(d_Y(b_1)).$$

By (c), to compute the cohomology class $[a]$ such that $\delta^p([c]) = [a]$ we can pick any $b \in Y^p$ such that $g^p(b) = c$, and since $c = g^p(d_Y(b_1))$ we can pick $b = d_Y(b_1)$ and then we obtain

$$d_Y(b) = d_Y \circ d_Y(b_1) = 0.$$

Since f^{p+1} is injective, if $a \in X^{p+1}$ is the unique element such that $f^{p+1}(a) = d_Y(b) = 0$, then $a = 0$, and thus $[a] = 0$.

It remains to prove that

$$\operatorname{Im}(g^p)^* = \operatorname{Ker}\delta^p \quad \text{and} \quad \operatorname{Im}\delta^p = \operatorname{Ker}(f^{p+1})^*.$$

For any cohomology class $[b] \in H^p(Y)$ for some $b \in Y^p$ such that $d_Y(b) = 0$ (b is a cocycle), since $(g^p)^*([b]) = [g^p(b)]$, if we write $c = g^p(b)$ then c is a cocycle in Z^p, and by definition of δ^p we have

$$\delta^p((g^p)^*([b])) = \delta^p([c]) = [(f^{p+1})^{-1}(d_Y(b))] = [(f^{p+1})^{-1}(0)] = 0.$$

Thus, $\operatorname{Im}(g^p)^* \subseteq \operatorname{Ker}\delta^p$.

Conversely, assume that $\delta^p([c]) = 0$, for some $c \in Z^p$ such that $d_Z(c) = 0$. By definition of δ^p, we have $\delta^p([c]) = [a]$ where $a \in X^{p+1}$ is given by $f^{p+1}(a) = d_Y(b)$ for any $b \in Y^p$ such that $g^p(b) = c$, and since $[a] = 0$ the element a must be a coboundary, which means that $a = d_X(a_1)$ for some $a_1 \in X^p$. Then by commutativity of the left middle square we have

$$d_Y(b) = f^{p+1}(a) = f^{p+1}(d_X(a_1)) = d_Y(f^p(a_1)),$$

so $d_Y(b - f^p(a_1)) = 0$, that is $b - f^p(a_1)$ is a cycle in Y^p. Since $\mathrm{Im}\, f^p = \mathrm{Ker}\, g^p$ we have $g^p \circ f^p = 0$, which implies that

$$c = g^p(b) = g^p(b - f^p(a)).$$

It follows that $(g^p)^*([b - f^p(a)]) = [c]$, proving that $\mathrm{Ker}\, \delta^p \subseteq \mathrm{Im}\, (g^p)^*$.

For any $[c] \in H^p(Z)$, since $\delta^p([c]) = [a]$ where $f^{p+1}(a) = d_Y(b)$ for any $b \in Y^p$ such that $g^p(b) = c$, as $d_Y(b)$ is a coboundary we have

$$(f^{p+1})^*(\delta^p([c])) = (f^{p+1})^*([a]) = [f^{p+1}(a)] = [d_Y(b)] = 0,$$

and thus $\mathrm{Im}\, \delta^p \subseteq \mathrm{Ker}\, (f^{p+1})^*$.

Conversely, assume that $(f^{p+1})^*([a]) = 0$, for some $a \in X^{p+1}$ with $d_X(a) = 0$, which means that $f^{p+1}(a) = d_Y(b)$ for some $b \in Y^p$. Since $\mathrm{Im}\, f^{p+1} = \mathrm{Ker}\, g^{p+1}$ we have $g^{p+1} \circ f^{p+1} = 0$, so by commutativity of the middle right square

$$d_Z(g^p(b)) = g^{p+1}(d_Y(b)) = g^{p+1}(f^{p+1}(a)) = 0,$$

which means that $g^p(b)$ is a cocycle in Z^p, and since $f^{p+1}(a) = d_Y(b)$ by definition of δ^p

$$\delta^p([g^p(b)]) = [a],$$

showing that $\mathrm{Ker}\, (f^{p+1})^* \subseteq \mathrm{Im}\, \delta^p$. $\square$

The maps $\delta^p \colon H^p(Z) \to H^{p+1}(X)$ are called *connecting homomorphisms*. The kind of argument used to prove Theorem 2.22 is known as *diagram chasing*.

Remark: The construction of the maps $\delta^p \colon H^p(Z) \to H^{p+1}(X)$ is often obtained as a corollary of the *snake lemma*. This is the approach followed in the classical texts by Mac Lane [Mac Lane (1975)] and Cartan–Eilenberg [Cartan and Eilenberg (1956)]. These books assume that the reader already has a fair amount of background in algebraic topology and the proofs are often rather terse or left to reader as “easy exercises” in diagram chasing.

Bott and Tu [Bott and Tu (1986)] refer to Mac Lane for help but as we just said Mac Lane leaves many details as exercises to the reader. More recent texts such as Munkres [Munkres (1984)], Rotman [Rotman (1988, 2009)], Madsen and Tornehave [Madsen and Tornehave (1998)], Tu [Tu (2008)] and Hatcher [Hatcher (2002)] show more compassion for the reader and provide much more details. Still, except for Hatcher and Munkres who give all the steps of the proof (for homology, and sometimes quickly) certain steps are left as "trivial" exercises (for example, step (e)). At the risk of annoying readers who have some familiarity with homological algebra we decided to provide all gory details of the proof so that readers who are novice in this area have a place to fall back if they get stuck, even if these proofs are not particularly illuminating (and rather tedious).

The assignment of a long exact sequence of cohomology to a short exact sequences of complexes is "natural" in the sense that it also applies to morphisms of short exact sequences of complexes.

Definition 2.14. Given two short exact sequences of cochain complexes

$$0 \longrightarrow X \xrightarrow{f} Y \xrightarrow{g} Z \longrightarrow 0$$

and

$$0 \longrightarrow X' \xrightarrow{f'} Y' \xrightarrow{g'} Z' \longrightarrow 0,$$

a *morphism* between these two exact sequences is a commutative diagram

$$\begin{array}{ccccccccc} 0 & \longrightarrow & X & \xrightarrow{f} & Y & \xrightarrow{g} & Z & \longrightarrow & 0 \\ & & \downarrow{\scriptstyle\alpha} & & \downarrow{\scriptstyle\beta} & & \downarrow{\scriptstyle\gamma} & & \\ 0 & \longrightarrow & X' & \xrightarrow[f']{} & Y' & \xrightarrow[g']{} & Z' & \longrightarrow & 0, \end{array}$$

where α, β, γ are chain maps.

The following proposition gives a precise meaning to the naturality of the assignment of a long exact sequence of cohomology to a short exact sequences of complexes.

Proposition 2.23. *For any morphism of exact sequences of cochain complexes*

$$\begin{array}{ccccccccc}
0 & \longrightarrow & X & \xrightarrow{f} & Y & \xrightarrow{g} & Z & \longrightarrow & 0 \\
 & & \downarrow{\scriptstyle\alpha} & & \downarrow{\scriptstyle\beta} & & \downarrow{\scriptstyle\gamma} & & \\
0 & \longrightarrow & X' & \xrightarrow[f']{} & Y' & \xrightarrow[g']{} & Z' & \longrightarrow & 0,
\end{array}$$

the following diagram of cohomology commutes.

$$\begin{array}{ccccccccc}
\longrightarrow & H^p(X) & \xrightarrow{f^*} & H^p(Y) & \xrightarrow{g^*} & H^p(Z) & \xrightarrow{\delta^p} & H^{p+1}(X) & \longrightarrow \\
 & \downarrow{\scriptstyle\alpha^*} & & \downarrow{\scriptstyle\beta^*} & & \downarrow{\scriptstyle\gamma^*} & & \downarrow{\scriptstyle\alpha^*} & \\
\longrightarrow & H^p(X') & \xrightarrow[(f')^*]{} & H^p(Y') & \xrightarrow[(g')^*]{} & H^p(Z') & \xrightarrow[(\delta')^p]{} & H^{p+1}(X') & \longrightarrow
\end{array}$$

Proof. A proof of Proposition 2.23 for homology can be found in Munkres [Munkres (1984)] (Chapter 3, Section 24, Theorem 24.2) and Hatcher [Hatcher (2002)] (Chapter 2, Section 2.1). The proof is a "diagram chasing" argument which can be modified to apply to cohomology as we now show. The first two squares commute because they already commute at the cochain level by definition of a morphism so we only have to prove that the third square commutes.

Recall how $\delta^p(\xi)$ is defined where $\xi = [c] \in H^p(Z)$ is represented by a cocycle $c \in Z^p$: pick any $b \in Y^p$ such that $g^p(b) = c$, push b down to Y^{p+1} by applying d_Y obtaining $d_Y(b)$, and then pull $d_Y(b)$ back to X^{p+1} by applying $(f^{p+1})^{-1}$, obtaining $a = (f^{p+1})^{-1}(d_Y(b))$; set $\delta^p([c]) = [a]$. Schematically,

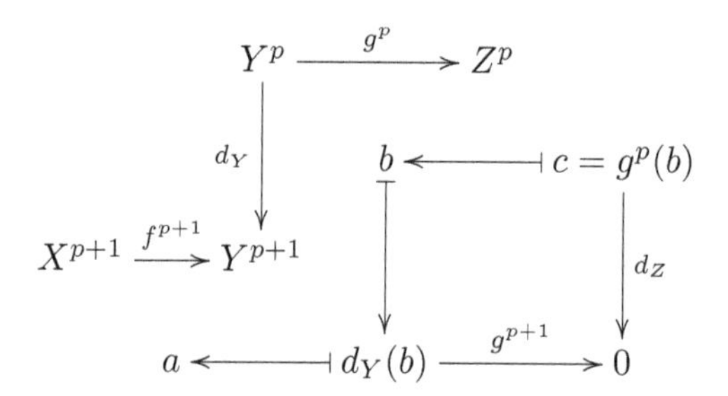

Since $a \in X^{p+1}$ is a cocycle and α is a chain map $\alpha(a) \in X'^{p+1}$ is a cocycle. Similarly $\gamma(c) \in Z'^{p+1}$ is a cocycle, and by definition $\gamma^*([c]) = [\gamma(c)]$. We

claim that

$$(\delta')^p([\gamma(c)]) = [\alpha(a)].$$

Since $c = g^p(b)$ we have $\gamma(c) = \gamma \circ g^p(b)$ and since the diagram

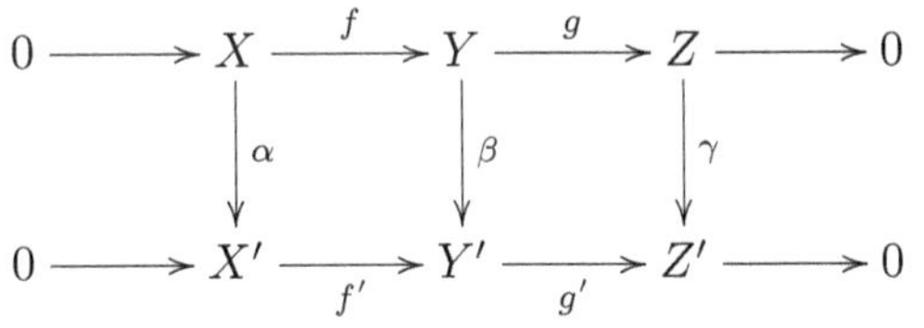

$$\begin{array}{ccccccccc} 0 & \longrightarrow & X & \xrightarrow{f} & Y & \xrightarrow{g} & Z & \longrightarrow & 0 \\ & & \downarrow{\scriptstyle \alpha} & & \downarrow{\scriptstyle \beta} & & \downarrow{\scriptstyle \gamma} & & \\ 0 & \longrightarrow & X' & \xrightarrow[f']{} & Y' & \xrightarrow[g']{} & Z' & \longrightarrow & 0 \end{array} \qquad (*)$$

commutes, we have $\gamma(c) = \gamma \circ g^p(b) = g'^p \circ \beta(b)$. Consider the following diagram:

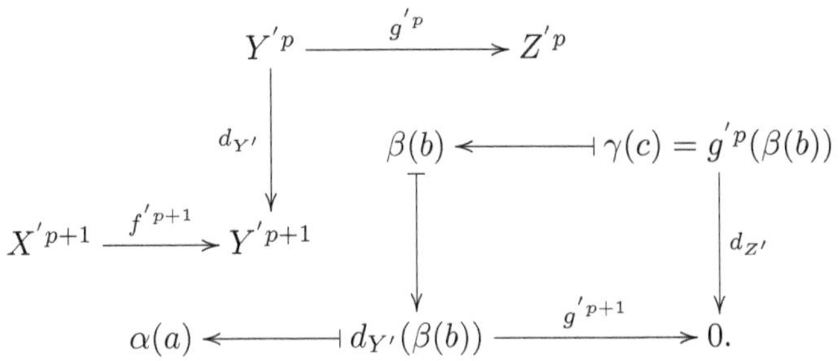

$$\begin{array}{ccccccc} & & Y'^p & \xrightarrow{g'^p} & & Z'^p & \\ & & \downarrow{\scriptstyle d_{Y'}} & & \beta(b) & \longleftarrow\!\shortmid & \gamma(c) = g'^p(\beta(b)) \\ X'^{p+1} & \xrightarrow{f'^{p+1}} & Y'^{p+1} & & \downarrow & & \downarrow{\scriptstyle d_{Z'}} \\ & \alpha(a) & \longleftarrow\!\shortmid & d_{Y'}(\beta(b)) & \xrightarrow{g'^{p+1}} & 0. & \end{array}$$

By commutativity of the diagram $(*)$, the fact that β is a chain map, and since $f^{p+1}(a) = d_Y(b)$, we have

$$f'^{p+1}(\alpha(a)) = \beta(f^{p+1}(a)) = \beta(d_Y(b)) = d_{Y'}(\beta(b)),$$

which shows that $(\delta')^p([\gamma(c)]) = [\alpha(a)]$. This part of the proof is illustrated in Figure 2.1.

But $\delta^p([c]) = [a]$, so we get

$$(\delta')^p(\gamma^*([c])) = (\delta')^p([\gamma(c)]) = [\alpha(a)] = \alpha^*([a]) = \alpha^*(\delta^p([c])),$$

namely

$$(\delta')^p \circ \gamma^* = \alpha^* \circ \delta^p,$$

as claimed. ☐

In the next chapter we discuss an example of a long exact sequence of cohomology arising from two open subsets U_1, U_2 of a manifold M that involves the cohomology space $H^p(U_1 \cup U_2)$ and the cohomology spaces $H^{p-1}(U_1 \cap U_2)$, $H^p(U_1)$ and $H^p(U_2)$. This long exact sequence is known as the *Mayer–Vietoris sequence*. If U is covered by a finite family $(U_i)_{i=1}^r$ of open sets and if this family is a "good cover," then by an inductive argument involving the *Mayer–Vietoris sequence* it is possible to prove that the cohomology spaces $H^p(U)$ are finite-dimensional.

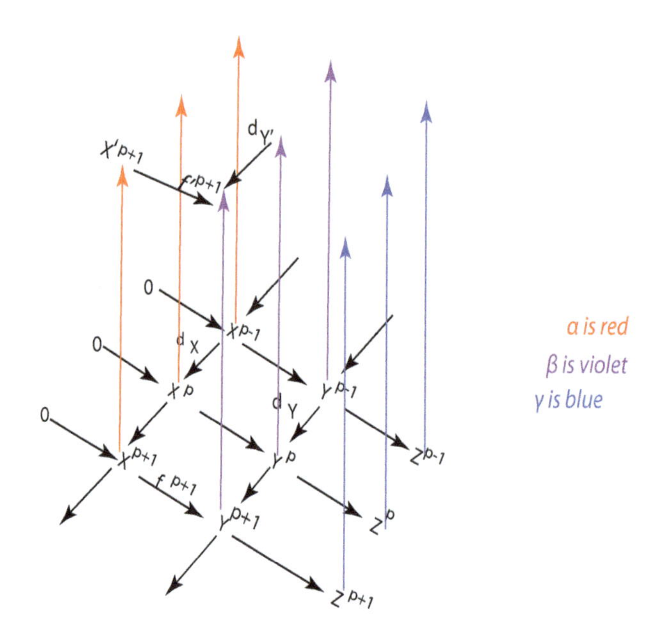

Fig. 2.1 Illustration for the proof of Proposition 2.23.

2.8 Problems

Problem 2.1. Prove Proposition 2.13.

Problem 2.2. Prove Proposition 2.14.

Problem 2.3. Prove Proposition 2.15.

Problem 2.4. Provide the details of Proposition 2.18.

Problem 2.5. (The snake lemma) Consider the commutative diagram shown below:

$$\begin{array}{ccccccc} & & A & \longrightarrow & B & \xrightarrow{\sigma} & C & \longrightarrow & 0 \\ & & \downarrow{\scriptstyle\alpha} & & \downarrow{\scriptstyle\beta} & & \downarrow{\scriptstyle\gamma} & & \\ 0 & \longrightarrow & A' & \xrightarrow[\kappa']{} & B' & \xrightarrow[\sigma']{} & C'. & & \end{array}$$

(1) Prove that if the rows are exact, then there is a map $D_*\colon \operatorname{Ker}\gamma \to \operatorname{Coker}\alpha$ such that the sequence

$$\operatorname{Ker}\alpha \longrightarrow \operatorname{Ker}\beta \longrightarrow \operatorname{Ker}\gamma \xrightarrow{D_*} \operatorname{Coker}\alpha \longrightarrow \operatorname{Coker}\beta \longrightarrow \operatorname{Coker}\gamma$$

is exact.

(2) Consider a short exact sequence

$$0 \longrightarrow X' \longrightarrow X \longrightarrow X'' \longrightarrow 0$$

of *chain* complexes X', X, X''. Prove that the commutative diagram

$$\begin{array}{ccccccc} & & X'_n/\operatorname{Im} d'_{n+1} & \longrightarrow & X_n/\operatorname{Im} d_{n+1} & \longrightarrow & X''_n/\operatorname{Im} d''_{n+1} & \longrightarrow 0 \\ & & \downarrow{\scriptstyle \Delta'} & & \downarrow{\scriptstyle \Delta} & & \downarrow{\scriptstyle \Delta''} & \\ 0 & \longrightarrow & Z_{n-1}(X') & \longrightarrow & Z_{n-1}(X) & \longrightarrow & Z_{n-1}(X'') & \end{array}$$

has exact rows, where the middle vertical map is defined such that $\Delta([x]) = d_n x$, for any $x \in X_n/\operatorname{Im} d_{n+1}$, and similarly for the other two maps.

Use (1) to prove the existence of the *long exact sequence of homology*, namely that there is an exact sequence

$$H_n(X') \longrightarrow H_n(X) \longrightarrow H_n(X'') \xrightarrow{\partial_n}$$
$$H_{n-1}(X') \longrightarrow H_{n-1}(X) \longrightarrow H_{n-1}(X'')$$

for all n.

Problem 2.6. Consider the following commutative diagram

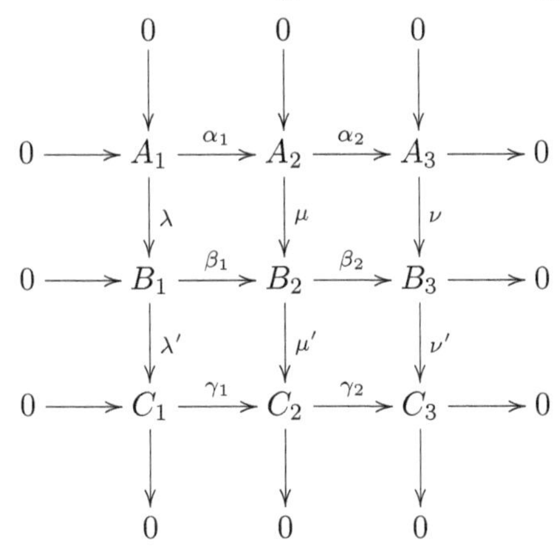

and suppose that all three columns and the first two rows are short exact. Prove that the third row is also short exact.

Hint. Use Problem 2.5.

Chapter 3

de Rham Cohomology

Differential forms offer a quick and rather easy approach to the cohomology groups (with real coefficients) of smooth manifolds. This approach was pioneered by Georges de Rham in the early 1930s. If M is a smooth manifold, then there is the *de Rham complex*

$$\mathcal{A}^0(M) \xrightarrow{d^0} \mathcal{A}^1(M) \xrightarrow{d^1} \mathcal{A}^2(M) \xrightarrow{d^2} \cdots \xrightarrow{d^{p-1}} \mathcal{A}^p(M) \xrightarrow{d^p} \mathcal{A}^{p+1}(M) \xrightarrow{d^{p+1}} \cdots$$

which uses the modules of smooth p-forms $\mathcal{A}^p(M)$ and the exterior derivatives $d^p \colon \mathcal{A}^p(M) \to \mathcal{A}^{p+1}(M)$. The corresponding cohomology groups are the *de Rham cohomology groups* $H^p_{\mathrm{dR}}(M)$. These are actually real vector spaces. This chapter offers a brief presentation of de Rham cohomology.

This chapter assumes a certain background in differential geometry, in particular, differential forms. However, although it gives a nice preview of some of the main themes of cohomology, such as Poincaré duality, it can be safely omitted. Readers who wish to review differential forms are referred to the excellent presentations in Tu [Tu (2008)], Morita [Morita (2001)], Madsen and Tornehave [Madsen and Tornehave (1998)], and Bott and Tu [Bott and Tu (1986)]. A detailed exposition, including an extensive review of tensor algebra, is also provided in Gallier and Quaintance [Gallier and Quaintance (2020b)].

In Section 3.1 we introduce the de Rham cohomology groups $H^p_{\mathrm{dR}}(M)$ and the de Rham cohomology groups with compact support $H^p_{\mathrm{dR,c}}(M)$. We state the Poincaré lemma which describes the de Rham cohomology of $\mathbb{R}^n$ (and the de Rham cohomology with compact support of $\mathbb{R}^n$).

In Section 3.2 we introduce an important tool, the Mayer–Vietoris argument. Let M be a smooth manifold and assume that $M = U_1 \cup U_2$ for two open subsets U_1 and U_2 of M. The Mayer–Vietoris argument makes use of an exact sequence which relates the cohomology of $M = U_1 \cup U_2$ to the cohomology of U_1, U_2 and $U_1 \cap U_2$. This method does not work of all covers, but it works for special covers called *good covers*. Every smooth manifold has a good cover, and a compact manifold has a finite good cover. We prove that if a manifold M has a finite good cover, then its cohomology groups are finite-dimensional vector spaces.

In Section 3.3 we discuss Poincaré duality for smooth orientable manifolds without boundary. Poincaré duality is a deep result which shows that the cohomology of a compact orientable manifold exhibits a fundamental symmetry. Technically, Poincaré duality states that if M is a smooth oriented manifold with a finite good cover, then we have isomorphisms

$$H^p(M) \cong (H_c^{n-p}(M))^*$$

for all p with $0 \leq p \leq n$, where $(H_c^{n-p}(M))^*$ is the algebraic dual of the vector space $H_c^{n-p}(M)$, the space of $\mathbb{R}$-linear forms on $H_c^{n-p}(M)$). In particular, if M is compact, then

$$H^p(M) \cong (H^{n-p}(M))^*$$

for all p with $0 \leq p \leq n$.

3.1 Review of de Rham Cohomology

Let M be a smooth manifold. The de Rham cohomology is based on differential forms. If $\mathcal{A}^p(M)$ denotes the real vector space of smooth p-forms on M, then we know that there is a mapping $d^p \colon \mathcal{A}^p(M) \to \mathcal{A}^{p+1}(M)$ called *exterior differentiation*, and d^p satisfies the crucial property

$$d^{p+1} \circ d^p = 0 \quad \text{for all } p \geq 0.$$

Recall that $\mathcal{A}^0(M) = C^\infty(M)$, the space of all smooth (real-valued) functions on M.

Definition 3.1. The sequence of vector spaces and linear maps between them satisfying $d^{p+1} \circ d^p = 0$ given by

$$\mathcal{A}^0(M) \xrightarrow{d^0} \mathcal{A}^1(M) \xrightarrow{d^1} \mathcal{A}^2(M) \xrightarrow{d^2} \cdots \xrightarrow{d^{p-1}} \mathcal{A}^p(M) \xrightarrow{d^p} \mathcal{A}^{p+1}(M) \xrightarrow{d^{p+1}} \cdots$$

is called a *differential complex*.

We can package together the vector spaces $\mathcal{A}^p(M)$ as the direct sum $\mathcal{A}^*(M)$ given by

$$\mathcal{A}^*(M) = \bigoplus_{p \geq 0} \mathcal{A}^p(M)$$

called the *de Rham complex* of M, and the family of maps (d^p) as the map

$$d\colon \mathcal{A}^*(M) \to \mathcal{A}^*(M),$$

where d on the pth summand $\mathcal{A}^p(M)$ is equal to d^p, so that

$$d \circ d = 0.$$

The direct sum $\mathcal{A}^*(M)$ is an example of the general concept of a graded vector space defined below.

Definition 3.2. A *gradation* of a vector space V is family (V_p) of subspaces $V_p \subseteq V$ such that

$$V = \bigoplus_{p \geq 0} V_p.$$

In this case, we say that V is a *graded vector space*.

The map d is an *anti-derivation*, which means that

$$d(\omega \wedge \tau) = d\omega \wedge \tau + (-1)^p \omega \wedge d\tau, \quad \omega \in \mathcal{A}^p(M), \tau \in \mathcal{A}^q(M).$$

For example, if $M = \mathbb{R}^3$, then

$$d^0\colon \mathcal{A}^0(M) \to \mathcal{A}^1(M)$$

correspond to grad,

$$d^1\colon \mathcal{A}^1(M) \to \mathcal{A}^2(M)$$

corresponds to curl, and

$$d^2\colon \mathcal{A}^2(M) \to \mathcal{A}^3(M)$$

corresponds to div.

In fact, $\mathcal{A}^*(U)$ is defined for every open subset U of M, and $\mathcal{A}^*$ is a sheaf of differential complexes.

Definition 3.3. A form $\omega \in \mathcal{A}^p(M)$ is *closed* if

$$d\omega = 0,$$

exact if

$$\omega = d\tau \quad \text{for some } \tau \in \mathcal{A}^{p-1}(M).$$

Let $Z^p(M)$ denote the subspace of $\mathcal{A}^p(M)$ consisting of closed p-forms, $B^p(M)$ denote the subspace of $\mathcal{A}^p(M)$ consisting of exact p-forms, with $B^0(M) = (0)$ (the trivial vector space), and let

$$Z^*(M) = \bigoplus_{p\geq 0} Z^p(M), \quad B^*(M) = \bigoplus_{p\geq 0} B^p(M).$$

Since $d \circ d = 0$, we have $B^p(M) \subseteq Z^p(M)$ for all $p \geq 0$ but the converse is generally false.

Definition 3.4. The *de Rham cohomology* of a smooth manifold M is the real vector space $H^*_{\mathrm{dR}}(M)$ given by the direct sum

$$H^*_{\mathrm{dR}}(M) = \bigoplus_{p\geq 0} H^p_{\mathrm{dR}}(M),$$

where the cohomology group (actually, real vector space) $H^p_{\mathrm{dR}}(M)$ is the quotient vector space

$$H^p_{\mathrm{dR}}(M) = Z^p(M)/B^p(M).$$

Thus, the cohomology group (vector space) $H^*_{\mathrm{dR}}(M)$ gives some measure of the failure of closed forms to be exact.

Note that by definition $H^*_{\mathrm{dR}}(M)$ is a graded vector space. Furthermore, exterior multiplication in $\mathcal{A}^*(M)$ induces a ring structure on the vector space $H^*_{\mathrm{dR}}(M)$. First it is clear by definition that

$$B^*(M) \subseteq Z^*(M) \subseteq \mathcal{A}^*(M).$$

Proposition 3.1. *The vector space $Z^*(M)$ is a subring of $\mathcal{A}^*(M)$ and $B^*(M)$ is an ideal in $Z^*(M)$.*

Proof. Assume that $d\omega = 0$ and $d\tau = 0$ for some $\omega \in Z^p(M)$ and some $\tau \in Z^q(M)$. Then since d is an anti-derivation, we have

$$d(\omega \wedge \tau) = d\omega \wedge \tau + (-1)^p \omega \wedge d\tau = 0 \wedge \tau + (-1)^p \omega \wedge 0 = 0,$$

which shows that $\omega \wedge \tau \in Z^*(M)$. Therefore, $Z^*(M)$ is a subring of $\mathcal{A}^*(M)$.

Next assume that $\omega \in Z^p(M)$ and $\tau \in B^q(M)$, so that $d\omega = 0$ and $\tau = d\alpha$ for some $\alpha \in \mathcal{A}^{q-1}(M)$. Then, we have

$$d(\omega \wedge (-1)^p \alpha) = d\omega \wedge (-1)^p \alpha + (-1)^p \omega \wedge (-1)^p d\alpha = 0 \wedge (-1)^p \alpha + \omega \wedge \tau = \omega \wedge \tau,$$

which shows that $\omega \wedge \tau \in B^*(M)$, so $B^*(M)$ is an ideal in $Z^*(M)$. □

Since $B^*(M)$ is an ideal in $Z^*(M)$, the quotient ring $Z^*(M)/B^*(M)$ is well-defined, and $H^*_{\mathrm{dR}}(M) = Z^*(M)/B^*(M)$ is a ring under the multiplication induced by $\wedge$. Therefore, $H^*_{\mathrm{dR}}(M)$ is an $\mathbb{R}$-algebra.

A variant of de Rham cohomology is *de Rham cohomology with compact support*.

Definition 3.5. The *de Rham cohomology with compact support* is obtained by considering the vector space $\mathcal{A}^*_c(M)$ of differential forms with compact support. As before, we have the subspaces $B^*_c(M) \subseteq Z^*_c(M)$, and we let

$$H^*_{\mathrm{dR},c}(M) = Z^*_c(M)/B^*_c(M).$$

The *Poincaré's Lemmas* are the following results:

Proposition 3.2. *The following facts hold:*

$$H^p_{\mathrm{dR}}(\mathbb{R}^n) = \begin{cases} 0 & \textit{if } p \neq 0 \\ \mathbb{R} & \textit{if } p = 0, \end{cases}$$

and

$$H^p_{\mathrm{dR},c}(\mathbb{R}^n) = \begin{cases} 0 & \textit{if } p \neq n \\ \mathbb{R} & \textit{if } p = n. \end{cases}$$

These facts also hold if $\mathbb{R}^n$ is replaced by any nonempty convex subset of $\mathbb{R}^n$ (or even a star-shaped subset of $\mathbb{R}^n$).

For a proof of Proposition 3.2, see Bott and Tu [Bott and Tu (1986)], Chapter 1.

3.2 The Mayer–Vietoris Argument

Let M be a smooth manifold and assume that $M = U_1 \cup U_2$ for two open subsets U_1 and U_2 of M. The Mayer–Vietoris argument makes use of an exact sequence which relates the cohomology of $M = U_1 \cup U_2$ to the cohomology of U_1, U_2 and $U_1 \cap U_2$. We obtain a method of proof which proceeds by induction on the size of the number of open subsets in an open cover. This method does not work of all covers, but it works for special covers called *good covers*. Fortunately, every smooth manifold has a good cover, and every compact (smooth) manifold has a finite good cover. The Mayer–Vietoris argument can be used to prove that the cohomology groups $H^p_{\mathrm{dR}}(M)$ of a manifold M with a finite good cover are finite-dimensional.

It can also be used to prove that the cohomology groups $H^p_{\mathrm{dR,c}}(M)$ with compact support of a smooth manifold M with a finite good cover are finite-dimensional. The Mayer–Vietoris argument can also be used to prove a version of Poincaré duality.

The inclusion maps $i_k\colon U_k \to M$ and $j_k\colon U_1 \cap U_2 \to U_k$ for $k = 1, 2$ induce a pullback map $f\colon \mathcal{A}^*(M) \to \mathcal{A}^*(U_1) \oplus \mathcal{A}^*(U_2)$ given by $f = (i_1^*, i_2^*)$ and a pullback map $g\colon \mathcal{A}^*(U_1) \oplus \mathcal{A}^*(U_2) \to \mathcal{A}^*(U_1 \cap U_2)$ given by $g = j_1^* - j_2^*$. We have the following short exact sequence.

Proposition 3.3. *For any smooth manifold M, if $M = U_1 \cup U_2$ for any two open subsets U_1 and U_2, then we have the short exact sequence*

$$0 \longrightarrow \mathcal{A}^*(M) \xrightarrow{f} \mathcal{A}^*(U_1) \oplus \mathcal{A}^*(U_2) \xrightarrow{g} \mathcal{A}^*(U_1 \cap U_2) \longrightarrow 0.$$

Proof. The proof is not really difficult. It involves the use of a partition of unity. For details, see Bott and Tu [Bott and Tu (1986)] (Chapter 1, Proposition 2.3) or Madsen and Tornehave [Madsen and Tornehave (1998)] (Chapter 5, Theorem 5.1). □

The short exact sequence given by Proposition 3.3 is called the *Mayer–Vietoris sequence*.

If we apply Theorem 2.22 to the Mayer–Vietoris sequence we obtain the long Mayer–Vietoris cohomology sequence shown below:

$$\cdots \longrightarrow H^{p-1}_{\mathrm{dR}}(U_1 \cap U_2) \xrightarrow{\delta^{p-1}} H^p_{\mathrm{dR}}(M) \xrightarrow{f^*} H^p_{\mathrm{dR}}(U_1) \oplus H^p_{\mathrm{dR}}(U_2) \xrightarrow{g^*} H^p_{\mathrm{dR}}(U_1 \cap U_2) \xrightarrow{\delta^p}$$
$$H^{p+1}_{\mathrm{dR}}(M) \xrightarrow{f^*} H^{p+1}_{\mathrm{dR}}(U_1) \oplus H^{p+1}_{\mathrm{dR}}(U_2) \xrightarrow{g^*} H^{p+1}_{\mathrm{dR}}(U_1 \cap U_2) \xrightarrow{\delta^{p+1}} H^{p+2}_{\mathrm{dR}}(M) \longrightarrow \cdots$$

(for all p).

This long exact sequence implies that

$$H^p_{\mathrm{dR}}(M) \cong \mathrm{Im}\,\delta^{p-1} \oplus \mathrm{Im}\,f^*;$$

see the paragraph in Section 2.1 just after Equation ($*_{\mathrm{cok}}$). It follows that if the spaces $H^{p-1}_{\mathrm{dR}}(U_1 \cap U_2)$, $H^{P}_{\mathrm{dR}}(U_1)$ and $H^{P}_{\mathrm{dR}}(U_2)$ are finite-dimensional, then so is $H^{p}_{\mathrm{dR}}(M)$. This suggests an inductive argument on the number of open subsets in a finite cover of M. For this argument to succeed, such covers must have some special properties about intersections of these open subsets; Bott and Tu call them *good covers*.

Definition 3.6. Given a smooth manifold M of dimension n, an open cover $\mathcal{U} = (U_\alpha)_{\alpha \in I}$ of M is called a *good cover* if all finite nonempty intersections $U_{\alpha_1} \cap \cdots \cap U_{\alpha_p}$ are diffeomorphic to $\mathbb{R}^n$. A manifold which has a finite good cover is said to be of *finite type*. See Figure 3.1.

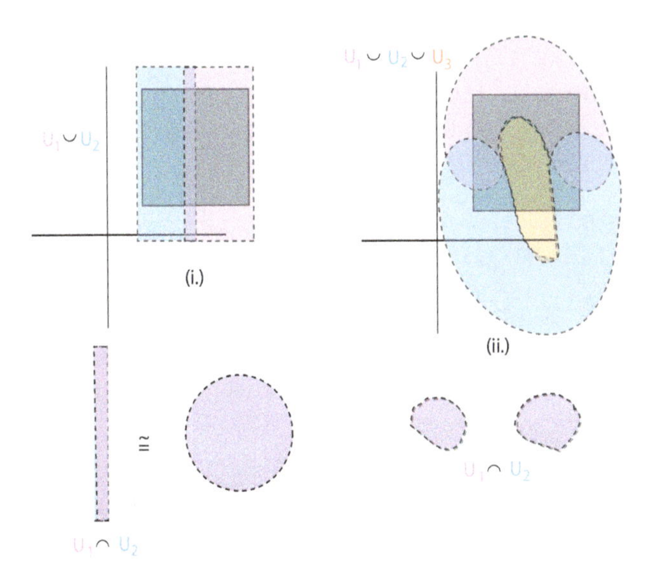

Fig. 3.1 The manifold M is an open unit square of $\mathbb{R}^2$. Figure (i.) is a good cover of M while Figure (ii.) is not a good cover of M since $U_1 \cap U_2$ is isomorphic to the disjoint union of two open disks.

Fortunately, every smooth manifold has a good cover.

Theorem 3.4. *Every smooth manifold M has a good cover. If M is a compact manifold, then M has a finite good cover.*

Proof Sketch. A detailed proof can be found in Bott and Tu [Bott and Tu (1986)], Chapter 1. The proof of Theorem 3.4 makes use of some differential geometry. First, using a partition of unity argument we can prove that every manifold has a Riemannian metric.

The second step uses the fact that in a Riemannian manifold, every point p has a geodesically convex neighborhood U, which means that any two points $p_1, p_2 \in U$ can be joined by a geodesic that stays inside U. Now any intersection of geodesically convex neighborhoods is still geodesically convex, and a geodesically convex neighborhood is diffeomorphic to $\mathbb{R}^n$, so any open cover consisting of geodesically convex open subsets is a good cover. $\square$

The above argument can be easily adapted to prove that every open cover of a manifold can be refined to a good open cover.

We can now prove that the de Rham cohomoloy spaces of a manifold endowed with a finite good cover are finite-dimensional. To simply notation, we write H^p instead of H^p_{dR}.

Theorem 3.5. *If a manifold M has a finite good cover, then the cohomology vector spaces $H^p(M)$ are finite-dimensional for all $p \geq 0$.*

Proof. We proceed by induction on the number of open sets in a good cover $(V_1, \ldots, V_p)$. If $p = 1$, then V_1 itself is diffeomorphic to $\mathbb{R}^n$, and by the Poincaré lemma (Proposition 3.2) the cohomology spaces are either (0) or $\mathbb{R}^n$. Thus, the base case holds.

For the induction step, assume that the cohomology of a manifold having a good cover with at most p open sets is finite-dimensional, and let $\mathcal{U} = (V_1, \ldots, V_{p+1})$ be a good cover with $p+1$ open subsets. The open subset $(V_1 \cup \cdots \cup V_p) \cap V_{p+1}$ has a good cover with p open subsets, namely $(V_1 \cap V_{p+1}, \ldots, V_p \cap V_{p+1})$. See Figures 3.2 and 3.3. By the induction hypothesis, the vector spaces $H^p(V_1 \cup \cdots \cup V_p)$, $H^p(V_{p+1})$ and $H^p((V_1 \cup \cdots \cup V_p) \cap V_{p+1})$ are finite-dimensional for all p, so by the consequence of the long Mayer–Vietoris cohomology sequence stated just before Definition 3.6, with $M = V_1 \cup \cdots \cup V_{p+1}$, $U_1 = V_1 \cup \cdots \cup V_p$, and $U_2 = V_{p+1}$, we conclude that the vector spaces $H^p(V_1 \cup \cdots \cup V_{p+1})$ are finite-dimensional for all p, which concludes the induction step. $\square$

As a special case of Theorem 3.5, we see that the cohomology of any compact manifold is finite-dimensional.

Fig. 3.2 A good cover of S^2 consisting of four open sets. Note $V_1 \cap V_2 = V_3 \cap V_4 = \emptyset$.

A similar result holds de Rham cohomology with compact support, but we have to be a little careful because in general, the pullback of a form with compact support by a smooth map may *not* have compact support. Fortunately, the Mayer–Vietoris sequence only needs inclusion maps between open sets.

Given any two open subsets U, V of M, if $U \subseteq V$ and $i\colon U \to V$ is the inclusion map, there is an induced map $i_*\colon \mathcal{A}^p_c(U) \to \mathcal{A}^p_c(V)$ defined such that

$$\begin{aligned} (i_*(\omega))(p) &= \omega(p) && \text{if } p \in U \\ (i_*(\omega))(p) &= 0 && \text{if } p \in V - \operatorname{supp} \omega. \end{aligned}$$

We say that ω has been extended to V by zero. Notice that unlike the definition of the pullback $f^*\omega$ of a form $\omega \in \mathcal{A}^p(V)$ by a smooth map $f\colon U \to V$ where $f^*\omega \in \mathcal{A}^p(U)$, the map i_* pushes a form $\omega \in \mathcal{A}^p_c(U)$ *forward* to a form $i_*\omega \in \mathcal{A}^p_c(V)$. If $i\colon U \to V$ and $j\colon V \to W$ are two inclusions, then $(j \circ i)_* = j_* \circ i_*$, with no reversal of the order of i_* and j_*.

Let M be a smooth manifold and assume that $M = U_1 \cup U_2$ for two open subsets U_1 and U_2 of M. The inclusion maps $i_k\colon U_k \to M$ and $j_k\colon U_1 \cap U_2 \to$

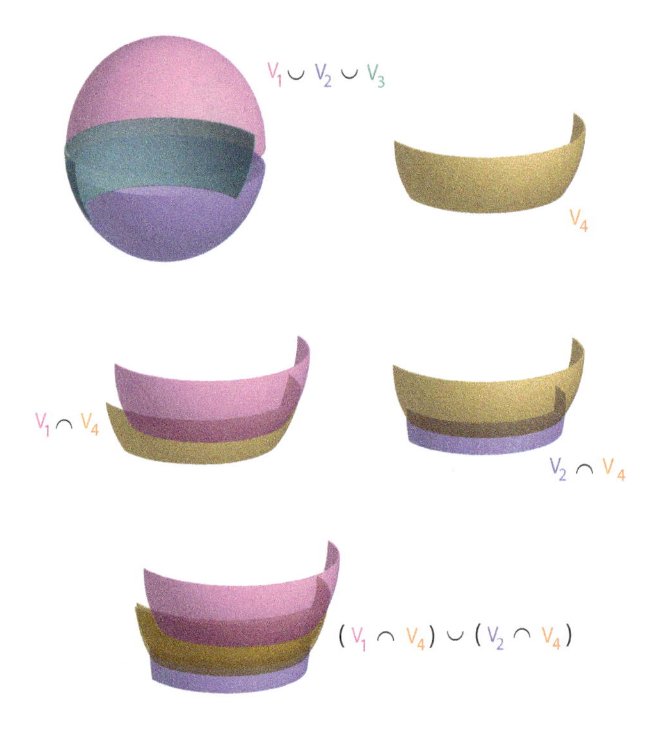

Fig. 3.3 The inductive good cover construction applied to $V_1 \cup V_2 \cup V_3 \cup V_4$, a good cover of S^2.

U_k for $k = 1, 2$ induce a map $s\colon \mathcal{A}_c^*(U_1) \oplus \mathcal{A}_c^*(U_2) \to \mathcal{A}_c^*(M)$ given by $s(\omega_1, \omega_2) = (i_1)_*(\omega_1) + (i_2)_*(\omega_2)$ and a map $j\colon \mathcal{A}_c^*(U_1 \cap U_2) \to \mathcal{A}_c^*(U_1) \oplus \mathcal{A}_c^*(U_2)$ given by $j(\omega) = ((j_1)_*(\omega), -(j_2)_*(\omega))$. We have the following short exact sequence called the *Mayer–Vietoris sequence for cohomology with compact support*.

Proposition 3.6. *For any smooth manifold M, if $M = U_1 \cup U_2$ for any two open subsets U_1 and U_2, then we have the short exact sequence*

$$0 \longrightarrow \mathcal{A}_c^*(U_1 \cap U_2) \stackrel{j}{\longrightarrow} \mathcal{A}_c^*(U_1) \oplus \mathcal{A}_c^*(U_2) \stackrel{s}{\longrightarrow} \mathcal{A}_c^*(M) \longrightarrow 0.$$

For a proof of Proposition 3.6, see Bott and Tu [Bott and Tu (1986)] (Chapter 1, Proposition 2.7). Observe that compared to the Mayer–Vietoris sequence of Proposition 3.3, the direction of the arrows is reversed.

If we apply Theorem 2.22 to the Mayer–Vietoris sequence of Proposition 3.6 we obtain the long Mayer–Vietoris sequence for cohomology with compact support shown below:

$$\cdots \longrightarrow H^{p-1}_{\mathrm{dR,c}}(M) \xrightarrow{\delta_c^{p-1}}$$

$$H^{p}_{\mathrm{dR,c}}(U_1 \cap U_2) \xrightarrow{j^*} H^{p}_{\mathrm{dR,c}}(U_1) \oplus H^{p}_{\mathrm{dR,c}}(U_2) \xrightarrow{s^*} H^{p}_{\mathrm{dR,c}}(M) \xrightarrow{\delta_c^{p}}$$

$$H^{p+1}_{\mathrm{dR,c}}(U_1 \cap U_2) \xrightarrow{j^*} H^{p+1}_{\mathrm{dR,c}}(U_1) \oplus H^{p+1}_{\mathrm{dR,c}}(U_2) \xrightarrow{s^*} H^{p+1}_{\mathrm{dR,c}}(M) \xrightarrow{\delta_c^{p+1}}$$

$$H^{p+2}_{\mathrm{dR,c}}(U_1 \cap U_2) \longrightarrow \cdots$$

(for all p). Then using the above sequence, the Poincaré lemma, and basically the same proof as in Theorem 3.5, we obtain the following result.

Theorem 3.7. *If a manifold M has a finite good cover, then the vector spaces $H^p_{\mathrm{dR,c}}(M)$ of cohomology with compact support are finite-dimensional for all $p \geq 0$.*

The long exact sequences of cohomology induced by Proposition 3.3 and Proposition 3.6 can be combined to prove a version of Poincaré duality. Following Bott and Tu [Bott and Tu (1986)] we give a brief presentation of this result.

3.3 Poincaré Duality on an Orientable Manifold

Let M be a smooth orientable manifold without boundary of dimension n. In this section, to simplify notation we write $H^p(M)$ for $H_{\mathrm{dR}}(M)$ and $H^p_c(M)$ for $H_{\mathrm{dR,c}}(M)$. For any form $\omega \in \mathcal{A}^p(M)$ and any form with compact support $\eta \in \mathcal{A}^{n-p}_c(M)$, the support of the n-form $\omega \wedge \eta$ is contained in both supports of ω and η, so $\omega \wedge \eta$ also has compact support and $\int_M \omega \wedge \eta$ makes sense. Since $B^*(M)$ is an ideal in $Z^*(M)$ and by Stokes' theorem $\int_M d\omega = 0$, we have a well-defined map

$$\langle -, - \rangle \colon H^p(M) \times H^{n-p}_c(M) \longrightarrow \mathbb{R}$$

defined by

$$\langle [\omega], [\eta] \rangle = \int_M \omega \wedge \eta,$$

for any closed form $\omega \in \mathcal{A}^p(M)$ and any closed form with compact support $\eta \in \mathcal{A}_c^{n-p}(M)$. The above map is clearly bilinear so it is a pairing. Recall that if the vector spaces $H^p(M)$ and $H_c^{n-p}(M)$ are finite-dimensional (which is the case if M has a finite good cover) and if the pairing is nondegenerate, then it induces a natural isomorphism between $H^p(M)$ and the dual space $(H_c^{n-p}(M))^*$ of $H_c^{n-p}(M)$.

Theorem 3.8. *(Poincaré duality) Let M be a smooth oriented n-dimensional manifold. If M has a finite good cover, then the map*

$$\langle -, - \rangle \colon H^p(M) \times H_c^{n-p}(M) \longrightarrow \mathbb{R}$$

is a nondegenerate pairing. This implies that we have isomorphisms

$$H^p(M) \cong (H_c^{n-p}(M))^*$$

for all p with $0 \leq p \leq n$. In particular, if M is compact then

$$H^p(M) \cong (H^{n-p}(M))^*$$

for all p with $0 \leq p \leq n$.

The proof of Theorem 3.8 uses induction on the size of a finite good cover for M. For the induction step, the long exact sequences of cohomology induced by Proposition 3.3 and Proposition 3.6 are combined in a clever way, and the five lemma (Proposition 2.5) is used. Proofs of Theorem 3.8 are given in Bott and Tu [Bott and Tu (1986)] (Chapter 1, Pages 44–46), and in more details in Madsen and Tornehave [Madsen and Tornehave (1998)] (Chapter 13).

The first step of the proof is to dualize the second long exact sequence of cohomology. It turns out that this yields an exact sequence, and for this we need the following proposition. This is actually a special case of Proposition 2.8, but it does not hurt to give a direct proof.

Proposition 3.9. *Let A, B, C be three vector spaces and let $\varphi \colon A \to B$ and $\psi \colon B \to C$ be two linear maps such that the sequence*

$$A \xrightarrow{\varphi} B \xrightarrow{\psi} C$$

is exact at B. Then the sequence

$$C^* \xrightarrow{\psi^\top} B^* \xrightarrow{\varphi^\top} A^*$$

is exact at B^*.

Proof. Recall that $\varphi^\top\colon B^* \to A^*$ is the linear map defined such that $\varphi^\top(f) = f \circ \varphi$ for every linear form $f \in B^*$ and similarly $\psi^\top\colon C^* \to B^*$ is given by $\psi^\top(g) = g \circ \psi$ for every linear form $g \in C^*$. The fact that the first sequence is exact at B means that $\operatorname{Im}\varphi = \operatorname{Ker}\psi$, which implies $\psi \circ \varphi = 0$, thus $\varphi^\top \circ \psi^\top = 0$, so $\operatorname{Im}\psi^\top \subseteq \operatorname{Ker}\varphi^\top$. Conversely, we need to prove that $\operatorname{Ker}\varphi^\top \subseteq \operatorname{Im}\psi^\top$.

Pick any $f \in \operatorname{Ker}\varphi^\top$, which means that $\varphi^\top(f) = 0$, that is $f \circ \varphi = 0$. Consequently $\operatorname{Im}\varphi \subseteq \operatorname{Ker} f$, and since $\operatorname{Im}\varphi = \operatorname{Ker}\psi$ we have

$$\operatorname{Ker}\psi \subseteq \operatorname{Ker} f.$$

We are going to construct a linear form $g \in C^*$ such that $f = g \circ \psi = \psi^\top(g)$. Observe that it suffices to construct such a linear form defined on $\operatorname{Im}\psi$, because such a linear form can then be extended to the whole of C.

Pick any basis $(v_i)_{i\in I}$ in $\operatorname{Im}\psi$, and let $(u_i)_{i\in I}$ be any family of vectors in B such that $\psi(u_i) = v_i$ for all $i \in I$. Then by a familiar argument $(u_i)_{i\in I}$ is linearly independent and it spans a subspace D of B such that

$$B = \operatorname{Ker}\psi \oplus D.$$

Define $g\colon C \to K$ such that

$$g(v_i) = f(u_i), \quad i \in I.$$

We claim that $f = g \circ \psi$.

Indeed, $f(u_i) = g(v_i) = (g \circ \psi)(u_i)$ for all $i \in I$, and if $w \in \operatorname{Ker}\psi$, since $\operatorname{Ker}\psi \subseteq \operatorname{Ker} f$, we have

$$f(w) = 0 = (g \circ \psi)(w) = 0.$$

Therefore, $f = g \circ \psi = \psi^\top(g)$, which shows that $f \in \operatorname{Im}\psi^\top$, as desired. $\square$

By applying Proposition 3.9 to the second long exact sequence of cohomology (of compact support), we obtain the following long exact sequence:

$$\cdots \longrightarrow H_c^{p+2}(U_1 \cap U_2)^* \xrightarrow{(\delta_c^{p+1})^\top} H_c^{p+1}(M)^* \xrightarrow{(s^*)^\top} H_c^{p+1}(U_1)^* \oplus H_c^{p+1}(U_2)^* \xrightarrow{(j^*)^\top} H_c^{p+1}(U_1 \cap U_2)^* \xrightarrow{(\delta_c^{p})^\top} H_c^{p}(M)^* \xrightarrow{(s^*)^\top} H_c^{p}(U_1)^* \oplus H_c^{p}(U_2)^* \xrightarrow{(j^*)^\top} H_c^{p}(U_1 \cap U_2)^* \xrightarrow{(\delta_c^{p-1})^\top} H_c^{p-1}(M)^* \longrightarrow \cdots$$

(for all p).

Let us denote by $\theta^p_M \colon H^p(M) \to (H^{n-p}_c(M))^*$ the isomorphism given by Theorem 3.8. The following propositions are shown in Bott and Tu [Bott and Tu (1986)] (Chapter 1, Lemma 5.6), and in Madsen and Tornehave [Madsen and Tornehave (1998)] (Chapter 13, Lemmas 13.6 and 13.7).

Proposition 3.10. *For any two open subsets U and V of a manifold M, if $U \subseteq V$ and $i \colon U \to V$ is the inclusion map, then the following diagrams commute for all p:*

$$\begin{array}{ccc} H^p(V) & \xrightarrow{\ i^*\ } & H^p(U) \\ \Big\downarrow{\scriptstyle \theta^p_V} & & \Big\downarrow{\scriptstyle \theta^p_U} \\ H^{n-p}_c(V)^* & \xrightarrow[\ i_*^\top\ ]{} & H^{n-p}_c(U)^*. \end{array}$$

Proposition 3.11. *For any two open subsets U_1 and U_2 of a manifold M, if $U = U_1 \cup U_2$ then the following diagrams commute for all p:*

$$\begin{array}{ccc} H^p(U_1 \cap U_2) & \xrightarrow{\qquad \delta^p \qquad} & H^{p+1}(U) \\ \Big\downarrow{\scriptstyle \theta^p_{U_1 \cap U_2}} & & \Big\downarrow{\scriptstyle \theta^{p+1}_U} \\ H^{n-p}_c(U_1 \cap U_2)^* & \xrightarrow[\ (-1)^{p+1}(\delta^{n-p-1}_c)^\top\ ]{} & H^{n-p-1}_c(U)^*. \end{array}$$

Using Proposition 3.10 and Proposition 3.11, we obtain a diagram in which the top and bottom rows are exact and every square commutes. Here is a fragment of this diagram in which we have omitted the labels of the horizontal arrows to unclutter this diagram. Due to space constraints we had to split the diagram into two parallel diagrams.

$$\begin{array}{ccccc} \longrightarrow & H^{p-1}(U_1) \oplus H^{p-1}(U_2) & \longrightarrow & H^{p-1}(U_1 \cap U_2) & \longrightarrow \\ & \Big\downarrow{\scriptstyle \theta^{p-1}_{U_1} \oplus\, \theta^{p-1}_{U_2}} & & \Big\downarrow{\scriptstyle \theta^{p-1}_{U_1 \cap U_2}} & \\ \longrightarrow & H^{n-p+1}_c(U_1)^* \oplus H^{n-p+1}_c(U_2)^* & \longrightarrow & H^{n-p+1}_c(U_1 \cap U_2)^* & \longrightarrow \end{array}$$

$$\begin{array}{ccccccc}
\longrightarrow H^p(U) & \longrightarrow & H^p(U_1) \oplus H^p(U_2) & \longrightarrow & H^p(U_1 \cap U_2) \longrightarrow \\
\downarrow \theta^p_U & & \downarrow \theta^p_{U_1} \oplus \theta^p_{U_2} & & \downarrow \theta^p_{U_1 \cap U_2} \\
\longrightarrow H^{n-p}_c(U)^* & \longrightarrow & H^{n-p}_c(U_1)^* \oplus H^{n-p}_c(U_2)^* & \longrightarrow & H^{n-p}_c(U_1 \cap U_2)^* \longrightarrow
\end{array}$$

Now here is the crucial step of the proof of Theorem 3.8. Suppose we can prove that the maps $\theta^p_{U_1}$, $\theta^p_{U_2}$ and $\theta^p_{U_1 \cap U_2}$ are isomorphisms for all p. Then by the five lemma (Proposition 2.5), we can conclude that the maps θ^p_U are also isomorphisms.

We can now give the main part of the proof of Theorem 3.8 using induction on the size of a finite good cover.

Proof sketch of Theorem 3.8. Let $\mathcal{U} = (V_1, \ldots, V_p)$ be a good cover for the orientable manifold M. We proceed by induction on p. If $p = 1$, then $M = V_1$ is diffeomorphic to $\mathbb{R}^n$ and by the Poincaré lemma (Proposition 3.2) we have

$$H^p_{\mathrm{dR}}(\mathbb{R}^n) = \begin{cases} 0 & \text{if } p \neq 0 \\ \mathbb{R} & \text{if } p = 0, \end{cases}$$

and

$$H^p_{\mathrm{dR},c}(\mathbb{R}^n) = \begin{cases} 0 & \text{if } p \neq n \\ \mathbb{R} & \text{if } p = n, \end{cases}$$

so we have the desired isomorphisms.

Assume inductively that Poincaré duality holds for any orientable manifold having a good cover with at most p open subsets, and let $(V_1, \ldots, V_{p+1})$ be a cover with $p+1$ open subsets. Observe that $(V_1 \cup \cdots \cup V_p) \cap V_{p+1}$ has a good cover with p open subsets, namely $(V_1 \cap V_{p+1}, \ldots, V_p \cap V_{p+1})$. By the induction hypothesis applied to $U_1 = V_1 \cup \cdots \cup V_p$, $U_2 = V_{p+1}$, and $U = M = V_1 \cup \cdots \cup V_{p+1}$, the maps $\theta^p_{U_1}$, $\theta^p_{U_2}$ and $\theta^p_{U_1 \cap U_2}$ in the diagram shown just after Proposition 3.11 are isomorphisms for all p, so by the five lemma (Proposition 2.5) we can conclude that the maps θ^p_U are also isomorphisms, establishing the induction step. ☐

Remark: The technique involving two Mayer–Vietoris sequences running in opposite direction (up on the top row, and down on the bottom row) is a preview of a similar technique used in the proof of the more general version of Poincaré duality stated in Theorem 7.16.

As a corollary of Poincaré duality, if M is an orientable and connected manifold, then $H^0(M) \cong \mathbb{R}$, and so $H_c^n(M) \cong \mathbb{R}$. In particular, if M is compact then $H^n(M) \cong \mathbb{R}$.

Remark: As explained in Bott and Tu [Bott and Tu (1986)], the assumption that the good cover is finite is not necessary. Then the statement of Poincaré duality is that if M is any orientable manifold of dimension n, then there are isomorphisms

$$H^p(M) \cong (H_c^{n-p}(M))^*$$

for all p with $0 \leq p \leq n$, even if $H^p(M)$ is infinite dimensional. However, the statement obtained by taking duals, namely

$$H_c^p(M) \cong (H^{n-p}(M))^*,$$

is generally false.

In Chapter 1 of their book, Bott and Tu derive more consequences of the Mayer–Vietoris method. The interested reader is referred to Bott and Tu [Bott and Tu (1986)].

The de Rham cohomology is a very effective tool to deal with manifolds but one of the drawbacks of using real coefficients is that torsion phenomena are overlooked. There are other cohomology theories of finer grain that use coefficients in rings such as $\mathbb{Z}$. One of the simplest uses singular chains, and we discuss it in the next chapter.

3.4 Problems

Problem 3.1. Let M be a connected n-dimensional compact smooth manifold. Let $\omega \in \mathcal{A}^k(M)$ and $\eta \in \mathcal{A}^{n-k}(M)$ be two closed forms on M. Prove that if $\omega \wedge \eta$ is not zero for any point of M, then the class $[\omega] \in H^k_{\mathrm{dR}}(M)$ represented by ω is nonzero.

Hint. Recall that if an n-dimensional smooth manifold has an n-form that it nowhere vanishing, then it is orientable, and that under any orientation, $\int_M \omega \wedge \eta \neq 0$.

Problem 3.2. Let $M = \mathbb{R}^2 - \{0\}$. Prove that

$$H^k_{\mathrm{dR}}(M) \cong H^k_{\mathrm{dR}}(S^1 \times \mathbb{R}).$$

Check that

$$\omega = \frac{xdy - ydx}{x^2 + y^2}$$

is a closed 1-form on M.

Problem 3.3. Consider the 2-form

$$\omega = x_1 dx_2 \wedge dx_3 - x_2 dx_1 \wedge dx_3 + x_3 dx_1 \wedge dx_2$$

on $\mathbb{R}^3$. Prove that

$$\int_{S^2} \omega = 4\pi.$$

Chapter 4

Singular Homology and Cohomology

Historically, singular homology and cohomology were developed in the 1940s, starting with a seminal paper of Eilenberg published in 1944 (building up on work by Alexander and Lefschetz among others). It was not the first homology theory. Indeed, simplicial homology emerged in the early 1920s.

One of the main differences between singular homology and simplicial homology is that singular homology groups can be assigned to *any* topological space X, but simplicial homology groups are only defined for certain combinatorial objects called *simplicial complexes*. In this respect, singular homology is superior to simplicial homology. It is also easier to prove that homeomorphic spaces, in fact, homotopy equivalent spaces, have isomorphic singular homology. The price to pay is that the singular homology groups have a more abstract definition than the simplicial homology groups, and their definition does not suggest methods to compute them.

Simplicial homology and singular homology agree, but it takes a lot of work to prove this fact (see Chapter 5). We feel that singular homology is less contrived than simplicial homology because it is defined directly for spaces, as opposed to simplicial homology which is defined for combinatorial objects that can be viewed as triangulations of spaces. Thus we will first define singular homology (and cohomology). Simplicial homology will be discussed in the next chapter (Chapter 5).

Roughly speaking, the singular homology groups are defined by chain complexes in which the modules in the chain complexes are built up from continuous maps from some simple geometric objects called simplices, which generalize line segments, triangles, tetrahedra, *etc.*

We begin by defining singular simplices and the chain complex $S_*(X;R)$ (consisting of a chain complex with modules $S_p(X;R)$ of singular p-chains) that gives rise to simplicial homology. For this we need to define the boundary $\partial\sigma$ of a singular simplex σ. Having assigned simplicial homology groups $H_p(X,R)$ to a topological space X (where R is a commutative ring with an identity element and $p \geq 0$), we show how a continuous map $f\colon X \to Y$ between two topological spaces X and Y induces homomorphisms $H_p(f)\colon H_p(X,R) \to H_p(Y,R)$ between homology groups.

Our next goal is to develop tools that will help us compute the singular homology groups of a space X. The first result is that homotopy equivalent spaces have isomorphic homology groups. This is called the *homotopy axiom*.

To compute singular homology groups it turns out that it is useful to define the singular homology groups $H_p(X,A;R)$ of a pair of spaces (X,A), where A is a subspace of X. The groups $H_p(X,A;R)$ are called the *relative singular homology groups*. The homotopy axiom also applies to relative singular homology. Using the zig-zag lemma (Theorem 2.22) we obtain the *long exact sequence of relative homology*, which plays a crucial role.

When (X,A) is a pair of spaces where A is a nonempty closed subspace that is a deformation retract of some neighborhood in X, the homology groups $H_p(X,A;R)$ are isomorphic to the groups $H_p(X/A;\{\text{pt}\};R)$, where X/A is the result of collapsing A to a single point. When the above condition holds for a pair (X,A), we say that (X,A) is a good pair. The groups $H_p(X/A;\{\text{pt}\};R)$ may be easier to compute that the groups $H_p(X,A;R)$ because X/A may be simpler than X. Technically, the groups $H_p(X/A;\{\text{pt}\};R)$ are isomorphic to some groups $\widetilde{H}_p(X;R)$, called *reduced homology groups*. The groups $\widetilde{H}_p(X;R)$ agree with the groups $H_p(X;R)$ for all $p \geq 1$, and $H_0(X;R) = \widetilde{H}_0(X;R) \oplus R$. Sometimes, the reduced homology groups are technically advantageous. When (X,A) is a good pair, there is a long exact sequence which involves the groups $\widetilde{H}_p(A;R)$, $\widetilde{H}_p(X;R)$ and $\widetilde{H}_p(X/A;R)$ which may be very helpful for computing $\widetilde{H}_p(X;R)$ in terms of the homology of the simpler spaces A and X/A.

The next tool for computing singular homology is the *excision axiom*. This axiom says that given a pair (X,A), if $Z \subseteq A \subseteq X$ is a subspace whose closure is contained in the interior of A, then we can carve out (excise) Z from both A and X and still have isomorphisms

$$H_p(X - Z, A - Z; R) \cong H_p(X,A;R), \quad p \geq 0.$$

The spaces $X - Z$ and $A - Z$ may be a lot simpler than the spaces X and A so it may be easier to compute the groups $H_p(X - Z, A - Z; R)$. There is also a very important long exact sequence, the *Mayer–Vietoris sequence*, which is useful for computing the homology $H_p(X; R)$ of a space X in terms of the homology of the simpler spaces $A \cap B$, A and B, where A and B are subspaces of X such that X is the union of the interiors of A and B. We also discuss the technical notion of a compact pair.

Next we apply the previous tools (homotopy equivalence, excision, long exact sequence of homology, long exact sequence of a good pair, Mayer–Vietoris sequence) to the computation of the homology groups of some classical spaces. We begin with the computation of the homology groups of the spheres S^n and of the discs D^n. Although very simple, these spaces occur frequently as building blocks for more complicated spaces. We also give (without proof) the homology groups of the real projective spaces, $\mathbb{RP}^n$, the complex projective spaces, $\mathbb{CP}^n$, and the n-tori T^n. We also indicate how homology can used to prove generalized versions of the Jordan curve theorem. We finish the section with a technical result about the homology groups $H_p(M, M - \{x\}; R)$, where M is a topological manifold and x is a point in M. This result will be needed later to define the notion of orientation.

Next we show how to generalize the homology groups of a space X so that they take values in an R-module G. These groups are denoted by $H_p(X; G)$. The process is algebraic and consists in tensoring the chain complex $S_*(X; R)$ with G. All previous results generalize to this situation. This section is quite technical and can be skipped upon first reading.

We then turn to singular cohomology. This is an algebraic process which consists in building a cochain complex basically by dualizing. The cochain group $S^p(X; R)$ is the space of R-linear maps from $S_p(X; R)$ to R (the dual of $S_p(X; R)$), and the coboundary map δ^p is the dual of ∂^{p+1}, namely

$$\delta^p f = f \circ \partial_{p+1}, \quad f \in S^p(X; R).$$

The only small issue is the sign assigned in front of the right-hand side in the above formula. We follow Bott and Tu and assign the sign +. We obtain the singular cohomology groups $H^p(X; R)$.

We explain how to define the singular cohomology groups $H^p(X; G)$ taking values in an R-module G. This is done by applying the dualization functor $\mathrm{Hom}_R(-, G)$. We state the *Mayer–Vietoris* long exact sequence in cohomology.

We show how to define relative singular cohomology groups $H^p(X, A, R)$ and $H^p(X, A; G)$ (where G is an R-module). We state the cohomology versions of the standard results that hold for homology:

(1) Homotopy axiom.
(2) Long exact sequence of relative cohomology.
(3) Excision axiom.

One of the technical advantages of cohomology over homology is that it is fairly easy to define multiplication operations on cohomology classes. This way the (graded) cohomology module $H^*(X; R)$ can be made into a ring, the *cohomology ring*. We define the *cup product* and states some of its basic properties.

4.1 Singular Homology

In this section we only assume that our space X is a Hausdorff topological space, and we consider continuous maps between such spaces. Singular homology (and cohomology) arises from chain complexes built from singular chains (and cochains). Singular chains are defined in terms of certain convex figures generalizing line segments, triangles, and tetrahedra called standard n-simplices. We adopt the definition from Milnor and Stasheff [Milnor and Stasheff (1974)].

Definition 4.1. For any integer $n \geq 0$, the *standard n-simplex* Δ^n is the convex subset of $\mathbb{R}^{n+1}$ consisting of the set of points

$$\Delta^n = \{(t_0, t_1, \dots, t_n) \in \mathbb{R}^{n+1} \mid t_0 + t_1 + \cdots + t_n = 1,\ t_i \geq 0\}.$$

The $n+1$ points corresponding to the canonical basis vectors $e_i^{n+1} = (0, \dots, 0, 1, 0, \dots, 0)$ $(1 \leq i \leq n+1)$ are called the *vertices* of the simplex Δ^n.

The simplex Δ^n is the convex hull of the $n+1$ points $(e_1^{n+1}, \dots, e_{n+1}^{n+1})$ since we can write

$$\Delta^n = \{t_0 e_1^{n+1} + t_1 e_2^{n+1} + \cdots + t_n e_{n+1}^{n+1} \mid t_0 + t_1 + \cdots + t_n = 1,\ t_i \geq 0\}.$$

Thus, Δ^n is a subset of $\mathbb{R}^{n+1}$. In particular, when $n = 0$, the 0-simplex Δ^0 consists of the single points $t_0 = 1$ on $\mathbb{R}$. Some simplices are illustrated in Figure 4.1.

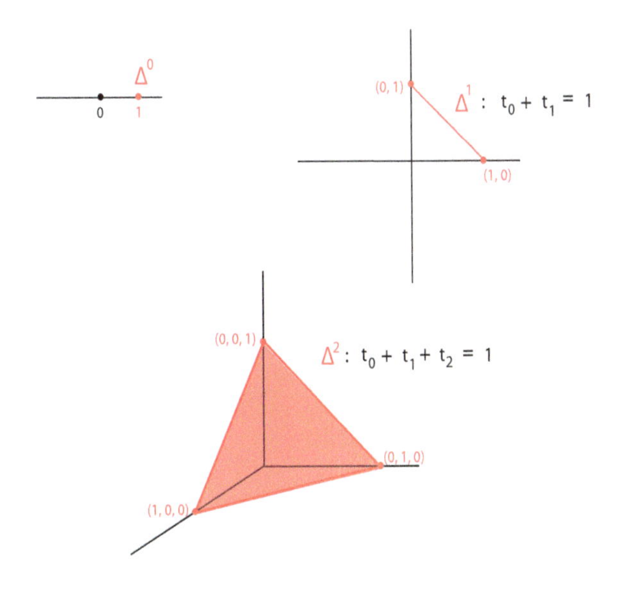

Fig. 4.1 The simplices $\Delta^0, \Delta^1, \Delta^2$.

Remark: Other authors such as Bott and Tu [Bott and Tu (1986)] and Warner [Warner (1983)] define the n-simplex Δ^n as a convex subset of $\mathbb{R}^n$. In their definition, if we denote the point corresponding to the origin of $\mathbb{R}^n$ as e_0^n, then

$$\begin{aligned}\Delta^n &= \{t_0 e_0^n + t_1 e_1^n + \cdots + t_n e_n^n \mid t_0 + t_1 + \cdots + t_n = 1,\ t_i \geq 0\}.\\ &= \{(t_1, \ldots, t_n) \in \mathbb{R}^n \mid t_1 + \cdots + t_n \leq 1,\ t_i \geq 0\}.\end{aligned}$$

Some of these simplices are illustrated in Figure 4.2.

These points of view are equivalent but one should be careful that the notion of face of a singular simplex (see below) is defined slightly differently.

Definition 4.2. Given a topological space X, a *singular p-simplex* is any continuous map $\sigma\colon \Delta^p \to X$ (with $p \geq 0$). If $p \geq 1$, the *ith face (map)* of the singular p-simplex σ is the $(p-1)$-singular simplex

$$\sigma \circ \phi_i^{p-1}\colon \Delta^{p-1} \to X, \quad 0 \leq i \leq p,$$

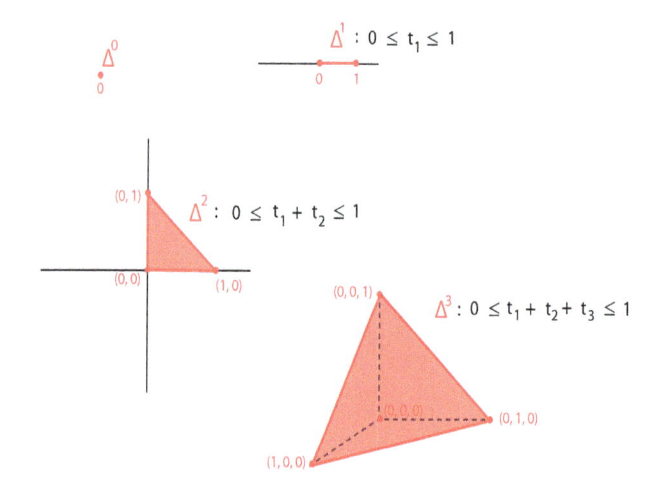

Fig. 4.2 Some simplices according to the second definition.

where $\phi_i^{p-1}\colon \Delta^{p-1} \to \Delta^p$ is the map given by

$$\phi_0^{p-1}(t_1, \ldots, t_p) = (0, t_1, \ldots, t_p)$$
$$\phi_i^{p-1}(t_0, \ldots, t_{i-1}, t_{i+1}, \ldots, t_p) = (t_0, \ldots, t_{i-1}, 0, t_{i+1}, \ldots, t_p), 1 \leq i \leq p-1$$
$$\phi_p^{p-1}(t_0, \ldots, t_{p-1}) = (t_0, \ldots, t_{p-1}, 0).$$

Some singular 1-simplices and singular 2-simplices are illustrated in Figure 4.3.

Note that a singular p-simplex σ has $p+1$ faces. The ith face $\sigma \circ \phi_i^{p-1}$ is sometimes denoted by σ^i. For example, if $p = 1$, since there is only one variable on $\mathbb{R}^1$ and $\Delta^0 = \{1\}$, the maps $\phi_0^0, \phi_1^0\colon \Delta^0 \to \Delta^1$ are given by

$$\phi_0^0(1) = (0, 1), \quad \phi_1^0(1) = (1, 0).$$

For $p = 2$, the maps $\phi_0^1, \phi_1^1, \phi_2^1\colon \Delta^1 \to \Delta^2$ are given by

$$\phi_0^1(t_1, t_2) = (0, t_1, t_2), \quad \phi_1^1(t_0, t_2) = (t_0, 0, t_2), \quad \phi_2^1(t_0, t_1) = (t_0, t_1, 0).$$

There does not seem to be any standard notation for the set of all singular p-simplices on X. We propose the notation $S_{\Delta^p}(X)$.

Remark: In Definition 4.2 we may replace X by any open subset U of X, in which case a continuous map $\sigma\colon \Delta^p \to U$ is called a *singular p-simplex in*

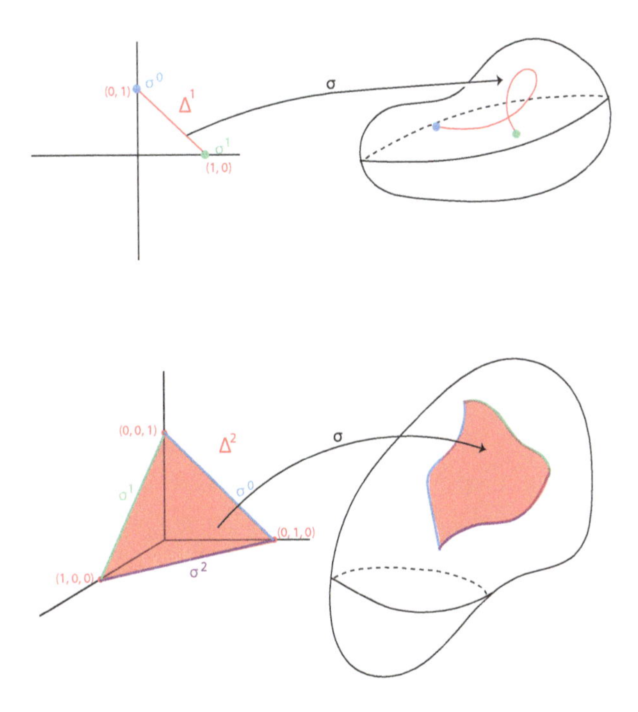

Fig. 4.3 Some singular simplices.

U. If X is a smooth manifold, following Warner [Warner (1983)], we define a *differentiable singular p-simplex* in U to be a singular p-simplex σ which can be extended to a smooth map of some open subset of $\mathbb{R}^{n+1}$ containing Δ^p into U.

We now come to the crucial definition of singular p-chains. In the framework of singular homology (and cohomology) we have the extra degree of freedom of choosing the coefficients. The set of coefficients will be a commutative ring with unit denoted by R. Better results are obtained if we assume that R is a PID. In most cases, we may assume that $R = \mathbb{Z}$.

Definition 4.3. Given a topological space X and a commutative ring R, a *singular p-chain with coefficients in R* is any formal linear combination $\alpha = \sum_{i=1}^{m} \lambda_i \sigma_i$ of singular p-simplices σ_i with coefficients $\lambda_i \in R$. The *singular chain group* $S_p(X; R)$ is the free R-module consisting of all singular p-chains; it is generated by the set $S_{\Delta^p}(X)$ of singular p-simplices. We set

$S_p(X;R) = (0)$ for $p < 0$. If $p \geq 1$, given any singular p-simplex σ, its *boundary* $\partial\sigma$ is the singular $(p-1)$-chain given by

$$\partial\sigma = \sigma \circ \phi_0^{p-1} - \sigma \circ \phi_1^{p-1} + \cdots + (-1)^p \sigma \circ \phi_p^{p-1}.$$

Extending the map ∂ to $S_p(X;R)$ by linearity, we obtain the *boundary homomorphism*

$$\partial\colon S_p(X;R) \to S_{p-1}(X;R).$$

When we want to be very precise, we write $\partial_p\colon S_p(X;R) \to S_{p-1}(X;R)$. We define $S_*(X;R)$ as the direct sum

$$S_*(X;R) = \bigoplus_{p\geq 0} S_p(X;R).$$

Then the boundary maps ∂_p yield the boundary map $\partial\colon S_*(X;R) \to S_*(X;R)$. For example, the boundary of a singular 1-simplex σ is $\sigma(0,1) - \sigma(1,0)$. The boundary of a singular 2-simplex σ is

$$\sigma^0 - \sigma^1 + \sigma^2,$$

where $\sigma^0, \sigma^1, \sigma^2$ are the faces of σ, in this case, three curves in X. For example, σ^0 is the curve given by the map

$$(t_1, t_2) \mapsto \sigma(0, t_1, t_2)$$

from Δ^1 to X, where $t_1 + t_2 = 1$ and $t_1, t_2 \geq 0$.

The following result is easy to check.

Proposition 4.1. *Given a topological space X and a commutative ring R, the boundary map $\partial\colon S_*(X;R) \to S_*(X;R)$ satisfies the equation*

$$\partial \circ \partial = 0.$$

We can put together the maps $\partial_p\colon S_p(X;R) \to S_{p-1}(X;R)$ to obtain the following chain complex of homology

$$0 \xleftarrow{\partial_0} S_0(X;R) \xleftarrow{\partial_1} S_1(X;R) \longleftarrow \cdots\ S_{p-1}(X;R) \xleftarrow{\partial_p} S_p(X;R) \xleftarrow{\partial_{p+1}}$$

in which the direction of the arrows is from right to left. Note that if we replace every nonnegative index p by $-p$ in ∂_p, $S_p(X;R)$ *etc.*, then we obtain a chain complex as defined in Section 2.5 and we now have all the ingredients to define homology groups. We have the familiar spaces $Z_p(X;R) = \operatorname{Ker}\partial_p$ of *singular p-cycles*, and $B_p(X;R) = \operatorname{Im}\partial_{p+1}$ of *singular p-boundaries*.

By Proposition 4.1, $B_p(X;R)$ is a submodule of $Z_p(X;R)$ so we obtain homology spaces.

Definition 4.4. Given a topological space X and a commutative ring R, for any $p \geq 0$, the module $Z_p(X;R) = \operatorname{Ker} \partial_p$ is the module of *singular p-cycles*, and the module $B_p(X;R) = \operatorname{Im} \partial_{p+1}$ is the module of *singular p-boundaries*. The *singular homology module* $H_p(X;R)$ is defined by

$$H_p(X;R) = \ker \partial_p / \operatorname{Im} \partial_{p+1} = Z_p(X;R)/B_p(X;R).$$

We set $H_p(X;R) = (0)$ for $p < 0$ and define $H_*(X;R)$ as the direct sum

$$H_*(X;R) = \bigoplus_{p \geq 0} H_p(X;R)$$

and call it the *singular homology of X with coefficients in R*.

The spaces $H_p(X;R)$ are R-modules but following common practice we often refer to them as groups.

A singular 0-chain is a linear combination $\sum_{i=1}^m \lambda_i P_i$ of points $P_i \in X$. Because the boundary of a singular 1-simplex is the difference of two points, if X is path-connected, it is easy to see that a singular 0-chain is the boundary of a singular 1-chain iff $\sum_{i=1}^m \lambda_i = 0$. *Thus, X is path connected iff*

$$H_0(X;R) = R.$$

More generally, we have the following proposition.

Proposition 4.2. *Given any topological space X, for any commutative ring R with an identity element, $H_0(X;R)$ is a free R-module. If $(X_\alpha)_{\alpha \in I}$ is the collection of path components of X and if σ_α is a singular 0-simplex whose image is in X_α, then the homology classes $[\sigma_\alpha]$ form a basis of $H_0(X;R)$.*

Proposition 4.2 is proven in Munkres [Munkres (1984)] (Chapter 4, Section 29, Theorem 29.2). In particular, if X has m path-connected components, then $H_0(X;R) \cong \underbrace{R \oplus \cdots \oplus R}_{m}$.

We leave it as an exercise (or look at Bott and Tu [Bott and Tu (1986)], Chapter III, §15) to show the following fact.

Proposition 4.3. *The homology groups of $\mathbb{R}^n$ are given by*

$$H_p(\mathbb{R}^n;R) = \begin{cases} (0) & \text{if } p \geq 1 \\ R & \text{if } p = 0. \end{cases}$$

The same result holds if $\mathbb{R}^n$ is replaced by any nonempty convex subset of $\mathbb{R}^n$, or a space consisting of a single point.

The homology groups (with coefficients in $\mathbb{Z}$) of the compact surfaces can be completely determined. Some of them, such as the projective plane $\mathbb{RP}^2$, have $\mathbb{Z}/2\mathbb{Z}$ as a homology group.

If X and Y are two topological spaces and if $f\colon X \to Y$ is a continuous function between them, then we have induced homomorphisms $H_p(f)\colon H_p(X;R) \to H_p(Y;R)$ between the homology groups of X and the homology groups of Y. We say that homology is functorial.

Proposition 4.4. *If X and Y are two topological spaces and if $f\colon X \to Y$ is a continuous function between them, then there are homomorphisms $H_p(f)\colon H_p(X;R) \to H_p(Y;R)$ for all $p \geq 0$.*

Proof. To prove the proposition we show that there is a chain map between the chain complexes associated with X and Y and apply Proposition 2.19. Given any singular p-simplex $\sigma\colon \Delta^p \to X$ we obtain a singular p-simplex $f\sigma\colon \Delta^p \to Y$ obtained by composing with f, namely $f\sigma = f \circ \sigma$. Since $S_p(X;R)$ is freely generated by $S_{\Delta^p}(X;R)$, the map $\sigma \mapsto f\sigma$ from $S_{\Delta^p}(X;R)$ to $S_p(Y;R)$ extends uniquely to a homomorphism $S_p(f)\colon S_p(X;R) \to S_p(Y;R)$. It is immediately verified that the following diagrams are commutative

$$\begin{array}{ccc} S_{p+1}(X;R) & \xrightarrow{\partial^X_{p+1}} & S_p(X;R) \\ {\scriptstyle S_{p+1}(f)}\downarrow & & \downarrow{\scriptstyle S_p(f)} \\ S_{p+1}(Y;R) & \xrightarrow[\partial^Y_{p+1}]{} & S_p(Y;R), \end{array}$$

which means that the maps $S_p(f)\colon S_p(X;R) \to S_p(Y;R)$ form a chain map $S(f)$. By Proposition 2.19, we obtain homomorphisms $S_p(f)^*\colon H_p(X;R) \to H_p(Y;R)$ for all p, which we denote by $H_p(f)$. $\square$

Following the convention that in homology subscripts are used to denote objects, the map $S_p(f)\colon S_p(X;R) \to S_p(Y;R)$ is also denoted $f_{\sharp,p}\colon S_p(X;R) \to S_p(Y;R)$, and the map $H_p(f)\colon H_p(X;R) \to H_p(Y;R)$ is also denoted $f_{*p}\colon H_p(X;R) \to H_p(Y;R)$ (or simply $f_*\colon H_p(X;R) \to H_p(Y;R)$).

Proposition 4.4 implies that if two spaces X and Y are homeomorphic, then X and Y have isomorphic homology. This gives us a way of showing that some spaces are not homeomorphic: if for some p the homology groups $H_p(X; R)$ and $H_p(Y; R)$ are not isomorphic, then X and Y are not homeomorphic.

4.2 Homotopy Equivalence and Homology

Actually, it turns out that the homology groups of two homotopy equivalent spaces are isomorphic. Intuitively, two continuous maps $f, g\colon X \to Y$ are homotopic is f can be continuously deformed into g, which means that there is a one-parameter family $F(-, t)$ of continuous maps $F(-, t)\colon X \to Y$ varying continuously in $t \in [0, 1]$ such that $F(x, 0) = f(x)$ and $F(x, 1) = g(x)$ for all $x \in X$. Here is the formal definition.

Definition 4.5. Two continuous maps $f, g\colon X \to Y$ (where X and Y are topological spaces) are *homotopic* if there is a continuous function $F\colon X \times [0, 1] \to Y$ (called a *homotopy with fixed ends*) such that

$$F(x, 0) = f(x), \quad F(x, 1) = g(x) \quad \text{for all } x \in X.$$

We write $f \simeq g$. See Figure 4.4.

Definition 4.6. A space X is said to be *contractible* if the identity map $\mathrm{id}_X\colon X \to X$ is homotopic to a constant function with domain X. For example, any convex subset of $\mathbb{R}^n$ is contractible.

Intuitively, a contractible space can be continuously deformed to a single point, so it is topologically trivial. In particular, it cannot contain holes. An example of a contractible set is shown in Figure 4.5.

Definition 4.7. A *deformation retraction* of a space X onto a subspace A is a homotopy $F\colon X \times [0, 1] \to X$ such that $F(x, 0) = x$ for all $x \in X$, $F(x, t) = x$ for all $x \in A$ and all $t \in (0, 1]$, and $F(X, 1) = A$. In this case, A is called a *deformation retract* of X.

An example of deformation retract is shown in Figure 4.6.

Topologically, homeomorphic spaces should be considered equivalent. From the point of view of homotopy, experience has shown that the more liberal notion of homotopy equivalence is the right notion of equivalence.

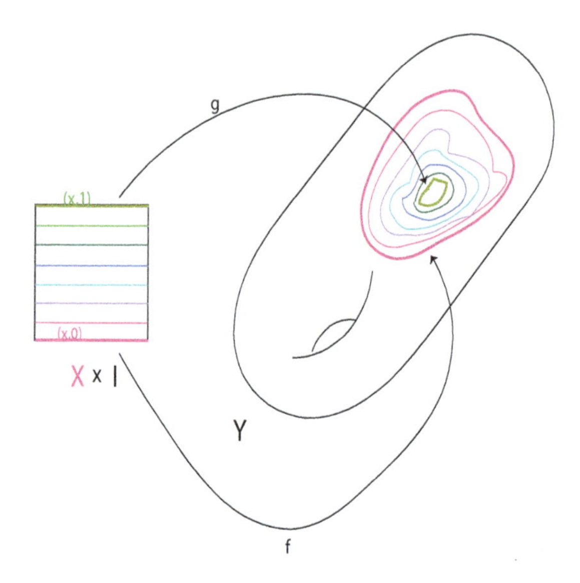

Fig. 4.4 The homotopy F between f and g, where $X = [0, 1]$ and Y is the torus.

Definition 4.8. Two topological spaces X and Y are *homotopy equivalent* if there are continuous functions $f\colon X \to Y$ and $g\colon Y \to X$ such that

$$g \circ f \simeq \mathrm{id}_X, \quad f \circ g \simeq \mathrm{id}_Y.$$

We write $X \simeq Y$. See Figure 4.7.

A great deal of homotopy theory has to do with developing tools to decide when two spaces are homotopy equivalent. It turns out that homotopy equivalent spaces have isomorphic homology. In this sense homology theory is cruder than homotopy theory. However, homotopy groups are generally more complicated and harder to compute than homology groups. For one thing, homotopy groups are generally nonabelian, whereas homology groups are abelian.

Proposition 4.5. *Given any two continuous maps $f, g\colon X \to Y$ (where X and Y are topological spaces), if f and g are homotopic, then the chain maps $S(f), S(g)\colon S_*(X; R) \to S_*(Y; R)$ are chain homotopic (see Definition 2.13).*

Proofs of Proposition 4.5 can be found in Mac Lane [Mac Lane (1975)] (Chapter II, Theorem 8.2) and Hatcher [Hatcher (2002)] (Chapter 2, Theorem 2.10). The idea is to reduce the proof to the case where the space Y

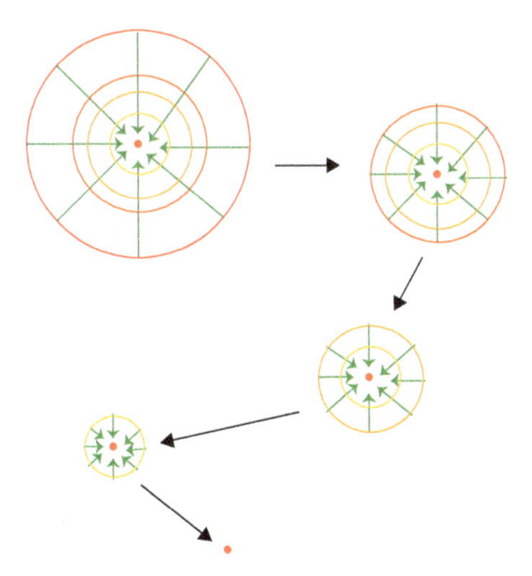

Fig. 4.5 A contractible set.

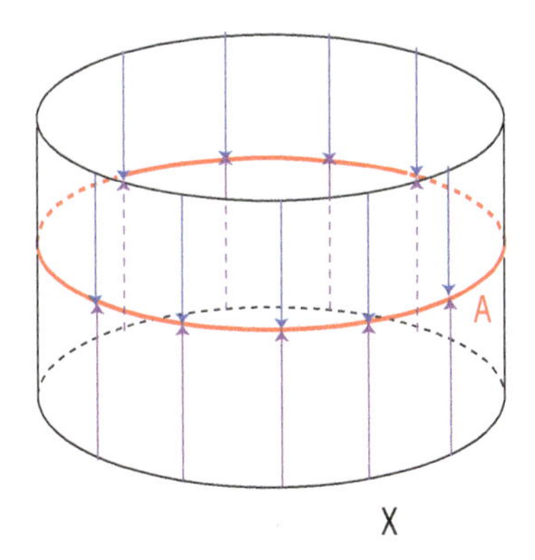

Fig. 4.6 A deformation retract of the cylinder X onto its median circle A.

is the cylinder $X \times [0, 1]$. In this case we have the two continuous maps $b, t \colon X \to X \times [0, 1]$ given by $b(x) = (x, 0)$ and $t(x) = (x, 1)$, which are clearly homotopic. Then one shows that a chain homotopy can be constructed between the chain maps $S(t)$ and $S(b)$.

As a corollary of Proposition 4.5, we obtain the following important result.

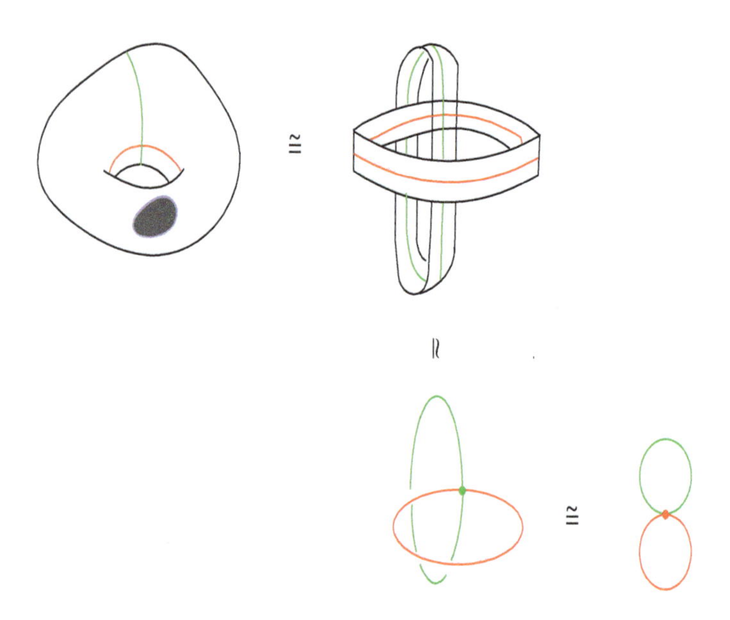

Fig. 4.7 The punctured torus is homotopically equivalent to the figure eight.

Proposition 4.6. *(Homotopy Axiom) Given any two continuous maps $f, g\colon X \to Y$ (where X and Y are topological spaces), if f and g are homotopic and $H_p(f), H_p(g)\colon H_p(X;R) \to H_p(Y;R)$ are the induced homomorphisms, then $H_p(f) = H_p(g)$ for all $p \geq 0$. As a consequence, if X and Y are homotopy equivalent, then the homology groups $H_p(X;R)$ and $H_p(Y;R)$ are isomorphic for all $p \geq 0$,*

Proof. By Proposition 4.5 there is a chain homotopy between $S(f)\colon S_*(X;R) \to S_*(Y;R)$ and $S(g)\colon S_*(X;R) \to S_*(Y;R)$, and by Proposition 2.20 the induced homomorphisms $H_p(f), H_p(g)\colon H_p(X;R) \to H_p(Y;R)$ are identical. If $f\colon X \to Y$ and $g\colon Y \to X$ are two maps making X and Y chain homotopic, we have $g \circ f \simeq \mathrm{id}_X$ and $f \circ g \simeq \mathrm{id}_Y$, so by the first part of the proposition

$$H_p(g \circ f) = H_p(g) \circ H_p(f) = H_p(\mathrm{id}_X) = \mathrm{id}_{H_p(X;R)}$$

and

$$H_p(f \circ g) = H_p(f) \circ H_p(g) = H_p(\mathrm{id}_Y) = \mathrm{id}_{H_p(Y;R)},$$

which shows that the maps $H_p(f)\colon H_p(X;R) \to H_p(Y;R)$ are isomorphisms with inverses $H_p(g)$. $\square$

4.3 Relative Singular Homology Groups

A more flexible theory is obtained if we consider homology groups $H_p(X, A)$ associated with pairs of spaces (X, A), where A is a subspace of X.

Since A is a subspace of X, each singular simplex $\sigma\colon \Delta^p \to A$ yields a singular simplex $\sigma\colon \Delta^p \to X$ by composing σ with the inclusion map from A to X, so the singular complex $S_*(A;R)$ is a subcomplex of the singular complex $S_*(X;R)$.

Definition 4.9. Let $S_p(X, A; R)$ be the quotient module

$$S_p(X, A; R) = S_p(X;R)/S_p(A;R)$$

and let $S_*(X, A; R)$ be the corresponding graded module (the direct sum of the $S_p(X, A; R)$).

The boundary map $\partial_{X,p}\colon S_p(X;R) \to S_{p-1}(X;R)$ of the original complex $S_*(X;R)$ restricts to the boundary map $\partial_{A,p}\colon S_p(A;R) \to S_{p-1}(A;R)$ of the complex $S_*(A;R)$ so the quotient map $\partial_p\colon S_p(X, A; R) \to S_{p-1}(X, A; R)$ induced by $\partial_{X,p}$ and given by

$$\partial_p(\sigma + S_p(A;R)) = \partial_{X,p}(\sigma) + S_{p-1}(A;R)$$

for every singular p-simplex σ is a boundary map for the chain complex $S_*(X, A; R)$.

Definition 4.10. The chain complex $S_*(X, A; R)$

$$\cdots \xleftarrow{\partial_{p-1}} S_{p-1}(X, A; R) \xleftarrow{\partial_p} S_p(X, A; R) \xleftarrow{\partial_{p+1}} \cdots$$

$$0 \xleftarrow{\partial_0} S_0(X, A; R) \xleftarrow{\partial_1} S_1(X, A; R) \longleftarrow \cdots$$

is called the *singular chain complex* of the pair (X, A).

We now have all the ingredients to define the singular relative homology groups.

Definition 4.11. Given a pair (X, A) where A is a subspace of X, the *singular relative homology groups* $H_p(X, A; R)$ of (X, A) are defined by

$$H_p(X, A; R) = H_p(S_*(X;R)/S_*(A;R)),$$

the singular homology groups of the chain complex $S_*(X, A; R)$. For short, we often drop the word "singular" in singular relative homology group.

Observe that the quotient module $S_p(X, A; R) = S_p(X; R)/S_p(A; R)$ is a free module. Indeed, the family of cosets of the form $\sigma + S_p(A; R)$ where the image of the singular p-simplex σ does not lie in A forms a basis of $S_p(X, A; R)$.

There is a useful alternative definition of the relative homology groups in terms of relative p-cycles and relative p-boundaries.

Definition 4.12. Given a pair of spaces (X, A), the group $Z_p(X, A; R)$ of *relative p-cycles*, consists of those chains $c \in S_p(X; R)$ such that $\partial_p c \in S_{p-1}(A; R)$, and the group $B_p(X, A; R)$ of *relative p-boundaries* consists of those chains $c \in S_p(X; R)$ such that $c = \partial_{p+1}\beta + \gamma$ with $\beta \in S_{p+1}(X; R)$ and $\gamma \in S_p(A; R)$.

Then the relative homology group $H_p(X, A; R)$ is also expressed as the quotient

$$H_p(X, A; R) = Z_p(X, A; R)/B_p(X, A; R).$$

An illustration of the notion of relative cycle is shown in Figure 4.8 and of a relative boundary in Figure 4.9.

A single space X may be regarded as the pair $(X, \emptyset)$, and so $H_p(X, \emptyset; R) = H_p(X; R)$.

Definition 4.13. Given two pairs (X, A) and (Y, B) with $A \subseteq X$ and $B \subseteq Y$, a map $f\colon (X, A) \to (Y, B)$ is a continuous function $f\colon X \to Y$ such that $f(A) \subseteq B$. A *homotopy* F between two maps $f, g\colon (X, A) \to (Y, B)$ is a homotopy between f and g such that $F(A \times [0, 1]) \subseteq B$; we write $f \simeq g$. Two pairs (X, A) and (Y, B) are *homotopy equivalent* if there exist maps $f\colon (X, A) \to (Y, B)$ and $g\colon (Y, B) \to (X, A)$ such that $g \circ f \simeq (\mathrm{id}_X, \mathrm{id}_A)$ and $f \circ g \simeq (\mathrm{id}_Y, \mathrm{id}_B)$.

Proposition 4.4 is easily generalized to pairs of spaces.

Proposition 4.7. *If (X, A) and (Y, A) are pairs of spaces and if $f\colon (X, A) \to (Y, B)$ is a continuous map between them, then there are homomorphisms $H_p(f)\colon H_p(X, A; R) \to H_p(Y, B; R)$ for all $p \geq 0$.*

Proof sketch. Given any singular p-simplex $\sigma\colon \Delta^p \to X$ by composition with f we obtain the singular p-simplex $f\sigma\colon \Delta^p \to Y$, and since $S_p(X; R)$ is freely generated by $S_{\Delta^p}(X; R)$ we get a homomorphism $S_p(f)\colon S_p(X; R) \to$

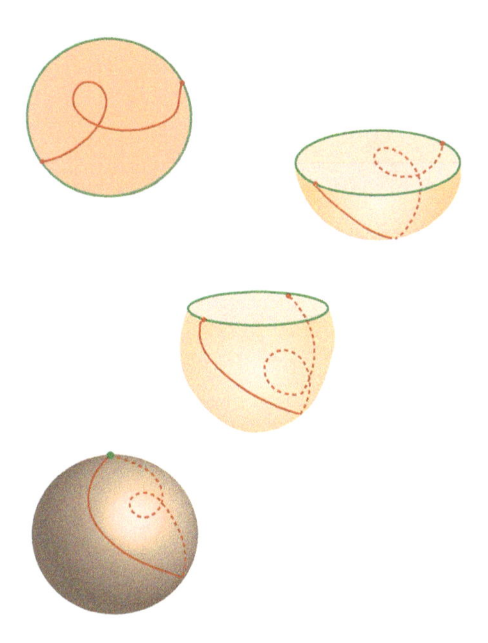

Fig. 4.8 Let X be the closed unit disk and A its circular boundary. Let $p = 1$. The red curve is a relative cycle since its boundary is in A. We show the effect of collapsing A to a point, namely transforming X into a unit sphere.

$S_p(Y;R)$. Consider the composite map $\varphi\colon S_p(X;R) \to S_p(Y;R)/S_p(B;R)$ given by

$$S_p(X;R) \xrightarrow{S_p(f)} S_p(Y;R) \xrightarrow{\pi_{Y,B}} S_p(Y;R)/S_p(B;R).$$

Since $f(A) \subseteq B$, the restriction of $S_p(f)$ to simplices in A yields a map $S_p(f)\colon S_p(A;R) \to S_p(B;R)$, so $S_p(f)(S_p(A;R)) \subseteq S_p(B;R)$, which implies that φ vanishes on $S_p(A;R)$. Thus $S_p(A;R) \subseteq \operatorname{Ker}\varphi$, which means that there is a unique homomorphism

$$f_{\sharp,p}\colon S_p(X;R)/S_p(A;R) \to S_p(Y;R)/S_p(B;R)$$

making the following diagram commute:

$$\begin{array}{ccc} S_p(X;R) & \xrightarrow{\pi_{X,A}} & S_p(X;R)/S_p(A;R) \\ & \searrow^{\varphi} & \downarrow{\scriptstyle f_{\sharp,p}} \\ & & S_p(Y;R)/S_p(B;R). \end{array}$$

Fig. 4.9 Let X be the closed unit disk and A its circular boundary. Let $p = 1$. The blue edges of the burnt orange triangle and the blue arc form a relative boundary.

One will verify that the maps

$$f_{\sharp,p}\colon S_p(X;R)/S_p(A;R) \to S_p(Y;R)/S_p(B;R)$$

define a chain map $f_\sharp$ from $S_*(X,A;R) = S_*(X;R)/S_*(A;R)$ to $S_*(Y,B;R) = S_*(Y;R)/S_*(B;R)$, and this chain map induces a homomorphism $H_p(f)\colon H_p(X,A;R) \to H_p(Y,B;R)$. ☐

The homomorphism $H_p(f)\colon H_p(X,A;R) \to H_p(Y,B;R)$ is also denoted by $f_{*p}\colon H_p(X,A;R) \to H_p(Y,B;R)$.

Proposition 4.6 is generalized to maps between pairs as follows.

Proposition 4.8. *(Homotopy Axiom) Given any two continuous maps $f,g\colon (X,A) \to (Y,B)$ if f and g are homotopic and $H_p(f), H_p(g)\colon H_p(X,A;R) \to H_p(Y,B;R)$ are the induced homomorphisms, then $H_p(f) = H_p(g)$ for all $p \geq 0$. As a consequence, if (X,A) and (Y,B) are homotopy equivalent, then the homology groups $H_p(X,A;R)$ and $H_p(Y,A;R)$ are isomorphic for all $p \geq 0$,*

Each pair (X, A) yields a short exact sequence of complexes

$$0 \longrightarrow S_*(A; R) \xrightarrow{i} S_*(X; R) \xrightarrow{j} S_*(X; R)/S_*(A; R) \longrightarrow 0,$$

where the second map is the inclusion map and the third map is the quotient map. Therefore, we can apply the zig-zag lemma (Theorem 2.22) to this short exact sequence. If we go back to the proof of this theorem and consider only spaces of index $p \leq 0$, then by changing each negative index p to $-p$ we obtain a diagram where the direction of the arrows is reversed and where each cohomology group H^p correspond to the homology group H_{-p} we obtain the "zig-zag lemma" for homology. Thus we obtain the following important result.

Theorem 4.9. *(Long Exact Sequence of Relative Homology) For every pair (X, A) of spaces, we have the following long exact sequence of homology groups*

$$\begin{array}{l} \cdots \longrightarrow H_{p+2}(X, A; R) \xrightarrow{\partial_{*p+2}} \\ H_{p+1}(A; R) \xrightarrow{i_*} H_{p+1}(X; R) \xrightarrow{j_*} H_{p+1}(X, A; R) \xrightarrow{\partial_{*p+1}} \\ H_p(A; R) \xrightarrow{i_*} H_p(X; R) \xrightarrow{j_*} H_p(X, A; R) \xrightarrow{\partial_{*p}} \\ H_{p-1}(A; R) \longrightarrow \cdots \end{array}$$

ending in

$$H_0(A; R) \longrightarrow H_0(X; R) \longrightarrow H_0(X, A; R) \longrightarrow 0.$$

It is actually possible to describe the boundary maps ∂_{*p} explicitly: for every relative cycle c, we have

$$\partial_{*p}([c]) = [\partial_p(c)].$$

4.4 Good Pairs and Reduced Homology

Pairs of spaces (X, A) where A is a nonempty closed subspace that is a deformation retract of some neighborhood in X occur naturally (for example if X is a cell complex and A is a nonempty subcomplex). Such pairs are called *good pairs*. In such a situation, it turns out that there are isomorphisms

$$H_p(X, A; R) \cong H_p(X/A, \{\text{pt}\}; R), \quad \text{for all } p \geq 0,$$

where the space X/A, *called a quotient space*, is obtained from X by identifying A with a single point, and where pt stands for any point in X (see Hatcher [Hatcher (2002)], Proposition 2.22).

It can also be shown that the homology groups $H_p(X, \{\text{pt}\}; R)$ are equal to some groups denoted by $\widetilde{H}_p(X)$, or more precisely by $\widetilde{H}_p(X; R)$, called *reduced homology groups of* X (see Hatcher [Hatcher (2002)], Proposition 2.22). The reduced homology groups $\widetilde{H}_p(X; R)$ agree with the homology groups $H_p(X; R)$ for all $p \geq 1$, and for $p = 0$, we have $H_0(X; R) = \widetilde{H}_0(X; R) \oplus R$. In particular, if X is path-connected, then $\widetilde{H}_0(X; R) = (0)$ (since $H_0(X; R) = R$). Technically, this is sometimes more convenient.

Definition 4.14. Given a nonempty space X, the *reduced homology groups*

$$\begin{aligned} \widetilde{H}_0(X; R) &= \operatorname{Ker} \epsilon / \operatorname{Im} \partial_1 \\ \widetilde{H}_p(X; R) &= \operatorname{Ker} \partial_p / \operatorname{Im} \partial_{p+1}, \quad p > 0 \end{aligned}$$

are defined by the *augmented chain complex*

$$\cdots \xleftarrow{\partial_{p-1}} S_{p-1}(X; R) \xleftarrow{\partial_p} S_p(X; R) \xleftarrow{\partial_{p+1}} \cdots$$

$$0 \longleftarrow R \xleftarrow{\epsilon} S_0(X; R) \xleftarrow{\partial_1} S_1(X; R) \longleftarrow \cdots$$

where $\epsilon \colon S_0(X; R) \to R$ is the unique R-linear map such that $\epsilon(\sigma) = 1$ for every singular 0-simplex $\sigma \colon \Delta^0 \to X$ in $S_{\Delta^0}(X)$, given by

$$\epsilon\left(\sum_i \lambda_i \sigma_i\right) = \sum_i \lambda_i.$$

It is immediate to see that $\epsilon \circ \partial_1 = 0$, so $\operatorname{Im} \partial_1 \subseteq \operatorname{Ker} \epsilon$. By definition $H_0(X; R) = S_0(X; R)/\operatorname{Im} \partial_1$. The module $S_0(X; R)$ is a free R-module isomorphic to the direct sum $\bigoplus_{\sigma \in S_{\Delta^0}(X)} R$ with one copy of R for every $\sigma \in S_{\Delta^0}(X)$, so by choosing one of the copies of R we can define an injective R-linear map $s \colon R \to S_0(X; R)$ such that $\epsilon \circ s = \text{id}$, and we obtain the following short split exact sequence:

$$0 \longrightarrow \operatorname{Ker} \epsilon \longrightarrow S_0(X; R) \underset{s}{\overset{\epsilon}{\rightleftarrows}} R \longrightarrow 0.$$

Thus

$$S_0(X; R) \cong \operatorname{Ker} \epsilon \oplus R,$$

and since $\operatorname{Im} \partial_1 \subseteq \operatorname{Ker} \epsilon$, we get

$$S_0(X; R)/\operatorname{Im} \partial_1 \cong (\operatorname{Ker} \epsilon/\operatorname{Im} \partial_1) \oplus R,$$

which yields

$$\begin{aligned} H_0(X; R) &= \widetilde{H}_0(X; R) \oplus R \\ H_p(X; R) &= \widetilde{H}_p(X; R), \quad p > 0. \end{aligned}$$

In the special case where $R = \mathbb{Z}$,

$$\begin{aligned} H_0(X) &= \widetilde{H}_0(X) \oplus \mathbb{Z} \\ H_p(X) &= \widetilde{H}_p(X), \quad p > 0. \end{aligned}$$

Since it is often used, we record the homology of a one-point space in the following proposition (see Proposition 4.3).

Proposition 4.10. *We have*

$$\begin{aligned} &H_0(\{\mathrm{pt}\}; R) = R \\ &\widetilde{H}_0(\{\mathrm{pt}\}; R) = (0) \\ &H_p(\{\mathrm{pt}\}; R) = \widetilde{H}_p(\{\mathrm{pt}\}; R) = (0), \quad \text{if } p > 0. \end{aligned}$$

One of the reasons for introducing the reduced homology groups is that

$$\widetilde{H}_p(\{\mathrm{pt}\}; R) = (0), \quad \text{for all } p \geq 0.$$

To define the *reduced singular relative homology groups* $\widetilde{H}_p(X, A; R)$ when $A \neq \emptyset$, we augment the singular chain complex

$$\cdots \xleftarrow{\partial_{p-1}} S_{p-1}(X, A; R) \xleftarrow{\partial_p} S_p(X, A; R) \xleftarrow{\partial_{p+1}} \cdots$$

$$0 \xleftarrow{\partial_0} S_0(X, A; R) \xleftarrow{\partial_1} S_1(X, A; R) \longleftarrow \cdots$$

of the pair (X, A) by adding one more 0 to the sequence:

$$0 \longleftarrow 0 \xleftarrow{\epsilon} S_0(X, A; R) \xleftarrow{\partial_1} S_1(X, A; R) \longleftarrow \cdots S_p(X, A; R) \xleftarrow{\partial_{p+1}} \cdots$$

Consequently, if $A \neq \emptyset$, we have

$$\widetilde{H}_p(X, A; R) = H_p(X, A; R) \quad \text{for all } p \geq 0.$$

In addition to the short exact sequence

$$0 \longrightarrow S_p(A;R) \longrightarrow S_p(X;R) \longrightarrow S_p(X;R)/S_p(A;R) \longrightarrow 0$$

that holds for all $p \geq 0$, we add the following exact sequence

$$0 \longrightarrow R \xrightarrow{\text{id}} R \longrightarrow 0 \longrightarrow 0$$

in dimension -1 and then we obtain a version of Theorem 4.9 for reduced homology.

Theorem 4.11. *(Long Exact Sequence of Reduced Relative Homology) For every pair (X, A) of spaces, we have the following long exact sequence of reduced homology groups*

$$\cdots \longrightarrow \widetilde{H}_{p+2}(X,A;R) \xrightarrow{\partial_{*p+2}} \widetilde{H}_{p+1}(A;R) \xrightarrow{i_*} \widetilde{H}_{p+1}(X;R) \xrightarrow{j_*} \widetilde{H}_{p+1}(X,A;R) \xrightarrow{\partial_{*p+1}} \widetilde{H}_p(A;R) \xrightarrow{i_*} \widetilde{H}_p(X;R) \xrightarrow{j_*} \widetilde{H}_p(X,A;R) \xrightarrow{\partial_{*p}} \widetilde{H}_{p-1}(A;R) \longrightarrow \cdots$$

ending in

$$\widetilde{H}_0(A;R) \longrightarrow \widetilde{H}_0(X;R) \longrightarrow \widetilde{H}_0(X,A;R) \longrightarrow 0.$$

If we apply Theorem 4.11 to the pair $(X, \{\text{pt}\})$ where $\text{pt} \in X$, since $\widetilde{H}_p(\{\text{pt}\};R) = (0)$ for all $p \geq 0$, we obtain the following isomorphisms as a corollary:

$$H_p(X, \{\text{pt}\}; R) \cong \widetilde{H}_p(X;R), \quad \text{for all } p \geq 0.$$

The following result is proven in Hatcher [Hatcher (2002)] (Proposition 2.22).

Proposition 4.12. *If (X, A) is a good pair, which means that A is a nonempty closed subspace that is a deformation retract of some neighborhood in X, then*

$$H_p(X,A;R) \cong H_p(X/A, \{\text{pt}\}; R) \cong \widetilde{H}_p(X/A;R), \quad \textit{for all } p \geq 0.$$

Using Proposition 4.12 we obtain the following theorem which can be used to compute the homology of a quotient space X/A from the homology of X and the homology of its subspace A (see Hatcher [Hatcher (2002)], Theorem 2.13).

Theorem 4.13. *For every pair of spaces (X, A), if (X, A) is a good pair, then we have the following long exact sequence of reduced homology groups*

$$\cdots \longrightarrow \widetilde{H}_{p+2}(X/A; R) \xrightarrow{\partial_{*p+2}}$$

$$\widetilde{H}_{p+1}(A; R) \xrightarrow{i_*} \widetilde{H}_{p+1}(X; R) \xrightarrow{j_*} \widetilde{H}_{p+1}(X/A; R) \xrightarrow{\partial_{*p+1}}$$

$$\widetilde{H}_{p}(A; R) \xrightarrow{i_*} \widetilde{H}_{p}(X; R) \xrightarrow{j_*} \widetilde{H}_{p}(X/A; R) \xrightarrow{\partial_{*p}}$$

$$\widetilde{H}_{p-1}(A; R) \longrightarrow \cdots$$

ending in

$$\widetilde{H}_0(A; R) \longrightarrow \widetilde{H}_0(X; R) \longrightarrow \widetilde{H}_0(X/A; R) \longrightarrow 0.$$

4.5 Excision and the Mayer–Vietoris Sequence

One of the main reasons why the relative homology groups are important is that they satisfy a property known as excision.

Theorem 4.14. *(Excision Axiom) Given subspaces $Z \subseteq A \subseteq X$ such that the closure of Z is contained in the interior of A, then the inclusion $(X - Z, A - Z) \longrightarrow (X, A)$ induces isomorphisms of singular homology*

$$H_p(X - Z, A - Z; R) \cong H_p(X, A; R), \quad \textit{for all } p \geq 0.$$

See Figure 4.10. Equivalently, for any subspaces $A, B \subseteq X$ whose interiors cover X, the inclusion map $(B, A \cap B) \longrightarrow (X, A)$ induces isomorphisms

$$H_p(B, A \cap B; R) \cong H_p(X, A; R), \quad \textit{for all } p \geq 0.$$

See Figure 4.11.

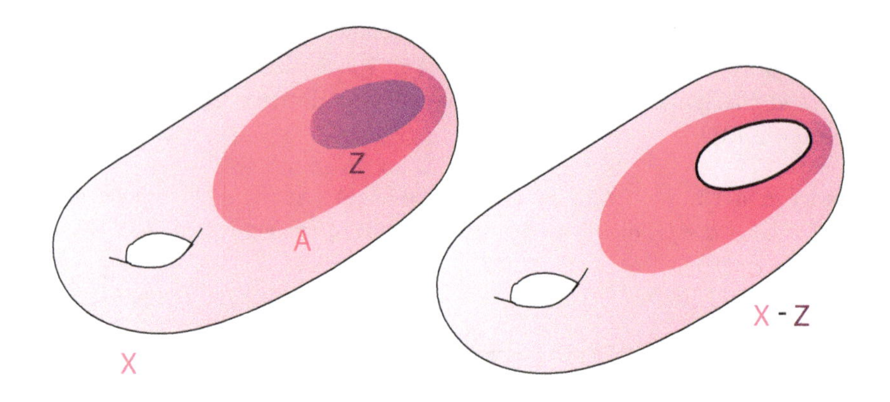

Fig. 4.10 Let X be the torus. This figure demonstrates the excision of the plum disk Z from X.

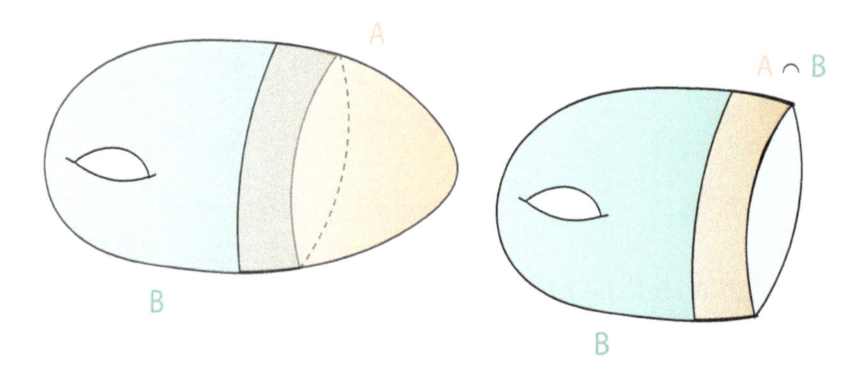

Fig. 4.11 Let X be the torus. This figure demonstrates the relationships between A, B and $A \cap B$.

The translation between the two versions is obtained by setting $B = X - Z$ and $Z = X - B$, in which case $A \cap B = A - Z$. The proof of Theorem 4.14 is rather technical and uses a technique known as barycentric subdivision. The reader is referred to Hatcher [Hatcher (2002)] (Chapter 2, Section 2.1) and Munkres [Munkres (1984)] (Chapter 4, Section 31).

Proposition 4.8, Theorem 4.9, and Theorem 4.14, state three of the properties that were singled out as characterizing homology theories by

Eilenberg and Steenrod [Eilenberg and Steenrod (1952)]. These properties hold for most of the known homology theories, and thus can be taken as axioms for homology theory; see Sato [Sato (1999)], Mac Lane [Mac Lane (1975)], Munkres [Munkres (1984)], or Hatcher [Hatcher (2002)].

The proof of Theorem 4.14 also relies on a technical lemma about the relationship between the chain complex $S_*(X;R)$ and the chain complex $S^{\mathcal{U}}_*(X;R)$ induced by a family $\mathcal{U}=(U_i)_{i\in I}$ of subsets of X whose interiors form an open cover of X.

Definition 4.15. Given a topological space X, for any family $\mathcal{U}=(U_i)_{i\in I}$ of subsets of X whose interiors form an open cover of X, we say that a singular p-simplex $\sigma\colon \Delta^p\to X$ is *$\mathcal{U}$-small* if its image is contained in one of the U_i. The submodule $S^{\mathcal{U}}_p(X;R)$ of $S_p(X;R)$ consists of all singular p-chains $\sum\lambda_k\sigma_k$ such that each p-simplex σ_k is $\mathcal{U}$-small. See Figure 4.12.

It is immediate that the boundary map $\partial_p\colon S_p(X;R)\to S_{p-1}(X;R)$ takes $S^{\mathcal{U}}_p(X;R)$ into $S^{\mathcal{U}}_{p-1}(X;R)$, so $S^{\mathcal{U}}_*(X;R)$ is a chain complex. The homology modules of the complex $S^{\mathcal{U}}_*(X;R)$ are denoted by $H^{\mathcal{U}}_p(X;R)$.

Proposition 4.15. *Given a topological space X, for any family $\mathcal{U}=(U_i)_{i\in I}$ of subsets of X whose interiors form an open cover of X, the inclusions $\iota_p\colon S^{\mathcal{U}}_p(X;R)\to S_p(X;R)$ induce a chain homotopy equivalence; that is, there is a family of chain maps $\rho_p\colon S_p(X;R)\to S^{\mathcal{U}}_p(X;R)$ such that $\rho\circ\iota$ is chain homotopic to the identity map of $S^{\mathcal{U}}_*(X;R)$ and $\iota\circ\rho$ is chain homotopic to the identity map of $S_*(X;R)$. As a consequence, we have isomorphisms $H^{\mathcal{U}}_p(X;R)\cong H_p(X;R)$ for all $p\geq 0$.*

The proof of Proposition 4.15 is quite involved. It uses barycentric subdivision; see Hatcher [Hatcher (2002)] (Chapter 2, Proposition 2.21) and Munkres [Munkres (1984)] (Chapter 4, Section 31, Theorem 31.5).

Besides playing a crucial role in proving the excision axiom, Proposition 4.15 yields a simple proof of the Mayer–Vietoris sequence in singular homology. For arbitrary topological spaces, partitions of unity are not available but the set-up of Proposition 4.15 yields an alternative method of proof.

Theorem 4.16. *(Mayer–Vietoris in singular homology) Given any topological space X, for any two subsets A, B of X such that $X=\mathrm{Int}(A)\cup\mathrm{Int}(B)$,*

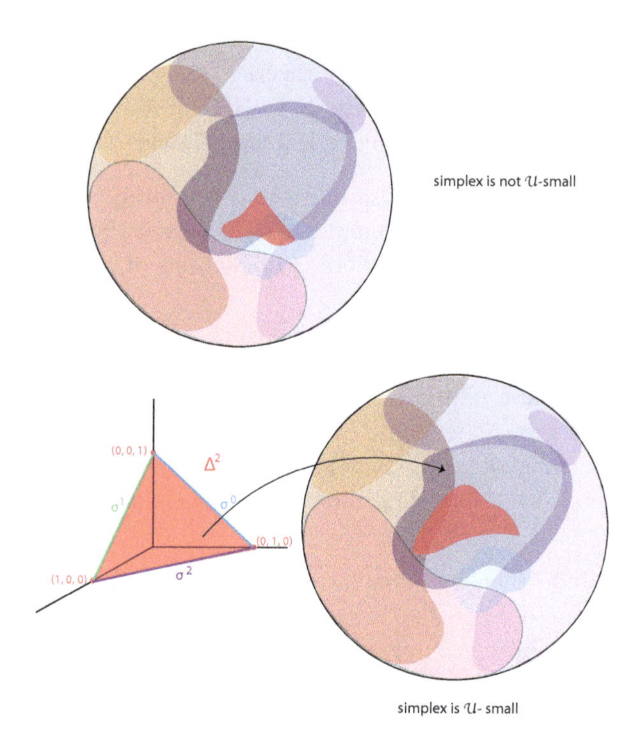

Fig. 4.12 Let X be the unit disk in $\mathbb{R}^2$. The colored patches represent the cover $\mathcal{U}$. In the top figure the 2-simplex is not $\mathcal{U}$-small since it not contained within any one U_i, while in the bottom figure, the 2-simplex is $\mathcal{U}$-small since it is contained within the "central" blue U_i.

there is a long exact sequence of homology

$$\longrightarrow H_p(A \cap B; R) \xrightarrow{\varphi_*} H_p(A; R) \oplus H_p(B; R) \xrightarrow{\psi_*} H_p(X; R) \xrightarrow{\partial_*}$$
$$H_{p-1}(A \cap B; R) \longrightarrow \cdots$$

where the maps φ and ψ are defined by

$$\varphi_*(c) = (i_*(c), -j_*(c))$$
$$\psi_*(a, b) = k_*(a) + l_*(b),$$

and where i, j, k, l are the inclusion maps shown in the diagram below:

$$\begin{array}{ccc} A \cap B & \xrightarrow{i} & A \\ {\scriptstyle j}\downarrow & & \downarrow{\scriptstyle k} \\ B & \xrightarrow[l]{} & X. \end{array}$$

If $A \cap B \neq \emptyset$, a similar sequence exists in reduced homology.

Proof. For simplicity of notation we suppress the ring R in writing $S_p(-, R)$ or $H_p(-, R)$. We define a sequence

$$0 \longrightarrow S_p(A \cap B) \xrightarrow{\varphi} S_p(A) \oplus S_p(B) \xrightarrow{\psi} S_p(A) + S_p(B) \longrightarrow 0$$

for every $p \geq 0$, where φ and ψ are given by

$$\varphi(c) = (i_\sharp(c), -j_\sharp(c))$$
$$\psi(a, b) = k_\sharp(a) + l_\sharp(b).$$

Observe that $\psi \circ \varphi = 0$. The map φ is injective, while ψ is surjective. We have $\operatorname{Im}\varphi \subseteq \operatorname{Ker}\psi$ since $\psi \circ \varphi = 0$. The kernel of ψ consists of all chains of the form $(c, -c)$ where $c \in S_p(A)$ and $-c \in S_p(B)$ so $c \in S_p(A \cap B)$ and $\varphi(c) = (c, -c)$, which shows that $\operatorname{Ker}\psi \subseteq \operatorname{Im}\varphi$. Therefore the sequence is exact, and we have a short exact sequence of chain complexes

$$0 \longrightarrow S_*(A \cap B) \xrightarrow{\varphi} S_*(A) \oplus S_*(B) \xrightarrow{\psi} S_*(A) + S_*(B) \longrightarrow 0. \tag{$*_{\mathrm{MV}}$}$$

By the long exact sequence of homology we have the long exact sequence

$$\longrightarrow H_p(A \cap B) \xrightarrow{\varphi_*} H_p(A) \oplus H_p(B) \xrightarrow{\psi_*} H_p(S_*(A) + S_*(B)) \xrightarrow{\partial_*}$$
$$H_{p-1}(A \cap B) \longrightarrow \cdots .$$

However, since $X = \operatorname{Int}(A) \cup \operatorname{Int}(B)$, Proposition 4.15 implies that

$$H_p(S_*(A) + S_*(B)) \cong H_p(X),$$

and we obtain the long exact sequence

$$\cdots \longrightarrow H_p(A \cap B) \xrightarrow{\varphi_*} H_p(A) \oplus H_p(B) \xrightarrow{\psi_*} H_p(X) \xrightarrow{\partial_*}$$
$$H_{p-1}(A \cap B) \longrightarrow \cdots ,$$

as desired. A similar argument applies to reduced homology by augmenting the complexes $S_*(A \cap B)$, $S_*(A) \oplus S_*(B)$, and $S_*(A) + S_*(B)$ using the maps $\epsilon\colon S_0(A \cap B) \to R$, $\epsilon \oplus \epsilon\colon S_0(A) \oplus S_0(B) \to R \oplus R$, and $\epsilon\colon S_0(A) + S_0(B) \to R$. ☐

Remark: The sequence $(*_{\text{MV}})$ is actually a split short exact sequence. This follows from Corollary 2.3, since $S_*(A) + S_*(B)$ is a free R-module.

The Mayer–Vietoris sequence can be used to compute the homology of spaces in terms of some of their pieces. For example, this is a way to compute the homology of the n-torus.

There are two more important properties of singular homology that should be mentioned:

(1) The axiom of compact support.
(2) The additivity axiom.

The axiom of compact support implies that the homology groups $H_p(X, A; R)$ are determined by the groups $H_p(C, D; R)$ where (C, D) is a compact pair in (X, A), which means that $D \subseteq C \subseteq X$, $D \subseteq A \subseteq X$, C is compact, and D is compact in C. See Figure 4.13.

Let $\mathcal{K}(X, A)$ be the sets of all compact pairs of (X, A) ordered by inclusion. It is a directed preorder.

Proposition 4.17. *For any pair (X, A) of topological spaces with $A \subseteq X$, the following properties hold:*

(1) Given any homology class $\alpha \in H_p(X, A)$, there is a compact pair (C, D) in (X, A) and a homology class $\beta \in H_p(C, D; R)$ such that $i_*(\beta) = \alpha$, where $i\colon (C, D) \to (X, A)$ is the inclusion map.
(2) Let (C, D) be any compact pair in (X, A), and let $\beta \in H_p(C, D; R)$ be any homology class such that $i_*(\beta) = 0$. Then there exists a compact pair (C', D') such that $(C, D) \subseteq (C', D') \subseteq (X, A)$ and $j_*(\beta) = 0$, where $j\colon (C, D) \to (C', D')$ is the inclusion map.
(3) The homology functor commutes with direct limits.
(4) The R-module $H_p(X, A; R)$ is isomorphic to the direct limit (see Section 8.3)

$$H_p(X, A; R) \cong \varinjlim_{(C,D) \in \mathcal{K}(X,A)} H_p(C, D; R).$$

Parts (1) and (2) of Proposition 4.17 are proven in Massey [Massey (1991)] (Chapter VIII, Section 6, Proposition 6.1) and Rotman [Rotman (1988)] (Chapter 4, Theorem 4.16). Part (3) is proven in Spanier [Spanier (1989)] (Chapter 4, Section 1, Theorem 7), as well as Part (4) (Chapter 4, Section 4, Theorem 6).

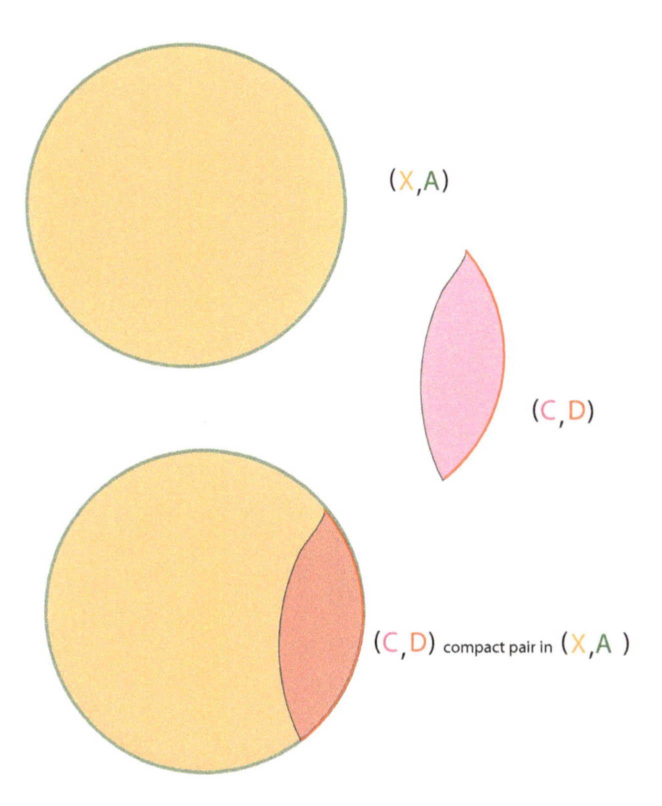

Fig. 4.13 Let X be the unit disk in $\mathbb{R}^2$ and A its green boundary, namely S^1. Then (C, D) is a compact pair of (X, A).

Sketch of proof. The proof of (1) is not difficult and relies on the fact that for any singular p-chain $a \in S_p(X; R)$ there is a compact subset C of X such that $a \in S_p(C; R)$. For simplicity of exposition assume that $A = \emptyset$. If $a = \sum_{i=1}^{k} \lambda_i \sigma_i \in S_p(X, R)$ is a cycle representing the homology class α, with $\lambda_i \in R$ and each σ_i a p-simplex $\sigma_i \colon \Delta^p \to X$, since Δ^p is compact and each σ_i is continuous, $C = \sigma_1(\Delta^p) \cup \cdots \cup \sigma_k(\Delta^p)$ is a compact subset of X and $a \in S_p(C; R)$. Let $b = \sum_{i=1}^{k} \lambda_i \sigma_i' \in S_p(C, R)$ be the p-chain in which $\sigma_i' \colon \Delta^p \to C$ is the corestriction of σ_i to C. We need to check that b is a p-cycle. By definition of the inclusion i we have $a = i_\sharp(b)$, and since a is a p-cycle we have

$$i_\sharp \circ \partial(b) = \partial \circ i_\sharp(b) = \partial a = 0.$$

Since i is an injection, $i_\sharp$ is also an injection, thus $\partial b = 0$, which means that $b \in S_p(C; R)$ is indeed a p-cycle, and if β denotes the homology class of b, we have $i_*(\beta) = \alpha$. The above argument is easily adapted to the case where $A \neq \emptyset$. The proof of (2) is similar and left as an exercise. $\square$

The above fact suggests the following axiom of homology.

Axiom of Compact Support.

Given any pair (X, A) with $A \subseteq X$ and given any homology class $\alpha \in H_p(X, A)$, there is a compact pair (C, D) in (X, A) and a homology class $\beta \in H_p(C, D; R)$ such that $i_*(\beta) = \alpha$, where $i\colon (C, D) \to (X, A)$ is the inclusion map.

This axiom is another of the axioms of a homology theory; see Munkres [Munkres (1984)] (Chapter 3, Section 26, Axiom 8), or Spanier [Spanier (1989)] (Chapter 4, Section 8, No. 11).

Remark: It turns out that Part (4) of Proposition 4.17 is equivalent to the fact that the axiom of compact support holds; see Spanier [Spanier (1989)] (Chapter 4, Section 8, Theorem 13).

To state the additivity axiom we need to define the topological sum of a family of spaces.

Definition 4.16. If $(X_i)_{i\in I}$ is a family of topological spaces we define the *topological sum* $\bigsqcup_{i\in I} X_i$ of the family $(X_i)_{i\in I}$ as the disjoint union of the spaces X_i, and we give it the topology for which a subset $Z \subseteq \bigsqcup_{i\in I} X_i$ is open iff $Z \cap X_i$ is open for all $i \in I$.

Additivity Axiom.

For any family of topological spaces $(X_i)_{i\in I}$ there is an isomorphism

$$H_p\left(\bigsqcup_{i\in I} X_i; R\right) \cong \bigoplus_{i\in I} H_p(X_i; R) \quad \text{for all } p \geq 0.$$

The above axiom introduced by Milnor is stated in Bredon [Bredon (1993)] (Chapter IV, Section 6), May [May (1999)] (Chapter 13, Section 1), and Hatcher [Hatcher (2002)] (Chapter 2, Section 2.3), where it is stated for relative homology and for a wedge sum of spaces.

The additivity axiom is a general property of chain complexes. Indeed, homology commutes with sums, products, and direct limits; see Spanier [Spanier (1989)] (Chapter 4, Section 1, Theorem 6 and Theorem 7). This axiom is only needed for infinite sums.

4.6 Some Applications of Singular Homology

It is remarkable that Proposition 4.8, Theorem 4.9, Theorem 4.11, Theorem 4.13, Theorem 4.14 and Theorem 4.16 can be used to compute the singular homology of some of the familiar simple spaces. The key idea is that the excision axiom, the homotopy axiom, and either the long exact sequence of relative homology (Theorem 4.9), or the long exact sequence of reduced relative homology (Theorem 4.11), or the long exact sequence of reduced homology for a good pair (Theorem 4.13), or the Mayer–Vietoris long exact sequence (Theorem 4.16), can be used to produce exact sequences in which two consecutive homology groups are "trapped" between zeros, and thus are isomorphic. Often we can obtain more isomorphisms by induction. For example, We show below how to compute the homology groups of the spheres.

Recall that the n-dimensional ball D^n and the n-dimensional sphere S^n are defined respectively as the subspaces of $\mathbb{R}^n$ and $\mathbb{R}^{n+1}$ given by

$$\begin{aligned} D^n &= \{x \in \mathbb{R}^n \mid \|x\|_2 \leq 1\} \\ S^n &= \{x \in \mathbb{R}^{n+1} \mid \|x\|_2 = 1\}. \end{aligned}$$

Observe that $D^0 = \{0\}$, a point-point space. Furthermore, $S^n = \partial D^{n+1}$, the boundary of D^{n+1}, and $D^n/\partial D^n$ is homeomorphic to S^n $(n \geq 1)$. We also know that D^n is convex for all $n \geq 0$, so by Proposition 4.3, its homology groups are given by

$$\begin{aligned} H_0(D^n; R) &= R \\ H_p(D^n; R) &= (0), \quad p \geq 0, \end{aligned}$$

or equivalently

$$\widetilde{H}_p(D^n; R) = (0), \quad p \geq 0.$$

Proposition 4.18. *The reduced homology of S^n is given by*

$$\widetilde{H}_p(S^n; R) = \begin{cases} R & \text{if } p = n \\ (0) & \text{if } p \neq n, \end{cases}$$

or equivalently the homology of S^n is given by

$$\begin{aligned} H_0(S^0; R) &= R \oplus R \\ H_p(S^0; R) &= (0), \quad p > 0, \end{aligned}$$

and for $n \geq 1$,

$$H_p(S^n; R) = \begin{cases} R & \text{if } p = 0, n \\ (0) & \text{if } p \neq 0, n. \end{cases}$$

Proof. For simplicity of notation, we drop the ring R in writing homology groups. Since $S^0 = \{-1, 1\}$, by the excision axiom (Theorem 4.14) with $X = S^0 = \{-1, 1\}$, $A = \{-1\}$ and $B = \{1\}$, we get

$$H_p(S^0, \{-1\}) \cong H_p(\{1\}, \emptyset) = H_p(\{1\})$$

for all $p \geq 0$. The long exact sequence of Theorem 4.9 for the pair $(S^0, \{-1\})$ gives the exact sequence

$$\longrightarrow H_p(\{-1\}) \longrightarrow H_p(S^0) \longrightarrow H_p(S^0, \{-1\}) \longrightarrow H_{p-1}(\{-1\}) \longrightarrow$$

If $p \geq 1$, since $H_p(\{-1\}) = H_p(\{1\}) = (0)$ and $H_p(S^0, \{-1\}) \cong H_p(\{1\})$, we get $H_p(S^0) = (0)$. If $p = 0$, since $H_0(\{-1\}) = H_0(\{1\}) = R$, we get $H_0(S^0) = R \oplus R$.

If $n \geq 1$, then since $D^n/\partial D^n$ is homeomorphic to S^n and $\partial D^n = S^{n-1}$ is a deformation retract of $D^n - \{x\}$, the long exact sequence of Theorem 4.13 for the good pair $(D^n, \partial D^n) = (D^n, S^{n-1})$ yields the exact sequence

$$\begin{array}{ccc} \longrightarrow \widetilde{H}_p(D^n) & \longrightarrow & \widetilde{H}_p(D^n/S^{n-1}) = \widetilde{H}_p(S^n) \longrightarrow \\ \widetilde{H}_{p-1}(S^{n-1}) & \longrightarrow & \widetilde{H}_{p-1}(D^{n-1}) \longrightarrow \end{array}$$

and if $p \geq 1$, since $\widetilde{H}_p(D^n) = \widetilde{H}_{p-1}(D^{n-1}) = (0)$, we get

$$\widetilde{H}_p(S^n) \cong \widetilde{H}_{p-1}(S^{n-1}) \quad p \geq 1.$$

We conclude by induction on $n \geq 1$. □

The most convenient setting to compute homology groups is the homology of cell complexes or simplicial homology; see Chapter 6. For example, cellular homology can used to compute the homology of the real and complex projective spaces $\mathbb{RP}^n$ and $\mathbb{CP}^n$; see Section 6.2, and also Hatcher [Hatcher (2002)], Munkres [Munkres (1984)], and Bredon [Bredon (1993)]. Even though we do not have the machinery to compute these homology groups, we believe that the reader will appreciate seeing concrete examples of homology groups, in particular for classical spaces such as the projective spaces and the tori.

Example 4.1. The real projective space $\mathbb{RP}^n$ is the quotient of $\mathbb{R}^{n+1} - \{0\}$ by the equivalence relation $\sim$ defined such that for all $(u_1, \ldots, u_{n+1}) \in \mathbb{R}^{n+1} - \{0\}$ and all $(v_1, \ldots, v_{n+1}) \in \mathbb{R}^{n+1} - \{0\}$,

$$(u_1, \ldots, u_{n+1}) \sim (v_1, \ldots, v_{n+1})$$

iff

$$(\exists \alpha \in \mathbb{R} - \{0\}) \quad (v_1, \ldots, v_{n+1}) = \alpha(u_1, \ldots, u_{n+1}).$$

Equivalently, $\mathbb{RP}^n$ is the quotient of the subset S^n of $\mathbb{R}^{n+1}$ defined by

$$S^n = \{(u_1, \ldots, u_{n+1}) \in \mathbb{R}^{n+1} \mid u_1^2 + \cdots + u_{n+1}^2 = 1\},$$

in other words, the n-sphere, by the equivalence relation $\sim$ on S^n defined so that for all $(u_1, \ldots, u_{n+1}) \in S^n$ and all $(v_1, \ldots, v_{n+1}) \in S^n$,

$$(u_1, \ldots, u_{n+1}) \sim (v_1, \ldots, v_{n+1}) \quad \text{iff} \quad (v_1, \ldots, v_{n+1}) = \pm(u_1, \ldots, u_{n+1}).$$

This says that two points on the sphere S^n are equivalent iff they are antipodal. See Figure 4.14. We have a quotient map $\pi\colon S^n \to \mathbb{RP}^n$.

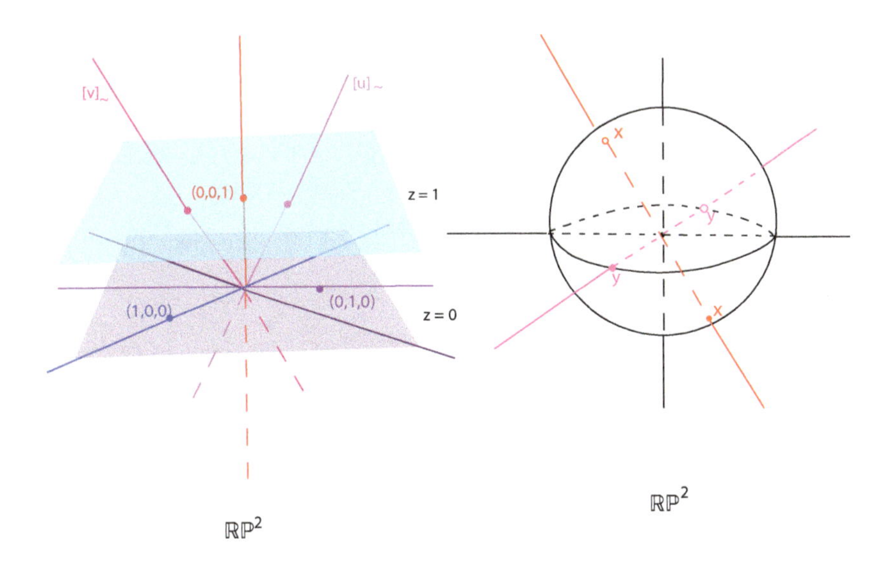

Fig. 4.14 Two representations of $\mathbb{RP}^2$. In the left representation, lines through the origin are "points". We can view $\mathbb{RP}^2$ as the union of the points in the blue plane $z = 1$ with the points at infinity corresponding to lines through the origin in the plane $z = 0$. In the right representation, $\mathbb{RP}^2$ is formed as quotient of S^2 via the antipodal equivalence.

The complex projective space $\mathbb{CP}^n$ is the quotient of $\mathbb{C}^{n+1} - \{0\}$ by the equivalence relation $\sim$ defined such that for all $(u_1, \ldots, u_{n+1}) \in \mathbb{C}^{n+1} - \{0\}$ and all $(v_1, \ldots, v_{n+1}) \in \mathbb{C}^{n+1} - \{0\}$,

$$(u_1, \ldots, u_{n+1}) \sim (v_1, \ldots, v_{n+1})$$

iff

$$(\exists \alpha \in \mathbb{C} - \{0\}) \quad (v_1, \ldots, v_{n+1}) = \alpha(u_1, \ldots, u_{n+1}).$$

Equivalently, $\mathbb{CP}^n$ is the quotient of the subset Σ^n of $\mathbb{C}^{n+1}$ defined by

$$\Sigma^n = \{(u_1, \ldots, u_{n+1}) \in \mathbb{C}^{n+1} \mid |u_1|^2 + \cdots + |u_{n+1}|^2 = 1\},$$

by the equivalence relation $\sim$ on Σ^n defined so that for all $(u_1, \ldots, u_{n+1}) \in \Sigma^n$ and all $(v_1, \ldots, v_{n+1}) \in \Sigma^n$,

$$(u_1, \ldots, u_{n+1}) \sim (v_1, \ldots, v_{n+1})$$

iff

$$(\exists \alpha \in \mathbb{C}, |\alpha| = 1) \quad (v_1, \ldots, v_{n+1}) = \alpha(u_1, \ldots, u_{n+1}).$$

If we write $u_j = x_j + iy_j$ with $x_j, y_j \in \mathbb{R}$, we have $(u_1, \ldots, u_{n+1}) \in \Sigma^n$ iff

$$x_1^2 + y_1^2 + \cdots + x_{n+1}^2 + y_{n+1}^2 = 1,$$

iff $(x_1, y_1, \ldots, x_{n+1}, y_{n+1}) \in S^{2n+1}$. Therefore we can identify Σ^n with S^{2n+1}, and we can view $\mathbb{CP}^n$ as the quotient of S^{2n+1} by the above equivalence relation. We have a quotient map $\pi \colon S^{2n+1} \to \mathbb{CP}^n$.

For $R = \mathbb{Z}$, the homology groups of $\mathbb{CP}^n$ and $\mathbb{RP}^n$ are given by

$$H_p(\mathbb{CP}^n; \mathbb{Z}) = \begin{cases} \mathbb{Z} & \text{for } p = 0, 2, 4, \ldots, 2n \\ (0) & \text{otherwise,} \end{cases}$$

and

$$H_p(\mathbb{RP}^n; \mathbb{Z}) = \begin{cases} \mathbb{Z} & \text{for } p = 0 \text{ and for } p = n \text{ odd} \\ \mathbb{Z}/2\mathbb{Z} & \text{for } p \text{ odd, } 0 < p < n \\ (0) & \text{otherwise.} \end{cases}$$

The homology of the n-torus $T^n = \underbrace{S^1 \times \cdots \times S^1}_{n}$ exhibits a remarkable symmetry:

$$H_p(T^n; R) = \underbrace{R \oplus \cdots \oplus R}_{\binom{n}{p}}.$$

The homology of the n-torus T^n can be computed by induction. Indeed, using the Mayer–Vietoris sequence (Theorem 4.16), it can be shown that

$$H_p(X \times S^1; \mathbb{Z}) \cong H_p(X; \mathbb{Z}) \oplus H_{p-1}(X; \mathbb{Z})$$

for any topological space X; see Exercise 36 in Hatcher [Hatcher (2002)].

Surprisingly, computing the homology groups $H_p(\mathbf{SO}(n); \mathbb{Z})$ of the rotation group $\mathbf{SO}(n)$ is more difficult. It can be shown that the groups $H_p(\mathbf{SO}(n); \mathbb{Z})$ are directs sums of copies of $\mathbb{Z}$ and $\mathbb{Z}/2\mathbb{Z}$, but their exact structure is harder to obtain. For more on this topic, we refer the reader to Hatcher [Hatcher (2002)] (Chapter 3, Sections 3.D and 3.E).

One of the most spectacular applications of homology is a proof of a generalized version of the Jordan curve theorem. First we need a bit of terminology.

Definition 4.17. Given two topological spaces X and Y, an *embedding* is a homeomorphism $f\colon X \to Y$ of X onto its image $f(X)$. A *m-cell* or *cell of dimension m* is any space B homeomorphic to the closed ball D^m. A subspace A of a space X *separates* X if $X - A$ is not connected.

Proposition 4.19. *Let B be a k-cell in S^n. Then $S^n - B$ is acyclic, which means that $H_p(S^n - B) = (0)$ for all $p \neq 0$. In particular B does not separate S^n.*

Proposition 4.19 is proven in Munkres [Munkres (1984)] (Chapter 4, Section 36, Theorem 36.1). See also Bredon [Bredon (1993)] (Chapter IV, Corollary 19.3).

Proposition 4.20. *Let $n > k \geq 0$. For any embedding $h\colon S^k \to S^n$ we have*

$$\widetilde{H}_p(S^n - h(S^k)) = \begin{cases} \mathbb{Z} & \text{if } p = n - k - 1 \\ 0 & \text{otherwise.} \end{cases}$$

This implies that $\widetilde{H}_p(S^n - h(S^k)) \cong \widetilde{H}_p(S^{n-k-1})$.

Proposition 4.20 is proven in Munkres [Munkres (1984)] (Chapter 4, Section 36, Theorem 36.2) and Bredon [Bredon (1993)] (Chapter IV, Theorem 19.4). The proof uses an induction on k and a Mayer-Vietoris sequence. Proposition 4.20 implies the following generalization of the Jordan curve theorem for $n \geq 1$.

Theorem 4.21. *(Generalized Jordan Curve Theorem in S^n) Let $n > 0$ and let C be any subset of S^n homeomorphic to S^{n-1}. Then $S^n - C$ has precisely two components, both acyclic, and C is their common topological boundary. See Figure 4.15.*

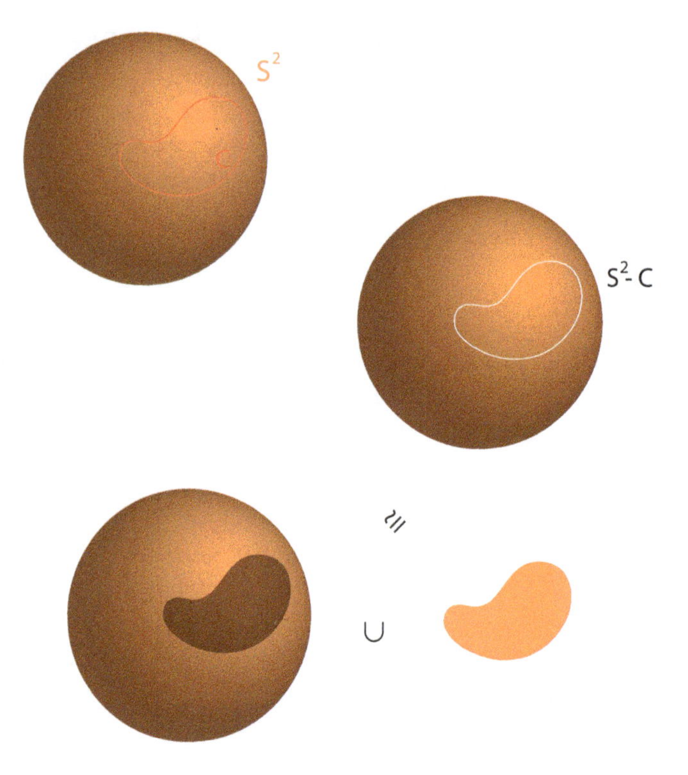

Fig. 4.15 An illustration of the Jordan curve theorem in S^2. Let C be the red curve which is homeomorphic to S^1. Then $S^2 - C$ has two component, which in this case, are each homeomorphic to D^2.

Theorem 4.21 is proved in Munkres [Munkres (1984)] (Chapter 4, Section 36, Theorem 36.3) and Bredon [Bredon (1993)] (Chapter IV, Theorem 19.5), in which it is called the *Jordan–Brouwer separation theorem*.

The first part of the theorem is obtained by applying Proposition 4.20 in the case where $k = n - 1$. In this case we see that $\widetilde{H}_0(S^n - C) = \mathbb{Z}$, so $H_0(S^n - C) = \mathbb{Z} \oplus \mathbb{Z}$ and this implies that $S^n - C$ has precisely two path components. The proof of the second part uses Proposition 4.19.

One might think that because C is homeomorphic to S^{n-1} the two components W_1 and W_2 of $S^n - C$ should be n-cells, but this is false in general. The problem is that an embedding of S^{n-1} into S^n can be very complicated. There is a famous embedding of S^2 into S^3 called the *Alexander horned sphere* for which the sets W_1 and W_2 are not even simply connected; see Bredon [Bredon (1993)] (Chapter IV, Page 232) and Hatcher [Hatcher (2002)] (Chapter 2, Example 2B.2). In the case $n = 2$, things are simpler; see Hatcher [Hatcher (2002)] (Chapter 2, Section 2.B) and Bredon [Bredon (1993)] (Chapter IV, Pages 235–236).

The classical version of the Jordan curve theorem is stated for embeddings of S^{n-1} into $\mathbb{R}^n$.

Theorem 4.22. *(Generalized Jordan curve theorem in $\mathbb{R}^n$) Let $n > 1$ and let C be any subset of $\mathbb{R}^n$ homeomorphic to S^{n-1}. Then $\mathbb{R}^n - C$ has precisely two components, one of which is bounded and the other one is not. The bounded component is acyclic and the other has the homology of S^{n-1}.*

Proof. Using the inverse stereographic projection from the north pole N we can embed C into S^n. See Figure 4.16. By Theorem 4.21 $S^n - C$ has two acyclic components. Let V be the component containing N. Obviously the other component U is bounded and acyclic. It follows that $S^n - U$ is homeomorphic to D^n so we can view V as being a subset of D^n. Next we follow Bredon [Bredon (1993)] (Chapter IV, Corollary 19.6). Consider the piece of the long exact sequence of the pair $(V, V - \{N\})$ given by Theorem 4.11 with $X = V$ and $A = V - \{N\}$:

$$\widetilde{H}_{p+1}(V) \longrightarrow H_{p+1}(V, V - \{N\}) \longrightarrow \widetilde{H}_p(V - \{N\}) \longrightarrow \widetilde{H}_p(V),$$

where we used the fact that $\widetilde{H}_{p+1}(V, V - \{N\}) = H_{p+1}(V, V - \{N\})$, since $p + 1 \geq 1$. By Theorem 4.21 the homology of V is acyclic, so we have the following isomorphisms

$$\begin{aligned}\widetilde{H}_p(V - \{N\}) &\cong H_{p+1}(V, V - \{N\}) \\ &\cong H_{p+1}(D^n, D^n - \{0\}) \\ &\cong \widetilde{H}_p(D^n - \{0\}) \\ &\cong \widetilde{H}_p(S^{n-1}),\end{aligned}$$

where the second isomorphism holds by excision since $V \subseteq D^n$, the third holds from the long exact sequence of $(D^n, D^n - \{0\})$, and the fourth by homotopy. $\square$

Later on, to define orientable manifolds we will need to compute the groups $H_p(M, M - \{x\}; R)$ where M is a topological manifold and x is any point in M.

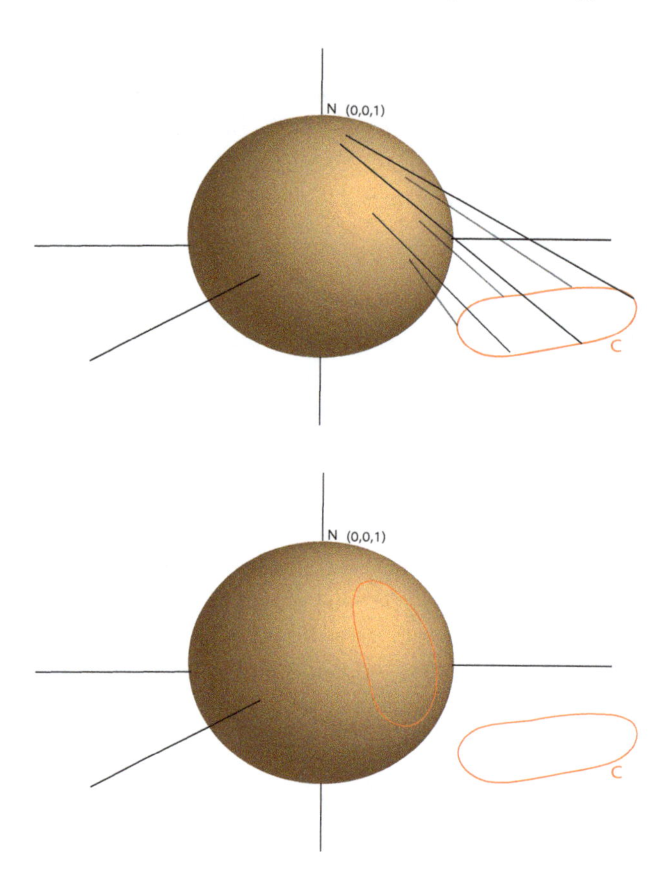

Fig. 4.16 Let C be the red curve in $\mathbb{R}^2$ which is homeomorphic to S^1. The top figure shows how to use the inverse stereographic projection to embed C into S^2. The embedded curve is illustrated in the bottom figure.

Recall the definition of a topological manifold.

Definition 4.18. A *topological manifold M of dimension n*, for short an *n-manifold*, is a topological space such that for every $x \in M$, there is some open subset U of M containing x and some homeomorphism $\varphi_U : U \to \Omega$ (called a *chart at x*) onto some open subset $\Omega \subseteq \mathbb{R}^n$. See Figure 4.17.

We have the following result.

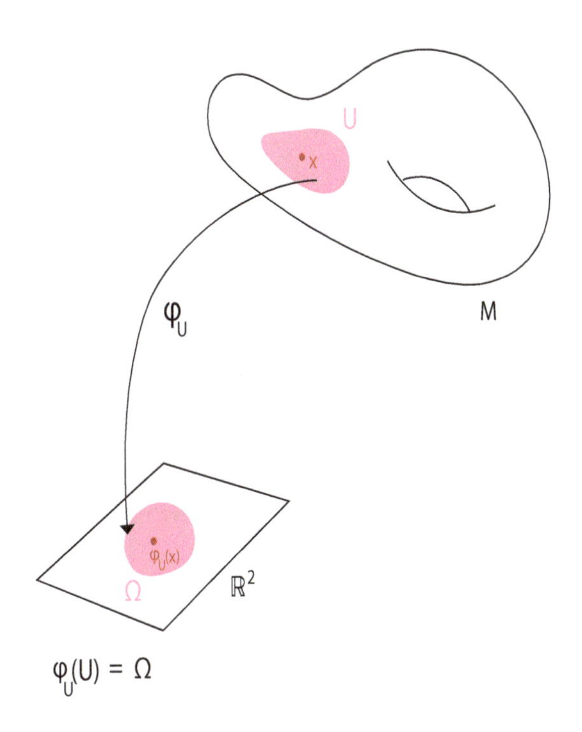

Fig. 4.17 A 2-dimensional manifold M and a chart at x. Note that M is homeomorphic to T^2.

Proposition 4.23. *If M is a topological manifold of dimension n and if R is any commutative ring with a multiplicative identity element, then*

$$\begin{aligned} H_p(M, M - \{x\}; R) &\cong H_p(\mathbb{R}^n, \mathbb{R}^n - \{x\}; R) \cong \widetilde{H}_{p-1}(\mathbb{R}^n - \{x\}; R) \\ &\cong \widetilde{H}_{p-1}(S^{n-1}) \end{aligned}$$

for all $p \geq 0$. Consequently

$$H_p(M, M - \{x\}; R) \cong \begin{cases} R & \text{if } p = n \\ (0) & \text{if } p \neq n. \end{cases}$$

Proof. By shrinking U is necessary we may assume that U is homeomorphic to $\mathbb{R}^n$, so by excision with $X = M, A = M - x$, and $Z = M - U$ (see Theorem 4.14), we obtain

$$H_p(M, M - \{x\}; R) \cong H_p(U, U - \{x\}; R) \cong H_p(\mathbb{R}^n, \mathbb{R}^n - \{x\}; R).$$

By Theorem 4.11 the long exact sequence of homology yields an exact sequence

$$\begin{aligned} &\widetilde{H}_{p+1}(\mathbb{R}^n;R) \longrightarrow \widetilde{H}_{p+1}(\mathbb{R}^n,\mathbb{R}^n-\{x\};R) \longrightarrow \\ &\widetilde{H}_p(\mathbb{R}^n-\{x\};R) \longrightarrow \widetilde{H}_p(\mathbb{R}^n;R). \end{aligned}$$

Since $\mathbb{R}^n$ is contractible, $\widetilde{H}_{p+1}(\mathbb{R}^n;R) = (0)$ and $\widetilde{H}_p(\mathbb{R}^n;R) = (0)$ so we have isomorphisms

$$\widetilde{H}_{p+1}(\mathbb{R}^n,\mathbb{R}^n-\{x\};R) \cong \widetilde{H}_p(\mathbb{R}^n-\{x\};R)$$

for all $p \geq 0$. Since $\widetilde{H}_{p+1}(\mathbb{R}^n,\mathbb{R}^n-\{x\};R) = H_{p+1}(\mathbb{R}^n,\mathbb{R}^n-\{x\};R)$ for $p \geq 1$, we get

$$H_p(\mathbb{R}^n,\mathbb{R}^n-\{x\};R) \cong \widetilde{H}_{p-1}(\mathbb{R}^n-\{x\};R)$$

for all $p \geq 1$. To finish the proof if $p \geq 1$, observe that S^{n-1} is a deformation retract of $\mathbb{R}^n-\{x\}$, so by the homotopy axiom (Proposition 4.8) we get

$$H_p(\mathbb{R}^n,\mathbb{R}^n-\{x\};R) \cong \widetilde{H}_{p-1}(S^{n-1};R)$$

for all $p \geq 1$. We conclude by using Proposition 4.18.

For $p = 0$, the end of the long exact sequence given by Theorem 4.9 yields

$$\begin{aligned} &H_1(\mathbb{R}^n,\mathbb{R}^n-\{x\}) \xrightarrow{f} H_0(\mathbb{R}^n-\{x\}) \xrightarrow{g} \\ &H_0(\mathbb{R}^n) \xrightarrow{h} H_0(\mathbb{R}^n,\mathbb{R}^n-\{x\}) \longrightarrow 0. \end{aligned}$$

If $n > 1$, then we just proved that $H_1(\mathbb{R}^n,\mathbb{R}^n-\{x\};R) = (0)$. In this case $H_0(\mathbb{R}^n;R) = R$ and $H_0(\mathbb{R}^n-\{x\};R) = R$ so we have the exact sequence

$$0 \xrightarrow{f} R \xrightarrow{g} R \xrightarrow{h} H_0(\mathbb{R}^n,\mathbb{R}^n-\{x\};R) \longrightarrow 0.$$

The map g is injective, so $R = \operatorname{Im} g = \operatorname{Ker} h$, and since h is also surjective, we conclude that $H_0(\mathbb{R}^n,\mathbb{R}^n-\{x\};R) = (0)$.

If $n = 1$, then we proved that $H_1(\mathbb{R}^1,\mathbb{R}^1-\{x\};R) = R$. In this case $H_0(\mathbb{R}^1;R) = R$ and $H_0(\mathbb{R}^1-\{x\};R) = R\oplus R$, so we have the exact sequence

$$R \xrightarrow{f} R\oplus R \xrightarrow{g} R \xrightarrow{h} H_0(\mathbb{R}^n,\mathbb{R}^n-\{x\};R) \longrightarrow 0.$$

We must have $\operatorname{Im} g = R$, because otherwise $\operatorname{Im} g = (0)$, so $\operatorname{Ker}\, g = R\oplus R$, and since the sequence is exact, $\operatorname{Im} f = \operatorname{Ker}\, g = R\oplus R$, which is impossible since the domain of f is R. By exactness, since $\operatorname{Ker}\, h = \operatorname{Im} g = R$ and since h is surjective, we conclude that $H_0(\mathbb{R}^1, \mathbb{R}^1 - \{x\}; R) = (0)$. Since homology (and reduced homology) of negative index are (0), we obtain the isomorphisms

$$H_p(\mathbb{R}^n, \mathbb{R}^n - \{x\}; R) \cong \widetilde{H}_{p-1}(\mathbb{R}^n - \{x\}; R) \cong \widetilde{H}_{p-1}(S^{n-1}; R)$$

for all $p \geq 0$. □

If M is an n-manifold, since the groups $H_n(M, M - \{x\}; R)$ are all isomorphic to R, a way to define a notion of orientation is to pick some generator μ_x from $H_n(M, M - \{x\}; R)$ for every $x \in M$. Since $H_n(M, M - \{x\}; R)$ is a ring with a unit, generators are just invertible elements. To say that M is orientable means that we can pick these $\mu_x \in H_n(M, M - \{x\}; R)$ in such a way that they "vary continuously" with x. We will show how to do this in Section 7.1.

In the next section we show how singular homology can be generalized to deal with more general coefficients.

4.7 Singular Homology with G-Coefficients

In the previous sections, given a commutative ring R with an identity element, we defined the singular chain group $S_p(X; R)$ as the free R-module generated by the set $S_{\Delta^p}(X)$ of singular p-simplices $\sigma\colon \Delta^p \to X$. Thus, a singular chain c can be expressed as a formal linear combination

$$c = \sum_{k=1}^{m} \lambda_i \sigma_i,$$

for some $\lambda_i \in R$ and some $\sigma_i \in S_{\Delta^p}(X)$.

If A is a subset of X, we defined the relative chain group $S_p(X, A; R)$ as the quotient $S_p(X; R)/S_p(A; R)$. We observed that $S_p(X, A; R)$ is also a free R-module, and a basis of $S_p(X, A; R)$ consists of the cosets $\sigma + S_p(A; R)$ where the image of the singular simplex $\sigma\colon \Delta^p \to X$ does not lie in A.

Experience shows that it is fruitful to generalize homology to allow coefficients in any R-module G. Intuitively, a chain with coefficients in G is a formal linear combination

$$c = \sum_{k=1}^{m} g_i \sigma_i,$$

where the g_i are elements of the module G. We may think of such chains as "vector-valued" as opposed to the original chains which are "scalar-valued."

As we will see shortly, the usual convention is to swap g_i and σ_i so that these formal sums are of the form $\sum \sigma_i g_i$.

A rigorous way to proceed is to define the following modules.

Definition 4.19. The module $S_p(X; G)$ of *singular p-chains with coefficients in G* is defined as the tensor product

$$S_p(X; G) = S_p(X; R) \otimes_R G.$$

It is an R-module.

Since the R-module $S_p(X; R)$ is freely generated by $S_{\Delta^p}(X)$, it is a standard result of linear algebra that we have an isomorphism

$$S_p(X; R) \otimes_R G \cong \bigoplus_{\sigma \in S_{\Delta^p}(X)} G,$$

the direct sum of copies of G, one for each $\sigma \in S_{\Delta^p}(X)$.

Recall that this direct sum is the R-module of all functions $c\colon S_{\Delta^p}(X) \to G$ that are zero except for finitely many σ. For any $g \neq 0$ and any $\sigma \in S_{\Delta^p}(X)$, if we denote by σg the function from $S_{\Delta^p}(X)$ to G which has the value 0 for all arguments except σ where its value is g, then every $c \in S_p(X; R) \otimes_R G = S_p(X; G)$ which is not identically 0 can be written in a unique way as a finite sum

$$c = \sum_{k=1}^{m} \sigma_i g_i$$

for some $\sigma_i \in S_{\Delta^p}(X)$ and some nonzero $g_i \in G$. Observe that in the above expression the "vector coefficient" g_i comes after σ_i, to conform with the fact that we tensor with G on the right.

Since we will always tensor over the ring R, for simplicity of notation we will drop the subscript R in $\otimes_R$. Now given the singular chain complex $(S_*(X; R), \partial_*)$ displayed below

$$0 \xleftarrow{\partial_0} S_0(X; R) \xleftarrow{\partial_1} S_1(X; R) \longleftarrow \cdots S_{p-1}(X; R) \xleftarrow{\partial_p} S_p(X; R) \cdots,$$

(recall that $\partial_i \circ \partial_{i+1} = 0$ for all $i \geq 0$) we can form the homology complex

$$\cdots \xleftarrow{\partial_p \otimes \mathrm{id}} S_p(X; R) \otimes G \longleftarrow \cdots$$

$$0 \xleftarrow{\partial_0 \otimes \mathrm{id}} S_0(X; R) \otimes G \xleftarrow{\partial_1 \otimes \mathrm{id}} S_1(X; R) \otimes G \longleftarrow \cdots$$

denoted $(S_*(X;R)\otimes G, \partial_*\otimes \mathrm{id})$ obtained by tensoring with G, and since by definition $S_p(X;G) = S_p(X;R)\otimes G$, we have the homology complex

$$\cdots \xleftarrow{\partial_p\otimes \mathrm{id}} S_p(X;G) \longleftarrow \cdots$$

$$0 \xleftarrow{\partial_0\otimes \mathrm{id}} S_0(X;G) \xleftarrow{\partial_1\otimes \mathrm{id}} S_1(X;G) \longleftarrow \cdots$$

denoted $(S_*(X;G), \partial_*\otimes \mathrm{id})$ (of course, $G_*(X;G) = S_*(X;R)\otimes G$).

Definition 4.20. Let R be a commutative ring with identity and let G be a R-module. The *singular homology modules $H_p(X;G)$ with coefficients in G* are the homology groups of the above complex; that is,

$$H_p(X;G) = H_p(S_*(X;G)) \quad p \geq 0.$$

It is easily checked that if $x\in X$ is a point then

$$H_p(\{x\};G) = \begin{cases} G & \text{if } p = 0 \\ (0) & \text{if } p \neq 0. \end{cases}$$

Similarly, if X is any contractible space then,

$$H_p(X;G) = \begin{cases} G & \text{if } p = 0 \\ (0) & \text{if } p \neq 0. \end{cases}$$

If $\epsilon\colon S_0(X;R)\to R$ is the map of Definition 4.14, then we obtain an augmentation map $\epsilon\otimes \mathrm{id}\colon S_0(X;R)\otimes G \to R\otimes G \cong G$, that is, a map $\epsilon\otimes \mathrm{id}\colon S_0(X;G)\to G$, and we obtain an augmented complex with G in dimension -1.

Definition 4.21. The corresponding homology groups are denoted $\widetilde{H}_p(X;G)$ and are called the *reduced singular homology groups with coefficients in G.*

As in Section 4.3 we can pick an injective map $s\colon R\to S_0(X;R)$ such that $\epsilon\circ s = \mathrm{id}$, and since $R\otimes G\cong G$ and the short exact sequence

$$0 \longrightarrow \mathrm{Ker}\,\epsilon \longrightarrow S_0(X;R) \underset{s}{\overset{\epsilon}{\rightleftarrows}} R \longrightarrow 0$$

splits, by tensoring with G we get the short split exact sequence

$$0 \longrightarrow (\mathrm{Ker}\,\epsilon)\otimes G \longrightarrow S_0(X;R)\otimes G \underset{s\otimes \mathrm{id}}{\overset{\epsilon\otimes \mathrm{id}}{\rightleftarrows}} R\otimes G \cong G \longrightarrow 0;$$

see Munkres [Munkres (1984)] (Chapter 6, Section 51, Exercise 1). Thus

$$S_0(X;G) = S_0(X;R) \otimes G \cong ((\mathrm{Ker}\,\epsilon) \otimes G) \oplus G,$$

and since $H_0(X;G) = S_0(X;G)/\mathrm{Im}(\partial_1 \otimes \mathrm{id})$, $\widetilde{H}_0(X;G) = (\mathrm{Ker}\,(\epsilon \otimes \mathrm{id}))/\mathrm{Im}(\partial_1 \otimes \mathrm{id}) \cong ((\mathrm{Ker}\,\epsilon) \otimes G)/\mathrm{Im}(\partial_1 \otimes \mathrm{id})$, and since $\mathrm{Im}\,\partial_1 \subseteq \mathrm{Ker}\,\epsilon$, we get

$$S_0(X;G)/\mathrm{Im}(\partial_1 \otimes \mathrm{id}) \cong (((\mathrm{Ker}\,\epsilon) \otimes G)/\mathrm{Im}(\partial_1 \otimes \mathrm{id})) \oplus G,$$

which shows that

$$\begin{aligned} H_0(X;G) &\cong \widetilde{H}_0(X;G) \oplus G \\ H_p(X;G) &\cong \widetilde{H}_p(X;G), \qquad p \geq 1. \end{aligned}$$

More generally, if A is a subset of X, we have the chain complex $(S_*(X,A;R), \partial_*)$ displayed below

$$\begin{array}{l} \cdots \xleftarrow{\partial_{p-1}} S_{p-1}(X,A;R) \xleftarrow{\partial_p} S_p(X,A;R) \longleftarrow \cdots \\ 0 \xleftarrow{\partial_0} S_0(X,A;R) \xleftarrow{\partial_1} S_1(X,A;R) \longleftarrow \cdots \end{array}$$

where $S_p(X,A;R) = S_p(X;R)/S_p(A;R)$, and by tensoring with G and writing

$$S_p(X,A;G) = S_p(X,A;R) \otimes G,$$

we obtain the chain complex $(S_*(X,A;R) \otimes G, \partial_* \otimes G)$

$$\begin{array}{l} \cdots \xleftarrow{\partial_p \otimes \mathrm{id}} S_p(X,A;G) \longleftarrow \cdots \\ 0 \xleftarrow{\partial_0 \otimes \mathrm{id}} S_0(X,A;G) \xleftarrow{\partial_1 \otimes \mathrm{id}} S_1(X,A;G) \longleftarrow \cdots \end{array}$$

denoted $(S_*(X,A;G), \partial_* \otimes G)$.

Definition 4.22. Let R be a commutative ring with identity and let G be a R-module. For any subset A of the space X, the *relative singular homology modules $H_p(X,A;G)$ with coefficients in G* are the homology groups of the above complex; that is,

$$H_p(X,A;G) = H_p(S_*(X,A;G)) \quad p \geq 0.$$

Similarly, the *reduced relative singular homology modules $\widetilde{H}_p(X,A;G)$ with coefficients in G* are the homology groups of the complex obtained by tensoring the reduced homology complex of (X,A) with G. As in Section 4.3, if $A \neq \emptyset$ we have

$$H_p(X,A;G) \cong \widetilde{H}_p(X,A;G), \qquad p \geq 0.$$

A continuous map $h\colon (X,A) \to (Y,B)$ gives rise to a chain map

$$h_\sharp \otimes \mathrm{id}\colon S_*(X,A;R) \otimes G \to S_*(Y,B;R) \otimes G$$

which induces a homology homomorphism

$$h_*\colon H_*(X,A;G) \to H_*(Y,B;G).$$

As we know (see the diagram just after Proposition 4.8), we have a short exact sequence

$$0 \longrightarrow S_p(A;R) \longrightarrow S_p(X;R) \longrightarrow S_p(X,A;R) \longrightarrow 0,$$

and since $S_p(X,A;R)$ is free, it is a split exact sequence. Therefore, by tensoring with G we obtain another short exact sequence

$$0 \longrightarrow S_p(A;R) \otimes G \longrightarrow S_p(X;R) \otimes G \longrightarrow S_p(X,A;R) \otimes G \longrightarrow 0;$$

that is, a short exact sequence

$$0 \longrightarrow S_p(A;G) \longrightarrow S_p(X;G) \longrightarrow S_p(X,A;G) \longrightarrow 0.$$

By Theorem 2.22, we obtain a long exact sequence of homology, as described in the following theorem which is the analog of Theorem 4.9.

Theorem 4.24. *(Long Exact Sequence of Relative Homology) For every pair (X,A) of spaces, for any R-module G, we have the following long exact sequence of homology groups*

$$\cdots \longrightarrow H_{p+2}(X,A;G) \xrightarrow{\partial_{*p+2}}$$

$$H_{p+1}(A;G) \xrightarrow{i_*} H_{p+1}(X;G) \xrightarrow{j_*} H_{p+1}(X,A;G) \xrightarrow{\partial_{*p+1}}$$

$$H_p(A;G) \xrightarrow{i_*} H_p(X;G) \xrightarrow{j_*} H_p(X,A;G) \xrightarrow{\partial_{*p}}$$

$$H_{p-1}(A;G) \longrightarrow \cdots$$

ending in

$$H_0(A;G) \longrightarrow H_0(X;G) \longrightarrow H_0(X,A;G) \longrightarrow 0.$$

The version of Theorem 4.24 for reduced homology also holds; it is the analog of Theorem 4.11.

It is quite easy to see that the homotopy axiom also holds for homology with coefficients in G (see Munkres [Munkres (1984)], Chapter 6, Section 51).

Proposition 4.25. *(Homotopy Axiom) Given any two continuous maps* $f, g\colon (X, A) \to (Y, B)$ *if* f *and* g *are homotopic and* $H_p(f), H_p(g)\colon H_p(X, A; G) \to H_p(Y, B; G)$ *are the induced homomorphisms, then* $H_p(f) = H_p(g)$ *for all* $p \geq 0$*. As a consequence, if* (X, A) *and* (Y, B) *are homotopy equivalent, then for any* R*-module* G *the homology groups* $H_p(X, A; G)$ *and* $H_p(Y, A; G)$ *are isomorphic for all* $p \geq 0$.

The excision axiom also holds but the proof requires a little more work (see Munkres [Munkres (1984)], Chapter 6, Section 51).

Theorem 4.26. *(Excision Axiom) Given subspaces* $Z \subseteq A \subseteq X$ *such that the closure of* Z *is contained in the interior of* A*, then for any* R*-module* G *the inclusion* $(X - Z, A - Z) \longrightarrow (X, A)$ *induces isomorphisms of singular homology*

$$H_p(X - Z, A - Z; G) \cong H_p(X, A; G), \quad \text{for all } p \geq 0.$$

Equivalently, for any subspaces $A, B \subseteq X$ *whose interiors cover* X*, the inclusion map* $(B, A \cap B) \longrightarrow (X, A)$ *induces isomorphisms*

$$H_p(B, A \cap B; G) \cong H_p(X, A; G), \quad \text{for all } p \geq 0.$$

Theorem 4.13 about good pairs also holds for coefficients in G. As a consequence, since the homotopy axiom, the excision axiom and the long exact sequence of homology exists, the proof of Proposition 4.18 goes through with G-coefficients. The homology of D^n is given by

$$\begin{aligned} H_0(D^n; G) &= G \\ H_p(D^n; G) &= (0), \quad p > 0, \end{aligned}$$

or equivalently

$$\widetilde{H}_p(D^n; G) = (0), \quad p \geq 0,$$

and we have the following result.

Proposition 4.27. *For any* R*-module* G *the reduced homology of* S^n *is given by*

$$\widetilde{H}_p(S^n; G) = \begin{cases} G & \text{if } p = n \\ (0) & \text{if } p \neq n, \end{cases}$$

or equivalently the homology of S^n is given by

$$H_0(S^0; G) = G \oplus G$$
$$H_p(S^0; G) = (0), \quad p > 0,$$

and for $n \geq 1$,

$$H_p(S^n; G) = \begin{cases} G & \text{if } p = 0, n \\ (0) & \text{if } p \neq 0, n. \end{cases}$$

Proposition 4.23 also extends to homology with G-coefficients.

Relative singular homology with coefficients in G satisfies the axioms of homology theory singled out by Eilenberg and Steenrod [Eilenberg and Steenrod (1952)]. The Mayer–Vietoris theorem (Theorem 4.16) also holds for homology with coefficients in G. The proof relies on the fact that the sequence ($*_{\mathrm{MV}}$) is actually a split short exact sequence, so by Proposition 2.17, tensoring with an R-module yields another split short exact sequence, and we can form the long exact sequence of homology. This version of the Mayer–Vietoris theorem is also discussed in Spanier [Spanier (1989)], Chapter 5, Section 1, Corollary 14.

A version of the Mayer–Vietoris sequence for relative singular homology will be needed to prove Poincaré duality. The version stated below is from May [May (1999)] (Chapter 14, Section 5).

Theorem 4.28. *(Mayer–Vietoris in relative singular homology) Given any two topological spaces X and Y with $Y \subseteq X$, for any two subsets A, B of X such that $Y = \mathrm{Int}(A) \cup \mathrm{Int}(B)$, there is a long exact sequence of relative homology*

$$\to H_{p+1}(X, A \cap B) \to H_{p+1}(X, A) \oplus H_{p+1}(X, B) \to H_{p+1}(X, Y) \to$$
$$\to H_p(X, A \cap B) \longrightarrow H_p(X, A) \oplus H_p(X, B) \longrightarrow H_p(X, Y) \longrightarrow$$

where we have omitted the module G for the lack of space.

The universal coefficient theorem for homology (Theorem 12.5) shows that if R is a PID, then the module $H_p(X, A; G)$ can be expressed in terms of the modules $H_p(X, A; R)$ and $H_{p-1}(X, A; R)$ for any R-module G.

For example, we find that the homology groups of the real projective space with values in an R-module G are given by

$$H_p(\mathbb{RP}^n; G) = \begin{cases} G & \text{for } p = 0, n \\ G/2G & \text{for } p \text{ odd, } 0 < p < n \\ \text{Ker}\,(G \stackrel{2}{\longrightarrow} G) & \text{for } p \text{ even } 0 < p < n \\ (0) & \text{otherwise} \end{cases}$$

if n is odd and

$$H_p(\mathbb{RP}^n; G) = \begin{cases} G & \text{for } p = 0 \\ G/2G & \text{for } p \text{ odd, } 0 < p < n \\ \text{Ker}\,(G \stackrel{2}{\longrightarrow} G) & \text{for } p \text{ even } 0 < p \leq n \\ (0) & \text{otherwise} \end{cases}$$

if n is even, where the map $G \stackrel{2}{\longrightarrow} G$ is the map $g \mapsto 2g$.

Although homology theory is a very interesting subject, we proceed with cohomology, which is our primary focus.

4.8 Singular Cohomology

Roughly, to obtain cohomology from homology we dualize everything.

Definition 4.23. Given a topological space X and a commutative ring R, for any $p \geq 0$ we define the *singular cochain group* $S^p(X; R)$ as the dual $\text{Hom}_R(S_p(X; R), R)$ of the R-module $S_p(X; R)$, namely the space of all R-linear maps from $S_p(X; R)$ to R. The elements of $S^p(X; R)$ are called *singular p-cochains*. We set $S^p(X; R) = (0)$ for $p < 0$.

Since $S_p(X; R)$ is the free R-module generated by the set $S_{\Delta^p}(X)$ of singular p-simplices, every linear map from $S_p(X; R)$ to R is completely determined by its restriction to $S_{\Delta^p}(X)$, so we may view an element of $S^p(X; R)$ as an *arbitrary function* $f\colon S_{\Delta^p}(X) \to R$ assigning some element of R to every singular p-simplex σ. Recall that the set of functions from $S_{\Delta^p}(X)$ to R forms an R-module under the operations of multiplication by a scalar and addition given by

$$(\lambda f)(\sigma) = \lambda(f(\sigma))$$
$$(f + g)(\sigma) = f(\sigma) + g(\sigma)$$

for any singular p-simplex $\sigma \in S_{\Delta^p}(X)$ and any scalar $\lambda \in R$. Any singular p-cochain $f\colon S_{\Delta^p}(X) \to R$ can be evaluated on any singular p-chain $\alpha = \sum_{i=1}^m \lambda_i \sigma_i$, where the σ_i are singular p-simplices in $S_{\Delta^p}(X)$, by

$$f(\sigma) = \sum_{i=1}^m \lambda_i f(\sigma_i).$$

We define the direct sum $S^*(X; R)$ as

$$S^*(X; R) = \bigoplus_{p \geq 0} S^p(X; R).$$

All we need to get a chain complex is to define the coboundary map $\delta^p\colon S^p(X; R) \to S^{p+1}(X; R)$.

It is quite natural to say that for any singular p-cochain $f\colon S_{\Delta^p}(X) \to R$, the value $\delta^p f$ should be the function whose value $(\delta^p f)(\alpha)$ on a singular $(p+1)$-chain α is given by

$$(\delta^p f)(\alpha) = \pm f(\partial_{p+1} \alpha).$$

If we write $\langle g, \beta \rangle = g(\beta)$ for the result of evaluating the singular p-cochain $g \in S^p(X; R)$ on the singular p-chain $\beta \in S_p(X; R)$, then the above is written as

$$\langle \delta^p f, \alpha \rangle = \pm \langle f, \partial_{p+1} \alpha \rangle,$$

which is reminiscent of an adjoint. It remains to pick the sign of the right-hand side. Bott and Tu [Bott and Tu (1986)], Greenberg and Harper [Greenberg and Harper (1981)], Hatcher [Hatcher (2002)], May [May (1999)], Munkres [Munkres (1984)] and Warner [Warner (1983)] pick the $+$ sign, whereas Bredon [Bredon (1993)], Mac Lane [Mac Lane (1975)] and Milnor and Stasheff [Milnor and Stasheff (1974)] pick the sign $(-1)^{p+1}$, so that

$$\langle \delta^p f, \alpha \rangle + (-1)^p \langle f, \partial_{p+1} \alpha \rangle = 0.$$

Milnor and Stasheff explain that their choice of sign agrees with the convention that whenever two symbols of dimension m and n are permuted, the sign $(-1)^{mn}$ should be introduced. Here δ is considered to have sign $+1$ and ∂ is considered to have sign -1. Mac Lane explains that the choice of the sign $(-1)^{p+1}$ is desirable if a generalization of cohomology is considered; see Mac Lane [Mac Lane (1975)] (Chapter II, Section 3).

Regardless of the choice of sign, $\delta^{p+1} \circ \delta^p = 0$. Since the $+$ sign is simpler, this is the one that we adopt. Thus, $\delta^p f$ is defined by

$$\delta^p f = f \circ \partial_{p+1} \quad \text{for all } f \in S^p(X; R).$$

If we let $A = S_{p+1}(X;R)$, $B = S_p(X;R)$ and $\varphi = \partial_{p+1}$, we see that the definition of δ^p is equivalent to

$$\delta^p = \partial_{p+1}^\top.$$

The cohomology complex is indeed obtained from the homology complex by dualizing spaces and maps.

Definition 4.24. Given a topological space X and a commutative ring R, for any $p \geq 0$, the *coboundary homomorphism*

$$\delta^p \colon S^p(X;R) \to S^{p+1}(X;R)$$

is defined by

$$\langle \delta^p f, \alpha \rangle = \langle f, \partial_{p+1}\alpha \rangle,$$

for every singular p-cochain $f \colon S_{\Delta^p}(X) \to R$ and every singular $(p+1)$-chain $\alpha \in S_{p+1}(X;R)$; equivalently,

$$\delta^p f = f \circ \partial_{p+1} \quad \text{for all } f \in S^p(X;R).$$

We obtain a coboundary map

$$\delta \colon S^*(X;R) \to S^*(X;R).$$

The following proposition is immediately obtained.

Proposition 4.29. *Given a topological space X and a commutative ring R, the coboundary map $\delta \colon S^*(X;R) \to S^*(X;R)$ satisfies the equation*

$$\delta \circ \delta = 0.$$

We now have all the ingredients to define cohomology groups. Since the $S^p(X;R)$ together with the coboundary maps δ^p form the chain complex

$$0 \xrightarrow{\delta^{-1}} S^0(X;R) \xrightarrow{\delta^0} S^1(X;R) \cdots \xrightarrow{\delta^{p-1}} S^p(X;R) \xrightarrow{\delta^p} S^{p+1}(X;R) \cdots$$

as in Section 2.5, we obtain familiar spaces.

Definition 4.25. Let $Z^p(X;R) = \operatorname{Ker} \delta^p$ be the space of *singular p-cocycles* and $B^p(X;R) = \operatorname{Im} \delta^{p-1}$ be the space of *singular p-coboundaries.*

By Proposition 4.29, $B^p(X; R)$ is a submodule of $Z^p(X; R)$ so we obtain cohomology spaces:

Definition 4.26. Given a topological space X and a commutative ring R, for any $p \geq 0$ the *singular cohomology module* $H^p(X; R)$ is defined by

$$H^p(X; R) = \ker \delta^p / \operatorname{Im} \delta^{p-1} = Z^p(X; R)/B^p(X; R).$$

We set $H^p(X; R) = (0)$ if $p < 0$ and define $H^*(X; R)$ as the direct sum

$$H^*(X; R) = \bigoplus_{p \geq 0} H^p(X; R)$$

and call it the *singular cohomology of X with coefficients in R.*

It is common practice to refer to the spaces $H^p(X; R)$ as groups even though they are R-modules.

Until now we have been very compulsive in adding the term *singular* in front of every notion (chain, cochain, cycle, cocycle, boundary, coboundary, *etc.*). From now on we will drop this term unless confusion may arise. We may also drop X or R in $H^p(X; R)$ *etc.* whenever possible (that is, not causing confusion).

At this stage, one may wonder if there is any connection between the homology groups $H_p(X; R)$ and the cohomology groups $H^p(X; R)$. The answer is yes and it is given by the *universal coefficient theorem*. However, even to state the universal coefficient theorem requires a fair amount of homological algebra, so we postpone this topic until Chapter 12. Let us just mention the following useful isomorphisms in dimension 0 and 1:

$$H^0(X; R) = \operatorname{Hom}_R(H_0(X; R), R)$$
$$H^1(X; R) = \operatorname{Hom}_R(H_1(X; R), R).$$

It is not hard to see that $H^0(X; R)$ consists of those functions from X to R that are constant on path-components. Readers who want to learn about universal coefficient theorems should consult Chapter 12. If R is a PID, then the following result proven in Milnor and Stasheff [Milnor and Stasheff (1974)] (Appendix A, Theorem A.1) gives a very clean answer.

Theorem 4.30. *Let X be a topological space X and let R be a PID. If the homology group $H_{p-1}(X; R)$ is a free R-module or (0), then the cohomology group $H^p(X; R)$ is canonically isomorphic to the dual $\operatorname{Hom}_R(H_p(X; R), R)$ of $H_p(X; R)$.*

In particular, Theorem 4.30 holds if R is a field.

There is a generalization of singular cohomology which is useful for certain applications. The idea is to use more general coefficients. We can use a R-module G as the set of coefficients.

Definition 4.27. Given a topological space X, a commutative ring R, and a R-module G, for any $p \geq 0$ the *singular cochain group* $S^p(X; G)$ is the R-module $\mathrm{Hom}_R(S_p(X; R), G)$ of R-linear maps from $S_p(X; R)$ to G. We set $S^p(X; G) = (0)$ for $p < 0$.

Following Warner [Warner (1983)], since $S_p(X; R)$ is the free R-module generated by the set $S_{\Delta^p}(X)$ of singular p-simplices, we can view $S^p(X; G)$ as the set of all functions $f \colon S_{\Delta^p}(X) \to G$. This is also a R-module. As a special case, if $R = \mathbb{Z}$, then G can be any abelian group. As before, we obtain R-modules $Z^p(X; G)$ and $B^p(X; G)$ and coboundary maps $\delta^p \colon S^p(X; G) \to S^{p+1}(X; G)$ defined by

$$\delta^p f = f \circ \partial_{p+1} \quad \text{for all } f \in S^p(X; G).$$

We get the chain complex

$$0 \xrightarrow{\delta^{-1}} S^0(X; G) \xrightarrow{\delta^0} S^1(X; G) \cdots \xrightarrow{\delta^{p-1}} S^p(X; G) \xrightarrow{\delta^p} S^{p+1}(X; G) \cdots$$

and we obtain cohomology groups.

Definition 4.28. Given a topological space X, a commutative ring R, and a R-module G, for any $p \geq 0$ the *singular cohomology module* $H^p(X; G)$ is defined by

$$H^p(X; G) = \ker \delta^p / \mathrm{Im}\, \delta^{p-1} = Z^p(X; G)/B^p(X; G).$$

We set $H^p(X; G) = (0)$ if $p < 0$ and define $H^*(X; G)$ as the direct sum

$$H^*(X; G) = \bigoplus_{p \geq 0} H^p(X; G)$$

and call it the *singular cohomology of X with coefficients in G*.

Warner uses the notation $H^p_\Delta(X; G)$ instead of $H^p(X; G)$. When more than one cohomology theory is used, this is a useful device to distinguish among the various cohomology groups.

Cohomology is also functorial. If $f \colon X \to Y$ is a continuous map, then we know from Proposition 4.4 that there is a chain map $f_{\sharp,p} \colon S_p(X; R) \to S_p(Y; R)$, so by applying $\mathrm{Hom}_R(-, G)$ we obtain a cochain map $f^{\sharp,p}\colon S^p(Y;G) \to S^p(X;G)$ which commutes with coboundaries, and thus a homomorphism $H^p(f)\colon H^p(Y;G) \to H^p(X;G)$. This fact is recorded as the following proposition.

Proposition 4.31. *If X and Y are two topological spaces and if $f\colon X \to Y$ is a continuous function between them, then there are homomorphisms $H^p(f)\colon H^p(Y;G) \to H^p(X;G)$ for all $p \geq 0$.*

The map $H^p(f)\colon H^p(Y;G) \to H^p(X;G)$ is also denoted by

$$f^{*p}\colon H^p(Y;G) \to H^p(X;G).$$

We also have the following version of Proposition 4.6 for cohomology.

Proposition 4.32. *Given any two continuous maps $f, g\colon X \to Y$ (where X and Y are topological spaces), if f and g are homotopic and $H^p(f), H^p(g)\colon H^p(Y;G) \to H^p(X;G)$ are the induced homomorphisms, then $H^p(f) = H^p(g)$ for all $p \geq 0$. As a consequence, if X and Y are homotopy equivalent, then the cohomology groups $H^p(X;G)$ and $H^p(Y;G)$ are isomorphic for all $p \geq 0$,*

For any PID R, there is a universal coefficient theorem for cohomology that yields an expression for $H^p(X;G)$ in terms of $H_{p-1}(X;R)$ and $H_p(X;R)$; see Theorem 12.11.

There is also a version of the Mayer–Vietoris exact sequence for singular cohomology. Given any topological space X, for any two subsets A, B of X such that $X = \mathrm{Int}(A) \cup \mathrm{Int}(B)$, recall from Theorem 4.16 that we have a short exact sequence

$$0 \longrightarrow S_p(A \cap B) \xrightarrow{\ \varphi\ } S_p(A) \oplus S_p(B) \xrightarrow{\ \psi\ } S_p(A) + S_p(B) \longrightarrow 0 \qquad (*_{\mathrm{MV}})$$

for every $p \geq 0$, where φ and ψ are given by

$$\begin{aligned} \varphi(c) &= (i_\sharp(c), -j_\sharp(c)) \\ \psi(a,b) &= k_\sharp(a) + l_\sharp(b). \end{aligned}$$

Because $S_p(A) \oplus S_p(B)$ is free and because $S_p(A \cap B)$ is a submodule of both $S_p(A)$ and $S_p(B)$, we can choose bases in $S_p(A)$ and $S_p(B)$ by completing a basis of $S_p(A \cap B)$, and as a consequence we can define a map $p\colon S_p(A) \oplus S_p(B) \to S_p(A \cap B)$ such that $p \circ \varphi = \mathrm{id}$. Therefore, by

Proposition 2.2, the above sequence splits, and if we apply $\mathrm{Hom}_R(-,R)$ to it, by Proposition 2.7, we obtain a split short exact sequence

$$0 \longrightarrow \mathrm{Hom}(S_p(A)+S_p(B),R) \xrightarrow{\psi^{\perp}} S^p(A)\oplus S^p(B) \xrightarrow{\varphi^{\perp}} S^p(A\cap B) \longrightarrow 0 \qquad (*)$$

where $\varphi^{\perp} = \mathrm{Hom}(\varphi, R)$ and $\psi^{\perp} = \mathrm{Hom}(\psi, R)$. Since the inclusions $\iota_p\colon S_p(A)+S_p(B) \to S_p(X)$ form a chain homotopy equivalence, which means that there are maps $\rho_p\colon S_p(X) \to S_p(A)+S_p(B)$ such that $\rho\circ\iota$ and $\iota\circ\rho$ are chain homotopic to id, by applying $\mathrm{Hom}_R(-,R)$ we see that there is also a chain homotopy equivalence between $\mathrm{Hom}(S_p(A)+S_p(B),R)$ and $\mathrm{Hom}(S_p(X),R) = S^p(X)$, so the long exact sequence associated with the short exact sequence $(*)$ yields the following result.

Theorem 4.33. *(Mayer–Vietoris in singular cohomology) Given any topological space X, for any two subsets A, B of X such that $X = \mathrm{Int}(A)\cup \mathrm{Int}(B)$, there is a long exact sequence of cohomology*

$$\longrightarrow H^p(X;G) \longrightarrow H^p(A;G)\oplus H^p(B;G) \longrightarrow H^p(A\cap B;G) \longrightarrow$$

$$H^{p+1}(X;G) \longrightarrow \cdots$$

If $A\cap B \neq \emptyset$, a similar sequence exists in reduced cohomology.

The Mayer-Vietoris theorem in cohomology (Theorem 4.33) also holds for cohomology with coefficients in G. This is because $(*_{\mathrm{MV}})$ is a split short exact sequence, and by Proposition 2.12, if we apply $\mathrm{Hom}(-,G)$ where G is an R-module, the analog of $(*)$ (with coefficients in G) also holds. This version of the Mayer–Vietoris theorem in cohomology is also discussed in Spanier [Spanier (1989)], Chapter 5, Section 4, Corollary 9.

There is a notion of singular cohomology with compact support and generalizations of Poincaré duality. Some of the steps still use the Mayer–Vietoris sequences and the five lemma, but the proof is harder and requires two kinds of induction. Basically, Poincaré duality asserts that for any orientable manifold M of dimension n and any commutative ring R with an identity element, there are isomorphisms

$$H^p_c(M;R) \cong H_{n-p}(M;R).$$

On left-hand side $H^p_c(M;R)$ denotes the pth singular cohomology group with compact support, and on the right-hand side $H_{n-p}(M;R)$ denotes the $(n-p)$th singular homology group. By manifold, we mean a topological manifold (thus, Hausdorff and paracompact), not necessarily a smooth manifold, so this is a very general theorem. For details, the interested reader is referred to Chapter 7 (Theorem 7.16), and for comprehensive presentations including proofs, to Milnor and Stasheff [Milnor and Stasheff (1974)] (Appendix A), Hatcher [Hatcher (2002)] (Chapter 3), and Munkres [Munkres (1984)] (Chapter 8).

If M is a smooth manifold and if $R = \mathbb{R}$, a famous result of de Rham states that *de Rham cohomology and singular cohomology are isomorphic*, that is

$$H_{\mathrm{dR}}(M) \cong H^*(M; \mathbb{R}).$$

This is a hard theorem to prove. A complete proof can be found Warner [Warner (1983)] (Chapter 5). Another proof can be found in Morita [Morita (2001)] (Chapter 3). These proofs use Čech cohomology, which will be discussed later. It should be pointed that Chapter 5 of Warner [Warner (1983)] covers far more than the de Rham theorem. It provides a very thorough presentation of sheaf cohomology from an axiomatic point of view and shows the equivalence of four "classical" cohomology theories for smooth manifolds: Alexander-Spanier, de Rham, singular, and Čech cohomology. Warner's presentation is based on an approach due to Henri Cartan written in the early 1950s and based on fine sheaves. In Chapter 13 we develop sheaf cohomology using a more general and more powerful approach due to Grothendieck based on derived functors and δ-functors. This material is very technical; don't give up, it will probably require many passes to be digested.

4.9 Relative Singular Cohomology Groups

In this section R is any commutative with unit 1 and G is any R-module.

Definition 4.29. The *reduced singular cohomology groups* $\widetilde{H}^p(X; G)$ are defined by dualizing the augmented chain complex

$$0 \longleftarrow R \xleftarrow{\epsilon} S_0(X;R) \xleftarrow{\partial_1} S_1(X;R) \cdots \xleftarrow{\partial_{p-1}} S_{p-1}(X;R) \xleftarrow{\partial_p} S_p(X;R) \cdots$$

by applying $\mathrm{Hom}_R(-, G)$. We have

$$\begin{aligned}\widetilde{H}^0(X;G) &= \mathrm{Hom}_R(\widetilde{H}_0(X;R), G)\\ \widetilde{H}^p(X;G) &= H^p(X;G) \quad p \geq 1.\end{aligned}$$

In fact, it can be shown that

$$H^0(X;G) \cong \widetilde{H}^0(X;G) \oplus G;$$

see Munkres [Munkres (1984)] (Chapter 5, Section 44).

To obtain the relative cohomology groups we dualize the chain complex of relative homology

$$\cdots \xleftarrow{\partial_{p-1}} S_{p-1}(X,A;R) \xleftarrow{\partial_p} S_p(X,A;R) \xleftarrow{\partial_{p+1}} \cdots$$

$$0 \xleftarrow{\partial_0} S_0(X,A;R) \xleftarrow{\partial_1} S_1(X,A;R) \longleftarrow \cdots$$

by applying $\mathrm{Hom}_R(-,G)$, where $S_p(X,A;R) = S_p(X,R)/S_p(A,R)$.

Definition 4.30. The chain complex $S^*(X,A;G)$ is the complex

$$0 \xrightarrow{\delta^{-1}} S^0(X,A;G) \xrightarrow{\delta^0} S^1(X,A;G) \longrightarrow \cdots$$

$$\cdots \xrightarrow{\delta^{p-1}} S^p(X,A;G) \xrightarrow{\delta^p} S^{p+1}(X,A;G) \xrightarrow{\delta^{p+1}} \cdots$$

with $S^p(X,A;G) = \mathrm{Hom}_R(S_p(X,A;R),G)$ and $\delta^p = \mathrm{Hom}_R(\partial_p,G)$ for all $p \geq 0$ (and δ^{-1} is the zero map). More explicitly

$$\delta^p(f) = f \circ \partial_{p+1} \quad \text{for all } f \in S^p(X,A;G);$$

that is,

$$\delta^p(f)(\sigma) = f(\partial_{p+1}(\sigma)) \quad \text{for all } f \in S^p(X,A;G) = \mathrm{Hom}_R(S_p(X,A;R),G) \text{ and all } \sigma \in S_{p+1}(X;A;R).$$

Note that

$$S^p(X,A;G) = \mathrm{Hom}_R(S_p(X,A;R),G) = \mathrm{Hom}_R(S_p(X;R)/S_p(A;R),G)$$

is isomorphic to the submodule of $S^p(X;G) = \mathrm{Hom}_R(S_p(X;R),G)$ consisting of all linear maps with values in G defined on singular simplices in $S_p(X;R)$ that vanish on singular simplices in $S_p(A;R)$. Consequently, the coboundary map

$$\delta^p \colon S^p(X,A;G) \to S^{p+1}(X,A;G)$$

is the restriction of $\delta^p_X \colon S^p(X;G) \to S^{p+1}(X;G)$ to $S^p(X,A;G)$ where δ^p_X is the coboundary map of absolute cohomology.

Definition 4.31. Given a pair of spaces (X,A), the *singular relative cohomology groups* $H^p(X,A;G)$ of (X,A) arise from the chain complex

$$0 \xrightarrow{\delta^{-1}} S^0(X,A;G) \xrightarrow{\delta^0} S^1(X,A;G) \longrightarrow \cdots$$

$$\cdots \xrightarrow{\delta^{p-1}} S^p(X,A;G) \xrightarrow{\delta^p} S^{p+1}(X,A;G) \xrightarrow{\delta^{p+1}} \cdots$$

and are given by

$$H^p(X,A;G) = \mathrm{Ker}\,\delta^p/\mathrm{Im}\,\delta^{p-1}, \quad p \geq 0.$$

As in the case of absolute singular cohomology, a continuous map $f\colon (X,A) \to (Y,B)$ induces a homomorphism of relative cohomology $f^*\colon H^*(Y,B) \to H^*(X,A)$. This is because by Proposition 4.7 the map f induces a chain map $f_\sharp\colon S_*(X,A;R) \to S_*(Y,B;R)$, and by applying $\mathrm{Hom}_R(-,G)$ we obtain a cochain map $f^\sharp\colon S^*(Y,B;G) \to S^*(X,A;G)$ which commutes with coboundaries, and thus induces homomorphisms $H^p(f)\colon H^p(Y,B;G) \to H^p(X,A;G)$.

Proposition 4.34. *If (X,A) and (Y,B) are pairs of topological spaces and if $f\colon (X,A) \to (Y,B)$ is a continuous function between them, then there are homomorphisms $H^p(f)\colon H^p(Y,B;G) \to H^p(X,A;G)$ for all $p \geq 0$.*

The map $H^p(f)\colon H^p(Y,B;G) \to H^p(X,A;G)$ is also denoted by $f^{*p}\colon H^p(Y,B;G) \to H^p(X,A;G)$.

We also have the following version of Proposition 4.6 for relative cohomology which is the cohomological version of Proposition 4.8.

Proposition 4.35. *(Homotopy Axiom) Given any two continuous maps $f,g\colon (X,A) \to (Y,B)$, if f and g are homotopic and $H^p(f), H^p(g)\colon H^p(Y,B;G) \to H^p(X,A;G)$ are the induced homomorphisms, then $H^p(f) = H^p(g)$ for all $p \geq 0$. As a consequence, if (X,A) and (Y,B) are homotopy equivalent then the cohomology groups $H^p(X,A;G)$ and $H^p(Y,B;G)$ are isomorphic for all $p \geq 0$,*

To obtain the long exact sequence of relative cohomology we dualize the short exact sequence

$$0 \longrightarrow S_*(A;R) \xrightarrow{\ i\ } S_*(X;R) \xrightarrow{\ j\ } S_*(X,A;R) \longrightarrow 0$$

where $S_*(X,A;R) = S_*(X,R)/S_*(A,R)$ by applying $\mathrm{Hom}(-,G)$ and we obtain the sequence

$$0 \longrightarrow S^*(X,A;G) \xrightarrow{\ j^\top\ } S^*(X;G) \xrightarrow{\ i^\top\ } S^*(A;G) \longrightarrow 0,$$

with $S^*(X,A;G) = \mathrm{Hom}_R(S_*(X;R)/S_*(A;R),G)$, and as before $S^*(A;G) = \mathrm{Hom}_R(S_*(A;R),G)$ and $S^*(X;G) = \mathrm{Hom}_G(S_*(X;R),G)$.

Since $S_p(X,A;R) = S_p(X,R)/S_p(A,R)$ is a free module for every p, by Proposition 2.8 the sequence of chain complexes

$$0 \longrightarrow S^*(X,A;G) \xrightarrow{\ j^\top\ } S^*(X;G) \xrightarrow{\ i^\top\ } S^*(A;G) \longrightarrow 0$$

is exact (this can also be verified directly; see Hatcher [Hatcher (2002)], Section 3.1). Therefore, we can apply the zig-zag lemma for cohomology (Theorem 2.22) to this short exact sequence and we obtain the following cohomological version of Theorem 4.9.

Theorem 4.36. *(Long Exact Sequence of Relative Cohomology) For every pair (X, A) of spaces, we have the following long exact sequence of cohomology groups*

$$\cdots \longrightarrow H^{p-1}(A;G) \xrightarrow{\delta^*_{p-1}} H^p(X,A;G) \xrightarrow{(j^\top)^*} H^p(X;G) \xrightarrow{(i^\top)^*} H^p(A;G) \xrightarrow{\delta^*_p} H^{p+1}(X,A;G) \xrightarrow{(j^\top)^*} H^{p+1}(X;G) \xrightarrow{(i^\top)^*} H^{p+1}(A;G) \xrightarrow{\delta^*_{p+1}} H^{p+2}(X,A;G) \longrightarrow \cdots$$

There is also a version of Theorem 4.36 for reduced relative cohomology with $A \neq \emptyset$. As in the case of reduced homology with $A \neq \emptyset$, we have

$$\widetilde{H}^p(X, A, G) = H^p(X, A, G) \quad \text{for all } p \geq 0.$$

By setting $A = \{\text{pt}\}$, the version of Theorem 4.36 for relative cohomology yields the isomorphisms

$$H^p(X, \{\text{pt}\}; G) \cong \widetilde{H}^p(X; G) \quad \text{for all } p \geq 0.$$

Finally, the excision property also holds for relative cohomology.

Theorem 4.37. *(Excision Axiom) Given subspaces $Z \subseteq A \subseteq X$ such that the closure of Z is contained in the interior of A, then the inclusion $(X - Z, A - Z) \longrightarrow (X, A)$ induces isomorphisms of singular cohomology*

$$H^p(X - Z, A - Z; G) \cong H^p(X, A; G), \quad \textit{for all } p \geq 0.$$

Equivalently, for any subspaces $A, B \subseteq X$ whose interiors cover X, the inclusion map $(B, A \cap B) \longrightarrow (X, A)$ induces isomorphisms

$$H^p(B, A \cap B; G) \cong H^p(X, A; G), \quad \textit{for all } p \geq 0.$$

The proof of Theorem 4.37 does not follow immediately by dualization of Theorem 4.14. For details the reader is referred to Munkres [Munkres (1984)] (Chapter 5, §44) or Hatcher [Hatcher (2002)] (Section 3.1).

Proposition 4.35, Theorem 4.36, and Theorem 4.37 state three of the properties that were singled out as characterizing cohomology theories by Eilenberg and Steenrod [Eilenberg and Steenrod (1952)]. As in the case of homology, these properties hold for most of the known cohomology theories, and thus can be taken as axioms for cohomology theory; see Sato [Sato (1999)], Mac Lane [Mac Lane (1975)], Munkres [Munkres (1984)], or Hatcher [Hatcher (2002)].

The axiom of compact support fails for cohomology.

A version of the Mayer–Vietoris sequence for relative singular cohomology will be needed to prove Poincaré duality. The version stated below is from May [May (1999)] (Chapter 19, Section 3).

Theorem 4.38. *(Mayer–Vietoris in relative singular cohomology) Given any two topological spaces X and Y with $Y \subseteq X$, for any two subsets A, B of X such that $Y = \mathrm{Int}(A) \cup \mathrm{Int}(B)$, there is a long exact sequence of relative cohomology*

$$\longrightarrow H^p(X,Y) \longrightarrow H^p(X,A) \oplus H^p(X,B) \longrightarrow H^p(X,A\cap B) \longrightarrow$$
$$\longrightarrow H^{p+1}(X,Y) \to H^{p+1}(X,A) \oplus H^{p+1}(X,B) \to H^{p+1}(X,A\cap B) \to$$

where we omitted the ring R due to lack of space.

There is an even more general version of Theorem 4.38 for pairs of spaces (X,Y), pairs of subsets (A,B) of X and pairs of subsets (C,D) of Y, with $Y \subseteq X$, $C \subseteq A$, $D \subseteq B$, $X = \mathrm{Int}(A) \cup \mathrm{Int}(B)$, and $Y = \mathrm{Int}(C) \cup \mathrm{Int}(D)$. We have the long exact sequence of relative cohomology

$$\longrightarrow H^p(X,Y) \longrightarrow H^p(A,C) \oplus H^p(B,D) \longrightarrow H^p(A\cap B, C\cap D) \longrightarrow$$
$$\longrightarrow H^{p+1}(X,Y) \to H^{p+1}(A,C) \oplus H^{p+1}(B,D) \to H^{p+1}(A\cap B, C\cap D) \to$$

where we omitted the ring R due to lack of space. See Hatcher [Hatcher (2002)] (Chapter 3, Section 3.1, Page 204). Theorem 4.38 corresponds to the special case $X = A = B$.

For any PID R, there is a universal coefficient theorem for cohomology that yields an expression for $H^p(X, A; G)$ in terms of $H_{p-1}(X, A; R)$ and $H_p(X, A; R)$; see Theorem 12.11.

4.10 The Cup Product and the Cohomology Ring

We will see later in Chapter 12 (the universal coefficient theorem for cohomology, Theorem 12.11) that the homology groups of a space with values in a PID R determine its cohomology groups with values in any R-module G. One might then think that cohomology groups are not useful, but this is far from the truth for several reasons.

First, cohomology groups arise naturally as various "obstructions," such as the Ext-groups discussed in Chapter 12, or in the problem of classifying, up to homotopy, maps from one space into another. We will also see that in some cases only cohomology can be defined, as in the case of sheaves. But another reason why cohomology is important is that there is a natural way to define a multiplication operation on cohomology classes that makes the direct sum of the cohomology modules into a (graded) algebra. This additional structure allows the distinction between spaces that would not otherwise be distinguished by their homology (and cohomology).

We would like to define an operation $\smile$ that takes two cochains $c \in S^p(X; R)$ and $d \in S^q(X; R)$ and produces a cochain $c \smile d \in S^{p+q}(X; R)$. For this we define two affine maps $\lambda_p \colon \Delta^p \to \Delta^{p+q}$ and $\rho_q \colon \Delta^q \to \Delta^{p+q}$ by

$$\lambda_p(e_i^{p+1}) = e_i^{p+q+1} \quad 1 \leq i \leq p+1$$
$$\rho_q(e_i^{q+1}) = e_{p+i}^{p+q+1}, \quad 1 \leq i \leq q+1.$$

For any singular $(p+q)$-simplex $\sigma \colon \Delta^{p+q} \to X$, observe that $\sigma \circ \lambda_p \colon \Delta^p \to X$ is a singular p-simplex and $\sigma \circ \rho_q \colon \Delta^q \to X$ is a singular q-simplex. See Figure 4.18.

Recall from Definition 4.23 that a singular p-cochain is a R-linear map from $S_p(X; R)$ to R, where $S_p(X; R)$ is the R-module of singular p-chains. Since $S_p(X; R)$ is the free R-module generated by the set $S_{\Delta^p}(X)$ of singular p-simplices, every singular p-cochain c is completely determined by its restriction to $S_{\Delta^p}(X)$, and thus can be viewed as a function from $S_{\Delta^p}(X)$ to R.

Definition 4.32. If $\sigma \colon \Delta^{p+q} \to X$ is a singular simplex, we call $\sigma \circ \lambda_p$ the *front p-face* of σ, and $\sigma \circ \rho_q$ the *back q-face* of σ. See Figure 4.19.

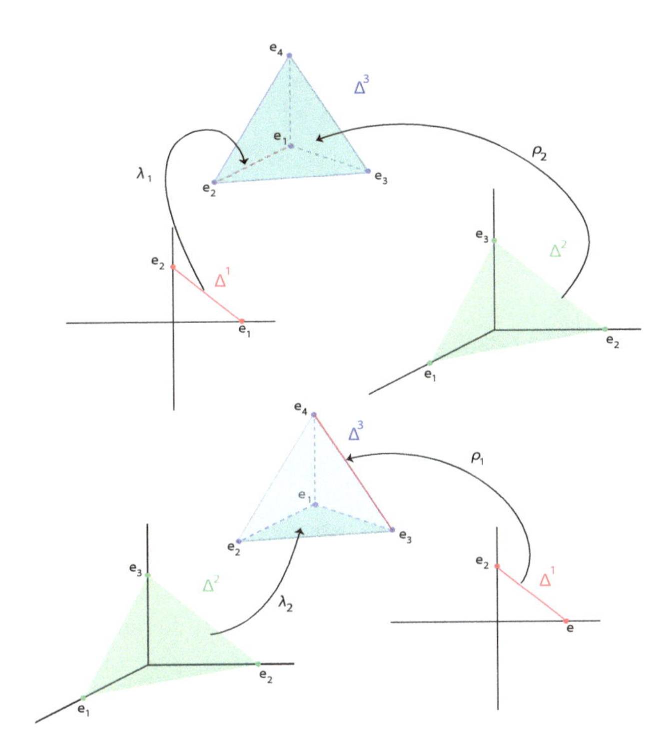

Fig. 4.18 Two ways of embedding a 1-simplex and a 2-simplex into a 3-simplex. For the top figure, $p = 1$ and $q = 2$, while for the bottom figure, $p = 2$ and $q = 1$.

Given any two cochains $c \in S^p(X; R)$ and $d \in S^q(X; R)$, their *cup product* $c \smile d \in S^{p+q}(X; R)$ is the cochain defined by

$$(c \smile d)(\sigma) = c(\sigma \circ \lambda_p)d(\sigma \circ \rho_q)$$

for all singular simplices $\sigma \in S_{\Delta^{p+q}}(X)$. The above defines a function

$$\smile : S^p(X; R) \times S^q(X; R) \to S^{p+q}(X; R).$$

Since $c(\sigma \circ \lambda_p) \in R$ and $d(\sigma \circ \rho_q) \in R$, we have $(c \smile d)(\sigma) \in R$, as desired.

Remark: Other authors, including Milnor and Stasheff [Milnor and Stasheff (1974)], add the sign $(-1)^{pq}$ to the formula in the definition of the cup product.

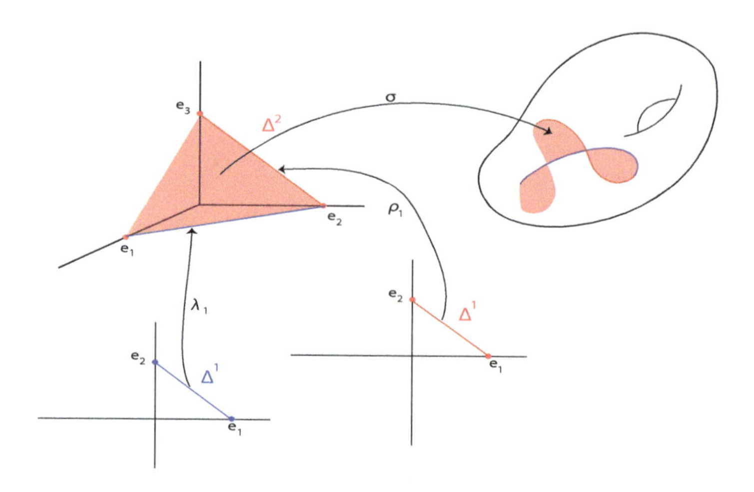

Fig. 4.19 A 2-simplex embedded in a torus, where $p = 1 = q$. The front 1-face is the blue edge while the back 1-face is the maroon edge.

The reader familiar with exterior algebra and differential forms will observe that the cup product can be viewed as a generalization of the wedge product.

Recall that $S^*(X;R)$ is the R-module $\bigoplus_{p\geq 0} S^p(X;R)$, and that $\epsilon\colon S_0(X;R) \to R$ is the unique homomorphism such that $\epsilon(x) = 1$ for every point $x \in S_0(X;R)$. Thus $\epsilon \in S^0(X;R)$ and since $\delta^0\epsilon = \epsilon \circ \partial_1 = 0$, the cochain ϵ is actually a cocycle and its cohomology class $[\epsilon] \in H^0(X;R)$ is denoted by 1.

The following proposition is immediate from the definition of the cup-product.

Proposition 4.39. *The cup product operation* $\smile$ *in* $S^*(X;R)$ *is bilinear, associative, and has the cocycle* ϵ *as identity element. Thus* $S^*(X;R)$ *is an associative graded ring with unit element.*

The following technical property implies that the cup product is well defined on cocycles.

Proposition 4.40. *For any two cochains* $c \in S^p(X;R)$ *and* $d \in S^q(X;R)$ *we have*

$$\delta(c \smile d) = (\delta c) \smile d + (-1)^p c \smile (\delta d).$$

Again note the analogy with the exterior derivative on differential forms. A proof of Proposition 4.39 can be found in Hatcher [Hatcher (2002)] (Chapter 3, Section 3.2, Lemma 3.6) and Munkres [Munkres (1984)] (Chapter 6, Theorem 48.1).

The formula of Proposition 4.40 implies that the cup product of cocycles is a cocycle, and that the cup product of a cocycle with a coboundary in either order is a coboundary, so we obtain an induced cup product on cohomology classes

$$\smile\colon H^p(X;R) \times H^q(X;R) \to H^{p+q}(X;R).$$

The cup product is bilinear, associative, and has 1 has identity element.

A continuous map $f\colon X \to Y$ induces a homomorphisms of cohomology $f^{p*}\colon H^p(Y;R) \to H^p(X;R)$ for all $p \geq 0$, and the cup product behaves well with respect to these maps.

Proposition 4.41. *Given any continuous map $f\colon X \to Y$, for all $\omega \in H^p(Y;R)$ and all $\eta \in H^q(Y;R)$, we have*

$$f^{(p+q)*}(\omega \smile_Y \eta) = f^{p*}(\omega) \smile_X f^{q*}(\eta).$$

Thus, $f^* = (f^{p*})_{p\geq 0}$ is a homomorphism between the graded rings $H^*(Y;R)$ (with the cup product $\smile_Y$) and $H^*(X;R)$ (with the cup product $\smile_X$).

Proposition 4.41 is proven in Hatcher [Hatcher (2002)] (Chapter 3, Section 3.2, Proposition 3.10) and Munkres [Munkres (1984)] (Chapter 6, Theorem 48.3).

Definition 4.33. Given a topological space X, its *cohomology ring* $H^*(X;R)$ is the graded ring $\bigoplus_{p\geq 0} H^p(X;R)$ equipped with the multiplication operation $\smile$ induced by the operations $\smile\colon H^p(X;R) \times H^q(X;R) \to H^{p+q}(X;R)$ for all $p, q \geq 0$.[1] An element $\omega \in H^p(X;R)$ is said to be of *degree* (or *dimension*) p, and we write $p = \deg(\omega)$.

Although the cup product is not commutative in general, it is skew-commutative in the following sense.

Proposition 4.42. *For all $\omega \in H^p(X;R)$ and all $\eta \in H^q(X;R)$, we have*

$$\omega \smile \eta = (-1)^{pq}(\eta \smile \omega).$$

[1]To be very precise, we have a family of multiplications $\smile_{p,q}\colon H^p(X;R) \times H^q(X;R) \to H^{p+q}(X;R)$, but this notation is too heavy and never used.

The proof of Proposition 4.42 is more complicated than the proofs of the previous propositions. It can be found in Hatcher [Hatcher (2002)] (Chapter 3, Section 2, Theorem 3.14). Another way to prove Proposition 4.42 is to first define the notion of cross-product and to define the cup product in terms of the cross-product. This is the approach followed by Bredon [Bredon (1993)] (Chapter VI, Sections 3 and 4), and Spanier [Spanier (1989)] (Chapter 5, Section 6).

The cohomology ring of most common spaces can be determined explicitly, but in some cases requires more machinery (such as Poincaré duality). Let us mention four examples.

Example 4.2. In the case of the sphere S^n, the cohomology ring $H^*(S^n; R)$ is the graded ring generated by one element α of degree n subject to the single relation $\alpha^2 = 0$.

The cohomology ring $H^*(T^n; R)$ of the n-torus T^n (with $T^n = S^1 \times \cdots \times S^1$ n times) is isomorphic to the exterior algebra $\bigwedge R^n$, with n-generators $\alpha_1, \ldots, \alpha_n$ of degree 1 satisfying the relations $\alpha_i \alpha_j = -\alpha_j \alpha_i$ for all $i \neq j$ and $\alpha_i^2 = 0$.

The cohomology ring $H^*(\mathbb{RP}^n, \mathbb{Z}/2\mathbb{Z})$ of real projective space $\mathbb{RP}^n$ with respect to $R = \mathbb{Z}/2\mathbb{Z}$ is isomorphic to the truncated polynomial ring $\mathbb{Z}/2\mathbb{Z}[\alpha]/(\alpha^{n+1})$, with α an element of degree 1. It is also possible to determine the cohomology ring $H^*(\mathbb{RP}^n, \mathbb{Z})$, but it is more complicated; see Hatcher [Hatcher (2002)] (Chapter 3, Theorem 3.12, and before Example 3.13).

The cohomology ring $H^*(\mathbb{CP}^n, \mathbb{Z})$ of complex projective space $\mathbb{CP}^n$ with respect to $R = \mathbb{Z}$ is isomorphic to the truncated polynomial ring $\mathbb{Z}[\alpha]/(\alpha^{n+1})$, with α an element of degree 2; see Hatcher [Hatcher (2002)] (Chapter 3, Theorem 3.12).

The cup product can be generalized in various ways. Here is a first generalization.

Definition 4.34. The cup product

$$\smile\colon S^p(X; R) \times S^q(X; G) \to S^{p+q}(X; G)$$

where G is any R-module, using the exact same formula

$$(c \smile d)(\sigma) = c(\sigma \circ \lambda_p) d(\sigma \circ \rho_q)$$

with $c \in S^p(X; R)$ and $d \in S^q(X; G)$, for all singular simplices $\sigma \in S_{\Delta^{p+q}}(X)$.

Since $c(\sigma \circ \lambda_p) \in R$ and $d(\sigma \circ \rho_q) \in G$, their product is in G so the above definition makes sense.

The formula

$$\delta(c \smile d) = (\delta c) \smile d + (-1)^p c \smile (\delta d)$$

of Proposition 4.40 still holds, but associativity only holds in a restricted fashion. Still, we obtain a cup product

$$\smile: H^p(X; R) \times H^q(X; G) \to H^{p+q}(X; G).$$

A second generalization if a version of the cup product for relative cohomology,

$$\smile: S^p(X, A; R) \times S^q(X; A, G) \to S^{p+q}(X; G)$$

where G is any R-module, using the exact same formula

$$(c \smile d)(\sigma) = c(\sigma \circ \lambda_p) d(\sigma \circ \rho_q)$$

as before. One has to check that the above formula yields an absolute cocycle in $S^{p+q}(X; G)$, which is left as an exercise. The above cup product induces a cup product

$$\smile: H^p(X, A; R) \times H^q(X; A, G) \to H^{p+q}(X; G).$$

Another generalization involves relative cohomology. For example, if A and B are open subset of a manifold X, there is a well-defined cup product

$$\smile: H^p(X, A; R) \times H^q(X, B; R) \to H^{p+q}(X, A \cup B; R);$$

see Hatcher [Hatcher (2002)] (Chapter 3, Section 3.2) and Milnor and Stasheff [Milnor and Stasheff (1974)] (Appendix A, Pages 264–265).

There are a number of interesting applications of the cup product but we will not go into this here, and instead refer the reader to Hatcher [Hatcher (2002)] (Chapter 3, Section 3.2), Bredon [Bredon (1993)] (Chapter VI), and Spanier [Spanier (1989)] (Chapter 5).

4.11 Problems

Problem 4.1. Prove Proposition 4.1.

Problem 4.2. Prove Proposition 4.3.

Problem 4.3. Prove the details of the proof of Proposition 4.7.

Problem 4.4. Prove that for any topological space X, we have

$$H_i(X \times S^n; R) \cong H_i(X; R) \oplus H_{i-n}(X; R)$$

for all $i, n \geq 0$ (recall that $H_j(X; R) = (0)$ for $j < 0$).

Hint. First prove that

$$H_i(X \times S^n; R) \cong H_i(X; R) \oplus H_i(X \times S^n, X \times \{x_0\}; R),$$
$$H_i(X \times S^n, X \times \{x_0\}; R) \cong H_{i-1}(X \times S^{n-1}, X \times \{x_0\}; R),$$

using Mayer–Vietoris for the second isomorphism.

Problem 4.5. Prove that for any n-torus T^n we have

$$H_p(T^n; R) = \underbrace{R \oplus \cdots \oplus R}_{\binom{n}{p}}.$$

Hint. Use Problem 4.4.

Problem 4.6. Prove Proposition 4.39.

Problem 4.7. Prove that the homology of $\mathbf{SO}(3)$ is given by

$$\begin{aligned}
H_0(\mathbf{SO}(3); \mathbb{Z}) &= \mathbb{Z} \\
H_1(\mathbf{SO}(3); \mathbb{Z}) &= \mathbb{Z}/2\mathbb{Z} \\
H_2(\mathbf{SO}(3); \mathbb{Z}) &= (0) \\
H_3(\mathbf{SO}(3); \mathbb{Z}) &= \mathbb{Z}.
\end{aligned}$$

Problem 4.8. (1) Prove that there is a homeomorphism between $\mathbf{SO}(4)$ and $\mathbf{SO}(3) \times S^3$.

(2) Prove that the homology of $\mathbf{SO}(4)$ is given by

$$\begin{aligned}
H_0(\mathbf{SO}(3); \mathbb{Z}) &= \mathbb{Z} \\
H_1(\mathbf{SO}(3); \mathbb{Z}) &= \mathbb{Z}/2\mathbb{Z} \\
H_2(\mathbf{SO}(3); \mathbb{Z}) &= (0) \\
H_3(\mathbf{SO}(3); \mathbb{Z}) &= \mathbb{Z} \oplus \mathbb{Z} \\
H_4(\mathbf{SO}(3); \mathbb{Z}) &= \mathbb{Z}/2\mathbb{Z} \\
H_5(\mathbf{SO}(3); \mathbb{Z}) &= (0) \\
H_6(\mathbf{SO}(3); \mathbb{Z}) &= \mathbb{Z}.
\end{aligned}$$

Hint. Use Problem 4.4.

Chapter 5

Simplicial Homology and Cohomology

In Chapter 4 we introduced the singular homology groups and the singular cohomology groups and presented some of their properties. Historically, singular homology and cohomology was developed in the 1940s, starting with a seminal paper of Eilenberg published in 1944 (building up on work by Alexander and Lefschetz among others), but it was not the first homology theory. Simplicial homology emerged in the early 1920s, more than thirty years after the publication of Poincaré's first seminal paper on "analysis situ" in 1892. Until the early 1930s, homology groups had not been defined and people worked with numerical invariants such as Betti numbers and torsion numbers. Emmy Noether played a significant role in introducing homology groups as the main objects of study.

One of the main differences between singular homology and simplicial homology is that singular homology groups can be assigned to *any* topological space X, but simplicial homology groups are defined for certain combinatorial objects called *simplicial complexes*. A simplicial complex is a combinatorial object that describes how to construct a space from simple building blocks generalizing points, line segments, triangles, and tetrahedra, called *simplices*. These building blocks are required to be glued in a "nice" way. Thus, simplicial homology is not as general as singular homology, but it is less abstract and more computational.

Given a simplicial complex K, we can associate to it a chain complex $C_*(K)$. In order to define the abelian groups $C_p(K)$ it is necessary to define the notion of oriented simplex. Then $C_p(K)$ is the free abelian group of oriented p-simplices. We can define boundary maps $\partial_p\colon C_p(K) \to C_{p-1}(K)$ to obtain a chain complex

$$0 \xleftarrow{\partial_0} C_0(K) \xleftarrow{\partial_1} C_1(K) \cdots \xleftarrow{\partial_{p-1}} C_{p-1}(K) \xleftarrow{\partial_p} C_p(K) \xleftarrow{\partial_{p+1}} \cdots$$

denoted $C_*(K)$. As usual, we let

$$Z_p(K) = \mathrm{Ker}\,\partial_p \quad \text{and} \quad B_p(K) = \mathrm{Im}\,\partial_{p+1},$$

and we define the *simplicial homology group* $H_p(K)$ as

$$H_p(K) = Z_p(K)/B_p(K).$$

In the construction above, it is implicitly assumed that the coefficients belong to $\mathbb{Z}$. We can generalize the construction to obtain simplicial homology modules $H_p(K;G)$ with coefficients in a module G over a commutative ring with unit R. Basically, the chain complex $C_*(K;G)$ is obtained by tensoring $C_*(K)$ with G.

We can also define relative simplicial homology groups $H_p(K,L;G)$, where L is a subcomplex of K. The zig-zag lemma yields the long exact sequence of relative simplicial homology.

The crucial connection between simplicial homology and singular homology is that the simplicial homology groups of a simplicial complex K are isomorphic to the singular homology groups of the space K_g built up from K, called its *geometric realization.*

Proving this result takes a fair amount of work and the introduction of various techniques (Mayer–Vietoris sequences, categories with models and acyclic models; see Spanier [Spanier (1989)] Chapter 4). As a consequence, if two simplicial complexes K and K' have homeomorphic geometric realizations K_g and K'_g, then the simplicial homology groups of K and K' are isomorphic. Thus, simplicial homology is subsumed by singular homology, but the more computational flavor of simplicial homology should not be overlooked as it provides techniques not offered by singular homology. In Chapter 6 we will present another homology theory based on spaces called CW complexes built up from spherical cells. This homology theory is also equivalent to singular homology but it is more computational.

The combinatorial nature of a simplicial complex K (of dimension m) allows the definition of the *Euler–Poincaré characteristic* $\chi(K)$ of K, namely

$$\chi(K) = \sum_{p=0}^{m} (-1)^p\, m_p,$$

where m_p is the number of p-simplices in K. The remarkable fact is that $\chi(K)$ depends only on the geometric realization of K. Indeed, it can be proven that

$$\chi(K) = \sum_{p=0}^{m} (-1)^p\, \mathrm{rank}(H_p(K)).$$

Here it is assumed that the homology groups $H_p(K)$ are defined with coefficients in $\mathbb{Z}$; that is, they are abelian groups. The above formula makes sense because it can be shown that the homology groups $H_p(K)$ are finitely generated abelian groups, so by the structure theorem for finitely abelian groups, the notion of rank is well-defined.

We conclude by defining simplicial cohomology and relative simplicial cohomology. The cohomology cochain complex

$$0 \xrightarrow{\delta^{-1}} C^0(K, L; G) \xrightarrow{\delta^0} C^1(K, L; G) \longrightarrow \cdots$$

$$\cdots \xrightarrow{\delta^{p-1}} C^p(K, L; G) \xrightarrow{\delta^p} C^{p+1}(K, L; G) \xrightarrow{\delta^{p+1}} \cdots$$

is obtained by applying the functor $\mathrm{Hom}_R(-, G)$ to the chain complex of relative simplicial homology

$$\cdots \xleftarrow{\partial_{p-1}} C_{p-1}(K, L; R) \xleftarrow{\partial_p} C_p(K, L; R) \xleftarrow{\partial_{p+1}} \cdots$$

$$0 \xleftarrow{\partial_0} C_0(K, L; R) \xleftarrow{\partial_1} C_1(K, L; R) \longleftarrow \cdots$$

As usual, the relative simplicial cohomology modules $H(K, L; G)$ are defined by

$$H^p(K, L; G) = \mathrm{Ker}\, \delta^p / \mathrm{Im}\, \delta^{p-1}, \quad p \geq 0.$$

If R is a PID, then for any R-module G we have isomorphisms

$$H^p(K, L; G) \cong H^p(K_g, L_g; G) \quad \text{for all } p \geq 0$$

between the relative simplicial cohomology of the pair of complexes (K, L) and the relative singular cohomology of the pair of geometric realizations (K_g, L_g).

In summary, simplicial cohomology is subsumed by singular cohomology (at least when R is a PID). Nevertheless, simplicial cohomology is much more amenable to computation than singular cohomology.

5.1 Simplices and Simplicial Complexes

In this section we define simplicial complexes. A simplicial complex is a combinatorial object which describes how to build a space by putting together some basic building blocks called simplices. The building blocks are required to be "glued" nicely, which means roughly that they can only be glued along faces (a notion to be define rigorously). The building blocks (simplices) are generalizations of points, line segments, triangles, tetrahedra. Simplices are very triangular in nature; in fact, they can be defined rigorously as convex hulls of affinely independent points.

To be on firm grounds we need to review some basics of affine geometry. For more comprehensive expositions the reader should consult Munkres [Munkres (1984)] (Chapter 1, Section 1), Rotman [Rotman (1988)] (Chapter 2), or Gallier [Gallier (2011)] (Chapter 2). The basic idea is that an affine space is a vector space without a prescribed origin. So properties of affine spaces are invariant not only under linear maps but also under translations. When we view $\mathbb{R}^n$ as an affine space we often refer to the vectors in $\mathbb{R}^n$ as *points*.

Definition 5.1. Given $n+1$ points, $a_0, a_1, \ldots, a_n \in \mathbb{R}^m$, these points are *affinely independent* iff the n vectors, $(a_1 - a_0 \ldots, a_n - a_0)$, are linearly independent.

Note that Munkres uses the terminology *geometrically independent* instead of affinely independent.

Definition 5.2. Given any sequence of n points $a_1, \ldots, a_n$ in $\mathbb{R}^m$, an *affine combination* of these points is a linear combination

$$\lambda_1 a_1 + \cdots + \lambda_n a_n,$$

with $\lambda_i \in \mathbb{R}$, and with the restriction that

$$\lambda_1 + \cdots + \lambda_n = 1. \tag{$*$}$$

Condition $(*)$ ensures that an affine combination does not depend on the choice of an origin.

Definition 5.3. An affine combination is a *convex combination* if the scalars λ_i satisfy the extra conditions $\lambda_i \geq 0$, in addition to $\lambda_1 + \cdots + \lambda_n = 1$.

Definition 5.4. A function $f\colon \mathbb{R}^n \to \mathbb{R}^m$ is *affine* if f preserves affine combinations, that is,

$$f(\lambda_1 a_1 + \cdots + \lambda_p a_p) = \lambda_1 f(a_1) + \cdots + \lambda_p f(a_p),$$

for all $a_1, \ldots, a_p \in \mathbb{R}^n$ and all $\lambda_1, \ldots, \lambda_p \in \mathbb{R}$ with $\lambda_1 + \cdots + \lambda_p = 1$.

A simplex is just the convex hull of a finite number of affinely independent points, but we also need to define faces, the boundary, and the interior of a simplex.

Definition 5.5. Given any $n+1$ affinely independent points, $a_0, \ldots, a_n$ in $\mathbb{R}^m$, the *n-simplex (or simplex) σ defined by* $a_0, \ldots, a_n$ is the convex hull of the points $a_0, \ldots, a_n$, that is, the set of all convex combinations $\lambda_0 a_0 + \cdots + \lambda_n a_n$, where $\lambda_0 + \cdots + \lambda_n = 1$, and $\lambda_i \geq 0$ for all i, $0 \leq i \leq n$. The scalars $\lambda_0, \ldots, \lambda_n$ are called *barycentric coordinates*. We call n the *dimension* of the n-simplex σ, and the points $a_0, \ldots, a_n$ are the *vertices* of σ.

Given any subset $\{a_{i_0}, \ldots, a_{i_k}\}$ of $\{a_0, \ldots, a_n\}$ (where $0 \leq k \leq n$), the k-simplex generated by $a_{i_0}, \ldots, a_{i_k}$ is called a *face* of σ. A face s of σ is a *proper face* if $s \neq \sigma$ (we agree that the empty set is a face of any simplex). For any vertex a_i, the face generated by $a_0, \ldots, a_{i-1}, a_{i+1}, \ldots, a_n$ (i.e., omitting a_i) is called the *face opposite* a_i. Every face which is a $(n-1)$-simplex is called a *boundary face*.

The union of the boundary faces is the *boundary of* σ, denoted as $\partial\sigma$, and the complement of $\partial\sigma$ in σ is the *interior* $\mathring{\sigma} = \sigma - \partial\sigma$ of σ. The interior $\mathring{\sigma}$ of σ is sometimes called an *open simplex*.

It should be noted that for a 0-simplex consisting of a single point $\{a_0\}$, $\partial\{a_0\} = \emptyset$, and $\mathring{\{a_0\}} = \{a_0\}$. Of course, a 0-simplex is a single point, a 1-simplex is the line segment (a_0, a_1), a 2-simplex is a triangle (a_0, a_1, a_2) (with its interior), and a 3-simplex is a tetrahedron (a_0, a_1, a_2, a_3) (with its interior), as illustrated in Figure 5.1.

We now state a number of properties of simplices whose proofs are left as an exercise. Clearly, a point x belongs to the boundary $\partial\sigma$ of σ iff at least one of its barycentric coordinates $(\lambda_0, \ldots, \lambda_n)$ is zero, and a point x belongs to the interior $\mathring{\sigma}$ of σ iff all of its barycentric coordinates $(\lambda_0, \ldots, \lambda_n)$ are positive, i.e., $\lambda_i > 0$ for all $i, 0 \leq i \leq n$. Then for every $x \in \sigma$, there is a unique face s such that $x \in \mathring{s}$, the face generated by those points a_i for which $\lambda_i > 0$, where $(\lambda_0, \ldots, \lambda_n)$ are the barycentric coordinates of x.

A simplex σ is convex, arcwise connected, compact, and closed. The interior $\mathring{\sigma}$ of a simplex is convex, arcwise connected, open, and σ is the closure of $\mathring{\sigma}$.

We now need to put simplices together to form more complex shapes. We define abstract simplicial complexes and their geometric realizations. This seems easier than defining simplicial complexes directly, as for example, in Munkres [Munkres (1984)].

Definition 5.6. An *abstract simplicial complex* (for short *simplicial complex*) is a pair, $K = (V, \mathcal{S})$, consisting of a (finite or infinite) nonempty set

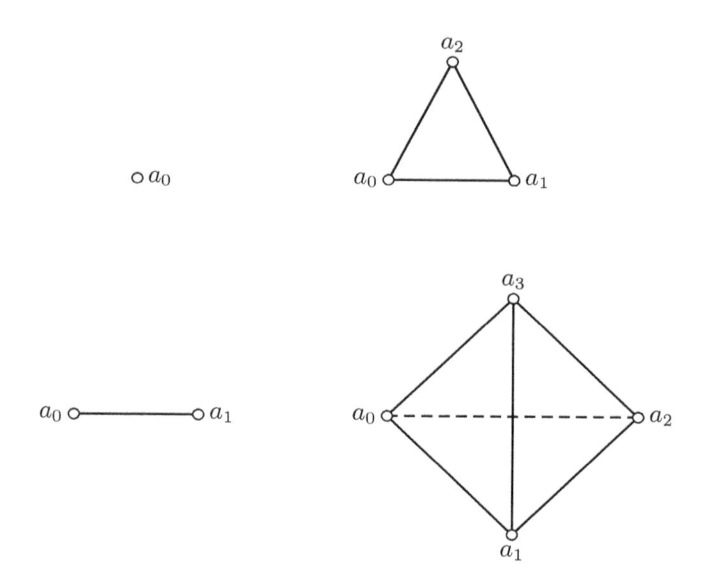

Fig. 5.1 Examples of simplices.

V of *vertices*, together with a family $\mathcal{S}$ of finite subsets of V called *abstract simplices* (for short *simplices*), and satisfying the following conditions:

(A1) Every $x \in V$ belongs to at least one and at most a finite number of simplices in $\mathcal{S}$.
(A2) Every subset of a simplex $\sigma \in \mathcal{S}$ is also a simplex in $\mathcal{S}$.

If $\sigma \in \mathcal{S}$ is a nonempty simplex of $n+1$ vertices, then its dimension is n, and it is called an *n-simplex*. A 0-simplex $\{x\}$ is identified with the vertex $x \in V$. The *dimension of an abstract complex* is the maximum dimension of its simplices if finite, and ∞ otherwise.

We will often use the abbreviation complex for abstract simplicial complex and simplex for abstract simplex. Also, given a simplex $s \in \mathcal{S}$, we will often use the notation $s \in K$.

The purpose of Condition (A1) is to insure that the geometric realization of a complex is locally compact. Condition (A2) is the technical way of defining faces.

Recall that given any set I, the real vector space $\mathbb{R}^{(I)}$ freely generated by I is defined as the subset of the Cartesian product $\mathbb{R}^I$ consisting of families $(\lambda_i)_{i\in I}$ of elements of $\mathbb{R}$ with finite support, which means that $\lambda_i = 0$ for all but finitely many indices $i \in I$ (where $\mathbb{R}^I$ denotes the set of all functions from I to $\mathbb{R}$). Then every abstract complex $(V, \mathcal{S})$ has a geometric realization as a topological subspace of the normed vector space $\mathbb{R}^{(V)}$ with the norm

$$\|(\lambda_v)_{v\in V}\| = \left(\sum_{v\in V} \lambda_v^2\right)^{1/2}.$$

Since $\lambda_v = 0$ for all but finitely many indices $v \in V$ this sum is well defined.

Definition 5.7. Given a simplicial complex, $K = (V, \mathcal{S})$, its *geometric realization* (also called the *polytope of* $K = (V, \mathcal{S})$) is the subspace K_g of $\mathbb{R}^{(V)}$ defined as follows: K_g is the set of all families $\lambda = (\lambda_a)_{a\in V}$ with finite support, such that:

(B1) $\lambda_a \geq 0$, for all $a \in V$;
(B2) The set $\{a \in V \mid \lambda_a > 0\}$ is a simplex in $\mathcal{S}$;
(B3) $\sum_{a\in V} \lambda_a = 1$.

The term *polyhedron* is sometimes used instead of polytope, and the notation $|K|$ is also used instead of K_g.

For every simplex $s \in \mathcal{S}$, we obtain a subset s_g of K_g by considering those families $\lambda = (\lambda_a)_{a\in V}$ in K_g such that $\lambda_a = 0$ for all $a \notin s$. In particular, every vertex $v \in V$ is realized as the point $v_g \in K_g$ whose coordinates $(\lambda_a)_{a\in V}$ are given by $\lambda_v = 1$ and $\lambda_a = 0$ for all $a \neq v$. We sometimes abuse notation and denote v_g by v. By (B2), we note that

$$K_g = \bigcup_{s\in\mathcal{S}} s_g.$$

It is also clear that for every n-simplex s, its geometric realization s_g can be identified with an n-simplex in $\mathbb{R}^n$.

Figure 5.2 illustrates the definition of a complex, where $V = \{v_1, v_2, v_3, v_4\}$ and

$$\mathcal{S} = \{\emptyset, \{v_1\}, \{v_2\}, \{v_3\}, \{v_4\}, \{v_1, v_2\}, \{v_1, v_3\}, \{v_2, v_3\}, \{v_3, v_4\}, \{v_2, v_4\}, \{v_1, v_2, v_3\}, \{v_2, v_3, v_4\}\}.$$

For clarity, the two triangles (2-simplices) are drawn as disjoint objects even though they share the common edge, (v_2, v_3) (a 1-simplex) and similarly for the edges that meet at some common vertex.

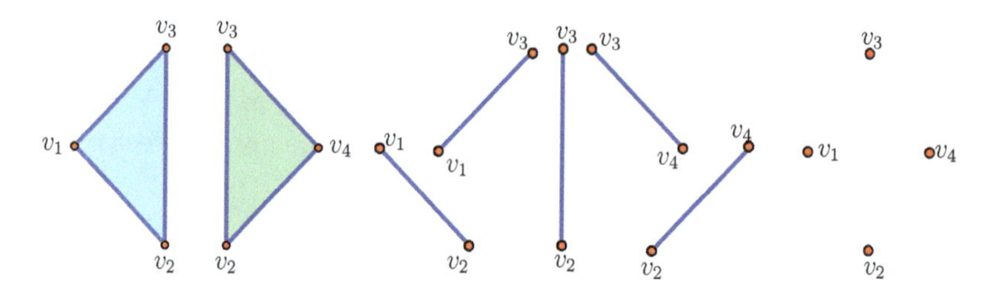

Fig. 5.2 A set of simplices forming a complex.

The geometric realization of the complex from Figure 5.2 is shown in Figure 5.3. Note that technically these polyhedra live in $\mathbb{R}^4$, so we are displaying homeomorphic copies. The same is true for the figures shown below.

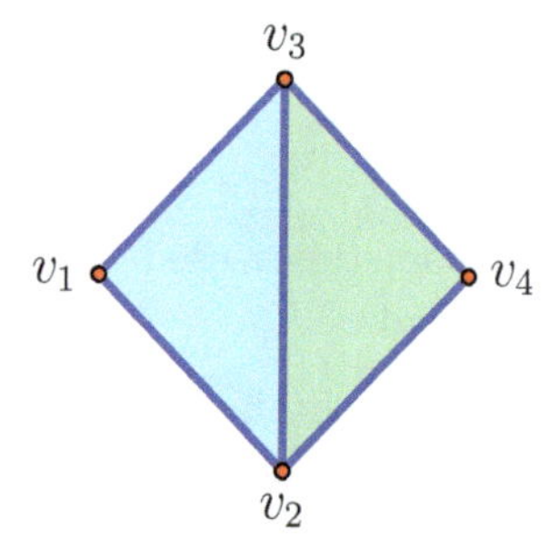

Fig. 5.3 The geometric realization of the complex of Figure 5.2.

Some collections of simplices violating Condition (A2) of Definition 5.6 are shown in Figure 5.4. In Figure 5.4(i),

$$V = \{v_1, v_2, v_3, v_4, v_5, v_6, w_1, w_2, w_3, w_4\}$$

and $\mathcal{S}$ contains the two 2-simplices $\{v_1, v_2, v_3\}$, $\{v_4, v_5, v_6\}$, neither of which intersect at along an edge or at a vertex of either triangle. In other words, $\mathcal{S}$ does not contain the 2-simplex $\{w_1, w_2, w_3\}$, a violation of Condition (A2). In Figure 5.4(ii), $V = \{v_1, v_2, v_3, v_4, v_5, v_6\}$ and

$$\mathcal{S} = \{\emptyset, \{v_1\}, \{v_2\}, \{v_3\}, \{v_4\}, \{v_5\}, \{v_6\}, \{v_1, v_2\}, \{v_2, v_3\}, \{v_1, v_3\}, \{v_4, v_5\}, \{v_5, v_6\}, \{v_4, v_6\}, \{v_1, v_2, v_3\}, \{v_4, v_5, v_6\}\}.$$

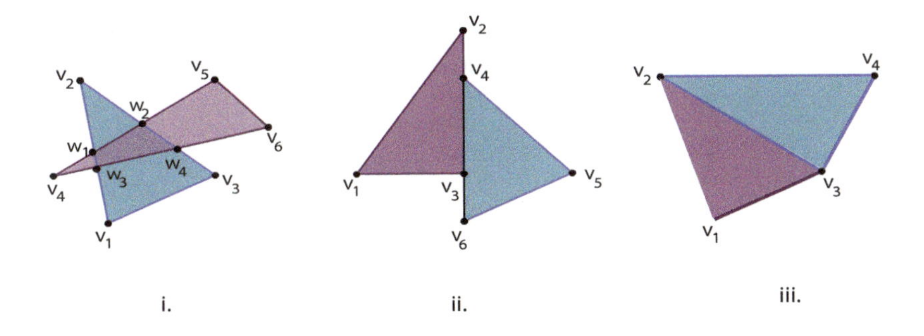

Fig. 5.4 Collections of simplices not forming a complex.

Note that the two 2-simplices meet along an edge $\{v_3, v_4\}$ which is not contained in $\mathcal{S}$, another violation of Condition (A2). In Figure 5.4(iii), $V = \{v_1, v_2, v_3, v_4\}$ and $\mathcal{S}$ contains the two 2-simplices $\{v_1, v_2, v_3\}$, $\{v_2, v_3, v_4\}$ but does not contain the edge $\{v_1, v_2\}$ and the vertex v_1.

Some geometric realizations of "legal" complexes are shown in Figure 5.5.

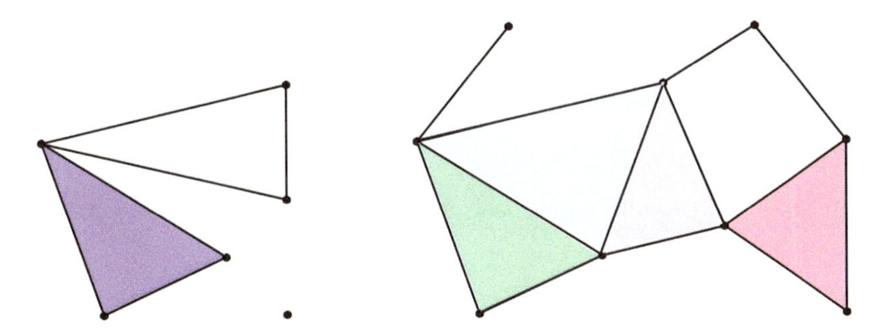

Fig. 5.5 Examples of geometric realizations of complexes.

Note that distinct complexes may have the same geometric realization. In fact, all the complexes obtained by subdividing the simplices of a given complex yield the same geometric realization (more exactly, homeomorphic copies).

Definition 5.8. Given a vertex $a \in V$, we define the *star of* a, denoted as $\mathrm{St}\, a$, as the finite union of the interiors $\overset{\circ}{s_g}$ of the geometric simplices s_g such that $a \in s$. Clearly, $a \in \mathrm{St}\, a$. The *closed star of* a, denoted as $\overline{\mathrm{St}}\, a$, is the finite union of the geometric simplices s_g such that $a \in s$. See Figure 5.6

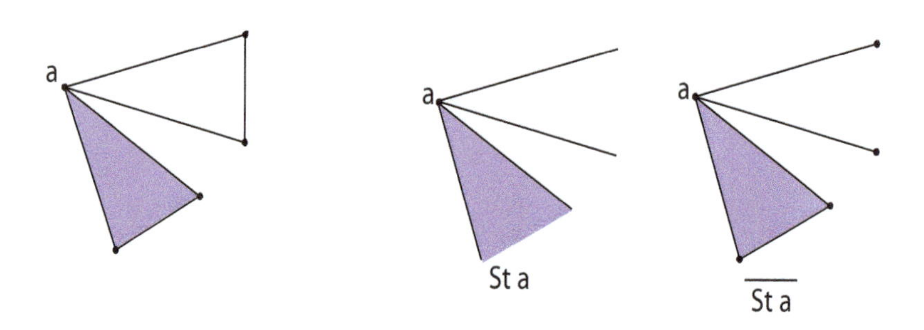

Fig. 5.6 Illustrations of $\mathrm{St}\, a$ and $\overline{\mathrm{St}}\, a$.

We define a topology on K_g by defining a subset F of K_g to be closed if $F \cap s_g$ is closed in s_g for all $s \in \mathcal{S}$ for the topology induced by $\mathbb{R}^{(V)}$. It is immediately verified that the axioms of a topological space hold.

Definition 5.9. A topological space X is *triangulable* if it is homeomorphic to the geometric realization K_g (with the above topology) of some simplicial complex K.

Actually, we can find a nice basis for this topology, as shown in the next proposition.

Proposition 5.1. *The family of subsets U of K_g such that $U \cap s_g = \emptyset$ for all but finitely many $s \in \mathcal{S}$, and such that $U \cap s_g$ is open in s_g when $U \cap s_g \neq \emptyset$, forms a basis of open sets for the topology of K_g. For any $a \in V$, the star* $\mathrm{St}\, a$ *of a is open, the closed star* $\overline{\mathrm{St}}\, a$ *is the closure of* $\mathrm{St}\, a$ *and is compact, and both* $\mathrm{St}\, a$ *and* $\overline{\mathrm{St}}\, a$ *are arcwise connected. The space K_g is locally compact, locally arcwise connected, and Hausdorff.*

We also observe that for any two simplices s_1, s_2 of $\mathcal{S}$, we have

$$(s_1 \cap s_2)_g = (s_1)_g \cap (s_2)_g.$$

We say that a complex $K = (V, \mathcal{S})$ is connected if it is not the union of two complexes $(V_1, \mathcal{S}_1)$ and $(V_2, \mathcal{S}_2)$, where $V = V_1 \cup V_2$ with V_1 and V_2

disjoint, and $\mathcal{S} = \mathcal{S}_1 \cup \mathcal{S}_2$ with $\mathcal{S}_1$ and $\mathcal{S}_2$ disjoint. The next proposition shows that a connected complex contains countably many simplices.

Proposition 5.2. *If $K = (V, \mathcal{S})$ is a connected complex, then $\mathcal{S}$ and V are countable.*

Next we give several examples of simplicial complexes whose geometric realizations are classical surfaces. These complexes have additional properties that make them triangulations but we will not discuss triangulations here. Figure 5.7 shows a triangulation of the *sphere*.

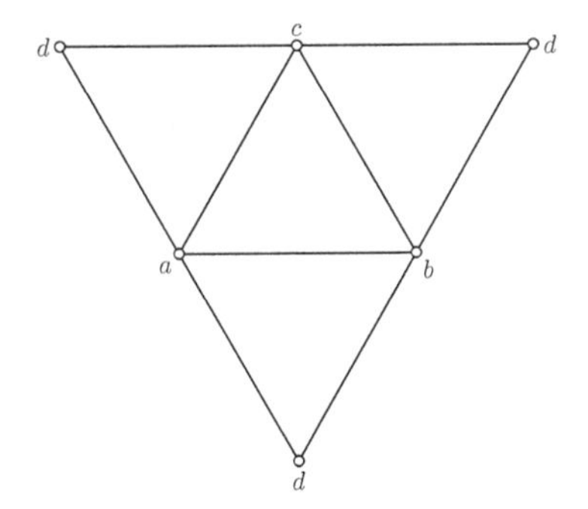

Fig. 5.7 A triangulation of the sphere.

The geometric realization of the above triangulation is obtained by pasting together the pairs of edges labeled (a, d), (b, d), (c, d). The geometric realization is a tetrahedron. See Figure 5.8.

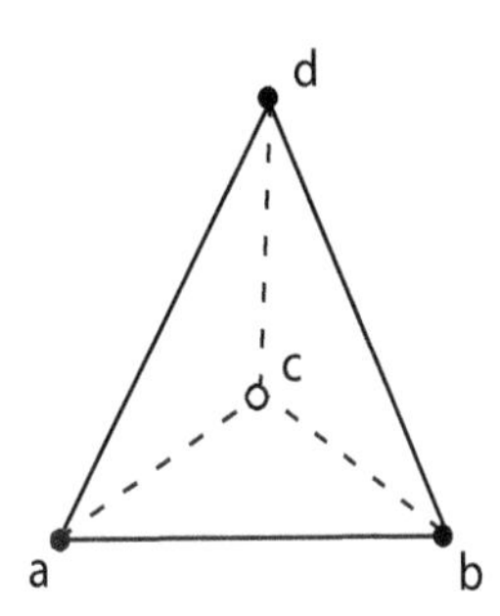

Fig. 5.8 The geometric realization of Figure 5.7.

Figure 5.9 shows a triangulation of a surface called a *torus*. The geometric realization of this triangulation is obtained by pasting together the pairs of edges labeled (a, d), (d, e), (e, a), and the pairs of edges labeled (a, b), (b, c), (c, a). See Figure 5.10.

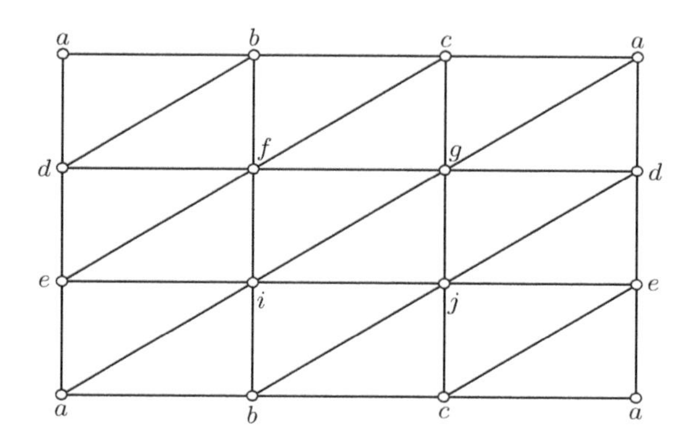

Fig. 5.9 A triangulation of the torus.

Figure 5.11 shows a triangulation of a surface called the *projective plane* and denoted by $\mathbb{RP}^2$. The geometric realization of the above triangulation is obtained by pasting together the pairs of edges labeled (a, f), (f, e), (e, d), and the pairs of edges labeled (a, b), (b, c), (c, d). This time, the gluing requires a "twist", since the paired edges have opposite orientation. Visualizing this surface in $\mathbb{R}^3$ is actually nontrivial.

Figure 5.12 shows a triangulation of a surface called the *Klein bottle*. The geometric realization of the above triangulation is obtained by pasting together the pairs of edges labeled (a, d), (d, e), (e, a), and the pairs of edges labeled (a, b), (b, c), (c, a). Again, some of the gluing requires a "twist", since some paired edges have opposite orientation. Visualizing this surface in $\mathbb{R}^3$ is not too difficult, but self-intersection cannot be avoided. See Figure 5.13.

The notion of subcomplex is defined as follows.

Definition 5.10. Given a simplicial complex $K = (V, \mathcal{S})$, a *subcomplex* L of K is a simplicial complex $L = (V_L, \mathcal{S}_L)$ such that $V_L \subseteq V$ and $\mathcal{S}_L \subseteq \mathcal{S}$.

Finally, the notion of map between simplicial complexes is defined as follows.

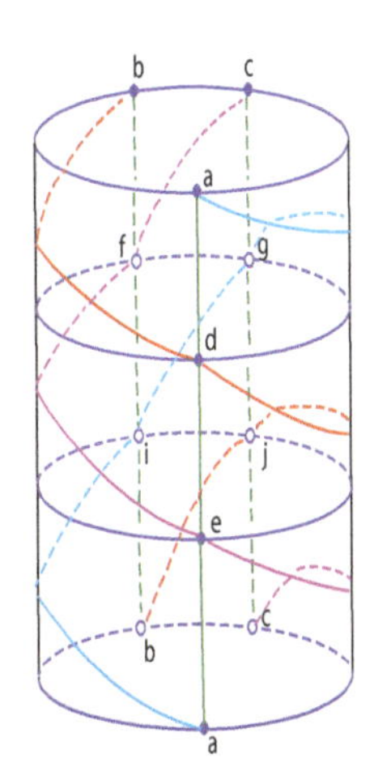

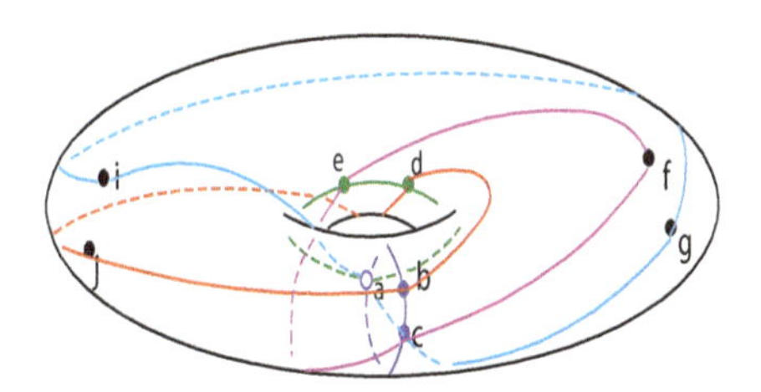

Fig. 5.10 Visualization of the torus with three spiral curves.

Definition 5.11. Given two simplicial complexes and $K_1 = (V_1, \mathcal{S}_1)$ and $K_2 = (V_2, \mathcal{S}_2)$, a *simplicial map* $f\colon K_1 \to K_2$ is a function $f\colon V_1 \to V_2$ such that whenever $\{v_1, \ldots, v_k\}$ is a simplex in $\mathcal{S}_1$, then $\{f(v_1), \ldots, f(v_k)\}$ is simplex in $\mathcal{S}_2$. Note that the $f(v_i)$ are not necessarily distinct. If L_1 is a subcomplex of K_1 and L_2 is a subcomplex of K_2, a *simplicial map* $f\colon (K_1, L_1) \to (K_2, L_2)$ is a simplicial map $f\colon K_1 \to K_2$ which carries every simplex of L_1 to a simplex of L_2.

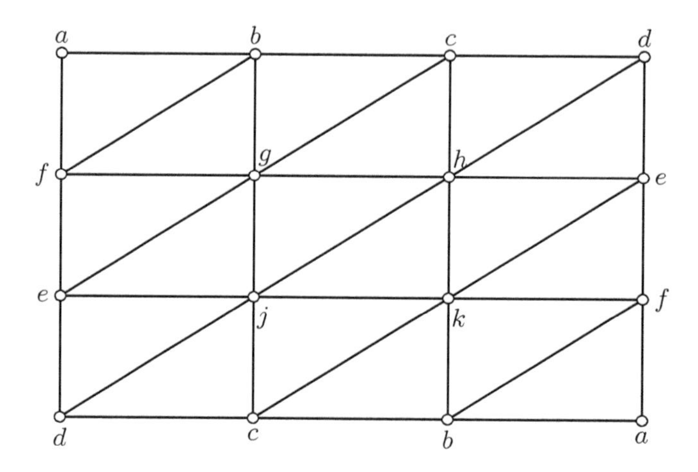

Fig. 5.11 A triangulation of the projective plane.

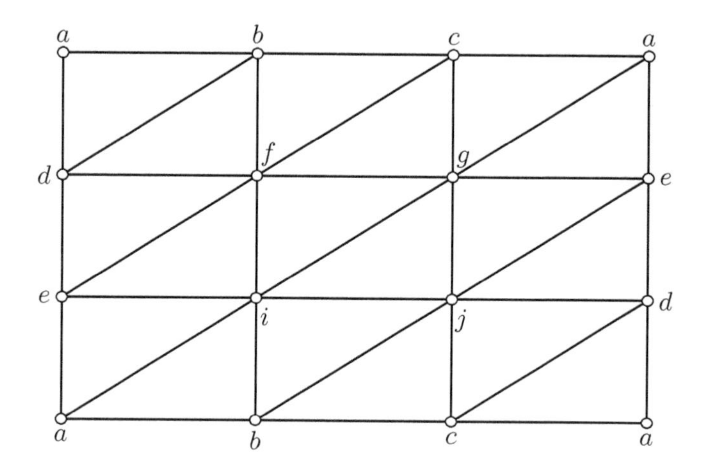

Fig. 5.12 A triangulation of the Klein bottle.

A simplicial map $f\colon K_1 \to K_2$ induces a continuous map $\widehat{f}\colon (K_1)_g \to (K_2)_g$, namely the function $\widehat{f}$ whose restriction to every simplex $s_g \in (K_1)_g$ is the unique affine map mapping v_i to $f(v_i)$ in $(K_2)_g$, where $s = \{v_1, \ldots, v_k\} \in \mathcal{S}_1$.

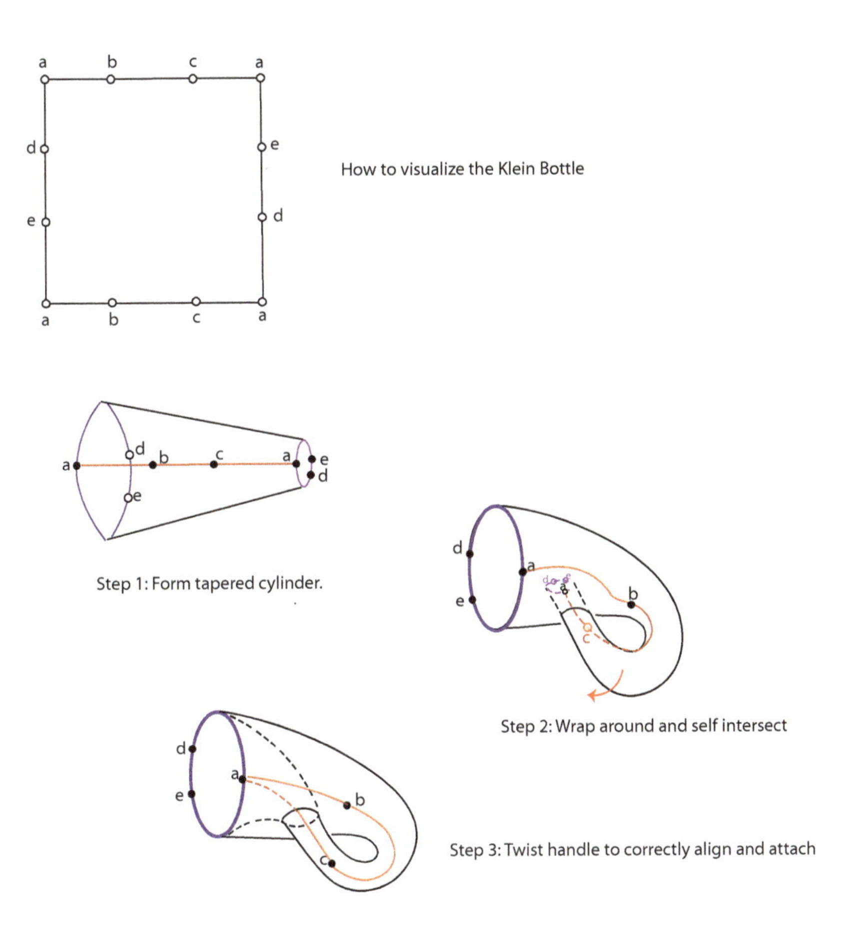

Fig. 5.13 Visualization of a Klein Bottle in $\mathbb{R}^3$.

5.2 Simplicial Homology Groups

In order to define the simplicial homology groups we need to describe how a chain complex $C_*(K)$, called a simplicial chain complex, is associated to a simplicial complex K. First we assume that the ring of homology coefficients is $R = \mathbb{Z}$.

Let $K = (V, \mathcal{S})$ be a simplicial complex, for short a complex. The chain complex $C_*(K)$ associated with K consists of free abelian groups $C_p(K)$ made out of oriented p-simplices. Every oriented p-simplex σ is assigned a *boundary* $\partial_p \sigma$. Technically, this is achieved by defining homomorphisms,

$$\partial_p \colon C_p(K) \to C_{p-1}(K),$$

with the property that $\partial_{p-1} \circ \partial_p = 0$. As in the case of singular homology, if we let $Z_p(K)$ be the kernel of ∂_p and

$$B_p(K) = \partial_{p+1}(C_{p+1}(K))$$

be the image of ∂_{p+1} in $C_p(K)$, since $\partial_p \circ \partial_{p+1} = 0$, the group $B_p(K)$ is a subgroup of the group $Z_p(K)$, and we define the simplicial homology group $H_p(K)$ as the quotient group

$$H_p(K) = Z_p(K)/B_p(K).$$

What makes the homology groups of a complex interesting is that they only depend on the geometric realization K_g of the complex K and not on the various complexes representing K_g. We will return to this point later.

The first step is to define oriented simplices. Given a complex $K = (V, \mathcal{S})$, recall that an n-simplex is a subset $\sigma = \{\alpha_0, \ldots, \alpha_n\}$ of V that belongs to the family $\mathcal{S}$. Thus, the set σ corresponds to $(n+1)!$ linearly ordered sequences $s\colon \{1, 2, \ldots, n+1\} \to \sigma$, where each s is a bijection. We define an equivalence relation on these sequences by saying that two sequences $s_1\colon \{1, 2, \ldots, n+1\} \to \sigma$ and $s_2\colon \{1, 2, \ldots, n+1\} \to \sigma$ are *equivalent* iff $\pi = s_2^{-1} \circ s_1$ is a permutation of even signature (π is the product of an even number of transpositions).

Definition 5.12. The two equivalence classes associated with a simplex σ are called *oriented simplices*, and if $\sigma = \{\alpha_0, \ldots, \alpha_n\}$, we denote the equivalence class of s as $[s(1), \ldots, s(n+1)]$, where s is one of the sequences $s\colon \{1, 2, \ldots, n+1\} \to \sigma$. We also say that the two classes associated with σ are the *orientations of* σ.

Two oriented simplices σ_1 and σ_2 are said to have *opposite orientation* if they are the two classes associated with some simplex σ. Given an oriented simplex, σ, we denote the oriented simplex having the opposite orientation by $-\sigma$, with the convention that $-(-\sigma) = \sigma$.

For example, if $\sigma = \{a_0, a_1, a_2\}$ is a 2-simplex (a triangle), there are six ordered sequences, the sequences $\langle a_2, a_1, a_0\rangle$, $\langle a_1, a_0, a_2\rangle$, and $\langle a_0, a_2, a_1\rangle$, are equivalent, and the sequences $\langle a_0, a_1, a_2\rangle$, $\langle a_1, a_2, a_0\rangle$, and $\langle a_2, a_0, a_1\rangle$, are also equivalent. Thus, we have the two oriented simplices, $[a_0, a_1, a_2]$ and $[a_2, a_1, a_0]$. We now define p-chains.

Definition 5.13. Given a complex, $K = (V, \mathcal{S})$, a *simplicial p-chain* on K is a function c from the set of oriented p-simplices to $\mathbb{Z}$, such that

(1) $c(-\sigma) = -c(\sigma)$, iff σ and $-\sigma$ have opposite orientation;
(2) $c(\sigma) = 0$, for all but finitely many simplices σ.

We define addition of p-chains pointwise, i.e., $c_1 + c_2$ is the p-chain such that $(c_1+c_2)(\sigma) = c_1(\sigma)+c_2(\sigma)$, for every oriented p-simplex σ. The group of simplicial p-chains is denoted by $C_p(K)$. If $p < 0$ or $p > \dim(K)$, we set $C_p(K) = \{0\}$.

To every oriented p-simplex σ is associated an *elementary p-chain c*, defined such that

$c(\sigma) = 1$,

$c(-\sigma) = -1$, where $-\sigma$ is the opposite orientation of σ, and

$c(\sigma') = 0$, for all other oriented simplices σ'.

We will often denote the elementary p-chain associated with the oriented p-simplex σ also by σ.

The following proposition is obvious, and simply confirms the fact that $C_p(K)$ is indeed a free abelian group.

Proposition 5.3. *For every complex, $K = (V, \mathcal{S})$, for every p, the group $C_p(K)$ is a free abelian group. For every choice of an orientation for every p-simplex, the corresponding elementary chains form a basis for $C_p(K)$.*

The only point worth elaborating is that except for $C_0(K)$, where no choice is involved, there is no canonical basis for $C_p(K)$ for $p \geq 1$, since different choices for the orientations of the simplices yield different bases.

If there are m_p p-simplices in K, the above proposition shows that $C_p(K) = \mathbb{Z}^{m_p}$.

As an immediate consequence of Proposition 5.3, for any abelian group G and any function f mapping the oriented p-simplices of a complex K to G and such that $f(-\sigma) = -f(\sigma)$ for every oriented p-simplex σ, there is a unique homomorphism, $\widehat{f}\colon C_p(K) \to G$, extending f.

We now define the boundary maps $\partial_p\colon C_p(K) \to C_{p-1}(K)$.

Definition 5.14. Given a complex, $K = (V, \mathcal{S})$, for every oriented p-simplex,

$$\sigma = [\alpha_0, \ldots, \alpha_p],$$

we define the *boundary*, $\partial_p \sigma$, *of* σ by

$$\partial_p \sigma = \sum_{i=0}^{p} (-1)^i [\alpha_0, \ldots, \widehat{\alpha_i}, \ldots, \alpha_p],$$

where $[\alpha_0, \ldots, \widehat{\alpha_i}, \ldots, \alpha_p]$ denotes the oriented $(p-1)$-simplex obtained by deleting vertex α_i. The *boundary map*, $\partial_p \colon C_p(K) \to C_{p-1}(K)$, is the unique homomorphism extending ∂_p on oriented p-simplices. For $p \leq 0$, ∂_p is the null homomorphism.

One must verify that $\partial_p(-\sigma) = -\partial_p \sigma$, but this is immediate. If $\sigma = [\alpha_0, \alpha_1]$, then

$$\partial_1 \sigma = \alpha_1 - \alpha_0.$$

If $\sigma = [\alpha_0, \alpha_1, \alpha_2]$, then

$$\partial_2 \sigma = [\alpha_1, \alpha_2] - [\alpha_0, \alpha_2] + [\alpha_0, \alpha_1] = [\alpha_1, \alpha_2] + [\alpha_2, \alpha_0] + [\alpha_0, \alpha_1].$$

If $\sigma = [\alpha_0, \alpha_1, \alpha_2, \alpha_3]$, then

$$\partial_3 \sigma = [\alpha_1, \alpha_2, \alpha_3] - [\alpha_0, \alpha_2, \alpha_3] + [\alpha_0, \alpha_1, \alpha_3] - [\alpha_0, \alpha_1, \alpha_2].$$

If σ is the chain

$$\sigma = [\alpha_0, \alpha_1] + [\alpha_1, \alpha_2] + [\alpha_2, \alpha_3],$$

shown in Figure 5.14(a), then

$$\begin{aligned} \partial_1 \sigma &= \partial_1 [\alpha_0, \alpha_1] + \partial_1 [\alpha_1, \alpha_2] + \partial_1 [\alpha_2, \alpha_3] \\ &= \alpha_1 - \alpha_0 + \alpha_2 - \alpha_1 + \alpha_3 - \alpha_2 \\ &= \alpha_3 - \alpha_0. \end{aligned}$$

On the other hand, if σ is the closed cycle,

$$\sigma = [\alpha_0, \alpha_1] + [\alpha_1, \alpha_2] + [\alpha_2, \alpha_0],$$

shown in Figure 5.14(b), then

$$\begin{aligned} \partial_1 \sigma &= \partial_1 [\alpha_0, \alpha_1] + \partial_1 [\alpha_1, \alpha_2] + \partial_1 [\alpha_2, \alpha_0] \\ &= \alpha_1 - \alpha_0 + \alpha_2 - \alpha_1 + \alpha_0 - \alpha_2 \\ &= 0. \end{aligned}$$

We have the following fundamental property:

Proposition 5.4. *For every complex,* $K = (V, \mathcal{S})$, *for every* p, *we have* $\partial_{p-1} \circ \partial_p = 0$.

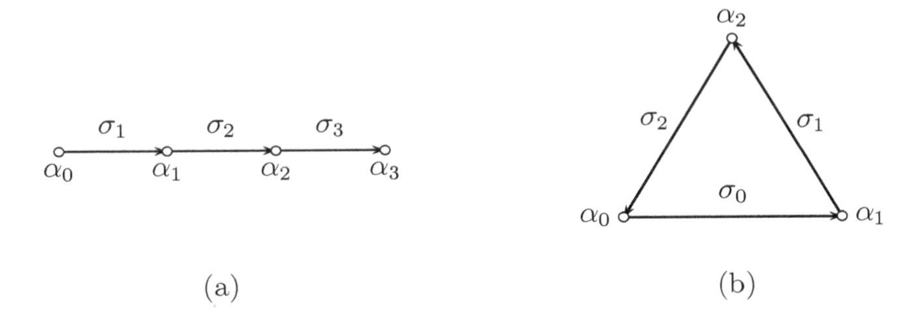

Fig. 5.14 (a) A chain with boundary $\alpha_3 - \alpha_0$. (b) A chain with 0 boundary.

Proof. For any oriented p-simplex, $\sigma = [\alpha_0, \ldots, \alpha_p]$, we have

$$\begin{aligned}
\partial_{p-1} \circ \partial_p \sigma &= \sum_{i=0}^{p} (-1)^i \partial_{p-1}[\alpha_0, \ldots, \widehat{\alpha_i}, \ldots, \alpha_p], \\
&= \sum_{i=0}^{p} \sum_{j=0}^{i-1} (-1)^i (-1)^j [\alpha_0, \ldots, \widehat{\alpha_j}, \ldots, \widehat{\alpha_i}, \ldots, \alpha_p] \\
&\quad + \sum_{i=0}^{p} \sum_{j=i+1}^{p} (-1)^i (-1)^{j-1} [\alpha_0, \ldots, \widehat{\alpha_i}, \ldots, \widehat{\alpha_j}, \ldots, \alpha_p] \\
&= 0.
\end{aligned}$$

The rest of the proof follows from the fact that $\partial_p \colon C_p(K) \to C_{p-1}(K)$ is the unique homomorphism extending ∂_p on oriented p-simplices. $\square$

Proposition 5.4 shows that the family $(C_p(K))_{p \geq 0}$ together with the boundary maps $\partial_p \colon C_p(K) \to C_{p-1}(K)$ form a chain complex

$$0 \xleftarrow{\partial_0} C_0(K) \xleftarrow{\partial_1} C_1(K) \cdots \xleftarrow{\partial_{p-1}} C_{p-1}(K) \xleftarrow{\partial_p} C_p(K) \xleftarrow{\partial_{p+1}} \cdots$$

denoted $C_*(K)$ called the *(oriented) simplicial chain complex* associated with the complex K.

Definition 5.15. Given a complex, $K = (V, \mathcal{S})$, the kernel $\operatorname{Ker} \partial_p$ of the homomorphism $\partial_p \colon C_p(K) \to C_{p-1}(K)$ is denoted by $Z_p(K)$, and the elements of $Z_p(K)$ are called *p-cycles*. The image $\partial_{p+1}(C_{p+1})$ of the homomorphism $\partial_{p+1} \colon C_{p+1}(K) \to C_p(K)$ is denoted by $B_p(K)$, and the elements of $B_p(K)$ are called *p-boundaries*. The *pth (oriented) simplicial homology group* $H_p(K)$ is the quotient group

$$H_p(K) = Z_p(K)/B_p(K).$$

Two p-chains c, c' are said to be *homologous* if there is some $(p+1)$-chain d such that $c = c' + \partial_{p+1} d$.

We will often omit the subscript p in ∂_p.

If $K = (V, \mathcal{S})$ is a finite-dimensional complex, as each group $C_p(K)$ is free and finitely generated, the homology groups $H_p(K)$ are all finitely generated.

Example 5.1. Consider the simplicial complex K_1 displayed in Figure 5.15. This complex consists of 6 vertices $\{v_1, \ldots, v_6\}$ and 8 oriented edges (1-simplices)

$$\begin{array}{llll} a_1 = [v_2, v_1] & a_2 = [v_1, v_4] & b_1 = [v_2, v_3] & b_2 = [v_3, v_4] \\ c_1 = [v_2, v_5] & c_2 = [v_5, v_4] & d_1 = [v_2, v_6] & d_2 = [v_6, v_4]. \end{array}$$

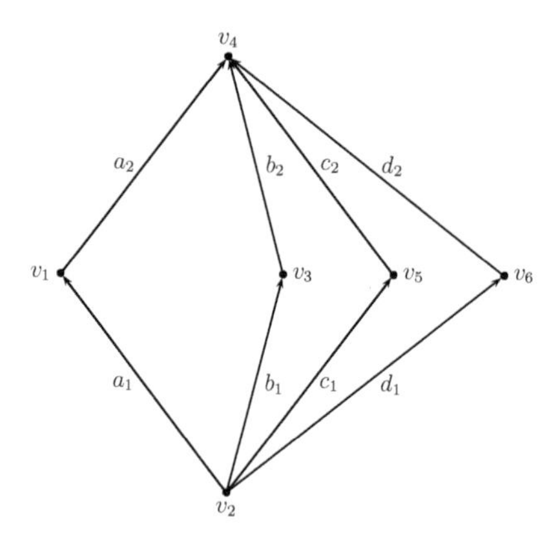

Fig. 5.15 A 1-dimensional simplicial complex.

Since this complex is connected, we claim that

$$H_0(K_1) = \mathbb{Z}.$$

Indeed, given any two vertices, u, u' in K_1, there is a path

$$\pi = [u_0, u_1], [u_1, u_2], \ldots, [u_{n-1}, u_n],$$

where each u_i is a vertex in K_1, with $u_0 = u$ and $u_n = u'$, and we have

$$\partial_1(\pi) = u_n - u_0 = u' - u,$$

which shows that u and u' are equivalent. Consequently, any 0-chain $\sum n_i v_i$ is equivalent to $\left(\sum n_i\right) v_0$, which proves that

$$H_0(K_1) = \mathbb{Z}.$$

If we look at the 1-cycles in $C_1(K_1)$, we observe that they are not all independent, but it is not hard to see that the three cycles

$$a_1 + a_2 - b_1 - b_2 \qquad b_1 + b_2 - c_1 - c_2 \qquad c_1 + c_2 - d_1 - d_2$$

form a basis of $C_1(K_1)$. It follows that

$$H_1(K_1) = \mathrm{Ker}\, \partial_1 / \mathrm{Im}\, \partial_2 = \mathrm{Ker}\, \partial_1 \cong \mathbb{Z} \oplus \mathbb{Z} \oplus \mathbb{Z}.$$

This reflects the fact that K_1 has three 1-dimensional holes.

Example 5.2. Next consider the 2-dimensional simplicial complex K_2 displayed in Figure 5.16. This complex consists of 6 vertices $\{v_1, \ldots, v_6\}$, 9 oriented edges (1-simplices)

$$\begin{aligned} a_1 &= [v_2, v_1] & a_2 &= [v_1, v_4] & b_1 &= [v_2, v_3] & b_2 &= [v_3, v_4] \\ c_1 &= [v_2, v_5] & c_2 &= [v_5, v_4] & d_1 &= [v_2, v_6] & d_2 &= [v_6, v_4] \\ e_1 &= [v_1, v_3], \end{aligned}$$

and two oriented triangles (2-simplices)

$$A_1 = [v_2, v_1, v_3] \qquad A_2 = [v_1, v_4, v_3].$$

We have

$$\partial_2 A_1 = a_1 + e_1 - b_1 \qquad \partial_2 A_2 = a_2 - b_2 - e_1.$$

It follows that

$$\partial_2(A_1 + A_2) = a_1 + a_2 - b_1 - b_2,$$

and $A_1 + A_2$ is a diamond with boundary $a_1 + a_2 - b_1 - b_2$. Since there are no 2-cycles,

$$H_2(K_2) = 0.$$

In order to compute

$$H_1(K_2) = \mathrm{Ker}\, \partial_1 / \mathrm{Im}\, \partial_2,$$

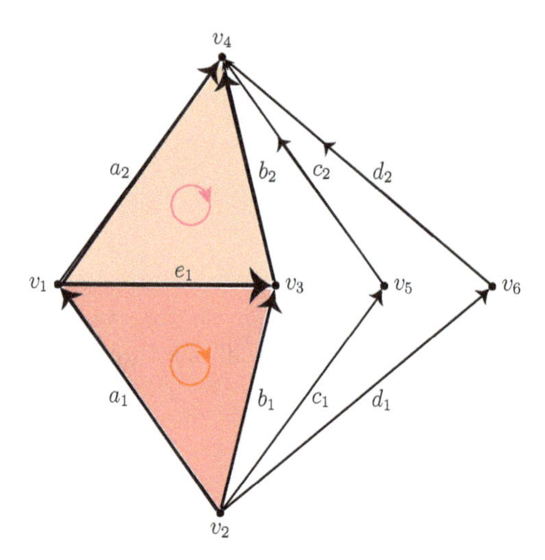

Fig. 5.16 A 2-dimensional simplicial complex with a diamond.

we observe that the cycles in $\operatorname{Im} \partial_2$ belong to the diamond $A_1 + A_2$, and so the only cycles in $C_1(K_2)$ whose equivalence class is nonzero must contain either $c_1 + c_2$ or $d_1 + d_2$. Then, any two cycles containing $c_1 + c_2$ (resp. $d_1 + d_2$) and passing through $A_1 + A_2$ are equivalent. For example, the cycles $a_1 + a_2 - c_1 - c_2$ and $b_1 + b_2 - c_1 - c_2$ are equivalent since their difference

$$a_1 + a_2 - c_1 - c_2 - (b_1 + b_2 - c_1 - c_2) = a_1 + a_2 - b_1 - b_2$$

is the boundary $\partial_2(A_1 + A_2)$. Similarly, the cycles $a_1 + e_1 + b_2 - c_1 - c_2$ and $a_1 + a_2 - c_1 - c_2$ are equivalent since their difference is

$$a_1 + e_1 + b_2 - c_1 - c_2 - (a_1 + a_2 - c_1 - c_2) = e_1 + b_2 - a_2 = \partial_2(-A_2).$$

Generalizing this argument, we can show that every cycle is equivalent to either a multiple of $a_1 + a_2 - c_1 - c_2$ or a multiple of $a_1 + a_2 - d_1 - d_2$, and thus

$$H_1(K_2) \cong \mathbb{Z} \oplus \mathbb{Z},$$

which reflects the fact that K_2 has two 1-dimensional holes. Observe that one of the three holes of the complex K_1 has been filled in by the diamond $A_1 + A_2$. Since K_2 is connected, $H_0(K_2) = \mathbb{Z}$.

Example 5.3. Now consider the 2-dimensional simplicial complex K_3 displayed in Figure 5.17. This complex consists of 8 vertices $\{v_1, \ldots, v_8\}$, 16 oriented edges (1-simplices)

$$\begin{array}{llll} a_1 = [v_5, v_1] & a_2 = [v_1, v_6] & b_1 = [v_5, v_3] & b_2 = [v_3, v_6] \\ c_1 = [v_5, v_7] & c_2 = [v_7, v_6] & d_1 = [v_5, v_8] & d_2 = [v_8, v_6] \\ e_1 = [v_1, v_2] & e_2 = [v_2, v_3] & f_1 = [v_1, v_4] & f_2 = [v_4, v_3] \\ g_1 = [v_5, v_2] & g_2 = [v_2, v_6] & h_1 = [v_5, v_4] & h_2 = [v_4, v_6], \end{array}$$

and 8 oriented triangles (2-simplices)

$$\begin{array}{llll} A_1 = [v_5, v_1, v_2] & A_2 = [v_5, v_2, v_3] & A_3 = [v_1, v_6, v_2] & A_4 = [v_2, v_6, v_3] \\ B_1 = [v_5, v_1, v_4] & B_2 = [v_5, v_4, v_3] & B_3 = [v_1, v_6, v_4] & B_4 = [v_4, v_6, v_3]. \end{array}$$

It is easy to check that

$$\begin{array}{ll} \partial_2 A_1 = a_1 + e_1 - g_1 & \partial_2 A_2 = g_1 + e_2 - b_1 \\ \partial_2 A_3 = a_2 - g_2 - e_1 & \partial_2 A_4 = g_2 - b_2 - e_2 \\ \partial_2 B_1 = a_1 + f_1 - h_1 & \partial_2 B_2 = h_1 + f_2 - b_1 \\ \partial_2 B_3 = a_2 - h_2 - f_1 & \partial_2 B_4 = h_2 - b_2 - f_2. \end{array}$$

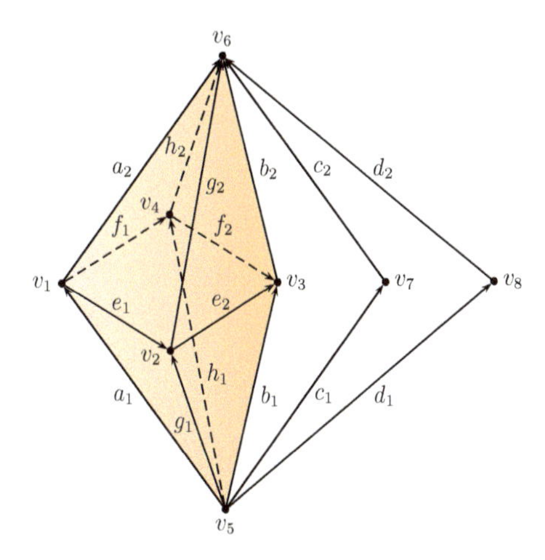

Fig. 5.17 A 2-dimensional simplicial complex with an octahedron.

If we let

$$A = A_1 + A_2 + A_3 + A_4 \quad \text{and} \quad B = B_1 + B_2 + B_3 + B_4,$$

then we get

$$\partial_2 A = \partial_2 B = a_1 + a_2 - b_1 - b_2,$$

and thus,

$$\partial_2(B - A) = 0.$$

Thus, $D = B - A$ is a 2-chain, and as we can see, it represents an octahedron. Observe that the chain group $C_2(K_3)$ is the eight-dimensional abelian group consisting of all linear combinations of A_is and B_js, and the fact that $\partial_2(B - A) = 0$ means that the kernel of the boundary map

$$\partial_2 \colon C_2(K_3) \to C_1(K_3)$$

is nontrivial. It follows that $B - A$ generates the homology group

$$H_2(K_3) = \mathrm{Ker}\, \partial_2 \cong \mathbb{Z}.$$

This reflects the fact that K_3 has a single 2-dimensional hole. The reader should check that as before,

$$H_1(K_3) = \mathrm{Ker}\, \partial_1 / \mathrm{Im}\, \partial_2 \cong \mathbb{Z} \oplus \mathbb{Z}.$$

Intuitively, this is because every cycle outside of the ocahedron D must contain either $c_1 + c_2$ or $d_1 + d_2$, and the "rest" of the cycle belongs to D. It follows that any two distinct cycles involving $c_1 + c_2$ (resp. $d_1 + d_2$) can be deformed into each other by "sliding" over D. The complex K_3 also has two 1-dimensional holes. Since K_3 is connected, $H_0(K_3) = \mathbb{Z}$.

Example 5.4. Finally consider the 3-dimensional simplicial complex K_4 displayed in Figure 5.18 obtained from K_3 by adding the oriented edge

$$k = [v_2, v_4]$$

and the four oriented tetrahedra (3-simplices)

$$T_1 = [v_1, v_2, v_4, v_6] \qquad T_2 = [v_3, v_4, v_2, v_6]$$
$$T_3 = [v_1, v_4, v_2, v_5] \qquad T_4 = [v_3, v_2, v_4, v_5].$$

We get

$$\begin{aligned}
\partial_3 T_1 &= [v_2, v_4, v_6] - [v_1, v_4, v_6] + [v_1, v_2, v_6] - [v_1, v_2, v_4] \\
\partial_3 T_2 &= [v_4, v_2, v_6] - [v_3, v_2, v_6] + [v_3, v_4, v_6] - [v_3, v_4, v_2] \\
\partial_3 T_3 &= [v_4, v_2, v_5] - [v_1, v_2, v_5] + [v_1, v_4, v_5] - [v_1, v_4, v_2] \\
\partial_3 T_4 &= [v_2, v_4, v_5] - [v_3, v_4, v_5] + [v_3, v_2, v_5] - [v_3, v_2, v_4].
\end{aligned}$$

Observe that

$$\begin{aligned}
\partial(T_1 + T_2 + T_3 + T_4) &= -[v_1, v_4, v_6] + [v_1, v_2, v_6] - [v_3, v_2, v_6] + [v_3, v_4, v_6] \\
&\quad - [v_1, v_2, v_5] + [v_1, v_4, v_5] - [v_3, v_4, v_5] + [v_3, v_2, v_5] \\
&= B_3 - A_3 - A_4 + B_4 - A_1 + B_1 + B_2 - A_2 \\
&= B_1 + B_2 + B_3 + B_4 - (A_1 + A_2 + A_3 + A_4) \\
&= B - A.
\end{aligned}$$

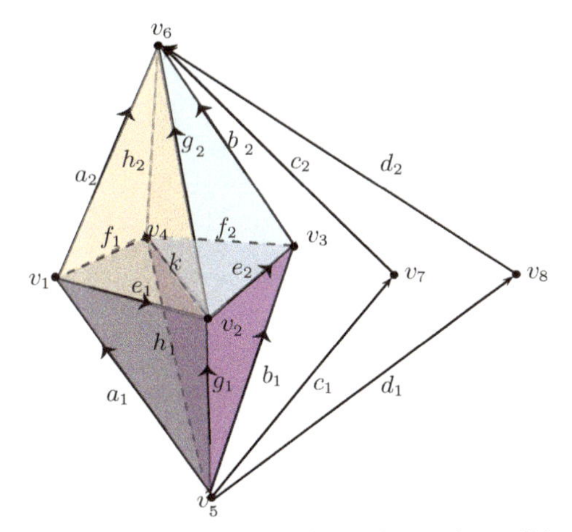

Fig. 5.18 A 3-dimensional simplicial complex with a solid octahedron.

It follows that

$$\partial_3 \colon C_3(K_4) \to C_2(K_4)$$

maps the solid octahedron $T = T_1 + T_2 + T_3 + T_4$ to $B - A$, and since $\operatorname{Ker} \partial_2$ is generated by $B - A$, we get

$$H_2(K_4) = \operatorname{Ker} \partial_2 / \operatorname{Im} \partial_3 = 0.$$

We also have

$$H_3(K_4) = \operatorname{Ker} \partial_3 / \operatorname{Im} \partial_3 = \operatorname{Ker} \partial_3 = 0,$$

and as before,

$$H_0(K_4) = \mathbb{Z} \quad \text{and} \quad H_1(K_4) = \mathbb{Z} \oplus \mathbb{Z}.$$

The complex K_4 still has two 1-dimensional holes but the 2-dimensional hole of K_3 has been filled up by the solid octahedron.

Example 5.5. For another example of a 2-dimensional simplicial complex with a hole (an annulus in the plane) consider the complex K_5 shown in Figure 5.19. This complex consists of 16 vertices, 32 edges (1-simplicies) oriented as shown in the Figure 5.19, and 16 triangles (2-simplicies) oriented according to the direction of their boundary edges. The boundary of K_5 is

$$\partial_2(K_5) = a_1 + a_2 + a_3 + b_1 + b_2 + b_3 + c_1 + c_2 + c_3 + d_1 + d_2 + d_3 + e + f + g + h.$$

As a consequence, the outer boundary $a_1 + a_2 + a_3 + b_1 + b_2 + b_3 + c_1 + c_2 + c_3 + d_1 + d_2 + d_3$ is equivalent to the inner boundary $-(e + f + g + h)$.

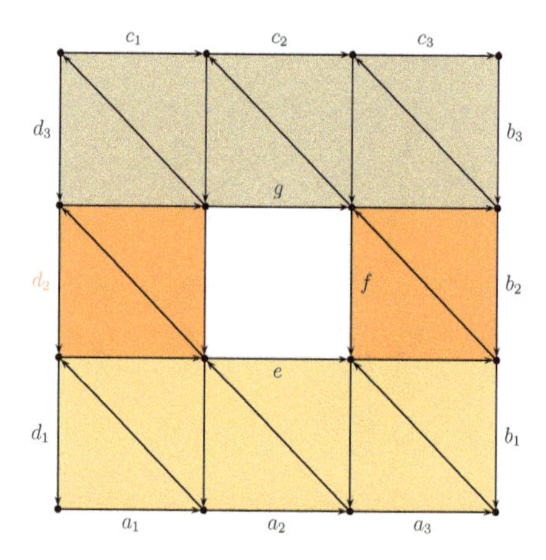

Fig. 5.19 A 2-dimensional simplicial complex with a hole.

It follows that all cycles in $C_2(K_5)$ not equivalent to zero are equivalent to a multiple of $e + f + g + h$, and thus

$$H_1(K_5) = \mathbb{Z},$$

indicating that K_5 has a single 1-dimensional hole. Since K_5 is connected, $H_0(K_5) = \mathbb{Z}$, and $H_2(K_5) = 0$ since $\mathrm{Ker}\, \partial_2 = 0$.

As we said in the introduction, the simplicial homology groups have a computational flavor, and this is one of the main reasons why they are attractive and useful. In fact, if K is any finite simplicial complex, there is an algorithm for computing the simplicial homology groups of K. This algorithm relies on a matrix reduction method (The Smith normal form) involving some simple row operations reminiscent of row-echelon reduction. This algorithm is described in detail in Munkres [Munkres (1984)] (Chapter 1, Section 11) and Rotman [Rotman (1988)] (Chapter 7).

5.3 Simplicial and Relative Homology with G-Coefficients

The generalization of simplicial homology to coefficients in any R-module G is immediate, where R is any commutative ring with an identity element. Simply define the chain group $C_p(K; G)$ as the R-module of functions c from the set of oriented p-simplices to G, such that

(1) $c(-\sigma) = -c(\sigma)$, iff σ and $-\sigma$ have opposite orientation;

(2) $c(\sigma) = 0$, for all but finitely many simplices σ.

A p-chain in $C_p(K; G)$ is a "vector-valued" formal finite linear combination

$$\sum_i \sigma_i g_i,$$

with $g_i \in G$ and σ_i an oriented p-simplex. Since by Proposition 5.3, the abelian group $C_p(K)$ (a $\mathbb{Z}$-module) is free with basis any choice of a set of oriented p-simplices, we have an isomorphism

$$C_p(K) \otimes_{\mathbb{Z}} G \cong \bigoplus_{[\sigma] \in C_{\Delta^p}(K)} G \cong C_p(K; G),$$

where $C_{\Delta^p}(K)$ denotes the set of equivalence classes of oriented p-simplices. The $\mathbb{Z}$-module $C_p(K; G)$ is made into an R-module by setting

$$\alpha \cdot \left(\sum_i \sigma_i g_i, \right) = \sum_i \sigma_i (\alpha g_i), \quad \alpha \in R.$$

Consequently, we can define the complex $C_*(K; G)$ as the complex $C_*(K) \otimes_{\mathbb{Z}} G$ obtained by tensoring the complex $C_*(K)$ with G (over the ring $\mathbb{Z}$) shown below:

$$0 \xleftarrow{\partial_0 \otimes \mathrm{id}} C_0(K) \otimes_{\mathbb{Z}} G \xleftarrow{\partial_1 \otimes \mathrm{id}} C_1(K) \otimes_{\mathbb{Z}} G \cdots \xleftarrow{\partial_p \otimes \mathrm{id}} C_p(K) \otimes_{\mathbb{Z}} G \cdots .$$

Since by definition, $C_p(K; G) = C_p(K) \otimes_{\mathbb{Z}} G$, we have the homology complex

$$0 \xleftarrow{\partial_0 \otimes \mathrm{id}} C_0(K; G) \xleftarrow{\partial_1 \otimes \mathrm{id}} C_1(K; G) \cdots \xleftarrow{\partial_p \otimes \mathrm{id}} C_p(K; G) \longleftarrow \cdots .$$

denoted $(C_*(K; G), \partial_* \otimes \mathrm{id})$. When $G = R$, each module $C_p(K; R)$ is a free R-module.

Definition 5.16. The simplicial homology groups $H_p(K; G)$ are the homology groups (really R-modules) of the simplicial chain complex $(C_*(K; G), \partial_* \otimes \mathrm{id})$.

Definition 5.17. Given two simplicial complexes K_1 and K_2, a simplicial map $f \colon K_1 \to K_2$ induces a homomorphism $f_{\sharp,p} \colon C_p(K_1; G) \to C_p(K_2; G)$ between the modules of oriented p-chains defined as follows: For any p-simplex $\{v_0, \ldots, v_p\}$ in K_1, we set

$$f_\sharp([v_0, \ldots, v_p]) = \begin{cases} [f(v_0), \ldots, f(v_p)] & \text{if the } f(v_i) \text{ are pairwise distinct} \\ 0 & \text{otherwise.} \end{cases}$$

It is easy to check that the $f_{\sharp,p}$ commute with the boundary maps, so $f_\sharp = (f_{\sharp,p})_{p\geq 0}$ is a chain map between the chain complexes $C_*(K_1;G)$ and $C_*(K_2;G)$ which induces homomorphisms

$$f_{*,p}\colon H_p(K_1;G) \to H_p(K_2;G) \quad \text{for all } p \geq 0.$$

This assignment is functorial; see Munkres [Munkres (1984)] (Chapter I, Section 12).

The relative simplicial homology groups are also easily defined (by analogy with relative singular homology).

Definition 5.18. Given a complex K and a subcomplex L of K, we define the *relative simplicial chain complex* $C_*(K, L; \mathbb{Z})$ by

$$C_p(K, L; \mathbb{Z}) = C_p(K; \mathbb{Z})/C_p(L; \mathbb{Z}).$$

As in the case of singular homology, $C_p(K, L; \mathbb{Z})$ is a free abelian group, because it has a basis consisting of the cosets of the form

$$\sigma + C_p(L; \mathbb{Z}),$$

where σ is an oriented p-simplex of K that is *not* in L. We obtain the *relative simplicial homology groups* $H_p(K, L; \mathbb{Z})$. We define the chain complex $C_*(K, L; G)$ as $C_*(K, L; \mathbb{Z}) \otimes_{\mathbb{Z}} G$, and we obtain *relative simplicial homology groups* $H_p(K, L; G)$ with coefficients in G.

Given two pairs of simplicial complexes (K_1, L_1) and (K_2, L_2), where L_1 is a subcomplex of K_1 and L_2 is a subcomplex of K_2, as in the absolute case a simplicial map $f\colon (K_1, L_1) \to (K_2, L_2)$ induces a homomorphism $f_{\sharp,p}\colon C_p(K_1, L_1; G) \to C_p(K_2, L_2; G)$ between the modules of oriented p-chains, and thus homomorphisms

$$f_{*,p}\colon H_p(K_1, L_1; G) \to H_p(K_2, L_2; G) \quad \text{for all } p \geq 0.$$

Again, this assignment is functorial.

A version of the excision axiom holds for relative simplicial homology. The following result is Theorem 9.1 in Munkres [Munkres (1984)] (Chapter I, Section 9).

Theorem 5.5. *Let K be a complex. Let K_0 be a subcomplex of K and let U be an open subset contained in $(K_0)_g$, such that $K_g - U$ is the geometric realization of a subcomplex L of K. Then inclusion induces isomorphisms*

$$H_p(L, L_0) \cong H_p(K, K_0), \quad p \geq 0.$$

A slightly more general version of Theorem 5.5 holds for triangulable spaces; see Theorem 27.2 in Munkres [Munkres (1984)] (Chapter III, Section 27).

The following version of the homotopy axiom holds, combining Theorem 19.2 and Theorem 19.5 in Munkres [Munkres (1984)] (Chapter II, Section 19).

Theorem 5.6. *Let K and L be two complexes. If $f, g\colon K_g \to L_g$ are homotopic maps and if $H_p(f)\colon H_p(K) \to H_p(L)$ and $H_p(g)\colon H_p(K) \to H_p(L)$ are the induced homomorphisms, then $H_p(f) = H_p(g)$ for all $p \geq 0$. In particular, if K_g and L_g are homotopy equivalent, then $H_p(K) \cong H_p(L)$ for all $p \geq 0$.*

Theorem 5.6 also holds for reduced simplicial homology and for relative simplicial homology; see Theorem 19.3 in Munkres [Munkres (1984)] (Chapter II, Section 19).

We also have a long exact sequence of homology of a pair (K_0, K); see Theorem 23.3 in Munkres [Munkres (1984)] (Chapter III, Section 23).

Theorem 5.7. *(Long Exact Sequence of Relative Simplicial Homology) For any pair (K_0, K) of complexes with K_0 a subcomplex of K, we have the following long exact sequence of homology groups*

$$\cdots \longrightarrow H_{p+2}(K, K_0) \xrightarrow{\partial_{*p+2}} H_{p+1}(K_0) \xrightarrow{i_*} H_{p+1}(K) \xrightarrow{j_*} H_{p+1}(K, K_0) \xrightarrow{\partial_{*p+1}} H_p(K_0) \xrightarrow{i_*} H_p(K) \xrightarrow{j_*} H_p(K, K_0) \xrightarrow{\partial_{*p}} H_{p-1}(K_0) \longrightarrow \cdots$$

ending in

$$H_0(K_0) \longrightarrow H_0(K) \longrightarrow H_0(K, K_0) \longrightarrow 0.$$

Theorem 5.5, Theorem 5.6 and Theorem 5.7 also hold for simplicial homology with coefficients in an R-module G; see Munkres [Munkres (1984)], Chapter 6, Section 51.

5.4 Equivalence of Simplicial and Singular Homology

Simplicial homology assigns homology groups to a simplicial complex K, not to a topological space. We can view the groups $H_p(K)$ as groups assigned to the geometric realization K_g of K, which is a space. Let us temporarily denote these groups by $H_p^{\Delta}(K_g)$. Now the following question arises.

If K and K' are two simplicial complexes whose geometric realizations K_g and K'_g are homeomorphic, are the groups $H_p^{\Delta}(K_g)$ and $H_p^{\Delta}(K'_g)$ isomorphic, that is, are the groups $H_p(K)$ and $H_p(K')$ isomorphic?

If the answer to this question was no, then the simplicial homology groups would not be useful objects for classifying spaces up to homeomorphism, but fortunately the answer is yes. However, the proof of this fact is quite involved. This can be proven directly as in Munkres [Munkres (1984)] (Chapter II), or by proving that the simplicial homology group $H_p(K)$ is isomorphic to the singular homology group $H_p(K_g)$ of the geometric realization of K. We will sketch this second approach. Unfortunately, the proof of this isomorphism also requires a lot of work.

In order to prove the equivalence of simplicial homology with singular homology we introduce a variant of the simplicial homology groups called *ordered simplicial homology groups*.

Definition 5.19. Let $K = (V, \mathcal{S})$ be a simplicial complex. An *ordered p-simplex* of K is a $(p+1)$-tuple $(v_0, \ldots, v_p)$ of vertices in V, where the v_i are vertices of some simplex σ of K *but need not be distinct*.

For example, if $\{v, w\}$ is a 1-simplex, then (v, w, w, v) is an ordered 3-simplex.

Let $C'_p(K; R)$ be the free R-module generated by the ordered p-simplices, called the group of *ordered p-chains*, and define the boundary map $\partial'_p \colon C'_p(K; R) \to C'_{p-1}(K; R)$ by

$$\partial'_p(v_0, \ldots, v_p) = \sum_{i=0}^{p} (-1)^i (v_0, \ldots, \widehat{v_i}, \ldots, v_p),$$

where $(v_0, \ldots, \widehat{v_i}, \ldots, v_p)$ denotes the ordered $(p-1)$-simplex obtained by deleting vertex v_i.

It is easily checked that $\partial'_p \circ \partial'_{p+1} = 0$, so we obtain a chain complex $C'_*(K; R)$ called the *ordered simplicial chain complex* of K. This is a huge and redundant complex, but it is useful to prove the equivalence of simplicial homology and singular homology.

Given a simplicial complex K and a subcomplex L, the *relative ordered simplicial chain complex* $C'_*(K, L; R)$ of (K, L) is defined by

$$C'_*(K, L; R) = C'_*(K; R)/C'_*(L; R).$$

We obtain the *ordered relative simplicial homology groups* $H'_p(K, L; R)$.

Theorem 5.8 below is proven in Munkres [Munkres (1984)] (Chapter I, Section 13, Theorem 13.6) and in Spanier [Spanier (1989)] (Chapter 4, Section 3, Theorem 8, and Section 5, Corollary 12). The proof uses a techniques known as "categories with models" and "acyclic models." These results are proven for $R = \mathbb{Z}$, but because the oriented chain modules $C_p(K, L; R)$ and the ordered chain modules $C'_p(K, L; R)$ are free R-modules, it can be checked that the constructions and the proofs go through for any commutative ring with an identity element 1.

Assuming for simplicity that $L = \emptyset$, the idea is to define two chain maps $\varphi\colon C_p(K; R) \to C'_p(K; R)$ and $\psi\colon C'_p(K; R) \to C_p(K; R)$ that are chain homotopy inverses. To achieve this, pick a partial order $\leq$ of the vertices of $K = (V, \mathcal{S})$ that induces a total order on the vertices of every simplex in $\mathcal{S}$, and define φ by

$$\varphi([v_0, \ldots, v_p]) = (v_0, \ldots, v_p) \quad \text{if } v_0 < v_1 < \cdots < v_p,$$

and ψ by

$$\psi((w_0, \ldots, w_p)) = \begin{cases} [w_0, \ldots, w_p] & \text{if the } w_i \text{ are pairwise distinct} \\ 0 & \text{otherwise.} \end{cases}$$

Then it can be shown that φ and ψ are natural transformations (with respect to simplicial maps) and that they are chain homotopy inverses. The maps φ and ψ can also be defined for pairs of complexes (K, L), as chain maps $\varphi\colon C_p(K, L; R) \to C'_p(K, L; R)$ and $\psi\colon C'_p(K, L; R) \to C_p(K, L; R)$ which are chain homotopic.

Theorem 5.8. *For any simplicial complex K and any subcomplex L of K, there are (natural) isomorphisms*

$$H_p(K, L; R) \cong H'_p(K, L; R) \quad \text{for all } p \geq 0$$

between the relative simplicial homology groups and the ordered relative simplicial homology groups.

Theorem 5.8 follows from the special case of the theorem in which $L = \emptyset$ by the five lemma (Proposition 2.5). This is a common trick in the subject which is used over and over again (see the proof of Theorem 5.9).

By naturality of the long exact sequence of homology of the pair (K, L), the chain map $\varphi\colon C_*(K, L; R) \to C'_*(K, L; R)$ yields the following commutative diagram:

$$\begin{array}{ccccccccc}
\cdots H_p(L;R) & \to & H_p(K;R) & \to & H_p(K,L;R) & \to & H_{p-1}(L;R) & \to & H_{p-1}(K;R)\cdots \\
\downarrow & & \downarrow & & \downarrow & & \downarrow & & \downarrow \\
\cdots H'_p(L;R) & \to & H'_p(K;R) & \to & H'_p(K,L;R) & \to & H'_{p-1}(L;R) & \to & H'_{p-1}(K;R)\cdots
\end{array}$$

in which the horizontal rows are exact. If we assume that the isomorphisms of the theorem hold in the absolute case, then all vertical arrows except the middle one are isomorphisms, and by the five lemma (Proposition 2.5), the middle arrow is also an isomorphism.

The proof that the simplicial homology group $H_p(K; \mathbb{Z})$ is isomorphic to the singular homology group $H_p(K_g; \mathbb{Z})$ is nontrivial. Proofs can be found in Munkres [Munkres (1984)] (Chapter 4, Section 34), Spanier [Spanier (1989)] (Chapter 4, Sections 4 and 6), Hatcher [Hatcher (2002)] (Chapter II, Section 2.1), and Rotman [Rotman (1988)] (Chapter 7). These proofs use variants of acyclic models, Mayer–Vietoris sequences, and the five lemma.

Given a simplicial complex K, the idea is to define a chain map $\theta\colon C'_*(K; \mathbb{Z}) \to S_*(K_g; \mathbb{Z})$ that induces isomorphisms $\theta_{*,p}\colon H'_p(K; \mathbb{Z}) \to H_p(K_g; \mathbb{Z})$ for all $p \geq 0$. Since K_g is a topological space, the only homology that applies is singular homology, and $S_*(K_g; \mathbb{Z})$ denotes the singular chain complex of singular homology; see Definition 4.3.

This can be done as follows: let $\ell(e_1, \ldots, e_{p+1})$ be the unique affine map from Δ^p (recall Definition 4.1) to K_g such that $\ell(e_{i+1}) = (v_i)_g$ for $i = 0, \ldots, p$. Then let

$$\theta((v_0, \ldots, v_p)) = \ell(e_1, \ldots, e_{p+1}).$$

It is also easy to define $\theta\colon C'_*(K, L; \mathbb{Z}) \to S_*(K_g, L_g; \mathbb{Z})$ for pairs of complexes (K, L) with L a subcomplex of K. Then we define the chain map $\eta\colon C_*(K, L; \mathbb{Z}) \to S_*(K_g, L_g; \mathbb{Z})$ as the composition $\eta = \theta \circ \varphi$, where $\varphi\colon C_*(K, L; \mathbb{Z}) \to C'_*(K, L; \mathbb{Z})$ is the chain map between oriented and ordered homology discussed earlier. The following important theorem shows that η induces an isomorphism between simplicial homology and singular homology.

Theorem 5.9. *Given any pair of simplicial complexes (K, L), where L is a subcomplex of K, the chain map $\eta\colon C_*(K, L; \mathbb{Z}) \to S_*(K_g, L_g; \mathbb{Z})$ induces isomorphisms*

$$H_p(K, L; \mathbb{Z}) \cong H_p(K_g, L_g; \mathbb{Z}) \quad \textit{for all } p \geq 0.$$

Proof sketch. By Theorem 5.8 it suffices to prove that the homology groups $H'_p(K, L; \mathbb{Z})$ and the singular homology groups $H_p(K_g, L_g; \mathbb{Z})$ are isomorphic. Again, we use the trick which consists in showing that Theorem 5.9 follows from the special case of the theorem in which $L = \emptyset$ by the five lemma (Proposition 2.5). Indeed, by naturality of the long exact sequence of homology of the pair (K, L), the chain map $\theta\colon C'_*(K, L; \mathbb{Z}) \to S_*(K_g, L_g; \mathbb{Z})$ yields the following commutative diagram

$$\begin{array}{ccccccccc}
H'_p(L;\mathbb{Z}) & \to & H'_p(K;\mathbb{Z}) & \to & H'_p(K,L;\mathbb{Z}) & \to & H'_{p-1}(L;\mathbb{Z}) & \to & H'_{p-1}(K;\mathbb{Z}) \cdots \\
\downarrow & & \downarrow & & \downarrow & & \downarrow & & \downarrow \\
H_p(L_g;\mathbb{Z}) & \to & H_p(K_g;\mathbb{Z}) & \to & H_p(K_g,L_g;\mathbb{Z}) & \to & H_{p-1}(L_g;\mathbb{Z}) & \to & H_{p-1}(K_g;\mathbb{Z}) \cdots
\end{array}$$

in which the horizontal rows are exact. If we assume that the isomorphisms of the theorem hold in the absolute case, then all vertical arrows except the middle one are isomorphisms, and by the five lemma (Proposition 2.5), the middle arrow is also an isomorphism.

The proof of the isomorphisms $H'_p(K; \mathbb{Z}) \cong H_p(K_g; \mathbb{Z})$ proceeds in two steps. We follow Spanier's proof Spanier [Spanier (1989)] (Theorem 8, Chapter 4, Section 6). Rotman's proof is nearly the same; see Rotman [Rotman (1988)] (Chapter 7), but beware that there appears to be some typos at the bottom of Page 151.

Step 1. We prove our result for a finite simplicial complex K by induction on the number n of simplices on K.

Base case, $n = 1$. For any abstract simplex s, let $\overline{s}$ be the simplicial complex consisting of all the faces of s (including s itself). The following result will be needed.

Proposition 5.10. *Given any abstract simplex s, there are isomorphisms*

$$H'_p(\overline{s}; \mathbb{Z}) \cong H_p(\overline{s}_g; \mathbb{Z}) \quad \textit{for all } p \geq 0.$$

Proposition 5.10 is Corollary 4.4.2 in Spanier [Spanier (1989)] (Chapter 4, Section 4). Intuitively, Proposition 5.10 is kind of obvious, since $\overline{s}$ corresponds to the combinatorial decomposition of a simplex, and $\overline{s}_g$ is a convex body homeomorphic to some ball D^m. Their corresponding homology should be (0) for $p > 0$ and $\mathbb{Z}$ for $p = 0$.

A rigorous proof of Proposition 5.10 uses the following results:

(1) We have the following isomorphisms between unreduced and reduced homology:

$$H'_0(K;\mathbb{Z}) \cong \widetilde{H'_0}(K;\mathbb{Z}) \oplus \mathbb{Z}$$
$$H'_p(K;\mathbb{Z}) \cong \widetilde{H'_p}(K;\mathbb{Z}) \quad p \geq 1$$

in ordered homology, and

$$H_0(K_g;\mathbb{Z}) \cong \widetilde{H}_0(K_g;\mathbb{Z}) \oplus \mathbb{Z}$$
$$H_p(K_g;\mathbb{Z}) \cong \widetilde{H}_p(K_g;\mathbb{Z}) \quad p \geq 1$$

in singular homology. This is Lemma 4.3.1 in Spanier [Spanier (1989)] (Chapter 4, Section 3).

(2) For any abstract simplex s, the reduced chain complex of ordered homology of $\overline{s}$ is acyclic; that is,

$$\widetilde{H'_p}(\overline{s};\mathbb{Z}) = (0) \quad \text{for all } p \geq 0.$$

This is Corollary 4.3.7 in Spanier [Spanier (1989)] (Chapter 4, Section 3). A more direct proof of the second fact (oriented simplicial homology) is given in Rotman [Rotman (1988)] (Chapter 7, Corollary 7.18). It is easily adapted to ordered homology.

(3) A chain complex C is said to be *contractible* if there is a chain homotopy between the identity chain map id_C of C and the zero chain map 0_C of C. Then a contractible chain complex is acyclic; that is, $H_p(C) = (0)$ for all $p \geq 0$. This is Corollary 4.2.3 in Spanier [Spanier (1989)] (Chapter 4, Section 2).

(4) Let X be any star-shaped subset of $\mathbb{R}^n$. Then the reduced singular complex of X is chain contractible. This is Lemma 4.4.1 in Spanier [Spanier (1989)] (Chapter 4, Section 4).

Induction step, $n > 1$. We will need the following facts:

(1) The Mayer–Vietoris sequence holds in ordered homology. This is not hard to prove; see Spanier [Spanier (1989)] (Chapter 4, Section 6).

(2) The Mayer–Vietoris sequence holds in reduced singular homology; this is Theorem 4.16.
(3) If K_1 and K_2 are subcomplexes of a simplicial complex K, then the Mayer–Vietoris sequence of singular homology holds for $(K_1)_g$ and $(K_2)_g$. This is Lemma 4.6.7 in Spanier [Spanier (1989)] (Chapter 4, Section 6). Actually, the above result is only needed in the following situation: if s is any simplex of K of highest dimension, then $K_1 = K - \{s\}$ and $K_2 = \overline{s}$; this is Lemma 7.20 in Rotman [Rotman (1988)] (Chapter 7). Since a Mayer–Vietoris sequence arises from a long exact sequence of homology, the chain map $\theta\colon C'_*(K, L; \mathbb{Z}) \to S_*(K_g, L_g; \mathbb{Z})$ induces a commutative diagram in which the top and bottom arrows are Mayer–Vietoris sequences and the vertical maps are induced by θ; see below.

Assume inductively that our result holds for any simplicial complex with less than $n > 1$ simplices. Pick any simplex s of maximal dimension, and let $K_1 = K - \{s\}$ and $K_2 = \overline{s}$, so that $K = K_1 \cup K_2$. Since $n > 1$ and s has maximal dimension, both K_1 and $K_1 \cap K_2$ are complexes (Condition (A2) is satisfied) and have less than n simplices, so by the induction hypothesis

$$H'_p(K_1; \mathbb{Z}) \cong H_p((K_1)_g; \mathbb{Z}) \quad \text{for all } p \geq 0$$

and

$$H'_p(K_1 \cap K_2; \mathbb{Z}) \cong H_p((K_1 \cap K_2)_g; \mathbb{Z}) \quad \text{for all } p \geq 0.$$

By Proposition 5.10 we also have

$$H'_p(K_2; R) = H'_p(\overline{s}; \mathbb{Z}) \cong H_p(\overline{s}_g; \mathbb{Z}) = H_p((K_2)_g; R) \quad \text{for all } p \geq 0.$$

Now Fact (3) (of the induction step) implies that we have the following diagram in which the horizontal rows are exact Mayer–Vietoris sequences (for a more direct argument, see Rotman [Rotman (1988)] (Chapter 7, Proposition 7.21)), and where we have suppressed the ring $\mathbb{Z}$ to simplify notation.

$$\begin{array}{ccccccccc}
H'_p(K_1 \cap \overline{s}) & \longrightarrow & H'_p(K_1) \oplus H'_p(\overline{s}) & \longrightarrow & H'_p(K) & \longrightarrow & H'_{p-1}(K_1 \cap \overline{s}) & \longrightarrow & Y \\
\downarrow & & \downarrow & & \downarrow & & \downarrow & & \downarrow \\
H_p((K_1 \cap \overline{s})_g) & \to & H_p((K_1)_g) \oplus H_p(\overline{s}_g) & \to & H_p(K_g) & \to & H_{p-1}((K_1 \cap \overline{s})_g) & \to & Z,
\end{array}$$

where $Y = H'_{p-1}(K_1) \oplus H'_{p-1}(\overline{s})$ and $Z = H_{p-1}((K_1)_g) \oplus H_{p-1}(\overline{s}_g)$. Since all vertical arrows except the middle one are isomorphisms, by the five lemma

(Proposition 2.5) the middle vertical arrow is also an isomorphism, which establishes the induction hypothesis. Therefore, we proved Theorem 5.9 for finite simplicial complexes.

Step 2. We prove our result for an infinite simplicial complex K. We resort to a direct limit argument (see Section 8.3). Let (K_α) be the family of finite subcomplexes of K under the inclusion ordering. It is a directed family. A version of this argument is given in Munkres [Munkres (1984)] (Chapter 4, Section 34, Lemma 44.2). Spanier proves that

$$H'_p(K;\mathbb{Z}) \cong \varinjlim H'_p(K_\alpha;\mathbb{Z})$$

and that

$$H_p(K_g;\mathbb{Z}) \cong \varinjlim H_p((K_\alpha)_g;\mathbb{Z}).$$

The first result is Theorem 4.3.11 in Spanier [Spanier (1989)] (Chapter 4, Section 3). This is an immediate consequence of the fact that homology commutes with direct limits; see Spanier [Spanier (1989)] (Theorem 4.1.7, Chapter 4, Section 1). The second result is the axiom of compact support for singular homology (Theorem 4.17). This completes the proof. $\square$

Theorem 5.9 proves the claim we made earlier that any two complexes K and K' that have homeomorphic geometric realizations have isomorphic simplicial homology groups, a result first proved by Alexander and Veblen.

The proofs of Theorem 5.9 found in the references cited earlier all assume that the ring of coefficients is $R = \mathbb{Z}$. However, close examination of Spanier's proof shows that the only result that makes use of the fact that $R = \mathbb{Z}$ is Proposition 5.10. If Proposition 5.10 holds for any commutative ring R with an identity element, then so does the theorem.

Fact (1) of Step 1 holds for any ring, in fact for any R-module G.

Fact (2) of Step 1 is a corollary of Theorem 4.3.6, which itself depends on Lemma 4.3.2; see Spanier [Spanier (1989)] (Chapter 4, Section 3). One needs to find right inverses to the augmentation maps $\epsilon\colon C'_0(K;R) \to R$ and $\epsilon\colon C'_0(K * w;R) \to R$, where $K * w$ is the cone with base K and vertex w; see Spanier [Spanier (1989)] (Chapter 3, Section 2). This is essentially the argument we gave in Section 4.4 just after Definition 4.14.

Actually, this argument can be generalized to any R-module G, as explained in Section 4.7 just after Definition 4.21, so we have the following generalization of Proposition 5.10: For any abstract simplex s and any R-module G, we have

$$H'_p(\overline{s}; G) \cong H_p(\overline{s}_g; G) \quad \text{for all } p \geq 0.$$

By tensoring with G, the chain map θ yields a chain map (also denoted θ) $\theta\colon C'_*(K, L; G) \to S_*(K_g, L_g; G)$. The chain map $\varphi\colon C_*(K, L; R) \to C'_*(K, L; R)$ can also be generalized to a chain map (also denoted φ) $\varphi\colon C_*(K, L; G) \to C'_*(K, L; G)$ by tensoring with G. We define $\varphi\colon C_*(K, L; G) \to S_*(K_g, L_g; G)$ as $\eta = \theta \circ \varphi$. Then we obtain a more general version of the isomorphism between simplicial homology and singular homology.

Theorem 5.11. *For any commutative ring R with an identity element 1 and for any R-module G, given any pair of simplicial complexes (K, L), where L is a subcomplex of K, the chain map $\eta\colon C_*(K, L; G) \to S_*(K_g, L_g; G)$ induces isomorphisms*

$$H_p(K, L; G) \cong H_p(K_g, L_g; G) \quad \text{for all } p \geq 0.$$

In summary, singular homology subsumes simplicial homology. Still, simplicial homology is much more computational.

5.5 The Euler–Poincaré Characteristic of a Simplicial Complex

In this section we assume that we are considering simplicial homology groups with coefficients in $\mathbb{Z}$. A fundamental invariant of finite complexes is the Euler–Poincaré characteristic. We saw earlier that the simplicial homology groups of a finite simplicial complex K are finitely generated abelian groups. We can assign a number $\chi(K)$ to K by making use of the fact that the structure of finitely generated abelian groups can be completely described. It turns out that every finitely generated abelian group can be expressed as the sum of the special abelian groups $\mathbb{Z}^r$ and $\mathbb{Z}/m\mathbb{Z}$. The crucial result is the following.

Proposition 5.12. *Let G be a free abelian group finitely generated by $(a_1, \ldots, a_n)$ and let H be any subgroup of G. Then H is a free abelian group and there is a basis, $(e_1, ..., e_n)$, of G, some $q \leq n$, and some positive natural numbers, $n_1, \ldots, n_q$, such that $(n_1 e_1, \ldots, n_q e_q)$ is a basis of H and n_i divides n_{i+1} for all i, with $1 \leq i \leq q - 1$.*

A neat proof of Proposition 5.12 can be found in Samuel [Samuel (2008)]; see also Dummit and Foote [Dummit and Foote (1999)] (Chapter 12, Theorem 4).

Remark: Actually, Proposition 5.12 is a special case of the structure theorem for finitely generated modules over a principal ring. Recall that $\mathbb{Z}$ is a principal ring, which means that every ideal $\mathcal{I}$ in $\mathbb{Z}$ is of the form $d\mathbb{Z}$, for some $d \in \mathbb{N}$.

We abbreviate the direct sum $\underbrace{\mathbb{Z} \oplus \cdots \oplus \mathbb{Z}}_{m}$ of m copies of $\mathbb{Z}$ as $\mathbb{Z}^m$. Using Proposition 5.12, we can also show the following useful result.

Theorem 5.13. *(Structure theorem for finitely generated abelian groups) Let G be a finitely generated abelian group. There is some natural number, $m \geq 0$, and some natural numbers $n_1, \ldots, n_q \geq 2$, such that G is isomorphic to the direct sum*

$$\mathbb{Z}^m \oplus \mathbb{Z}/n_1\mathbb{Z} \oplus \cdots \oplus \mathbb{Z}/n_q\mathbb{Z},$$

and where n_i divides n_{i+1} for all i, with $1 \leq i \leq q-1$.

Proof. Assume that G is generated by $A = (a_1, \ldots, a_n)$ and let $F(A)$ be the free abelian group generated by A. The inclusion map $i \colon A \to G$ can be extended to a unique homomorphism $f \colon F(A) \to G$ which is surjective since A generates G, and thus G is isomorphic to $F(A)/f^{-1}(0)$. By Proposition 5.12, $H = f^{-1}(0)$ is a free abelian group and there is a basis $(e_1, ..., e_n)$ of G, some $p \leq n$, and some positive natural numbers $k_1, \ldots, k_p$, such that $(k_1e_1, \ldots, k_pe_p)$ is a basis of H, and k_i divides k_{i+1} for all i, with $1 \leq i \leq p-1$. Let r, $0 \leq r \leq p$, be the largest natural number such that $k_1 = \ldots = k_r = 1$, rename k_{r+i} as n_i, where $1 \leq i \leq p-r$, and let $q = p-r$. Then we can write

$$H = \mathbb{Z}^{p-q} \oplus n_1\mathbb{Z} \oplus \cdots \oplus n_q\mathbb{Z},$$

and since $F(A)$ is isomorphic to $\mathbb{Z}^n$, it is easy to verify that $F(A)/H$ is isomorphic to

$$Z^{n-p} \oplus \mathbb{Z}/n_1\mathbb{Z} \oplus \cdots \oplus \mathbb{Z}/n_q\mathbb{Z},$$

which proves the proposition. $\square$

Observe that G is a free abelian group iff $q = 0$, and otherwise $\mathbb{Z}/n_1\mathbb{Z} \oplus \cdots \oplus \mathbb{Z}/n_q\mathbb{Z}$ is the torsion subgroup of G. Thus, as a corollary of

Proposition 5.13, we obtain the fact that every finitely generated abelian group G is a direct sum, $G = Z^m \oplus T$, where T is the torsion subgroup of G and Z^m is the free abelian group of dimension m.

One verifies that m is the rank (the maximal dimension of linearly independent sets in G) of G, denoted $\text{rank}(G)$.

Definition 5.20. The number $m = \text{rank}(G)$ is called the *Betti number* of G and the numbers $n_1, \ldots, n_q$ are the *torsion numbers* of G.

It can also be shown that q and the n_i only depend on G.

In the early days of algebraic topology (between the late 1890s and the early 1930s), an area of mathematics started by Henri Poincaré in the late 1890s, homology groups had not been defined and people worked with Betti numbers and torsion coefficients. Emmy Noether played a crucial role in introducing homology groups into the field.

Fig. 5.20 Leonhard Euler, 1707–1783 (left), and Henri Poincaré, 1854–1912 (right).

Definition 5.21. Given a finite complex $K = (V, \mathcal{S})$ of dimension m, if we let m_p be the number of p-simplices in K, we define the *Euler–Poincaré characteristic* $\chi(K)$ *of* K by

$$\chi(K) = \sum_{p=0}^{m} (-1)^p \, m_p.$$

In order to prove Theorem 5.15 we make use of Proposition 5.14 stated below.

Proposition 5.14. *If*

$$0 \longrightarrow E \longrightarrow F \longrightarrow G \longrightarrow 0$$

is a short exact sequence of homomorphisms of abelian groups and if F has finite rank, then

$$\mathrm{rank}(F) = \mathrm{rank}(E) + \mathrm{rank}(G).$$

In particular, if G is an abelian group of finite rank and if H is a subgroup of G, then $\mathrm{rank}(G) = \mathrm{rank}(H) + \mathrm{rank}(G/H)$.

Proposition 5.14 follows from the fact that $\mathbb{Q}$ is a flat $\mathbb{Z}$-module (see Definition 11.1 and Proposition 11.12). By tensoring with $\mathbb{Q}$ with obtain an exact sequence in which the spaces $E \otimes_{\mathbb{Z}} \mathbb{Q}$, $F \otimes_{\mathbb{Z}} \mathbb{Q}$, and $G \otimes_{\mathbb{Z}} \mathbb{Q}$, are vector spaces over $\mathbb{Q}$ whose dimensions are equal to the ranks of the abelian groups being tensored with; see Proposition 11.13. A proof of Proposition 5.14 is also given in Greenberg and Harper [Greenberg and Harper (1981)] (Chapter 20, Lemma 20.7 and Lemma 20.8).

The following remarkable theorem holds:

Theorem 5.15. *Given a finite complex $K = (V, \mathcal{S})$ of dimension m, we have*

$$\chi(K) = \sum_{p=0}^{m} (-1)^p \, \mathrm{rank}(H_p(K)),$$

the alternating sum of the Betti numbers (the ranks) of the homology groups of K.

Proof. We know that $C_p(K)$ is a free group of rank m_p. Since $H_p(K) = Z_p(K)/B_p(K)$, by Proposition 5.14, we have

$$\mathrm{rank}(H_p(K)) = \mathrm{rank}(Z_p(K)) - \mathrm{rank}(B_p(K)).$$

Since we have a short exact sequence

$$0 \longrightarrow Z_p(K) \longrightarrow C_p(K) \xrightarrow{\partial_p} B_{p-1}(K) \longrightarrow 0,$$

again, by Proposition 5.14, we have

$$\mathrm{rank}(C_p(K)) = m_p = \mathrm{rank}(Z_p(K)) + \mathrm{rank}(B_{p-1}(K)).$$

Also, note that $B_m(K) = 0$, and $B_{-1}(K) = 0$. Then, we have

$$\begin{aligned}
\chi(K) &= \sum_{p=0}^{m} (-1)^p \, m_p \\
&= \sum_{p=0}^{m} (-1)^p \, (\mathrm{rank}(Z_p(K)) + \mathrm{rank}(B_{p-1}(K))) \\
&= \sum_{p=0}^{m} (-1)^p \, \mathrm{rank}(Z_p(K)) + \sum_{p=0}^{m} (-1)^p \, \mathrm{rank}(B_{p-1}(K)).
\end{aligned}$$

Using the fact that $B_m(K) = 0$, and $B_{-1}(K) = 0$, we get

$$\begin{aligned}\chi(K) &= \sum_{p=0}^{m}(-1)^p \operatorname{rank}(Z_p(K)) + \sum_{p=0}^{m}(-1)^{p+1} \operatorname{rank}(B_p(K)) \\ &= \sum_{p=0}^{m}(-1)^p \left(\operatorname{rank}(Z_p(K)) - \operatorname{rank}(B_p(K))\right) \\ &= \sum_{p=0}^{m}(-1)^p \operatorname{rank}(H_p(K)).\end{aligned}$$

□

A striking corollary of Theorem 5.15 (together with Theorem 5.9) is that the Euler–Poincaré characteristic, $\chi(K)$, of a complex of finite dimension m only depends on the geometric realization K_g of K, since it only depends on the homology groups $H_p(K) = H_p(K_g)$ of the polytope K_g. Thus, the Euler–Poincaré characteristic is an invariant of all the finite complexes corresponding to the same polytope, $X = K_g$. We can say that it is *the* Euler–Poincaré characteristic of the polytope $X = K_g$, and denote it by $\chi(X)$. In particular, this is true of surfaces that admit a triangulation. The Euler–Poincaré characteristic in one of the major ingredients in the classification of the compact surfaces. In this case, $\chi(K) = m_0 - m_1 + m_2$, where m_0 is the number of vertices, m_1 the number of edges, and m_2 the number of triangles in K.

Going back to the triangulations of the sphere, the torus, the projective space, and the Klein bottle, we find that they have Euler–Poincaré characteristics 2 (sphere), 0 (torus), 1 (projective space), and 0 (Klein bottle).

5.6 Simplicial Cohomology

In this section G is any R-module over a commutative ring R with an identity element 1. The relative (and absolute) simplicial cohomology groups of a pair of simplicial complexes (K, L) (where L is a subcomplex of K) are defined the same way that the singular relative cohomology groups are defined from the singular homology groups by applying $\mathrm{Hom}_R(-; G)$, as in Section 4.9.

Given the chain complex of relative simplicial homology

$$0 \xleftarrow{\partial_0} C_0(K, L; R) \xleftarrow{\partial_1} C_1(K, L; R) \longleftarrow \cdots C_{p-1}(K, L; R) \xleftarrow{\partial_p} C_p(K, L; R) \xleftarrow{\partial_{p+1}}$$

by applying $\mathrm{Hom}_R(-, G)$, where $C_p(K, L; R) = C_p(K, R)/C_p(L, R)$, we obtain the chain complex

$$0 \xrightarrow{\delta^{-1}} C^0(K, L; G) \xrightarrow{\delta^0} C^1(K, L; G) \to \cdots C^p(K, L; G) \xrightarrow{\delta^p} C^{p+1}(K, L; G) \xrightarrow{\delta^{p+1}}$$

with $C^p(K, L; G) = \mathrm{Hom}_R(C_p(K, L; R), G)$ and $\delta^p = \mathrm{Hom}_R(\partial_p, G)$ for all $p \geq 0$ (and δ^{-1} is the zero map). More explicitly

$$\delta^p(f) = f \circ \partial_{p+1} \quad \text{for all } f \in C^p(K, L; G);$$

that is

$$\delta^p(f)(\sigma) = f(\partial_{p+1}(\sigma)) \quad \text{for all } f \in C^p(K, L; G) = \mathrm{Hom}_R(C_p(K, L; R), G)$$
$$\text{and all } \sigma \in C_{p+1}(K; L; R).$$

Definition 5.22. Given a pair of complexes (K, L) with L a subcomplex of K, the *simplicial relative cohomology groups* $H^p(K, L; G)$ of (K, L) arise from the chain complex

$$0 \xrightarrow{\delta^{-1}} C^0(K, L; G) \xrightarrow{\delta^0} C^1(K, L; G) \to \cdots C^p(K, L; G) \xrightarrow{\delta^p} C^{p+1}(K, L; G) \xrightarrow{\delta^{p+1}}$$

with

$$\delta^p(f) = f \circ \partial_{p+1} \quad \text{for all } f \in C^p(K, L; G),$$

and are given by

$$H^p(K, L; G) = \mathrm{Ker}\, \delta^p / \mathrm{Im}\, \delta^{p-1}, \quad p \geq 0.$$

To obtain the long exact sequence of relative simplicial cohomology we dualize the short exact sequence

$$0 \longrightarrow C_*(L; R) \xrightarrow{i} C_*(K; R) \xrightarrow{j} C_*(K, L; R) \longrightarrow 0$$

where $C_*(K, L; R) = C_*(K, R)/C_*(L, R)$ by applying $\mathrm{Hom}(-, G)$ and we obtain the sequence

$$0 \longrightarrow C^*(K, L; G) \xrightarrow{j^\top} C^*(K; G) \xrightarrow{i^\top} C^*(L; G) \longrightarrow 0,$$

with $C^*(K, L; G) = \mathrm{Hom}_R(C_*(K; R)/C_*(L; R), G)$, and as before $C^*(L; G) = \mathrm{Hom}_R(C_*(L; R), G)$ and $C^*(K; G) = \mathrm{Hom}_G(C_*(K; R), G)$.

Since $C_p(K, L; R) = C_p(K, R)/C_p(L, R)$ is a free module for every p, by Proposition 2.8 the sequence of chain complexes

$$0 \longrightarrow C^*(K, L; G) \xrightarrow{j^\top} C^*(K; G) \xrightarrow{i^\top} C^*(L; G) \longrightarrow 0$$

is exact.

Given two pairs of simplicial complexes (K_1, L_1) and (K_2, L_2), where L_1 is a subcomplex of K_1 and L_2 is a subcomplex of K_2, a simplicial map $f\colon (K_1, L_1) \to (K_2, L_2)$ induces a homomorphism $f_{\sharp,p}\colon C_p(K_1, L_1; R) \to C_p(K_2, L_2; R)$ between the modules of oriented p-chains, and thus by applying $\mathrm{Hom}_R(-, G)$ we get a homomorphism $f^{\sharp,p}\colon C^p(K_2, L_2; G) \to C^p(K_1, L_1; G)$ commuting with coboundaries which induces homomorphisms

$$f^{*,p}\colon H^p(K_2, L_2; G) \to H^p(K_1, L_1; G) \quad \text{for all } p \geq 0.$$

Again, this assignment is functorial. The above fact is the simplicial analog of Proposition 4.36.

If R is a PID, then the simplicial cohomology group $H^p(K, L; G)$ is isomorphic to the singular cohomology group $H^p(K_g, L_g; G)$ for every $p \geq 0$. This result is easily obtained from the universal coefficient theorem for cohomology, or by an argument about free chain complexes; see Munkres [Munkres (1984)] (Chapter 5, Section 45, Theorem 45.5).

Theorem 5.16. *Let (K, L) be any pair of simplicial complexes with L a subcomplex of K. If R is a PID, then for any R-module G we have isomorphisms*

$$H^p(K, L; G) \cong H^p(K_g, L_g; G) \quad \textit{for all } p \geq 0$$

between the relative simplicial cohomology of the pair of complexes (K, L) and the relative singular cohomology of the pair of geometric realizations (K_g, L_g).

Proof. Let $\eta\colon C_*(K, L; R) \to S_*(K_g, L_g; R)$ be the chain map of Theorem 5.9. By the naturality part of universal coefficient theorem for cohomology (Theorem 12.6, and see Example 11.1 for the definition of Ext^1_R), we have the commutative diagram

$$\begin{array}{ccccccccc}
0 & \longrightarrow & Y & \longrightarrow & H^p(K_g, L_g; G) & \longrightarrow & \mathrm{Hom}_R(H_p(K_g, L_g; R), G) & \longrightarrow & 0 \\
 & & \big\downarrow{\scriptstyle \mathrm{Ext}^R_1(\eta_*)} & & \big\downarrow{\scriptstyle (\mathrm{Hom}_R(\eta, G))^*} & & \big\downarrow{\scriptstyle \mathrm{Hom}_R(\eta_*, \mathrm{id})} & & \\
0 & \longrightarrow & Z & \longrightarrow & H^p(K, L; G) & \longrightarrow & \mathrm{Hom}_R(H_p(K, L; R), G) & \longrightarrow & 0,
\end{array}$$

where $Y = \mathrm{Ext}^1_R(H_{p-1}(K_g, L_g; R), G)$ and $Z = \mathrm{Ext}^1_R(H_{p-1}(K, L; R), G)$. By Theorem 5.9 the chain map η induces isomorphisms $H_{p-1}(K, L; R) \cong H_{p-1}(K_g, L_g; R)$ and $H_p(K, L; R) \cong H_p(K_g, L_g; R)$, so the first and the third map in the above diagram are isomorphisms. By the short five lemma (Proposition 2.4) we conclude that the middle map is an isomorphism. $\square$

The above proof shows the stronger result that if $H_{p-1}(K, L; R) \cong H_{p-1}(K_g, L_g; R)$ and $H_p(K, L; R) \cong H_p(K_g, L_g; R)$, then $H^p(K, L; G) \cong H^p(K_g, L_g; G)$.

In summary, simplicial cohomology is subsumed by singular cohomology (at least when R is a PID). Nevertheless, simplicial cohomology is much more amenable to computation than singular cohomology. In particular, simplicial cohomology can be used to compute the cohomology ring of various spaces; see Munkres [Munkres (1984)] (Chapter 5, Section 49).

Indeed, it is possible to define a cup product on the simplicial cohomology of a complex. If $K = (V, \mathcal{S})$ is a simplicial complex, let $\leq$ be a partial order of the vertices of K that induces a total order on the vertices of every simplex in $\mathcal{S}$.

Definition 5.23. Given a simplicial complex $K = (V, \mathcal{S})$ and a partial order of its vertices as above, define a map

$$\smile^{\Delta}: C^p(K; R) \times C^q(K; R) \to C^{p+q}(K; R)$$

by

$$(c \smile^{\Delta} d)([v_0, \ldots, v_{p+q}]) = c([v_0, \ldots, v_p])\, d([v_p, \ldots, v_{p+q}])$$

iff $v_0 < v_1 < \cdots < v_{p+q}$, for all simplicial p-cochains $c \in C^p(K; R)$ and all simplicial q-cochains $d \in C^q(K; R)$.

It can be shown that the map $\smile^{\Delta}: C^p(K; R) \times C^q(K; R) \to C^{p+q}(K; R)$ induces a *cup product*

$$\smile^{\Delta}: H^p(K; R) \times H^q(K; R) \to H^{p+q}(K; R)$$

which is bilinear and associative and independent of the partial order $\leq$ chosen on V; see Munkres [Munkres (1984)] (Chapter 5, Section 49, Theorem 49.1 and Theorem 49.2).

It can also be shown that if $\eta: C_*(K; R) \to S_*(K_g; R)$ is the chain map of Theorem 5.9, then $\eta^* = \mathrm{Hom}_R(\eta, R)$ carries the cup product $\smile$ of singular cohomology to the cup product $\smile^{\Delta}$ of simplicial cohomology of Definition 4.32. If $h: K_1 \to K_2$ is a simplicial map between two simplicial complexes, then h^* preserves cup products; see Munkres [Munkres (1984)] (Chapter 5, Section 49, Theorem 49.1 and Theorem 49.2).

5.7 Problems

Problem 5.1. Prove that homology (with integer coefficients) of the torus T specified by the complex of Figure 5.9 is given by

$$\begin{aligned} H_0(T) &= \mathbb{Z} \\ H_1(T) &= \mathbb{Z} \oplus \mathbb{Z} \\ H_2(T) &= \mathbb{Z}. \end{aligned}$$

Problem 5.2. Prove that homology (with integer coefficients) of the projective plane $\mathbb{RP}^2$ specified by the complex of Figure 5.11 is given by

$$\begin{aligned} H_0(\mathbb{RP}^2) &= \mathbb{Z} \\ H_1(\mathbb{RP}^2) &= \mathbb{Z}/2\mathbb{Z} \\ H_2(\mathbb{RP}^2) &= (0). \end{aligned}$$

Problem 5.3. Prove that homology (with integer coefficients) of the Klein bottle K specified by the complex of Figure 5.12 is given by

$$\begin{aligned} H_0(K) &= \mathbb{Z} \\ H_1(K) &= \mathbb{Z} \oplus (\mathbb{Z}/2\mathbb{Z}) \\ H_2(K) &= (0). \end{aligned}$$

Problem 5.4. Given two (2-dimensional) tori T, their connected sum $T \sharp T$ is the space (a surface) obtained by deleting a region homeomorphic to an open disc (in the plane) from each torus and gluing together the pieces that remain. Such a space is the geometric realization of a complex shown in Figure 6.11 of Munkres [Munkres (1984)], Page 40. Prove that the homology of $T \sharp T$ is given by

$$\begin{aligned} H_0(T \sharp T) &= \mathbb{Z} \\ H_1(T \sharp T) &= \mathbb{Z}^4 \\ H_2(T \sharp T) &= \mathbb{Z}. \end{aligned}$$

Problem 5.5. Given a (2-dimensional) torus T, we can form its g-fold connected sum

$$X_g = \underbrace{T \sharp \cdots \sharp T}_{g}$$

by gluing together $g \geq 2$ tori. Prove that the homology of X_g is given by

$$\begin{aligned} H_0(X_g) &= \mathbb{Z} \\ H_1(X_g) &= \mathbb{Z}^{2g} \\ H_2(X_g) &= \mathbb{Z}. \end{aligned}$$

You will have to figure out a complex whose geometric realization is X_g.

We let $X_0 = S^2$ and $X_1 = T$. Check that the Euler characteristic of X_g is given by

$$\chi(X_g) = 2 - 2g, \quad g \geq 0.$$

The number g is called the *genus* of the surface X_g.

Problem 5.6. Given two projective planes $\mathbb{RP}^2$, their connected sum $\mathbb{RP}^2 \sharp \mathbb{RP}^2$ is the space (a surface) obtained by deleting a region homeomorphic to an open disc (in the plane) from each projective plane and gluing together the pieces that remain. Such a space is the geometric realization of a complex shown in Figure 6.8 of Munkres [Munkres (1984)], Page 38. Prove that the homology of $\mathbb{RP}^2 \sharp \mathbb{RP}^2$ is given by

$$\begin{aligned} H_0(\mathbb{RP}^2 \sharp \mathbb{RP}^2) &= \mathbb{Z} \\ H_1(\mathbb{RP}^2 \sharp \mathbb{RP}^2) &= \mathbb{Z} \oplus (\mathbb{Z}2/\mathbb{Z}) \\ H_2(\mathbb{RP}^2 \sharp \mathbb{RP}^2) &= (0). \end{aligned}$$

Prove that $\mathbb{RP}^2 \sharp \mathbb{RP}^2$ is homeomorphic to the Klein bottle.

Problem 5.7. Given a projective plane $\mathbb{RP}^2$, we can form its g-fold connected sum

$$Y_g = \underbrace{\mathbb{RP}^2 \sharp \cdots \sharp \mathbb{RP}^2}_{g}$$

by gluing together $g \geq 2$ projective planes. Prove that the homology of Y_g is given by

$$\begin{aligned} H_0(Y_g) &= \mathbb{Z} \\ H_1(Y_g) &= \mathbb{Z}^{g-1} \oplus (\mathbb{Z}/2\mathbb{Z}) \\ H_2(Y_g) &= (0). \end{aligned}$$

We let $Y_1 = \mathbb{RP}^2$. Check that the Euler characteristic of Y_g is given by

$$\chi(Y_g) = 2 - g, \quad g \geq 1.$$

The number g is called the *genus* of the surface Y_g.

Remark: A classical theorem of topology (the classification theorem for compact surfaces) states that any compact surface (a topological 2-dimensional manifold without boundary which is compact) is homeomorphic to one of the spaces (surfaces)

$$S^2,\, X_g,\, Y_g, \quad g \geq 1.$$

This is a nontrivial theorem and it takes a lot of work to prove it in full. Problems 5.5 and 5.7 imply that no two of these surfaces are homeomorphic. The surfaces S^2, X_g are orientable, and the surfaces Y_g are nonorientable. For comprehensive expositions, see Massey [Massey (1987)], Alhfors and Sario [Ahlfors and Sario (1960)], or Gallier and Xu [Gallier and Xu (2013)].

Chapter 6

Homology and Cohomology of CW Complexes

Computing the singular homology (or cohomology) groups of a space X is generally very difficult. J.H.C. Whitehead invented a class of spaces called CW complexes for which the computation of the singular homology groups is much more tractable. Roughly speaking, a CW complex X is built up inductively starting with a collection of points, in such a way that if the space X^p has been obtained at stage p, then the space X^{p+1} is obtained from X^p by gluing, or as it is customary to say attaching, a collection of closed balls whose boundaries are glued to X^p in a specific fashion. Each space X^p is called a p-skeleton of the space X. Every compact manifold is homotopy equivalent to a CW complex, so the class of CW complexes is quite rich. It also plays an important role in homotopy theory. In this short chapter we describe CW complexes and explain how their homology and cohomology can be computed.

One of the nice features of CW complexes is the fact that it is possible to assign to each CW complex X a chain complex $S_*^{\mathrm{CW}}(X;R)$ called its cellular chain complex, where

$$S_p^{\mathrm{CW}}(X;R) = H_p(X^p, X^{p-1};R),$$

the relative pth singular homology group of the pair (X^p, X^{p-1}), where X^p is the p-skeleton of X. We will show that the module $H_p(X^p, X^{p-1};R)$ is a free R-module whose dimension (when finite) is equal to the number of p-cells in X. Furthermore, the homology of the cellular complex agrees with the singular homology. That is, if we write $H_p^{\mathrm{CW}}(X;R) = H_p(S_*^{\mathrm{CW}}(X;R))$, then

$$H_p^{\mathrm{CW}}(X;R) \cong H_p(X;R) \quad \text{for all } p \geq 0,$$

where $H_p(X;R)$ is the pth singular homology module of X; see Theorem 6.8. In many practical cases, the number of p-cells is quite small so the cellular complex $S_*^{\mathrm{CW}}(X;R)$ is much more manageable than the singular complex $S_*(X;R)$.

The formula for the Euler–Poincaré characteristic given for simplicial complexes can be generalized to a finite CW complex X of dimension n. We have

$$\chi(X) = \sum_p (-1)^p \,\mathrm{rank}\, H_p(X;\mathbb{Z}) = \sum_{p=0}^{n} (-1)^p a_p,$$

where a_p is the number of p-cells in X (here we use singular homology with coefficients in $\mathbb{Z}$). We discuss what happens when we replace $\mathbb{Z}$ by a more general ring R. It turns out that $\chi(X)$ is independent of R if R is a PID.

To define cohomology module we consider the dual modules

$$H^k(X^p, X^{p-1}; G) = H^k(\mathrm{Hom}_R(S_k(X^p, X^{p-1}; R), G),$$

where X is a CW complex. If we assume that R is a PID, then

$$H^p(X^p, X^{p-1}; G) = \mathrm{Hom}_R(H_p(X^p, X^{p-1}; R), G),$$

and it is possible to define a cochain complex $S^*_{\mathrm{CW}}(X;G)$ whose modules are the cohomology modules $H^p(X^p, X^{p-1}; G)$, which defines cellular cohomology modules

$$H^p_{\mathrm{CW}}(X;G) = H^p(S^*_{\mathrm{CW}}(X;G)).$$

It can be shown that for any PID R, there are isomorphisms

$$H^p_{\mathrm{CW}}(X;G) \cong H^p(X;G) \quad \text{for all } p \geq 0$$

between the cellular cohomology modules and the singular cohomology modules of X. Furthermore, the cellular cochain complex $S^*_{\mathrm{CW}}(X;G)$ is isomorphic to the cochain complex $\mathrm{Hom}_R(S_*^{\mathrm{CW}}(X;R), G)$ (the dual of the cellular chain complex $S_*^{\mathrm{CW}}(X;R)$ with respect to G); see Theorem 6.16.

6.1 CW Complexes

First we define closed and open cells, and then we describe the process of attaching space (or adjunction space). Recall that the n-dimensional ball D^n, the n-dimensional open ball Int D^n, and the n-dimensional sphere S^n, are defined by

$$\begin{aligned} D^n &= \{x \in \mathbb{R}^n \mid \|x\|_2 \leq 1\} \\ \mathrm{Int}\, D^{n+1} &= \{x \in \mathbb{R}^{n+1} \mid \|x\|_2 < 1\} \\ S^n &= \{x \in \mathbb{R}^{n+1} \mid \|x\|_2 = 1\}. \end{aligned}$$

Furthermore, $S^n = \partial D^{n+1} = D^{n+1} - \mathrm{Int}\, D^{n+1}$, the boundary of D^{n+1}, and $D^n/\partial D^n$ is homeomorphic to S^n $(n \geq 1)$. When $n = 0$, we set Int $D^0 = D^0 = \{0\}$, and $\partial D^0 = S^{-1} = \emptyset$.

Definition 6.1. A *(closed) cell of dimension* $m \geq 0$ (or *closed* m*-cell*) is a space homeomorphic to D^m, and an *open cell of dimension* $m \geq 0$ (or *open* m*-cell*) is a space homeomorphic to Int D^m. We will usually denote an open m-cell by e^m (or simply e), and its closure by $\overline{e^m}$ (or simply $\overline{e}$). The set $\overline{e} - e$ is denoted by $\dot{e}$.

Observe that an open or closed 0-cell is a point.

Given two topological spaces X and Y, given a closed subset A of X, and given a continuous map $f\colon A \to Y$, we would like to define the space $X \cup_f Y$ obtained by gluing X and Y "along A." We will define $X \cup_f Y$ as a quotient space of the disjoint union $X \sqcup Y$ of X and Y with the topology in which a subset $Z \in X \sqcup Y$ is open iff $Z \cap X$ is open in X and $Z \cap Y$ is open in Y. See Figure 6.3. More generally, recall the definition of the topological sum of a family of spaces (Definition 4.16).

Definition 6.2. If $(X_i)_{i\in I}$ is a family of topological spaces we define the *topological sum* $\bigsqcup_{i\in I} X_i$ of the family $(X_i)_{i\in I}$ as the disjoint union of the spaces X_i, and we give it the topology for which a subset $Z \subseteq \bigsqcup_{i\in I} X_i$ is open iff $Z \cap X_i$ is open for all $i \in I$.

We will also need the notion of coherent union.

Definition 6.3. Given a topological space X, if $(X_i)_{i\in I}$ is a family of subspaces of X such that $X = \bigcup_{i\in I} X_i$, we say that the topology of X is *coherent* with the family $(X_i)_{i\in I}$ if a subset $A \subseteq X$ is open in X iff $A \cap X_i$ is open in X_i for all $i \in I$. We say that X is the *coherent union* of the family $(X_i)_{i\in I}$. See Figures 6.1 and 6.2.

Given X, Y, A, and $f\colon A \to Y$ as above, we form the quotient space of $X \sqcup Y$ by identifying each set

$$f^{-1}(y) \cup \{y\}$$

for each $y \in Y$ to a point. This means that we form the quotient set corresponding to the partition of $X \sqcup Y$ into the subsets of the form $f^{-1}(y) \cup \{y\}$ for all $y \in Y$, and all singleton sets $\{x\}$ for all $x \in X - A$. Observe

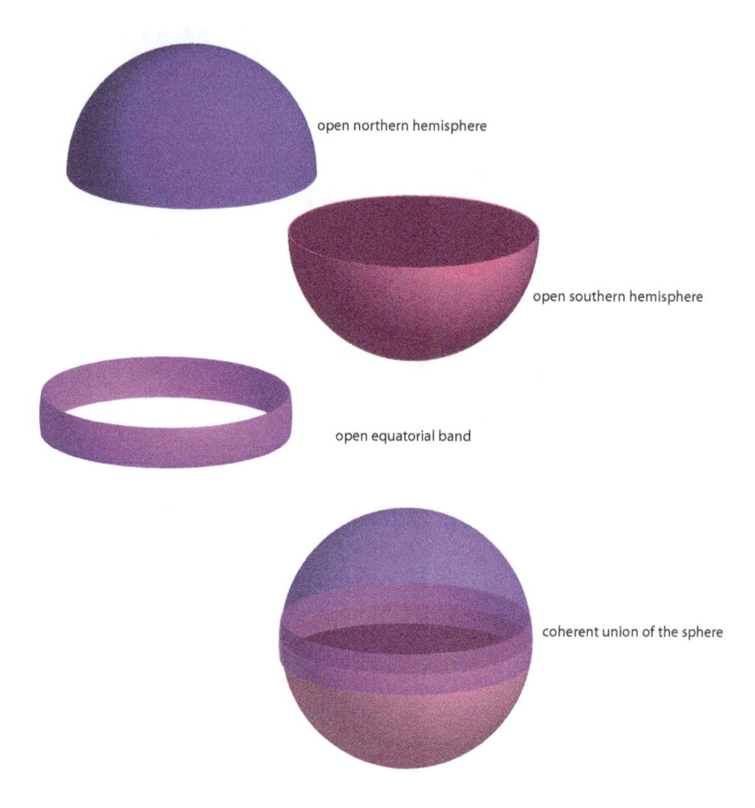

Fig. 6.1 A coherent union of the sphere composed of the union of two open hemispheres and an open equatorial cylinder.

that if $y \notin f(A)$, then $f^{-1}(y) = \emptyset$, so in this case the subset $f^{-1}(y) \cup \{y\}$ reduces to $\{y\}$.

Definition 6.4. Given two topological spaces X and Y, given a closed subset A of X, and given a continuous map $f\colon A \to Y$, the *adjunction space determined by* f (or *attaching space determined by* f), denoted by $X \cup_f Y$, is the quotient space of the disjoint sum $X \sqcup Y$ corresponding to the partition of $X \sqcup Y$ into the subsets of the form $f^{-1}(y) \cup \{y\}$ for all $y \in Y$, and all singleton sets $\{x\}$ for all $x \in X - A$. The map f is called the *adjunction map* (or *attaching map*). See Figure 6.3. Let $\pi\colon X \sqcup Y \to X \cup_f Y$ be the quotient map. The space $X \cup_f Y$ is given the quotient topology induced by π; that is, $Z \subseteq X \cup_f Y$ is open iff $\pi^{-1}(Z)$ is open in $X \sqcup Y$.

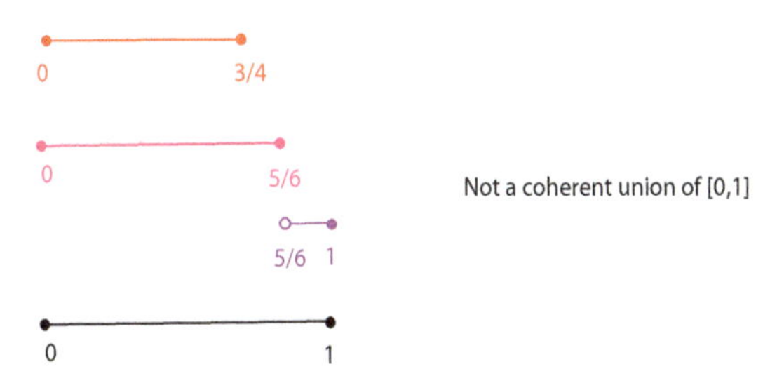

Fig. 6.2 Let $X = [0, 1]$, where $X = [0, 3/4] \cup [0, 5/6] \cup (5/6, 1] = A \cup B \cup C$. This is not coherent union since under the induced topology $(3/4, 5/6]$ is open B and trivially open in A and C, yet $(3/4, 5/6]$ is not open in $[0, 1]$.

Observe that the adjunction map $f\colon A \to Y$ needs not be injective, that is, it could cause some collapsing of parts of A. For example, if $X = D^1$, $A = S^1$, $Y = \{0\}$ and $f\colon A \to Y$ is the constant function that "collapses" S^1 onto $\{0\}$, then the adjunction space $X \cup_f Y$ is homeomorphic to the sphere S^2. See Figure 6.4.

It is easy to show that the quotient map $\pi\colon X \sqcup Y \to X \cup_f Y$ maps Y homeomorphically onto a closed subspace of $X \cup_f Y$.

Definition 6.5. A topological space X is *normal* if the singleton subset $\{x\}$ is closed for all $x \in X$, and if for any two closed disjoint subsets A and B of X there exist two disjoint open subsets U and V of X such that $A \subseteq U$ and $B \subseteq V$.

Since every singleton subset is closed, a normal space is Hausdorff.

The following result is shown in Munkres [Munkres (1984)] (Chapter 4, Theorem 37.2).

Proposition 6.1. *Given X, Y, A, and $f\colon A \to Y$ as in Definition 6.4, if X and Y are normal, then $X \cup_f Y$ is also normal, and in particular Hausdorff.*

A CW complex can be defined intrinsically or by an inductive definition involving the process of attaching cells. We begin with the second method since it is easier to grasp. To simplify matters we begin with the notion of a finite CW complex.

Definition 6.6. A finite *CW complex X of dimension n* is defined inductively as follows:

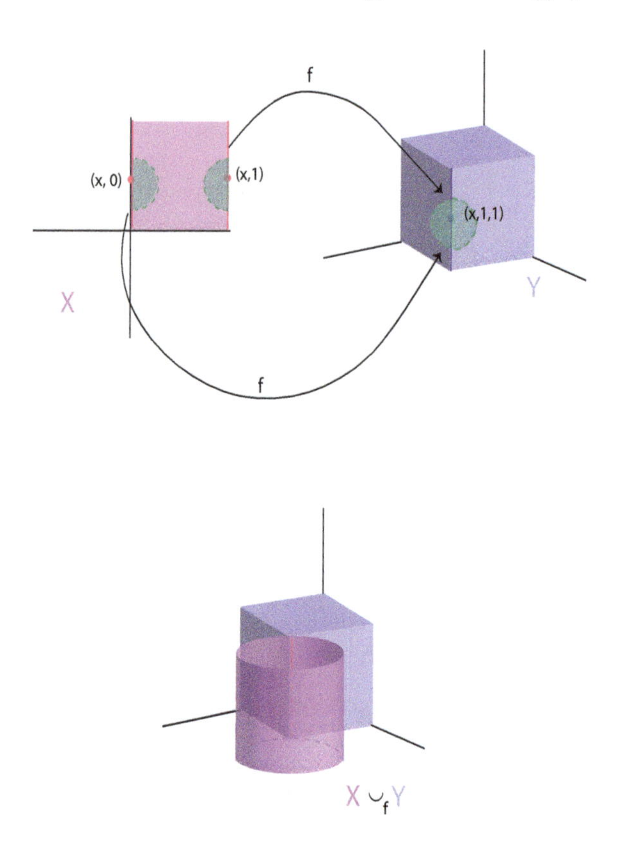

Fig. 6.3 Let X be the unit square in $\mathbb{R}^2$ and Y be the boundary of the unit cube in $\mathbb{R}^3$. Let A be the union of the vertical line segments corresponding to $x = 0$ and $x = 1$. The attaching map $f\colon A \to Y$ is defined via $f(x, 0) = (1, 1, x) = f(x, 1)$. The upper figure shows an open set in $X \cup_f Y$ as defined in Definition 6.4. The lower figure shows the 3-dimensional rendering of the quotient space $X \cup_f Y$.

(1) Let X^0 be a finite set of points (0-cells) with the discrete topology.
(2) If $p < n$ and if X^p has been constructed, let I_{p+1} be a finite (possibly empty) index set, let $\bigsqcup_{i \in I_{p+1}} D_i^{p+1}$ be the disjoint union of closed $(p+1)$-balls, and if we write $S_i^p = \partial D_i^{p+1}$ let $g_{p+1}\colon \bigsqcup_{i \in I_{p+1}} S_i^p \to X^p$ be a continuous map (an *attaching map*). Then X^{p+1} is the adjunction space

$$X^{p+1} = \left(\bigsqcup_{i \in I_{p+1}} D_i^{p+1} \right) \cup_{g_{p+1}} X^p.$$

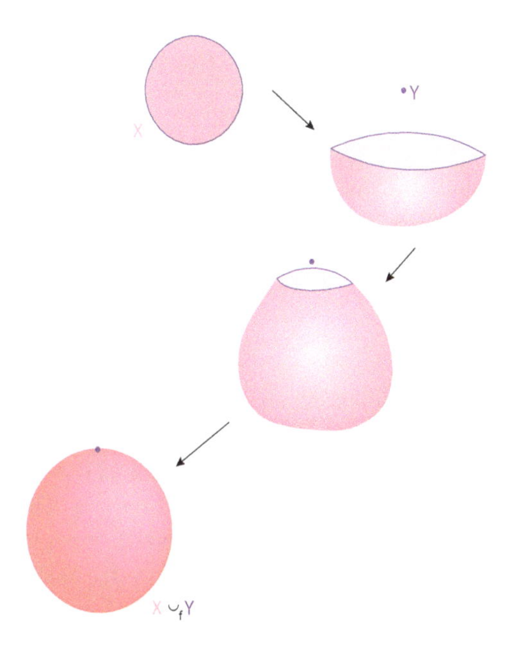

Fig. 6.4 Let X be the unit disk in $\mathbb{R}^2$ and Y be a point. Let A be the circular boundary of X. The attaching map $f\colon A \to Y$ collapses A to a point and wraps the disk into a sphere as depicted by the four stage rendering of $X \cup_f Y$.

Either $n = 0$ and $X = X^0$, or $n \geq 1$ in which case $X^0 \neq \emptyset$ and $I_n \neq \emptyset$, that is, there is some open n-cell, and we let $X = X^n$. The subspace X^p is called the *p-skeleton* of X.

If π_{p+1}^{CW} is the quotient map

$$\pi_{p+1}^{\mathrm{CW}}\colon \left(\coprod_{i \in I_{p+1}} D_i^{p+1}\right) \sqcup X^p \to \left(\coprod_{i \in I_{p+1}} D_i^{p+1}\right) \cup_{g_{p+1}} X^p = X^{p+1},$$

then we write $e_i^{p+1} = \pi_{p+1}^{\mathrm{CW}}(\mathrm{Int}\, D_i^{p+1})$.

It is not hard to see that e_i^{p+1} is an open $(p+1)$-cell (*i.e.* π_{p+1}^{CW} maps $\mathrm{Int}\, D_i^{p+1}$ homeomorphically onto e_i^{p+1}). Furthermore, since π_{p+1}^{CW} maps X^p homeomorphically onto a subspace of X^{p+1}, we can view π_{p+1}^{CW} as the inclusion on X^p and as g_{p+1} on $\coprod_{i \in I_p} D_i^{p+1}$. It follows that the open $(p+1)$-cells e_i^{p+1} are disjoint from all the open cells in X^p. Since π_{p+1}^{CW} is a homeomorphism on each $\mathrm{Int}\, D_i^{p+1}$, we have $e_i^{p+1} \cap e_j^{p+1} = \emptyset$ for all $i \neq j$. It follows by induction that $X = X^n$ is the disjoint union of all the open cells e_i^p for $p = 0, \ldots, n$ and all $i \in I_p$. The topology of the X^p, in particular $X = X^n$, is the quotient topology of an adjunction space, as in Definition 6.4.

Since X^0 is normal, by Proposition 6.1 we conclude that $X = X^n$ is normal, thus Hausdorff. It is also clear that a finite CW complex is compact.

Example 6.1.

(1) A 0-dimensional CW complex is simply a discrete set of points. A 1-dimensional CW complex X consists of 0-cells and 1-cells, where each 1-cell e_i^1 is homeomorphic to the open line segment $(-1, 1)$, whose boundaries are attached to some 0-cells x and y, possibly identical. If we view each 1-cell as a directed edge and each 0-cell as a node (or vertex), then the CW complex X is a (directed) *graph* in which several edges may have the same endpoints and an edge may have identical endpoints (self-loops). See Figure 6.5.

(2) The n-sphere S^n $(n \geq 1)$ is homeomorphic to the CW complex with one 0-cell e^0, one n-cell e^n, and with the attaching map $g_n \colon S^{n-1} \to e^0$, the constant map, with $S^n = X^n$. See Figure 6.4. This is equivalent to viewing S^n as the quotient $D^n/\partial D^n = D^n/S^{n-1}$. When $n = 0$, S^0 is the CW complex consisting of two disjoint 0-cells.

(3) The n-ball D^n $(n \geq 1)$ is homeomorphic to the CW complex X with one 0-cell e^0, one $(n-1)$-cell e^{n-1}, and one n-cell e^n. First, $X^{n-1} = S^{n-1}$ as explained in (2), and then $D^n = X^n$ is obtained using as attaching map the identity map $g_n \colon S^{n-1} \to S^{n-1}$. See Figure 6.6.

(4) The real projective space $\mathbb{RP}^2$ is homeomorphic to the CW complex X with one 0-cell e^0, one 1-cell e^1, and one 2-cell e^2. First, X^1 is obtained by using the constant map $g_1 \colon S^0 \to e^0$ as attaching map, and then X^2 is obtained by using as attaching map the map $g_2 \colon S^1 \to S^1$ that sends S^1 around S^1 twice $(g_2(e^{i\theta}) = e^{2i\theta})$. Observe that $X^1 = \mathbb{RP}^1$. See Figure 6.7. This suggest a recursive method for obtaining a cell structure for $\mathbb{RP}^n$.

(5) The projective space $\mathbb{RP}^n (n \geq 0)$ is homeomorphic to the CW complex X with exactly one p-cell e^p for $p = 0, \ldots, n$; that is, the set of cells $\{e^0, e^1, \ldots, e^n\}$. We have $X^0 = \{e^0\}$, and assuming that $X^{n-1} = \mathbb{RP}^{n-1}$ has been constructed, $X^n = \mathbb{RP}^n$ is obtained by using the quotient map $g_n \colon S^{n-1} \to \mathbb{RP}^{n-1}$ that identifies two antipodal points as attaching map; see Example 4.1.

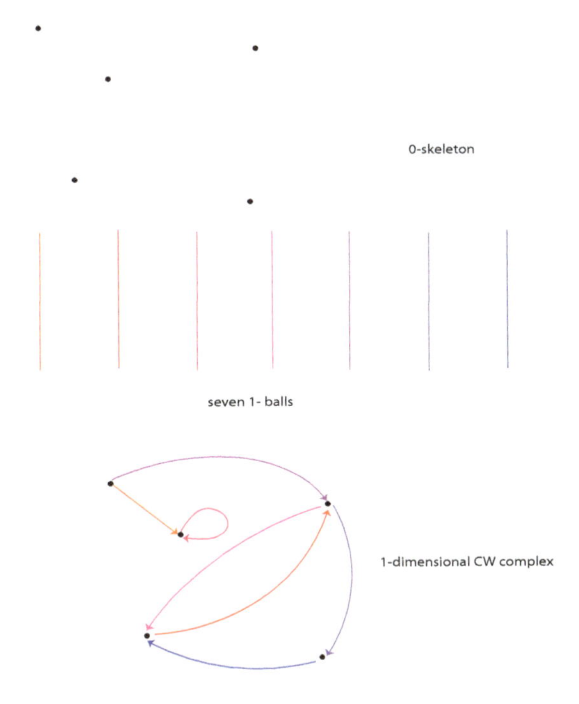

Fig. 6.5 A directed graph is a 1-dimensional CW complex.

(6) The complex projective space $\mathbb{CP}^n (n \geq 0)$ is homeomorphic to the CW complex X with exactly one $2p$-cell e^{2p} for $p = 0, \ldots, n$; that is, the set of cells $\{e^0, e^2, \ldots, e^{2n}\}$. We have $X^0 = \{e^0\}$, and assuming that $X^{2n-2} = \mathbb{CP}^{n-1}$ has been constructed, $X^{2n} = \mathbb{CP}^n$ is obtained by using the quotient map $g_{2n}\colon S^{2n-1} \to \mathbb{CP}^{n-1}$ as attaching map; see Example 4.1.

(7) The 2-torus $T^2 = S^1 \times S^1$ is homeomorphic to the CW complex X with one 0-cell e^0, two 1-cells e_1^1, e_2^1, and one 2-cell e^2. First X^1 is obtained by using the constant map $g_1\colon S^0 \sqcup S^0 \to e^0$ as attaching map. The space X^1 consists of two circles on a torus in $\mathbb{R}^3$ (in orthogonal planes) intersecting in a common point. Then $T^2 = X^2$ is obtained by using the map $g_2\colon S^1 \to X^1$ that "wraps" S^1 around the two circles of X^1, as attaching map; think of the construction of a torus from a square in which opposite sides are glued in two steps. See Figure 6.8.

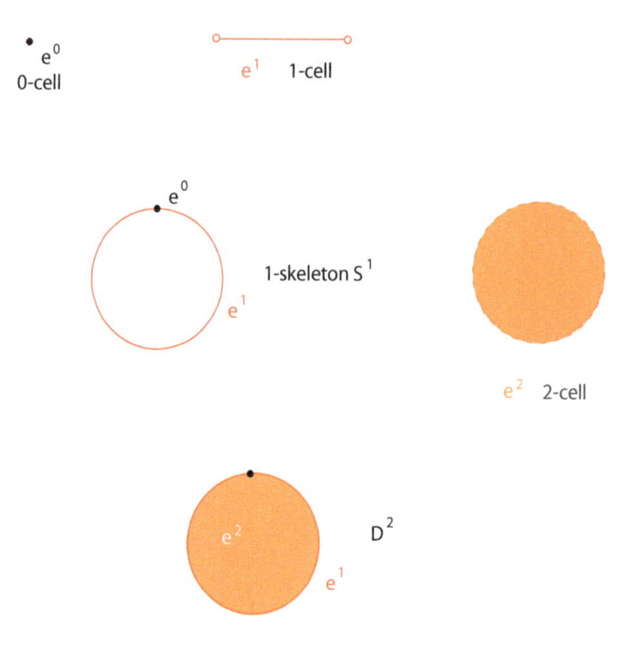

Fig. 6.6 D^2 as a 2-dimensional CW complex.

Remark: Ambitious readers should read Chapter 6 of Milnor and Stasheff [Milnor and Stasheff (1974)], where a cell structure for the Grassmann manifolds is described. This is a generalization of the cell structure for $\mathbb{RP}^n$.

The definition of a CW complex can be generalized by allowing the index sets I_p to be infinite and by allowing the sequence of p-skeleta X^p to be infinite.

Definition 6.7. A *CW complex* X is defined inductively as follows:

(1) Let X^0 be a set of points (0-cells) with the discrete topology. If $X^0 = \emptyset$, then let $X = \emptyset$.
(2) If X^p has been constructed ($p \geq 0$) and if $X^p \neq \emptyset$, let I_{p+1} be a (possibly empty) index set, let $\bigsqcup_{i \in I_{p+1}} D_i^{p+1}$ be the disjoint union of closed $(p+1)$-balls, and if we write $S_i^p = \partial D_i^{p+1}$ let $g_{p+1} \colon \bigsqcup_{i \in I_{p+1}} S_i^p \to X^p$ be

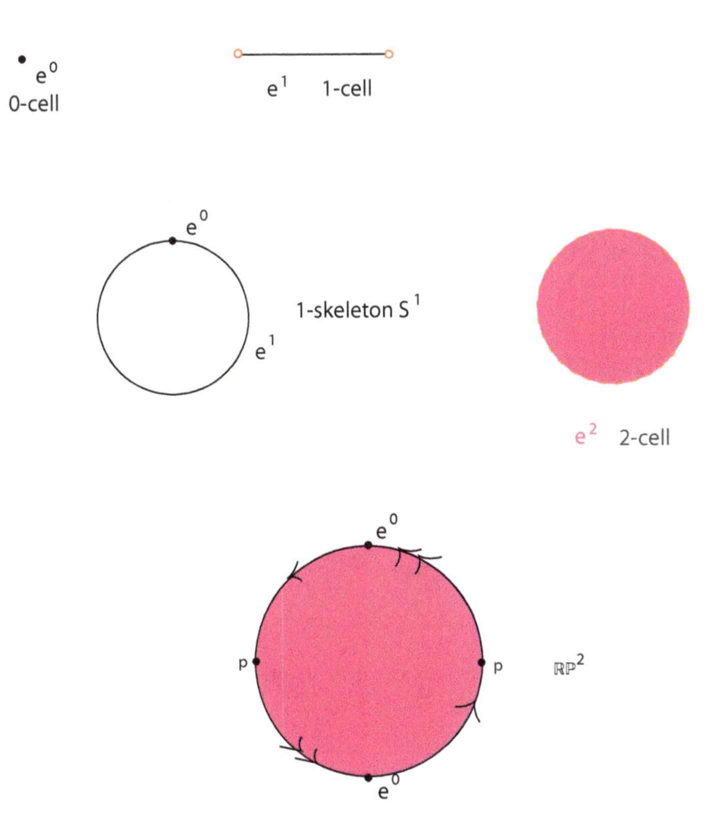

Fig. 6.7 $\mathbb{RP}^2$ as a 2-dimensional CW complex, where antipodal points of e^1 are glued together.

a continuous map (an *attaching map*). Then X^{p+1} is the adjunction space

$$X^{p+1} = \left(\coprod_{i \in I_{p+1}} D_i^{p+1} \right) \cup_{g_{p+1}} X^p.$$

Suppose $X^0 \neq \emptyset$. If there is a smallest $n \geq 0$ such that $I_p = \emptyset$ for all $p \geq n+1$, then we let $X = X^n$ and we say that X has *dimension* n. In this case, note that X^n must have some open n-cell. Otherwise we let $X = \bigcup_{p \geq 0} X^p$, and we give X the topology for which X is the coherent union of the family $(X^p)_{p \geq 0}$; that is, a subset Z of X is open iff $Z \cap X^p$ is open in X^p for all $p \geq 0$. Each subspace X^p is called a *p-skeleton* of X.

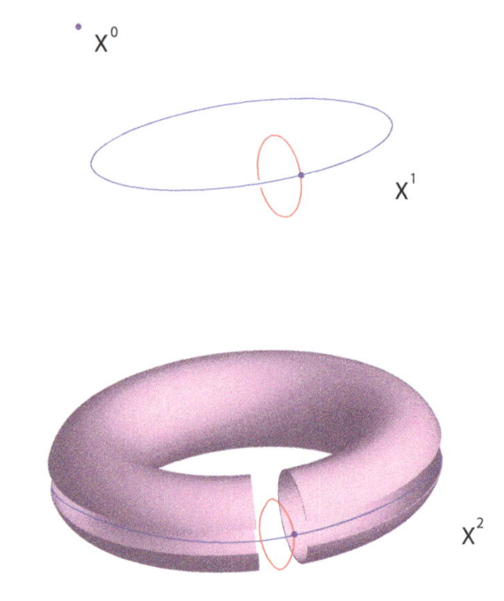

Fig. 6.8 The CW complex construction of the torus T^2.

As before if π^{CW}_{p+1} is the quotient map $\pi^{\mathrm{CW}}_{p+1}\colon \left(\coprod_{i\in I_{p+1}} D^{p+1}_i\right) \sqcup X^p \to X^{p+1}$, then we write

$$e^{p+1}_i = \pi^{\mathrm{CW}}_{p+1}(\mathrm{Int}\, D^{p+1}_i),$$

and it is not hard to see that e^{p+1}_i is an open $(p+1)$-cell (*i.e.* π^{CW}_{p+1} maps $\mathrm{Int}\, D^{p+1}_i$ homeomorphically onto e^{p+1}_i). It follows that X is the disjoint union of the cells e^p_i for all $p \geq 0$ and all $i \in I_p$. The topology of the X^p is the quotient topology of an adjunction space, as in Definition 6.4.

Definition 6.8. For every p-ball D^p_i, the restriction to D^p_i of the composition of the quotient map π^{CW}_p from $\left(\coprod_{i\in I_p} D^p_i\right) \sqcup X^{p-1}$ to X^p with the inclusion $X^p \longrightarrow X$ is a map from D^p_i to X denoted by f_i (or f^p_i if we want to be very precise) and called the *characteristic map* of $e^p_i = \pi^{\mathrm{CW}}_p(\mathrm{Int}\, D^p_i)$.

It is not hard to show that $f_i(D^p_i) = \overline{e^p_i}$, $f_i(S^{p-1}_i) = \dot{e}^p_i$, and f_i is a homeomorphism of $\mathrm{Int}\, D^p_i$ onto e^p_i.

Remark: One should be careful that the terminology "open cell" is slightly misleading. Although an open cell e^p_i is open in X^p, it may *not* be open in

X. Consider the example of the torus T^2 from Example 6.1(7). The open cell $e_1^1 = \pi_1(\text{Int } D_1^1)$ of X^1 is not open in T^2.

Example 6.2. The infinite union $X = \mathbb{RP}^\infty = \bigcup_{n\geq 0} \mathbb{RP}^n$ is an infinite CW complex whose n-skeleton X^n is $\mathbb{RP}^n$. The CW complex $\mathbb{RP}^\infty$ has infinitely many n-cells e^n, one for each dimension.

Similarly, the infinite union $X = \mathbb{CP}^\infty = \bigcup_{n\geq 0} \mathbb{CP}^n$ is an infinite CW complex whose $2n$-skeleta X^{2n} and X^{2n+1} are both $\mathbb{CP}^n$. The CW complex $\mathbb{CP}^\infty$ has infinitely many n-cells e^{2n}, one for each even dimension.

Definition 6.9. A *subcomplex* of a CW complex X is a subspace A of X which is a union of open cells e_i of X such that the closure $\overline{e_i}$ of each open cell e_i in A is also in A.

Note that each p-skeleton X^p is a subcomplex of X. It is easy to show by induction over skeleta that a subcomplex is a closed subspace; see Munkres [Munkres (1984)] (Chapter 4, Section 38, Page 217).

The following proposition states a crucial compactness property of CW complexes.

Proposition 6.2. *If X is a CW complex, then the following properties hold and are all equivalent.*

(1) *If a subspace A of X has no two points in the same open cell, then A is closed and discrete.*
(2) *If a subspace C of X is compact, then C is contained in a finite union of open cells.*
(3) *Each open cell of X is contained in a finite subcomplex of X.*

Proposition 6.2 is proven in Bredon [Bredon (1993)] (Chapter IV, Section 8, Proposition 8.1). As a corollary we have the following result.

Proposition 6.3. *If X is a CW complex, then any compact subset C of X is contained in a finite subcomplex.*

Proof. By Proposition 6.2(2) the compact subset C is contained in a union of a finite number of open cells of X. By Proposition 6.2(3) each of these open cells is contained in a finite subcomplex. But the union of this finite number of finite subcomplexes is a finite subcomplex which contains C. $\square$

It can be shown that a CW complex X is normal; see Munkres [Munkres (1984)] (Chapter 4, Section 38, Theorem 38.2 and Theorem 38.3). In fact, more can be proved.

Proposition 6.4. *Let X be a CW complex as defined in Definition 6.7. Then the following properties hold:*

(1) The space X is the disjoint union of a collection of open cells.
(2) X is Hausdorff.
(3) For each open p-cell e_i of the collection, there is a continuous map $f_i\colon D^p \to X$ that maps $\mathrm{Int}\, D^p$ homeomorphically onto e_i and carries $S^{p-1} = \partial D^p$ into a finite union of open cells e_j^k, each of dimension $k < p$.
(4) A set Z is closed in X iff $Z \cap \overline{e_i}$ is closed in $\overline{e_i}$ for all open cells e_i.

Proposition 6.4 is proven in Hatcher [Hatcher (2002)] (Appendix, Topology of cell complexes, Proposition A2).

Property (3) is what is referred to as "closure-finiteness" by J.H.C. Whitehead. Property (4) expresses the fact that X has the "weak topology." This explains the CW in CW complexes!

It is easy to see that Properties (3) and (4) imply that $f_i(D^p) = \overline{e_i}$ and $f_i(S^{p-1}) = \dot{e}_i$. The map f_i is called a *characteristic map* for the open cell e_i.

The properties of Proposition 6.4 can be taken as the definition of a CW complex. This is what J.H.C. Whitehead did originally, and this is the definition used by Munkres [Munkres (1984)] and Milnor and Stasheff [Milnor and Stasheff (1974)]. Then it can be shown that this alternate definition is equivalent to our previous definition (Definition 6.7). This is proven in Munkres [Munkres (1984)] (Chapter 4, Section 38, Theorem 38.2 and Theorem 38.3).

Since our primary goal is to determine the homology (and cohomology) groups of CW complexes, we will not go into a more detailed study of these spaces. Let us just mention that every CW complex X is normal, paracompact, compactly generated (which means that X is the union of its compact subsets and that a set $A \subseteq X$ is closed in X iff $A \cap C$ is closed in C for every compact subset C of X), and a finite CW complex is an ENR (Euclidean neighborhood retract).

We will also need the fact that a subcomplex A of a CW complex is a deformation retract of a neighborhood of X. The following result is proven in Hatcher [Hatcher (2002)] (Appendix, Proposition A.5).

Proposition 6.5. *For any CW complex X and any subcomplex A of X, there is a neighborhood $N(A)$ of X that deformation retracts onto A. In other words, (X, A) is a good pair.*

In particular, if X is a CW complex, then (X^p, X^{p-1}) is a good pair.

For a more comprehensive exposition of CW complexes we refer the interested reader to Hatcher [Hatcher (2002)] (Appendix, Topology of cell complexes), Bredon [Bredon (1993)] (Chapter IV, Sections 8–14), and Massey [Massey (1991)] (Chapter IX). Rotman [Rotman (1988)] also contains a rather thorough yet elementary treatment.

6.2 Homology of CW Complexes

Given a CW complexes X, it is possible to assign to X a chain complex $S_*^{\mathrm{CW}}(X; R)$ called its cellular chain complex, where

$$S_p^{\mathrm{CW}}(X; R) = H_p(X^p, X^{p-1}; R),$$

the relative pth singular homology group of the pair (X^p, X^{p-1}), where X^p is the p-skeleton of X (by convention $X^{-1} = \emptyset$). The module $H_p(X^p, X^{p-1}; R)$ is a free R-module whose dimension (when finite) is equal to the number of p-cells in X. This means that we can view $H_p(X^p, X^{p-1}; R)$ as the set of formal linear combinations $\sum_i \lambda_i e_i^p$, where $\lambda_i \in R$ and the e_i^p are open p-cells. Furthermore, the homology of the cellular complex agrees with the singular homology. That is, if we write $H_p^{\mathrm{CW}}(X; R) = H_p(S_*^{\mathrm{CW}}(X; R))$, then

$$H_p^{\mathrm{CW}}(X; R) \cong H_p(X; R) \quad \text{for all } p \geq 0,$$

where $H_p(X; R)$ is the pth singular homology module of X. In many practical cases, the number of p-cells is quite small so the cellular complex $S_*^{\mathrm{CW}}(X; R)$ is much more manageable than the singular complex $S_*(X; R)$.

We will need of few properties of the modules $H_k(X^p, X^{p-1}; R)$. By convention, if X is a CW complex we set $X^{-1} = \emptyset$. Then $H_0(X^0, X^{-1}; R) = H_0(X^0; R)$.

Proposition 6.6. *If X is a CW complex, then the following properties hold.*

(a) *We have* $H_k(X^p, X^{p-1}; R) = (0)$ *for* $k \neq p$ *and* $H_p(X^p, X^{p-1}; R)$ *is a free* R*-module with a basis in one-to-one correspondence with the* p*-cells of* X.

(b) $H_k(X^p; R) = (0)$ *for all* $k > p$. *In particular, if* X *has finite dimension* n *then* $H_p(X; R) = (0)$ *for all* $p > n$.

Sketch of proof. To prove (a) we use Proposition 6.5 which says that (X^p, X^{p-1}) is a good pair. By Proposition 4.12

$$H_k(X^p, X^{p-1}; R) \cong H_k(X^p/X^{p-1}, \{\mathrm{pt}\}; R) \cong \widetilde{H}_k(X^p/X^{p-1}; R).$$

Then we use Corollary 2.25 from Hatcher [Hatcher (2002)] (Chapter 2, Section 2.1), the fact that X^p/X^{p-1} is the wedge sum of p-spheres (the disjoint sum of p-spheres glued at the south pole, the *basepoint*), and Proposition 4.18.

To prove (b) first observe that $H_k(X^0; R) = (0)$ for all $k > 0$. Next consider the following piece of the long exact sequence of homology of the pair (X^p, X^{p-1}) (see Theorem 4.9):

$$H_{k+1}(X^p, X^{p-1}; R) \longrightarrow H_k(X^{p-1}; R) \longrightarrow H_k(X^p; R) \longrightarrow H_k(X^p, X^{p-1}; R).$$

If $k \neq p, p-1$, then the first and the fourth groups are zero by (a), so we have isomorphisms

$$H_k(X^p; R) \cong H_k(X^{p-1}; R) \quad k \neq p, p-1.$$

Thus if $k > p$, by induction we get

$$H_k(X^p) \cong H_k(X^0) = (0),$$

proving (b). □

Proposition 6.6(a) implies that we can view $H_p(X^p, X^{p-1}; R)$ as the set of formal linear combinations $\sum_i \lambda_i e_i^p$, where $\lambda_i \in R$ and the e_i^p are open p-cells.

Proposition 6.7. *If* X *is a CW complex, then we have* $H_k(X^p; R) \cong H_k(X; R)$ *for all* $k < p$.

Sketch of proof. Consider the following piece of the long exact sequence of homology of the pair (X^p, X^{p-1}):

$$H_{k+1}(X^{p+1}, X^p; R) \longrightarrow H_k(X^p; R) \longrightarrow H_k(X^{p+1}; R) \longrightarrow H_k(X^{p+1}, X^p; R).$$

If $k < p$ then $k+1 < p+1$ so the first and fourth groups are zero and we have isomorphisms

$$H_k(X^p; R) \cong H_k(X^{p+1}; R) \quad k < p.$$

By induction, if $k < p$ then

$$H_k(X^p; R) \cong H_k(X^{p+m}; R) \quad \text{for all } m \geq 0.$$

If X is finite-dimensional, we are done. Otherwise, following Milnor and Stasheff [Milnor and Stasheff (1974)] (Appendix A, Corollary A.3), we use the fact that

$$H_k(X; R) \cong \varinjlim_{r \geq 0} H_k(X^r; R),$$

because every singular simplex of X is contained in a compact subset, and hence in some X^r. A similar proof is given in Hatcher [Hatcher (2002)] (Chapter 2, Lemma 2.34). $\square$

We now show that we can form a chain complex with the modules $H_p(X^p, X^{p-1}; R)$.

Recall that $S_k(X^p, X^{p-1}; G) = S_k(X^p; G)/S_k(X^{p-1}; G)$, so we have the quotient map $\pi_k\colon S_k(X^p; G) \to S_k(X^p, X^{p-1}; G)$ which yields the map $j_k\colon H_k(X^p; G) \to H_k(X^p, X^{p-1}; G)$. Consider the following pieces of the long exact sequence of homology of the pairs (X^{p+1}, X^p), (X^p, X^{p-1}), and (X^{p-1}, X^{p-2}):

$$H_{p+1}(X^{p+1}, X^p; R) \xrightarrow{\partial_{p+1}} H_p(X^p; R) \longrightarrow H_p(X^{p+1}; R) \longrightarrow$$

$$H_p(X^{p+1}, X^p; R) \longrightarrow \cdots$$

$$H_p(X^{p-1}; R) \longrightarrow H_p(X^p; R) \xrightarrow{j_p} H_p(X^p, X^{p-1}; R) \xrightarrow{\partial_p}$$

$$H_{p-1}(X^{p-1}; R) \longrightarrow \cdots$$

$$H_{p-1}(X^{p-2}; R) \longrightarrow H_{p-1}(X^{p-1}; R) \xrightarrow{j_{p-1}} H_{p-1}(X^{p-1}, X^{p-2}; R) \longrightarrow$$

$$H_{p-2}(X^{p-2}; R) \longrightarrow \cdots.$$

Observe that by Proposition 6.6 the modules showed in red are (0); that is, we have

$$H_p(X^{p+1}, X^p; R) = H_p(X^{p-1}; R) = H_{p-1}(X^{p-2}; R) = (0),$$

and by Proposition 6.7 we have $H_p(X^{p+1}; R) \cong H_p(X; R)$. We form the following diagram

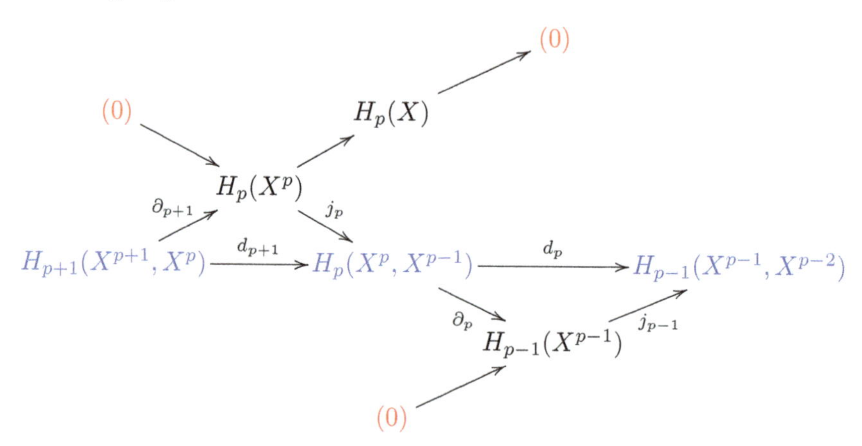

in which for simplicity of notation we omitted the ring R, and where $d_{p+1} = j_p \circ \partial_{p+1}$ and $d_p = j_{p-1} \circ \partial_p$. Since $\partial_p \circ j_p = 0$ (because the sequence on that descending diagonal is exact), we have

$$d_p \circ d_{p+1} = j_{p-1} \circ \partial_p \circ j_p \circ \partial_{p+1} = 0.$$

Therefore, the modules $H_p(X^p, X^{p-1}; R)$ together with the boundary maps $d_p \colon H_p(X^p, X^{p-1}; R) \to H_{p-1}(X^{p-1}, X^{p-2}; R)$ form a chain complex. Recall that we set $X^{-1} = \emptyset$.

Definition 6.10. Given a CW complex X, the *cellular chain complex* $S_*^{\mathrm{CW}}(X; R)$ associated with X is the chain complex where $S_p^{\mathrm{CW}}(X; R) = H_p(X^p, X^{p-1}; R)$ and the boundary maps $d_p \colon H_p(X^p, X^{p-1}; R) \to H_{p-1}(X^{p-1}, X^{p-2}; R)$ are given by $d_p = j_{p-1} \circ \partial_p$ as in the diagram above. We denote the *cellular homology module* $H_p(S_*^{\mathrm{CW}}(X; R))$ of the chain complex $S_*^{\mathrm{CW}}(X; R)$ by $H_p^{\mathrm{CW}}(X; R)$.

The reason for introducing the modules $H_p^{\mathrm{CW}}(X; R)$ is that they are isomorphic to the singular homology modules $H_p(X; R)$, and in practice they are usually much easier to compute.

Theorem 6.8. *Let X be a CW complex. There are isomorphisms*

$$H_p^{\mathrm{CW}}(X; R) \cong H_p(X; R) \quad \text{for all } p \geq 0$$

between the cellular homology modules and the singular homology modules of X.

Proof. Exactness of the left ascending diagonal sequence in the diagram above (and the first isomorphism theorem) shows that

$$H_p(X;R) \cong H_p(X^p;R)/\mathrm{Im}\,\partial_{p+1}.$$

Since j_p is injective, it maps $\mathrm{Im}\,\partial_{p+1}$ isomorphically onto $\mathrm{Im}\,j_p \circ \partial_{p+1} = \mathrm{Im}\,d_{p+1}$ and it maps $H_p(X^p;R)$ isomorphically onto $\mathrm{Im}\,j_p = \mathrm{Ker}\,\partial_p$, so

$$H_p(X;R) \cong \mathrm{Ker}\,\partial_p/\mathrm{Im}\,d_{p+1}.$$

Since j_{p-1} is injective, $\mathrm{Ker}\,\partial_p = \mathrm{Ker}\,d_p$, thus we obtain an isomorphism

$$H_p(X;R) \cong \mathrm{Ker}\,d_p/\mathrm{Im}\,d_{p+1} = H_p^{\mathrm{CW}}(X;R),$$

as claimed. □

Theorem 6.8 has the following immediate corollaries:

(1) If the CW complex X has no p-cells, then $H_p(X;R) = (0)$.
(2) If the CW complex X has k p-cells, then $H_p(X;R)$ is generated by at most k elements.
(3) If the CW complex X has no two of its cells in adjacent dimensions, then $H_p(X;R)$ is a free R-module with a basis in one-to-one correspondence with the p-cells in X. This is because whenever there is some p-cell, then there are no $(p-1)$-cells and no $(p+1)$-cells so $X^{p-2} = X^{p-1}$ and $X^p = X^{p+1}$, which implies that $H_{p-1}(X^{p-1}, X^{p-2}) = H_{p+1}(X^{p+1}, X^p) = (0)$ and then we have the piece of the cellular chain complex

$$H_{p+1}(X^{p+1}, X^p) = (0) \xrightarrow{0} H_p(X^p, X^{p-1}) \xrightarrow{d_p} (0) = H_{p-1}(X^{p-1}, X^{p-2}),$$

and $H_p(X;R) = \mathrm{Ker}\,d_p = H_p(X^p, X^{p-1})$.

Example 6.3. Property (3) immediately yields the homology of $\mathbb{CP}^n$. Indeed, recall from Example 6.1 that as a CW complex $\mathbb{CP}^n$ has $n+1$ cells

$$e^0, e^2, e^4, \ldots, e^{2n}.$$

Therefore, we get

$$H_p(\mathbb{CP}^n;R) = \begin{cases} R & \text{for } p = 0, 2, 4, \ldots, 2n \\ (0) & \text{otherwise.} \end{cases}$$

We also get the homology of $\mathbb{CP}^\infty$:

$$H_p(\mathbb{CP}^\infty;R) = \begin{cases} R & \text{for } p \text{ even} \\ (0) & \text{otherwise.} \end{cases}$$

Computing the homology of $\mathbb{RP}^n$ is more difficult. The problem is to figure out what are the boundary maps d_p.

Generally, in order to be able to compute the cellular homology groups, we need a method to "compute" the boundary maps d_p. This can indeed be done in principle, and often in practice although this can be tricky, using the notion of degree of a map of the sphere to itself. To simplify matters assume that $R = \mathbb{Z}$, although any abelian group G will do.

Definition 6.11. Let $f\colon S^n \to S^n$ be a continuous map. We have the homomorphism $f_*\colon \widetilde{H}_n(S^n;\mathbb{Z}) \to \widetilde{H}_n(S^n;\mathbb{Z})$, and since $\widetilde{H}_n(S^n;\mathbb{Z}) \cong \mathbb{Z}$ (see Proposition 4.18), the homomorphism f_* must be of the form $f_*(\alpha) = d\alpha$ for some $d \in \mathbb{Z}$. The integer d is called the *degree* of f and is denoted by $\deg f$.

The degree is an important invariant of a map $f\colon S^n \to S^n$. Intuitively, the degree $d = \deg f$ measures how many times f wraps around S^n (and preserves or reverses direction). For example, it can be shown that the degree of the antipodal map $-\mathbb{1}\colon S^n \to S^n$ given by $-\mathbb{1}(x) = -x$ is $(-1)^{n+1}$.

Our intention is not to discuss degree theory, but simply to point out that this notion can be used to determine the boundary maps d_n. Detailed expositions about degrees of maps can be found in Hatcher [Hatcher (2002)] (Chapter 2, Section 2.2), Bredon [Bredon (1993)] (Chapter IV, Sections 6 and 7), and Rotman [Rotman (1988)] (Chapter 6).

To compute d_p, for every open p-cell $e_i^p \in X$ considered as a chain in $H_p(X^p, X^{p-1}; R)$ and for any open $(p-1)$-cell $e_j^{p-1} \in X$ considered as a chain in $H_{p-1}(X^{p-1}, X^{p-2}; R)$, we define a map $f_{ij}\colon S^{p-1} \to S^{p-1}$ as follows:

(1) Let $q\colon X^{p-1} \to X^{p-1}/X^{p-2}$ be the quotient map.
(2) Recall that X^{p-1}/X^{p-2} is homeomorphic to the wedge sum of $(p-1)$-spheres S^{p-1}, one for each $j \in I_{p-1}$ (this is the disjoint sum of $(p-1)$-spheres with their south pole identified). See Figure 6.9. Let $q_j\colon X^{p-1}/X^{p-2} \to S^{p-1}$ be the projection onto the jth sphere. It is the map that collapses all the other spheres in the wedge sum except the jth one onto a point (the south pole). Then we let
$$f_{ij} = q_j \circ q \circ f_i|S^{p-1},$$
where $f_i\colon D_i^p \to X$ is the characteristic map of the cell e_i^p and $f_i|S^{p-1}$ is the restriction of f_i to S^{p-1}.

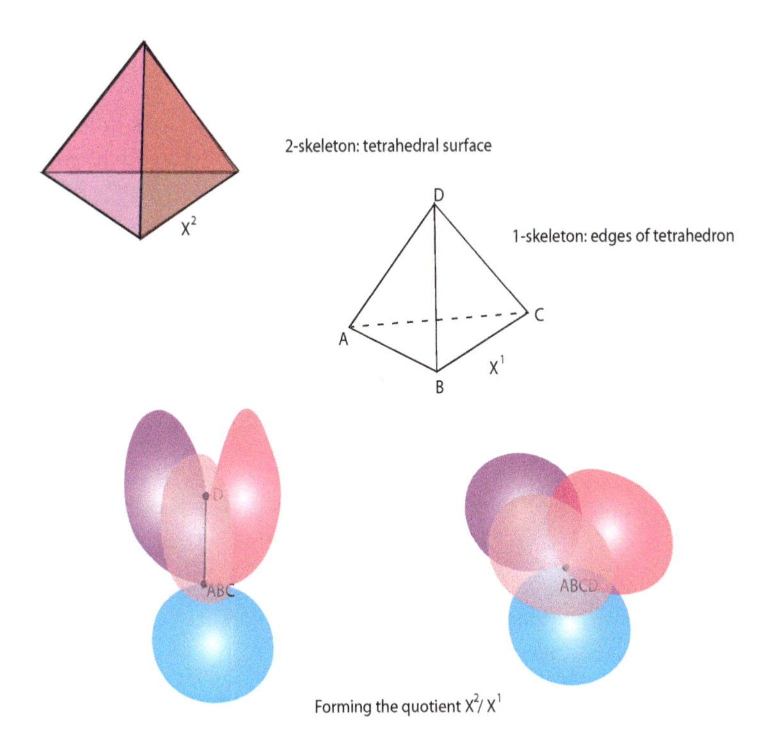

Fig. 6.9 Let X^3 be the solid tetrahedron. Then X^2 is the surface comprised of four 2-cells while X^1 is the union of the edges. If we collapse all the edges to a single point, we obtain four spheres joined to a single point, one sphere for each face of the tetrahedron.

The following proposition is proven in Hatcher [Hatcher (2002)] (Chapter 2, Section 2.2, after Theorem 2.35) and in Bredon [Bredon (1993)] (Chapter IV, Section 10, Theorem 10.3).

Proposition 6.9. *Let X be a CW complex. Then the boundary map $d_p\colon H_p(X^p, X^{p-1}; \mathbb{Z}) \to H_{p-1}(X^{p-1}, X^{p-2}; \mathbb{Z})$ of the cellular complex $S_*^{\mathrm{CW}}(X; \mathbb{Z})$ associated with X is given by*

$$d_p(e_i^p) = \sum_j d_{ij} e_j^{p-1}$$

where $d_{ij} = \deg f_{ij}$ is the degree of the map $f_{ij}\colon S^{p-1} \to S^{p-1}$ defined above as the composition $f_{ij} = q_j \circ q \circ f_i | S^{p-1}$.

The sum in Proposition 6.9 is finite because f_i maps S^{p-1} into a the union of a finite number of cells of dimension at most $p-1$ (by Proposition 6.4(3)). The degrees d_{ij} are often called *incidence numbers*.

The boundary map $d_1 \colon H_1(X^1, X^0; \mathbb{Z}) \to H_0(X^0; \mathbb{Z})$ is much easier to compute than it appears. Recall that X^1 is a graph in which every 1-cell e_i (an edge) is attached to some 0-cells (nodes) x and y, with x attached to -1 and y attached to $+1$ (x and y may be identical). Then it is not hard to show that

$$d_1(e) = y - x.$$

Details of this computation are given in Bredon [Bredon (1993)] (Chapter IV, Section 10).

As an illustration of Proposition 6.9 we can compute the homology groups of $\mathbb{RP}^n$.

Example 6.4. Recall that as a CW complex $\mathbb{RP}^n$ has a cell structure with $n+1$ cells

$$e^0, e^1, e^2, \ldots, e^n.$$

It follows that the cellular cell complex is of the form

$$0 \longrightarrow \mathbb{Z} \xrightarrow{d_n} \mathbb{Z} \xrightarrow{d_{n-1}} \cdots \longrightarrow \mathbb{Z} \xrightarrow{d_3} \mathbb{Z} \xrightarrow{d_2} \mathbb{Z} \xrightarrow{d_1} \mathbb{Z} \longrightarrow 0.$$

In this case there is only one cell of dimension $k-1$ so $q_1 = \mathrm{id}$ and we just have to find the degree of the map $q \circ f_k | S^{k-1}$ (from S^{k-1} to itself). This map is a homeomorphism when restricted to the two components of $S^{k-1} - S^{k-2}$, and these two homeomorphisms are obtained from each other by precomposing with the antipodal map of S^{k-1}, which has degree $(-1)^k$. Then one finds that the degree of the map $q \circ f_k | S^{k-1}$ is $1 + (-1)^k$; see Hatcher [Hatcher (2002)] (Chapter 2, Example 2.42). It follows that d_k is either 0 or multiplication by 2, according to the parity of k. Thus if n is even we have the chain complex

$$0 \longrightarrow \mathbb{Z} \xrightarrow{2} \mathbb{Z} \xrightarrow{0} \cdots \xrightarrow{2} \mathbb{Z} \xrightarrow{0} \mathbb{Z} \xrightarrow{2} \mathbb{Z} \xrightarrow{0} \mathbb{Z} \longrightarrow 0,$$

and if n is odd we have the chain complex

$$0 \longrightarrow \mathbb{Z} \xrightarrow{0} \mathbb{Z} \xrightarrow{2} \cdots \xrightarrow{2} \mathbb{Z} \xrightarrow{0} \mathbb{Z} \xrightarrow{2} \mathbb{Z} \xrightarrow{0} \mathbb{Z} \longrightarrow 0.$$

From this we get

$$H_p(\mathbb{RP}^n; \mathbb{Z}) = \begin{cases} \mathbb{Z} & \text{for } p = 0 \text{ and for } p = n \text{ odd} \\ \mathbb{Z}/2\mathbb{Z} & \text{for } p \text{ odd}, 0 < p < n \\ (0) & \text{otherwise,} \end{cases}$$

as stated in Section 4.6.

Similarly we find that the homology of $\mathbb{RP}^\infty$ is given by

$$H_p(\mathbb{RP}^\infty; \mathbb{Z}) = \begin{cases} \mathbb{Z} & \text{for } p = 0 \\ \mathbb{Z}/2\mathbb{Z} & \text{for } p \text{ odd} \\ (0) & \text{otherwise.} \end{cases}$$

Other examples are given in Hatcher [Hatcher (2002)] (Chapter 2, Section 2.2). A slightly different approach to incidence numbers is presented in Massey [Massey (1991)] (Chapter IX, Sections 5–7). Massey shows that for special types of CW complexes called *regular complexes* there is a procedure for computing the incidence numbers (see Massey [Massey (1991)], Chapter IX, Section 7).

The generalization of cellular homology to coefficients in an R-module G is immediate. We define the R-modules $S_p^{\mathrm{CW}}(X; G)$ by

$$S_p^{\mathrm{CW}}(X; G) = H_p(X^p, X^{p-1}; G),$$

where as before we set $X^{-1} = \emptyset$. The only change in Proposition 6.6 is that

$$H_p(X^p, X^{p-1}; G) \cong \bigoplus_{e_i^p \mid i \in I_p} G$$

is the direct sum of copies of G, one for each open p-cell of X. This means that we can view $H_p(X^p, X^{p-1}; R)$ as the set of formal "vector-valued" linear combinations $\sum_i e_i^p g_i$, where $g_i \in G$ and the e_i^p are open p-cells. Then Proposition 6.7 goes through, the boundary maps are defined as before and we get the following theorem.

Theorem 6.10. *Let X be a CW complex. For any R-module G there are isomorphisms*

$$H_p^{\mathrm{CW}}(X; G) \cong H_p(X; G) \quad \textit{for all } p \geq 0$$

between the cellular homology modules and the singular homology modules of X.

6.3 The Euler–Poincaré Characteristic of a CW Complex

In this section we generalize the Euler–Poincaré formula obtained for simplicial complexes in Section 5.5 to CW complexes. Let us assume that our ring R is $R = \mathbb{Z}$ and that $G = \mathbb{Z}$. In this case we abbreviate $H_p(X; \mathbb{Z})$ as $H_p(X)$ (since cellular homology agrees with singular homology we may assume that we are using singular homology). We know that if X is a finite CW complex then its homology groups $H_p(X;\mathbb{Z})$ are finitely generated abelian groups. More generally we have the following definition.

Definition 6.12. Let X be a topological space. We say that X is of *finite type* if $H_p(X)$ if a finitely generated abelian group for all $p \geq 0$, and X is of *bounded finite type* if it is of finite type and $H_p(X) = 0$ for all but a finite number of indices p.

We can now define a famous invariant of a space.

Definition 6.13. If X is a space of bounded finite type, then its *Euler–Poincaré characteristic* $\chi(X)$ is defined as

$$\chi(X) = \sum_p (-1)^p \operatorname{rank} H_p(X).$$

Since X is of finite bounded type the above sum contains only finitely many nonzero terms. The natural number $\operatorname{rank} H_p(X) = \operatorname{rank} H_p(X;\mathbb{Z})$ is called the *pth Betti number of* X and is denoted by b_p.

If X is a finite CW complex of dimension n, then each p-skeleton has a finite number of p-cells, say a_p. Remarkably $\chi(X) = \sum_{p=0}^{n}(-1)^p a_p$, a formula generalizing Euler's formula in the case of a convex polyhedron. We can now prove the following beautiful result generalizing Theorem 5.15 to CW complexes.

Theorem 6.11. *(Euler–Poincaré) Let X be a finite CW complex of dimension n and let a_p be the number of p-cells in X. We have*

$$\chi(X) = \sum_p (-1)^p \operatorname{rank} H_p(X) = \sum_{p=0}^{n} (-1)^p a_p.$$

Proof. As usual let $B_p = \operatorname{Im} d_{p+1} \subseteq S_p^{\mathrm{CW}}(X)$ be the group of p-boundaries and let $Z_p = \operatorname{Ker} d_p \subseteq S_p^{\mathrm{CW}}(X)$ be the group of p-cycles. By definition $H_p^{\mathrm{CW}}(X) = Z_p/B_p$, by Theorem 6.8 we have $H_p^{\mathrm{CW}}(X) \cong H_p(X)$, and $S_p^{\mathrm{CW}}(X)$ is a free abelian group of rank a_p (the number of p-cells). Observe that $B_n = B_{-1} = (0)$. We have the exact sequence

$$0 \longrightarrow Z_p \xrightarrow{\ i\ } S_p^{\mathrm{CW}}(X) \xrightarrow{\ d_p\ } B_{p-1} \longrightarrow 0$$

which (by Proposition 5.14) shows that

$$a_p = \mathrm{rank}(S_p^{\mathrm{CW}}(X)) = \mathrm{rank}(Z_p) + \mathrm{rank}(B_{p-1}), \tag{$*$}$$

and the exact sequence

$$0 \longrightarrow B_p \longrightarrow Z_p \longrightarrow H_p(X) \longrightarrow 0$$

which (by Proposition 5.14) shows that

$$\mathrm{rank}(Z_p) = \mathrm{rank}(B_p) + \mathrm{rank}(H_p(X)). \tag{$**$}$$

From Equation ($**$) we obtain

$$\sum_p (-1)^p (\mathrm{rank}(B_p) + \mathrm{rank}(H_p(X))) = \sum_p (-1)^p \,\mathrm{rank}(Z_p),$$

and from Equation ($*$) we obtain

$$\sum_p (-1)^p \,\mathrm{rank}(Z_p) = \sum_p (-1)^p (a_p - \mathrm{rank}(B_{p-1})),$$

so we obtain

$$\begin{aligned}\sum_p (-1)^p \,\mathrm{rank}(B_p) &+ \sum_p (-1)^p \,\mathrm{rank}(H_p(X)) \\ &= \sum_p (-1)^p a_p + \sum_p (-1)^{p-1} \,\mathrm{rank}(B_{p-1}).\end{aligned}$$

The sums involving the B_* cancel out because $B_n = B_{-1} = (0)$, and we obtain

$$\sum_p (-1)^p a_p = \sum_p (-1)^p \,\mathrm{rank}(H_p(X))) = \chi(X),$$

as claimed. ◻

Theorem 6.11 proves that the number $\sum_{p=0}^{n} (-1)^p a_p$ is the same for *all* cell structures (of CW complexes) defining a given space X. It is a topological invariant.

Example 6.5. For example, if $X = S^2$, we know that as a CW complex S^2 has two cells e^0 and e^2, so we get

$$\chi(S^2) = 1 + (-1)^2 \times 1 = 2.$$

As a consequence, if X is any CW complex homeomorphic to S^2 with V 0-cells, E 1-cells and F 2-cells, we must have

$$F - E + V = 2,$$

a famous equation due to Euler (for convex polyhedra in $\mathbb{R}^3$).

Example 6.6. More generally, since the n-sphere S^n has a structure with one 0-cell and one n-cell, we see that

$$\chi(S^n) = 1 + (-1)^n.$$

This is the Euler–Poincaré characteristic of any convex polytope in $\mathbb{R}^{n+1}$, a formula proven by Poincaré.

Example 6.7. For the real projective plane $\mathbb{RP}^2$ we have a CW cell structure with three cells e^0, e^1, e^2, so we get

$$\chi(\mathbb{RP}^2) = 1.$$

In general

$$\chi(\mathbb{RP}^{2n}) = 1 \quad \text{and} \quad \chi(\mathbb{RP}^{2n+1}) = 0.$$

Example 6.8. For the torus T^2, we have a CW cell structure with four cells e^0, e_1^1, e_2^1, e^2, so we get

$$\chi(T^2) = 0.$$

More generally, since the homology groups of the n-torus T^n are given by

$$H_p(T^n) = \mathbb{Z}^{\binom{n}{p}},$$

using the fact that $0 = (1-1)^n = \sum_{p=0}^n (-1)^p \binom{n}{p}$, we have

$$\chi(T^n) = \sum_{p=0}^{n} (-1)^p \binom{n}{p} = 0.$$

Definition 6.14. If R is any ring and if X is a space of bounded finite type, then its *Euler–Poincaré characteristic* $\chi_R(X)$ is defined as

$$\chi_R(X) = \sum_p (-1)^p \operatorname{rank} H_p(X; R),$$

where $\operatorname{rank} H_p(X; R)$ is the rank of R-module $H_p(X; R)$.

Since Proposition 5.14 actually holds for finitely generated modules over an integral domain R (see Proposition 11.13), and since the rest of the proof of Theorem 6.11 does not depend on the ring R, we have the following slight generalization of Theorem 6.11.

Theorem 6.12. *(Euler–Poincaré) Let X be a finite CW complex of dimension n and let a_p be the number of p-cells in X. For any integral domain R, we have*

$$\chi_R(X) = \sum_p (-1)^p \operatorname{rank} H_p(X; R) = \sum_{p=0}^{n} (-1)^p a_p.$$

Thus, for finite CW complexes, the Euler–Poincaré characteristic

$$\chi_R(X) = \sum_p (-1)^p \operatorname{rank} H_p(X; R)$$

is independent of the ring R, as long as it is an integral domain. This fact is also noted in Greenberg and Harper in the special case where R is a PID; see [Greenberg and Harper (1981)] (Chapter 20, Remark 20.19).

We also have the following proposition showing that for any space X of bounded finite type, the Euler–Poincaré characteristic $\chi_R(X) = \sum_p (-1)^p$ rank $H_p(X; R)$ is independent the ring R, provided that it is a PID.

Proposition 6.13. *Let X be any space of bounded finite type and let R be any PID. Then we have*

$$\chi_R(X) = \sum_p (-1)^p \operatorname{rank} H_p(X; R) = \chi(X) = \sum_p (-1)^p \operatorname{rank} H_p(X; \mathbb{Z}).$$

Proof. We use the universal coefficient theorem for homology (Theorem 12.1) and some properties of $\operatorname{Tor}_1^{\mathbb{Z}}$ stated after Theorem 12.5, including the following facts:

$$\begin{aligned} \operatorname{Tor}_1^{\mathbb{Z}}(\mathbb{Z}/m\mathbb{Z}, A) &\cong \operatorname{Ker}(A \stackrel{m}{\longrightarrow} A) \\ \mathbb{Z}/m\mathbb{Z} \otimes_{\mathbb{Z}} A &\cong A/mA \\ \operatorname{Tor}_1^{\mathbb{Z}}(\mathbb{Z}, A) &\cong (0), \end{aligned}$$

where A is any abelian group and the map $A \stackrel{m}{\longrightarrow} A$ is multiplication by m. Also, the Tor^R functor is defined in Example 11.1. Recall that the homology groups are finitely generated abelian groups of the form

$$H_p(X; \mathbb{Z}) = \mathbb{Z}^k \oplus \mathbb{Z}/m_1\mathbb{Z} \oplus \cdots \oplus \mathbb{Z}/m_q\mathbb{Z},$$

with $k \geq 0$ and $m_1, \ldots, m_q \geq 2$. Since (Theorem 12.1)

$$H_p(X; R) \cong (H_p(X; \mathbb{Z}) \otimes_{\mathbb{Z}} R) \oplus \operatorname{Tor}_1^{\mathbb{Z}}(H_{p-1}(X; \mathbb{Z}), R),$$

the term $\mathbb{Z}^k$ in $H_p(X; \mathbb{Z})$ after being tensored with R yields the term R^k in $H_p(X; R)$, and every term $\mathbb{Z}/m\mathbb{Z}$ in $H_p(X; \mathbb{Z})$ after being tensored with R yields the term $\mathbb{Z}/m\mathbb{Z} \otimes_{\mathbb{Z}} R \cong R/mR$ in $H_p(X; R)$. Since

$$H_{p+1}(X; R) \cong (H_{p+1}(X; \mathbb{Z}) \otimes_{\mathbb{Z}} R) \oplus \operatorname{Tor}_1^{\mathbb{Z}}(H_p(X; \mathbb{Z}), R),$$

every term $\mathbb{Z}/m\mathbb{Z}$ in $H_p(X; \mathbb{Z})$ yields the term $\operatorname{Tor}_1^{\mathbb{Z}}(\mathbb{Z}/m\mathbb{Z}, R) \cong \operatorname{Ker}(R \stackrel{m}{\longrightarrow} R)$ in $H_{p+1}(X; R)$. Since R is a PID, we have $\operatorname{Ker}(R \stackrel{m}{\longrightarrow} R) = sR$ for some natural number s, so we have the exact sequence

$$0 \longrightarrow sR \stackrel{i}{\longrightarrow} R \stackrel{m}{\longrightarrow} mR \longrightarrow 0,$$

and since R is a PID it is an integral domain so the module mR is free over R and the above sequence splits, which implies that

$$R \cong sR \oplus mR,$$

and thus

$$R/mR \cong sR.$$

Either $sR \ncong R$, in which case $R/mR \cong sR$ is a torsion term that does not contribute to the sum $\sum_p (-1)^p \operatorname{rank} H_p(X; R)$, or $R/mR \cong sR \cong R$, in which case the contributions of the term $\mathbb{Z}/m\mathbb{Z} \otimes_{\mathbb{Z}} R \cong R$ in $H_p(X; R)$ and of the term $\operatorname{Tor}_1^{\mathbb{Z}}(\mathbb{Z}/m\mathbb{Z}, R) \cong R$ in $H_{p+1}(X; R)$ to the sum $\sum_p (-1)^p \operatorname{rank} H_p(X; R)$ cancel out since they have the signs $(-1)^p$ and $(-1)^{p+1}$, which proves that

$$\sum_p (-1)^p \operatorname{rank} H_p(X; R) = \sum_p (-1)^p \operatorname{rank} H_p(X; \mathbb{Z}),$$

as claimed. Properties of $\operatorname{Tor}_1^{\mathbb{Z}}$ stated just after Theorem 12.5 are heavily used. We leave the details as an exercise. ☐

Proposition 6.13 justifies using the ring $\mathbb{Z}$ in the definition of the Euler–Poincaré characteristic. This remark is also made in Greenberg and Harper; see [Greenberg and Harper (1981)] (Chapter 20, Remark 20.19).

In the next section we take a quick look at cellular cohomology.

6.4 Cohomology of CW Complexes

Recall that by Definition 4.30 that the cochain groups $S^*(X, A; G)$ of a pair (X, A) with coefficients in an R-module G are given by $S^p(X, A; G) = \operatorname{Hom}_R(S_p(X, A; R), G)$, so that the cohomology modules $H^p(X, A; G)$ are given by

$$H^p(X, A; G) = H^p(\operatorname{Hom}_R(S_p(X, A; R), G)).$$

If we specialize X and A to X^p and X^{p-1}, where X is a CW complex, we obtain

$$H^k(X^p, X^{p-1}; G) = H^k(\operatorname{Hom}_R(S_k(X^p, X^{p-1}; R), G).$$

By Proposition 6.6(a), $H_p(X^p, X^{p-1}; R)$ is a free R-module and we have $H_k(X^p, X^{p-1}; R) = (0)$ if $k \neq p$, so this suggests using the universal coefficient theorem for cohomology to compute $H^k(X^p, X^{p-1}; G)$. In order to be able to do this we assume that R is a PID, and we let G be any R-module.

By the version of the universal coefficient theorem for cohomology given by Proposition 12.8 (with $X^{-1} = \emptyset$ as before) with $C = S_*(X^p, X^{p-1}; R)$, a chain complex of free R-modules, we have

$$\begin{aligned} H^p(X^p, X^{p-1}; G) &= H^k(\mathrm{Hom}_R(S_*(X^p, X^{p-1}; R), G)) \\ &\cong \mathrm{Hom}_R(H_k(S_*(X^p, X^{p-1}R), G) \\ &= \mathrm{Hom}_R(H_k(X^p, X^{p-1}; R), G), \end{aligned}$$

so we obtain

$$\begin{aligned} &H^p(X^p, X^{p-1}; G) \cong \mathrm{Hom}_R(H_p(X^p, X^{p-1}; R), G) \\ &H^k(X^p, X^{p-1}; G) = (0) \qquad k \neq p. \end{aligned}$$

Proposition 6.14. *If X is a CW complex, then the following properties hold.*

(a) We have $H^k(X^p, X^{p-1}; G) = (0)$ for all $k \neq p$, and
$H^p(X^p, X^{p-1}; G) \cong \mathrm{Hom}_R(H_p(X^p, X^{p-1}; R), G)$.
(b) We have $H^k(X^p; G) \cong (0)$ for all $k > p$.
(c) We have $H^k(X^p; G) \cong H^k(X; G)$ for all $k < p$.

Proof. (a) has already been proven.

(b) We have the following piece of the long exact sequence of cohomology (see Theorem 4.36) for the pair (X^p, X^{p-1}):

$$H^k(X^p, X^{p-1}; G) \to H^k(X^p; G) \to H^k(X^{p-1}; G) \to H^{k+1}(X^p, X^{p-1}; G),$$

and if $k \neq p-1, p$ we know that $H^k(X^p, X^{p-1}; G) = H^{k+1}(X^p, X^{p-1}; G) \cong (0)$, so we have isomorphisms

$$H^k(X^p; G) \cong H^k(X^{p-1}; G) \quad \text{for all } k \neq p-1, p.$$

If we assume that $k > p$, then by induction on p we get

$$H^k(X^p; G) \cong H^k(X^0; G) \cong (0).$$

(c) To prove (c) we will use the fact that $H_k(X, X^p; R) = (0)$ for all $k \leq p$. This is proven in Hatcher [Hatcher (2002)] (Chapter 2, Lemma 2.34) using a construction known as the "mapping telescope." In Milnor and Stasheff [Milnor and Stasheff (1974)] (Page 262) it is shown that $H_k(X, X^p; R) \cong H_k(X^{p+1}, X^p; R)$, and since $H_k(X^{p+1}, X^p; R) = (0)$ for all $k \neq p+1$ we conclude that $H_k(X, X^p; R) = (0)$ for all $k \leq p$.

By the universal coefficient theorem for cohomology (Proposition 12.8) we deduce that

$$H^k(X, X^p; G) = (0) \quad \text{for all } k \leq p.$$

Consider the following piece of the long exact sequence of cohomology of the pair (X, X^p):

$$H^k(X, X^p; G) \longrightarrow H^k(X; G) \longrightarrow H^k(X^p; G) \longrightarrow H^{k+1}(X, X^p; G).$$

If $k < p$ then $k + 1 \leq p$ and we know that $H^k(X, X^p; G) = H^{k+1}(X, X^p; G) = (0)$, so we get isomorphisms

$$H^k(X; G) \cong H^k(X^p; G) \quad \text{for all } k < p,$$

as claimed. □

In particular, Proposition 6.14 implies that $H^p(X; G) \cong H^p(X^{p+1}; G)$.

Recall that $S_k(X^p, X^{p-1}; G) = S_k(X^p; G)/S_k(X^{p-1}; G)$, so we have the quotient map $\pi_k \colon S_k(X^p; G) \to S_k(X^p, X^{p-1}; G)$ which yields the map $j^k \colon H^k(X^p, X^{p-1}; G) \to H^k(X^p; G)$. Consider the following pieces of the long exact sequences of cohomology for the pairs (X^{p-1}, X^{p-2}), (X^p, X^{p-1}), and (X^{p+1}, X^p) (see Theorem 4.36):

$$H^{p-2}(X^{p-2}; G) \longrightarrow H^{p-1}(X^{p-1}, X^{p-2}; G) \xrightarrow{j^{p-1}} H^{p-1}(X^{p-1}; G) \to$$

$$H^{p-1}(X^{p-2}; G) \longrightarrow \cdots$$

$$H^{p-1}(X^{p-1}; G) \xrightarrow{\delta^{p-1}} H^p(X^p, X^{p-1}; G) \xrightarrow{j^p} H^p(X^p; G) \longrightarrow$$

$$H^p(X^{p-1}; G) \longrightarrow \cdots$$

$$H^p(X^{p+1}, X^p; G) \longrightarrow H^p(X^{p+1}; G) \longrightarrow H^p(X^p; G) \xrightarrow{\delta^p}$$

$$H^{p+1}(X^{p+1}, X^p; G) \longrightarrow \cdots.$$

Since by Proposition 6.14 we also have

$$H^{p-1}(X^{p-2}; G) = H^p(X^{p-1}; G) = H^p(X^{p+1}, X^p; G) = (0),$$

and $H^p(X;G) \cong H^p(X^{p+1};G)$, we have the following diagram:

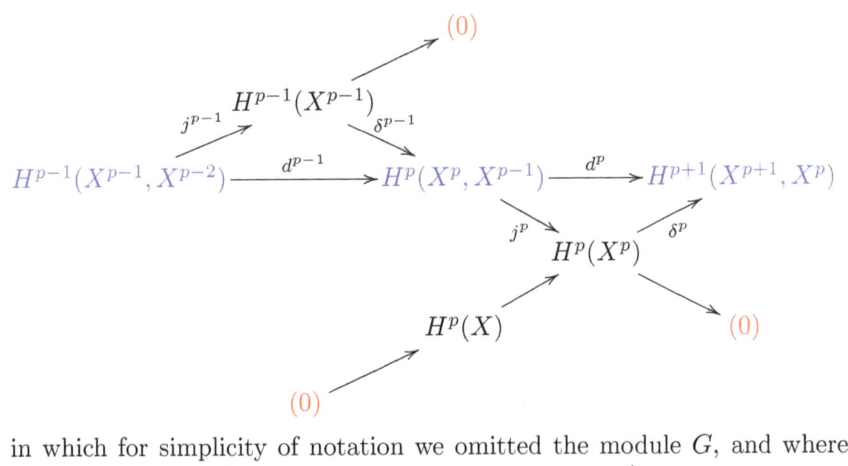

in which for simplicity of notation we omitted the module G, and where $d^{p-1} = \delta^{p-1} \circ j^{p-1}$ and $d^p = \delta^p \circ j^p$. Since $j^p \circ \delta^{p-1} = 0$ (because the sequence on that diagonal is exact), we have

$$d^p \circ d^{p-1} = \delta^p \circ j^p \circ \delta^{p-1} \circ j^{p-1} = 0.$$

Definition 6.15. Given a CW complex X, the modules $H^p(X^p, X^{p-1}; G)$ together with the coboundary maps

$$d^p \colon H^p(X^p, X^{p-1}; G) \to H^{p+1}(X^{p+1}, X^p; G)$$

defined above form a cochain complex $S^*_{\mathrm{CW}}(X;G)$ called the *cellular cochain complex* associated with X. The cohomology modules associated with the cochain complex $S^*_{\mathrm{CW}}(X;G)$ are denoted by

$$H^p_{\mathrm{CW}}(X;G) = H^p(S^*_{\mathrm{CW}}(X;G))$$

and called the *cellular cohomology modules* of the cochain complex $S^*_{\mathrm{CW}}(X;G)$.

The following simple proposition will needed to prove Theorem 6.16.

Proposition 6.15. *If the following diagram is commutative and if $j \colon A \to B$ is surjective*

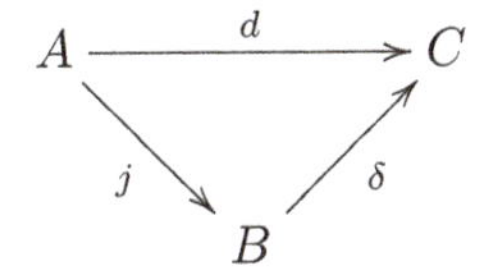

then

$$\mathrm{Ker}\, \delta \cong \mathrm{Ker}\, d / \mathrm{Ker}\, j.$$

Proof. Define a map $\varphi\colon \operatorname{Ker}\delta \to \operatorname{Ker} d/\operatorname{Ker} j$ as follows: for any $b \in \operatorname{Ker}\delta$, let

$$\varphi(b) = a + \operatorname{Ker} j$$

for any $a \in \operatorname{Ker} d$ such that $j(a) = b$. Since j is surjective, there is some $a \in A$ such that $j(a) = b$. Furthermore, for any $a \in A$ such that $j(a) = b \in \operatorname{Ker}\delta$, since $d = \delta \circ j$ we have $d(a) = \delta(j(a)) = \delta(b) = 0$, so $a \in \operatorname{Ker} d$. This map is well defined because if another $a' \in \operatorname{Ker} d$ is chosen such that $j(a') = b$, then $j(a') = j(a)$ so $j(a' - a) = 0$, that is, $a' - a \in \operatorname{Ker} j$, so $a + \operatorname{Ker} j = a' + \operatorname{Ker} j$.

The map φ is injective because if $\varphi(b) = \operatorname{Ker} j$, since $\varphi(b) = a + \operatorname{Ker} j$ for any $a \in \operatorname{Ker} d$ such that $j(a) = b$, we have $a + \operatorname{Ker} j = \operatorname{Ker} j$, which implies that $a \in \operatorname{Ker} j$ so $b = j(a) = 0$. The map φ is surjective because for any $a + \operatorname{Ker} j$ with $a \in \operatorname{Ker} d$, by definition of φ we have $\varphi(j(a)) = a + \operatorname{Ker} j$. Therefore $\varphi\colon \operatorname{Ker}\delta \to \operatorname{Ker} d/\operatorname{Ker} j$ is an isomorphism. ☐

Theorem 6.16. *Let X be a CW complex. For any PID R and any R-module G there are isomorphisms*

$$H^p_{\mathrm{CW}}(X;G) \cong H^p(X;G) \quad \textit{for all } p \geq 0$$

between the cellular cohomology modules and the singular cohomology modules of X. Furthermore, the cellular cochain complex $S^*_{\mathrm{CW}}(X;G)$ is isomorphic to the cochain complex $\operatorname{Hom}_R(S^{\mathrm{CW}}_*(X;R), G)$ (the dual of the cellular chain complex $S^{\mathrm{CW}}_*(X;R)$ with respect to G).

Proof. The above diagram shows that

$$H^p(X;G) \cong \operatorname{Ker}\delta^p.$$

Since j^p is surjective, Proposition 6.15 (with $A = H^p(X^p, X^{p-1}), B = H^p(X^p)$ and $C = H^{p+1}(X^{p+1}, X^p)$) shows that

$$\operatorname{Ker}\delta^p = \operatorname{Ker} d^p/\operatorname{Ker} j^p,$$

which yields $H^p(X;G) \cong \operatorname{Ker} d^p/\operatorname{Ker} j^p$. But $\operatorname{Ker} j^p = \operatorname{Im}\delta^{p-1}$ so

$$H^p(X;G) \cong \operatorname{Ker} d^p/\operatorname{Im}\delta^{p-1}.$$

Since j^{p-1} is surjective, $\operatorname{Im}\delta^{p-1} = \operatorname{Im} d^{p-1}$, and finally we obtain

$$H^p(X;G) \cong \operatorname{Ker} d^p/\operatorname{Im} d^{p-1} = H^p_{\mathrm{CW}}(X;G),$$

as claimed.

We now construct two commutative diagrams as follows.

Step 1. The first one is obtained from the naturality part of the universal coefficient theorem for cohomology (Theorem 12.6) applied to a chain map $\theta\colon C \to C'$ of chain complexes C and C' by retaining only the rightmost of the two squares in the diagram of Theorem 12.6, and by flipping it about its diagonal. We obtain the diagram

$$\begin{array}{ccc} H^p(\mathrm{Hom}_R(C', G)) & \xrightarrow{(\mathrm{Hom}_R(\theta,\mathrm{id}))^*} & H^p(\mathrm{Hom}_R(C, G)) \\ \Big\downarrow h' & & \Big\downarrow h \\ \mathrm{Hom}_R(H_p(C'), G) & \xrightarrow[\mathrm{Hom}_R(\theta_*,\mathrm{id})]{} & \mathrm{Hom}_R(H_p(C), G). \end{array}$$

If we specialize the above diagram to the chain map $\pi\colon S_*(X^p; R) \to S_*(X^p, X^{p-1}; R)$ (with $C = S_*(X^p, R)$, $C' = S_*(X^p, X^{p-1}; R)$), we find that $(\theta_*)_p = j_p$, so the lower arrow is $j_p^* = \mathrm{Hom}_R(\theta_*, \mathrm{id})$, and $j^p = (\mathrm{Hom}_R(\theta, \mathrm{id}))^*$, and since by definition

$$\begin{aligned} H_p(X^p; R) &= H_p(S_*(X^p; R)) = H_p(C), \;\; H_p(X^p, X^{p-1}; R) \\ &= H_p(S_*(X^p, X^{p-1}; R)) = H_p(C'), \\ S^*(X^p; G) &= \mathrm{Hom}_R(S_*(X^p; R), G) = \mathrm{Hom}_R(C; G), \\ S^*(X^p, X^{p-1}; G) &= \mathrm{Hom}_R(S_*(X^p, X^{p-1}; R), G) = \mathrm{Hom}_R(C'; G), \\ H^p(X^p; G) &= H^p(S^*(X^p; G)) \\ H^p(X^p, X^{p-1}; G) &= H^p(S^*(X^p, X^{p-1}; G)), \end{aligned}$$

we obtain the following commutative diagram:

$$\begin{array}{ccc} H^p(X^p, X^{p-1}; G) & \xrightarrow{\quad j^p \quad} & H^p(X^p; G) \\ \Big\downarrow h^p & & \Big\downarrow h \\ \mathrm{Hom}_R(H_p(X^p, X^{p-1}; R), G) & \xrightarrow[j_p^*]{} & \mathrm{Hom}_R(H_p(X^p; R), G). \end{array}$$

Step 2. The second commutative diagram expresses a duality relationship between the connecting homomorphisms $\delta^p\colon H^p(A; G) \to H^{p+1}(X, A; G)$ and $\partial_{p+1}\colon H_{p+1}(X, A; G) \to H_p(A; G)$ arising in the long exact sequences of relative cohomology and homology of a pair (X, A). The reason why the diagram shown below commutes is explained in Hatcher [Hatcher (2002)] (Chapter 3, Section 3.1, Pages 200–201).

$$\begin{array}{ccc} H^p(A;G) & \xrightarrow{\delta^p} & H^{p+1}(X,A;G) \\ \Big\downarrow h & & \Big\downarrow h' \\ \mathrm{Hom}_R(H_p(A;R);G) & \xrightarrow[\partial^*_{p+1}]{} & \mathrm{Hom}_R(H_{p+1}(X,A;R);G). \end{array}$$

If we specialize to $X = X^{p+1}, A = X^p$, we obtain the commutative diagram

$$\begin{array}{ccc} H^p(X^p;G) & \xrightarrow{\delta^p} & H^{p+1}(X^{p+1},X^p;G) \\ \Big\downarrow h & & \Big\downarrow h^{p+1} \\ \mathrm{Hom}_R(H_p(X^p;R);G) & \xrightarrow[\partial^*_{p+1}]{} & \mathrm{Hom}_R(H_{p+1}(X^{p+1},X^p;R);G). \end{array}$$

Next if we concatenate the (last) commutative diagrams obtained in Step 1 and Step 2, we obtain the commutative diagram shown below.

$$\begin{array}{ccccc} H^p(X^p,X^{p-1};G) & \xrightarrow{j^p} & H^p(X^p;G) & \xrightarrow{\delta^p} & H^{p+1}(X^{p+1},X^p;G) \\ \Big\downarrow h^p & & \Big\downarrow h & & \Big\downarrow h^{p+1} \\ \mathrm{Hom}_R(H_p(X^p,X^{p-1};R),G) & \xrightarrow[j^*_p]{} & Z & \xrightarrow[\partial^*_{p+1}]{} & \mathrm{Hom}_R(H_{p+1}(X^{p+1},X^p;R),G), \end{array}$$

where $Z = \mathrm{Hom}_R(H_p(X^p;R),G)$. Thus the big rectangle commutes. Furthermore, by Proposition 12.8, the maps h^p and h^{p+1} are isomorphisms. But the composition of the two maps on the top row is d^p, the cellular coboundary map, and the composition of the two maps on the bottom row is $d^*_{p+1} = \mathrm{Hom}_R(d_{p+1};G)$ since $d_{p+1} = j_p \circ \partial_{p+1}$ which implies that $d^*_{p+1} = \partial^*_{p+1} \circ j^*_p$, so we have the commutative diagram

$$\begin{array}{ccc} H^p(X^p,X^{p-1};G) & \xrightarrow{d^p} & H^{p+1}(X^{p+1},X^p;G) \\ \Big\downarrow h^p & & \Big\downarrow h^{p+1} \\ \mathrm{Hom}_R(H_p(X^p,X^{p-1};R),G) & \xrightarrow[d^*_{p+1}]{} & \mathrm{Hom}_R(H_{p+1}(X^{p+1},X^p;R),G), \end{array}$$

which shows that the cellular cochain complex $S^*_{\mathrm{CW}}(X;G)$ is isomorphic to the cochain complex $\mathrm{Hom}_R(S^{\mathrm{CW}}_*(X;R),G)$. $\square$

As a consequence, although this is not obvious *a priori*, the cellular cochain complex $S^*_{\mathrm{CW}}(X;G)$ is isomorphic to the cochain complex obtained by applying $\mathrm{Hom}_R(-,G)$ to the cellular chain complex $S^{\mathrm{CW}}_*(X;R)$. Also, the cellular cohomology modules "compute" the singular cohomology modules.

6.5 Problems

Problem 6.1. Prove that the Klein bottle can be expressed as a CW-complex with one 0-cell, two 2-cells, and one 2-cell, just as the torus T, but with different attaching maps.

Problem 6.2. Refer back to Problem 5.5 in which, given a (2-dimensional) torus T, we formed the g-fold connected sum

$$X_g = \underbrace{T \sharp \cdots \sharp T}_{g}$$

by gluing together $g \geq 2$ tori. Prove that X_g is obtained as a CW-complex having one 0-cell, $2n$ 1-cells, and one 2-cell.

Problem 6.3. Refer back to Problem 5.7 in which, given a projective plane $\mathbb{RP}^2$, we formed its g-fold connected sum

$$Y_g = \underbrace{\mathbb{RP}^2 \sharp \cdots \sharp \mathbb{RP}^2}_{g}$$

by gluing together $g \geq 2$ projective planes. Prove that Y_g is obtained as a CW-complex having one 0-cell, n 1-cells, and one 2-cell.

Problem 6.4. Prove Proposition 6.2.

Problem 6.5. Another way to compute the boundary maps of the cellular chain complex associated with $\mathbb{RP}^n$ is to triangulate S^n and $\mathbb{RP}^n$ and to use simplicial homology. This approach is discussed in Munkres [Munkres (1984)], see Lemma 40.4, Lemma 40.7, Theorem 40.6, and Theorem 40.7. Study this approach carefully.

Problem 6.6. (1) Prove that the index of the identity map $I\colon S^n \to S^n$ is $+1$.

(2) Prove that if $f\colon S^n \to S^n$ is not surjective, then $\deg f = 0$.

Problem 6.7. (1) Prove that if $f, g\colon S^n \to S^n$ are homotopic maps, then $\deg f = \deg g$.

Prove that $\deg(g \circ f) = (\deg f)(\deg g)$. Deduce that if $f\colon S^n \to S^n$ is a homotopy equivalence, then $\deg f = \pm 1$.

Problem 6.8. (1) Prove that if $f\colon S^n \to S^n$ is a reflection of S^n, which means that f fixes a subspace homeomorphic to S^{n-1} and exchanges the two complementary hemispheres, then $\deg f = -1$.

(2) Prove that the antipodal map $-\mathbb{1}\colon S^n \to S^n$ given by $-\mathbb{1}(x) = -x$ has degree $(-1)^{n+1}$.

Problem 6.9. Prove that if $f\colon S^n \to S^n$ has no fixed point, then $\deg f = (-1)^{n+1}$.

Problem 6.10. Prove that the cellular chain complex of the CW-complex for X_g determined in Problem 6.2 is

$$0 \longrightarrow \mathbb{Z} \xrightarrow{d_2} \mathbb{Z}^{2g} \xrightarrow{d_1} \mathbb{Z} \longrightarrow 0,$$

and that $d_1 = d_2 = 0$. Use this to compute the homology of X_g.

Problem 6.11. Prove that the cellular chain complex of the CW-complex for Y_g determined in Problem 6.3 is

$$0 \longrightarrow \mathbb{Z} \xrightarrow{d_2} \mathbb{Z}^{g} \xrightarrow{d_1} \mathbb{Z} \longrightarrow 0.$$

Prove that $d_1 = 0$ and that d_2 is defined by its action on 1 by $d_2(1) = (2, \ldots, 2)$. Deduce that $H_2(Y_g) = (0)$.

By replacing the last vector $(0, \ldots, 0, 1)$ in the canonical basis for $\mathbb{Z}^g$ by $(1, \ldots, 1)$, prove that $H_1(Y) = \mathbb{Z}^{g-1} \oplus (\mathbb{Z}/2\mathbb{Z})$.

Problem 6.12. Prove that

$$\chi(\mathbb{RP}^{2n}) = 1 \quad \text{and} \quad \chi(\mathbb{RP}^{2n+1}) = 0.$$

Problem 6.13. Prove that if $f\colon S^n \to S^n$ has degree d, then the homomorphism induced on cohomology, $f^*\colon H^n(S^n; G) \to H^n(S^n; G)$, is multiplication by d.

Problem 6.14. Prove that if A is a closed subspace of X that is a deformation retract of some neighborhood of X, then the quotient map $\pi\colon X \to X/A$ induces isomorphisms $H^p(X, A; G) \cong H^p(X/A; G)$, for all $p \geq 0$.

Chapter 7

Poincaré Duality

Our goal is to state a version of the Poincaré duality for singular homology and cohomology, one of the most important results about the topology of manifolds. The basic version is that if M is a "nice" n-manifold, then there are isomorphisms

$$H^p(M; \mathbb{Z}) \cong H_{n-p}(M; \mathbb{Z}) \qquad (*)$$

for all $p \in \mathbb{Z}$. Here nice means compact and orientable, a notion that will be defined in Section 7.1.

The isomorphisms $(*)$ are actually induced by an operation

$$\frown : S^p(M; \mathbb{Z}) \times S_n(M; \mathbb{Z}) \to S_{n-p}(M; \mathbb{Z})$$

combining a chain and a cochain to make a chain, called *cap product*, which induces an operation

$$\frown : H^p(M; \mathbb{Z}) \times H_n(M; \mathbb{Z}) \to H_{n-p}(M; \mathbb{Z})$$

combining a homology class and a cohomology class to make a homology class. Furthermore, if M is orientable, then there is a unique special homology class $\mu_M \in H_n(M; \mathbb{Z})$ called the *fundamental class* of M, and Poincaré duality means that the map

$$c \mapsto c \frown \mu_M$$

is an isomorphism between $H^p(M; \mathbb{Z})$ and $H_{n-p}(M; \mathbb{Z})$.

All this can be generalized to coefficients in any commutative ring R with an identity element and to compact manifolds that are R-orientable, a notion defined in Section 7.1.

It is even possible to generalize Poincaré duality to noncompact R-orientable manifolds, by replacing singular cohomology by the more general notion of singular cohomology with compact support. We will explain all this in the following sections. We begin with the notion of orientation.

7.1 Orientations of a Manifold

Since 0-dimensional manifolds constitute a degenerate case of little interest (discrete sets of points), we assume that our manifolds have dimension $n > 0$.

If M is a topological manifold of dimension $n > 0$ and if R is any commutative ring with multiplicative unit, we saw in Proposition 4.23 that

$$H_p(M, M - \{x\}; R) \cong \begin{cases} R & \text{if } p = n \\ (0) & \text{if } p \neq n. \end{cases}$$

Since the groups $H_n(M, M - \{x\}; R)$ are all isomorphic to R, a way to define a notion of orientation is to pick some generator μ_x from $H_n(M, M - \{x\}; R)$, for every $x \in M$. Here we view R as a free R-module so a generator of R is an element $s \in R$ such that the map $r \mapsto rs$ ($r \in R$) is surjective. Since R is a ring with an identity element, a generator of R is just an invertible element. To say that M is orientable means that we can pick these invertible elements $\mu_x \in H_n(M, M - \{x\}; R)$ in such a way that they "vary continuously" with x.

A way to achieve this is to introduce the notion of fundamental class of M at a subspace A.

Definition 7.1. Given an n-manifold M and any subset A of M, an *R-fundamental (homology) class of M at the subspace A* is a homology class $\mu_A \in H_n(M, M - A; R)$ such that for every $x \in A$,

$$\rho_x^A(\mu_A) = \mu_x \in H_n(M, M - \{x\}; R)$$

is a generator of $H_n(M, M - \{x\}; R)$, where $\rho_x^A \colon H_n(M, M - A; R) \to H_n(M, M - \{x\}; R)$ is the homomorphism induced by the inclusion $M - A \subseteq M - \{x\}$. If $A = M$, we call μ_M an *R-fundamental (homology) class of M.*

An *R-orientation* of M is an open cover $\mathcal{U} = (U_i)_{i \in I}$ together with a family $(\mu_{U_i})_{i \in I}$ of fundamental classes of M at U_i such that whenever $U_i \cap U_j \neq \emptyset$, then

$$\rho_{U_i \cap U_j}^{U_i}(\mu_{U_i}) = \rho_{U_i \cap U_j}^{U_j}(\mu_{U_j}), \tag{$\dagger$}$$

where $\rho_{U_i \cap U_j}^{U_i} \colon H_n(M, M - U_i; R) \to H_n(M, M - U_i \cap U_j; R)$ and $\rho_{U_i \cap U_j}^{U_j} \colon H_n(M, M - U_j; R) \to H_n(M, M - U_i \cap U_j; R)$ are the homomorphisms induced by the inclusions $U_i \cap U_j \subseteq U_i$ and $U_i \cap U_j \subseteq U_j$. A manifold M is *R-orientable* if it has an R-orientation.

When $R = \mathbb{Z}$, we use the terminology fundamental classes and orientations (we drop the prefix R). For simplicity of notation, we write μ_i instead of μ_{U_i}.

Observe that if $(\mathcal{U} = (U_i)_{i\in I}, (\mu_i)_{i\in I})$ is an R-orientation of M, since $\rho_x^{U_i} = \rho_x^{U_i\cap U_j} \circ \rho_{U_i\cap U_j}^{U_i}$ and $\rho_x^{U_j} = \rho_x^{U_i\cap U_j} \circ \rho_{U_i\cap U_j}^{U_j}$, the condition $\rho_{U_i\cap U_j}^{U_i}(\mu_i) = \rho_{U_i\cap U_j}^{U_j}(\mu_j)$ implies that

$$\rho_x^{U_i}(\mu_i) = \rho_x^{U_j}(\mu_j) \quad \text{for all } x \in U_i \cap U_j,$$

that is, the R-orientation is indeed consistent. Consequently, if we set $\mu_x = \rho_x^{U_i}(\mu_i)$ for every $i \in U$ and for every $x \in U_i$, we obtain a well-defined family $(\mu_x)_{x\in M}$ of generators $\mu_x \in H_n(M, M - x; R)$. We call $(\mu_x)_{x\in M}$ the *family of generators* induced by the orientation.

Remark: Readers familiar with differential geometry will observe the analogy between a fundamental class and a (global) volume form in the case where the n-manifold is smooth. In the smooth case, there is a tangent space at every point $x \in M$, and an orientation is given by a nonzero global section ω of the bundle $\bigwedge^n T^*M$. In the absence of the tangent bundle, the substitute is the orientation bundle whose fibres are the homology rings $H_n(M, M - \{x\}; R)$.

Definition 7.2. For any chart $\varphi_U \colon U \to \Omega$ where U is an open subset of M, if D is a closed ball contained in $\Omega \subseteq \mathbb{R}^n$, then $B = \varphi_U^{-1}(D)$ is a compact subset of M and we call it a *compact and convex subset of* M. See Figure 7.1.

Then a minor modification of Proposition 4.23 can be used to show the following fact (see Bredon [Bredon (1993)], Chapter VI, Proposition 7.1).

Proposition 7.1. *Given a topological n-manifold M, for any compact and convex subset B, for any point $x \in B$, the homomorphism $\rho_x^B \colon H_p(M, M - B; R) \to H_p(M, M - x; R)$ induced by the inclusion $M - B \subseteq M - x$ is an isomorphism for all $p \geq 0$. We have*

$$H_p(M, M - B; R) \cong \begin{cases} R & \text{if } p = n \\ (0) & \text{if } p \neq n. \end{cases}$$

Proof. By shrinking the domain U of the chart $\varphi_U \colon U \to \Omega$ such that $B = \varphi_U^{-1}(D)$ as in Definition 7.2 and Figure 7.1, we may assume that U is

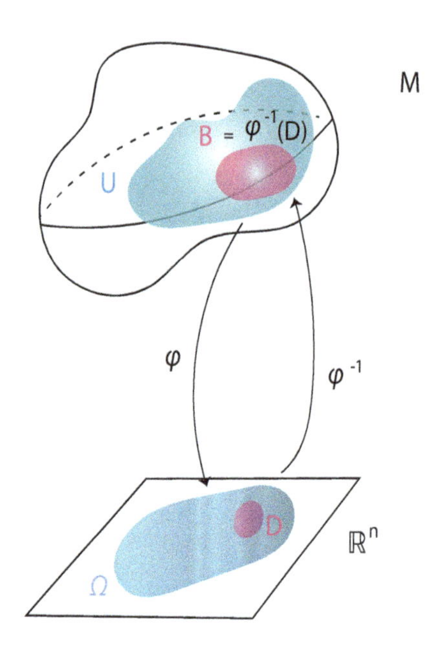

Fig. 7.1 An illustration of a compact and convex subset of the 2-manifold M.

homeomorphic to $\mathbb{R}^n$. As in the proof of Proposition 4.23, by excision with $X = M, A = M - x$, and $Z = M - U$ (see Theorem 4.14), we obtain

$$H_p(M, M - \{x\}; R) \cong H_p(U, U - \{x\}; R) \cong H_p(\mathbb{R}^n, \mathbb{R}^n - \{x\}; R),$$

and by excision with $X = M, A = M - B$, and $Z = M - U$, we obtain

$$H_p(M, M - B; R) \cong H_p(U, U - B; R) \cong H_p(\mathbb{R}^n, \mathbb{R}^n - B; R).$$

By the proof of Proposition 4.23, we have

$$H_p(\mathbb{R}^n, \mathbb{R}^n - \{x\}; R) \cong \widetilde{H}_{p-1}(S^{n-1}; R)$$

for all $p \geq 0$. The exact same proof with $\mathbb{R}^n - \{x\}$ replaced by $\mathbb{R}^n - B$ also shows that

$$H_p(\mathbb{R}^n, \mathbb{R}^n - B; R) \cong \widetilde{H}_{p-1}(S^{n-1}; R)$$

for all $p \geq 0$. We also know from Proposition 4.23 that the only nonzero modules occur when $p = n$, in which case they are isomorphic to R. We can check that we have the commutative diagram

$$\begin{array}{ccc} H_p(M, M - B; R) & \xrightarrow{\rho_x^B} & H_p(M, M - \{x\}; R) \\ \cong \downarrow & & \uparrow \cong \\ H_p(\mathbb{R}^n, \mathbb{R}^n - B; R) & \longrightarrow & H_p(\mathbb{R}^n, \mathbb{R}^n - \{x\}; R) \\ \downarrow \cong & & \uparrow \cong \\ \widetilde{H}_{n-1}(S^{n-1}) & \xrightarrow[=]{} & \widetilde{H}_{n-1}(S^{p-1}), \end{array}$$

which implies that the second horizontal arrow and the first horizontal arrow labeled by ρ_x^B are isomorphisms. □

Proposition 7.1 shows that for any small enough compact subset B, the manifold M has an R-fundamental class at B. Indeed, we can pick the fundamental class $\mu_B \in H_n(M, M - B; R) \cong R$ as any generator of R, and since ρ_x^B is an isomorphism, each $\mu_x = \rho_x^B(\mu_B)$ is a generator of $H_n(M, M - \{x\}; R)$. It is also easy to show that Proposition 7.1 implies that Condition (†) in Definition 7.1 can be replaced by the condition

$$\rho_x^{U_i}(\mu_i) = \rho_x^{U_j}(\mu_j) \quad \text{for all } x \in U_i \cap U_j.$$

Some textbooks use this condition instead of (†).

In the special case where $R = \mathbb{Z}/2\mathbb{Z}$, since $\mathbb{Z}/2\mathbb{Z} = \{0, 1\}$, the only generator of $H_n(M, M - \{x\}, \mathbb{Z}/2\mathbb{Z}) \cong \mathbb{Z}/2\mathbb{Z}$ is 1.

Proposition 7.2. *Every manifold has a $\mathbb{Z}/2\mathbb{Z}$-orientation.*

Proof. Since $H_n(M, M - \{x\}, \mathbb{Z}/2\mathbb{Z}) \cong \mathbb{Z}/2\mathbb{Z}$ only has 1 as generator, the consistency conditions are trivial. We can use Proposition 7.1 to create a $\mathbb{Z}/2\mathbb{Z}$-orientation of M by covering M with open balls and making a fundamental class associated with the closure of that ball. □

Definition 7.3. A $\mathbb{Z}/2\mathbb{Z}$-orientation is also called a *mod 2 orientation.*

Proposition 7.3. *If a manifold M has an R-fundamental class, then it has an R-orientation.*

Proof. Since for any open cover $\mathcal{U} = (U_i)_{i \in I}$ of M we have $\rho_x^M = \rho_x^{U_i} \circ \rho_{U_i}^M$, we can take $\mu_i = \rho_{U_i}^M(\mu_M) \in H_n(M, M - U_i; R)$ as fundamental class of U_i. The consistency conditions are immediately verified. □

The converse of Proposition 7.3 holds if M is compact. This is a non-trivial and deep fact whose proof is difficult (see Theorem 7.7, which relies on Theorem 7.4).

Remark: There are other ways of defining R-orientability. One can define the *orientation bundle* M_R of M by taking the disjoint union of the groups $H_n(M, M - \{x\}; R)$ where x ranges over M, and giving it a suitable topology that amounts to a local consistency condition for R-orientability. Then an R-orientation is a continuous section $s\colon M \to M_R$ that picks a generator of $H_n(M, M - \{x\}; R)$ for every $x \in M$. We refer the reader to Hatcher [Hatcher (2002)] (Chapter 3, Section 3.3), Bredon [Bredon (1993)] (Chapter VI, Section 7), and Spanier [Spanier (1989)] (Chapter 6, Sections 2 and 3). The notion of R-orientation in Definition 7.1 corresponds to the notion of a $\mathcal{U}$-compatible family in Spanier [Spanier (1989)] (Chapter 6, Section 3). Milnor and Stasheff [Milnor and Stasheff (1974)] use a condition using the notion of a small cell, as defined in Spanier [Spanier (1989)] (Chapter 6, Section 3). The equivalence of the condition of Definition 7.1 with the orientation bundle condition amounts to the proof of Theorem 4 in Spanier [Spanier (1989)] (Chapter 6, Section 3); see also Proposition 7.3 in Bredon [Bredon (1993)] (Chapter VI, Section 7).

It can also be shown that a connected nonorientable n-manifold has a two-sheeted connected covering space which is orientable. This implies that every simply connected manifold is orientable; see Hatcher [Hatcher (2002)] (Chapter 3, Section 3.3, Proposition 3.25).

We see that we are naturally led to the study of the groups $H_n(M, M - K; R)$, where K is a compact subset of M. We have the following theorem which is the key to the existence of an R-fundamental class $\mu_K \in H_n(M, M - K; R)$ if K is compact and M is R-orientable (see Theorem 7.7). It is also the key to the vanishing theorem, which may be considered as a prelude to Poincaré duality.

Theorem 7.4. *Let M be an n-manifold.*

(i) For any compact subset K, if $p > n$, then $H_p(M, M - K; R) = (0)$.
(ii) For any homology class $\alpha \in H_n(M, M - K; R)$, we have $\alpha = 0$ iff $\rho_x^K(\alpha) = 0$ for all $x \in K$, where $\rho_x^K\colon H_n(M, M - K; R) \to H_n(M, M - x; R)$ is the homomorphism induced by the inclusion $M - K \subseteq M - x$.

Theorem 7.4 is proven in Milnor and Stasheff [Milnor and Stasheff (1974)] (Appendix A, Lemma A.7), Hatcher [Hatcher (2002)] (Chapter 3, Lemma 3.27), May [May (1999)] (Chapter 20, Section 3), Massey [Massey (1991)] (Chapter XIV, Lemma 2.3), and the first statement of the theorem is proven in Bredon [Bredon (1993)] (Chapter VI, Theorem 7.8(a)).

The following notation will be used in the proofs below.

Definition 7.4. For any two compact subsets L_1 and L_2 such that $L_1 \subseteq L_2 \subseteq M$, the map $\rho^{L_2}_{L_1}\colon H_p(M, M - L_2; R) \to H_p(M, M - L_1; R)$ is the homomorphism induced by the inclusion $M - L_2 \subseteq M - L_1$. To simplify notation, we often write ρ_{L_1} instead of $\rho^{L_2}_{L_1}$. When $L_1 = \{x\}$ (a single point), we write $\rho^{L_2}_x$.

For any three compact subsets such that $L_1 \subseteq L_2 \subseteq L_3 \subseteq M$, the composition

$$H_n(M, M - L_3; R) \xrightarrow{\rho^{L_3}_{L_2}} H_n(M, M - L_2; R) \xrightarrow{\rho^{L_2}_{L_1}} H_n(M, M - L_1; R)$$

is equal to the map $\rho^{L_3}_{L_1}\colon H_n(M, M - L_3; R) \to H_n(M, M - L_1; R)$, that is

$$\rho^{L_3}_{L_1} = \rho^{L_2}_{L_1} \circ \rho^{L_3}_{L_2}. \tag{$\dagger_1$}$$

In particular, if $L_1 = \{x\}$, then

$$\rho^{L_3}_{x} = \rho^{L_2}_{x} \circ \rho^{L_3}_{L_2}. \tag{$\dagger_2$}$$

See Figure 7.2.

Sketch of proof. We prove (i) and some cases of (ii) following Milnor and Stasheff [Milnor and Stasheff (1974)] (Appendix A, Lemma A.7) with some help from Massey [Massey (1991)] (Chapter XIV, Lemma 2.3). In (ii), since ρ^K_x is a homomorphism, if $\alpha = 0$, then obviously $\rho^K_x(\alpha) = 0$ so we will focus on proving that if $\rho^K_x(\alpha) = 0$ for all $x \in K$, then $\alpha = 0$. The proof is divided in six steps.

Case 1. Suppose $M = \mathbb{R}^n$ and K is a compact and convex subset.

(i) By Proposition 7.1, we have isomorphisms

$$H_p(\mathbb{R}^n, \mathbb{R}^n - K; R) \cong H_p(\mathbb{R}^n, \mathbb{R}^n - \{x\}; R) \cong \widetilde{H}(S^{n-1}; R)$$

for all p, and again by Proposition 7.1, $H_p(\mathbb{R}^n, \mathbb{R}^n - K; R) \cong (0)$ for $p > n$.

(ii) By Proposition 7.1, the map $\rho^K_x\colon H_n(\mathbb{R}^n, \mathbb{R}^n - K; R) \to H_n(\mathbb{R}^n, \mathbb{R}^n - \{x\}; R)$ is an isomorphism.

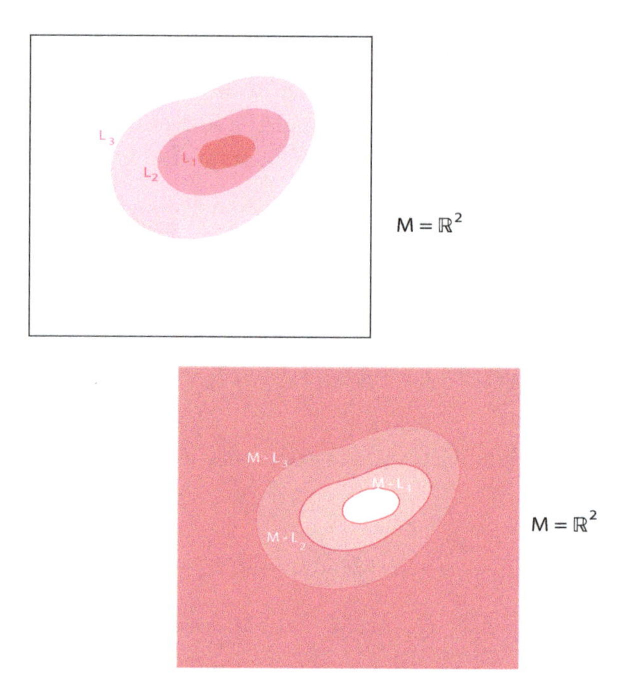

Fig. 7.2 An illustration of the inclusions $L_1 \subseteq L_2 \subseteq L_3 \subseteq M$ and the inclusions $M - L_3 \subseteq M - L_2 \subseteq M - L_1$.

Case 2. Let $K = K_1 \cup K_2$, with K_1 and K_2 compact, and assume that the theorem holds for K_1 and K_2 and $K_1 \cap K_2$.

(i) If we construct the Mayer–Vietoris homology long exact sequence given by Theorem 4.28 applied to $X = M$, $Y = M-(K_1 \cap K_2)$, $A = M-K_1$, $B = M - K_2$, since $A \cap B = M - (K_1 \cup K_2) = M - K$, (see Figure 7.3), we obtain the long exact sequence of relative homology

$$\longrightarrow H_{p+1}(M, M-(K_1 \cap K_2)) \xrightarrow{\partial} H_p(M, M-K) \xrightarrow{\varphi} H_p(X, M-K_1) \oplus H_p(M, M-K_2) \xrightarrow{\psi} H_p(M, M-(K_1 \cap K_2)) \to \cdots$$

where

$$\varphi(\alpha) = \rho^K_{K_1}(\alpha) \oplus \rho^K_{K_2}(\alpha) \quad \text{and} \quad \psi(\beta \oplus \gamma) = \rho^{K_1}_{K_1 \cap K_2}(\beta) - \rho^{K_2}_{K_1 \cap K_2}(\gamma).$$

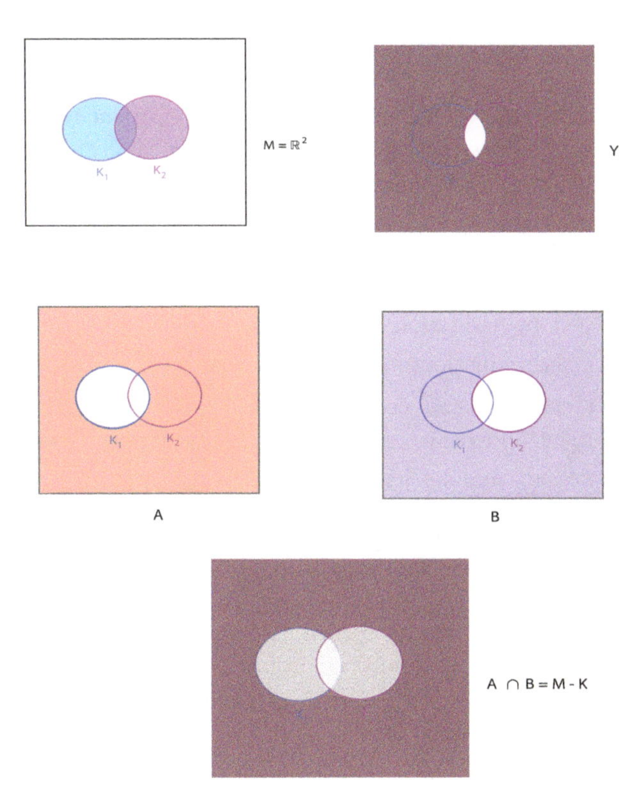

Fig. 7.3 An illustration of the spaces $K = K_1 \cup K_2$, Y, A, B, and $A \cap B$ utilized in the Mayer–Vietoris of Case 2.

Since by hypothesis $H_{p+1}(M, M - (K_1 \cap K_2); R) \cong (0)$, $H_p(M, M - K_1; R) \cong (0)$ and $H_p(M, M - K_2; R) \cong (0)$ for $p > n$, and since we have an exact sequence, we conclude that $H_p(M, M - K) = (0)$ for $p > n$.

(ii) If $p = n$, since we just showed that $H_{n+1}(M, M - (K_1 \cap K_2)) \cong (0)$, we have the piece of exact sequence

$$\longrightarrow (0) \xrightarrow{\partial} H_p(M, M - K) \xrightarrow{\varphi} H_p(X, M - K_1) \oplus H_p(M, M - K_2) \xrightarrow{\psi} H_p(M, M - (K_1 \cap K_2)) \longrightarrow \cdots$$

which shows that φ is injective. For any $\alpha \in H_p(M, M - K; R)$, we have $\alpha = 0$ iff $\varphi(\alpha) = 0$ (since φ is injective) iff $\rho^K_{K_1}(\alpha) \oplus \rho^K_{K_2}(\alpha) = 0$ iff $\rho^K_{K_1}(\alpha) =$

0 and $\rho^K_{K_2}(\alpha) = 0$. Since by hypothesis (ii) holds for K_1 and K_2, we have $\rho^K_{K_1}(\alpha) = 0$ iff $\rho^{K_1}_x(\rho^K_{K_1}(\alpha)) = 0$ for all $x \in K_1$ and $\rho^K_{K_2}(\alpha) = 0$ iff $\rho^{K_2}_y(\rho^K_{K_2}(\alpha)) = 0$ for all $y \in K_2$, which by ($\dagger_2$) is equivalent to $\rho^K_x(\alpha) = 0$ for all $x \in K_1$ and $\rho^K_y(\alpha) = 0$ for all $y \in K_2$, and finally equivalent to $\rho^K_z(\alpha) = 0$ for all $z \in K = K_1 \cup K_2$.

Case 3. Suppose $K = K_1 \cup \cdots \cup K_m$, the union of compact and convex subsets of $\mathbb{R}^n$.

(i) We proceed by induction on m. The base case $m = 1$ follows by Case 1. For the induction step, observe that $\left(\bigcup_{j=1}^m K_j\right) \cap K_{m+1}$ is the union of m compact and convex subsets, and $\bigcup_{j=1}^m K_j$ is also the union of m compact and convex subsets, so the induction hypothesis applies to these two sets and we have $H_p\left(M, M - \bigcup_{j=1}^m K_j; R\right) = (0)$ and $H_p\left(M, M - \left(\bigcup_{j=1}^m K_j\right) \cap K_{m+1}; R\right) = (0)$ for all $p > n$. By Case 1, $H_p(M, M - K_{m+1}; R) = (0)$ for all $p > n$, so using Case 2 (applied to $\bigcup_{j=1}^m K_j$ and K_{m+1}), we deduce that $H_p\left(M, M - \bigcup_{j=1}^{m+1} K_j; R\right) = (0)$ for all $p > n$.

(ii) This is also proven by induction on m using Case 1 and Case 2. The details are left as an exercise.

Case 4. Assume that K is an arbitrary compact subset of $\mathbb{R}^n$. This case is technically more difficult than the others. We reproduce Minor and Stasheff's proof supplemented by Massey [Massey (1991)] (Chapter XIV, Lemma 2.3).

(i) Given a class $\alpha \in H_p(\mathbb{R}^n, \mathbb{R}^n - K; R)$, we would like to find an open subset $N \subseteq \mathbb{R}^n$ containing K small enough so that $C \cap N = \emptyset$, and some class $\alpha' \in H_p(\mathbb{R}^n, \mathbb{R}^n - N; R)$ such that $\alpha = \rho^N_K(\alpha')$. For this, we use the fact that homology has compact support; see just before and just after Proposition 4.17. Since homology has compact support, by Proposition 4.17(1), there is a compact pair $(B, C) \subseteq (\mathbb{R}^n, \mathbb{R}^n - K)$ and some homology class $\beta \in H_p(B, C; \mathbb{R})$ such that $\rho^{B,C}_K(\beta) = \alpha$, where $\rho^{B,C}_K \colon H_p(B, C; R) \to H_p(\mathbb{R}^n, \mathbb{R}^n - K; R)$ is the homomorphism induced by the inclusion $(B, C) \subseteq (\mathbb{R}^n, \mathbb{R}^n - K)$.

Next we can pick an open subset $N \subseteq \mathbb{R}^n$ containing K small enough so that $C \cap N = \emptyset$. Since

$$\rho^{B,C}_K = \rho^N_K \circ (i_{B,C})_*,$$

where $(i_{B,C})_*\colon H_p(B,C;R) \to H_p(\mathbb{R}^n, \mathbb{R}^n - N; R)$ is the map induced by the inclusion $i_{B,C}\colon (B,C) \to (\mathbb{R}^n, \mathbb{R}^n - N)$ and $\rho^N_K\colon H_p(\mathbb{R}^n - N; R) \to H_p(\mathbb{R}^n, \mathbb{R}^n - K; R)$ is the map induced by the inclusion $\mathbb{R}^n - N \subseteq \mathbb{R}^n - K$, if we pick $\alpha' = (i_{B,C})_*(\beta)$, then $\alpha' \in H_p(\mathbb{R}^n, \mathbb{R}^n - N; R)$ is a homology class such that $\alpha = \rho^{B,C}_K(\beta) = \rho^N_K \circ (i_{B,C})_*(\beta) = \rho^N_K(\alpha')$. See Figure 7.4.

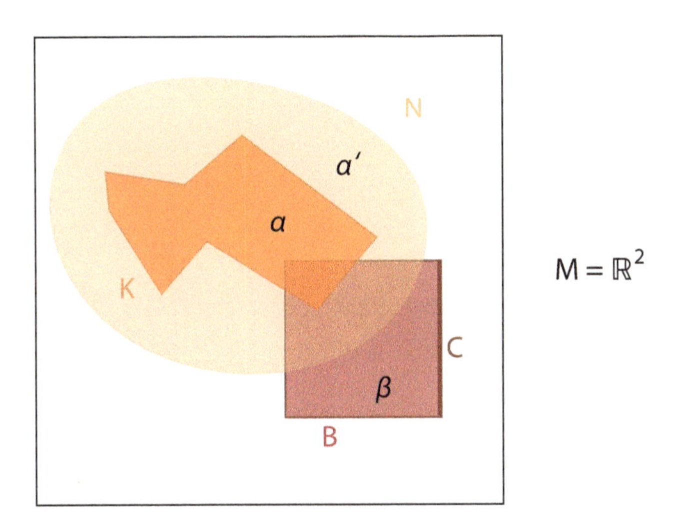

Fig. 7.4 A schematic illustration of the relationships between K, N, B, C and the associated homology classes necessary for Case 4(i).

Cover K by finitely many closed balls $B_1, \ldots, B_m$ such that $B_i \subseteq N$ and $B_i \cap K \neq \emptyset$ and write $\mathcal{B} = B_1 \cup \cdots \cup B_m$. Then $K \subseteq \mathcal{B} \subseteq N$ and we have the following commutative diagram.

$$\begin{array}{ccc} \alpha' \in H_p(\mathbb{R}^n, \mathbb{R}^n - N; R) & \xrightarrow{\rho^N_{\mathcal{B}}} & H_p(\mathbb{R}^n, \mathbb{R}^n - \mathcal{B}; R) \ni \alpha'' \\ & \searrow^{\rho^N_K} & \downarrow \rho^{\mathcal{B}}_K \\ & & \alpha \in H_p(\mathbb{R}^n, \mathbb{R}^n - K; R) \end{array}$$

where $\rho^N_{\mathcal{B}}$ is the map induced by the inclusion $\mathbb{R}^n - N \subseteq \mathbb{R}^n - \mathcal{B}$, $\rho^{\mathcal{B}}_K$ is the map induced by the inclusion $\mathbb{R}^n - \mathcal{B} \subseteq \mathbb{R}^n - K$, and ρ^N_K is the map induced by the inclusion $\mathbb{R}^n - N \subseteq \mathbb{R}^n - K$. See Figure 7.5.

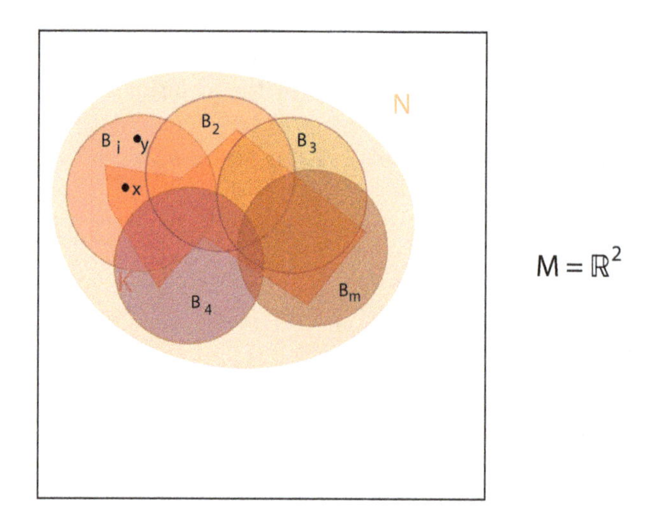

Fig. 7.5 A schematic illustration of the covering $\mathcal{B}$ used in the proof of Case 4(i).

If $p > n$, then $H_p(\mathbb{R}^n, \mathbb{R}^n - \mathcal{B}; R) = (0)$ by Case 3, hence $\rho^N_{\mathcal{B}}(\alpha') = 0$ and by ($\dagger_1$) we have $\alpha = \rho^N_K(\alpha') = \rho^{\mathcal{B}}_K(\rho^N_{\mathcal{B}}(\alpha')) = \rho^{\mathcal{B}}_K(0) = 0$.

(ii) Now consider the situation where $p = n$. Assume that $\rho^K_x(\alpha) = 0$ for all $x \in K$. As in (i), we can find an open subset $N \subseteq \mathbb{R}^n$ containing K and some $\alpha' \in H_n(\mathbb{R}^n, \mathbb{R}^n - N; R)$ such that $\alpha = \rho^N_K(\alpha')$. Let $\alpha'' = \rho^N_{\mathcal{B}}(\alpha')$ so that by ($\dagger_1$) $\alpha = \rho^N_K(\alpha') = \rho^{\mathcal{B}}_K(\rho^N_{\mathcal{B}}(\alpha')) = \rho^{\mathcal{B}}_K(\alpha'')$.

We claim that $\rho^{\mathcal{B}}_y(\alpha'') = 0$ for all $y \in \mathcal{B}$.

To show this, assume that $y \in B_i$ and pick some $x \in B_i \cap K$. Refer to Figure 7.5. Consider the following commutative diagram due to Massey.

$$\begin{array}{ccccc}
 & & H_n(\mathbb{R}^n, \mathbb{R}^n - \{y\}; R) & & \\
 & \nearrow^{\rho^{\mathcal{B}}_y} & & \nwarrow^{\rho^{B_i}_y} & \\
\alpha'' \in H_n(\mathbb{R}^n, \mathbb{R}^n - \mathcal{B}; R) & & \xrightarrow{\rho^{\mathcal{B}}_{B_i}} & & H_n(\mathbb{R}^n, \mathbb{R}^n - B_i; R) \\
\downarrow \rho^{\mathcal{B}}_K & & & & \downarrow \rho^{B_i}_x \\
\alpha \in H_n(\mathbb{R}^n, \mathbb{R}^n - K; R) & & \xrightarrow[\rho^K_x]{} & & H_n(\mathbb{R}^n, \mathbb{R}^n - \{x\}; R)
\end{array}$$

All homomorphisms are induced by inclusions. Since B_i is a closed ball, by Proposition 7.1, the maps $\rho_x^{B_i}$ and $\rho_y^{B_i}$ are isomorphisms. Since $\rho_x^K(\alpha) = 0$ and $\alpha = \rho_K^{\mathcal{B}}(\alpha'')$, using the commutative square we have

$$0 = \rho_x^K(\alpha) = \rho_x^K(\rho_K^{\mathcal{B}}(\alpha'')) = \rho_x^{B_i}(\rho_{B_i}^{\mathcal{B}}(\alpha'')).$$

Since $\rho_x^{B_i}$ is an isomorphism, we have

$$\rho_{B_i}^{\mathcal{B}}(\alpha'') = 0.$$

Using the commutative triangle we have

$$\rho_y^{\mathcal{B}}(\alpha'') = \rho_y^{B_i}(\rho_{B_i}^{\mathcal{B}}(\alpha'')) = \rho_y^{B_i}(0) = 0,$$

that is, $\rho_y^{\mathcal{B}}(\alpha'') = 0$ for all $y \in \mathcal{B}$. By Case 3, we must have $\alpha'' = 0$, and thus $\alpha = \rho_K^{\mathcal{B}}(\alpha'') = 0$.

Case 5. Suppose K is small enough so that $K \subseteq U$ for some open subset $U \subseteq M$ homeomorphic to $\mathbb{R}^n$.

(i) By excision (Theorem 4.14), $H_p(M, M - K; R) \cong H_p(U, U - K; R)$, and since U is homeomorphic to $\mathbb{R}^n$,

$$H_p(M, M - K; R) \cong H_p(U, U - K; R) \cong H_p(\mathbb{R}^n, \mathbb{R}^n - K; R),$$

so by Case 4, $H_p(\mathbb{R}^n, \mathbb{R}^n - K; R) \cong (0)$ for $p > n$ and thus $H_p(M, M - K; R) \cong (0)$ for $p > n$.

(ii) We use the isomorphism $\varphi\colon H_n(M, M-K; R) \to H_n(\mathbb{R}^n, \mathbb{R}^n - K; R)$ and Case 4. Details are left as an exercise.

Case 6. Suppose K is an arbitrary compact subset of M. Since M is a manifold and K is compact, there are finitely many charts covering K so we can write $K = K_1 \cup \cdots \cup K_m$, a union of small compact subsets in the sense of Case 5.

(i) By induction on m essentially as in Case 3, we prove using Case 2 and Case 5 that $H_p(M, M - K; R) \cong (0)$ for $p > n$.

(ii) The proof is also by induction on m. Details are left as an exercise.

□

Remark: It can be shown that if A is a closed subset of M, then $H_p(M, M - A; R) = (0)$ for all $p > n$; see Greenberg and Harper [Greenberg and Harper (1981)] (Section 22, Theorem 22.24).

Theorem 7.5. *(Vanishing) Let M be an n-manifold. We have $H_p(M; R) = (0)$ if $p > n$. If M is connected and noncompact, then $H_n(M; R) = (0)$.*

Theorem 7.5 is proven in Hatcher [Hatcher (2002)] (Chapter 3, Theorem 3.26(c) and Proposition 3.29), May [May (1999)] (Chapter 20, Section 4), and Bredon [Bredon (1993)] (Chapter VI, Corollary 7.12).

Sketch of proof. If $M = K$ is a compact n-manifold, by Theorem 7.4, we have $H_p(K; R) = (0)$ for all $p > n$. Since homology satisfies the axiom of compact support, by Theorem 4.17(4), we get

$$H_p(M; R) = \varinjlim_{K \in \mathcal{K}(M)} H_p(K; R) = (0)$$

for all $p > n$ (where $\mathcal{K}(M)$ denotes the set of compact subsets of M), as claimed. $\square$

Remark: The proof technique used to prove Theorem 7.4, as well as a number of other results, is a type of induction on compact subsets involving some limit argument. It is nicely presented in Bredon [Bredon (1993)] (Chapter VI, Section 7), where it is called the *Bootstrap Lemma*. Omitting proofs, here is a presentation of this method.

The Bootstrap Method

Given an n-manifold M, we would like to prove some property $P_M(A)$ about closed subsets A of M. Consider the following five properties:

(i) If A is a compact and convex subset of M, then $P_M(A)$ holds.
(ii) If $P_M(A), P_M(B)$ and $P_M(A \cap B)$ hold for some closed subsets A and B, then $P_M(A \cup B)$ holds.
(iii) if $A_1 \supseteq A_2 \supseteq \cdots A_i \supseteq A_{i+1} \supseteq \cdots$ is a sequence of compact subsets and if $P_M(A_i)$ holds for all i, then $P\left(\bigcap_i A_i\right)$ holds.
(iv) If $(A_i)_{i \in I}$ is a family of disjoint compact subsets with disjoint neighborhoods and if $P_M(A_i)$ holds for all i, then $P\left(\bigcup_i A_i\right)$ holds.
(v) For any closed subset A, if $P_M(A \cap W)$ holds for all open subsets W of M such that the closure of W is compact, then $P_M(A)$ holds.

We have the following proposition shown in Bredon [Bredon (1993)] (Chapter VI, Section 7, Lemma 7.9) and called the Bootstrap Lemma.

Proposition 7.6. *(Bootstrap Lemma) Let M be any n-manifold.*

(1) Let $P_M(A)$ be a property about compact subsets A of M. If (i), (ii), and (iii) hold, then $P_M(A)$ holds for all compact subsets A of M.

(2) *If M is a separable metric space, $P_M(A)$ is a property about closed subsets A of M, and all four statements (i)–(iv) hold, then $P_M(A)$ holds for all closed subsets A of M.*
(3) *Let $P_M(A)$ be a property about closed subsets A of M. If all five statements (i)–(v) hold, then $P_M(A)$ holds for all closed subsets A of M.*

Finally, we have our major result.

Theorem 7.7. *(Existence of a fundamental class) Let M be an n-manifold. For any compact subset K of M, for any R-orientation of M, there is a unique R-fundamental class μ_K of M at K which determines the same family of generators as the family $(\mu_x)_{x\in K}$ induced by the R-orientation of M. If M is compact, then M has a unique R-fundamental class μ_M corresponding to the family of generators $(\mu_x)_{x\in M}$ induced by the R-orientation.*

Theorem 7.7 is proven in Hatcher [Hatcher (2002)] (Chapter 3, Theorem 3.26), May [May (1999)] (Chapter 20, Section 3), Massey [Massey (1991)] (Chapter XIV, Theorem 2.2) and Milnor and Stasheff [Milnor and Stasheff (1974)] (Appendix A, Theorem A.8).

Sketch of proof. We follow Milnor and Stasheff's proof [Milnor and Stasheff (1974)] (Appendix A, Theorem A.8).

First we prove uniqueness. Recall that if K and L are two compact subsets of M such that $K \subseteq L$, we write $\rho^L_K\colon H_p(M, M-L; R) \to H_p(M, M-K; R)$ for the homomorphism induced by the inclusion $M - L \subseteq M - K$. Assume that there are two fundamental classes μ^1_K and μ^2_K. Then we have $\rho^K_x(\mu^1_K) = \mu_x$ and $\rho^K_x(\mu^2_K) = \mu_x$ for all $x \in K$, so $\rho^K_x(\mu^1_K - \mu^2_K) = 0$ for all $x \in K$. By Theorem 7.4(ii), we deduce that $\mu^1_K - \mu^2_K = 0$, that is, $\mu^1_K = \mu^2_K$.

The existence proof is divided in three steps.

Case 1. Since M is orientable, it has some open cover $(U_i)_{i\in I}$ such that each U_i has a fundamental class μ_i satisfying the consistency property of an orientation. Assume that K is contained in some U_i, and denote U_i as U and μ_i as μ_U. Then for all $x \in U$, the map $\rho^U_x : H_n(M, M-U; R) \to H_n(M, M - \{x\})$ is equal to the composition

$$H_n(M, M-U; R) \xrightarrow{\rho^U_K} H_n(M, M-K; R) \xrightarrow{\rho^K_x} H_n(M, M-\{x\}; R).$$

Then if we let $\mu_K = \rho_K^U(\mu_U)$, because by (†2) $\rho_x^U = \rho_x^K \circ \rho_K^U$, we have $\rho_x^K(\mu_K) = \rho_x^K(\rho_K^U(\mu_U)) = \rho_x^U(\mu_U) = \mu_x$, where $\mu_x = \rho_x^U(\mu_U)$ is a generator of $H_n(M, M - \{x\}; R)$ since μ_U is a fundamental class for U, and the consistency properties hold for μ_K since they hold for μ_U, so μ_K is a fundamental class for K. By construction, μ_U and μ_K induce the same set of generators.

Case 2. Suppose that $K = K_1 \cup K_2$, where K_1 and K_2 are two compact subsets and assume that μ_{K_1}, μ_{K_2} and $\mu_{K_1 \cap K_2}$ exist. As in Case 2 of the proof of Theorem 7.4, there is a Mayer–Vietoris long exact sequence of relative homology

$$\longrightarrow H_{p+1}(M, M - (K_1 \cap K_2)) \xrightarrow{\partial} H_p(M, M - K) \xrightarrow{\varphi} H_p(X, M - K_1) \oplus H_p(M, M - K_2) \xrightarrow{\psi} H_p(M, M - (K_1 \cap K_2)) \to \cdots$$

where

$$\varphi(\alpha) = \rho_{K_1}^K(\alpha) \oplus \rho_{K_2}^K(\alpha) \quad \text{and} \quad \psi(\beta \oplus \gamma) = \rho_{K_1 \cap K_2}^{K_1}(\beta) - \rho_{K_1 \cap K_2}^{K_2}(\gamma).$$

If $p = n$, by Theorem 7.4(i), we have $H_{n+1}(M, M - (K_1 \cap K_2)) \cong (0)$, so we have the piece of exact sequence

$$\longrightarrow (0) \xrightarrow{\partial} H_n(M, M - K) \xrightarrow{\varphi} H_n(X, M - K_1) \oplus H_n(M, M - K_2) \xrightarrow{\psi} H_n(M, M - (K_1 \cap K_2)) \to \cdots$$

which shows that φ is injective. Since by hypothesis μ_{K_1} and μ_{K_2} exist, by (†2) we have

$$\rho_x^{K_1 \cap K_2} \circ \rho_{K_1 \cap K_2}^{K_i}(\mu_{K_i}) = \rho_x^{K_i}(\mu_{K_i}) = \mu_x \quad \text{for all } x \in K_1 \cap K_2,\ i = 1, 2,$$

so

$$\rho_x^{K_1 \cap K_2}(\rho_{K_1 \cap K_2}^{K_1}(\mu_{K_1}) - \rho_{K_1 \cap K_2}^{K_2}(\mu_{K_2})) = 0 \quad \text{for all } x \in K_1 \cap K_2.$$

By Theorem 7.4(ii) applied to $K_1 \cap K_2$, we have

$$\rho_{K_1 \cap K_2}^{K_1}(\mu_{K_1}) - \rho_{K_1 \cap K_2}^{K_2}(\mu_{K_2}) = 0,$$

and since $\psi(\mu_{K_1} \oplus \mu_{K_2}) = \rho_{K_1 \cap K_2}^{K_1}(\mu_{K_1}) - \rho_{K_1 \cap K_2}^{K^2}(\mu_{K_2})$, we get

$$\psi(\mu_{K_1} \oplus \mu_{K_2}) = 0.$$

Since our sequence is exact, $\text{Im}\varphi = \text{Ker}\,\psi$, and since φ is injective, there is a unique $\mu_K \in H_n(M, M - K; R)$ such that

$$\varphi(\mu_K) = \mu_{K_1} \oplus \mu_{K_2}.$$

It remains to check that μ_K has properties required of a fundamental class, which is left as an exercise.

Case 3. Assume that K is an arbitrary compact subset of M. Since M is an R-orientable manifold and K is compact, K is covered by finitely many oriented charts, so we can express K as $K = K_1 \cup \cdots \cup K_m$, where each K_i is a compact subset of some open subset U_i of M that has some fundamental class μ_i as in Case 1. Using Case 1 and Case 2, we construct μ_K by induction on m. □

Definition 7.5. The fundamental class of a compact orientable manifold M is denoted by $[M]$.

If M is *any* manifold, not necessarily compact, then we know that M is $\mathbb{Z}/2\mathbb{Z}$-orientable and we have the following version of Theorem 7.7.

Theorem 7.8. *(Existence of a fundamental class, mod 2 case) Let M be any n-manifold (not necessarily orientable). For any compact subset K of M, there is a unique fundamental class μ_K of M at K such that $\rho_x^K(\mu_K) = \mu_x$ for all $x \in K$, where μ_x is the unique nonzero element of $H_n(M, M - K, \mathbb{Z}/2\mathbb{Z}) \cong \mathbb{Z}/2\mathbb{Z}$.*

The proof of Theorem 7.8 is essentially the same as the proof of Theorem 7.7.

The next theorem tells us what the group $H_n(M; R)$ looks like.

Theorem 7.9. *Let M be an n-manifold. If M is connected, then*

$$H_n(M;R) = \begin{cases} R & \text{if } M \text{ is compact and orientable} \\ \text{Ker}\,(R \xrightarrow{2} R) & \text{if } M \text{ is compact and not orientable} \\ (0) & \text{if } M \text{ is not compact.} \end{cases}$$

Here, the map $R \xrightarrow{2} R$ is the map $r \mapsto 2r$.

Theorem 7.9 is proven in Bredon [Bredon (1993)] (Chapter VI, Corollary 7.12).

In particular, Theorem 7.9 shows that if $R = \mathbb{Z}$ and if M is compact and not orientable then $H_n(M; R) = (0)$, and that if M is compact then $H_n(M; \mathbb{Z}/2\mathbb{Z}) = \mathbb{Z}/2\mathbb{Z}$.

Theorem 7.9 yields a crisp characterization of the orientability of a compact n-manifold (when $R = \mathbb{Z}$) in terms of the vanishing of $H_n(M; \mathbb{Z})$.

Proposition 7.10. *If M is a connected and compact n-manifold, then either $H_n(M; \mathbb{Z}) = (0)$ and M is not orientable, or $H_n(M; \mathbb{Z}) \cong \mathbb{Z}$, M is orientable, and the homomorphisms $H_n(M; \mathbb{Z}) = H_n(M, \emptyset; \mathbb{Z}) \longrightarrow H_n(M, M - \{x\}; \mathbb{Z})$ are isomorphisms for all $x \in M$.*

Proposition 7.10 is a special case of Corollary 8 in Spanier [Spanier (1989)] (Chapter 6, Section 3). It is also proven in May [May (1999)] (Chapter 20, Section 3). This second proof only uses Theorem 7.9 together with the universal coefficient theorem for homology (Theorem 12.5), but it is a nice proof worth presenting.

Proof. Since M is a compact manifold, for any $x \in M$, the manifold $M - \{x\}$ is not compact. By Theorem 7.9, we have $H_n(M - \{x\}; R) = (0)$. The long exact sequence of relative homology of the pair $(M, M - \{x\})$ (Theorem 4.9) yields the exact sequence

$$H_n(M - \{x\}; R) \longrightarrow H_n(M; R) \longrightarrow H_n(M, M - \{x\}; R),$$

and since $H_n(M - \{x\}; R) = (0)$ we deduce that

$$H_n(M; R) \longrightarrow H_n(M, M - \{x\}; R) \cong R$$

is an injective homomorphism for every ring R. We would like to conclude that if $R = \mathbb{Z}$ and if $H_n(M; \mathbb{Z}) \neq (0)$, then $H_n(M; \mathbb{Z}) \cong \mathbb{Z}$ and the above map is an isomorphism.

Since $\mathrm{Tor}_1^{\mathbb{Z}}(\mathbb{Z}, \mathbb{Z}/p\mathbb{Z}) = (0)$ (see the discussion after Theorem 12.5), by Theorem 12.4 we have

$$H_n(M; \mathbb{Z}/p\mathbb{Z}) \cong H_n(M; \mathbb{Z}) \otimes \mathbb{Z}/p\mathbb{Z}$$

and similarly

$$H_n(M, M - \{x\}; \mathbb{Z}/p\mathbb{Z}) \cong H_n(M, M - \{x\}; \mathbb{Z}) \otimes \mathbb{Z}/p\mathbb{Z}$$

for all $p > 0$. Since $H_n(M; R) \longrightarrow H_n(M, M - \{x\}; R) \cong R$ is an injective homomorphism for every ring R, for $R = \mathbb{Z}$ the homomorphism

$$H_n(M; \mathbb{Z}) \otimes \mathbb{Z}/p\mathbb{Z} \longrightarrow H_n(M, M - \{x\}; \mathbb{Z}) \otimes \mathbb{Z}/p\mathbb{Z} \cong \mathbb{Z} \otimes \mathbb{Z}/p\mathbb{Z} \cong \mathbb{Z}/p\mathbb{Z} \quad (*)$$

is injective for all $p > 0$. If $H_n(M;\mathbb{Z}) \neq (0)$, then we leave it as an exercise to prove that $H_n(M;\mathbb{Z}) \cong \mathbb{Z}$. Finally, the map $H_n(M;\mathbb{Z}) \longrightarrow H(M, M - \{x\};\mathbb{Z})$ must be an isomorphism since otherwise 1 would be mapped to some $m > 1$, but then the map $(*)$ would not be injective for $p = m$ (since $m \otimes z = 0$ for all $z \in \mathbb{Z}/m\mathbb{Z}$). $\square$

An important (and deep fact) about a compact manifold M is that its homology groups are finitely generated. This is not easy to prove; see Bredon [Bredon (1993)] (Appendix E, Corollary E.5), and Hatcher [Hatcher (2002)] (Appendix, Topology of Cell Complexes, Corollaries A.8 and A.9). As a consequence, using the universal coefficient theorem for cohomology (Theorem 12.11) we have the following result about the cohomology group $H^n(M;R)$ (see Bredon [Bredon (1993)], Chapter VI, Section 7, Corollary 7.14).

Proposition 7.11. *For any n-manifold M, if M is compact and connected, then*

$$H^n(M;R) = \begin{cases} R & \text{if } M \text{ is orientable} \\ R/2R & \text{if } M \text{ is not orientable.} \end{cases}$$

It should also be noted that if M is a smooth manifold, then the notion of orientability in terms of Jacobians of transition functions or the existence of a volume form, as defined for instance in Warner [Warner (1983)] or Tu [Tu (2008)], is equivalent to the notion of orientability given in Definition 7.1. This is proven (with a bit of handwaving) in Bredon [Bredon (1993)] (Chapter VI, Section 7, Theorem 7.15).

The second step to state the Poincaré duality theorem is to define the cap-product.

7.2 The Cap Product

Recall the definition of the maps $\lambda_p \colon \Delta^p \to \Delta^{p+q}$ and $\rho_q \colon \Delta^q \to \Delta^{p+q}$ defined in Section 4.10 and the definition of the cup product $\smile$; see Definition 4.32. In what follows, we write $n = p + q$, so $q = n - p$.

Definition 7.6. Given a cochain $c \in S^p(X;R)$ and a chain $\sigma \in S_n(X;R)$ (with $n \geq p \geq 0$), define the *cap product* $c \frown \sigma$ as the chain in $S_{n-p}(X;R)$ given by

$$c \frown \sigma = c(\sigma \circ \rho_p)(\sigma \circ \lambda_{n-p})$$

where $\sigma \circ \lambda_{n-p}$ is the front $(n-p)$-face of Δ^n and $\sigma \circ \rho_p$ is the back p-face of Δ^n.

Since $\sigma \circ \rho_p \in S_p(X; R)$ and $\sigma \circ \lambda_{n-p} \in S_{n-p}(X; R)$ we have $c(\sigma \circ \rho_p) \in R$, and indeed $c(\sigma \circ \rho_p)(\sigma \circ \lambda_{n-p}) \in S_{n-p}(X; R)$.

Definition 7.6 is designed so that

$$a(b \frown \sigma) = (a \smile b)(\sigma)$$

for all $a \in S^{n-p}(X; R)$, all $b \in S^p(X; R)$, and all $\sigma \in S_n(X; R)$, or equivalently using the bracket notation for evaluation as

$$\langle a, b \frown \sigma \rangle = \langle a \smile b, \sigma \rangle, \qquad (*)$$

which shows that $\frown$ is the adjoint of $\smile$ with respect to the evaluation pairing $\langle -, - \rangle$. Recall that the evaluation pairing $\langle -, - \rangle$ is defined by

$$\langle c, \tau \rangle = c(\tau), \quad c \in S^p(X; R), \tau \in S_p(X; R).$$

Indeed, since $a \in S^{n-p}(X; R)$ and $b \in S^p(X; R)$, by Definition 4.32, $a \smile b \in S^n(X; R)$ is the cocycle such that

$$(a \smile b)(\sigma) = a(\sigma \circ \lambda_{n-p}) b(\sigma \circ \rho_p), \quad \sigma \in S_n(X; R),$$

so $b \frown \sigma \in S_{n-p}(X; R)$ should be the cycle satisfying the equation

$$a(b \frown \sigma) = (a \smile b)(\sigma) = a(\sigma \circ \lambda_{n-p}) b(\sigma \circ \rho_p) = a(b(\sigma \circ \rho_p)(\sigma \circ \lambda_{n-p}))$$

for all $a \in S^{n-p}(X; R)$ and all $\sigma \in S_n(X; R)$, which implies

$$b \frown \sigma = b(\sigma \circ \rho_p)(\sigma \circ \lambda_{n-p}),$$

confirming that Definition 7.6 is forced by Condition $(*)$.

The reader familiar with exterior algebra and differential forms will observe that the cap product is a type of contraction (or hook).

Remark: There are several variants of Definition 7.6. Our version is the one adopted by Munkres [Munkres (1984)] (Chapter 8, Section 66). Milnor and Stasheff [Milnor and Stasheff (1974)] use the same formula except for the presence of the sign $(-1)^{p(n-p)}$ (also recall their sign convention for the coboundary operator). Hatcher [Hatcher (2002)] (Chapter 3, Section 3.3) uses the formula

$$\sigma \frown c = c(\sigma \circ \lambda_p)(\sigma \circ \rho_{n-p}),$$

with the order of σ and c switched, which forces λ and ρ to be switched.

Beware that Greenberg and Harper [Greenberg and Harper (1981)] (Part III, Section 24, Page 205) also switch the order of the arguments in the evaluation bracket: they write $\langle \tau, c\rangle = c(\tau)$; see Section 23, Page 174. They also define the cap product $\sigma \frown c$ as in Hatcher, by

$$\sigma \frown c = c(\sigma \circ \lambda_p)(\sigma \circ \rho_{n-p}).$$

Their pairing relation between the cup and the cap product is

$$(a \smile b)(\sigma) = \langle \sigma, a \smile b\rangle = \langle \sigma \frown a, b\rangle = b(\sigma \frown a).$$

In the end, this makes no difference but one has to be very careful about signs when stating the formula for $\partial(c \frown \sigma)$.

Recall that $\epsilon\colon S_0(X;R) \to R$ (the augmentation map; see Definition 4.14) is the unique homomorphism such that $\epsilon(x) = 1$ for every point $x \in S_0(X;R)$. The cohomology class of the cocycle ϵ (in $H^0(X;R)$) is denoted by 1.

Also recall (see Proposition 4.4) that if $f\colon X \to Y$ is a continuous map between two topological spaces X and Y, then there are induced homomorphisms $f_{\sharp,p}\colon S_p(X;R) \to S_p(Y;R)$ and $f_{*p}\colon H_p(X;R) \to H_p(Y;R)$. By applying $\mathrm{Hom}_R(-,R)$, we obtain homomorphisms $f^{\sharp,p}\colon S^p(Y;R) \to S^p(X;R)$ and $f^{*p}\colon H^p(Y;R) \to H^p(X;R)$ (see Proposition 4.31).

Proposition 7.12. *For any $c \in S^p(X;R)$ and any $\sigma \in S_n(X;R)$, the cap product*
$\frown\colon S^p(X;R) \times S_n(X;R) \to S_{n-p}(X;R)$ is bilinear and we have

$$\partial(c \frown \sigma) = (-1)^{n-p}(\delta c \frown \sigma) + c \frown \partial\sigma.$$

Furthermore, we have

$$\epsilon \frown \sigma = \sigma$$

for all $\sigma \in S_n(X;R)$, and

$$c \frown (d \frown \sigma) = (c \smile d) \frown \sigma,$$

for all $c \in S^p(X;R)$, all $d \in S^q(X;R)$, and all $\sigma \in S_{p+q+r}(X;R)$.

The cap product is natural with respect to continuous maps $f\colon X \to Y$, which means that for all $c \in S^p(Y;R)$ and all $\sigma \in S_n(X;R)$, we have

$$f_\sharp(f^\sharp(c) \frown \sigma) = c \frown f_\sharp(\sigma).$$

Proposition 7.12 is from Munkres [Munkres (1984)] (Chapter 8, Section 66, Theorem 66.1). As a consequence of the first formula, we see that the cap product induces a bilinear operation on cohomology and homology classes

$$\frown : H^p(X;R) \times H_n(X;R) \to H_{n-p}(X;R)$$

(if $0 \leq p \leq n$), also called *cap product.*

Remark: Using Milnor and Stasheff's sign convention both for δ and for the cap product, the formula for $\partial(c \frown \sigma)$ is

$$\partial(c \frown \sigma) = \delta c \frown \sigma + (-1)^p (c \frown \partial\sigma);$$

see Milnor and Stasheff [Milnor and Stasheff (1974)], Appendix A Formula (4), Page 276. The virtue of this formula is that there is a + sign in front of the term $\delta c \frown \sigma$, so in the proof of Poincaré duality the diagram in Case 2 of the proof commutes, not just up to sign. This sign issue is discussed in Hatcher [Hatcher (2002)] (Chapter 3, Section 3.3, Lemma 3.36) and Massey [Massey (1991)] (Chapter XIV, Section 8).

The following properties are immediate consequences of Proposition 7.12.

Proposition 7.13. *For any $a \in H_n(X;R)$ we have*

$$1 \frown a = a,$$

and

$$\omega \frown (\eta \frown a) = (\omega \smile \eta) \frown a,$$

for all $\omega \in H^p(X;R)$, all $\eta \in H^q(X;R)$, and all $a \in H_{p+q+r}(X;R)$.

The cap product is natural with respect to continuous maps $f \colon X \to Y$, which means that for all $[c] \in H^p(Y;R)$ and all $[\sigma] \in H_n(X;R)$, we have

$$f_*(f^*([c]) \frown [\sigma]) = [c] \frown f_*([\sigma]).$$

Given any cochain $c \in S^p(X;R)$ and any chain $\sigma \in S_p(X;R)$, the operation (evaluation) $(c,\sigma) \mapsto c(\sigma)$ is bilinear, and it is easy to check that it induces a bilinear map $\langle -,- \rangle \colon H^p(X;R) \times H_p(X;R) \longrightarrow R$ called the *Kronecker index*. The map $\epsilon \colon S_0(X;R) \to R$ carries boundaries to zero, hence it induces a homomorphism $\epsilon_* \colon H_0(X;R) \to R$; see just after Definition 4.14.

If X is path connected, then this homomorphism is an isomorphism (since $H_0(X; R) \cong R$). The following result shows how ϵ_* and the Kronecker index are related in terms of the cap product.

Proposition 7.14. *Let M be an n-manifold. For all $\omega \in H^p(X; R)$ and all $a \in H_p(X; R)$, we have*

$$\epsilon_*(\omega \frown a) = \langle \omega, a \rangle,$$

with $0 \leq p \leq n$.

Proposition 7.14 is proven in Munkres [Munkres (1984)] (Chapter 8, Section 66, Theorem 66.3).

There is also a version of the cap product for relative homology and cohomology,

$$\frown: H^p(X, A; R) \times H_n(X, A \cup B; R) \to H_{n-p}(X, B; R),$$

where A and B are open in X.

We will need the version where $B = \emptyset$ in the proof of the Poincaré duality theorem, namely

$$\frown: H^p(X, A; R) \times H_n(X, A; R) \to H_{n-p}(X; R).$$

First we define the cap product

$$\frown: S^p(X, A; R) \times S_n(X, A; R) \to S_{n-p}(X; R)$$

using the formula of Definition 7.6. To show that this definition makes sense at the level of relative cochains and chains, we need to check that for any cochain $\omega \in S^p(X; R)$ and any chain $\sigma \in S_n(X; R)$, if ω vanishes on all chains carried by A, and if σ is carried by A, then $\omega \frown \sigma = 0$, which is left as an exercise. Proposition 7.12 holds for this cap product so we can define a cap product

$$\frown: H^p(X, A; R) \times H_n(X, A; R) \to H_{n-p}(X; R).$$

Proposition 7.14 also holds for this relative version of the cap product.

For any continuous map $f\colon (X, A) \to (Y, B)$ (see Definition 4.13), we have induced homorphisms $f_\sharp\colon S_*(X, A; R) \to S_*(Y, B; R)$ and $f_*\colon H_*(X, A; R) \to H_*(Y, B; R)$ (see Proposition 4.7), and induced homorphisms $f^\sharp\colon S^*(Y, B; R) \to S^*(X, A; R)$ and $f^*\colon H^*(Y, B; R) \to H^*(X, A; R)$ (see Proposition 4.34). The cap product defined above is natural with respect to continuous map $f\colon (X, A) \to (Y, B)$, which means that for all $c \in S^p(Y, B; R)$ and all $\sigma \in S_n(X, A; R)$, we have

$$f_\sharp(f^\sharp(c) \frown \sigma) = c \frown f_\sharp(\sigma).$$

We leave it as an exercise to prove that Proposition 7.13 also holds for this relative version of the cap product.

7.3 Cohomology with Compact Support

We define a subcomplex $S_c^*(X;R)$ of $S^*(X;R)$ where each module $S_c^p(X;R)$ consists of cochains with compact support as follows.

Definition 7.7. A cochain $c \in S^p(X;R)$ is said to have *compact support* if there is some compact subset $K \subseteq X$ such that such that $c \in S^p(X, X - K;R)$, or equivalently if c has value zero on every singular simplex in $X - K$. For such a cochain c we see that δc also vanishes on all singular simplices in $X - K$, so the modules $S_c^p(X;R)$ of cochains with compact support form a subcomplex $S_c^*(X;R)$ of $S^*(X;R)$ whose cohomology modules are denoted $H_c^p(X;R)$ and called *cohomology groups with compact support.*

It turns out that the group $H_c^p(X;R)$ can be conveniently expressed as the direct limit of the groups of the form $H^p(X, X - K;R)$ where K is compact. Observe that if K and L are any two compact subsets of X and if $K \subseteq L$, then $S^p(X;X - K;R) \subseteq S^p(X, X - L;R)$, so we have a module homomorphism $\rho_L^K : H^p(X, X - K;R) \to H^p(X, X - L;R)$. The family $\mathcal{K}$ of all compact subsets of X ordered by inclusion is a directed set since the union of two compact sets is compact, so the direct limit

$$\varinjlim_{K \in \mathcal{K}} H^p(X, X - K;R)$$

of the mapping family $(H^p(X, X - K;R)_{K\in\mathcal{K}}, (\rho_L^K)_{K\subseteq L})$ is well-defined; see Section 8.3.

Proposition 7.15. *We have isomorphisms*

$$H_c^p(X;R) \cong \varinjlim_{K \in \mathcal{K}} H^p(X, X - K;R)$$

for all $p \geq 0$. Furthermore, if X is compact, then $H_c^p(X;R) \cong H^p(X;R)$.

Proposition 7.15 is actually not hard to prove; see Hatcher [Hatcher (2002)] (Chapter 3, Section 3.3, just after Proposition 3.33). Intuitively, X is approximated by larger and larger compact subsets K. If K is very large, $X - K$ is very small, so the group $H^p(X, X - K;R)$ is a "good" approximation of $H_c^p(X;R)$.

Remark: Unlike the case for ordinary singular cohomology, if $f : X \to Y$ is a continuous map, there is not necessarily an induced map $f^* : H_c^p(Y;R) \to H_c^p(X;R)$. The problem is that if K is a compact subset of Y, then $f^{-1}(K)$

is not necessarily compact. However, *proper* maps have this property and induce a corresponding map between cohomology groups with compact support. If f is proper, for any compact subset K of Y, $f^{-1}(K)$ is compact in X, so f maps $X - f^{-1}(K)$ into $Y - K$ and there is an induced homomorphism from $H^p(Y, Y - K; R)$ to $H^p(X, X - f^{-1}(K); R)$. Since $H_c^p(X; R)$ is the direct limit of the $H^p(X, X - L; R)$ where L ranges over compact subsets in X, there is a homomorphism from $H^p(X, X - f^{-1}(K); R)$ to $H_c^p(X; R)$, so we have a homomorphism from $H^p(Y, Y - K; R)$ to $H_c^p(X; R)$. We leave it as an exercise that these homomorphisms induce a homomorphism $f^* \colon H_c^p(Y; R) \to H_c^p(X; R)$.

Fortunately, the maps involved in Poincaré duality are inclusions and they are proper.

We know from Theorem 7.7 that if K is compact and if the n-manifold M is R-orientable, then there is a unique R-fundamental class $\mu_K \in H_n(M, M - K; R)$ of M at K. In particular, if M itself is compact and R-orientable, then there is a R-fundamental class μ_M.

Definition 7.8. Let M be a compact and R-orientable manifold. The map

$$D_M \colon H^p(M; R) \to H_{n-p}(M; R) \quad (0 \le p \le n)$$

is defined by $D_M(\omega) = \omega \frown \mu_M$.

Poincaré duality asserts that the map D_M is an isomorphism. To extend this isomorphism to cohomology with compact support when M is R-orientable we need to define D_M for noncompact spaces. We do this as follows.

Recall that there is a cap product

$$\frown \colon H^p(M, M - K; R) \times H_n(M, M - K; R) \to H_{n-p}(M; R).$$

Since there is an isomorphism

$$H_c^p(M; R) \cong \varinjlim_{K \in \mathcal{K}} H^p(M, M - K; R),$$

we generalize Definition 7.8 as follows.

Definition 7.9. Let M be an R-orientable manifold. For any $\omega \in H_c^p(M; R)$ $(0 \le p \le n)$, pick some representative ω' in the equivalence class defining ω in $\varinjlim H^p(M, M - K; R)$, namely some $\omega' \in H^p(M, M - K; R)$ for some compact subset K, and set

$$D_M(\omega) = \omega' \frown \mu_K \in H_{n-p}(M; R),$$

where $\mu_K \in H_n(M, M - K; R)$.

We need to prove that the above definition does not depend on the choice of the representative $\omega' \in H^p(M, M - K; R)$. If $\omega'' \in H^p(M, M - L; R)$ is another representative for some compact subset such that $K \subseteq L$, then it is easy to show that the diagram

$$\begin{array}{ccc} H^p(M, M-K; R) & \longrightarrow & H^p(M, M-L; R) \\ & {\scriptstyle - \frown \mu_K} \searrow \quad \swarrow {\scriptstyle - \frown \mu_L} & \\ & H_{n-p}(M; R) & \end{array}$$

is commutative, and thus

$$D_M \colon H^p_c(M; R) \to H_{n-p}(M; R)$$

as specified above is indeed well-defined.

7.4 The Poincaré Duality Theorem

The following theorem is a very general version of Poincaré duality applying to compact as well as noncompact manifolds.

Theorem 7.16. *(Poincaré Duality Theorem) Let M be an n-manifold and let R be a PID. If M is R-orientable, then the map*

$$D_M \colon H^p_c(M; R) \to H_{n-p}(M; R)$$

given in Definition 7.9 is an isomorphism for all $p \in \mathbb{Z}$. Furthermore, $H_p(M; R) = (0)$ and $H^p_c(M; R) = (0)$ for all p such that $p < 0$ or $p > n$. If $R = \mathbb{Z}/2\mathbb{Z}$, the above map is an isomorphism whether M is orientable or not.

If M is compact and R-orientable, then the map

$$D_M \colon \omega \mapsto \omega \frown \mu_M$$

is an isomorphism between $H^p(M; R)$ and $H_{n-p}(M; R)$ for all $p \in \mathbb{Z}$. Furthermore, $H_p(M; R) = (0)$ and $H^p(M; R) = (0)$ for all p such that $p < 0$ or $p > n$.

The "canonical" proof of Theorem 7.4 can be found in Milnor and Stasheff [Milnor and Stasheff (1974)] (Appendix A, Pages 277–279). This is a very elegant proof but some of the details are not worked out explicitly. Hatcher [Hatcher (2002)] (Chapter 3, Theorem 3.35), Greenberg and Harper [Greenberg and Harper (1981)] (Part III, Section 26, Theorem 26.6), Massey [Massey (1991)] (Chapter XIV, Theorem 4.1), and May

[May (1999)] (Chapter 20, Section 5) give more detailed and slightly more general proofs (it is not assumed that R is a PID).

A sticky point in the proof is the commutativity of a certain diagram in which the top row is a Mayer–Vietoris sequence of cohomology and the bottom row is a Mayer–Vietoris sequence of homology (see Case 2 of the proof in Milnor and Stasheff [Milnor and Stasheff (1974)], Appendix A). With the sign convention for δ and the definition of the cap product used by Hatcher and Greenberg and Harper, this diagram only commutes up to sign. This fact is carefully proven by these authors. The diagram commutes with the definitions used by Massey [Massey (1991)] (Chapter XIV, Lemma 8.2). May claims that the diagram commutes with his definition of the cap product but leaves this fact as an exercise to the reader.

On the other hand, and this is where the sign conventions used by Milnor and Stasheff pay off, it can be checked that the diagram in Milnor and Stasheff's proof commutes (not just up to sign). In the end, this is a technical point that does not affect the final result, but we felt that the reader should be warned.

The proof of Poincaré duality can be viewed as a sophisticated type of induction making use of Mayer–Vietoris sequences. The technical difficulty is that the induction step applies to open subsets. Cohomology with compact support comes to the rescue since we can take limits (really colimits) involving compact subsets. It turns out that we also need to use two kinds of induction: usual (finitary) induction, and transfinite induction in the form of Zorn's lemma.

Proof. We now present Milnor and Stasheff's proof, occasionally elaborated as in Hatcher, Greenberg and Harper, and Massey. The proof consists of five steps. Step 2 is one of the technically most involved.

Case 1. Assume that $M = \mathbb{R}^n$.

For any closed ball B, we know by Proposition 7.1 that $H_n(\mathbb{R}^n, \mathbb{R}^n - B; R) \cong R$ with generator μ_B, and $H_p(\mathbb{R}^n, \mathbb{R}^n - B; R) = (0)$ for all $p \neq n$. By Theorem 12.11 or Theorem 4.30 we have

$$H^n(\mathbb{R}^n, \mathbb{R}^n - B; R) \cong \mathrm{Hom}_R(H_n(\mathbb{R}^n, \mathbb{R}^n - B; R), R) \cong \mathrm{Hom}_R(R, R) \cong R$$

with a generator a such that $\langle a, \mu_B \rangle = 1$. Now Proposition 7.13 and Property $(*)$ applied to the cap product

$$\frown : H^n(\mathbb{R}^n, \mathbb{R}^n - B; R) \times H_n(\mathbb{R}^n, \mathbb{R}^n - B; R) \to H_0(\mathbb{R}^n; R)$$

imply that

$$1 = \langle a, \mu_B \rangle = \langle 1 \smile a, \mu_B \rangle = \langle 1, a \frown \mu_B \rangle,$$

and by definition of 1 (as the cohomology class of ϵ), $a \frown \mu_B$ is a generator of $H_0(\mathbb{R}^n; R) \cong R$. Thus $- \frown \mu_B$ maps $H^n(\mathbb{R}^n, \mathbb{R}^n - B; R)$ isomorphically to $H_0(\mathbb{R}^n; R)$, and since all the other modules are zero for $p \neq n$, by passing to the direct limit over the balls B as they become larger it follows that D_M maps $H_c^*(\mathbb{R}^n; R)$ isomorphically onto $H_*(\mathbb{R}^n; R)$.

Case 2. Suppose that $M = U \cup V$, where U and V are two open subsets of M, and assume that Poincaré duality holds for U, V, and $U \cap V$. See Figure 7.6.

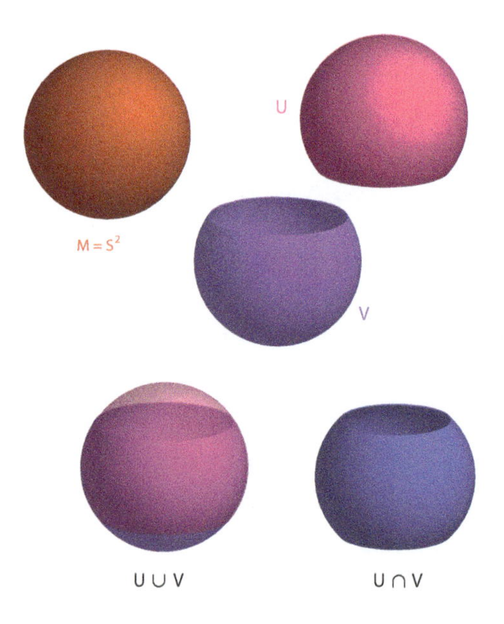

Fig. 7.6 An illustration of Case 2 where $M = S^2 = U \cup V$.

The goal is to construct the following commutative diagram involving a homological exact Mayer–Vietoris sequence on the bottom row and a cohomological Mayer–Vietoris sequence on the top row.

$$
\begin{array}{ccccccc}
\longrightarrow H_c^p(U\cap V) & \longrightarrow & H_c^p(U)\oplus H_c^p(V) & \longrightarrow & H_c^p(M) & \longrightarrow \\
\downarrow D_{U\cap V} & & \downarrow D_U\oplus D_V & & \downarrow D_M & \\
\longrightarrow H_{n-p}(U\cap V) & \longrightarrow & H_{n-p}(U)\oplus H_{n-p}(V) & \longrightarrow & H_{n-p}(M) & \longrightarrow
\end{array}
$$

Due to the lack of space we could not show more modules but the above diagram continues as

$$
\begin{array}{ccccccc}
\longrightarrow H_c^p(M) & \longrightarrow & H_c^{p+1}(U\cap V) & \longrightarrow & H_c^{p+1}(U)\oplus H_c^{p+1}(V) & \longrightarrow \\
\downarrow D_M & & \downarrow D_{U\cap V} & & \downarrow D_U\oplus D_V & \\
\longrightarrow H_{n-p}(M) & \longrightarrow & H_{n-p-1}(U\cap V) & \longrightarrow & H_{n-p-1}(U)\oplus H_{n-p-1}(V) & \longrightarrow
\end{array}
$$

The homological Mayer–Vietoris sequence on the bottom row is exact and we will prove that the cohomological sequence on the top row is also exact. Since by hypothesis (applied to U, V and $U \cap V$) the two leftmost and the two rightmost vertical arrows are isomorphisms, by the five lemma, the middle map $D_M\colon H_c^p(M) \to H_{n-p}(M)$ is an isomorphism.

It remains to prove that the above diagram commutes (at least, up to signs).

The bottom row is obtained using the Mayer–Vietoris sequence for homology for $X = M, A = U, B = V$; see Theorem 4.16.

To obtain the top row, pick some compact subsets $K \subseteq U$ and $L \subseteq V$ and apply Mayer–Vietoris in relative singular cohomology (Theorem 4.38) to $X = M, Y = M - (K \cap L), A = M - K, B = M - L$. See Figures 7.7 and 7.8.

Since $A \cap B = (M - K) \cap (M - L) = M - (K \cup L)$, we obtain the long exact sequence

$$
\begin{array}{l}
\longrightarrow H^p(M, M-(K\cap L)) \longrightarrow H^p(M, M-K)\oplus H^p(M, M-L) \longrightarrow \\
\longrightarrow H^p(M, M-(K\cup L)) \longrightarrow H^{p+1}(M, M-(K\cap L)) \longrightarrow
\end{array}
$$

By excision (Theorem 4.37), deleting $M - (U \cap V)$ from M and $M - (K \cap L)$ we obtain

$$
\begin{aligned}
H^p(M, M-(K\cap L)) &\cong H^p(U\cap V, (U\cap V)-(K\cap L)) \\
H^{p+1}(M, M-(K\cap L)) &\cong H^{p+1}(U\cap V, (U\cap V)-(K\cap L)),
\end{aligned}
$$

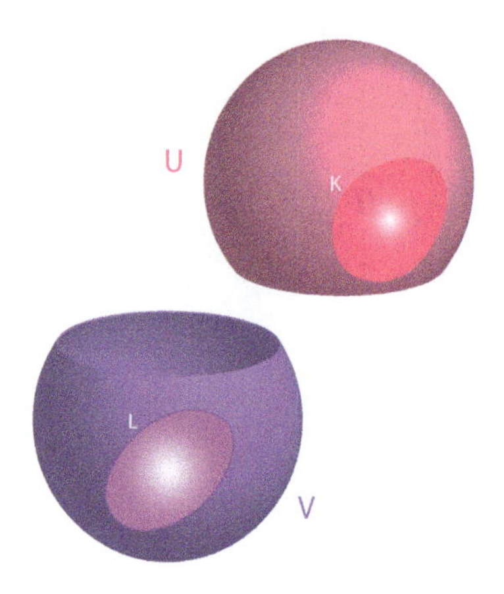

Fig. 7.7 Illustrations of the compact subsets K and L, where $S^2 = U \cup V$.

deleting $M - U$ from M and K we obtain

$$H^p(M, M - K) \cong H^p(U, U - K),$$

and deleting $M - V$ from M and L we obtain

$$H^p(M, M - L) \cong H^p(V, V - K).$$

Thus we obtain the exact sequence

$$\to H^p(U \cap V, (U \cap V) - (K \cap L)) \to H^p(U, U - K) \oplus H^p(V, V - L) \to H^p(M, M - (K \cup L)) \longrightarrow H^{p+1}(U \cap V, (U \cap V) - (K \cap L)) \to$$

Abbreviating $H^p(U\cap V, (U\cap V)-(K\cap L))$ as $H^p(U\cap V \mid K\cap L)$, $H^p(U, U - K)$ as $H^p(U \mid K)$, $H^p(V, V - L)$ as $H^p(V \mid L)$, $H^p(M, M - (K \cup L))$ as $H^p(M \mid K \cup L)$, *etc.*, we obtain the following diagram which is shown in two pieces since it does not fit on one line.

$$\begin{array}{ccccccc}
\to H^p(U \cap V \mid K \cap L) & \to & H^p(U \mid K) \oplus H^p(V \mid L) & \to & H^p_c(M \mid K \cup L) & \to \\
\downarrow D_{U\cap V} & & \downarrow D_U \oplus D_V & & \downarrow D_M & \\
\longrightarrow H_{n-p}(U \cap V) & \longrightarrow & H_{n-p}(U) \oplus H_{n-p}(V) & \longrightarrow & H_{n-p}(M) & \longrightarrow
\end{array}$$

$$\begin{array}{ccccccc}
H^p(M \mid K \cup L) & \xrightarrow{\delta} & H^{p+1}(U \cap V \mid K \cap L) & \to & H^{p+1}(U \mid K) \oplus H^{p+1}(V \mid L) & \to \\
\downarrow D_M & & \downarrow D_{U\cap V} & & \downarrow D_U \oplus D_V & \\
H_{n-p}(M) & \xrightarrow[\partial]{} & H_{n-p-1}(U \cap V) & \longrightarrow & H_{n-p-1}(U) \oplus H_{n-p-1}(V) & \longrightarrow
\end{array}$$

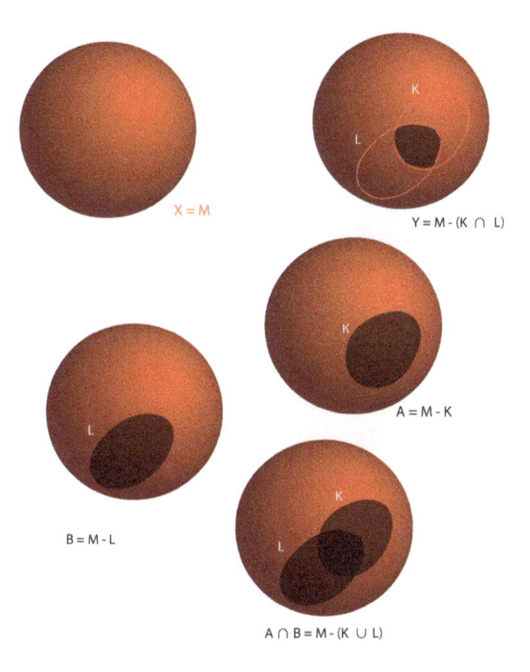

Fig. 7.8 The spaces associated with the Mayer–Vietoris sequence of Case 2.

We now come to the most tedious part of the proof where it is necessary to prove that all the squares commute. The commutativity of the squares not involving the coboundary map δ and the boundary map ∂ are a consequence of the naturality of the map D which follows immediately from the naturality of the relative cap product (Proposition 7.13 generalized to the relative cap product). The commutativity of the diagram involving δ and ∂ is tedious and a bit tricky. With our sign convention (which is the same as Bott and Tu), this diagram commutes only up to sign. The same thing holds with Hatcher and Greenberg and Harper's convention. These facts are proven in gory details by these authors. Massey proves that with his definitions, the diagram commutes. In fact, he devotes an entire appendix to this proof. It is a little tricky to follow his proof because the cap product is defined in terms of the slant product (see Massey [Massey (1991)] (Chapter XIII, Section 3). With Milnor and Stasheff's convention, the diagram also commutes. The verification is left as an exercise.

To finish the proof, we pass to the limit over the compact subsets $K \subseteq U$ and $L \subseteq V$, so in the limit we obtain the commutative diagram with cohomology with compact support on the top row discussed earlier and repeated below:

$$\begin{array}{ccccccc}
\longrightarrow H_c^p(U\cap V) & \longrightarrow & H_c^p(U)\oplus H_c^p(V) & \longrightarrow & H_c^p(M) & \longrightarrow \\
\downarrow D_{U\cap V} & & \downarrow D_U\oplus D_V & & \downarrow D_M & \\
\longrightarrow H_{n-p}(U\cap V) & \longrightarrow & H_{n-p}(U)\oplus H_{n-p}(V) & \longrightarrow & H_{n-p}(M) & \longrightarrow
\end{array}$$

$$\begin{array}{ccccccc}
\longrightarrow H_c^p(M) & \longrightarrow & H_c^{p+1}(U\cap V) & \longrightarrow & H_c^{p+1}(U)\oplus H_c^{p+1}(V) & \longrightarrow \\
\downarrow D_M & & \downarrow D_{U\cap V} & & \downarrow D_U\oplus D_V & \\
\longrightarrow H_{n-p}(M) & \longrightarrow & H_{n-p-1}(U\cap V) & \longrightarrow & H_{n-p-1}(U)\oplus H_{n-p-1}(V) & \longrightarrow
\end{array}$$

Since colimits preserve exactness, the top row is exact. Then we finish the proof using the five lemma as we did before.

Case 3. Suppose that $M = U = \bigcup_{i\in I} U_i$, for some family of open subsets U_i such that $U_i \subseteq U_{i+1}$ for all $i \geq 0$, and assume that Poincaré duality holds for each U_i, namely $H_c^p(U_i; R) \cong H_{n-p}(U_i; R)$. Since direct limits (colimits) preserve isomorphisms, we obtain an isomorphism

$$\varinjlim_{i\in I} H_c^p(U_i; R) \cong \varinjlim_{i\in I} H_{n-p}(U_i; R).$$

We need to prove that

$$H_c^p(U; R) \cong \varinjlim_{i\in I} H_c^p(U_i; R) \quad \text{and} \quad H_{n-p}(U, R) \cong \varinjlim_{i\in I} H_{n-p}(U_i; R)$$

to finish the proof.

Any compact subset K of U is contained in some U_i. Since homology is compactly supported (Proposition 4.17), we conclude that

$$H_{n-p}(U, R) \cong \varinjlim_{i\in I} H_{n-p}(U_i; R).$$

To prove the other isomorphisms, observe that

$$\begin{aligned}
H_c^p(U; R) &\cong \varinjlim_{i\in I} \varinjlim_{K \mid K\subseteq U_i} H^p(U_i, U_i - K; R) \\
&\cong \varinjlim_{K\subseteq U} \varinjlim_{i \mid K\subseteq U_i} H^p(U_i, U_i - K; R) \\
&\cong \varinjlim_{K\subseteq U} H^p(U, U - K; R).
\end{aligned}$$

The first isomorphism is an exchange of colimits, and the second isomorphism is a consequence of the fact that by excision, for i large enough, $H^p(U_i, U_i - K; R) \cong H^p(U, U - K; R)$; the details are left as an exercise. Finally, by definition,

$$\varinjlim_{K \subseteq U} H^p(U, U - K; R) \cong H^p_c(U; R).$$

Case 4. M is an open subset of $\mathbb{R}^n$. We can write M as a countable union of convex subsets, in fact, open balls U_i, say $M = \bigcup_{i \geq 1} U_i$. See Figure 7.9.

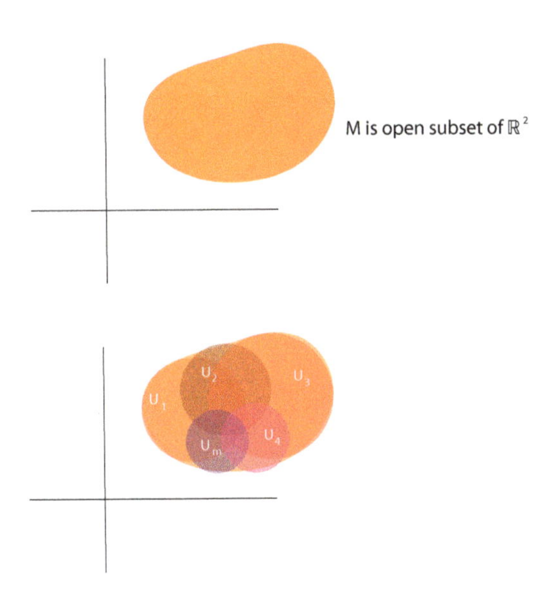

Fig. 7.9 Let M be an open subset of $\mathbb{R}^2$ with an associated open cover $\bigcup_{i \geq 1} U_i$.

Also recall that a convex open subset of $\mathbb{R}^n$ is homeomorphic to $\mathbb{R}^n$. Define the sequence (V_i) of open sets V_i given by

$$V_1 = U_1$$
$$V_{i+1} = V_i \cup U_{i+1}.$$

See Figure 7.10.

Both V_i and $V_i \cap U_{i+1}$ are unions of i convex open sets so we are reduced to proving that if $(W_j)_{j=1}^m$ is any finite family of convex opens sets, then Poincaré duality holds for their union $\bigcup_{j=1}^m W_j$.

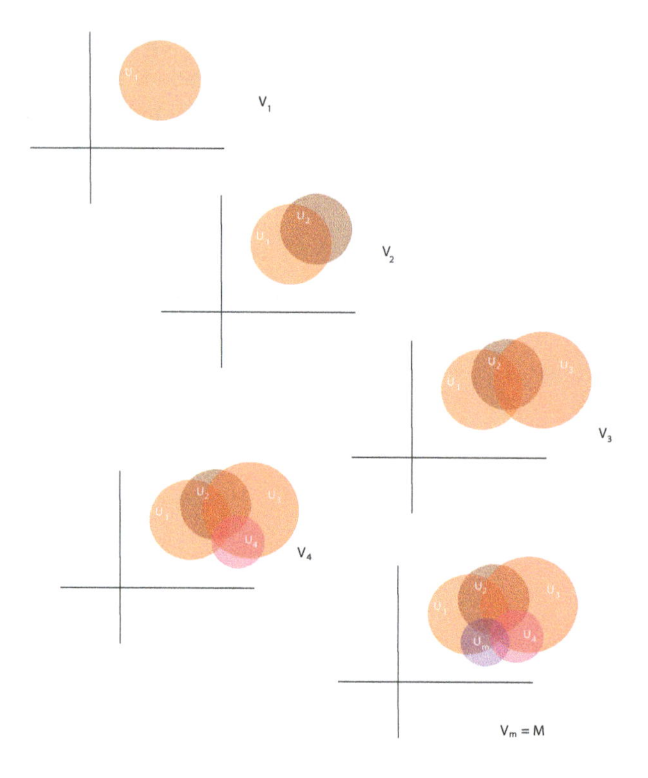

Fig. 7.10 The construction of V_i from the open cover $\bigcup_{i\geq 1} U_i$. Note $V_i \subseteq V_{i+1}$ for $1 \leq i \leq m-1$.

We proceed by induction on m. The base case $m = 1$ follows by Case 1. If we assume inductively that Poincaré duality holds for any family of m convex open sets, since all the W_j are convex and open for $j = 1, \ldots, m+1$, the intersection $\left(\bigcup_{j=1}^m W_j\right) \cap W_{m+1}$ is the union of m convex open sets, so by the induction hypothesis, Poincaré duality holds for $\bigcup_{j=1}^m W_j$ and $\left(\bigcup_{j=1}^m W_j\right) \cap W_{m+1}$, but it also holds for W_{m+1} by the base case, so by Case 2 it holds for $\bigcup_{j=1}^{m+1} W_j$.

Applying the above with $W_j = U_j$, we conclude that Poincaré duality holds for $\bigcup_{j=1}^m U_j = V_m$ for all $m \geq 1$. By Case 3, Poincaré duality holds for $M = \bigcup_{i\geq 1} V_i = \bigcup_{i \geq 1} U_i$.

Case 5. If M is covered by a finite or a countable family of open subsets U_i (domains of charts) homeomorphic to $\mathbb{R}^n$, then we can repeat the argument in Case 4 to conclude that Poincaré duality holds for M.

Otherwise, we use transfinite induction. This is what Milnor and Stasheff do. An alternative is to use Zorn's lemma. Consider the family of open subsets $U \subseteq M$ for which Poincaré duality holds. We can check that this family has the property that for every totally ordered sequence (under inclusion) $(U_i)_{i \in I}$ of such open subsets, by Case 3, Poincaré duality holds for $\bigcup_{i \in I} U_i$. By Zorn's lemma, there is a maximal open set $V \subseteq M$ such that Poincaré duality holds for V. If $V \neq M$, then pick some $x \in M - V$ and some coordinate chart W with $x \in W$. Since W is homeomorphic to an open subset of $\mathbb{R}^n$ and since V is open, $V \cap W$ is also homeomorphic to an open subset of $\mathbb{R}^n$, so Poincaré duality holds for W and $V \cap W$ by Case 4, and also for V by hypothesis, so by Case 2 Poincaré duality holds for $V \cup W$. But $V \neq V \cup W$ since $x \in W$ and $x \notin V$, contradicting the maximality of V. □

Theorem 7.16 actually holds for any commutative ring R with an identity element, not necessarily a PID. The only change in the proof occurs in Case 1; see Hatcher [Hatcher (2002)] (Chapter 3, Case (1) in the proof of Theorem 3.35), May [May (1999)] (Chapter 20, Section 5, Page 159), Greenberg and Harper [Greenberg and Harper (1981)] (Part III, Section 26, Step 3, Pages 220–221), and Massey [Massey (1991)] (Chapter XIV, Theorem 4.1).

Sketch of proof of Case (1) for any commutative ring. We follow Massey [Massey (1991)] (Chapter XIV, Theorem 4.1) with a twist of Greenberg and Harper [Greenberg and Harper (1981)]. In this case $M = \mathbb{R}^n$ and since every compact subset of $\mathbb{R}^n$ is contained in some closed ball B, we have

$$H^p_c(\mathbb{R}^n; R) \cong \varinjlim_{B} H^p(\mathbb{R}^n, \mathbb{R}^n - B; R)$$

with the limit taken over all closed balls in $\mathbb{R}^n$.

Using a translation of a homological argument into a cohomological argument (as in Proposition 4.18) it can be shown that

$$\widetilde{H}^p(S^n; R) = \begin{cases} R & \text{if } p = n \\ (0) & \text{if } p \neq n; \end{cases}$$

see Greenberg and Harper [Greenberg and Harper (1981)] (Part II, Section 15, Page 84, and Part III, Section 23, Page 183).

It can also be shown (by adapting the homological proof of Proposition 4.23 into a cohomological proof) that for any $x \in B$,

$$H^p(\mathbb{R}^n, \mathbb{R}^n - B; R) \cong H^p(\mathbb{R}^n, \mathbb{R}^n - \{x\}; R) \cong \widetilde{H}^{p-1}(S^{n-1}; R) \cong \widetilde{H}^p(S^n; R);$$

see Greenberg and Harper [Greenberg and Harper (1981)] (Part III, Section 26, Page 216). Therefore, we have

$$H_c^p(\mathbb{R}^n; R) \cong \widetilde{H}^p(S^n; R),$$

and the only nonzero module occurs for $p = n$. Since $\mathbb{R}^n$ is path connected, $\epsilon_*\colon H_0(\mathbb{R}^n; R) \to R$ is an isomorphism, and by Proposition 7.14, for all $\omega \in H^n(\mathbb{R}^n; \mathbb{R}^n - B; R)$ and all $a \in H_n(\mathbb{R}^n, \mathbb{R}^n - B; R)$, we have

$$\epsilon_*(\omega \frown a) = \langle \omega, a \rangle = \omega(a).$$

It follows that the map D_B given by $D_B(\omega) = \omega \frown \mu_B$ is an isomorphism

$$D_B\colon H^n(\mathbb{R}^n, \mathbb{R}^n - B; R) \to H_0(\mathbb{R}^n; R)$$

for every closed ball B, so by passing to the limit, as isomorphisms are preserved, we get an isomorphism

$$D_{\mathbb{R}^n}\colon H_c^n(\mathbb{R}^n; R) \to H_0(\mathbb{R}^n; R) \cong R,$$

proving Poincaré duality for $\mathbb{R}^n$. □

Example 7.1. Since the sphere S^n is compact and orientable, we can obtain its cohomology from its homology. Recall from Proposition 4.18 that for $n \geq 1$ we have

$$H_p(S^n; R) = \begin{cases} R & \text{if } p = 0, n \\ (0) & \text{if } p \neq 0, n. \end{cases}$$

Thus we obtain

$$H^p(S^n; R) = \begin{cases} R & \text{if } p = 0, n \\ (0) & \text{if } p \neq 0, n. \end{cases}$$

Similarly, since the n-torus $T^n = \underbrace{S^1 \times \cdots \times S^1}_{n}$ is compact and orientable, its cohomology is given by

$$H^p(T^n; R) = \underbrace{R \oplus \cdots \oplus R}_{\binom{n}{p}}.$$

As in the case of the sphere, it is identical to its homology, which reconfirms that these spaces are very symmetric.

Applications of Poincaré duality often involve the universal coefficient theorems (see Chapter 12). The reader is referred to Hatcher [Hatcher (2002)] (Chapter 3) for some of these applications. In particular, one will find a proof of the fact that the cohomology ring $H^*(\mathbb{CP}^n; \mathbb{Z})$ is isomorphic to $\mathbb{Z}[\alpha]/(\alpha^{n+1})$, with α of degree 2. As an application of Poincaré duality, we prove an important fact about compact manifolds of odd dimension.

Recall from Section 6.3 that the *Euler–Poincaré characteristic* $\chi(M)$ of a compact n-dimensional manifold is defined by

$$\chi(M) = \sum_{p=0}^{n} (-1)^p \operatorname{rank} H_p(M; \mathbb{Z}).$$

The natural numbers $\operatorname{rank} H_p(M; \mathbb{Z})$ are the *Betti numbers* of M and are denoted by b_p.

Proposition 7.17. *If M is a compact topological manifold (orientable or not) of odd dimension, then its Euler–Poincaré characteristic is zero, that is, $\chi(M) = 0$.*

Proof. Let $\dim M = 2m+1$. If M is orientable, by Poincaré duality $H^{2m+1-p}(M; \mathbb{Z}) \cong H_p(M; \mathbb{Z})$ for $p = 0, \ldots, 2m+1$, so $\operatorname{rank} H_p(M; \mathbb{Z}) = \operatorname{rank} H^{2m+1-p}(M; \mathbb{Z})$, but by Proposition 12.12, we have

$$H^n(M; \mathbb{Z}) \cong F_n \oplus T_{n-1}$$

where $H_n(M; \mathbb{Z}) = F_n \oplus T_n$ with F_n free and T_n a torsion abelian group, so $\operatorname{rank} H^{2m+1-p}(M; \mathbb{Z}) = \operatorname{rank} H_{2m+1-p}(M; \mathbb{Z})$. Therefore,

$$\operatorname{rank} H_p(M; \mathbb{Z}) = \operatorname{rank} H_{2m+1-p}(M; \mathbb{Z}),$$

and since $2m+1$ is odd we get

$$\begin{aligned} \chi(M) &= \sum_{p=0}^{2m+1} (-1)^p \operatorname{rank} H_p(M; \mathbb{Z}) \\ &= \sum_{p=0}^{2m+1} (-1)^p \operatorname{rank} H_{2m+1-p}(M; \mathbb{Z}) \\ &= -\sum_{p=0}^{2m+1} (-1)^{2m+1-p} \operatorname{rank} H_{2m+1-p}(M; \mathbb{Z}) \\ &= -\chi(M), \end{aligned}$$

so $\chi(M) = 0$.

If M is not orientable we apply Poincaré duality with $R = \mathbb{Z}/2\mathbb{Z}$. In this case each $H_p(M; \mathbb{Z}/2\mathbb{Z})$ and each $H^{2m+1-p}(M; \mathbb{Z}/2\mathbb{Z})$ is a vector space and their rank is just their dimension. Because $\mathbb{Z}/2\mathbb{Z}$ is a field, $H^{2m+1-p}(M; \mathbb{Z}/2\mathbb{Z})$ and $H_{2m+1-p}(M; \mathbb{Z}/2\mathbb{Z})$ are dual spaces of the same dimension, and as above we conclude that

$$\sum_{p=0}^{2m+1} (-1)^p \dim H_p(M; \mathbb{Z}/2\mathbb{Z}) = 0.$$

If we can show that

$$\chi(M) = \sum_{p=0}^{2m+1} (-1)^p \dim H_p(M; \mathbb{Z}/2\mathbb{Z}),$$

we are done. Since $\mathbb{Z}/2\mathbb{Z}$ is a field it is a PID, and the above equation follows from Proposition 6.13. For the sake of those readers who have not read Chapter 6 we provide the proof in the special case $R = \mathbb{Z}/2\mathbb{Z}$.

By the universal coefficient theorem for homology (Theorem 12.1) and the fact that

$$\mathbb{Z}/m\mathbb{Z} \otimes_{\mathbb{Z}} \mathbb{Z}/n\mathbb{Z} \cong \mathrm{Tor}_1^{\mathbb{Z}}(\mathbb{Z}/m\mathbb{Z}, \mathbb{Z}/n\mathbb{Z}) \cong \mathbb{Z}/\mathrm{gcd}(m,n)\mathbb{Z},$$

every term $\mathbb{Z}^k$ in $H_p(M; \mathbb{Z})$ when tensored with $\mathbb{Z}/2\mathbb{Z}$ gives a term $(\mathbb{Z}/2\mathbb{Z})^k$ in $H_p(M; \mathbb{Z}/2\mathbb{Z})$, every term $\mathbb{Z}/q\mathbb{Z}$ in $H_p(M; \mathbb{Z})$ with $q > 2$ when tensored with $\mathbb{Z}/2\mathbb{Z}$ yields (0), and every term $(\mathbb{Z}/2\mathbb{Z})^h$ in $H_p(M; \mathbb{Z})$ when tensored with $\mathbb{Z}/2\mathbb{Z}$ yields a term $(\mathbb{Z}/2\mathbb{Z})^h$ in $H_p(M; \mathbb{Z}/2\mathbb{Z})$, and the same term $(\mathbb{Z}/2\mathbb{Z})^h$ in $H_{p+1}(M; \mathbb{Z}/2\mathbb{Z})$ as the contribution of $\mathrm{Tor}_1^{\mathbb{Z}}((\mathbb{Z}/2\mathbb{Z})^h, \mathbb{Z}/2\mathbb{Z})$. The contribution of the two terms $(\mathbb{Z}/2\mathbb{Z})^h$ to the sum $\sum_{p=0}^{2m+1} (-1)^p \dim H_p(M; \mathbb{Z}/2\mathbb{Z})$ cancel out since their respective signs are $(-1)^p$ and $(-1)^{p+1}$, so

$$\chi(M) = \sum_{p=0}^{2m+1} (-1)^p \dim H_p(M; \mathbb{Z}/2\mathbb{Z}),$$

which concludes the proof. ☐

In the next section we present an even more general version of Poincaré Duality for cohomology and homology with coefficients in any R-module G and any commutative ring with identity element 1.

7.5 The Poincaré Duality Theorem with Coefficients in G

The first step is to define a version of the cap product that accommodates coefficients in G.

Definition 7.10. Define the cap product

$$\frown : S^p(X;G) \times S_n(X;R) \to S_{n-p}(X;G)$$

using a variant of the formula of Definition 7.6, namely for any cochain $c \in S^p(X;G)$ and any chain $\sigma \in S_n(X;R)$,

$$c \frown \sigma = (\sigma \circ \lambda_{n-p})c(\sigma \circ \rho_p),$$

where we switched the order of the two expressions on the right-hand side to conform with the convention that a chain in $S_{n-p}(X;G)$ is a formal combination of the form $\sum \sigma_i g_i$ with $g_i \in G$ and σ_i a $(n-p)$-simplex.

Since $\sigma \circ \rho_p \in S_p(X;R)$, $\sigma \circ \lambda_{n-p} \in S_{n-p}(X;R)$, and $c \in S^p(X;G)$, we have $c(\sigma \circ \rho_p) \in G$, and indeed $(\sigma \circ \lambda_{n-p})c(\sigma \circ \rho_p) \in S_{n-p}(X;G)$ so the above definition makes sense.

If $a \in S^{n-p}(X;R)$, $b \in S^p(X;G)$ and $\sigma \in S_n(X;R)$, by Definition 4.34, we have

$$\langle a \smile b, \sigma \rangle = a(\sigma \circ \lambda_{n-p})b(\sigma \circ \rho_p),$$

and

$$b \frown \sigma = (\sigma \circ \lambda_{n-p})b(\sigma \circ \rho_p),$$

so if $\langle f, s \rangle$ with $f \in S^p(X;R)$ and $s \in S_p(X;G)$ is defined the right way, the identity

$$\langle a, b \frown \sigma \rangle = \langle a \smile b, \sigma \rangle$$

will hold. But the definition of a pairing $\langle -,- \rangle : S^q(X;R) \times S_q(X;G) \to G$ is standard, namely

$$\left\langle f, \sum \sigma_i g_i \right\rangle = \sum f(\sigma_i) g_i,$$

where $f \in S^q(X;R)$ and $\sum \sigma_i g_i$ is a singular q-simplex in $S_q(X;G)$ (where the σ_i are q-simplices). In the above situation, $q = n - p$.

It is even possible to define a pairing $\langle -,- \rangle : S^q(X;G) \times S_q(X;G') \to G \otimes G'$, where G and G' are two different R-modules; see Spanier [Spanier (1989)] (Chapter 5, Section 5, Page 243). In summary, the equation

$$\langle a, b \frown \sigma \rangle = \langle a \smile b, \sigma \rangle$$

holds for this more general version of cup products and cap products.

The formula

$$\partial(c \frown \sigma) = (-1)^{n-p}(\delta c \frown \sigma) + c \frown \partial\sigma$$

of Proposition 7.12 still holds for any $c \in S^p(X;G)$ and any $\sigma \in S_n(X;R)$, so we obtain a cap product

$$\frown\colon H^p(X;G) \times H_n(X;R) \to H_{n-p}(X;G).$$

There is also a relative version of the cap product

$$\frown\colon H^p(X,A;G) \times H_n(X,A;R) \to H_{n-p}(X;G)$$

which will be used in the version of Poincaré duality with coefficients in G; see May [May (1999)] (Chapter 20, Section 2), but beware that this definition is very abstract. Actually it is possible to use Definition 7.10 to define the above relative cap product and to justify this definition using the reasoning at the end of Section 7.2. We leave this verification as an exercise.

The most general relative cap product is a bilinear map

$$\frown\colon H^p(X,A;G) \times H_n(X,A \cup B;R) \to H_{n-p}(X;B,G)$$

where A and B are subsets of X such that $\mathrm{Int}(A) \cup \mathrm{Int}(B) = A \cup B$, where these interiors are defined with respect to the subspace topology on $A \cup B$ induced by X; see Spanier [Spanier (1989)] (Chapter 5, Section 7, Page 254), Munkres [Munkres (1984)] (Chapter 8, Section 66, Page 392) and Hatcher [Hatcher (2002)] (Chapter 3, Section 3.3). This version of the cap product will be used in Section 14.5 on Alexander–Lefschetz duality.

Next we promote singular cohomology with coefficients in G to cohomology with compact support. All one has to do is replace R by G everywhere. We obtain the *cohomology groups with compact support* $H^p_c(X;G)$. It is easy to verify that Proposition 7.15 also holds.

Proposition 7.18. *We have isomorphisms*

$$H^p_c(X;G) \cong \varinjlim_{K \in \mathcal{K}} H^p(X, X-K;G)$$

for all $p \geq 0$. *Furthermore, if* X *is compact, then* $H^p_c(X;G) \cong H^p(X;G)$.

Given an R-orientable manifold M we also have to generalize the mapping
$D_M : H_c^p(M; R) \to H_{n-p}(M; R)$ to a mapping

$$D_M : H_c^p(M; G) \to H_{n-p}(M; G),$$

and for this we use the cup product

$$\frown : H^p(M, M - K; G) \times H_n(M, M - K; R) \to H_{n-p}(X; G)$$

defined above. Since there is an isomorphism

$$H_c^p(M; G) \cong \varinjlim_{K \in \mathcal{K}} H^p(M, M - K; G),$$

for any $\omega \in H_c^p(M; G)$ we pick some representative ω' in the equivalence class defining ω in $\varinjlim H^p(M, M - K; G)$, namely some $\omega' \in H^p(M, M - K; G)$ for some compact subset K, and since $\mu_K \in H_n(M, M - K; R)$ we set

$$D_M(\omega) = \omega' \frown \mu_K \in H_{n-p}(M; G).$$

Then we prove that the above definition does not depend on the choice of the representative $\omega' \in H^p(M, M - K; G)$ just as in the case where $G = R$. In conclusion, we obtain our map

$$D_M : H_c^p(M; G) \to H_{n-p}(M; G).$$

Using this map, the following version of Poincaré duality can be proven.

Theorem 7.19. *(Poincaré Duality Theorem for Coefficients in a Module) Let M be an n-manifold, let R be any commutative ring with unit 1, and let G be any R-module. If M is R-orientable, then the map*

$$D_M : H_c^p(M; G) \to H_{n-p}(M; G)$$

defined above is an isomorphism for all $p \in \mathbb{Z}$. Furthermore, $H_p(M; G) = (0)$ and $H_c^p(M; G) = (0)$ for all p such that $p < 0$ or $p > n$. If $R = \mathbb{Z}/2\mathbb{Z}$, the above map is an isomorphism whether M is orientable or not.

If M is compact and R-orientable, then the map

$$D_M : \omega \mapsto \omega \frown \mu_M$$

is an isomorphism between $H^p(M; G)$ and $H_{n-p}(M; G)$ for all $p \in \mathbb{Z}$. Furthermore, we have $H_p(M; G) = (0)$ and $H^p(M; G) = (0)$ for all p such that $p < 0$ or $p > n$.

Theorem 7.19 is proven in May [May (1999)] (Chapter 20, Section 5) and Massey [Massey (1991)] (Chapter XIV, Theorem 4.1). Except for Case 1, the proof is basically identical to the proof of Theorem 7.16.

The proof of Case 1 is modified as follows. By tensoring with G, the map $\epsilon\colon S_0(X;R) \to R$ yields a map $S_0(X;R) \otimes G \to R \otimes G \cong G$ which induces a homomorphism which we also denote $\epsilon_*\colon H_0(X;G) \to G$. This homomorphism is an isomorphism if X is path connected, and Proposition 7.14 holds, namely

$$\epsilon_*(\omega \frown a) = \langle \omega, a \rangle,$$

for all $\omega \in H^p(X;G)$ and all $a \in H_p(X;G)$ $(0 \leq p \leq n)$. The rest of the proof is analogous to the proof given in Section 7.4.

We will see later on that there is an even more general version of duality known as Alexander–Lefschetz duality; see Chapter 14.

7.6 Problems

Problem 7.1. Prove that Condition (†) in Definition 7.1 can be replaced by the condition

$$\rho_x^{U_i}(\mu_i) = \rho_x^{U_j}(\mu_j) \quad \text{for all } x \in U_i \cap U_j.$$

Hint. Use Proposition 7.1.

Problem 7.2. Prove the identities from Proposition 7.12 listed below. For any $c \in S^p(X;R)$ and any $\sigma \in S_n(X;R)$, we have

$$\partial(c \frown \sigma) = (-1)^{n-p}(\delta c \frown \sigma) + c \frown \partial\sigma.$$

Furthermore, we have

$$\epsilon \frown \sigma = \sigma$$

for all $\sigma \in S_n(X;R)$, and

$$c \frown (d \frown \sigma) = (c \smile d) \frown \sigma,$$

for all $c \in S^p(X;R)$, all $d \in S^q(X;R)$, and all $\sigma \in S_{p+q+r}(X;R)$.

Problem 7.3. Prove Proposition 7.14.

Problem 7.4. Define the cap product

$$\frown : S^p(X, A; R) \times S_n(X, A; R) \to S_{n-p}(X; R)$$

using the formula of Definition 7.6.

(1) Check carefully that this cap product is well defined.

(2) Prove Proposition 7.12 for this cap product.

(3) Prove Proposition 7.14 for this cap product.

Problem 7.5. Prove that the diagram

$$\begin{array}{ccc} H^p(M, M-K; R) & \longrightarrow & H^p(M, M-L; R) \\ & {\scriptstyle -\frown \mu_K} \searrow \quad \swarrow {\scriptstyle -\frown \mu_L} & \\ & H_{n-p}(M; R) & \end{array}$$

where $K \subseteq L$ are compact subsets of M is commutative, and thus

$$D_M : H^p_c(M; R) \to H_{n-p}(M; R)$$

as specified in Definition 7.9 at the end of Section 7.3 is indeed well-defined.

Problem 7.6. It can be shown that projective space $\mathbb{RP}^n$ with n odd is orientable. Use this fact to prove that

$$H^p(\mathbb{RP}^n; \mathbb{Z}) = \begin{cases} \mathbb{Z} & \text{for } p = 0 \text{ and for } p = n \text{ odd} \\ \mathbb{Z}/2\mathbb{Z} & \text{for } p \text{ even, } 0 < p \leq n \\ (0) & \text{otherwise.} \end{cases}$$

Problem 7.7. Recall from Problem 5.5 that the space X_g (a surface) is obtained by forming the g-fold connected sum

$$X_g = \underbrace{T \sharp \cdots \sharp T}_{g}$$

by gluing together $g \geq 2$ tori. It can be shown that these surfaces are orientable. Prove that the cohomology of X_g is given by

$$H^0(X_g) = \mathbb{Z}$$
$$H^1(X_g) = \mathbb{Z}^{2g}$$
$$H^2(X_g) = \mathbb{Z}.$$

Problem 7.8. Prove that if M is a compact orientable manifold of dimension $n = 4k + 2$, then $\chi(M)$ is even.

Hint. This is a hard problem. For help, see Greenberg and Harper [Greenberg and Harper (1981)] (Corollary 26.11).

Chapter 8

Presheaves and Sheaves; Basics

Presheaves and sheaves are two of the indispensable tools used in some of the more advanced parts of algebraic topology and algebraic geometry. Therefore it is important for the reader to be exposed to these concepts as soon as possible. Unfortunately, many presentations of these concepts quickly take a very abstract turn, especially when explaining the process of converting a presheaf into a sheaf.

We believe that it is best to proceed in two stages. In the first stage, the concepts of preseaves and sheaves are defined as concretely as possible, using familiar examples as illustrations. This is what we do in this chapter. In Chapter 9 we show how the notion of presheaf can be used to define a very general kind of cohomology based on open covers, called Čech cohomology. In the second stage, we discuss more sophisticated aspects of sheaves, including the process of converting a presheaf into a sheaf (sheafification) and exact sequences of sheaves. This second stage is presented in Chapter 10.

According to Dieudonné the origin of the notions of presehaves and sheaves can be traced to papers of Jean Leray published in 1945–1947. In his 1945 paper, Leray's goal was to define a cohomology theory on an arbitrary topological space X, starting from some (almost) arbitrary cochain complex C^*. In this theory, every cochain $c \in C^*$ is assigned a *support* $S(c) \subseteq X$ subject to certain axioms. A pair of the form (C^*, S) is called a *concrete complex*. The central objects in Leray's theory are special kinds of concrete complexes that he called *couvertures*. The English translation of "couverture" is "cover" (it could also be "blanket"). In his 1947 paper, Leray introduced the notion of *fine complex* and *fine couverture*. Using these notions he was able to establish the equivalence of his notion of cohomology with the Alexander–Spanier cohomology discussed later in this book.

The notion of couverture was abandoned shortly after its creation but there is little doubt that it was one of the intermediate steps that led Leray to the much more flexible notion of sheaf.

Another motivation for the notion of sheaf was the desire to define cohomology theories with varying coefficients (as opposed to using a fixed abelian group G, use a family of abelian group G_α). Reidemeister came up with such a theory in 1935 to study the homology of a covering space. Steenrod in his work on fibre bundles in 1942 considered homology and cohomology with local coefficients.

Presheaves and sheaves were introduced for the first time in a paper of Jean Leray published in 1946. One of Leray's main motivations was the following problem: given a continuous map $f\colon Y \to X$ between two topological spaces Y and X, find a relationship between the cohomology of Y and the cohomology of X using properties of f. In particular, assuming f surjective, is it possible to reconstruct the cohomology $H^*(Y)$ of Y from the cohomology $H^*(X)$ of X and the cohomology $H^*(f^{-1}(x))$ of each fibre $f^{-1}(x)$ $(x \in X)$?

The above question suggests considering the assignment of some module $\mathcal{F}(x)$ to $x \in X$, and more generally of some module $\mathcal{F}(E)$ to each subset E in some designated family of subsets of X. Leray picked the closed subsets. Such assignments $E \mapsto \mathcal{F}(E)$ must satisfy certain properties which allow the "passage from local to global information." These are sheaves in the sense of Leray.

Leray's paper and subsequent lectures on the subject triggered some major activity on the subject in the years 1947–1951. Henri Cartan became very active on this topic, as well as two of his students, Jean-Pierre Serre and Jean-Louis Koszul. Armand Borel also played a key role in these developments. Henri Cartan who had worked on some problems in complex analysis was very familiar with the passage from local to global and realized that it was preferable to define a sheaf as an assignment of a module $\mathcal{F}(U)$ to an *open* set U. His definition is essentially the definition we adopted, as presented by Godement [Godement (1958)]; see also Serre [Serre (1955)]. Another definition of a sheaf was given by Henri Cartan and Michel Lazard in 1951 based on the notion of "espace étalé." The Cartan–Lazard notion of a sheaf is equivalent to the previous definition of a sheaf. We will discuss this equivalence in Chapter 10.

After this historical introduction, let us return to the topics covered in this chapter. In Section 8.1 we define presheaves and maps (morphisms) of presheaves.

Presheaves are typically used to keep track of local information assigned to a global object (the space X). It is usually desirable to use consistent local information to recover some global information, but this requires a sharper notion, that of a sheaf. Section 8.2 is devoted to an elementary presentation of sheaves. A deeper study is conducted in Chapter 10.

In general, a presheaf fails to satisfy the consistency conditions of a sheaf but there is a procedure (known as sheafification) for converting a presheaf into a sheaf (see Chapter 10). This method uses a notion of limit of a family of modules called a direct mapping family. This notion of limit is also needed in defining the Čech cohomology modules $\check{H}^p(X, \mathcal{F})$ of a space X with values in a presheaf $\mathcal{F}$ from the family of Čech cohomology modules $\check{H}^p(\mathcal{U}, \mathcal{F})$ associated with open covers $\mathcal{U}$ of X (see Chapter 9). In preparation for the topics mentioned above we carefully discuss direct mapping families and direct limits in Section 8.3.

8.1 Presheaves

Roughly speaking, presheaves (and sheaves) are a way of packaging local information about a topological space X in a way that is mathematically useful. We can imagine that above every open subset U of X there is a "balloon" $\mathcal{F}(U)$ of information about U, often a set of functions, and that this information is compatible with restriction; namely if V is another open set contained in U, then the balloon of information $\mathcal{F}(V)$ is obtained from $\mathcal{F}(U)$ by some restriction function ρ^U_V. See Figure 8.1.

The typical example of a presheaf (in this case, actually sheaf) is as follows: given a topological space X (for simplicity, you may assume that $X = \mathbb{R}$, or $X = \mathbb{R}^n$), for every (nonempty) open subset U of X, let $C^0(U)$ be the set of all real-valued continuous functions $f\colon U \to \mathbb{R}$. For any open subset $V \subseteq U$, we obtain a function $\rho^U_V\colon C^0(U) \to C^0(V)$ by restricting any function $f\colon U \to \mathbb{R}$ to V. See Figure 8.1.

Observe that if $W \subseteq V \subseteq U$, then

$$\rho^U_W = \rho^V_W \circ \rho^U_V$$

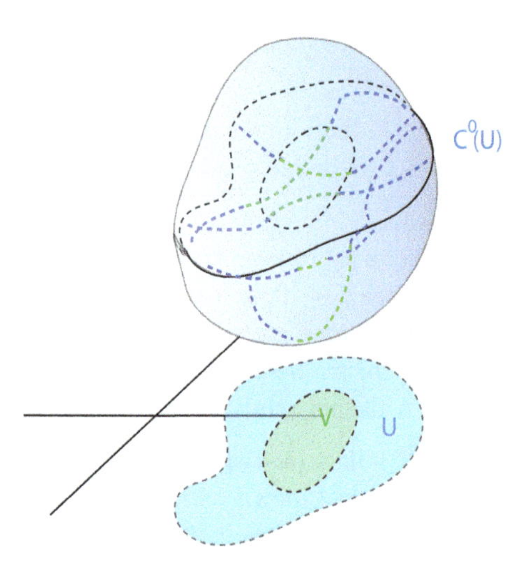

Fig. 8.1 The elevated blue balloon is a schematic representation of a presheaf of real valued functions over the open set $U \subseteq \mathbb{R}^2$. Each "function" is represented as blue and green dotted lines, where the green dash is the restriction of the function on V.

and

$$\rho_U^U = \mathrm{id}_U.$$

See Figure 8.2.

The assignment $U \mapsto C^0(U)$ is a presheaf on X. In the above example each $C^0(U)$ can be viewed as a set, but also as a real vector space, or a ring, or even as an algebra, since functions can be added, rescaled, and multiplied pointwise.

More generally, we can pick a class of structures, say sets, vector spaces, R-modules (where R is a commutative ring with a multiplicative identity), groups, commutative rings, R-algebra, *etc.*, and assign an object $\mathcal{F}(U)$ in this class to every open subset U of X, and for every pair of open subsets U, V such that $V \subseteq U$, if we write $i\colon V \to U$ for the inclusion map from V to U, then we assign to i a map $\mathcal{F}(i)\colon \mathcal{F}(U) \to \mathcal{F}(V)$ which is a morphism of the class of objects under consideration. This means that if the $\mathcal{F}(U)$ are sets, then the $\mathcal{F}(i)$ are just functions; if the $\mathcal{F}(U)$ are R-modules then the $\mathcal{F}(i)$ are R-linear maps; if the $\mathcal{F}(U)$ are groups then the

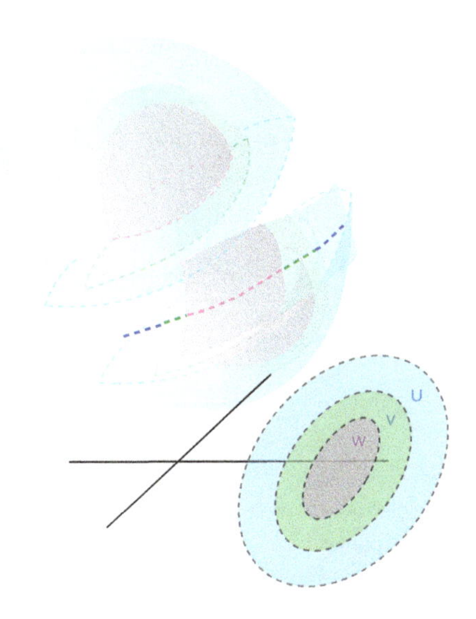

Fig. 8.2 A schematic representation of the nested presheaves of continuous functions associated with the open subsets $W \subseteq V \subseteq U \subseteq \mathbb{R}^2$. The wavy plane with the bold dashed line represents the graph of a continuous real-valued function with domain in U. If this function is restricted to the different colored "balloons," (which have been opened to show the graph of the continuous function), the domain is restricted appropriately, namely to either V or W, as evidenced by the color change.

$\mathcal{F}(i)$ are group homomorphisms; if the $\mathcal{F}(U)$ are rings then the $\mathcal{F}(i)$ are ring homomorphisms, *etc.*

A fancy way to proceed would be assume that we have a category $\mathbf{C}$ and that objects of $\mathbf{C}$ are assigned to open subsets of X and morphisms of $\mathbf{C}$ are assigned to inclusion maps, so that a presheaf is a contravariant functor. For our purposes it is not necessary to assume such generality.

Definition 8.1. Given a topological space X and a class $\mathbf{C}$ of structures (a category), say sets, vector spaces, R-modules, groups, commutative rings, *etc.*, a *presheaf on X with values in* $\mathbf{C}$ consists of an assignment of some object $\mathcal{F}(U)$ in $\mathbf{C}$ to every open subset U of X and of a map $\mathcal{F}(i)\colon \mathcal{F}(U) \to \mathcal{F}(V)$ of the class of structures in $\mathbf{C}$ to every inclusion $i\colon V \to U$ of open subsets $V \subseteq U \subseteq X$, such that

$$\mathcal{F}(i \circ j) = \mathcal{F}(j) \circ \mathcal{F}(i)$$
$$\mathcal{F}(\mathrm{id}_U) = \mathrm{id}_{\mathcal{F}(U)},$$

for any two inclusions $i\colon V \to U$ and $j\colon W \to V$, with $W \subseteq V \subseteq U$.

Note that the order of composition is switched in $\mathcal{F}(i \circ j) = \mathcal{F}(j) \circ \mathcal{F}(i)$.

Intuitively, the map $\mathcal{F}(i)\colon \mathcal{F}(U) \to \mathcal{F}(V)$ is a restriction map if we think of $\mathcal{F}(U)$ and $\mathcal{F}(V)$ as sets of functions (which is often the case). For this reason, the map $\mathcal{F}(i)\colon \mathcal{F}(U) \to \mathcal{F}(V)$ is also denoted by $\rho^U_V\colon \mathcal{F}(U) \to \mathcal{F}(V)$, and the first equation of Definition 8.1 is expressed by

$$\rho^U_W = \rho^V_W \circ \rho^U_V.$$

See Figures 8.1 and 8.2. Here are some examples of presheaves.

Example 8.1.

(1) The *constant presheaf* G_X with values in $G \in \mathbf{C}$, defined such that $G_X(U) = G$ for all open subsets U of X, and ρ^U_V is the identity function of G for all open subsets U, V with $V \subseteq U$. A variant of the constant presheaf which comes up in cohomology has $G_X(\emptyset) = (0)$ instead of $G_X(\emptyset) = G$ when G is an algebraic structure with an identity element 0.
(2) If Y is another topological space, then $\mathcal{C}^0_Y$ is the presheaf defined so that $\mathcal{C}^0_Y(U)$ is the set of all continuous functions $f\colon U \to Y$ from the open subset U of X to Y.
(3) If $Y = (\mathbb{R}, +, \text{usual metric topology})$, then $\mathcal{C}^0_Y$ is the presheaf of real-valued continuous functions on X. It is presheaf of $\mathbb{R}$-algebras.
(4) If $Y = (\mathbb{R}, +, \text{trivial topology})$, then $\mathcal{C}^0_Y$ is the presheaf of all real-valued functions on X. It is presheaf of $\mathbb{R}$-algebras.
(5) If M is a smooth manifold, then $\mathcal{C}^\infty$ is the presheaf defined so that $\mathcal{C}^\infty(U)$ is the set of all smooth real-valued functions $f\colon U \to \mathbb{R}$ from the open subset U of M.

A map between two presheaves is defined as follows.

Definition 8.2. Given a topological space X and a fixed class $\mathbf{C}$ of structures (a category), say sets, vector spaces, R-modules, groups, commutative rings, *etc.*, a *map (or morphism)* $\varphi\colon \mathcal{F} \to \mathcal{G}$ of presheaves $\mathcal{F}$ and $\mathcal{G}$ on X consists of a family of maps $\varphi_U\colon \mathcal{F}(U) \to \mathcal{G}(U)$ of the class of structures in $\mathbf{C}$, such that

$$\varphi_V \circ (\rho_{\mathcal{F}})^U_V = (\rho_{\mathcal{G}})^U_V \circ \varphi_U$$

for every pair of open subsets U, V such that $V \subseteq U \subseteq X$. Equivalently, the following diagrams commute for every pair of open subsets U, V such that $V \subseteq U \subseteq X$ (and $i\colon V \to U$ is the corresponding inclusion map):

$$\begin{array}{ccc} \mathcal{F}(U) & \xrightarrow{\varphi_U} & \mathcal{G}(U) \\ {\scriptstyle \mathcal{F}(i)}\downarrow & & \downarrow{\scriptstyle \mathcal{G}(i)} \\ \mathcal{F}(V) & \xrightarrow[\varphi_V]{} & \mathcal{G}(V), \end{array}$$

or using the restriction notation $(\rho_{\mathcal{F}})^U_V$ for $\mathcal{F}(i)$ and $(\rho_{\mathcal{G}})^U_V$ for $\mathcal{G}(i)$,

$$\begin{array}{ccc} \mathcal{F}(U) & \xrightarrow{\varphi_U} & \mathcal{G}(U) \\ {\scriptstyle (\rho_{\mathcal{F}})^U_V}\downarrow & & \downarrow{\scriptstyle (\rho_{\mathcal{G}})^U_V} \\ \mathcal{F}(V) & \xrightarrow[\varphi_V]{} & \mathcal{G}(V). \end{array}$$

See Figure 8.3.

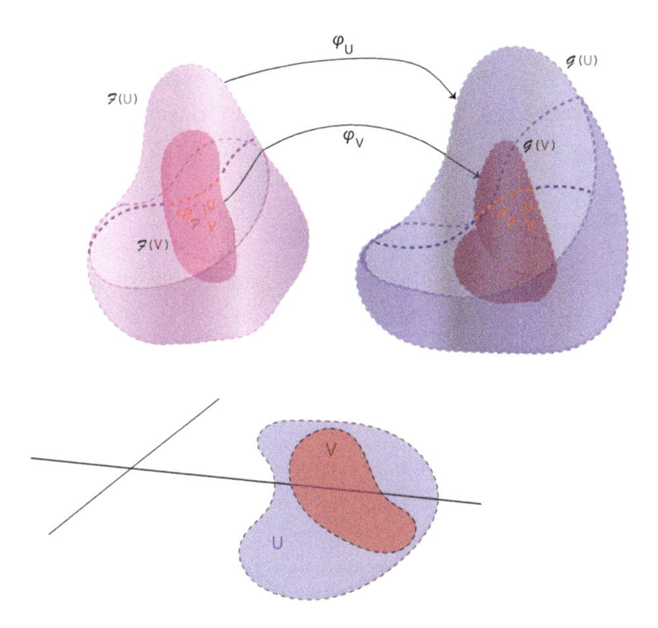

Fig. 8.3 The two purple "eggplants" represent the elements of the presheaves $\mathcal{F}$ and $\mathcal{G}$. The presheaf map $\varphi_U\colon \mathcal{F}(U) \to \mathcal{G}(U)$ maps the left "eggplant" to the right "eggplant" in a manner which preserves restrictions associated with the inclusion $V \subseteq U \subseteq \mathbb{R}^2$.

Remark: In fancy terms, a map of presheaves is a natural transformation.

Definition 8.3. Given three presheaves $\mathcal{F}, \mathcal{G}, \mathcal{H}$ on X and two maps of presheaves $\varphi\colon \mathcal{F} \to \mathcal{G}$ and $\psi\colon \mathcal{G} \to \mathcal{H}$, the *composition* $\psi \circ \varphi$ of φ and ψ is defined by the family of maps

$$(\psi \circ \varphi)_U = \psi_U \circ \varphi_U$$

for all open subsets U of X.

It is easily checked that $\psi \circ \varphi$ is indeed a map of presheaves from $\mathcal{F}$ to $\mathcal{H}$.

Definition 8.4. Given two presheaves $\mathcal{F}$ and $\mathcal{G}$ on X, a presheaf map $\varphi\colon \mathcal{F} \to \mathcal{G}$ is *injective* if every map $\varphi_U\colon \mathcal{F}(U) \to \mathcal{G}(U)$ is injective, *surjective* if every map $\varphi_U\colon \mathcal{F}(U) \to \mathcal{G}(U)$ is surjective (for each open subset U of X). Two presheaves $\mathcal{F}$ and $\mathcal{G}$ are *isomorphic* if there exists some presheaf map $\varphi\colon \mathcal{F} \to \mathcal{G}$ and $\psi\colon \mathcal{G} \to \mathcal{F}$ such that $\psi \circ \varphi = \mathrm{id}$ and $\varphi \circ \psi = \mathrm{id}$.

It is not hard to see that a presheaf map is an isomorphism iff it is injective and surjective.

If $\mathcal{F}$ and $\mathcal{G}$ are presheaves of algebraic structures (modules, groups, commutative rings, *etc.*) then there is a notion of kernel, image, and cokernel of a map of presheaves. This allows the definition of exact sequences of presheaves. We will come back to this point later on.

8.2 Sheaves

In Section 8.1 we defined the notion of a presheaf. Presheaves are typically used to keep track of local information assigned to a global object (the space X). It is usually desirable to use consistent local information to recover some global information, but this requires a sharper notion, that of a sheaf.

Expositions on the subject of sheaves tend to be rather abstract and assume a significant amount of background. Our goal is to provide just enough background to have a good understanding of the sheafification process and of the subtleties involving exact sequences of presheaves and sheaves. These topics will be discussed in Chapter 10.

We should mention some of the classics, including (in alphabetic order) Bredon [Bredon (1993)], Eisenbud and Harris [Eisenbud and Harris (1992)], Forster [Forster (1981)], Godement [Godement (1958)], Griffiths and Harris [Griffiths and Harris (1978)], Gunning [Gunning (1990)],

Hartshorne [Hartshorne (1977)], Hirzebruch [Hirzebruch (1978)], Kashiwara and Shapira [Kashiwara and Schapira (1994)], Mac Lane and Moerdijk [MacLane and Moerdijk (1992)], Mumford [Mumford (1999)], Narasimham [Narasimham (1992)], Rotman [Rotman (2009)], Serre FAC [Serre (1955)], Shafarevich [Shafarevich (1994)], Spanier [Spanier (1989)]. One of the most accessible (and quite thorough) presentations is found in Tennison [Tennison (1975)].

The motivation for the extra condition that a sheaf should satisfy is this. Suppose we consider the presheaf of continuous functions on a topological space X. If U is any open subset of X and if $(U_i)_{i \in I}$ is an open cover of U, for any family $(f_i)_{i \in I}$ of continuous functions $f_i \colon U_i \to \mathbb{R}$, if f_i and f_j agree on the overlap $U_i \cap U_j$, then the f_i patch to a *unique* continuous function $f \colon U \to \mathbb{R}$ whose restriction to U_i is f_i.

Definition 8.5. Given a topological space X and a class $\mathbf{C}$ of structures (a category), say sets, vector spaces, R-modules, groups, commutative rings, *etc.*, a *sheaf on X with values in* $\mathbf{C}$ is a presheaf $\mathcal{F}$ on X such that for any open subset U of X, for every open cover $(U_i)_{i \in I}$ of U (that is, $U = \bigcup_{i \in I} U_i$ for some open subsets $U_i \subseteq U$ of X), the following conditions hold:

(G) (Gluing condition) For every family $(f_i)_{i \in I}$ with $f_i \in \mathcal{F}(U_i)$, if the f_i are consistent, which means that
$$\rho^{U_i}_{U_i \cap U_j}(f_i) = \rho^{U_j}_{U_i \cap U_j}(f_j) \quad \text{for all } i, j \in I,$$
then there is some $f \in \mathcal{F}(U)$ such that $\rho^U_{U_i}(f) = f_i$ for all $i \in I$. See Figure 8.4.

(M) (Monopresheaf condition) For any two elements $f, g \in \mathcal{F}(U)$, if f and g agree on all the U_i, which means that
$$\rho^U_{U_i}(f) = \rho^U_{U_i}(g) \quad \text{for all } i \in I,$$
then $f = g$.

Obviously, Condition (M) implies that in Condition (G) the element f obtained by patching the f_i is unique.

Another notation often used for $\mathcal{F}(U)$ is $\Gamma(U, \mathcal{F})$. An element of $\Gamma(U, \mathcal{F})$ is called a *section above* U, and elements of $\Gamma(X, \mathcal{F}) = \mathcal{F}(X)$ are called *global sections*. This terminology is justified by the fact that many sheaves arise as continuous sections of some surjective continuous map $p \colon E \to X$; that is, continuous functions $s \colon U \to E$ such that $p \circ s = \mathrm{id}_U$; see Example 8.2(1).

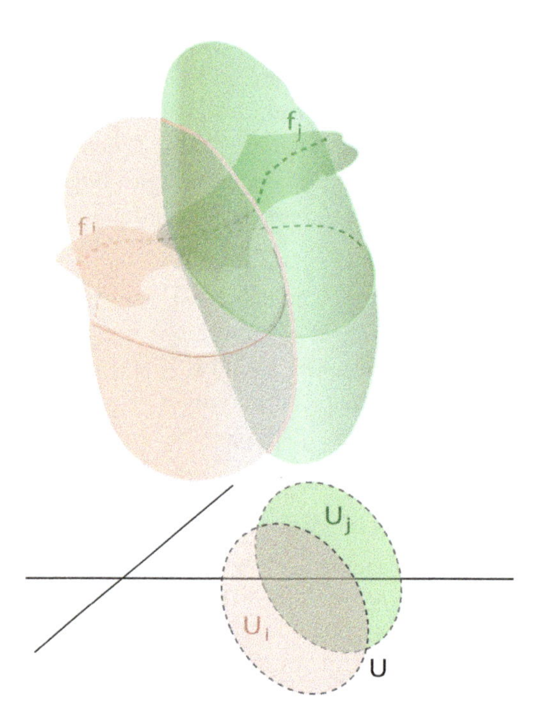

Fig. 8.4 A schematic representation Condition (G) for the set $U = U_i \cup U_j \subseteq \mathbb{R}^2$. The element $f_i \in \mathcal{F}(U_i)$ is represented by the wavy peach plane with the bold peach dotted line in the peach "balloon" while the element $f_j \in \mathcal{F}(U_j)$ is represented by the wavy green plane with the bold green dotted line. Where the two "balloons" intersect, the peach plane overlaps the green plane. In other words $\rho^{U_i}_{U_i \cap U_j}(f_i) = \rho^{U_j}_{U_i \cap U_j}(f_j)$.

For any two open subsets U and V with $V \subseteq U$, for any $s \in \Gamma(U, \mathcal{F}) = \mathcal{F}(U)$, it is often convenient to abbreviate $\rho^U_V(s)$ by $s|V$.

Remarks:

(1) If $\mathcal{F}(U) = \emptyset$ for some open subset U of X, then $\mathcal{F}$ is the trivial sheaf such that $\mathcal{F}(V) = \emptyset$ for all open subsets V of X. This is because there is a restriction function $\rho^X_U : \mathcal{F}(X) \to \emptyset$, but the only function with range $\emptyset$ is the empty function with domain $\emptyset$ so $\mathcal{F}(X) = \emptyset$. Since there is restriction function $\rho^X_V : \mathcal{F}(X) \to \mathcal{F}(V)$ for every open subset V of X, we deduce that $\mathcal{F}(V) = \emptyset$ for all open subsets of X. This observation is due to Godement [Godement (1958)]. From now on, we rule out the above possibility. Note that it is ruled out automatically for sheaves of algebraic structures having an identity element.

(2) Assuming that $\mathcal{F}$ is not the trivial sheaf, then Conditions (G) and (M) apply to all open subsets U of X and all families of open covers $(U_i)_{i\in I}$ of U, including the case where $U = \emptyset$ and $I = \emptyset$. In this case, Conditions (G) and (M) implies that $\mathcal{F}(\emptyset)$ is a one-element set. In the case of groups, modules, groups, commutative rings, *etc.*, we have $\mathcal{F}(\emptyset) = \{0\}$.

(3) Condition (G) applies to open subsets U that are the *disjoint union* of open subsets $U_i \subseteq U$. In this case, every family $(f_i)_{i\in I}$ with $f_i \in \mathcal{F}(U_i)$ must patch to yield some global element $f \in \mathcal{F}(U)$ such that $\rho^U_{U_i}(f) = f_i$. Thus, the gluing condition imposes some *consistency* among the local pieces $f_i \in \mathcal{F}(U_i)$, even if the U_i are pairwise disjoint. This is a major difference with presheaves, where unrelated and inconsistent objects may be assigned to disjoint open subsets.

(4) If $\mathcal{F}$ is a sheaf of R-modules or commutative rings, then Condition (M) can be replaced by the following condition which is often more convenient:

(M) (Monopresheaf condition) For any element $f \in \mathcal{F}(U)$, if f is zero on the U_i, which means that

$$\rho^U_{U_i}(f) = 0 \quad \text{for all } i \in I,$$

then $f = 0$.

(5) If $\mathcal{F}$ is a sheaf of R-modules or commutative rings, then Conditions (M) and (G) can be stated as an exactness condition. For any nonempty subset U of X, for any open cover $(U_i)_{i\in I}$ of U, define the maps $f\colon \mathcal{F}(U) \to \prod_{i\in I} \mathcal{F}(U_i)$ and $g\colon \prod_{i\in I} \mathcal{F}(U_i) \to \prod_{i,j\in I} \mathcal{F}(U_i \cap U_j)$ by

$$\begin{aligned} f(s) &= (\rho^U_{U_i}(s))_{i\in I} \\ g((s_i)_{i\in I}) &= (\rho^{U_i}_{U_i\cap U_j}(s_i) - \rho^{U_j}_{U_i\cap U_j}(s_j))_{(i,j)\in I\times I}. \end{aligned}$$

Then Conditions (M) and (G) are equivalent to the hypothesis that the sequence

$$0 \longrightarrow \mathcal{F}(U) \xrightarrow{\ f\ } \prod_{i\in I} \mathcal{F}(U_i) \xrightarrow{\ g\ } \prod_{(i,j)\in I\times I} \mathcal{F}(U_i \cap U_j)$$

is exact.

(6) Intuitively, we may think of the elements $f \in \mathcal{F}(U)$ (the sections above U) as abstract functions. In fact, this point of view can be justified rigorously. For every sheaf $\mathcal{F}$ on a space X, we can construct a "big" space E with a continuous projection function $p\colon E \to X$ so that for

every open subset U of X, every $s \in \mathcal{F}(U)$ can be viewed as a function $\widetilde{s}\colon U \to E$ (a section of p, see Example 8.2(1) below). In fact, p is a local homeomorphism. We will investigate the construction of E in Section 10.1.

Here are some examples of sheaves.

Example 8.2.

(1) Let $p\colon E \to X$ be a surjective continuous map between two topological spaces E and X. We define the *sheaf* $\Gamma[E,p]$ *of (continuous) sections of* p on X as follows: for every open subset U of X,

$$\begin{aligned}\Gamma[E,p](U) &= \Gamma(U,\Gamma[E,p]) \\ &= \{s\colon U \to E \mid p \circ s = \mathrm{id} \text{ and } s \text{ is continuous}\};\end{aligned}$$

equivalently, the following diagram commutes:

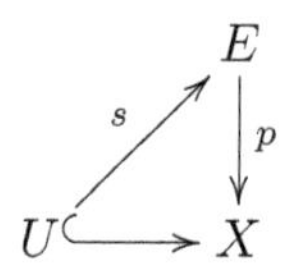

where the horizontal arrow is inclusion; see Figure 8.5. For the sake of notational simplicity, the sheaf $\Gamma[E,p]$ is often denoted by ΓE.

(2) If Y is another topological space, $E = X \times Y$, and $p\colon X \times Y \to X$ is the first projection, then the sheaf $\Gamma[E,p]$ in (1) corresponds to the presheaf on X of Example 8.1(2–4), which is actually a sheaf. Indeed, since p is the map $(x,y) \mapsto x$, every continuous section s of p above U is a function of the form $x \mapsto (x, f(x))$, where $f\colon U \to Y$ is a continuous function. Therefore, *there is a bijection between the set of continuous sections of* p *above* U *and the set of continuous functions from* U *to* Y. See Figure 8.6.

(3) If Y is given the discrete topology, $E = X \times Y$, and $p\colon X \times Y \to X$ is the first projection, then the sheaf $\Gamma[E,p]$ in (1) corresponds to the sheaf of locally constant functions with values in Y, because every continuous section s of p above U is a function of the form $x \mapsto (x, f(x))$, where $f\colon U \to Y$ is a locally constant function. Recall that a function $f\colon U \to Y$ is locally constant if for every $x \in U$ there is some open subset V of U containing x such that f is constant on V. For any $x \in U$, since Y is discrete the set $\{f(x)\}$ is open, and since f is continuous $V = f^{-1}(f(x))$ is some open subset of U containing x and f is constant on V (with

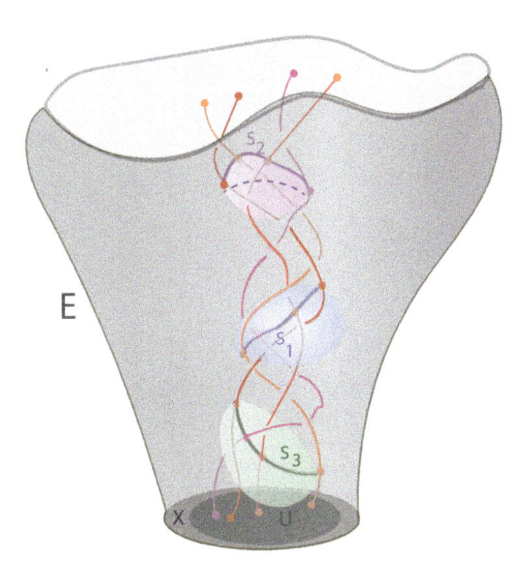

Fig. 8.5 A schematic representation of the sheaf of sections $\Gamma[E,p]$ where X is the circular base and E be the solid gray upside-down "lamp shade". Note $p^{-1}(x)$ for $x \in X$ is a twisted orange "spaghetti strand." We illustrate three elements of $\Gamma[E,p](U)$ as blue, green, and purple wavy disks which connect the various preimages in a continuous manner, namely $p \circ s(u) = u$ for $u \in U$.

value $f(x)$). A locally constant function must have a constant value on a connected open subset. See Figure 8.7.

The sheaf of locally constant functions on X with values in Y is denoted $\widetilde{Y}_X$ (or Y_X^+ if the "tilde" notation is already used). Beware that in general this is *not* the constant presheaf Y_X with values in Y. Indeed if X is the union of two disjoint open subsets U_1 and U_2 and if Y has at least two distinct elements y_1, y_2, then we can pick the family (y_1, y_2) with $y_1 \in Y_X(U_1) = Y$ and $y_2 \in Y_X(U_2) = Y$, and since $U_1 \cap U_2 = \emptyset$, by Condition (G) there should be some element $y \in Y_X(X) = Y$ such that $\rho^X_{U_1}(y) = y_1$ and $\rho^X_{U_2}(y) = y_2$. But since Y_X is the constant presheaf, $\rho^X_{U_1} = \rho^X_{U_2} = \mathrm{id}$, so we should have $y = y_1 = y_2$, which is impossible since $y_1 \neq y_2$. The sheaf $\widetilde{Y}_X$ of locally constant functions with values in Y is usually called (confusingly) the *constant sheaf with values in* Y.

(4) Given a smooth manifold M, the smooth real-valued functions on M form a sheaf $\mathcal{C}^\infty$. For every open subset U of M, let $\mathcal{C}^\infty(U)$ be the $\mathbb{R}$-algebra of smooth functions on U.

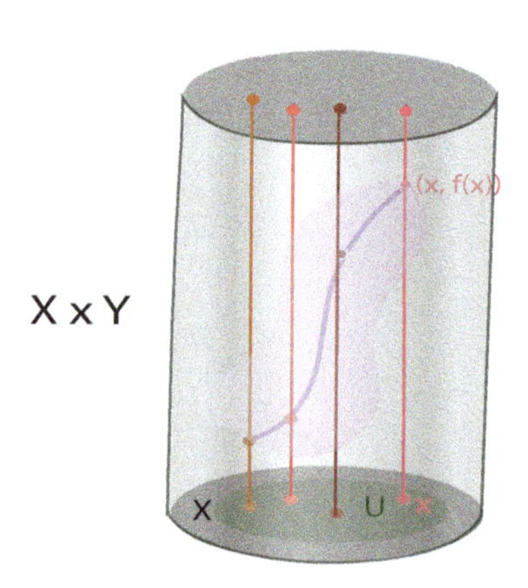

Fig. 8.6 Let X be the closed unit disk, $Y = [0, 1]$, and $E = X \times Y$ be the solid grey cylinder. Each $p^{-1}(x)$ is straight orange "spaghetti strand." We illustrate an element of $\Gamma[E, p](U)$ associated with the continuous function $f: U \to Y$ as a wavy purple disk.

(5) Given a smooth manifold M, the differential forms on M form a sheaf $\mathcal{A}_X^*$. For every open subset U of M, let $\mathcal{A}^p(U)$ be the vector space of p-forms on U, and let $\mathcal{A}_X^*(U) = \mathcal{A}^p(U)$. Then it is easy to check that we obtain a sheaf of vector spaces; the restriction maps are the pullbacks of forms.

We just observed that in general the constant presheaf with values in Y in not a sheaf. Here is another example of a presheaf which is not a sheaf.

Example 8.3. Let X be any topological space with at least two points (for example, $X = \{0, 1\}$), and let $\mathcal{F}_1$ be the presheaf given by

$$\mathcal{F}_1(U) = \begin{cases} \mathbb{Z} & \text{if } U = X \\ (0) & \text{if } U \neq X \text{ is an open subset,} \end{cases}$$

with all ρ_V^U equal to the zero map except if $U = V = X$ (in which case it is the identity). It is easy to check that Condition (M) fails. In particular if $X = \{0, 1\}$ with the discrete topology, then $X = \{0\} \cup \{1\}$, where $\{0\}$ and $\{1\}$ are open sets in X. Let $f \in \mathcal{F}_1(X)$ be $f = 1$, while $g \in \mathcal{F}_1(X)$ is $g = -1$. Then

$$\rho_{\{0\}}^X(f) = 0 = \rho_{\{1\}}^X(g),$$

where $f \neq g$.

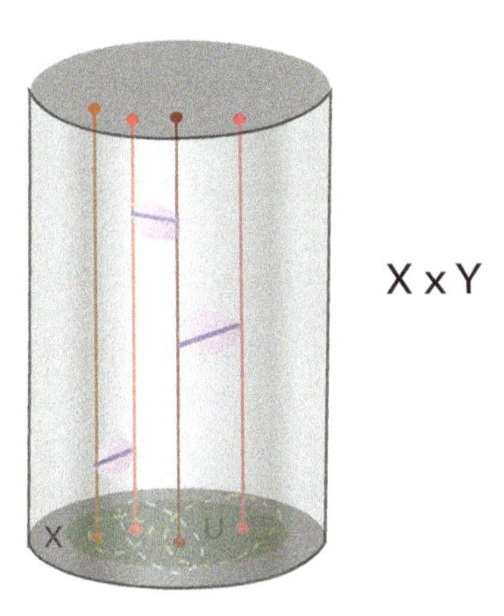

Fig. 8.7 Let X be the closed unit disk, $Y = [0,1]$, and $E = X \times Y$ be the solid grey cylinder. Each $p^{-1}(x)$ is straight orange "spaghetti strand" composed of disjoint open points. Since Y has the discrete topology, an element of $\Gamma[E,p](U)$ is illustrated as the purple "jump" function.

The notion of a map $\varphi\colon \mathcal{F} \to \mathcal{G}$ between two sheaves $\mathcal{F}$ and $\mathcal{G}$ is exactly as in Definition 8.2.

Definition 8.6. Two sheaves $\mathcal{F}$ and $\mathcal{G}$ are *isomorphic* if there exist some sheaf morphisms $\varphi\colon \mathcal{F} \to \mathcal{G}$ and $\psi\colon \mathcal{G} \to \mathcal{F}$ such that $\psi \circ \varphi = \mathrm{id}$ and $\varphi \circ \psi = \mathrm{id}$.

It turns out that every sheaf is isomorphic to a sheaf of sections as in Example 8.2(1), but to prove this we need the notion of direct limit; see Section 8.8.

Definition 8.7. Given a topological space X, for every (nonempty) open subset U of X, for every presheaf (or sheaf) $\mathcal{F}$ on X, the *restriction* $\mathcal{F}|U$ of $\mathcal{F}$ to U is defined so that for every open subset V of U,

$$(\mathcal{F}|U)(V) = \mathcal{F}(V).$$

If $\mathcal{F}$ is a sheaf, it is immediate that $\mathcal{F}|U$ is a also a sheaf. Given two preshaves (or sheaves) $\mathcal{F}$ and $\mathcal{G}$ on X, the presheaf $\mathcal{H}om(\mathcal{F},\mathcal{G})$ is defined by

$$\mathcal{H}om(\mathcal{F},\mathcal{G})(U) = \mathrm{Hom}(\mathcal{F}|U, \mathcal{G}|U)$$

for every open subset U of X. If $\mathcal{F}$ and $\mathcal{G}$ are sheaves, it is easy to see that $\mathcal{H}om(\mathcal{F},\mathcal{G})$ is also a sheaf.

The next section is devoted to direct limits, an indispensable tool in sheaf theory and the cohomology of sheaves.

8.3 Direct Mapping Families and Direct Limits

We begin our study of direct limits with the following two definitions.

Definition 8.8. A *directed set* is a set I equipped with a preorder $\leq$ (where $\leq$ is a reflexive and transitive relation) such that for all $i, j \in I$, there is some $k \in I$ such that $i \leq k$ and $j \leq k$. A subset J of I is said to be *cofinal* in I if for every $i \in I$ there is some $j \in J$ such that $i \leq j$. For example, $2\mathbb{Z}$ is cofinal in $\mathbb{Z}$, where $2\mathbb{Z} = \{2x \mid x \in \mathbb{Z}\}$.

Definition 8.9. A *direct mapping family* of sets (or R-modules, or commutative rings, *etc.*) is a pair $((F_i)_{i \in I}, (\rho^i_j)_{i \leq j})$ where $(F_i)_{i \in I}$ is a family of sets (R-modules, commutative rings, *etc.*) F_i whose index set I is a directed set, and for all $i, j \in I$ with $i \leq j$, $\rho^i_j \colon F_i \to F_j$ is a map (R-linear, ring homomorphism, *etc.*) so that

$$\rho^i_i = \mathrm{id}$$
$$\rho^i_k = \rho^j_k \circ \rho^i_j$$

for all $i, j, k \in I$ with $i \leq j \leq k$, as illustrated below

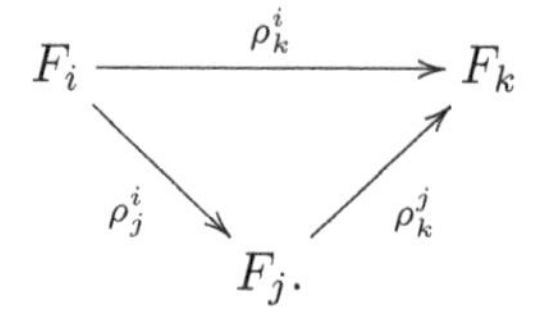

Here are two examples of direct mapping families.

Example 8.4.

(1) Let X be a topological space and pick any point $x \in X$. Then the family of open subsets U of X such that $x \in U$ forms a directed set under the preorder $U \prec V$ iff $V \subseteq U$. If $C^0(U)$ is the set of continuous $\mathbb{R}$-valued functions defined in U and if $\rho^U_V \colon C^0(U) \to C^0(V)$ is the restriction map, then the family of sets (rings) $(C^0(U))_{U \ni x}$ (for all open subsets U of X containing x) forms a direct mapping family.

(2) More generally, if $\mathcal{F}$ is a presheaf on X, then the family of sets (R-modules, *etc.*) $(\mathcal{F}(U))_{U \ni x}$ forms a direct mapping family, with $\rho^U_V \colon \mathcal{F}(U) \to \mathcal{F}(V)$ whenever $V \subseteq U$, the presheaf restriction map.

The direct limit of a direct mapping family $((F_i)_{i\in I}, (\rho^i_j)_{i\leq j})$ is obtained as a quotient of a disjoint union of the F_i.

Definition 8.10. The *direct limit* (or *inductive limit*) $\varinjlim F_i$ of the direct mapping family $((F_i)_{i\in I}, (\rho^i_j)_{i\leq j})$ of sets (R-modules, commutative rings, *etc.*) is defined as follows:

First form the disjoint union $\coprod_{i\in I} F_i$. Next let $\sim$ be the equivalence relation on $\coprod_{i\in I} F_i$ defined by:

$$f_i \sim f_j \quad \text{iff} \quad \rho^i_k(f_i) = \rho^j_k(f_j) \quad \text{for some } k \in I \text{ with } k \geq i, j,$$

for any $f_i \in F_i$ and any $f_j \in F_j$; see Figure 8.8. Finally the direct limit $\varinjlim F_i$ is given by

$$\varinjlim_{i\in I} F_i = \left(\coprod_{i\in I} F_i\right) / \sim .$$

It is clear that $\sim$ is reflexive and symmetric but we need to check transitivity. This is where the fact that I is a directed set is used. If $f_i \sim f_j$ and $f_j \sim f_k$, then there exist $p, q \in I$ such that $i, j \leq p$, $j, k \leq q$, $\rho^i_p(f_i) = \rho^j_p(f_j)$ and $\rho^j_q(f_j) = \rho^k_q(f_k)$. Since I is a directed preorder there is some $r \in I$ such that $p, q \leq r$. We claim that

$$\rho^i_r(f_i) = \rho^k_r(f_k),$$

showing that $f_i \sim f_k$. This is because

$$\rho^i_r(f_i) = \rho^p_r \circ \rho^i_p(f_i) = \rho^p_r \circ \rho^j_p(f_j) = \rho^j_r(f_j) = \rho^q_r \circ \rho^j_q(f_j) = \rho^q_r \circ \rho^k_q(f_k) = \rho^k_r(f_k),$$

as illustrated by the following diagram

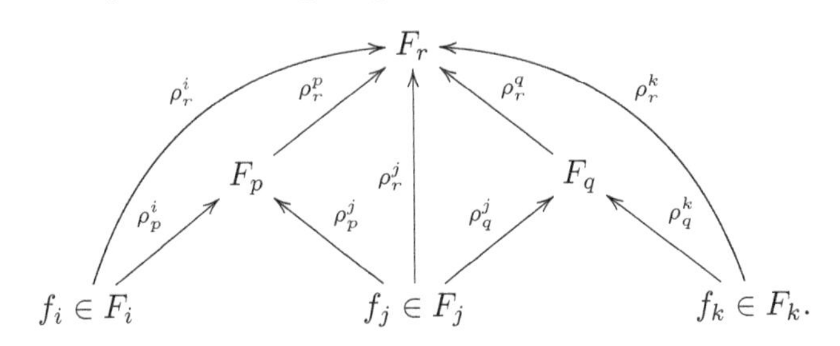

For every index $i \in I$, we have the canonical injection $\epsilon_i \colon F_i \to \coprod_{i\in I} F_i$, and thus, a canonical map $\pi_i \colon F_i \longrightarrow \varinjlim F_i$, namely

$$\pi_i \colon f \mapsto [\epsilon_i(f)]_\sim = [\epsilon_i(f)].$$

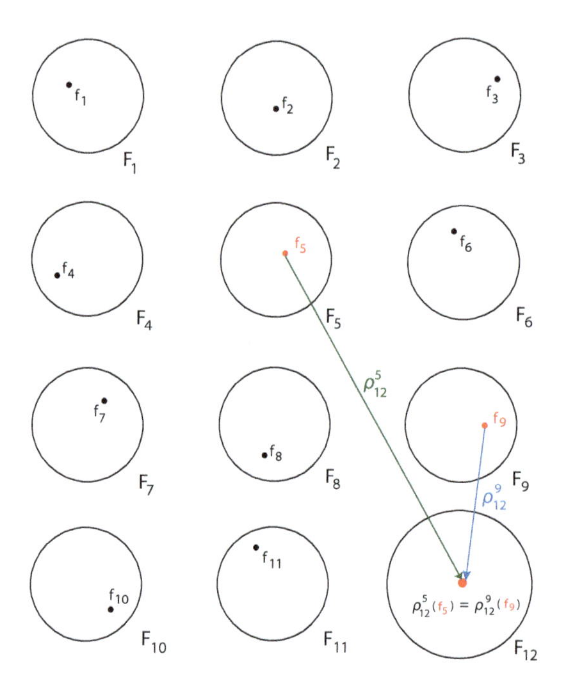

Fig. 8.8 An illustration of the equivalence relation $\sim$ used in the direct limit construction. Since $\rho^5_{12}(f_5) = \rho^9_{12}(f_9)$, $f_5 \sim f_9$.

(Here, $[x]_\sim = [x]$ means equivalence class of x modulo $\sim$.) It is obvious that $\pi_i = \pi_j \circ \rho^i_j$ for all $i, j \in I$ with $i \leq j$ as illustrated in the diagram below

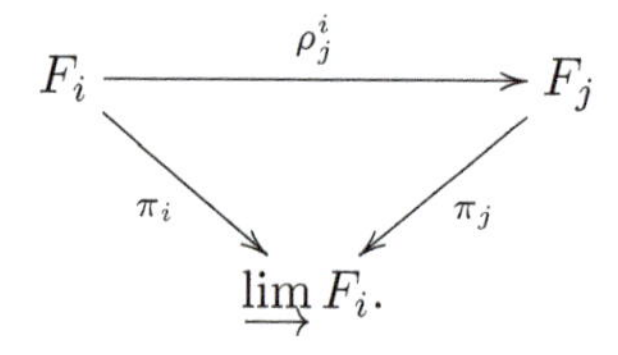

If each F_i is a R-module, then $\varinjlim F_i$ is also a R-module (a ring, *etc.*). We define addition by

$$[f_i] + [f_j] = [\rho^i_k(f_i) + \rho^j_k(f_j)], \quad \text{for any } k \in I \text{ with } k \geq i, j,$$

and multiplication by a scalar as

$$\lambda[f_i] = [\lambda f_i].$$

If the F_i are rings, then we define multiplication by

$$[f_i] \cdot [f_j] = [\rho^i_k(f_i) \cdot \rho^j_k(f_j)], \quad \text{for any } k \in I \text{ with } k \geq i, j.$$

The direct limit $(\varinjlim F_i, (\pi_i)_{i\in I})$ is characterized by the important *universal mapping property*: for every set (R-module, commutative ring, *etc.*) G and every family of maps $\theta_i \colon F_i \to G$ such that $\theta_i = \theta_j \circ \rho^i_j$, for all $i, j \in I$ with $i \leq j$ as in the diagram below

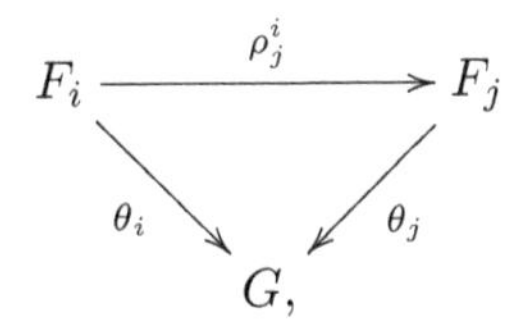

there is a unique map $\varphi\colon \varinjlim F_i \to G$, so that

$$\theta_i = \varphi \circ \pi_i, \quad \text{for all } i \in I$$

as illustrated in the diagram below

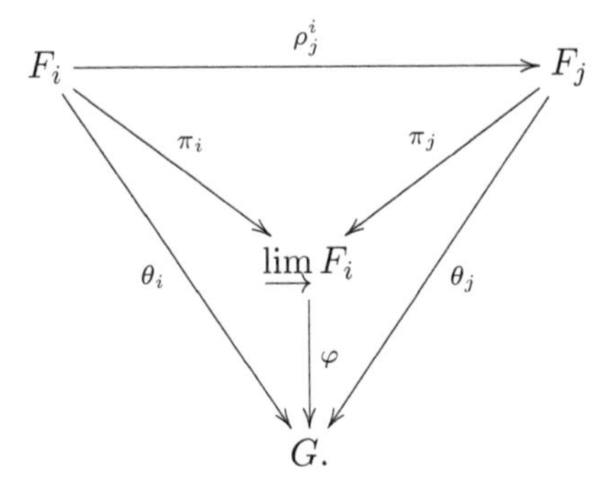

The universal mapping property of the direct limit implies that it is unique up to isomorphism.

Remark: The direct limit $\varinjlim F_i$ is actually a *colimit*; it is an initial object in a suitably defined category. Unfortunately, following common practice (probably due to some obscure historical tradition) it is called a *direct* limit.

The following proposition gives a useful criterion to show that an object is a direct limit.

Proposition 8.1. *Given a direct mapping family $((F_i)_{i\in I}, (\rho^i_j)_{i\leq j})$ of sets (R-modules, commutative rings, etc.), suppose G is a set (R-module, ring,* *etc.) and* $(\theta_i)_{i\in I}$ *is a family of maps* $\theta_i\colon F_i \to G$ *such that* $\theta_i = \theta_j \circ \rho^i_j$, *for all* $i, j \in I$ *with* $i \leq j$ *as in the diagram below*

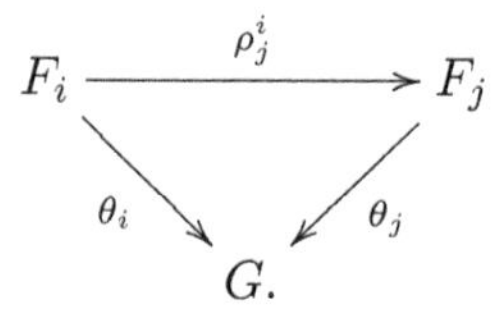

If the following two conditions are satisfied

(a) *For every* $g \in G$, *there is some* $i \in I$ *and some* $f_i \in F_i$ *such that* $g = \theta_i(f_i)$.

(b) *For all* $i, j \in I$, *for any* $f_i \in F_i$ *and any* $f_j \in F_j$,
$$\theta_i(f_i) = \theta_j(f_j) \quad \text{iff} \quad \exists k \text{ such that } i \leq k, j \leq k \text{ and } \rho^i_k(f_i) = \rho^j_k(f_j),$$

then $(G, (\theta_i)_{i\in I})$ *is a direct limit of the direct mapping family* $((F_i)_{i\in I}, (\rho^i_j)_{i\leq j})$.

Proof. It suffices to prove that $(G, (\theta_i)_{i\in I})$ satisfies the universal mapping family. Let H be a set (R-module, commutative ring, *etc.*) and $(\eta_i)_{i\in I}$ is a family of maps $\eta_i\colon F_i \to H$ such that $\eta_i = \eta_j \circ \rho^i_j$, for all $i, j \in I$ with $i \leq j$ as in the diagram below

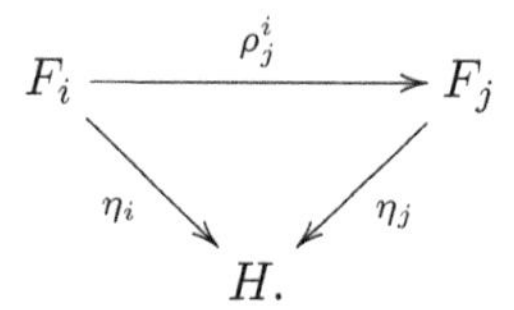

We need to prove that there is a unique map $\varphi\colon G \to H$ such that the following diagram commutes for all i, j

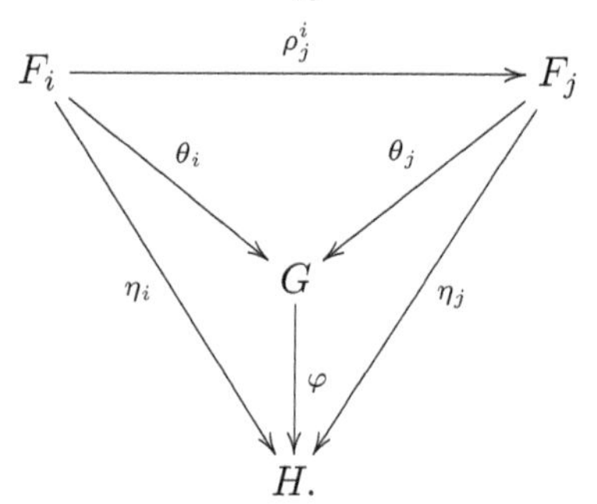

By (a), since every $g \in G$ is of the form $g = \theta_i(f_i)$ for some $f_i \in F_i$, then we must have
$$\varphi(g) = \varphi(\theta_i(f_i)) = \eta_i(f_i).$$

Thus, if φ exists, it is unique. It remains to show that the definition of $\varphi(g)$ as $\eta_i(f_i)$ does not depend on the choice of f_i. If $f_j \in F_j$ is another element such that $\theta_j(f_j) = g$, then $\theta_i(f_i) = \theta_j(f_j)$, which by (b) means that there is some $k \in I$ such that, $i \leq k$, $j \leq k$ and $\rho^i_k(f_i) = \rho^j_k(f_j)$. But then since the following diagrams commute

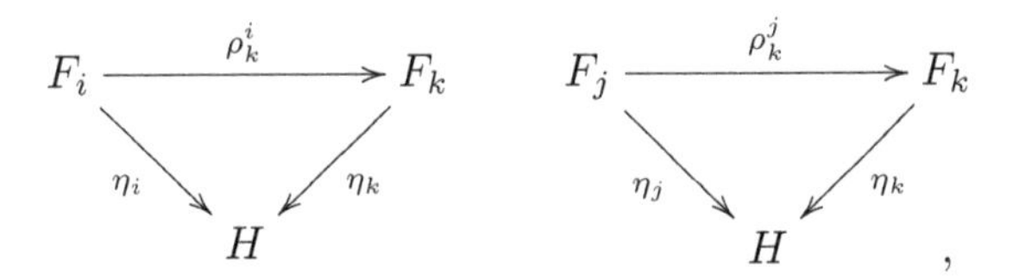

we have

$$\eta_i(f_i) = \eta_k(\rho^i_k(f_i)) = \eta_k(\rho^j_k(f_j)) = \eta_j(f_j),$$

which shows that $\varphi(g)$ is well defined. $\square$

We will also need the notion of map between two direct mapping families and of the direct limit of such a map.

Definition 8.11. Given two direct mapping families $((F_i)_{i\in I}, ((\rho_F)^i_j)_{i\leq j})$ and $((G_i)_{i\in I}, ((\rho_G)^i_j)_{i\leq j})$ of sets (R-modules, commutative rings, *etc.*) over the same directed preorder I, a *map* from $((F_i)_{i\in I}, ((\rho_F)^i_j)_{i\leq j})$ to $((G_i)_{i\in I}, ((\rho_G)^i_j)_{i\leq j})$ is a family $\varphi = (\varphi_i)_{\in I}$ of maps $\varphi_i \colon F_i \to G_i$ (of sets, of R-modules, commutative rings, *etc.*) such that the following diagram commutes for all $i \leq j$:

$$\begin{array}{ccc} F_i & \xrightarrow{(\rho_F)^i_j} & F_j \\ {\scriptstyle \varphi_i}\downarrow & & \downarrow{\scriptstyle \varphi_j} \\ G_i & \xrightarrow[(\rho_G)^i_j]{} & G_j. \end{array}$$

Let $\varphi = (\varphi_i)_{\in I}$ be a map between two direct mapping families $((F_i)_{i\in I}, ((\rho_F)^i_j)_{i\leq j})$ and $((G_i)_{i\in I}, ((\rho_G)^i_j)_{i\leq j})$. If we write $(F = \varinjlim F_i, \theta_i \colon F_i \to F)$ for the direct limit of the first family and $(G = \varinjlim G_i, \eta_i \colon G_i \to G)$ for the direct limit of the second family, the

commutativity of the following diagram

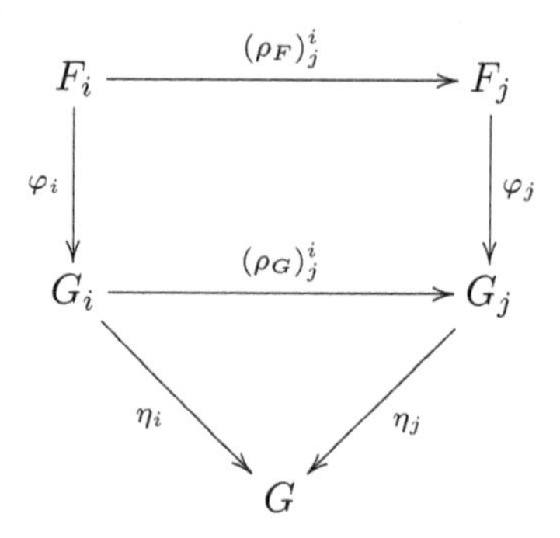

shows that if we write $\psi_i = \eta_i \circ \varphi_i$, then following diagram commutes

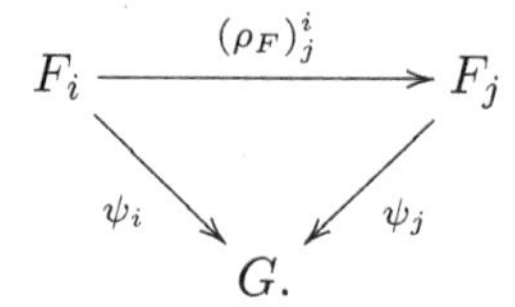

Therefore by the universal mapping property of the direct limit ($F = \varinjlim F_i, \theta_i \colon F_i \to F$), there is a unique map $\Phi \colon F \to G$ such that the following diagram commutes

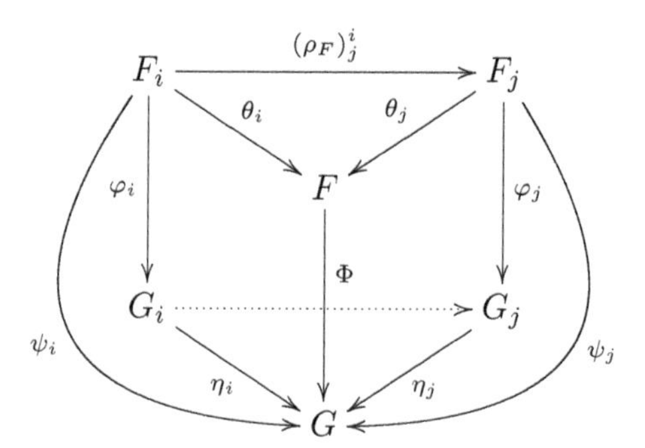

and so the following diagram commutes for all $i \in I$:

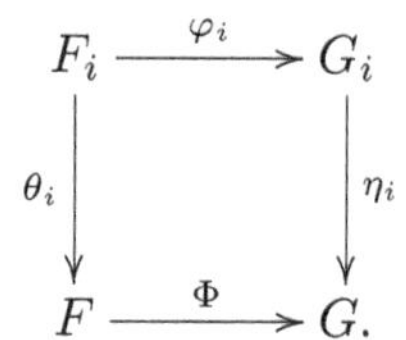

Definition 8.12. Let $\varphi = (\varphi_i)_{\in I}$ be a map between two direct mapping families $((F_i)_{i\in I}, ((\rho_F)^i_j)_{i\leq j})$ and $((G_i)_{i\in I}, ((\rho_G)^i_j)_{i\leq j})$. If we write $(F = \varinjlim F_i, \theta_i\colon F_i \to F)$ for the direct limit of the first family and $(G = \varinjlim G_i, \eta_i\colon G_i \to G)$ for the direct limit of the second family, the *direct limit* $\Phi = \varinjlim \varphi_i$ is the unique map $\Phi\colon \varinjlim F_i \to \varinjlim G_i$such that the diagram below commutes for all $i \in I$

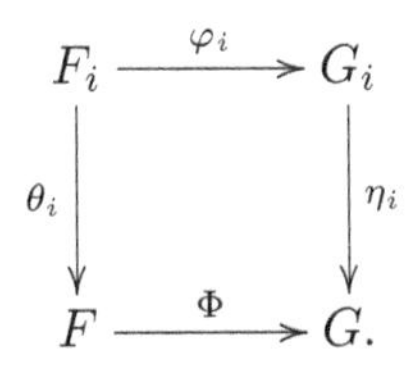

We will also need a generalization of the notion of map of direct mapping families for families indexed by different index sets. Such maps will be needed to define the notion of homomorphism induced by a continuous map in Čech cohomology.

Definition 8.13. Given two direct mapping families $((F_i)_{i\in I}, ((\rho_F)^i_k)_{i\leq k})$ and $((G_j)_{j\in J}, ((\rho_G)^j_l)_{j\leq l})$ of sets (R-modules, commutative rings, *etc.*) over the directed preorders I and J, a *map* from $((F_i)_{i\in I}, ((\rho_F)^i_k)_{i\leq k})$ to $((G_j)_{j\in J}, ((\rho_G)^j_l)_{j\leq l})$ is pair (τ, φ), where $\tau\colon I \to J$ is an order-preserving map and φ is a family $\varphi = (\varphi_i)_{\in I}$ of maps $\varphi_i\colon F_i \to G_{\tau(i)}$ (of sets, of R-modules, commutative rings, *etc.*) such that the following diagram commutes for all $i \leq k$

$$\begin{array}{ccc} F_i & \xrightarrow{(\rho_F)^i_k} & F_k \\ {\scriptstyle \varphi_i}\downarrow & & \downarrow{\scriptstyle \varphi_k} \\ G_{\tau(i)} & \xrightarrow[(\rho_G)^{\tau(i)}_{\tau(k)}]{} & G_{\tau(k)}. \end{array}$$

Let $(\tau, \varphi = (\varphi_i)_{\in I})$ be a map between two direct mapping families $((F_i)_{i\in I}, ((\rho_F)^i_k)_{i\leq k})$ and $((G_j)_{j\in J}, ((\rho_G)^j_l)_{j\leq l})$. If we write $(F = \varinjlim F_i, \theta_i\colon F_i \to F)$ for the direct limit of the first family and $(G = \varinjlim G_j, \eta_j\colon G_j \to G)$ for the direct limit of the second family, the

commutativity of the following diagram

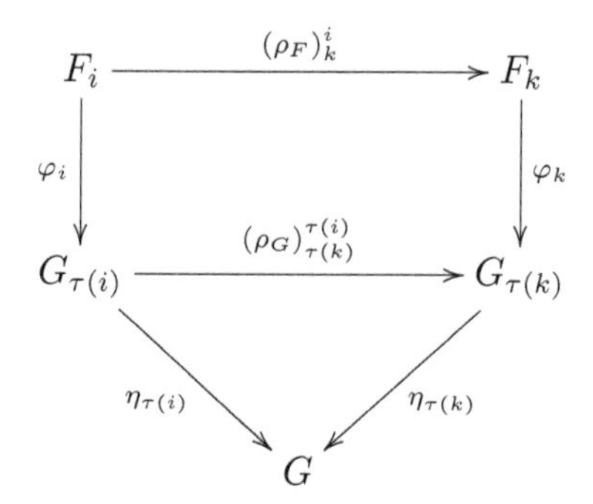

shows that if we write $\psi_i = \eta_{\tau(i)} \circ \varphi_i$, then the following diagram commutes for all i, k

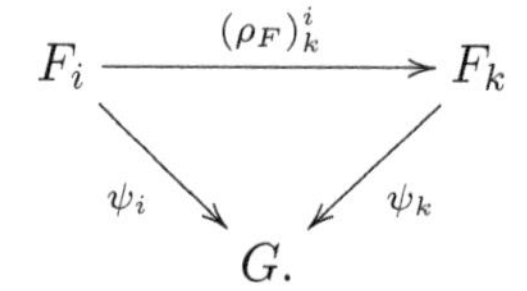

Therefore by the universal mapping property of the direct limit ($F = \varinjlim F_i, \theta_i \colon F_i \to F$), there is a unique map $\Phi \colon F \to G$ such that the following diagram commutes for all $i \in I$

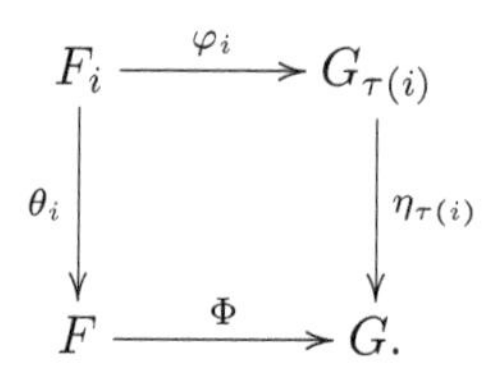

Definition 8.14. Let $(\tau, \varphi = (\varphi_i)_{\in I})$ be a map between two direct mapping families $((F_i)_{i \in I}, ((\rho_F)^i_k)_{i \leq k})$ and $((G_j)_{j \in J}, ((\rho_G)^j_l)_{j \leq l})$. If we write $(F = \varinjlim F_i, \theta_i \colon F_i \to F)$ for the direct limit of the first family and $(G = \varinjlim G_j, \eta_j \colon G_j \to G)$ for the direct limit of the second family, the *direct limit* $\Phi = \varinjlim \varphi_i$ is the unique map $\Phi \colon \varinjlim F_i \to \varinjlim G_j$ such that the diagram below commutes for all $i \in I$

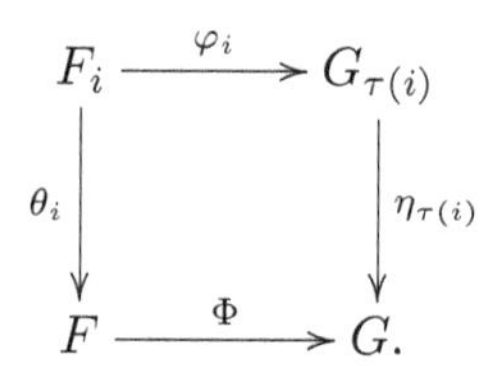

8.4 Problems

Problem 8.1. Prove that a presehaf map is an isomorphism iff it is injective and surjective.

Problem 8.2. Given a smooth manifold M, check that the assignment of the space of differential p-forms $\mathcal{A}^p(U)$ to an open subset U of M is a sheaf.

Problem 8.3. Check that if $\mathcal{F}$ and $\mathcal{G}$ are sheaves on a space X, then $\mathcal{H}om(\mathcal{F}, \mathcal{G})$ is also a sheaf.

Problem 8.4. The notion of kernel $\mathrm{Ker}\,\varphi$ and image $\mathrm{Im}\,\varphi$ of a presheaf or sheaf map $\varphi\colon \mathcal{F} \to \mathcal{G}$ of presheaves of R-modules over a space X is defined as follows. The presheaf $\mathrm{Ker}\,\varphi$ is defined by $(\mathrm{Ker}\,\varphi)(U) = \mathrm{Ker}\,\varphi_U$, and the presheaf $\mathrm{Im}\,\varphi$ is defined by $(\mathrm{Im}\,\varphi)(U) = \mathrm{Im}\,\varphi_U$ (where U is any open subset of X).

(1) Prove that if $\mathcal{F}$ and $\mathcal{G}$ are presheaves, then $\mathrm{Ker}\,\varphi$ and $\mathrm{Im}\,\varphi$ are also presheaves.

(2) Prove that if $\mathcal{F}$ and $\mathcal{G}$ are sheaves, then $\mathrm{Ker}\,\varphi$ is a sheaf. Give an example where $\mathcal{F}$ and $\mathcal{G}$ are sheaves but $\mathrm{Im}\,\varphi$ is not a sheaf.

Problem 8.5. Check the universal mapping property of the direct limit of a direct mapping family $((F_i)_{i\in I}, (\rho^i_j)_{i\leq j})$. Prove that any two direct limits of a direct mapping family $((F_i)_{i\in I}, (\rho^i_j)_{i\leq j})$ are isomorphic.

Chapter 9

Čech Cohomology with Values in a Presheaf

Given a topological space X and a presehaf $\mathcal{F}$, there is a way of defining cohomology groups $\check{H}^p(X, \mathcal{F})$ as a limit process involving the definition of some cohomology groups $\check{H}^p(\mathcal{U}, \mathcal{F})$ associated with open covers $\mathcal{U} = (U_j)_{j \in J}$ of the space X. Given two open covers $\mathcal{U}$ and $\mathcal{V}$, we can define when $\mathcal{V}$ is a refinement of $\mathcal{U}$, and then we define the cohomology group $\check{H}^p(X, \mathcal{F})$ as the direct limit of the directed system of groups $\check{H}^p(\mathcal{U}, \mathcal{F})$. When the presheaf $\mathcal{F}$ has some special properties and when nice covers exist, the limit process can be bypassed.

Because it can be defined for any presheaf and for any topological space, Čech cohomology is a very powerful and most valuable tool. It plays a major role in algebraic topology (duality) and algebraic geometry (derived functors cohomology).

From a historical perspective, Čech defined certain kinds of *homology* groups (with coefficients in $\mathbb{Q}$) in a paper published in 1932. The definition of these homology groups involved *finite* covers $\mathcal{U}$ of a topological space X, and a notion of refinement. Then the Čech homology groups are defined by taking an *inverse limit* rather than a direct limit. Roughly at the same time, Alexandroff extended this concept to coefficients in any commutative ring. A few years later (1936), Steenrod made an extensive study of these Čech homology groups.

The Čech cohomology groups defined in terms of covers seem to have been first introduced and studied by Spanier (1948), Dowker (1950), and Eilenberg and Steenrod (1952). At first, finite covers were used, but Dowker realized that this led to some pathologies and switched to arbitrary covers, arriving at the definition given in this chapter. Čech cohomology is given a very thorough treatment in Eilenberg and Steenrod's famous book

[Eilenberg and Steenrod (1952)]. The generalization of Čech cohomology to presheaves is probably due to the "Cartan school." It is quite an obvious step for someone familiar with sheaves. An early occurrence of this definition appears in Serre [Serre (1955)].

After this historical digression we return to the topics discussed in this chapter. In Section 9.1 we define the cohomology modules $\check{H}^p(\mathcal{U}, \mathcal{F})$ associated with a cover $\mathcal{U}$ of a topological space X. The definition of the modules $C^p(\mathcal{U}, \mathcal{F})$ of Čech cochains with values in a presheaf $\mathcal{F}$ requires considering intersections

$$U_{i_0} \cap \cdots \cap U_{i_p},$$

where the U_k are open subsets in the cover $\mathcal{U}$. Technically, it is simpler to consider sequences with possible repetitions to deal correctly with the passage to a finer cover. It is also possible to use alternating cochains, which are more economical. We state a result of Serre [Serre (1955)] which shows that both approaches are equivalent.

In Section 9.2 we define the Čech cohomology modules $\check{H}^p(X, \mathcal{F})$ associated with a topological space X. The module $\check{H}^p(X, \mathcal{F})$ is obtained as the direct limit of the mapping family $(\check{H}^p(\mathcal{U}, \mathcal{F}))_{\mathcal{U}}$ with respect to the directed set of open covers under the notion of refinement. Here one has to be careful to avoid set theoretic pitfalls (the family of all open covers of a given space is **not** a set). This difficulty can be avoided using a device due to Serre [Serre (1955)]. In writing this section we have greatly benefited from Serre's classical exposition of Čech cohomology in one of his landmark papers, *Faisceaux algébriques cohérents* [Serre (1955)], abbreviated as *FAC*, and published in 1955.

In Section 9.3 we investigate the relationship between de Rham cohomology and classical Čech cohomology for the constant sheaf $\widetilde{\mathbb{R}}_X$ (corresponding to coefficients in $\mathbb{R}$). If M is a smooth manifold and if $\mathcal{U}$ is a good cover of M (as in Definition 3.6), then the de Rham cohomology and the Čech cohomology modules are isomorphic, that is,

$$H^p_{\mathrm{dR}}(M) \cong \check{H}^p(M, \widetilde{\mathbb{R}}_M) \cong \check{H}^p(\mathcal{U}, \widetilde{\mathbb{R}}_M),$$

for all $p \geq 0$. The main technical tool to prove the above equivalence is a double complex known as the *Čech–de Rham complex*. This elegant proof method is due to André Weil and we follow closely Bott and Tu's exposition [Bott and Tu (1986)].

If X is a paracompact manifold (see Definition 13.7), then singular cohomology and classical Čech cohomology for the constant sheaf $\widetilde{\mathbb{Z}}_X$ (corresponding to coefficients in $\mathbb{Z}$) are isomorphic. More can be said if X has a good cover.

9.1 Čech Cohomology of a Cover

Throughout this chapter, R will denote a fixed commutative ring with unit. Let $\mathcal{F}$ be a presheaf of R-modules on X. We always assume that $\mathcal{F}(\emptyset) = (0)$, as in the case of a sheaf. Our first goal is to define R-modules of cochains, $C^p(\mathcal{U}, \mathcal{F})$. Here a decision must be made, namely whether we use sequences of indices with or without repetitions allowed. This is one of the confusing aspects of the set up of Čech cohomology, as the literature uses both approaches typically without any motivation. The crucial point is to deal correctly with the passage to a finer cover. The proof is simpler if we allow repetitions of indices, and we will follow this approach. However, it can also be shown that using special kinds of cochains called alternating cochains, isomorphic cohomology R-modules are obtained. As a corollary, one may indeed assume that sequences without repetitions are used.

Let X be a topological space and let $\mathcal{U}$ be an open cover for X. Given any finite sequence $I = (i_0, \ldots, i_p)$ of elements of some index set J (where $p \geq 0$ and the i_j are not necessarily distinct), we let

$$U_I = U_{i_0 \cdots i_p} = U_{i_0} \cap \cdots \cap U_{i_p}.$$

Note that it may happen that $U_I = \emptyset$ (this is another confusing point: some authors only consider sequences $I = (i_0, \ldots, i_p)$ for which $U_{i_0 \cdots i_p} \neq \emptyset$). We denote by $U_{i_0 \cdots \widehat{i_j} \cdots i_p}$ the intersection

$$U_{i_0 \cdots \widehat{i_j} \cdots i_p} = U_{i_0} \cap \cdots \cap \widehat{U_{i_j}} \cap \cdots \cap U_{i_p}$$

of the p subsets obtained by omitting U_{i_j} from $U_{i_0 \cdots i_p} = U_{i_0} \cap \cdots \cap U_{i_p}$ (the intersection of the $p+1$ subsets). See Figure 9.1. Then we have $p+1$ inclusion maps

$$\delta_j^p \colon U_{i_0 \cdots i_p} \longrightarrow U_{i_0 \cdots \widehat{i_j} \cdots i_p}, \quad 0 \leq j \leq p.$$

For example, if $p = 0$ we have the map

$$\delta_0^0 \colon U_{i_0} \longrightarrow X;$$

for $p = 1$, we have the two maps

$$\delta_0^1 \colon U_{i_0} \cap U_{i_1} \longrightarrow U_{i_1}, \quad \delta_1^1 \colon U_{i_0} \cap U_{i_1} \longrightarrow U_{i_0};$$

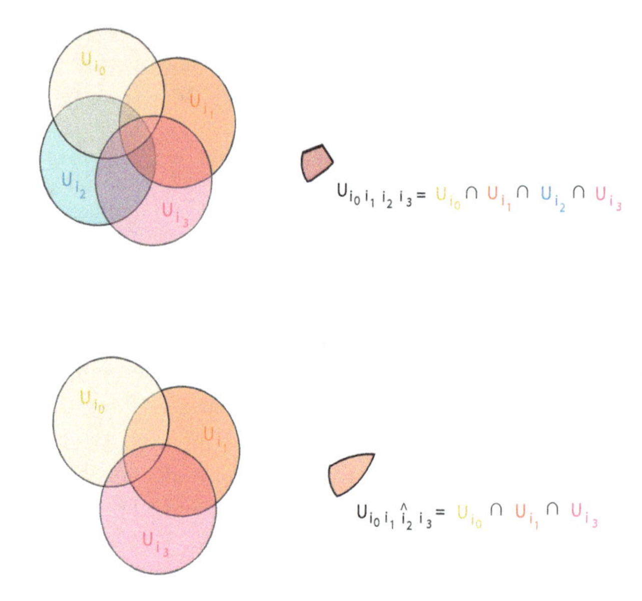

Fig. 9.1 An illustration of $U_{i_0 i_1 i_2 i_3}$ and $U_{i_0 i_1 \widehat{i_2} i_3}$.

for $p = 2$, we have the three maps

$$\begin{aligned}
\delta_0^2 &: U_{i_0} \cap U_{i_1} \cap U_{i_2} \longrightarrow U_{i_1} \cap U_{i_2}, \\
\delta_1^2 &: U_{i_0} \cap U_{i_1} \cap U_{i_2} \longrightarrow U_{i_0} \cap U_{i_2}, \\
\delta_2^2 &: U_{i_0} \cap U_{i_1} \cap U_{i_2} \longrightarrow U_{i_1} \cap U_{i_2}.
\end{aligned}$$

Definition 9.1. Given a topological space X, an open cover $\mathcal{U} = (U_j)_{j \in J}$ of X, and a presheaf of abelian groups $\mathcal{F}$ on X, for any $p \geq 0$, the R-module of *Čech p-cochains* $C^p(\mathcal{U}, \mathcal{F})$ is the set of all functions f with domain J^{p+1} such that $f(i_0, \ldots, i_p) \in \mathcal{F}(U_{i_0 \cdots i_p})$; in other words,

$$C^p(\mathcal{U}, \mathcal{F}) = \prod_{(i_0, \ldots, i_p) \in J^{p+1}} \mathcal{F}(U_{i_0 \cdots i_p}),$$

the set of all J^{p+1}-indexed families $(f_{i_0, \ldots, i_p})_{(i_0, \ldots, i_p) \in J^{p+1}}$ with $f_{i_0, \ldots, i_p} \in \mathcal{F}(U_{i_0 \cdots i_p})$.

In particular, for $p = 0$ we have

$$C^0(\mathcal{U}, \mathcal{F}) = \prod_{j \in J} \mathcal{F}(U_j),$$

so a 0-cochain is a J-indexed family $f = (f_j)_{j\in J}$ with $f_j \in \mathcal{F}(U_j)$, and for $p = 1$ we have

$$C^1(\mathcal{U}, \mathcal{F}) = \prod_{(i,j)\in J^2} \mathcal{F}(U_i \cap U_j),$$

so a 1-cochain is a J^2-indexed family $f = (f_{i,j})_{(i,j)\in J^2}$ with $f_{i,j} \in \mathcal{F}(U_i \cap U_j)$.

Remark: Since $\mathcal{F}(\emptyset) = (0)$, for any cochain $f \in C^p(\mathcal{U}, \mathcal{F})$, if $U_{i_0\cdots i_p} = \emptyset$, then $f_{i_0\cdots i_p} = 0$. Therefore, we could define $C^p(\mathcal{U}, \mathcal{F})$ as the set of families $f_{i_0\cdots i_p} \in \mathcal{F}(U_{i_0\cdots i_p})$ corresponding to tuples $(i_0, \ldots, i_p) \in J^{p+1}$ such that $U_{i_0\cdots i_p} \neq \emptyset$. This is the definition adopted by several authors, including Warner [Warner (1983)] (Chapter 5, Section 5.33).

Each inclusion map $\delta^p_j \colon U_{i_0\cdots i_p} \longrightarrow U_{i_0\cdots \widehat{i_j}\cdots i_p}$ induces a map

$$\mathcal{F}(\delta^p_j) \colon \mathcal{F}(U_{i_0\cdots \widehat{i_j}\cdots i_p}) \longrightarrow \mathcal{F}(U_{i_0\cdots i_p})$$

which is none other that the restriction map $\rho^{U_{i_0\cdots \widehat{i_j}\cdots i_p}}_{U_{i_0\cdots i_p}}$ which, for the sake of notational simplicity, we denote by $\rho^j_{i_0\cdots i_p}$.

Definition 9.2. Given a topological space X, an open cover $\mathcal{U} = (U_j)_{j\in J}$ of X, and a presheaf of R-modules $\mathcal{F}$ on X, the *coboundary maps* $\delta^p_{\mathcal{F}} \colon C^p(\mathcal{U}, \mathcal{F}) \to C^{p+1}(\mathcal{U}, \mathcal{F})$ are given by

$$\delta^p_{\mathcal{F}} = \sum_{j=0}^{p+1} (-1)^j \mathcal{F}(\delta^{p+1}_j), \quad p \geq 0.$$

More explicitly, for any p-cochain $f \in C^p(\mathcal{U}, \mathcal{F})$, for any sequence $(i_0, \ldots, i_{p+1}) \in J^{p+2}$, we have

$$(\delta^p_{\mathcal{F}} f)_{i_0,\ldots,i_{p+1}} = \sum_{j=0}^{p+1} (-1)^j \rho^j_{i_0\cdots i_{p+1}}(f_{i_0,\ldots,\widehat{i_j},\ldots,i_{p+1}}).$$

Note that the definition of $(\delta^p_{\mathcal{F}} f)_{i_0,\ldots,i_{p+1}}$ is reminiscent of the definition of the boundary map $\partial\sigma$ given in Definition 4.3, but here we are dealing with cohomology.

Unravelling Definition 9.2, for $p = 0$ we have

$$(\delta^0_{\mathcal{F}} f)_{i,j} = \rho^0_{ij}(f_j) - \rho^1_{ij}(f_i), \tag{δ^0}$$

and for $p = 1$ we have

$$(\delta^1_{\mathcal{F}} f)_{i,j,k} = \rho^0_{ijk}(f_{j,k}) - \rho^1_{ijk}(f_{i,k}) + \rho^2_{ijk}(f_{i,j}). \tag{δ^1}$$

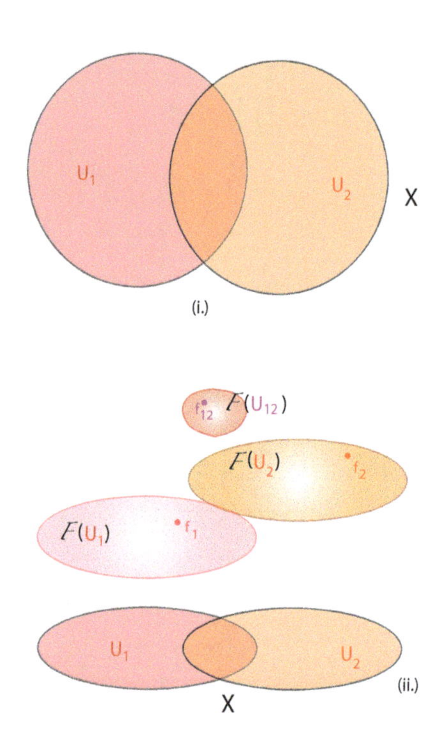

Fig. 9.2 An illustration of X in Example 9.1. Figure (ii.) illustrates the associated presheaf $\mathcal{F}$.

Example 9.1. As an explicit example of Definitions 9.1 and 9.2, let X be the union of two open sets, namely $X = U_1 \cup U_2$. See Figure 9.2. Then

$$\begin{aligned} C^0(\mathcal{U}, \mathcal{F}) &= \mathcal{F}(U_1) \times \mathcal{F}(U_2) \\ C^1(\mathcal{U}, \mathcal{F}) &= \mathcal{F}(U_{11}) \times \mathcal{F}(U_{12}) \times \mathcal{F}(U_{21}) \times \mathcal{F}(U_{22}) \\ C^2(\mathcal{U}, \mathcal{F}) &= \mathcal{F}(U_{111}) \times \mathcal{F}(U_{112}) \times \mathcal{F}(U_{121}) \times \mathcal{F}(U_{122}) \\ &\qquad \times \mathcal{F}(U_{211}) \times \mathcal{F}(U_{212}) \times \mathcal{F}(U_{221}) \times \mathcal{F}(U_{222}), \end{aligned}$$

where

$$U_{11} = U_1 \cap U_1 = U_1, \qquad U_{12} = U_1 \cap U_2 = U_{21}, \qquad U_{22} = U_2 \cap U_2 = U_2$$

$$U_{111} = U_1 \cap U_1 \cap U_1 = U_1, \qquad U_{222} = U_2 \cap U_2 \cap U_2 = U_2$$

$$U_{112} = U_{121} = U_{211} = U_1 \cap U_1 \cap U_2 = U_1 \cap U_2 \cap U_2 = U_{221} = U_{212} = U_{122}.$$

In general $C^p(\mathcal{U}, \mathcal{F})$ is a product with 2^{p+1} factors. A typical element of $C^0(\mathcal{U}, \mathcal{F})$ has the form (f_1, f_2) where f_1 is an element of the group associated with U_1 and f_2 is an element of the group associated with U_2. A typical element of $C^1(\mathcal{U},\mathcal{F})$ has the form $(f_{1,1}, f_{1,2}, f_{2,1}, f_{2,2})$ where $f_{1,1}$ is an element of the group associated with $U_{11} = U_1$, $f_{1,2}$ is an element of the group associated with $U_{12} = U_1 \cap U_2$, $f_{2,1}$ is another element of the group associated with $U_{21} = U_{12}$, and $f_{2,2}$ is an element of the group associated with U_{22}. In general $f_{1,2} \neq f_{2,1}$. A typical element of $C^2(\mathcal{U},\mathcal{F})$ has the form

$$(f_{1,1,1}, f_{1,1,2}, f_{1,2,1}, f_{1,2,2}, f_{2,1,1}, f_{2,1,2}, f_{2,2,1}, f_{2,2,2}),$$

where $f_{1,1,1}$ is an element of the group associated with $U_{111} = U_1$, $f_{1,1,2}$ is an element of the group associated with $U_{112} = U_1 \cap U_2$, $f_{1,2,1}$ is an element of the group associated with $U_{121} = U_{112}$, $f_{1,2,2}$ is an element of the group associated with $U_{122} = U_{112}$, $f_{2,1,1}$ is an element of the group associated with $U_{211} = U_{112}$, $f_{2,1,2}$ is an element of the group associated with $U_{212} = U_{112}$, $f_{2,2,1}$ is an element of the group associated with $U_{221} = U_{112}$, and $f_{2,2,2}$ is an element of the group associated with $U_{222} = U_2$. In general, a typical element of $C^p(\mathcal{U},\mathcal{F})$ is a 2^{p+1}-tuple.

The coboundary map $\delta^0_{\mathcal{F}} \colon C^0(\mathcal{U},\mathcal{F}) \to C^1(\mathcal{U},\mathcal{F})$ takes $f \in C^0(\mathcal{U},\mathcal{F})$, say $f = (f_1, f_2)$, and makes it into element of $C^1(\mathcal{U},\mathcal{F})$ by calculating

$$\begin{aligned}
(\delta^0_{\mathcal{F}} f)_{1,1} &= \rho^0_{11}(f_1) - \rho^1_{11}(f_1) = 0\\
(\delta^0_{\mathcal{F}} f)_{1,2} &= \rho^0_{12}(f_2) - \rho^1_{12}(f_1)\\
(\delta^0_{\mathcal{F}} f)_{2,1} &= \rho^0_{21}(f_1) - \rho^1_{21}(f_2)\\
(\delta^0_{\mathcal{F}} f)_{2,2} &= \rho^0_{22}(f_2) - \rho^1_{22}(f_2) = 0.
\end{aligned}$$

In other words

$$\delta^0_{\mathcal{F}}(f_1, f_2) = (0, \rho^0_{12}(f_2) - \rho^1_{12}(f_1), \rho^0_{21}(f_1) - \rho^1_{21}(f_2), 0) \in C^1(\mathcal{U},\mathcal{F}).$$

The coboundary map $\delta^1_{\mathcal{F}} \colon C^1(\mathcal{U},\mathcal{F}) \to C^2(\mathcal{U},\mathcal{F})$ takes $f \in C^1(\mathcal{U},\mathcal{F})$, say $f = (f_{1,1}, f_{1,2}, f_{2,1}, f_{2,2})$, and makes it into element of $C^2(\mathcal{U},\mathcal{F})$ by calculating

$$\begin{aligned}
(\delta^1_{\mathcal{F}} f)_{1,1,1} &= \rho^0_{111}(f_{1,1}) - \rho^1_{111}(f_{1,1}) + \rho^2_{111}(f_{1,1}) = \rho^2_{111}(f_{1,1})\\
(\delta^1_{\mathcal{F}} f)_{1,1,2} &= \rho^0_{112}(f_{1,2}) - \rho^1_{112}(f_{1,2}) + \rho^2_{112}(f_{1,1}) = \rho^2_{112}(f_{1,1})\\
(\delta^1_{\mathcal{F}} f)_{1,2,1} &= \rho^0_{121}(f_{2,1}) - \rho^1_{121}(f_{1,1}) + \rho^2_{121}(f_{1,2})\\
(\delta^1_{\mathcal{F}} f)_{1,2,2} &= \rho^0_{122}(f_{2,2}) - \rho^1_{122}(f_{1,2}) + \rho^2_{122}(f_{1,2}) = \rho^0_{122}(f_{2,2})\\
(\delta^1_{\mathcal{F}} f)_{2,1,1} &= \rho^0_{211}(f_{1,1}) - \rho^1_{211}(f_{2,1}) + \rho^2_{211}(f_{2,1}) = \rho^0_{211}(f_{1,1})\\
(\delta^1_{\mathcal{F}} f)_{2,1,2} &= \rho^0_{212}(f_{1,2}) - \rho^1_{212}(f_{2,2}) + \rho^2_{212}(f_{2,1})\\
(\delta^1_{\mathcal{F}} f)_{2,2,1} &= \rho^0_{221}(f_{2,1}) - \rho^1_{221}(f_{2,1}) + \rho^2_{221}(f_{2,2}) = \rho^2_{221}(f_{2,2})\\
(\delta^1_{\mathcal{F}} f)_{2,2,2} &= \rho^0_{222}(f_{2,2}) - \rho^1_{222}(f_{2,2}) + \rho^2_{222}(f_{2,2}) = \rho^2_{222}(f_{2,2}).
\end{aligned}$$

In other words

$$\begin{aligned}\delta^1_{\mathcal{F}}(f_{1,1}, f_{1,2}, f_{2,1}, f_{2,2}) = (&\rho^2_{111}(f_{1,1}), \rho^2_{112}(f_{1,1}),\\ &\rho^0_{121}(f_{2,1}) - \rho^1_{121}(f_{1,1}) + \rho^2_{121}(f_{1,2}), \rho^0_{122}(f_{2,2}),\\ &\rho^0_{211}(f_{1,1}), \rho^0_{212}(f_{1,2}) - \rho^1_{212}(f_{2,2}) + \rho^2_{212}(f_{2,1}),\\ &\rho^2_{221}(f_{2,2}), \rho^2_{222}(f_{2,2})).\end{aligned}$$

Families of the form $(\delta^0_{\mathcal{F}} f)_{i,j}$ form the group (R-module) $B^1(\mathcal{U}, \mathcal{F})$ of Čech coboundaries, and the group (R-module) $Z^0(\mathcal{U}, \mathcal{F})$ of Čech cocycles consists of the families $(f_j)_{j\in J} \in C^0(\mathcal{U}, \mathcal{F})$ such that $(\delta^0_{\mathcal{F}} f) = 0$; that is, families $(f_j)_{j\in J} \in C^0(\mathcal{U}, \mathcal{F})$ such that

$$\rho^0_{ij}(f_j) = \rho^1_{ij}(f_i)$$

for all $i, j \in J$.

Families of the form $(\delta^1_{\mathcal{F}} f)_{i,jk}$ form the group (R-module) $B^2(\mathcal{U}, \mathcal{F})$ of Čech coboundaries, and the group (R-module) $Z^1(\mathcal{U}, \mathcal{F})$ of Čech cocycles consists of the families $(f_{ij})_{(i,j)\in J^2} \in C^1(\mathcal{U}, \mathcal{F})$ such that $(\delta^1_{\mathcal{F}} f) = 0$; that is, families $(f_{i,j})_{(i,j)\in J^2} \in C^1(\mathcal{U}, \mathcal{F})$ such that

$$\rho^1_{ijk}(f_{i,k}) = \rho^2_{ijk}(f_{i,j}) + \rho^0_{ijk}(f_{j,k})$$

for all $i, j, k \in J$.

In general the definition of $B^p(\mathcal{U}, \mathcal{F})$ and $Z^p(\mathcal{U}, \mathcal{F})$ is as follows.

Definition 9.3. Given a topological space X, an open cover $\mathcal{U} = (U_j)_{j\in J}$ of X, and a presheaf of R-modules $\mathcal{F}$ on X, the R-module $B^p(\mathcal{U}, \mathcal{F})$ of *Čech p-boundaries* is given by $B^p(\mathcal{U}, \mathcal{F}) = \operatorname{Im} \delta^{p-1}_{\mathcal{F}}$ for $p \geq 1$ with $B^0(\mathcal{U}, \mathcal{F}) = (0)$, and the R-module $Z^p(\mathcal{U}, \mathcal{F})$ of *Čech p-cocycles* is given by $Z^p(\mathcal{U}, \mathcal{F}) = \operatorname{Ker} \delta^p_{\mathcal{F}}$, for $p \geq 0$.

It is easy to check that $\delta^{p+1}_{\mathcal{F}} \circ \delta^p_{\mathcal{F}} = 0$ for all $p \geq 0$, so we have a chain complex $C^*(\mathcal{U}, \mathcal{F})$ of cohomology

$$0 \xrightarrow{\delta^{-1}_{\mathcal{F}}} C^0(\mathcal{U}, \mathcal{F}) \xrightarrow{\delta^0_{\mathcal{F}}} C^1(\mathcal{U}, \mathcal{F}) \cdots \xrightarrow{\delta^{p-1}_{\mathcal{F}}} C^p(\mathcal{U}, \mathcal{F}) \xrightarrow{\delta^p_{\mathcal{F}}} C^{p+1}(\mathcal{U}, \mathcal{F}) \xrightarrow{\delta^{p+1}_{\mathcal{F}}} \cdots$$

called the *Čech complex*, and we can define the Čech cohomology groups as follows. Let G be an R-module, and write G_X for the constant presheaf on X such that $G_X(U) = G$ for every nonempty open subset $U \subseteq X$ (with $G_X(\emptyset) = (0)$).

Definition 9.4. Given a topological space X, an open cover $\mathcal{U} = (U_j)_{j\in J}$ of X, and a presheaf of R-modules $\mathcal{F}$ on X, the *Čech cohomology groups*

$\check{H}^p(\mathcal{U}, \mathcal{F})$ of the cover $\mathcal{U}$ with values in $\mathcal{F}$ are defined by

$$\check{H}^p(\mathcal{U}, \mathcal{F}) = Z^p(\mathcal{U}, \mathcal{F})/B^p(\mathcal{U}, \mathcal{F}), \quad p \geq 0.$$

The *classical Čech cohomology groups* $\check{H}^p(\mathcal{U}; G)$ *of the cover* $\mathcal{U}$ *with coefficients in the* R*-module* G are the groups $\check{H}^p(\mathcal{U}, G_X)$.

The groups $\check{H}^p(\mathcal{U}, \mathcal{F})$ and $\check{H}^p(\mathcal{U}, G_X)$ are in fact R-modules.

If $\mathcal{F}$ is a sheaf, then $\check{H}^0(\mathcal{U}, \mathcal{F})$ is independent of the cover $\mathcal{U}$.

Proposition 9.1. *Given a topological space X, an open cover $\mathcal{U} = (U_j)_{j \in J}$ of X, and a presheaf of R-modules $\mathcal{F}$ on X, if $\mathcal{F}$ is a sheaf, then*

$$\check{H}^0(\mathcal{U}, \mathcal{F}) = \mathcal{F}(X) = \Gamma(X, \mathcal{F}),$$

the module of global sections of $\mathcal{F}$.

Proof. We saw earlier that a 0-cocycle is a family $(f_j)_{j \in J} \in C^0(\mathcal{U}, \mathcal{F})$ such that

$$\rho^0_{ij}(f_j) = \rho^1_{ij}(f_i)$$

for all $i, j \in J$. Since $\mathcal{F}$ is a sheaf, the f_i patch to a global section $f \in \mathcal{F}(X)$ such that $\rho^X_{U_i}(f) = f_i$ for all $i \in I$. ☐

The module of p-cochains $C^p(\mathcal{U}, \mathcal{F})$ consists of the set of all families $(f_{i_0,\ldots,i_p})_{(i_0,\ldots,i_p) \in J^{p+1}}$ with $f_{i_0,\ldots,i_p} \in \mathcal{F}(U_{i_0 \cdots i_p})$. This is not a very economical definition. It turns out that the same Čech cohomology groups are obtained using the more economical notion of alternating cochain.

Definition 9.5. Given a topological space X, an open cover $\mathcal{U} = (U_j)_{j \in J}$ of X, and a presheaf of R-modules $\mathcal{F}$ on X, a cochain $f \in C^p(\mathcal{U}, \mathcal{F})$ is *alternating* if it satisfies the following conditions:

(a) $f_{i_0,\ldots,i_p} = 0$ whenever two of the indices $i_0, \ldots, i_p$ are equal.
(b) $f_{\sigma(i_0),\ldots,\sigma(i_p)} = \text{sign}(\sigma) f_{i_0,\ldots,i_p}$, for every permutation σ of the set $\{0, \ldots, p\}$ (where $\text{sign}(\sigma)$ denotes the sign of the permutation σ).

The set of alternating p-cochains forms a submodule $C'^p(\mathcal{U}, \mathcal{F})$ of $C^p(\mathcal{U}, \mathcal{F})$.

It is easily checked that $\delta^p_{\mathcal{F}} f$ is alternating if f is alternating. As a consequence the alternating cochains yield a chain complex $(C'^*(\mathcal{U}, \mathcal{F}), \delta_{\mathcal{F}})$. The corresponding cohomology groups are denoted by $\check{H}'^p(\mathcal{U}, \mathcal{F})$. The following proposition is shown in FAC [Serre (1955)] (Chapter 1, §3, Subsection 20).

Proposition 9.2. *Given a topological space X, an open cover $\mathcal{U} = (U_j)_{j \in J}$ of X, and a presheaf of R-modules $\mathcal{F}$ on X, the Čech cohomology groups $\check{H}^p(\mathcal{U}, \mathcal{F})$ and $\check{H}'^p(\mathcal{U}, \mathcal{F})$ are isomorphic for all $p \geq 0$.*

The proof of Proposition 9.2 consists in defining a suitable chain homotopy. It also justifies the fact that we may assume that the index set J is totally ordered (say by $\leq$), and using cochains $f_{i_0,\ldots,i_p}$ where the indices form a strictly increasing sequence $i_0 < i_1 < \cdots < i_p$; Bott and Tu [Bott and Tu (1986)] use this approach (Chapter II, §8).

Our next goal is to define Čech cohomology groups $\check{H}^p(X, \mathcal{F})$ that are independent of the open cover $\mathcal{U}$ chosen for X. Such groups are obtained as direct limits of direct mapping families of modules, as defined in Section 8.8. The direct limit construction is applied to the preorder of refinement among open coverings.

9.2 Čech Cohomology with Values in a Presheaf

First we need to define the notion of refinement of a cover.

Definition 9.6. Given two covers $\mathcal{U} = (U_i)_{i \in I}$ and $\mathcal{V} = (V_j)_{j \in J}$ of a space X, we say that $\mathcal{V}$ *is a refinement of* $\mathcal{U}$, denoted $\mathcal{U} \prec \mathcal{V}$,[1] if there is a function $\tau \colon J \to I$ such that

$$V_j \subseteq U_{\tau(j)} \quad \text{for all } j \in J.$$

See Figure 9.3. We say that two covers $\mathcal{U}$ and $\mathcal{V}$ are *equivalent* if $\mathcal{V} \prec \mathcal{U}$ and $\mathcal{U} \prec \mathcal{V}$.

Let $\tau \colon J \to I$ be a function such that

$$V_j \subseteq U_{\tau(j)} \quad \text{for all } j \in J$$

as above.

Definition 9.7. The homomorphism τ^p from $C^p(\mathcal{U}, \mathcal{F})$ to $C^p(\mathcal{V}, \mathcal{F})$ is defined as follows: for every p-cochain $f \in C^p(\mathcal{U}, \mathcal{F})$, let $\tau^p f \in C^p(\mathcal{V}, \mathcal{F})$ be the p-cochain given by

$$(\tau^p f)_{j_0 \cdots j_p} = \rho^U_V (f_{\tau(j_0) \cdots \tau(j_p)})$$

[1]This is the notation used by Bott and Tu [Bott and Tu (1986)]. Serre uses the opposite notation $\mathcal{V} \prec \mathcal{U}$, in FAC [Serre (1955)] (Chapter 1, §3, Subsection 22).

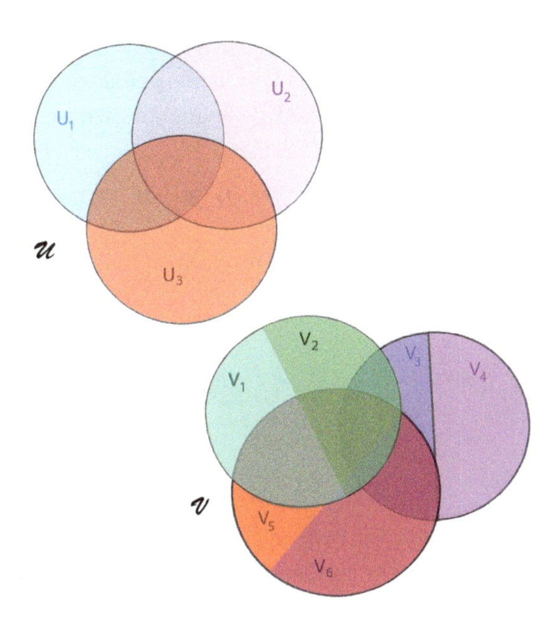

Fig. 9.3 Let $\mathcal{U} = U_1 \cup U_2 \cup U_3$. Let $\mathcal{V} = V_1 \cup V_2 \cup V_3 \cup V_4 \cup V_5 \cup V_6$. Then $\mathcal{U} \prec \mathcal{V}$ with $\tau\colon \{1,2,3,4,5,6\} \to \{1,2,3\}$ where $\tau(1) = 1$, $\tau(2) = 1$, $\tau(3) = 2$, $\tau(4) = 2$, $\tau(5) = 3$, $\tau(6) = 3$ since $V_1 \subseteq U_1$, $V_2 \subseteq U_1$, $V_3 \subseteq U_2$, $V_4 \subseteq U_2$, $V_5 \subseteq U_3$, $V_6 \subseteq U_3$.

for all $(j_0, \ldots, j_p) \in J^{p+1}$, where ρ_V^U denotes the restriction map associated with the inclusion of $V_{j_0\cdots j_p}$ into $U_{\tau(j_0)\cdots\tau(j_p)}$.

Example 9.2. For example, if we take the refinement $\mathcal{U} \prec \mathcal{V}$ illustrated by Figure 9.3, set $p = 0$, and take a cochain $f = (f_1, f_2, f_3) \in C^0(\mathcal{U}, \mathcal{F})$, where f_1 is an element of the group associated with U_1, f_2 is an element of the group associated with U_2, and f_3 is an element of the group associated with U_3, we calculate $\tau^0 f \in C^0(\mathcal{V}, \mathcal{F})$ as

$$
\begin{aligned}
(\tau^0 f)_1 &= \rho_{V_1}^{U_1}(f_{\tau(1)}) = \rho_{V_1}^{U_1}(f_1)\\
(\tau^0 f)_2 &= \rho_{V_2}^{U_1}(f_{\tau(2)}) = \rho_{V_2}^{U_1}(f_1)\\
(\tau^0 f)_3 &= \rho_{V_3}^{U_2}(f_{\tau(3)}) = \rho_{V_3}^{U_2}(f_2)\\
(\tau^0 f)_4 &= \rho_{V_4}^{U_2}(f_{\tau(4)}) = \rho_{V_4}^{U_2}(f_2)\\
(\tau^0 f)_5 &= \rho_{V_5}^{U_3}(f_{\tau(5)}) = \rho_{V_5}^{U_3}(f_3)\\
(\tau^0 f)_6 &= \rho_{V_6}^{U_3}(f_{\tau(6)}) = \rho_{V_6}^{U_3}(f_3).
\end{aligned}
$$

In other words

$$\tau^0(f_1, f_2, f_3) = (\rho_{V_1}^{U_1}(f_1), \rho_{V_2}^{U_1}(f_1), \rho_{V_3}^{U_2}(f_2), \rho_{V_4}^{U_2}(f_2), \rho_{V_5}^{U_3}(f_3), \rho_{V_6}^{U_3}(f_3)).$$

Note that even if the j_k's are distinct, τ may not be injective so the $\tau(j_k)$'s may not be distinct. This is why it is necessary to define the modules $C^p(\mathcal{U}, \mathcal{F})$ using families indexed by sequences whose elements are not necessarily distinct.

It is easy to see that the map $\tau^p\colon C^p(\mathcal{U}, \mathcal{F}) \to C^p(\mathcal{V}, \mathcal{F})$ commutes with $\delta_{\mathcal{F}}$ so we obtain homomorphisms

$$\tau^{*p}\colon \check{H}^p(\mathcal{U}, \mathcal{F}) \to \check{H}^p(\mathcal{V}, \mathcal{F}).$$

Proposition 9.3. *Given any two open covers $\mathcal{U}$ and $\mathcal{V}$ of a space X, if $\mathcal{U} \prec \mathcal{V}$ and if $\tau_1\colon J \to I$ and $\tau_2\colon J \to I$ are functions such that*

$$V_j \subseteq U_{\tau_1(j)} \quad \textit{and} \quad V_j \subseteq U_{\tau_2(j)} \quad \textit{for all } j \in J,$$

then $\tau_1^{*p} = \tau_2^{*p}$ for all $p \geq 0$.

Proof Sketch. Following Serre (see FAC [Serre (1955)], Chapter 1, §3, Subsection 21), we define the maps $k^p\colon C^p(\mathcal{U}, \mathcal{F}) \to C^{p-1}(\mathcal{V}, \mathcal{F})$ such that given any $f \in C^p(\mathcal{U}, \mathcal{F})$,

$$(k^p f)_{j_0 \cdots j_{p-1}} = \sum_{h=0}^{p-1} (-1)^h \rho_h(f_{\tau_1(j_0) \cdots \tau_1(j_h) \tau_2(j_h) \cdots \tau_2(j_{p-1})})$$

for all $(j_0, \ldots, j_{p-1}) \in J^p$, where ρ_h denotes the restriction map associated with the inclusion of $V_{j_0 \cdots j_{p-1}}$ into $U_{\tau_1(j_0) \cdots \tau_1(j_h) \tau_2(j_h) \cdots \tau_2(j_{p-1})}$. Then it can be verified that

$$\delta_{\mathcal{F}} \circ k^p(f) + k^{p+1} \circ \delta_{\mathcal{F}}(f) = \tau_2^p(f) - \tau_1^p(f).$$

It follows that the maps $k^p\colon C^p(\mathcal{U}, \mathcal{F}) \to C^{p-1}(\mathcal{V}, \mathcal{F})$ define a chain homotopy, and by Proposition 2.20, we have $\tau_1^{*p} = \tau_2^{*p}$ for all $p \geq 0$. ☐

Proposition 9.3 implies that if $\mathcal{U} \prec \mathcal{V}$, then there is a homomorphism

$$\rho^{\mathcal{U}}_{\mathcal{V}}\colon \check{H}^p(\mathcal{U}, \mathcal{F}) \to \check{H}^p(\mathcal{V}, \mathcal{F}).$$

It is easy to check that the relation $\mathcal{U} \prec \mathcal{V}$ among covers is a directed preorder; indeed, given any two covers $\mathcal{U} = (U_i)_{i \in I}$ and $\mathcal{V} = (V_j)_{j \in J}$, the cover $\mathcal{W} = (U_i \cap V_j)_{(i,j) \in I \times J}$ is a common refinement of both $\mathcal{U}$ and $\mathcal{V}$, so $\mathcal{U} \prec \mathcal{W}$ and $\mathcal{V} \prec \mathcal{W}$. See Figure 9.4.

It is also immediately verified that if $\mathcal{U} \prec \mathcal{V} \prec \mathcal{W}$, then

$$\rho^{\mathcal{U}}_{\mathcal{W}} = \rho^{\mathcal{V}}_{\mathcal{W}} \circ \rho^{\mathcal{U}}_{\mathcal{V}}$$

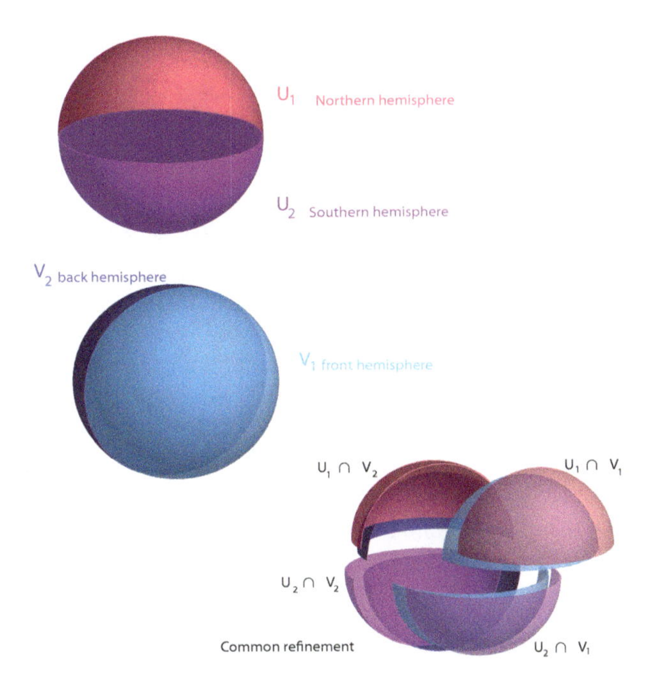

Fig. 9.4 Let $X = S^2$ with two covers $\mathcal{U} = U_1 \cup U_2$ and $\mathcal{V} = V_1 \cup V_2$. The common refinement $\mathcal{W}$ is an open cover consisting four sets.

and that

$$\rho_{\mathcal{U}}^{\mathcal{U}} = \mathrm{id}.$$

Furthermore, if $\mathcal{U}$ and $\mathcal{V}$ are equivalent, then because

$$\rho_{\mathcal{U}}^{\mathcal{V}} \circ \rho_{\mathcal{V}}^{\mathcal{U}} = \mathrm{id} \quad \text{and} \quad \rho_{\mathcal{V}}^{\mathcal{U}} \circ \rho_{\mathcal{U}}^{\mathcal{V}} = \mathrm{id},$$

we see that

$$\rho_{\mathcal{V}}^{\mathcal{U}} \colon \check{H}^p(\mathcal{U}, \mathcal{F}) \to \check{H}^p(\mathcal{V}, \mathcal{F})$$

is an isomorphism.

Consequently, it appears that the family $(\check{H}^p(\mathcal{U}, \mathcal{F}))_{\mathcal{U}}$ is a direct mapping family of modules indexed by the directed set of open covers of X.

However, there is a set-theoretic difficulty, which is that the family of open covers of X is **not** a set because it allows arbitrary index sets.[2]

[2]Most textbook presentations of Čech cohomology ignore this subtle point.

A way to circumvent this difficulty is provided by Serre (see FAC [Serre (1955)], Chapter 1, §3, Subsection 22). The key observation is that any covering $(U_i)_{i\in I}$ is equivalent to a covering $(U'_\lambda)_{\lambda\in L}$ whose index set L is a subset of 2^X. Indeed, we can take for $(U'_\lambda)_{\lambda\in L}$ the *set* of all open subsets of X that belong to the family $(U_i)_{i\in I}$.

As we noted earlier, if $\mathcal{U} = (U_i)_{i\in I}$ and $\mathcal{V} = (V_j)_{j\in J}$ are equivalent, then there is an isomorphism between $\check{H}^p(\mathcal{U}, \mathcal{F})$ and $\check{H}^p(\mathcal{V}, \mathcal{F})$, so we can define

$$\check{H}^p(X, \mathcal{F}) = \varinjlim_{\mathcal{U}} \check{H}^p(\mathcal{U}, \mathcal{F})$$

with respect to coverings $\mathcal{U} = (U_i)_{i\in I}$ whose index set I is a subset of 2^X. Another way to circumvent the set theoretic difficulty is to use a device due to Godement ([Godement (1958)], Chapter 5, Section 5.8).

In summary, we have the following definition.

Definition 9.8. Given a topological space X and a presheaf $\mathcal{F}$ of R-modules on X, the *Čech cohomology groups* $\check{H}^p(X, \mathcal{F})$ *with values in* $\mathcal{F}$ are defined by

$$\check{H}^p(X, \mathcal{F}) = \varinjlim_{\mathcal{U}} \check{H}^p(\mathcal{U}, \mathcal{F})$$

with respect to coverings $\mathcal{U} = (U_i)_{i\in I}$ whose index set I is a subset of 2^X. The *classical Čech cohomology groups* $\check{H}^p(X; G)$ *with coefficients in the R-module* G are the groups $\check{H}^p(X, G_X)$ where G_X is the constant presheaf with value G.

Remark: Warner [Warner (1983)] and Bott and Tu [Bott and Tu (1986)] (second edition) define the classical Čech cohomology groups $\check{H}^p(X; G)$ as the groups $\check{H}^p(X, \widetilde{G}_X)$, where $\widetilde{G}_X$ is the sheaf of locally constant functions with values in G. Although this is not obvious, if X is paracompact (see Definition 13.7), then the groups $\check{H}^p(X, G_X)$ are $\check{H}^p(X, \widetilde{G}_X)$ are isomorphic; this is proven in Proposition 13.16. As a consequence, for manifolds (which by definition are paracompact), this makes no difference. However, Alexander–Lefschetz duality is proven for the classical definition of Čech cohomology corresponding to the case where the constant presheaf G_X is used, and this is why we used it in our definition.

9.3 Equivalence of Čech Cohomology to Other Cohomologies

Next we will investigate the relationship between de Rham cohomology and classical Čech cohomology for the constant sheaf $\widetilde{\mathbb{R}}_X$ (corresponding to coefficients in $\mathbb{R}$), and singular cohomology and classical Čech cohomology for the constant sheaf $\widetilde{\mathbb{Z}}_X$ (corresponding to coefficients in $\mathbb{Z}$). For manifolds, the de Rham cohomology and the classical Čech cohomology for the constant sheaf $\widetilde{\mathbb{R}}_X$ are isomorphic, and the singular cohomology and the classical Čech cohomology for the constant sheaf $\widetilde{\mathbb{Z}}_X$ are also isomorphic. Furthermore, we will see that if our spaces have a good cover $\mathcal{U}$ (recall Definition 3.6), then the Čech cohomology groups $\check{H}^p(\mathcal{U}, \widetilde{\mathbb{R}}_X)$ are independent of $\mathcal{U}$ and in fact isomorphic to the de Rham cohomology groups $H^p_{\mathrm{dR}}(X)$. The main technical tool to prove the above equivalence is a double complex known as the *Čech–de Rham complex*.

Similarly, the Čech cohomology groups $\check{H}^p(\mathcal{U}, \widetilde{\mathbb{Z}}_X)$ are independent of $\mathcal{U}$ and in fact isomorphic to the singular cohomology groups $H^p(X; \mathbb{Z})$ (if X is triangularizable).

Theorem 9.4. *Let M be a smooth manifold. The de Rham cohomology groups are isomorphic to the Čech cohomology groups with values in the sheaf $\widetilde{\mathbb{R}}_M$, and also isomorphic to the Čech cohomology groups associated with good covers (with values in the sheaf $\widetilde{\mathbb{R}}_M$):*

$$H^p_{\mathrm{dR}}(M) \cong \check{H}^p(\mathcal{U}, \widetilde{\mathbb{R}}_M) \cong \check{H}^p(M, \widetilde{\mathbb{R}}_M),$$

for all $p \geq 0$ and all good covers $\mathcal{U}$ of M.

By a previous remark, since manifolds are paracompact, the above theorem also holds with the constant presheaf $\mathbb{R}_M$ instead of the sheaf $\widetilde{\mathbb{R}}_M$.

Theorem 9.4 is proven in Bott and Tu [Bott and Tu (1986)] (Theorem 8.9 and Proposition 10.6). The technique used for proving the first isomorphism is based on an idea of André Weil. The idea is to use a double complex known as the *Čech–de Rham complex*. A complete exposition is given in Chapter 2, Section 8, of Bott and Tu [Bott and Tu (1986)], and we provide most of the proof.

Let M be a smooth manifold. The differential p-forms on M form a sheaf $\mathcal{A}^p_M$ with $\Gamma(U, \mathcal{A}^p_M) = \mathcal{A}^p(U)$, the vector space of p-forms on the open subset $U \subseteq M$.

Given an open cover $\mathcal{U} = (U_j)_{j \in J}$ of M we define the double complex $\mathcal{A}C^{*,*}(\mathcal{U})$ by

$$\mathcal{A}C^{p,q}(\mathcal{U}) = \prod_{(i_0, \ldots, i_p) \in J^{p+1}} \Gamma(U_{i_0 \cdots i_p}, \mathcal{A}^q_M) = \prod_{(i_0, \ldots, i_p) \in J^{p+1}} \mathcal{A}^q(U_{i_0 \cdots i_p}).$$

There are two differentials

$$\delta^{p,q}\colon \mathcal{A}C^{p,q}(\mathcal{U}) \to \mathcal{A}C^{p+1,q}(\mathcal{U}) \quad \text{and} \quad d^{p,q}\colon \mathcal{A}C^{p,q}(\mathcal{U}) \to \mathcal{A}C^{p,q+1}(\mathcal{U})$$

and we have $\delta^{p+1,q} \circ \delta^{p,q} = 0$, $d^{p,q+1} \circ d^{p,q} = 0$, and $\delta^{p,q}$ and $d^{p,q}$ obviously commute. To reduce the amount of notation we often write δ^p (or even δ) instead of $\delta^{p,q}$ and d^q (or even d) instead of $d^{p,q}$.

We also define $D^{p,q}\colon \mathcal{A}C^{p,q}(\mathcal{U}) \to \mathcal{A}C^{p+1,q}(\mathcal{U}) \oplus \mathcal{A}C^{p,q+1}(\mathcal{U})$ by

$$D^{p,q} = \delta^p + (-1)^p d^q.$$

We associate to the double complex $\mathcal{A}C^{*,*}(\mathcal{U})$ the single complex $\mathcal{A}C^*(\mathcal{U})$ defined by

$$\mathcal{A}C^n(\mathcal{U}) = \bigoplus_{p+q=n} \mathcal{A}C^{p,q}(\mathcal{U}),$$

with the differential $D^n\colon \mathcal{A}C^n(\mathcal{U}) \to \mathcal{A}C^{n+1}(\mathcal{U})$ given by

$$D^n = \sum_{p+q=n} D^{p,q}.$$

For any $\omega = \omega_0 + \cdots + \omega_n$ with $\omega_p \in \mathcal{A}C^{p,n-p}(\mathcal{U})$, we have

$$D^n(\omega) = \sum_{p=0}^{n} D^{p,n-p}(\omega_p).$$

It is easily verified that

$$D^{n+1} \circ D^n = 0.$$

It suffices to verify that $D^{n+1}(D^n\omega) = 0$ for $\omega \in \mathcal{A}C^{p,q}(\mathcal{U})$ with $p+q = n$. Since $D^{p,q}\omega = \delta^p\omega + (-1)^p d^q\omega$ with $\delta^p\omega \in \mathcal{A}C^{p+1,q}(\mathcal{U})$ and $d^q\omega \in \mathcal{A}C^{p,q+1}(\mathcal{U})$, only $D^{p+1,q} = \delta^{p+1} + (-1)^{p+1}d^q$ and $D^{p,q+1} = \delta^p + (-1)^p d^{q+1}$ apply, and we get

$$\begin{aligned}
(D^{p+1,q} + D^{p,q+1})(D^{p,q}(\omega)) &= D^{p+1,q}(\delta^p\omega) + D^{p,q+1}((-1)^p d^q\omega) \\
&= (\delta^{p+1} + (-1)^{p+1}d^q)(\delta^p\omega) \\
&\quad + (\delta^p + (-1)^p d^{q+1})((-1)^p d^q\omega) \\
&= \delta^{p+1}(\delta^p\omega) + (-1)^{p+1}d^q(\delta^p\omega) \\
&\quad + (-1)^p\delta^p(d^q\omega) + d^{q+1}(d^q\omega) \\
&= (-1)^{p+1}d^q(\delta^p\omega) + (-1)^p d^q(\delta^p\omega) = 0
\end{aligned}$$

since δ^p and d^q commute. See the diagram below for a graphical illustration.

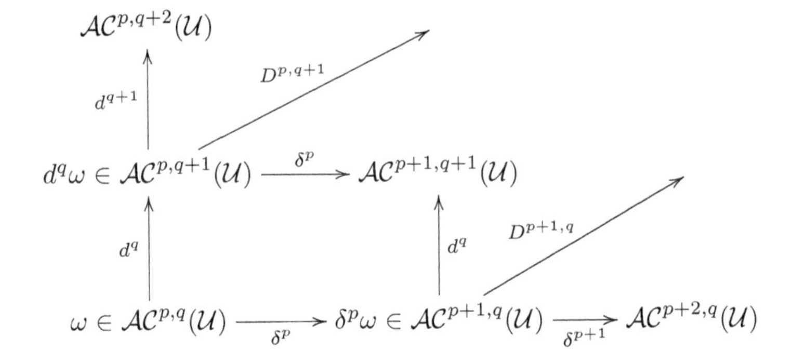

For simplicity of notation we often write D instead of $D^{p,q}$ or D^n.

The double complex $\mathcal{A}C^{*,*}(\mathcal{U})$ can be displayed as a 2-dimensional array (infinite in both dimensions) with the x-axis corresponding to the index p and the y-axis corresponding to the index q as follows.

q	$\vdots$	$\vdots$	$\vdots$	
$\mathcal{A}^2(M)$	$\mathcal{A}C^{0,2}(\mathcal{U})$	$\mathcal{A}C^{1,2}(\mathcal{U})$	$\mathcal{A}C^{2,2}(\mathcal{U})$	$\cdots$
$\mathcal{A}^1(M)$	$\mathcal{A}C^{0,1}(\mathcal{U})$	$\mathcal{A}C^{1,1}(\mathcal{U})$	$\mathcal{A}C^{2,1}(\mathcal{U})$	$\cdots$
$\mathcal{A}^0(M)$	$\mathcal{A}C^{0,0}(\mathcal{U})$	$\mathcal{A}C^{1,0}(\mathcal{U})$	$\mathcal{A}C^{2,0}(\mathcal{U})$	$\cdots$
	$C^0(\mathcal{U},\widetilde{\mathbb{R}}_M)$	$C^1(\mathcal{U},\widetilde{\mathbb{R}}_M)$	$C^2(\mathcal{U},\widetilde{\mathbb{R}}_M)$	p

We added an extra column consisting of the spaces of differential forms $\mathcal{A}^q(M)$ and an extra row consisting of the Čech cochain modules $C^p(\mathcal{U},\widetilde{\mathbb{R}}_M)$ associated with the cover $\mathcal{U}$ and the constant sheaf $\widetilde{\mathbb{R}}_M$. Each $C^p(\mathcal{U},\widetilde{\mathbb{R}}_M)$ is the kernel of the lowest d in the pth column, so $C^p(\mathcal{U},\widetilde{\mathbb{R}}_M)$ consists of locally constant functions on the open subsets $U_{i_0\cdots i_p}$ with $(i_0,\ldots,i_p)\in J^{p+1}$. The reason for doing so is that this extra column is the de Rham complex (with differential d) whose cohomology is the de Rham cohomology $H^*_{\mathrm{dR}}(M)$, and this extra row is the Čech complex (with differential δ) whose cohomology is the Čech cohomology $\check{H}^*(\mathcal{U},\widetilde{\mathbb{R}}_M)$. The module

$$\mathcal{A}C^n(\mathcal{U}) = \bigoplus_{p+q=n} \mathcal{A}C^{p,q}(\mathcal{U})$$

is obtained by summing along the diagonal line $p+q=n$. The cohomology of the complex $(\mathcal{A}C^*(\mathcal{U}), D)$ is denoted by $H_D^*(\mathcal{A}C^*(\mathcal{U}))$. In Bott and Tu [Bott and Tu (1986)] it is denoted $H_D\{\mathcal{A}C^*(\mathcal{U}, \mathcal{A}_M^*)\}$.

The following result is shown in Bott and Tu [Bott and Tu (1986)].

Theorem 9.5. *For any smooth manifold M and any open cover $\mathcal{U}$, there is an isomorphism*

$$H_{\mathrm{dR}}^*(M) \cong H_D^*(\mathcal{A}C^*(\mathcal{U})).$$

Theorem 9.5 follows from the following facts:

(1) Each (augmented) row

$$0 \longrightarrow \mathcal{A}^q(M) \xrightarrow{r^q} \mathcal{A}C^{0,q}(\mathcal{U}) \xrightarrow{\delta^0} \mathcal{A}C^{1,q}(\mathcal{U}) \xrightarrow{\delta^1} \mathcal{A}C^{2,q}(\mathcal{U}) \xrightarrow{\delta^2} \cdots$$

of the double complex is exact, where the map r^q is the restriction from $\mathcal{A}^q(M)$ to $\mathcal{A}C^{0,q}(\mathcal{U}) = \prod_{j\in J} \mathcal{A}^q(U_j)$ given by $r^q(\omega) = (\omega \mid U_j)_{j\in J}$. This is proven in Proposition 8.5 of Bott and Tu [Bott and Tu (1986)]. The argument uses a partition of unity.

(2) The fact that the rows of the double complex are exact implies that the cohomology of the complex $(\mathcal{A}C^*(\mathcal{U}), D)$ is equal to the cohomology of the first column (consisting of $\mathcal{A}C^{0,0}(\mathcal{U}), \mathcal{A}C^{0,1}(\mathcal{U}), \ldots, \mathcal{A}C^{0,q}(\mathcal{U}), \ldots$) of the double complex.

(3) The de Rham cohomology $H_{\mathrm{dR}}^*(M)$ is isomorphic to the cohomology of the first column (consisting of $\mathcal{A}C^{0,0}(\mathcal{U}), \mathcal{A}C^{0,1}(\mathcal{U}), \ldots, \mathcal{A}C^{0,q}(\mathcal{U}), \ldots$) of the double complex.

Both (2) and (3) are proven in Proposition 8.8 of Bott and Tu [Bott and Tu (1986)].

Proof of (2). Every cocycle in $\mathcal{A}C^n(\mathcal{U}) = \bigoplus_{p+q=n} \mathcal{A}C^{p,q}(\mathcal{U})$ is of the form

$$\omega = \omega_0 + \omega_1 + \cdots + \omega_n,$$

with ω_p a D-cocycle in $\mathcal{A}C^{p,n-p}(\mathcal{U})$ for $p = 0, \ldots, n$ (that is, $D^{p,n-p}\omega_p = 0$). Since $\omega_n \in \mathcal{A}C^{n,0}(\mathcal{U})$, $D^{n,0} = \delta^n + (-1)^n d^0$, $\delta^n\omega_n \in \mathcal{A}C^{n+1,0}(\mathcal{U})$ and $d^0\omega_n \in \mathcal{A}C^{n,1}(\mathcal{U})$, we see that $D^{n,0}\omega_n = 0$ iff $\delta^n\omega_n = 0$ and $d^0\omega_n = 0$. Since $\omega_n \in \mathcal{A}C^{n,0}(\mathcal{U})$, $\delta^n\omega_n = 0$, and since every row of the double complex is exact, for $q = 0$ the diagram

$$\cdots \longrightarrow \mathcal{A}C^{n-1,0}(\mathcal{U}) \xrightarrow{\delta^{n-1}} \mathcal{A}C^{n,0}(\mathcal{U}) \xrightarrow{\delta^n} \mathcal{A}C^{n+1,0}(\mathcal{U}) \longrightarrow \cdots$$

is exact, we have $\omega_n \in \operatorname{Ker} \delta^n = \operatorname{Im} \delta^{n-1}$, which means that $\omega_n = \delta^{n-1}\beta$ with $\beta \in \mathcal{AC}^{n-1,0}(\mathcal{U})$. Now $D^{n-1,0}\beta \in \mathcal{AC}^n(\mathcal{U})$ and

$$D^{n-1,0}\beta = \delta^{n-1}\beta + (-1)^{n-1}d^0\beta = \omega_n + (-1)^{n-1}d^0\beta$$

with $d^0\beta \in \mathcal{AC}^{n-1,1}(\mathcal{U})$, so $\omega_n - D^{n-1,0}\beta = (-1)^n d^0\beta \in \mathcal{AC}^{n-1,1}(\mathcal{U})$ with $D^{n,0}(\omega_n - D^{n-1,0}\beta) = D^{n,0}\omega_n - D^{n,0}D^{n-1,0}\beta = 0$, so the cohomology class $[\omega]$ is also represented by the cocycle

$$\omega_0 + \omega_1 + \cdots + \omega_{n-2} + \omega_{n-1} + \omega_n - D^{n-1,0}\beta$$

with $\omega_{n-1} + \omega_n - D^{n-1,0}\beta = \omega_{n-1} + (-1)^n d^0\beta$ a cocycle in $\mathcal{AC}^{n-1,1}(\mathcal{U})$. A graphical illustration of this argument is provided by Figure 9.5. The idea is to climb up the diagonal starting from the $(n,0)$ slot up to the $(0,n)$ slot, each time subtracting $D^{p-1,n-p}\beta$ from ω_p in the $(p, n-p)$ slot for some $\beta \in \mathcal{AC}^{p-1,n-p}(\mathcal{U})$ such that $\delta^{p-1}\beta = \omega_p$. Such a form β exists because $\delta^p\omega_p = 0$ (since $D^{p,n-p}\omega_p = 0$) and because the pth row of the double complex is exact.

By induction we can prove that the cohomology class $[\omega]$ with $\omega \in \mathcal{AC}^n(\mathcal{U})$ is represented by a cocycle of the form

$$\omega = \omega_0, \quad \omega_0 \in \mathcal{AC}^{0,n}(\mathcal{U}),$$

proving (2). □

Proof of (3). We need to show that the restriction map r from $\mathcal{A}^n(M)$ to $\mathcal{AC}^{0,n}(\mathcal{U}) \subseteq \mathcal{AC}^n(\mathcal{U}) = \bigoplus_{p+q=n} \mathcal{AC}^{p,q}(\mathcal{U})$ induces an isomorphism r^* between $H^n_{\mathrm{dR}}(M)$ and $H^n_D(\mathcal{AC}^n(\mathcal{U}))$. In all rigor we should write r^n but the context makes it clear which r^n applies.

First we check that $r = (r^n)$ is chain map, and for this we need to check that for every n the following diagram is commutative.

$$\begin{array}{ccc} \mathcal{A}^n(M) & \xrightarrow{r^n} & \mathcal{AC}^n(\mathcal{U}) \\ {\scriptstyle d}\downarrow & & \downarrow{\scriptstyle D} \\ \mathcal{A}^{n+1}(M) & \xrightarrow[r^{n+1}]{} & \mathcal{AC}^{n+1}(\mathcal{U}). \end{array}$$

Since the rows of the double complex are exact and $\operatorname{Im} r^n \subseteq \mathcal{AC}^{0,n}(\mathcal{U})$, only $D^{0,n}$ applies and we have $\delta^0 \circ r^n = 0$, so (as $p = 0$)

$$\begin{aligned} D^{0,n} \circ r^n &= (\delta^0 + d^n) \circ r^n \\ &= d^n \circ r^n \\ &= r^{n+1} \circ d^n \end{aligned}$$

by definition of r^n and r^{n+1}.

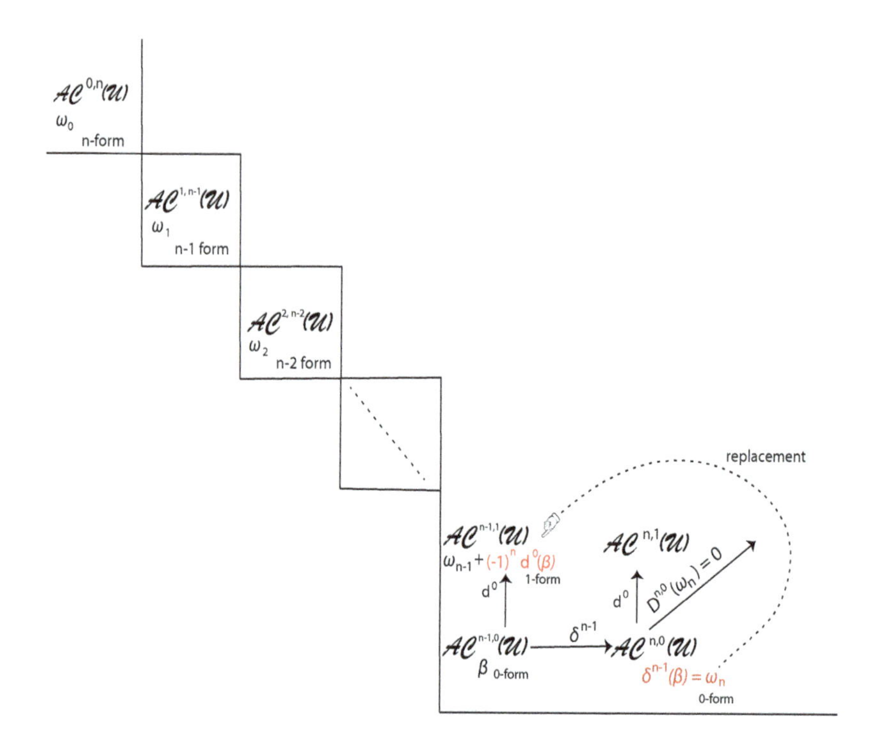

Fig. 9.5 A graphical illustration of the replacement induction of Case 2.

Next we need to prove that $r^*: H^n_{\mathrm{dR}}(M) \to H^n_D(\mathcal{AC}^n(\mathcal{U}))$ is an isomorphism.

First we prove that r^* is surjective. By the reasoning in Part (2), a cohomology class $[\omega] \in H^n_D(\mathcal{AC}^n(\mathcal{U}))$ is represented by a cocycle $\omega \in \mathcal{AC}^{0,n}(\mathcal{U})$ such that $D^{0,n}\omega = 0$. Since $\omega \in \mathcal{AC}^{0,n}(\mathcal{U})$, we have

$$D^{0,n}\omega = \delta^0\omega + d^n\omega$$

with $\delta^0\omega \in \mathcal{AC}^{1,n}(\mathcal{U})$ and $d^n\omega \in \mathcal{AC}^{0,n+1}(\mathcal{U})$ so $D^{0,n}\omega = 0$ iff $\delta^0\omega = 0$ and $d^n\omega = 0$. By definition of $\delta^0: \mathcal{AC}^{0,n}(\mathcal{U}) \to \mathcal{AC}^{1,n}(\mathcal{U})$ and Equation (δ^0), $\delta^0\omega = 0$ means that the forms $\omega \mid U_j$ agree on overlaps $U_j \cap U_k$, so ω is actually a closed form in $\mathcal{A}^n(M)$, which shows that r^* is surjective. The following diagram is an illustration of the proof.

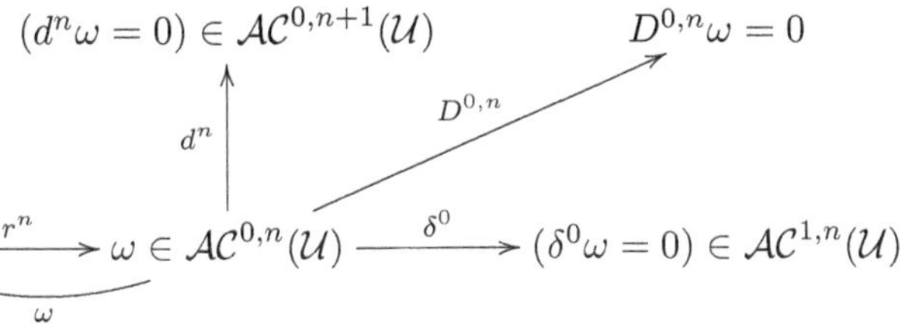

$$\begin{array}{ccccc} & & (d^n\omega = 0) \in \mathcal{AC}^{0,n+1}(\mathcal{U}) & & D^{0,n}\omega = 0 \\ & & \uparrow d^n & \nearrow D^{0,n} & \\ \mathcal{A}^n(M) & \xrightarrow{r^n} & \omega \in \mathcal{AC}^{0,n}(\mathcal{U}) & \xrightarrow{\delta^0} & (\delta^0\omega = 0) \in \mathcal{AC}^{1,n}(\mathcal{U}) \\ & \xleftarrow[\omega]{} & & & \end{array}$$

Second we prove that r^* is injective. If $r^*([\omega_1]) = r^*([\omega_2])$, then $r^*([\omega_1] - [\omega_2]) = 0$, which means that $r^n(\omega_1 - \omega_2) = D^{n-1}\beta$ for some cochain $\beta \in \mathcal{AC}^{n-1}(\mathcal{U})$. Using the reasoning in Part (2), we may assume that $\beta \in \mathcal{AC}^{0,n-1}(\mathcal{U})$. Since $\omega_1 - \omega_2 \in \mathcal{A}^n(M)$, we have $r^n(\omega_1 - \omega_2) \in \mathcal{AC}^{0,n}(\mathcal{U})$, and we also have

$$D^{0,n-1}\beta = \delta^0\beta + d^{n-1}\beta$$

with $\delta^0\beta \in \mathcal{AC}^{1,n-1}(\mathcal{U})$ and $d^{n-1}\beta \in \mathcal{AC}^{0,n}(\mathcal{U})$, so $r^n(\omega_1 - \omega_2) = D^{0,n-1}\beta = \delta^0\beta + d^{n-1}\beta$ implies that

$$\delta^0\beta = 0.$$

Therefore, β is a global form in $\mathcal{A}^{n-1}(M)$, so $r^n(\omega_1 - \omega_2) = D^{0,n-1}\beta = d^{n-1}\beta$ implies that $\omega_1 - \omega_2 = d^{n-1}\beta$, that is $[\omega_1] = [\omega_2]$, proving that r^* is injective. The following diagram is an illustration of the proof.

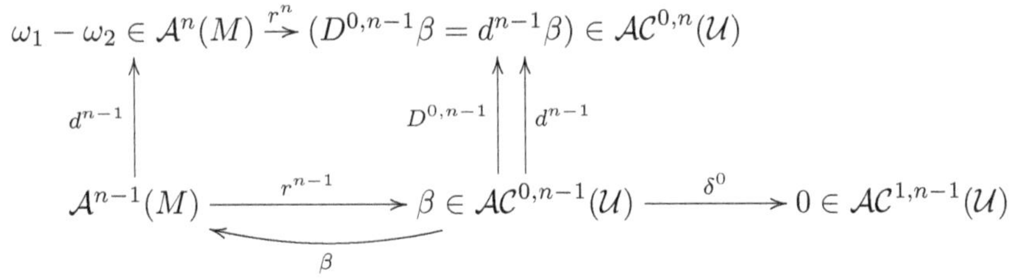

$$\begin{array}{ccccc} \omega_1 - \omega_2 \in \mathcal{A}^n(M) & \xrightarrow{r^n} & (D^{0,n-1}\beta = d^{n-1}\beta) \in \mathcal{AC}^{0,n}(\mathcal{U}) & & \\ \uparrow d^{n-1} & & \uparrow D^{0,n-1} \ \uparrow d^{n-1} & & \\ \mathcal{A}^{n-1}(M) & \xrightarrow{r^{n-1}} & \beta \in \mathcal{AC}^{0,n-1}(\mathcal{U}) & \xrightarrow{\delta^0} & 0 \in \mathcal{AC}^{1,n-1}(\mathcal{U}) \\ & \xleftarrow[\beta]{} & & & \end{array}$$

This concludes the proof that r^* is an isomorphism. ☐

Furthermore, if $\mathcal{U}$ is a good cover, the following result is shown in Bott and Tu [Bott and Tu (1986)] (before Theorem 8.9).

Theorem 9.6. *For any smooth manifold M and any good cover $\mathcal{U}$, there is an isomorphism*

$$\check{H}^p(\mathcal{U}, \widetilde{\mathbb{R}}_M) \cong H^*_D(\mathcal{AC}^*(\mathcal{U})).$$

The reason is that if $\mathcal{U}$ is a good cover, then the augmented columns (consisting of $C^p(\mathcal{U}, \widetilde{\mathbb{R}}_M)$, $\mathcal{A}C^{p,0}(\mathcal{U}), \mathcal{A}C^{p,1}(\mathcal{U}), \ldots, \mathcal{A}C^{p,q}(\mathcal{U}), \ldots$) of the double complex are exact. Here the first map $i^p\colon C^p(\mathcal{U}, \widetilde{\mathbb{R}}_M) \to \mathcal{A}C^{p,0}(\mathcal{U})$ is the inclusion map (recall that $C^p(\mathcal{U}, \widetilde{\mathbb{R}}_M)$ is the kernel of $d^0\colon \mathcal{A}C^{p,0}(\mathcal{U}) \to \mathcal{A}C^{p,1}(\mathcal{U})$). Indeed, the qth cohomology group of the pth column ($q \geq 1$) is

$$\prod_{(i_0,\ldots,i_p)\in J^{p+1}} H^q(U_{i_0\cdots i_p}),$$

but if $\mathcal{U}$ is a good cover, the open subsets $U_{i_0\cdots i_p}$ are contractible,[3] so by the Poincaré lemma, $H^q(U_{i_0\cdots i_p}) = (0)$ for all $q \geq 1$, thus $\prod_{(i_0,\ldots,i_p)\in J^{p+1}} H^q(U_{i_0\cdots i_p}) = (0)$, and since $C^p(\mathcal{U}, \widetilde{\mathbb{R}}_M)$ is the kernel of the lowest d in the column, the pth column is exact.

By an argument analogous to (2) we can prove that the cohomology of the complex $(\mathcal{A}C^*, D)$ is equal to the cohomology of the first row (consisting of $\mathcal{A}C^{0,0}(\mathcal{U}), \mathcal{A}C^{1,0}(\mathcal{U}), \cdots, \mathcal{A}C^{p,0}(\mathcal{U}), \ldots$) of the double complex. By an argument analogous to (3) we can prove that the Čech cohomology $\check{H}^*(\mathcal{U}, \widetilde{\mathbb{R}}_M)$ is isomorphic to the cohomology of the first row (consisting of $\mathcal{A}C^{0,0}(\mathcal{U}), \mathcal{A}C^{1,0}(\mathcal{U}), \ldots, \mathcal{A}C^{p,0}(\mathcal{U}), \ldots$) of the double complex.

Consequently, by Theorem 9.5 and Theorem 9.6, we obtain an isomorphism

$$H^p_{\mathrm{dR}}(M) \cong \check{H}^p(\mathcal{U}, \widetilde{\mathbb{R}}_M)$$

for all good covers $\mathcal{U}$ and all $p \geq 0$.

Since every smooth manifold has a good cover (see Theorem 3.4), and since the good covers are cofinal in the set of all covers of M (with index set in 2^M), following Bott and Tu [Bott and Tu (1986)] (Proposition 10.6), we obtain the isomorphism

$$\check{H}^p(M, \widetilde{\mathbb{R}}_M) \cong \check{H}^p(\mathcal{U}, \widetilde{\mathbb{R}}_M)$$

for all good covers $\mathcal{U}$ and all $p \geq 0$.

Remark: Morita also proves Theorem 9.4 using the double complex $\mathcal{A}C^{*,*}(\mathcal{U})$, but without introducing the single complex $\mathcal{A}C^*(\mathcal{U})$; see Morita [Morita (2001)] (Chapter 3). Morita does not prove Theorem 9.5 and Theorem 9.6.

[3]In fact, diffeomorphic to $\mathbb{R}^n$. A more general notion of a good cover $\mathcal{U}$ on a topological space (not necessarily a manifold) is that all finite intersections are contractible; see Bott and Tu [Bott and Tu (1986)], Chapter II, Section 13.

We now turn to singular cohomology.

Theorem 9.7. *If M is a paracompact topological manifold and if G is a R-module over a commutative ring R with a unit, then the singular cohomology groups $H^p(M;G)$ are isomorphic to the Čech cohomology groups $\check{H}^p(M,\widetilde{G}_M)$:*

$$H^p(M;G) \cong \check{H}^p(M,\widetilde{G}_M) \quad \textit{for all } p \geq 0.$$

If X is a topological space and if $\mathcal{U}$ is a good cover of X, then we have isomorphisms between the singular cohomology groups $H^p(X;\mathbb{Z})$ and the Čech cohomology groups $\check{H}^p(X,\widetilde{\mathbb{Z}}_X)$ and $\check{H}^p(\mathcal{U},\widetilde{\mathbb{Z}}_X)$:

$$H^p(X,\mathbb{Z}) \cong \check{H}^p(\mathcal{U},\widetilde{\mathbb{Z}}_X) \cong \check{H}^p(X,\widetilde{\mathbb{Z}}_X) \quad \textit{for all } p \geq 0.$$

In particular, the above holds if X is a smooth manifold.

By a previous remark, since our spaces are paracompact, the above theorem also holds with the constant presheaf G_X (or $\mathbb{Z}_X$) instead of the sheaf $\widetilde{G}_X$ (or $\widetilde{\mathbb{Z}}_X$)).

The proof of the isomorphism $H^p(X;G) \cong \check{H}^p(X,\widetilde{G}_X)$ takes a lot of work. A version of this proof can be found in Warner [Warner (1983)] (Chapter 5). Another type of cohomology known as sheaf cohomology is introduced, and it is shown that both singular cohomology and Čech cohomology agree with sheaf cohomology if X is paracompact and locally Euclidean. Sheaf cohomology is a special case of Grothendieck's approach to cohomology using derived functors. This is a very general and powerful approach which is discussed thoroughly in Chapter 13.

The other isomorphisms involving good covers are proven in Bott and Tu [Bott and Tu (1986)] using double complexes; see Chapter III, §15, Theorem 15.8.

It should be noted that if the space X is not well-behaved, then singular cohomology and Čech cohomology may differ. For example, if X is the *topologist's sine curve* (a space which is connected but neither locally connected nor path connected), it can be shown that

$$\begin{aligned} H^1(X;\mathbb{Z}) &= (0) \\ \check{H}^1(X;\mathbb{Z}) &= \mathbb{Z}; \end{aligned}$$

see Munkres [Munkres (1984)] (Chapter 8, §73).

9.4 Problems

Problem 9.1. Prove Proposition 9.2.

Hint. For help, see Serre FAC [Serre (1955)] (Chapter 1, §3, Subsection 20).

Problem 9.2. Provide the details of the proof of Proposition 9.3.

Hint. For help, see Serre FAC [Serre (1955)] (Chapter 1, §3, Subsection 20).

Problem 9.3. Provide the details of the induction step in the proof of (2) in Theorem 9.5.

Problem 9.4. Provide the details of the proof of (2) in Theorem 9.6.

Problem 9.5. Provide the details of the proof of (3) in Theorem 9.6.

Problem 9.6. Prove that if a smooth manifold M has a finite good cover, then its Čech cohomology is finite-dimensional, and thus its de Rham cohomology $H^*_{\mathrm{dR}}(M)$ is finite-dimensional. Deduce that if M is compact, then its de Rham cohomology $H^*_{\mathrm{dR}}(M)$ is finite-dimensional.

Chapter 10

Presheaves and Sheaves; A Deeper Look

One of the main goals of this chapter is to define the notion of exact sequence of sheaves

$$\cdots \longrightarrow \mathcal{F} \xrightarrow{\varphi} \mathcal{G} \xrightarrow{\psi} \mathcal{H} \longrightarrow \cdots$$

where φ and ψ are maps of sheaves. The obvious definition is $\mathrm{Im}\,\varphi = \mathrm{Ker}\,\psi$, and this requires defining the kernel and the image of a map of sheaves.

The notion of kernel $\mathrm{Ker}\,\varphi$ and image $\mathrm{Im}\,\varphi$ of a presheaf or sheaf map $\varphi\colon \mathcal{F} \to \mathcal{G}$ is easily defined. The presheaf $\mathrm{Ker}\,\varphi$ is defined by $(\mathrm{Ker}\,\varphi)(U) = \mathrm{Ker}\,\varphi_U$, and the presheaf $\mathrm{Im}\,\varphi$ is defined by $(\mathrm{Im}\,\varphi)(U) = \mathrm{Im}\,\varphi_U$. In the case or presheaves, they are also presheaves, but in the case of sheaves, the kernel $\mathrm{Ker}\,\varphi$ is indeed a sheaf, but the image $\mathrm{Im}\,\varphi$ is **not** a sheaf in general.

This failure of the image of a sheaf map to be a sheaf is a problem that causes significant technical complications. In particular, it is not clear what it means for a sheaf map to be surjective, and a "good" definition of the notion of an exact sequence of sheaves is also unclear.

Fortunately, there is a procedure for converting a presheaf $\mathcal{F}$ into a sheaf $\widetilde{\mathcal{F}}$ which is reasonably well-behaved. This procedure is called *sheafification*. There is a sheaf map $\eta_{\mathcal{F}}\colon \mathcal{F} \to \widetilde{F}$ which is generally not injective.

The *sheafification* process is universal in the sense that given any presheaf $\mathcal{F}$ and any sheaf $\mathcal{G}$, for any presheaf map $\varphi\colon \mathcal{F} \to \mathcal{G}$, there is a unique sheaf map $\widehat{\varphi}\colon \widetilde{\mathcal{F}} \to \mathcal{G}$ such that

$$\varphi = \widehat{\varphi} \circ \eta_{\mathcal{F}}$$

as illustrated by the following commutative diagram

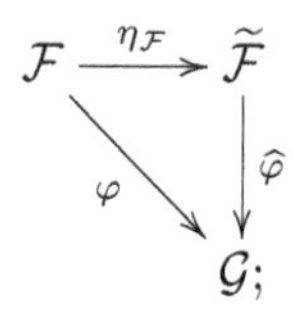

see Theorem 10.12.

The *sheafification* process involves constructing a topological space $S\mathcal{F}$ from the presheaf $\mathcal{F}$ that we call the *stalk space* of $\mathcal{F}$. Godement calls it the *espace étalé*. The stalk space is the disjoint union of sets (modules) $\mathcal{F}_x$ called *stalks*. Each stalk $\mathcal{F}_x$ is the direct limit $\varinjlim(\mathcal{F}(U))_{U\ni x}$ of the family of modules $\mathcal{F}(U)$ for all "small" open sets U containing x (see Definition 10.1). There is a surjective map $p\colon S\mathcal{F} \to X$ which, under the topology given to $S\mathcal{F}$, is a local homeomorphism, which means that for every $y \in S\mathcal{F}$, there is some open subset V of $S\mathcal{F}$ containing y such that the restriction of p to V is a homeomorphism. The sheaf $\widetilde{\mathcal{F}}$ consists of the continuous sections of p, that is, the continuous functions $s\colon U \to S\mathcal{F}$ such that $p \circ s = \mathrm{id}_U$, for any open subset U of X. This construction is presented in detail in Section 10.1, Section 10.2, and Section 10.4.

The construction of the pair $(S\mathcal{F}, p)$ from a presheaf $\mathcal{F}$ suggests another definition of a sheaf as a pair (E, p), where E is a topological space and $p\colon E \to X$ is a surjective local homeomorphism onto another space X. Such a pair (E, p) is often called a *sheaf space*, but we prefer to call it a *stalk space*. This is the definition that was given by H. Cartan and M. Lazard around 1950. The sheaf ΓE associated with the stalk space (E, p) is defined as follows: for any open subset U or X, the *sections* of ΓE are the continuous sections $s\colon U \to E$, that is, the continuous functions such that $p \circ s = \mathrm{id}$. We can also define a notion of map between two stalk spaces. Stalk spaces are discussed in Section 10.3.

As this stage, given a topological space X we have three categories:

(1) The category $\mathbf{Psh}(X)$ of presheaves and their morphisms.
(2) The category $\mathbf{Sh}(X)$ of sheaves and their morphisms.
(3) The category $\mathbf{StalkS}(X)$ of stalk spaces and their morphisms.

There is also a functor

$$S\colon \mathbf{PSh}(X) \to \mathbf{StalkS}(X)$$

from the category $\mathbf{PSh}(X)$ to the category $\mathbf{StalkS}(X)$ given by the construction of a stalk space $S\mathcal{F}$ from a presheaf $\mathcal{F}$, and a functor

$$\Gamma\colon \mathbf{StalkS}(X) \to \mathbf{Sh}(X)$$

from the category $\mathbf{StalkS}(X)$ to the category $\mathbf{Sh}(X)$, given by the sheaf ΓE of continuous sections of E. Here, we are using the term functor in an informal way. A more precise definition is given in Section 10.10.

Note that every sheaf $\mathcal{F}$ is also a presheaf, and that every map $\varphi\colon \mathcal{F} \to \mathcal{G}$ of sheaves is also a map of presheaves. Therefore, we have an inclusion map

$$i\colon \mathbf{Sh}(X) \to \mathbf{PSh}(X),$$

which is a functor. As a consequence, S restricts to an operation (functor)

$$S\colon \mathbf{Sh}(X) \to \mathbf{StalkS}(X).$$

There is also a map η which maps a presheaf $\mathcal{F}$ to the sheaf $\Gamma S(\mathcal{F}) = \widetilde{\mathcal{F}}$. This map η is a natural isomorphism between the functors id (the identity functor) and ΓS from $\mathbf{Sh}(X)$ to itself.

We can also define a map ϵ which takes a stalk space (E, p) and makes the stalk space $S\Gamma E$. The map ϵ is a natural isomorphism between the functors id (the identity functor) and $S\Gamma$ from $\mathbf{StalkS}(X)$ to itself.

Then we see that the two operations (functors)

$$S\colon \mathbf{Sh}(X) \to \mathbf{StalkS}(X) \quad \text{and} \quad \Gamma\colon \mathbf{StalkS}(X) \to \mathbf{Sh}(X)$$

are almost mutual inverses, in the sense that there is a natural isomorphism η between ΓS and id and a natural isomorphism ϵ between $S\Gamma$ and id. In such a situation, we say that the classes (categories) $\mathbf{Sh}(X)$ and $\mathbf{StalkS}(X)$ are *equivalent*. The upshot is that it is basically a matter of taste (or convenience) whether we decide to work with sheaves or stalk spaces. All this is explained in Sections 10.3 and 10.4. We also discuss stalk spaces of rings and modules in Section 10.5.

We still need to define the image of a sheaf map in such a way that the notion of exact sequence of sheaves makes sense. Recall that if $f\colon A \to B$ is a homomorphism of modules, the *cokernel* $\mathrm{Coker} f$ of f is defined by $B/\mathrm{Im} f$. It is a measure of the surjectivity of f. We also have the projection map $\mathrm{coker}(f)\colon B \to \mathrm{Coker}\, f$, and observe that

$$\mathrm{Im}\, f = \mathrm{Ker}\, \mathrm{coker}(f).$$

The above suggests defining notions of cokernels of presheaf maps and sheaf maps. For a presheaf map $\varphi\colon \mathcal{F} \to \mathcal{G}$ this is easy, and we can define the *presheaf cokernel* $\mathrm{PCoker}(\varphi)$. It comes with a presheaf map $\mathrm{pcoker}(\varphi)\colon \mathcal{G} \to \mathrm{PCoker}(\varphi)$.

If $\mathcal{F}$ and $\mathcal{G}$ are sheaves, we define the *sheaf cokernel* $\mathrm{SCoker}(\varphi)$ as the sheafification of $\mathrm{PCoker}(\varphi)$. It also comes with a presheaf map $\mathrm{scoker}(\varphi)\colon \mathcal{G} \to \mathrm{SCoker}(\varphi)$.

Then it can be shown that if $\varphi\colon \mathcal{F} \to \mathcal{G}$ is a sheaf map, $\mathrm{SCoker}(\varphi) = (0)$ iff the stalk maps $\varphi_x\colon \mathcal{F}_x \to \mathcal{G}_x$ are surjective for all $x \in X$; see Proposition 10.19.

It follows that the "correct" definition for the image $\mathrm{SIm}\,\varphi$ of a sheaf map $\varphi\colon \mathcal{F} \to \mathcal{G}$ is

$$\mathrm{SIm}\,\varphi = \mathrm{Ker}\,\mathrm{scoker}(\varphi).$$

With this definition, a sequence of sheaves

$$\mathcal{F} \xrightarrow{\ \varphi\ } \mathcal{G} \xrightarrow{\ \psi\ } \mathcal{H}$$

is said to be exact if $\mathrm{SIm}\,\varphi = \mathrm{Ker}\,\psi$. Then it can be shown that

$$\mathcal{F} \xrightarrow{\ \varphi\ } \mathcal{G} \xrightarrow{\ \psi\ } \mathcal{H}$$

is an exact sequence of sheaves iff the sequence

$$\mathcal{F}_x \xrightarrow{\ \varphi_x\ } \mathcal{G}_x \xrightarrow{\ \psi_x\ } \mathcal{H}_x$$

is an exact sequence of R-modules (or rings) for all $x \in X$; see Proposition 10.24. This second characterization of exactness (for sheaves) is usually much more convenient than the first condition.

The definitions of cokernels and images of presheaves and sheaves as well as the notion of exact sequences of presheaves and sheaves are discussed in Sections 10.6, 10.7, 10.8, 10.9, 10.10, and 10.11.

In Section 10.12 we introduce ring spaces which generalize significantly the notion of manifold.

10.1 Stalks and Maps of Stalks

In the case where $\mathcal{F}$ is a presheaf on a topological space X and x is any given point in X, the direct limit $\varinjlim(\mathcal{F}(U))_{U \ni x}$ of the direct mapping family $(\mathcal{F}(U))_{U \ni x}$ plays an important role (where U is any open subset of X). In particular, these limits, called stalks, can be used to construct a sheaf $\widetilde{\mathcal{F}}$ from a presheaf $\mathcal{F}$; furthermore, the sheaf $\widetilde{\mathcal{F}}$ is the "smallest" sheaf extending $\mathcal{F}$, in a technical sense that will be explained later. If $\mathcal{F}$ is already a sheaf, then $\widetilde{\mathcal{F}}$ is isomorphic to $\mathcal{F}$.

Definition 10.1. If $\mathcal{F}$ is a presheaf on a topological space X and x is any given point in X, the direct limit $\varinjlim(\mathcal{F}(U))_{U \ni x}$ of the direct mapping family $(\mathcal{F}(U))_{U \ni x}$, as defined in Example 8.4(2), is called the *stalk* of $\mathcal{F}$ at x, and is denoted by $\mathcal{F}_x$. For every open subset U such that $x \in U$, we have a projection map $\rho_{U,x} \colon \mathcal{F}(U) \to \mathcal{F}_x$, and we write $s_x = \rho_{U,x}(s)$ for every $s \in \mathcal{F}(U)$. One calls s_x the *germ* of s at x. See Figure 10.1.

If $\mathcal{F}$ is the presheaf (actually a sheaf) of continuous functions given by $\mathcal{F}(U) = C^0(U)$, the set of continuous functions defined on an open subset U containing x, then $\mathcal{F}_x$ is just the set of *germs* of locally defined functions near x. Indeed, two locally defined functions $f \in C^0(U)$ and $g \in C^0(V)$ near x are equivalent iff their restrictions to $U \cap V$ agree. In general the stalks are characterized as follows.

Definition 10.2. For an arbitrary presheaf $\mathcal{F}$ on a topological space X, for any $x \in X$, the stalk $\mathcal{F}_x$ is the set of equivalence classes defined such that for any two open subsets U and V both containing x, the "local" sections $f \in \mathcal{F}(U)$ and $g \in \mathcal{F}(V)$ are equivalent, written $f \sim_{\mathcal{F}} g$ or simply $f \sim g$, iff there is some open subset W containing x such that $W \subseteq U \cap V$ and $\rho^U_W(f) = \rho^V_W(g)$.

So we can also think of the elements of $\mathcal{F}_x$ are "abstract germs" of local sections near x. Observe that any element γ of the stalk $\mathcal{F}_x$ is the equivalence class of some section $s \in \mathcal{F}(U)$ for some open subset U of X containing x, namely $\gamma = s_x$, where s_x is the germ of s at x. For any smaller open subset $V \subseteq U$ containing x, the sections $s \in \mathcal{F}(U)$ and $\rho^U_V(s) \in \mathcal{F}(V)$ are obviously equivalent, so we also have $\gamma = (\rho^U_V(s))_x$. We will use this fact all the time.

For a constant presheaf G_X on X with values in G, we have $G_{X,x} = G$ for all $x \in X$. Beware that for some pathological presheaves $\mathcal{F}$ (for example, of abelian groups), it is possible that $\mathcal{F}_x = (0)$ for all $x \in X$, even though $\mathcal{F}$ is not the constant presheaf with value 0. An example is given by the following presheaf from Example 8.3. Let X be any topological space with at least two points (for example, $X = \{0, 1\}$), and let $\mathcal{F}_1$ be the presheaf

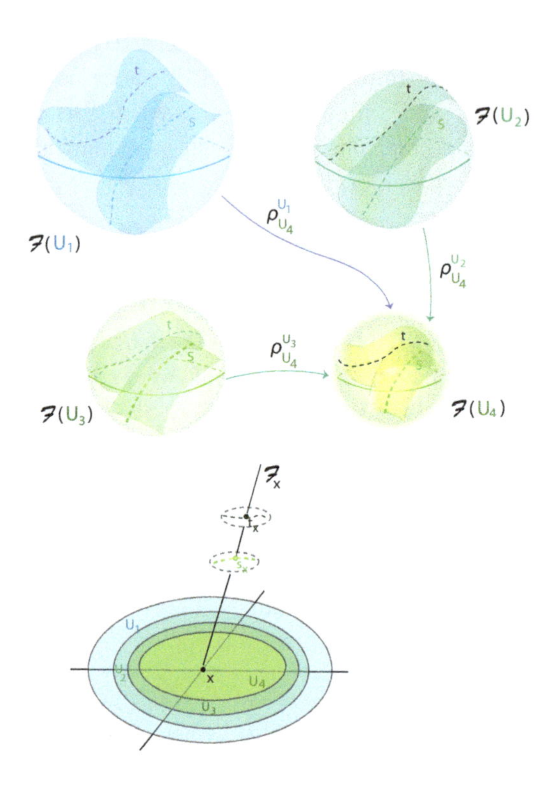

Fig. 10.1 A schematic representation of $\mathcal{F}_x$ for $x \in \mathbb{R}^2$. We illustrate the direct limit construction for two germs, s_x and t_x. Elements of the presheaf $\mathcal{F}$ are the spherical balloons. Since $U_4 \subseteq U_3 \subseteq U_2 \subseteq U_1$, the presheaf restriction maps imply that all images of s are equivalent to the image of s in U_4, and all the images of t are equivalent to the image of t in U_4. By continuing this process, we form the equivalence classes s_x and t_x, which we illustrate as little disks centered on the radial stalk extending from $x \in \mathbb{R}^2$.

given by

$$\mathcal{F}_1(U) = \begin{cases} \mathbb{Z} & \text{if } U = X \\ (0) & \text{if } U \neq X \text{ is an open subset,} \end{cases}$$

with all ρ_V^U equal to the zero map except if $U = V = X$ (in which case it is the identity). It is easy to check that $\mathcal{F}_{1,x} = (0)$ for all $x \in X$.

The following result will be needed in Section 10.6.

Proposition 10.1. *Let $\mathcal{F}$ be a presheaf on a topological space X. If $\mathcal{F}$ satisfies Condition (M), then for any open subset U of X, for any sections*

$s, t \in \mathcal{F}(U)$, we have

$$s = t \quad \textit{iff} \quad s_x = t_x \quad \textit{for all } x \in U.$$

Proof. Obviously if $s = t$, then $s_x = t_x$ for all $x \in U$. Conversely, if $s_x = t_x$ for all $x \in U$, then for each $x \in U$ there is some open subset $U_x \subseteq U$ containing x such that $\rho^U_{U_x}(s) = \rho^U_{U_x}(t)$, and since the family $(U_x)_{x \in U}$ is an open cover of U, Condition (M) implies that $s = t$. $\square$

A map $\varphi\colon \mathcal{F} \to \mathcal{G}$ between two presheaves $\mathcal{F}$ and $\mathcal{G}$ on a topological space X induces maps of stalks $\varphi_x\colon \mathcal{F}_x \to \mathcal{G}_x$ for all $x \in X$. When $\mathcal{F}$ and $\mathcal{G}$ are sheaves, these maps carry a lot of information about φ.

To define $\varphi_x\colon \mathcal{F}_x \to \mathcal{G}_x$ we proceed as follows. Any element $\gamma \in \mathcal{F}_x$ is an equivalence class $\gamma = s_x$ for some section $s \in \mathcal{F}(U)$ and some open subset U of X containing x. Let

$$\varphi_x(s_x) = (\varphi_U(s))_x,$$

where $\varphi_U\colon \mathcal{F}(U) \to \mathcal{G}(U)$ is the map defining φ on U. We need to prove that this definition does not depend on the choice of the representative in the equivalence class γ. If $t \in \mathcal{F}(V)$ is another section such that $s \sim_{\mathcal{F}} t$, then there is some open subset W such that $W \subseteq U \cap V$ and $(\rho_{\mathcal{F}})^U_W(s) = (\rho_{\mathcal{F}})^V_W(t)$. Since φ is a map of presheaves, the following diagrams commute

$$\begin{array}{ccc} \mathcal{F}(U) & \xrightarrow{\varphi_U} & \mathcal{G}(U) \\ {\scriptstyle (\rho_{\mathcal{F}})^U_W}\downarrow & & \downarrow{\scriptstyle (\rho_{\mathcal{G}})^U_W} \\ \mathcal{F}(W) & \xrightarrow[\varphi_W]{} & \mathcal{G}(W) \end{array} \qquad \begin{array}{ccc} \mathcal{F}(V) & \xrightarrow{\varphi_V} & \mathcal{G}(V) \\ {\scriptstyle (\rho_{\mathcal{F}})^V_W}\downarrow & & \downarrow{\scriptstyle (\rho_{\mathcal{G}})^V_W} \\ \mathcal{F}(W) & \xrightarrow[\varphi_W]{} & \mathcal{G}(W), \end{array}$$

and we get

$$(\rho_{\mathcal{G}})^U_W(\varphi_U(s)) = \varphi_W((\rho_{\mathcal{F}})^U_W(s)) = \varphi_W((\rho_{\mathcal{F}})^V_W(t)) = (\rho_{\mathcal{G}})^V_W(\varphi_V(t)),$$

which shows that $\varphi_U(s) \sim_{\mathcal{G}} \varphi_V(t)$, thus $(\varphi_U(s))_x = (\varphi_V(t))_x$. Therefore, φ_x is well defined and suggests the following definition of a map of stalks, which a special instance of Definition 8.12.

Definition 10.3. A map $\varphi\colon \mathcal{F} \to \mathcal{G}$ between two presheaves $\mathcal{F}$ and $\mathcal{G}$ on a topological space X induces *maps of stalks* $\varphi_x\colon \mathcal{F}_x \to \mathcal{G}_x$ for all $x \in X$ defined as follows: for every $\gamma \in \mathcal{F}_x$, if $\gamma = s_x$ for some section $s \in \mathcal{F}(U)$ and some open subset U of X containing x, set

$$\varphi_x(s_x) = (\varphi_U(s))_x.$$

See Figure 10.2. By the above argument this definition does not depend on the choice of the representative chosen in the equivalence class γ.

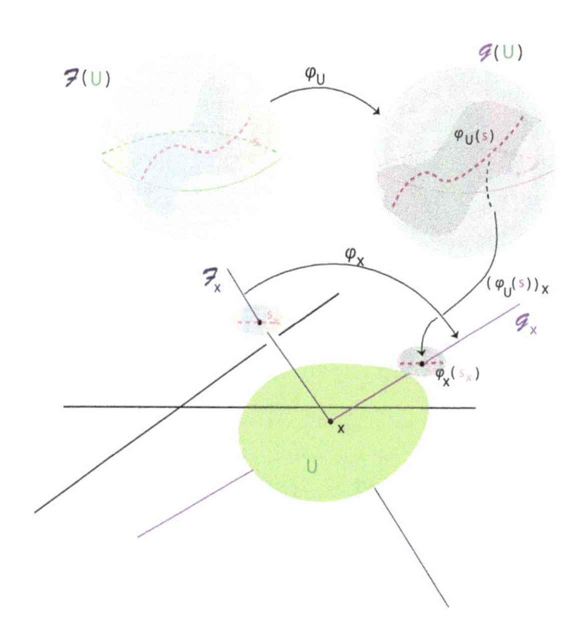

Fig. 10.2 A schematic representation of $\varphi_x\colon \mathcal{F}_x \to \mathcal{G}_x$ which maps the dark purple "stick" onto the plum "stick". The result of this stalk mapping is the same as first mapping the presheaf element $\mathcal{F}(U)$ onto $\mathcal{G}(U)$ and then using the direct limiting procedure to compute the stalk of $\varphi_U(s)$ where $s \in \mathcal{F}(U)$.

If $\varphi\colon \mathcal{F} \to \mathcal{G}$ and $\psi\colon \mathcal{G} \to \mathcal{H}$ are two maps of presheaves, it is immediately verified that

$$(\psi \circ \varphi)_x = \psi_x \circ \varphi_x$$

and

$$(\mathrm{id}_{\mathcal{F}})_x = \mathrm{id}_{\mathcal{F}_x},$$

for all $x \in X$ (where $\mathrm{id}_{\mathcal{F}}$ denotes the identity map of the presheaf $\mathcal{F}$).

Proposition 10.2. *Let $\mathcal{F}$ and $\mathcal{G}$ be two presheaves on a topological space X, and let $\varphi\colon \mathcal{F} \to \mathcal{G}$ and $\psi\colon \mathcal{F} \to \mathcal{G}$ be two maps of presheaves. If $\mathcal{G}$ satisfies Condition (M) (in particular, if $\mathcal{G}$ is a sheaf) and if $\varphi_x = \psi_x$ for all $x \in X$, then $\varphi = \psi$.*

Proof. We need to prove that $\varphi_U(s) = \psi_U(s)$ for any open subset U of X and any $s \in \mathcal{F}(U)$. Since $\varphi_x = \psi_x$ for every $x \in X$, for every $x \in U$ we have

$$\varphi_x(s_x) = \psi_x(s_x),$$

that is,

$$(\varphi_U(s))_x = (\psi_U(s))_x.$$

The above equations means that there is some open subset U_x of X such that $U_x \subseteq U$ and

$$(\rho_{\mathcal{G}})^U_{U_x}(\varphi_U(s)) = (\rho_{\mathcal{G}})^U_{U_x}(\psi_U(s)).$$

Since the family $(U_x)_{x \in U}$ is an open cover of U, Condition (M) implies that $\varphi_U(s) = \psi_U(s)$, and so $\varphi = \psi$. $\square$

Proposition 10.2 has the following corollary.

Corollary 10.3. *If $\varphi\colon \mathcal{F} \to \mathcal{G}$ is a map of sheaves, then φ is uniquely determined by the family of stalk maps $\varphi_x\colon \mathcal{F}_x \to \mathcal{G}_x$.*

Next, given a presheaf $\mathcal{F}$ on X, we construct a sheaf $\widetilde{F}$ and a presheaf map $\eta\colon \mathcal{F} \to \widetilde{\mathcal{F}}$ such that $\mathcal{F}$ satisfies Condition (M) iff η is injective, and $\mathcal{F}$ is a sheaf iff η is an isomorphism.

10.2 Sheafification of a Presheaf

We follow Godement's exposition [Godement (1958)] (Chapter II, Section 1.2), which we find to be one of the most lucid.

The key idea is to make the disjoint union $\coprod_{x \in X} \mathcal{F}_x$ of all the stalks into a topological space denoted $S\mathcal{F}$, with a projection function $p\colon S\mathcal{F} \to X$, and to let $\widetilde{\mathcal{F}}$ be the sheaf $\Gamma[S\mathcal{F}, p]$ of continuous sections of p, as in Example 8.2(1). See Figures 10.3 and 10.4.

If we let $S\mathcal{F} = \coprod_{x \in X} \mathcal{F}_x$ be the disjoint union of all the stalks, we denote by p the function $p\colon S\mathcal{F} \to X$ given by $p(\gamma) = x$ for all $\gamma \in \mathcal{F}_x$. For every (nonempty) open subset U of X, we view each "abstract" section $s \in \mathcal{F}(U)$ as the actual function $\widetilde{s}\colon U \to S\mathcal{F}$ given by

$$\widetilde{s}(x) = s_x, \quad x \in U.$$

By definition, $\widetilde{s}$ is a section of p. The final step is to give $S\mathcal{F}$ the coarsest topology (the topology with the least amount of open sets) which makes all the functions $\widetilde{s}$ continuous. Consequently, a subset Ω of $S\mathcal{F}$ is open iff for every open subset U of X and every $s \in \mathcal{F}(U)$, the subset

$$\{x \in U \mid \widetilde{s}(x) \in \Omega\}$$

is open in X.

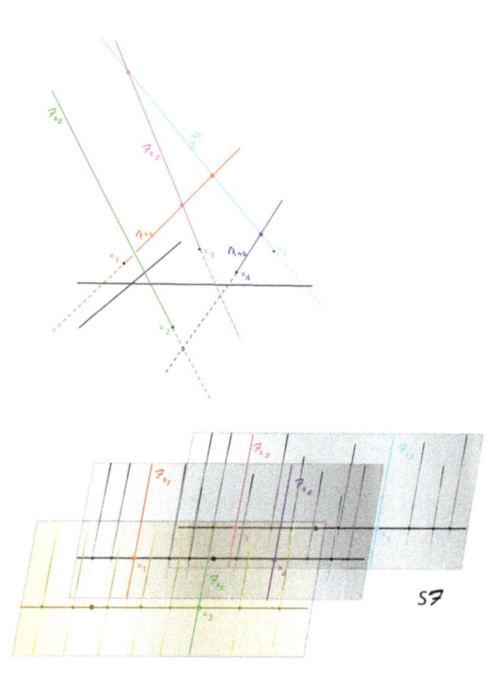

Fig. 10.3 A schematic representation of $\coprod_{x\in X}\mathcal{F}_x$ for $X=\mathbb{R}^2$. The top picture illustrates five "stalks" before taking the disjoint union. Once the disjoint union is formed, the "stalks" are lined up in parallel planes.

Definition 10.4. The space $S\mathcal{F}$ endowed with the above topology is called the *stalk space* of the presheaf $\mathcal{F}$, and we let $\widetilde{\mathcal{F}}$ be the sheaf $\Gamma[S\mathcal{F},p]$ of continuous sections of p. See Figure 10.4.

We claim that $\widetilde{s}(U)$ is open in $S\mathcal{F}$ for every open subset U and every $s\in\mathcal{F}(U)$.

Proof. We need to prove that for every open subset V and every $t\in\mathcal{F}(V)$ the set

$$\{y\in V\mid \widetilde{t}(y)\in\widetilde{s}(U)\}$$

is open. See Figure 10.5.

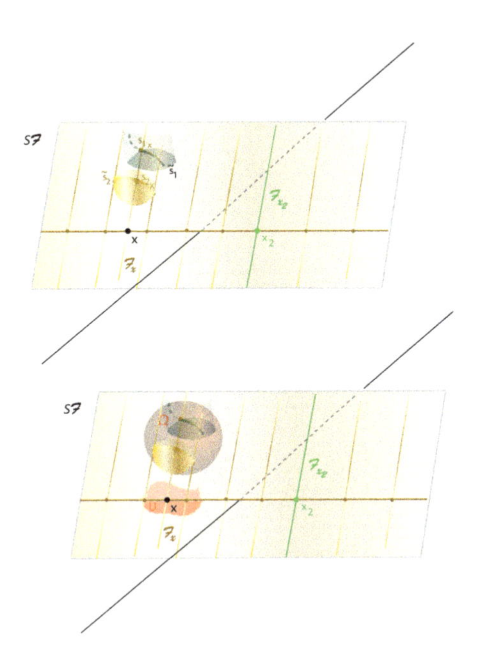

Fig. 10.4 A schematic representation of the region around the "first plane" in $S\mathcal{F}$ where $X = \mathbb{R}^2$. The top picture illustrates two sections $\tilde{s}_1$ and $\tilde{s}_2$. The bottom picture illustrates the relationship between Ω, an open "spherical" set of $S\mathcal{F}$, and U, an open set of X containing x. Both $\tilde{s}_1(U)$ and $\tilde{s}_2(U)$ are open sets in $S\mathcal{F}$.

We have

$$\begin{aligned}\{y \in V \mid \widetilde{t}(y) \in \widetilde{s}(U)\} &= \bigcup_{x \in U} \{y \in V \mid \widetilde{t}(y) = \widetilde{s}(x)\} \\ &= \{x \in U \cap V \mid s_x = t_x\},\end{aligned}$$

since $\widetilde{s}(x) = s_x$, $\widetilde{t}(y) = t_y$, and the stalks are pairwise disjoint. It suffices to show that the subset $\{x \in U \cap V \mid s_x = t_x\}$ is open in X. However, $s_x = t_x$ means that there is some open subset $W \subseteq U \cap V$ containing x such that $\rho^U_W(s) = \rho^V_W(t)$ on W, which means that $\{x \in U \cap V \mid s_x = t_x\}$ is indeed open in X. $\square$

We now show that the function p is continuous. As in the previous argument it suffices to prove that for any open subset V and any $t \in \mathcal{F}(V)$, the set

$$\{y \in V \mid \widetilde{t}(y) \in p^{-1}(U)\}$$

is open for any open subset U. We have

$$\begin{aligned}\{y \in V \mid \widetilde{t}(y) \in p^{-1}(U)\} &= \bigcup_{\substack{W \subseteq U \\ s \in \mathcal{F}(W) \\ x \in W}} \{y \in V \mid \widetilde{t}(y) = s_x\} \\ &= \bigcup_{\substack{W \subseteq U \\ s \in \mathcal{F}(W)}} \{x \in V \cap W \mid s_x = t_x\}.\end{aligned}$$

By the previous argument, $\{x \in V \cap W \mid s_x = t_x\}$ is open in X, so we are done.

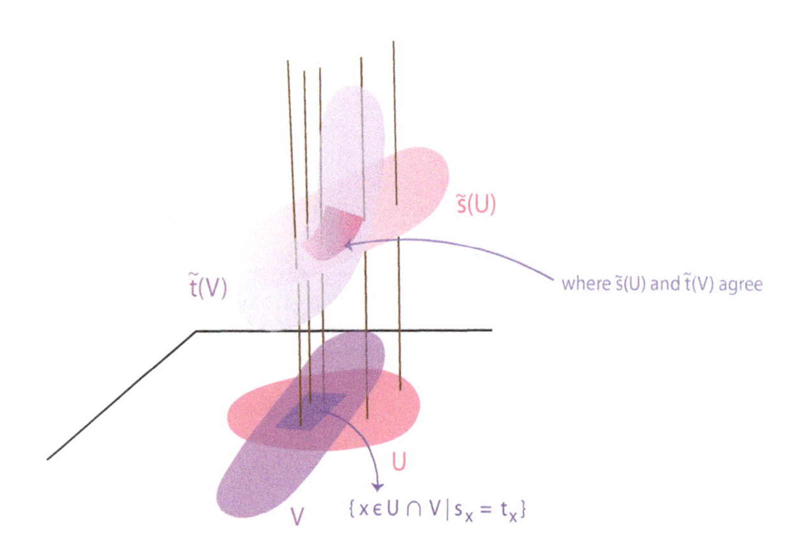

Fig. 10.5 A schematic illustration of $\tilde{s}(U)$ in $S\mathcal{F}$ when $X = \mathbb{R}^2$. The overlap between the sections $\tilde{s}(U)$ and $\tilde{s}(V)$ is the open purple rhombus in $\mathbb{R}^2$.

Any element of $\mathcal{F}_x$ is of the form s_x for some open subset U of X containing x and some $s \in \mathcal{F}(U)$. Observe that $\widetilde{s}$ is the inverse of the restriction of p to $\widetilde{s}(U)$, and since $\Omega = \widetilde{s}(U)$ is open, the map p is a homeomorphism from Ω to U. Therefore p is a local homeomorphism.

Consider any open subset Ω of $S\mathcal{F}$ and pick any $s_x \in \Omega$. As above we may assume that s_x is the germ of some section $s \in \mathcal{F}(U)$ for some open subset U containing x. Since p is a local homeomorphism, there is an open subset Ω_1 containing s_x such that the restriction of p to Ω_1 is a homeomorphism. Consequently p is homeomorphism from $\Omega \cap \Omega_1$ to some open subset V of X containing x. But then $\rho^U_{U\cap V}(s)$ being a section in $\mathcal{F}(U \cap V)$, we know that $\widetilde{\rho^U_{U\cap V}(s)}(U \cap V)$ is an open subset of Ω containing s_x.

Therefore Ω is the union of open subsets of the form $\widetilde{t}(W)$ with $t \in \mathcal{F}(W)$, where W is some open subset of X. It is a standard fact of topology that this condition implies that the sets of form $\widetilde{s}(U)$ with $s \in \mathcal{F}(U)$ form a basis of the topology.

In summary, we proved the following proposition.

Proposition 10.4. *Let $\mathcal{F}$ be a presheaf on a topological space X. The stalk space $S\mathcal{F}$ together with the coarsest topology that makes all the maps $\widetilde{s}\colon U \to S\mathcal{F}$ continuous has a basis for its topology consisting of the subsets of the form $\widetilde{s}(U)$, for all open subsets U of X and all $s \in \mathcal{F}(U)$. Furthermore, the projection map $p\colon S\mathcal{F} \to X$ is a local homeomorphism.*

It should be noted that the topology of $S\mathcal{F}$ is not assumed to be Hausdorff. In fact, in many interesting examples it is not. We called the space $S\mathcal{F}$ the *stalk space* of $\mathcal{F}$. In Godement [Godement (1958)] and most of the French literature, the space $S\mathcal{F}$ is called "espace étalé." A rough translation is "spread over space" or "laid over space."

Definition 10.5. Given any presheaf $\mathcal{F}$ on a topological space X, the map $\eta\colon \mathcal{F} \to \widetilde{\mathcal{F}}$ is defined such that for every open subset U of X, for every $s \in \mathcal{F}(U)$,

$$\eta_U(s) = \widetilde{s}.$$

If we need to very precise, we use the notation $\eta_{\mathcal{F}}$ instead of η.

It is easily checked that $\eta = (\eta_U)$ is indeed a map of presheaves. We now take a closer look at the map $\eta\colon \mathcal{F} \to \widetilde{\mathcal{F}}$.

Proposition 10.5. *Let $\mathcal{F}$ be a presheaf on a topological space X. The presheaf $\mathcal{F}$ satisfies Condition (M) iff the presheaf map $\eta\colon \mathcal{F} \to \widetilde{\mathcal{F}}$ is injective.*

Proof. We follow Serre's proof in FAC [Serre (1955)] (Chapter I, Section 3). First assume that $\mathcal{F}$ satisfies Condition (M). First we prove that η is injective. We have to prove that for every open subset U of X, for any two elements $s, t \in \mathcal{F}$, if $\widetilde{s} = \widetilde{t}$, then $s = t$. Now, $\widetilde{s} = \widetilde{t}$ iff $s_x = t_x$ for all $x \in U$, which means that there is some open subset U_x of U containing x such that

$$\rho^U_{U_x}(s) = \rho^U_{U_x}(t).$$

Since the family $(U_x)_{x \in U}$ is an open cover of U, by Condition (M) we must have $s = t$.

Conversely, assume that $\eta_U \colon \mathcal{F}(U) \to \widetilde{\mathcal{F}}(U)$ is injective. Pick any $s, t \in \mathcal{F}(U)$, and assume there is some open cover $(U_i)_{i \in I}$ of U such that $\rho^U_{U_i}(s) = \rho^U_{U_i}(t)$ for all $i \in I$. By definition of a direct limit, for any $x \in U$,

$$\widetilde{s}(x) = s_x = (\rho^U_{U_i}(s))_x \quad \text{and} \quad \widetilde{t}(x) = t_x = (\rho^U_{U_i}(t))_x,$$

so if $\rho^U_{U_i}(s) = \rho^U_{U_i}(t)$ then $\widetilde{s}(x) = \widetilde{t}(x)$ for all $x \in U$; that is, $\widetilde{s} = \widetilde{t}$. Since η_U is injective, we conclude that $s = t$, which means that Condition (M) holds. $\square$

The next proposition characterizes when η is an isomorphism.

Proposition 10.6. *Let $\mathcal{F}$ be a presheaf on a topological space X and assume that $\mathcal{F}$ satisfies Condition (M). The presheaf map $\eta \colon \mathcal{F} \to \widetilde{\mathcal{F}}$ is surjective iff Condition (G) holds. As a consequence, η is an isomorphism iff $\mathcal{F}$ is a sheaf.*

Proof. Again, we follow Serre's proof in FAC [Serre (1955)] (Chapter I, Section 3). By Proposition 10.5 Condition (M) holds iff η is injective, so we may assume that η is injective.

First assume that $\mathcal{F}$ satisfies Condition (G). We wish to prove that η_U is surjective for every open subset U. For any open subset U of X, for any continuous section $f \colon U \to S\mathcal{F}$, for any $x \in U$, we claim that there is some open subset U_x of U containing x and some $s^x \in \mathcal{F}(U_x)$ such that the restriction of f to U_x agrees with $\widetilde{s^x}$.

Since $f(x) \in \mathcal{F}_x$, there is some open subset U_x of U containing x and some $s^x \in \mathcal{F}(U_x)$ such that $f(x) = (s^x)_x$. Since $\widetilde{s^x}$ and f both invert p on U_x, the restriction of f to U_x agrees with $\widetilde{s^x}$.

The same argument holds for any $y \in U$ so there is some open subset U_y of U containing y and some $s^y \in \mathcal{F}(U_y)$ such that the restriction of f

to U_y agrees with $\widetilde{s^y}$. It follows that $(\widetilde{\rho^U_{U_x\cap U_y}(s^x)})_z = (\widetilde{\rho^U_{U_x\cap U_y}(s^y)})_z = f(z)$ for all $x, y \in U$ and all $z \in U_x \cap U_y$, that is, $\widetilde{\rho^U_{U_x\cap U_y}(s^x)} = \widetilde{\rho^U_{U_x\cap U_y}(s^y)}$. Since η is injective, we get

$$\rho^U_{U_x\cap U_y}(s^x) = \rho^U_{U_x\cap U_y}(s^y).$$

But then, by Condition (G), the s^x patch to some $s \in \mathcal{F}(U)$ such that $\rho^U_{U_x}(s) = s^x$, thus $\eta_U(s) = \widetilde{s}$ agrees with $\widetilde{s^x} = f|U_x$ on each U_x, which means that $\eta_U(s) = f$. See Figure 10.6.

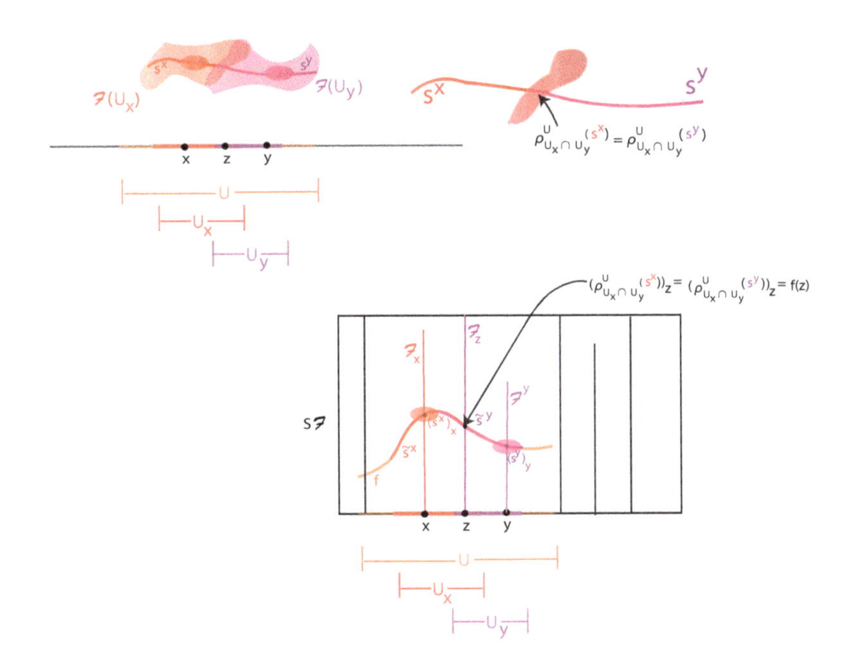

Fig. 10.6 A schematic representation of the proof that a presheaf $\mathcal{F}$ which satisfies Conditions (M) and (G) implies η is surjective. The top two diagrams are related to $\mathcal{F}$ while the bottom diagram is related to $\widetilde{\mathcal{F}}$. Note that X is $\mathbb{R}$.

Conversely, assume that η_U is surjective (and injective) for every open subset U of X. Let $(U_i)_{i\in I}$ be some open cover of U and let $(s_i)_{i\in I}$ be a family of elements $s_i \in \mathcal{F}(U_i)$ such that

$$\rho^U_{U_i\cap U_j}(s_i) = \rho^U_{U_i\cap U_j}(s_j)$$

for all i, j. It follows that the sections $f_i = \widetilde{s_i}$ and $f_j = \widetilde{s_j}$ agree on $U_i \cap U_j$, so they patch to a continuous section $f\colon U \to S\mathcal{F}$ which agrees with f_i on each U_i. Since η_U is assumed to be surjective, there is some $s \in \mathcal{F}(U)$ such that $\eta_U(s) = f$. Then, if we write $s'_i = \rho^U_{U_i}(s)$, we see that $\widetilde{s'_i} = f_i$. Since $f_i = \widetilde{s_i} = \widetilde{s'_i}$ for all i and since η is injective, we conclude that $s_i = s'_i$; that is, $\rho^U_{U_i}(s) = s_i$, which shows that Condition (G) holds. $\square$

Propositions 10.5 and 10.6 show that the Conditions (M) and (G) in the definition of a sheaf (Definition 8.5) are not as arbitrary as they might seem. They are just the conditions needed to ensure that a sheaf is isomorphic to a sheaf of sections of a certain space.

Remark: We proved earlier that for any open subset U of X, for any two continuous sections f and g in $\Gamma(U, S\mathcal{F})$, the subset $W = \{x \in U \mid f(x) = g(x)\}$ is open. If the stalk space $S\mathcal{F}$ is Hausdorff, then W is also closed (because the diagonal $\{(\gamma, \gamma) \mid \gamma \in S\mathcal{F}\}$ is closed). In this case it follows that if U is a connected open subset of X, if two continuous sections f and g in $\Gamma(U, S\mathcal{F})$ agree at some point, then $f = g$. In other words, the principle of analytic continuation holds. If $\mathcal{F}$ is the sheaf of continuous functions on $\mathbb{R}^n$, the principle of analytic continuation fails so $S\mathcal{F}$ is not Hausdorff. However, if $\mathcal{F}$ is the sheaf of holomorphic functions on a complex analytic manifold, then $S\mathcal{F}$ is Hausdorff.

If we examine more closely the construction of the sheaf $\widetilde{\mathcal{F}}$ from a presheaf $\mathcal{F}$, we see that we actually used two constructions:

(1) Given a presheaf $\mathcal{F}$, we constructed the stalk space $S\mathcal{F}$ and we gave it a topology that made the projection $p\colon S\mathcal{F} \to X$ into a local homeomorphism. This is the construction S ("stalkification"), which constructs the stalk space $(S\mathcal{F}, p)$ from a presheaf $\mathcal{F}$.
(2) Given a pair (E, p), where $p\colon E \to X$ is a local homeomorphism, we constructed the sheaf $\Gamma[E, p]$ (abbreviated as ΓE) of continuous sections of p.

Observe that the construction $\mathcal{F} \mapsto \widetilde{\mathcal{F}}$ is the composition of S and Γ, that is, $\widetilde{\mathcal{F}} = \Gamma S\mathcal{F}$, and Proposition 10.6 shows that if $\mathcal{F}$ is a sheaf, then $\Gamma S\mathcal{F}$ is isomorphic to $\mathcal{F}$.

Remark: If $\mathcal{F}$ is a presheaf on a space X, we define the presheaf $\mathcal{F}^{(+)}$ as follows: for every open subset U of X,

$$\mathcal{F}^{(+)}(U) = \check{H}^0(U, \mathcal{F}|U),$$

where $\check{H}^0(U, \mathcal{F}|U)$ is a Čech cohomology groups defined in Section 9.1. Then it can be shown that $\mathcal{F}^{(+)}$ satisfies Condition (M), and that $\mathcal{F}^{(+)(+)}$ is isomorphic to the sheafification $\widetilde{\mathcal{F}}$ of $\mathcal{F}$.

It is natural to take a closer look at the properties of a pair (E, p), where $p\colon E \to X$ is a local homeomorphism, and to ask what is the effect of applying the operations Γ and S to the space E. We will see that the stalk space $S\Gamma E$ is isomorphic to the original space E.

The upshot of all this is that the constructions S and Γ are essentially inverse of each other, modulo some isomorphisms. To make this more precise we need to define what kind of objects are in the domain of Γ, and what are the maps between such objects.[1]

10.3 Stalk Spaces (or Sheaf Spaces)

As we just explained, given a presheaf $\mathcal{F}$, the construction of the stalk space $S\mathcal{F}$ yields a pair $(S\mathcal{F}, p)$, where $p\colon S\mathcal{F} \to X$ is the projection, and by Proposition 10.4 the map p is a local homeomorphism. This suggests the following definition.

Definition 10.6. A pair (E, p) where E is a topological space and $p\colon E \to X$ is a surjective local homeomorphism is called a *stalk space* (or *sheaf space*[2]). A *map* (or *morphism*) of stalk spaces (E_1, p_1) and (E_2, p_2) is a continuous map $\varphi\colon E_1 \to E_2$ such that the following diagram commutes:

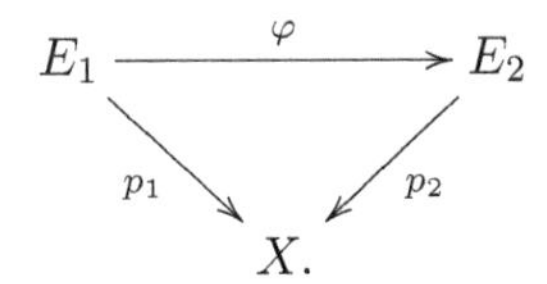

See Figure 10.7.

Observe that the commutativity of the diagram implies that φ maps fibres of E_1 to fibres of E_2.

The main construction on a stalk space (E, p) is the construction Γ described in Example 8.2(1), which yields the sheaf $\Gamma[E, p]$ (abbreviated ΓE) of continuous sections of p, with

$$\Gamma[E, p](U) = \Gamma(U, \Gamma[E, p]) = \{s\colon U \to E \mid p \circ s = \mathrm{id} \text{ and } s \text{ is continuous}\}$$

[1] Actually, S and Γ are functors between certain categories.

[2] The terminology "sheaf space" is used by Tennison [Tennison (1975)]. Godement uses the terminology "espace étalé."

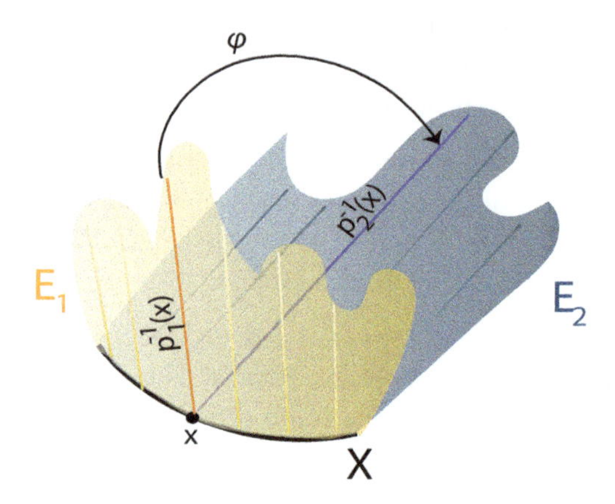

Fig. 10.7 A schematic representation of two stalk spaces E_1 and E_2, along with the map $\varphi\colon E_1 \to E_2$ which maps the fibres of E_1 to the fibres of E_2.

for any open subset U of X. This construction also applies to maps of stalk spaces (it is functorial).

Definition 10.7. Given a map $\varphi\colon E_1 \to E_2$ of stalk spaces (E_1, p_1) and (E_2, p_2) we obtain a map of sheaves $\Gamma\varphi\colon \Gamma E_1 \to \Gamma E_2$ defined as follows: for every open subset U of X, the map $(\Gamma\varphi)_U \colon \Gamma(U, E_1) \to \Gamma(U, E_2)$ is given by

$$(\Gamma\varphi)_U(f) = \varphi \circ f,$$

as illustrated by the diagram below:

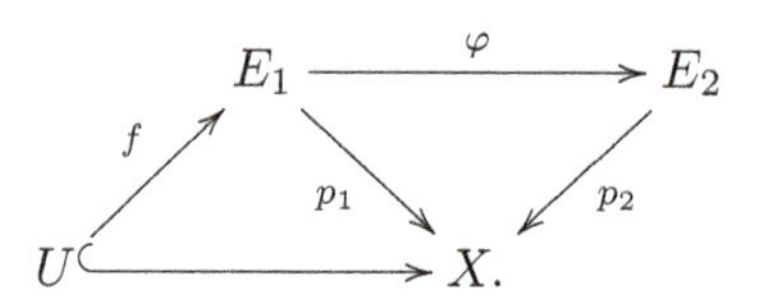

See Figure 10.8.

It is immediately checked that $\Gamma\varphi$ is a map of sheaves. Also, if $\varphi\colon E_1 \to E_2$ and $\psi\colon E_2 \to E_3$ are two maps of stalk spaces, then

$$\Gamma(\psi \circ \varphi) = \Gamma\psi \circ \Gamma\varphi,$$

and $\Gamma\mathrm{id}_E = \mathrm{id}_{\Gamma E}$. This means that the construction Γ is functorial.

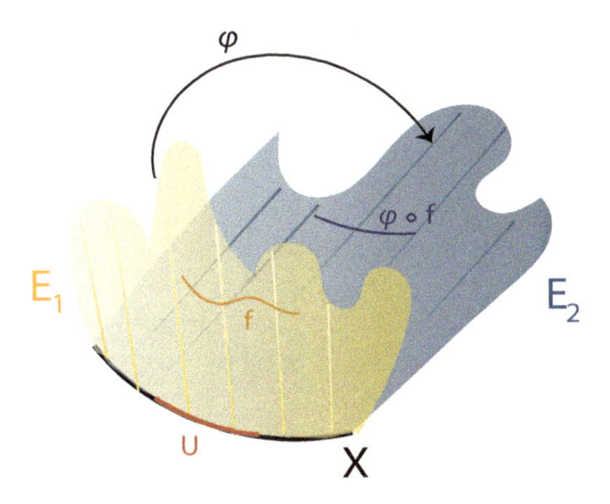

Fig. 10.8 A schematic representation of two stalk spaces E_1 and E_2, along with the sheaf map $(\Gamma\varphi)_U\colon \Gamma(U, E_1) \to \Gamma(U, E_2)$ which maps the ochre section of E_1 to the blue section of E_2.

Here are a few useful properties of stalk spaces. In particular, we will see that the fibres of a stalk space are isomorphic to the stalks of the sheaf ΓE of continuous sections.

Proposition 10.7. *Let (E, p) be a stalk space. Then the following properties hold:*

(a) The map p is an open map.
(b) For any open subset U of X and any continuous section $f \in \Gamma(U, E)$, the subset $f(U)$ is open in E; such open subsets form a basis for the topology of E.
(c) For any commutative diagram

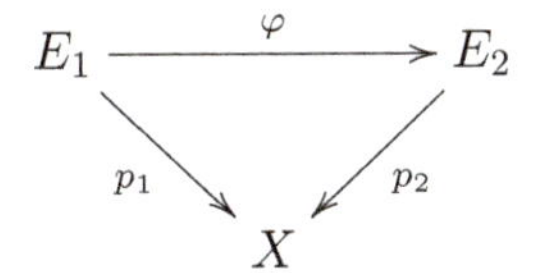

$$\begin{array}{ccc} E_1 & \xrightarrow{\varphi} & E_2 \\ & {\scriptstyle p_1}\searrow \quad \swarrow{\scriptstyle p_2} & \\ & X & \end{array}$$

where (E_1, p_2) and (E_2, p_2) are stalk spaces, the map φ is continuous iff it is an open map iff it is a local homeomorphism.

Proof. (a) Let V be any nonempty open subset in E. For any $x \in p(V)$ let $e \in E$ be any point in E such that $p(e) = x$. Since p is a local homeomorphism, there is some open subset W of E containing e such that $p(W)$ is open in X. Then $p(W)$ is some open subset of $p(V)$ containing x, so $p(V)$ is open. See Figure 10.9.

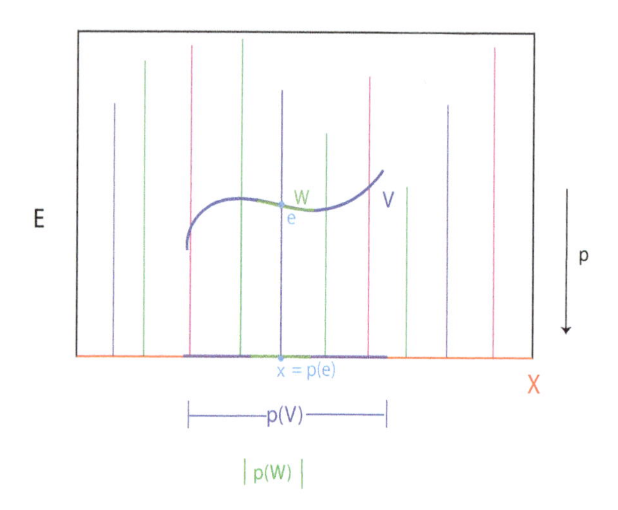

Fig. 10.9 A schematic representation of the stalk space (E, p) where E is the rectangle and X its red edge. The open set V may be thought of as a section $f \in \Gamma(p(V), E)$.

(b) For any $e \in f(U)$, since p is a local homeomorphism there is some open subset W of E containing e such that $p(W)$ is open in X and p maps W homeomorphically onto $p(W)$. It follows that p maps $f(U) \cap W$ homeomorphically onto $U \cap V$, where $V = p(W)$ (since f is a section of p). Since $U \cap V$ is open in X and p is a homeomorphism between $f(U) \cap W$ and $U \cap V$, the subset $f(U) \cap W$ is an open subset of $f(U)$ containing e, which shows that $f(U)$ is open. Using (a), it is easy to see that open subsets of the form $f(U)$ form a basis for the topology of E.

(c) A proof can be found in Tennison [Tennison (1975)] (see Chapter 2, Lemma 3.5). $\square$

The construction of the stalk space $S\mathcal{F}$ (and of the sheaf $\widetilde{\mathcal{F}}$) from a presheaf $\mathcal{F}$ is functorial in the following sense.

Proposition 10.8. *Given any map of preshaves $\varphi \colon \mathcal{F} \to \mathcal{G}$, there is a map of stalk spaces $S\varphi \colon S\mathcal{F} \to S\mathcal{G}$ induced by the stalk maps $\varphi_x \colon \mathcal{F}_x \to \mathcal{G}_x$ for all $x \in X$, and a map of sheaves $\widetilde{\varphi} \colon \widetilde{\mathcal{F}} \to \widetilde{\mathcal{G}}$.*

Proof. Since $S\mathcal{F}$ is the disjoint union of the stalks $\mathcal{F}_x$ of $\mathcal{F}$ and $S\mathcal{G}$ is the disjoint union of the stalks $\mathcal{G}_x$ of $\mathcal{G}$, the stalk maps $\varphi_x \colon \mathcal{F}_x \to \mathcal{G}_x$ define a map $S\varphi \colon S\mathcal{F} \to S\mathcal{G}$ given by

$$S\varphi(\gamma) = \varphi_x(\gamma), \quad \gamma \in \mathcal{F}_x,\, x \in X.$$

It is immediately verified that the following diagram commutes

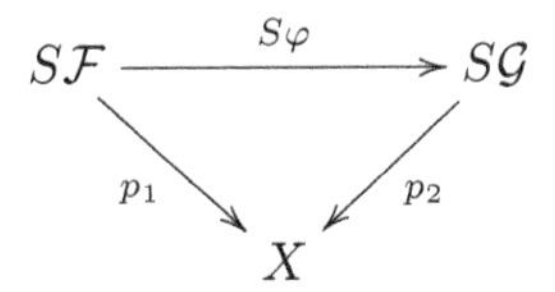

and that $S\varphi$ is continuous using Proposition 10.7(c). The map $\widetilde{\varphi} \colon \widetilde{\mathcal{F}} \to \widetilde{\mathcal{G}}$ is obtained from $S\varphi \colon S\mathcal{F} \to S\mathcal{G}$ by applying Γ as in Definition 10.7. $\square$

It is easy to check that if $\varphi \colon \mathcal{F} \to \mathcal{G}$ and $\psi \colon \mathcal{G} \to \mathcal{H}$ are maps of presheaves, then $S(\psi \circ \varphi) = S\psi \circ S\varphi$ and $S\mathrm{id}_{\mathcal{F}} = \mathrm{id}_{S\mathcal{F}}$. Similarly $\widetilde{\psi \circ \varphi} = \widetilde{\psi} \circ \widetilde{\varphi}$ and $\widetilde{\mathrm{id}_{\mathcal{F}}} = \mathrm{id}_{\widetilde{\mathcal{F}}}$.

Strictly speaking the map $\eta \colon \mathcal{F} \to \widetilde{\mathcal{F}}$ depend on $\mathcal{F}$, so it should really be denoted by $\eta_{\mathcal{F}} \colon \mathcal{F} \to \widetilde{\mathcal{F}}$. It is easy to check that the family η of maps $\eta_{\mathcal{F}}$ is natural in the following sense: given any presheaf map $\varphi \colon \mathcal{F} \to \mathcal{G}$, the following diagram commutes:

$$\begin{array}{ccc} \mathcal{F} & \xrightarrow{\eta_{\mathcal{F}}} & \widetilde{\mathcal{F}} \\ {\scriptstyle\varphi}\downarrow & & \downarrow{\scriptstyle\widetilde{\varphi}} \\ \mathcal{G} & \xrightarrow[\eta_{\mathcal{G}}]{} & \widetilde{\mathcal{G}}. \end{array}$$

The next proposition tells us that the fibres of a stalk space are stalks of the sheaf ΓE.

Proposition 10.9. *Let (E, p) be a stalk space. For any $x \in X$, the stalk $(\Gamma E)_x$ of the sheaf ΓE of continuous sections of p is isomorphic to the fibre $p^{-1}(x)$ at x. Furthermore, as a subspace of E, the fibre $p^{-1}(x)$ has the discrete topology.*

Proof. Pick any $x \in X$. For any open subset U of X with $x \in U$ we have a map $\mathrm{Eval}_{U,x} \colon \Gamma(U, E) \to p^{-1}(x)$ given by

$$\mathrm{Eval}_{U,x}(f) = f(x)$$

for any continuous section $f\colon U \to E$ of p. For any open subset V such that $V \subseteq U$ and $x \in V$ the following diagram commutes

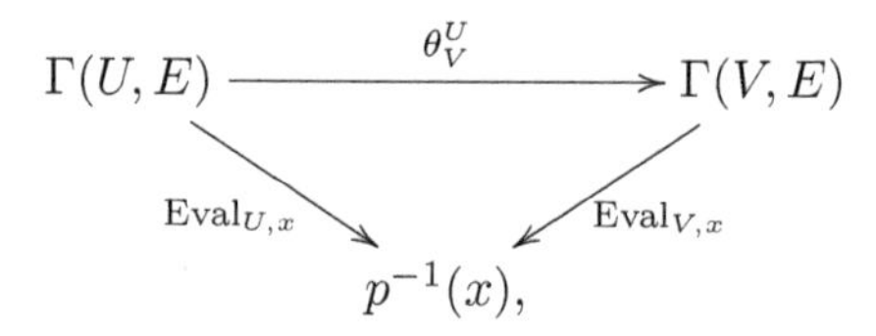

where the map $\theta_V^U\colon \Gamma(U,E) \to \Gamma(V,E)$ is the restriction map. We use Proposition 8.1 to prove that $(p^{-1}(x), \mathrm{Eval}_{U,x})$ is a direct limit. By the universal mapping property, $p^{-1}(x)$ is isomorphic to the direct limit $(\Gamma E)_x$ of the direct mapping family $((\Gamma(U,E))_U, (\theta_V^U))$.

(a) We need to show that for every $e \in p^{-1}(x)$, there is some open subset U of X and some section $f \in \Gamma(U,E)$ such that $f(x) = e$. Since p is a local homeomorphism, there is some open subset W of E such that $e \in W$ and the restriction $p|W$ maps W homeomorphically onto an open subset $U = p(W)$ of X. Then the inverse f of $p|W$ is a continuous section in $\Gamma(U,E)$ such that $f(x) = e$. Observe that $p^{-1}(x) \cap W = \{e\}$, which shows that the fibre $p^{-1}(x)$ has the discrete topology.

(b) For any $x \in X$, suppose that $\mathrm{Eval}_{U,x}(f) = f(x) = g(x) = \mathrm{Eval}_{V,x}(g)$ where $f \in \Gamma(U,E)$ and $g \in \Gamma(V,E)$, with $x \in U \cap V$. Then by Proposition 10.7 both $f(U)$ and $g(V)$ are open in E so $W = f(U) \cap g(U)$ is open and f and g agree on $p(W)$ (since they are both the inverse of p on $U \cap V$), which by Proposition 10.7 is open. This means that

$$\theta_{p(W)}^U(f) = \theta_{p(W)}^U(g),$$

which shows that Condition (b) of Proposition 8.1 is also satisfied.

Therefore, the stalk $(\Gamma E)_x$ of the sheaf ΓE is isomorphic to the fibre $p^{-1}(x)$ at x. ☐

Proposition 10.9, when combined with Definition 10.5, has the following corollaries.

Proposition 10.10. *For any presheaf $\mathcal{F}$ on a space X, the map $\eta\colon \mathcal{F} \to \widetilde{\mathcal{F}}$ induces isomorphisms of stalks $\eta_x\colon \mathcal{F}_x \to \widetilde{\mathcal{F}}_x$ for all $x \in X$.*

Proof. By construction the stalk $\mathcal{F}_x$ of $\mathcal{F}$ at x is equal to the fibre $p^{-1}(x)$ of the stalk space $S\mathcal{F}$, and $\widetilde{\mathcal{F}} = \Gamma S\mathcal{F}$, the sheaf of continuous sections of p, so $\widetilde{\mathcal{F}}_x = (\Gamma S\mathcal{F})_x$. By Proposition 10.9, we have $\mathcal{F}_x \cong \widetilde{\mathcal{F}}_x$. It remains to show that η_x is a stalk isomorphism. The stalk map $\eta_x : \mathcal{F}_x \to \widetilde{\mathcal{F}}_x$ as given by Definition 8.12 is the unique map that makes the following diagram commute

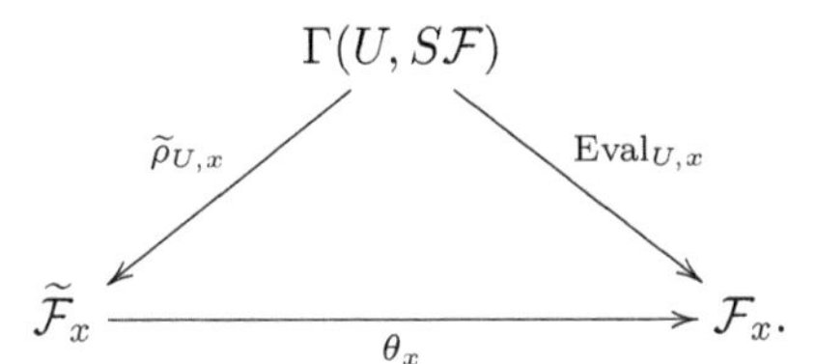

for all open subsets U of X with $x \in U$. Since $p^{-1}(x) = \mathcal{F}_x$, by Proposition 10.9, there are isomorphisms $\theta_x : \widetilde{\mathcal{F}}_x \to p^{-1}(x)$ and thus $\theta_x : \widetilde{\mathcal{F}}_x \to \mathcal{F}_x$ such that the following diagram commutes:

$$\begin{array}{ccc} & \Gamma(U, S\mathcal{F}) & \\ {}^{\widetilde{\rho}_{U,x}}\swarrow & & \searrow^{\mathrm{Eval}_{U,x}} \\ \widetilde{\mathcal{F}}_x & \xrightarrow[\theta_x]{} & \mathcal{F}_x. \end{array}$$

Consequently, the diagrams

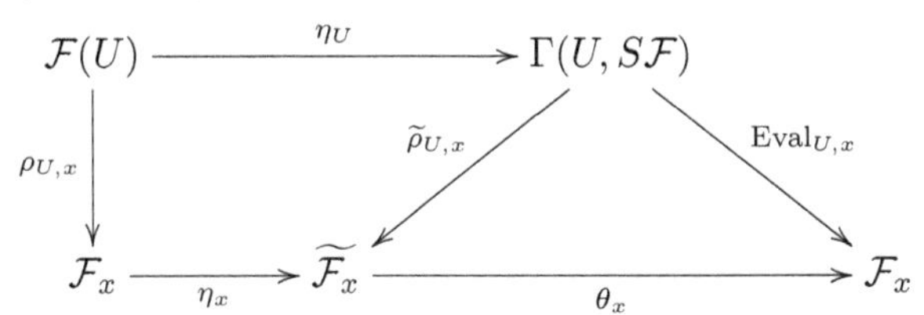

all commute. However, for all $s \in \mathcal{F}(U)$, we have

$$\rho_{U,x}(s) = s_x = \widetilde{s}(x) = \mathrm{Eval}_{U,x}(\eta_U(s)) = (\mathrm{Eval}_{U,x} \circ \eta_U)(s),$$

so the diagrams

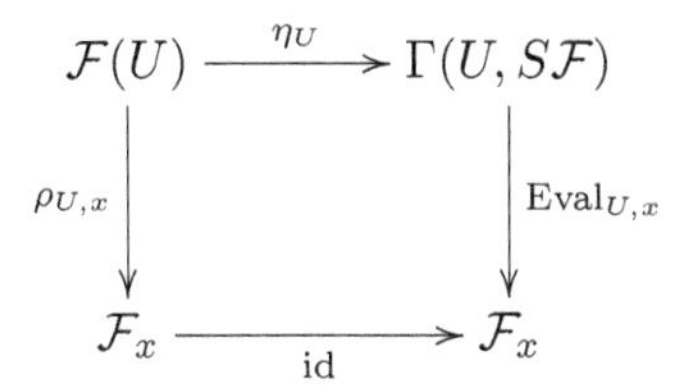

also commute, and by uniqueness of the bottom map making all these diagrams commute, we must have

$$\theta_x \circ \eta_x = \mathrm{id}.$$

Since θ_x is an isomorphism, so must be η_x. $\square$

Proposition 10.11. *For any stalk space (E, p), there is a stalk space isomorphism $\epsilon\colon E \to S\Gamma E$.*

Proof sketch. For every $x \in X$, by Proposition 10.9 there are isomorphisms $\epsilon_x\colon p^{-1}(x) \to (\Gamma E)_x$. Since the fibre of $S\Gamma E$ at x is equal to $(\Gamma E)_x$, the bijections ϵ_x define a bijection $\epsilon\colon E \to S\Gamma E$ defined by

$$\epsilon(e) = \epsilon_x(e), \quad e \in p^{-1}(x),\, x \in X,$$

such that $p = \Gamma p \circ \epsilon$, where $\Gamma p\colon S\Gamma E \to X$ is the projection associated with the stalk space $S\Gamma E$, as illustrated in the following diagram

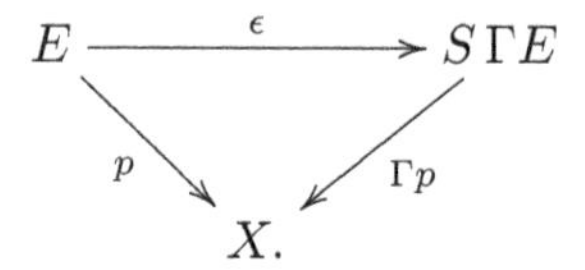

It remains to check that ϵ is continuous, which is shown in Tennison [Tennison (1975)] (Chapter II, Theorem 3.10). ☐

Strictly speaking the map $\epsilon\colon E \to S\Gamma E$ depends on E, so it should really be denoted by ϵ_E.

Definition 10.8. Given a stalk space (E, p), the stalk map $\epsilon_E\colon E \to S\Gamma E$ is defined by

$$\epsilon_E(e) = \epsilon_x(e), \quad e \in p^{-1}(x),\, x \in X,$$

where the map $\epsilon_x\colon p^{-1}(x) \to (\Gamma E)_x$ (an isomorphism) is given by Proposition 10.9.

It can be shown that the family ϵ of maps ϵ_E is natural in the following sense: for every map $\varphi\colon E_1 \to E_2$ of stalk spaces (E_1, p_1) and (E_2, p_2), the following diagram commutes:

$$\begin{array}{ccc} E_1 & \xrightarrow{\epsilon_{E_1}} & S\Gamma E_1 \\ {\scriptstyle\varphi}\downarrow & & \downarrow{\scriptstyle S\Gamma\varphi} \\ E_2 & \xrightarrow[\epsilon_{E_2}]{} & S\Gamma E_2. \end{array}$$

The results of the previous sections can be put together to show that the construction $\mathcal{F} \mapsto \widetilde{\mathcal{F}} = \Gamma S \mathcal{F}$ of a sheaf from a presheaf (the *sheafification* of $\mathcal{F}$) is universal, and that the constructions S and Γ are essentially mutual inverses.

10.4 The Equivalence of Sheaves and Stalk Spaces

The following theorem shows the universality of the sheafification construction $\mathcal{F} \mapsto \widetilde{\mathcal{F}}$.

Theorem 10.12. *Given any presheaf $\mathcal{F}$ and any sheaf $\mathcal{G}$, for any presheaf map $\varphi\colon \mathcal{F} \to \mathcal{G}$, there is a unique sheaf map $\widehat{\varphi}\colon \widetilde{\mathcal{F}} \to \mathcal{G}$ such that*

$$\varphi = \widehat{\varphi} \circ \eta_{\mathcal{F}}$$

as illustrated by the following commutative diagram

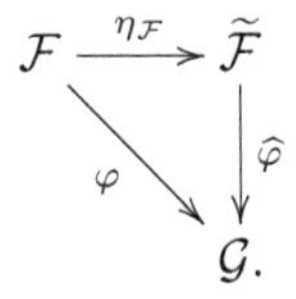

Proof. First we prove that if $\widehat{\varphi}\colon \widetilde{\mathcal{F}} \to \mathcal{G}$ exists, then it is unique. Since $\varphi = \widehat{\varphi} \circ \eta_{\mathcal{F}}$, for every $x \in X$, by considering the stalk maps we must have

$$\varphi_x = \widehat{\varphi}_x \circ \eta_x.$$

However, by Proposition 10.10, the map η_x is an isomorphism, which shows that $\widehat{\varphi}_x = \varphi_x \circ \eta_x^{-1}$ is uniquely defined. Since $\mathcal{G}$ is a sheaf, by Proposition 10.2 the map $\widehat{\varphi}$ is uniquely determined.

We now show the existence of the map $\widehat{\varphi}$. By Proposition 10.8, the presheaf map $\varphi\colon \mathcal{F} \to \mathcal{G}$ yields the sheaf map $\widetilde{\varphi}\colon \widetilde{\mathcal{F}} \to \widetilde{\mathcal{G}}$. Furthermore, since $\mathcal{G}$ is a sheaf, by Proposition 10.6, the map $\eta_{\mathcal{G}}\colon \mathcal{G} \to \widetilde{\mathcal{G}}$ is an isomorphism. Therefore, we get the sheaf map $\widehat{\varphi} = \widetilde{\eta_{\mathcal{G}}}^{-1} \circ \widetilde{\varphi}$ from $\widetilde{\mathcal{F}}$ to $\mathcal{G}$ as illustrated in the following diagram.

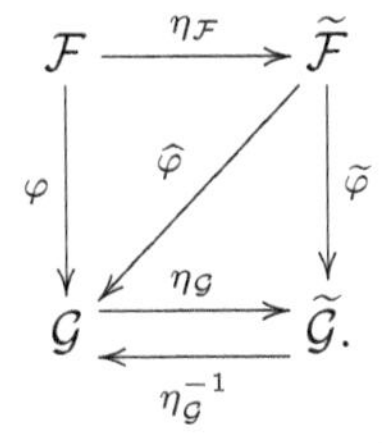

Using the naturality of η we see that $\varphi = \widetilde{\eta_{\mathcal{G}}}^{-1} \circ \widetilde{\varphi} \circ \eta_{\mathcal{F}} = \widehat{\varphi} \circ \eta_{\mathcal{F}}$. ☐

We now go back to the constructions S and Γ to make the equivalence of sheaves and stalk spaces more precise. The "right" framework to do so is category theory, but we prefer to remain more informal.

The situation is that we have three kinds of objects and maps between these objects (categories):

(1) The class (category) $\mathbf{PSh}(X)$ whose objects are presheaves over a topological space X and whose maps (morphisms) are maps of presheaves.
(2) The class (category) $\mathbf{Sh}(X)$ whose objects are sheaves over a topological space X and whose maps (morphisms) are maps of sheaves.
(3) The class (category) $\mathbf{StalkS}(X)$ whose objects are stalk spaces over a topological space X and whose maps (morphisms) are maps of stalk spaces.

Definition 10.3 implies that the operation S maps an object $\mathcal{F}$ of $\mathbf{PSh}(X)$ to an object $(S\mathcal{F}, p\colon S\mathcal{F} \to X)$ of $\mathbf{StalkS}(X)$, and a map $\varphi\colon \mathcal{F} \to \mathcal{G}$ between objects of $\mathbf{PSh}(X)$ to a map $S\varphi\colon S\mathcal{F} \to S\mathcal{G}$ between objects in $\mathbf{StalkS}(X)$, in such that a way that $S(\psi \circ \varphi) = S\psi \circ S\varphi$ and $S\mathrm{id}_{\mathcal{F}} = \mathrm{id}_{S\mathcal{F}}$. In sophisticated terms,

$$S\colon \mathbf{PSh}(X) \to \mathbf{StalkS}(X)$$

is a functor from the category $\mathbf{PSh}(X)$ to the category $\mathbf{StalkS}(X)$.

Definition 10.7 implies that the operation Γ maps an object (E, p) from $\mathbf{StalkS}(X)$ to an object ΓE in $\mathbf{Sh}(X)$, and a map $\varphi\colon E_1 \to E_2$ between two objects (E_1, p_1) and (E_2, p_2) in $\mathbf{StalkS}(X)$ to a map $\Gamma\varphi\colon \Gamma E_1 \to \Gamma E_2$ between objects in $\mathbf{Sh}(X)$, in such a way that $\Gamma(\psi \circ \varphi) = \Gamma\psi \circ \Gamma\varphi$ and $\Gamma\mathrm{id}_E = \mathrm{id}_{\Gamma E}$. In sophisticated terms,

$$\Gamma\colon \mathbf{StalkS}(X) \to \mathbf{Sh}(X)$$

is a functor from the category $\mathbf{StalkS}(X)$ to the category $\mathbf{Sh}(X)$.

Note that every sheaf $\mathcal{F}$ is also a presheaf, and that every map $\varphi\colon \mathcal{F} \to \mathcal{G}$ of sheaves is also a map of presheaves. Therefore, we have an inclusion map

$$i\colon \mathbf{Sh}(X) \to \mathbf{PSh}(X),$$

which is a functor. As a consequence, S restricts to an operation (functor)

$$S\colon \mathbf{Sh}(X) \to \mathbf{StalkS}(X).$$

We also defined the map η which maps a presheaf $\mathcal{F}$ to the sheaf $\Gamma S(\mathcal{F}) = \widetilde{\mathcal{F}}$, and showed in Proposition 10.6 that this map is an isomorphism iff $\mathcal{F}$ is a sheaf. We also showed that η is natural. This can be restated as saying that η is a natural isomorphism between the functors id (the identity functor) and ΓS from $\mathbf{Sh}(X)$ to itself.

We also defined the map ϵ which takes a stalk space (E, p) and makes the stalk space $S\Gamma E$, and proved in Proposition 10.11 that $\epsilon\colon E \to S\Gamma E$

is an isomorphism. This can be restated as saying that ϵ is a natural isomorphism between the functors id (the identity functor) and $S\Gamma$ from $\mathbf{StalkS}(X)$ to itself. Then, we see that the two operations (functors)

$$S\colon \mathbf{Sh}(X) \to \mathbf{StalkS}(X) \quad \text{and} \quad \Gamma\colon \mathbf{StalkS}(X) \to \mathbf{Sh}(X)$$

are almost mutual inverses, in the sense that there is a natural isomorphism η between ΓS and id and a natural isomorphism ϵ between $S\Gamma$ and id. In such a situation, we say that the classes (categories) $\mathbf{Sh}(X)$ and $\mathbf{StalkS}(X)$ are *equivalent*. The upshot is that it is basically a matter of taste (or convenience) whether we decide to work with sheaves or stalk spaces.[3]

We also have the operator (functor)

$$\Gamma S\colon \mathbf{PSh}(X) \to \mathbf{Sh}(X)$$

which "sheafifies" a presheaf $\mathcal{F}$ into the sheaf $\widetilde{\mathcal{F}}$. Theorem 10.12 can be restated as saying that there is an isomorphism

$$\mathrm{Hom}_{\mathbf{PSh}(X)}(\mathcal{F}, i(\mathcal{G})) \cong \mathrm{Hom}_{\mathbf{Sh}(X)}(\widetilde{\mathcal{F}}, \mathcal{G}),$$

between the set (category) of maps between the presheaves $\mathcal{F}$ and $i(\mathcal{G})$ and the set (category) of maps between the sheaves $\widetilde{\mathcal{F}}$ and $\mathcal{G}$. In fact, such an isomorphism is natural, so in categorical terms i and $\widetilde{\ } = \Gamma S$ are *adjoint functors*. This is as far as we will go with our excursion into category theory. The reader should consult Tennison [Tennison (1975)] for a comprehensive treatment of a preshaves and sheaves in the framework of *abelian categories*.

10.5 Stalk Spaces of Modules or Rings

In Sections 10.3 and 10.4 we have considered presheaves and sheaves of sets. If $\mathcal{F}$ is a sheaf of R-modules, then it is immediately verified that for every $x \in X$ the stalk $\mathcal{F}_x$ at x is an R-module, and similarly if $\mathcal{F}$ is a sheaf of rings, then $\mathcal{F}_x$ is a ring.

Minor modifications need to be made to the notion of a stalk space to extend the equivalence between sheaves of R-modules, rings, *etc.* and stalk spaces. We simply need to assume that every fibre $p^{-1}(x)$ (with $x \in X$) is a R-module, ring, *etc.*, and that the R- module operations, ring operations, *etc.*, are continuous.

[3]Actually, if we deal with sheaves of modules or rings, it turns out that stalk spaces have a better behavior when it comes to images of morphisms, or quotients.

More precisely, we have the following definitions taken from Serre [Serre (1955)] (Chapter I, Sections 1 and 6).

Definition 10.9. A *stalk space of R-modules* is a pair $(E, p\colon E \to X)$ where p is a surjective local homeomorphism, and the following conditions hold:

(1) Every fibre $p^{-1}(x)$ (with $x \in X$) is an R-module.
(2) There is a continuous function $\cdot_R\colon R \times E \to E$ such that for all $\lambda \in R$ and all $e \in E$, $\lambda \cdot_R e = \lambda \cdot e$, where $\cdot$ is scalar multiplication in the fibre $p^{-1}(p(e))$.
(3) There is a continuous function $-_E\colon E \to E$ such that for all $e \in E$, $-_E e = -e$, where $-e$ is the additive inverse of e in the fibre $p^{-1}(p(e))$.
(4) If we set $E \sqcap E = \{(e_1, e_2) \in E \times E \mid p(e_1) = p(e_2)\}$, then there is a continuous function $+_E\colon E \sqcap E \to E$ such that $e_1 +_E e_2 = e_1 + e_2$, where $+$ is addition in the fibre $p^{-1}(p(e_1))$ $(= p^{-1}(p(e_2)))$.

Definition 10.10. A map of stalk spaces of R-modules (E_1, p_1) and (E_2, p_2) is a map $\varphi\colon (E_1, p_1) \to (E_2, p_2)$ of stalk spaces such that for every $x \in X$, the restriction of φ to the fibre $p_1^{-1}(x)$ is a R-linear map between $p_1^{-1}(x)$ and $p_2^{-1}(x)$.

Here is the definition of a stalk space when the fibres are commutative rings.

Definition 10.11. A *stalk space of commutative rings* is a pair $(E, p\colon E \to X)$ where p is a surjective local homeomorphism, and the following conditions hold:

(1) Every fibre $p^{-1}(x)$ (with $x \in X$) is a commutative ring.
(2) There is a continuous function $-_E\colon E \to E$ such that for all $e \in E$, $-_E e = -e$, where $-e$ is the additive inverse of e in the fibre $p^{-1}(p(e))$.
(3) If we set $E \sqcap E = \{(e_1, e_2) \in E \times E \mid p(e_1) = p(e_2)\}$, then there is a continuous function $+_E\colon E \sqcap E \to E$ such that $e_1 +_E e_2 = e_1 + e_2$, where $+$ is addition in the fibre $p^{-1}(p(e_1))$ $(= p^{-1}(p(e_2)))$.
(4) There is a continuous function $*_E\colon E \sqcap E \to E$ such that $e_1 *_E e_2 = e_1 * e_2$, where $*$ is multiplication in the fibre $p^{-1}(p(e_1))$ $(= p^{-1}(p(e_2)))$.

Definition 10.12. A map of stalk spaces of rings (E_1, p_1) and (E_2, p_2) is a map $\varphi\colon (E_1, p_1) \to (E_2, p_2)$ of stalk spaces such that for every $x \in X$, the restriction of φ to the fibre $p_1^{-1}(x)$ is a ring homomorphism between $p_1^{-1}(x)$ and $p_2^{-1}(x)$.

Having the notion of stalk space of commutative rings we can generalize the notion of stalk space of R-modules to allow the ring R to vary. This notion plays an important role in algebraic geometry. In the following definition, if $(\mathcal{A}, p_{\mathcal{A}}\colon \mathcal{A} \to X)$ is a stalk space of commutative rings, the fibre $p_{\mathcal{A}}^{-1}(x)$ above $x \in X$ (a commutative ring) is denoted $\mathcal{A}_x$.

Definition 10.13. Given a stalk space of commutative rings $(\mathcal{A}, p_{\mathcal{A}}\colon \mathcal{A} \to X)$, a *stalk space of $\mathcal{A}$-modules* is a pair $(E, p\colon E \to X)$ where p is a surjective local homeomorphism, and the following conditions hold:

(1) Every fibre $p^{-1}(x)$ (with $x \in X$) is an $\mathcal{A}_x$-module.
(2) If we set $\mathcal{A} \sqcap E = \{(\lambda, e) \in \mathcal{A} \times E \mid p_{\mathcal{A}}(\lambda) = p(e)\}$, then there is a continuous function $\cdot_E\colon \mathcal{A} \sqcap E \to E$ such that $\lambda \cdot_E e = \lambda \cdot e$, where $\cdot$ is the scalar multiplication on $\mathcal{A}_{p(e)} \times p^{-1}(p(e))$.
(3) There is a continuous function $-_E\colon E \to E$ such that for all $e \in E$, $-_E e = -e$, where $-e$ is the additive inverse of e in the fibre $p^{-1}(p(e))$.
(4) If we set $E \sqcap E = \{(e_1, e_2) \in E \times E \mid p(e_1) = p(e_2)\}$, then there is a continuous function $+_E\colon E \sqcap E \to E$ such that $e_1 +_E e_2 = e_1 + e_2$, where $+$ is addition in the fibre $p^{-1}(p(e_1))\ (= p^{-1}(p(e_2)))$.

Definition 10.10 is modified in the obvious way. That is, a map of stalk spaces of $\mathcal{A}$-modules (E_1, p_1) and (E_2, p_2) is a map $\varphi\colon (E_1, p_1) \to (E_2, p_2)$ of stalk spaces such that for every $x \in X$, the restriction of φ to the fibre $p_1^{-1}(x)$ is an $\mathcal{A}_x$-linear map between $p_1^{-1}(x)$ and $p_2^{-1}(x)$.

Finally, the restriction of a stalk space to a subspace is defined as follows.

Definition 10.14. Given a stalk space of R-modules (or rings) (E, p) over a space X, for every subset Y of X, we define the *restriction* $(E, p)|Y$ of (E, p) to Y as the stalk space $(p^{-1}(Y), p|p^{-1}(Y))$.

The reader is referred to Tennison [Tennison (1975)] for more details on the equivalence between sheaves with an algebraic structure and stalk spaces with the same algebraic structure on the fibres.

10.6 Kernels of Presheaves and Sheaves

If $f\colon A \to B$ is a homomorphism between two R-modules A and B, recall that the *kernel* $\mathrm{Ker}\,(f)$ of f is defined by

$$\mathrm{Ker}\,(f) = \{u \in A \mid f(u) = 0\},$$

the *image* $\mathrm{Im}(f)$ of f is defined by
$$\mathrm{Im}(f) = \{v \in B \mid (\exists u \in A)(v = f(u))\},$$
the *cokernel* $\mathrm{Coker}(f)$ of f is defined by
$$\mathrm{Coker}(f) = B/\mathrm{Im}(f),$$
and the *coimage* $\mathrm{Coim}(f)$ of f is defined by
$$\mathrm{Coim}(f) = A/\mathrm{Ker}\,(f).$$
Furthermore, f is injective iff $\mathrm{Ker}\,(f) = (0)$, f is surjective iff $\mathrm{Coker}(f) = (0)$, and there is an isomorphism $\mathrm{Coim}(f) \cong \mathrm{Im}(f)$. A sequence of R-modules
$$A \xrightarrow{f} B \xrightarrow{g} C$$
is *exact* at B if $\mathrm{Im}(f) = \mathrm{Ker}\,(g)$.

We would like to generalize the above notions to maps of presheaves and sheaves of R-modules or commutative rings. In the case of presheaves, everything works perfectly, but in the case of sheaves, there are two problems:

(1) In general, the presheaf image of a sheaf is not a sheaf.
(2) In general, the presheaf quotient of two sheaves is not a sheaf.

A way to fix these problems is to apply the sheafification process to the presheaf, but in the case of the image of a sheaf morphism $\varphi\colon \mathcal{F} \to \mathcal{G}$, this has the slightly unpleasant consequence that $\widetilde{\mathrm{Im}(\varphi)}$ is a not a subsheaf of $\mathcal{G}$. This small problem can be avoided by defining the image of a sheaf morphism as the kernel of its cokernel map (as this would be the case in an abelian category).

From now on in this section we assume that we are dealing with presheaves and sheaves of R-modules or commutative rings. We follow closely Tennison [Tennison (1975)], so many proof are omitted.

We begin with kernels. If $\varphi\colon \mathcal{F} \to \mathcal{G}$ is a map of presheaves on a space X, then for every open subset U of X, define $(\mathrm{Ker}\,\varphi)_U$ by
$$(\mathrm{Ker}\,\varphi)_U = \mathrm{Ker}\,\varphi_U = \{s \in \mathcal{F}(U) \mid \varphi_U(s) = 0\}.$$
If V is some open subset of U, the commutativity of the diagram
$$\begin{array}{ccc} \mathcal{F}(U) & \xrightarrow{\varphi_U} & \mathcal{G}(U) \\ {\scriptstyle (\rho_{\mathcal{F}})^U_V}\downarrow & & \downarrow{\scriptstyle (\rho_{\mathcal{G}})^U_V} \\ \mathcal{F}(V) & \xrightarrow[\varphi_V]{} & \mathcal{G}(V) \end{array}$$

implies that if $s \in (\mathrm{Ker}\,\varphi)_U$, that is, $\varphi_U(s) = 0$, then

$$\varphi_V((\rho_{\mathcal{F}})^U_V(s)) = (\rho_{\mathcal{G}})^U_V(\varphi_U(s)) = (\rho_{\mathcal{G}})^U_V(0) = 0,$$

so $(\rho_{\mathcal{F}})^U_V(s) \in (\mathrm{Ker}\,\varphi)_V$. This shows that the $(\mathrm{Ker}\,\varphi)_U$ together with the restriction functions ρ^U_V (as a function from $(\mathrm{Ker}\,\varphi)_U$ to $(\mathrm{Ker}\,\varphi)_V$) is a presheaf on X.

Definition 10.15. If $\varphi\colon \mathcal{F} \to \mathcal{G}$ is a map of presheaves on a space X, then for every open subset U of X, define $(\mathrm{Ker}\,\varphi)_U$ by

$$(\mathrm{Ker}\,\varphi)_U = \mathrm{Ker}\,\varphi_U = \{s \in \mathcal{F}(U) \mid \varphi_U(s) = 0\}.$$

Then the $(\mathrm{Ker}\,\varphi)_U$ together with the restriction functions ρ^U_V (as a function from $(\mathrm{Ker}\,\varphi)_U$ to $(\mathrm{Ker}\,\varphi)_V$) is a presheaf called the *presheaf kernel* of φ and denoted $\mathrm{Ker}\,\varphi$.

If $\mathcal{F}$ and $\mathcal{G}$ are sheaves, then $\mathrm{Ker}\,\varphi$ is a sheaf.

Proposition 10.13. *If $\mathcal{F}$ is a sheaf and $\mathcal{G}$ satisfies Condition (M), then* $\mathrm{Ker}\,\varphi$ *is a sheaf. In particular, if $\mathcal{F}$ and $\mathcal{G}$ are sheaves, then* $\mathrm{Ker}\,\varphi$ *is a sheaf.*

Proof. We need to check Conditions (M) and (G). Since $\mathcal{F}$ is a sheaf, it satisfies Condition (M), and it is easy to show that $\mathrm{Ker}\,\varphi$ also satisfies Condition (M).

To check Condition (G), let U be any open subset of X, let $(U_i)_{i\in I}$ be any open cover of U, and let $(s_i)_{i\in I}$ be a family of sections $s_i \in (\mathrm{Ker}\,\varphi)_{U_i}$ such that $(\rho_{\mathcal{F}})^U_{U_i\cap U_j}(s_i) = (\rho_{\mathcal{F}})^U_{U_i\cap U_j}(s_j)$ for all i, j. Since $\mathcal{F}$ is a sheaf, there is some $s \in \mathcal{F}(U)$ such that $(\rho_{\mathcal{F}})^U_{U_i}(s) = s_i$ for all $i \in I$. Since $\varphi_{U_i}(s_i) = 0$, the commutativity of the diagram

$$\begin{array}{ccc} \mathcal{F}(U) & \xrightarrow{\varphi_U} & \mathcal{G}(U) \\ {\scriptstyle (\rho_{\mathcal{F}})^U_{U_i}}\downarrow & & \downarrow{\scriptstyle (\rho_{\mathcal{G}})^U_{U_i}} \\ \mathcal{F}(U_i) & \xrightarrow[\varphi_{U_i}]{} & \mathcal{G}(U_i) \end{array}$$

implies that

$$0 = \varphi_{U_i}(s_i) = \varphi_{U_i}((\rho_{\mathcal{F}})^U_{U_i}(s)) = (\rho_{\mathcal{G}})^U_{U_i}(\varphi_U(s))$$

for all $i \in I$. Since $\mathcal{G}$ satisfies Condition (M) (as formulated in Remark 4 after Definition 8.5), $\varphi_U(s) = 0$, which means that $s \in (\mathrm{Ker}\,\varphi)_U$. ☐

The next proposition generalizes the property that a module or (ring) map $f\colon A \to B$ is injective iff $\mathrm{Ker}\, f = (0)$.

Proposition 10.14. *Let $\varphi\colon \mathcal{F} \to \mathcal{G}$ be a map of presheaves. Conditions (i) and (ii) are equivalent.*

(i) $\mathrm{Ker}\, \varphi = (0)$ *(the trivial zero sheaf).*
(ii) *φ_U is injective for all open subsets U of X.*
(iii) *If (ii) (equivalently (i)) holds, then φ_x is injective for all $x \in X$. If φ_x is injective for all $x \in X$ and if $\mathcal{F}$ satisfies Condition (M), then (ii) (equivalently (i)) holds.*

Proof. The equivalence of (i) and (ii) is immediate by definition of $(\mathrm{Ker}\, \varphi)_U$.

Assume that (ii) holds, and suppose that $\varphi_x(\gamma) = 0$ for some $\gamma \in \mathcal{F}_x$ (with $x \in X$). This means that there is some open subset U of X containing x and some $s \in \mathcal{F}(U)$ such that $s_x = \gamma$ and since by Definition 10.3, $\varphi_x(\gamma) = \varphi_x(s_x) = (\varphi_U(s))_x$, that $(\varphi_U(s))_x = 0$, which in turn means that there is some open subset $V \subseteq U$ containing x such that

$$(\rho_{\mathcal{G}})^U_V(\varphi_U(s)) = 0.$$

Since φ is a map of presheaves,

$$(\rho_{\mathcal{G}})^U_V(\varphi_U(s)) = \varphi_V((\rho_{\mathcal{F}})^U_V(s)),$$

we get $\varphi_V((\rho_{\mathcal{F}})^U_V(s)) = 0$, and since φ_V is injective, $(\rho_{\mathcal{F}})^U_V(s) = 0$. But $(\rho_{\mathcal{F}})^U_V(s) = 0$ implies that $\gamma = ((\rho_{\mathcal{F}})^U_V(s))_x = 0$, so φ_x is injective.

Conversely, assume that φ_x is injective for all $x \in X$ and that $\mathcal{F}$ satisfies Condition (M). Suppose $\varphi_U(s) = 0$ for some $s \in \mathcal{F}(U)$ (where U is any open subset of X). We need to prove that $s = 0$. Then by Definition 10.3,

$$\varphi_x(s_x) = (\varphi_U(s))_x = 0$$

for all $x \in U$, and since φ_x is injective for all x, we deduce that $s_x = 0$ for all $x \in U$. Since $\mathcal{F}$ satisfies Condition (M), by Proposition 10.1 (with $t = 0$), we conclude that $s = 0$, which shows that φ_U is injective. $\square$

Definition 10.16. A map of presheaves $\varphi\colon \mathcal{F} \to \mathcal{G}$ is *injective* if any of the Conditions (i) and (ii) of Proposition 10.14 holds. A map of sheaves $\varphi\colon \mathcal{F} \to \mathcal{G}$ is *injective* if any of the Conditions (i)–(iii) of Proposition 10.14 holds.

Remark: A presheaf or sheaf map $\varphi\colon \mathcal{F} \to \mathcal{G}$ is said to a *monic* if for every presheaf $\mathcal{H}$ any two presheaf maps $\psi_1, \psi_2\colon \mathcal{H} \to \mathcal{F}$, if $\varphi \circ \psi_1 = \varphi \circ \psi_2$, then $\psi_1 = \psi_2$. It can be shown that being a monic is equivalent to any of the conditions of Proposition 10.14; see Tennison [Tennison (1975)] (Chapter III, Theorem 3.5).

The following two propositions are stated without proof; see Tennison [Tennison (1975)] (Chapter III) for details.

Proposition 10.15. *If $\varphi\colon \mathcal{F} \to \mathcal{G}$ is a map of presheaves, then*

$$(\mathrm{Ker}\,\varphi)_x = \mathrm{Ker}\,\varphi_x$$

for all $x \in X$.

Proposition 10.16. *If $\varphi\colon (E_1, p_1) \to (E_2, p_2)$ is a map of stalk spaces, then $\Gamma\varphi\colon \Gamma E_1 \to \Gamma E_2$ is an injective map of sheaves iff φ is injective iff φ is a homeomorphism onto an open subspace of E_2.*

10.7 Cokernels of Presheaves and Sheaves

The notions of subpresheaves and subsheaves are defined as follows.

Definition 10.17. Given two presheaves $\mathcal{F}$ and $\mathcal{G}$ on a space X, we say that *$\mathcal{F}$ is a subpresheaf of $\mathcal{G}$* if for every open subset U of X, the R-module (resp. ring) $\mathcal{F}(U)$ is a submodule (resp. subring) of $\mathcal{G}(U)$, and the restriction functions of $\mathcal{F}$ are induced by the restriction functions of $\mathcal{G}$ ($(\rho_{\mathcal{F}})^U_V$ is the restriction of $(\rho_{\mathcal{G}})^U_V$ for any two open subsets $V \subseteq U$). If $\mathcal{F}$ and $\mathcal{G}$ are sheaves and the above condition hold, we say that *$\mathcal{F}$ is a subsheaf of $\mathcal{G}$*.

Remark: In terms of stalk spaces, in view of Proposition 10.16, we say that (E_1, p_1) is a *substalk space* of (E_2, p_2) if E_1 is an open subset of E_2, p_1 is the restriction of p_2 to E_1, and the fibre $p_1^{-1}(x)$ is a submodule (resp. subring) of the fibre $p_2^{-1}(x)$ for all $x \in X$.

The following proposition will be needed.

Proposition 10.17. *Let $\mathcal{G}$ be a sheaf and assume that $\mathcal{F}$ and $\mathcal{F}'$ are two subsheaves of $\mathcal{G}$. Then $\mathcal{F} = \mathcal{F}'$ iff $\mathcal{F}_x = \mathcal{F}'_x$ for all $x \in X$ (as submodules or subrings).*

Proof. First we prove that if $\mathcal{F}_x \subseteq \mathcal{F}'_x$ for all $x \in X$ (as submodules or subrings) then $\mathcal{F}$ is a subsheaf of $\mathcal{F}'$. We claim that for any open subset U for X, for any section $s \in \mathcal{F}(U)$, there is a unique section $t \in \mathcal{F}'(U)$ such that $s_x = t_x$ for all $x \in U$. Since $\mathcal{F}_x \subseteq \mathcal{F}'_x$ for all $x \in U$, there is an open cover $(U_x)_{x \in U}$ of U and a family of sections $t^x \in \mathcal{F}'(U_x)$ such that $(\rho_{\mathcal{F}})^U_{U_x}(s) = (\rho_{\mathcal{F}'})^U_{U_x}(t^x)$ for all $x \in U$. It follows that $(\rho_{\mathcal{F}'})^{U_x}_{U_x \cap U_y}(t^x) = (\rho_{\mathcal{F}'})^{U_y}_{U_x \cap U_y}(t^y)$ for all x, y and since $\mathcal{F}'$ is a sheaf there is a unique section $t \in \mathcal{F}'(U)$ such that $(\rho_{\mathcal{F}'})^U_{U_x}(t) = t^x$ for all $x \in U$. Observe that $(\rho_{\mathcal{F}})^U_{U_x}(s) = (\rho_{\mathcal{F}'})^U_{U_x}(t^x)$ and $(\rho_{\mathcal{F}'})^U_{U_x}(t) = t^x$ imply that $s_x = t^x_x = t_x$. Now the construction of t depends on the open cover (U_x), but since $\mathcal{F}'$ is a sheaf, by Proposition 10.1, there is a unique section $t \in \mathcal{F}'(U)$ with prescribed germs s_x for all $x \in U$. Therefore, we obtain a map $\varphi_U \colon \mathcal{F}(U) \to \mathcal{F}'(U)$ by setting $\varphi_U(s) = t$, and it is easy to see that these maps define a sheaf map $\varphi \colon \mathcal{F} \to \mathcal{F}'$. At first glance it is not obvious that φ is an inclusion map, but it is as the following argument shows. Recall that $s_x = t_x$ for all $x \in U$. But by Definition 10.3, we also have $\varphi_x(s_x) = (\varphi_U(s))_x = t_x$, so the composition $i' \circ \varphi$ where i' is the inclusion of $\mathcal{F}'$ in $\mathcal{G}$ agrees on stalks with the inclusion i of $\mathcal{F}$ in $\mathcal{G}$. By Proposition 10.2, we have $i' \circ \varphi = i$, so φ is an inclusion.

Now, if $\mathcal{F}_x = \mathcal{F}'_x$ for all $x \in X$, by the above $\mathcal{F}$ is a subsheaf of $\mathcal{F}'$ and $\mathcal{F}'$ is a subsheaf of $\mathcal{F}$ so $\mathcal{F} = \mathcal{F}'$.

If $\mathcal{F} = \mathcal{F}'$, then obviously $\mathcal{F}_x = \mathcal{F}'_x$ for all $x \in X$. ☐

Let us now consider cokernels and images. Let $\varphi \colon \mathcal{F} \to \mathcal{G}$ be a map of presheaves. For every open subset U of X, define PCoker_U by

$$\mathrm{PCoker}_U = \mathcal{G}(U)/\varphi_U(\mathcal{F}(U)) = \mathcal{G}(U)/\mathrm{Im}\,\varphi_U,$$

the quotient module (resp. quotient ring) of $\mathcal{G}(U)$ modulo $\varphi_U(\mathcal{F}(U))$, which is well defined since $\varphi_U(\mathcal{F}(U))$ is a submodule (resp. subring) of $\mathcal{G}(U)$ because φ_U is a homomorphism.

For any open subset $V \subseteq U$, the commutativity of the diagram

$$\begin{array}{ccc} \mathcal{F}(U) & \xrightarrow{\varphi_U} & \mathcal{G}(U) \\ {\scriptstyle (\rho_{\mathcal{F}})^U_V}\downarrow & & \downarrow{\scriptstyle (\rho_{\mathcal{G}})^U_V} \\ \mathcal{F}(V) & \xrightarrow[\varphi_V]{} & \mathcal{G}(V) \end{array}$$

implies that for any $s \in \mathcal{F}(U)$, we have

$$(\rho_{\mathcal{G}})^U_V(\varphi_U(s)) = \varphi_V((\rho_{\mathcal{F}})^U_V(s)),$$

which shows that $(\rho_{\mathcal{G}})^U_V(\varphi_U(s)) \in \mathrm{Im}(\varphi_V)$, that is, $(\rho_{\mathcal{G}})^U_V(\mathrm{Im}(\varphi_U)) \subseteq \mathrm{Im}(\varphi_V)$. If we let $\mathrm{pcoker}_U \colon \mathcal{G}(U) \to \mathcal{G}(U)/\mathrm{Im}(\varphi_U)$ be the projection map, then $\mathrm{pcoker}_V \circ (\rho_{\mathcal{G}})^U_V \colon \mathcal{G}(U) \to \mathcal{G}(V)/\mathrm{Im}(\varphi_V)$ vanishes on $\mathrm{Im}(\varphi_U)$, which implies that there is a unique map $(\overline{\rho}_{\mathcal{G}})^U_V \colon \mathcal{G}(U)/\mathrm{Im}(\varphi_U) \to \mathcal{G}(V)/\mathrm{Im}(\varphi_V)$ making the following diagram commute

$$\begin{array}{ccc} \mathcal{G}(U) & \xrightarrow{\mathrm{pcoker}_U} & \mathcal{G}(U)/\mathrm{Im}(\varphi_U) \\ {\scriptstyle (\rho_G)^U_V}\downarrow & & \downarrow{\scriptstyle (\overline{\rho}_{\mathcal{G}})^U_V} \\ \mathcal{G}(V) & \xrightarrow[\mathrm{pcoker}_V]{} & \mathcal{G}(V)/\mathrm{Im}(\varphi_V). \end{array}$$

Therefore, the Pcoker_U together with the restriction functions $(\overline{\rho}_{\mathcal{G}})^U_V$ define a presheaf on X.

Definition 10.18. If $\varphi \colon \mathcal{F} \to \mathcal{G}$ is a map of presheaves on a space X, then for every open subset U of X, define PCoker_U by

$$\mathrm{PCoker}_U = \mathcal{G}(U)/\varphi_U(\mathcal{F}(U)) = \mathcal{G}(U)/\mathrm{Im}\,\varphi_U.$$

Then the PCoker_U together with the restriction functions $(\overline{\rho}_{\mathcal{G}})^U_V$ define a presheaf called the *presheaf cokernel* of φ, and denoted $\mathrm{PCoker}(\varphi)$. The projection maps $\mathrm{pcoker}_U \colon \mathcal{G}(U) \to \mathcal{G}(U)/\mathrm{Im}(\varphi_U)$ define a presheaf map $\mathrm{pcoker}(\varphi) \colon \mathcal{G} \to \mathrm{PCoker}(\varphi)$.

Obviously, $\mathrm{pcoker}(\varphi) \circ \varphi = 0$ as illustrated in the diagram below

$$\mathcal{F} \xrightarrow{\ \varphi\ } \mathcal{G} \xrightarrow{\mathrm{pcoker}(\varphi)} \mathrm{PCoker}(\varphi).$$

In fact, $\mathrm{pcoker}(\varphi)$ is characterized by a universal property of this kind; see Tennison [Tennison (1975)] (Chapter III) for details.

If $\varphi \colon \mathcal{F} \to \mathcal{G}$ is a map of sheaves, in general the presheaf cokernel $\mathrm{PCoker}(\varphi)$ is not a sheaf. To obtain a sheaf, we sheafify it.

Definition 10.19. If $\varphi \colon \mathcal{F} \to \mathcal{G}$ is a map of sheaves on a space X, then the *sheaf cokernel* of φ, denoted $\mathrm{SCoker}(\varphi)$, is the sheafification $\widetilde{\mathrm{PCoker}(\varphi)}$ of the presheaf cokernel $\mathrm{PCoker}(\varphi)$ of φ. The presheaf map $\mathrm{scoker}(\varphi) \colon \mathcal{G} \to \mathrm{SCoker}(\varphi)$ is defined as the composition

$$\mathcal{G} \xrightarrow{\mathrm{pcoker}(\varphi)} \mathrm{PCoker}(\varphi) \xrightarrow{\eta_{\mathrm{PCoker}(\varphi)}} \widetilde{\mathrm{PCoker}(\varphi)} = \mathrm{SCoker}(\varphi),$$

where $\eta_{\mathrm{PCoker}(\varphi)} \colon \mathrm{PCoker}(\varphi) \to \widetilde{\mathrm{PCoker}(\varphi)}$ is the canonical map of Definition 10.5.

Again, $\text{scoker}(\varphi) \circ \varphi = 0$ as illustrated in the diagram below

$$\mathcal{F} \xrightarrow{\varphi} \mathcal{G} \xrightarrow{\text{scoker}(\varphi)} \text{SCoker}(\varphi).$$

In fact, $\text{scoker}(\varphi)$ is characterized by a universal property of this kind; see Tennison [Tennison (1975)] (Chapter III) for details.

The following propositions generalize the characterization of the surjectivity of a module (resp. ring) homomorphism $f\colon A \to B$ in terms of its cokernel to presheaves and sheaves.

Proposition 10.18. *Let $\varphi\colon \mathcal{F} \to \mathcal{G}$ be a map of presheaves on a space X. Then the following conditions are equivalent:*

(i) $\text{PCoker}(\varphi) = (0)$.
(ii) For every open subset U of X, the map φ_U is surjective.

Proof. The equivalence of (i) and (ii) follows immediately from the definitions. ☐

Proposition 10.19. *Let $\varphi\colon \mathcal{F} \to \mathcal{G}$ be a map of sheaves on a space X. Then the following conditions are equivalent:*

(i) $\text{SCoker}(\varphi) = (0)$.
(ii) For every $x \in X$, $(\text{PCoker}(\varphi))_x = (0)$.
(iii) For every $x \in X$, φ_x is surjective.
(iv) For every open subset U of X, for every $t \in \mathcal{G}(U)$, there is some open cover $(U_i)_{i\in I}$ of U and a family $(s_i)_{i\in I}$ of sections $s_i \in \mathcal{F}(U_i)$ such that $\varphi_{U_i}(s_i) = (\rho_{\mathcal{G}})^U_{U_i}(t)$ for all $i \in I$.

Any of the conditions of Proposition 10.18 implies the above conditions.

Proof. The equivalence of (i) and (ii) goes as follows. Since $\text{SCoker}(\varphi)$ is a sheaf, by Proposition 10.2 (with ψ the zero map), $\text{SCoker}(\varphi) = (0)$ iff $(\text{SCoker}(\varphi))_x = (0)$ for all $x \in X$. But by Proposition 10.10 the stalks $(\text{SCoker}(\varphi))_x$ and $(\text{PCoker}(\varphi))_x$ are isomorphic, so $(\text{SCoker}(\varphi))_x = (0)$ iff $(\text{PCoker}(\varphi))_x = (0)$ for all $x \in X$.

To prove the equivalence of (ii) and (iii) we need to unwind the definitions. We have $(\text{PCoker}(\varphi))_x = (0)$ iff for every open subset U of containing x and any $s \in \text{PCoker}(\varphi)(U) = \text{PCoker}_U$ there is some open subset $V \subseteq U$ containing x such that $(\overline{\rho}_{\mathcal{G}})^U_V(s) = 0$ iff (since $(\overline{\rho}_{\mathcal{G}})^U_V(s) \in \text{PCoker}_V$ and $\text{PCoker}_V = \mathcal{G}(V)/\varphi_U(\mathcal{F}(V)))$ for every open subset U of containing x and any $t \in \mathcal{G}(U)$ there is some open subset $V \subseteq U$ containing x

such that $(\rho_{\mathcal{G}})^U_V(t) \in \varphi_V(\mathcal{F}(V))$ iff there is some $s_1 \in \mathcal{F}(V)$ such that $\varphi_V(s_1) = (\rho_{\mathcal{G}})^U_V(t)$ (so $\varphi_x((s_1)_x) = ((\varphi_V(s_1))_x = t_x)$ iff φ_x is surjective.

Next we prove that (iii) $\Longrightarrow$ (iv). Assume (iii) holds. For any open subset U of X and for any $t \in \mathcal{G}(U)$, for any $x \in U$, since φ_x is surjective, there is some $\alpha \in \mathcal{F}_x$ such that $\varphi_x(\alpha) = t_x$. If α is represented by some $f^x \in \mathcal{F}(V_x)$ for some open subset V_x of U containing x, to say that $\varphi_x(\alpha) = t_x$ means that there is some open subset U_x of V_x containing x such that $(\rho_{\mathcal{G}})^{V_x}_{U_x}(\varphi_{V_x}(f^x)) = (\rho_{\mathcal{G}})^U_{U_x}(t)$. However, the commutativity of the diagram

$$\begin{array}{ccc} \mathcal{F}(V_x) & \xrightarrow{\varphi_{V_x}} & \mathcal{G}(V_x) \\ {\scriptstyle (\rho_{\mathcal{F}})^{V_x}_{U_x}} \downarrow & & \downarrow {\scriptstyle (\rho_{\mathcal{G}})^{V_x}_{U_x}} \\ \mathcal{F}(U_x) & \xrightarrow[\varphi_{U_x}]{} & \mathcal{G}(U_x) \end{array}$$

shows that $(\rho_{\mathcal{G}})^{V_x}_{U_x}(\varphi_{V_x}(f^x)) = \varphi_{U_x}((\rho_{\mathcal{F}})^{V_x}_{U_x}(f^x))$, and thus

$$\varphi_{U_x}((\rho_{\mathcal{F}})^{V_x}_{U_x}(f^x)) = (\rho_{\mathcal{G}})^U_{U_x}(t).$$

If we let $s^x = (\rho_{\mathcal{F}})^{V_x}_{U_x}(f^x)$, then we have a family $(s^x)_{x \in U}$ of sections $s^x \in \mathcal{F}(U_x)$ such that the U_x form an open cover of U and $\varphi_{U_x}(s^x) = (\rho_{\mathcal{G}})^U_{U_x}(t)$ for all $x \in U$, which is (iv).

The implication (iv) $\Longrightarrow$ (iii) is immediate. Indeed, any $\gamma \in \mathcal{G}_x$ is represented by some section $t \in \mathcal{G}(U)$ for some open subset U containing x, and by (iv), we have $\varphi_x((s_i)_x) = (\varphi_{U_i}(s_i))_x = t_x$ for any of the $s_i \in \mathcal{F}(U_i)$ given by (iv) since $\varphi_{U_i}(s_i) = (\rho_{\mathcal{G}})^U_{U_i}(t)$ for all $i \in I$. ☐

It is important to note that in the case of a map of sheaves $\varphi\colon \mathcal{F} \to \mathcal{G}$, unlike the case of presheaves, Condition (i) ($\mathrm{SCoker}(\varphi) = (0)$) *does not* imply that the maps φ_U are surjective for all open subsets U. We can only assert a *local version* of the surjectivity of the φ_U, as in Condition (iv).

An example of the failure of surjectivity of the φ_U is provided by $X = \mathbb{C}$ (the complex numbers), the sheaf of holomorphic functions $\mathcal{F} = \mathcal{C}^\omega$, and $\varphi = d$, the differentiation operator on $\mathcal{F}$ (here, $\mathcal{G} = \mathcal{F}$). For any $x \in \mathbb{C}$, locally near x a holomorphic function f can be integrated as a holomorphic function g such that $d/dz(g) = f$, but if U is not simply connected there are holomorphic functions which cannot be expressed as $d/dz(g)$ for some holomorphic function g, for example $f = 1/z$ on $U = \{z \in \mathbb{C} \mid z \neq 0\}$.

Definition 10.20. A map of presheaves $\varphi\colon \mathcal{F} \to \mathcal{G}$ is *surjective* if any of the Conditions (i) and (ii) of Proposition 10.18 holds. A map of sheaves

$\varphi\colon \mathcal{F} \to \mathcal{G}$ is *surjective* if any of the Conditions (i)–(iv) of Proposition 10.19 holds.

Remark: A presheaf map $\varphi\colon \mathcal{F} \to \mathcal{G}$ is said to be an *epic* if for every presheaf $\mathcal{H}$ any two presheaf maps $\psi_1, \psi_2\colon \mathcal{G} \to \mathcal{H}$, if $\psi_1 \circ \varphi = \psi_2 \circ \varphi$, then $\psi_1 = \psi_2$. Similarly, a sheaf map $\varphi\colon \mathcal{F} \to \mathcal{G}$ is said to be an *epic* if for every sheaf $\mathcal{H}$ any two sheaf maps $\psi_1, \psi_2\colon \mathcal{G} \to \mathcal{H}$, if $\psi_1 \circ \varphi = \psi_2 \circ \varphi$, then $\psi_1 = \psi_2$. It can be shown that being a presheaf epic is equivalent to any of the conditions of Proposition 10.18, and being a sheaf epic is equivalent to any of the conditions of Proposition 10.19; see Tennison [Tennison (1975)] (Chapter III, Theorems 4.7 and 4.8). Technically, Definition 10.20 defines the notions of presheaf epic and sheaf epic. A presheaf morphism is surjective on sections (*i.e.* all φ_U are surjective). The failure of a sheaf morphism to be a surjection on sections is closely related to sheaf cohomology.

10.8 Presheaf and Sheaf Isomorphisms

We can combine Propositions 10.14, 10.18, and 10.19 to obtain the following criteria for a map of presheaves or a map of sheaves to be an isomorphism.

Proposition 10.20. *Let $\varphi\colon \mathcal{F} \to \mathcal{G}$ be a map of presheaves on a space X. Then the following conditions are equivalent:*

(i) φ is an isomorphism.
(ii) For every open subset U of X, φ_U is bijective.

If $\mathcal{F}$ and $\mathcal{G}$ are sheaves, then we have the further equivalent condition:

(iii) φ_x is bijective for all $x \in X$.

Proof. By definition φ is a presheaf isomorphism iff there is some presheaf morphism $\psi\colon \mathcal{G} \to \mathcal{F}$ such that $\psi \circ \varphi = \mathrm{id}_{\mathcal{F}}$ and $\varphi \circ \psi = \mathrm{id}_{\mathcal{G}}$ iff there is some $\psi\colon \mathcal{G} \to \mathcal{F}$ such that $\psi_U \circ \varphi_U = \mathrm{id}_{\mathcal{F}(U)}$ and $\varphi_U \circ \psi_U = \mathrm{id}_{\mathcal{G}(\mathcal{U})}$ for all open subsets U iff φ_U is an isomorphism for all open subsets U. It remains to check that the inverses $\psi_U : \mathcal{G}(U) \to \mathcal{F}(U)$ are compatible with the restriction functions, which is easy to do. This proves that (i) and (ii) are equivalent.

It is clear that (i) implies (iii). Now assume that $\mathcal{F}$ and $\mathcal{G}$ are sheaves and that the φ_x are bijective. Since each φ_x is injective, we know from Proposition 10.14 that φ_U is injective for every open subset U. We now prove that because the φ_x are surjective, each φ_U is also surjective.

By Proposition 10.19(iv), for every open subset U of X, for every $t \in \mathcal{G}(U)$, there is some open cover $(U_i)_{i\in I}$ of U and a family $(s_i)_{i\in I}$ of sections $s_i \in \mathcal{F}(U_i)$ such that $\varphi_{U_i}(s_i) = (\rho_{\mathcal{G}})^U_{U_i}(t)$ for all $i \in I$. By applying $\rho^{U_i}_{U_i\cap U_j}$ to both sides of the equation $\varphi_{U_i}(s_i) = (\rho_{\mathcal{G}})^U_{U_i}(t)$ and $\rho^{U_j}_{U_i\cap U_j}$ to both sides of the equation $\varphi_{U_j}(s_j) = (\rho_{\mathcal{G}})^U_{U_j}(t)$ and using the fact that

$$(\rho_{\mathcal{G}})^{U_i}_{U_i\cap U_j}(\varphi_{U_i}(s_i)) = \varphi_{U_i\cap U_j}((\rho_{\mathcal{F}})^{U_i}_{U_i\cap U_j}(s_i))$$
$$(\rho_{\mathcal{G}})^{U_j}_{U_i\cap U_j}(\varphi_{U_j}(s_j)) = \varphi_{U_i\cap U_j}((\rho_{\mathcal{F}})^{U_j}_{U_i\cap U_j}(s_j))$$

as shown by the commutativity of the diagrams

$$\begin{array}{ccc} \mathcal{F}(U_i) & \xrightarrow{\varphi_{U_i}} & \mathcal{G}(U_i) \\ \downarrow (\rho_{\mathcal{F}})^{U_i}_{U_i\cap U_j} & & \downarrow (\rho_{\mathcal{G}})^{U_i}_{U_i\cap U_j} \\ \mathcal{F}(U_i\cap U_j) & \xrightarrow[\varphi_{U_i\cap U_j}]{} & \mathcal{G}(U_i\cap U_j) \end{array} \qquad \begin{array}{ccc} \mathcal{F}(U_j) & \xrightarrow{\varphi_{U_j}} & \mathcal{G}(U_j) \\ \downarrow (\rho_{\mathcal{F}})^{U_j}_{U_i\cap U_j} & & \downarrow (\rho_{\mathcal{G}})^{U_j}_{U_i\cap U_j} \\ \mathcal{F}(U_i\cap U_j) & \xrightarrow[\varphi_{U_i\cap U_j}]{} & \mathcal{G}(U_i\cap U_j), \end{array}$$

we get

$$\varphi_{U_i\cap U_j}((\rho_{\mathcal{F}})^{U_i}_{U_i\cap U_j}(s_i)) = \varphi_{U_i\cap U_j}((\rho_{\mathcal{F}})^{U_j}_{U_i\cap U_j}(s_j)) = (\rho_{\mathcal{G}})^{U}_{U_i\cap U_j}(t),$$

and since $\varphi_{U_i\cap U_j}$ is injective, we conclude that

$$(\rho_{\mathcal{F}})^{U_i}_{U_i\cap U_j}(s_i) = (\rho_{\mathcal{F}})^{U_j}_{U_i\cap U_j}(s_j)$$

for all i, j. Since $\mathcal{F}$ is a sheaf, by Condition (G), there is some $s \in \mathcal{F}(U)$ such that $(\rho_{\mathcal{F}})^U_{U_i}(s) = s_i$ for all i. We claim that $\varphi_U(s) = t$. For this, observe that

$$(\rho_{\mathcal{G}})^U_{U_i}(\varphi_U(s)) = \varphi_{U_i}((\rho_{\mathcal{F}})^U_{U_i}(s)) = \varphi_{U_i}(s_i) = (\rho_{\mathcal{G}})^U_{U_i}(t)$$

for all i, and since $\mathcal{G}$ is a sheaf, by Condition (M) we get

$$\varphi_U(s) = t,$$

as claimed. Therefore, φ_U is surjective. ☐

We also have the following result that we state without proof. The proof consists in unwinding the definitions; see Tennison [Tennison (1975)] (Chapter III, Proposition 4.11).

Proposition 10.21. *Let $\varphi\colon \mathcal{F} \to \mathcal{G}$ be a map of presheaves on a space X. Then*

$$(\text{PCoker}\,\varphi)_x = \text{Coker}\,\varphi_x = \mathcal{G}_x/\text{Im}\,\varphi_x$$

for all $x \in X$. If $\mathcal{F}$ and $\mathcal{G}$ are sheaves, then

$$(\text{SCoker}\,\varphi)_x = \text{Coker}\,\varphi_x$$

for all $x \in X$.

In general, if $\varphi\colon \mathcal{F} \to \mathcal{G}$ is a presheaf morphism, even if φ is surjective and $\mathcal{F}$ is a sheaf $\mathcal{G}$ need not be a sheaf. However, it is under the following conditions.

Proposition 10.22. *Let $\mathcal{F}$ be a sheaf and $\mathcal{G}$ be a presheaf. If $\varphi\colon \mathcal{F} \to \mathcal{G}$ is a presheaf isomorphism, then $\mathcal{G}$ is a sheaf.*

Proof. Let $\psi\colon \mathcal{G} \to \mathcal{F}$ be the inverse of φ. For any open subset U of X and any open cover $(U_i)_{i\in I}$ of U, let $s, t \in \mathcal{G}(U)$ be such that $(\rho_{\mathcal{G}})^U_{U_i}(s) = (\rho_{\mathcal{G}})^U_{U_i}(t)$ for all i. Since ψ is a presheaf map, the commutativity of the diagram

$$\begin{array}{ccc} \mathcal{G}(U) & \xrightarrow{\psi_U} & \mathcal{F}(U) \\ {\scriptstyle (\rho_{\mathcal{G}})^U_{U_i}}\downarrow & & \downarrow{\scriptstyle (\rho_{\mathcal{F}})^U_{U_i}} \\ \mathcal{G}(U_i) & \xrightarrow[\psi_{U_i}]{} & \mathcal{F}(U_i) \end{array}$$

yields

$$\begin{aligned} \psi_{U_i}((\rho_{\mathcal{G}})^U_{U_i}(s)) &= (\rho_{\mathcal{F}})^U_{U_i}(\psi_U(s)) \\ \psi_{U_i}((\rho_{\mathcal{G}})^U_{U_i}(t)) &= (\rho_{\mathcal{F}})^U_{U_i}(\psi_U(t)), \end{aligned}$$

and since $(\rho_{\mathcal{G}})^U_{U_i}(s) = (\rho_{\mathcal{G}})^U_{U_i}(t)$, we get

$$(\rho_{\mathcal{F}})^U_{U_i}(\psi_U(s)) = (\rho_{\mathcal{F}})^U_{U_i}(\psi_U(t))$$

for all i. Since $\mathcal{F}$ is a sheaf, by Condition (M), we must have $\psi_U(s) = \psi_U(t)$. Since ψ_U is injective, $s = t$; that is, $\mathcal{G}$ satisfies Condition (M).

Next let $(t_i)_{\in I}$ be a family with $t_i \in \mathcal{G}(U_i)$ such that $(\rho_{\mathcal{G}})^{U_i}_{U_i\cap U_j}(t_i) = (\rho_{\mathcal{G}})^{U_j}_{U_i\cap U_j}(t_j)$ for all i, j. Since ψ is a presheaf map, the commutativity of the diagrams

$$\begin{array}{ccc} \mathcal{G}(U_i) & \xrightarrow{\psi_{U_i}} & \mathcal{F}(U_i) \\ {\scriptstyle (\rho_{\mathcal{G}})^{U_i}_{U_i\cap U_j}}\downarrow & & \downarrow{\scriptstyle (\rho_{\mathcal{F}})^{U_i}_{U_i\cap U_j}} \\ \mathcal{G}(U_i\cap U_j) & \xrightarrow[\psi_{U_i\cap U_j}]{} & \mathcal{F}(U_i\cap U_j) \end{array} \qquad \begin{array}{ccc} \mathcal{G}(U_j) & \xrightarrow{\psi_{U_j}} & \mathcal{F}(U_j) \\ {\scriptstyle (\rho_{\mathcal{G}})^{U_j}_{U_i\cap U_j}}\downarrow & & \downarrow{\scriptstyle (\rho_{\mathcal{F}})^{U_j}_{U_i\cap U_j}} \\ \mathcal{G}(U_i\cap U_j) & \xrightarrow[\psi_{U_i\cap U_j}]{} & \mathcal{F}(U_i\cap U_j) \end{array}$$

yields

$$\begin{aligned} \psi_{U_i\cap U_j}((\rho_{\mathcal{G}})^{U_i}_{U_i\cap U_j}(t_i)) &= (\rho_{\mathcal{F}})^{U_i}_{U_i\cap U_j}(\psi_{U_i}(t_i)) \\ \psi_{U_i\cap U_j}((\rho_{\mathcal{G}})^{U_j}_{U_i\cap U_j}(t_j)) &= (\rho_{\mathcal{F}})^{U_j}_{U_i\cap U_j}(\psi_{U_j}(t_j)), \end{aligned}$$

and since $(\rho_\mathcal{G})^{U_i}_{U_i \cap U_j}(t_i) = (\rho_\mathcal{G})^{U_j}_{U_i \cap U_j}(t_j)$, we get

$$(\rho_\mathcal{F})^{U_i}_{U_i \cap U_j}(\psi_{U_i}(t_i)) = (\rho_\mathcal{F})^{U_j}_{U_i \cap U_j}(\psi_{U_j}(t_j))$$

for all i, j. Since $\mathcal{F}$ is a sheaf, by Condition (G), there is some $s \in \mathcal{F}(U)$ such that

$$(\rho_\mathcal{F})^U_{U_i}(s) = \psi_{U_i}(t_i)$$

for all $i \in I$. Now since φ_{U_i} and ψ_{U_i} are mutual inverses, we get

$$(\rho_\mathcal{G})^U_{U_i}(\varphi_U(s)) = \varphi_{U_i}((\rho_\mathcal{F})^U_{U_i}(s)) = \varphi_{U_i}(\psi_{U_i}(t_i)) = t_i$$

for all $i \in I$, which shows that Condition (G) holds with $\varphi_U(s) \in \mathcal{G}(U)$. Therefore, $\mathcal{G}$ is a sheaf. □

Remark: The notions of image and quotient of a map of stalk spaces do not present the difficulties encountered with sheaves. If $\varphi\colon (E_1, p_1) \to (E_2, p_2)$ is a map of stalk spaces, because φ is a local homeomorphism (see Proposition 10.7(c)), the subspace $\varphi(E_1)$ is open in E_2, and so it is a substalk space of (E_2, p_2). Similarly, if (E_1, p_1) is a substalk space of (E_2, p_2), then for every $x \in X$ we can form the quotient $H_x = p_2^{-1}(x)/p_1^{-1}(x)$ and make the disjoint union of the H_x into a stalk space by giving it the quotient topology of the topology of E_2. This is what Serre does in FAC [Serre (1955)] (Chapter 1, Section 7.1).

10.9 Exact Sequences of Presheaves and Sheaves

The key to the "correct" definition of an exact sequence of sheaves is the appropriate notion of image of a sheaf morphism.

Definition 10.21. If $\varphi\colon \mathcal{F} \to \mathcal{G}$ is map of presheaves on a space X, then the *(presheaf) image* of φ, denoted $\mathrm{PIm}\,\varphi$, is the kernel $\mathrm{Ker}\,\mathrm{pcoker}(\varphi)$ of the cokernel map $\mathrm{pcoker}(\varphi)\colon \mathcal{G} \to \mathrm{PCoker}(\varphi)$ (with $\mathrm{PCoker}_U = \mathcal{G}(U)/\varphi_U(\mathcal{F}(U))$). If $\varphi\colon \mathcal{F} \to \mathcal{G}$ is map of sheaves on a space X, then the *(sheaf) image* of φ, denoted $\mathrm{SIm}\,\varphi$, is the kernel $\mathrm{Ker}\,\mathrm{scoker}(\varphi)$ of the cokernel map $\mathrm{scoker}(\varphi)\colon \mathcal{G} \to \mathrm{SCoker}(\varphi)$.

It is not hard to check that if $\varphi\colon \mathcal{F} \to \mathcal{G}$ is a map of presheaves, then $(\mathrm{PIm}\varphi)(U) = \mathrm{Im}\varphi_U$, while if $\varphi\colon \mathcal{F} \to \mathcal{G}$ is map of sheaves, then $(\mathrm{SIm}\varphi)_x = \mathrm{Im}\,\varphi_x$ for all $x \in X$.

Remark: The image $\operatorname{Im}\varphi$ of a map of sheaves $\varphi\colon \mathcal{F} \to \mathcal{G}$ is often defined as the sheafification $\widetilde{\mathrm{PIm}\,\varphi}$ of the presheaf $\mathrm{PIm}\,\varphi$. The small problem with this approach is that this sheaf is not a subsheaf of $\mathcal{G}$. There is an injective morphism from $\widetilde{\mathrm{PIm}\,\varphi}$ into $\mathcal{G}$ so the image of φ should really be the image of $\widetilde{\mathrm{PIm}\,\varphi}$ by that morphism. It seems to us that using $\mathrm{SIm}\,\varphi$ for the image of φ is a cleaner approach (which agrees with the definition of image in an abelian category).

If $\varphi\colon \mathcal{F} \to \mathcal{G}$ is map of sheaves and $\mathrm{PIm}\varphi$ is a sheaf, then $\mathrm{SIm}\varphi = \mathrm{PIm}\varphi$. Indeed, both are subsheaves of $\mathcal{G}$ and their stalks are equal to $\operatorname{Im}\varphi_x$ for all x, so by Proposition 10.17 they are equal. As a consequence, we obtain the following result.

Proposition 10.23. *If $\varphi\colon \mathcal{F} \to \mathcal{G}$ is an injective map of sheaves, then* $\mathrm{SIm}\,\varphi = \mathrm{PIm}\,\varphi$.

Proof. Indeed, since φ is injective there is a presheaf isomorphism from $\mathcal{F}$ to $\mathrm{PIm}\,\varphi$, and by Proposition 10.22 we conclude that $\mathrm{PIm}\,\varphi$ is sheaf, so by the fact stated just before this proposition $\mathrm{SIm}\,\varphi = \mathrm{PIm}\,\varphi$. ☐

Definition 10.22. Let

$$\cdots \longrightarrow \mathcal{F} \xrightarrow{\varphi} \mathcal{G} \xrightarrow{\psi} \mathcal{H} \longrightarrow \cdots$$

be a sequence of maps of preshaves (over a space X). We say that the sequence is *exact at $\mathcal{G}$ as a sequence of presheaves* if

$$\mathrm{PIm}\,\varphi = \operatorname{Ker}\psi.$$

We say that it is an *exact sequence of presheaves* if it is exact at each point where it makes sense.

If the sequence consists of sheaves, then we say that it is *exact at $\mathcal{G}$ as a sequence of sheaves* if

$$\mathrm{SIm}\,\varphi = \operatorname{Ker}\psi.$$

It is an *exact sequence of sheaves* if it is exact at each point where it makes sense.

We have the following result stating more convenient conditions for checking that a sequence is an exact sequence of presheaves or an exact sequence of sheaves.

Proposition 10.24. *The following facts hold:*

(i) *If the sequence*

$$\mathcal{F} \xrightarrow{\varphi} \mathcal{G} \xrightarrow{\psi} \mathcal{H}$$

is an exact sequence of presheaves, then for every open subset U of X

$$\mathcal{F}(U) \xrightarrow{\varphi_U} \mathcal{G}(U) \xrightarrow{\psi_U} \mathcal{H}(U)$$

is an exact sequence of R-modules (or rings).

(ii) *The sequence*

$$\mathcal{F} \xrightarrow{\varphi} \mathcal{G} \xrightarrow{\psi} \mathcal{H}$$

is an exact sequence of sheaves iff the sequence

$$\mathcal{F}_x \xrightarrow{\varphi_x} \mathcal{G}_x \xrightarrow{\psi_x} \mathcal{H}_x$$

is an exact sequence of R-modules (or rings) for all $x \in X$.

(iii) *If the sequence of sheaves*

$$\mathcal{F} \xrightarrow{\varphi} \mathcal{G} \xrightarrow{\psi} \mathcal{H}$$

is exact as a sequence of presheaves, then it is exact as a sequence of sheaves.

Proof. A complete proof is given in Tennison [Tennison (1975)] (Chapter III, Theorem 6.5). We only give the proof of (ii). By definition, the sequence is exact iff $\mathrm{SIm}\,\varphi = \mathrm{Ker}\,\psi$ iff by Proposition 10.17

$$(\mathrm{SIm}\,\varphi)_x = (\mathrm{Ker}\,\psi)_x$$

for all $x \in X$. But by definition

$$\begin{aligned}
(\mathrm{SIm}\,\varphi)_x &= (\mathrm{Ker}\,(\mathrm{scoker}\,\varphi))_x \\
&= \mathrm{Ker}\,((\mathrm{scoker}\,\varphi)_x \colon \mathcal{G}_x \longrightarrow (\mathrm{SCoker}\,\varphi)_x) \quad \text{by Proposition 10.15} \\
&= \mathrm{Ker}\,((\mathrm{scoker}\,\varphi)_x \colon \mathcal{G}_x \longrightarrow (\mathcal{G}_x/\mathrm{Im}\,\varphi_x)) \quad \text{by Proposition 10.21} \\
&= \mathrm{Im}\,\varphi_x.
\end{aligned}$$

Therefore, $\mathrm{SIm}\,\varphi = \mathrm{Ker}\,\psi$ iff (by Proposition 10.15) $\mathrm{Im}\,\varphi_x = (\mathrm{Ker}\,\psi)_x = \mathrm{Ker}\,\psi_x$, as claimed. ☐

As a corollary of Proposition 10.24, we have the following result.

Proposition 10.25. *The following facts hold as sequences of preseaves or sheaves.*

(*i*) *The sequence*

$$0 \longrightarrow \mathcal{F} \stackrel{\varphi}{\longrightarrow} \mathcal{G}$$

is exact iff φ *is injective (a monic); see Proposition 10.14 and Definition 10.16.*

(*ii*) *The sequence*

$$\mathcal{F} \stackrel{\varphi}{\longrightarrow} \mathcal{G} \longrightarrow 0$$

is exact iff φ *is surjective (an epic; see Proposition 10.18, Proposition 10.19, and Definition 10.20.*

(*iii*) *For any map* $\varphi\colon \mathcal{F} \to \mathcal{G}$ *of preshaves the sequence*

$$0 \longrightarrow \mathrm{Ker}\,\varphi \longrightarrow \mathcal{F} \stackrel{\varphi}{\longrightarrow} \mathcal{G} \longrightarrow \mathrm{PCoker}\,\varphi \longrightarrow 0$$

is exact, and for any map $\varphi\colon \mathcal{F} \to \mathcal{G}$ *of sheaves the sequence*

$$0 \longrightarrow \mathrm{Ker}\,\varphi \longrightarrow \mathcal{F} \stackrel{\varphi}{\longrightarrow} \mathcal{G} \longrightarrow \mathrm{SCoker}\,\varphi \longrightarrow 0$$

is exact.

10.10 Categories, Functors, Additive Categories

We now want to discuss the preservation of exactness by various operations (functors). Some examples of these operations are:

(1) The inclusion map $i\colon \mathbf{Sh}(X) \to \mathbf{PSh}(X)$ which maps a sheaf to the corresponding presheaf, and a morphism $\varphi\colon \mathcal{F} \to \mathcal{G}$ to the corresponding presheaf morphism.

(2) The sheafification operation $\Gamma S\colon \mathbf{PSh}(X) \to \mathbf{Sh}(X)$ which maps a presheaf $\mathcal{F}$ to its sheafification $\widetilde{\mathcal{F}}$, and a map of preshaves $\varphi\colon \mathcal{F} \to \mathcal{G}$ to the map of sheaves $\widetilde{\varphi}\colon \widetilde{\mathcal{F}} \to \widetilde{\mathcal{G}}$ (see Proposition 10.8).

(3) For every open subset U of X, for every presheaf $\mathcal{F} \in \mathbf{PSh}(X)$, we have the operation $\Gamma(U, -)$, "sections over U," given by

$$\Gamma(U, \mathcal{F}) = \mathcal{F}(U),$$

which yields an R-module (or a ring). Any presheaf morphism $\varphi\colon \mathcal{F} \to \mathcal{G}$ is mapped to the R-module (or ring) homomorphism $\varphi_U\colon \mathcal{F}(U) \to \mathcal{G}(U)$.

(4) For every open subset U of X, for every sheaf $\mathcal{F} \in \mathbf{Sh}(X)$, we have the operation $\Gamma(U, -)$, "sections over U," given by

$$\Gamma(U, \mathcal{F}) = \mathcal{F}(U),$$

which yields an R-module (or a ring). Any sheaf morphism $\varphi\colon \mathcal{F} \to \mathcal{G}$ is mapped to the R-module (or ring) homomorphism $\varphi_U\colon \mathcal{F}(U) \to \mathcal{G}(U)$. This functor is crucial in sheaf cohomology.

All the concepts we have discussed so far, R-modules, (commutative) rings, abelian groups, presheaves, sheaves, share a common abstract structure, that of a category. We have used the term category informally many times, and we finally define it precisely.

Definition 10.23. A *category* $\mathbf{C}$ consists of

(1) A class $\mathrm{Ob}_{\mathbf{C}}$ of *objects*.
(2) A family $\mathrm{Ar}_{\mathbf{C}}$ of pairwise disjoint sets $\mathrm{Hom}_{\mathbf{C}}(A, B)$ of elements called *morphisms* (or *arrows*), for any pair (A, B) of objects $A, B \in \mathrm{Ob}_{\mathbf{C}}$. Each set $\mathrm{Hom}_{\mathbf{C}}(A, B)$ is called a *Hom-set*. To simplify notation, a morphism $f \in \mathrm{Hom}_{\mathbf{C}}(A, B)$ is also denoted by $A \xrightarrow{f} B$ or $f \colon A \to B$. The object A is called the *domain* of f and the object B is called the *range* (or *codomain*) of f. A morphism $f \colon A \to B$ is also called a *map*.
(3) For any triple of objects $A, B, C \in \mathrm{Ob}_{\mathbf{C}}$, an operation
$$\circ_{A,B,C} \colon \mathrm{Hom}_{\mathbf{C}}(B, C) \times \mathrm{Hom}_{\mathbf{C}}(A, B) \to \mathrm{Hom}_{\mathbf{C}}(A, C)$$
called *composition*, which assigns a morphism $g \circ_{A,B,C} f$ to any pair of morphisms $f \in \mathrm{Hom}_{\mathbf{C}}(A, B)$ and $g \in \mathrm{Hom}_{\mathbf{C}}(B, C)$.
(4) A function which assigns to each object $A \in \mathrm{Ob}_{\mathbf{C}}$ a morphism $\mathrm{id}_A \in \mathrm{Hom}_{\mathbf{C}}(A, A)$.

The above data satisfies the following axioms:

(i) (*Associativity*) For all objects $A, B, C, D \in \mathrm{Ob}_{\mathbf{C}}$, for all morphisms $f \colon A \to B$, $g \colon B \to C$, $h \colon C \to D$,
$$h \circ_{A,C,D} (g \circ_{A,B,C} f) = (h \circ_{B,C,D} g) \circ_{A,B,D} f.$$
(ii) (*Identity*) For any two objects $A, B \in \mathrm{Ob}_{\mathbf{C}}$, for any morphism $f \colon A \to B$,
$$f \circ_{A,A,B} \mathrm{id}_A = f = \mathrm{id}_B \circ_{A,B,B} f.$$

Informally, we can think of a category as a graph with vertices $A \in \mathrm{Ob}_{\mathbf{C}}$ and all morphisms $f \in \mathrm{Hom}_{\mathbf{C}}(A, B)$ as "parallel" edges between A and B. There is also a way of composing the edges which makes a category into a kind of generalized monoid. Since the amount of notation is quite formidable, we often abuse it. For example, we drop the subscripts in the composition operations $\circ_{A,B,C}$ and simply write $\circ$. We also write $A \in \mathbf{C}$ instead of $A \in \mathrm{Ob}_{\mathbf{C}}$ and $\mathrm{Hom}(A, B)$ instead of $\mathrm{Hom}_{\mathbf{C}}(A, B)$.

The notion of isomorphism in a category is the obvious one.

Definition 10.24. Given a category $\mathbf{C}$, a morphism $\alpha\colon A \to B$ is an *isomorphism* (some authors say an *equivalence*) if there is a morphism $\beta\colon B \to A$ such that $\beta \circ \alpha = \mathrm{id}_A$ and $\alpha \circ \beta = \mathrm{id}_B$.

The common thread between the previous examples is that we have two categories $\mathbf{C}$ and $\mathbf{D}$, and we have a transformation T (a functor) which works as follows:

(i) Each object A of $\mathbf{C}$ is mapped to some object $T(A)$ of $\mathbf{D}$.

(ii) Each map $A \xrightarrow{f} B$ between two objects A and B in $\mathbf{C}$ is mapped to some map $T(A) \xrightarrow{T(f)} T(B)$ between the objects $T(A)$ and $T(B)$ in $\mathbf{D}$ in such a way that the following properties hold:

(a) Given two maps $A \xrightarrow{f} B$ and $B \xrightarrow{g} C$ between objects A, B, C in $\mathbf{C}$ such that the composition $A \xrightarrow{g\circ f} C \;=\; A \xrightarrow{f} B \xrightarrow{g} C$ makes sense, the composition $T(A) \xrightarrow{T(f)} T(B) \xrightarrow{T(g)} T(C)$ makes sense in $\mathbf{D}$, and

$$T(g \circ f) = T(g) \circ T(f).$$

(b) If $A \xrightarrow{\mathrm{id}_A} A$ is the identity map of the object A in $\mathbf{C}$, then $T(A) \xrightarrow{T(\mathrm{id}_A)} T(A)$ is the identity map of $T(A)$ in $\mathbf{D}$; that is,

$$T(\mathrm{id}_A) = \mathrm{id}_{T(A)}.$$

Definition 10.25. Whenever a transformation $T\colon \mathbf{C} \to \mathbf{D}$ satisfies the Properties (i), (ii) (a), (b), we call it a *(covariant) functor* from $\mathbf{C}$ to $\mathbf{D}$.

If $T\colon \mathbf{C} \to \mathbf{D}$ satisfies Properties (i), (b), and if Properties (ii) and (a) are replaced by the Properties (ii') and (a') below

(ii') Each map $A \xrightarrow{f} B$ between two objects A and B in $\mathbf{C}$ is mapped to some map $T(B) \xrightarrow{T(f)} T(A)$ between the objects $T(B)$ and $T(A)$ in $\mathbf{D}$ in such a way that the following properties hold:

(a') Given two maps $A \xrightarrow{f} B$ and $B \xrightarrow{g} C$ between objects A, B, C in $\mathbf{C}$ such that the composition $A \xrightarrow{g\circ f} C \;=\; A \xrightarrow{f} B \xrightarrow{g} C$

makes sense, the composition $T(C) \xrightarrow{T(g)} T(B) \xrightarrow{T(f)} T(A)$ makes sense in $\mathbf{D}$, and

$$T(g \circ f) = T(f) \circ T(g),$$

then T is called a *contravariant functor*.

Definition 10.26. Whenever a transformation $T \colon \mathbf{C} \to \mathbf{D}$ satisfies the Properties (i), (ii') (a'), (b), we call it a *contravariant functor* from $\mathbf{C}$ to $\mathbf{D}$.

Example 10.1. The four functors defined at the beginning of this section are covariant functors. Another example of a covariant functor is the functor $\mathrm{Hom}(A, -)$ (for a fixed R-module A) from the category of R-modules to itself (the category of abelian groups if R is not commutative) which maps a module B to the module $\mathrm{Hom}(A, B)$ and a module homomorphism $f \colon B \to C$ to the module homomorphism $\mathrm{Hom}(A, f)$ from $\mathrm{Hom}(A, B)$ to $\mathrm{Hom}(A, C)$ given by

$$\mathrm{Hom}(A, f)(\varphi) = f \circ \varphi \quad \text{for all } \varphi \in \mathrm{Hom}(A, B);$$

see Section 2.4.

The tensor product $- \otimes_R B$ is another example of covariant functor; see Example 10.5.

Example 10.2. An example of a contravariant functor is the functor $\mathrm{Hom}(-, A)$ (for a fixed R-module A) from the category of R-modules to itself (the category of abelian groups if R is not commutative) which maps a module B to the module $\mathrm{Hom}(B, A)$ and a module homomorphism $f \colon B \to C$ to the module homomorphism $\mathrm{Hom}(f, A)$ from $\mathrm{Hom}(C, A)$ to $\mathrm{Hom}(B, A)$ given by

$$\mathrm{Hom}(f, A)(\varphi) = \varphi \circ f \quad \text{for all } \varphi \in \mathrm{Hom}(C, A);$$

see Section 2.4.

Let us not forget that our main goal is to generalize the notion of exact sequence to structures more general than R-modules. This means generalizing notions such as

(1) Injectivity and surjectivity.
(2) Kernels of maps.
(3) Images of maps.

It is also desirable to define quotient objects and direct sums. Since in a category objects may not possess members, we have to define the above concepts in terms of maps. For injectivity and surjectivity, this is achieved as follows.

Definition 10.27. Given any category $\mathbf{C}$, a map $\varphi\colon A \to B$ is a *monic* if for any two maps $\psi_1, \psi_2\colon C \to A$, if $\varphi \circ \psi_1 = \varphi \circ \psi_2$, then $\psi_1 = \psi_2$.[4] A map $\varphi\colon A \to B$ is an *epic* if for any two maps $\psi_1, \psi_2\colon B \to C$, if $\psi_1 \circ \varphi = \psi_2 \circ \varphi$, then $\psi_1 = \psi_2$.[5]

We leave it as an exercise to check that for sets and functions, for R-modules and R-linear maps, and for commutative rings and ring homomorphisms, a map is injective iff it is monic and a map is surjective iff it is epic. However, in the category of sheaves, epic is not equivalent to surjective.

In order to define kernels and images, we need to impose more structure on our category. In particular, we need the notion of a zero. To achieve this we give the Hom-sets $\mathrm{Hom}_{\mathbf{C}}(A, B)$ the additional structure of an abelian group. This way, we have a zero map $0_{A,B}$ between any two objects A and B, and maps $f \in \mathrm{Hom}_{\mathbf{C}}(A, B)$ can be added or subtracted, as if they were linear maps.

Given a category $\mathbf{C}$, recall that for notational convenience we use the notations $f \in \mathrm{Hom}_{\mathbf{C}}(A, B)$ and $f\colon A \to B$ interchangeably.

Definition 10.28. A category $\mathbf{C}$ is an **Ab**-*category* (or a *pre-additive category*) if for all $A, B \in \mathbf{C}$ the set of maps $\mathrm{Hom}_{\mathbf{C}}(A, B)$ is an abelian group (with addition operation $+_{A,B}$ and a zero map $0_{A,B}$), and if the following distributivity axioms hold: for all $A, B, C, D \in \mathbf{C}$, for all maps $f \in \mathrm{Hom}_{\mathbf{C}}(A, B)$, $g_1, g_2 \in \mathrm{Hom}_{\mathbf{C}}(B, C)$, and $h \in \mathrm{Hom}_{\mathbf{C}}(C, D)$,

$$
\begin{aligned}
h \circ (g_1 + g_2) &= h \circ g_1 + h \circ g_2 \\
(g_1 + g_2) \circ f &= g_1 \circ f + g_2 \circ f.
\end{aligned}
$$

If $\mathbf{C}$ and $\mathbf{D}$ are two **Ab**-categories, a functor $T\colon \mathbf{C} \to \mathbf{D}$ is *additive* if for all $A, B \in \mathbf{C}$ and all $f, g \in \mathrm{Hom}_{\mathbf{C}}(A, B)$,

$$T(f + g) = T(f) + T(g).$$

[4] Some authors use the terminology *monomorphism*. However, the term monomorphism refers to an injective homomorphism. The notion of monic is more general. In the category of R-modules or commutative rings, the notions of monomorphism and monic are equivalent.

[5] Some authors use the terminology *epimorphism*. However, the term epimorphism refers to a surjective homomorphism. The notion of epic is more general. In the category of R-modules or commutative rings, the notions of epimorphism and epic are equivalent.

Observe that if T is an additive functor, then $T(0_{A,B}) = 0_{T(A),T(B)}$.

Proposition 10.26. *In an* **Ab**-*category we have* $0_{B,C} \circ f = 0_{A,C}$ *for all* $f \colon A \to B$ *and* $g \circ 0_{A,B} = 0_{A,C}$ *for all* $g \colon B \to C$.

Proof. This is because by distributivity,

$$0_{B,C} \circ f = (0_{B,C} + 0_{B,C}) \circ f = 0_{B,C} \circ f + 0_{B,C} \circ f,$$

and similarly,

$$g \circ 0_{A,B} = g \circ (0_{A,B} + 0_{A,B}) = g \circ 0_{A,B} + g \circ 0_{A,B}. \qquad \square$$

For simplicity of notation we usually drop the subscript A, B in $+_{A,B}$ and $0_{A,B}$.

Example 10.3. The category of R-modules is an **Ab**-category. The category of sheaves (or presheaves) of R-modules or rings is also an **Ab**-category. The functors $\mathrm{Hom}_R(A, -)$, $\mathrm{Hom}_R(-, A)$, $- \otimes B$, and $\Gamma(U, -)$ are additive.

As in the case of R-modules, in an **Ab**-category, there is a nicer way to characterize monics and epics.

Proposition 10.27. *In an* **Ab**-*category, a map* $\varphi \colon A \to B$ *is a monic iff for every* $\psi \colon C \to A$, $\varphi \circ \psi = 0$ *implies that* $\psi = 0$. *Similarly, a map* $\varphi \colon A \to B$ *is an epic iff for every* $\psi \colon B \to C$, $\psi \circ \varphi = 0$ *implies that* $\psi = 0$.

Proof. Indeed, if φ is monic, since by Proposition 10.26, $\varphi \circ 0_{C,A} = 0_{C,B}$, if $\varphi \circ \psi = 0_{C,B}$, then $\psi = 0_{C,A}$.

Conversely, if $\varphi \circ \psi_1 = \varphi \circ \psi_2$, then $\varphi \circ (\psi_2 - \psi_1) = 0_{C,B}$, and by hypothesis we must have $\psi_2 - \psi_1 = 0_{C,A}$, that is, $\psi_1 = \psi_2$, and φ is monic.

The statement about epic maps is left as an exercise. $\square$

The question of determining when a zero map $0_{A,B}$ is monic or epic arises naturally.

Proposition 10.28. *Let* $\mathbf{C}$ *be an* **Ab**-*category. If the zero map* $0_{A,B} \colon A \to B$ *is monic, then* $\mathrm{Hom}(A, A) = \{0_{A,A}\}$ *and* $\mathrm{id}_A = 0_{A,A}$. *If the zero map* $0_{A,B} \colon A \to B$ *is epic, then* $\mathrm{Hom}(B, B) = \{0_{B,B}\}$ *and* $\mathrm{id}_B = 0_{B,B}$.

Proof. If the zero map $0_{A,B} \colon A \to B$ is monic, since by Proposition 10.26, $0_{A,B} \circ \psi = 0_{C,B}$ for any $\psi \colon C \to A$, for $C = A$ and $\psi = \mathrm{id}_A$, we should have $\mathrm{id}_A = 0_{A,A}$, which implies that $\mathrm{Hom}(A, A) = \{0_{A,A}\}$.

The proof that if the zero map $0_{A,B} \colon A \to B$ is epic, then $\mathrm{Hom}(B, B) = \{0_{B,B}\}$ and $\mathrm{id}_B = 0_{B,B}$ is dual to the previous proof. $\square$

An object such that $\mathrm{Hom}(A, A) = \{0_{A,A}\}$ is called a zero object. As we just saw, in order to deal with monics and epics as we would in the case of R-modules, it is desirable to assume their existence. Although we will not use them in this section, direct sums are also desirable. This suggests the following definition.

Definition 10.29. Let $\mathbf{C}$ be an $\mathbf{Ab}$-category. An object $A \in \mathbf{C}$ is called a *zero object* if $\mathrm{Hom}(A, A) = \{0_{A,A} = \mathrm{id}_A\}$, a one-element group. An $\mathbf{Ab}$-category $\mathbf{C}$ is an *additive category* if there is a zero object $\mathbf{0}$ in $\mathbf{C}$ and if the notion of *direct sum* makes sense for any two objects $A, B \in \mathbf{C}$. More precisely, this means that for any two objects $A_1, A_2 \in \mathbf{C}$, there is an object $A_1 \oplus A_2 \in \mathbf{C}$ and four morphisms $i_1 \colon A_1 \to A_1 \oplus A_2$, $i_2 \colon A_2 \to A_1 \oplus A_2$, $\pi_1 \colon A_1 \oplus A_2 \to A_1$, and $\pi_2 \colon A_1 \oplus A_2 \to A_2$ as in the following diagram

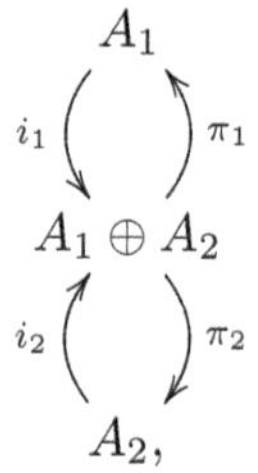

$$\begin{array}{c} A_1 \\ {\scriptstyle i_1}\big\downarrow\big\uparrow{\scriptstyle \pi_1} \\ A_1 \oplus A_2 \\ {\scriptstyle i_2}\big\uparrow\big\downarrow{\scriptstyle \pi_2} \\ A_2, \end{array}$$

such that

$$\pi_1 \circ i_1 = \mathrm{id}_{A_1} \qquad \pi_2 \circ i_2 = \mathrm{id}_{A_2} \qquad i_1 \circ \pi_1 + i_2 \circ \pi_2 = \mathrm{id}_{A_1 \oplus A_2}.$$

For any object $A \in \mathbf{C}$ and any zero object $\mathbf{0}$, since

$$f = \mathrm{id}_{\mathbf{0}} \circ f = 0_{\mathbf{0},\mathbf{0}} \circ f = 0_{A,\mathbf{0}}$$

for all $f \colon A \to \mathbf{0}$, we deduce that $\mathrm{Hom}(A, \mathbf{0}) = \{0_{A,\mathbf{0}}\}$, a one-element group. Similarly, for any object $B \in \mathbf{C}$, since

$$g = g \circ \mathrm{id}_{\mathbf{0}} = g \circ 0_{\mathbf{0},\mathbf{0}} = 0_{\mathbf{0},B}$$

for all $g \colon \mathbf{0} \to B$, we deduce that $\mathrm{Hom}(\mathbf{0}, B) = \{0_{\mathbf{0},B}\}$, a one-element group. We record the above facts as the following proposition.

Proposition 10.29. *Let* $\mathbf{C}$ *be an* $\mathbf{Ab}$*-category with a zero object* $\mathbf{0}$*. For any object* $A \in \mathbf{C}$*, we have* $\mathrm{Hom}(A, \mathbf{0}) = \{0_{A,\mathbf{0}}\}$*, and for any object* $B \in \mathbf{C}$*, we have* $\mathrm{Hom}(\mathbf{0}, B) = \{0_{\mathbf{0},B}\}$.

Proposition 10.28 can be sharpened as follows,

Proposition 10.30. *Let* $\mathbf{C}$ *be an* $\mathbf{Ab}$*-category with a zero object. The zero map* $0_{A,B} \colon A \to B$ *is monic iff* A *is a zero object. The zero map* $0_{A,B} \colon A \to B$ *is epic iff* B *is a zero object.*

Proof. One direction of the proposition was proven in Proposition 10.28. For the other direction, if A is a zero object, by Proposition 10.29 we have $\mathrm{Hom}(C, A) = \{0_{C,A}\}$, a one element group, so the zero map $0_{A,B}\colon A \to B$ is monic.

Dually, if B is a zero object, then $\mathrm{Hom}(B, C) = \{0_{B,C}\}$, so the zero map $0_{A,B}\colon A \to B$ is epic. □

Let us now consider direct sums. The equations

$$\pi_1 \circ i_1 = \mathrm{id}_{A_1} \qquad \pi_2 \circ i_2 = \mathrm{id}_{A_2} \qquad i_1 \circ \pi_1 + i_2 \circ \pi_2 = \mathrm{id}_{A_1 \oplus A_2}$$

imply that

$$\begin{aligned}\pi_1 \circ i_2 &= \pi_1 \circ (i_1 \circ \pi_1 + i_2 \circ \pi_2) \circ i_2 \\ &= \pi_1 \circ i_1 \circ \pi_1 \circ i_2 + \pi_1 \circ i_2 \circ \pi_2 \circ i_2 \\ &= \mathrm{id}_{A_1} \circ \pi_1 \circ i_2 + \pi_1 \circ i_2 \circ \mathrm{id}_{A_2} = \pi_1 \circ i_2 + \pi_1 \circ i_2,\end{aligned}$$

so we deduce that $\pi_1 \circ i_2 = 0$. Similarly, we have $\pi_2 \circ i_1 = 0$. The equations $\pi_1 \circ i_1 = \mathrm{id}_{A_1}$ and $\pi_2 \circ i_2 = \mathrm{id}_{A_2}$ imply that i_1, i_2 are monic and π_1, π_2 are epic.

Suppose we have two maps $f\colon A_1 \to C$ and $g\colon A_2 \to C$. Let $h\colon A_1 \oplus A_2 \to C$ be the map defined by

$$h = f \circ \pi_1 + g \circ \pi_2.$$

We have

$$\begin{aligned}h \circ i_1 &= (f \circ \pi_1 + g \circ \pi_2) \circ i_1 = f \circ \pi_1 \circ i_1 + g \circ \pi_2 \circ i_1 = f \\ h \circ i_2 &= (f \circ \pi_1 + g \circ \pi_2) \circ i_2 = f \circ \pi_1 \circ i_2 + g \circ \pi_2 \circ i_2 = g.\end{aligned}$$

If h' is any other map such that $f = h' \circ i_1$ and $g = h' \circ i_2$, then

$$h' = h' \circ (i_1 \circ \pi_1 + i_2 \circ \pi_2) = h' \circ i_1 \circ \pi_1 + h' \circ i_2 \circ \pi_2 = f \circ \pi_1 + g \circ \pi_2 = h.$$

Therefore, we proved that $h = f \circ \pi_1 + g \circ \pi_2$ is the unique map such that $f = h \circ i_1$ and $g = h \circ i_2$, as illustrated in the diagram below

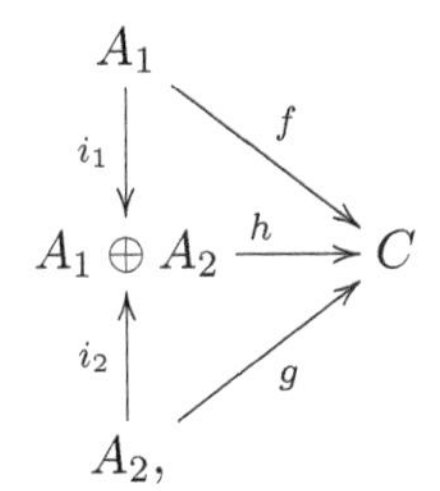

which expresses that $(A_1 \oplus A_2, i_1, i_2)$ is a *coproduct* (in the sense of category theory). We showed that the map $h \mapsto h \circ i_1 + h \circ i_2$ with inverse $f + g \mapsto f \circ \pi_1 + g \circ \pi_2$ is an isomorphism

$$\mathrm{Hom}(A_1 \oplus A_2, C) \cong \mathrm{Hom}(A_1, C) \oplus \mathrm{Hom}(A_2, C).$$

Dually, given any two maps $f\colon C \to A_1$ and $g\colon C \to A_2$, the map $h\colon C \to A_1 \oplus A_2$ defined by

$$h = i_1 \circ f + i_2 \circ g$$

is the unique map such that $f = \pi_1 \circ h$ and $g = \pi_2 \circ h$, as illustrated in the diagram below

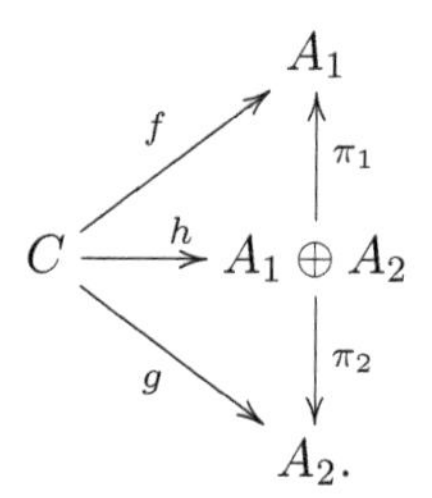

The proof of the above fact is left as an exercise. This fact shows that $(A_1 \oplus A_2, \pi_1, \pi_2)$ is a *product* (in the sense of category theory). The map $h \mapsto \pi_1 \circ h + \pi_2 \circ h$ with inverse $f + g \mapsto i_1 \circ f + i_2 \circ g$ is an isomorphism

$$\mathrm{Hom}(C, A_1 \oplus A_2) \cong \mathrm{Hom}(C, A_1) \oplus \mathrm{Hom}(C, A_2).$$

Arbitrary finite direct sums and finite products are readily defined. The fact that in any additive category (finite) direct sums and products are isomorphic generalizes a well-known fact about R-modules. For more on products and coproducts in additive categories, see Mac Lane [Mac Lane (1975)] (Chapter IX, Section 1).

Having zero maps and zero objects we can define kernels and cokernels. Cokernels will play the role of quotient modules. Also, note that the existence of direct sum is not needed.

Definition 10.30. Let $\mathbf{C}$ be additive category and let $\alpha \in \mathrm{Hom}_{\mathbf{C}}(A, B)$ be any map.

(1) A map $k\colon K \to A$ is a *kernel of* α if k is a monic such that for any map $\varphi\colon C \to A$,

$$\alpha \circ k = 0, \quad \text{and} \quad \alpha \circ \varphi = 0 \text{ implies } \varphi = k \circ \varphi'$$

for some unique map $\varphi' : C \to K$, as illustrated in the following commutative diagram:

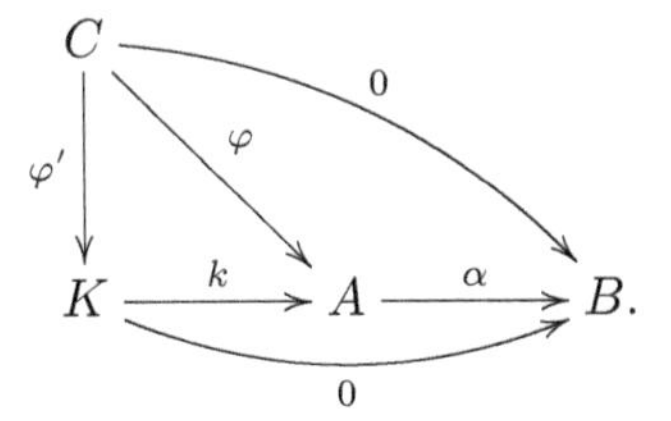

We write $k \in \ker(\alpha)$ or $k \in \ker \alpha$, where $\ker \alpha$ denotes the set of kernels of α.

(2) A map $\sigma : B \to C$ is a *cokernel of* α if σ is an epic such that for any map $\psi : B \to D$,

$$\sigma \circ \alpha = 0, \quad \text{and} \quad \psi \circ \alpha = 0 \text{ implies } \psi = \psi' \circ \sigma$$

for some unique map $\psi' : C \to D$, as illustrated in the following commutative diagram:

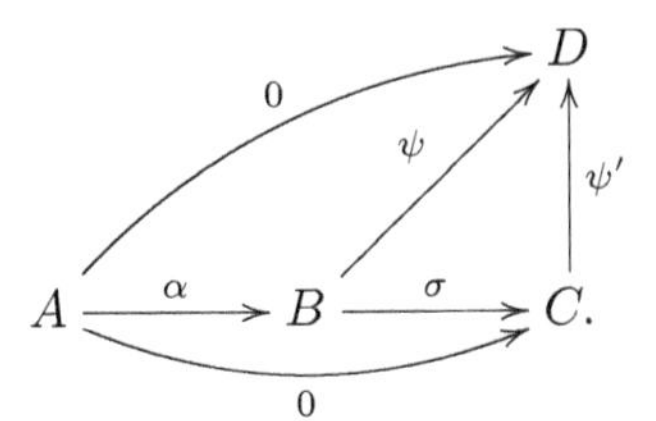

We write $\sigma \in \operatorname{coker}(\alpha)$ or $\sigma \in \operatorname{coker} \alpha$, where $\operatorname{coker} \alpha$ denotes the set of cokernels of α.

We see immediately from the definitions that if $k_1 : K_1 \to A$ and $k_2 : K_2 \to A$ are two kernels for $\alpha : A \to B$, then there are two (unique) maps $k_1' : K_1 \to K_2$ and $k_2' : K_2 \to K_1$ such that $k_2' \circ k_1' = \mathrm{id}_{K_1}$, $k_1' \circ k_2' = \mathrm{id}_{K_2}$, $k_2 = k_1 \circ k_2'$, and $k_1 = k_2 \circ k_1'$.

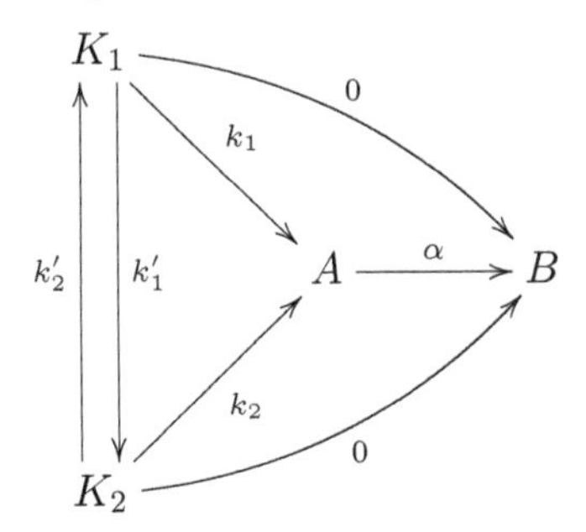

In other words, there is an isomorphism (unique) $k_2'\colon K_2 \to K_1$ such that $k_2 = k_1 \circ k_2'$.

Definition 10.31. Given any category $\mathbf{C}$, the relation defined on maps $\beta_1\colon K_1 \to A$ and $\beta_2\colon K_2 \to A$ by requiring that there is an isomorphism $\beta_2'\colon K_2 \to K_1$ such that $\beta_2 = \beta_1 \circ \beta_2'$ is an equivalence relation on maps with range A called *right equivalence*.

It is immediately verified that right equivalence is indeed an equivalence relation.

If α is a monic, then its right equivalence class consists of monics. This fact shows that any two kernels of α are right equivalent, and that the equivalence class of any kernel of α under right equivalence is ker α.

In particular, if $0_{K,A} \in \ker \alpha$ for some zero object K, then the maps right equivalent to $0_{K,A}$ are all the zero maps with domain a zero object and range A, so ker α is the right equivalence class of zero maps with domain a zero object and range A. By abuse of notation we write ker $\alpha = 0$.

The (right) equivalence class of a monic β with range A is called a *subobject of* A. If $\ker \alpha \neq \emptyset$, then the (right) equivalence class $\ker \alpha$ of all kernels of α is a subobject of A.

Similarly, if $\sigma_1\colon B \to C_1$ and $\sigma_2\colon B \to C_2$ are two cokernels for $\alpha\colon A \to B$, then there are two (unique) maps $\sigma_1'\colon C_2 \to C_1$ and $\sigma_2'\colon C_1 \to C_2$ such that $\sigma_2' \circ \sigma_1' = \mathrm{id}_{C_2}$, $\sigma_1' \circ \sigma_2' = \mathrm{id}_{C_1}$, $\sigma_2 = \sigma_2' \circ \sigma_1$, and $\sigma_1 = \sigma_1' \circ \sigma_2$.

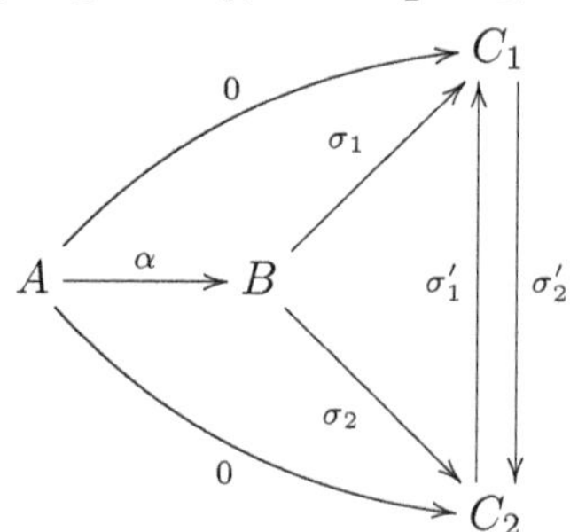

In other words, there is an isomorphism (unique) $\sigma_2'\colon C_1 \to C_2$ such that $\sigma_2 = \sigma_2' \circ \sigma_1$.

Definition 10.32. Given any category $\mathbf{C}$, the relation defined on maps $\beta_1\colon B \to C_1$ and $\beta_2\colon B \to C_2$ by requiring that there is an isomorphism $\beta_2'\colon C_1 \to C_2$ such that $\beta_2 = \beta_2' \circ \beta_1$ is an equivalence relation on maps with domain B called *left equivalence*.

It is immediately verified that left equivalence is indeed an equivalence relation.

If α is an epic, then its left equivalence class consists of epics. This fact shows that any two cokernels of α are left equivalent, and that the equivalence class of any cokernel of α under left equivalence is $\operatorname{coker} \alpha$.

In particular, if $0_{B,C} \in \operatorname{coker} \alpha$ for some zero object C, then the maps left equivalent to $0_{B,C}$ are all the zero maps with domain B and range a zero object, so $\operatorname{coker} \alpha$ is the left equivalence class of zero maps with domain B and range a zero object. By abuse of notation we write $\operatorname{coker} \alpha = 0$.

The (left) equivalence class of an epic β with domain B is called a *quotient object of* B. If $\operatorname{coker} \alpha \neq \emptyset$, then the (left) equivalence class $\operatorname{coker} \alpha$ of all cokernels of α is a quotient object of B.

The above definitions have been designed so that the following desirable facts hold. See Mac Lane [Mac Lane (1975)] (Chapter IX, Sections 1 and 2).

Proposition 10.31. *Let* $\mathbf{C}$ *be an additive category.*

(1) A map $\alpha\colon A \to B$ *is a monic iff* $\ker \alpha = 0$, *and an epic iff* $\operatorname{coker} \alpha = 0$.
(2) For any maps $\alpha\colon A \to B$, $\beta\colon B \to C$ *and* $\sigma\colon D \to A$ *if* $\beta\colon B \to C$ *is monic and* $\sigma\colon D \to A$ *is epic, then*

$$\ker(\beta \circ \alpha) = \ker\, \alpha, \quad \operatorname{coker}(\alpha \circ \sigma) = \operatorname{coker} \alpha.$$

Sketch of proof. (1) Suppose $0_{K,A} \in \ker\, \alpha$ for some zero object K. To prove that α is monic it suffices to show that for any $\psi\colon C \to A$, if $\alpha \circ \psi = 0$, then $\psi = 0_{C,A}$. Since $0_{K,A}$ is a kernel of α and $\alpha \circ \psi = 0$, there is a unique map $\psi'\colon C \to K$ such that $\psi = 0_{K,A} \circ \psi' = 0_{C,A}$, as claimed.

Conversely, assume α is monic. If $k\colon K \to A$ is a kernel of α, we have $\alpha \circ k = 0$, but since α is monic, $k = 0_{K,A}$, and so $0_{K,A} \in \ker\, \alpha$ (this also implies that K is a zero object). The statement about epics is left as an exercise.

(2) Assume that $k\colon K \to A$ is a kernel of $\alpha\colon A \to B$. We have $\alpha \circ k = 0$, so $(\beta \circ \alpha) \circ k = 0$. If $\beta \circ \alpha \circ \psi = 0$ for some $\psi\colon K_2 \to A$, since β is monic, we have $\alpha \circ \psi = 0$, and since k is a kernel of α, there is an isomorphism ψ' such that $\psi = k \circ \psi'$. This shows that k is a kernel of $\beta \circ \alpha$.

Conversely, let k be a kernel of $\beta \circ \alpha$. We have $\beta \circ \alpha \circ k = 0$. Since β is monic, $\alpha \circ k = 0$. If $\alpha \circ \psi = 0$ for some $\psi\colon K_2 \to A$, then $\beta \circ \alpha \circ \psi = 0$, and since k is a kernel of $\beta \circ \alpha$, there is an isomorphism ψ' such that $\psi = k \circ \psi'$. This shows that k is a kernel of α.

The proof in the cokernel case is dual and left to the reader. $\square$

Intuitively speaking an *abelian category* is an additive category in which the notion of kernel and cokernel of a map makes sense. Then we can define the notion of image of a map f as the kernel of the cokernel of f, so the notion of exact sequence makes sense.

10.11 Abelian Categories and Exactness

Definition 10.33. An *abelian category* $\mathbf{C}$ is an additive category such that the following three properties hold:

(1) Every map $\alpha \in \mathrm{Hom}_{\mathbf{C}}(A, B)$ has a kernel and a cokernel.
(2) For any monic k and any epic σ, we have $k \in \ker(\sigma)$ iff $\sigma \in \mathrm{coker}(k)$.
(3) Every map $\alpha \in \mathrm{Hom}_{\mathbf{C}}(A, B)$ can be factored as $\alpha = \lambda \circ \sigma$, with λ monic and σ epic.

Let $\sigma_1 \colon B \to C_1$ and $\sigma_2 \colon B \to C_2$ be two cokernels for $\alpha \colon A \to B$. Consequently there is an isomorphism (unique) $\sigma_2' \colon C_1 \to C_2$ (with inverse σ_1') such that $\sigma_2 = \sigma_2' \circ \sigma_1$. If follows that for any map $k \colon K \to B$ we have $\sigma_1 \circ k = 0$ iff $\sigma_2 \circ k = 0$.

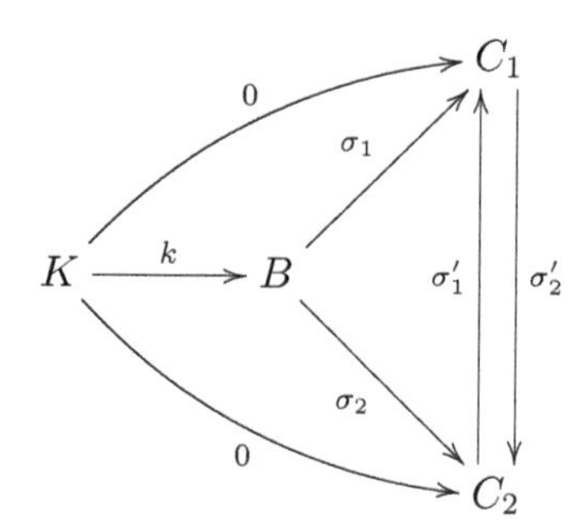

Therefore, $\ker \sigma_1 = \ker \sigma_2$ (as right equivalence classes), and the notation $\ker \mathrm{coker}\, \alpha$ makes sense. It is the (right) equivalence class of kernels of any cokernel of α.

Similarly, if $k_1 \colon K_1 \to A$ and $k_2 \colon K_2 \to A$ are two kernels for $\alpha \colon A \to B$, there is an isomorphism (unique) $k_2' \colon K_2 \to K_1$ (with inverse k_1') such that

$k_2 = k_1 \circ k_2'$. Consequently, for any map $\sigma\colon A \to C$ we have $\sigma \circ k_1 = 0$ iff $\sigma \circ k_2 = 0$.

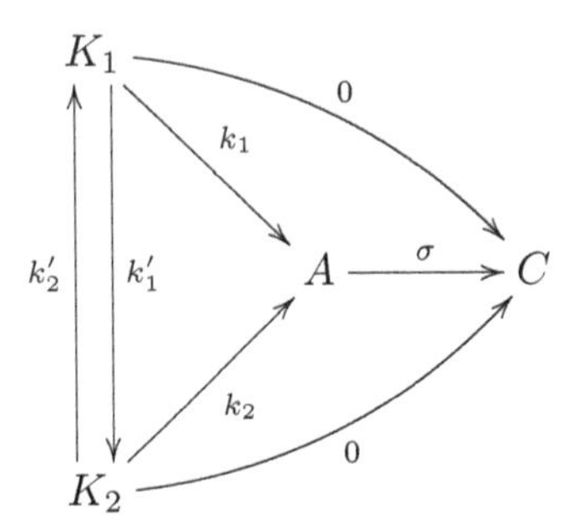

Therefore, $\operatorname{coker} k_1 = \operatorname{coker} k_2$ (as left equivalence classes), and the notation $\operatorname{coker}\ker\alpha$ makes sense. It is the (left) equivalence class of cokernels of any kernels of α.

Definition 10.34. Let $\mathbf{C}$ be an abelian category. The *image* $\operatorname{im}\alpha$ of a map $\alpha\colon A \to B$, is defined as $\operatorname{im}\alpha = \ker(\operatorname{coker}(\alpha))$, the (right) equivalence class of kernels of any cokernel of α. The *coimage* $\operatorname{coim}\alpha$ of a map $\alpha\colon A \to B$, is defined as $\operatorname{coim}\alpha = \operatorname{coker}(\ker(\alpha))$, the (left) equivalence class of cokernels of any kernel of α.

Observe that $\operatorname{im}\alpha$ consists of monics and is a subobject B and $\operatorname{coim}\alpha$ consists of epics and is a quotient object of A

Using Definition 10.34 we define exactness as follows.

Definition 10.35. Given two maps $\alpha\colon A \to B$ and $\beta\colon C \to D$ in an abelian category, the sequence

$$A \xrightarrow{\alpha} B \xrightarrow{\beta} C$$

is *exact* if $\operatorname{im}\alpha = \ker\beta$ (as right equivalence classes), which means that some kernel of $\operatorname{coker}\alpha$ is a kernel of β. The sequence

$$0 \longrightarrow A \xrightarrow{\alpha} B \xrightarrow{\beta} C \longrightarrow 0$$

is *short exact sequence* if α is a monic, β is an epic, and $\alpha \in \ker\beta$ (equivalently $\beta \in \operatorname{coker}\alpha$).

By Axiom (2) of an abelian category, $\alpha \in \ker\beta$ if $\beta \in \operatorname{coker}\alpha$, and we leave it as an exercise to prove that in a short exact sequence, $\operatorname{im}\alpha = \ker\beta$. In other words, a short exact sequence is exact at B, as it should be.

Proposition 10.32. *Given any map $\alpha\colon A \to B$, there are maps $k\colon K \to A$, $\sigma\colon A \to C$, $\lambda\colon C \to B$, $\tau\colon B \to D$, with k, λ monic, σ, τ epic, $\alpha = \lambda \circ \sigma$, $k \in \ker \alpha$, $\sigma \in \operatorname{coim} \alpha$, $\lambda \in \operatorname{im} \alpha$, and $\tau \in \operatorname{coker} \alpha$, as illustrated in the following diagram in which the horizontal row and the vertical row are exact.*

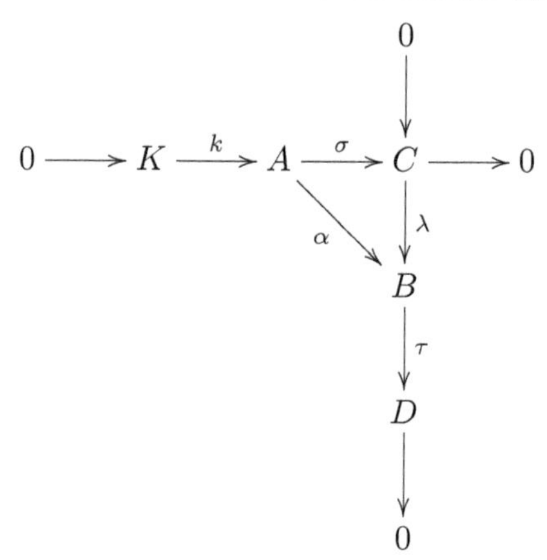

If $\sigma_1\colon A \to C_1$ and $\lambda_1\colon C_1 \to B$ are maps such that $\alpha = \lambda_1 \circ \sigma_1$, with σ_1 epic and λ_1 monic, then there is an isomorphism $\eta\colon C \to C_1$ such that $\sigma_1 = \eta \circ \sigma$ and $\lambda_1 = \lambda \circ \eta^{-1}$.

Proof. Using Axiom (3) of abelian categories, there is an epic $\sigma\colon A \to C$ and a monic $\lambda\colon C \to B$ such that $\alpha = \lambda \circ \sigma$. By Axiom (1), there is some monic $k\colon K \to A$ in $\ker \alpha$ and some epic $\tau\colon B \to D$ in $\operatorname{coker} \alpha$. Since λ is monic, by Proposition 10.31(2), $\ker\, \alpha = \ker\, \sigma$, and since τ is epic, $\operatorname{coker} \alpha = \operatorname{coker} \lambda$. By Axiom (2), $k \in \ker\, \sigma$ iff $\sigma \in \operatorname{coker} k$, and $\tau \in \operatorname{coker} \lambda$ iff $\lambda \in \ker\, \tau$. Since $k \in \ker\, \alpha$ and $\sigma \in \operatorname{coker} k$, we have $\sigma \in \operatorname{coker}(\ker(\alpha)) = \operatorname{coim} \alpha$, and since $\lambda \in \ker\, \tau$ and $\tau \in \operatorname{coker} \alpha$, we have $\lambda \in \ker(\operatorname{coker}(\alpha)) = \operatorname{im} \alpha$.

Assume that $\alpha = \lambda \circ \sigma = \lambda_1 \circ \sigma_1$, with λ, λ_1 monic and σ, σ_1 epic. Since λ and λ_1 are monic, by Proposition 10.31(2),

$$\ker\, \alpha = \ker\, \sigma = \ker\, \sigma_1.$$

It follows that there is some monic β such that $\beta \in \ker\, \sigma$ and $\beta \in \ker\, \sigma_1$, but by Axiom (2) of abelian categories, we have $\sigma \in \operatorname{coker} \beta$ and $\sigma_1 \in \operatorname{coker} \beta$, so σ and σ_1 are left equivalent, which means that there is an isomorphism $\eta\colon C \to C_1$ such that $\sigma_1 = \eta \circ \sigma$. Then we have

$$\alpha = \lambda \circ \sigma = \lambda_1 \circ \sigma_1 = \lambda_1 \circ \eta \circ \sigma,$$

and since σ is epic, we deduce that $\lambda = \lambda_1 \circ \eta$, or equivalently $\lambda_1 = \lambda \circ \eta^{-1}$. ☐

The diagram of Proposition 10.32 is called an *analysis* of α and the factorization $\alpha = \lambda \circ \sigma$ a *standard factorization* of α. The maps λ and σ are unique up to an isomorphism, in the sense that any other standard factorization of α consists of maps $\lambda \circ \eta^{-1}$ and $\eta \circ \sigma$ for some isomorphism η. The right equivalence class of λ is $\operatorname{im} \alpha$ and the left equivalence class of σ is $\operatorname{coim} \alpha$.

In the category of R-modules, if $\alpha\colon A \to B$ is an R-linear map, recall that $\operatorname{Coker} \alpha = B/\operatorname{Im} \alpha$ and $\operatorname{Coim} \alpha = A/\operatorname{Ker} \alpha$. We have the maps $k\colon \operatorname{Ker} \alpha \to A$ (the inclusion map), $p\colon A \to A/\operatorname{Ker} \alpha$ (the projection onto the quotient), $i\colon \operatorname{Im} \alpha \to B$ (the inclusion map), and $\tau\colon B \to B/\operatorname{Im} \alpha$ (the projection onto the quotient). By the first isomorphism theorem, there is an isomorphism $\overline{\alpha}\colon \operatorname{Coim} \alpha \to \operatorname{Im} \alpha$ (recall that $\operatorname{Coim} \alpha = A/\operatorname{Ker} \alpha$), and we have $\alpha = i \circ \overline{\alpha} \circ p$.

The correspondence with the previous proposition is that $K = \operatorname{Ker} \alpha$ and $D = B/\operatorname{Im} \alpha = \operatorname{Coker} \alpha$, but there are two possible choices for C.

(1) Pick $C = \operatorname{Im} \alpha$ and then $\sigma = \overline{\alpha} \circ p$ and $\lambda = i$, obtaining the following diagram

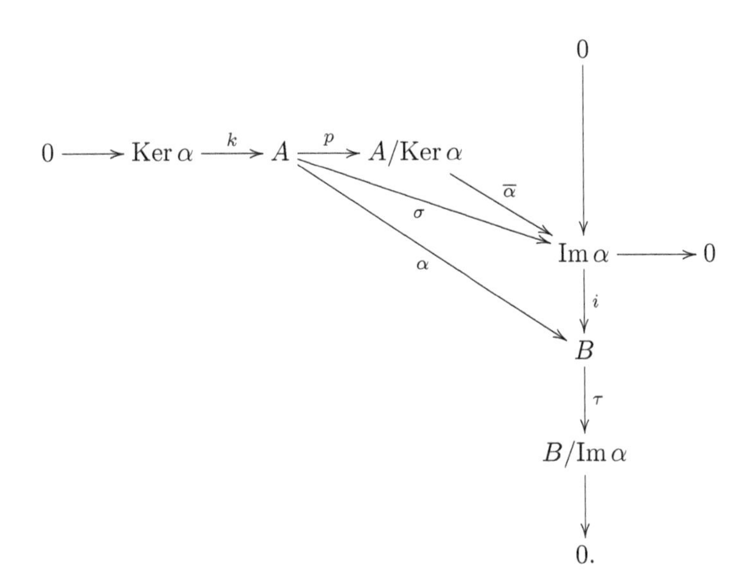

(2) Pick $C = A/\operatorname{Ker} \alpha = \operatorname{Coim} \alpha$, and then $\sigma = p$ and $\lambda = i \circ \overline{\alpha}$, obtaining

the following diagram

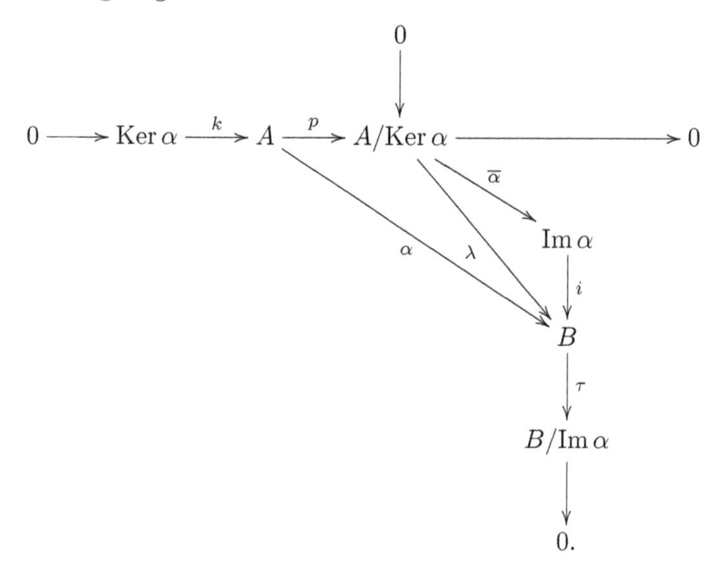

There are some useful identities relating $\ker, \operatorname{coker}, \operatorname{im}$ and coim. For example,

$$\ker \alpha = \ker(\operatorname{coim} \alpha)$$
$$\operatorname{coker} \alpha = \operatorname{coker}(\operatorname{im} \alpha).$$

Using the above equations we find that the condition $\operatorname{im} \alpha = \ker \beta$ for exactness of (α, β) (see Definition 10.35) is equivalent to $\operatorname{coim} \beta = \operatorname{coker} \alpha$.

A thorough treatment can be found in Mac Lane [Mac Lane (1975)] (Chapter IX). Proving that the five lemma holds in any abelian category is an informative exercise.

There is an aspect of abelian categories which is puzzling, not to say disturbing. Even though all the abelian categories that we will encounter (R-modules, presheaves, sheaves) have objects with members, the definition of an abelian category is so general that it allows categories where an object A many not have members, and maps $\alpha\colon A \to B$ are not linear. The whole idea still is to generalize the algebra of modules. Many arguments are indeed purely arrow-theoretic (do not refer to members of objects), but some are not, so how are we supposed to carry them out?

Fortunately, there is a deep theorem due to Freyd and Mitchell that asserts that every abelian category is" equivalent" to some category of

R-modules over some suitable ring R. To make this precise, we need to define the notion of full and faithful functor.

Definition 10.36. Given two categories $\mathbf{C}$ and $\mathbf{D}$, let $F\colon \mathbf{C} \to \mathbf{D}$ be functor. We say that the functor F is *faithful* if the map $f \mapsto F(f)$ from $\mathrm{Hom}_{\mathbf{C}}(A, B)$ to $\mathrm{Hom}_{\mathbf{D}}(F(A), F(B))$ is injective. We say that the functor F is *full* if the map $f \mapsto F(f)$ from $\mathrm{Hom}_{\mathbf{C}}(A, B)$ to $\mathrm{Hom}_{\mathbf{D}}(F(A), F(B))$ is surjective. A functor F which is full and faithful is called *fully faithful*.

Here is the precise statement of the Freyd–Mitchell embedding theorem (1964).

Theorem 10.33. *(Freyd–Mitchell embedding theorem) If $\mathbf{C}$ is a small*[6] *abelian category, then there is a (generally noncommutative) ring R (with multiplicative unit) and a fully faithful and exact functor F from $\mathbf{C}$ into the (abelian) category of R-modules.*

Note that the ring R is not commutative in general. Thus technically we have to deal with left R-modules and we need to be more careful (for example, when taking tensors and Hom). The good news is that each object $A \in \mathbf{C}$ can be viewed as an R-modules $F(A)$, and that the set of maps $\mathrm{Hom}_{\mathbf{C}}(A, B)$ can be viewed as the space of *all* linear maps from $F(A)$ to $F(B)$. The Freyd–Mitchell embedding theorem allows us to prove theorems about abelian categories using standard methods of linear algebra (for R-modules).

For more details about abelian categories, see Weibel [Weibel (1994)], Rotman [Rotman (2009)] and Mac Lane [Mac Lane (1975)]. For our purposes it is enough to think of an abelian category as a category of modules over a suitable ring. The categories of R-modules and the categories of sheaves (or presheaves) are abelian categories.

Definition 10.37. Given two abelian categories $\mathbf{C}$ and $\mathbf{D}$, an additive functor $T\colon \mathbf{C} \to \mathbf{D}$ is said to be *exact* if whenever the sequence

$$0 \longrightarrow A \longrightarrow B \longrightarrow C \longrightarrow 0$$

is exact in $\mathbf{C}$, then the sequence

$$0 \longrightarrow T(A) \longrightarrow T(B) \longrightarrow T(C) \longrightarrow 0$$

[6] This means that the class of objects of $\mathbf{C}$ is actually a set. This is a technical condition needed to avoid set-theoretic paradoxes.

is exact in $\mathbf{D}$; *left exact* if whenever the sequence

$$0 \longrightarrow A \longrightarrow B \longrightarrow C$$

is exact in $\mathbf{C}$, then the sequence

$$0 \longrightarrow T(A) \longrightarrow T(B) \longrightarrow T(C)$$

is exact; *right exact* if whenever the sequence

$$A \longrightarrow B \longrightarrow C \longrightarrow 0$$

is exact in $\mathbf{C}$, then the sequence

$$T(A) \longrightarrow T(B) \longrightarrow T(C) \longrightarrow 0$$

is exact. If $T\colon \mathbf{C} \to \mathbf{D}$ is a contravariant additive functor, then T is said to be *exact* if whenever the sequence

$$0 \longrightarrow A \longrightarrow B \longrightarrow C \longrightarrow 0$$

is exact in $\mathbf{C}$, then the sequence

$$0 \longrightarrow T(C) \longrightarrow T(B) \longrightarrow T(A) \longrightarrow 0$$

is exact in $\mathbf{D}$; *left exact* if whenever the sequence

$$A \longrightarrow B \longrightarrow C \longrightarrow 0$$

is exact in $\mathbf{C}$, then the sequence

$$0 \longrightarrow T(C) \longrightarrow T(B) \longrightarrow T(A)$$

is exact; *right exact* if whenever the sequence

$$0 \longrightarrow A \longrightarrow B \longrightarrow C$$

is exact in $\mathbf{C}$, then the sequence

$$T(C) \longrightarrow T(B) \longrightarrow T(A) \longrightarrow 0$$

is exact.

Example 10.4. For example, the (contravariant) functor $\mathrm{Hom}(-, A)$ is left-exact but not exact in general. The proof that $\mathrm{Hom}(-, A)$ is left-exact is identical to the proof of Proposition 2.7 except that R is replaced by any R-module A and $f^\top$ is replaced by $\mathrm{Hom}(f, A)$. Similarly, the functor $\mathrm{Hom}(A, -)$ is left-exact but not exact in general.

Modules for which the functor $\mathrm{Hom}(A, -)$ is exact play an important role. They are called *projective modules*. Similarly, modules for which the functor $\mathrm{Hom}(-, A)$ is exact are called *injective modules*.

Another important functor is given by the tensor product of modules.

Example 10.5. Given a fixed R-module M, we have a functor T from R-modules to R-modules such that $T(A) = A \otimes_R M$ for any R-module A, and $T(f) = f \otimes_R \mathrm{id}_M$ for any R-linear map $f\colon A \to B$. This functor usually denoted $- \otimes_R M$ is right-exact; see Section 2.4.

Modules M for which the functor $- \otimes_R M$ is exact are called *flat*.

Here is a result giving us more exact or left exact functors.

Proposition 10.34. *The following results hold:*

(1) The inclusion functor $i\colon \mathbf{Sh}(X) \to \mathbf{PSh}(X)$ is left-exact.
(2) The sheafification functor $\Gamma S\colon \mathbf{PSh}(X) \to \mathbf{Sh}(X)$ which maps a presheaf $\mathcal{F}$ to its sheafification $\widetilde{\mathcal{F}}$, is exact.
(3) For every open subset U of X, the functor $\Gamma(U, -)$ (sections over U) from $\mathbf{PSh}(X)$ to abelian groups is exact.
(4) For every open subset U of X, the functor $\Gamma(U, -)$, (sections over U) from $\mathbf{Sh}(X)$ to abelian groups is left-exact.

Proof. A proof of Proposition 10.34 can be found in Tennison [Tennison (1975)] (Chapter III, Theorem 6.9). We simply indicate how to prove (1) and (4).

(1) If

$$0 \longrightarrow \mathcal{F} \xrightarrow{\varphi} \mathcal{G} \xrightarrow{\psi} \mathcal{H} \longrightarrow 0$$

is exact as sheaves, then by Proposition 10.25 φ is injective. It follows from Proposition 10.23 that $\mathrm{PIm}\,\varphi = \mathrm{SIm}\,\varphi$, and then exactness at $\mathcal{G}$ (in the sense of sheaves) means that

$$\mathrm{PIm}\,\varphi = \mathrm{SIm}\,\varphi = \mathrm{Ker}\,\psi,$$

which is exactness in the sense of presheaves.

(4) By (1), if

$$0 \longrightarrow \mathcal{F} \xrightarrow{\varphi} \mathcal{G} \xrightarrow{\psi} \mathcal{H} \longrightarrow 0$$

is exact as sheaves, then

$$0 \longrightarrow \mathcal{F} \xrightarrow{\varphi} \mathcal{G} \xrightarrow{\psi} \mathcal{H}$$

is exact as presheaves. By Proposition 10.24 we deduce that the sequence

$$0 \longrightarrow \mathcal{F}(U) \xrightarrow{\varphi_U} \mathcal{G}(U) \xrightarrow{\psi_U} \mathcal{H}(U)$$

is exact for all open subsets of X. □

One of the most useful applications of sheaves is that they can be used to generalize the notion of manifold. In the next section we sketch this approach.

10.12 Ringed Spaces

The notion of a manifold X captures the intuition that many spaces look locally like familiar spaces, such as $\mathbb{R}^n$ (which means that for every point $x \in X$ there is some open subset U containing x which "looks" like $\mathbb{R}^n$, more precisely U is homeomorphic to $\mathbb{R}^n$), and that certain types of functions can be defined on them; for example continuous functions, smooth functions, analytic functions, *etc.* The notion of a ringed space provides an abstract way of specifying which are the "nice" functions on a space.

Definition 10.38. A *ringed space* is a pair $(X, \mathcal{O}_X)$ where X is a topological space and $\mathcal{O}_X$ is a sheaf of commutative rings called the *structure sheaf*.

The next step is to define the notion of map between two ringed spaces $(X, \mathcal{O}_X)$ and $(Y, \mathcal{O}_Y)$. The basic idea is that such a map f is a continuous map between the underlying spaces X and Y that pulls back the sheaf of functions on Y to the sheaf of functions on X. For simplicity, let us first assume that $\mathcal{O}_X$ and $\mathcal{O}_Y$ are both sheaves of functions respectively on X and Y. Let $f \colon X \to Y$ be a continuous function. Given any function $h \in \mathcal{O}_Y(V)$ (where V is some open subset of Y), denote the restriction of $h \circ f$ to $f^{-1}(V)$ by f^*h. Then f should be a map of ringed spaces if the following condition holds: for every open subset V of Y,

$$\text{if } h \in \mathcal{O}_Y(V) \text{ then } f^*h \in \mathcal{O}_X(f^{-1}(V)).$$

See Figure 10.10.

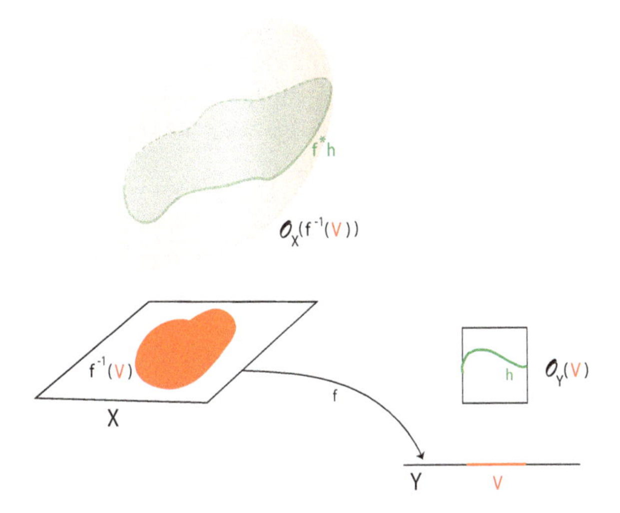

Fig. 10.10 A schematic illustration of f^*h where $X = \mathbb{R}^2$ and $Y = \mathbb{R}$. The green plane in the peach balloon is the pull back of the section $h \in \mathcal{O}_Y(V)$.

Observe that the assignment $h \mapsto f^*h$ defines a map

$$f_V^* \colon \mathcal{O}_Y(V) \to \mathcal{O}_X(f^{-1}(V))$$

which is a ring homomorphism. Thus, to define the notion of map of ringed spaces, it seems natural to require that there is a map of sheaves between $\mathcal{O}_Y$ and some sheaf over the base space Y whose sections over any open subset V of Y come from sections of $\mathcal{O}_X$ over $f^{-1}(V)$. Such a sheaf corresponds to the notion of direct image of a sheaf.

Definition 10.39. Given any continuous function $f \colon X \to Y$ between two topological spaces X and Y, for any sheaf $\mathcal{F}$ on X, define the presheaf $f_*\mathcal{F}$ on Y by

$$f_*\mathcal{F}(V) = \mathcal{F}(f^{-1}(V))$$

for all open subsets V of Y. It is easily verified that $f_*\mathcal{F}$ is a sheaf on Y called the *direct image* of $\mathcal{F}$ under f.

We can now define the notion of morphism of ringed spaces $(X, \mathcal{O}_X)$ and $(Y, \mathcal{O}_Y)$ even if $\mathcal{O}_X$ and $\mathcal{O}_Y$ are not sheaves of functions.

Definition 10.40. A *map* (or *morphism*) between two ringed spaces $(X, \mathcal{O}_X)$ and $(Y, \mathcal{O}_Y)$ is a pair (f, g), where $f\colon X \to Y$ is a continuous function and $g\colon \mathcal{O}_Y \to f_*\mathcal{O}_X$ is a map of sheaves, with each $g_V : \mathcal{O}_Y(V) \to f_*\mathcal{O}_X(V)$ a ring homomorphism for every open subset V of Y.

Given two maps of ringed spaces $(f_1, g_1)\colon (X, \mathcal{O}_X) \to (Y, \mathcal{O}_Y)$ and $(f_2, g_2)\colon (Y, \mathcal{O}_Y) \to (Z, \mathcal{O}_Z)$, their composition is the ring space map $(f_2, g_2) \circ (f_1, g_1)\colon (X, \mathcal{O}_X) \to (Z, \mathcal{O}_Z)$ given by the pair of maps

$$(f_2, g_2) \circ (f_1, g_1) = (f_2 \circ f_1, g_2 \circ g_1).$$

Definition 10.41. A map of ringed spaces $(f, g)\colon (X, \mathcal{O}_X) \to (Y, \mathcal{O}_Y)$ is an *isomorphism* iff there is some ring map $(f', g')\colon (Y, \mathcal{O}_Y) \to (X, \mathcal{O}_X)$ such that $(f, g) \circ (f', g') = (\mathrm{id}, \mathrm{id})$ and $(f', g') \circ (f, g) = (\mathrm{id}, \mathrm{id})$.

Given a ringed space $(X, \mathcal{O}_X)$, for every open subset U of X it is clear that $(U, \mathcal{O}_X|U)$ is a ringed space on U.

We can now use the above notions to define a far reaching definition of the notion of a manifold. The idea is that a ringed space $(X, \mathcal{O}_X)$ is a certain type of manifold (also called a variety in the algebraic case) if it is locally isomorphic to some other ringed space of the required type. The sheaf $\mathcal{O}_X$ specifies the "nice" functions on X.

Definition 10.42. Given two ringed spaces $(X, \mathcal{O}_X)$ and $(Y, \mathcal{O}_Y)$, we say that $(X, \mathcal{O}_X)$ is *locally isomorphic to* $(Y, \mathcal{O}_Y)$ if for every $x \in X$ there is some open subset U of X containing x and some open subset V of Y such that the ringed spaces $(U, \mathcal{O}_X|U)$ and $(V, \mathcal{O}_Y|V)$ are isomorphic.

Here are some examples illustrating that familiar types of manifolds can be cast in the framework of ringed spaces.

Example 10.6.

(1) A *topological (or continuous) manifold* M is a ringed space which is locally isomorphic to $(\mathbb{R}^n, \mathcal{C}(\mathbb{R}^n))$, where $\mathcal{C}(\mathbb{R}^n)$ is the sheaf of algebras of continuous (real-valued) functions on $\mathbb{R}^n$.

(2) A *smooth manifold* M is a ringed space which is locally isomorphic to $(\mathbb{R}^n, \mathcal{C}^\infty(\mathbb{R}^n))$, where $\mathcal{C}^\infty(\mathbb{R}^n)$ is the sheaf of algebras of smooth (real-valued) functions on $\mathbb{R}^n$.

(3) A *complex analytic manifold* M is a ringed space which is locally isomorphic to $(\mathbb{C}^n, \mathrm{Hol}(\mathbb{C}^n))$, where $\mathrm{Hol}(\mathbb{C}^n)$ is the sheaf of smooth (complex-valued) functions on $\mathbb{C}^n$.

To illustrate the power of the notion of ringed space, if we had defined the notion of affine variety (where the functions are given by ratios of polynomials), then an *algebraic variety* is a ringed space which is locally isomorphic to an affine variety.

More generally, in algebraic geometry the central notion is that of a *scheme*, which is a ringed space locally isomorphic to an *affine scheme* (an affine scheme is a ringed space locally isomorphic to the "spectrum" of a ring, whatever that is). Ambitious readers are referred to Hartshorne [Hartshorne (1977)] for an advanced treatment of algebraic geometry based on schemes.

10.13 Problems

Problem 10.1. Check that the map $\eta\colon \mathcal{F} \to \widetilde{\mathcal{F}}$ of Definition 10.5 is indeed a map of presheaves, where $\mathcal{F}$ is a presheaf on a topological space X.

Problem 10.2. Prove Part (c) of Proposition 10.7.

Problem 10.3. Check that the family η of maps $\eta_{\mathcal{F}}$ is natural in the following sense: given any presheaf map $\varphi\colon \mathcal{F} \to \mathcal{G}$, the following diagram commutes:

$$\begin{array}{ccc} \mathcal{F} & \xrightarrow{\eta_{\mathcal{F}}} & \widetilde{\mathcal{F}} \\ {\scriptstyle\varphi}\big\downarrow & & \big\downarrow{\scriptstyle\widetilde{\varphi}} \\ \mathcal{G} & \xrightarrow[\eta_{\mathcal{G}}]{} & \widetilde{\mathcal{G}}. \end{array}$$

Problem 10.4. Prove that the stalk map $\epsilon_E\colon E \to S\Gamma E$ given in Definition 10.8 is continuous.

Problem 10.5. Prove that the family ϵ of maps ϵ_E is natural in the following sense: for every map $\varphi\colon E_1 \to E_2$ of stalk spaces (E_1, p_1) and (E_2, p_2), the following diagram commutes:

$$\begin{array}{ccc} E_1 & \xrightarrow{\epsilon_{E_1}} & S\Gamma E_1 \\ {\scriptstyle\varphi}\big\downarrow & & \big\downarrow{\scriptstyle S\Gamma\varphi} \\ E_2 & \xrightarrow[\epsilon_{E_2}]{} & S\Gamma E_2. \end{array}$$

Problem 10.6. Prove Proposition 10.15.

Problem 10.7. Prove Proposition 10.16.

Problem 10.8. Prove Proposition 10.21.

Problem 10.9. Prove (i) and (iii) of Proposition 10.24.

Problem 10.10. Prove the characterization of epics given in Proposition 10.27.

Problem 10.11. Given any two maps $f\colon C \to A_1$ and $g\colon C \to A_2$, prove that the map $h\colon C \to A_1 \oplus A_2$ defined by

$$h = i_1 \circ f + i_2 \circ g$$

is the unique map such that $f = \pi_1 \circ h$ and $g = \pi_2 \circ h$, as illustrated in the diagram below

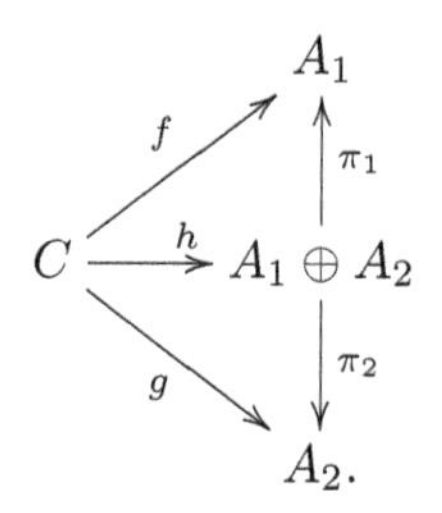

Problem 10.12. Check that right-equivalence (see Definition 10.31) is indeed an equivalence relation.

Problem 10.13. Check that left-equivalence (see Definition 10.32) is indeed an equivalence relation.

Problem 10.14. Check the identity

$$\mathrm{coker}\,(\alpha \circ \sigma) = \mathrm{coker}\,\alpha$$

from Proposition 10.27.

Problem 10.15. Prove (2) and (3) of Proposition 10.34.

Problem 10.16. Check that the presheaf $f_*\mathcal{F}$ of Definition 10.39 is indeed a sheaf.

Chapter 11

Derived Functors, δ-Functors, and ∂-Functors

The main goal of this chapter is to define the notions of derived functors, δ-functors, and ∂-functors. This machinery plays a crucial role in the definition of sheaf cohomology, an indispensable tool in advanced algebraic geometry (based on schemes) and algebraic topology.

Roughly speaking, derived functors provide a way of "measuring" how much a functor fails to be exact by computing certain homology and cohomology groups.

Recall that a functor $T\colon \mathbf{C} \to \mathbf{D}$ (where $\mathbf{C}$ and $\mathbf{D}$ are abelian categories) is said to be *exact* if whenever the sequence

$$0 \longrightarrow A \longrightarrow B \longrightarrow C \longrightarrow 0$$

is exact in $\mathbf{C}$, then the sequence

$$0 \longrightarrow T(A) \longrightarrow T(B) \longrightarrow T(C) \longrightarrow 0$$

is exact in $\mathbf{D}$, *left exact* if whenever the sequence

$$0 \longrightarrow A \longrightarrow B \longrightarrow C$$

is exact in $\mathbf{C}$, then the sequence

$$0 \longrightarrow T(A) \longrightarrow T(B) \longrightarrow T(C)$$

is exact, *right exact* if whenever the sequence

$$A \longrightarrow B \longrightarrow C \longrightarrow 0$$

is exact in $\mathbf{C}$, then the sequence

$$T(A) \longrightarrow T(B) \longrightarrow T(C) \longrightarrow 0$$

is exact. A similar definition can be given for a contravariant functor but some of the arrows are turned around.

For example, the functor $\mathrm{Hom}(-, A)$ is *left-exact* but not exact in general (see Section 2.1). Similarly, the functor $\mathrm{Hom}(A, -)$ is *left-exact* but not exact in general (see Section 2.4).

Modules for which the functor $\mathrm{Hom}(A, -)$ is exact play an important role. They are called *projective modules*. Similarly, modules for which the functor $\mathrm{Hom}(-, A)$ is exact are called *injective modules*.

The functor $- \otimes_R M$ is *right-exact* but not exact in general (see Section 2.4). Modules M for which the functor $- \otimes_R M$ is exact are called *flat*.

A good deal of homological algebra has to do with understanding how much a module fails to be projective or injective (or flat). As we will see in Section 11.1, injective and projective modules are also characterized by extension properties.

Injective modules were introduced by Baer in 1940 and projective modules by Cartan and Eilenberg in the early 1950s. Every free module is projective. Injective modules are more elusive. If the ring R is a PID an R-module M is injective iff it is divisible (which means that for every nonzero $\lambda \in R$, the map given by $u \mapsto \lambda u$ for $u \in M$ is surjective).

One of the most useful properties of projective modules is that every module M is the image of some projective (even free) module P, which means that there is a surjective homomorphism $\rho\colon P \to M$. Similarly, every module M can be embedded in an injective module I, which means that there is an injective homomorphism $i\colon M \to I$. This second fact is harder to prove (see Baer's embedding theorem, Theorem 11.6).

The above properties can be used to construct inductively projective and injective resolutions of a module M, a process that is the key to the definition of derived functors. Intuitively, projective resolutions measure how much a module deviates from being projective, and injective resolutions measure how much a module deviates from being injective.

Given any R-module A, a *projective resolution* of A is any exact sequence

$$\cdots \longrightarrow P_n \xrightarrow{d_n} P_{n-1} \xrightarrow{d_{n-1}} \cdots \longrightarrow P_1 \xrightarrow{d_1} P_0 \xrightarrow{p_0} A \longrightarrow 0 \tag{$*_1$}$$

in which every P_n is a projective module. The exact sequence

$$\cdots \longrightarrow P_n \xrightarrow{d_n} P_{n-1} \xrightarrow{d_{n-1}} \cdots \longrightarrow P_1 \xrightarrow{d_1} P_0$$

obtained by truncating the projective resolution of A after P_0 is denoted by $\mathbf{P}^A$, and the projective resolution $(*_1)$ is denoted by

$$\mathbf{P}^A \xrightarrow{p_0} A \longrightarrow 0.$$

Given any R-module A, an *injective resolution* of A is any exact sequence

$$0 \longrightarrow A \xrightarrow{i_0} I^0 \xrightarrow{d^0} I^1 \xrightarrow{d^1} \cdots \longrightarrow I^n \xrightarrow{d^n} I^{n+1} \longrightarrow \cdots \qquad (**_1)$$

in which every I^n is an injective module. The exact sequence

$$I^0 \xrightarrow{d^0} I^1 \xrightarrow{d^1} \cdots \longrightarrow I^n \xrightarrow{d^n} I^{n+1} \longrightarrow \cdots$$

obtained by truncating the injective resolution of A before I^0 is denoted by $\mathbf{I}_A$, and the injective resolution $(**_1)$ is denoted by

$$0 \longrightarrow A \xrightarrow{i_0} \mathbf{I}_A.$$

Now suppose that we have a functor $T\colon \mathbf{C} \to \mathbf{D}$, where $\mathbf{C}$ is the category of R-modules and $\mathbf{D}$ is the category of abelian groups. If we apply T to $\mathbf{P}^A$ we obtain the chain complex

$$0 \longleftarrow T(P_0) \xleftarrow{T(d_1)} T(P_1) \xleftarrow{T(d_2)} \cdots \longleftarrow T(P_{n-1}) \xleftarrow{T(d_n)} T(P_n) \longleftarrow \cdots, \qquad \text{(Lp)}$$

denoted $T(\mathbf{P}^A)$. The above is no longer exact in general but it defines homology groups $H_p(T(\mathbf{P}^A))$.

Similarly, if we apply T to $\mathbf{I}_A$ we obtain the cochain complex

$$0 \longrightarrow T(I^0) \xrightarrow{T(d^0)} T(I^1) \xrightarrow{T(d^1)} \cdots \longrightarrow T(I^n) \xrightarrow{T(d^n)} T(I^{n+1}) \longrightarrow \cdots, \qquad \text{(Ri)}$$

denoted $T(\mathbf{I}_A)$. The above is no longer exact in general but it defines cohomology groups $H^p(T(\mathbf{I}_A))$.

The reason why projective resolutions are so special is that even though the homology groups $H_p(T(\mathbf{P}^A))$ appear to depend on the projective resolution $\mathbf{P}^A$, in fact they don't; *the groups $H_p(T(\mathbf{P}^A))$ only depend on A and T*. This is proven in Theorem 11.28.

Similarly, the reason why injective resolutions are so special is that even though the cohomology groups $H^p(T(\mathbf{I}_A))$ appear to depend on the injective resolution $\mathbf{I}_A$, in fact they don't; *the groups $H^p(T(\mathbf{I}_A))$ only depend on A and T*. This is proven in Theorem 11.27.

Proving the above facts takes some work; we make use of the *comparison theorems*; see Section 11.2, Theorem 11.17 and Theorem 11.21. In view of the above results, given a functor T as above, Cartan and Eilenberg were led to define the *left derived functors* L_nT of T by

$$L_nT(A) = H_n(T(\mathbf{P}^A)),$$

for any projective resolution $\mathbf{P}^A$ of A, and the *right derived functors* R^nT of T by

$$R^nT(A) = H^n(T(\mathbf{I}_A)),$$

for any injective resolution $\mathbf{I}_A$ of A. The functors L_nT and R^nT can also be defined on maps. If T is right-exact, then L_0T is isomorphic to T (as a functor), and if T is left-exact, then R^0T is isomorphic to T (as a functor).

For example, the left derived functors of the right-exact functor $T_B(A) = A \otimes B$ (with B fixed) are the "Tor" functors. We have $\mathrm{Tor}_0^R(A, B) \cong A \otimes B$, and the functor $\mathrm{Tor}_1^R(-, G)$ plays an important role in comparing the homology of a chain complex C and the homology of the complex $C \otimes_R G$; see Chapter 12. Čech introduced the functor $\mathrm{Tor}_1^R(-, G)$ in 1935 in terms of generators and relations. It is only after Whitney defined tensor products of arbitrary $\mathbb{Z}$-modules in 1938 that the definition of Tor was expressed in the intrinsic form that we are now familiar with.

There are also versions of left and right derived functors for contravariant functors. For example, the right derived functors of the contravariant left-exact functor $T_B(A) = \mathrm{Hom}_R(A, B)$ (with B fixed) are the "Ext" functors. We have $\mathrm{Ext}_R^0(A, B) \cong \mathrm{Hom}_R(A, B)$, and the functor $\mathrm{Ext}_R^1(-, G)$ plays an important role in comparing the homology of a chain complex C and the cohomology of the complex $\mathrm{Hom}_R(C, G)$; see Chapter 12. The Ext functors were introduced in the context of algebraic topology by Eilenberg and Mac Lane (1942).

Everything we discussed so far is presented in Cartan and Eilenberg's groundbreaking book, *Homological Algebra* [Cartan and Eilenberg (1956)], published in 1956. It is in this book that the name *homological algebra* is introduced. Mac Lane [Mac Lane (1975)] (1975) and Rotman [Rotman (1979, 2009)] give more "gentle" presentations (see also Weibel [Weibel (1994)] and Eisenbud [Eisenbud (1995)]).

Derived functors can be defined for functors $T\colon \mathbf{C} \to \mathbf{D}$ where $\mathbf{C}$ or $\mathbf{D}$ is a more general category than the category of R-modules or the category of abelian groups. For example, in sheaf cohomology, the category $\mathbf{C}$ is the category of sheaves of rings. In general, it suffices that **C** and **D** are abelian categories.

We say that **C** *has enough projectives* if every object in **C** is the image of some projective object in **C**, and that **C** *has enough injectives* if every object in **C** can be embedded (injectively) into some injective object in **C**.

The most important property of derived functors is that short exact sequences yield long exact sequences of homology or cohomology. This property was proven by Cartan and Eilenberg, but Grothendieck realized how crucial it was and this led him to the fundamental concept of a *universal δ-functor*. Since we will be using right derived functors much more than left derived functors we state the existence of the long exact sequences of cohomology for right derived functors.

Theorem *Assume the abelian category* **C** *has enough injectives, let* $0 \longrightarrow A' \longrightarrow A \longrightarrow A'' \longrightarrow 0$ *be an exact sequence in* **C**, *and let* $T\colon \mathbf{C} \to \mathbf{D}$ *be a left-exact (additive) functor.*

(1) Then for every $n \geq 0$, *there is a map*

$$(R^nT)(A'') \xrightarrow{\delta^n} (R^{n+1}T)(A'),$$

and the sequence

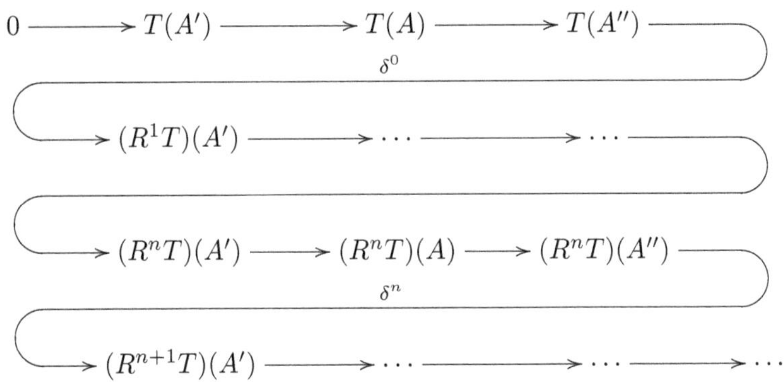

is exact. This property is similar to the property of the zig-zag lemma from Section 1.2.

(2) If $0 \longrightarrow B' \longrightarrow B \longrightarrow B'' \longrightarrow 0$ *is another exact sequence in* **C**, *and if there is a commutative diagram*

$$\begin{array}{ccccccccc}
0 & \longrightarrow & A' & \longrightarrow & A & \longrightarrow & A'' & \longrightarrow & 0 \\
 & & \downarrow & & \downarrow & & \downarrow & & \\
0 & \longrightarrow & B' & \longrightarrow & B & \longrightarrow & B'' & \longrightarrow & 0,
\end{array}$$

then the induced diagram beginning with

$$\begin{array}{ccccccccc} 0 & \longrightarrow & T(A') & \longrightarrow & T(A) & \longrightarrow & T(A'') & \xrightarrow{\delta^0_A} & \\ & & \downarrow & & \downarrow & & \downarrow & & \\ 0 & \longrightarrow & T(B') & \longrightarrow & T(B) & \longrightarrow & T(B'') & \xrightarrow[\delta^0_B]{} & \end{array}$$

and continuing with

$$\begin{array}{ccccccccccc} \cdots & \longrightarrow & R^nT(A') & \longrightarrow & R^nT(A) & \longrightarrow & R^nT(A'') & \xrightarrow{\delta^n_A} & (R^{n+1}T)(A') & \longrightarrow & \cdots \\ & & \downarrow & & \downarrow & & \downarrow & & \downarrow & & \\ \cdots & \longrightarrow & R^nT(B') & \longrightarrow & R^nT(B) & \longrightarrow & R^nT(B'') & \xrightarrow[\delta^n_B]{} & (R^{n+1}T)(B') & \longrightarrow & \cdots \end{array}$$

is also commutative.

The proof of this result (Theorem 11.31) is fairly involved and makes use of the horseshoe lemma (Theorem 11.25).

The previous theorem suggests the definition of families of functors originally proposed by Cartan and Eilenberg [Cartan and Eilenberg (1956)] and then investigated by Grothendieck in his legendary "Tohoku" paper [Grothendieck (1957)] (1957).

A *δ-functors* consists of a countable family $T = (T^n)_{n\geq 0}$ of functors $T^n\colon \mathbf{C} \to \mathbf{D}$ that satisfy the two conditions of the previous theorem. There is a notion of map, also called morphism, between δ-functors.

Given two δ-functors $S = (S^n)_{n\geq 0}$ and $T = (T^n)_{n\geq 0}$, a *morphism* $\eta\colon S \to T$ between S and T is a family $\eta = (\eta^n)_{n\geq 0}$ of natural transformations $\eta^n\colon S^n \to T^n$ such that a certain diagram commutes; see Definition 11.21.

Grothendieck also introduced the key notion of universal δ-functor; see Grothendieck [Grothendieck (1957)] (Chapter II, Section 2.2) and Definition 11.22.

A δ-functor $T = (T^n)_{n\geq 0}$ is *universal* if for every δ-functor $S = (S^n)_{n\geq 0}$ and every natural transformation $\varphi\colon T^0 \to S^0$ there is a *unique* morphism $\eta\colon T \to S$ such that $\eta^0 = \varphi$; we say that η *lifts* φ;

The reason why universal δ-functors are important is a kind of uniqueness property that shows that a universal δ-functor *is completely determined by the component* T^0; see Proposition 11.38.

One might wonder whether (universal) δ-functors exist. Indeed there are plenty of them; see Theorem 11.39.

Theorem *Assume the abelian category* $\mathbf{C}$ *has enough injectives. For every additive left-exact functor* $T\colon \mathbf{C} \to \mathbf{D}$, *the family* $(R^nT)_{n\geq 0}$ *of right derived functors of* T *is a* δ*-functor. Furthermore* T *is isomorphic to* R^0T.

In fact, the δ-functors $(R^nT)_{n\geq 0}$ are universal.

Grothendieck came up with an ingenious sufficient condition for a δ-functor to be universal: the notion of an *erasable* functor. Since Grothendieck's paper is written in French, this notion defined in Section 2.2 (Page 141) of [Grothendieck (1957)] is called *effaçable*, and many books and paper use it. Since the English translation of "effaçable" is "erasable," as advocated by Lang we will use the English word.

A functor $T\colon \mathbf{C} \to \mathbf{D}$ is *erasable* (or *effaçable*) (see Definition 11.26) if for every object $A \in \mathbf{C}$ there is some object M_A and an injection $u\colon A \to M_A$ such that $T(u) = 0$. In particular this will be the case if $T(M_A)$ is the zero object of $\mathbf{D}$. If the category $\mathbf{C}$ has enough injectives, it can be shown that T is erasable iff $T(I) = (0)$ for all injectives I.

Our favorite functors, namely the right derived functors R^nT, are erasable by injectives for all $n \geq 1$. The following result due to Grothendieck is crucial: see Theorem 11.46.

Theorem *Assume the abelian category* $\mathbf{C}$ *has enough injectives. Let* $T = (T^n)_{n\geq 0}$ *be a* δ*-functor between two abelian categories* $\mathbf{C}$ *and* $\mathbf{D}$. *If* $T^n(I) = (0)$ *for every injective* I, *for all* $n \geq 1$, *then* T *is a universal* δ*-functor.*

Actually, using the notion of injective erasing (see Definition 11.27), Grothendieck proved a more general result; see Theorem 11.44.

Finally, by combining the previous results, we obtain the most important theorem about universal δ-functors: see Theorem 11.47.

Theorem *Assume the abelian category* $\mathbf{C}$ *has enough injectives. For every left-exact functor* $T\colon \mathbf{C} \to \mathbf{D}$, *the right derived functors* $(R^nT)_{n\geq 0}$ *form a universal* δ*-functor such that* T *is isomorphic to* R^0T. *Conversely, every universal* δ*-functor* $T = (T^n)_{n\geq 0}$ *is isomorphic to the right derived* δ*-functor* $(R^nT^0)_{n\geq 0}$.

After all, the mysterious universal δ-functors are just the right derived functors of left-exact functors. As an example, the functors $\mathrm{Ext}^n_R(A, -)$ constitute a universal δ-functor (for any fixed R-module A).

The machinery of universal δ-functors can be used to prove that different kinds of cohomology theories yield isomorphic groups. If two cohomology theories $(H_S^n(-))_{n\geq 0}$ and $(H_T^n(-))_{n\geq 0}$ defined for objects in a category $\mathbf{C}$ (say, topological spaces) are given by universal δ-functors S and T in the sense that the cohomology groups $H_S^n(A)$ and $H_T^n(A)$ are given by $H_S^n(A) = S^n(A)$ and $H_T^n(A) = T^n(A)$ for all objects $A \in \mathbf{C}$, and if $H_S^0(A)$ and $H_T^0(A)$ are isomorphic, then $H_S^n(A)$ and $H_T^n(A)$ are isomorphic for all $n \geq 0$. This technique will be used in Chapter 13 to prove that sheaf cohomology and Čech cohomology are isomorphic for paracompact spaces.

Later we will see how the machinery of right derived functors can be used to define sheaf cohomology (where the category $\mathbf{C}$ is the category of sheaves of R-modules, the category $\mathbf{D}$ is the category of abelian groups, and T is the "global section functor").

11.1 Projective, Injective, and Flat Modules

We saw in Section 2.4 that the functors $\mathrm{Hom}(M,-)$ and $\mathrm{Hom}(-,M)$ are left-exact but not exact in general, and that the functor $-\otimes M$ is right-exact but not exact in general. Thus it is natural to take a closer look at the modules for which these functors are exact.

Definition 11.1. An R-module M is *projective* if the functor $\mathrm{Hom}(M,-)$ is exact, *injective* if the functor $\mathrm{Hom}(-,M)$ is exact, and *flat* if the functor $-\otimes M$ is exact.

Observe that the trivial module (0) is injective, projective, and flat. The above definition does not tell us what kind of animals these modules are. The propositions of this section give somewhat more illuminating characterizations. Recall that for any linear map $h\colon A \to B$, we have $\mathrm{Hom}(M,h)(\varphi) = h \circ \varphi$ for all $\varphi \in \mathrm{Hom}(M,A)$; see Definition 2.6.

Proposition 11.1. *Let P be an R-module. Then the following properties are equivalent:*

(1) P is projective.
(2) For any surjective linear map $h\colon A \to B$ and any linear map $f\colon P \to B$, there is some linear map $\widehat{f}\colon P \to A$ lifting $f\colon P \to B$ in the sense that

$f = h \circ \widehat{f}$, as in the following commutative diagram:

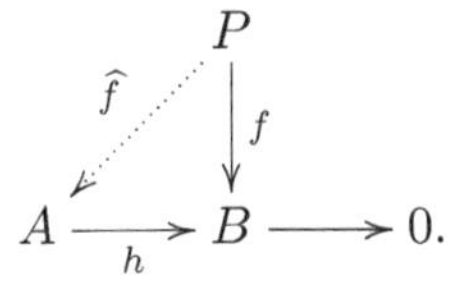

(3) *Any exact sequence*

$$0 \longrightarrow A \longrightarrow B \longrightarrow P \longrightarrow 0$$

splits.

(4) *There is a free module F and some other module Q such that $F \cong P \oplus Q$.*

Proof. This is a standard result of commutative algebra. Proofs can be found in Dummit and Foote [Dummit and Foote (1999)], Rotman [Rotman (1979, 2009)], Mac Lane [Mac Lane (1975)], Cartan–Eilenberg [Cartan and Eilenberg (1956)], and Weibel [Weibel (1994)], among others. We only show that (1) is equivalent to (2) and that (2) implies (3).

Since $\text{Hom}(P, -)$ is left exact, to say that it is exact means that if

$$A \xrightarrow{h} B \longrightarrow 0$$

is exact, then the sequence

$$\text{Hom}(P, A) \xrightarrow{\text{Hom}(P,h)} \text{Hom}(P, B) \longrightarrow 0$$

is also exact. This is equivalent to saying that if $h\colon A \to B$ is surjective, then the map $\text{Hom}(P, h)\colon \text{Hom}(P, A) \to \text{Hom}(P, B)$ is surjective, which by definition of $\text{Hom}(P, h)$ means that for any linear map $f \in \text{Hom}(P, B)$ there is some $\widehat{f} \in \text{Hom}(P, A)$ such that $f = h \circ \widehat{f}$ as in

$$\begin{array}{ccccc} & & P & & \\ & \swarrow^{\widehat{f}} & \downarrow f & & \\ A & \xrightarrow[h]{} & B & \longrightarrow & 0, \end{array}$$

which is exactly (2).

Suppose

$$0 \longrightarrow A \xrightarrow{f} B \xrightarrow{g} P \longrightarrow 0$$

is an exact sequence. We have the diagram

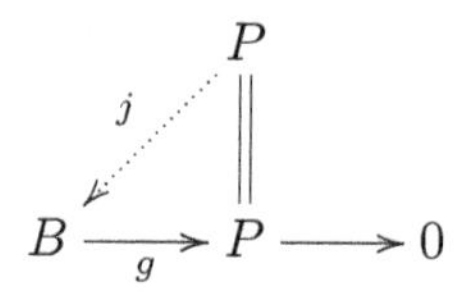

and since P is projective the lifting property gives a map $j\colon P \to B$ such that $g \circ j = \mathrm{id}_P$, which by Proposition 2.2(3) shows that (3) holds. ☐

Proposition 11.1(4) shows that projective modules are almost free, in the sense that they are a summand of a free module. It also shows that *free modules are projective*, an invaluable fact.

Another fact that we will need later is that every module is the image of some projective module.

Proposition 11.2. *For every R-module M, there is some projective (in fact, free) module P and a surjective homomorphism $\rho\colon P \to M$.*

Proof. Pick any set S of generators for M (possibly M itself) and let $P = R^{(S)}$ be the free R-module generated by S. The inclusion map $i\colon S \to M$ extends to a surjective linear map $\rho\colon P \to M$. ☐

The notion of projective module is generalized to abelian categories as follows (see Mac Lane [Mac Lane (1975)], Chapter IX, Section 4).

Definition 11.2. Let $\mathbf{C}$ be an abelian category. An object $P \in \mathbf{C}$ is a *projective object* if for any epic $h\colon A \to B$ and any map $f\colon P \to B$, there is some map $\widehat{f}\colon P \to A$ lifting $f\colon P \to B$ in the sense that $f = h \circ \widehat{f}$, as in the following commutative diagram:

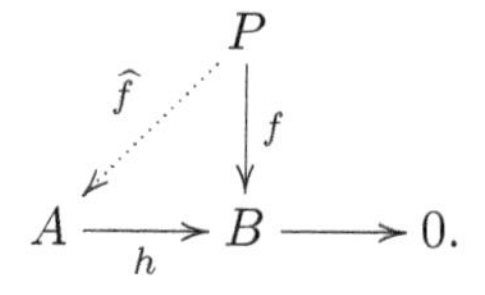

Parts (1), (2), (3) of Proposition 11.1 generalize to abelian categories; see Mac Lane [Mac Lane (1975)] (Chapter IX, Section 4, Proposition 4.2).

Injective modules are more elusive, although the diagram in Proposition 11.1(2) dualizes. The simplest characterization of injective modules is probably the condition given by Theorem 11.10, namely that an R-module

E is injective iff every injection $f: E \to M$ has a retraction $r: M \to E$ (that is, $r \circ f = \mathrm{id}_E$).

Recall that for any linear map $h: A \to B$, we have $\mathrm{Hom}(h, M)(\varphi) = \varphi \circ h$ for all $\varphi \in \mathrm{Hom}(B, M)$; see Definition 2.5.

Proposition 11.3. *Let I be an R-module. Then the following properties are equivalent:*

(1) I is injective.

(2) For any injective linear map $h: A \to B$ and any linear map $f: A \to I$, there is some linear map $\widehat{f}: B \to I$ extending $f: A \to I$ in the sense that $f = \widehat{f} \circ h$, as in the following commutative diagram:

$$\begin{array}{ccccc} 0 & \longrightarrow & A & \xrightarrow{h} & B \\ & & {\scriptstyle f}\downarrow & \swarrow {\scriptstyle \widehat{f}} & \\ & & I. & & \end{array}$$

(3) Any exact sequence

$$0 \longrightarrow I \longrightarrow B \longrightarrow C \longrightarrow 0$$

splits.

Proof. This is also a standard result of commutative algebra. Proofs can be found in Dummit and Foote [Dummit and Foote (1999)], Rotman [Rotman (1979, 2009)], Mac Lane [Mac Lane (1975)], Cartan–Eilenberg [Cartan and Eilenberg (1956)], and Weibel [Weibel (1994)], among others. We only show that (1) is equivalent to (2) and that (2) implies (3). Since $\mathrm{Hom}(-, I)$ is left exact, to say that it is exact means that if

$$0 \longrightarrow A \xrightarrow{h} B$$

is exact, then the sequence

$$\mathrm{Hom}(B, I) \xrightarrow{\mathrm{Hom}(h,I)} \mathrm{Hom}(A, I) \longrightarrow 0$$

is also exact. This is equivalent to saying that if $h: A \to B$ is injective, then the map $\mathrm{Hom}(h, I): \mathrm{Hom}(B, I) \to \mathrm{Hom}(A, I)$ is surjective, which by definition of $\mathrm{Hom}(h, I)$ means that for any linear map $f \in \mathrm{Hom}(A, I)$ there is some $\widehat{f} \in \mathrm{Hom}(B, I)$ such that $f = \widehat{f} \circ h$ as in

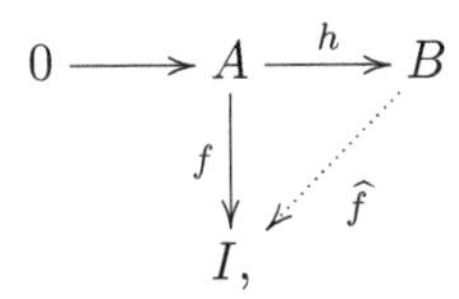

which is exactly (2).

Suppose

$$0 \longrightarrow I \xrightarrow{f} B \xrightarrow{g} C \longrightarrow 0$$

is an exact sequence. We have the diagram

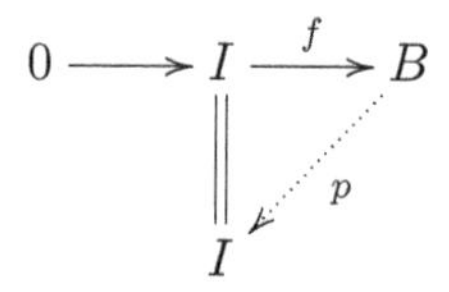

and since I is injective the lifting property gives a map $p\colon B \to I$ such that $p \circ f = \mathrm{id}_I$, which by Proposition 2.2(2) is (3). □

The notion of injective module being dual to the notion of projective module is generalized to abelian categories as follows (see Mac Lane [Mac Lane (1975)], Chapter IX, Section 4).

Definition 11.3. Let $\mathbf{C}$ be an abelian category. An object $I \in \mathbf{C}$ is an *injective object* if for any monic $h\colon A \to B$ and any map $f\colon A \to I$, there is some map $\widehat{f}\colon B \to I$ extending $f\colon A \to I$ in the sense that $f = \widehat{f} \circ h$, as in the following commutative diagram:

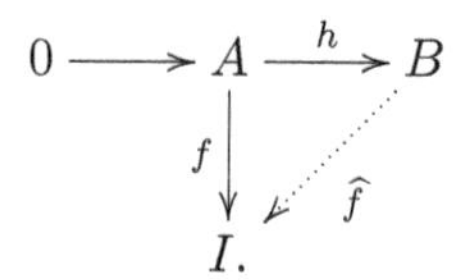

Proposition 11.3 also generalizes to abelian categories.

The following theorem due to Baer shows that to test whether a module is injective it is enough to check the extension property (Proposition 11.3(2)) for sequences $0 \longrightarrow \mathfrak{A} \longrightarrow R$ for all ideals $\mathfrak{A}$ of the ring R.

Theorem 11.4. *(Baer Representation Theorem) An R-module I is injective iff it has the extension property with respect to all sequences $0 \longrightarrow \mathfrak{A} \longrightarrow R$ where $\mathfrak{A}$ is an ideal of the ring R.*

The proof is a gem. Versions of the proof can be found in Dummit and Foote [Dummit and Foote (1999)], Rotman [Rotman (1979, 2009)], Mac Lane [Mac Lane (1975)], Cartan–Eilenberg [Cartan and Eilenberg (1956)], and Weibel [Weibel (1994)], among others.

Proof. We follow Rotman. If I is injective, the extension property w.r.t. sequences of the form $0 \longrightarrow \mathfrak{A} \longrightarrow R$ is a special case of the condition for being injective.

Conversely, assume that the extension property holds for sequences $0 \longrightarrow \mathfrak{A} \longrightarrow R$, where $\mathfrak{A}$ is an ideal in R. What does this mean? We have the diagram

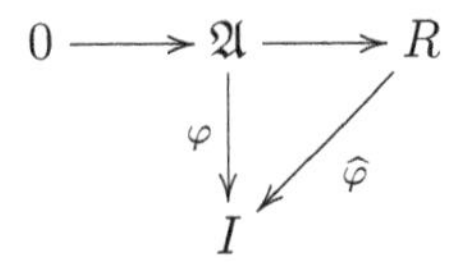

in which $\widehat{\varphi}$ extends φ. So for all $\xi \in \mathfrak{A}$, we have $\varphi(\xi) = (\widehat{\varphi} \restriction \mathfrak{A})(\xi)$. In particular, $\widehat{\varphi}(1) \in I$ exists, say $q = \widehat{\varphi}(1)$. Since $\xi \cdot 1 = \xi$ for all $\xi \in \mathfrak{A}$, we have

$$\varphi(\xi) = \widehat{\varphi}(\xi) = \xi\widehat{\varphi}(1) = \xi q.$$

Define $\mathcal{S}$ by

$$\mathcal{S} = \left\{ (N, \psi) \;\middle|\; \begin{array}{ll} (1)\ N \text{ is a submodule of } B, & (2)\ A \subseteq N, \\ (3)\ \psi\colon N \to I \text{ extends } \varphi \text{ to } N & \end{array} \right\},$$

as illustrated in the following diagram

$$\begin{array}{ccccccc} 0 & \longrightarrow & A & \longrightarrow & N & \longrightarrow & B \\ & & {\scriptstyle\varphi}\big\downarrow & \swarrow{\scriptstyle\psi} & & & \\ & & I. & & & & \end{array}$$

Partially order $\mathcal{S}$ by inclusion and agreement of extensions. Then we easily check that $\mathcal{S}$ is inductive (which means that every totally ordered subset of $\mathcal{S}$ has an upper bound). By Zorn's lemma, there is a maximal element (N_0, ψ_0) in $\mathcal{S}$. We claim that $N_0 = B$.

If $N_0 \neq B$, there is some $m \in B - N_0$, and let $\mathfrak{A}$ be the ideal given by

$$\mathfrak{A} = \{\rho \in R \mid \rho m \in N_0\}.$$

Observe that $0 \in \mathfrak{A}$.

Remark: If you know some algebra, $\mathfrak{A} = (m \longrightarrow N_0)$ is the *transporter* of m into N_0.

Define the R-module map $\theta\colon \mathfrak{A} \to I$ by

$$\theta(\rho) = \psi_0(\rho m).$$

Since we have the diagram

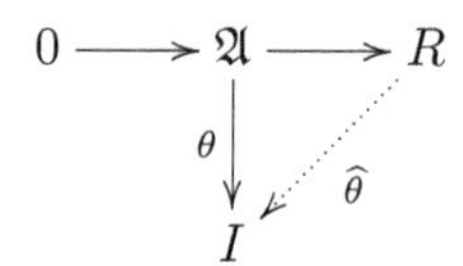

in which the top row is exact, by hypothesis, there is a map $\widehat{\theta}\colon R \to I$ extending θ.

Consider the module $N_0 + Rm$, which strictly contains N_0. If $z \in N_0 + Rm$, then $z = z_0 + \rho m$ for some $z_0 \in N_0$ and some $\rho \in R$. Set

$$\psi(z) = \psi_0(z_0) + \rho q, \quad z = z_0 + \rho m \in N_0 + Rm,$$

where $q = \widehat{\theta}(1)$. We must prove that $\psi\colon N_0 + Rm \to I$ is a well-defined map, that is, if $z = z_0 + \rho m = \widetilde{z}_0 + \widetilde{\rho} m$, then

$$\psi_0(z_0) + \rho q = \psi_0(\widetilde{z}_0) + \widetilde{\rho} q.$$

Now if we can prove that $\psi\colon N_0 + Rm \to I$ is indeed well-defined, then it is an extension of ψ_0 to the new module $N_0 + Rm$ strictly containing N_0, contradicting the maximality of N_0. Therefore, $N_0 = B$, and we are done.

If $z = z_0 + \rho m = \widetilde{z}_0 + \widetilde{\rho} m$, then $z_0 - \widetilde{z}_0 = (\widetilde{\rho} - \rho)m$; so $\widetilde{\rho} - \rho \in \mathfrak{A}$. Consequently,

$$\theta(\widetilde{\rho} - \rho) = \psi_0((\widetilde{\rho} - \rho)m).$$

Yet,

$$\theta(\widetilde{\rho} - \rho) = \widehat{\theta}(\widetilde{\rho} - \rho) = (\widetilde{\rho} - \rho)\widehat{\theta}(1) = (\widetilde{\rho} - \rho)q,$$

and so we get

$$\psi_0(z_0 - \widetilde{z}_0) = \psi_0((\widetilde{\rho} - \rho)m) = \theta(\widetilde{\rho} - \rho) = (\widetilde{\rho} - \rho)q.$$

Therefore, we deduce that

$$\psi_0(z_0) + \rho q = \psi_0(\widetilde{z}_0) + \widetilde{\rho} q,$$

establishing that ψ is well-defined. ☐

As a corollary of Theorem 11.4, it is possible to characterize injective modules when the ring R is a PID.

Definition 11.4. An R-module M is *divisible* if for every nonzero $\lambda \in R$, the multiplication map given by $u \mapsto \lambda u$ for all $u \in M$ is surjective.

Proposition 11.5. *If the ring R has no zero divisors, then any injective module is divisible. Furthermore, if R is a PID, then a module is injective iff it is divisible.*

Proof. Assume I is an injective R-module and pick any nonzero $\lambda \in R$. We wish to prove that the map $u \mapsto \lambda u$ ($u \in I$) from I to itself is surjective. This means that for any element $m \in I$, we can find some $u \in I$ such that $m = \lambda u$. Since R has no zero divisors, the map $\alpha \mapsto \alpha\lambda$ from R to itself is injective, so the map $f\colon R\lambda \to I$ given by $f(\alpha\lambda) = \alpha m$ is well-defined. Obviously $R\lambda$ is an ideal in R so we have the diagram

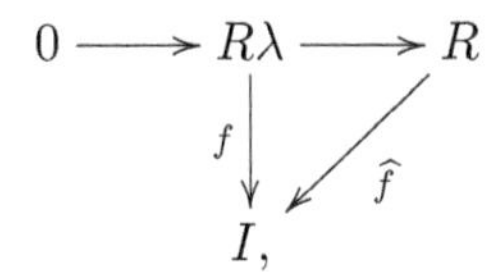

and since I is injective, there is a map $\widehat{f}\colon R \to I$ extending f. Then we have

$$m = f(\lambda) = \widehat{f}(\lambda) = \lambda \widehat{f}(1),$$

with $\widehat{f}(1) \in I$, which shows that the map $u \mapsto \lambda u$ ($u \in I$) is surjective. Therefore, I is divisible.

Now assume that R is a PID and that I is divisible. By Theorem 11.4, to prove that I is injective it suffices to prove that the extension property holds for sequences of the form $0 \longrightarrow \mathfrak{A} \longrightarrow R$, where $\mathfrak{A}$ is an ideal in R. Since R is a PID, any (left) ideal $\mathfrak{A}$ in R is of the form $R\lambda$ for some $\lambda \in R$. Consider any linear map $f\colon R\lambda \to I$. We wish to extend f to R. The case $\lambda = 0$ is trivial, so assume $\lambda \neq 0$. Since I is divisible, there is some $m \in I$ such that

$$f(\lambda) = \lambda m.$$

Define the map $\widehat{f}\colon R \to I$ by

$$\widehat{f}(\alpha) = \alpha m, \quad \alpha \in R.$$

Since for any $\beta \in R$,

$$\widehat{f}(\beta\lambda) = \beta\lambda m = \beta f(\lambda) = f(\beta\lambda),$$

we see that $\widehat{f}\colon R \to I$ extends $f\colon R\lambda \to I$, as desired. $\square$

The reader should check that the $\mathbb{Z}$-module $\mathbb{Q}/\mathbb{Z}$ is injective. More generally, if R is a PID and if K is the fraction field of R, then K/R is an injective R-module.

A result dual to the statement of Proposition 11.2 holds for injective modules but is harder to prove.

Theorem 11.6. *(Baer Embedding Theorem) For every R-module M, there is some injective module I and an injection $i\colon M \to I$.*

A particularly short proof of Theorem 11.6 can be found in Godement [Godement (1958)]. It uses the fact that if M is a projective $\mathbb{Z}$-module, then $\mathrm{Hom}_{\mathbb{Z}}(M, \mathbb{Q}/\mathbb{Z})$ is an injective $\mathbb{Z}$-module. We have to make $\mathrm{Hom}_{\mathbb{Z}}(M, \mathbb{Q}/\mathbb{Z})$ into an R-module but we will deal with this technical issue later. Observe that an R-module is automatically a $\mathbb{Z}$-module (since it is an abelian group).

The first step is to show that any $\mathbb{Z}$-module M can be embedded into M^{DD}, where $M^D = \mathrm{Hom}_{\mathbb{Z}}(M, \mathbb{Q}/\mathbb{Z})$. Given a $\mathbb{Z}$-module M, we define a natural $\mathbb{Z}$-linear map $m \mapsto \widehat{m}$ from M to M^{DD} in the usual way: for every $m \in M$ and every $f \in \mathrm{Hom}_{\mathbb{Z}}(M, \mathbb{Q}/\mathbb{Z})$,

$$\widehat{m}(f) = f(m).$$

It is clear that such a map is $\mathbb{Z}$-linear.

Proposition 11.7. *For every $\mathbb{Z}$-module M, the natural map $M \longrightarrow M^{DD}$ is injective.*

Proof. It is enough to show that $m \neq 0$ implies that $\widehat{m} \neq 0$, *i.e.*, there is some $f \in M^D = \mathrm{Hom}_{\mathbb{Z}}(M, \mathbb{Q}/\mathbb{Z})$ so that $f(m) \neq 0$.

Consider the cyclic subgroup $\mathbb{Z}m$ of M generated by m. If m has finite order $n \geq 1$, then $\mathbb{Z}m \cong \mathbb{Z}/n\mathbb{Z}$. The $\mathbb{Z}$-linear map $f\colon \mathbb{Z}/n\mathbb{Z} \to \mathbb{Q}/\mathbb{Z}$ given by $f(1) = 1/n \pmod{\mathbb{Z}}$ is obviously an injection. Since $0 \longrightarrow \mathbb{Z}/n\mathbb{Z} \longrightarrow M$ is exact and $\mathbb{Q}/\mathbb{Z}$ is injective, the map $f\colon \mathbb{Z}/n\mathbb{Z} \to \mathbb{Q}/\mathbb{Z}$ extends to a $\mathbb{Z}$-linear map $\widehat{f}\colon M \to \mathbb{Q}/\mathbb{Z}$ with $\widehat{f}(m) \neq 0$, as claimed.

If $\mathbb{Z}m$ is infinite (m has infinite order), then we have the $\mathbb{Z}$-linear surjection $g\colon \mathbb{Z}m \to \mathbb{Z}/2\mathbb{Z}$ given by $g(m) = 1 \pmod 2$. We also have the injective $\mathbb{Z}$-linear map $f_2\colon \mathbb{Z}/2\mathbb{Z} \to \mathbb{Q}/\mathbb{Z}$ given by $f_2(1) = 1/2 \pmod{\mathbb{Z}}$, and since $\mathbb{Q}/\mathbb{Z}$ is injective, the $\mathbb{Z}$-linear map $f_2\colon \mathbb{Z}/2\mathbb{Z} \to \mathbb{Q}/\mathbb{Z}$ extends to a $\mathbb{Z}$-linear map $\widehat{f_2}\colon M \to \mathbb{Q}/\mathbb{Z}$, with $\widehat{f_2}(1) \neq 0$. Then the composition $\widehat{f} = \widehat{f_2} \circ g$ is a $\mathbb{Z}$-linear map $\widehat{f}\colon M \to \mathbb{Q}/\mathbb{Z}$ such that $\widehat{f}(m) = \widehat{f_2}(g(m)) = \widehat{f_2}(1) \neq 0$. $\square$

Remark: Godement [Godement (1958)] claims that an infinite cyclic group $\mathbb{Z}m$ embeds in $\mathbb{Q}/\mathbb{Z}$ (see Page 7). This is false, but this also does not matter since the crucial point is that there is a surjection of $\mathbb{Z}m$ onto $\mathbb{Z}/2\mathbb{Z}$. This fact is used in the proof given in Bourbaki [Bourbaki (1980)] (Section, No. 8, Proposition 12).

As constructed, $\mathrm{Hom}_{\mathbb{Z}}(M, \mathbb{Q}/\mathbb{Z})$ is a $\mathbb{Z}$-module but we need to make it into an R-module. Theorem 11.6 is actually valid for modules over a noncommutative ring but we need to be careful how we define the action of the ring R. Since R is not necessarily commutative, if M is an R-module, it turns out that R acts on $\mathrm{Hom}_{\mathbb{Z}}(M, \mathbb{Q}/\mathbb{Z})$ on the right, or equivalently that the ring R^{op} (see below) acts on $\mathrm{Hom}_{\mathbb{Z}}(M, \mathbb{Q}/\mathbb{Z})$ (on the left).

Recall that given a ring R, the ring R^{op} is the ring with the same underlying set R, the same addition operation $+$, and the multiplication operation $*^{\mathrm{op}}$ given by $\lambda *^{\mathrm{op}} \mu = \mu * \lambda$ for all $\lambda, \mu \in R$. If M is an R-module and N is any $\mathbb{Z}$-module, then we can define a map from $R \times \mathrm{Hom}_{\mathbb{Z}}(M, N)$ to $\mathrm{Hom}_{\mathbb{Z}}(M, N)$ as follows: for all $\alpha \in R$ and all $f \in \mathrm{Hom}_{\mathbb{Z}}(M, N)$,

$$(\alpha f)(m) = f(\alpha m), \quad \text{for all } m \in M. \tag{$*_R$}$$

Since $\alpha *^{\mathrm{op}} \beta = \beta * \alpha$, we have

$$\begin{aligned}(\alpha(\beta f))(m) &= (\beta f)(\alpha m) = f(\beta(\alpha m)) = f((\beta * \alpha)m) \\ &= ((\beta * \alpha) f)(m) = ((\alpha *^{\mathrm{op}} \beta) f)(m).\end{aligned}$$

The equation

$$(\alpha(\beta f))(m) = f(\beta(\alpha m)) = ((\alpha *^{\mathrm{op}} \beta) f)(m)$$

shows that $(*_R)$ defines a left action of R^{op} on $\mathrm{Hom}_{\mathbb{Z}}(M, N)$ which makes $\mathrm{Hom}_{\mathbb{Z}}(M, N)$ into a R^{op}-module.

Similarly, if M is an R^{op}-module and N is any $\mathbb{Z}$-module, then $(*_R)$ defines a left action of R on $\mathrm{Hom}_{\mathbb{Z}}(M, N)$ which makes $\mathrm{Hom}_{\mathbb{Z}}(M, N)$ into an R-module, since

$$\begin{aligned}(\alpha(\beta f))(m) &= (\beta f)(\alpha m) = f(\beta(\alpha m)) = f((\beta *^{\mathrm{op}} \alpha)m) \\ &= f((\alpha * \beta)m) = ((\alpha * \beta) f)(m).\end{aligned}$$

Then $M^D = \mathrm{Hom}_{\mathbb{Z}}(M, \mathbb{Q}/\mathbb{Z})$ is an R^{op}-module if M is an R-module (resp. an R-module if M is an R^{op}-module). Furthermore, the $\mathbb{Z}$-injection, $M \longrightarrow M^{DD}$, is an R-injection. The crux of Godement's proof is the following proposition.

Proposition 11.8. *If M is a projective R^{op}-module, then M^D is an injective R-module.*

Proof. Consider the diagram

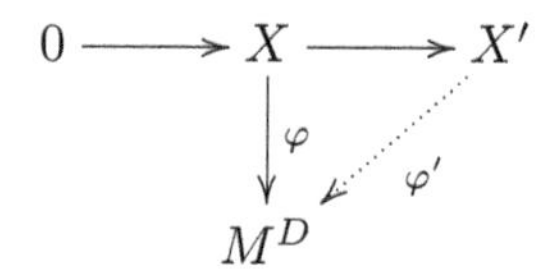

where the upper row is exact. To prove that M^D is injective, we need to prove that φ extends to a map $\varphi'\colon X' \to M^D$. By applying $\mathrm{Hom}_{\mathbb{Z}}(-, \mathbb{Q}/\mathbb{Z})$ to φ we obtain the map $\mathrm{Hom}_{\mathbb{Z}}(\varphi, \mathbb{Q}/\mathbb{Z})\colon \mathrm{Hom}_{\mathbb{Z}}(M^D, \mathbb{Q}/\mathbb{Z}) \longrightarrow \mathrm{Hom}_{\mathbb{Z}}(X, \mathbb{Q}/\mathbb{Z})$, *i.e.*, a map $M^{DD} \longrightarrow X^D$, and since we have an injection $M \longrightarrow M^{DD}$, by composition we get a map $\theta\colon M \to X^D$. Now since $\mathbb{Q}/\mathbb{Z}$ is injective, $\mathrm{Hom}_{\mathbb{Z}}(-, \mathbb{Q}/\mathbb{Z})$ maps the exact sequence

$$0 \longrightarrow X \longrightarrow X'$$

to the exact sequence

$$\mathrm{Hom}_{\mathbb{Z}}(X', \mathbb{Q}/\mathbb{Z}) \longrightarrow \mathrm{Hom}_{\mathbb{Z}}(X, \mathbb{Q}/\mathbb{Z}) \longrightarrow 0,$$

i.e., $X'^D \longrightarrow X^D \longrightarrow 0$. So we have the diagram

$$\begin{array}{ccccc} & & M & & \\ & \overset{\theta'}{\swarrow} & \downarrow{\scriptstyle \theta} & & \\ X'^D & \longrightarrow & X^D & \longrightarrow & 0, \end{array}$$

where the lower row is exact, and since M is projective, the map θ lifts to a map $\theta'\colon M \to X'^D$. Consequently, by applying $\mathrm{Hom}_{\mathbb{Z}}(-, \mathbb{Q}/\mathbb{Z})$ we get a map $X'^{DD} \longrightarrow M^D$, and since we have an injection $X' \longrightarrow X'^{DD}$, by composition we get a map $X' \longrightarrow M^D$ extending φ, as desired. Therefore, M^D is injective. ◻

We can now prove Theorem 11.6.

Proof of Theorem 11.6. Consider the R^{op}-module M^D. By Proposition 11.2 we know that there is a free R^{op}-module F such that

$$F \longrightarrow M^D \longrightarrow 0 \quad \text{is exact.}$$

But F being free is projective, and since $\mathrm{Hom}_{\mathbb{Z}}(-, \mathbb{Q}/\mathbb{Z})$ is left-exact, we get the exact sequence

$$0 \longrightarrow M^{DD} \longrightarrow F^D.$$

By Proposition 11.8, the module F^D is injective. Composing the natural injection $M \longrightarrow M^{DD}$ with the injection $M^{DD} \longrightarrow F^D$, we obtain our injection $M \longrightarrow F^D$ of M into an injective. ◻

Theorem 11.6 can be used to give an interesting characterization of injective modules. The following auxiliary result is needed.

Proposition 11.9. *Let I be an injective R-module. If C is any R-module such that there are R-linear maps $i\colon C \to I$ and $p\colon I \to C$ such that $p \circ i = \mathrm{id}_C$ (so C is a retract of I), then C is also injective.*

Proof. Let $h\colon A \to B$ be any injection and let $f\colon A \to C$ be any linear map. We need to show that f extends to a map $\widehat{f}\colon B \to C$ such that $f = \widehat{f} \circ h$. We have the map $i \circ f\colon A \to I$, and since I is injective, there is some linear map $g\colon B \to I$ such that $i \circ f = g \circ h$, as shown in the diagram below.

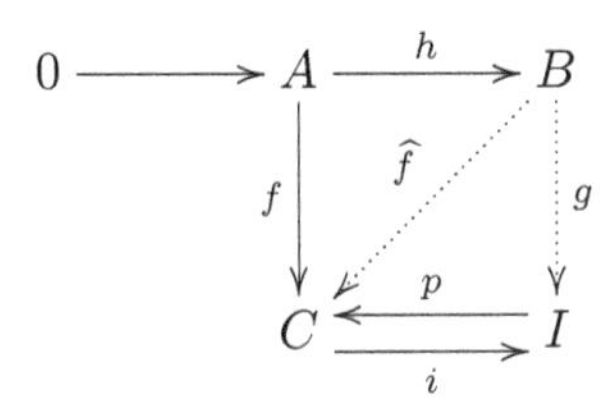

If we let $\widehat{f} = p \circ g$, then since $i \circ f = g \circ h$ and $p \circ i = \mathrm{id}_D$, we have

$$\widehat{f} \circ h = p \circ g \circ h = p \circ i \circ f = \mathrm{id}_D \circ f = f,$$

so $\widehat{f}$ is the required extension of f. ☐

The hypothesis of Proposition 11.9 is equivalent to $I = i(C) \oplus D$ with $D = \mathrm{Ker}\, p$. As a corollary, if I is an injective module and if $I = C \oplus D$, then C and D are injective.

The following theorem provides an interesting characterization of injective modules.

Theorem 11.10. *An R-module E is injective iff every injection $f\colon E \to M$ has a retraction $r\colon M \to E$, that is, $r \circ f = \mathrm{id}_E$ (both f and r are R-linear maps).*

Sketch of proof. We leave it as an exercise to prove that if E is injective, then every injection $f\colon E \to M$ has a retraction.

Conversely, by Theorem 11.6, there is some injective module I and some injection $i\colon E \to I$. By hypothesis, there is a retraction $r\colon I \to E$ such that $r \circ i = \mathrm{id}_E$, and by Proposition 11.9, E is injective. ☐

Finally, we come to flat modules.

Proposition 11.11. *Let M and N be any two R-modules. If $M \oplus N$ is flat, then M and N are flat. Every projective module is flat. Direct sums of flat modules are flat.*

A proof of Proposition 11.11 can be found in Rotman [Rotman (1988, 2009)]. The following result gives us a precise idea of what a flat module is when the ring R is a PID.

Proposition 11.12. *If the ring R has nonzero divisors, then any flat module is torsion-free. Furthermore, if R is a PID then a module is a flat module iff it is torsion-free.*

A proof of Proposition 11.12 can be found in Weibel [Weibel (1994)] (Chapter 3, Section 3.2), Bourbaki [Bourbaki (1989)] (Chapter I, §2, Section 4, Proposition 3), and as a exercise in Dummit and Foote [Dummit and Foote (1999)]. In particular, $\mathbb{Q}$ is a flat $\mathbb{Z}$-module.

More generally, if R is an integral domain and if K is its fraction field, then K is a flat R-module; see Atiyah and MacDonald [Atiyah and Macdonald (1969)] (Chapter 3, Corollary 3.6) or Bourbaki [Bourbaki (1989)] (Chapter II, §2, Section 4, Theorem 1). This last result has an interesting application.

If M is a finitely generated R-module where R is an integral domain, recall that the *rank* $\operatorname{rank} M$ of M is the largest number of linearly independent vectors in M. Since the fraction field K of R is a field, the tensor product $M \otimes_R K$ is a vector space, and it is easy to see that the dimension of the vector space $M \otimes_R K$ is equal to the rank of M; see Matsumura [Matsumura (1989)] (Chapter 4, Section 11, Page 84).

Proposition 11.13. *Let R be an integral domain. For any finitely generated R-module A, B, C, if there is a short exact sequence*

$$0 \longrightarrow A \longrightarrow B \longrightarrow C \longrightarrow 0,$$

then

$$\operatorname{rank} B = \operatorname{rank} A + \operatorname{rank} C.$$

Proof. Since the fraction field K of R is a flat R-module, if we tensor with K we get the short exact sequence

$$0 \longrightarrow A \otimes_R K \longrightarrow B \otimes_R K \longrightarrow C \otimes_R K \longrightarrow 0,$$

in which all the modules involved are vector spaces over K. But then this is a split exact sequence and we have

$$\dim B \otimes_R K = \dim A \otimes_R K + \dim C \otimes_R K.$$

By a previous remark, $\operatorname{rank} A = \dim A \otimes_R K$ and similarly with B and C, so we obtain

$$\operatorname{rank} B = \operatorname{rank} A + \operatorname{rank} C,$$

as claimed. □

In the special case where $R = \mathbb{Z}$ and A, B, C are finitely generated abelian groups, the equation of Proposition 11.13 is obtained by tensoring with $\mathbb{Q}$. Another proof of this formula (for abelian groups) is given in Greenberg and Harper [Greenberg and Harper (1981)] (Chapter 20, Lemma 20.7 and Lemma 20.8).

This is an equation which is used in proving the Euler–Poincaré formula; see Theorem 6.11.

It can be shown that $\mathbb{Q}/\mathbb{Z}$ is an injective $\mathbb{Z}$-module which is not flat and the $\mathbb{Z}$-module $\mathbb{Q} \oplus \mathbb{Z}$ is flat but neither projective nor injective. See Figure 11.1.

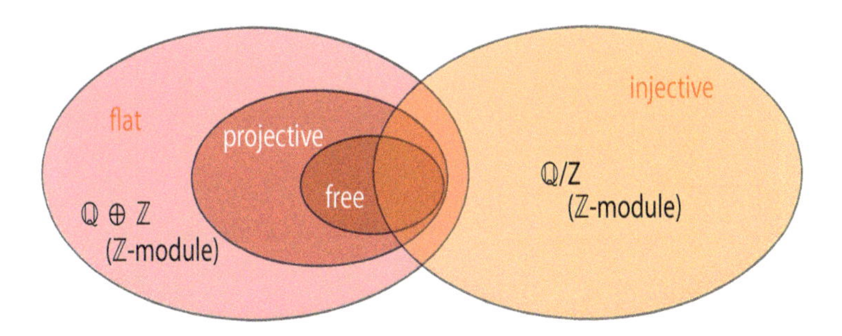

Fig. 11.1 A Venn diagram representing the containment structure of free, flat, projective, and injective modules.

We are now ready to discuss (projective and injective) resolutions, one of the most important technical tools in homological algebra.

11.2 Projective and Injective Resolutions

We saw in Section 11.1 that in general there are modules that are not projective or not injective (or neither). Then it is natural to ask whether it is possible to quantify how much a module deviates from being projective or injective. Let us first consider the projective case.

We know from Proposition 11.2 that given any module M, there is some projective (in fact, free) module P_0 and a surjection $p_0\colon P_0 \to M$. It follows that M is isomorphic to $P_0/\mathrm{Ker}\, p_0$, but the module $K_0 = \mathrm{Ker}\, p_0$ may not be projective, so we repeat the process. There is some projective module P_1 and a surjection $p_1\colon P_1 \to K_0$. Again K_0 is isomorphic to $P_1/\mathrm{Ker}\, p_1$, but $K_1 = \mathrm{Ker}\, p_1$ may not be projective. We repeat the process.

By induction, we obtain exact sequences

$$0 \longrightarrow K_n \xrightarrow{i_n} P_n \xrightarrow{p_n} K_{n-1} \longrightarrow 0$$

with P_n projective, $K_n = \mathrm{Ker}\, p_n$, and i_n the inclusion map for all $n \geq 1$, and the starting sequence

$$0 \longrightarrow K_0 \xrightarrow{i_0} P_0 \xrightarrow{p_0} M \longrightarrow 0,$$

as illustrated by the following diagram.

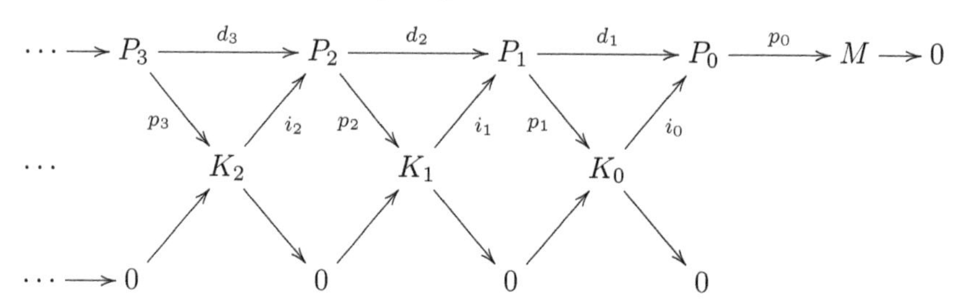

If we define $d_n\colon P_n \to P_{n-1}$ by

$$d_n = i_{n-1} \circ p_n \quad (n \geq 1),$$

then since i_{n-1} is injective we have

$$\mathrm{Ker}\, d_n = \mathrm{Ker}\, p_n = K_n,$$

and since p_n is surjective we have

$$\mathrm{Im}\, d_n = \mathrm{Im}\, i_{n-1} = K_{n-1}.$$

Therefore, $\mathrm{Im}\, d_{n+1} = \mathrm{Ker}\, d_n$ for all $n \geq 1$. See Figure 11.2.

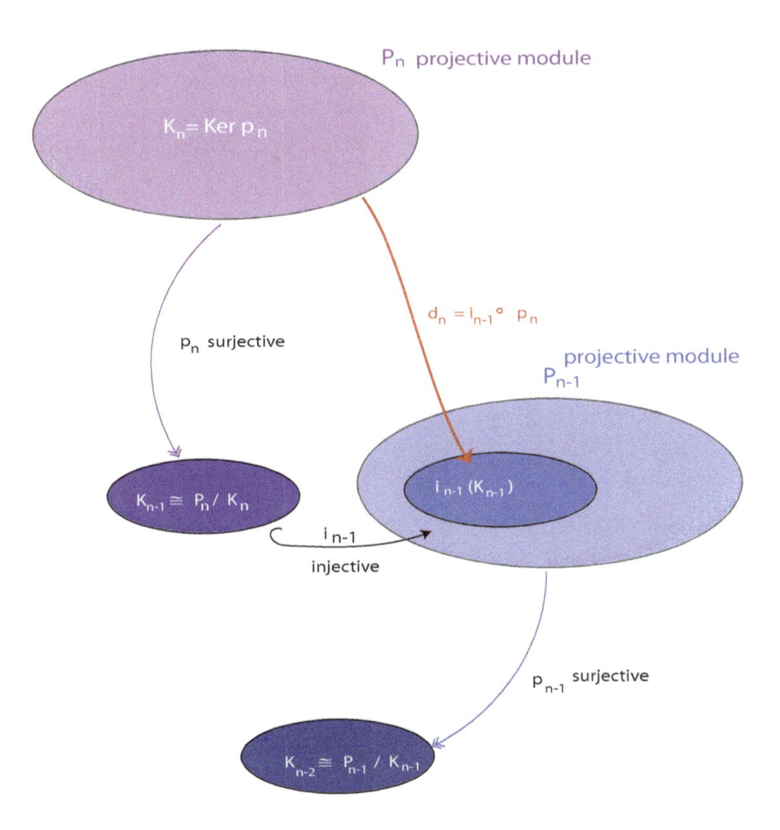

Fig. 11.2 A schematic illustration of the map $d_n = i_{n-1} \circ p_n$ used for the construction of a projective resolution.

We also have $\operatorname{Im} d_1 = K_0 = \operatorname{Ker} p_0$ and p_0 is surjective, therefore the top row is an exact sequence. In summary, we proved the following result.

Proposition 11.14. *For every R-module M, there is some exact sequence*

$$\cdots \longrightarrow P_n \xrightarrow{d_n} P_{n-1} \xrightarrow{d_{n-1}} \cdots \longrightarrow P_1 \xrightarrow{d_1} P_0 \xrightarrow{p_0} M \longrightarrow 0$$

in which every P_n is a projective module. Furthermore, we may assume that the P_n are free.

Exact sequences of the above from are called resolutions.

Definition 11.5. Given any R-module M, a *projective* (resp. *free*, resp. *flat*) *resolution* of M is any exact sequence

$$\cdots \longrightarrow P_n \xrightarrow{d_n} P_{n-1} \xrightarrow{d_{n-1}} \cdots \longrightarrow P_1 \xrightarrow{d_1} P_0 \xrightarrow{p_0} M \longrightarrow 0 \qquad (*)$$

in which every P_n is a projective (resp. free, resp. flat) module. The exact sequence

$$\cdots \longrightarrow P_n \xrightarrow{d_n} P_{n-1} \xrightarrow{d_{n-1}} \cdots \longrightarrow P_1 \xrightarrow{d_1} P_0$$

obtained by truncating the projective resolution of M after P_0 is denoted by $\mathbf{P}^M$ or $\mathbf{P}_\bullet$, and the projective resolution $(*)$ is denoted by

$$\mathbf{P}^M \xrightarrow{p_0} M \longrightarrow 0.$$

An exact sequence $(*)$ where the P_i are not necessarily projective (nor free, nor flat) is called a *left acyclic resolution* of M.

Remark: Following the convention for writing complexes with lower indices discussed in Section 2.5, the exact sequence $(*)$ of Definition 11.5 can also be written as

$$0 \longleftarrow M \xleftarrow{p_0} P_0 \xleftarrow{d_1} P_1 \longleftarrow \cdots \xleftarrow{d_{n-1}} P_{n-1} \xleftarrow{d_n} P_n \longleftarrow \cdots \qquad (**)$$

and the truncated sequence

$$P_0 \xleftarrow{d_1} P_1 \longleftarrow \cdots \xleftarrow{d_{n-1}} P_{n-1} \xleftarrow{d_n} P_n \longleftarrow \cdots$$

is still denoted by $\mathbf{P}^M$ or $\mathbf{P}_\bullet$. The projective resolution $(**)$ is denoted by

$$0 \longleftarrow M \xleftarrow{p_0} \mathbf{P}^M.$$

Proposition 11.14 shows that every module has some projective (resp. free, resp. flat) resolution. A projective resolution may stop after finitely many steps, which means that there is some m such that $P_n = (0)$ for all $n \geq m$. For example, if the ring R is a PID, since every submodule of a free module is free, every R-module has a free resolution with two steps:

$$0 \longrightarrow P_1 \xrightarrow{d_1} P_0 \xrightarrow{p_0} M \longrightarrow 0,$$

with $P_1 = K_0 = \operatorname{Ker} p_0$, a free submodule of the free module P_0.

If we apply the functor $\mathrm{Hom}(-, B)$ to the exact sequence $\mathbf{P}^A$

$$P_0 \xleftarrow{d_1} P_1 \longleftarrow \cdots \xleftarrow{d_{n-1}} P_{n-1} \xleftarrow{d_n} P_n \longleftarrow \cdots$$

obtained from a projective resolution of a module A by dropping the term A, exactness is usually lost but we still obtain the chain complex $\mathrm{Hom}(\mathbf{P}^A, B)$ given by

$$0 \longrightarrow \mathrm{Hom}(P_0, B) \longrightarrow \cdots \longrightarrow \mathrm{Hom}(P_{n-1}, B) \longrightarrow \mathrm{Hom}(P_n, B) \longrightarrow \cdots,$$

with the maps $\mathrm{Hom}(P_{n-1}, B) \xrightarrow{\mathrm{Hom}(d_n, B)} \mathrm{Hom}(P_n, B)$.

Consequently, we have the cohomology groups $H^p(\mathrm{Hom}(\mathbf{P}^A, B))$ of the cohomology complex $\mathrm{Hom}(\mathbf{P}^A, B)$.

These cohomology modules seem to depend of the choice of the projective resolution $\mathbf{P}^A$. However, the remarkable fact about projective resolutions is that these cohomology groups *are independent* of the projective resolution chosen. This is what makes projective resolutions so special. In our case where we applied the functor $\mathrm{Hom}(-, B)$, the cohomology groups are denoted by $\mathrm{Ext}^n_R(A, B)$ (the "Ext" groups).

Definition 11.6. For any two R-modules A and B, the cohomology groups $\mathrm{Ext}^n_R(A, B)$, the *Ext groups*, are the cohomology groups obtained by applying the functor $\mathrm{Hom}(-, B)$ to the exact sequence $\mathbf{P}^A$

$$P_0 \xleftarrow{d_1} P_1 \longleftarrow \cdots \xleftarrow{d_{n-1}} P_{n-1} \xleftarrow{d_n} P_n \longleftarrow \cdots$$

obtained from any projective resolution of a module A by dropping the term A.

Since $\mathrm{Hom}(-, B)$, is left exact, the exact sequence

$$P_1 \xrightarrow{d_1} P_0 \xrightarrow{p_0} A \longrightarrow 0$$

yields the exact sequence

$$0 \longrightarrow \mathrm{Hom}(A, B) \xrightarrow{\mathrm{Hom}(p_0, B)} \mathrm{Hom}(P_0, B) \xrightarrow{\mathrm{Hom}(d_1, B)} \mathrm{Hom}(P_1, B).$$

This above implies that $\mathrm{Hom}(A, B)$ is isomorphic to $\mathrm{Ker}\,\mathrm{Hom}(d_1, B) = H^0(\mathrm{Hom}(\mathbf{P}^A, B))$ that is,

$$\mathrm{Ext}^0_R(A, B) \cong \mathrm{Hom}(A, B).$$

If A is a projective module, then we have the trivial resolution $0 \longrightarrow A \xrightarrow{\mathrm{id}} A \longrightarrow 0$, and $\mathrm{Ext}^n_R(A, B) = (0)$ for all $n \geq 1$.

If the ring R is a PID, then every module A has a free resolution

$$0 \longrightarrow P_1 \xrightarrow{d_1} P_0 \xrightarrow{p_0} A \longrightarrow 0\,,$$

so $\mathrm{Ext}^n_R(A,B) = (0)$ for all $n \geq 2$. The group $\mathrm{Ext}^1_R(A,B)$ plays a crucial role in the universal coefficient theorem for cohomology which expresses the cohomology groups of a complex in terms of its cohomology. The cohomology complex is obtained from the homology complex by applying the functor $\mathrm{Hom}(-,R)$.

If we apply the functor $-\otimes B$ to the exact sequence $\mathbf{P}^A$

$$P_0 \xleftarrow{d_1} P_1 \longleftarrow \cdots \xleftarrow{d_{n-1}} P_{n-1} \xleftarrow{d_n} P_n \longleftarrow \cdots$$

obtained from a projective resolution of a module A by dropping the term A, exactness is usually lost but we still obtain the chain complex $\mathbf{P}^A \otimes B$ given by

$$0 \longleftarrow P_0 \otimes B \longleftarrow \cdots \longleftarrow P_{n-1} \otimes B \longleftarrow P_n \otimes B \longleftarrow \cdots$$

with maps $P_n \otimes B \xrightarrow{d_n \otimes \mathrm{id}_B} P_{n-1} \otimes B$.

This time we have the homology groups $H_p(\mathbf{P}^A \otimes B)$ of the homology complex $\mathbf{P}^A \otimes B$.

As before, these homology groups *are independent* of the resolution chosen. These homology groups are denoted by $\mathrm{Tor}^R_n(A,B)$ (the "Tor" groups).

Definition 11.7. For any two R-modules A and B, the homology groups $\mathrm{Tor}^R_n(A,B)$, called the *Tor groups*, are the homology groups obtained by applying the functor $-\otimes B$ to the exact sequence $\mathbf{P}^A$

$$P_0 \xleftarrow{d_1} P_1 \longleftarrow \cdots \xleftarrow{d_{n-1}} P_{n-1} \xleftarrow{d_n} P_n \longleftarrow \cdots$$

obtained from any projective resolution of a module A by dropping the term A.

Because $-\otimes B$ is right-exact, we have an isomorphism

$$\mathrm{Tor}^R_0(A,B) \cong A \otimes B.$$

If the ring R is a PID, then $\mathrm{Tor}^R_0(A,B) = (0)$ for all $n \geq 2$. The group $\mathrm{Tor}^R_1(A,B)$ plays a crucial role in the universal coefficient theorem that expresses the homology groups with coefficients in an R-module B in terms of the homology groups with coefficients in R.

Using Theorem 11.6, we can dualize the construction of Proposition 11.14 to show that every module has an injective resolution, a notion defined below.

Definition 11.8. Given any R-module M, an *injective resolution* of M is any exact sequence

$$0 \longrightarrow M \xrightarrow{i_0} I^0 \xrightarrow{d^0} I^1 \xrightarrow{d^1} \cdots \longrightarrow I^n \xrightarrow{d^n} I^{n+1} \longrightarrow \cdots \tag{$*$}$$

in which every I^n is an injective module. The exact sequence

$$I^0 \xrightarrow{d^0} I^1 \xrightarrow{d^1} \cdots \longrightarrow I^n \xrightarrow{d^n} I^{n+1} \longrightarrow \cdots$$

obtained by truncating the injective resolution of M before I^0 is denoted by $\mathbf{I}_M$ or $\mathbf{I}^\bullet$, and the injective resolution $(*)$ is denoted by

$$0 \longrightarrow M \xrightarrow{i_0} \mathbf{I}_M.$$

An exact sequence $(*)$ where the I^i are not necessarily injective is called a *right acyclic resolution* of M.

Proposition 11.15. *Every R-module M has some injective resolution.*

Proof. Using Theorem 11.6 we can find an injective module I^0 and an injection $i^0 \colon M \to I^0$. Let $C^1 = \mathrm{Coker}\, i^0$ be the cokernel of i^0. If C^1 is not injective then by Theorem 11.6 we can find an injective module I^1 and an injection $i^1 \colon C^1 \to I^1$. Let $C^2 = \mathrm{Coker}\, i^1$. If C^2 is not injective we repeat the process. By induction we obtain exact sequences

$$0 \longrightarrow C^n \xrightarrow{i^n} I^n \xrightarrow{p^n} C^{n+1} \longrightarrow 0,$$

where $C^{n+1} = \mathrm{Coker}\, i^n = I^n/\mathrm{Im}\, i^n$ and p^n is the projection map for all $n \geq 0$, starting with

$$0 \longrightarrow M \xrightarrow{i^0} I^0 \xrightarrow{p^0} C^1 \longrightarrow 0,$$

as illustrated by the following diagram.

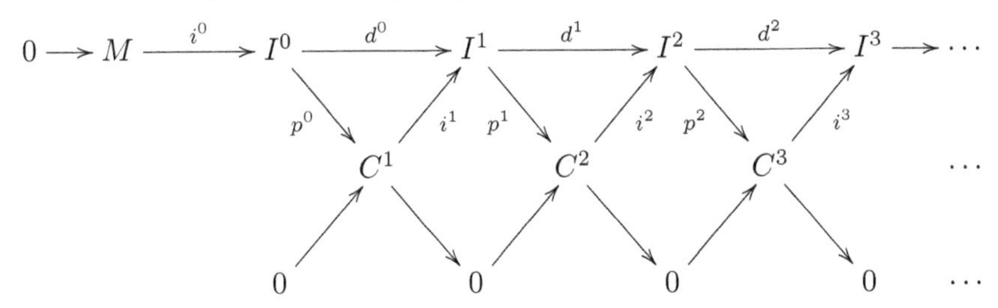

If we define $d^n \colon I^n \to I^{n+1}$ by

$$d^n = i^{n+1} \circ p^n \quad (n \geq 0),$$

then we immediately check $\operatorname{Ker} d^n = \operatorname{Ker} p^n = \operatorname{Im} i^n$ and $\operatorname{Im} d^n = \operatorname{Im} i^{n+1}$, so the top row is exact; that is, it is an injective resolution of M. See Figure 11.3. $\square$

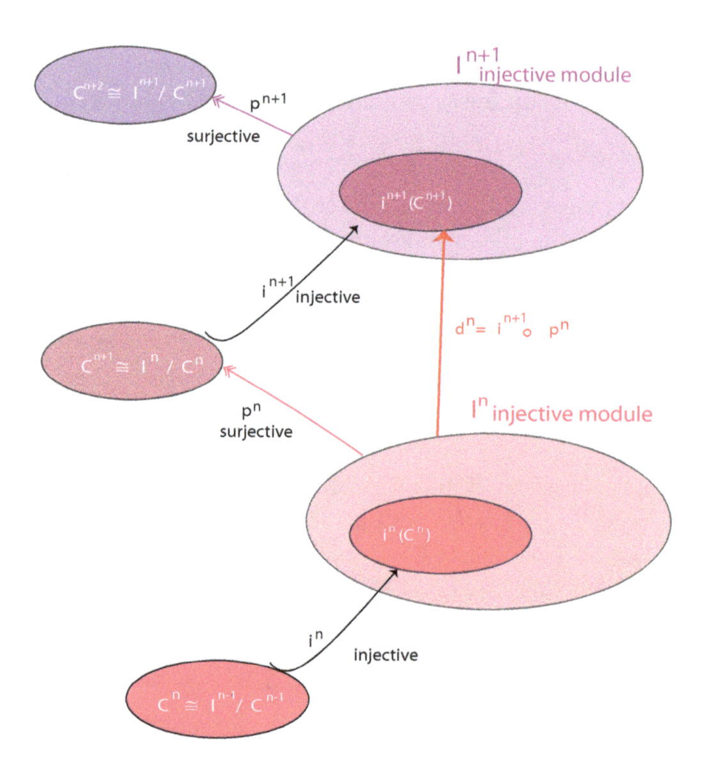

Fig. 11.3 A schematic illustration of the map $d^n = i^{n+1} \circ p^n$ used for the construction of an injective resolution.

If we apply the functor $\operatorname{Hom}(A, -)$ to the exact sequence

$$I^0 \xrightarrow{d^0} I^1 \xrightarrow{d^1} \cdots \longrightarrow I^n \xrightarrow{d^n} I^{n+1} \longrightarrow \cdots$$

obtained by truncating the injective resolution of B before I^0 we obtain the complex

$$\operatorname{Hom}(A, I^0) \longrightarrow \operatorname{Hom}(A, I^1) \cdots \longrightarrow \operatorname{Hom}(A, I^n) \longrightarrow \operatorname{Hom}(A, I^{n+1}) \longrightarrow \cdots$$

with maps $\operatorname{Hom}(A, I^n) \xrightarrow{\operatorname{Hom}(A, d^n)} \operatorname{Hom}(A, I^{n+1})$.

We have the cohomology groups $H^p(\mathrm{Hom}(A, \mathbf{I}_B))$ of the complex $\mathrm{Hom}(A, \mathbf{I}_B)$. Remarkably, as in the case of projective resolutions, these cohomology groups are *independent* of the injective resolution chosen. This is what makes injective resolutions so special. In our case where we applied the functor $\mathrm{Hom}(A, -)$ we obtain some cohomology modules $\mathrm{Ext}'^p_R(A, B)$.

Definition 11.9. For any two R-modules A and B, the cohomology groups $\mathrm{Ext}'^n_R(A, B)$, the Ext' *groups*, are the cohomology groups obtained by applying the functor $\mathrm{Hom}(A, -)$ to the exact sequence $\mathbf{I}^A$

$$I^0 \xrightarrow{d^0} I^1 \xrightarrow{d^1} \cdots \longrightarrow I^n \xrightarrow{d^n} I^{n+1} \longrightarrow \cdots$$

obtained by truncating any injective resolution of B before I^0.

It is natural to ask whether the modules $\mathrm{Ext}'^p_R(A, B)$ are related to the cohomology modules $\mathrm{Ext}^p_R(A, B)$ induced by the functor $\mathrm{Hom}(-, B)$ and defined in terms of projective resolutions. The answer is that they are *isomorphic*; see Rotman [Rotman (1988, 2009)] or Weibel [Weibel (1994)] for a thorough exposition.

11.3 Comparison Theorems for Resolutions

We now return to the fundamental property of projective and injective resolutions, a kind of quasi-uniqueness. To be more precise, there is a chain homotopy equivalence between the complexes $\mathbf{P}^A$ and $\mathbf{P}'^A$ arising from any two projective resolutions of a module A (a similar result holds for injective resolutions). To understand this, let us review the notions of chain map and chain homotopy from Section 2.6 in the context of projective and injective resolutions.

Definition 11.10. Let A and B be two R-modules, let

$$\mathbf{P}^A \xrightarrow{\epsilon} A \longrightarrow 0 \qquad (*)$$

and

$$\mathbf{P}'^B \xrightarrow{\epsilon'} B \longrightarrow 0 \qquad (**)$$

be two complexes, and let $f \colon A \to B$ be a map of R-modules. A *map* (or *morphism*) *from* $\mathbf{P}^A$ *to* $\mathbf{P}'^B$ *over* f (or *lifting* f) is a family $g = (g_n)_{n \geq 0}$

of maps $g_n\colon P_n \to P'_n$ such that the following diagram commutes for all $n \geq 1$:

$$\begin{array}{ccc} P_n & \xrightarrow{d_n^P} & P_{n-1} \\ {\scriptstyle g_n}\downarrow & & \downarrow{\scriptstyle g_{n-1}} \\ P'_n & \xrightarrow[d_n^{P'}]{} & P'_{n-1} \end{array} \qquad\qquad \begin{array}{ccc} P_0 & \xrightarrow{\epsilon} & A \\ {\scriptstyle g_0}\downarrow & & \downarrow{\scriptstyle f} \\ P'_0 & \xrightarrow[\epsilon']{} & B. \end{array}$$

Given two morphisms g and h from $\mathbf{P}^A$ to $\mathbf{P}'^B$ over f, a *chain homotopy* between g and h is a family $s = (s_n)_{n\geq 0}$ of maps $s_n\colon P_n \to P'_{n+1}$ for $n \geq 0$, such that

$$g_n - h_n = s_{n-1} \circ d_n^P + d_{n+1}^{P'} \circ s_n, \qquad n \geq 1$$

and

$$g_0 - h_0 = d_1^{P'} \circ s_0,$$

as illustrated in the diagrams

$$\begin{array}{ccccccccc} \cdots & \longrightarrow & P_{n+1} & \xrightarrow{d_{n+1}^P} & P_n & \xrightarrow{d_n^P} & P_{n-1} & \xrightarrow{d_{n-1}^P} & \cdots \\ & & {\scriptstyle \Delta^{n+1}}\downarrow & \swarrow{\scriptstyle s_n} & \downarrow{\scriptstyle \Delta_n} & \swarrow{\scriptstyle s_{n-1}} & \downarrow{\scriptstyle \Delta_{n-1}} & & \\ \cdots & \longrightarrow & P'_{n+1} & \xrightarrow[d_{n+1}^{P'}]{} & P'_n & \xrightarrow[d_n^{P'}]{} & P'_{n-1} & \xrightarrow[d_{n-1}^{P'}]{} & \cdots \end{array}$$

and

$$\begin{array}{ccccc} \cdots & \longrightarrow & P_1 & \xrightarrow{d_1^P} & P_0 \\ & & {\scriptstyle \Delta^1}\downarrow & \swarrow{\scriptstyle s_0} & \downarrow{\scriptstyle \Delta_0} \\ \cdots & \longrightarrow & P'_1 & \xrightarrow[d_1^{P'}]{} & P'_0 \end{array}$$

where $\Delta_n = g_n - h_n$.

In particular, a special case of Definition 11.10 is the case where $(*)$ and $(**)$ are projective resolutions. Dually, we have a definition that specializes to injective resolutions.

Definition 11.11. Let A and B be two R-modules, let

$$0 \longrightarrow A \xrightarrow{\epsilon} \mathbf{I}_A \qquad (*)$$

and

$$0 \longrightarrow B \xrightarrow{\epsilon'} \mathbf{I}'_B \qquad (**)$$

be two complexes, and let $f\colon A \to B$ be a map of R-modules. A *map* (or *morphism*) *from* $\mathbf{I}_A$ *to* $\mathbf{I}'_B$ *over* f (or *lifting* f) is a family $g = (g^n)_{n\geq 0}$ of maps $g^n\colon I^n \to I'^n$ such that the following diagram commutes for all $n \geq 0$:

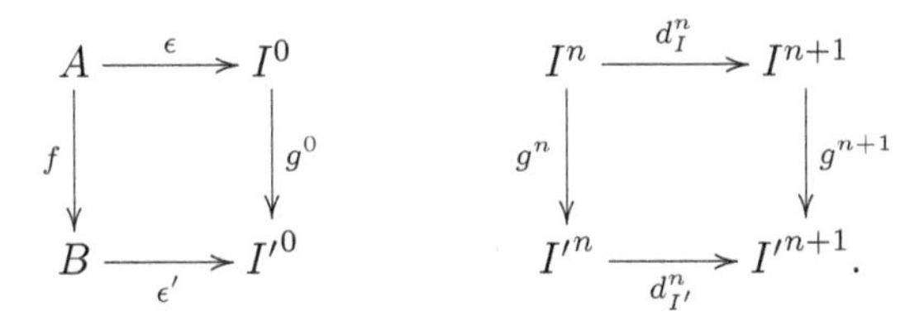

Given two morphisms g and h from $\mathbf{I}_A$ to $\mathbf{I}'_B$ over f, a *chain homotopy* between g and h is a family $s = (s^n)_{n\geq 1}$ of maps $s^n\colon I^n \to I'^{n-1}$ for $n \geq 1$, such that

$$g^n - h^n = s^{n+1} \circ d_I^n + d_{I'}^{n-1} \circ s^n, \qquad n \geq 1$$

and

$$g^0 - h^0 = s^1 \circ d_I^0,$$

as illustrated in the diagrams

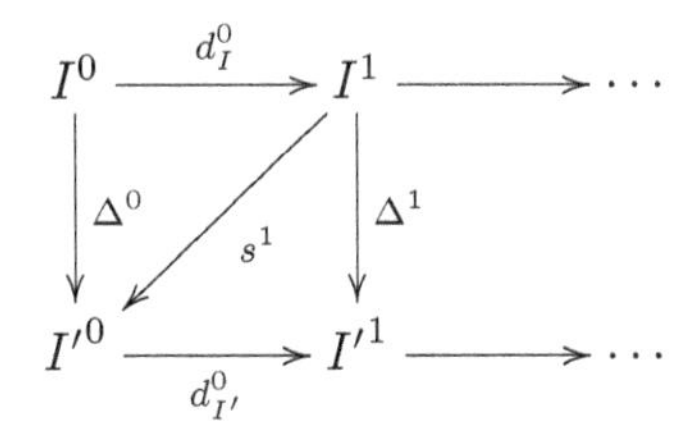

and

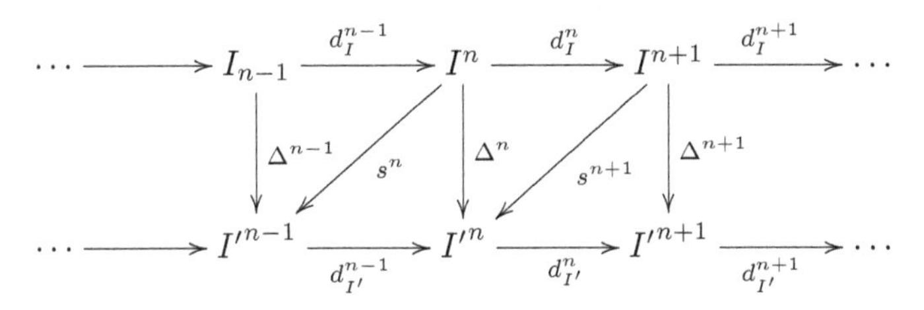

where $\Delta^n = g^n - h^n$.

We now come to the small miracle about projective resolutions. We begin with a crucial observation.

Proposition 11.16. *If we have a diagram*

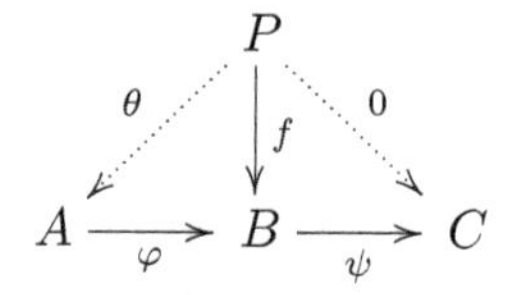

in which

(1) P is projective.
(2) The lower sequence is exact (i.e., $\mathrm{Im}\,\varphi = \mathrm{Ker}\,\psi$).
(3) $\psi \circ f = 0$,

then there is a map $\theta\colon P \to A$ lifting f (as shown by the dotted arrow above).

Proof. Indeed, $\psi \circ f = 0$ implies that $\mathrm{Im}\, f \subseteq \mathrm{Ker}\,\psi = \mathrm{Im}\,\varphi$; so, we have $\mathrm{Im}\, f \subseteq \mathrm{Im}\,\varphi$, and we are reduced to the usual diagram

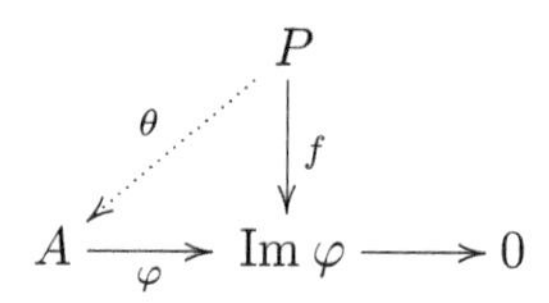

where φ is surjective. $\square$

Theorem 11.17. *(Comparison Theorem, Projective Case) Let A and B be R-modules. If $\mathbf{P}^A \xrightarrow{\epsilon} A \longrightarrow 0$ is a chain complex with all P_n in $\mathbf{P}^A$ projective and if $\mathbf{X}^B \xrightarrow{\epsilon'} B \longrightarrow 0$ is an exact sequence (a left resolution of B), then any R-linear map $f\colon A \to B$ lifts to a morphism g from $\mathbf{P}^A$ to $\mathbf{X}^B$ as illustrated by the following commutative diagram:*

$$
\begin{array}{ccccccccccc}
\cdots & \longrightarrow & P_2 & \xrightarrow{d_2^P} & P_1 & \xrightarrow{d_1^P} & P_0 & \xrightarrow{\epsilon} & A & \longrightarrow & 0 \\
 & & \Big\downarrow{\scriptstyle g_2} & & \Big\downarrow{\scriptstyle g_1} & & \Big\downarrow{\scriptstyle g_0} & & \Big\downarrow{\scriptstyle f} & & \\
\cdots & \longrightarrow & X_2 & \xrightarrow[d_2^X]{} & X_1 & \xrightarrow[d_1^X]{} & X_0 & \xrightarrow[\epsilon']{} & B & \longrightarrow & 0.
\end{array}
$$

Any two morphisms from $\mathbf{P}^A$ to $\mathbf{X}^B$ lifting f are chain homotopic.

Proof. Here is a slightly expanded version of the classical proof from Cartan–Eilenberg [Cartan and Eilenberg (1956)] (Chapter V, Proposition 1.1).

We begin by proving the existence of the lift, stepwise, by induction. Since we have morphisms $\epsilon\colon P_0 \to A$ and $f\colon A \to B$, we get a morphism $f \circ \epsilon\colon P_0 \to B$ and we have the diagram

$$\begin{array}{ccccc} & & P_0 & & \\ & \swarrow^{g_0} & \downarrow f\circ\epsilon & & \\ X_0 & \xrightarrow[\epsilon']{} & B & \longrightarrow & 0. \end{array}$$

As P_0 is projective, the map $g_0\colon P_0 \to X_0$ exists and makes the diagram commute. Assume the lift exists up to level n. We have the diagram

$$\begin{array}{ccccccc} P_{n+1} & \xrightarrow{d^P_{n+1}} & P_n & \xrightarrow{d^P_n} & P_{n-1} & \longrightarrow & \cdots \\ & & \downarrow g_n & & \downarrow g_{n-1} & & \\ X_{n+1} & \xrightarrow[d^X_{n+1}]{} & X_n & \xrightarrow[d^X_n]{} & X_{n-1} & \longrightarrow & \cdots, \end{array} \tag{$\dagger$}$$

so we get a map $g_n \circ d^P_{n+1}\colon P_{n+1} \to X_n$ and a diagram

$$\begin{array}{ccccc} & & P_{n+1} & & \\ & \swarrow^{g_{n+1}} & \downarrow g_n\circ d^P_{n+1} & & \\ X_{n+1} & \xrightarrow[d^X_{n+1}]{} & X_n & \xrightarrow[d^X_n]{} & X_{n-1}. \end{array}$$

But, by commutativity in ($\dagger$), we get

$$d^X_n \circ g_n \circ d^P_{n+1} = g_{n-1} \circ d^P_n \circ d^P_{n+1} = 0.$$

Observe that in the above step we only use the fact that the first sequence is a chain complex. Now P_{n+1} is projective and the lower row in the above diagram is exact, so by Proposition 11.16, there is a lifting $g_{n+1}\colon P_{n+1} \to X_{n+1}$, as required.

Say we have two lifts $g = (g_n)$ and $h = (h_n)$. We construct the chain homotopy (s_n), by induction on $n \geq 0$.

For the base case, we have the diagram

$$\begin{array}{ccccccc} & & P_0 & \xrightarrow{\epsilon} & A & \longrightarrow & 0 \\ & \swarrow^{s_0} & g_0\downarrow\ \downarrow h_0 & & \downarrow f & & \\ X_1 & \xrightarrow[d^X_1]{} & X_0 & \xrightarrow[\epsilon']{} & B & \longrightarrow & 0. \end{array}$$

As $\epsilon'(g_0 - h_0) = (f - f)\epsilon = 0$, the lower row is exact and P_0 is projective, we get our lifting $s_0 \colon P_0 \to X_1$ with $g_0 - h_0 = d_1^X \circ s_0$.

Assume, for the induction step, that we already have $s_0, \ldots, s_{n-1}$. Write $\Delta_n = g_n - h_n$, then we get the diagram

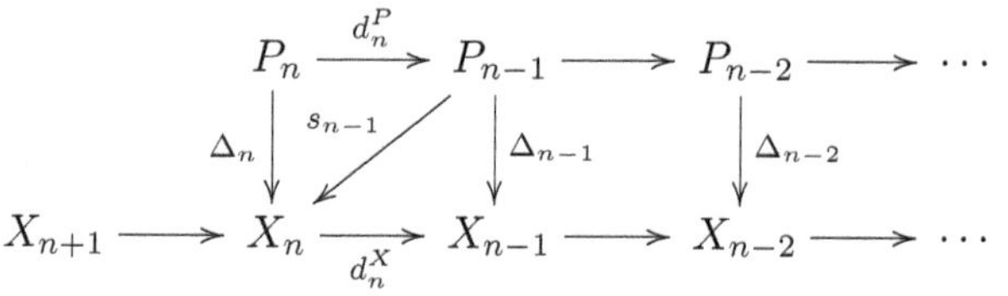

$$\begin{array}{ccccccccc} & & P_n & \xrightarrow{d_n^P} & P_{n-1} & \longrightarrow & P_{n-2} & \longrightarrow & \cdots \\ & & {\scriptstyle \Delta_n}\downarrow & \swarrow{\scriptstyle s_{n-1}} & \downarrow{\scriptstyle \Delta_{n-1}} & & \downarrow{\scriptstyle \Delta_{n-2}} & & \\ X_{n+1} & \longrightarrow & X_n & \xrightarrow[d_n^X]{} & X_{n-1} & \longrightarrow & X_{n-2} & \longrightarrow & \cdots \end{array} \qquad (\dagger\dagger)$$

There results a map $\Delta_n - s_{n-1} \circ d_n^P \colon P_n \longrightarrow X_n$ and a diagram

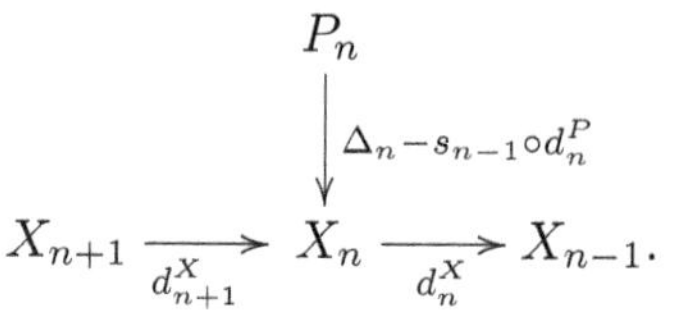

$$\begin{array}{ccccc} & & P_n & & \\ & & \downarrow{\scriptstyle \Delta_n - s_{n-1} \circ d_n^P} & & \\ X_{n+1} & \xrightarrow[d_{n+1}^X]{} & X_n & \xrightarrow[d_n^X]{} & X_{n-1}. \end{array}$$

As usual, if we show that $d_n^X \circ (\Delta_n - s_{n-1} \circ d_n^P) = 0$, then there will be a lift $s_n \colon P_n \to X_{n+1}$ making the diagram

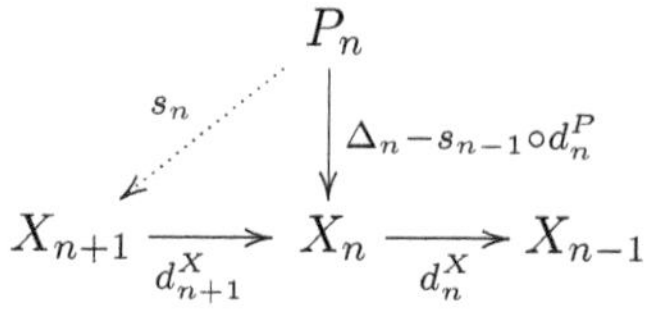

$$\begin{array}{ccccc} & & P_n & & \\ & \swarrow{\scriptstyle s_n} & \downarrow{\scriptstyle \Delta_n - s_{n-1} \circ d_n^P} & & \\ X_{n+1} & \xrightarrow[d_{n+1}^X]{} & X_n & \xrightarrow[d_n^X]{} & X_{n-1} \end{array}$$

commute. Now by the commutativity of $(\dagger\dagger)$, we have $d_n^X \circ \Delta_n = \Delta_{n-1} \circ d_n^P$; so

$$d_n^X \circ (\Delta_n - s_{n-1} \circ d_n^P) = \Delta_{n-1} \circ d_n^P - d_n^X \circ s_{n-1} \circ d_n^P.$$

By the induction hypothesis, $\Delta_{n-1} = g_{n-1} - h_{n-1} = s_{n-2} \circ d_{n-1}^P + d_n^X \circ s_{n-1}$, and therefore

$$\begin{aligned} \Delta_{n-1} \circ d_n^P - d_n^X \circ s_{n-1} \circ d_n^P &= s_{n-2} \circ d_{n-1}^P \circ d_n^P + d_n^X \circ s_{n-1} \circ d_n^P \\ &\quad - d_n^X \circ s_{n-1} \circ d_n^P = 0. \end{aligned}$$

Hence, by Proposition 11.16, s_n exists and we are done. □

Note that Theorem 11.17 holds under hypotheses weaker than the assumption that both $\mathbf{P}^A \xrightarrow{\epsilon} A \longrightarrow 0$ and $\mathbf{X}^B \xrightarrow{\epsilon'} B \longrightarrow 0$ are projective resolutions. It suffices that the first sequence is a chain complex with all P_n projective and that the second sequence is exact (with arbitrary X_n).

There are two important corollaries of the comparison theorem.

Proposition 11.18. *Given any R-linear map $f\colon A \to B$ between some R-modules A and B, if $\mathbf{P}^A \xrightarrow{\epsilon} A \longrightarrow 0$ and $\mathbf{P}'^B \xrightarrow{\epsilon'} B \longrightarrow 0$ are any two projective resolutions of A and B, then f has a lift g from $\mathbf{P}^A$ to $\mathbf{P}'^B$. Furthermore, any two lifts of f are chain homotopic.*

Recall that a *homotopy equivalence* between two chain complexes C and D consists of a pair (g, h) of chain maps $g\colon C \to D$ and $h\colon D \to C$ such that $h \circ g$ is chain homotopic to id_C and $g \circ h$ is chain homotopic to id_D.

We have the following important result which plays a key role in showing that the notion of derived functor does not depend on the choice of a projective resolution.

Theorem 11.19. *Given any R-module A, if $\mathbf{P}^A \xrightarrow{\epsilon} A \longrightarrow 0$ and $\mathbf{P}'^A \xrightarrow{\epsilon'} A \longrightarrow 0$ are any two projective resolutions of A, then $\mathbf{P}^A$ and $\mathbf{P}'^A$ are homotopy equivalent.*

Proof. By Proposition 11.18, the identity map $\mathrm{id}_A\colon A \to A$ has a lift g from $\mathbf{P}^A$ and $\mathbf{P}'^A$ and a lift h from $\mathbf{P}'^A$ and $\mathbf{P}^A$. Then $h \circ g$ is a lift of id_A from $\mathbf{P}^A$ to $\mathbf{P}^A$, and since the identity map $\mathrm{id}_\mathbf{P}$ of the complex $\mathbf{P}^A$ is also a lift of id_A, by Proposition 11.18 there is a chain homotopy from $h \circ g$ to $\mathrm{id}_{\mathbf{P}^A}$. Similarly, $g \circ h$ is a lift of id_A from $\mathbf{P}'^A$ to $\mathbf{P}'^A$, and since the identity map $\mathrm{id}_{\mathbf{P}'}$ of the complex $\mathbf{P}'^A$ is also a lift of id_A, by Proposition 11.18 there is a chain homotopy from $g \circ h$ to $\mathrm{id}_{\mathbf{P}'^A}$. Therefore, g and h define a homotopy equivalence between $\mathbf{P}^A$ and $\mathbf{P}'^A$. $\square$

Since the definition of an injective module is obtained from the definition of a projective module by changing the direction of the arrows it is not unreasonable to expect that a version of Theorem 11.17 holds for injective resolutions. The proof is basically obtained by changing the direction of the arrows, but it takes a little more than that. Indeed, some quotients show up in the proof. Paraphrazing Lang [Lang (1993)]: "The books on homological algebra that I know of in fact carry out the projective case, and leave the injective case to the reader."

We begin with a crucial observation dual to the crucial observation in Proposition 11.16.

Proposition 11.20. *If we have a diagram*

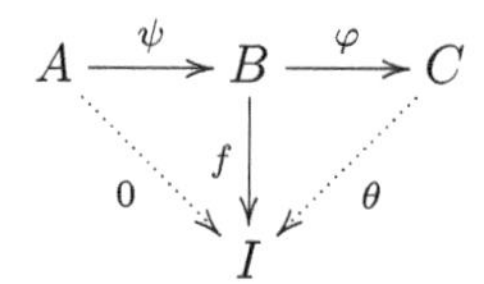

in which

(1) I is injective.
(2) The upper sequence is exact (i.e., $\mathrm{Im}\ \psi = \mathrm{Ker}\,\varphi$*).*
(3) $f \circ \psi = 0$,

then there is a map $\theta\colon C \to I$ *lifting* f *(as shown by the dotted arrow above).*

Proof. Indeed, $f \circ \psi = 0$ implies that $\mathrm{Im}\ \psi \subseteq \mathrm{Ker}\, f$; so we have $\mathrm{Ker}\,\varphi = \mathrm{Im}\ \psi \subseteq \mathrm{Ker}\, f$, that is $\mathrm{Ker}\,\varphi \subseteq \mathrm{Ker}\, f$. It follows that there is a unique map $\overline{f}\colon B/\mathrm{Ker}\,\varphi \to I$ such that the following diagram commutes:

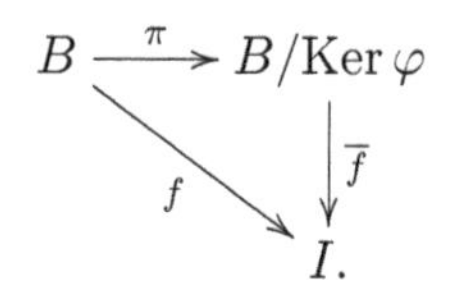

The map $\varphi\colon B \to C$ factors through the quotient map $\overline{\varphi}\colon B/\mathrm{Ker}\,\varphi \to C$ as $\varphi = \overline{\varphi} \circ \pi$ so we have the commutative diagram

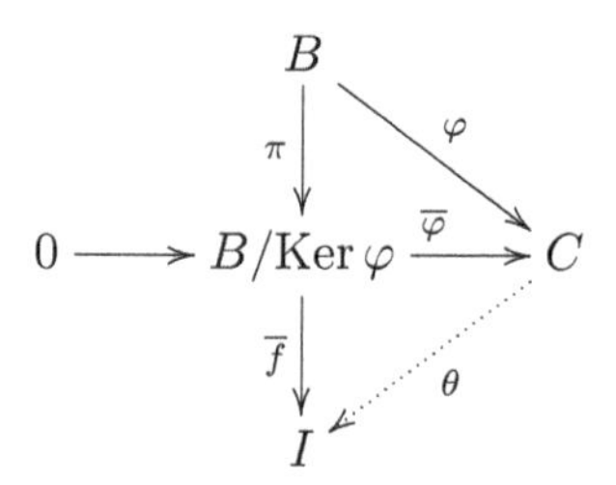

and since I is injective there is a map $\theta\colon C \to I$ lifting $\overline{f}$ as shown in the diagram above. Since $f = \overline{f} \circ \pi$, the commutativity of the above diagram yields $f = \overline{f} \circ \pi = \theta \circ \varphi$, which shows that θ lifts f, as claimed. ☐

Theorem 11.21. *(Comparison Theorem, Injective Case) Let A and B be R-modules. If $0 \longrightarrow A \xrightarrow{\epsilon} \mathbf{X}_A$ is an exact sequence (a right resolution of A) and if $0 \longrightarrow B \xrightarrow{\epsilon'} \mathbf{I}_B$ is a chain complex with all I^n in $\mathbf{I}_B$ injective, then any R-linear map $f\colon A \to B$ lifts to a morphism g from $\mathbf{X}_A$ to $\mathbf{I}_B$ as illustrated by the following commutative diagram:*

$$\begin{array}{ccccccccccc}
0 & \longrightarrow & A & \xrightarrow{\epsilon} & X^0 & \xrightarrow{d_X^0} & X^1 & \xrightarrow{d_X^1} & X^2 & \xrightarrow{d_X^2} & \cdots \\
 & & \downarrow f & & \downarrow g^0 & & \downarrow g^1 & & \downarrow g^2 & & \\
0 & \longrightarrow & B & \xrightarrow[\epsilon']{} & I^0 & \xrightarrow[d_I^0]{} & I^1 & \xrightarrow[d_I^1]{} & I^2 & \xrightarrow[d_I^2]{} & \cdots .
\end{array}$$

Any two morphisms from $\mathbf{X}_A$ to $\mathbf{I}_B$ lifting f are chain homotopic.

Proof. Using Proposition 11.20, the proof of the theorem proceeds by induction and is very similar to the proof of Theorem 11.17. Lang [Lang (1993)] gives most of the details. ☐

Note that Theorem 11.21 holds under hypotheses weaker than the assumption that both $0 \longrightarrow A \xrightarrow{\epsilon} \mathbf{X}_A$ and $0 \longrightarrow B \xrightarrow{\epsilon'} \mathbf{I}_B$ are injective resolutions. It suffices that the first sequence is exact (with arbitrary X^n) and that the second sequence is a chain complex with all I^n injective.

Analogously to the projective case we have the following important corollaries.

Proposition 11.22. *Given any R-linear map $f\colon A \to B$ between some R-modules A and B, if $0 \longrightarrow A \xrightarrow{\epsilon} \mathbf{I}_A$ and $0 \longrightarrow B \xrightarrow{\epsilon'} \mathbf{I}'_B$ are any two injective resolutions of A and B, then f has a lift g from $\mathbf{I}_A$ to $\mathbf{I}'_B$. Furthermore, any two lifts of f are chain homotopic.*

The following result plays a key role in showing that the notion of derived functor does not depend on the choice of an injective resolution.

Theorem 11.23. *Given any R-module A, if $0 \longrightarrow A \xrightarrow{\epsilon} \mathbf{I}_A$ and $0 \longrightarrow A \xrightarrow{\epsilon'} \mathbf{I}'_A$ are any two injective resolutions of A, then $\mathbf{I}_A$ and $\mathbf{I}'_A$ are homotopy equivalent.*

At this stage we are ready to define the central concept of this chapter, the notion of derived functor. A key observation is that the existence of projective resolutions or injective resolutions depends only on the fact that for every object A there is some projective object P and a surjection $\rho\colon P \to A$, and there is some injective object I and an injection $\epsilon\colon A \to I$.

If $\mathbf{C}$ is an abelian category then the notions of projective and injective objects make sense since they are defined purely in terms of conditions on maps; see Definition 11.2 and Definition 11.3. The notions of projective and injective resolutions are also defined by replacing projective modules by projective objects and injective modules by injective objects.

Definition 11.12. Given an abelian category $\mathbf{C}$, we say that $\mathbf{C}$ has *enough injectives* if for every object $A \in \mathbf{C}$ there is some injective object $I \in \mathbf{C}$ and a monic $\epsilon\colon A \to I$ (which means that $\ker \epsilon = 0$) (resp. *enough projectives* if for every $A \in \mathbf{C}$ there is some projective object $P \in \mathbf{C}$ and an epic $\rho\colon P \to A$ (which means that $\operatorname{coker} \rho = 0$)).

If can be shown that if an abelian category $\mathbf{C}$ has enough projectives, then the results of this section (in particular Propositions 11.16, 11.18, and Theorem 11.19) hold. Similarly, if an abelian category $\mathbf{C}$ has enough injectives, then the results of this section (in particular Propositions 11.20, 11.22, and Theorem 11.23) hold.

As we saw, the category of R-modules has enough injectives and projectives. Now it turns out that the category of sheaves (which is abelian) has enough injectives, but does not have enough projectives (as we saw, cokernels and quotients are problematic).

Derived functors have the property that any short exact sequence yields a long cohomology (or homology) exact sequence, and that it is so naturally (as in Theorem 2.22 and Proposition 2.23). To prove these facts requires some rather technical propositions involving projective and injective resolutions. We content ourselves with stating these results. Furthermore, since our ultimate goal is to apply derived functors to the category of sheaves to obtain sheaf cohomology, and since the category of sheaves does not have enough projectives but has enough injectives, we will focus our attention on results involving injectives.

We need to define what we mean by an exact sequence of chain complexes.

Definition 11.13. If $\mathfrak{A} = (A, d_A)$, $\mathfrak{B} = (B, d_B)$ and $\mathfrak{C} = (C, d_C)$ are three chain complexes and $f\colon \mathfrak{A} \to \mathfrak{B}$ and $g\colon \mathfrak{B} \to \mathfrak{C}$ are two chain maps with

$f = (f^n)$ and $g = (g^n)$, we say that the sequence of complexes

$$0 \longrightarrow \mathfrak{A} \xrightarrow{f} \mathfrak{B} \xrightarrow{g} \mathfrak{C} \longrightarrow 0$$

is *exact* iff the sequence

$$0 \longrightarrow A^n \xrightarrow{f^n} B^n \xrightarrow{g^n} C^n \longrightarrow 0$$

is exact for every n.

Proposition 11.24. *(Horseshoe Lemma, Projective Case) Consider the diagram (in some abelian category* $\mathbf{C}$*)*

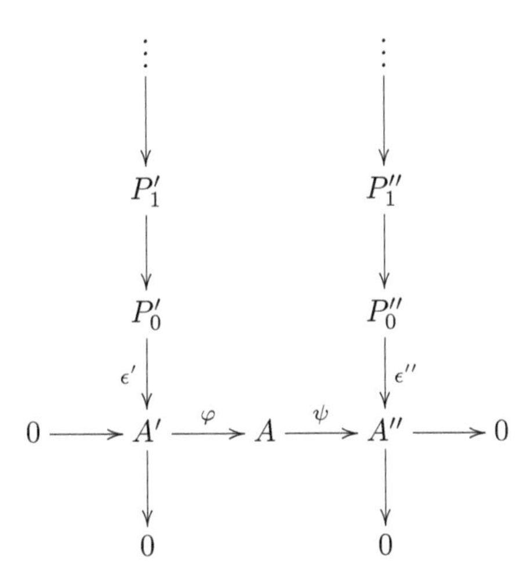

where the left column is a projective resolution $\mathfrak{P}'\colon\ \mathbf{P}^{A'} \xrightarrow{\epsilon'} A' \longrightarrow 0$ *of* A'*, the right column* $\mathfrak{P}''\colon\ \mathbf{P}^{A''} \xrightarrow{\epsilon''} A'' \longrightarrow 0$ *is a projective resolution of* A''*, and the row is an exact sequence. Then there is a projective resolution* $\mathfrak{P}\colon\ \mathbf{P}^{A} \xrightarrow{\epsilon} A \longrightarrow 0$ *of* A *and chain maps* $f\colon \mathfrak{P}' \to \mathfrak{P}$ *and* $g\colon \mathfrak{P} \to \mathfrak{P}''$ *such that the sequence*

$$0 \longrightarrow \mathfrak{P}' \xrightarrow{f} \mathfrak{P} \xrightarrow{g} \mathfrak{P}'' \longrightarrow 0$$

is exact.

A proof of Proposition 11.24 can be found in Rotman [Rotman (1979, 2009)] (Chapter 6, Lemma 6.20).

Proposition 11.25. *(Horseshoe Lemma, Injective Case) Consider the diagram (in some abelian category* $\mathbf{C}$*)*

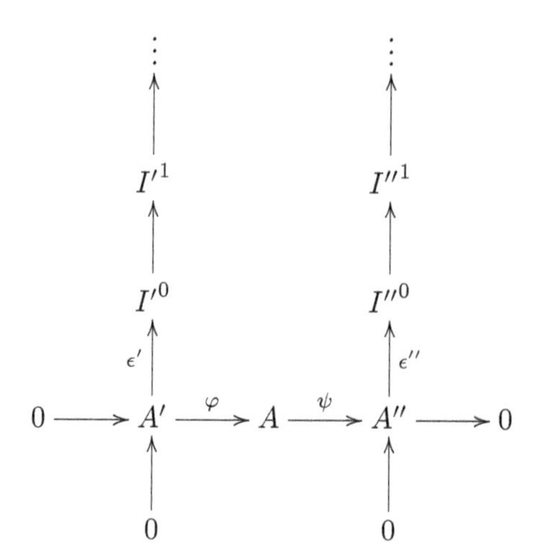

where the left column is an injective resolution $\mathfrak{I}'\colon 0 \longrightarrow A' \xrightarrow{\epsilon'} \mathbf{I}_{A'}$ *of* A', *the right column* $\mathfrak{I}''\colon 0 \longrightarrow A'' \xrightarrow{\epsilon''} \mathbf{I}_{A''}$ *is an injective resolution of* A'', *and the row is an exact sequence. Then there is an injective resolution* $\mathfrak{I}\colon 0 \longrightarrow A \xrightarrow{\epsilon} \mathbf{I}_A$ *of* A *and chain maps* $f\colon \mathfrak{I}' \to \mathfrak{I}$ *and* $g\colon \mathfrak{I} \to \mathfrak{I}''$ *such that the sequence*

$$0 \longrightarrow \mathfrak{I}' \xrightarrow{f} \mathfrak{I} \xrightarrow{g} \mathfrak{I}'' \longrightarrow 0$$

is exact.

We will also need a generalization of the Horseshoe Lemma for chain maps of exact sequences.

Proposition 11.26. *Suppose we have a map of exact sequences (in some abelian category* $\mathbf{C}$*)*

$$\begin{array}{ccccccccc} 0 & \longrightarrow & A' & \xrightarrow{\varphi} & A & \xrightarrow{\psi} & A'' & \longrightarrow & 0 \\ & & \downarrow{\scriptstyle f'} & & \downarrow{\scriptstyle f} & & \downarrow{\scriptstyle f''} & & \\ 0 & \longrightarrow & B' & \xrightarrow[\varphi']{} & B & \xrightarrow[\psi']{} & B'' & \longrightarrow & 0 \end{array}$$

and that we have some injective resolutions $0 \longrightarrow A' \xrightarrow{\epsilon^{A'}} \mathbf{I}_{A'}$, $0 \longrightarrow A'' \xrightarrow{\epsilon^{A''}} \mathbf{I}_{A''}$, $0 \longrightarrow B' \xrightarrow{\epsilon^{B'}} \mathbf{I}_{B'}$ *and* $0 \longrightarrow B'' \xrightarrow{\epsilon^{B''}} \mathbf{I}_{B''}$

of the corners A', A'', B', B'', and chain maps $F' \colon \mathbf{I}_{A'} \to \mathbf{I}_{B'}$ over f' and $F'' \colon \mathbf{I}_{A''} \to \mathbf{I}_{B''}$ over f''. Then there exist injective resolutions $0 \longrightarrow A \xrightarrow{\epsilon^A} \mathbf{I}_A$ of A and $0 \longrightarrow B \xrightarrow{\epsilon^B} \mathbf{I}_B$ of B and a chain map $F \colon \mathbf{I}_A \to \mathbf{I}_B$ over f such that the following diagram commutes

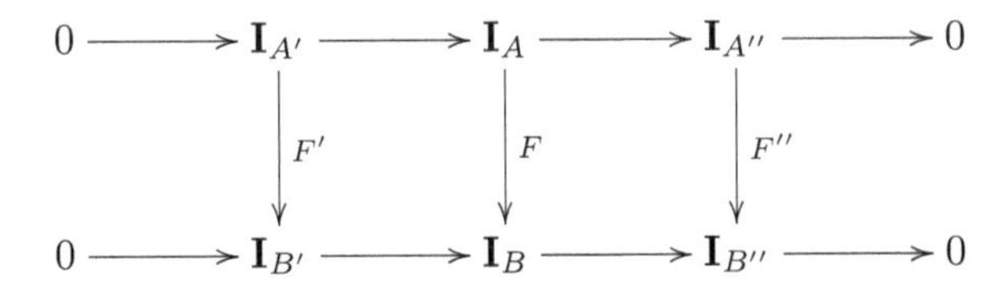

and has exact rows.

There is also a version of Proposition 11.26 for projective resolutions; see Rotman [Rotman (1979, 2009)] (Chapter 6, Lemma 6.24). The reader should enjoy the use of 3-dimensional diagrams involving cubes.

11.4 Left and Right Derived Functors

Let $\mathbf{C}$ and $\mathbf{D}$ be two abelian categories, and let $T \colon \mathbf{C} \to \mathbf{D}$ be an additive functor. Actually, in all our examples $\mathbf{C}$ is either the category of R-modules, the category of presheaves, or the category or sheaves, and $\mathbf{D}$ is either the category of R-modules or the category of abelian groups, so the reader may assume this if the abstract nature of abelian categories makes her/him uncomfortable.

Assume that $\mathbf{C}$ has enough injectives. For any $A \in \mathbf{C}$, if $0 \longrightarrow A \xrightarrow{\epsilon} \mathbf{I}_A$ is an injective resolution of A, then if we apply T to $\mathbf{I}_A$ we obtain the cochain complex

$$0 \longrightarrow T(I^0) \xrightarrow{T(d^0)} T(I^1) \xrightarrow{T(d^1)} \cdots \longrightarrow T(I^n) \xrightarrow{T(d^n)} T(I^{n+1}) \longrightarrow \cdots, \tag{Ri}$$

denoted $T(\mathbf{I}_A)$. If $T \colon \mathbf{C} \to \mathbf{D}$ is a contravariant functor and if we apply T to $\mathbf{I}_A$ we obtain the chain complex

$$0 \longleftarrow T(I^0) \xleftarrow{T(d^0)} T(I^1) \xleftarrow{T(d^1)} \cdots \longleftarrow T(I^n) \xleftarrow{T(d^n)} T(I^{n+1}) \longleftarrow \cdots, \tag{Li}$$

also denoted $T(\mathbf{I}_A)$.

Now assume that $\mathbf{C}$ has enough projectives. For any $A \in \mathbf{C}$, if $\mathbf{P}^A \xrightarrow{\epsilon} A \longrightarrow 0$ is a projective resolution of A, then if we apply T to $\mathbf{P}^A$ we obtain the chain complex

$$0 \longleftarrow T(P_0) \xleftarrow{T(d_1)} T(P_1) \xleftarrow{T(d_2)} \cdots \longleftarrow T(P_{n-1}) \xleftarrow{T(d_n)} T(P_n) \longleftarrow \cdots, \tag{Lp}$$

denoted $T(\mathbf{P}^A)$. If $T\colon \mathbf{C} \to \mathbf{D}$ is a contravariant functor and if we apply T to $\mathbf{P}^A$ we obtain the cochain complex

$$0 \longrightarrow T(P_0) \xrightarrow{T(d_1)} T(P_1) \xrightarrow{T(d_2)} \cdots \longrightarrow T(P_{n-1}) \xrightarrow{T(d_n)} T(P_n) \longrightarrow \cdots, \tag{Rp}$$

also denoted $T(\mathbf{P}^A)$. The above four complexes have (co)homology that defines the left and right derived functors of T.

Definition 11.14. Let $\mathbf{C}$ and $\mathbf{D}$ be two abelian categories, and let $T\colon \mathbf{C} \to \mathbf{D}$ be an additive functor.

(Ri) Assume that $\mathbf{C}$ has enough injectives. For any $A \in \mathbf{C}$, if $0 \longrightarrow A \xrightarrow{\epsilon} \mathbf{I}_A$ is an injective resolution of A, then the cohomology groups of the complex $T(\mathbf{I}_A)$ are denoted by

$$R^nT(\mathbf{I}_A) = H^n(T(\mathbf{I}_A)), \qquad n \geq 0.$$

(Li) If $T\colon \mathbf{C} \to \mathbf{D}$ is a contravariant functor, then the homology groups of the complex $T(\mathbf{I}_A)$ are denoted by

$$L_nT(\mathbf{I}_A) = H_n(T(\mathbf{I}_A)), \qquad n \geq 0.$$

(Lp) Now assume that $\mathbf{C}$ has enough projectives. For any $A \in \mathbf{C}$, if $\mathbf{P}^A \xrightarrow{\epsilon} A \longrightarrow 0$ is a projective resolution of A, then the homology groups of the complex $T(\mathbf{P}^A)$ are denoted by

$$L_nT(\mathbf{P}^A) = H_n(T(\mathbf{P}^A)), \qquad n \geq 0.$$

(Rp) If $T\colon \mathbf{C} \to \mathbf{D}$ is a contravariant functor, then the cohomology groups of the complex $T(\mathbf{P}^A)$ are denoted by

$$R^nT(\mathbf{P}^A) = H^n(T(\mathbf{P}^A)), \qquad n \geq 0.$$

The reason for using R^nT or L_nT is that the chain complexes $T(\mathbf{I}_A)$ in (Ri) and $T(\mathbf{P}^A)$ in (Rp) have arrows going to the right since they are cohomology complexes so the corresponding functors are R^nT, and the chain complexes $T(\mathbf{I}_A)$ in (Li) and $T(\mathbf{P}^A)$ in (Lp) have arrows going to the left since they are homology complexes so the corresponding functors are L_nT.

We also follow the (almost) universally adopted convention that superscripts are used for denoting objects involving cohomology and subscripts are used for denoting objects involving homology.

In the rest of this chapter we always assume that $\mathbf{C}$ and $\mathbf{D}$ are abelian categories and that $\mathbf{C}$ has enough injectives or projectives, as needed.

All the operators introduced in Definition 11.14 are actually functors so let us clarify what are the categories involved. In Cases (Li) and (Ri) the domain category is the set of all injective resolutions $0 \longrightarrow A \xrightarrow{\epsilon} \mathbf{I}_A$ for all $A \in \mathbf{C}$, and a morphism from $0 \longrightarrow A \xrightarrow{\epsilon} \mathbf{I}_A$ to $0 \longrightarrow B \xrightarrow{\epsilon'} \mathbf{I}'_B$ is simply a map $f\colon A \to B$. To be absolutely precise $R^nT(\mathbf{I}_A)$ and $L_nT(\mathbf{I}_A)$ should be denoted $R^nT(\, 0 \longrightarrow A \xrightarrow{\epsilon} \mathbf{I}_A\,)$ and $L_nT(\, 0 \longrightarrow A \xrightarrow{\epsilon} \mathbf{I}_A\,)$ but for the sake of notational simplicity we use the former notation.

In Cases (Lp) and (Rp) the domain category is the set of all projective resolutions $\mathbf{P}^A \xrightarrow{\epsilon} A \longrightarrow 0$ $(A \in \mathbf{C})$, and a morphism from $\mathbf{P}^A \xrightarrow{\epsilon} A \longrightarrow 0$ to $\mathbf{P}'^B \xrightarrow{\epsilon'} B \longrightarrow 0$ is simply a map $f\colon A \to B$. Again, to be absolutely precise $L_nT(\mathbf{P}^A)$ and $R^nT(\mathbf{P}^A)$ should be denoted $L_nT(\, \mathbf{P}^A \xrightarrow{\epsilon} A \longrightarrow 0\,)$ $R^nT(\, \mathbf{P}^A \xrightarrow{\epsilon} A \longrightarrow 0\,)$ but we use the simpler notation.

In both cases the codomain category is $\mathbf{D}$. Definition 11.14 describes how R^nT and L_nT act on objects. We also have to explain how they act on maps $f\colon A \to B$. First consider Case (Ri).

If $0 \longrightarrow A \xrightarrow{\epsilon} \mathbf{I}_A$ is any injective resolution of A and $0 \longrightarrow B \xrightarrow{\epsilon'} \mathbf{I}'_B$ is any injective resolution of B, then by Proposition 11.22 the map f has a lift g from $\mathbf{I}_A$ to $\mathbf{I}'_B$ as illustrated by the following commutative diagram

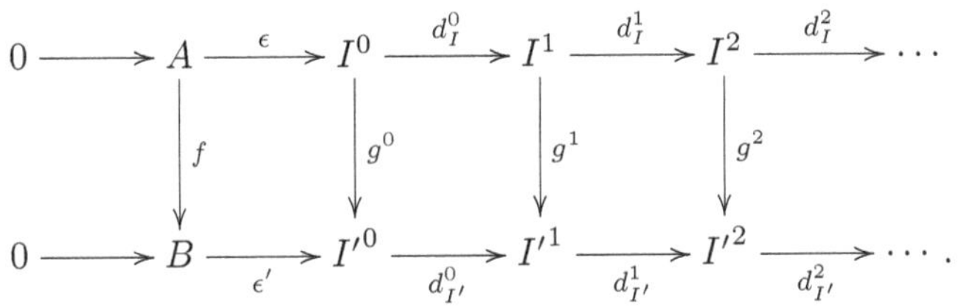

$$\begin{array}{ccccccccccc}
0 & \longrightarrow & A & \xrightarrow{\epsilon} & I^0 & \xrightarrow{d_I^0} & I^1 & \xrightarrow{d_I^1} & I^2 & \xrightarrow{d_I^2} & \cdots \\
 & & \downarrow f & & \downarrow g^0 & & \downarrow g^1 & & \downarrow g^2 & & \\
0 & \longrightarrow & B & \xrightarrow[\epsilon']{} & I'^0 & \xrightarrow[d_{I'}^0]{} & I'^1 & \xrightarrow[d_{I'}^1]{} & I'^2 & \xrightarrow[d_{I'}^2]{} & \cdots .
\end{array}$$

Since T is a functor, $T(g)$ is a chain map from $T(\mathbf{I}_A)$ to $T(\mathbf{I}'_B)$ lifting $T(f)$

as illustrated by the following commutative diagram

$$\begin{array}{ccccccccc}
0 \longrightarrow & T(A) & \xrightarrow{T(\epsilon)} & T(I^0) & \xrightarrow{T(d_I^0)} & T(I^1) & \xrightarrow{T(d_I^1)} & T(I^2) & \xrightarrow{T(d_I^2)} \cdots \\
& \downarrow{\scriptstyle T(f)} & & \downarrow{\scriptstyle T(g^0)} & & \downarrow{\scriptstyle T(g^1)} & & \downarrow{\scriptstyle T(g^2)} & \\
0 \longrightarrow & T(B) & \xrightarrow[T(\epsilon')]{} & T(I'^0) & \xrightarrow[T(d_{I'}^0)]{} & T(I'^1) & \xrightarrow[T(d_{I'}^1)]{} & T(I'^2) & \xrightarrow[T(d_{I'}^2)]{} \cdots .
\end{array}$$

By Proposition 2.19, $T(g)$ induces a homomorphism of cohomology $T(g^n)^*\colon H^n(T(\mathbf{I}_A)) \to H^n(T(\mathbf{I}'_B))$ for all $n \geq 0$. Furthermore, if h is another lift of f, since by Proposition 11.22 any two lifts of f are chain homotopic say by the chain homotopy $(s^n)_{n\geq 0}$, since T is additive, by applying T to the equations

$$g^n - h^n = s^{n+1} \circ d_I^n + d_{I'}^{n-1} \circ s^n$$

we obtain

$$T(g^n) - T(h^n) = T(s^{n+1}) \circ T(d_I^n) + T(d_{I'}^{n-1}) \circ T(s^n),$$

which shows that $(T(s^n))_{n\geq 0}$ is a chain homotopy between $T(g)$ and $T(h)$, and by Proposition 2.20 we have $T(g^n)^* = T(h^n)^*$. Therefore, the homomorphism $T(g^n)^*\colon H^n(T(\mathbf{I}_A)) \to H^n(T(\mathbf{I}'_B))$ is independent of the lift g of f, and we define $R^nT(\mathbf{I}_A, \mathbf{I}'_B)(f)\colon R^nT(\mathbf{I}_A) \to R^nT(\mathbf{I}'_B)$ by

$$R^nT(\mathbf{I}_A, \mathbf{I}'_B)(f) = T(g^n)^*.$$

In Case (Li), since T is a contravariant functor, a lift g of f induces a chain map $T(g)$ between the homology complexes $T(\mathbf{I}'_B)$ and $T(\mathbf{I}_A)$ lifting $T(f)$ as illustrated by the following commutative diagram

$$\begin{array}{ccccccccc}
0 \longleftarrow & T(B) & \xleftarrow{T(\epsilon')} & T(I'^0) & \xleftarrow{T(d_{I'}^0)} & T(I'^1) & \xleftarrow{T(d_{I'}^1)} & T(I'^2) & \xleftarrow{T(d_{I'}^2)} \cdots \\
& \downarrow{\scriptstyle T(f)} & & \downarrow{\scriptstyle T(g^0)} & & \downarrow{\scriptstyle T(g^1)} & & \downarrow{\scriptstyle T(g^2)} & \\
0 \longleftarrow & T(A) & \xleftarrow[T(\epsilon)]{} & T(I^0) & \xleftarrow[T(d_I^0)]{} & T(I^1) & \xleftarrow[T(d_I^1)]{} & T(I^2) & \xleftarrow[T(d_I^2)]{} \cdots .
\end{array}$$

The map $T(g_n)_*\colon H_n(T(\mathbf{I}'_B)) \to H_n(T(\mathbf{I}_A))$ is a homomorphism of homology and we obtain a well-defined map $L_nT(\mathbf{I}'_B, \mathbf{I}_A)(f)\colon L_nT(\mathbf{I}'_B) \to L_nT(\mathbf{I}_A)$ (independent of the lifting g) given by

$$L_nT(\mathbf{I}'_B, \mathbf{I}_A)(f) = T(g_n)_*.$$

In Case (Lp) we use projective resolutions $\mathbf{P}^A \xrightarrow{\epsilon} A \longrightarrow 0$ and $\mathbf{P}'^B \xrightarrow{\epsilon'} B \longrightarrow 0$. By Proposition 11.18, the map f has a lift g from $\mathbf{P}^A$ to $\mathbf{P}'^B$ as illustrated by the following commutative diagram

$$\begin{array}{ccccccccccc}
0 & \longleftarrow & A & \xleftarrow{\epsilon} & P_0 & \xleftarrow{d_0^P} & P_1 & \xleftarrow{d_1^P} & P^2 & \xleftarrow{d_2^P} & \cdots \\
 & & \downarrow f & & \downarrow g_0 & & \downarrow g_1 & & \downarrow g_2 & & \\
0 & \longleftarrow & B & \xleftarrow[\epsilon']{} & P'_0 & \xleftarrow[d_0^{P'}]{} & P'_1 & \xleftarrow[d_1^{P'}]{} & P'_2 & \xleftarrow[d_2^{P'}]{} & \cdots .
\end{array}$$

Since T is a functor, $T(g)$ is a chain map from $T(\mathbf{P}^A)$ to $T(\mathbf{P}'^B)$ lifting $T(f)$ as illustrated by the following commutative diagram

$$\begin{array}{ccccccccccc}
0 & \longleftarrow & T(A) & \xleftarrow{T(\epsilon)} & T(P_0) & \xleftarrow{T(d_0^P)} & T(P_1) & \xleftarrow{T(d_1^P)} & T(P^2) & \xleftarrow{T(d_2^P)} & \cdots \\
 & & \downarrow T(f) & & \downarrow T(g_0) & & \downarrow T(g_1) & & \downarrow T(g_2) & & \\
0 & \longleftarrow & T(B) & \xleftarrow[T(\epsilon')]{} & T(P'_0) & \xleftarrow[T(d_0^{P'})]{} & T(P'_1) & \xleftarrow[T(d_1^{P'})]{} & T(P'_2) & \xleftarrow[T(d_2^{P'})]{} & \cdots .
\end{array}$$

Then $T(g)$ is a chain map of homology from $T(\mathbf{P}^A)$ to $T(\mathbf{P}'^B)$, and $T(g_n)_*\colon H_n(T(\mathbf{P}^A)) \to H_n(T(\mathbf{P}'^B))$ is the induced map of homology. We obtain a well-defined map of homology (independent of the lifting g) $L_nT(\mathbf{P}^A, \mathbf{P}'^B)(f)\colon L_nT(\mathbf{P}^A) \to L_nT(\mathbf{P}'^B)$ given by

$$L_nT(\mathbf{P}^A, \mathbf{P}'^B)(f) = T(g_n)_*.$$

In Case (Rp), we use projective resolutions and Proposition 11.18. Since T is a contravariant functor, $T(g)$ is a chain map from $T(\mathbf{P}'^B)$ to $T(\mathbf{P}^A)$ lifting $T(f)$ as illustrated by the following commutative diagram

$$\begin{array}{ccccccccccc}
0 & \longrightarrow & T(B) & \xrightarrow{T(\epsilon')} & T(P'_0) & \xrightarrow{T(d_0^{P'})} & T(P'_1) & \xrightarrow{T(d_1^{P'})} & T(P'_2) & \xrightarrow{T(d_2^{P'})} & \cdots \\
 & & \downarrow T(f) & & \downarrow T(g_0) & & \downarrow T(g_1) & & \downarrow T(g_2) & & \\
0 & \longrightarrow & T(A) & \xrightarrow[T(\epsilon)]{} & T(P_0) & \xrightarrow[T(d_0^P)]{} & T(P_1) & \xrightarrow[T(d_1^P)]{} & T(P_2) & \xrightarrow[T(d_2^P)]{} & \cdots .
\end{array}$$

Then $T(g)$ is a chain map of cohomology from $T(\mathbf{P}'^B)$ to $T(\mathbf{P}^A)$ and $T(g^n)^*\colon H^n(T(\mathbf{P}'^B)) \to H^n(T(\mathbf{P}^A))$ is the induced map of cohomology.

We obtain a well-defined map of cohomology (independent of the lifting g) $R^nT(\mathbf{P}'^B, \mathbf{P}^A)(f)\colon R^nT(\mathbf{P}'^B) \to R^nT(\mathbf{P}^A)$ given by

$$R^nT(\mathbf{P}'^B, \mathbf{P}^A)(f) = T(g^n)^*.$$

In summary we make the following definition.

Definition 11.15. Let $A, B \in \mathbf{C}$ be objects in $\mathbf{C}$ and let $f\colon A \to B$ be any map.

(Ri) If $0 \longrightarrow A \xrightarrow{\epsilon} \mathbf{I}_A$ is any injective resolution of A and $0 \longrightarrow B \xrightarrow{\epsilon'} \mathbf{I}'_B$ is any injective resolution of B, then we define $R^nT(\mathbf{I}_A, \mathbf{I}'_B)(f)\colon R^nT(\mathbf{I}_A) \to R^nT(\mathbf{I}'_B)$ by

$$R^nT(\mathbf{I}_A, \mathbf{I}'_B)(f) = T(g^n)^*$$

for any lift g of f. The map $T(g^n)^*\colon H^n(T(\mathbf{I}_A)) \to H^n(T(\mathbf{I}'_B))$ is independent of the lift g.

(Li) We define $L_nT(\mathbf{I}'_B, \mathbf{I}_A)(f)\colon L_nT(\mathbf{I}'_B) \to L_nT(\mathbf{I}_A)$ by

$$L_nT(\mathbf{I}'_B, \mathbf{I}_A)(f) = T(g_n)_*$$

for any lift g of f. The map $T(g_n)_*\colon H_n(T(\mathbf{I}'_B)) \to H_n(T(\mathbf{I}_A))$ is independent of the lift g.

(Lp) If $\mathbf{P}^A \xrightarrow{\epsilon} A \longrightarrow 0$ is any projective resolution of A and $\mathbf{P}'^B \xrightarrow{\epsilon'} B \longrightarrow 0$ is any projective resolution of B, then we define $L_nT(\mathbf{P}^A, \mathbf{P}'^B)(f)\colon L_nT(\mathbf{P}^A) \to L_nT(\mathbf{P}'^B)$ by

$$L_nT(\mathbf{P}^A, \mathbf{P}'^B)(f) = T(g_n)_*$$

for any lift g of f. The map $T(g_n)_*\colon H_n(T(\mathbf{P}^A)) \to H_n(T(\mathbf{P}'^B))$ is independent of the lift g.

(Rp) We define $R^nT(\mathbf{P}'^B, \mathbf{P}^A)(f)\colon R^nT(\mathbf{P}'^B) \to R^nT(\mathbf{P}^A)$ by

$$R^nT(\mathbf{P}'^B, \mathbf{P}^A)(f) = T(g^n)^*$$

for any lift g of f. The map $T(g^n)^*\colon H^n(T(\mathbf{P}'^B)) \to H^n(T(\mathbf{P}^A))$ is independent of the lift g.

It is an easy exercise to check that R^nT and L_nT are additive functors, contravariant in Cases (Li) and (Rp).

The next two theorems are absolutely crucial results. Indeed, they show that even though the objects $R^nT(\mathbf{I}_A)$ (and $L_nT(\mathbf{I}_A)$) depend on the injective resolution $\mathbf{I}_A$ chosen for A, this dependency is inessential because any other resolution $\mathbf{I}'_A$ for A yields an object $R^nT(\mathbf{I}'_A)$ *isomorphic* to $R^nT(\mathbf{I}_A)$. Similarly if $\mathbf{P}^A$ and $\mathbf{P}'^A$ are two different resolutions for A then $L_nT(\mathbf{P}^A)$ and $L_nT(\mathbf{P}'^A)$ are isomorphic. The key to these isomorphisms are the comparison theorems. These isomorphisms are actually isomorphisms of functors known as natural transformations that we now define. A natural transformation is a simple generalization of the notion of morphism of presheaves.

Definition 11.16. Given two categories $\mathbf{C}$ and $\mathbf{D}$ and two functors $F, G\colon \mathbf{C} \to \mathbf{D}$ between them, a *natural transformation* $\eta\colon F \to G$ is a family $\eta = (\eta_A)_{A\in\mathbf{C}}$ of maps $\eta_A\colon F(A) \to G(A)$ in $\mathbf{D}$ such that the following diagram commutes for all maps $f\colon A \to B$ between objects $A, B \in \mathbf{C}$:

$$\begin{array}{ccc} F(A) & \xrightarrow{\eta_A} & G(A) \\ {\scriptstyle F(f)}\downarrow & & \downarrow{\scriptstyle G(f)} \\ F(B) & \xrightarrow[\eta_B]{} & G(B). \end{array}$$

We are now ready to state and prove our crucial theorems.

Theorem 11.27. *Let* $0 \longrightarrow A \xrightarrow{\epsilon_A} \mathbf{I}_A$ *and* $0 \longrightarrow A \xrightarrow{\epsilon'_A} \mathbf{I}'_A$ *be any two injective resolutions for any* $A \in \mathbf{C}$. *If* $T\colon \mathbf{C} \to \mathbf{D}$ *is any additive functor, then there are isomorphisms*

$$\eta^n_A\colon R^nT(\mathbf{I}_A) \to R^nT(\mathbf{I}'_A)$$

for all $n \geq 0$ *that depend only on* A *and* T. *Furthermore, for any map* $f\colon A \to B$, *for any injective resolutions* $0 \longrightarrow B \xrightarrow{\epsilon_B} \mathbf{I}_B$ *and*

$0 \longrightarrow B \xrightarrow{\epsilon'_B} \mathbf{I}'_B$ *of* B *the following diagram*

$$\begin{array}{ccc} R^nT(\mathbf{I}_A) & \xrightarrow{\eta^n_A} & R^nT(\mathbf{I}'_A) \\ {\scriptstyle R^nT(\mathbf{I}_A,\mathbf{I}_B)(f)}\downarrow & & \downarrow{\scriptstyle R^nT(\mathbf{I}'_A,\mathbf{I}'_B)(f)} \\ R^nT(\mathbf{I}_B) & \xrightarrow[\eta^n_B]{} & R^nT(\mathbf{I}'_B) \end{array}$$

commutes for all $n \geq 0$.

If $T\colon \mathbf{C} \to \mathbf{D}$ *is a contravariant additive functor, then there are isomorphisms*

$$\eta^A_n\colon L_nT(\mathbf{I}_A) \to L_nT(\mathbf{I}'_A)$$

for all $n \geq 0$ that depend only on A and T. Furthermore, the following diagram

$$\begin{array}{ccc} L_nT(\mathbf{I}_B) & \xrightarrow{\eta_n^B} & L_nT(\mathbf{I}'_B) \\ {\scriptstyle L_nT(\mathbf{I}_B,\mathbf{I}_A)(f)}\downarrow & & \downarrow{\scriptstyle L_nT(\mathbf{I}'_B,\mathbf{I}'_A)(f)} \\ L_nT(\mathbf{I}_A) & \xrightarrow[\eta_n^A]{} & L_nT(\mathbf{I}'_A) \end{array}$$

commutes for all $n \geq 0$.

Proof. By Theorem 11.23 the complexes $\mathbf{I}_A$ and $\mathbf{I}'_A$ are homotopy equivalent, which means that there are chain maps $g\colon \mathbf{I}_A \to \mathbf{I}'_A$ and $h\colon \mathbf{I}'_A \to \mathbf{I}_A$ both lifting id_A such that $h \circ g$ is chain homotopic to $\mathrm{id}_{\mathbf{I}_A}$ and $g \circ h$ is chain homotopic to $\mathrm{id}_{\mathbf{I}'_A}$. Since T is additive, $T(h) \circ T(g)$ is chain homotopic to $\mathrm{id}_{T(\mathbf{I}_A)}$ and $T(g) \circ T(h)$ is chain homotopic to $\mathrm{id}_{T(\mathbf{I}'_A)}$. These chain maps induce cohomology homomorphisms for all $n \geq 0$ and by Proposition 2.20, we obtain

$$\begin{aligned} T(h^n)^* \circ T(g^n)^* &= \mathrm{id}_{T(\mathbf{I}_A)} \\ T(g^n)^* \circ T(h^n)^* &= \mathrm{id}_{T(\mathbf{I}'_A)}. \end{aligned}$$

Therefore, $T(g^n)^*\colon H^n(T(\mathbf{I}_A)) \to H^n(T(\mathbf{I}'_A))$ is an isomorphism of cohomology.

We still have to show that this map depends only on T and A. This is because by Proposition 11.22, any two lifts g and g' of id_A are chain homotopic, so $T(g)$ and $T(g')$ are chain homotopic, and by Proposition 2.20 we have $T(g^n)^* = T(g'^n)^*$. As a consequence, it is legitimate to set $\eta_A^n = T(g^n)^*$, a well-defined isomorphism $\eta_A^n\colon R^nT(\mathbf{I}_A) \to R^nT(\mathbf{I}'_A)$.

Finally, we need to check that the η_A^n yield a natural transformation. For any map $f\colon A \to B$ we need to show that the following diagram commutes:

$$\begin{array}{ccc} R^nT(\mathbf{I}_A) & \xrightarrow{\eta_A^n} & R^nT(\mathbf{I}'_A) \\ {\scriptstyle R^nT(\mathbf{I}_A,\mathbf{I}_B)(f)}\downarrow & & \downarrow{\scriptstyle R^nT(\mathbf{I}'_A,\mathbf{I}'_B)(f)} \\ R^nT(\mathbf{I}_B) & \xrightarrow[\eta_B^n]{} & R^nT(\mathbf{I}'_B). \end{array}$$

The map η_A^n is given by a lifting g_A of id_A from $\mathbf{I}_A$ to $\mathbf{I}'_A$, and the map $R^nT(\mathbf{I}'_A, \mathbf{I}'_B)(f)$ is given by a lifting h' of f from $\mathbf{I}'_A$ to $\mathbf{I}'_B$. Thus $h' \circ g_A$

is a lifting of $f \circ \mathrm{id}_A = f$ from $\mathbf{I}_A$ to $\mathbf{I}'_B$, as illustrated in the following commutative diagram

$$\begin{array}{ccccccccccc}
0 & \longrightarrow & A & \longrightarrow & I_A^0 & \longrightarrow & I_A^1 & \longrightarrow & I_A^2 & \longrightarrow & I_A^3 & \longrightarrow \cdots \\
 & & \downarrow{\scriptstyle \mathrm{id}_A} & & \downarrow{\scriptstyle g_A^0} & & \downarrow{\scriptstyle g_A^1} & & \downarrow{\scriptstyle g_A^2} & & \downarrow{\scriptstyle g_A^3} & \\
0 & \longrightarrow & A & \longrightarrow & I'^0_A & \longrightarrow & I'^1_A & \longrightarrow & I'^2_A & \longrightarrow & I'^3_A & \longrightarrow \cdots \\
 & & \downarrow{\scriptstyle f} & & \downarrow{\scriptstyle h'^0} & & \downarrow{\scriptstyle h'^1} & & \downarrow{\scriptstyle h'^2} & & \downarrow{\scriptstyle h'^3} & \\
0 & \longrightarrow & B & \longrightarrow & I'^0_B & \longrightarrow & I'^1_B & \longrightarrow & I'^2_B & \longrightarrow & I'^3_B & \longrightarrow \cdots
\end{array}$$

Similarly the map η_B^n is given by a lifting g_B of id_B from $\mathbf{I}_B$ to $\mathbf{I}'_B$, and the map $R^nT(\mathbf{I}_A, \mathbf{I}_B)(f)$ is given by a lifting h of f from $\mathbf{I}_A$ to $\mathbf{I}_B$. Thus $g_B \circ h$ is a lifting of $\mathrm{id}_B \circ f = f$ from $\mathbf{I}_A$ to $\mathbf{I}'_B$, as illustrated in the following commutative diagram

$$\begin{array}{ccccccccccc}
0 & \longrightarrow & A & \longrightarrow & I_A^0 & \longrightarrow & I_A^1 & \longrightarrow & I_A^2 & \longrightarrow & I_A^3 & \longrightarrow \cdots \\
 & & \downarrow{\scriptstyle f} & & \downarrow{\scriptstyle h^0} & & \downarrow{\scriptstyle h^1} & & \downarrow{\scriptstyle h^2} & & \downarrow{\scriptstyle h^3} & \\
0 & \longrightarrow & B & \longrightarrow & I_B^0 & \longrightarrow & I_B^1 & \longrightarrow & I_B^2 & \longrightarrow & I_B^3 & \longrightarrow \cdots \\
 & & \downarrow{\scriptstyle \mathrm{id}_B} & & \downarrow{\scriptstyle g_B^0} & & \downarrow{\scriptstyle g_B^1} & & \downarrow{\scriptstyle g_B^2} & & \downarrow{\scriptstyle g_B^3} & \\
0 & \longrightarrow & B & \longrightarrow & I'^0_B & \longrightarrow & I'^1_B & \longrightarrow & I'^2_B & \longrightarrow & I'^3_B & \longrightarrow \cdots
\end{array}$$

Since T is a functor, $T(h') \circ T(g_A)$ and $T(g_B) \circ T(h)$ both lift $T(f)$, and by Proposition 11.22 they are chain homotopic, so

$$T(h'^n)^* \circ T(g_A^n)^* = T(g_B^n)^* \circ T(h^n)^*$$

or equivalently

$$R^nT(\mathbf{I}'_A, \mathbf{I}'_B)(f) \circ \eta_A^n = \eta_B^n \circ R^nT(\mathbf{I}_A, \mathbf{I}_B)(f)$$

as desired. The proof in the case of a contravariant functor is similar. ☐

We have a similar theorem for projective resolutions using Proposition 11.18 and Theorem 11.19 instead of Proposition 11.22 and Theorem 11.23.

Theorem 11.28. *Let* $\mathbf{P}^A \xrightarrow{\epsilon^A} A \longrightarrow 0$ *and* $\mathbf{P}'^A \xrightarrow{\epsilon'^A} A \longrightarrow 0$ *be any two projective resolutions for any* $A \in \mathbf{C}$. *If* $T \colon \mathbf{C} \to \mathbf{D}$ *is any additive functor, then there are isomorphisms*

$$\eta_n^A \colon L_nT(\mathbf{P}^A) \to L_nT(\mathbf{P}'^A)$$

for all $n \geq 0$ that depend only on A and T. Furthermore, for any map $f\colon A \to B$, for any projective resolutions $\mathbf{P}^B \xrightarrow{\epsilon^B} B \longrightarrow 0$ and $\mathbf{P}'^B \xrightarrow{\epsilon'^B} B \longrightarrow 0$ of B, the following diagram

$$\begin{array}{ccc} L_nT(\mathbf{P}^A) & \xrightarrow{\eta_n^A} & L_nT(\mathbf{P}'^A) \\ {\scriptstyle L_nT(\mathbf{P}^A,\mathbf{P}^B)(f)}\downarrow & & \downarrow{\scriptstyle L_nT(\mathbf{P}'^A,\mathbf{P}'^B)(f)} \\ L_nT(\mathbf{P}^B) & \xrightarrow[\eta_n^B]{} & L_nT(\mathbf{P}'^B) \end{array}$$

commutes for all $n \geq 0$.

If $T\colon \mathbf{C} \to \mathbf{D}$ is a contravariant additive functor, then there are isomorphisms

$$\eta_A^n\colon R^nT(\mathbf{P}^A) \to R^nT(\mathbf{P}'^A)$$

for all $n \geq 0$ that depend only on A and T. Furthermore, the following diagram

$$\begin{array}{ccc} R^nT(\mathbf{P}^B) & \xrightarrow{\eta_B^n} & R^nT(\mathbf{P}'^B) \\ {\scriptstyle R^nT(\mathbf{P}^B,\mathbf{P}^A)(f)}\downarrow & & \downarrow{\scriptstyle R^nT(\mathbf{P}'^B,\mathbf{P}'^A)(f)} \\ R^nT(\mathbf{P}^A) & \xrightarrow[\eta_A^n]{} & R^nT(\mathbf{P}'^A) \end{array}$$

commutes for all $n \geq 0$.

Theorem 11.27 and Theorem 11.28 suggest defining R^nT and L_nT as functors with domain $\mathbf{C}$ rather than projective or injective resolutions.

Definition 11.17. Let $\mathbf{C}$ and $\mathbf{D}$ be two abelian categories, and let $T\colon \mathbf{C} \to \mathbf{D}$ be an additive functor.

(Ri) Assume that $\mathbf{C}$ has enough injectives and for every object A in $\mathbf{C}$ choose (once and for all) some injective resolution $0 \longrightarrow A \xrightarrow{\epsilon} \mathbf{I}_A$. The *right derived functors* R^nT of T are defined for every $A \in \mathbf{C}$ by

$$R^nT(A) = R^nT(\mathbf{I}_A) = H^n(T(\mathbf{I}_A)), \qquad n \geq 0,$$

and for every map $f\colon A \to B$, by

$$R^nT(f) = R^nT(\mathbf{I}_A, \mathbf{I}'_B)(f), \qquad n \geq 0.$$

(Li) If $T\colon \mathbf{C} \to \mathbf{D}$ is a contravariant functor, then the *left derived functors* L_nT of T are defined for every $A \in \mathbf{C}$ by

$$L_nT(A) = L_nT(\mathbf{I}_A) = H_n(T(\mathbf{I}_A)), \qquad n \geq 0,$$

and for every map $f\colon A \to B$, by

$$L_nT(f) = L_nT(\mathbf{I}'_B, \mathbf{I}_A)(f), \qquad n \geq 0.$$

(Lp) Now assume that $\mathbf{C}$ has enough projectives and for every object A in $\mathbf{C}$ choose (once and for all) some projective resolution $\mathbf{P}^A \xrightarrow{\epsilon} A \longrightarrow 0$. The *left derived functors* L_nT of T are defined for every $A \in \mathbf{C}$ by

$$L_nT(A) = L_nT(\mathbf{P}^A) = H_n(T(\mathbf{P}^A)), \qquad n \geq 0,$$

and for every map $f\colon A \to B$, by

$$L_nT(f) = L_nT(\mathbf{P}^A, \mathbf{P'}^B)(f), \qquad n \geq 0.$$

(Rp) If $T\colon \mathbf{C} \to \mathbf{D}$ is a contravariant functor, then the *right derived functors* R^nT of T are defined for every $A \in \mathbf{C}$ by

$$R^nT(A) = R^nT(\mathbf{P}^A) = H^n(T(\mathbf{P}^A)), \qquad n \geq 0,$$

and for every map $f\colon A \to B$, by

$$R^nT(f) = R^nT(\mathbf{P'}^B, \mathbf{P}^A)(f), \qquad n \geq 0.$$

Observe that in (Li) and (Rp) the derived functors are contravariant. Any other choice of injective resolutions or projective resolutions yields derived functors $(\widehat{R}^nT)_{n\geq 0}$ and $(\widehat{L}_nT)_{n\geq 0}$ that are naturally isomorphic to the derived functors $(R^nT)_{n\geq 0}$ and $(L_nT)_{n\geq 0}$ associated to the original fixed choice of resolutions (in the sense that the $(\eta^n_A)_{A\in\mathbf{C}}$ and $(\eta^A_n)_{A\in\mathbf{C}}$ in Theorems 11.27 and 11.28 are natural transformations with all η^n_A and all η^A_n isomorphisms). For example, in Case (Ri), for all maps $f\colon A \to B$, we have the commutative diagram

$$\begin{array}{ccc} R^nT(A) & \xrightarrow{\eta^n_A} & \widehat{R}^nT(A) \\ \downarrow{\scriptstyle R^nT(f)} & & \downarrow{\scriptstyle \widehat{R}^nT(f)} \\ R^nT(B) & \xrightarrow[\eta^n_B]{} & \widehat{R}^nT(B) \end{array}$$

for every $n \geq 0$.

11.5 Left-Exact and Right-Exact Derived Functors

One of the main reasons for defining the derived functors $(R^nT)_{n\geq 0}$ and $(L_nT)_{n\geq 0}$ is to investigate properties of T, in particular how much does T preserve exactness. For T fixed, the objects $R^nT(A)$ (or $L_nT(A)$) (groups if $\mathbf{D}$ is the category of abelian groups) are important invariants of the object A.

It turns out that more useful information is obtained if either R^0T is isomorphic to T or L_0T is isomorphic to T. The following proposition gives sufficient conditions for this to happen.

Proposition 11.29. *Let $\mathbf{C}$ and $\mathbf{D}$ be two abelian categories, and let $T\colon \mathbf{C}\to\mathbf{D}$ be an additive functor.*

(1) If T is left-exact then R^0T is naturally isomorphic to T. If T is right-exact and contravariant then L_0T is naturally isomorphic to T.
(2) If T is right-exact then L_0T is naturally isomorphic to T. If T is left-exact and contravariant then R^0T is naturally isomorphic to T.

Proof. (1) Let $0 \longrightarrow A \xrightarrow{\epsilon} \mathbf{I}_A$ be an injective resolution of A. Since T is left-exact we have the exact sequence

$$0 \longrightarrow T(A) \xrightarrow{T(\epsilon)} T(I_0) \xrightarrow{T(d^0)} T(I_1).$$

Since $T(\epsilon)$ is injective, it follows that $T(A)$ is isomorphic to $\mathrm{Im}\, T(\epsilon) = \mathrm{Ker}\, T(d^0)$. The chain complex $T(\mathbf{I}_A)$ given by

$$0 \longrightarrow T(I_0) \xrightarrow{T(d^0)} T(I_1) \xrightarrow{T(d^1)} T(I^2) \longrightarrow \cdots$$

yields $R^0T(A) = H^0(T(\mathbf{I}_A)) = \mathrm{Ker}\, T(d^0)$, so $T(A)$ is isomorphic to $R^0T(A)$. We leave it as an exercise to show that these isomorphisms constitute a natural transformation. The case where T is right-exact and contravariant is left as an exercise.

(2) Let $\mathbf{P}^A \xrightarrow{\epsilon} A \longrightarrow 0$ be a projective resolution of A. Since T is right-exact we have the exact sequence

$$0 \longleftarrow T(A) \xleftarrow{T(\epsilon)} T(P^0) \xleftarrow{T(d_1)} T(P^1).$$

Since the map $T(\epsilon)$ is surjective $T(A)$ is isomorphic to $T(P^0)/\mathrm{Ker}\, T(\epsilon) = T(P^0)/\mathrm{Im}\, T(d_1)$. The chain complex $T(\mathbf{P}^A)$ given by

$$0 \longleftarrow T(P^0) \xleftarrow{T(d_1)} T(P^1) \xleftarrow{T(d_2)} T(P^2) \longleftarrow \cdots$$

yields $L_0T(A) = H_0(T(\mathbf{P}^A)) = T(P^0)/\mathrm{Im}\, T(d_1)$, so $T(A)$ is isomorphic to $L_0T(A)$. We leave it as an exercise to show that these isomorphisms constitute a natural transformation. The case where T is left-exact and contravariant is also left as an exercise. $\square$

Remark: We will show later that in Case (Ri) R^0T is left-exact, in Case (Li) L_0T is right-exact, in Case (Lp) L_0T is right-exact, and in Case (Rp) R^0T is left-exact. These properties also proven in Rotman [Rotman (1979, 2009)]. As a consequence, the conditions of Proposition 11.29 are necessary and sufficient.

Example 11.1. We know that the contravariant functor $T_B(A) = \mathrm{Hom}(A, B)$ with B fixed is left-exact. Its right derived functors are the "Ext" functors (see Definition 11.6)

$$\mathrm{Ext}^n_R(A, B) = (R^nT_B)(A),$$

with

$$\mathrm{Ext}^0_R(A, B) \cong \mathrm{Hom}(A, B).$$

This corresponds to Case (Rp).

We also know that the functor $T'_A(B) = \mathrm{Hom}(A, B)$ with A fixed is left-exact. Its right derived functors are also "Ext" functors (see Definition 11.9)

$$\mathrm{Ext}'^n_R(A, B) = (R^nT'_A)(B),$$

with

$$\mathrm{Ext}'^0_R(A, B) \cong \mathrm{Hom}(A, B).$$

This corresponds to Case (Ri). It turns out that $\mathrm{Ext}^n_R(A, B)$ and $\mathrm{Ext}'^n_R(A, B)$ are isomorphic; see Rotman [Rotman (1979, 2009)] (Chapter 7, Theorem 7.8).

The functor $T_B(A) = A \otimes B$ with B fixed is right-exact. Its left derived functors are the "Tor" functors (see Definition 11.7)

$$\mathrm{Tor}^R_n(A, B) = (L_nT_B)(A),$$

with

$$\mathrm{Tor}^R_0(A, B) \cong A \otimes B.$$

This corresponds to Case (Lp).

Similarly the functor $T_A(B) = A \otimes B$ with A fixed is right-exact. Its left derived functors are also the "Tor" functors

$$\mathrm{Tor}'^R_n(A, B) = (L_n T_A)(B),$$

with

$$\mathrm{Tor}'^R_0(A, B) \cong A \otimes B.$$

This also corresponds to Case (Lp). It turns out that $\mathrm{Tor}^R_n(A, B)$ and $\mathrm{Tor}'^R_n(A, B)$ are isomorphic; see Rotman [Rotman (1979, 2009)] (Chapter 7, Theorem 7.9). It can be shown that for all R-modules A and B, the R-module $\mathrm{Tor}^R_n(A, B)$ is a torsion module for all $n \geq 1$; see Rotman [Rotman (1979, 2009)] (Chapter 8, Theorem 8.21).

Since Hom is not right-exact, its left derived functors convey no obvious information about Hom. Similarly, since $\otimes$ is not left-exact, its right derived functors convey no obvious information about it.

Although quite trivial the following proposition has significant implications, namely that the family of right derived functors $(R^n T)_{n \geq 0}$ are universal δ-functors, and that the family of left derived functors $(L_n T)_{n \geq 0}$ are universal ∂-functors; see Section 11.8.

Proposition 11.30. *Let* $\mathbf{C}$ *and* $\mathbf{D}$ *be two abelian categories, and let* $T \colon \mathbf{C} \to \mathbf{D}$ *be an additive functor.*

(1) For every injective object I*, we have* $R^n T(I) = (0)$ *for all* $n \geq 1$*, and* $T(I)$ *is isomorphic to* $R^0 T(I)$*. If* T *is contravariant we have* $L_n T(I) = (0)$ *for all* $n \geq 1$*, and* $T(I)$ *is isomorphic to* $L^0 T(I)$*.*
(2) For every projective object P*, we have* $L_n T(P) = (0)$ *for all* $n \geq 1$*, and* $T(P)$ *is isomorphic to* $L_0 T(P)$*. If* T *is contravariant we have* $R^n T(P) = (0)$ *for all* $n \geq 1$*, and* $T(P)$ *is isomorphic to* $R^0 T(P)$*.*

Proof. (1) If I is injective we can pick the resolution

$$0 \longrightarrow I \xrightarrow{\mathrm{id}} I \longrightarrow 0,$$

which yields the complex $T(\mathbf{I})$ given by

$$0 \longrightarrow T(I) \longrightarrow 0,$$

and obviously $R^0 T(I) = H^0(T(\mathbf{I})) = T(I)$ and $H^n(T(\mathbf{I})) = (0)$ for all $n \geq 1$. The proof for the other cases is similar and left as an exercise. ☐

It should also be noted that if T is an exact functor then $R^nT = (0)$ and $L_nT = (0)$ for all $n \geq 1$.

Proposition 11.30 implies that if A or B is a projective R-module (in particular, a free module), then

$$\mathrm{Tor}_n^R(A, B) = (0) \quad \text{for all } n \geq 1.$$

It can also be shown that the above property holds if A or B is a flat R-module; see Rotman [Rotman (1979, 2009)] (Chapter 8, Theorem 8.7). Proposition 11.30 also implies that if A is a projective R-module (in particular, a free module) or if B is an injective R-module then

$$\mathrm{Ext}_R^n(A, B) = (0) \quad \text{for all } n \geq 1.$$

11.6 Long Exact Sequences Induced by Derived Functors

We now come to the most important properties of derived functors, that short-exact sequences yield long exact sequences of cohomology or homology.

Theorem 11.31. *(Long exact sequence, Case (Ri)) Assume the abelian category* $\mathbf{C}$ *has enough injectives, let* $0 \longrightarrow A' \longrightarrow A \longrightarrow A'' \longrightarrow 0$ *be an exact sequence in* $\mathbf{C}$*, and let* $T\colon \mathbf{C} \to \mathbf{D}$ *be an additive left-exact functor.*

(1) Then for every $n \geq 0$*, there is a map*

$$(R^nT)(A'') \xrightarrow{\delta^n} (R^{n+1}T)(A'),$$

and the sequence

$$\begin{array}{l}
0 \longrightarrow T(A') \longrightarrow T(A) \longrightarrow T(A'') \xrightarrow{\delta^0} \\
\longrightarrow (R^1T)(A') \longrightarrow \cdots \longrightarrow \cdots \longrightarrow \\
\longrightarrow (R^nT)(A') \longrightarrow (R^nT)(A) \longrightarrow (R^nT)(A'') \xrightarrow{\delta^n} \\
\longrightarrow (R^{n+1}T)(A') \longrightarrow \cdots \longrightarrow \cdots \longrightarrow \cdots
\end{array}$$

is exact.

(2) If $0 \longrightarrow B' \longrightarrow B \longrightarrow B'' \longrightarrow 0$ *is another exact sequence in* $\mathbf{C}$, *and if there is a commutative diagram*

$$\begin{array}{ccccccccc} 0 & \longrightarrow & A' & \longrightarrow & A & \longrightarrow & A'' & \longrightarrow & 0 \\ & & \downarrow & & \downarrow & & \downarrow & & \\ 0 & \longrightarrow & B' & \longrightarrow & B & \longrightarrow & B'' & \longrightarrow & 0, \end{array}$$

then the induced diagram (shown in two pieces for the lack of space)

$$\begin{array}{ccccccccc} 0 & \longrightarrow & T(A') & \longrightarrow & T(A) & \longrightarrow & T(A'') & \xrightarrow{\delta_A^0} & \\ & & \downarrow & & \downarrow & & \downarrow & & \\ 0 & \longrightarrow & T(B') & \longrightarrow & T(B) & \longrightarrow & T(B'') & \xrightarrow[\delta_B^0]{} & \end{array}$$

$$\begin{array}{ccccccccccc} \cdots & \longrightarrow & R^nT(A') & \longrightarrow & R^nT(A) & \longrightarrow & R^nT(A'') & \xrightarrow{\delta_A^n} & (R^{n+1}T)(A') & \longrightarrow & \cdots \\ & & \downarrow & & \downarrow & & \downarrow & & \downarrow & & \\ \cdots & \longrightarrow & R^nT(B') & \longrightarrow & R^nT(B) & \longrightarrow & R^nT(B'') & \xrightarrow[\delta_B^n]{} & (R^{n+1}T)(B') & \longrightarrow & \cdots \end{array}$$

is also commutative.

Proof. We have injective resolutions (from the collection of resolutions picked once and for all) $0 \longrightarrow A' \xrightarrow{\epsilon'} \mathbf{I}_{A'}$ and $0 \longrightarrow A'' \xrightarrow{\epsilon''} \mathbf{I}_{A''}$ for A' and A''. We are in the situation where we can apply the horseshoe lemma (Proposition 11.25) to obtain an injective resolution $0 \longrightarrow A \xrightarrow{\epsilon} \widehat{\mathbf{I}}_A$ for A as illustrated in the following diagram in which all rows and columns are exact

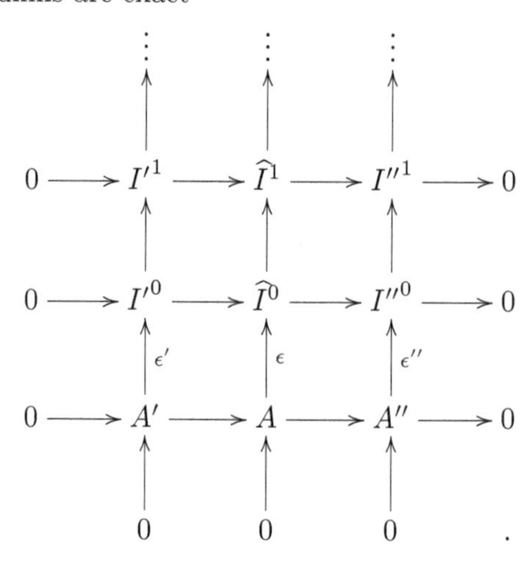

Since all the rows are exact we obtain an exact sequence of complexes

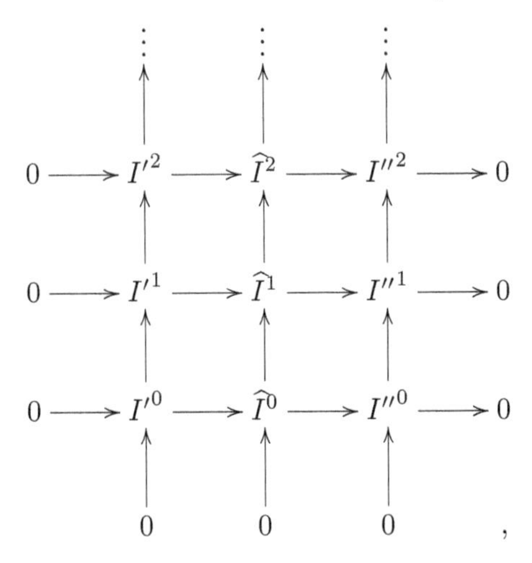

denoted by

$$0 \longrightarrow \mathbf{I}_{A'} \longrightarrow \widehat{\mathbf{I}}_A \longrightarrow \mathbf{I}_{A''} \longrightarrow 0.$$

Observe that the injective resolution $\widehat{\mathbf{I}}_A$ for A given by the Horseshoe Lemma may not be the original resolution that was picked originally and this is why it is denoted with hats. In the end, we will see that Theorem 11.27 implies that this does not matter.

If we apply T to this complex we obtain another sequence of complexes

$$0 \longrightarrow T(\mathbf{I}_{A'}) \longrightarrow T(\widehat{\mathbf{I}}_A) \longrightarrow T(\mathbf{I}_{A''}) \longrightarrow 0$$

as illustrated below

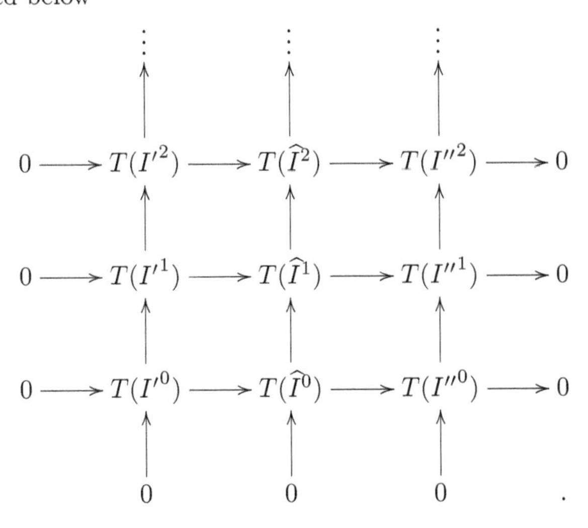

Because the I'^n are injective and the rows

$$0 \longrightarrow I'^n \longrightarrow \widehat{I}^n \longrightarrow I''^n \longrightarrow 0$$

are exact, by Proposition 11.3 these sequence split and since T is an additive functor the sequences

$$0 \longrightarrow T(I'^n) \longrightarrow T(\widehat{I}^n) \longrightarrow T(I'^n) \longrightarrow 0$$

also split and thus are exact. Therefore the sequence

$$0 \longrightarrow T(\mathbf{I}_{A'}) \longrightarrow T(\widehat{\mathbf{I}}_A) \longrightarrow T(\mathbf{I}_{A''}) \longrightarrow 0$$

is a short exact sequence, so our fundamental theorem applies (the zig-zag lemma for cohomology, Theorem 2.22) and we obtain a long exact sequence of cohomology

$$\begin{array}{l}
0 \longrightarrow H^0(T(\mathbf{I}_{A'})) \longrightarrow H^0(T(\widehat{\mathbf{I}}_A)) \longrightarrow H^0(T(\mathbf{I}_{A''})) \xrightarrow{\delta^0} \\
\longrightarrow H^1(T(\mathbf{I}_{A'})) \longrightarrow \cdots \longrightarrow \cdots \longrightarrow \\
\longrightarrow H^n(T(\mathbf{I}_{A'})) \longrightarrow H^n(T(\widehat{\mathbf{I}}_A)) \longrightarrow H^n(T(\mathbf{I}_{A''})) \xrightarrow{\delta^n} \\
\longrightarrow H^{n+1}(T(\mathbf{I}_{A'})) \longrightarrow \cdots \longrightarrow \cdots \longrightarrow \cdots
\end{array}$$

namely the following long exact sequence

$$\begin{array}{l}
0 \longrightarrow R^0T(A') \longrightarrow \widehat{R}^0T(A) \longrightarrow R^0T(A'') \xrightarrow{\delta^0} \\
\longrightarrow (R^1T)(A') \longrightarrow \cdots \longrightarrow \cdots \longrightarrow \\
\longrightarrow (R^nT)(A') \longrightarrow (\widehat{R}^nT)(A) \longrightarrow (R^nT)(A'') \xrightarrow{\delta^n} \\
\longrightarrow (R^{n+1}T)(A') \longrightarrow \cdots \longrightarrow \cdots \longrightarrow \cdots .
\end{array}$$

The right derived functors $\widehat{R}^nT$ may not be those corresponding to the original choice of injective resolutions but we can use Theorem 11.27 to replace it by the isomorphic derived functors R^nT corresponding to the original choice of injective resolutions and adjust the isomorphisms. Since T is left-exact, by Proposition 11.29 we may also replace the R^0T terms (as well as the $\widehat{R}^0T$ terms) by T and adjust the isomorphisms. After all this, we do obtain the promised long exact sequence.

To prove naturality we use Proposition 11.26. Assume we have a commutative diagram

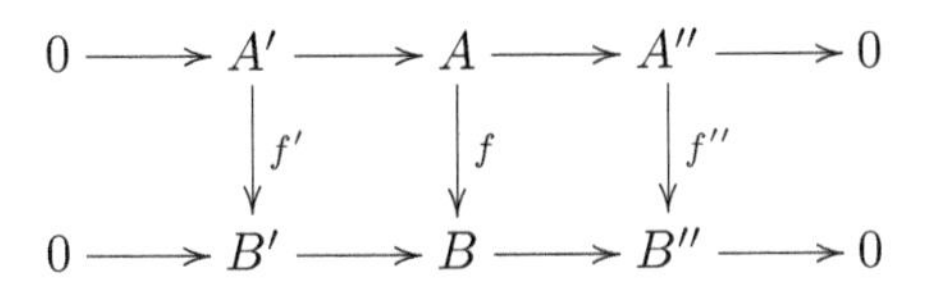

with exact rows. We have the injective resolutions $0 \longrightarrow A' \xrightarrow{\epsilon^{A'}} \mathbf{I}_{A'}$, $0 \longrightarrow A'' \xrightarrow{\epsilon^{A''}} \mathbf{I}_{A''}$, $0 \longrightarrow B' \xrightarrow{\epsilon^{B'}} \mathbf{I}_{B'}$ and $0 \longrightarrow B'' \xrightarrow{\epsilon^{B''}} \mathbf{I}_{B''}$ of the corners A', A'', B', B'', and chain maps $F' \colon \mathbf{I}_{A'} \to \mathbf{I}_{B'}$ over f' and $F'' \colon \mathbf{I}_{B''} \to \mathbf{I}_{B''}$ over f''. Then there exist injective resolutions $0 \longrightarrow A \xrightarrow{\epsilon^{A}} \widehat{\mathbf{I}}_A$ of A and $0 \longrightarrow B \xrightarrow{\epsilon^{B}} \widehat{\mathbf{I}}_B$ of B and a chain map $F \colon \widehat{\mathbf{I}}_A \to \widehat{\mathbf{I}}_B$ over f such that the following diagram commutes

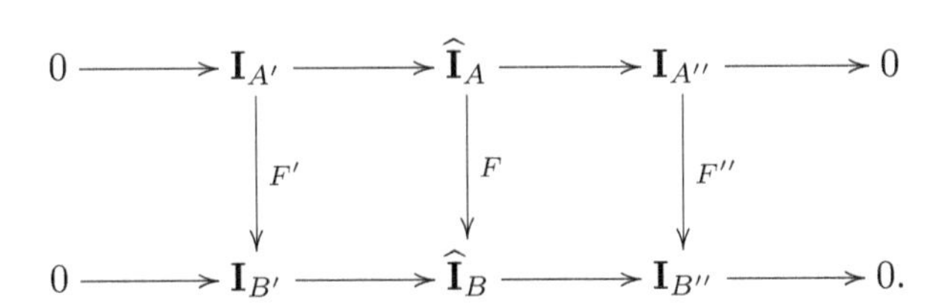

Since the $I^n_{A'}$ and the $I^n_{B'}$ are injective, every row of the diagram above splits, thus after applying T we obtain a commutative diagram with exact rows

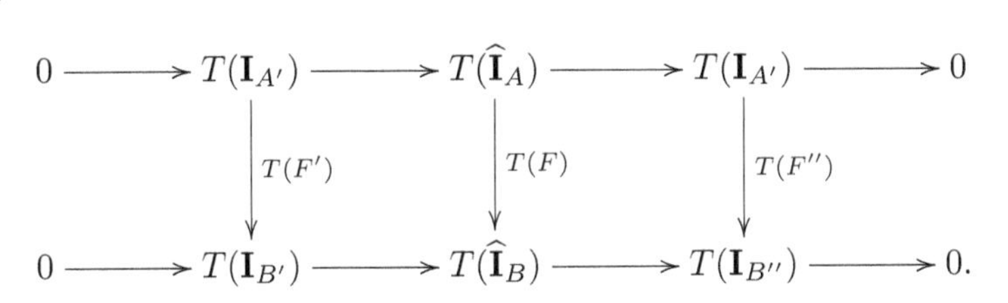

We now conclude by applying Proposition 2.23 and replacing the terms $\widehat{R}^nT$ by R^nT as we did before. $\square$

Remark: If T is not left-exact, the proof of Theorem 11.31 shows that R^0T is left-exact.

A similar theorem holds for the left derived functors L_nT of a (right-exact) functor; we obtain a long exact sequence of homology type involving the L_nT applied to A', A, A'', and L_0T is right-exact.

Theorem 11.32. *(Long exact sequence, Case (Lp)) Assume the abelian category* $\mathbf{C}$ *has enough projectives, let* $0 \longrightarrow A' \longrightarrow A \longrightarrow A'' \longrightarrow 0$ *be an exact sequence in* $\mathbf{C}$*, and let* $T\colon \mathbf{C} \to \mathbf{D}$ *be an additive right-exact functor.*

(1) Then for every $n \geq 1$*, there is a map*

$$(L_nT)(A'') \xrightarrow{\partial_n} (L_{n-1}T)(A'),$$

and the sequence

$$\cdots \longrightarrow L_nT(A') \longrightarrow L_nT(A) \longrightarrow L_nT(A'') \xrightarrow{\partial_n} L_{n-1}T(A') \longrightarrow \cdots$$

$$\cdots \longrightarrow L_1T(A'') \xrightarrow{\partial_1} T(A') \longrightarrow T(A) \longrightarrow T(A'') \longrightarrow 0$$

is exact.

(2) If $0 \longrightarrow B' \longrightarrow B \longrightarrow B'' \longrightarrow 0$ *is another exact sequence in* $\mathbf{C}$*, and if there is a commutative diagram*

$$\begin{array}{ccccccccc} 0 & \longrightarrow & A' & \longrightarrow & A & \longrightarrow & A'' & \longrightarrow & 0 \\ & & \downarrow & & \downarrow & & \downarrow & & \\ 0 & \longrightarrow & B' & \longrightarrow & B & \longrightarrow & B'' & \longrightarrow & 0, \end{array}$$

then the induced diagram

$$\begin{array}{ccccccccc} \cdots (L_nT)(A') & \longrightarrow & (L_nT)(A) & \longrightarrow & (L_nT)(A'') & \xrightarrow{\partial_n^A} & (L_{n-1}T)(A') & \longrightarrow & \cdots \\ \downarrow & & \downarrow & & \downarrow & & \downarrow & & \\ \cdots (L_nT)(B') & \longrightarrow & (L_nT)(B) & \longrightarrow & (L_nT)(B'') & \xrightarrow[\partial_n^B]{} & (L_{n-1}T)(B') & \longrightarrow & \cdots \end{array}$$

and ending with

$$\begin{array}{ccccccccc}
\cdots \longrightarrow & L_1T(A'') & \xrightarrow{\partial_1^A} & T(A') & \longrightarrow & T(A) & \longrightarrow & T(A'') & \longrightarrow 0 \\
& \downarrow & & \downarrow & & \downarrow & & \downarrow & \\
\cdots \longrightarrow & L_1T(B'') & \xrightarrow[\partial_1^B]{} & T(B') & \longrightarrow & T(B) & \longrightarrow & T(B'') & \longrightarrow 0
\end{array}$$

is also commutative.

Remark: If T is not right-exact, the proof of Theorem 11.32 shows that L_0T is right-exact.

If **C** has enough injectives and T is a contravariant (right-exact) functor, we have a version of Theorem 11.32 showing that there is a long-exact sequence of homology type involving the L_nT applied to A', A, A'', with the terms A', A, A'' appearing in reverse order (Case (Li)). As a consequence, L_0T is right-exact. This case does not seem to arise in practice.

If **C** has enough projectives and T is a contravariant (left-exact) functor, we have a version of Theorem 11.31 showing that there is a long-exact sequence of cohomology type involving the R^nT applied to A', A, A'' with the terms A', A, A'' appearing in reverse order (Case (Rp)). As a consequence, R_0T is left-exact.

Remember: *Right derived functors go with left-exact functors; left derived functors go with right-exact functors.*

11.7 T-Acyclic Resolutions

There are situations (for example, when dealing with sheaves) where it is useful to know that right derived functors can be computed by resolutions involving objects that are not necessarily injective, but T-acyclic, as defined below. Assume that **C** is an abelian category that has enough injectives.

Definition 11.18. Given an additive left-exact functor $T\colon \mathbf{C} \to \mathbf{D}$, an object $J \in \mathbf{C}$ is *(right) T-acyclic* if $R^nT(J) = (0)$ for all $n \geq 1$ (see Definition 11.17, Case (Ri)).

The following proposition shows that right derived functors can be computed using T-acyclic resolutions. The following auxiliary result is needed.

Proposition 11.33. *If the sequence*

$$0 \longrightarrow A \xrightarrow{f} B \xrightarrow{g} C$$

is exact and if T is left-exact, then $\mathrm{Ker}\, T(g) \cong T(\mathrm{Ker}\, g)$.

Proof. Since the above is exact

$$A \cong \operatorname{Im} f = \operatorname{Ker} g,$$

and as T is a functor

$$T(A) \cong T(\operatorname{Ker} g).$$

Since T is left-exact we obtain the exact sequence

$$0 \longrightarrow T(A) \xrightarrow{T(f)} T(B) \xrightarrow{T(g)} T(C),$$

so

$$T(A) \cong \operatorname{Im} T(f) = \operatorname{Ker} T(g),$$

and thus

$$\operatorname{Ker} T(g) \cong T(A) \cong T(\operatorname{Ker} g),$$

as claimed. □

Proposition 11.34. *Given an additive left-exact functor $T\colon \mathbf{C} \to \mathbf{D}$, for any $A \in \mathbf{C}$ suppose there is an exact sequence*

$$0 \longrightarrow A \xrightarrow{\epsilon} J^0 \xrightarrow{d^0} J^1 \xrightarrow{d^1} J^2 \xrightarrow{d^2} \cdots \tag{†}$$

in which every J^n is right T-acyclic (a right T-acyclic resolution $\mathbf{J}_A$). Then for every $n \geq 0$ we have an isomorphism between $R^nT(A)$ and $H^n(T(\mathbf{J}_A))$, where $T(\mathbf{J}_A)$ is the cochain complex

$$0 \longrightarrow T(J^0) \xrightarrow{T(d^0)} T(J^1) \xrightarrow{T(d^1)} T(J^2) \xrightarrow{T(d^2)} \cdots .$$

Proof. The proof is a good illustration of the use of the long exact sequence given by Theorem 11.31. Since (†) is exact and T is left-exact we obtain the exact sequence

$$0 \longrightarrow T(A) \xrightarrow{T(\epsilon)} T(J^0) \xrightarrow{T(d^0)} T(J^1),$$

which (see the proof of Proposition 11.29(1)) implies that

$$R^0T(A) \cong T(A) \cong \operatorname{Ker} T(d^0) = H^0(T(\mathbf{J}_A)).$$

Let $K^n = \operatorname{Ker} d^n$ for all $n \geq 1$. The exact sequence (†) implies that $\operatorname{Im} d^n = \operatorname{Ker} d^{n+1} = K^{n+1}$ and the surjection $p^n\colon J^n \to K^{n+1}$ has kernel K^n so we have the short exact sequence

$$0 \longrightarrow K^n \longrightarrow J^n \xrightarrow{p^n} K^{n+1} \longrightarrow 0 \tag{$*$}$$

for all $n \geq 1$. We also have the short exact sequence

$$0 \longrightarrow A \longrightarrow J^0 \xrightarrow{p^0} K^1 \longrightarrow 0. \qquad (**)$$

If we denote the injection of K^{n+1} into J^{n+1} by ϵ^{n+1}, then we can factor d^n as

$$d^n = \epsilon^{n+1} \circ p^n.$$

We have the following commutative diagram

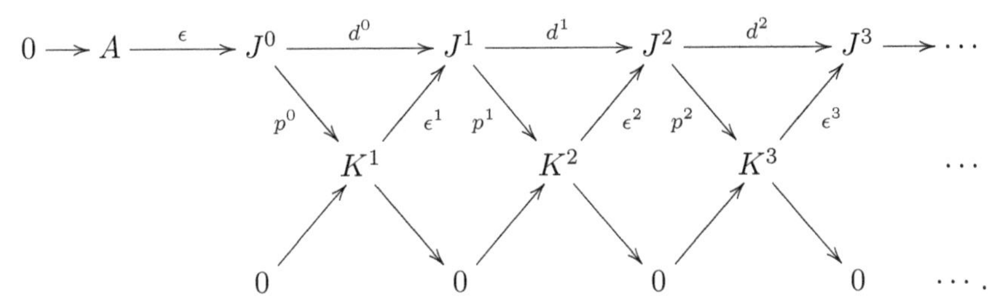

If we apply T we get

$$T(d^n) = T(\epsilon^{n+1}) \circ T(p^n).$$

Since ϵ^{n+1} is injective, the sequence $0 \longrightarrow K^{n+1} \xrightarrow{\epsilon^{n+1}} J^{n+1} \xrightarrow{d^{n+1}} J^{n+2}$ is exact, and since T is left exact we see that

$0 \longrightarrow T(K^{n+1}) \xrightarrow{T(\epsilon^{n+1})} T(J^{n+1}) \xrightarrow{T(d^{n+1})} T(J^{n+2})$ is also exact, so $T(\epsilon^{n+1})$ is injective. It follows that the restriction of $T(\epsilon^{n+1})$ to $\operatorname{Im} T(p^n)$ is an isomorphism onto the image of $T(d^n)$, which implies that

$$\operatorname{Im} T(d^n) \cong \operatorname{Im} T(p^n), \qquad n \geq 0.$$

By definition of $K^n = \operatorname{Ker} d^n$, we have the exact sequence

$$0 \longrightarrow K^n \longrightarrow J^n \xrightarrow{d^n} J^{n+1},$$

so by Proposition 11.33 we get

$$\operatorname{Ker} T(d^n) \cong T(\operatorname{Ker} d^n). \qquad (*_{\text{Ker}})$$

If we apply Theorem 11.31 to (**), the long exact sequence begins with

$$0 \longrightarrow T(A) \longrightarrow T(J^0) \xrightarrow{T(p^0)} T(K^1) \longrightarrow R^1T(A) \longrightarrow R^1T(J^0) = (0),$$

which yields

$$\begin{aligned} R^1T(A) &\cong T(K^1)/\operatorname{Im} T(p^0) = T(\operatorname{Ker} d^1)/\operatorname{Im} T(p^0) \\ &\cong \operatorname{Ker} T(d^1)/\operatorname{Im} T(d^0) = H^1(T(\mathbf{J}_A)). \end{aligned}$$

So far, we proved that $R^0T(A) \cong H^0(T(\mathbf{J}_A))$ and $R^1T(A) \cong H^1(T(\mathbf{J}_A))$. To prove that $R^nT(A) \cong H^n(T(\mathbf{J}_A))$ for $n \geq 2$ again we use the long exact sequence applied to $(**)$, which gives

$$R^{n-1}T(J^0) \longrightarrow R^{n-1}T(K^1) \longrightarrow R^nT(A) \longrightarrow R^nT(J^0),$$

and since J^0 is T-acyclic $R^{n-1}T(J^0) = R^nT(J^0) = (0)$ for $n \geq 2$, so we obtain isomorphisms

$$R^{n-1}T(K^1) \cong R^nT(A), \qquad n \geq 2.$$

The long exact sequence applied to $(*)$ yields

$$R^{n-i-1}T(J^i) \longrightarrow R^{n-i-1}T(K^{i+1}) \longrightarrow R^{n-i}T(K^i) \longrightarrow R^{n-i}T(J^i),$$

and since J^i is T-acyclic $R^{n-i-1}T(J^i) = R^{n-i}T(J^i) = (0)$ so we have the isomorphisms

$$R^{n-i-1}T(K^{i+1}) \cong R^{n-i}T(K^i), \qquad 1 \leq i \leq n-2.$$

By induction we obtain

$$R^{n-1}T(K^1) \cong R^1T(K^{n-1}), \qquad n \geq 2.$$

However, we showed that $R^{n-1}T(K^1) \cong R^nT(A)$, so we obtain

$$R^nT(A) \cong R^{n-1}T(K^1) \cong R^1T(K^{n-1}).$$

The long exact sequence applied to $(*)$ yields

$$T(J^{n-1}) \xrightarrow{T(p^{n-1})} T(K^n) \longrightarrow R^1T(K^{n-1}) \longrightarrow R^1T(J^{n-1}) = (0)$$

which by $(*_{\mathrm{Ker}})$ and the first isomorphism theorem implies that

$$\begin{aligned} R^nT(A) &\cong R^1T(K^{n-1}) \\ &\cong T(K^n)/\mathrm{Im}\, T(p^{n-1}) \\ &= T(\mathrm{Ker}\, d^n)/\mathrm{Im}\, T(p^{n-1}) \\ &\cong \mathrm{Ker}\, T(d^n)/\mathrm{Im}\, T(d^{n-1}) = H^n(T(\mathbf{J}_A)). \end{aligned}$$

Therefore we proved that $R^nT(A) \cong H^n(T(\mathbf{J}_A))$ for all $n \geq 0$. $\square$

Another proof of Proposition 11.34 can be found in Lang [Lang (1993)] (Chapter XX, §6, Theorem 6.2). Actually, Lang proves a stronger result. This result is that for any injective resolution $0 \longrightarrow A \xrightarrow{\epsilon'} \mathbf{I}_A$, the morphism from the complex $\mathbf{J}_A$ to the complex $\mathbf{I}_A$ lifting id_A given by Proposition 11.21 induces isomorphisms $H^n(T(\mathbf{J}_A)) \cong R^nT(A)$ for all $n \geq 0$.

Lang's proof makes use of a result of independent interest that we discuss below.

Proposition 11.35. *Let $T\colon \mathbf{C} \to \mathbf{D}$ be an additive left-exact functor. For any exact sequence*

$$0 \longrightarrow X^0 \xrightarrow{d^0} X^1 \xrightarrow{d^1} X^2 \xrightarrow{d^2} X^3 \xrightarrow{d^3} \cdots, \tag{$\dagger$}$$

if the X^i are T-acyclic for all $i \geq 0$, then

$$0 \longrightarrow T(X^0) \xrightarrow{T(d^0)} T(X^1) \xrightarrow{T(d^1)} T(X^2) \xrightarrow{T(d^2)} T(X^3) \cdots$$

is also an exact sequence.

Proof. The proof uses an inductive process involving the cokernels $C^n = \operatorname{Im} d^n$ $(n \geq 1)$. Since $\operatorname{Im} d^n = \operatorname{Ker} d^{n+1}$, by the first isomorphism theorem

$$C^n = \operatorname{Im} d^n \cong X^n/\operatorname{Ker} d^n \cong X^n/\operatorname{Im} d^{n-1} = \operatorname{Coker} d^{n-1}, \quad n \geq 1,$$

so $C^n = \operatorname{Im} d^n$ is indeed isomorphic to the cokernel of d^{n-1}. We can factor $d^n\colon X^n \to X^{n+1}$ as

$$d^n = \epsilon^n \circ p^n,$$

where $p^n\colon X^n \to C^n$ is a surjection and $\epsilon^n\colon C^n \to X^{n+1}$ is an injection. It follows that $\operatorname{Ker} p^{n+1} = \operatorname{Ker} d^{n+1} = \operatorname{Im} d^n = \operatorname{Im} \epsilon^n$ and $\operatorname{Im} \epsilon^n = \operatorname{Im} d^n = \operatorname{Ker} d^{n+1}$ for all $n \geq 1$, so we have the exact sequences

$$0 \longrightarrow C^n \xrightarrow{\epsilon^n} X^{n+1} \xrightarrow{p^{n+1}} C^{n+1} \longrightarrow 0 \tag{$\dagger_n$}$$

and

$$0 \longrightarrow C^n \xrightarrow{\epsilon^n} X^{n+1} \xrightarrow{d^{n+1}} X^{n+2}. \tag{$\dagger\dagger_n$}$$

We wish to prove by induction on n that exactness holds up to $T(X^{n+2})$ and that C^{n+1} is T-acyclic.

Let us consider the case $n = 0$ (base step). Since T is left exact, we have an exact sequence

$$0 \longrightarrow T(X^0) \xrightarrow{T(d^0)} T(X^1) \xrightarrow{T(d^1)} T(X^2),$$

which shows that we have exactness at $T(X^0)$ and $T(X^1)$. We prove that we also have exactness at $T(X^2)$.

If we let $C^1 = \operatorname{Im} d^1$, since $\operatorname{Ker} p^1 = \operatorname{Ker} d^1$, we have the exact sequence

$$0 \longrightarrow X^0 \xrightarrow{d^0} X^1 \xrightarrow{p^1} C^1 \longrightarrow 0. \tag{$\dagger_0$}$$

If we apply Theorem 11.31 to the above exact sequence, the long exact sequence begins with

$$0 \longrightarrow T(X^0) \xrightarrow{T(d^0)} T(X^1) \xrightarrow{T(p^1)} T(C^1) \longrightarrow RT^1(X^0),$$

but since X^1 is T-acyclic, $RT^1(X^0) = (0)$, so we have the exact sequence

$$0 \longrightarrow T(X^0) \xrightarrow{T(d^0)} T(X^1) \xrightarrow{T(p^1)} T(C^1) \longrightarrow 0. \qquad (*_1)$$

As we just showed, we have an exact sequence

$$0 \longrightarrow C^1 \xrightarrow{\epsilon^1} X^2 \xrightarrow{d^2} X^3. \qquad (\dagger\dagger_1)$$

Since T is left exact, we obtain the exact sequence

$$0 \longrightarrow T(C^1) \xrightarrow{T(\epsilon^1)} T(X^2) \xrightarrow{T(d^2)} T(X^3). \qquad (*_2)$$

We can splice the sequences $(*_1)$ and $(*_2)$ to obtain the sequence

$$0 \to T(X^0) \xrightarrow{T(d^0)} T(X^1) \xrightarrow{T(p^1)} T(C^1) \xrightarrow{T(\epsilon^1)} T(X^2) \xrightarrow{T(d^2)} T(X^3)$$

which is exact except at $T^1(C)$, but since $d^1 = \epsilon^1 \circ p^1$ and $T(p^1)$ is surjective,

$$\operatorname{Im} T(d^1) = \operatorname{Im} T(\epsilon^1) \circ T(p^1) = \operatorname{Im} T(\epsilon^1) = \operatorname{Ker} T(d^2),$$

the sequence

$$0 \longrightarrow T(X^0) \xrightarrow{T(d^0)} T(X^1) \xrightarrow{T(d^1)} T(X^2) \xrightarrow{T(d^2)} T(X^3)$$

is exact at $T(X^2)$.

We prove that C^1 is T-acyclic as follows. If we apply Theorem 11.31 to the exact sequence

$$0 \longrightarrow X^0 \xrightarrow{d^0} X^1 \xrightarrow{p^1} C^1 \longrightarrow 0, \qquad (\dagger_0)$$

we obtain the piece of exact sequence

$$R^pT(X^1) \longrightarrow R^pT(C^1) \longrightarrow R^{p+1}T(X^0),$$

and since X^0 and X^1 are acyclic, $R^pT(X^1) = R^{p+1}T(X^0) = (0)$ for all $p \geq 1$, so $R^pT(C^1) = (0)$ for all $p \geq 1$.

The induction step is to prove that exactness holds at $T(X^{n+2})$ and that C^{n+1} is T-acyclic for $n \geq 1$, assuming that C^n and X^{n+1} are T-acyclic.

We have the exact sequence

$$0 \longrightarrow C^n \xrightarrow{\epsilon^n} X^{n+1} \xrightarrow{p^{n+1}} C^{n+1} \longrightarrow 0, \qquad (\dagger_n)$$

where C^n, X^{n+1} are T-acyclic, and the exact sequence

$$0 \longrightarrow C^{n+1} \xrightarrow{\epsilon^{n+1}} X^{n+2} \xrightarrow{d^{n+2}} X^{n+3}, \qquad (\dagger\dagger_{n+1})$$

so we can repeat the argument used for the exact sequences

$$0 \longrightarrow X^0 \xrightarrow{d^0} X^1 \xrightarrow{p^1} C^1 \longrightarrow 0$$

and

$$0 \longrightarrow C^1 \xrightarrow{\epsilon^1} X^2 \xrightarrow{d^2} X^3$$

to prove that exactness holds at X^2 to prove that exactness holds at X^{n+2} and that C^{n+1} is T-acyclic, which establishes the induction step. □

A proposition analogous to Proposition 11.34 holds for left T-acyclic resolutions and the left derived functors L_nT. This time we assume that the abelian category $\mathbf{C}$ has enough projectives.

Definition 11.19. Given an additive left-exact functor $T\colon \mathbf{C} \to \mathbf{D}$, an object $J \in \mathbf{C}$ is *(left) T-acyclic* if $L_nT(J) = (0)$ for all $n \geq 1$ (see Definition 11.17, Case (Lp)).

Proposition 11.36. *Given an additive left-exact functor $T\colon \mathbf{C} \to \mathbf{D}$, for any $A \in \mathbf{C}$ suppose there is an exact sequence*

$$0 \longleftarrow A \xleftarrow{\epsilon} P_0 \xleftarrow{d_0} P_1 \xleftarrow{d_1} P_2 \xleftarrow{d_2} \cdots \qquad (\dagger)$$

in which every P^n is left T-acyclic (a left T-acyclic resolution $\mathbf{P}^A$). Then for every $n \geq 0$ we have an isomorphism between $L_nT(A)$ and $H_n(T(\mathbf{P}^A))$, where $T(\mathbf{P}^A)$ is the chain complex

$$0 \longleftarrow T(P_0) \xleftarrow{T(d_0)} T(P_1) \xleftarrow{T(d_1)} T(P_2) \xleftarrow{T(d_2)} \cdots$$

Proposition 11.34 has an interesting application to de Rham cohomology. Say M is a smooth manifold. Recall that for every $p \geq 0$ we have the sheaf $\mathcal{A}^p_M$ of differential forms on M (where for every open subset U of M, $\mathcal{A}^p_M(U) = \mathcal{A}^p(U)$ is the vector space of smooth p-forms on U).

Proposition 11.37. *If $\widetilde{\mathbb{R}}_M$ denotes the sheaf of locally constant real-valued functions on a smooth manifold M, then*

$$0 \longrightarrow \widetilde{\mathbb{R}}_M \xrightarrow{\epsilon} \mathcal{A}^0_M \xrightarrow{d} \mathcal{A}^1_M \xrightarrow{d} \cdots \xrightarrow{d} \mathcal{A}^p_M \xrightarrow{d} \mathcal{A}^{p+1}_M \xrightarrow{d} \cdots$$

is a resolution of $\widetilde{\mathbb{R}}_M$, where ϵ is the inclusion map.

Proof. The above fact is proved using Proposition 10.24(ii) by showing that for every $x \in M$, the stalk complex

$$0 \longrightarrow \mathbb{R} \longrightarrow \mathcal{A}^0_{M,x} \longrightarrow \mathcal{A}^1_{M,x} \cdots \longrightarrow \mathcal{A}^p_{M,x} \longrightarrow \mathcal{A}^{p+1}_{M,x} \longrightarrow \cdots$$

is exact. Since M is a smooth manifold, we may assume that M is an open subset of $\mathbb{R}^n$, and use a fundamental system of convex open neighborhoods of x to compute the direct limit $\mathcal{A}^p_{M,x} = \varinjlim(\mathcal{A}^p(U))_{U \ni x}$. If U is convex, the complex

$$0 \longrightarrow \mathbb{R} \longrightarrow \mathcal{A}^0(U) \longrightarrow \mathcal{A}^1(U) \cdots \longrightarrow \mathcal{A}^p(U) \longrightarrow \mathcal{A}^{p+1}(U) \longrightarrow \cdots$$

is exact by the Poincaré lemma (Proposition 3.2). Since a direct limit of exact sequences is exact, we conclude that

$$0 \longrightarrow \mathbb{R} \longrightarrow \mathcal{A}^0_{M,x} \longrightarrow \mathcal{A}^1_{M,x} \cdots \longrightarrow \mathcal{A}^p_{M,x} \longrightarrow \mathcal{A}^{p+1}_{M,x} \longrightarrow \cdots$$

is exact. For details, see Brylinski [Brylinski (1993)] (Section 1.4, Proposition 1.4.3). ☐

If $\Gamma(M, -)$ is the global section functor with $\Gamma(M, \mathcal{A}^p_M) = \mathcal{A}^p(M)$, then it can also be shown that the sheaves $\mathcal{A}^p_M$ are $\Gamma(M, -)$-acyclic. This is because the sheaves $\mathcal{A}^p_M$ are soft, and soft sheaves on a paracompact space are $\Gamma(M, -)$-acyclic; see Godement [Godement (1958)] (Chapter 3, Section 3.9), or Brylinski [Brylinski (1993)] (Section 1.4, Theorem 1.4.6 and Proposition 1.4.9), or Section 13.5.

Now, it is also true that sheaves have enough injectives (we will see this in Chapter 13). Therefore, we conclude that the cohomology groups $R^p\Gamma(M, -)(\widetilde{\mathbb{R}}_M)$ and the de Rham cohomology groups $H^p_{\mathrm{dR}}(M)$ are isomorphic. The groups $R^p\Gamma(M, -)(\widetilde{\mathbb{R}}_M)$ are called the *sheaf cohomology groups* of the sheaf $\widetilde{\mathbb{R}}_M$ and are denoted by $H^p(M, \widetilde{\mathbb{R}}_M)$. We will also show in the next chapter that for a paracompact space M, the Čech cohomology groups $\check{H}^p(M, \mathcal{F})$ and the sheaf cohomology groups $H^p(M, \mathcal{F}) = R^p\Gamma(M, -)(\mathcal{F})$ are isomorphic (where $\Gamma(M, -)$ is the global section functor, $\Gamma(M, \mathcal{F}) = \mathcal{F}(M)$); thus, for smooth manifolds we have isomorphisms

$$H^p(M, \widetilde{\mathbb{R}}_M) \cong \check{H}^p(M, \widetilde{\mathbb{R}}_M) \cong H^p_{\mathrm{dR}}(M),$$

proving part of Theorem 9.4.

Theorems 11.31 and 11.32 suggest the definition of families of functors originally proposed by Cartan and Eilenberg [Cartan and Eilenberg (1956)] and then investigated by Grothendieck in his legendary "Tohoku" paper [Grothendieck (1957)] (1957).

11.8 Universal δ-Functors and ∂-Functors

In his famous Tohoku paper [Grothendieck (1957)] Grothendieck introduced the terminology "∂-functor" and "∂^*-functor;" see Chapter II, Section 2.1. The notion of ∂-functor is a slight generalization of the notion of "connected sequence of functors" introduced earlier by Cartan and Eilenberg [Cartan and Eilenberg (1956)] (Chapter 3). Since ∂-functor have a cohomological flavor and ∂^*-functor have a homological flavor, everybody now appears to use the terminology δ-functor instead of ∂-functor and ∂-functor for ∂^*-functor.

Definition 11.20. Given two abelian categories $\mathbf{C}$ and $\mathbf{D}$, a *δ-functor* consists of a countable family $T = (T^n)_{n\geq 0}$ of additive functors $T^n\colon \mathbf{C} \to \mathbf{D}$, and for every short exact sequence $0 \longrightarrow A' \longrightarrow A \longrightarrow A'' \longrightarrow 0$ in the abelian category $\mathbf{C}$ and every $n \geq 0$ of a map

$$T^n(A'') \xrightarrow{\delta^n} T^{n+1}(A')$$

such that the following two properties hold:

(i) The sequence

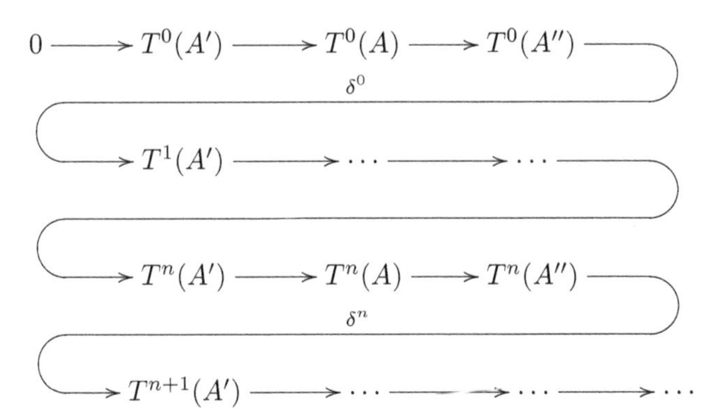

is exact (a long exact sequence).

(ii) If $0 \longrightarrow B' \longrightarrow B \longrightarrow B'' \longrightarrow 0$ is another exact sequence in $\mathbf{C}$, and if there is a commutative diagram

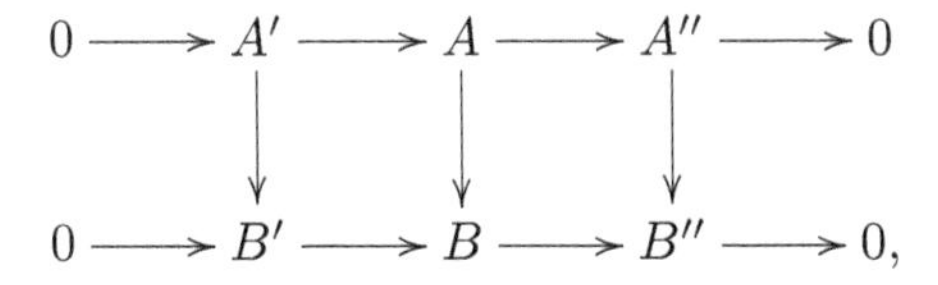

then the induced diagram beginning with

$$\begin{array}{ccccccc} 0 \longrightarrow & T^0(A') & \longrightarrow & T^0(A) & \longrightarrow & T^0(A'') & \xrightarrow{\delta_A^0} \\ & \downarrow & & \downarrow & & \downarrow & \\ 0 \longrightarrow & T^0(B') & \longrightarrow & T^0(B) & \longrightarrow & T^0(B'') & \xrightarrow[\delta_B^0]{} \end{array}$$

and continuing with

$$\begin{array}{ccccccccc} \cdots \longrightarrow & T^n(A') & \longrightarrow & T^n(A) & \longrightarrow & T^n(A'') & \xrightarrow{\delta_A^n} & T^{n+1}(A') & \longrightarrow \cdots \\ & \downarrow & & \downarrow & & \downarrow & & \downarrow & \\ \cdots \longrightarrow & T^n(B') & \longrightarrow & T^n(B) & \longrightarrow & T^n(B'') & \xrightarrow[\delta_B^n]{} & T^{n+1}(B') & \longrightarrow \cdots \end{array}$$

is also commutative.

In particular, T^0 is left-exact.

The notion of morphism of δ-functors is defined as follows.

Definition 11.21. Given two δ-functors $S = (S^n)_{n \geq 0}$ and $T = (T^n)_{n \geq 0}$, a *morphism* $\eta\colon S \to T$ between S and T is a family $\eta = (\eta^n)_{n \geq 0}$ of natural transformations $\eta^n\colon S^n \to T^n$ such that the following diagram commutes

$$\begin{array}{ccc} S^n(A'') & \xrightarrow{\delta_S^n} & S^{n+1}(A') \\ {\scriptstyle (\eta^n)_{A''}} \downarrow & & \downarrow {\scriptstyle (\eta^{n+1})_{A'}} \\ T^n(A'') & \xrightarrow[\delta_T^n]{} & T^{n+1}(A') \end{array}$$

for all $n \geq 0$ and for every short exact sequence $0 \longrightarrow A' \longrightarrow A \longrightarrow A'' \longrightarrow 0$.

Morphisms of δ-functors are composed in the obvious way. The notion of isomorphism is also obvious (each η^n is an isomorphism).

Grothendieck introduced the important notion of universal δ-functor; see Grothendieck [Grothendieck (1957)] (Chapter II, Section 2.2).

Definition 11.22. A δ-functor $T = (T^n)_{n \geq 0}$ is *universal* if for every δ-functor $S = (S^n)_{n \geq 0}$ and every natural transformation $\varphi\colon T^0 \to S^0$ there is a *unique* morphism $\eta\colon T \to S$ such that $\eta^0 = \varphi$ as illustrated in the commutative diagram below; we say that η *lifts* φ,

$$\begin{array}{ccccccccccc}
\cdots T^0(A'') & \xrightarrow{\delta_T^0} & T^1(A') & \longrightarrow & T^1(A) & \longrightarrow & T^1(A'') & \xrightarrow{\delta_T^1} & T^2(A') & \longrightarrow & \cdots \\
\downarrow{\scriptstyle \varphi_{A''}} & & \downarrow{\scriptstyle \eta^1_{A'}} & & \downarrow{\scriptstyle \eta^1_{A}} & & \downarrow{\scriptstyle \eta^1_{A''}} & & \downarrow{\scriptstyle \eta^2_{A'}} & & \\
\cdots S^0(A'') & \xrightarrow[\delta_S^0]{} & S^1(A') & \longrightarrow & S^1(A) & \longrightarrow & S^1(A'') & \xrightarrow[\delta_S^1]{} & S^2(A') & \longrightarrow & \cdots
\end{array}$$

for every short exact sequence $0 \longrightarrow A' \longrightarrow A \longrightarrow A'' \longrightarrow 0$.

Proposition 11.38. *Suppose $S = (S^n)_{n\geq 0}$ and $T = (T^n)_{n\geq 0}$ are both universal δ-functors and there is an isomorphism $\varphi\colon S^0 \to T^0$ (a natural transformation φ which is an isomorphism). Then there is a unique isomorphism $\eta\colon S \to T$ lifting φ.*

Proof. Since φ is an isomorphism, it has an inverse $\psi\colon T^0 \to S^0$, that is, we have $\psi \circ \varphi = \mathrm{id}_{S^0}$ and $\varphi \circ \psi = \mathrm{id}_{T^0}$. Since S is universal there is a unique lift $\eta\colon S \to T$ of φ and since T is universal there is a unique lift $\theta\colon T \to S$ of ψ. But $\theta \circ \eta$ lifts $\psi \circ \varphi = \mathrm{id}_{S^0}$ and $\eta \circ \theta$ lifts $\varphi \circ \psi = \mathrm{id}_{T^0}$. However, id_S is a lift of id_{S^0} and id_T is a lift of id_{T^0}, so by uniqueness of lifts we must have $\theta \circ \eta = \mathrm{id}_S$ and $\eta \circ \theta = \mathrm{id}_T$, which shows that η is an isomorphism. ☐

Proposition 11.38 shows a significant property of a universal δ-functor T: *it is completely determined by the component T^0.*

One might wonder whether (universal) δ-functors exist. Indeed there are plenty of them.

Theorem 11.39. *Assume the abelian category* **C** *has enough injectives. For every additive left-exact functor $T\colon \mathbf{C} \to \mathbf{D}$, the family $(R^nT)_{n\geq 0}$ of right derived functors of T is a δ-functor. Furthermore T is isomorphic to R^0T.*

Proof. Now that we have done all the hard work the proof is short: apply Theorem 11.31. The second property follows from Proposition 11.29. ☐

In fact, the δ-functors $(R^nT)_{n\geq 0}$ are universal. Before explaining the technique due to Grothendieck for proving this fact, let us take a quick look at ∂-functors.

Definition 11.23. Given two abelian categories $\mathbf{C}$ and $\mathbf{D}$, a *∂-functor* consists of a countable family $T = (T_n)_{n \geq 0}$ of additive functors $T_n \colon \mathbf{C} \to \mathbf{D}$, and for every short exact sequence $0 \longrightarrow A' \longrightarrow A \longrightarrow A'' \longrightarrow 0$ in the abelian category $\mathbf{C}$ and every $n \geq 1$ of a map

$$T_n(A'') \xrightarrow{\partial_n} T_{n-1}(A')$$

such that the following two properties hold:

(i) The sequence

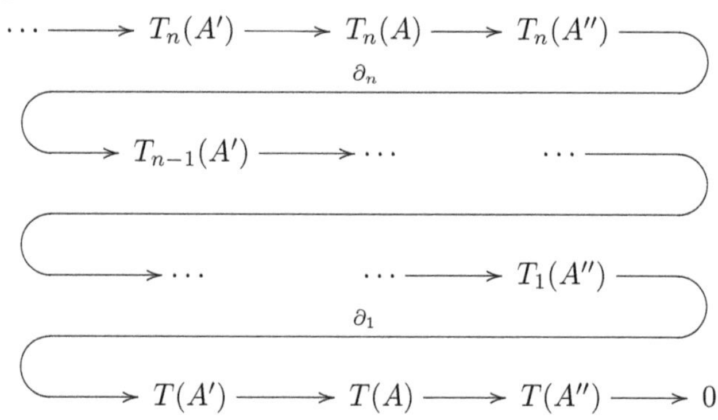

is exact.

(ii) If $0 \longrightarrow B' \longrightarrow B \longrightarrow B'' \longrightarrow 0$ is another exact sequence in $\mathbf{C}$, and if there is a commutative diagram

$$\begin{array}{ccccccccc} 0 & \longrightarrow & A' & \longrightarrow & A & \longrightarrow & A'' & \longrightarrow & 0 \\ & & \downarrow & & \downarrow & & \downarrow & & \\ 0 & \longrightarrow & B' & \longrightarrow & B & \longrightarrow & B'' & \longrightarrow & 0, \end{array}$$

then the induced diagram

$$\begin{array}{ccccccccccc} \cdots & \longrightarrow & T_n(A') & \longrightarrow & T_n(A) & \longrightarrow & T_n(A'') & \xrightarrow{\partial_n^A} & T_{n-1}(A') & \longrightarrow & \cdots \\ & & \downarrow & & \downarrow & & \downarrow & & \downarrow & & \\ \cdots & \longrightarrow & T_n(B') & \longrightarrow & T_n(B) & \longrightarrow & T_n(B'') & \xrightarrow[\partial_n^B]{} & T_{n-1}(B') & \longrightarrow & \cdots \end{array}$$

and ending with

$$\begin{array}{ccccccccccc} \cdots & \longrightarrow & T_1(A'') & \xrightarrow{\partial_1^A} & T(A') & \longrightarrow & T(A) & \longrightarrow & T(A'') & \longrightarrow & 0 \\ & & \downarrow & & \downarrow & & \downarrow & & \downarrow & & \\ \cdots & \longrightarrow & T_1(B'') & \xrightarrow[\partial_1^B]{} & T(B') & \longrightarrow & T(B) & \longrightarrow & T(B'') & \longrightarrow & 0 \end{array}$$

is also commutative.

In particular, T_0 is right-exact.

Definition 11.24. Given two ∂-functors $S = (S_n)_{n \geq 0}$ and $T = (T_n)_{n \geq 0}$, a *morphism* $\eta\colon S \to T$ between S and T is a family $\eta = (\eta_n)_{n \geq 0}$ of natural transformations $\eta_n : S_n \to T_n$ such that the following diagram commutes

$$\begin{array}{ccc} S_n(A'') & \xrightarrow{\partial_n^S} & S_{n-1}(A') \\ {\scriptstyle (\eta_n)_{A''}}\downarrow & & \downarrow{\scriptstyle (\eta_{n-1})_{A'}} \\ T_n(A'') & \xrightarrow[\partial_n^T]{} & T_{n-1}(A') \end{array}$$

for all $n \geq 1$ and for every short exact sequence $0 \longrightarrow A' \longrightarrow A \longrightarrow A'' \longrightarrow 0$.

Morphisms of ∂-functors are composed in the obvious way. The notion of isomorphism is clear (each η_n is an isomorphism).

Grothendieck introduced the important notion of universal ∂-functor; see Grothendieck [Grothendieck (1957)] (Chapter II, Section 2.2).

Definition 11.25. A ∂-functor $T = (T_n)_{n \geq 0}$ is *universal* if for every ∂-functor $S = (S_n)_{n \geq 0}$ and every natural transformation $\varphi\colon S_0 \to T_0$ there is a *unique* morphism $\eta\colon S \to T$ such that $\eta_0 = \varphi$ as illustrated in the commutative diagram below; we say that η *lifts* φ.

$$\begin{array}{ccccccccccc} \cdots T_2(A'') & \xrightarrow{\partial_2^T} & T_1(A') & \longrightarrow & T_1(A) & \longrightarrow & T_1(A'') & \xrightarrow{\partial_1^T} & T_0(A') & \longrightarrow & \cdots \\ \downarrow{\scriptstyle \eta^2_{A''}} & & \downarrow{\scriptstyle \eta^1_{A'}} & & \downarrow{\scriptstyle \eta^1_{A}} & & \downarrow{\scriptstyle \eta^1_{A''}} & & \downarrow{\scriptstyle \varphi_{A'}} & & \\ \cdots S_2(A'') & \xrightarrow[\partial_2^S]{} & S_1(A') & \longrightarrow & S_1(A) & \longrightarrow & S_1(A'') & \xrightarrow[\partial_1^S]{} & S_0(A') & \longrightarrow & \cdots \end{array}$$

for every short exact sequence $0 \longrightarrow A' \longrightarrow A \longrightarrow A'' \longrightarrow 0$.

Proposition 11.40. *Suppose $S = (S_n)_{n \geq 0}$ and $T = (T_n)_{n \geq 0}$ are both universal ∂-functors and there is an isomorphism $\varphi\colon S_0 \to T_0$ (a natural transformation φ which is an isomorphism). Then there is a unique isomorphism $\eta\colon S \to T$ lifting φ.*

The proof of Proposition 11.40 is the same as the proof of Proposition 11.38. Proposition 11.40 shows a significant property of a universal ∂-functor T: *it is completely determined by the component* T_0.

There are plenty of (universal) ∂-functors.

Theorem 11.41. *Assume the abelian category* $\mathbf{C}$ *has enough projectives. For every additive right-exact functor* $T\colon \mathbf{C} \to \mathbf{D}$, *the family* $(L_nT)_{n\geq 0}$ *of left derived functors of* T *is a* ∂*-functor. Furthermore* T *is isomorphic to* L_0T.

Proof. Now that we have done all the hard work the proof is short: apply Theorem 11.32. The second property follows from Proposition 11.29. □

Grothendieck came up with an ingenious sufficient condition for a δ-functor to be universal: the notion of an *erasable* functor. Since Grothendieck's paper is written in French, this notion defined in Section 2.2 (Page 141) of [Grothendieck (1957)] is called *effaçable*, and many books and paper use it. Since the English translation of "effaçable" is "erasable," as advocated by Lang we will use the English word.

Definition 11.26. An additive functor $T\colon \mathbf{C} \to \mathbf{D}$ is *erasable* (or *effaçable*) if for every object $A \in \mathbf{C}$ there is some object $M_A \in \mathbf{C}$ and a monic $u\colon A \to M_A$ such that $T(u) = 0$. In particular this will be the case if $T(M_A)$ is the zero object of $\mathbf{D}$. We say that T *coerasable* (or *coeffaçable*) if for every object $A \in \mathbf{C}$ there is some object $M_A \in \mathbf{C}$ and an epic $u\colon M_A \to A$ such that $T(u) = 0$.

In many cases T is erasable by injectives (which means that M_A can be chosen to be injective) and T is coerasable by projectives (which means that M_A can be chosen to be projective). However, this is not always desirable.

The following proposition shows that our favorite functors, namely right derived functors, are erasable functors (and left derived functors are coerasable by projectives).

Proposition 11.42. *Assume the abelian category* $\mathbf{C}$ *has enough injectives. For every additive (left-exact) functor* $T\colon \mathbf{C} \to \mathbf{D}$, *the right derived functors* R^nT *are erasable by injectives for all* $n \geq 1$. *Assume the abelian category* $\mathbf{C}$ *has enough projectives. For every additive (right-exact) functor* $T\colon \mathbf{C} \to \mathbf{D}$, *the left derived functors* L_nT *are coerasable by projectives for all* $n \geq 1$.

Proof. For every $A \in \mathbf{C}$ there is a monic $u\colon A \to I$ into some injective I. Applying R^nT we get a map $R^nT(u)\colon R^nT(A) \to R^nT(I)$, but by Proposition 11.30 we have $R^nT(I) = (0)$ for all $n \geq 1$. The proof in the projective case is similar and left as an exercise. □

In order to state Grothendieck's theorem (Theorem 11.44), we need the notion of injective erasing of an object, due to Grothendieck; see Grothendieck's Tohoku [Grothendieck (1957)], Section 1.10.

Definition 11.27. Let $\mathbf{C}$ be an abelian category. For any object $A \in \mathbf{C}$, an *injective erasing* of A is a monic $u\colon A \to M$ such that for every monic $g\colon B \to C$ and any map $f\colon B \to A$, there is some map $\widetilde{f}\colon C \to M$ making the following diagram commute

$$\begin{array}{ccccc} 0 & \longrightarrow & B & \xrightarrow{g} & C \\ & & \downarrow{\scriptstyle f} & & \downarrow{\scriptstyle \widetilde{f}} \\ 0 & \longrightarrow & A & \xrightarrow[u]{} & M. \end{array}$$

If $\mathbf{C}$ has enough injectives, then any monic $u\colon A \to I$ where I is injective is an injective erasing. Definition 11.27 allows more general kinds of erasing.

The following proposition reveals some relationships between the notion of erasability and the notion of injective erasing.

Proposition 11.43. *Suppose that T is an additive functor from $\mathbf{C}$ to some other abelian category $\mathbf{D}$.*

(1) If T is erasable, then for any injective erasing $u\colon A \to M$, we have $T(u) = 0$.
(2) If every object $A \in \mathbf{C}$ has an injective erasing, then T is erasable iff $T(u) = (0)$ for every injective erasing $u\colon A \to M$.
(3) If T is erasable, then $T(I) = (0)$ for every injective object I.
(4) If $\mathbf{C}$ has enough injectives, then T is erasable iff $T(I) = (0)$ for every injective object I.

Proof. (1) Suppose that A is erased by some (monic) map $v\colon A \to M_A$ ($T(v) = 0$). Since $u\colon A \to M$ is an injective erasing, we have the following commutative diagram

$$\begin{array}{ccccc} 0 & \longrightarrow & A & \xrightarrow{v} & M_A \\ & & \downarrow{\scriptstyle \mathrm{id}} & & \downarrow{\scriptstyle \widetilde{\mathrm{id}}} \\ 0 & \longrightarrow & A & \xrightarrow[u]{} & M, \end{array}$$

and if apply the functor T we get the following commutative diagram

$$\begin{array}{ccc} T(A) & \xrightarrow{T(v)} & T(M_A) \\ {\scriptstyle T(\mathrm{id})}\downarrow & & \downarrow{\scriptstyle T(\widetilde{\mathrm{id}})} \\ T(A) & \xrightarrow[T(u)]{} & T(M), \end{array}$$

and since $T(v) = 0$ and $T(\mathrm{id}) = \mathrm{id}_{T(A)}$, we obtain

$$T(u) = T(u) \circ \mathrm{id}_{T(A)} = T(\widetilde{\mathrm{id}}) \circ 0 = 0.$$

(2) Assume that every object $A \in \mathbf{C}$ has an injective erasing $u \colon A \to M$. If $T(u) = (0)$ for every such injective erasing, then A is erased by $u \colon A \to M$, so T is erasable. The converse is given by (1).

(3) Assume that T is erasable and let I be some injective object. Since T is erasable, we have monic $v \colon I \to M_I$ such that $T(v) = 0$. Since I is injective, we have the following commutative diagram:

$$\begin{array}{ccccc} 0 & \longrightarrow & I & \xrightarrow{v} & M_I \\ & & {\scriptstyle \mathrm{id}}\downarrow & & \downarrow{\scriptstyle \widetilde{\mathrm{id}}} \\ 0 & \longrightarrow & I & \xrightarrow[\mathrm{id}]{} & I. \end{array}$$

If we apply T to the above diagram, since $T(v) = 0$ we get the commutative diagram

$$\begin{array}{ccc} T(I) & \xrightarrow{0} & T(M_I) \\ {\scriptstyle T(\mathrm{id})}\downarrow & & \downarrow{\scriptstyle T(\widetilde{\mathrm{id}})} \\ T(I) & \xrightarrow[T(\mathrm{id})]{} & T(I), \end{array}$$

so we get

$$\mathrm{id}_{T(I)} = T(\mathrm{id}) = T(\widetilde{\mathrm{id}}) \circ 0 = 0,$$

which implies $T(I) = (0)$.

(4) Assume $\mathbf{C}$ has enough injectives. First assume that $T(I) = (0)$ for every injective I. For any object A there is a monic $u \colon A \to I$ with I injective, so by applying T there is a map $T(u) \colon T(A) \to T(I)$. Since I is injective, $T(I) = (0)$, so $T(u) = 0$ and u erases A. Therefore, T is erasable. The converse has been proven in (3). $\square$

The following theorem shows the significance of the seemingly strange notion of injective erasability.

Theorem 11.44. *(Grothendieck) Let $T = (T^n)_{n\geq 0}$ be a δ-functor between two abelian categories* **C** *and* **D**. *If every object $A \in$* **C** *has an injective erasing $v\colon A \to M_A$ such that $T^n(v) = 0$ for all $n \geq 1$, then T is a universal δ-functor.*

Proof. Theorem 11.44 is essentially Proposition 2.2.1 on Page 141 of Grothendieck's Tohoku [Grothendieck (1957)], with the slightly stronger hypothesis of injective erasability because the proof is simpler. Grothendieck's version requiring only erasability will be discussed after the proof of this theorem.

The proof takes two thirds of a page. Even if you read French, you are likely to be frustrated. All the pieces are there but as Grothendieck says

"Des raisonnements standarts montrent que le morphisme ainsi défini ne dépend pas du choix particulier de la suite exacte $0 \longrightarrow A \longrightarrow M \longrightarrow A' \longrightarrow 0$, puis le fait que ce morphisme est fonctoriel, et permute à ∂."

Roughly translated, the above says that the details constitute "standard reasoning." No doubt that experts in the field will have no trouble supplying the details but for the rest of us, where is a complete proof?

The proof that we present consists of four steps. It is essentially due to Steve Shatz, except that we use injective erasings, which makes it a little more general.

Let us begin by explaining the main construction in the proof. The proof is by induction on n; we shall treat only the case $n = 1$; the other cases are very similar.

Step 1. Construction of the lift map u_1.

Let $S = (S^n)_{n\geq 0}$ be another δ-functor and let $u_0\colon T^0 \to S^0$ be a given map of functors. If A is an object of **C**, injective erasability of A for T^1 shows that there is an exact sequence

$$0 \longrightarrow A \xrightarrow{\ v\ } M_A \xrightarrow{\ p\ } A'' \longrightarrow 0, \qquad (\dagger)$$

with $A'' = \mathrm{Coker}(v)$, such that the map $\delta^0_{T^0}$ in the induced sequence

$$T^0(M_A) \xrightarrow{T^0(p)} T^0(A'') \xrightarrow{\delta^0_{T^0}} T^1(A) \xrightarrow{\ 0\ } T^1(M_A)$$

is surjective (since $T^1(v) = 0$). Since T is a δ-functor we have the commutative diagram

$$\begin{array}{ccccccc} T^0(M_A) & \xrightarrow{T^0(p)} & T^0(A'') & \xrightarrow{\delta^0_{T^0}} & T^1(A) & \xrightarrow{0} & T^1(M_A) \\ \downarrow{\scriptstyle u_0(M_A)} & & \downarrow{\scriptstyle u_0(A'')} & & \vdots{\scriptstyle u_1} & & \\ S^0(M_A) & \xrightarrow[S^0(p)]{} & S^0(A'') & \xrightarrow[\delta^0_{S^0}]{} & S^1(A). & & \end{array}$$

Since $\mathrm{Ker}\, \delta^0_{T^0} = \mathrm{Im}\, T^0(p)$, since the left square commutes

$$u_0(A'') \circ T^0(p) = S^0(p) \circ u_0(M_A),$$

and since the bottom row is exact, we get

$$\delta^0_{S^0} \circ u_0(A'') \circ T^0(p) = \delta^0_{S^0} \circ S^0(p) \circ u_0(M_A) = 0,$$

which proves that

$$\mathrm{Ker}\, \delta^0_{T^0} \subseteq \mathrm{Ker}\, (\delta^0_{S^0} \circ u_0(A'')).$$

Since $\delta^0_{T^0}$ is surjective we define $u_1 \colon T^1(A) \to S^1(A)$ as follows: for any $a \in T^1(A)$, pick any $b \in T^0(A'')$ such that $a = \delta^0_{T^0}(b)$, and set

$$u_1(a) = (\delta^0_{S^0} \circ u_0(A''))(b). \qquad (*)$$

This map is well-defined, because if $a = \delta^0_{T^0}(b')$ for some other $b' \in T^0(A'')$, then $\delta^0_{T^0}(b) = \delta^0_{T^0}(b')$, so $\delta^0_{T^0}(b' - b) = 0$, that is $b' - b = c$ for some $c \in \mathrm{Ker}\, \delta^0_{T^0}$, and since $\mathrm{Ker}\, \delta^0_{T^0} \subseteq \mathrm{Ker}\, (\delta^0_{S^0} \circ u_0(A''))$, we have $b' = b + c$ with $c \in \mathrm{Ker}\, (\delta^0_{S^0} \circ u_0(A''))$, which implies that

$$\begin{aligned} (\delta^0_{S^0} \circ u_0(A''))(b') &= (\delta^0_{S^0} \circ u_0(A''))(b + c) \\ &= (\delta^0_{S^0} \circ u_0(A''))(b) + (\delta^0_{S^0} \circ u_0(A''))(c) \\ &= (\delta^0_{S^0} \circ u_0(A''))(b) + 0 = (\delta^0_{S^0} \circ u_0(A''))(b). \end{aligned}$$

Therefore the map $u_1 \colon T^1(A) \to S^1(A)$ making the second square commute is uniquely defined. It remains to check that u_1 has the required properties and that it does not depend on the choice of the exact sequence (†). Lang [Lang (1993)] actually spells out most of the details but leaves out the verification that the argument does not depend on the choice of the short exact sequence defining M_A; see Chapter XX, §7, Theorem 7.1. This is where the assumption that injective erasings exist is needed.

Step 2. The proof of independence from the choice of the injective erasing $A \xrightarrow{v} M_A$ is a nice illustration of the extension property of injective erasings. Suppose we have another exact sequence

$$0 \longrightarrow \widetilde{A} \xrightarrow{\widetilde{v}} \widetilde{M_A} \xrightarrow{\widetilde{p}} \widetilde{A''} \longrightarrow 0,$$

where $\widetilde{v}\colon \widetilde{A} \to \widetilde{M_A}$ is an injective erasing of $\widetilde{A}$ (which exists, by hypothesis), with $\widetilde{A''} = \mathrm{Coker}(\widetilde{v})$. By hypothesis, we have $T^1(\widetilde{v}) = 0$. Assume we have a map $g\colon A \to \widetilde{A}$. Since $\widetilde{v}\colon \widetilde{A} \to \widetilde{M_A}$ is an injective erasing and v is a monic, there is a map θ extending $\widetilde{v} \circ g$ making the following diagram commute:

$$\begin{array}{ccccccccc}
0 & \longrightarrow & A & \xrightarrow{v} & M_A & \xrightarrow{p} & A'' & \longrightarrow & 0 \\
 & & \downarrow{\scriptstyle g} & & \downarrow{\scriptstyle \theta} & & & & \\
0 & \longrightarrow & \widetilde{A} & \xrightarrow[\widetilde{v}]{} & \widetilde{M_A} & \xrightarrow[\widetilde{p}]{} & \widetilde{A''} & \longrightarrow & 0.
\end{array}$$

Now the diagram

$$\begin{array}{ccccccc}
A & \xrightarrow{v} & M_A & \xrightarrow{p} & A'' & \longrightarrow & 0 \\
\downarrow{\scriptstyle g} & & \downarrow{\scriptstyle \theta} & & \vdots{\scriptstyle \overline{\theta}} & & \\
\widetilde{A} & \xrightarrow[\widetilde{v}]{} & \widetilde{M_A} & \xrightarrow[\widetilde{p}]{} & \widetilde{A''} & &
\end{array}$$

is similar to the commutative diagram used in the construction of u_1 in Step 1, and it has exact rows, so the same argument shows that there is a map $\overline{\theta}$ making the diagram

$$\begin{array}{ccccccccc}
0 & \longrightarrow & A & \xrightarrow{v} & M_A & \xrightarrow{p} & A'' & \longrightarrow & 0 \\
 & & \downarrow{\scriptstyle g} & & \downarrow{\scriptstyle \theta} & & \downarrow{\scriptstyle \overline{\theta}} & & \\
0 & \longrightarrow & \widetilde{A} & \xrightarrow[\widetilde{v}]{} & \widetilde{M_A} & \xrightarrow[\widetilde{p}]{} & \widetilde{A''} & \longrightarrow & 0
\end{array}$$

commute.

Theorem 11.31 applied to the above diagram with T and S yields the two commutative diagrams

$$\begin{array}{ccccccc}
T^0(M_A) & \longrightarrow & T^0(A'') & \longrightarrow & T^1(A) & \longrightarrow & 0 \\
\downarrow & & \downarrow & & \downarrow{\scriptstyle T^1(g)} & & \\
T^0(\widetilde{M_A}) & \longrightarrow & T^0(\widetilde{A''}) & \longrightarrow & T^1(\widetilde{A}) & \longrightarrow & 0,
\end{array}$$

since $T^1(v) = 0$ and $T^1(\widetilde{v}) = 0$, and

$$\begin{array}{ccccc}
S^0(M_A) & \longrightarrow & S^0(A'') & \longrightarrow & S^1(A) \\
\downarrow & & \downarrow & & \downarrow{\scriptstyle S^1(g)} \\
S^0(\widetilde{M_A}) & \longrightarrow & S^0(\widetilde{A''}) & \longrightarrow & S^1(\widetilde{A}),
\end{array}$$

and the two commutative diagrams involved in the construction of u_1 and $\widetilde{u_1}$ in Step 1,

$$\begin{array}{ccccccc}
T^0(M_A) & \longrightarrow & T^0(A'') & \longrightarrow & T^1(A) & \xrightarrow{0} & 0 \\
\downarrow{\scriptstyle u_0(M_A)} & & \downarrow{\scriptstyle u_0(A'')} & & \downarrow{\scriptstyle u_1} & & \\
S^0(M_A) & \longrightarrow & S^0(A'') & \longrightarrow & S^1(A) & &
\end{array}$$

and

$$\begin{array}{ccccccc}
T^0(\widetilde{M_A}) & \longrightarrow & T^0(\widetilde{A''}) & \longrightarrow & T^1(\widetilde{A}) & \xrightarrow{0} & 0 \\
\downarrow{\scriptstyle u_0(\widetilde{M_A})} & & \downarrow{\scriptstyle u_0(\widetilde{A''})} & & \downarrow{\scriptstyle \widetilde{u_1}} & & \\
S^0(\widetilde{M_A}) & \longrightarrow & S^0(\widetilde{A''}) & \longrightarrow & S^1(\widetilde{A}). & &
\end{array}$$

We can combine these four diagrams into the following diagram.

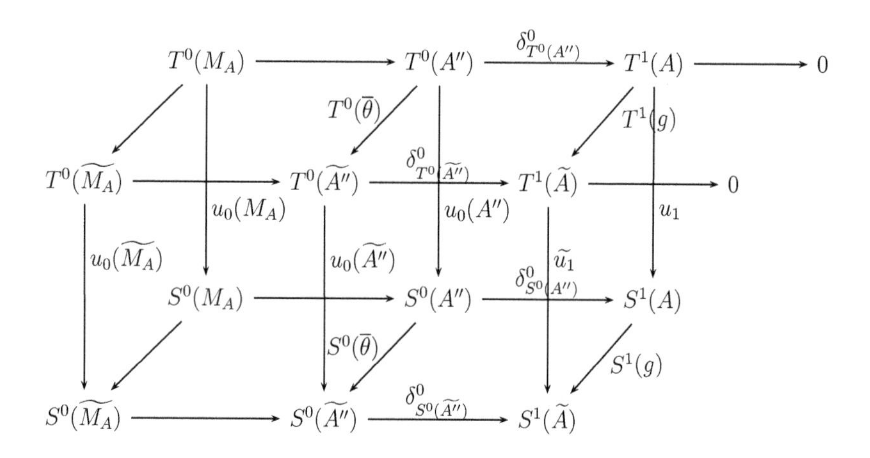

All squares at top, bottom, front, and back commute, and the two left hand vertical squares also commute (by naturality of u_0). Since $\delta^0_{T^0(A'')}\colon T^0(A'') \to T^1(A)$ is surjective, if we prove that

$$\widetilde{u_1} \circ T^1(g) \circ \delta^0_{T^0(A'')} = S^1(g) \circ u_1 \circ \delta^0_{T^0(A'')},$$

then we can conclude that

$$\widetilde{u_1} \circ T^1(g) = S^1(g) \circ u_1,$$

which is the commutativity of the righthand vertical square. For this we use the commutativity of the other five faces of the rightmost cube, in the order

top, front, left, bottom, back.

The details are left as an exercise.

If we set $A = \widetilde{A}$ and $g = \mathrm{id}$ (perhaps for different M_A and $\widetilde{M_A}$), we see that

$$\widetilde{u_1} = u_1,$$

so u_1 is independent of M_A.

Step 3. To prove that the construction of u_1 given in Step 1 is functorial, we need to show that for any map $g\colon A \to \widetilde{A}$, if u_1 and $\widetilde{u_1}$ are obtained using the construction in Step 1 involving the two diagrams just before the big diagram, then the following diagram commutes

$$\begin{array}{ccc} T^1(A) & \xrightarrow{T^1(g)} & T^1(\widetilde{A}) \\ {\scriptstyle u_1}\downarrow & & \downarrow{\scriptstyle \widetilde{u_1}} \\ S^1(A) & \xrightarrow[S^1(g)]{} & S^1(\widetilde{A}). \end{array}$$

However, this is just the diagram corresponding to the right face of the right cube, and we just proved that the construction makes it commute.

Step 4. Finally, we need to prove that for any short exact sequence

$$0 \longrightarrow A' \xrightarrow{\varphi} A \xrightarrow{\psi} A'' \longrightarrow 0,$$

the diagram

$$\begin{array}{ccc} T^0(A'') & \xrightarrow{\delta^0_{T^0}} & T^1(A') \\ {\scriptstyle u_0(A'')}\downarrow & & \downarrow{\scriptstyle u_1} \\ S^0(A'') & \xrightarrow[\delta^0_{S^0}]{} & S^1(A') \end{array}$$

commutes, where u_1 is constructed in Step 1 (see Definition 11.22). Here we have to be careful because ψ is not necessarily erased, so the previous construction does not work. However, there is an injective erasing

$$0 \longrightarrow A' \xrightarrow{v} M_{A'} \xrightarrow{p} X \longrightarrow 0,$$

and as before we obtain a commutative diagram

$$\begin{array}{ccccccccc} 0 & \longrightarrow & A' & \xrightarrow{\varphi} & A & \xrightarrow{\psi} & A'' & \longrightarrow & 0 \\ & & \downarrow{\scriptstyle \mathrm{id}_{A'}} & & \downarrow{\scriptstyle \theta} & & \downarrow{\scriptstyle \overline{\theta}} & & \\ 0 & \longrightarrow & A' & \xrightarrow[v]{} & M_{A'} & \xrightarrow[p]{} & X & \longrightarrow & 0. \end{array}$$

Since T is a δ-functor, we obtain the commutative diagram

$$\begin{array}{ccc} T^0(A'') & \xrightarrow{\delta^0_{T^0(A'')}} & T^1(A') \\ {\scriptstyle T^0(\overline{\theta})}\downarrow & & \downarrow{\scriptstyle \mathrm{id}} \\ T^0(X) & \xrightarrow[\delta^0_{T^0(X)}]{} & T^1(A'). \end{array}$$

Similarly, since S is a δ-functor, we obtain the commutative diagram

$$\begin{array}{ccc} S^0(A'') & \xrightarrow{\delta^0_{S^0(A'')}} & S^1(A') \\ {\scriptstyle S^0(\overline{\theta})}\downarrow & & \downarrow{\scriptstyle \mathrm{id}} \\ S^0(X) & \xrightarrow[\delta^0_{S^0(X)}]{} & S^1(A'). \end{array}$$

Since u_0 is a natural transformation, we have the commutative diagram

$$\begin{array}{ccc} T^0(A'') & \xrightarrow{T^0(\overline{\theta})} & T^0(X) \\ {\scriptstyle u_0(A'')}\downarrow & & \downarrow{\scriptstyle u_0(X)} \\ S^0(A'') & \xrightarrow[S^0(\overline{\theta})]{} & S^0(X). \end{array}$$

The construction of u_1 in Step 1 (with X instead of A'' and A' instead of A) yields the following commutative diagram

$$\begin{array}{ccc} T^0(X) & \xrightarrow{\delta^0_{T^0}} & T^0(A') \\ {\scriptstyle u_0(X)}\downarrow & & \downarrow{\scriptstyle u_1} \\ S^0(X) & \xrightarrow[\delta^0_{S^0}]{} & S^0(A'). \end{array}$$

We leave it as an exercise to put the four diagrams above as four faces of a prism whose top and bottom faces are the triangles corresponding to the first two diagrams (because of the edge id, the vertices corresponding to $T^1(A')$ can be merged and similarly the vertices corresponding to $S^1(A')$ can be merged), the left-hand square face corresponds to the third diagram, and the front square face corresponds to the fourth diagram. They all commute, and one can deduce that the right-hand square face also commutes, which is the desired commutative diagram that we are seeking. To prove that

$$u_1 \circ \delta^0_{T^0} = \delta^0_{S^0} \circ u_0(A''),$$

one should use commutations in the order

top triangular face, front face, left face, bottom triangular face. $\square$

Observe that Theorem 11.44 does not require that $\mathbf{C}$ has enough injectives. The hypothesis of the theorem relies on the condition of erasability of the functors T^n given by Proposition 11.43(3).

Actually, the weaker hypothesis that the functors T^n are erasable for all $n \geq 1$ is enough to prove that the functors (T^n) constitute a universal δ-functor. We thank Steve Shatz for communicating the following clever argument.

Observe that if

$$0 \longrightarrow A \longrightarrow \widetilde{M_A} \longrightarrow \widetilde{A''} \longrightarrow 0$$

is another exact sequence, and *if this sequence dominates the former* in the sense that there is a commutative diagram

$$\begin{array}{ccccccccc} 0 & \longrightarrow & A & \longrightarrow & M_A & \longrightarrow & A'' & \longrightarrow & 0 \\ & & \downarrow & & \downarrow & & \downarrow & & \\ 0 & \longrightarrow & A & \longrightarrow & \widetilde{M_A} & \longrightarrow & \widetilde{A''} & \longrightarrow & 0, \end{array}$$

then the proof of Step 2 shows that the maps u_1 and $\widetilde{u_1}$ induced by these sequences are the same. From this it follows that given two sequences

$$0 \longrightarrow A \longrightarrow M_A \longrightarrow A'' \longrightarrow 0$$

and

$$0 \longrightarrow A \longrightarrow \widetilde{M_A} \longrightarrow \widetilde{A''} \longrightarrow 0,$$

we need only find a common dominant.

If ξ is the composed map $A \longrightarrow M_A \longrightarrow M_A \oplus \widetilde{M_A}$ and η is the composed map $A \longrightarrow \widetilde{M_A} \longrightarrow M_A \oplus \widetilde{M_A}$, then $\xi - \eta$ is an injection of A into $M_A \oplus \widetilde{M_A}$. Let M be the cokernel of $\xi - \eta$, then we leave it as an exercise to prove that the exact sequence

$$0 \longrightarrow A \longrightarrow M \longrightarrow A'' \oplus \widetilde{A''} \longrightarrow 0,$$

is the required dominant. Therefore we have the following theorem as stated by Grothendieck in [Grothendieck (1957)] (Section 2.2, Proposition 2.2.1).

Theorem 11.45. *(Grothendieck) Let $T = (T^n)_{n\geq 0}$ be a δ-functor between two abelian categories $\mathbf{C}$ and $\mathbf{D}$. If the functors T^n are erasable for all $n \geq 1$, then T is a universal δ-functor.*

If $\mathbf{C}$ has enough injectives, then by Proposition 11.43(4), the functors T^n are erasable iff $T^n(I) = (0)$ for all injective objects I, for every $n \geq 1$. This is the situation generally encountered. In this case we have the following corollary.

Theorem 11.46. *(Grothendieck) Let $T = (T^n)_{n\geq 0}$ be a δ-functor between two abelian categories $\mathbf{C}$ and $\mathbf{D}$. Suppose $\mathbf{C}$ has enough injectives. If $T^n(I) = (0)$ for all injective I, for all $n \geq 1$, then T is a universal δ-functor.*

There is also a version of Theorem 11.44 for a contravariant ∂-functor which is erasable.

Combining Theorem 11.44 and Theorem 11.39 we obtain the most important result of this chapter.

Theorem 11.47. *Assume the abelian category $\mathbf{C}$ has enough injectives. For every additive left-exact functor $T\colon \mathbf{C} \to \mathbf{D}$, the right derived functors $(R^nT)_{n\geq 0}$ form a universal δ-functor such that T is isomorphic to R^0T. Conversely, every universal δ-functor $T = (T^n)_{n\geq 0}$ is isomorphic to the right derived δ-functor $(R^nT^0)_{n\geq 0}$.*

Proof. The first statement is obtained by combining Theorem 11.44, Proposition 11.42, and Theorem 11.39. Conversely, if $T = (T^n)_{n\geq 0}$ is a universal δ-functor, then T^0 is left-exact, so by the first part of the theorem applied to T^0, $(R^nT^0)_{n\geq 0}$ is a universal δ-functor with R^0T^0 isomorphic to T^0, thus T and $(R^nT^0)_{n\geq 0}$ are isomorphic by Proposition 11.38. ☐

After all, the mysterious universal δ-functors are just the right derived functors of left-exact functors. As an example, the functors $\mathrm{Ext}^n_R(A, -)$ constitute a universal δ-functor (for any fixed R-module A). For every sheaf $\mathcal{F}$ on a topological space X, the global section functor $\Gamma(X, -)$ is left-exact, so its right derived functors $R^p\Gamma(X, -)$ form a universal δ-functor. The corresponding cohomology groups $R^p\Gamma(X, -)(\mathcal{F})$, denoted $H^p(X, \mathcal{F})$, are called the *cohomology groups of the sheaf* $\mathcal{F}$. The cohomology of sheaves will be thoroughly investigated in Chapter 13. It is one of the most sophisticated (and powerful) tools discussed in this book.

Of course there is a version of Theorem 11.44 for coerasable ∂-functors. We leave to reader the task of stating the dual notion of Definition 11.27, which should be called *projective coerasing*, and to formulate the dual of Proposition 11.43. We state a version using coerasability by projectives.

Theorem 11.48. *(Grothendieck) Let $T = (T_n)_{n\geq 0}$ be a ∂-functor between two abelian categories* **C** *and* **D**. *If T_n is coerasable by projectives for all $n \geq 1$, then T is a universal ∂-functor.*

Remark: As the case of δ-functors, there are versions of Theorem 11.48 using coerasability criteria not requiring coerasability by projectives.

There is a version of Theorem 11.48 for a contravariant δ-functor which is coerasable.

Combining Theorem 11.48 and Theorem 11.41 we obtain the other most important result of this section.

Theorem 11.49. *Assume the abelian category* **C** *has enough projectives. For every additive right-exact functor $T\colon \mathbf{C} \to \mathbf{D}$ the left derived functors $(L_nT)_{n\geq 0}$ form a universal ∂-functor such that T is isomorphic to L_0T. Conversely, every universal ∂-functor $T = (T_n)_{n\geq 0}$ is isomorphic to the left derived ∂-functor $(L_nT_0)_{n\geq 0}$.*

After all, the mysterious universal ∂-functors are just the left derived functors of right-exact functors. For example, the functors $\mathrm{Tor}_n^R(A, -)$ and $\mathrm{Tor}_n^R(-, B)$ constitute universal ∂-functors.

Remark: Theorem 11.47 corresponds to Case (Ri). If **C** has enough injectives there is also a version of Theorem 11.47 for a contravariant right-exact functor T saying that $(L_nT)_{n\geq 0}$ is a contravariant universal ∂-functor (Case (Li)). There doesn't seem to be any practical example of this case.

Theorem 11.49 corresponds to Case (Lp). If **C** has enough projectives there is a version of Theorem 11.49 for a contravariant left-exact functor T saying that $(R^nT)_{n\geq 0}$ is a contravariant universal δ-functor (Case (Rp)). As an example, the functors $\mathrm{Ext}_R^n(-, B)$ constitute a contravariant universal δ-functor (for any fixed R-module B).

11.9 Problems

Problem 11.1. In Proposition 11.1, prove that (1) (P is projective) is equivalent to (3) and also equivalent to (4).

Problem 11.2. Check that in the direct sum $\mathbb{Z}/6\mathbb{Z} \cong \mathbb{Z}/2\mathbb{Z} \oplus \mathbb{Z}/3\mathbb{Z}$, the $\mathbb{Z}/6\mathbb{Z}$-module $\mathbb{Z}/2\mathbb{Z}$ is projective but not free.

Problem 11.3. Prove that an R-module M is projective iff there is some family $(u_i)_{i\in I}$ of elements of M and a family of R-linear forms $(\varphi_i \colon M \to R)_{i\in I}$ such that:

(1) For every $x \in M$, we have $\varphi_i(x) = 0$ for all but finitely many $i \in I$.
(2) For every $x \in M$, we have $x = \sum_{i\in I}(\varphi_i(x))u_i$.

Furthermore, M is spanned by $(u_i)_{i\in I}$.

Problem 11.4. In Proposition 11.3, prove that (1) (I is injective) is equivalent to (3).

Problem 11.5. In Theorem 11.10, prove that if E is injective, then every injection $f \colon E \to M$ has a retraction.

Problem 11.6. Prove that if R is a PID and if K is the fraction field of R, then K/R is an injective R-module.

Problem 11.7. For any commutative ring R with a unit element, prove that R is a flat R-module.

Problem 11.8. For any commutative ring R with a unit element, prove that for any family $(M_i)_{i\in I}$ of R-modules, $\bigoplus_{i\in I} M_i$ is flat iff each M_i is flat. Prove that every projective module is flat.

Problem 11.9. Prove Proposition 11.12.

Problem 11.10. Prove that if R is an integral domain and if K is its fraction field, then K is a flat R-module.

Problem 11.11. Prove that $\mathbb{Q}/\mathbb{Z}$ is an injective $\mathbb{Z}$-module which is not flat and the $\mathbb{Z}$-module $\mathbb{Q} \oplus \mathbb{Z}$ is flat but neither projective nor injective.

Problem 11.12. Provide a detailed proof of Theorem 11.21. Use Proposition 11.20 and proceed by induction mimicking the proof of Theorem 11.17.

Problem 11.13. Prove the remaining cases of Proposition 11.30.

Problem 11.14. In the proof of Theorem 11.44, complete the verification of Step 2, namely that the righthand vertical square of the cube commutes.

Problem 11.15. In the proof of Theorem 11.44, complete the verification of Step 4, namely that the righthand square face of the prism commutes.

Problem 11.16. Prove that given two sequences

$$0 \longrightarrow A \longrightarrow M_A \longrightarrow A'' \longrightarrow 0$$

and

$$0 \longrightarrow A \longrightarrow \widetilde{M_A} \longrightarrow \widetilde{A''} \longrightarrow 0,$$

the sequence

$$0 \longrightarrow A \longrightarrow M \longrightarrow A'' \oplus \widetilde{A''} \longrightarrow 0$$

is a common dominant (where M is defined in the paragraph just before Theorem 11.45).

Chapter 12

Universal Coefficient Theorems

Suppose we have a homology chain complex

$$0 \xleftarrow{d_0} C_0 \xleftarrow{d_1} C_1 \cdots \xleftarrow{d_{p-1}} C_{p-1} \xleftarrow{d_p} C_p \xleftarrow{d_{p+1}} C_{p+1} \longleftarrow \cdots,$$

where the C_i are R-modules over some commutative ring R with a multiplicative identity element (recall that $d_i \circ d_{i+1} = 0$ for all $i \geq 0$). Given another R-module G we can form the homology complex

$$0 \xleftarrow{d_0 \otimes \mathrm{id}} C_0 \otimes_R G \xleftarrow{d_1 \otimes \mathrm{id}} C_1 \otimes_R G \cdots \xleftarrow{d_p \otimes \mathrm{id}} C_p \otimes_R G \longleftarrow \cdots,$$

obtained by tensoring with G, denoted $C \otimes_R G$, and the cohomology complex

$$0 \xrightarrow{\mathrm{Hom}_R(d_0, G)} \mathrm{Hom}_R(C_0, G) \longrightarrow \cdots$$

$$\cdots \longrightarrow \mathrm{Hom}_R(C_p, G) \xrightarrow{\mathrm{Hom}_R(d_{p+1}, G)} \mathrm{Hom}_R(C_{p+1}, G) \longrightarrow \cdots$$

obtained by applying $\mathrm{Hom}_R(-, G)$, and denoted $\mathrm{Hom}_R(C, G)$.

The question is: what is the relationship between the homology groups $H_p(C \otimes_R G)$ and the original homology groups $H_p(C)$ in the first case, and what is the relationship between the cohomology groups $H^p(\mathrm{Hom}_R(C, G))$ and the original homology groups $H_p(C)$ in the second case?

The ideal situation would be that

$$H_p(C \otimes_R G) \cong H_p(C) \otimes_R G \quad \text{and} \quad H^p(\mathrm{Hom}_R(C, G)) \cong \mathrm{Hom}_R(H_p(C), G),$$

but this is generally not the case. If the ring R is nice enough and if the modules C_p are nice enough, then $H_p(C \otimes_R G)$ can be expressed in terms of $H_p(C) \otimes_R G$ and $\mathrm{Tor}_1^R(H_{p-1}(C), G)$, where $\mathrm{Tor}_1^R(-, G)$ is a one of the left-derived functors of $- \otimes_R G$, and $H^p(\mathrm{Hom}_R(C, G))$ can be expressed in terms of $\mathrm{Hom}_R(H_p(C), G))$ and $\mathrm{Ext}^1_R(H_{p-1}(C), G)$, where $\mathrm{Ext}^1_R(-, G)$ is one of the right-derived functors of $\mathrm{Hom}_R(-, G)$; both derived functors are defined in Section 11.2 and further discussed in Example 11.1. These formulae are known as universal coefficient theorems.

12.1 Universal Coefficient Theorems for Homology

Following Rotman [Rotman (1988, 2009)] (Chapter 8), we give universal coefficients formulae that are general enough to cover all the cases of interest in singular homology and singular cohomology, for (commutative) rings that are hereditary and modules that are projective.

Definition 12.1. A commutative ring R (with an identity element) is *hereditary* if every ideal in R is a projective module.

Every PID is hereditary (and every semisimple ring is hereditary). The reason why hereditary rings are interesting is that if R is hereditary, then every submodule of a projective R-module is also projective. In fact, a theorem of Cartan and Eilenberg states that *a ring is hereditary iff every submodule of a projective R-module is also projective*; see Rotman [Rotman (1988, 2009)] (Chapter 4, Theorem 4.23).

The next theorem is a universal coefficient theorem for homology.

Theorem 12.1. *(Universal Coefficient Theorem for Homology) Let R be a commutative hereditary ring, G be any R-module, and let C be a chain complex of projective R-modules. Then there is a split exact sequence*

$$0 \longrightarrow H_n(C) \otimes_R G \xrightarrow{\mu} H_n(C \otimes_R G) \xrightarrow{p} \mathrm{Tor}^R_1(H_{n-1}(C), G) \longrightarrow 0$$

for all $n \geq 0$. (It is assumed that $H_n(C) = (0)$ for all $n < 0$.) Thus, we have an isomorphism

$$H_n(C \otimes_R G) \cong (H_n(C) \otimes_R G) \oplus \mathrm{Tor}^R_1(H_{n-1}(C), G)$$

for all $n \geq 0$. Furthermore, the maps involved in the exact sequence of the theorem are natural, which means that for any chain map $\varphi\colon C \to C'$ between two chain complexes C and C' the following diagram commutes:

$$\begin{array}{ccccccccc} 0 & \longrightarrow & H_n(C) \otimes_R G & \xrightarrow{\mu} & H_n(C \otimes_R G) & \xrightarrow{p} & \mathrm{Tor}^R_1(H_{n-1}(C), G) & \longrightarrow & 0 \\ & & \downarrow{\scriptstyle \varphi_* \otimes \mathrm{id}} & & \downarrow{\scriptstyle (\varphi \otimes \mathrm{id})_*} & & \downarrow{\scriptstyle \mathrm{Tor}^R_1(\varphi_*)} & & \\ 0 & \longrightarrow & H_n(C') \otimes_R G & \xrightarrow[\mu']{} & H_n(C' \otimes_R G) & \xrightarrow[p']{} & \mathrm{Tor}^R_1(H_{n-1}(C'), G) & \longrightarrow & 0. \end{array} \quad (\dagger)$$

Theorem 12.1 is proven in Rotman [Rotman (1988, 2009)] and we follow this proof (Chapter 8, Theorem 8.22). We warn the reader that in all the proofs that we are aware of (including Rotman's proof), the details involved in verifying that the maps μ and p are natural are omitted (or sketched). We decided to provide complete details (with a little help from Spanier [Spanier (1989)]), which makes the proof quite long. The reader is advised to skip such details upon first reading.

Before launching into the detailed proof we provide an outline.

Proof outline. There are two parts to the theorem.

(A) Derive the desired split exact sequence.
(B) Prove the naturality of the exact sequence.

Part A: Derive the split exact sequence

$$0 \longrightarrow H_n(C) \otimes_R G \xrightarrow{\mu} H_n(C \otimes_R G) \xrightarrow{p} \mathrm{Tor}_1^R(H_{n-1}(C), G) \longrightarrow 0.$$

Step A1: The first challenge in deriving this sequence is to get a grip on the term $\mathrm{Tor}_1^R(H_{n-1}(C), G)$. By definition, this means we must develop a projective resolution for $H_{n-1}(C)$. The desired projective resolution, namely

$$0 \longrightarrow Z_n \xrightarrow{i_n} C_n \xrightarrow{\widetilde{d}_n} Z_{n-1} \longrightarrow H_{n-1} \longrightarrow 0 \qquad (**)$$

is obtained by splicing together two short exact sequences

$$0 \longrightarrow Z_n(C) \xrightarrow{i_n} C_n \xrightarrow{d_n^B} B_{n-1}(C) \longrightarrow 0 \qquad (*)$$

and

$$0 \longrightarrow B_{n-1}(C) \longrightarrow Z_{n-1}(C) \longrightarrow H_{n-1}(C) \longrightarrow 0. \qquad (\dagger)$$

These sequences are "pictured" below in Figure 12.1, Figure 12.2, and Figure 12.3, where $Z_n = \mathrm{Ker}\, d_n$, $B_n = \mathrm{Im}\, d_{n+1}$ and $H_n = Z_n/B_n$. Note that we drop the argument (C) in $Z_n(C), B_n(C), H_n(C)$ since it is clear from the context. We use the projective resolution to calculate $\mathrm{Tor}_1^R(H_{n-1}(C), G)$ by tensoring with G to form the homology chain complex

$$0 \longrightarrow Z_n \otimes G \xrightarrow{i_n \otimes \mathrm{id}} C_n \otimes G \xrightarrow{\widetilde{d}_n \otimes \mathrm{id}} Z_{n-1} \otimes G \longrightarrow 0\,,$$

and discover that

$$\begin{aligned}\mathrm{Tor}_1^R(H_{n-1}, G) &= \mathrm{Ker}\,(\widetilde{d}_n \otimes \mathrm{id})/\mathrm{Im}\,(i_n \otimes \mathrm{id})\\ &\cong \mathrm{Ker}\,(\widetilde{d}_n \otimes \mathrm{id})/(Z_n \otimes G) \cong \mathrm{Ker}\,(d_n \otimes \mathrm{id})/(Z_n \otimes G).\end{aligned}$$

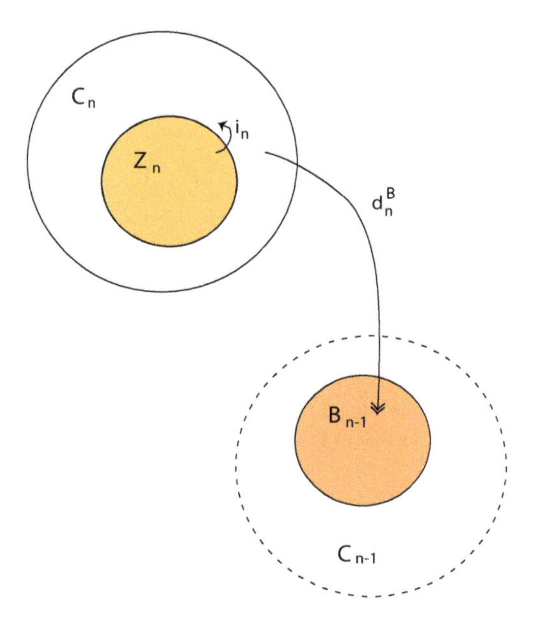

Fig. 12.1 A schematic representation of the exact sequence (∗).

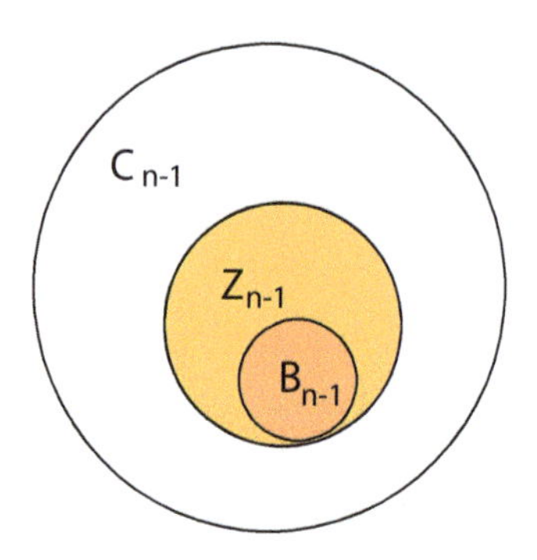

Fig. 12.2 A schematic representation of the exact sequence (†).

Step A2: Actually obtaining the short exact sequence of the theorem. First we verify that

$$\operatorname{Im}(d_{n+1} \otimes \mathrm{id}) \subseteq Z_n \otimes G \subseteq \operatorname{Ker}(d_n \otimes \mathrm{id}) \subseteq C_n \otimes G.$$

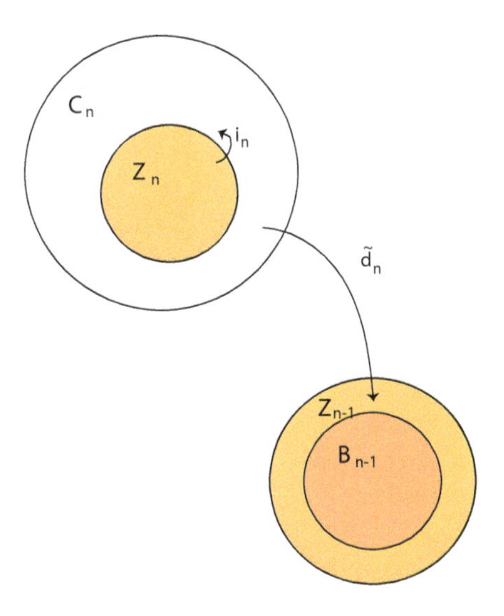

Fig. 12.3 A schematic representation of the projective resolution (∗∗).

See Figure 12.4. Then we apply the third isomorphism theorem to the containment identity, and the result directly follows after we observe that

$$\mathrm{Im}(d_{n+1} \otimes \mathrm{id}) = \{d_{n+1}(c) \otimes g \in C_n \otimes G \mid c \in C_{n+1}, g \in G\} = B_n \otimes G.$$

Step A3: Showing that the short exact sequence of Step A2 actually splits. This follows from the fact that the exact sequence (∗) used to build the projective resolution is in fact a short split exact sequence.

Part B: Prove the naturality of the exact sequence

Step B1: Show that the left square commutes. The slightly tricky part is we don't have a "nice" closed form for $\mathrm{Ker}\,(d_n \otimes \mathrm{id})$, but since

$$H_n(C) \otimes G \cong (Z_n \otimes G)/(B_n \otimes G),$$

and

$$H_n(C \otimes G) \cong (\mathrm{Ker}\,(d_n \otimes \mathrm{id}))/(B_n \otimes G),$$

and since we have the correct containment $Z_n \otimes G \subseteq \mathrm{Ker}\,(d_n \otimes \mathrm{id})$, the diagram chasing goes through.

Step B2: Show that the right square commutes. Here the tricky part is to define the map $\mathrm{Tor}_1(\varphi_*)$. To define the existence of $\mathrm{Tor}_1(\varphi_*)$ we

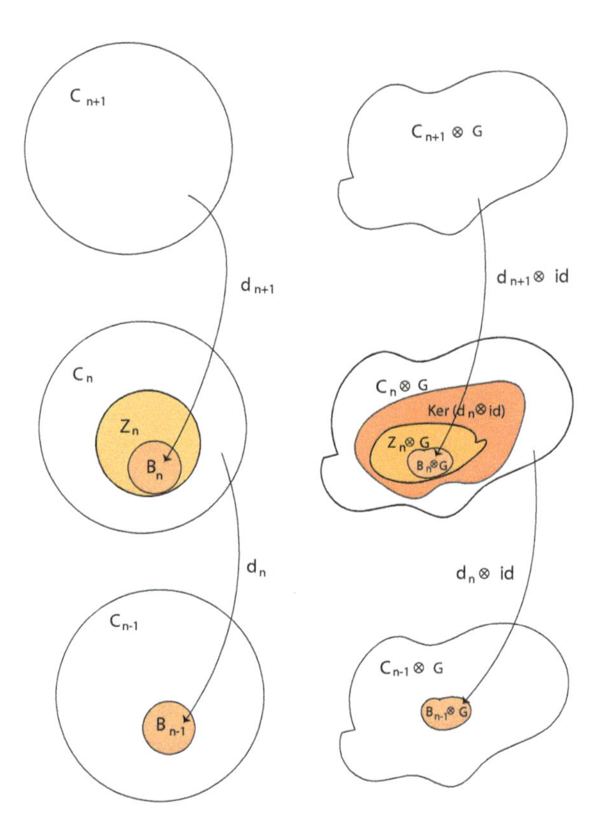

Fig. 12.4 A schematic representation of the containment identity of Step A2.

form a lift of the projective resolution of H_{n-1}, and then tensor this lift appropriately to define the correct push down of homology classes. Then we use the fact that

$$H_n(C \otimes G) \cong (\mathrm{Ker}\,(d_n \otimes \mathrm{id}))/(B_n \otimes G),$$

and

$$\mathrm{Tor}_1^R(H_{n-1}, G) \cong (\mathrm{Ker}\,(d_n \otimes \mathrm{id}))/(Z_n \otimes G)$$

to define the map p as a "modified" inclusion (you have to take equivalence classes over $B_n \otimes G$ instead of $Z_n \otimes G$). Then commutativity follows as desired. □

Proof of Theorem 12.1. We begin by observing that we have some exact sequences

$$0 \longrightarrow Z_n(C) \xrightarrow{i_n} C_n \xrightarrow{d_n^B} B_{n-1}(C) \longrightarrow 0 \qquad (*)$$

and

$$0 \longrightarrow B_{n-1}(C) \xrightarrow{\iota_{n-1}} Z_{n-1}(C) \longrightarrow H_{n-1}(C) \longrightarrow 0. \qquad (*')$$

The first sequence $(*)$ is exact by definition of $Z_n(C)$ as $Z_n(C) = \mathrm{Ker}\, d_n$ and $B_{n-1}(C)$ as $B_{n-1}(C) = \mathrm{Im}\, d_n$, where the map $d_n^B \colon C_n \to B_{n-1}(C)$ is the corestriction of $d_n \colon C_n \to C_{n-1}$ to $B_{n-1}(C)$. The second sequence $(*')$ is exact by definition of $H_{n-1}(C)$, as $H_{n-1}(C) = Z_{n-1}(C)/B_{n-1}(C) = \mathrm{Ker}\, d_{n-1}/\mathrm{Im}\, d_n$. From now on, to simplify notation we drop the argument (C) in $Z_n(C), B_n(C), H_n(C)$. These can be spliced using the diagram of exact sequences

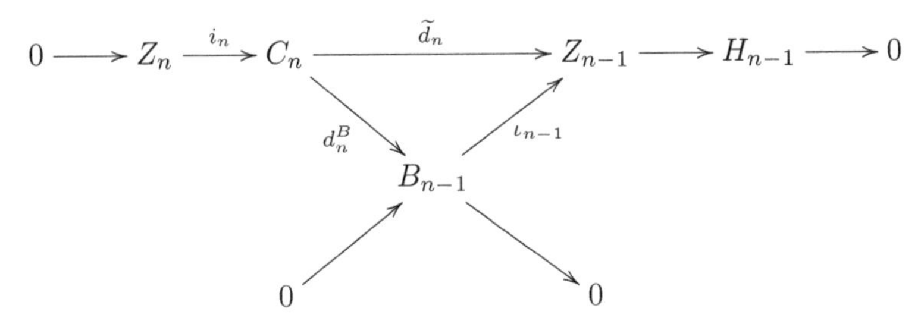

to form an exact sequence

$$0 \longrightarrow Z_n \xrightarrow{i_n} C_n \xrightarrow{\widetilde{d}_n} Z_{n-1} \longrightarrow H_{n-1} \longrightarrow 0. \qquad (**)$$

Here ι_{n-1} is the inclusion map of B_{n-1} into Z_{n-1} and $\widetilde{d}_n \colon C_n \to Z_{n-1}$ is the corestriction of $d_n \colon C_n \to C_{n-1}$ to Z_{n-1}. Since every C_n is projective and R is hereditary, the submodules Z_{n-1} and B_{n-1} of C_{n-1} are also projective. This implies that the short exact sequence $(*)$ splits (by Proposition 11.1(3)) and that the exact sequence $(**)$ is a projective resolution of H_{n-1}. If we tensor $(**)$ with G and drop the term H_{n-1} we obtain the homology chain complex

$$0 \longrightarrow Z_n \otimes G \xrightarrow{i_n \otimes \mathrm{id}} C_n \otimes G \xrightarrow{\widetilde{d}_n \otimes \mathrm{id}} Z_{n-1} \otimes G \longrightarrow 0 \qquad (L)$$

denoted L, and by definition of $\mathrm{Tor}^R(-, G)$, we have

$$\mathrm{Tor}_j^R(H_{n-1}, G) = H_j(L), \quad j \geq 0.$$

Because $(*)$ is a split exact sequence, the sequence obtained by tensoring $(*)$ with G is also exact, so $i_n \otimes \mathrm{id}$ is injective. This implies that $\mathrm{Tor}_2^R(H_{n-1}, G) = (0)$. We can compute $\mathrm{Tor}_j^R(H_{n-1}, G)$ for $j = 0, 1$ as follows:

$$\begin{aligned}\operatorname{Tor}_1^R(H_{n-1}, G) &= H_1(L) = \operatorname{Ker}(\widetilde{d_n} \otimes \mathrm{id})/\operatorname{Im}(i_n \otimes \mathrm{id}) \\ &\cong \operatorname{Ker}(d_n \otimes \mathrm{id})/(Z_n \otimes G) \\ H_{n-1} \otimes G \cong \operatorname{Tor}_0^R(H_{n-1}, G) &= H_0(L) = (Z_{n-1} \otimes G)/\operatorname{Im}(\widetilde{d_n} \otimes \mathrm{id}) \\ &\cong (Z_{n-1} \otimes G)/(B_{n-1} \otimes G).\end{aligned}$$

These equations are justified as follows. The maps d_n and $\widetilde{d_n}$ only differ in their codomain so they have the same value on all $c \in C_n$, and we have

$$\begin{aligned}\operatorname{Im}(d_n \otimes \mathrm{id}) &= \operatorname{Im}(\widetilde{d_n} \otimes \mathrm{id}) \\ &= \{d_n(c) \otimes g \in C_{n-1} \otimes G \mid c \in C_n, g \in G\} = B_{n-1} \otimes G,\end{aligned}$$

which justifies the equation $(Z_{n-1} \otimes G)/\operatorname{Im}(\widetilde{d_n} \otimes \mathrm{id}) = (Z_{n-1} \otimes G)/(B_{n-1} \otimes G)$. Since $d_n = i_{n-1} \circ \widetilde{d_n}$, with $d_n \colon C_n \to C_{n-1}$, $\widetilde{d_n} \colon C_n \to Z_{n-1}$, and $i_{n-1} \colon Z_{n-1} \to C_{n-1}$, we have

$$d_n \otimes \mathrm{id} = (i_{n-1} \circ \widetilde{d_n}) \otimes \mathrm{id} = (i_{n-1} \otimes \mathrm{id}) \circ (\widetilde{d_n} \otimes \mathrm{id}),$$

and since $i_{n-1} \otimes \mathrm{id}$ is injective, $\operatorname{Ker}(d_n \otimes \mathrm{id}) = \operatorname{Ker}(\widetilde{d_n} \otimes \mathrm{id})$, which implies that

$$\operatorname{Tor}_1^R(H_{n-1}, G) = \operatorname{Ker}(\widetilde{d_n} \otimes \mathrm{id})/\operatorname{Im}(i_n \otimes \mathrm{id}) \cong \operatorname{Ker}(d_n \otimes \mathrm{id})/(Z_n \otimes G),$$

which justifies the last equation on the first line. In summary,

$$\operatorname{Tor}_1^R(H_{n-1}, G) \cong \operatorname{Ker}(d_n \otimes \mathrm{id})/(Z_n \otimes G) \tag{T}$$

and

$$H_{n-1} \otimes G \cong (Z_{n-1} \otimes G)/\operatorname{Im}(d_n \otimes \mathrm{id}) = (Z_{n-1} \otimes G)/(B_{n-1} \otimes G). \tag{H}$$

Now look at the sequence

$$C_{n+1} \xrightarrow{d_{n+1}} C_n \xrightarrow{d_n} C_{n-1}$$

and tensor it with G to obtain the sequence

$$C_{n+1} \otimes G \xrightarrow{d_{n+1} \otimes \mathrm{id}} C_n \otimes G \xrightarrow{d_n \otimes \mathrm{id}} C_{n-1} \otimes G.$$

One verifies that

$$\operatorname{Im}(d_{n+1} \otimes \mathrm{id}) = B_n \otimes G \subseteq Z_n \otimes G \subseteq \operatorname{Ker}(d_n \otimes \mathrm{id}) \subseteq C_n \otimes G.$$

By the third isomorphism theorem, we have

$$\begin{aligned}(\operatorname{Ker}(d_n \otimes \mathrm{id})/\operatorname{Im}(d_{n+1} \otimes \mathrm{id}))/[(Z_n \otimes G)/\operatorname{Im}(d_{n+1} \otimes \mathrm{id})] \\ \cong \operatorname{Ker}(d_n \otimes \mathrm{id})/(Z_n \otimes G),\end{aligned}$$

which may be rewritten as an exact sequence

$$0 \longrightarrow (Z_n \otimes G)/\mathrm{Im}(d_{n+1} \otimes \mathrm{id}) \longrightarrow \mathrm{Ker}\,(d_n \otimes \mathrm{id})/\mathrm{Im}(d_{n+1} \otimes \mathrm{id}) \longrightarrow \mathrm{Ker}\,(d_n \otimes \mathrm{id})/(Z_n \otimes G) \longrightarrow 0.$$

The middle term is just $H_n(C \otimes G)$, while by (H) the first term is isomorphic to $H_n(C) \otimes G$ and by (T) the third term is equal to $\mathrm{Tor}_1^R(H_{n-1}, G)$, so we obtain the exact sequence of the theorem.

It remains to prove that this sequence splits. Since (*) splits, we have an isomorphism

$$C_n \cong Z_n \oplus B_{n-1}$$

and by tensoring with G we obtain

$$C_n \otimes G \cong (Z_n \otimes G) \oplus (B_{n-1} \otimes G).$$

The reader should check that this implies that $Z_n \otimes G$ is a summand of $\mathrm{Ker}\,(d_n \otimes \mathrm{id})$. It follows from this that $(Z_n \otimes G)/(B_n \otimes G)$ is a summand of $\mathrm{Ker}\,(d_n \otimes \mathrm{id})/(B_n \otimes G)$, and the sequence of the theorem splits.

Suppose we have a chain map $\varphi\colon C \to C'$ between two chain complexes C and C'. First we prove that the left square of the diagram (†) commutes, that is the following diagram commutes:

$$\begin{array}{ccc} H_n(C) \otimes_R G & \xrightarrow{\mu} & H_n(C \otimes_R G) \\ {\scriptstyle \varphi_* \otimes \mathrm{id}} \downarrow & & \downarrow {\scriptstyle (\varphi \otimes \mathrm{id})_*} \\ H_n(C') \otimes_R G & \xrightarrow[\mu']{} & H_n(C' \otimes_R G). \end{array}$$

Since by (H) (with n instead of $n-1$)

$$H_n(C) \otimes G \cong (Z_n \otimes G)/(B_n \otimes G)$$

and

$$H_n(C \otimes_R G) = \mathrm{Ker}\,(d_n \otimes \mathrm{id})/\mathrm{Im}(d_{n+1} \otimes \mathrm{id}) = \mathrm{Ker}\,(d_n \otimes \mathrm{id})/(B_n \otimes G),$$

the commutativity of the above diagram is equivalent to the commutativity of the following diagram:

$$\begin{array}{ccc} (Z_n \otimes_R G)/(B_n \otimes_R G) & \xrightarrow{\mu} & \mathrm{Ker}\,(d_n \otimes \mathrm{id})/(B_n \otimes G) \\ {\scriptstyle \varphi_* \otimes \mathrm{id}} \downarrow & & \downarrow {\scriptstyle (\varphi \otimes \mathrm{id})_*} \\ (Z'_n \otimes_R G)/(B'_n \otimes_R G) & \xrightarrow[\mu']{} & \mathrm{Ker}\,(d'_n \otimes \mathrm{id})/(B'_n \otimes G). \end{array} \qquad (\dagger_1)$$

Since

$$H_n(C) \otimes G \cong (Z_n \otimes G)/(B_n \otimes G),$$

the linear map $\varphi_* \otimes \mathrm{id}\colon H_n \otimes G \to H'_n \otimes G$ is given by

$$(\varphi_* \otimes \mathrm{id})([c \otimes g]_{B_n \otimes G}) = [\varphi(c) \otimes g]_{B'_n \otimes G}, \qquad (*_1)$$

where $[c \otimes g]_{B_n \otimes G}$ is the equivalence class of $c \otimes g \in Z_n \otimes G$ modulo $B_n \otimes G$ and $[\varphi(c) \otimes g]_{B'_n \otimes G}$ is the equivalence class of $\varphi(c) \otimes g \in Z'_n \otimes G$ modulo $B'_n \otimes G$. Since φ is a chain map, $\varphi(B_n) \subseteq B'_n$ and $\varphi(Z_n) \subseteq Z'_n$, so for any $d \otimes g' \in B_n \otimes G$ we have

$$\begin{aligned}(\varphi_* \otimes \mathrm{id})([c \otimes g + d \otimes g']_{B_n \otimes G}) &= [\varphi(c) \otimes g]_{B'_n \otimes G} + [\varphi(d) \otimes g']_{B'_n \otimes G} \\ &= [\varphi(c) \otimes g]_{B'_n \otimes G},\end{aligned}$$

since $\varphi(d) \otimes g' \in B'_n \otimes G$, and $\varphi(c) \otimes g \in Z'_n \otimes G$. Thus, the map $\varphi_* \otimes \mathrm{id}$ is well defined.

Since

$$H_n(C \otimes_R G) = \mathrm{Ker}\,(d_n \otimes \mathrm{id})/(B_n \otimes G)$$

and

$$H_n(C) \otimes G \cong (Z_n \otimes G)/(B_n \otimes G),$$

the linear map $\mu\colon H_n(C) \otimes_R G \to H_n(C \otimes_R G)$ is given by

$$\mu([c \otimes g]_{B_n \otimes G}) = [c \otimes g]_{B_n \otimes G}, \qquad (*_2)$$

where $c \in Z_n$ is a cycle and g is any element in G, with $Z_n \otimes G \subseteq \mathrm{Ker}\,(d_n \otimes \mathrm{id})$ and where equivalence classes are taken modulo $B_n \otimes G$. If $c \in Z_n$ is a cycle, then $d_n(c) = 0$ so $(d_n \otimes \mathrm{id})(c \otimes g) = d_n(c) \otimes g = 0$, which implies that $c \otimes g \in \mathrm{Ker}\,(d_n \otimes \mathrm{id})$. If $d \otimes g' \in B_n \otimes G$, then

$$\mu([c \otimes g + d \otimes g']_{B_n \otimes G}) = [c \otimes g]_{B_n \otimes G} + [d \otimes g']_{B_n \otimes G} = [c \otimes g]_{B_n \otimes G}$$

because $d \otimes g' \in B_n \otimes G$, so the map μ is well defined. The map $\mu'\colon H_n(C') \otimes_R G \to H_n(C' \otimes_R G)$ is given by

$$\mu'([c' \otimes g]_{B'_n \otimes G}) = [c' \otimes g]_{B'_n \otimes G}, \qquad (*_3)$$

where $c' \in Z'_n$ is a cycle and g is any element in G, and where the equivalence classes are taken modulo $B'_n \otimes G$.

The linear map $(\varphi \otimes \mathrm{id})_*\colon H_n(C \otimes G) \to H_n(C' \otimes G)$ is given by

$$(\varphi \otimes \mathrm{id})_*([c \otimes g]_{B_n \otimes G}) = [\varphi(c) \otimes g]_{B'_n \otimes G}, \qquad (*_4)$$

where $[c \otimes g]_{B_n \otimes G}$ is the equivalence class of $c \otimes g \in \mathrm{Ker}\,(d_n \otimes \mathrm{id})$ modulo $B_n \otimes G$ and $[\varphi(c) \otimes g]_{B'_n \otimes G} \in \mathrm{Ker}\,(d'_n \otimes \mathrm{id})$ is the equivalence class of $\varphi(c) \otimes g \in \mathrm{Ker}\,(d'_n \otimes \mathrm{id})$ modulo $B'_n \otimes G$. Since φ is a chain map, we have $\varphi \circ d_n = d'_n \circ \varphi$, so

$$\begin{aligned}(d'_n \otimes \mathrm{id})(\varphi(c) \otimes g) &= d'_n(\varphi(c)) \otimes g = \varphi(d_n(c)) \otimes g \\ &= (\varphi \otimes \mathrm{id})((d_n \otimes \mathrm{id})(c \otimes g)) = 0\end{aligned}$$

so $\varphi(c) \otimes g \in \mathrm{Ker}\,(d'_n \otimes \mathrm{id})$. Since φ is a chain map $\varphi(B_n) \subseteq B'_n$, and for any $d \otimes g' \in B_n \otimes G$

$$\begin{aligned}(\varphi \otimes \mathrm{id})_*([c \otimes g + d \otimes g']_{B_n \otimes G}) &= [\varphi(c) \otimes g]_{B'_n \otimes G} + [\varphi(d) \otimes g']_{B'_n \otimes G} \\ &= [\varphi(c) \otimes g]_{B'_n \otimes G}\end{aligned}$$

since $\varphi(d) \otimes g' \in B'_n \otimes G$. Therefore, $(\varphi \otimes \mathrm{id})_*$ is well defined. Then we have

$$\begin{aligned}(\varphi \otimes \mathrm{id})_*(\mu([c \otimes g]_{B_n \otimes G})) &= (\varphi \otimes \mathrm{id})_*([c \otimes g]_{B_n \otimes G}), && \text{by } (*_2) \\ &= [\varphi(c) \otimes g]_{B'_n \otimes G}, && \text{by } (*_4) \\ &= \mu'([\varphi(c) \otimes g]_{B'_n \otimes G}), && \text{by } (*_3) \\ &= \mu'((\varphi_* \otimes \mathrm{id})([c \otimes g]_{B_n \otimes G})), && \text{by } (*_1),\end{aligned}$$

which shows that

$$(\varphi \otimes \mathrm{id})_* \circ \mu = \mu' \circ (\varphi_* \otimes \mathrm{id}),$$

so the left square of the diagram (†) commutes.

Next we prove that the right square of the diagram (†) commutes, that is, the following diagram commutes:

$$\begin{array}{ccc} H_n(C \otimes_R G) & \xrightarrow{\ p\ } & \mathrm{Tor}_1^R(H_{n-1}(C), G) \\ {\scriptstyle (\varphi \otimes \mathrm{id})_*}\downarrow & & \downarrow{\scriptstyle \mathrm{Tor}_1^R(\varphi_*)} \\ H_n(C' \otimes_R G) & \xrightarrow[p']{} & \mathrm{Tor}_1^R(H_{n-1}(C'), G). \end{array}$$

Since

$$H_n(C \otimes_R G) = \mathrm{Ker}\,(d_n \otimes \mathrm{id})/(B_n \otimes G)$$

and

$$\mathrm{Tor}_1^R(H_{n-1}, G) \cong \mathrm{Ker}\,(d_n \otimes \mathrm{id})/(Z_n \otimes G),$$

the commutativity of the above diagram is equivalent to the commutativity of the following diagram:

$$\begin{array}{ccc} \mathrm{Ker}\,(d_n \otimes_R \mathrm{id})/(B_n \otimes_R G) & \xrightarrow{p} & \mathrm{Ker}\,(d_n \otimes \mathrm{id})/(Z_n \otimes G) \\ \Big\downarrow (\varphi\otimes \mathrm{id})_* & & \Big\downarrow \mathrm{Tor}_1^R(\varphi_*) \\ \mathrm{Ker}\,(d'_n \otimes_R \mathrm{id})/(B'_n \otimes_R G) & \xrightarrow[p']{} & \mathrm{Ker}\,(d'_n \otimes \mathrm{id})/(Z'_n \otimes G). \end{array} \qquad (\dagger_2)$$

To figure out what $\mathrm{Tor}_1(\varphi_*)$ is we go back to the projective resolution $(**)$ of H_{n-1}

$$0 \longrightarrow Z_n \xrightarrow{i_n} C_n \xrightarrow{\widetilde{d}_n} Z_{n-1} \longrightarrow H_{n-1} \longrightarrow 0. \qquad (**)$$

If $\varphi\colon C_n \to C'_n$ is a chain map, we claim that the following diagram commutes:

$$\begin{array}{ccccccc} Z_n & \xrightarrow{i_n} & C_n & \xrightarrow{\widetilde{d}_n} & Z_{n-1} & \longrightarrow & H_{n-1} \\ \Big\downarrow \varphi|Z_n & & \Big\downarrow \varphi & & \Big\downarrow \varphi|Z_{n-1} & & \Big\downarrow \varphi_* \\ Z'_n & \xrightarrow[i'_n]{} & C'_n & \xrightarrow[\widetilde{d}'_n]{} & Z'_{n-1} & \longrightarrow & H'_{n-1}. \end{array} \qquad (**_1)$$

The leftmost square commutes because i_n and i'_n are inclusions, the middle square commutes because φ is a chain map, and the rightmost square commutes because $H_{n-1} = Z_{n-1}/B_{n-1}$ and $H'_{n-1} = Z'_{n-1}/B'_{n-1}$ and by the definition of $\varphi_*\colon H_{n-1} \to H'_{n-1}$, namely $\varphi_*([c]) = [\varphi(c)]$, for any $c \in Z_n$. Therefore we obtain a lifting of φ_* between two projective resolutions of H_{n-1} and H'_{n-1}, so by applying $- \otimes G$ we obtain

$$\begin{array}{ccccccc} Z_n \otimes G & \xrightarrow{i_n\otimes \mathrm{id}} & C_n \otimes G & \xrightarrow{\widetilde{d}_n\otimes \mathrm{id}} & Z_{n-1} \otimes G & \longrightarrow & 0 \\ \Big\downarrow (\varphi|Z_n)\otimes \mathrm{id} & & \Big\downarrow \varphi\otimes \mathrm{id} & & \Big\downarrow (\varphi|Z_{n-1})\otimes \mathrm{id} & & \\ Z'_n \otimes G & \xrightarrow[i'_n\otimes \mathrm{id}]{} & C'_n \otimes G & \xrightarrow[\widetilde{d}'_n\otimes \mathrm{id}]{} & Z'_{n-1} \otimes G & \longrightarrow & 0, \end{array} \qquad (**_2)$$

and if we denote the upper row by $\mathcal{C}$ and the lower row by $\mathcal{C}'$, as explained just after Definition 11.14, the maps $\mathrm{Tor}_j^R(\varphi_*)\colon \mathrm{Tor}_j^R(H_{n-1}, G) \to \mathrm{Tor}_j^R(H'_{n-1}, G)$ are the maps of homology $\mathrm{Tor}_j^R(\varphi_*)\colon H_j(\mathcal{C}) \to H_j(\mathcal{C}')$ induced by the chain map of the diagram $(**_2)$ and are independent of the lifting of φ_* in $(**_1)$. Since

$$\mathrm{Tor}_1^R(H_{n-1}(C), G) \cong \mathrm{Ker}\,(d_n \otimes \mathrm{id})/(Z_n \otimes G)$$

and

$$\mathrm{Tor}_1^R(H_{n-1}(C'), G) \cong \mathrm{Ker}\,(d'_n \otimes \mathrm{id})/(Z'_n \otimes G),$$

the map $\mathrm{Tor}_1^R(\varphi_*)\colon \mathrm{Tor}_1^R(H_{n-1}(C), G) \to \mathrm{Tor}_1^R(H_{n-1}(C'), G)$ is the unique linear map given by

$$\mathrm{Tor}_1^R(\varphi_*)([c \otimes g]_{Z_n \otimes G}) = [\varphi(c) \otimes g]_{Z'_n \otimes G} \qquad (*_5)$$

for any $c \in C_n$ and any $g \in G$ such that $c \otimes g \in \mathrm{Ker}\,(d_n \otimes \mathrm{id})$. The subscript $Z_n \otimes G$ indicates that the equivalence class is taken modulo $Z_n \otimes G$ and the subscript $Z'_n \otimes G$ indicates that the equivalence class is taken modulo $Z'_n \otimes G$. If $(d_n \otimes \mathrm{id})(c \otimes g) = 0$, that is, $d_n(c) \otimes g = 0$, since φ is a chain map

$$(d'_n \otimes \mathrm{id})(\varphi(c) \otimes \mathrm{id}) = d'_n(\varphi(c)) \otimes g = \varphi(d_n(c)) \otimes g = (\varphi \otimes \mathrm{id})(d_n(c) \otimes g) = 0.$$

Also, for any $d \otimes g' \in Z_n \otimes G$, since φ is a chain map $\varphi(Z_n) \subseteq Z'_n$, and we have

$$\begin{aligned}\mathrm{Tor}_1^R(\varphi_*)([c \otimes g + d \otimes g']_{Z_n \otimes G}) &= [\varphi(c) \otimes g]_{Z'_n \otimes G} + [\varphi(d) \otimes g']_{Z'_n \otimes G}\\ &= [\varphi(c) \otimes g]_{Z'_n \otimes G},\end{aligned}$$

so $\mathrm{Tor}_1^R(\varphi_*)$ is well defined. Since

$$H_n(C \otimes_R G) = \mathrm{Ker}\,(d_n \otimes \mathrm{id})/(B_n \otimes G)$$

the map $p\colon H_n(C \otimes_R G) \to \mathrm{Tor}_1^R(H_{n-1}(C), G)$ is given by

$$p([c \otimes g]_{B_n \otimes G}) = [c \otimes g]_{Z_n \otimes G} \qquad (*_6)$$

for any $c \otimes g \in \mathrm{Ker}\,(d_n \otimes \mathrm{id})$. Since $B_n \otimes G \subseteq Z_n \otimes G$, this map is well defined. Similarly, the map $p'\colon H_n(C' \otimes_R G) \to \mathrm{Tor}_1^R(H_{n-1}(C'), G)$ is given by

$$p'([c' \otimes g]_{B'_n \otimes G}) = [c' \otimes g]_{Z'_n \otimes G} \qquad (*_7)$$

for any $c' \otimes g \in \mathrm{Ker}\,(d'_n \otimes \mathrm{id})$. Then we have

$$\begin{aligned}\mathrm{Tor}_1^R(\varphi_*)(p([c \otimes g]_{B_n \otimes G})) &= \mathrm{Tor}_1^R(\varphi_*)([c \otimes g]_{Z_n \otimes G}), && \text{by } (*_6),\\ &= [\varphi(c) \otimes g]_{Z'_n \otimes G}, && \text{by } (*_5),\end{aligned}$$

and

$$\begin{aligned}p'((\varphi \otimes \mathrm{id})_*([c \otimes g]_{B_n \otimes G})) &= p'([\varphi(c) \otimes g]_{B'_n \otimes G}), && \text{by } (*_1),\\ &= [\varphi(c) \otimes g]_{Z'_n \otimes G}, && \text{by } (*_7).\end{aligned}$$

Therefore

$$\mathrm{Tor}_1^R(\varphi_*) \circ p = p' \circ (\varphi \otimes \mathrm{id})_*,$$

which proves that the second square of the diagram (†) commutes. ☐

However, the splitting is not natural. This means that a splitting of the upper row may not map to a splitting of the lower row. Also, the theorem holds if the C_n are flat; what is needed is that if R is hereditary, then any submodule of a flat R-module is flat (see Rotman [Rotman (1988, 2009)], Theorem 9.25 and Theorem 11.31).

A weaker version of Theorem 12.1 is proven in Munkres for $R = \mathbb{Z}$ and where the C_n are free abelian groups; see Munkres [Munkres (1984)] (Chapter 7, Theorem 55.1). This version of Theorem 12.1 is also proved in Hatcher; see Hatcher [Hatcher (2002)] (Chapter 3, Appendix 3.A, Theorem 3.A.3). Theorem 12.1 is proven in Spanier for free modules over a PID; see Spanier [Spanier (1989)] (Chapter 5, Section 2, Theorem 8).

Remark: The injective map $\mu\colon H_n(C) \otimes G \to H_n(C \otimes G)$ is given by $\mu([c \otimes g]) = [c \otimes g]$ if we view $H_n(C)$ as isomorphic to $(Z_n \otimes G)/(B_n \otimes G)$, or by $\mu([c] \otimes g) = [c \otimes g]$ if we don't use this isomorphism; see Spanier [Spanier (1989)] (Chapter 5, Section 1, Page 214).

Whenever $\mathrm{Tor}_1^R(H_{n-1}(C), G)$ vanishes we obtain the "ideal result." This happens in the following two cases.

Proposition 12.2. *If C is a complex of vector spaces and if V is a vector space over the same field K, then we have*

$$H_n(C \otimes_K V) \cong H_n(C) \otimes_K V$$

for all $n \geq 0$.

Proposition 12.3. *If C is a complex of free abelian groups, G is an abelian group, and if either $H_{n-1}(C)$ or G is torsion-free, then we have*

$$H_n(C \otimes_{\mathbb{Z}} G) \cong H_n(C) \otimes_{\mathbb{Z}} G$$

for all $n \geq 0$.

As a corollary of Theorem 12.1, we obtain the following result about singular homology, since $\mathbb{Z}$ is a PID, and the abelian groups in the complex $S_*(X, A; \mathbb{Z})$ are free.

Theorem 12.4. *If X is a topological space, A is a subset of X, and G is any abelian group, then we have the following isomorphism of relative singular homology:*

$$H_n(X, A; G) \cong (H_n(X, A; \mathbb{Z}) \otimes_{\mathbb{Z}} G) \oplus \mathrm{Tor}_1^{\mathbb{Z}}(H_{n-1}(X, A; \mathbb{Z}), G)$$

for all $n \geq 0$.

Proof. By definition $H_n(X, A; \mathbb{Z}) = H_n(S_*(X, A; \mathbb{Z}))$ and $H_n(X, A; G) = H_n(S_*(X, A; G))$. But by definition $S_*(X, A; G) \cong S_*(X, A; \mathbb{Z}) \otimes_{\mathbb{Z}} G$, and the $S_n(X, A; \mathbb{Z})$ are free abelian groups, and thus projective. $\square$

Theorem 12.4 shows that the singular homology groups with coefficients in an abelian group G are determined by the singular homology groups with integer coefficients.

Since the modules in the relative chain complex $S_*(X, A; R)$ are free, and thus projective, and a PID is hereditary, Theorem 12.1 has the following corollary.

Theorem 12.5. *If X is a topological space, A is a subset of X, R is a PID, and G is any R-module, then we have the following isomorphism of relative singular homology:*

$$H_n(X, A; G) \cong (H_n(X, A; R) \otimes_R G) \oplus \mathrm{Tor}_1^R(H_{n-1}(X, A; R), G)$$

for all $n \geq 0$.

Theorem 12.5 is also proven in Spanier [Spanier (1989)] (Chapter 5, Section 2, Theorem 8). The reader should be warned that the assumption that R is a PID is missing in the statement of his Theorem 8. This is because Spanier reminds the reader earlier on Page 220 that R is a PID. Spanier also proves a more general theorem similar to Theorem 12.1 but applying to a chain complex C such that $C \otimes G$ is acyclic and with R a PID; see Theorem 14 in Spanier [Spanier (1989)] (Chapter 5, Section 2).

12.2 Computing Tor

If G is a finitely generated abelian group and A is any abelian group, then $\mathrm{Tor}_1^{\mathbb{Z}}(A, G)$ can be computed recursively using some simple rules. It is customary to drop the subscript 1 in $\mathrm{Tor}_1^R(-,-)$.

The main rules that allow us to use a recursive method are

$$\mathrm{Tor}^R\left(\bigoplus_{i \in I} A_i, B\right) \cong \bigoplus_{i \in I} \mathrm{Tor}^R(A_i, B)$$

$$\mathrm{Tor}^R\left(A, \bigoplus_{i \in I} B_i\right) \cong \bigoplus_{i \in I} \mathrm{Tor}^R(A, B_i)$$

$$\mathrm{Tor}^R(A, B) \cong \mathrm{Tor}^R(B, A)$$

$$\mathrm{Tor}^R(A, B) \cong (0) \text{ if } A \text{ or } B \text{ is flat (in particular, projective, or free),}$$

which hold for any commutative ring R (with an identity element) any R-modules, and any index set I; see Munkres [Munkres (1984)] (Chapter 7, Section 54) and Rotman [Rotman (1988, 2009)] (Chapter 8). When $R = \mathbb{Z}$, we also have

$$\mathrm{Tor}^{\mathbb{Z}}(\mathbb{Z}, A) = (0)$$

and

$$\mathrm{Tor}^{\mathbb{Z}}(\mathbb{Z}/m\mathbb{Z}, A) \cong \mathrm{Ker}\,(A \xrightarrow{m} A),$$

where A is an abelian group and the map $A \xrightarrow{m} A$ is multiplication by m. The proof of this last equation involves a clever use of a free resolution.

Proof. It is immediately checked that the sequence

$$0 \longrightarrow \mathbb{Z} \xrightarrow{m} \mathbb{Z} \longrightarrow \mathbb{Z}/m\mathbb{Z} \longrightarrow 0$$

is exact, and since $\mathbb{Z}$ is a free abelian group, the above sequence is a free resolution of $\mathbb{Z}/m\mathbb{Z}$. Then since $\mathrm{Tor}^{\mathbb{Z}}(-, A)$ is the left derived functor of $- \otimes A$, we deduce that $\mathrm{Tor}_j^{\mathbb{Z}}(\mathbb{Z}/m\mathbb{Z}, A) = (0)$ for all $j \geq 2$, and the long exact sequence given by Theorem 11.32 yields the exact sequence

$$0 \longrightarrow \mathrm{Tor}_1^{\mathbb{Z}}(\mathbb{Z}/m\mathbb{Z}, A) \longrightarrow \mathbb{Z} \otimes_{\mathbb{Z}} A \xrightarrow{m \otimes \mathrm{id}} \mathbb{Z} \otimes_{\mathbb{Z}} A \longrightarrow (\mathbb{Z}/m\mathbb{Z}) \otimes_{\mathbb{Z}} A \longrightarrow 0.$$

But $\mathbb{Z} \otimes_{\mathbb{Z}} A \cong A$, so we obtain an exact sequence

$$0 \longrightarrow \mathrm{Tor}_1^{\mathbb{Z}}(\mathbb{Z}/m\mathbb{Z}, A) \xrightarrow{j} A \xrightarrow{m} A \longrightarrow (\mathbb{Z}/m\mathbb{Z}) \otimes_{\mathbb{Z}} A \longrightarrow 0,$$

and since j is injective and $\mathrm{Im}\, j = \mathrm{Ker}\, m$, we get $\mathrm{Tor}^{\mathbb{Z}}(\mathbb{Z}/m\mathbb{Z}, A) \cong \mathrm{Ker}\,(A \xrightarrow{m} A)$, as claimed. ☐

We also use the following identities about tensor products:

$$\begin{aligned} \Big(\bigoplus_{i \in I} A_i\Big) \otimes_R B &\cong \bigoplus_{i \in I} A_i \otimes_R B \\ A \otimes_R B &\cong B \otimes_R A \\ R \otimes_R A &\cong A, \end{aligned}$$

which hold for any commutative ring R (with an identity element), any R-modules and any index set I; see Rotman [Rotman (1988, 2009)] (Theorems 1.12, 1.13 and 2.8) and Munkres [Munkres (1984)] (Chapter 6, Section 50). When $R = \mathbb{Z}$, we also have

$$\mathbb{Z}/m\mathbb{Z} \otimes_{\mathbb{Z}} A \cong A/mA$$

where A is an abelian group; see Munkres [Munkres (1984)] (Chapter 6, Corollary 50.5). These rules imply that

$$\mathrm{Tor}^{\mathbb{Z}}(\mathbb{Z}/m\mathbb{Z}, \mathbb{Z}) = (0)$$

and

$$\mathbb{Z}/m\mathbb{Z} \otimes_{\mathbb{Z}} \mathbb{Z}/n\mathbb{Z} \cong \mathrm{Tor}^{\mathbb{Z}}(\mathbb{Z}/m\mathbb{Z}, \mathbb{Z}/n\mathbb{Z}) \cong \mathbb{Z}/\gcd(m,n)\mathbb{Z}.$$

For details, see Munkres [Munkres (1984)] (Chapter 7, Section 54), Rotman [Rotman (1988, 2009)] (Chapter 8), and Hatcher [Hatcher (2002)] (Chapter 3, Appendix 3.A, Proposition 3.A.5).

12.3 Universal Coefficient Theorems for Cohomology

Regarding the cohomology complex obtained by using $\mathrm{Hom}_R(-, G)$, we have the following theorem.

Theorem 12.6. *(Universal Coefficient Theorem for Cohomology) Let R be a commutative hereditary ring, G be any R-module, and let C be a chain complex of projective R-modules. Then there is a split exact sequence*

$$0 \longrightarrow \mathrm{Ext}^1_R(H_{n-1}(C), G) \xrightarrow{j} H^n(\mathrm{Hom}_R(C, G)) \xrightarrow{h} \mathrm{Hom}_R(H_n(C), G) \longrightarrow 0$$

for all $n \geq 0$. (It is assumed that $H_n(C) = (0)$ for all $n < 0$.) Thus, we have an isomorphism

$$H^n(\mathrm{Hom}_R(C, G)) \cong \mathrm{Hom}_R(H_n(C), G) \oplus \mathrm{Ext}^1_R(H_{n-1}(C), G)$$

for all $n \geq 0$. Furthermore, the maps in the exact sequence of the theorem are natural, which means that for any chain map $\theta\colon C \to C'$ between two chain complexes C and C' we have the following commutative diagram

$$\begin{array}{ccccccccc} 0 \longrightarrow & \mathrm{Ext}^1_R(H_{n-1}(C'), G) & \xrightarrow{j'} & H^n(\mathrm{Hom}_R(C', G)) & \xrightarrow{h'} & \mathrm{Hom}_R(H_n(C'), G) & \longrightarrow 0 \\ & \Big\downarrow {\scriptstyle \mathrm{Ext}^1_R(\theta_*)} & & \Big\downarrow {\scriptstyle (\mathrm{Hom}_R(\theta, \mathrm{id}))^*} & & \Big\downarrow {\scriptstyle \mathrm{Hom}_R(\theta_*, \mathrm{id})} & \\ 0 \longrightarrow & \mathrm{Ext}^1_R(H_{n-1}(C), G) & \xrightarrow[j]{} & H^n(\mathrm{Hom}_R(C, G)) & \xrightarrow[h]{} & \mathrm{Hom}_R(H_n(C), G) & \longrightarrow 0. \end{array}$$

Theorem 12.6 is proven by modifying the proof of Theorem 12.1 by replacing the functor $- \otimes_R G$ by the functor $\mathrm{Hom}_R(-, G)$. Again, we warn the reader that in all the proofs that we are aware of (Rotman leaves the entire proof to the reader), the details involved in verifying that the maps j and h are natural are omitted (or sketched). The dualization process (applying $\mathrm{Hom}(-, G)$) also causes technical complications that do not come up when tensoring with G. In particular it is no longer obvious how to identify $\mathrm{Hom}(H_n(C), G)$, and some auxiliary proposition is needed (Proposition 2.9). We decided to provide complete details (with a little help from Spanier [Spanier (1989)]), which makes the proof quite long. The reader is advised to skip such details upon first reading. We begin with an outline of the proof.

Proof outline. There are two parts to the proof.

(A) Derive the desired split exact sequence.
(B) Prove the naturality of the exact sequence.

Part A: Derive the split exact sequence

$$0 \to \mathrm{Ext}^1_R(H_{n-1}(C), G) \xrightarrow{j} H^n(\mathrm{Hom}_R(C, G)) \xrightarrow{h} \mathrm{Hom}_R(H_n(C), G) \to 0.$$

In the above sequence, there are two terms that need to be "properly" understood (defined in a concrete manner), namely $\mathrm{Ext}^1_R(H_{n-1}(C), G$ and $\mathrm{Hom}_R(H_n(C), G)$.

Step A1: Calculating $\mathrm{Ext}^1_R(H_{n-1}(C), G)$.

By definition this requires calculating a projective resolution for $H_{n-1}(C)$. Fortunately we can use the projective resolution we derived for Theorem 12.1, namely

$$0 \longrightarrow Z_n \xrightarrow{i_n} C_n \xrightarrow{\widetilde{d}_n} Z_{n-1} \longrightarrow H_{n-1} \longrightarrow 0. \qquad (**)$$

If we apply $\mathrm{Hom}(-, G)$ to $(**)$, we find that

$$\begin{aligned}\mathrm{Ext}^1_R(H_{n-1}, G) = H^1(\mathcal{C}) &= (\mathrm{Ker}\,\mathrm{Hom}(i_n, \mathrm{id}))/(\mathrm{Im}\,\mathrm{Hom}(\widetilde{d}_n, \mathrm{id})) \\ &= (\mathrm{Ker}\,\mathrm{Hom}(i_n, \mathrm{id}))/(\mathrm{Im}\,\mathrm{Hom}(d_n, \mathrm{id})).\end{aligned}$$

Step A2: Verifying the containment identity

$$\mathrm{Im}\,\mathrm{Hom}(d_n, \mathrm{id}) \subseteq \mathrm{Ker}\,\mathrm{Hom}(i_n, \mathrm{id}) \subseteq \mathrm{Ker}\,\mathrm{Hom}(d_{n+1}, \mathrm{id}) \qquad (*_5)$$

which when combined with the third isomorphism theorem gives desired exact sequence of the theorem.

To actually derive $(*_5)$ and to help prove naturality part of Theorem 12.6, we write the expressions which appear in the numerators and denominators as follows

$$\begin{aligned}\mathrm{Im}\,\mathrm{Hom}(d_n, \mathrm{id}) &= \{\psi \circ d_n \in \mathrm{Hom}(C_n, G) \mid \psi \in \mathrm{Hom}(C_{n-1}, G)\} \\ \mathrm{Ker}\,\mathrm{Hom}(i_n, \mathrm{id}) &= \{\varphi \in \mathrm{Hom}(C_n, G) \mid \varphi(c) = 0 \quad \text{for all } c \in Z_n\} \\ \mathrm{Ker}\,\mathrm{Hom}(d_{n+1}, \mathrm{id}) &= \{\varphi \in \mathrm{Hom}(C_n, G) \mid \varphi(c) = 0 \quad \text{for all } c \in B_n\}.\end{aligned}$$

See Figures 12.5 and 12.6.

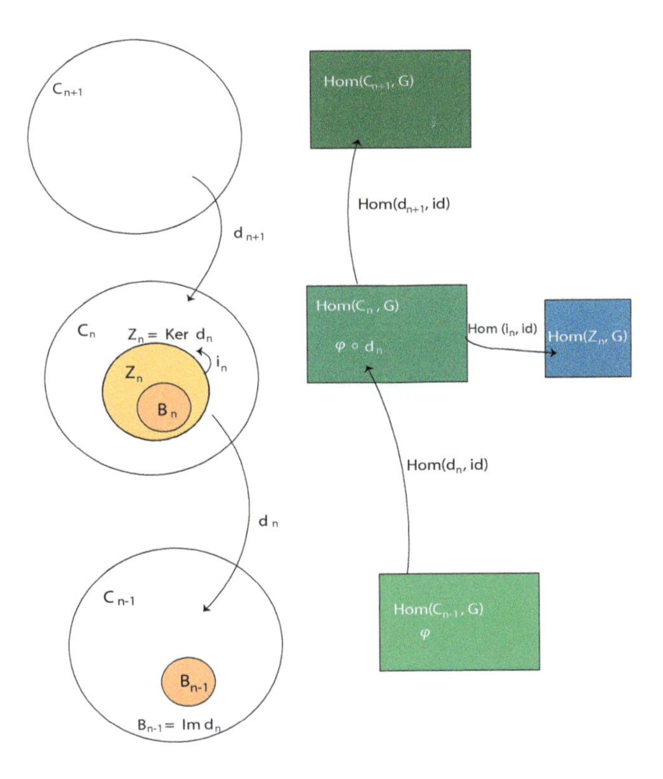

Fig. 12.5 A schematic representation of the containment identity $(*_5)$.

Then we apply the third isomorphism theorem to $(*_5)$ to obtain the following exact sequence

$$\begin{aligned} &0 \longrightarrow \mathrm{Ext}^1_R(H_{n-1}, G) \longrightarrow H^n(\mathrm{Hom}(C, G)) \to \\ &\quad \mathrm{Ker}\,\mathrm{Hom}(d_{n+1}, \mathrm{id})/\mathrm{Ker}\,\mathrm{Hom}(i_n, \mathrm{id}) \longrightarrow 0. \end{aligned} \tag{$\dagger$}$$

Step A3: Show that

$$\mathrm{Ker}\,\mathrm{Hom}(d_{n+1}, \mathrm{id})/\mathrm{Ker}\,\mathrm{Hom}(i_n, \mathrm{id}) \cong \mathrm{Hom}(H_n(C), G).$$

This is where we use the set theoretic descriptions of $\mathrm{Ker}\,\mathrm{Hom}(i_n, \mathrm{id})$ and $\mathrm{Ker}\,\mathrm{Hom}(d_{n+1}, \mathrm{id})$ from Part A2 along with Proposition 2.9. Once this is complete, the exact sequence of $(\dagger)$ becomes the desired exact sequence.

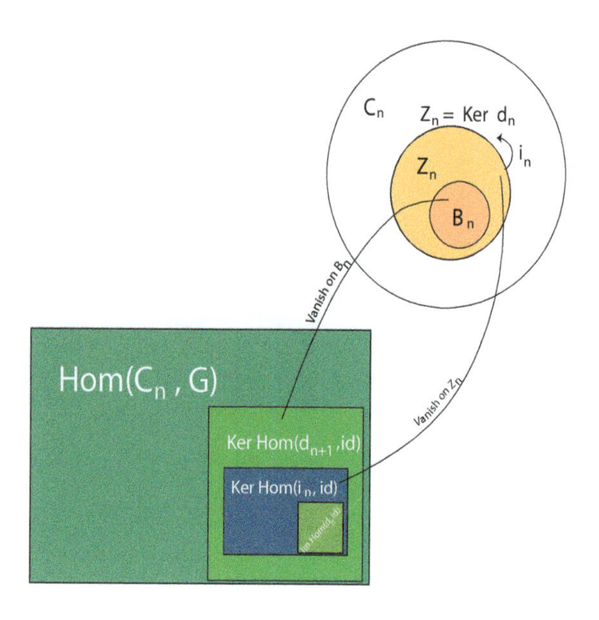

Fig. 12.6 A "close up" view of the containment identity $(*_5)$.

Step A4: Show that (†) is a split exact sequence.

When doing this calculation we use the fact that $(*)$ is a split exact sequence, find that

$$\operatorname{Ext}^1_R(H_{n-1}, G) = \operatorname{Hom}(B_{n-1}, G)/\operatorname{Im}\operatorname{Hom}(d_n, \mathrm{id}), \qquad (*_8)$$

and apply Proposition 2.10. Note that $(*_8)$ is not used again.

Step B: Verifying the naturality part of the theorem.

Step B1: Showing that the right square commutes. For this we need the auxiliary result

$$\operatorname{Hom}(H_n, G) = \operatorname{Hom}(Z_n/B_n, G) \cong \operatorname{Ker}\operatorname{Hom}(\gamma_n, \mathrm{id}), \qquad (*_{11})$$

where $\gamma_n \colon B_n \to Z_n$ is the inclusion map illustrated by Figure 12.7.

Since

$$\operatorname{Ker}\operatorname{Hom}(\gamma_n, \mathrm{id}) = \{\varphi \in \operatorname{Hom}(Z_n, G) \mid \varphi|B_n \equiv 0\},$$

an application of Proposition 2.10 provides the desired isomorphism of

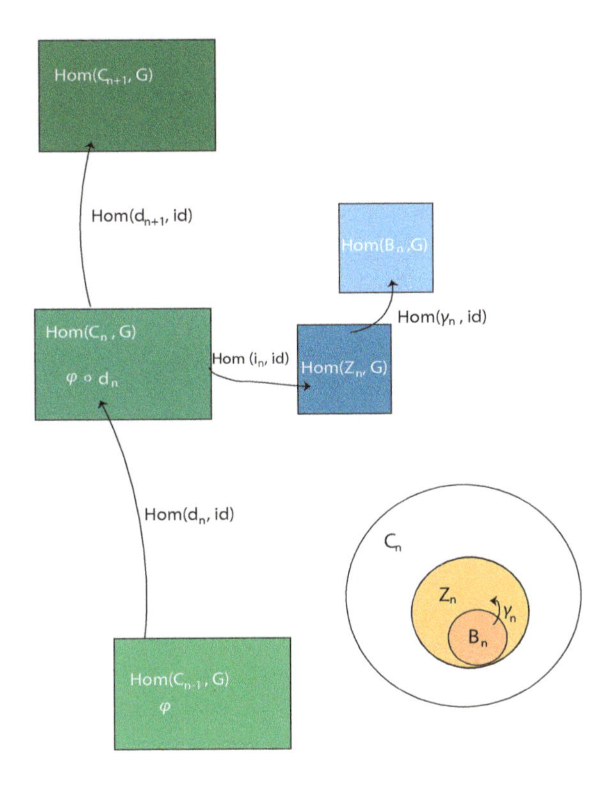

Fig. 12.7 Expanding Figure 12.5 to include $(*_{11})$.

$(*_{11})$. This means we can rewrite the right hand square as

$$\begin{array}{ccc} \operatorname{Ker}\operatorname{Hom}(d'_{n+1},\mathrm{id})/\operatorname{Im}\operatorname{Hom}(d'_n,\mathrm{id}) & \xrightarrow{h'} & \operatorname{Ker}\operatorname{Hom}(\gamma'_n,\mathrm{id}) \\ \Big\downarrow{\scriptstyle (\operatorname{Hom}(\theta,\mathrm{id}))^*} & & \Big\downarrow{\scriptstyle \operatorname{Hom}(\theta_*,\mathrm{id})} \\ \operatorname{Ker}\operatorname{Hom}(d_{n+1},\mathrm{id})/\operatorname{Im}\operatorname{Hom}(d_n,\mathrm{id}) & \xrightarrow[h]{} & \operatorname{Ker}\operatorname{Hom}(\gamma_n,\mathrm{id}). \end{array} \qquad (\dagger_2)$$

Then intuitively $(\operatorname{Hom}(\theta,\mathrm{id}))^*$ (after precomposition with the chain map) pushes down a cohomology class, $\operatorname{Hom}(\theta_*,\mathrm{id})$ pushes down a restricted domain version of map, and $h's$ shift across a cohomology class with domain restricted to Z_n. When the aforementioned maps are rigorously defined, it is easy to show the commutativity of $(\dagger_2)$.

Step 2B: Show the commutativity of the left square.

The minor issue in this situation is to figure out the meaning of $\mathrm{Ext}^1_R(\theta_*)$. To show the existence of $\mathrm{Ext}^1_R(\theta_*)$, we use the same lift of $(*)$ we developed for Theorem 12.1 and then apply $\mathrm{Hom}(-, G)$ to this lift. We find that $\mathrm{Ext}^1_R(\theta_*)$ is once again (after precomposition with the chain map) a push down of a cohomology class while the $j's$ are a modified inclusion maps, all of which make the right square commute as desired. □

Proof of Theorem 12.6. Recall from the beginning of the proof of Theorem 12.1 that we have the split short exact sequence

$$0 \longrightarrow Z_n(C) \xrightarrow{i_n} C_n \xrightarrow{d_n^B} B_{n-1}(C) \longrightarrow 0 \qquad (*)$$

and the exact sequence

$$0 \longrightarrow Z_n \xrightarrow{i_n} C_n \xrightarrow{\widetilde{d}_n} Z_{n-1} \longrightarrow H_{n-1} \longrightarrow 0 \qquad (**)$$

where $\widetilde{d}_n : C_n \to Z_{n-1}$ is the corestriction of $d_n : C_n \to C_{n-1}$ to Z_{n-1} and $d_n^B : C_n \to B_{n-1}(C)$ is the corestriction of $d_n : C_n \to C_{n-1}$ to $B_{n-1}(C)$. Since every C_n is projective and R is hereditary, the exact sequence $(**)$ is a projective resolution of H_{n-1}. If we apply $\mathrm{Hom}(-, G)$ to $(**)$ and drop the term H_{n-1} we obtain the cohomology chain complex

$$0 \longrightarrow \mathrm{Hom}(Z_{n-1}, G) \xrightarrow{\mathrm{Hom}(\widetilde{d}_n, \mathrm{id})} \mathrm{Hom}(C_n, G) \xrightarrow{\mathrm{Hom}(i_n, \mathrm{id})} \mathrm{Hom}(Z_n, G) \longrightarrow 0$$

denoted $\mathcal{C}$, and by definition of $\mathrm{Ext}^j_R(-, G)$, we have

$$\mathrm{Ext}^j_R(H_{n-1}, G) = H^j(\mathcal{C}).$$

Since the sequence $(*)$ is a split exact sequence and i_n is injective, $\mathrm{Hom}(i_n, \mathrm{id})$ is surjective, and this implies that

$$\begin{aligned}\mathrm{Ext}^2_R(H_{n-1}, G) &= H^2(\mathcal{C}) = \mathrm{Hom}(Z_n, G)/\mathrm{Im}\,\mathrm{Hom}(i_n, \mathrm{id}) \\ &= \mathrm{Hom}(Z_n, G)/\mathrm{Hom}(Z_n, G) = (0).\end{aligned}$$

We also have

$$\mathrm{Ext}^1_R(H_{n-1}, G) = H^1(\mathcal{C}) = \mathrm{Ker}\,\mathrm{Hom}(i_n, \mathrm{id})/\mathrm{Im}\,\mathrm{Hom}(\widetilde{d}_n, \mathrm{id}).$$

From the original chain complex

$$0 \xleftarrow{d_0} C_0 \xleftarrow{d_1} C_1 \longleftarrow \cdots \xleftarrow{d_{n-1}} C_{n-1} \xleftarrow{d_n} C_n \xleftarrow{d_{n+1}} C_{n+1} \longleftarrow \cdots$$

we have

$$H_n = \mathrm{Ker}\, d_n/\mathrm{Im}\, d_{n+1} = Z_n/B_n, \qquad (*_1)$$

and from the complex

$$0 \xrightarrow{\mathrm{Hom}_R(d_0,\mathrm{id})} \mathrm{Hom}_R(C_0, G) \longrightarrow \cdots$$

$$\cdots \longrightarrow \mathrm{Hom}_R(C_{n-1}, G) \xrightarrow{\mathrm{Hom}_R(d_n,\mathrm{id})} \mathrm{Hom}_R(C_n, G) \longrightarrow \cdots$$

we have

$$H^n(\mathrm{Hom}(C, G)) = \mathrm{Ker}\,\mathrm{Hom}(d_{n+1}, \mathrm{id})/\mathrm{Im}\,\mathrm{Hom}(d_n, \mathrm{id}). \qquad (*_2)$$

Since $d_n = i_{n-1} \circ \widetilde{d}_n$, with $d_n \colon C_n \to C_{n-1}$, $\widetilde{d}_n \colon C_n \to Z_{n-1}$, and $i_{n-1} \colon Z_{n-1} \to C_{n-1}$ we have

$$\mathrm{Hom}(d_n, \mathrm{id}) = \mathrm{Hom}(\widetilde{d}_n, \mathrm{id}) \circ \mathrm{Hom}(i_{n-1}, \mathrm{id}).$$

Since $\mathrm{Hom}(C_{n-1}, G) \xrightarrow{\mathrm{Hom}(i_{n-1},G)} \mathrm{Hom}(Z_{n-1}, G)$ is a surjection, we have

$$\mathrm{Im}\,\mathrm{Hom}(\widetilde{d}_n, \mathrm{id}) = \mathrm{Im}\,\mathrm{Hom}(d_n, \mathrm{id}). \qquad (*_3)$$

Consequently

$$\mathrm{Ext}^1_R(H_{n-1}, G) = \mathrm{Ker}\,\mathrm{Hom}(i_n, \mathrm{id})/\mathrm{Im}\,\mathrm{Hom}(d_n, \mathrm{id}). \qquad (*_4)$$

We claim that

$$\mathrm{Im}\,\mathrm{Hom}(d_n, \mathrm{id}) \subseteq \mathrm{Ker}\,\mathrm{Hom}(i_n, \mathrm{id}) \subseteq \mathrm{Ker}\,\mathrm{Hom}(d_{n+1}, \mathrm{id}). \qquad (*_5)$$

Since $\mathrm{Hom}(d_n, \mathrm{id}) \colon \mathrm{Hom}(C_{n-1}, G) \to \mathrm{Hom}(C_n, G)$ is given by $\varphi \mapsto \varphi \circ d_n$ for all $\varphi \in \mathrm{Hom}(C_{n-1}, G)$, we have

$$\mathrm{Im}\,\mathrm{Hom}(d_n, \mathrm{id}) = \{\psi \circ d_n \in \mathrm{Hom}(C_n, G) \mid \psi \in \mathrm{Hom}(C_{n-1}, G)\}.$$

Also, since $\mathrm{Hom}(d_{n+1}, \mathrm{id}) \colon \mathrm{Hom}(C_n, G) \to \mathrm{Hom}(C_{n+1}, G)$ is given by $\varphi \mapsto \varphi \circ d_{n+1}$ for all $\varphi \in \mathrm{Hom}(C_n, G)$, and $\mathrm{Hom}(i_n, \mathrm{id}) \colon \mathrm{Hom}(C_n, G) \to \mathrm{Hom}(Z_n, G)$ is given by $\varphi \mapsto \varphi \circ i_n$ for all $\varphi \in \mathrm{Hom}(C_n, G)$, we see that $\varphi \in \mathrm{Ker}\,\mathrm{Hom}(d_{n+1}, \mathrm{id})$ iff $\varphi \circ d_{n+1} = 0$ iff φ vanishes on $B_n = \mathrm{Im}\, d_{n+1}$, and $\varphi \in \mathrm{Ker}\,\mathrm{Hom}(i_n, \mathrm{id})$ iff $\varphi \circ i_n = 0$ iff φ vanishes on $Z_n = \mathrm{Im}\, i_n$. Therefore

$$\begin{aligned}
\mathrm{Im}\,\mathrm{Hom}(d_n, \mathrm{id}) &= \{\psi \circ d_n \in \mathrm{Hom}(C_n, G) \mid \psi \in \mathrm{Hom}(C_{n-1}, G)\} \\
\mathrm{Ker}\,\mathrm{Hom}(i_n, \mathrm{id}) &= \{\varphi \in \mathrm{Hom}(C_n, G) \mid \varphi(c) = 0 \quad \text{for all } c \in Z_n\} \\
\mathrm{Ker}\,\mathrm{Hom}(d_{n+1}, \mathrm{id}) &= \{\varphi \in \mathrm{Hom}(C_n, G) \mid \varphi(c) = 0 \quad \text{for all } c \in B_n\}.
\end{aligned}$$

The above equations will be used to prove $(*_6)$ below and to prove naturality.

Since $Z_n = \mathrm{Ker}\, d_n$, any function $\psi \circ d_n \in \mathrm{Im}\,\mathrm{Hom}(d_n, \mathrm{id})$ vanishes on Z_n, so $\mathrm{Im}\,\mathrm{Hom}(d_n, \mathrm{id}) \subseteq \mathrm{Ker}\,\mathrm{Hom}(i_n, \mathrm{id})$, and since $B_n \subseteq Z_n$, any function $\varphi \in \mathrm{Hom}(C_n, G)$ that vanishes on Z_n also vanishes on B_n, so $\mathrm{Ker}\,\mathrm{Hom}(i_n, \mathrm{id}) \subseteq \mathrm{Ker}\,\mathrm{Hom}(d_{n+1}, \mathrm{id})$.

Then we can apply the third isomorphism theorem and we get

$$\begin{aligned}(\mathrm{Ker}\,\mathrm{Hom}(d_{n+1}, \mathrm{id})/\mathrm{Im}\,\mathrm{Hom}(d_n, \mathrm{id}))/(\mathrm{Ker}\,\mathrm{Hom}(i_n, \mathrm{id})/\mathrm{Im}\,\mathrm{Hom}(d_n, \mathrm{id}))\\ \cong \mathrm{Ker}\,\mathrm{Hom}(d_{n+1}, \mathrm{id})/\mathrm{Ker}\,\mathrm{Hom}(i_n, \mathrm{id}),\end{aligned}$$

and this can be rewritten as the exact sequence

$$\begin{aligned}0 \longrightarrow \mathrm{Ker}\,\mathrm{Hom}(i_n, \mathrm{id})/\mathrm{Im}\,\mathrm{Hom}(d_n, \mathrm{id}) \longrightarrow\\ \mathrm{Ker}\,\mathrm{Hom}(d_{n+1}, \mathrm{id})/\mathrm{Im}\,\mathrm{Hom}(d_n, \mathrm{id}) \longrightarrow\\ \mathrm{Ker}\,\mathrm{Hom}(d_{n+1}, \mathrm{id})/\mathrm{Ker}\,\mathrm{Hom}(i_n, \mathrm{id}) \longrightarrow 0.\end{aligned}$$

Since

$$\mathrm{Ext}^1_R(H_{n-1}, G) = \mathrm{Ker}\,\mathrm{Hom}(i_n, \mathrm{id})/\mathrm{Im}\,\mathrm{Hom}(d_n, \mathrm{id}),$$

the first term in the exact sequence is $\mathrm{Ext}^1_R(H_{n-1}, G)$, and the second term is $H^n(\mathrm{Hom}(C, G))$, so our exact sequence can be written as

$$\begin{aligned}0 \longrightarrow \mathrm{Ext}^1_R(H_{n-1}, G) \longrightarrow H^n(\mathrm{Hom}(C, G)) \longrightarrow\\ \mathrm{Ker}\,\mathrm{Hom}(d_{n+1}, \mathrm{id})/\mathrm{Ker}\,\mathrm{Hom}(i_n, \mathrm{id}) \longrightarrow 0.\end{aligned} \qquad (\dagger)$$

It remains to figure out what is $\mathrm{Ker}\,\mathrm{Hom}(d_{n+1}, \mathrm{id})/\mathrm{Ker}\,\mathrm{Hom}(i_n, \mathrm{id})$. We will show that this term is isomorphic to $\mathrm{Hom}(H_n, G)$.

We proved earlier that

$$\begin{aligned}\mathrm{Ker}\,\mathrm{Hom}(i_n, \mathrm{id}) &= \{\varphi \in \mathrm{Hom}(C_n, G) \mid \varphi(c) = 0 \quad \text{for all } c \in Z_n\}\\ \mathrm{Ker}\,\mathrm{Hom}(d_{n+1}, \mathrm{id}) &= \{\varphi \in \mathrm{Hom}(C_n, G) \mid \varphi(c) = 0 \quad \text{for all } c \in B_n\},\end{aligned}$$

so

$$\begin{aligned}&\mathrm{Ker}\,\mathrm{Hom}(d_{n+1}, \mathrm{id})/\mathrm{Ker}\,\mathrm{Hom}(i_n, \mathrm{id})\\ &\quad = \{\varphi \in \mathrm{Hom}(C_n, G) \mid \varphi|B_n \equiv 0\}/\{\varphi \in \mathrm{Hom}(C_n, G) \mid \varphi|Z_n \equiv 0\}.\end{aligned}$$

We use Proposition 2.9 to conclude that

$$\begin{aligned}&\mathrm{Ker}\,\mathrm{Hom}(d_{n+1}, \mathrm{id})/\mathrm{Ker}\,\mathrm{Hom}(i_n, \mathrm{id})\\ &\quad = \{\varphi \in \mathrm{Hom}(C_n, G) \mid \varphi|B_n \equiv 0\}/\{\varphi \in \mathrm{Hom}(C_n, G) \mid \varphi|Z_n \equiv 0\}\\ &\quad = B^0_n/Z^0_n \cong \mathrm{Hom}(Z_n/B_n, G) = \mathrm{Hom}(H_n, G),\end{aligned}$$

where

$$B_n^0 = \{\varphi \in \mathrm{Hom}(C_n, G) \mid \varphi(b) = 0 \quad \text{for all } b \in B_n\}$$
$$Z_n^0 = \{\varphi \in \mathrm{Hom}(C_n, G) \mid \varphi(z) = 0 \quad \text{for all } z \in Z_n\}.$$

Since the exact sequence $(*)$ splits, we have $C_n = Z_n \oplus Z_n'$ for some submodule Z_n' of C_n, and we can apply Proposition 2.9 to $M = C_n$, $Z = Z_n$, and $B = B_n$. Therefore, the exact sequence $(\dagger)$ yields

$$0 \longrightarrow \mathrm{Ext}_R^1(H_{n-1}, G) \longrightarrow H^n(\mathrm{Hom}(C, G)) \longrightarrow \mathrm{Hom}(H_n, G) \longrightarrow 0. \quad (\dagger\dagger)$$

We now prove that the exact sequence $(\dagger\dagger)$ splits. For this we use the fact that since the exact sequence $(*)$ splits we have an isomorphism

$$C_n \cong Z_n \oplus B_{n-1}.$$

Applying $\mathrm{Hom}(-, G)$, we get

$$\mathrm{Hom}(C_n, G) \cong \mathrm{Hom}(Z_n, G) \oplus \mathrm{Hom}(B_{n-1}, G). \quad (*_6)$$

Recall that

$$\mathrm{Ker}\,\mathrm{Hom}(i_n, \mathrm{id}) = \{\varphi \in \mathrm{Hom}(C_n, G) \mid \varphi|Z_n \equiv 0\}$$
$$\mathrm{Ker}\,\mathrm{Hom}(d_{n+1}, \mathrm{id}) = \{\varphi \in \mathrm{Hom}(C_n, G) \mid \varphi|B_n \equiv 0\}.$$

We deduce from the above that

$$\mathrm{Ker}\,\mathrm{Hom}(i_n, \mathrm{id}) \cong \mathrm{Hom}(B_{n-1}, G), \quad (*_7)$$

so by $(*_4)$ we obtain

$$\mathrm{Ext}_R^1(H_{n-1}, G) \cong \mathrm{Hom}(B_{n-1}, G)/\mathrm{Im}\,\mathrm{Hom}(d_n, \mathrm{id}). \quad (*_8)$$

Since $(*_5)$ implies that $\mathrm{Ker}\,\mathrm{Hom}(i_n, \mathrm{id}) \subseteq \mathrm{Ker}\,\mathrm{Hom}(d_{n+1}, \mathrm{id})$, by $(*_6)$ we have

$$\mathrm{Ker}\,\mathrm{Hom}(d_{n+1}, \mathrm{id}) \cong \{\varphi \in \mathrm{Hom}(Z_n, G) \mid \varphi|B_n \equiv 0\} \oplus \mathrm{Hom}(B_{n-1}, G).$$

Now by Proposition 2.10 there is an isomorphism

$$\kappa\colon \{\varphi \in \mathrm{Hom}(Z_n, G) \mid \varphi|B_n \equiv 0\} \to \mathrm{Hom}(Z_n/B_n, G), \quad (*_9)$$

where κ is given by

$$(\kappa(\varphi))([z]) = \varphi(z) \quad \text{for all } [z] \in Z_n/B_n. \quad (*_\kappa)$$

Since $Z_n/B_n = H_n$, we obtain

$$\mathrm{Ker}\,\mathrm{Hom}(d_{n+1}, \mathrm{id}) \cong \mathrm{Hom}(H_n, G) \oplus \mathrm{Hom}(B_{n-1}, G). \quad (*_{10})$$

We now take the quotient modulo $\mathrm{Im}\,\mathrm{Hom}(d_n, \mathrm{id})$. Since we showed that $\mathrm{Im}\,\mathrm{Hom}(d_n, \mathrm{id}) \subseteq \mathrm{Ker}\,\mathrm{Hom}(i_n, \mathrm{id}) \cong \mathrm{Hom}(B_{n-1}, G)$, we get

$$\begin{aligned}&\mathrm{Ker}\,\mathrm{Hom}(d_{n+1}, \mathrm{id})/\mathrm{Im}\,\mathrm{Hom}(d_n, \mathrm{id})\\ &\qquad\cong \mathrm{Hom}(H_n, G) \oplus (\mathrm{Hom}(B_{n-1}, G)/\mathrm{Im}\,\mathrm{Hom}(d_n, \mathrm{id})),\end{aligned}$$

and by $(*_8)$ this means that

$$H^n(\mathrm{Hom}(C, G)) \cong \mathrm{Hom}(H_n, G) \oplus \mathrm{Ext}^1_R(H_{n-1}, G),$$

which proves that the exact sequence (††) splits.

To prove naturality of the exact sequence (††) we first give another expression for $\mathrm{Hom}(Z_n/B_n, G) = \mathrm{Hom}(H_n, G)$ in terms of the inclusion map $\gamma_n \colon B_n \to Z_n$ as in Spanier [Spanier (1989)] (Chapter 5, Section 5, Theorem 3). We claim that

$$\mathrm{Hom}(H_n, G) = \mathrm{Hom}(Z_n/B_n, G) \cong \mathrm{Ker}\,\mathrm{Hom}(\gamma_n, \mathrm{id}). \qquad (*_{11})$$

Indeed, since $\gamma_n \colon B_n \to Z_n$ we have $\mathrm{Hom}(\gamma_n, \mathrm{id}) \colon \mathrm{Hom}(Z_n, G) \to \mathrm{Hom}(B_n, G)$, and we have $\varphi \in \mathrm{Ker}\,\mathrm{Hom}(\gamma_n, \mathrm{id})$ iff $\varphi \circ \gamma_n = 0$ iff φ vanishes on B_n, thus

$$\mathrm{Ker}\,\mathrm{Hom}(\gamma_n, \mathrm{id}) = \{\varphi \in \mathrm{Hom}(Z_n, G) \mid \varphi | B_n \equiv 0\},$$

but we know $(*_9)$ that this last term is isomorphic to $\mathrm{Hom}(Z_n/B_n, G) = \mathrm{Hom}(H_n, G)$. We now prove the naturality of (††).

Let $\theta \colon C \to C'$ be a chain map. First we prove that the diagram

$$\begin{array}{ccc} H^n(\mathrm{Hom}_R(C', G)) & \xrightarrow{h'} & \mathrm{Hom}_R(H_n(C'), G) \\ \Big\downarrow{\scriptstyle (\mathrm{Hom}_R(\theta, \mathrm{id}))^*} & & \Big\downarrow{\scriptstyle \mathrm{Hom}_R(\theta_*, \mathrm{id})} \\ H^n(\mathrm{Hom}_R(C, G)) & \xrightarrow[h]{} & \mathrm{Hom}_R(H_n(C), G) \end{array} \qquad (\dagger_1)$$

commutes, which in view of $(*_2)$ and $(*_{11})$ is equivalent to the commutativity of the following diagram

$$\begin{array}{ccc} \mathrm{Ker}\,\mathrm{Hom}(d'_{n+1}, \mathrm{id})/\mathrm{Im}\,\mathrm{Hom}(d'_n, \mathrm{id}) & \xrightarrow{h'} & \mathrm{Ker}\,\mathrm{Hom}(\gamma'_n, \mathrm{id}) \\ \Big\downarrow{\scriptstyle (\mathrm{Hom}(\theta, \mathrm{id}))^*} & & \Big\downarrow{\scriptstyle \mathrm{Hom}(\theta_*, \mathrm{id})} \\ \mathrm{Ker}\,\mathrm{Hom}(d_{n+1}, \mathrm{id})/\mathrm{Im}\,\mathrm{Hom}(d_n, \mathrm{id}) & \xrightarrow[h]{} & \mathrm{Ker}\,\mathrm{Hom}(\gamma_n, \mathrm{id}), \end{array} \qquad (\dagger_2)$$

where the various maps involved are defined below. Recall that

$$\begin{aligned} \mathrm{Ker}\,\mathrm{Hom}(d_{n+1}, \mathrm{id}) &= \{\varphi \in \mathrm{Hom}(C_n, G) \mid \varphi | B_n \equiv 0\} \\ \mathrm{Im}\,\mathrm{Hom}(d_n, \mathrm{id}) &= \{\psi \circ d_n \in \mathrm{Hom}(C_n, G) \mid \psi \in \mathrm{Hom}(C_{n-1}, G)\} \\ \mathrm{Ker}\,\mathrm{Hom}(\gamma_n, \mathrm{id}) &= \{\varphi \in \mathrm{Hom}(Z_n, G) \mid \varphi | B_n \equiv 0\}. \end{aligned}$$

The map $(\mathrm{Hom}(\theta, \mathrm{id}))^*$ is given by

$$(\mathrm{Hom}(\theta, \mathrm{id}))^*([\varphi']) = [\varphi' \circ \theta] \qquad (*_{12})$$

for any $\varphi' \in \mathrm{Hom}(C'_n, G)$ such that $\varphi'|B'_n \equiv 0$. Technically, the above should be written as

$$(\mathrm{Hom}(\theta, \mathrm{id}))^*([\varphi']_{\mathrm{Im}\,\mathrm{Hom}(d'_n, \mathrm{id})}) = [\varphi' \circ \theta]_{\mathrm{Im}\,\mathrm{Hom}(d_n, \mathrm{id})},$$

where the modulus of the equivalence class is indicated as a subscript. But since we used this kind of notation in our proof of Theorem 12.1, to alleviate notation we omit these subscripts. The reader should have no difficulty in determining the modulus of the equivalence class.

The map $\mathrm{Hom}(\theta_*, \mathrm{id})$ is given by

$$\mathrm{Hom}(\theta_*, \mathrm{id})(\varphi') = \varphi' \circ (\theta|Z_n) \qquad (*_{13})$$

for any $\varphi' \in \mathrm{Hom}(Z'_n, G)$ such that $\varphi'|B'_n \equiv 0$, the map h is given by

$$h([\varphi]) = \varphi|Z_n \qquad (*_{14})$$

for any $\varphi \in \mathrm{Hom}(C_n, G)$ such that $\varphi|B_n \equiv 0$, and the map h' is given by

$$h'([\varphi']) = \varphi'|Z'_n \qquad (*_{15})$$

for any $\varphi' \in \mathrm{Hom}(C'_n, G)$ such that $\varphi'|B'_n \equiv 0$. To be very precise, the equivalence classes $[\varphi']$ of maps $\varphi' \in \mathrm{Hom}(Z'_n, G)$ such that $\varphi'|B'_n \equiv 0$ should be denoted $[\varphi']_{\mathrm{Im}\,\mathrm{Hom}(d'_n, \mathrm{id})}$, but by now the reader should be used to this kind of notational abuse. The map $(\mathrm{Hom}(\theta, \mathrm{id}))^*$ is well defined because θ is a chain map so for any $\psi' \circ d'_n \in \mathrm{Im}\,\mathrm{Hom}(d'_n, \mathrm{id})$ we have

$$(\mathrm{Hom}(\theta, \mathrm{id}))^*([\varphi' + \psi' \circ d'_n]) = [\varphi' \circ \theta + \psi' \circ d'_n \circ \theta] = [\varphi' \circ \theta + \psi' \circ \theta \circ d_n] = [\varphi' \circ \theta].$$

If $\varphi'|B'_n \equiv 0$, then because θ is a chain map, for any $c \in C_{n+1}$

$$(\varphi' \circ \theta)(d_{n+1}(c)) = \varphi'(d'_{n+1}(\theta(c))) = 0$$

so $(\varphi' \circ \theta)|B_n \equiv 0$. The map $\mathrm{Hom}(\theta_*, \mathrm{id})$ is well defined because $\theta(Z_n) \subseteq Z'_n$ since θ is a chain map, and if $\varphi'|B'_n \equiv 0$ for any $\varphi' \in \mathrm{Hom}(Z'_n, G)$, then using the same reasoning as above $(\varphi' \circ \theta)|B_n \equiv 0$. The map h is well defined because if $\varphi \in \mathrm{Hom}(C_n, G)$ with $\varphi|B_n \equiv 0$ then $\varphi|Z_n$ vanishes on B_n since $B_n \subseteq Z_n$, and for any $\psi \circ d_n \in \mathrm{Im}\,\mathrm{Hom}(d_n, \mathrm{id})$, we have

$$(\varphi + \psi \circ d_n)|Z_n = \varphi|Z_n + (\psi \circ d_n)|Z_n = \varphi|Z_n,$$

since $d_n|Z_n \equiv 0$ ($Z_n = \mathrm{Ker}\, d_n$). Similarly the map h' is well defined.

Then by $(*_{15})$ an $(*_{13})$ we have

$$\mathrm{Hom}(\theta_*, \mathrm{id})(h'([\varphi'])) = \mathrm{Hom}(\theta_*, \mathrm{id})(\varphi'|Z'_n) = (\varphi'|Z'_n) \circ (\theta|Z_n),$$

and by $(*_{12})$ and $(*_{14})$

$$h((\mathrm{Hom}(\theta,\mathrm{id}))^*([\varphi']) = h([\varphi' \circ \theta]) = (\varphi' \circ \theta)|Z_n.$$

Since $\theta(Z_n) \subseteq Z'_n$, we have

$$(\varphi'|Z'_n) \circ (\theta|Z_n) = (\varphi' \circ \theta)|Z_n,$$

which proves that the diagram $(\dagger_2)$ commutes.

We now prove that the diagram

$$\begin{array}{ccc} \mathrm{Ext}^1_R(H_{n-1}(C'),G) & \xrightarrow{j'} & H^n(\mathrm{Hom}_R(C',G)) \\ \Big\downarrow{\scriptstyle \mathrm{Ext}^1_R(\theta_*)} & & \Big\downarrow{\scriptstyle (\mathrm{Hom}_R(\theta,\mathrm{id}))^*} \\ \mathrm{Ext}^1_R(H_{n-1}(C),G) & \xrightarrow[j]{} & H^n(\mathrm{Hom}_R(C,G)) \end{array} \qquad (\dagger_3)$$

commutes, which in view of $(*_2)$ and $(*_4)$ is equivalent to the commutativity of the following diagram

$$\begin{array}{ccc} \mathrm{Ker}\,\mathrm{Hom}(i'_n,\mathrm{id})/\mathrm{Im}\,\mathrm{Hom}(d'_n,\mathrm{id}) & \xrightarrow{j'} & \mathrm{Ker}\,\mathrm{Hom}(d'_{n+1},\mathrm{id})/\mathrm{Im}\,\mathrm{Hom}(d'_n,\mathrm{id}) \\ \Big\downarrow{\scriptstyle \mathrm{Ext}^1(\theta_*)} & & \Big\downarrow{\scriptstyle (\mathrm{Hom}(\theta,\mathrm{id}))^*} \\ \mathrm{Ker}\,\mathrm{Hom}(i_n,\mathrm{id})/\mathrm{Im}\,\mathrm{Hom}(d_n,\mathrm{id}) & \xrightarrow[j]{} & \mathrm{Ker}\,\mathrm{Hom}(d_{n+1},\mathrm{id})/\mathrm{Im}\,\mathrm{Hom}(d_n,\mathrm{id}), \end{array} \qquad (\dagger_4)$$

where the maps involved (besides the right vertical map) are defined below.

To figure out what $\mathrm{Ext}^1(\theta_*)$ is we go back to the projective resolution $(**)$ of H_{n-1}

$$0 \longrightarrow Z_n \xrightarrow{i_n} C_n \xrightarrow{\widetilde{d}_n} Z_{n-1} \longrightarrow H_{n-1} \longrightarrow 0. \qquad (**)$$

If $\theta\colon C_n \to C'_n$ is a chain map, we showed during the proof of Theorem 12.1 that the following diagram commutes:

$$\begin{array}{ccccccc} Z_n & \xrightarrow{i_n} & C_n & \xrightarrow{\widetilde{d}_n} & Z_{n-1} & \longrightarrow & H_{n-1} \\ \Big\downarrow{\scriptstyle \theta|Z_n} & & \Big\downarrow{\scriptstyle \theta} & & \Big\downarrow{\scriptstyle \theta|Z_{n-1}} & & \Big\downarrow{\scriptstyle \theta_*} \\ Z'_n & \xrightarrow[i'_n]{} & C'_n & \xrightarrow[\widetilde{d}'_n]{} & Z'_{n-1} & \longrightarrow & H'_{n-1}. \end{array} \qquad (**_1)$$

Therefore we obtain a lifting of θ_* between two projective resolutions of H_{n-1} and H'_{n-1} so by applying $\mathrm{Hom}(-,G)$ we obtain the commutative diagram

$$\begin{array}{ccccccc} 0 \longrightarrow \mathrm{Hom}(Z'_{n-1}, G) & \xrightarrow{\mathrm{Hom}(\widetilde{d}'_n, G)} & \mathrm{Hom}(C'_n, G) & \xrightarrow{\mathrm{Hom}(i'_n, \mathrm{id})} & \mathrm{Hom}(Z'_n, \mathrm{id}) \\ \Big\downarrow {\scriptstyle \mathrm{Hom}(\theta|Z_{n-1}, \mathrm{id})} & & \Big\downarrow {\scriptstyle \mathrm{Hom}(\theta|C_n, \mathrm{id})} & & \Big\downarrow {\scriptstyle \mathrm{Hom}(\theta|Z_n, \mathrm{id})} \\ 0 \longrightarrow \mathrm{Hom}(Z_{n-1}, G) & \xrightarrow{\mathrm{Hom}(\widetilde{d}_n, G)} & \mathrm{Hom}(C_n, G) & \xrightarrow{\mathrm{Hom}(i_n, \mathrm{id})} & \mathrm{Hom}(Z_n, \mathrm{id}), \end{array} \qquad (**_2)$$

and if we denote the upper row by $\mathcal{C}'$ and the lower row by $\mathcal{C}$, as explained just after Definition 11.14, the maps $\mathrm{Ext}^j_R(\theta_*)\colon \mathrm{Ext}^j_R(H'_{n-1}, G) \to \mathrm{Ext}^j_R(H_{n-1}, G)$ are the maps of cohomology $\mathrm{Ext}^j_R(\theta_*)\colon H^j(\mathcal{C}') \to H^j(\mathcal{C})$ induced by the chain map of the diagram $(**_2)$ and are independent of the lifting of θ_* in $(**_1)$.

Recall that

$$\begin{aligned} \mathrm{Ker}\,\mathrm{Hom}(d_{n+1}, \mathrm{id}) &= \{\varphi \in \mathrm{Hom}(C_n, G) \mid \varphi|B_n \equiv 0\} \\ \mathrm{Im}\,\mathrm{Hom}(d_n, \mathrm{id}) &= \{\psi \circ d_n \in \mathrm{Hom}(C_n, G) \mid \psi \in \mathrm{Hom}(C_{n-1}, G)\} \\ \mathrm{Ker}\,\mathrm{Hom}(i_n, \mathrm{id}) &= \{\varphi \in \mathrm{Hom}(C_n, G) \mid \varphi|Z_n \equiv 0\}. \end{aligned}$$

Since by $(*_4)$

$$\begin{aligned} \mathrm{Ext}^1_R(H_{n-1}, G) &= \mathrm{Ker}\,\mathrm{Hom}(i_n, \mathrm{id})/\mathrm{Im}\,\mathrm{Hom}(\widetilde{d}_n, \mathrm{id}) \\ &= \mathrm{Ker}\,\mathrm{Hom}(i_n, \mathrm{id})/\mathrm{Im}\,\mathrm{Hom}(d_n, \mathrm{id}) \end{aligned}$$

and similarly for $\mathrm{Ext}^1_R(H'_{n-1}, G)$, the cohomology map $\mathrm{Ext}^1_R(\theta_*)$ is given by

$$\mathrm{Ext}^1_R(\theta_*)([\varphi']) = [\varphi' \circ \theta], \qquad (*_{16})$$

for all $\varphi' \in \mathrm{Hom}(C'_n, G)$ such that $\varphi'|Z'_n \equiv 0$. It is well defined because θ is a chain map and for any $\psi' \circ d'_n \in \mathrm{Im}\,\mathrm{Hom}(d'_n, \mathrm{id})$ we have

$$\mathrm{Ext}^1_R(\theta_*)([\varphi' + \psi' \circ d'_n]) = [\varphi' \circ \theta + \psi' \circ d'_n \circ \theta] = [\varphi' \circ \theta + \psi' \circ \theta \circ d_n] = [\varphi' \circ \theta].$$

The map
$j\colon \mathrm{Ker}\,\mathrm{Hom}(i_n, \mathrm{id})/\mathrm{Im}\,\mathrm{Hom}(d_n, \mathrm{id}) \to \mathrm{Ker}\,\mathrm{Hom}(d_{n+1}, \mathrm{id})/\mathrm{Im}\,\mathrm{Hom}(d_n, \mathrm{id})$ is the quotient of the inclusion map $\mathrm{Ker}\,\mathrm{Hom}(i_n, \mathrm{id}) \longrightarrow \mathrm{Ker}\,\mathrm{Hom}(d_{n+1}, \mathrm{id})$ given by

$$j([\varphi]) = [\varphi], \qquad (*_{17})$$

for any $\varphi \in \mathrm{Hom}(C_n, G)$ such that $\varphi|Z_n \equiv 0$. This map is well defined because for any $\psi \circ d_n \in \mathrm{Im}\,\mathrm{Hom}(d_n, \mathrm{id})$ we have

$$j([\varphi + \psi \circ d_n]) = [\varphi + \psi \circ d_n] = [\varphi],$$

because $B_n \subseteq Z_n$ and $Z_n = \operatorname{Ker} d_n$ so $\psi \circ d_n$ vanishes on B_n. The map j' is defined analogously as

$$j'([\varphi']) = [\varphi'], \tag{$*_{18}$}$$

for any $\varphi' \in \operatorname{Hom}(C'_n, G)$ such that $\varphi|Z'_n \equiv 0$. By $(*_{12})$ and $(*_{18})$ we have

$$(\operatorname{Hom}(\theta, \mathrm{id}))^*(j'([\varphi']) = (\operatorname{Hom}(\theta, \mathrm{id}))^*([\varphi']) = [\varphi' \circ \theta]$$

for any $\varphi' \in \operatorname{Hom}(C'_n, G)$ such that $\varphi'|Z'_n \equiv 0$, and by $(*_{16})$ and $(*_{17})$ we have

$$j(\operatorname{Ext}^1_R(\theta_*)([\varphi'])) = j([\varphi' \circ \theta]) = [\varphi' \circ \theta].$$

Therefore,

$$(\operatorname{Hom}(\theta, \mathrm{id}))^* \circ j' = j \circ \operatorname{Ext}^1(\theta_*),$$

which proves that $(\dagger_4)$ commutes, and finishes the proof of naturality. □

As in the case of homology, the splitting is not natural.

Spanier proves a version of Theorem 12.6 for a chain complex C such that $\operatorname{Ext}_R(C, G)$ is acyclic and with R a PID; see Theorem 3 in Spanier [Spanier (1989)] (Chapter 5, Section 5).

Remarks:

(1) Under the isomorphism $\kappa\colon \{\varphi \in \operatorname{Hom}(Z_n, G) \mid \varphi|B_n \equiv 0\} \to \operatorname{Hom}(Z_n/B_n, G)$, the map

$$h\colon H^n(\operatorname{Hom}(C, G)) \to \{\varphi \in \operatorname{Hom}(Z_n, G) \mid \varphi|B_n \equiv 0\}$$

is given by $h([\varphi]) = \varphi|Z_n$ for any $[\varphi] \in H^n(\operatorname{Hom}(C, G))$. Composing with the isomorphism κ, we obtain the surjection (also denoted h)

$$h\colon H^n(\operatorname{Hom}(C, G)) \to \operatorname{Hom}(H_n(C), G)$$

given by

$$(h([\varphi]))([z]) = \varphi(z),$$

for any $[\varphi] \in H^n(\operatorname{Hom}(C, G))$ and any $[z] \in H_n(C)$; this matches Spanier's definition; see Spanier [Spanier (1989)] (Chapter 5, Section 5, Page 242). In Munkres, the map $h\colon H^n(\operatorname{Hom}(C, G)) \to \operatorname{Hom}(H_n(C), G)$ is defined on Page 276 ([Munkres (1984)], Section 45), and called the *Kronecker map* (it is denoted by κ rather than h).

(2) We can prove that

$$\mathrm{Ext}^1_R(H_{n-1}, G) \cong \mathrm{Coker}\,\mathrm{Hom}(\gamma_{n-1}, \mathrm{id}) = \mathrm{Hom}(B_{n-1}, G)/\mathrm{Im}\,\mathrm{Hom}(\gamma_{n-1}, \mathrm{id}). \quad (*_{19})$$

This will establish a connection with Spanier's proof of the naturality of the exact sequence (††); see Spanier [Spanier (1989)] (Chapter 5, Section 5).

Recall from $(*_4)$ that

$$\mathrm{Ext}^1_R(H_{n-1}, G) = \mathrm{Ker}\,\mathrm{Hom}(i_n, \mathrm{id})/\mathrm{Im}\,\mathrm{Hom}(d_n, \mathrm{id}).$$

We already showed in $(*_7)$ that $\mathrm{Ker}\,\mathrm{Hom}(i_n, \mathrm{id}) \cong \mathrm{Hom}(B_{n-1}, G)$ so we just have to prove that

$$\mathrm{Im}\,\mathrm{Hom}(d_n, \mathrm{id}) \cong \mathrm{Im}\,\mathrm{Hom}(\gamma_{n-1}, \mathrm{id}). \quad (*_{20})$$

This is because

$$\mathrm{Im}\,\mathrm{Hom}(d_n, \mathrm{id}) = \{\psi \circ d_n \in \mathrm{Hom}(C_n, G) \mid \psi \in \mathrm{Hom}(C_{n-1}, G)\}$$
$$\mathrm{Im}\,\mathrm{Hom}(\gamma_{n-1}, \mathrm{id}) = \{\psi \circ \gamma_{n-1} \in \mathrm{Hom}(B_{n-1}, G) \mid \psi \in \mathrm{Hom}(Z_{n-1}, G)\}$$

and since $d_n \colon C_n \to B_{n-1}$ is a surjection and $\gamma_n \colon B_n \to Z_n$ is an injection,

$$\{\psi \circ d_n \in \mathrm{Hom}(C_n, G) \mid \psi \in \mathrm{Hom}(C_{n-1}, G)\} \cong \{\psi | B_{n-1} \in \mathrm{Hom}(B_{n-1}, G) \mid \psi \in \mathrm{Hom}(C_{n-1}, G)\}$$

and

$$\{\psi \circ \gamma_{n-1} \in \mathrm{Hom}(B_{n-1}, G) \mid \psi \in \mathrm{Hom}(Z_{n-1}, G)\} \cong \{\psi | B_{n-1} \in \mathrm{Hom}(B_{n-1}, G) \mid \psi \in \mathrm{Hom}(Z_{n-1}, G)\},$$

but since $B_{n-1} \subseteq Z_{n-1} \subseteq C_{n-1}$, the sets of the right-hand sides of the two equations above are identical.

Therefore, we proved that the exact sequence

$$0 \to \mathrm{Ext}^1_R(H_{n-1}, G) \to H^n(\mathrm{Hom}(C, G)) \to \mathrm{Hom}(H_n, G) \to 0 \quad (\dagger\dagger)$$

is equivalent to the exact sequence

$$0 \to \mathrm{Coker}\,\mathrm{Hom}(\gamma_{n-1}, \mathrm{id}) \to H^n(\mathrm{Hom}(C, G)) \to \mathrm{Ker}\,\mathrm{Hom}(\gamma_n, \mathrm{id}) \to 0, \quad (\dagger\dagger_2)$$

which is the exact sequence found in the middle of Page 243 in Spanier (and others, such as Munkres and Hatcher); see Spanier [Spanier (1989)] (Chapter 5, Section 5). We can now refer to Spanier's proof of naturality of this sequence.

Whenever $\mathrm{Ext}^1_R(H_{n-1}(C), G)$ vanishes, we obtain the "ideal result."

Recall form Definition 11.4 that an R-module M is divisible if for every nonzero $\lambda \in R$, the multiplication map given by $u \mapsto \lambda u$ for all $u \in M$ is surjective. Here we let $R = \mathbb{Z}$ and M be an abelian group.

Proposition 12.7. *If C is a complex of free abelian groups, G is an abelian group, and if either $H_{n-1}(C)$ or G is divisible, then we have an isomorphism*

$$H^n(\mathrm{Hom}_{\mathbb{Z}}(C, G)) \cong \mathrm{Hom}_{\mathbb{Z}}(H_n(C), G)$$

for all $n \geq 0$.

We also have the following generalization of Theorem 4.30 to G-coefficients.

Proposition 12.8. *If R is a PID, G is an R-module, C is a complex of free R-modules, and if $H_{n-1}(C)$ is a free R-module or (0), then we have an isomorphism*

$$H^n(\mathrm{Hom}_R(C, G)) \cong \mathrm{Hom}_R(H_n(C), G)$$

for all $n \geq 0$.

Proposition 12.9. *If C is a complex of vector spaces and V is a vector space, both over the same field K, then we have an isomorphism*

$$H^n(\mathrm{Hom}_K(C, V)) \cong \mathrm{Hom}_K(H_n(C), V)$$

for all $n \geq 0$. In particular, for $V = K$, we have isomorphisms

$$H^n(\mathrm{Hom}_K(C, K)) \cong \mathrm{Hom}_K(H_n(C), K) = H_n(C)^*,$$

where $H_n(C)^*$ is the dual of the vector space $H_n(C)$, for all $n \geq 0$.

Since the modules $S_*(X, A; \mathbb{Z})$ are free abelian groups, Theorem 12.6 yields the following result showing that the singular cohomology groups with coefficients in an abelian group G are determined by the singular homology groups with coefficients in $\mathbb{Z}$.

Theorem 12.10. *If X is a topological space, A is a subset of X, and G is any abelian group, then there is an isomorphism relative singular cohomology*

$$H^n(X, A; G) \cong \mathrm{Hom}_{\mathbb{Z}}(H_n(X, A; \mathbb{Z}), G) \oplus \mathrm{Ext}^1_{\mathbb{Z}}(H_{n-1}(X, A; \mathbb{Z}), G)$$

for all $n \geq 0$.

Theorem 12.10 is also proven in Munkres [Munkres (1984)] (Chapter 7, Section 53, Theorem 53.1) and in Hatcher [Hatcher (2002)] (Chapter 3, Section 3.1, Theorem 3.2).

Since the modules $S_*(X, A; R)$ are free, Theorem 12.6 has the following corollary.

Theorem 12.11. *If X is a topological space, A is a subset of X, R is any PID, and G is any R-module, then there is an isomorphism of relative singular cohomology*

$$H^n(X, A; G) \cong \mathrm{Hom}_R(H_n(X, A; R), G) \oplus \mathrm{Ext}^1_R(H_{n-1}(X, A; R), G)$$

for all $n \geq 0$.

12.4 Computing Ext

If A is a finitely generated abelian group and G is any abelian group, then $\mathrm{Ext}^1_{\mathbb{Z}}(A, G)$ can be computed recursively. It is customary to drop the superscript 1 in $\mathrm{Ext}^1_R(-,-)$. We have the identities

$$\begin{aligned}
\mathrm{Ext}_R\left(\bigoplus_{i\in I} A_i, B\right) &\cong \prod_{i\in I} \mathrm{Ext}_R(A_i, B) \\
\mathrm{Ext}_R\left(A, \prod_{i\in I} B_i\right) &\cong \prod_{i\in I} \mathrm{Ext}_R(A, B_i) \\
\mathrm{Ext}_R(A, B) &\cong (0) \quad \text{if } A \text{ is projective or } B \text{ is injective,}
\end{aligned}$$

for any commutative ring R and any R-modules; see Munkres [Munkres (1984)] (Chapter 7, Section 52) and Rotman [Rotman (1988, 2009)] (Chapter 7). If the index set I is finite, we can replace $\prod$ by $\bigoplus$. When $R = \mathbb{Z}$ we also have

$$\begin{aligned}
\mathrm{Ext}_{\mathbb{Z}}(\mathbb{Z}, G) &\cong (0) \\
\mathrm{Ext}_{\mathbb{Z}}(\mathbb{Z}/m\mathbb{Z}, G) &\cong G/mG,
\end{aligned}$$

where G is an abelian group. This last equation is proven as follows.

Proof. We know that the sequence

$$0 \longrightarrow \mathbb{Z} \xrightarrow{\ m\ } \mathbb{Z} \longrightarrow \mathbb{Z}/m\mathbb{Z} \longrightarrow 0$$

is a free resolution of $\mathbb{Z}/m\mathbb{Z}$. Since $\mathrm{Ext}_{\mathbb{Z}}(-, G)$ is the right derived functor of $\mathrm{Hom}_{\mathbb{Z}}(-, G)$, we deduce that $\mathrm{Ext}^j_{\mathbb{Z}}(\mathbb{Z}/m\mathbb{Z}, G) = (0)$ for all $j \geq 2$, and the long exact sequence given by Theorem 11.31 yields the exact sequence

$$0 \longrightarrow \mathrm{Hom}(\mathbb{Z}/m\mathbb{Z}, G) \longrightarrow \mathrm{Hom}(\mathbb{Z}, G) \xrightarrow{\mathrm{Hom}(m,G)}$$

$$\mathrm{Hom}(\mathbb{Z}, G) \longrightarrow \mathrm{Ext}^1_{\mathbb{Z}}(\mathbb{Z}/m\mathbb{Z}, G) \longrightarrow 0.$$

Since $\mathrm{Hom}(\mathbb{Z}, G) \cong G$, we obtain an exact sequence

$$0 \longrightarrow \mathrm{Hom}(\mathbb{Z}/m\mathbb{Z}, G) \longrightarrow G \xrightarrow{m} G \xrightarrow{p} \mathrm{Ext}^1_{\mathbb{Z}}(\mathbb{Z}/m\mathbb{Z}, G) \longrightarrow 0,$$

and since p is surjective and $\mathrm{Im}\, m = \mathrm{Ker}\, p$, we have

$$\mathrm{Ext}^1_{\mathbb{Z}}(\mathbb{Z}/m\mathbb{Z}, G) \cong G/\mathrm{Ker}\, p \cong G/mG,$$

as claimed. ☐

We also use the following rules for $\mathrm{Hom}_R(-,-)$:

$$\mathrm{Hom}_R\left(\bigoplus_{i\in I} A_i, B\right) \cong \prod_{i\in I} \mathrm{Hom}_R(A_i, B)$$

$$\mathrm{Hom}_R\left(A, \prod_{i\in I} B_i\right) \cong \prod_{i\in I} \mathrm{Hom}_R(A, B_i)$$

for any commutative ring and any R-modules; see Rotman [Rotman (1988, 2009)] (Theorem 2.4 and Theorem 2.6). If the index set I is finite, we can replace $\prod$ by $\bigoplus$. When $R = \mathbb{Z}$, we also have

$$\mathrm{Hom}_{\mathbb{Z}}(\mathbb{Z}, G) \cong G$$

$$\mathrm{Hom}_{\mathbb{Z}}(\mathbb{Z}/m\mathbb{Z}, G) \cong \mathrm{Ker}\,(G \xrightarrow{m} G),$$

where G is an abelian group. The above formula is proven as follows.

Proof. We have the exact sequence

$$0 \longrightarrow \mathbb{Z} \xrightarrow{m} \mathbb{Z} \longrightarrow \mathbb{Z}/m\mathbb{Z} \longrightarrow 0.$$

Since $\mathrm{Hom}_{\mathbb{Z}}(-, G)$ is right-exact, we obtain the exact sequence

$$0 \longrightarrow \mathrm{Hom}_{\mathbb{Z}}(\mathbb{Z}/m\mathbb{Z}, G) \longrightarrow \mathrm{Hom}_{\mathbb{Z}}(\mathbb{Z}, G) \xrightarrow{\mathrm{Hom}_{\mathbb{Z}}(m,G)} \mathrm{Hom}_{\mathbb{Z}}(\mathbb{Z}, G).$$

Since $\mathrm{Hom}(\mathbb{Z}, G) \cong G$, we obtain an exact sequence

$$0 \longrightarrow \mathrm{Hom}_{\mathbb{Z}}(\mathbb{Z}/m\mathbb{Z}, G) \longrightarrow G \xrightarrow{m} G,$$

which yields $\mathrm{Hom}_{\mathbb{Z}}(\mathbb{Z}/m\mathbb{Z}, G) \cong \mathrm{Ker}\,(G \xrightarrow{m} G)$, as claimed. ☐

These rules imply that

$$\mathrm{Hom}_{\mathbb{Z}}(\mathbb{Z}/m\mathbb{Z}, \mathbb{Z}) \cong (0)$$

and

$$\mathrm{Hom}_{\mathbb{Z}}(\mathbb{Z}/m\mathbb{Z}, \mathbb{Z}/n\mathbb{Z}) \cong \mathrm{Ext}_{\mathbb{Z}}(\mathbb{Z}/m\mathbb{Z}, \mathbb{Z}/n\mathbb{Z}) \cong \mathbb{Z}/\gcd(m,n)\mathbb{Z}.$$

For details, see Munkres [Munkres (1984)] (Chapter 7, Section 52), Rotman [Rotman (1988, 2009)] (Chapter 7), and Hatcher [Hatcher (2002)] (Chapter 3, Section 3.1).

If A is a finitely generated abelian group, we know that A can be written (uniquely) as a direct sum

$$A = F \oplus T$$

where A is a free abelian group and F is a torsion abelian group. Then the above rules imply the following useful result that allows to compute integral cohomology from integral homology.

Proposition 12.12. *Let C be a chain complex of free abelian groups. If $H_{n-1}(C)$ and $H_n(C)$ are finitely generated and if we write $H_n(C) = F_n \oplus T_n$ where F_n is the free part of $H_n(C)$ and T_n is the torsion part of $H_n(C)$ (and similarly $H_{n-1}(C) = F_{n-1} \oplus T_{n-1}$), then we have an isomorphism*

$$H^n(\mathrm{Hom}_{\mathbb{Z}}(C, \mathbb{Z})) \cong F_n \oplus T_{n-1}.$$

In particular, the above holds for the singular homology groups $H_n(X;\mathbb{Z})$ and the singular cohomology groups $H^n(X;\mathbb{Z})$ of a topological space X; that is,

$$H^n(X;\mathbb{Z}) \cong F_n \oplus T_{n-1}$$

where $H_n(X;\mathbb{Z}) = F_n \oplus T_n$ with F_n free and T_n a torsion abelian group.

Proof. Using the above rules, since T_n is a finitely generated torsion abelian group it is a direct sum of abelian groups of the form $\mathbb{Z}/m\mathbb{Z}$, and since F_n is a finitely generated free abelian group it is of the form $\mathbb{Z}^n$, so we have

$$\begin{aligned}\mathrm{Hom}_{\mathbb{Z}}(H_n(C), \mathbb{Z}) &= \mathrm{Hom}_{\mathbb{Z}}(F_n \oplus T_n, \mathbb{Z}) \cong \mathrm{Hom}_{\mathbb{Z}}(F_n, \mathbb{Z}) \oplus \mathrm{Hom}_{\mathbb{Z}}(T_n, \mathbb{Z}) \\ &\cong \mathrm{Hom}_{\mathbb{Z}}(F_n, \mathbb{Z}) \cong F_n,\end{aligned}$$

and

$$\begin{aligned}\mathrm{Ext}_{\mathbb{Z}}(H_{n-1}(C), \mathbb{Z}) &= \mathrm{Ext}_{\mathbb{Z}}(F_{n-1} \oplus T_{n-1}, \mathbb{Z}) \\ &\cong \mathrm{Ext}_{\mathbb{Z}}(F_{n-1}, \mathbb{Z}) \oplus \mathrm{Ext}_{\mathbb{Z}}(T_{n-1}, \mathbb{Z}) \\ &\cong \mathrm{Ext}_{\mathbb{Z}}(T_{n-1}, \mathbb{Z}) \cong T_{n-1}.\end{aligned}$$

By Theorem 12.6, we conclude that $H^n(\mathrm{Hom}_{\mathbb{Z}}(C, \mathbb{Z})) \cong F_n \oplus T_{n-1}$. $\square$

Proposition 12.12 is found in Bott and Tu [Bott and Tu (1986)] (Chapter III, Corollary 15.14.1), Hatcher [Hatcher (2002)] (Chapter 3, Corollary 3.3), and Spanier [Spanier (1989)] (Chapter 5, Section 5, Corollary 4). As an application of Proposition 12.12, we can compute the cohomology groups of the real projective spaces $\mathbb{RP}^n$ and of the complex projective space $\mathbb{CP}^n$. Recall from Section 4.3 that the homology groups of $\mathbb{CP}^n$ and $\mathbb{RP}^n$ are given by

$$H_p(\mathbb{CP}^n; \mathbb{Z}) = \begin{cases} \mathbb{Z} & \text{for } p = 0, 2, 4, \ldots, 2n \\ (0) & \text{otherwise,} \end{cases}$$

and

$$H_p(\mathbb{RP}^n; \mathbb{Z}) = \begin{cases} \mathbb{Z} & \text{for } p = 0 \text{ and for } p = n \text{ odd} \\ \mathbb{Z}/2\mathbb{Z} & \text{for } p \text{ odd, } 0 < p < n \\ (0) & \text{otherwise.} \end{cases}$$

Using Proposition 12.12, we obtain

$$H^p(\mathbb{CP}^n; \mathbb{Z}) = \begin{cases} \mathbb{Z} & \text{for } p = 0, 2, 4, \ldots, 2n \\ (0) & \text{otherwise,} \end{cases}$$

and

$$H^p(\mathbb{RP}^n; \mathbb{Z}) = \begin{cases} \mathbb{Z} & \text{for } p = 0 \text{ and for } p = n \text{ odd} \\ \mathbb{Z}/2\mathbb{Z} & \text{for } p \text{ even, } 0 < p \leq n \\ (0) & \text{otherwise.} \end{cases}$$

Spanier [Spanier (1989)] (Chapter 5, Sections 2 and 5) and Munkres [Munkres (1984)] (Chapter 7, Section 56) discuss other types of universal coefficient theorems.

In the next section we discuss briefly some generalizations of the universal coefficient theorems known as the *Künneth Theorems* or *Künneth Formulae*.

12.5 Künneth Formulae

In order to state the Künneth formulae we need to generalize the notion of tensor product and the Hom functor to complexes. Here it is technically important to spell out the index conventions used to denote chain complexes and cochain complexes and to allow negative indices. Following Rotman

[Rotman (1988, 2009)], as in Section 2.5, a chain complex $\mathcal{C}_* = (C_p)_{p\in\mathbb{Z}}$ is denoted by

$$\cdots \longleftarrow C_{p-2} \xleftarrow{d_{p-1}} C_{p-1} \xleftarrow{d_p} C_p \xleftarrow{d_{p+1}} C_{p+1} \longleftarrow \cdots,$$

using increasing subscripts, with the arrows going from right to left, and a cochain complex $\mathcal{C}^* = (C^p)_{p\in\mathbb{Z}}$ is denoted by

$$\cdots \longrightarrow C^{p-1} \xrightarrow{d^{p-1}} C^p \xrightarrow{d^p} C^{p+1} \xrightarrow{d^{p+1}} C^{p+2} \longrightarrow \cdots,$$

using increasing superscripts, with the arrows going from left to right.

As we explained in Section 2.5, a cochain complex can be converted to a chain complex, and conversely, by changing C^p to C_{-p} and d^p to d_{-p} and changing the direction of the arrows. The cochain complex

$$\cdots \longrightarrow C^{p-1} \xrightarrow{d^{p-1}} C^p \xrightarrow{d^p} C^{p+1} \xrightarrow{d^{p+1}} C^{p+2} \longrightarrow \cdots$$

becomes the chain complex

$$\cdots \longleftarrow C_{-(p+2)} \xleftarrow{d_{-(p+1)}} C_{-(p+1)} \xleftarrow{d_{-p}} C_{-p} \xleftarrow{d_{-(p-1)}} C_{-(p-1)} \longleftarrow \cdots.$$

Conversely we get a cochain complex from a chain complex by changing C_p to C^{-p} and d_p to d^{-p} and changing the direction of the arrows. In most cases, given a chain complex $\mathcal{C}_*$ we have $C_p = (0)$ for all $p < 0$. We call such a complex a *positive chain complex*. Similarly, in most cases, given a cochain complex $\mathcal{C}^*$ we have $C_p = (0)$ for all $p < 0$. We call such a complex a *positive cochain complex*. If we convert a positive cochain complex $(C^p)_{p\in\mathbb{N}}$ into a chain complex $(C_{-p})_{-p\in\mathbb{N}}$, then we obtain a *negative chain complex*. This trick allows us to view a positive cochain complex as a negative chain complex. By symmetry, a negative cochain complex is converted to a positive chain complex.

It is usually more pleasant to avoid negative subscripts in negative chain complexes by turning them into cochain complexes by switching signs and raising indices but there are constructions (for example, Hom functors) for which it is more convenient to use complexes with negative and positive indices. Whether we pick chain complexes or cochain complexes is a matter of taste. Rotman favors chain complexes, but Weibel favors cochain complexes. We follow Rotman and use chain complexes.

Our first construction is the tensor product $\mathcal{C} \otimes \mathcal{D}$ of chain complexes $\mathcal{C}$ and $\mathcal{D}$. Then we will state a formula relating the homology of $\mathcal{C} \otimes \mathcal{D}$ to the homology of $\mathcal{C}$ and the homology of $\mathcal{D}$. Such a formula is known as *Künneth theorem* (or *Künneth formula*).

Definition 12.2. Given two chain complexes $\mathcal{C} = (C_p)_{p\in\mathbb{Z}}$ and $\mathcal{D} = (D_q)_{q\in\mathbb{Z}}$ where the C_p an D_q are R-modules, the *tensor product of the complexes $\mathcal{C}$ and $\mathcal{D}$* is the chain complex $\mathcal{C}\otimes\mathcal{D} = ((\mathcal{C}\otimes\mathcal{D})_n)_{n\in\mathbb{Z}}$ defined such that

$$(\mathcal{C}\otimes\mathcal{D})_n = \bigoplus_{p+q=n} C_p\otimes D_q, \quad n\in\mathbb{Z},$$

with differential $\partial_n\colon (\mathcal{C}\otimes\mathcal{D})_n \to (\mathcal{C}\otimes\mathcal{D})_{n-1}$ given by

$$\partial_n(c_p\otimes d_q) = (\partial^C c_p)\otimes d_q + (-1)^p c_p\otimes(\partial^D d_q), \quad c_p\in C_p, d_q\in D_q.$$

Clearly, if $\mathcal{C}$ an $\mathcal{D}$ are both positive chain complexes or both negative chain complexes, then there are only finitely many indices p,q such that $p+q=n$. In the first case $\mathcal{C}\otimes\mathcal{D}$ is a positive chain complex and in the second case it is a negative chain complex (equivalent to a positive cochain complex). The following remarkable theorem holds.

Theorem 12.13. *(Künneth formula) Let R be a hereditary ring, and let $\mathcal{C}$ and $\mathcal{D}$ be two chain complexes with all C_p flat. Then for every $n\in\mathbb{Z}$ there is natural sequence*

$$0 \longrightarrow \bigoplus_{p+q=n} H_p(\mathcal{C})\otimes H_q(\mathcal{D}) \xrightarrow{\ \alpha\ } H_n(\mathcal{C}\otimes\mathcal{D}) \longrightarrow$$

$$\bigoplus_{p+q=n} \mathrm{Tor}_1^R(H_{p-1}(\mathcal{C}), H_q(\mathcal{D})) \longrightarrow 0$$

that splits.

The splitting need not be natural.

Theorem 12.13 is proven in Rotman [Rotman (1988, 2009)] (Chapter 11, Theorem 11.31). The proof is hard and long. There is also a more sophisticated proof using spectral sequences.

Theorem 12.13 is very general. It yields the (strong) universal coefficient theorem for homology (Theorem 12.1) as a corollary with $\mathcal{D}$ the chain complex consisting of the single nonzero module $\mathcal{D}_0 = G$ and $\mathcal{C}$ a positive chain complex whose modules C_p are flat.

When both $\mathcal{C}$ and $\mathcal{D}$ are positive chain complexes, Theorem 12.13 yields what is usually known as the *Künneth formula for chain complexes*; see

Munkres [Munkres (1984)] (Chapter 7, Section 58). In this case, since $p, q \geq 0$, the direct sums are finite.

When both $\mathcal{C}$ and $\mathcal{D}$ are negative chain complexes, in other words, positive cochain complexes, Theorem 12.13 yields a *Künneth formula for cochain complexes*. In this case, since $p, q < 0$, $(\mathcal{C} \otimes \mathcal{D})_n = (0)$ for all $n > 0$, and for each $n \leq 0$, there are only finitely many p, q such that $p + q = n$. The homology groups H_{-k} with $k \geq 0$ become cohomology groups H^k, and we obtain the following exact sequences

$$0 \longrightarrow \bigoplus_{p+q=n} H^p(\mathcal{C}) \otimes H^q(\mathcal{D}) \xrightarrow{\ \alpha\ } H^n(\mathcal{C} \otimes \mathcal{D}) \longrightarrow$$

$$\bigoplus_{p+q=n} \mathrm{Tor}_1^R(H^{p+1}(\mathcal{C}), H^q(\mathcal{D})) \longrightarrow 0,$$

that split. When $\mathcal{D}$ consists of a single nonzero module $D_0 = G$, we obtain a universal coefficient theorem for computing the cohomology modules $H^n(\mathcal{C} \otimes G)$ in terms of $H^n(\mathcal{C})$ and $H^{n+1}(\mathcal{C})$ (and some Tor module), namely the exact sequence

$$0 \longrightarrow H^n(\mathcal{C}) \otimes G \longrightarrow H^n(\mathcal{C} \otimes G) \longrightarrow \mathrm{Tor}_1^R(H^{n+1}(\mathcal{C}), G) \longrightarrow 0$$

splits; see Munkres [Munkres (1984)] (Chapter 7, Corollary 56.4).

Another application of Theorem 12.13 is a formula for computing the homology of the product of two topological spaces. For this we need to state the Eilenberg–Zilber theorem.

Theorem 12.14. *(Eilenberg–Zilber theorem) Given any two topological spaces X and Y, there are chain homotopies $\mu\colon S_*(X) \otimes S_*(Y) \to S_*(X \times Y)$ and $\nu\colon S_*(X \times Y) \to S_*(X) \otimes S_*(Y)$ (in singular homology with coefficients in $\mathbb{Z}$) that are mutual inverses. These chain homotopies are natural with respect to chain maps induced by continuous maps.*

For a proof of Theorem 12.14, see Munkres [Munkres (1984)] (Chapter 7, Sections 59). The Eilenberg–Zilber theorem immediately implies that

$$H_m(X \times Y) \cong H_m(S_*(X) \otimes S_*(Y)), \quad m \geq 0.$$

As a corollary we obtain the following result.

Theorem 12.15. *(Künneth formula for topological spaces) Given any topological spaces X and Y, for every $n \in \mathbb{N}$ there is natural sequence*

$$0 \longrightarrow \bigoplus_{p+q=n} H_p(X) \otimes H_q(Y) \xrightarrow{\alpha} H_n(X \times Y) \longrightarrow$$

$$\bigoplus_{p+q=n} \mathrm{Tor}_1^R(H_{p-1}(X), H_q(Y)) \longrightarrow 0$$

that splits. Here we are dealing with singular homology with coefficients in $\mathbb{Z}$.

Theorem 12.15 is proven in Munkres [Munkres (1984)] (Chapter 7, Section 59). As an application of Theorem 12.15, it is easy to compute the homology groups $H_p(T^n)$ of the n-torus $T^n = (S^1)^n$ by induction and to confirm that

$$H_p(T^n) = \mathbb{Z}^{\binom{n}{p}}.$$

We now consider the generalization of Hom to complexes.

Definition 12.3. Given two chain complexes $\mathcal{C} = (C_p)_{p\in\mathbb{Z}}$ and $\mathcal{D} = (D_q)_{q\in\mathbb{Z}}$ where the C_p an D_q are R-modules, the chain complex $\mathrm{Hom}(\mathcal{C},\mathcal{D}) = (\mathrm{Hom}(\mathcal{C},\mathcal{D})_n)_{\in\mathbb{Z}}$ is defined by

$$\mathrm{Hom}(\mathcal{C},\mathcal{D})_n = \prod_{p+q=n} \mathrm{Hom}(C_{-p}, D_q), \quad n \in \mathbb{Z},$$

with differential $\partial_n \colon \mathrm{Hom}(\mathcal{C},\mathcal{D})_n \to \mathrm{Hom}(\mathcal{C},\mathcal{D})_{n-1}$ given by

$$\partial_n = \prod_{p+q=n-1} \partial_{p,q},$$

with

$$\partial_{p,q}\Big((f_{ij})_{i+j=n}\Big) = (-1)^{p+q} f_{p+1,q} \circ \partial^C_{-p} + \partial^D_{q+1} \circ f_{p,q+1}, \quad p+q = n-1,$$

where $f_{i,j} \in \mathrm{Hom}(C_{-i}, D_j)$, $i + j = n$.

Observe that

$$(\partial^C_{-p})^*(f_{p+1,q}) = \mathrm{Hom}(\partial^C_{-p}, \mathrm{id})(f_{p+1,q}) = f_{p+1,q} \circ \partial^C_{-p}$$

and

$$(\partial^D_{q+1})_*(f_{p,q+1}) = \mathrm{Hom}(\mathrm{id}, \partial^D_{q+1})(f_{p,q+1}) = \partial^D_{q+1} \circ f_{p,q+1},$$

so we can also write

$$\partial_n = \prod_{p+q=n-1} ((-1)^{p+q}(\partial^C_{-p})^* + (\partial^D_{q+1})_*).$$

If $\mathcal{C}$ is a positive chain complex (C_p) (with $C_p = (0)$ for $p < 0$) and if $\mathcal{D}$ is a negative chain complex (D_q) (with $D_q = (0)$ for $q > 0$), then $\mathrm{Hom}(C_{-p}, D_q)$ is nonzero only if $p \leq 0$ and $q \leq 0$, so for $n \leq 0$ there are only finitely many $p, q \leq 0$ such that $p + q = n$ and we have

$$\mathrm{Hom}(\mathcal{C}, \mathcal{D})_n = \bigoplus_{p+q=n, p,q\leq 0} \mathrm{Hom}(C_{-p}, D_q).$$

If we let $p' = -p$, $q' = -q$, and $n' = -n$, by switching signs and raising the indices, for $n' \geq 0$ we have

$$\mathrm{Hom}(\mathcal{C}, \mathcal{D})^{n'} = \bigoplus_{p'+q'=n', p',q'\geq 0} \mathrm{Hom}(C_{p'}, D^{q'}).$$

This is the case that occurs most of the time. For this reason, some authors define $\mathrm{Hom}(\mathcal{C}, \mathcal{D})$ directly as a cochain complex.

Remark: As in the case of tensor products of modules, for any three chain complexes $\mathcal{C}, \mathcal{D}, \mathcal{E}$, we have an isomorphism

$$\mathrm{Hom}(\mathcal{C} \otimes \mathcal{D}, \mathcal{E}) \cong \mathrm{Hom}(\mathcal{C}, \mathrm{Hom}(\mathcal{D}, \mathcal{E})).$$

We have the following Künneth formula for $\mathrm{Hom}(\mathcal{C}, \mathcal{D})$; see Rotman [Rotman (1988)] (Chapter 11, Theorem 11.32), Rotman [Rotman (2009)] (Chapter 10, Theorem 10.85), and Hilton and Stammbach [Hilton and Stammbach (1996)] (Chapter V, Theorem 3.1).

Theorem 12.16. *(Künneth formula for* Hom*) Let R be a hereditary ring, and let $\mathcal{C}$ and $\mathcal{D}$ be two chain complexes with all C_p projective. Then for every $n \in \mathbb{Z}$ there is natural sequence*

$$0 \longrightarrow \prod_{p-q=n-1} \mathrm{Ext}^1_R(H_p(\mathcal{C}), H_q(\mathcal{D})) \longrightarrow H^n(\mathrm{Hom}(\mathcal{C}, \mathcal{D})) \longrightarrow \prod_{p-q=n} \mathrm{Hom}(H_p(\mathcal{C}), H_q(\mathcal{D})) \longrightarrow 0$$

that splits. Here the chain complex $\mathrm{Hom}(\mathcal{C}, \mathcal{D}) = (\mathrm{Hom}(\mathcal{C}, \mathcal{D})_{-n})$ *is turned into a cochain complex* $(\mathrm{Hom}(\mathcal{C}, \mathcal{D})^n)$ *as explained earlier so that the cohomology groups are well-defined.*

The splitting need not be natural.

In the formula in Rotman [Rotman (1988)], H^n should be H_n, and in Rotman [Rotman (2009)], $p - q$ should be $p + q$. The formula in Hilton and Stammbach [Hilton and Stammbach (1996)] is stated with $\prod_{q-p=n+1}$ in the first product, $\prod_{q-p=n}$ in the second product, and H_n instead of H^n. This equivalent to the formula in our statement since $q-p=n+1$ in the first case and $q-p=n$ in the second case are respectively equivalent to $p-q=-n-1$ and $p-q=-n$, and we changed $-n$ to n.

We prefer our version for the following reason. If $\mathcal{C}$ is a positive chain complex and $\mathcal{D}$ is a negative chain complex (a positive cochain complex), we saw that $\mathrm{Hom}(\mathcal{C},\mathcal{D})$ is a negative chain complex with negative indices $-n$ (with $n \geq 0$), so $\mathrm{Hom}(\mathcal{C},\mathcal{D})$ is a positive cochain complex, we have $p \geq 0$, $q \leq 0$, and the H^{-q} and H^n are cohomology groups. In this special case, changing q to $-q$ (with $q \geq 0$), the Künneth formula says that the exact sequence

$$0 \longrightarrow \prod_{p+q=n-1} \mathrm{Ext}^1_R(H_p(\mathcal{C}), H^q(\mathcal{D})) \longrightarrow H^n(\mathrm{Hom}(\mathcal{C},\mathcal{D})) \longrightarrow$$

$$\prod_{p+q=n} \mathrm{Hom}(H_p(\mathcal{C}), H^q(\mathcal{D})) \longrightarrow 0$$

splits. See Weibel [Weibel (1994)] (Exercise 3.6.1, Page 90).

In the special case where $\mathcal{C}$ is a positive chain complex of projectives and $\mathcal{D}$ consists of the single nonzero module $D_0 = G$, we obtain the universal coefficient theorem for cohomology (Theorem 12.6) as a corollary.

For more on Künneth formulae we refer the reader to Rotman [Rotman (1988, 2009)], Munkres [Munkres (1984)] (Chapter 7, Sections 58 and 60), Hatcher [Hatcher (2002)] (Chapter 3, Sections 3.2 and 3.B) and Spanier [Spanier (1989)] (Chapter 5).

12.6 Problems

Problem 12.1. Prove the identities

$$\mathrm{Tor}^R\left(\bigoplus_{i\in I} A_i, B\right) \cong \bigoplus_{i\in I} \mathrm{Tor}^R(A_i, B)$$

$$\mathrm{Tor}^R\left(A, \bigoplus_{i\in I} B_i\right) \cong \bigoplus_{i\in I} \mathrm{Tor}^R(A, B_i)$$

$$\mathrm{Tor}^R(A,B) \cong \mathrm{Tor}^R(B,A),$$

where A, A_i, B_i, B are R-modules for a commutative ring R (with an identity element), and I is an arbitrary index set.

Problem 12.2. Prove that

$$\mathrm{Tor}^R(A,B) \cong (0) \quad \text{if } A \text{ or } B \text{ is flat (in particular, projective, or free)},$$

for any commutative ring R.

Problem 12.3. Compute $A \otimes B$ and $\mathrm{Tor}^{\mathbb{Z}}(A,B)$ for

$$A = \mathbb{Z} \oplus \mathbb{Z}/2\mathbb{Z} \oplus \mathbb{Z}/4\mathbb{Z} \oplus \mathbb{Z}/6\mathbb{Z}, \qquad B = \mathbb{Z} \oplus \mathbb{Z} \oplus \mathbb{Z}/9\mathbb{Z} \oplus \mathbb{Z}/12\mathbb{Z}.$$

Problem 12.4. Let A and B be two finitely generated abelian groups and let T_A and T_B be their torsion groups. Prove that $\mathrm{Tor}^R(A,B) \cong \mathrm{Tor}^R(T_A,T_B)$.

Problem 12.5. Prove the identities

$$\mathrm{Ext}_R\left(\bigoplus_{i\in I} A_i, B\right) \cong \prod_{i\in I} \mathrm{Ext}_R(A_i,B)$$

$$\mathrm{Ext}_R\left(A, \prod_{i\in I} B_i\right) \cong \prod_{i\in I} \mathrm{Ext}_R(A,B_i),$$

where A, A_i, B_i, B are R-modules for a commutative ring R (with an identity element), and I is an arbitrary index set.

Problem 12.6. Prove that

$$\mathrm{Ext}_R(A,B) \cong (0) \quad \text{if } A \text{ is projective or } B \text{ is injective},$$

for any commutative ring R.

Problem 12.7. Compute $\mathrm{Hom}(A,B)$ and $\mathrm{Ext}_{\mathbb{Z}}(A,B)$ for

$$A = \mathbb{Z} \oplus \mathbb{Z}/2\mathbb{Z} \oplus \mathbb{Z}/4\mathbb{Z} \oplus \mathbb{Z}/6\mathbb{Z}, \qquad B = \mathbb{Z} \oplus \mathbb{Z} \oplus \mathbb{Z}/9\mathbb{Z} \oplus \mathbb{Z}/12\mathbb{Z}.$$

Problem 12.8. Consider the spheres S^r and S^s. Prove that

$$H_m(S^r \times S^r;\mathbb{Z}) = \begin{cases} \mathbb{Z} & \text{if } m = 0, 2r \\ \mathbb{Z} \oplus \mathbb{Z} & \text{if } m = 2r \\ (0) & \text{otherwise,} \end{cases}$$

and if $r \neq s$, then

$$H_m(S^r \times S^s;\mathbb{Z}) = \begin{cases} \mathbb{Z} & \text{if } m = 0, r, s, r+s \\ (0) & \text{otherwise.} \end{cases}$$

Problem 12.9. Check that in Definition 12.2, the formula $\partial_n : (\mathcal{C} \otimes \mathcal{D})_n \to (\mathcal{C} \otimes \mathcal{D})_{n-1}$ given by

$$\partial_n(c_p \otimes d_q) = (\partial^C c_p) \otimes d_q + (-1)^p c_p \otimes (\partial^D d_q), \quad c_p \in C_p, d_q \in D_q$$

defines a differential ($\partial_{n-1} \circ \partial_n = 0$).

Problem 12.10. Prove using the Künneth formula that the homology groups $H_p(T^n)$ of the n-torus $T^n = (S^1)^n$ are given by

$$H_p(T^n) = \mathbb{Z}^{\binom{n}{p}}.$$

Problem 12.11. Check that in Definition 12.3, the formula $\partial_n : \mathrm{Hom}(\mathcal{C}, \mathcal{D})_n \to \mathrm{Hom}(\mathcal{C}, \mathcal{D})_{n-1}$ given by

$$\partial_n = \prod_{p+q=n-1} \partial_{p,q},$$

with

$$\partial_{p,q}\Big((f_{ij})_{i+j=n}\Big) = (-1)^{p+q} f_{p+1,q} \circ \partial^C_{-p} + \partial^D_{q+1} \circ f_{p,q+1}, \quad p+q = n-1,$$

where $f_{i,j} \in \mathrm{Hom}(C_{-i}, D_j)$, $i+j = n$, defines a differential ($\partial_{n-1} \circ \partial_n = 0$).

Chapter 13

Cohomology of Sheaves

In this chapter we apply the results of Sections 11.4 and 11.8 to the case where $\mathbf{C}$ is the abelian category of sheaves of R-modules on a topological space X, $\mathbf{D}$ is the (abelian) category of abelian groups, and T is the left-exact global section functor $\Gamma(X, -)$, with $\Gamma(X, \mathcal{F}) = \mathcal{F}(X)$ for every sheaf $\mathcal{F}$ on X. It turns out that the category of sheaves has enough injectives, thus the right derived functors $R^p\Gamma(X, -)$ exist, and for every sheaf $\mathcal{F}$ on X, the cohomology groups $R^p\Gamma(X, -)(\mathcal{F})$ are defined. These groups, denoted by $H^p(X, \mathcal{F})$, are called the *cohomology groups* of the sheaf $\mathcal{F}$ (or the *cohomology groups of X with values in $\mathcal{F}$*).

In principle, computing the cohomology groups $H^p(X, \mathcal{F})$ requires finding injective resolutions of sheaves. However injective sheaves are very big and hard to deal with. Fortunately, there is a class of sheaves known as *flasque* sheaves (due to Godement) which are $\Gamma(X, -)$-acyclic, and every sheaf has a resolution by flasque sheaves. Therefore, by Proposition 11.34, the cohomology groups $H^p(X, \mathcal{F})$ can be computed using flasque resolutions.

If the space X is paracompact (see Definition 13.7), then it turns out that for any sheaf $\mathcal{F}$, the Čech cohomology groups $\check{H}^p(X, \mathcal{F})$ are isomorphic to the cohomology groups $H^p(X, \mathcal{F})$. Furthermore, if $\mathcal{F}$ is a presheaf, then the Čech cohomology groups $\check{H}^p(X, \mathcal{F})$ and $\check{H}^p(X, \widetilde{\mathcal{F}})$ are isomorphic, where $\widetilde{\mathcal{F}}$ is the sheafification of $\mathcal{F}$. Several other results (due to Leray and Henri Cartan) about the relationship between Čech cohomology and sheaf cohomology will be stated.

When X is a topological manifold (thus paracompact), for every R-module G, we will show that the singular cohomology groups $H^p(X; G)$ are isomorphic to the cohomology groups $H^p(X, \widetilde{G}_X)$ of the constant sheaf $\widetilde{G}_X$. Technically, we will need to define *soft* and *fine* sheaves.

We will also define Alexander–Spanier cohomology and prove that it is equivalent to sheaf cohomology (and Čech cohomology) for paracompact spaces and for the constant sheaf $\widetilde{G}_X$.

In summary, if X is a paracompact topological space (for example, a topological manifold) and if G is any R-module, then singular cohomology, Čech cohomology, Alexander–Spanier cohomology, and sheaf cohomology for the constant sheaf $\widetilde{G}_X$ or the presheaf G_X are all equivalent; there are isomorphisms

$$H^p(X, G) \cong \check{H}^p(X, G) \cong H^p_{\text{A-S}}(X; G) \cong H^p(X, \widetilde{G}_X) \cong H^p(X, G_X)$$

for all $p \geq 0$. If X is a smooth manifold and $R = \mathbb{R}$, we also have the de Rham isomorphisms

$$H^p_{\text{dR}}(X) \cong H^p(X, \widetilde{\mathbb{R}}_X)$$

for all $p \geq 0$.

13.1 Cohomology Groups of a Sheaf of Modules

It is convenient to use for a definition of an injective sheaf the condition of Proposition 11.3 which applies to abelian categories. Recall the definition of an injective, or monic, sheaf map from Definition 10.16.

Definition 13.1. A sheaf $\mathcal{I}$ is *injective* if for any injective (monic) sheaf map $h\colon \mathcal{F} \to \mathcal{G}$ and any sheaf map $f\colon \mathcal{F} \to \mathcal{I}$, there is some sheaf map $\widehat{f}\colon \mathcal{G} \to \mathcal{I}$ extending $f\colon \mathcal{F} \to \mathcal{I}$ in the sense that $f = \widehat{f} \circ h$, as in the following commutative diagram:

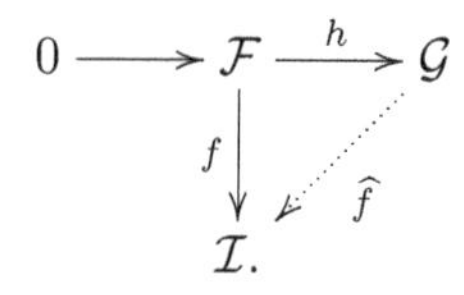

We need to prove that the category of sheaves of R-modules has enough injectives.

Proposition 13.1. *For any sheaf $\mathcal{F}$ of R-modules, there is an injective sheaf $\mathcal{I}$ and an injective sheaf homomorphism $\varphi\colon \mathcal{F} \to \mathcal{I}$.*

Proof. We know that the category of R-modules has enough injectives (see Theorem 11.6). For every fixed $x \in X$, pick some injection $\mathcal{F}_x \longrightarrow I^x$ with I^x an injective R-module, which always exists by Theorem 11.6 (recall that $\mathcal{F}_x$ is also an R-module). Define the "skyscraper sheaf" $\mathcal{I}^x$ as the sheaf given by

$$\mathcal{I}^x(U) = \begin{cases} I^x & \text{if } x \in U \\ (0) & \text{if } x \notin U \end{cases}$$

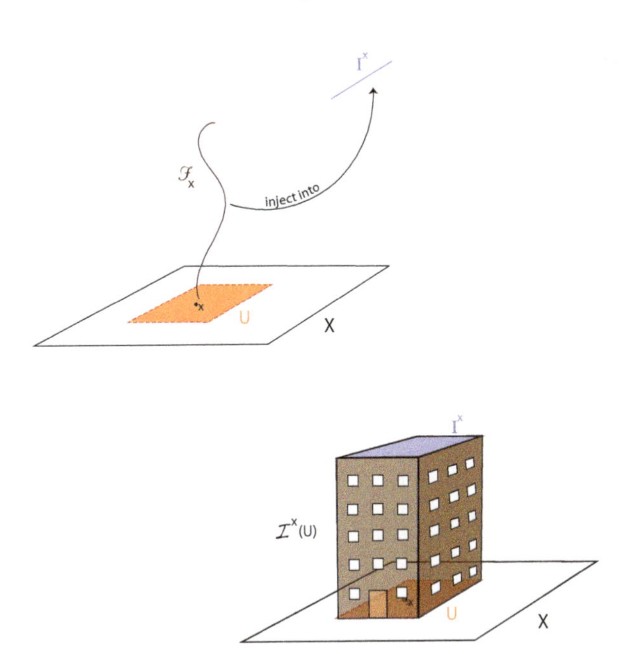

Fig. 13.1 A schematic representation of the "skyscraper sheaf" $\mathcal{I}^x$. We represent X as a white plane. The top figure injects the maroon stalk $\mathcal{F}_x$ into the blue R-module I^x. The bottom figure shows that for any open set U of X, $\mathcal{I}^x(U)$ is a "skyscraper" whose blue "roof" is the fixed module I^x.

for every open subset U of X (we use a superscript in $\mathcal{I}^x$ to avoid the potential confusion with the stalk at x). See Figure 13.1. It is easy to check that there is an isomorphism

$$\mathrm{Hom}_{\mathbf{Sh}(X)}(\mathcal{F}, \mathcal{I}^x) \cong \mathrm{Hom}_R(\mathcal{F}_x, I^x)$$

for any sheaf $\mathcal{F}$ given by $\varphi \mapsto \varphi_x$ with $\varphi \in \mathrm{Hom}_{\mathbf{Sh}(X)}(\mathcal{F}, \mathcal{I}^x)$ (see Definition 10.3 and Corollary 10.3), and this implies that $\mathcal{I}^x$ is an injective sheaf.

We also have a sheaf map from $\mathcal{F}$ to $\mathcal{I}^x$ given by the injection $\mathcal{F}_x \longrightarrow I^x$. Consequently we obtain an injective sheaf map

$$\mathcal{F} \longrightarrow \prod_{x \in X} \mathcal{I}^x.$$

Since a product of injective sheaves is injective, $\mathcal{F}$ is embedded into an injective sheaves. $\square$

Remark: The category of sheaves does not have enough projectives. This is the reason why projective resolutions of sheaves are of little interest.

As we explained in Section 11.2, since the category of sheaves is an abelian category and since it has enough injectives, Proposition 11.15 holds for sheaves; that is, every sheaf has some injective resolution. Since the global section functor on sheaves is left-exact (see Proposition 10.34(4)), as a corollary of Theorem 11.27 we make the following definition.

Definition 13.2. Let X be a topological space, and let $\Gamma(X, -)$ be the global section functor from the abelian category $\mathbf{Sh}(X)$ of sheaves of R-modules to the category of abelian groups. The *cohomology groups* of the sheaf $\mathcal{F}$ (or the *cohomology groups of X with values in $\mathcal{F}$*), denoted by $H^p(X, \mathcal{F})$, are the groups $R^p\Gamma(X, -)(\mathcal{F})$ induced by the right derived functor $R^p\Gamma(X, -)$ (with $p \geq 0$).

To compute the sheaf cohomology groups $H^p(X, \mathcal{F})$, pick any resolution of $\mathcal{F}$

$$0 \longrightarrow \mathcal{F} \longrightarrow \mathcal{I}^0 \xrightarrow{d^0} \mathcal{I}^1 \xrightarrow{d^1} \mathcal{I}^2 \xrightarrow{d^2} \cdots$$

by injective sheaves $\mathcal{I}^n$, apply the global section functor $\Gamma(X, -)$ to obtain the complex of R-modules

$$0 \xrightarrow{\delta^{-1}} \mathcal{I}^0(X) \xrightarrow{\delta^0} \mathcal{I}^1(X) \xrightarrow{\delta^1} \mathcal{I}^2(X) \xrightarrow{\delta^2} \cdots,$$

and then

$$H^p(X, \mathcal{F}) = \mathrm{Ker}\, \delta^p / \mathrm{Im}\, \delta^{p-1}.$$

By Theorem 11.47 the right derived functors $R^p\Gamma(X, -)$ constitute a universal δ-functor, so all the properties of δ-functors apply.

In algebraic geometry it is useful to consider sheaves defined on a ringed space generalizing modules. Roughly speaking, we consider sheaves of modules for which we allow the ring of coefficients $\mathcal{O}_X(U)$ to vary with U.

Definition 13.3. Given a ringed space $(X, \mathcal{O}_X)$, an $\mathcal{O}_X$*-module* (or *sheaf of modules over* X) is a sheaf $\mathcal{F}$ of abelian groups on X such that for every open subset U, the group $\mathcal{F}(U)$ is an $\mathcal{O}_X(U)$-module, and the following conditions hold for all open subsets $V \subseteq U$:

$$\begin{array}{ccc} \mathcal{O}_X(U) \times \mathcal{F}(U) & \longrightarrow & \mathcal{F}(U) \\ \Big\downarrow {\scriptstyle (\rho\mathcal{O}^U_V, \rho\mathcal{F}^U_V)} & & \Big\downarrow {\scriptstyle \rho\mathcal{F}^U_V} \\ \mathcal{O}_X(V) \times \mathcal{F}(V) & \longrightarrow & \mathcal{F}(V). \end{array}$$

Any sheaf of R-modules on X can be viewed as an $\mathcal{O}_X$-module with respect to the constant sheaf $\widetilde{R}_X$. There is an obvious notion of morphism of $\mathcal{O}_X$-modules induced by the notion of morphism of sheaves. The category of $\mathcal{O}_X$-modules on a ringed space $(X, \mathcal{O}_X)$ is denoted by $\mathfrak{Mod}(X, \mathcal{O}_X)$. Proposition 13.1 has the following generalization.

Proposition 13.2. *For any sheaf $\mathcal{F}$ of $\mathcal{O}_X$-modules, there is an injective $\mathcal{O}_X$-module $\mathcal{I}$ and an injective morphism $\varphi\colon \mathcal{F} \to \mathcal{I}$.*

A proof of Proposition 13.2 can be found in Hartshorne [Hartshorne (1977)] (Chapter III, Section 2, Proposition 2.2). As a consequence, we can define the cohomology groups $H^p(X, \mathcal{F})$ of the $\mathcal{O}_X$-module $\mathcal{F}$ over the ringed space $(X, \mathcal{O}_X)$ as the groups induced by the right derived functors $R^p\Gamma(X, -)$ of the functor $\Gamma(X, -)$ from the category $\mathfrak{Mod}(X, \mathcal{O}_X)$ of $\mathcal{O}_X$-modules to the category of abelian groups (with $p \geq 0$).

We now turn to flasque sheaves.

13.2 Flasque Sheaves

The notion of flasque sheaf is due to Godement (see [Godement (1958)], Chapter 3). The word *flasque* is French and it is hard to find an accurate English translation for it. The closest approximations we can think of are *flabby*, *limp*, or *soggy*; a good example of a "flasque" object is a slab of jello or a jellyfish. Most authors use the French word "flasque" so we will use it too.

Definition 13.4. A sheaf $\mathcal{F}$ on a topological space X is *flasque* if for every open subset U of X the restriction map $\rho^X_U : \mathcal{F}(X) \to \mathcal{F}(U)$ is surjective.

We will see shortly that injective sheaves are flasque. Although this is not obvious from the definition, the notion of being flasque is local.

Proposition 13.3. *Let $\mathcal{F}$ be an $\mathcal{O}_X$-module. If $\mathcal{F}$ is flasque, so is $\mathcal{F} \upharpoonright U$ for every open subset U of X. Conversely, if for every $x \in X$, there is a neighborhood U such that $\mathcal{F} \upharpoonright U$ is flasque, then $\mathcal{F}$ is flasque.*

Proof. The first statement is trivial, so let us prove the converse. Given any open set V of X, let s be a section of $\mathcal{F}$ over V. Let T be the set of all pairs (U, σ), where U is an open in X containing V, and σ is an extension of s to U. Partially order T by saying that $(U_1, \sigma_1) \leq (U_2, \sigma_2)$ if $U_1 \subseteq U_2$ and σ_2 extends σ_1, and observe that T is inductive, which means that every chain has an upper bound. Zorn's lemma provides us with a maximal extension of s to a section σ over an open set U_0. Were U_0 not X, there would exist an open set W in X not contained in U_0 such that $\mathcal{F} \upharpoonright W$ is flasque. Thus we could extend the section $\rho^{U_0}_{U_0 \cap W}(\sigma)$ to a section σ' of $\mathcal{F} \upharpoonright W$. Since σ and σ' agree on $U_0 \cap W$ by construction, their common extension to $U_0 \cup W$ extends s, a contradiction; see Figure 13.2. ☐

Proposition 13.4. *Every $\mathcal{O}_X$-module may be embedded in a canonical functorial way into a flasque $\mathcal{O}_X$-module. Every $\mathcal{O}_X$-module has a canonical flasque resolution (i.e., a resolution by flasque $\mathcal{O}_X$-modules.)*

Proof. Let $\mathcal{F}$ be an $\mathcal{O}_X$-module, and define a presheaf $\mathcal{C}^0(X, \mathcal{F})$ by

$$U \mapsto \prod_{x \in U} \mathcal{F}_x.$$

It is immediate that $\mathcal{C}^0(X, \mathcal{F})$ is actually a sheaf and that we have an *injection* of $\mathcal{O}_X$-modules $j \colon \mathcal{F} \to \mathcal{C}^0(X, \mathcal{F})$. An element of $\mathcal{C}^0(X, \mathcal{F})$ over any open set U is a collection (s_x) of elements indexed by U, each s_x lying over the $\mathcal{O}_{X,x}$-module $\mathcal{F}_x$. Such a sheaf is flasque because every U-indexed sequence s_x can be extended to an X-indexed sequence by assigning any arbitrary element of $\mathcal{F}_x$ to any $x \in X - U$. Hence $\mathfrak{Mod}(X, \mathcal{O}_X)$ possesses enough flasque sheaves.

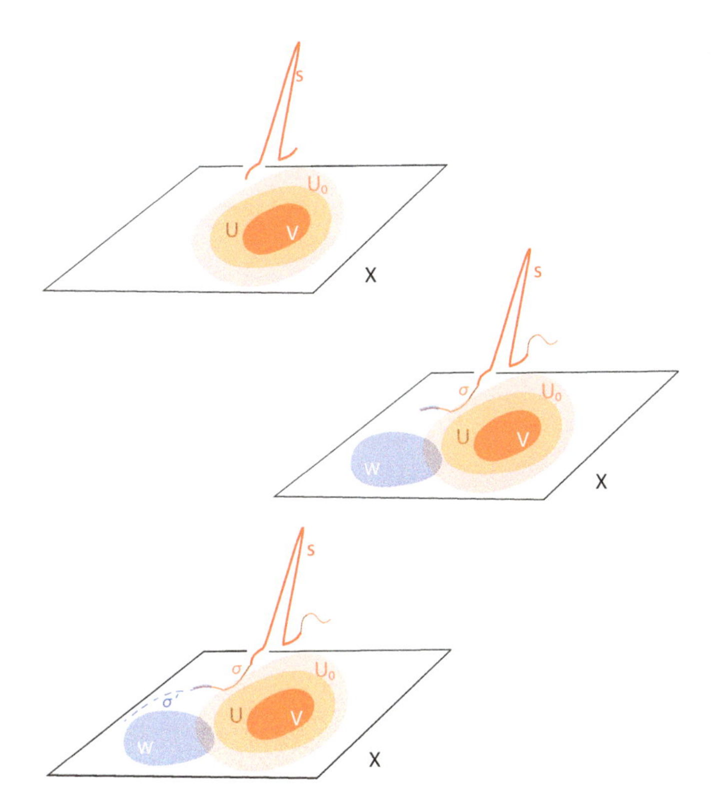

Fig. 13.2 A schematic illustration of the proof of Proposition 13.3. The space X is represented by the white plane. The top figure shows the red section s over the open set V. The middle figure shows the "supposed" maximal extension of s as the section σ. But if U_0 is not X, the bottom figure illustrates how to extend σ, thus contradicting maximality.

If $\mathcal{Z}_1$ is the (sheaf) cokernel $\mathcal{C}^0(X,\mathcal{F})/j(\mathcal{F})$ of the canonical injection $j\colon \mathcal{F} \to \mathcal{C}^0(X,\mathcal{F})$, we define $\mathcal{C}^1(X,\mathcal{F})$ to be the flasque sheaf $\mathcal{C}^0(X,\mathcal{Z}_1)$, and d^0 is the composite map

$$d^0\colon \mathcal{C}^0(X,\mathcal{F}) \longrightarrow \mathcal{C}^0(X,\mathcal{F})/j(\mathcal{F}) = \mathcal{Z}^1 \longrightarrow \mathcal{C}^0(X,\mathcal{Z}_1) = \mathcal{C}^1(X,\mathcal{F}).$$

In general,

$$\mathcal{Z}_n = \mathcal{C}^{n-1}(X,\mathcal{F})/d^{n-2}\mathcal{C}^{n-2}(X,\mathcal{F}),$$

the (sheaf) cokernel of the map $d^{n-2}\colon \mathcal{C}^{n-2}(X,\mathcal{F}) \to \mathcal{C}^{n-1}(X,\mathcal{F})$, and

$$\mathcal{C}^n(X,\mathcal{F}) = \mathcal{C}^0(X,\mathcal{Z}_n),$$

a flasque sheaf. The map $d^{n-1}\colon \mathcal{C}^{n-1}(X,\mathcal{F}) \to \mathcal{C}^n(X,\mathcal{F})$ is the composite

$$d^{n-1}\colon \mathcal{C}^{n-1}(X,\mathcal{F}) \longrightarrow \mathcal{C}^{n-1}(X,\mathcal{F})/d^{n-2}\mathcal{C}^{n-2}(X,\mathcal{F}) = \mathcal{Z}_n \longrightarrow$$
$$\mathcal{C}^0(X,\mathcal{Z}_n) = \mathcal{C}^n(X,\mathcal{F}).$$

Observe that $\mathrm{Im}\, d^{n-1} \cong \mathcal{Z}_n$, so we could define $\mathcal{Z}_n$ as

$$\mathcal{Z}_n = \mathcal{C}^{n-1}(X,\mathcal{F})/\mathcal{Z}_{n-1},$$

as in Godement [Godement (1958)] (Chapter 4, Section 4.2). Putting all this information together, we obtain the desired flasque resolution of $\mathcal{F}$

$$0 \longrightarrow \mathcal{F} \xrightarrow{j} \mathcal{C}^0(X,\mathcal{F}) \xrightarrow{d^0} \mathcal{C}^1(X,\mathcal{F}) \xrightarrow{d^1} \mathcal{C}^2(X,\mathcal{F}) \xrightarrow{d^1} \cdots$$

as claimed. □

Definition 13.5. The resolution of $\mathcal{F}$ constructed in Proposition 13.4 is called the *canonical flasque resolution of* $\mathcal{F}$ or the *Godement resolution of* $\mathcal{F}$. We define the R-modules $C^n(X,\mathcal{F})$ as

$$C^n(X,\mathcal{F}) = \Gamma(X,\mathcal{C}^n(X,\mathcal{F})) = \mathcal{C}^n(X,\mathcal{F})(X),$$

where $\Gamma(X,-)$ is the global section functor.

By applying the global section functor $\Gamma(X,-)$ (which is exact) to the canonical resolution

$$0 \longrightarrow \mathcal{F} \xrightarrow{j} \mathcal{C}^0(X,\mathcal{F}) \xrightarrow{d^0} \mathcal{C}^1(X,\mathcal{F}) \xrightarrow{d^1} \mathcal{C}^2(X,\mathcal{F}) \xrightarrow{d^2} \cdots$$

of a sheaf $\mathcal{F}$ yields the cochain complex

$$0 \longrightarrow \Gamma(X,\mathcal{F}), \xrightarrow{j^*} C^0(X,\mathcal{F}) \xrightarrow{(d^0)^*} C^1(X,\mathcal{F}) \xrightarrow{(d^1)^*} C^2(X,\mathcal{F}) \xrightarrow{(d^2)^*} \cdots$$

denoted $C(X,\mathcal{F})$, and we will see in Proposition 13.7 that the cohomology of the above complex computes the sheaf cohomology modules $H^p(X,\mathcal{F})$ (defined in terms of injective resolutions), that is,

$$H^p(X,\mathcal{F}) \cong H^p(C(X,\mathcal{F}); R).$$

Also recall that $H^0(X,\mathcal{F}) \cong \Gamma(X,\mathcal{F})$.

Definition 13.6. Given two sheaves of R-modules $\mathcal{F}'$ and $\mathcal{F}''$, we obtain a presheaf $\mathcal{F} = \mathcal{F}' \oplus \mathcal{F}''$ by setting

$$\mathcal{F}(U) = (\mathcal{F}' \oplus \mathcal{F}'')(U) = \mathcal{F}'(U) \oplus \mathcal{F}''(U)$$

for every open subset U of X. Actually, $\mathcal{F}' \oplus \mathcal{F}''$ is a sheaf. We call $\mathcal{F}'$ and $\mathcal{F}''$ *direct factors* of $\mathcal{F}$.

Here is the principal property of flasque sheaves.

Theorem 13.5. *Let $0 \longrightarrow \mathcal{F}' \longrightarrow \mathcal{F} \longrightarrow \mathcal{F}'' \longrightarrow 0$ be an exact sequence of $\mathcal{O}_X$-modules, and assume $\mathcal{F}'$ is flasque. Then this sequence is exact as a sequence of presheaves. If both $\mathcal{F}'$ and $\mathcal{F}$ are flasque, so is $\mathcal{F}''$. Finally, any direct factor of a flasque sheaf is flasque.*

Proof. Given any open set U, we must prove that

$$0 \longrightarrow \mathcal{F}'(U) \stackrel{\varphi}{\longrightarrow} \mathcal{F}(U) \stackrel{\psi}{\longrightarrow} \mathcal{F}''(U) \longrightarrow 0$$

is exact. By Proposition 10.34(4), the sole problem is to prove that $\mathcal{F}(U) \longrightarrow \mathcal{F}''(U)$ is surjective. By restricting attention to U, we may assume $U = X$; hence, we are going to prove that a global section of $\mathcal{F}''$ may be lifted to a global section of $\mathcal{F}$. Let s'' be a global section of $\mathcal{F}''$, then by Proposition 10.19(iv), locally s'' may be lifted to sections of $\mathcal{F}$. Let T be the family of all pairs (U, σ) where U is an open in X, and σ is a section of $\mathcal{F}$ over U whose image σ'' in $\mathcal{F}''(U)$ is equal to $\rho^X_{\mathcal{F}''(U)}(s'')$. Partially order T as in the proof of Proposition 13.3 and observe that T is inductive. Zorn's lemma provides us with a maximal lifting of s'' to a section $\sigma \in \mathcal{F}(U_0)$.

Were U_0 not X, there would exist $x \in X - U_0$, a neighborhood V of x, and a section τ of $\mathcal{F}$ over V which is a local lifting of $\rho^X_V(s'')$. The sections $\rho^{U_0}_{U_0 \cap V}(\sigma)$, $\rho^V_{U_0 \cap V}(\tau)$ have the same image under ψ in $\mathcal{F}''(U_0 \cap V)$ so their difference maps to 0. Since $\mathrm{SIm}\,\varphi = \mathrm{Ker}\,\psi$, there is a section t of $\mathcal{F}'(U_0 \cap V)$ such that

$$\rho^{U_0}_{U_0 \cap V}(\sigma) = \rho^V_{U_0 \cap V}(\tau) + \varphi(t).$$

Since $\mathcal{F}'$ is flasque, the section t is the restriction of a section $t' \in \mathcal{F}'(V)$. Upon replacing τ by $\tau + \varphi(t')$ (which does not affect the image in $\mathcal{F}''(V)$ since by definition $\varphi(t') = \varphi(t) = \rho^{U_0}_{U_0 \cap V}(\sigma) - \rho^V_{U_0 \cap V}(\tau)$ is in the kernel of ψ), we may assume that $\rho^{U_0}_{U_0 \cap V}(\sigma) = \rho^V_{U_0 \cap V}(\tau)$; that is, that σ and τ agree on the overlap $U_0 \cap V$. Clearly, we may extend σ (*via* τ) to $U_0 \cup V$, contradicting the maximality of (U_0, σ); hence, $U_0 = X$; see Figure 13.3.

Now suppose that $\mathcal{F}'$ and $\mathcal{F}$ are flasque. If $s'' \in \mathcal{F}''(U)$, then by the above, there is a section $s \in \mathcal{F}(U)$ mapping onto s''. Since $\mathcal{F}$ is also flasque, we may lift s to a global section t of $\mathcal{F}$. The image t'' of t in $\mathcal{F}''(X)$ is the required extension of s'' to a global section of $\mathcal{F}''$.

Finally, assume that $\mathcal{F}$ is flasque, and that $\mathcal{F} = \mathcal{F}' \oplus \mathcal{F}''$ for some sheaf $\mathcal{F}''$. For any open subset U of X and any section $s \in \mathcal{F}'(U)$, we can make

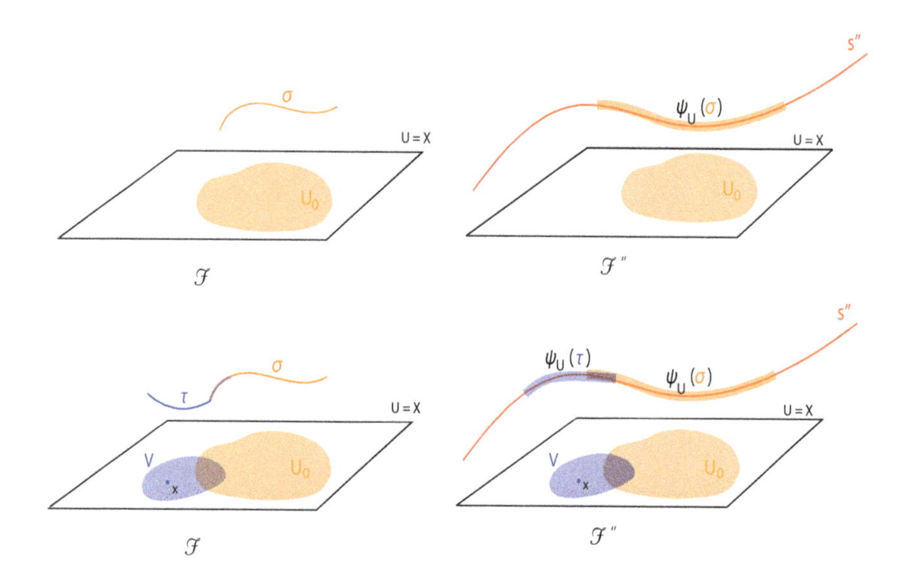

Fig. 13.3 A schematic illustration of the proof of Theorem 13.5. The space X is represented by the white plane. The top figure shows the "supposed" maximal local lifting σ of the global section $s'' \in \mathcal{F}''$. But if U_0 is not X, the bottom figure illustrates how to extend σ, thus contradicting maximality. Since $\mathcal{F}'$ is flasque, we could illustrate σ and τ as agreeing on $U_0 \cap V$.

s into a section $\widetilde{s} \in \mathcal{F}(U)$ by setting the component of $\widetilde{s}(U)$ in $\mathcal{F}''(U)$ equal to the zero section. Since $\mathcal{F}$ is flasque, there is some section $t \in \mathcal{F}(X)$ such that $\rho^X_U(t) = \widetilde{s}$. But $t = t_1 + t_2$ for some unique $t_1 \in \mathcal{F}'(X)$ and $t_2 \in \mathcal{F}''(X)$, and since ρ^X_U is linear,

$$s + 0 = \widetilde{s} = \rho^X_U(t) = \rho^X_U(t_1) + \rho^X_U(t_2)$$

with $\rho^X_U(t_1) \in \mathcal{F}'(U)$ and $\rho^X_U(t_2) \in \mathcal{F}''(U)$, so $s = \rho^X_U(t_1)$ with $t_1 \in \mathcal{F}'(X)$, which shows that $\mathcal{F}'$ is flasque. □

The following general proposition from Tohoku ([Grothendieck (1957)], Section 3.3) implies that flasque sheaves are $\Gamma(X, -)$-acyclic. It also implies that soft sheaves over a paracompact space are $\Gamma(X, -)$-acyclic (see Section 13.5). Since the only functor involved is the global section functor, it is customary to abbreviate $\Gamma(X, -)$-acyclic to acyclic.

Proposition 13.6. *Let T be an additive functor from the abelian category $\mathbf{C}$ to the abelian category $\mathbf{C}'$, and suppose that $\mathbf{C}$ has enough injectives. Let X be a class of objects in $\mathbf{C}$ which satisfies the following conditions:*

(i) *$\mathbf{C}$ possesses enough X-objects, which means that for every object $A \in \mathbf{C}$, there is a monic map from A to some object in X.*
(ii) *If A is an object of $\mathbf{C}$ and A is a direct factor of some object in X, then A belongs to X.*
(iii) *If $0 \longrightarrow A' \longrightarrow A \longrightarrow A'' \longrightarrow 0$ is exact and if A' belongs to X, then $0 \longrightarrow T(A') \longrightarrow T(A) \longrightarrow T(A'') \longrightarrow 0$ is exact, and if A also belongs to X, then A'' belongs to X.*

Under these conditions, every injective object belongs to X, for each M in X we have $R^nT(M) = (0)$ for $n > 0$, and finally the functors R^nT may be computed by taking X-resolutions.

Proof. The following proof is due to Steve Shatz. Let I be an injective of $\mathbf{C}$. By (i), I admits a monic into some object M of the class X. We have an exact sequence

$$0 \longrightarrow I \xrightarrow{\varphi} M \longrightarrow \mathrm{Coker}\,\varphi \longrightarrow 0,$$

and as I is injective and $\varphi\colon I \to M$ is a monic map, there is a map $p\colon M \to I$ such that $p \circ \varphi = \mathrm{id}$ as in the following diagram

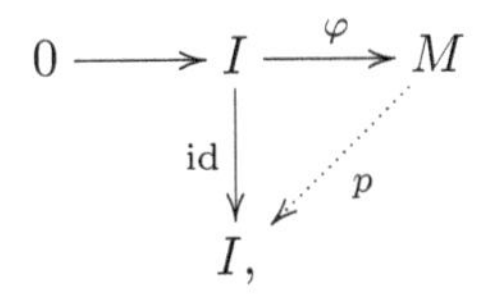

and by Proposition 2.2(2) (which generalizes to abelian categories) the above sequence is split so I is a direct factor of M (this is the generalization of the proof of Proposition 11.3(3) to abelian categories); hence (ii) implies I lies in X. Let us now show that $R^nT(M) = (0)$ for $n > 0$ if M lies in X. Now $\mathbf{C}$ possesses enough injectives, so if we set $C_0 = \mathrm{Coker}(M \longrightarrow I_0)$ and inductively $C_{i+1} = \mathrm{Coker}(C_i \longrightarrow I_{i+1})$ where the maps $M \longrightarrow I_0$ and $C_i \longrightarrow I_{i+1}$ are injections and the I_i are injective, we have the exact sequences

$$\begin{gathered}0 \longrightarrow M \longrightarrow I_0 \longrightarrow C_0 \longrightarrow 0\\ 0 \longrightarrow C_0 \longrightarrow I_1 \longrightarrow C_1 \longrightarrow 0\\ 0 \longrightarrow C_1 \longrightarrow I_2 \longrightarrow C_2 \longrightarrow 0\\ \cdots\cdots\cdots\cdots\cdots\cdots\end{gathered}$$

$$0 \longrightarrow C_n \longrightarrow I_{n+1} \longrightarrow C_{n+1} \longrightarrow 0$$

$$\cdots\cdots\cdots\cdots\cdots\cdots\cdots$$

Here each I_i is injective, so lies in X. As M belongs to X, (iii) shows that C_0 lies in X. By induction, C_i belongs to X for every $i \geq 0$. Again, by (iii), the sequences

$$0 \longrightarrow T(M) \longrightarrow T(I_0) \longrightarrow T(C_0) \longrightarrow 0$$

$$\cdots\cdots\cdots\cdots\cdots\cdots\cdots\cdots\cdots\cdots$$

$$0 \longrightarrow T(C_n) \longrightarrow T(I_{n+1}) \longrightarrow T(C_{n+1}) \longrightarrow 0$$

$$\cdots\cdots\cdots\cdots\cdots\cdots\cdots\cdots\cdots\cdots$$

are exact. Then, as in the proof of Proposition 11.15, we obtain the exact sequence

$$0 \longrightarrow T(M) \longrightarrow T(I_0) \longrightarrow T(I_1) \longrightarrow T(I_2) \longrightarrow \cdots$$

and this proves that $R^nT(M) = (0)$ for positive n. Finally, by Proposition 11.34, the functors R^nT may be computed from arbitrary X-resolutions (which exist by (i)). □

Using Proposition 13.4 and Theorem 13.5, Proposition 13.6 applied with $\mathbf{C}$ the abelian category of sheaves, X the family of flasque sheaves, and T the global section functor, yields the following result.

Proposition 13.7. *Flasque sheaves are acyclic, that is $H^p(X, \mathcal{F}) = (0)$ for every flasque sheaf $\mathcal{F}$ and all $p \geq 1$, and the cohomology groups $H^p(X, \mathcal{F})$ of any arbitrary sheaf $\mathcal{F}$ can be computed using flasque resolutions.*

In view of Proposition 13.2, we also have the following result.

Proposition 13.8. *If $(X, \mathcal{O}_X)$ is a ringed space, then the right derived functors of the functor $\Gamma(X, -)$ from the category $\mathfrak{Mod}(X, \mathcal{O}_X)$ of $\mathcal{O}_X$-modules to the category of abelian groups coincide with the sheaf cohomology functors $H^p(X, -)$.*

Proof. The right derived functors of the functor $\Gamma(X, -)$ from the category $\mathfrak{Mod}(X, \mathcal{O}_X)$ of $\mathcal{O}_X$-modules to the category of abelian groups is computed using resolutions of injectives in the category $\mathfrak{Mod}(X, \mathcal{O}_X)$. But injective sheaves are flasque, and flasque sheaves are acyclic, so by Proposition 11.34 these resolutions compute sheaf cohomology. □

In the rest of this chapter we restrict our attention to *presheaves and sheaves of R-modules*. Our next goal is to compare Čech cohomology and sheaf cohomology.

13.3 Comparison of Čech Cohomology and Sheaf Cohomology

The reader may want to review Sections 9.1 and 9.2 before reading this section. We begin by proving that for every space X, every open cover $\mathcal{U}$ of X, every sheaf $\mathcal{F}$ of R-modules on X, and every $p \geq 0$, there is a homomorphism

$$\check{H}^p(\mathcal{U}, \mathcal{F}) \longrightarrow H^p(X, \mathcal{F}).$$

For every open subset U of X let $\mathcal{U}_{/U}$ denote the induced covering of U consisting of all open subsets of the form $U_i \cap U$, with $U_i \in \mathcal{U}$. Then it is immediately verified that the presheaf $\mathcal{C}^p(\mathcal{U}, \mathcal{F})$ defined by

$$\mathcal{C}^p(\mathcal{U}, \mathcal{F})(U) = C^p(\mathcal{U}_{/U}, \mathcal{F})$$

for any open subset U of X is a sheaf. The crucial property of the sheaves $\mathcal{C}^p(\mathcal{U}, \mathcal{F})$ is that the complex

$$0 \longrightarrow \mathcal{F} \longrightarrow \mathcal{C}^0(\mathcal{U}, \mathcal{F}) \xrightarrow{\delta} \mathcal{C}^1(\mathcal{U}, \mathcal{F}) \xrightarrow{\delta} \cdots \mathcal{C}^p(\mathcal{U}, \mathcal{F}) \xrightarrow{\delta} \mathcal{C}^{p+1}(\mathcal{U}, \mathcal{F}) \cdots$$

is a resolution of the sheaf $\mathcal{F}$.

Proposition 13.9. *For every open cover $\mathcal{U}$ of the space X, for every $\mathcal{F}$ of R-modules on X, the complex*

$$0 \longrightarrow \mathcal{F} \longrightarrow \mathcal{C}^0(\mathcal{U}, \mathcal{F}) \xrightarrow{\delta} \mathcal{C}^1(\mathcal{U}, \mathcal{F}) \xrightarrow{\delta} \cdots \mathcal{C}^p(\mathcal{U}, \mathcal{F}) \xrightarrow{\delta} \mathcal{C}^{p+1}(\mathcal{U}, \mathcal{F}) \cdots$$

is a resolution of the sheaf $\mathcal{F}$.

Sketch of proof. We follow Brylinski [Brylinski (1993)] (Section 1.3, Proposition 1.3.3). By Proposition 10.24(ii) it suffices to show that the stalk sequence

$$0 \longrightarrow \mathcal{F}_x \longrightarrow \mathcal{C}^0(\mathcal{U}, \mathcal{F})_x \xrightarrow{\delta} \cdots \mathcal{C}^p(\mathcal{U}, \mathcal{F})_x \xrightarrow{\delta} \mathcal{C}^{p+1}(\mathcal{U}, \mathcal{F})_x \xrightarrow{\delta} \cdots$$

is exact for every $x \in X$, and since direct limits of exact sequences are still exact it suffice to show that for every $x \in X$, there is some open neighborhood V of x such that the sequence

$$0 \longrightarrow \mathcal{F}(W) \xrightarrow{\epsilon} C^0(\mathcal{U}_{/W}, \mathcal{F}) \xrightarrow{\delta} \cdots C^p(\mathcal{U}_{/W}, \mathcal{F}) \xrightarrow{\delta} C^{p+1}(\mathcal{U}_{/W}, \mathcal{F}) \cdots$$

is exact for every open subset W of V. Pick $V = U_{i_0}$ for some open subset U_{i_0} in such that $x \in U_{i_0}$. Then for $W \subseteq V = U_{i_0}$, the open cover $\{U_i \cap W \mid U_i \in \mathcal{U}\}$ contains $W = W \cap U_{i_0}$. The map ϵ with domain $\mathcal{F}(W)$ is clearly injective and we conclude by using the following simple proposition which is proven in Brylinski [Brylinski (1993)] (Section 1.3, Lemma 1.3.2) and Bredon [Bredon (1997)] (Chapter III, Lemma 4.8):

Proposition 13.10. *If $\mathcal{U} = (U_i)_{i \in I}$ is an open cover of X and if $U_i = X$ for some index i, then for any presheaf $\mathcal{F}$ of R-modules we have $\check{H}^p(\mathcal{U}, \mathcal{F}) = (0)$ for all $p > 0$.*

It follows that the above sequence is exact. ☐

Proposition 13.11. *For every space X, every open cover $\mathcal{U}$ of X, every sheaf $\mathcal{F}$ of R-modules on X, and every $p \geq 0$, there is a homomorphism*

$$\check{H}^p(\mathcal{U}, \mathcal{F}) \longrightarrow H^p(X, \mathcal{F})$$

from Čech cohomology to sheaf cohomology. Consequently there is also a homomorphism

$$\check{H}^p(X, \mathcal{F}) \longrightarrow H^p(X, \mathcal{F})$$

for every $p \geq 0$.

Proof. By Proposition 13.9 we have a resolution

$$0 \longrightarrow \mathcal{F} \longrightarrow \mathcal{C}^*(\mathcal{U}, \mathcal{F})$$

of the sheaf $\mathcal{F}$. For every injective resolution $0 \longrightarrow \mathcal{F} \longrightarrow \mathbf{I}$ of $\mathcal{F}$, by Theorem 11.21, there is a map of resolutions from $\mathcal{C}^*(\mathcal{U}, \mathcal{F})$ to $\mathbf{I}$ lifting the identity and unique up to homotopy. Thus, there is a homomorphism of cohomology $\check{H}^p(\mathcal{U}, \mathcal{F}) \longrightarrow H^p(X, \mathcal{F})$. Since $\check{H}^p(X, \mathcal{F})$ is a direct limit of the $\check{H}^p(\mathcal{U}, \mathcal{F})$, we obtain the homomorphism $\check{H}^p(X, \mathcal{F}) \longrightarrow H^p(X, \mathcal{F})$ by passing to the limit. ☐

In general, the homomorphism $\check{H}^p(X, \mathcal{F}) \longrightarrow H^p(X, \mathcal{F})$ of Proposition 13.11 is neither injective nor surjective. A sufficient condition for having an isomorphism is that X be a *paracompact* topological space (see Definition 13.7).

The strategy to prove that the maps $\check{H}^p(X, \mathcal{F}) \longrightarrow H^p(X, \mathcal{F})$ are isomorphisms is to prove that (under certain conditions) the family of functors $(\check{H}^p(X, -))_{p \geq 0}$ is a universal δ-functor. Indeed, if two cohomology theories $(H^n_S(-))_{n \geq 0}$ and $(H^n_T(-))_{n \geq 0}$ defined for objects in a category $\mathbf{C}$ (say, topological spaces) are given by universal δ-functors S and T in the sense that the cohomology groups $H^n_S(A)$ and $H^n_T(A)$ are given by $H^n_S(A) = S^n(A)$

and $H_T^n(A) = T^n(A)$ for all objects $A \in \mathbf{C}$, and if $H_S^0(A)$ and $H_T^0(A)$ are isomorphic, then $H_S^n(A)$ and $H_T^n(A)$ are isomorphic for all $n \geq 0$. Since sheaf cohomology is defined by right derived δ-functors, which are universal by Theorem 11.47, since for a sheaf, $\check{H}^0(X, \mathcal{F}) \cong \Gamma(X, \mathcal{F}) = \mathcal{F}(X)$, by Proposition 11.38 we obtain the desired isomorphisms.

To prove that the family of functors $(\check{H}^p(X, \mathcal{F}))_{p\geq 0}$ is a universal δ-functor we use Grothendieck's theorem (Theorem 11.45).

We begin by proving that the functors $\check{H}^p(\mathcal{U}, -)$ on sheaves are erasable. Next we will show that the family $(\check{H}^p(\mathcal{U}, -))_{p\geq 0}$ is a δ-functors on sheaves. To do this, we will first show that they constitute a δ-functor on preshaves and then use the fact that if X is paracompact and if $\mathcal{F}$ is a presheaf, then $\check{H}^p(X, \mathcal{F}) \cong \check{H}^p(X, \widetilde{\mathcal{F}})$ for all $p \geq 0$ (see Proposition 13.16).

Proposition 13.12. *For every space X, every open cover $\mathcal{U}$ of X, if the sheaf $\mathcal{F}$ is flasque then*

$$\check{H}^p(\mathcal{U}, \mathcal{F}) = (0) \quad p \geq 1.$$

Consequently the functors $\check{H}^p(\mathcal{U}, -)$ and the functors $\check{H}^p(X, -)$ on sheaves are erasable for all $p \geq 1$.

Proof. Proposition 13.12 is proven in Godement [Godement (1958)] (Chapter 5, Theorem 5.2.3), Hartshorne [Hartshorne (1977)] (Chapter III, Proposition 4.3), and Bredon [Bredon (1997)] (Chapter III, Corollary 4.10).

Observe that since $\mathcal{F}$ is assumed to be flasque, the sheaves $\mathcal{C}^p(\mathcal{U}, \mathcal{F})$ are also flasque because the restriction of $\mathcal{F}$ to any open subset $U_{i_0 \cdots i_p}$ is flasque and a product of flasque sheaves is flasque. Thus by Proposition 13.9 $0 \longrightarrow \mathcal{F} \longrightarrow \mathcal{C}^*(\mathcal{U}, \mathcal{F})$ is a resolution of $\mathcal{F}$ by flasque sheaves. By Proposition 13.7 the cohomology groups $H^p(X, \mathcal{F})$ can be computed using this resolution, but by definition this resolution computes the cohomology groups $\check{H}^p(\mathcal{U}, \mathcal{F})$, so we get

$$H^p(X, \mathcal{F}) = \check{H}^p(\mathcal{U}, \mathcal{F}), \quad \text{for all } p \geq 0.$$

However since $\mathcal{F}$ is flasque, by Proposition 13.7 we have $H^p(X, \mathcal{F}) = (0)$ for all $p \geq 1$, so $\check{H}^p(\mathcal{U}, \mathcal{F}) = (0)$ for all $p \geq 1$. Since every sheaf can be embedded in a flasque sheaf (Proposition 13.4), the functors $\check{H}^p(\mathcal{U}, -)$ are erasable for all $p \geq 1$. By passing to the limit over coverings we obtain the fact that the functors $\check{H}^p(X, -)$ are erasable for all $p \geq 1$. $\square$

The next important fact is that, on *presheaves*, the functors $C^p(\mathcal{U}, -)$ are exact.

Proposition 13.13. *For every space X and every open cover $\mathcal{U}$ of X, the functor $C^p(\mathcal{U}, -)$ from presheaves to abelian groups is exact for all $p \geq 0$.*

Proof. If

$$0 \longrightarrow \mathcal{F}' \longrightarrow \mathcal{F} \longrightarrow \mathcal{F}'' \longrightarrow 0 \qquad (*)$$

is an exact sequence of presheaves, then the sequence

$$0 \longrightarrow C^p(\mathcal{U}, \mathcal{F}') \longrightarrow C^p(\mathcal{U}, \mathcal{F}) \longrightarrow C^p(\mathcal{U}, \mathcal{F}'') \longrightarrow 0$$

is of the form

$$0 \longrightarrow \prod_{(i_0,\ldots,i_p)} \mathcal{F}'(U_{i_0\cdots i_p}) \longrightarrow \prod_{(i_0,\ldots,i_p)} \mathcal{F}(U_{i_0\cdots i_p}) \longrightarrow \prod_{(i_0,\ldots,i_p)} \mathcal{F}''(U_{i_0\cdots i_p}) \longrightarrow 0.$$

But since $(*)$ is an exact sequence of presheaves, by Proposition 10.24(i), every sequence

$$0 \longrightarrow \mathcal{F}'(U_{i_0\cdots i_p}) \longrightarrow \mathcal{F}(U_{i_0\cdots i_p}) \longrightarrow \mathcal{F}''(U_{i_0\cdots i_p}) \longrightarrow 0$$

is exact, and since exactness is preserved under direct products, the sequence

$$0 \longrightarrow \prod_{(i_0,\ldots,i_p)} \mathcal{F}'(U_{i_0\cdots i_p}) \longrightarrow \prod_{(i_0,\ldots,i_p)} \mathcal{F}(U_{i_0\cdots i_p}) \longrightarrow \prod_{(i_0,\ldots,i_p)} \mathcal{F}''(U_{i_0\cdots i_p}) \longrightarrow 0$$

is exact. ∎

As a corollary of Proposition 13.13 we have the next result.

Proposition 13.14. *Every exact sequence of presheaves*

$$0 \longrightarrow \mathcal{F}' \longrightarrow \mathcal{F} \longrightarrow \mathcal{F}'' \longrightarrow 0$$

yields the short exact sequence

$$0 \longrightarrow C^*(X, \mathcal{F}') \longrightarrow C^*(X, \mathcal{F}) \longrightarrow C^*(X, \mathcal{F}'') \longrightarrow 0,$$

which yields a long exact sequence of Čech cohomology groups.

Proof. Indeed, by Proposition 13.13, every exact sequence of presheaves

$$0 \longrightarrow \mathcal{F}' \longrightarrow \mathcal{F} \longrightarrow \mathcal{F}'' \longrightarrow 0$$

yields an exact sequence of Čech cohomology complexes

$$0 \longrightarrow C^*(\mathcal{U}, \mathcal{F}') \longrightarrow C^*(\mathcal{U}, \mathcal{F}) \longrightarrow C^*(\mathcal{U}, \mathcal{F}'') \longrightarrow 0,$$

and thus, by Theorem 2.22, a long exact sequence of Čech cohomology groups over the cover $\mathcal{U}$. By passing to the limit over covers, we obtain the short exact sequence

$$0 \longrightarrow C^*(X, \mathcal{F}') \longrightarrow C^*(X, \mathcal{F}) \longrightarrow C^*(X, \mathcal{F}'') \longrightarrow 0,$$

which yields a long exact sequence of Čech cohomology groups. □

Condition (ii) of Definition 11.20 is verified in a similar fashion (for preseaves).

Thus, for *presheaves*, the family of functors $(\check{H}^p(X, -))_{p \geq 0}$ is a δ-functor (and even a universal δ-functor, in view of a previous remark). The difficulty is that for sheaves, in general, it fails to be a δ-functor. If X is paracompact, Proposition 13.16 implies that the family of functors $(\check{H}^p(X, -))_{p \geq 0}$ is a δ-functor for sheaves.

Fortunately, since $\check{H}^0(X, \mathcal{F}) \cong \mathcal{F}$, by Proposition 10.34(4), the functors $\check{H}^0(X, -)$ are *left-exact on sheaves*. Given an exact sequence of sheaves

$$0 \longrightarrow \mathcal{F}' \xrightarrow{\varphi} \mathcal{F} \xrightarrow{\psi} \mathcal{F}'' \longrightarrow 0 \tag{$*$}$$

we can consider the exact sequence of presheaves

$$0 \longrightarrow \mathcal{F}' \longrightarrow \mathcal{F} \longrightarrow \mathcal{G} \longrightarrow 0$$

where $\mathcal{G} = \mathrm{PCoker}(\varphi)$, and by Proposition 13.14, we obtain a long exact sequence of cohomology whose rows

$$\longrightarrow \check{H}^p(X, \mathcal{F}') \longrightarrow \check{H}^p(X, \mathcal{F}) \longrightarrow \check{H}^p(X, \mathcal{G}) \longrightarrow \tag{$**$}$$

involve the Čech cohomology groups $\check{H}^p(X, \mathcal{F}'), \check{H}^p(X, \mathcal{F})$, and $\check{H}^p(X, \mathcal{G})$. The exactness of ($*$) means that $\mathcal{F}'' = \mathrm{SCoker}(\varphi)$, with $\mathrm{SCoker}(\varphi) = \widetilde{\mathrm{PCoker}}(\varphi)$, the sheafification of $\mathrm{PCoker}(\varphi) = \mathcal{G}$, so

$$\mathcal{F}'' = \widetilde{\mathcal{G}}.$$

Thus, if we can show that

$$\check{H}^p(X, \mathcal{G}) \cong \check{H}^p(X, \widetilde{\mathcal{G}}) \tag{$\dagger$}$$

for every presheaf $\mathcal{G}$, by replacing $\check{H}^p(X, \mathcal{G})$ by $\check{H}^p(X, \widetilde{\mathcal{G}}) = \check{H}^p(X, \mathcal{F}'')$ in $(**)$ we obtain a long exact sequence with rows

$$\longrightarrow \check{H}^p(X, \mathcal{F}') \longrightarrow \check{H}^p(X, \mathcal{F}) \longrightarrow \check{H}^p(X, \mathcal{F}'') \longrightarrow$$

which constitutes a long exact sequence (in the sense of presheaves) of cohomology associated with $(*)$, which by Proposition 10.24(iii) is also exact in the sense of sheaves, and the family $(\check{H}^p(X, -))_{p\geq 0}$ is a δ-functor. This is where the paracompactness condition comes in to save the day (see Proposition 13.16).

Definition 13.7. A space X is *paracompact* if it is Hausdorff and if every open cover has an open, locally finite, refinement. An open cover $\mathcal{U} = (U_i)_{i\in I}$ of X is *locally finite* if for every point $x \in X$, there is some open subset V containing x such that $V \cap U_i \neq \emptyset$ for only finitely many $i \in I$; see Figure 13.4.

Every metric space is paracompact and so is every locally compact and second-countable space.

Assume that X is paracompact. The key fact due to Godement is the following somewhat bizarre result which implies the crucial fact (†).

Proposition 13.15. *Assume the space X is paracompact. For any presheaf $\mathcal{F}$ on X, if $\widetilde{\mathcal{F}} = (0)$ (the sheafification of $\mathcal{F}$ is the zero sheaf), then*

$$\check{H}^p(X, \mathcal{F}) = (0), \quad \textit{for all } p \geq 0.$$

Proposition 13.15 is proven Godement [Godement (1958)] (Chapter 5, Theorem 5.10.2). Another proof can be found in Bredon [Bredon (1997)] (Chapter III, Theorem 4.4). See also Spanier [Spanier (1989)] (Chapter 6, Theorem 16). None of these proofs are particularly illuminating. The significance of Proposition 13.15 is that it implies (†).

The proof of the next proposition requires the notion of quotient of presheaves of R-modules defined below.

Definition 13.8. Given any two presheaves $\mathcal{F}$ and $\mathcal{G}$ of R-modules over a topological space X, if $\mathcal{G}$ is a subsheaf of $\mathcal{F}$ (in particular, $\mathcal{G}(U)$ is a submodule of $\mathcal{F}(U)$ for all open subsets U of X), then the *quotient presheaf* $\mathcal{F}/\mathcal{G}$ is the presheaf defined such that

$$(\mathcal{F}/\mathcal{G})(U) = \mathcal{F}(U)/\mathcal{G}(U)$$

for every open subset U of X.

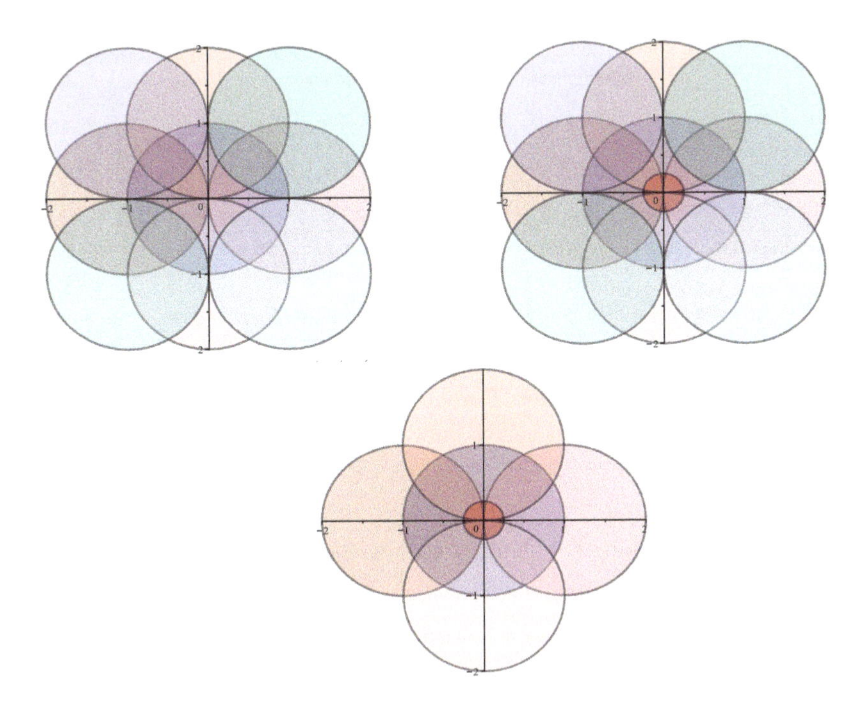

Fig. 13.4 Let X be $\mathbb{R}^2$ with $\mathcal{U}$ consisting of open unit disks centered at integer lattice points. The left figure shows $\mathcal{U}$ around the origin. Since $\mathcal{U}$ is homogenous in nature, the middle and right figures "demonstrate" the paracompactness of $\mathbb{R}^2$ by showing local finiteness at the origin. In particular if V is the open disk centered at the with radius $1/4$, V only intersects five elements of $\mathcal{U}$.

It is easily verified that $\mathcal{F}/\mathcal{G}$ is indeed a presheaf. However, if $\mathcal{F}$ and $\mathcal{G}$ are sheaves, $\mathcal{F}/\mathcal{G}$ may fail to be a sheaf. Thus for sheaves, the quotient sheaf $\mathcal{F}/\mathcal{G}$ is defined as the sheafification of the presheaf $\mathcal{F}/\mathcal{G}$.

Proposition 13.16. *Assume the space X is paracompact. For any presheaf $\mathcal{F}$ on X, we have isomorphisms*

$$\check{H}^p(X, \mathcal{F}) \cong \check{H}^p(X, \widetilde{\mathcal{F}}) \quad \textit{for all } p \geq 0.$$

Proof. We follow Godement [Godement (1958)] (Chapter 5, Page 230). Let $\eta\colon \mathcal{F} \to \widetilde{\mathcal{F}}$ be the morphism from $\mathcal{F}$ to its sheafification $\widetilde{\mathcal{F}}$ (see Definition 10.5), and let $\mathcal{K} = \operatorname{Ker} \eta$ and $\mathcal{I} = \operatorname{PIm} \eta$, as presheaves. We have the exact sequences of presheaves

$$0 \longrightarrow \mathcal{K} \longrightarrow \mathcal{F} \longrightarrow \mathcal{I} \longrightarrow 0$$

and

$$0 \longrightarrow \mathcal{I} \longrightarrow \widetilde{\mathcal{F}} \longrightarrow \widetilde{\mathcal{F}}/\mathcal{I} \longrightarrow 0.$$

Furthermore, we claim that

$$\widetilde{\mathcal{K}} = (0) \quad \text{and} \quad \widetilde{\widetilde{\mathcal{F}}/\mathcal{I}} = (0).$$

It suffices to prove that $\mathcal{K}_x = (0)$ and $(\widetilde{\mathcal{F}}/\mathcal{I})_x = (0)$ for all $x \in X$. In the first case, by definition of η, for every open subset U of X and every $s \in \mathcal{F}(U)$ we have $\eta_U(s) = \widetilde{s}$, with $\widetilde{s}(x) = s_x$ for all $x \in U$, so $s \in \operatorname{Ker} \eta_U = \mathcal{K}(U)$ iff $s_x = 0$ for all $x \in U$, which implies that $\mathcal{K}_x = (0)$.

To prove that $(\widetilde{\mathcal{F}}/\mathcal{I})_x = (0)$ we use the fact (which is not hard to prove) that for any two presheaves $\mathcal{F}$ and $\mathcal{G}$, we have $(\mathcal{F}/\mathcal{G})_x = \mathcal{F}_x/\mathcal{G}_x$. Then $(\widetilde{\mathcal{F}}/\mathcal{I})_x = \widetilde{\mathcal{F}}_x/\mathcal{I}_x$, but it is easily shown that $\mathcal{I}_x = \widetilde{\mathcal{F}}_x$ since any continuous section in $\widetilde{\mathcal{F}}(U)$ agrees locally with some section of the form $\widetilde{s} \in \mathcal{I}(V)$ for some $V \subseteq U$.

By taking the long cohomology sequence associated with the first exact sequence we obtain exact sequences

$$\check{H}^p(X, \mathcal{K}) \longrightarrow \check{H}^p(X, \mathcal{F}) \longrightarrow \check{H}^p(X, \mathcal{I}) \longrightarrow \check{H}^{p+1}(X, \mathcal{K})$$

for all $p \geq 0$, and since $\widetilde{\mathcal{K}} = (0)$, by Proposition 13.15, we have

$$\check{H}^p(X, \mathcal{K}) = \check{H}^{p+1}(X, \mathcal{K}) = (0),$$

which yields isomorphisms

$$\check{H}^p(X, \mathcal{F}) \cong \check{H}^p(X, \mathcal{I}), \quad p \geq 0.$$

Similarly, by taking the long cohomology sequence associated with the second exact sequence we obtain exact sequences

$$0 \longrightarrow \check{H}^0(X, \mathcal{I}) \longrightarrow \check{H}^0(X, \widetilde{\mathcal{F}}) \longrightarrow \check{H}^0(X, \widetilde{\mathcal{F}}/\mathcal{I})$$

and

$$\check{H}^p(X, \widetilde{\mathcal{F}}/\mathcal{I}) \longrightarrow \check{H}^{p+1}(X, \mathcal{I}) \longrightarrow \check{H}^{p+1}(X, \widetilde{\mathcal{F}}) \longrightarrow \check{H}^{p+1}(X, \widetilde{\mathcal{F}}/\mathcal{I})$$

for all $p \geq 0$, and since $\widetilde{\widetilde{\mathcal{F}}/\mathcal{I}} = (0)$, by Proposition 13.15, we have

$$\check{H}^p(X, \widetilde{\mathcal{F}}/\mathcal{I}) = \check{H}^{p+1}(X, \widetilde{\mathcal{F}}/\mathcal{I}) = (0),$$

so we obtain isomorphisms

$$\check{H}^p(X, \mathcal{I}) \cong \check{H}^p(X, \widetilde{\mathcal{F}}), \quad p \geq 0.$$

It follows that

$$\check{H}^p(X, \mathcal{F}) \cong \check{H}^p(X, \widetilde{\mathcal{F}}), \quad p \geq 0,$$

as claimed. ☐

By putting the previous results together, we proved the following important theorem.

Theorem 13.17. *Assume the space X is paracompact. For any sheaf $\mathcal{F}$ on X, we have isomorphisms*

$$\check{H}^p(X,\mathcal{F}) \cong H^p(X,\mathcal{F}) \quad \text{for all } p \geq 0$$

between Čech cohomology and sheaf cohomology. Furthermore, for every presheaf $\mathcal{F}$, we have isomorphisms

$$\check{H}^p(X,\mathcal{F}) \cong H^p(X,\widetilde{\mathcal{F}}) \quad \text{for all } p \geq 0.$$

Remark: The fact that for a paracompact space, every short exact sequence of sheaves yields a long exact sequence of cohomology is already proven in Serre's FAC [Serre (1955)] (Chapter 1, Section 25, Proposition 7).

Observe that all that is needed to prove Proposition 13.16 is the fact that for any presheaf $\mathcal{F}$, if $\widetilde{\mathcal{F}} = (0)$, then

$$\check{H}^p(X,\mathcal{F}) = (0), \quad \text{for all } p \geq 0.$$

This condition holds if X paracompact (this is the content of Proposition 13.15), but there are other situations where it holds (perhaps for specific values of p). For example, for any space X (not necessarily paracompact), it is shown in Godement ([Godement (1958)] Chapter 5, Lemma on Page 227) that for any presheaf $\mathcal{F}$, if $\widetilde{\mathcal{F}} = (0)$, then $\check{H}^0(X,\mathcal{F}) = (0)$. As a consequence, for any space X, for any sheaf $\mathcal{F}$ on X, we have isomorphisms

$$\check{H}^p(X,\mathcal{F}) \cong H^p(X,\mathcal{F}), \quad p = 0,1;$$

see Godement ([Godement (1958)] Chapter 5, Corollary of Theorem 5.9.1 on Page 227).

Grothendieck shows that the map $\check{H}^2(X,\mathcal{F}) \longrightarrow H^2(X,\mathcal{F})$ is injective and gives an example where is it not an isomorphism; see Tohoku [Grothendieck (1957)] (Section 3.8, Example, Pages 177–179).

We now briefly discuss conditions not involving the space X but instead the cover $\mathcal{U}$ that yield isomorphisms between the Čech cohomology groups $\check{H}^p(\mathcal{U},\mathcal{F})$ and the sheaf cohomology groups $H^p(X,\mathcal{F})$.

First we state a result due to Leray involving the vanishing of certain sheaf cohomology groups on various open sets.

Theorem 13.18. *(Leray) For any topological space X and any sheaf $\mathcal{F}$ on X, for any open cover $\mathcal{U}$, if $H^p(U_{i_0\cdots i_p}, \mathcal{F}) = (0)$ for all $p > 0$ and all $(i_0, \ldots, i_p)$, then*

$$\check{H}^p(X, \mathcal{F}) \cong H^p(X, \mathcal{F}), \quad \textit{for all} \quad p \geq 0.$$

A proof of Theorem 13.18 can be found in Bredon [Bredon (1997)] (Chapter III, Theorem 4.13). The proof involves a double complex. Leray's theorem is used in algebraic geometry where X is a scheme and $\mathcal{F}$ is a quasi-coherent sheaf; see Hartshorne [Hartshorne (1977)] (Chapter III, Section 4, Theorem 4.5), and EGA III [Grothendieck and Dieudonné (1961)] (1.4.1).

Next we state a result due to Henri Cartan involving the vanishing of certain Čech cohomology groups on various open sets.

Theorem 13.19. *(H. Cartan) For any topological space X and any sheaf $\mathcal{F}$ on X, for any open cover $\mathcal{U}$, if $\mathcal{U}$ is a basis for the topology of X closed under finite intersections and if $\check{H}^p(U_{i_0\cdots i_p}, \mathcal{F}) = (0)$ for all $p > 0$ and all $(i_0, \ldots, i_p)$, then*

$$\check{H}^p(X, \mathcal{F}) \cong H^p(X, \mathcal{F}), \quad \textit{for all} \quad p \geq 0.$$

A proof of Theorem 13.19 is given in Grothendieck [Grothendieck (1957)] (Section 3.8, Corollary 4), and in more details in Godement [Godement (1958)] (Chapter 5, Theorem 5.9.2).

We now compare singular cohomology and sheaf cohomology (for constant sheaves). To do so, we will need to introduce soft sheaves and fine sheaves.

13.4 Singular Cohomology and Sheaf Cohomology

If R is a commutative ring with an identity element and G is an R-module, how can we relate the singular cohomology groups $H^p(X; G)$ to some sheaf cohomology groups? The answer is to consider the cohomology groups $H^p(X, \widetilde{G}_X)$ of the constant sheaf $\widetilde{G}_X$ (the sheafification of the constant presheaf G; see Example 8.2(1)). The key idea is to consider some suitable resolution of $\widetilde{G}_X$ by acyclic sheaves such that the complex obtained by applying the global section functor to this resolution yields the singular cohomology groups, and to apply Proposition 11.34 to conclude that we have isomorphisms $H^p(X;G) \cong H^p(X, \widetilde{G}_X)$, provided some mild assumptions on X.

The natural candidate for the sheaves involved in a resolution of $\widetilde{G}_X$ are the presheaves $S^p(-;G)$ given by

$$U \mapsto S^p(U;G),$$

where $S^p(U;G)$ is the R-module of singular cochains on the open subset U, as defined in Definition 4.27, replacing X by U.

The first problem is that the presheaves $S^p(-;G)$ satisfy Axiom (G), but in general fail to satisfy Axiom (M). To fix this problem we consider the sheafification $\mathcal{S}^p(-;G)$ of $S^p(-,G)$ (see Definition 10.7 and Proposition 10.8). The coboundary maps $\delta^p\colon S^p(U;G) \to S^{p+1}(U;G)$ induce maps $\delta^p\colon \mathcal{S}^p(-;G) \to \mathcal{S}^{p+1}(-;G)$, where we wrote δ instead of $\widetilde{\delta}$ to simplify the notation. Then we obtain a complex

$$0 \longrightarrow \widetilde{G}_X \longrightarrow \mathcal{S}^0(-;G) \xrightarrow{\delta} \mathcal{S}^1(-;G) \xrightarrow{\delta} \mathcal{S}^2(-;G) \xrightarrow{\delta} \cdots . \tag{$*$}$$

When is this a resolution of $\widetilde{G}_X$ and when are the sheaves $\mathcal{S}^p(-;G)$ acyclic?

It turns out that if X is locally Euclidean, then the complex $(*)$ is exact; that is, a resolution. There is a more general condition implying that the complex $(*)$ is a resolution, namely that X is an HLC-space (X is homologically locally connected). Any locally contractible space, any manifold, or any CW-complex is HLC; for details, see Bredon [Bredon (1997)] (Chapter II, Section 1). For our purposes, it suffices to assume that X is a topological manifold. The proof that the complex $(*)$ is a resolution if M is a topological manifold can be found in Warner [Warner (1983)] (Chapter V, Section 5.31). It is very technical.

Furthermore, if X is paracompact, then the sheaves $\mathcal{S}^p(-;G)$ are acyclic. These sheaves are generally not flasque but they are soft sheaves. In fact, fine sheaves and soft sheaves are acyclic; we will see this in the next section. By Proposition 11.34, if we apply the global section functor $\Gamma(X,-)$ to the resolution $(*)$, we obtain the complex $\mathcal{S}^*(X;G)$ (of modules)

$$0 \longrightarrow \mathcal{S}^0(X;G) \xrightarrow{\delta} \mathcal{S}^1(X;G) \xrightarrow{\delta} \mathcal{S}^2(X;G) \xrightarrow{\delta} \cdots$$

whose cohomology is isomorphic to the sheaf cohomology $H^*(X, \widetilde{G}_X)$.

However, there is a new problem: the cohomology groups of the complex $\mathcal{S}^*(X;G)$ involve the modules $\mathcal{S}^p(X;G)$, but the singular cohomology groups involve the modules $S^p(X;G)$; how do we know that these groups are isomorphic? They are indeed isomorphic if X is paracompact.

Let us settle this point before dealing with soft sheaves. Recall that we are only considering presheaves and sheaves of R-modules. Assume that X is paracompact. If $\mathcal{F}$ is a presheaf on X and if $\widetilde{\mathcal{F}}$ is its sheafification, the natural map $\eta\colon \mathcal{F} \to \widetilde{\mathcal{F}}$ induces the map $\eta\colon \mathcal{F}(X) \to \widetilde{\mathcal{F}}(X)$ given by $\eta = \eta_X$ as in Definition 10.5; that is, for every $s \in \mathcal{F}(X)$,

$$\eta(s) = \widetilde{s}$$

with $\widetilde{s}(x) = s_x$ for all $x \in X$. Define the presheaf $\mathcal{F}(X)_0$ by

$$\mathcal{F}(X)_0 = \{s \in \mathcal{F}(X) \mid \eta(s) = 0\} = \operatorname{Ker} \eta.$$

Then we have the following result.

Proposition 13.20. *Assume that the space X is paracompact. For every presheaf $\mathcal{F}$, if $\mathcal{F}$ satisfies Condition (G), then the sequence*

$$0 \longrightarrow \mathcal{F}(X)_0 \longrightarrow \mathcal{F}(X) \xrightarrow{\ \theta\ } \widetilde{\mathcal{F}}(X) \longrightarrow 0$$

is exact.

The only thing that needs to be proven is that θ is surjective. This is proven in Warner [Warner (1983)] (Chapter V, Proposition 5.27) and in Bredon [Bredon (1997)] (Chapter I, Theorem 6.2). The proof relies heavily on the existence of a locally finite open cover (this is where paracompactness is used).

As a consequence of Proposition 13.20, we have an exact sequence of cochain complexes

$$0 \longrightarrow S^*(X;G)_0 \longrightarrow S^*(X;G) \longrightarrow \mathcal{S}^*(X;G) \longrightarrow 0. \qquad (\dagger)$$

We claim that if we can prove that

$$H^p(S^*(X;G)_0) = (0) \quad \text{for all } p \geq 0,$$

then we have isomorphisms

$$H^p(X;G) = H^p(S^*(X;G)) \cong H^p(\mathcal{S}^*(X;G)), \quad \text{for all } p \geq 0.$$

Proof. This follows easily by taking the long exact sequence of cohomology associated with the exact sequence (†). We have exact sequences

$$H^p(S^*(X;G)_0) \longrightarrow H^p(X;G) \longrightarrow H^p(\mathcal{S}^*(X;G)) \longrightarrow H^{p+1}(S^*(X;G)_0)$$

for all $p \geq 0$, and since by hypothesis $H^p(S^*(X;G)_0) = H^{p+1}(S^*(X;G)_0) = (0)$, we obtain the isomorphisms

$$H^p(X;G) = H^p(S^*(X;G)) \cong H^p(\mathcal{S}^*(X;G)), \quad \text{for all } p \geq 0,$$

as claimed. □

Now it is shown in Warner [Warner (1983)] (Chapter 5, Section 5.32) that indeed

$$H^p(S^*(X;G)_0) = (0) \quad \text{for all } p \geq 0.$$

This is a very technical argument involving barycentric subdivision and a bit of topology (but does not require X to be paracompact).

In summary, we have shown that if X is paracompact and a topological manifold, provided that the sheaves $\mathcal{S}^p(-;G)$ are acyclic, then we have isomorphisms

$$H^p(X;G) \cong H^p(X, \widetilde{G}_X), \quad \text{for all } p \geq 0$$

between singular cohomology and sheaf cohomology of the constant sheaf $\widetilde{G}_X$.

The sheaves $\mathcal{S}^p(-;G)$ are indeed acyclic because they are soft, and soft sheaves over a paracompact space are acyclic; this will be proven in Section 13.5. Assuming that this result has been proved, we have the following theorem showing the equivalence of singular cohomology and sheaf cohomology for the constant sheaf $\widetilde{G}_X$ and a (paracompact) topological manifold X.

Theorem 13.21. *Assume X is a paracompact topological manifold. For any R-module G, there are isomorphisms*

$$H^p(X;G) \cong H^p(X, \widetilde{G}_X), \quad \textit{for all } p \geq 0$$

between singular cohomology and sheaf cohomology of the constant sheaf $\widetilde{G}_X$.

Remark: There is a variant of singular cohomology that uses differentiable singular simplices instead of singular simplices as defined in Definition 4.2. Given a topological space X, if $p \geq 1$, a *differentiable singular p-simplex* is any map $\sigma\colon \Delta^p \to X$ that can be extended to a smooth map of a neighborhood of Δ^p. Then $S^p_\infty(U; G)$ denotes the R-module of functions which assign to each differentiable singular p-simplex an element of G (for $p \geq 1$), and $S^0_\infty(U; G) = S^0(X; G)$. Elements of $S^p_\infty(U; G)$ are called *differentiable singular p-cochains*. Then we obtain the cochain complex $S^*_\infty(X; G)$ and its cohomology groups denoted $H^p_{\Delta^\infty}(X; G)$ are called the *differentiable singular cohomology groups* of X with coefficients in G. Each $S^p_\infty(-; G)$ is a presheaf satisfying Condition (M), and we let $\mathcal{S}^p_\infty(-; G)$ be its sheafification. As in the continuous case, we obtain a version of Theorem 13.21.

Theorem 13.22. *Assume X is a paracompact topological manifold. For any R-module G, there are isomorphisms*

$$H^p_{\Delta^\infty}(X; G) \cong H^p(X, \widetilde{G}_X), \quad \textit{for all } p \geq 0$$

between differentiable singular cohomology and sheaf cohomology of the constant sheaf $\widetilde{G}_X$.

Details can be found in Warner [Warner (1983)] (Chapter 5, Sections 5.31, 5.32). The significance of differentiable singular cohomology is that it yields a stronger version of the equivalence with de Rham cohomology when $G = \mathbb{R}$ and X is a smooth manifold; see Section 13.7.

13.5 Soft Sheaves

Roughly speaking a sheaf is soft if it satisfies the condition for being flasque for closed subsets of X; that is, for every closed subset A of X, the restriction map from $\mathcal{F}(X)$ to $\mathcal{F}(A)$ is surjective. The problem is that sheaves are only defined over open subsets!

The remedy is to work with stalk spaces (E, p). Before proceeding the reader may want to review Sections 10.3 and 10.4. Since every sheaf $\mathcal{F}$ is isomorphic to the sheaf of sections $\widetilde{\mathcal{F}}$ associated with the stalk space $(S\mathcal{F}, \pi)$, this is not a problem, although at times it is a little awkward.

Definition 13.9. If (E, p) is a stalk space of R-modules on X with $p\colon E \to X$, and $\Gamma[E, p]$ is the sheaf of continuous sections associated with (E, p) (see Example 8.2(1)), following Godement [Godement (1958)] (Chapter 1, bottom of Page 110), for every subset Y of X (not necessarily open) we define

$$\Gamma(Y, \Gamma[E,p]) = \{s \colon Y \to E \mid p \circ s = \mathrm{id} \text{ and } s \text{ is continuous}\}$$

as the set of all continuous sections from Y viewed as a subspace of X; see Figure 13.5.

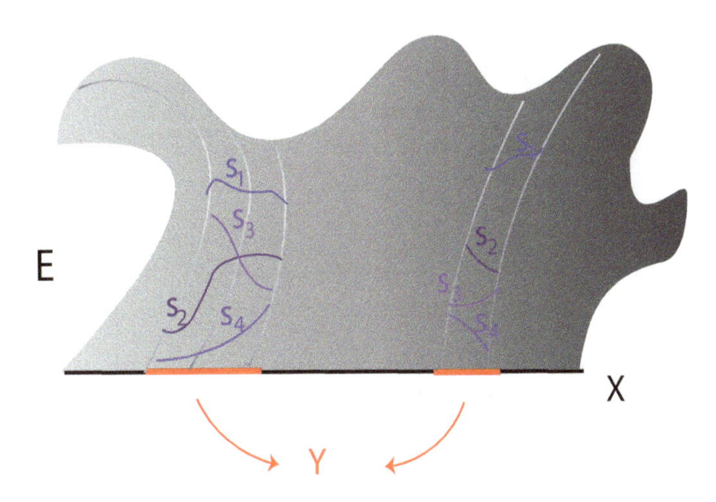

Fig. 13.5 A schematic representation of four sections of $\Gamma(Y, \Gamma[E,p])$.

We usually abuse notation a little and denote the sheaf $\Gamma[E,p]$ associated with the stalk space (E,p) by $\mathcal{F}$. We write $\Gamma(Y,\mathcal{F})$ for $\Gamma(Y,\Gamma[E,p])$. Then we can make the following definition.

Definition 13.10. If $\mathcal{F}$ is the sheaf induced by a stalk space (E,p) of R-modules on X, we say that the sheaf $\mathcal{F}$ is *soft* if the restriction map from $\Gamma(X,\mathcal{F})$ to $\Gamma(A,\mathcal{F})$ is surjective for every closed subset A of X.

In order to prove that soft sheaves are acyclic, which is one of our main goals, we need to assume that X is paracompact. Then we will see that every flasque sheaf is soft.

Given a sheaf $\mathcal{F}$ and its sheafification $\widetilde{\mathcal{F}}$, the sheaf isomorphism $\eta \colon \mathcal{F} \to \widetilde{\mathcal{F}}$ ensures that $\mathcal{F}$ is flasque iff $\widetilde{\mathcal{F}}$ is flasque, so there is no problem.

In this section we will content ourselves with stating the properties of soft sheaves that are needed to finish the proof of the equivalence of singular cohomology and sheaf cohomology (for the constant sheaves $\widetilde{G}_X$), and the proof of the equivalence of de Rham cohomology and sheaf cohomology (for the constant sheaves $\widetilde{\mathbb{R}}_X$). Details and proofs can be found in Bredon [Bredon (1997)] (Chapter II, Section 9) and Godement [Godement (1958)] (Chapters 3, 4, 5). Soft sheaves are also discussed in Brylinski [Brylinski (1993)] (Chapter I, Section 1,4), but a different definition is used.

Proposition 13.23. *Let $\mathcal{F}$ be the sheaf induced by a stalk space (E, p) of R-modules over a space X, let Y be any subset of X and let $s \in \Gamma(Y, \mathcal{F})$ be any section over Y. If Y admits a fundamental system of paracompact neighborhoods,[1] then s has an extension to some open neighborhood of Y in X.*

Proposition 13.23 is proven in Godement [Godement (1958)] (Chapter III, Theorem 3.3.1) and Bredon [Bredon (1997)] (Chapter I, Theorem 9.5). As an immediate corollary we obtain the following result.

Proposition 13.24. *Let $\mathcal{F}$ be the sheaf induced by a stalk space (E, p) of R-modules over a space X. If X is paracompact and $\mathcal{F}$ is flasque, then $\mathcal{F}$ is soft.*

Recall that we are only considering presheaves and sheaves of R-modules. To prove that soft sheaves on a paracompact space are acyclic, we need the following two propositions.

Proposition 13.25. *If X is paracompact, for any exact sequence of sheaves (induced by stalk spaces)*

$$0 \longrightarrow \mathcal{F}' \longrightarrow \mathcal{F} \longrightarrow \mathcal{F}'' \longrightarrow 0,$$

if $\mathcal{F}'$ is soft, then the sequence

$$0 \longrightarrow \Gamma(X, \mathcal{F}') \longrightarrow \Gamma(X, \mathcal{F}) \longrightarrow \Gamma(X, \mathcal{F}'') \longrightarrow 0,$$

is exact.

A proof of Proposition 13.25 is given in Bredon [Bredon (1997)] (Chapter II, Theorem 9.9); see also Godement [Godement (1958)] (Chapter 3, Theorem 3.5.2). The proof uses Zorn's lemma and is fairly involved.

[1] This means that there is a family $\mathcal{N}$ of paracompact neighborhoods of Y such that for every neighborhood V of Y there is some W in $\mathcal{N}$ such that $W \subseteq V$.

Proposition 13.26. *If X is paracompact, for any exact sequence of sheaves (induced by stalk spaces)*

$$0 \longrightarrow \mathcal{F}' \longrightarrow \mathcal{F} \longrightarrow \mathcal{F}'' \longrightarrow 0,$$

if $\mathcal{F}'$ and $\mathcal{F}$ are soft, then $\mathcal{F}''$ is also soft.

A proof of Proposition 13.26 is given in Bredon [Bredon (1997)] (Chapter II, Theorem 9.10); see also Godement [Godement (1958)] (Chapter 3, Theorem 3.5.3). The proof is analogous to the proof given for flasque sheaves in Theorem 13.5.

It is also easy to see that every direct factor of a soft sheaf is soft; the proof is analogous to the proof given for flasque sheaves in Theorem 13.5 with closed subsets playing the role of open subsets. But now (as in the case of flasque sheaves) the assumptions of Proposition 13.6 apply, and we immediately get the following result.

Proposition 13.27. *For any sheaf $\mathcal{F}$ induced by a stalk space (E, p), if X is paracompact and $\mathcal{F}$ is soft, then $\mathcal{F}$ is acyclic, that is*

$$H^p(X, \mathcal{F}) = (0) \quad \textit{for all } p \geq 1.$$

Neither Godement nor Bredon have Proposition 13.6 from Tohoku at their disposal, so they need to prove Proposition 13.27; see Godement [Godement (1958)] (Chapter 4, Theorem 4.4.3) and Bredon [Bredon (1997)] (Chapter II, Theorem 9.11).

Going back to singular cohomology, it remains to prove that the sheaves $\mathcal{S}^p(X; G)$ are soft.

Proposition 13.28. *If the space X is paracompact, then the sheaves (of singular cochains) $\mathcal{S}^p(X; G)$ are soft.*

A proof of Proposition 13.28 is given in Godement [Godement (1958)] (Chapter 3, Section 3.9, Example 3.9.1).

Propositions 13.27 and 13.28 conclude the proof of Theorem 13.21.

13.6 Fine Sheaves

Another way to prove Proposition 13.28 is to prove that the sheaves $\mathcal{S}^p(X;G)$ are fine and that fine sheaves are soft. Fine sheaves will also be needed in Section 13.7.

Definition 13.11. If $\mathcal{F}$ is the sheaf induced by a stalk space (E,p) where $p\colon E \to X$ is a continuous surjection, for any subset Y of X, the sheaf $\mathcal{F}|Y$ is the sheaf of continuous sections of the stalk space $(p^{-1}(Y), p|p^{-1}(Y))$, where Y is endowed with the subspace topology.

Observe that Definition 13.11 specifies what is the restriction of a sheaf $\mathcal{F}$ induced by a stalk space (E,p) with projection $p\colon E \to X$ to a subset Y of X, whose sections are continuous functions over open subsets of Y endowed with the subspace topology, but Definition 13.9 defines sections of $\mathcal{F}$ over the *fixed* subset Y. Recall that we are only considering presheaves and sheaves of R-modules.

Definition 13.12. Given two sheaves $\mathcal{F}$ and $\mathcal{G}$ induced by stalk spaces over the same space X, we have a definition of the presheaf $\mathcal{H}om(\mathcal{F},\mathcal{G})$ analogous to Definition 8.7:

$$\mathcal{H}om(\mathcal{F},\mathcal{G})(U) = \mathrm{Hom}(\mathcal{F}|U, \mathcal{G}|U)$$

for every open subset U of X, where $\mathrm{Hom}(\mathcal{F}|U, \mathcal{G}|U)$ denotes the set of maps between the sheaves $\mathcal{F}|U$ and $\mathcal{G}|U$.

Even though $\mathcal{H}om(\mathcal{F},\mathcal{G})$ is a sheaf if $\mathcal{F}$ and $\mathcal{G}$ are sheaves induced by stalk spaces, because we need to work with stalk spaces when dealing with soft sheaves, with some abuse of notation, we also denote the sheafification of the above presheaf by $\mathcal{H}om(\mathcal{F},\mathcal{G})$. Then we have the following definition due to Godement [Godement (1958)] (Chapter 3, Section 3.7).

Definition 13.13. For any sheaf $\mathcal{F}$ (of R-modules) on X induced by the stalk space (E,p), we say that $\mathcal{F}$ is *fine* if $\mathcal{H}om(\mathcal{F},\mathcal{F})$ is soft.

The following results about fine and soft sheaves are proven in Godement [Godement (1958)] (Chapter 3, Section 3.7) and in Bredon [Bredon (1997)] (Chapter II, Section 9).

Proposition 13.29. *Assume the space X is paracompact. If $\mathcal{O}_X$ is any sheaf of rings with unit induced by a stalk space and if $\mathcal{O}_X$ is soft, then any $\mathcal{O}_X$-module is soft.*

This is Theorem 3.7.1 in Godement [Godement (1958)].

Proposition 13.30. *Assume the space X is paracompact. If $\mathcal{O}_X$ is sheaf of rings with unit induced by a stalk space, then $\mathcal{O}_X$ is soft iff every $x \in X$ has some open neighborhood U such that for any two disjoint open subsets S, T contained in U, there is some section $s \in \mathcal{O}_X(U)$ such that $s \equiv 1$ on S and $s \equiv 0$ on T.*

This is Theorem 3.7.2 in Godement [Godement (1958)]. The proof uses Urysohn's theorem and a local characterization of soft sheaves, namely Theorem 3.4.1 in Godement [Godement (1958)]. We omitted Theorem 3.4.1 because of its technical nature (its proof uses Zorn's lemma).

Proposition 13.31. *Assume the space X is paracompact. A sheaf $\mathcal{F}$ (of R-modules) induced by a stalk space (E, p) is fine iff for any two disjoint open subsets S, T in X, there is a sheaf homomorphism $\varphi\colon \mathcal{F} \to \mathcal{F}$ such that $\varphi \equiv 1$ in a neighborhood of S and $\varphi \equiv 0$ in a neighborhood of T. Every fine sheaf is soft.*

See Godement [Godement (1958)] (Section 3.7, Page 157) and Bredon [Bredon (1997)] (Chapter II, Theorem 9.16). Since every soft sheaf is acyclic, so is every fine sheaf (over a paracompact space).

Remark: If X is paracompact, then any injective sheaf on X is fine; see Bredon [Bredon (1997)] (Chapter II, Exercise 17). The following diagram summarizes the relationships between injective, flasque, fine, and soft sheaves (assuming that X is paracompact):

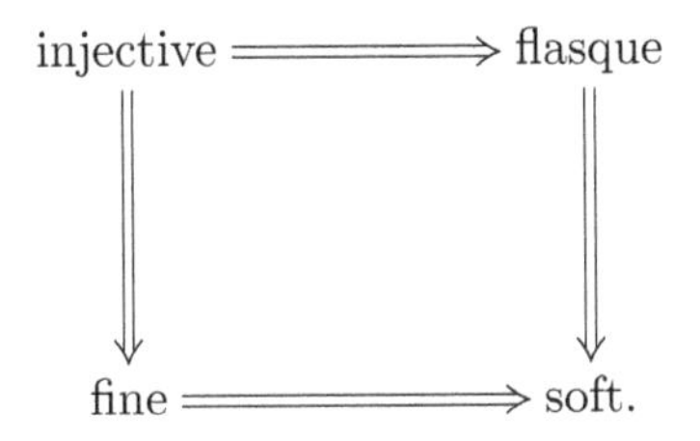

Godement proves that the sheaves $\mathcal{S}^p(-; G)$ are fine (Godement, Example 3.7.1, Page 161); see also Bredon [Bredon (1997)] (Chapter III, Page 180).

Besides being acyclic, fine sheaves behave well with respect to tensor products, which, historically motivated their introduction.

Definition 13.14. Given two sheaves $\mathcal{F}$ and $\mathcal{G}$ of R-modules, the presheaf $\mathcal{F} \otimes \mathcal{G}$ is defined by

$$(\mathcal{F} \otimes \mathcal{G})(U) = \mathcal{F}(U) \otimes \mathcal{G}(U)$$

for any open subset U of X.

Actually, the presheaf $\mathcal{F} \otimes \mathcal{G}$ is a sheaf. If $\mathcal{F}$ and $\mathcal{G}$ are induced by stalk spaces of R-modules, with a minor abuse of notation we let $\mathcal{F} \otimes \mathcal{G}$ be the sheafification of the above sheaf.

Proposition 13.32. *Assume the space X is paracompact. For any fine sheaf $\mathcal{F}$ and any sheaf $\mathcal{G}$ induced by stalk spaces on X, the sheaf $\mathcal{F} \otimes \mathcal{G}$ is fine.*

Proposition 13.32 is proven in Godement [Godement (1958)] (Chapter 3, Theorem 3.7.3), Bredon [Bredon (1997)] (Chapter II, Corollary 9.18), and Warner [Warner (1983)] (Chapter V, Section 5.10).

Proposition 13.32 can used to create resolutions. Indeed, suppose that we have a resolution

$$0 \longrightarrow \widetilde{R}_X \longrightarrow \mathcal{C}^0 \longrightarrow \mathcal{C}^1 \longrightarrow \mathcal{C}^2 \longrightarrow \cdots$$

of the locally constant sheaf $\widetilde{R}_X$ by fine and torsion-free sheaves $\mathcal{C}^p$ (which means that each stalk $\mathcal{C}^p_x$ is a torsion-free R-module, where by stalk we mean the fibre over $x \in X$ in the stalk space defining $\mathcal{C}^p$). Then it can be shown that for any sheaf $\mathcal{F}$ of R-modules, the complex

$$0 \longrightarrow \widetilde{R}_X \otimes \mathcal{F} \longrightarrow \mathcal{C}^0 \otimes \mathcal{F} \longrightarrow \mathcal{C}^1 \otimes \mathcal{F} \longrightarrow \mathcal{C}^2 \otimes \mathcal{F} \longrightarrow \cdots \quad (*)$$

is a resolution of $\mathcal{F} \cong \widetilde{R}_X \otimes \mathcal{F}$ by fine sheaves; see Warner [Warner (1983)] (Chapter V, Section 5.10), Theorem 5.15). Furthermore, if X is paracompact and if the ring R is a PID, resolutions of $\widetilde{R}_X$ by fine and torsion-free sheaves do exist; for example, the sheaves $\mathcal{S}^p(X; R)$ of singular cochains are fine and torsion-free; see Warner [Warner (1983)] (Chapter V, Section 5.31).

Thus, if X is paracompact and if R is a PID, we can define the sheaf cohomology groups $H^p(X, \mathcal{F})$ in terms of the resolution $(*)$ as

$$H^p(X, \mathcal{F}) = H^p(\Gamma(\mathcal{C}^* \otimes \mathcal{F})).$$

Since fine sheaves are acyclic, it follows that these groups are independent of the fine and torsion-free resolution of $\widetilde{R}_X$ chosen.

This method to define sheaf cohomology in terms of resolutions of fine sheaves is due to Henri Cartan and is presented in Chapter V of Warner [Warner (1983)]. It is also the approach used by Bredon [Bredon (1997)].

The advantage of this method is that it does not require the machinery of derived functors. The disadvantage is that it relies on fine sheaves, and thus on paracompactness, and assumes that the ring R is a PID. This makes it unsuitable for more general spaces and sheaves that arise naturally in algebraic geometry.

Fine sheaves are often defined in terms of partitions of unity, as in Warner [Warner (1983)] (Chapter V, Definition 5.10) or Spanier [Spanier (1989)] (Chapter 6, Section 8). Given a sheaf $\mathcal{F}$ induced by a stalk space (E, p), the *support* of a map $\varphi\colon \mathcal{F} \to \mathcal{F}$, denoted by $\mathrm{supp}(\varphi)$, is the closure of the set of elements $x \in X$ such that $\varphi(x)|\mathcal{F}_x \neq 0$ (where $\mathcal{F}_x = p^{-1}(x)$ denotes the stalk of $\mathcal{F}$ at x).

Definition 13.15. Given a sheaf $\mathcal{F}$ induced by a stalk space of rings (E, p) over X, we say that $\mathcal{F}$ is *p-fine* if for each locally finite open cover $\mathcal{U} = (U_i)_{i\in I}$ of X, for each $i \in I$ there is some map $\varphi_i\colon \mathcal{F} \to \mathcal{F}$ such that

(a) $\mathrm{supp}(\varphi_i) \subseteq U_i$.
(b) $\sum \varphi_i = \mathrm{id}$.

This sum makes sense because $\mathcal{U}$ is locally finite.

The family $(\varphi_i)_{i\in I}$ is called a *partition of unity for $\mathcal{F}$ subordinate to the cover $\mathcal{U}$*.

Then if X is paracompact, using a partition of unity, it is not hard to show to the sheaves $\mathcal{S}^p(-;G)$ and $\mathcal{S}^p_\infty(-;G)$ are p-fine; see Warner [Warner (1983)] (Chapter V, Sections 5.31 and 5.32, Pages 193–196).

It is not obvious that on a paracompact space, a sheaf is fine iff it is p-fine. It is shown in Brylinski [Brylinski (1993)] (Chapter 1, Proposition 1.4.9) that a p-fine sheaf is soft. It is shown in Warner that a p-fine sheaf is acyclic; see [Warner (1983)] (Chapter V, Section 5.20, Page 179). Therefore, *both fine sheaves and p-fine sheaves are acyclic*. It is also claimed in Exercise 13 in Bredon ([Bredon (1997)], Chapter II, Page 170) that Definition 13.13 is equivalent to Definition 13.15 for a paracompact space; thus, a sheaf is fine iff it is p-fine; see Figure 13.6.

Remark: There is a slight generalization of the various cohomology theories involving "families of support." A *family of support* on X is a family

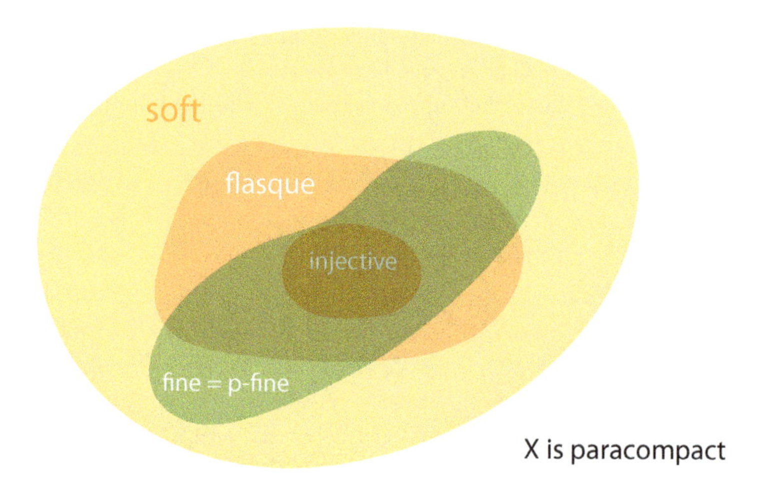

Fig. 13.6 Sheaf containment diagram for X paracompact.

Φ of closed subsets of X satisfying certain closure properties. Interesting families of support are also *paracompactifying*; see Godement [Godement (1958)] (Chapter 3, Section 3.2). Then given a sheaf $\mathcal{F}$ induced by a stalk space, for any section $s \in \Gamma(X, \mathcal{F})$, the *support* $|s|$ of s is the closed set of $x \in X$ such that $s(x) \neq 0$. We define Γ_Φ by

$$\Gamma_\Phi(X, \mathcal{F}) = \{s \in \Gamma(X, \mathcal{F}) \mid |s| \in \Phi\}.$$

Then we can define the cohomology groups $H^p_\Phi(X, \mathcal{F})$ by considering the (left-exact) functor Γ_Φ instead of Γ. We can also define Φ-soft and Φ-fine sheaves, and the results that we have presented generalize to paracompactifying families of support Φ. For details on this approach, see Godement [Godement (1958)] and Bredon [Bredon (1997)].

Another example of a p-fine sheaf is the sheaf $\mathcal{A}^p_X$ of differential forms on a smooth manifold X. Here, since we have to use stalk spaces, we are really dealing with the sheafification of the sheaf of differential forms, but we will use the same notation. This will allow us to finish the discussion of the comparison between the de Rham cohomology and sheaf cohomology started with Proposition 11.37.

13.7 de Rham Cohomology and Sheaf Cohomology

Let X be a smooth manifold. Recall that we proved in Proposition 11.37 that the sequence

$$0 \longrightarrow \widetilde{\mathbb{R}}_X \xrightarrow{\epsilon} \mathcal{A}_X^0 \xrightarrow{d} \mathcal{A}_X^1 \xrightarrow{d} \cdots \xrightarrow{d} \mathcal{A}_X^p \xrightarrow{d} \mathcal{A}_X^{p+1} \xrightarrow{d} \cdots$$

is a resolution of the locally constant sheaf $\widetilde{\mathbb{R}}_X$. As we stated in the previous section, we have the following result.

Proposition 13.33. *For any (paracompact) smooth manifold X, the sheaves $\mathcal{A}_X^p$ (actually, the sheafifications of the sheaves $\mathcal{A}_X^p$) are p-fine and fine sheaves.*

That the $\mathcal{A}_X^p$ are fine sheaves is proven in Godement [Godement (1958)] (Chapter 3, Example 3.7.1, Page 158). That the $\mathcal{A}_X^p$ are p-fine sheaves is proven in Warner [Warner (1983)] (Chapter V, Section 5.28) and Brylinski [Brylinski (1993)] (Section 1.4, Page 139). Since fine sheaves and p-fine sheaves are equivalent and thus acyclic, by Proposition 11.34 the sheaf cohomology groups of the sheaf $\widetilde{\mathbb{R}}_X$ are computed by the resolution of fine (and p-fine) sheaves

$$0 \longrightarrow \widetilde{\mathbb{R}}_X \xrightarrow{\epsilon} \mathcal{A}_X^0 \xrightarrow{d} \mathcal{A}_X^1 \xrightarrow{d} \cdots \xrightarrow{d} \mathcal{A}_X^p \xrightarrow{d} \mathcal{A}_X^{p+1} \xrightarrow{d} \cdots$$

Thus, in view of Theorem 13.17 and Theorem 13.21, we obtain the following version of the *de Rham theorem*:

Theorem 13.34. *Let X be a (paracompact) smooth manifold. There are isomorphisms*

$$H^p_{\mathrm{dR}}(X) \cong H^p(X, \widetilde{\mathbb{R}}_X) \cong \check{H}^p(X, \widetilde{\mathbb{R}}_X) \cong H^p(X; \mathbb{R})$$

between de Rham cohomology, the sheaf cohomology of the locally constant sheaf $\widetilde{\mathbb{R}}_X$, Čech cohomology of $\widetilde{\mathbb{R}}_X$, and singular cohomology over $\mathbb{R}$.

Theorem 13.22 also yields an isomorphism

$$H^p_{\mathrm{dR}}(X) \cong H^p_{\Delta^\infty}(X; \mathbb{R})$$

between de Rham cohomology and differentiable singular cohomology with coefficients in $\mathbb{R}$. It is possible to give a more explicit definition of the above isomorphism using integration.

For any $p \geq 1$, define the map $k_p \colon \mathcal{A}^p(X) \to S^p_\infty(X; \mathbb{R})$ by

$$k_p(\omega)(\sigma) = \int_\sigma \omega,$$

for any p-form $\omega \in \mathcal{A}^p(X)$ and any differentiable singular p-simplex σ in X. Using Stokes' theorem, it can be shown that the k_p induce a cochain map

$$k\colon \mathcal{A}^*(X) \to S^*_\infty(X;\mathbb{R}).$$

The above map induces a map of cohomology, and a strong version of the de Rham theorem is this:

Theorem 13.35. *For any smooth manifold X, the cochain map $k\colon \mathcal{A}^*(X) \to S^*_\infty(X;\mathbb{R})$ induces an isomorphism*

$$k^*_p\colon H^p_{\mathrm{dR}}(X) \to H^p_{\Delta^\infty}(X;\mathbb{R})$$

for every $p \geq 0$, between de Rham cohomology and differentiable singular cohomology.

For details, see Warner [Warner (1983)] (Chapter 5, Sections 5.35–5.37). Chapter 5 of Warner also contains a treatment of the multiplicative structure of cohomology.

There is yet another cohomology theory, *Alexander–Spanier cohomology*. It turns out to be equivalent to Čech cohomology, but it occurs naturally in a version of duality called Alexander–Lefschetz duality.

Alexander–Spanier cohomology is discussed extensively in Warner [Warner (1983)] (Chapter V), Bredon [Bredon (1997)] (Chapters I, II, III), and Spanier [Spanier (1989)] (Chapter 6).

13.8 Alexander–Spanier Cohomology and Sheaf Cohomology

Let X be a paracompact space and let G be an R-module.

Definition 13.16. For any open subset U of X, for any $p \geq 0$, let $A^p(U;G)$ denote the R-module of all functions $f\colon U^{p+1} \to G$. The homomorphism

$$\delta^p\colon A^p(U;G) \to A^{p+1}(U;G)$$

is defined by

$$(\delta^p f)(x_0,\ldots,x_{p+1}) = \sum_{i=0}^{p+1}(-1)^i f(x_0,\ldots,\widehat{x_i},\ldots,x_{p+1}),$$

for all $f \in A^p(U;G)$ and all $(x_0,\ldots,x_{p+1}) \in U^{p+2}$.

It is easily checked that $\delta^{p+1} \circ \delta^p = 0$ for all $p \geq 0$, so we obtain a cochain complex

$$0 \longrightarrow A^0(U;G) \xrightarrow{\delta^0} A^1(U;G) \xrightarrow{\delta^1} A^2(U;G) \xrightarrow{\delta^2} \cdots$$

denoted by $A^*(U;G)$. If $V \subseteq U$, then there is a restriction homomorphism

$$\rho^U_V \colon A^p(U;G) \to A^p(V;G),$$

so we obtain a presheaf $A^p(-;G)$ of R-modules called the *presheaf of Alexander–Spanier p-cochains*. The presheaf $A^p(-;G)$ satisfies Condition (G) for $p \geq 1$ but not Condition (M).

Let $\mathcal{A}^p_{\text{A-S}}(-;G)$ be the sheafification of $A^p(-;G)$. As in the case of singular cohomology we obtain a complex

$$0 \longrightarrow \widetilde{G}_X \longrightarrow \mathcal{A}^0_{\text{A-S}}(-;G) \xrightarrow{\delta} \mathcal{A}^1_{\text{A-S}}(-;G) \xrightarrow{\delta} \mathcal{A}^2_{\text{A-S}}(-;G) \xrightarrow{\delta} \cdots. \tag{$*$}$$

The following result is proven in Warner [Warner (1983)] (Chapter 5, Section 5.26).

Proposition 13.36. *The sheaves $\mathcal{A}^p_{\text{A-S}}(-;G)$ are fine and the complex $(*)$ is a resolution of $\widetilde{G}_X$.*

By Proposition 11.34, if we apply the global section functor $\Gamma(X,-)$ to the resolution $(*)$, we obtain the complex $\mathcal{A}^*_{\text{A-S}}(X;G)$ (of modules)

$$0 \longrightarrow \mathcal{A}^0_{\text{A-S}}(X;G) \xrightarrow{\delta^0} \mathcal{A}^1_{\text{A-S}}(X;G) \xrightarrow{\delta^1} \mathcal{A}^2_{\text{A-S}}(X;G) \xrightarrow{\delta^2} \cdots$$

whose cohomology is isomorphic to the sheaf cohomology $H^*(X, \widetilde{G}_X)$.

We can give an alternative and more direct definition of $\mathcal{A}^p_{\text{A-S}}(X;G)$. Since X is paracompact and since the presheaves $A^p(-;G)$ satisfy Condition (G), Proposition 13.20 implies that the sequence of cochain complexes

$$0 \longrightarrow A^*_0(X;G) \longrightarrow A^*(X;G) \longrightarrow \mathcal{A}^*_{\text{A-S}}(X;G) \longrightarrow 0$$

is exact, with

$$A^p_0(X;G) = \{f \in A^p(X;G) \mid f_x = 0 \text{ for all } x \in X\}.$$

Then we have isomorphisms

$$A^p(X;G)/A^p_0(X;G) \cong \mathcal{A}^p_{\text{A-S}}(X;G)$$

for all $p \geq 0$, and the sheaf cohomology groups $H^p(X; \widetilde{G}_X)$ are the cohomology groups of the complex

$$0 \longrightarrow A^0(X;G)/A^0_0(X;G) \xrightarrow{\delta^0} A^1(X;G)/A^1_0(X;G) \xrightarrow{\delta^1} A^2(X;G)/A^2_0(X;G) \xrightarrow{\delta^2} \cdots$$

Now, the elements of $A^p_0(X;G)$ can be described as functions $f \in A^p(X;G)$ that are *locally zero*.

Definition 13.17. A function $f \in A^p(X;G)$ is *locally zero* if there is some open cover $\mathcal{U} = (U_i)_{i \in I}$ of X such that $f(x_0, \ldots x_p) = 0$ for all $(x_0, \ldots, x_p) \in U_i^{p+1}$ in some $U_i \in \mathcal{U}$.

Equivalently, if we write

$$\mathcal{U}^{p+1} = \bigcup_{i \in I} U_i^{p+1} \subseteq X^{p+1},$$

then $f \in A^p(X;G)$ is locally zero if there is some open cover $\mathcal{U} = (U_i)_{i \in I}$ of X such that f vanishes on $\mathcal{U}^{p+1}$.

It follows that the restriction of δ to $A^p_0(X;G)$ has its image in $A^{p+1}_0(X;G)$, because if f vanishes on $\mathcal{U}^{p+1}$, then δf vanishes on $\mathcal{U}^{p+2}$. It follows that we obtain the quotient complex

$$0 \longrightarrow A^0(X;G)/A^0_0(X;G) \xrightarrow{\delta^0} A^1(X;G)/A^1_0(X;G) \xrightarrow{\delta^1} A^2(X;G)/A^2_0(X;G) \xrightarrow{\delta^2} \cdots$$

as above. By definition, its cohomology groups are the Alexander–Spanier cohomology groups.

Definition 13.18. For any topological space X, the *Alexander–Spanier* complex is the complex

$$0 \longrightarrow A^0(X;G)/A^0_0(X;G) \xrightarrow{\delta^0} A^1(X;G)/A^1_0(X;G) \xrightarrow{\delta^1} A^2(X;G)/A^2_0(X;G) \xrightarrow{\delta^2} \cdots$$

where the $A^p(-;G)$ are the Alexander–Spanier presheaves and $A^p_0(X;G)$ consists of the functions in $A^p(X;G)$ that are locally zero. The cohomology groups of the above complex are the *Alexander–Spanier* cohomology groups and are denoted by $H^p_{\text{A-S}}(X;G)$.

Observe that the Alexander–Spanier cohomology groups are defined for all topological spaces, not necessarily paracompact. However, we proved that if X is paracompact, then they agree with the sheaf cohomology groups of the sheaf $\widetilde{G}_X$.

Theorem 13.37. *If the space X is paracompact, then we have isomorphisms*

$$H^p_{A\text{-}S}(X;G) \cong H^p(X;\widetilde{G}_X) \quad \textit{for all } p \geq 0$$

between Alexander–Spanier cohomology and the sheaf cohomology of the constant sheaf $\widetilde{G}_X$.

In view of Theorem 13.17, we also have the following theorem (proven in full in Warner [Warner (1983)], Chapter 5, Section 5.26, Pages 187–188).

Theorem 13.38. *If the space X is paracompact, then we have isomorphisms*

$$H^p_{A\text{-}S}(X;G) \cong \check{H}^p(X;\widetilde{G}_X) \quad \textit{for all } p \geq 0$$

between Alexander–Spanier cohomology and the Čech cohomology of the constant sheaf $\widetilde{G}_X$ (classical Čech cohomology).

Theorem 13.38 is also proven in Spanier [Spanier (1989)] (Chapter 6, Section 8, Corollary 8). In fact, the above isomorphisms hold even if X is not paracompact, a theorem due to Dowker; see Theorem 14.5, and also Spanier [Spanier (1989)] (Chapter 6, Exercise 6.D.3).

Remark: The cohomology of the complex

$$0 \longrightarrow A^0(X;G) \xrightarrow{\delta^0} A^1(X;G) \xrightarrow{\delta^1} A^2(X;G) \xrightarrow{\delta^2} \cdots$$

is trivial; that is, its cohomology groups are all equal to G; see Spanier [Spanier (1989)] (Chapter 6, Section 4, Lemma 1).

13.9 Problems

Problem 13.1. Prove that for any sheaf $\mathcal{F}$, the map

$$\mathrm{Hom}_{\mathbf{Sh}(X)}(\mathcal{F}, \mathcal{I}^x) \cong \mathrm{Hom}_R(\mathcal{F}_x, I^x)$$

given by $\varphi \mapsto \varphi_x$, with $\varphi \in \mathrm{Hom}_{\mathbf{Sh}(X)}(\mathcal{F}, \mathcal{I}^x)$, is an isomorphism.

Problem 13.2. Prove Proposition 13.10.

Problem 13.3. Prove that for any two presheaves $\mathcal{F}$ and $\mathcal{G}$, we have $(\mathcal{F}/\mathcal{G})_x = \mathcal{F}_x/\mathcal{G}_x$.

Problem 13.4. Prove that for any space X and for any sheaf $\mathcal{F}$ on X, we have isomorphisms

$$\check{H}^p(X,\mathcal{F}) \cong H^p(X,\mathcal{F}), \quad p = 0,1.$$

Problem 13.5. Prove Proposition 13.26.

Problem 13.6. Consider the homomorphism

$$\delta^p \colon A^p(U;G) \to A^{p+1}(U;G)$$

defined by

$$(\delta^p f)(x_0,\ldots,x_{p+1}) = \sum_{i=0}^{p+1} (-1)^i f(x_0,\ldots,\widehat{x_i},\ldots,x_{p+1}),$$

for all $f \in A^p(U;G)$ and all $(x_0,\ldots,x_{p+1}) \in U^{p+2}$. Check that $\delta^{p+1} \circ \delta^p = 0$ for all $p \geq 0$.

Chapter 14

Alexander and Alexander–Lefschetz Duality

Our goal is to present various generalizations of Poincaré duality. These versions of duality involve taking direct limits of direct mapping families of singular cohomology groups which, in general, are not singular cohomology groups. However, such limits are isomorphic to Alexander–Spanier cohomology groups, and thus to Čech cohomology groups. These duality results also require relative versions of homology and cohomology. Thus, in preparation for Alexander–Lefschetz duality we need to define relative Alexander–Spanier cohomology and relative Čech cohomology.

14.1 Relative Alexander–Spanier Cohomology

Given any topological space X (not necessarily paracompact), let us denote by $A^p_{\text{A-S}}(X;G)$[1] the Alexander–Spanier cochain modules

$$A^p_{\text{A-S}}(X;G) = A^p(X;G)/A^p_0(X;G),$$

where $A^p_0(X;G)$ is the set of functions in $A^p(X;G)$ that are locally zero (which means that there is some open cover $\mathcal{U} = (U_i)_{i\in I}$ of X such that $f(x_0,\ldots x_p) = 0$ for all $(x_0,\ldots,x_p) \in U_i^{p+1}$ in some $U_i \in \mathcal{U}$). Recall that if we write

$$\mathcal{U}^{p+1} = \bigcup_{i\in I} U_i^{p+1} \subseteq X^{p+1},$$

then $f \in A^p(X;G)$ is locally zero if there is some open cover $\mathcal{U} = (U_i)_{i\in I}$ of X such that f vanishes on $\mathcal{U}^{p+1}$.

We are going to provide three equivalent definitions of relative Alexander–Spanier cohomology. The first two definitions parallel the technique used in Section 13.8. The first definition (Definition 14.1) uses an

[1]In Section 13.8 we used the notation $\mathcal{A}^p_{\text{A-S}}(X;G)$, but for the sake of simplicity we will use the notation $A^p_{\text{A-S}}(X;G)$.

abstract complex. The second definition (see Proposition 14.1) uses a concrete quotient module definition. The third definition involves a direct limit over open covers. Since Čech cohomology is also defined in terms of open covers, this third definition provides the link between Alexander–Spanier cohomology and Čech cohomology. In fact, they are isomorphic.

If $h\colon X \to Y$ is a continuous map, then we have an induced cochain maps

$$h^{p\sharp}\colon A^p(Y;G) \to A^p(X;G)$$

given by

$$h^{p\sharp}(\varphi)(x_0,\ldots,x_p) = \varphi(h(x_0),\ldots,h(x_p))$$

for all $(x_0,\ldots,x_p) \in X^{p+1}$ and all $\varphi \in A^p(Y;G)$.

If φ vanishes on $\mathcal{V}^{p+1}$, where $\mathcal{V}$ is some open cover of Y, since h is continuous we see that $h^{-1}(\mathcal{V})$ is an open cover of X and then $h^{p\sharp}$ vanishes on $(h^{-1}(\mathcal{V}))^{p+1}$. It follows that $h^{p\sharp}$ maps $A_0^p(Y;G)$ into $A_0^p(X;G)$, so there is an induced map

$$h^{p\sharp}\colon A^p_{\text{A-S}}(Y;G) \to A^p_{\text{A-S}}(X;G),$$

and thus a module homomorphism

$$h^{p*}\colon H^p_{\text{A-S}}(Y;G) \to H^p_{\text{A-S}}(X;G).$$

If A is a subspace of X and $i\colon A \to X$ is the inclusion map, then the homomorphisms $i^{p\sharp}\colon A^p_{\text{A-S}}(X;G) \to A^p_{\text{A-S}}(A;G)$ are surjective (see the proof of Proposition 14.1 for an explicit definition of $i^{p\sharp}$). Therefore

$$A^p_{\text{A-S}}(X,A;G) = \operatorname{Ker} i^{p\sharp}$$

is a submodule of $A^p_{\text{A-S}}(X;G)$ called the module of *relative Alexander–Spanier p-cochains*, and by restriction we obtain a cochain complex

$$0 \longrightarrow A^0_{\text{A-S}}(X,A;G) \xrightarrow{\delta^0} A^1_{\text{A-S}}(X,A;G) \xrightarrow{\delta^1} A^2_{\text{A-S}}(X,A;G) \xrightarrow{\delta^2} \cdots. \tag{$*$}$$

Definition 14.1. If X is a topological space and if A is a subspace of X, the *relative Alexander–Spanier* cohomology groups $H^p_{\text{A-S}}(X,A;G)$ are the cohomology groups of the complex $(*)$.

Observe that by definition the sequence

$$0 \longrightarrow A^*_{\text{A-S}}(X, A; G) \longrightarrow A^*_{\text{A-S}}(X; G) \longrightarrow A^*_{\text{A-S}}(A; G) \longrightarrow 0$$

is an exact sequence of cochain complexes. Therefore by Theorem 2.22 we have the following long exact sequence of cohomology:

$$\cdots \longrightarrow H^{p-1}_{\text{A-S}}(A; G) \xrightarrow{\delta^*_{p-1}} H^p_{\text{A-S}}(X, A; G) \longrightarrow H^p_{\text{A-S}}(X; G) \longrightarrow H^p_{\text{A-S}}(A; G) \xrightarrow{\delta^*_p} H^{p+1}_{\text{A-S}}(X, A; G) \longrightarrow H^{p+1}_{\text{A-S}}(X; G) \longrightarrow H^{p+1}_{\text{A-S}}(A; G) \xrightarrow{\delta^*_{p+1}} H^{p+2}_{\text{A-S}}(X, A; G) \longrightarrow \cdots$$

A continuous map $h\colon (X, A) \to (Y, B)$ (with $h(A) \subseteq B$) also yields the commutative diagram

$$\begin{array}{ccccccccc}
0 & \longrightarrow & A^*_{\text{A-S}}(Y, B; G) & \longrightarrow & A^*_{\text{A-S}}(Y; G) & \longrightarrow & A^*_{\text{A-S}}(B; G) & \longrightarrow & 0 \\
 & & \Big\downarrow{\scriptstyle h^\sharp} & & \Big\downarrow{\scriptstyle (h|X)^\sharp} & & \Big\downarrow{\scriptstyle (h|A)^\sharp} & & \\
0 & \longrightarrow & A^*_{\text{A-S}}(X, A; G) & \longrightarrow & A^*_{\text{A-S}}(X; G) & \longrightarrow & A^*_{\text{A-S}}(A; G) & \longrightarrow & 0
\end{array}$$

in which the rows are exact, and a diagram chasing argument proves the existence of a map $h^\sharp$ making the left square commute. We define the homomorphism

$$h^*\colon H^*_{\text{A-S}}(Y, B; G) \to H^*_{\text{A-S}}(X, A; G)$$

induced by $h\colon (X, A) \to (Y, B)$ as the homomorphism induced by the cochain homomorphism

$$h^\sharp\colon A^*_{\text{A-S}}(Y, B; G) \to A^*_{\text{A-S}}(X, A; G)$$

given by the above commutative diagram.

The Alexander–Spanier relative cohomology modules are also limits of certain cohomology groups defined in terms of open covers. This characterization is needed to prove that relative Alexander–Spanier cohomology satisfies the homotopy axiom, and also to prove later on its equivalence with relative classical Čech cohomology defined in Section 14.4. We now sketch this development.

The first step is to give another characterization of $A^*_{\text{A-S}}(X, A; G)$ in terms of $A^*_0(X; G)$ and a certain submodule of $A^*(X; G)$.

Definition 14.2. For any space X and any subspace A of X, we define $A^p(X, A; G)$ as the submodule of $A^p(X; G)$ consisting of all functions in $A^p(X; G)$ which are locally zero on A. More precisely, there is some open cover $\mathcal{U}$ of X such that $f \in A^p(X; G)$ vanishes on $\mathcal{U}^{p+1} \cap A^{p+1}$.

It is immediate that $\delta\colon A^*(X; G) \to A^*(X; G)$ restricts to $A^*(X, A; G)$ so $A^*(X, A; G)$ is a cochain complex. Observe that $A^*(X, \emptyset; G) = A^*(X; G)$.

Proposition 14.1. *Let (X, A) be a pair of spaces with $A \subseteq X$. There is an isomorphism*

$$A^*_{A\text{-}S}(X, A; G) \cong A^*(X, A; G)/A^*_0(X; G).$$

Proof. The surjective homomorphism $i^{p\sharp}\colon A^p_{\text{A-S}}(X; G) \to A^p_{\text{A-S}}(A; G)$ induced by the inclusion $i\colon A \to X$ is defined by

$$i^\sharp([f]) = [f|A],$$

where on the left-hand side $[f]$ is the equivalence class of $f \in A^p(X; G)$ modulo $A^p_0(X; G)$, and on the right-hand side $[f|A]$ is the equivalence modulo $A^p_0(A; G)$ of the restriction of f to A^p. If $f' = f + g$ where g is locally zero on X, there is some open cover $\mathcal{U}$ of X such that g vanishes on $\mathcal{U}^{p+1}$, and $g|A$ vanishes on $\mathcal{U}^{p+1} \cap A^{p+1}$. Since $f'|A = f|A + g|A$ this shows that $[f'|A] = [f|A]$ and the above map is well defined. This reasoning also shows that the map φ given by the composition

$$A^*(X; G) \xrightarrow{\pi} A^*(X; G)/A^*_0(X; G) \xrightarrow{i^\sharp} A^*(A; G)/A^*_0(A; G) = A^*_{\text{A-S}}(A; G)$$

is given by

$$\varphi(f) = [f|A],$$

and that the kernel of φ is equal to $A^*(X, A; G)$, so we have an exact sequence

$$0 \longrightarrow A^*(X, A; G) \xrightarrow{\iota} A^*(X; G) \xrightarrow{\varphi} A^*_{\text{A-S}}(A; G) \longrightarrow 0,$$

and $A^*_0(X;G) \subseteq A^*(X,A;G)$. Since $A^*_0(X;G) \subseteq A^*(X,A;G)$, $\mathrm{Ker}\,\varphi = A^*(X,A;G)$ (since $\mathrm{Im}\,\iota = \mathrm{Ker}\,\varphi$), and the following diagram commutes

$$\begin{array}{ccc} A^*(X;G) & \xrightarrow{\pi} & A^*(X;G)/A^*_0(X;G) \\ & \searrow^{\varphi} & \downarrow i^\sharp \\ & & A^*_{\text{A-S}}(A;G), \end{array}$$

we have $\mathrm{Ker}\, i^\sharp \cong A^*(X,A;G)/A^*_0(X;G)$, and we conclude that we have the isomorphism

$$A^*_{\text{A-S}}(X,A;G) \cong A^*(X,A;G)/A^*_0(X;G),$$

as claimed. □

Observe that $A^*_{\text{A-S}}(X,\emptyset;G) = A^*_{\text{A-S}}(X;G)$.

14.2 Alexander–Spanier Cohomology as a Direct Limit

The next step is to define some cohomology groups based on open covers of (X,A), and for this we need a few facts about open covers.

Definition 14.3. Given a pair of topological spaces (X,A) where A is a subset of X, a pair $(\mathcal{U},\mathcal{U}^A)$ is an *open cover of* (X,A) if $\mathcal{U} = (U_i)_{i\in I}$ is an open cover of X and $\mathcal{U}^A = (U_i)_{i\in I^A}$ is a subcover of $\mathcal{U}$ which is a cover of A; that is, $I^A \subseteq I$ and $A \subseteq \bigcup_{i\in I^A} U_i$; see Figure 14.1.

Recall from Definition 9.6 that given two covers $\mathcal{U} = (U_i)_{i\in I}$ and $\mathcal{V} = (V_j)_{j\in J}$ of a space X, we say that $\mathcal{V}$ *is a refinement of* $\mathcal{U}$, denoted $\mathcal{U} \prec \mathcal{V}$, if there is a function $\tau\colon J \to I$ (sometimes called a *projection*) such that

$$V_j \subseteq U_{\tau(j)} \quad \text{for all } j \in J.$$

Definition 14.4. Given a pair of topological spaces (X,A) where A is a subset of X, for any two open covers $(\mathcal{U},\mathcal{U}^A)$ and $(\mathcal{V},\mathcal{V}^A)$ of (X,A), with $\mathcal{U} = (U_i)_{i\in I}$, $I^A \subseteq I$, $\mathcal{V} = (V_j)_{j\in J}$, $J^A \subseteq J$, we say that $(\mathcal{V},\mathcal{V}^A)$ is a *refinement* of $(\mathcal{U},\mathcal{U}^A)$, written $(\mathcal{U},\mathcal{U}^A) \prec (\mathcal{V},\mathcal{V}^A)$, if there is a function $\tau\colon J \to I$ (sometimes called a *projection*) such that $\tau(J^A) \subseteq I^A$ and

$$V_j \subseteq U_{\tau(j)} \quad \text{for all } j \in J\text{; see Figure 14.2.}$$

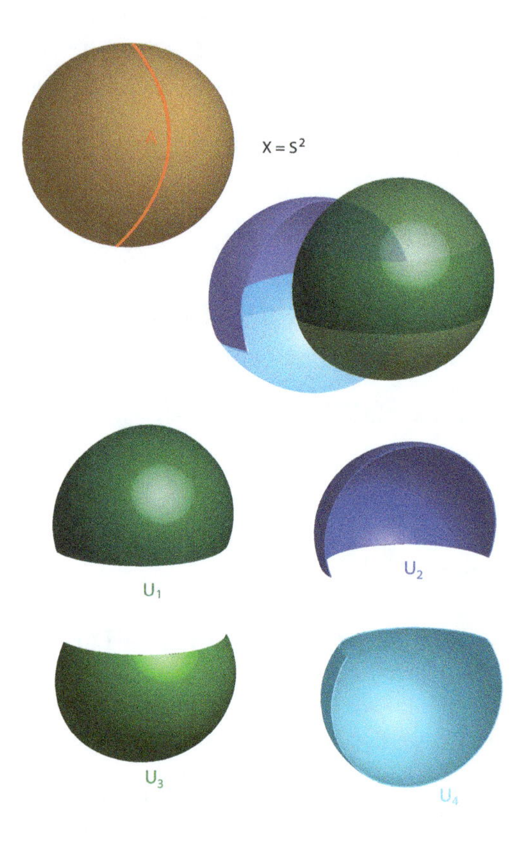

Fig. 14.1 Let $X = S^2$ and A be the orange semicircle. Let $\mathcal{U} = \{U_1, U_2, U_3, U_4\}$ with $I = \{1, 2, 3, 4\}$. Then $\mathcal{U}^A = \{U_1, U_3\}$ with $I_A = \{1, 3\}$.

Let $\mathrm{Cov}(X, A)$ be the preorder of open covers $(\mathcal{U}, \mathcal{U}^A)$ of (X, A) under refinement. If $(\mathcal{U}, \mathcal{U}^A)$ and $(\mathcal{V}, \mathcal{V}^A)$ are two open covers of (X, A), if we let

$$\mathcal{W} = \{U_i \cap V_j \mid (i, j) \in I \times J\}$$

and

$$\mathcal{W}^A = \{U_i \cap V_j \mid (i, j) \in I^A \times J^A\},$$

we see that $(\mathcal{W}, \mathcal{W}^A)$ is an open cover of (X, A) that refines both $(\mathcal{U}, \mathcal{U}^A)$ and $(\mathcal{V}, \mathcal{V}^A)$; see Figure 14.3. Therefore, $\mathrm{Cov}(X, A)$ is a directed preorder.

We also define $\mathrm{Cov}(X)$ as the preorder of open covers of X under refinement; it is a directed preorder. However, observe that $\mathrm{Cov}(X)$ is not equal to $\mathrm{Cov}(X, \emptyset)$, because even if $A = \emptyset$, a cover of $(X, \emptyset)$ consists of a

Fig. 14.2 Let $X = S^2$ and A be the orange semicircle. Let $\mathcal{V} = \{V_1, V_2, V_3, V_4, V_5, V_6\}$ with $J = \{1, 2, 3, 4, 5, 6\}$, and $\mathcal{V}^A = \{V_1, V_4\}$ with $J_A = \{1, 4\}$. Then $(\mathcal{V}, \mathcal{V}_A)$ is a refinement of $(\mathcal{U}, \mathcal{U}_A)$ of Figure 14.1 since $V_1 \subseteq U_1$, $V_2 \subseteq U_1$, $V_3 = U_2$, $V_4 \subseteq U_3$, $V_5 \subseteq U_3$, and $V_6 = U_4$.

pair $(\mathcal{U}, \mathcal{U}^A)$ where $\mathcal{U}^A$ is a subcover of $\mathcal{U}$ associated with some index set $I^A \subseteq I$ which is not necessarily empty. Covers in $\mathrm{Cov}(X)$ correspond to those covers $(\mathcal{U}, \emptyset)$ in $\mathrm{Cov}(X, \emptyset)$ for which $I^A = \emptyset$. In the end this will not matter but this a subtle point that should not be overlooked.

We are ready to show that $A^*_{\text{A-S}}(X, A; G)$ is the limit of cochain complexes associated with covers $(\mathcal{U}, \mathcal{U}^A)$ of (X, A).

Definition 14.5. Let (X, A) be a pair of topological spaces with $A \subseteq X$. For any open cover $(\mathcal{U}, \mathcal{U}^A)$ of (X, A), let $A^p(\mathcal{U}, \mathcal{U}^A; G)$ be the submodule of $A^p(X; G)$ given by

$$A^p(\mathcal{U}, \mathcal{U}^A; G) = \{f \colon \mathcal{U}^{p+1} \to G \mid f(x_0, \ldots, x_p) = 0 \\ \text{if } (x_0, \ldots, x_p) \in (\mathcal{U}^A)^{p+1} \cap A^{p+1}\}.$$

The homomorphism

$$\delta^p \colon A^p(\mathcal{U}, \mathcal{U}^A; G) \to A^{p+1}(\mathcal{U}, \mathcal{U}^A; G)$$

Fig. 14.3 Let X be the unit disk and A the red boundary arc. The top level shows $(\mathcal{U}, \mathcal{U}^A)$ where $\mathcal{U} = \{U_1, U_2\}$ and $\mathcal{U}_A = \{U_1\}$. The middle level shows $(\mathcal{V}, \mathcal{V}^A)$ where $\mathcal{V} = \{V_1, V_2, V_3\}$ and $\mathcal{V}_A = \{V_1, V_2\}$. The bottom level shows the common refinement $(\mathcal{W}, \mathcal{W}^A)$ where $\mathcal{W} = \{U_1 \cap V_1 = V_1, U_1 \cap V_2 = V_2, U_1 \cap V_3, U_2 \cap V_1, U_2 \cap V_2, V_3\}$ and $\mathcal{W}_A = \{U_1 \cap V_1 = V_1, U_1 \cap V_2 = V_2\}$.

is defined as in Definition 13.16 by

$$(\delta^p f)(x_0, \ldots, x_{p+1}) = \sum_{i=0}^{p+1} (-1)^i f(x_0, \ldots, \widehat{x_i}, \ldots, x_{p+1}).$$

It is easily checked that $\delta^{p+1} \circ \delta^p = 0$ for all $p \geq 0$, so the modules $A^p(\mathcal{U}, \mathcal{U}^A; G)$ form a cochain complex.

Remark: The module $A^p(\mathcal{U}, \mathcal{U}^A; G)$ can be viewed as an ordered simplicial cochain complex; see Spanier [Spanier (1989)] (Chapter 6, Section 5).

If $(\mathcal{V}, \mathcal{V}^A)$ is a refinement of $(\mathcal{U}, \mathcal{U}^A)$, then the restriction map is a cochain map

$$\rho^{\mathcal{U}, \mathcal{U}^A}_{\mathcal{V}, \mathcal{V}^A} \colon A^p(\mathcal{U}, \mathcal{U}^A; G) \to A^p(\mathcal{V}, \mathcal{V}^A; G),$$

so the directed family $(A^p(\mathcal{U},\mathcal{U}^A;G))_{(\mathcal{U},\mathcal{U}^A)\in \mathrm{Cov}(X,A)}$ together with the family of maps $\rho^{\mathcal{U},\mathcal{U}^A}_{\mathcal{V},\mathcal{V}^A}$ with $(\mathcal{U},\mathcal{U}^A) \prec (\mathcal{V},\mathcal{V}^A)$ is a direct mapping family.

Remark: As usual, one has to exercise some care because the set of all covers of (X,A) is not a set. This can be dealt with as in Serre's FAC [Serre (1955)] or as in Eilenberg and Steenrod [Eilenberg and Steenrod (1952)] (Chapter IX, Page 238).

The remarkable fact is that if $A \neq \emptyset$, then we have an isomorphism

$$A^p_{\text{A-S}}(X,A;G) \cong \varinjlim_{(\mathcal{U},\mathcal{U}^A)\in \mathrm{Cov}(X,A)} A^p(\mathcal{U},\mathcal{U}^A;G),$$

and if $A = \emptyset$, we have an isomorphism

$$A^p_{\text{A-S}}(X;G) \cong \varinjlim_{\mathcal{U}\in \mathrm{Cov}(X)} A^p(\mathcal{U},\emptyset;G).$$

To prove the above isomorphism, first if $A \neq \emptyset$, we will define a map

$$\lambda\colon A^p(X,A;G) \to \varinjlim_{(\mathcal{U},\mathcal{U}^A)\in \mathrm{Cov}(X,A)} A^p(\mathcal{U},\mathcal{U}^A;G),$$

where $A^*(X,A;G)$ is the module defined in Definition 14.2, and if $A = \emptyset$, we will define a map

$$\lambda\colon A^p(X;G) \to \varinjlim_{\mathcal{U}\in \mathrm{Cov}(X)} A^p(\mathcal{U},\emptyset;G).$$

Assume $A \neq \emptyset$. For any $f \in A^p(X,A;G)$, there is some open cover $\mathcal{U}^A$ of A consisting of open subsets of X such that f vanishes on $(\mathcal{U}^A)^{p+1} \cap A^{p+1}$, and we let $\mathcal{U}$ be the open cover of X obtained by adding X itself to the cover $\mathcal{U}^A$ and giving it some new index, say k (we need to do this to obey the indexing convention of Definition 14.3). Then $(\mathcal{U},\mathcal{U}^A)$ is an open cover of (X,A) and by restriction f determines an element $f|(\mathcal{U},\mathcal{U}^A) \in A^p(\mathcal{U},\mathcal{U}^A;G)$. Passing to the limit, we obtain a homomorphism

$$\lambda^p\colon A^p(X,A;G) \to \varinjlim_{(\mathcal{U},\mathcal{U}^A)\in \mathrm{Cov}(X,A)} A^p(\mathcal{U},\mathcal{U}^A;G).$$

Theorem 14.2. *If $A \neq \emptyset$, then the map*

$$\lambda\colon A^*(X,A;G) \to \varinjlim_{(\mathcal{U},\mathcal{U}^A)\in \mathrm{Cov}(X,A)} A^*(\mathcal{U},\mathcal{U}^A;G)$$

as defined above is surjective and its kernel is given by $\mathrm{Ker}\,\lambda = A_0^*(X;G)$. *Consequently, we have an isomorphism*

$$A^*_{A\text{-}S}(X,A;G) \cong \varinjlim_{(\mathcal{U},\mathcal{U}^A)\in \mathrm{Cov}(X,A)} A^*(\mathcal{U},\mathcal{U}^A;G).$$

If $A=\emptyset$, *then the map*

$$\lambda\colon A^*(X;G) \to \varinjlim_{\mathcal{U}\in \mathrm{Cov}(X)} A^*(\mathcal{U},\emptyset;G)$$

is surjective and its kernel is given by $\mathrm{Ker}\,\lambda = A_0^*(X;G)$. *Consequently, we have an isomorphism*

$$A^*_{A\text{-}S}(X;G) \cong \varinjlim_{\mathcal{U}\in \mathrm{Cov}(X)} A^*(\mathcal{U},\emptyset;G).$$

Proof. We follow Spanier's proof, see Spanier [Spanier (1989)] (Chapter 6, Section 4, Theorem 1). Assume that $A\neq\emptyset$. First we prove that λ is surjective. Pick any $u\in A^p(\mathcal{U},\mathcal{U}^A;G)$, and define $f_u\in A^p(X,A;G)$ by

$$f_u(x_0,\ldots,x_p) = \begin{cases} u(x_0,\ldots,x_p) & \text{if } (x_0,\ldots,x_p)\in \mathcal{U}^{p+1} \\ 0 & \text{otherwise.}\end{cases}$$

Then f_u vanishes on $(\mathcal{U}^A)^{p+1}\cap A^{p+1}$, and therefore $f_u|(\mathcal{U},\mathcal{U}^A)\in A^p(\mathcal{U},\mathcal{U}^A;G)$. By definition, we have $f_u| = u$, so λ is surjective.

Next we prove that $\mathrm{Ker}\,\lambda = A_0^*(X;G)$. A function $f\in A^p(X,A;G)$ is in the kernel of λ iff there is some open cover $(\mathcal{U},\mathcal{U}^A)$ such that $f|(\mathcal{U},\mathcal{U}^A)=0$. Thus, $\lambda(f)=0$ iff there is some open covering $\mathcal{U}$ such that f vanishes on $\mathcal{U}^{p+1}$. By the definition of $A_0^*(X;G)$, we have $\lambda(f)=0$ iff $f\in A_0^*(X;G)$.

The case where $A=\emptyset$ is similar but slightly simpler. $\square$

An important corollary of Theorem 14.2 is the following characterization of the relative Alexander–Spanier cohomology groups as certain limits of simpler cohomology groups (in fact, simplicial cohomology).

Theorem 14.3. *Let* (X,A) *be a pair of spaces with* $A\subseteq X$. *If* $A\neq\emptyset$, *then we have an isomorphism*

$$H^p_{A\text{-}S}(X,A;G) \cong \varinjlim_{(\mathcal{U},\mathcal{U}^A)\in \mathrm{Cov}(X,A)} H^p(\mathcal{U},\mathcal{U}^A;G), \quad \textit{for all } p\geq 0.$$

If $A=\emptyset$, *then we have an isomorphism*

$$H^p_{A\text{-}S}(X;G) \cong \varinjlim_{\mathcal{U}\in \mathrm{Cov}(X)} H^p(\mathcal{U},\emptyset;G), \quad \textit{for all } p\geq 0.$$

Proof. It is shown in Spanier [Spanier (1989)] (Chapter 4) that cohomology commutes with direct limits (this is a general categorical fact about direct limits). Using Theorem 14.2 we obtain our result. □

Spanier uses Theorem 14.3 to prove that Alexander–Spanier cohomology satisfies the homotopy axiom; see Spanier [Spanier (1989)] (Chapter 6, Section 5). Actually, Spanier proves that Alexander–Spanier cohomology satisfies all of the Eilenberg–Steenrod axioms. A more detailed treatment of Alexander–Spanier cohomology is found in Spanier [Spanier (1989)] (Chapter 6, Sections 4–9).

14.3 Alexander–Spanier Cohomology with Compact Support

In order to state the most general version of Alexander–Lefschetz duality (not restricted to the compact case), it is necessary to introduce Alexander–Spanier cohomology with compact support

Definition 14.6. A subset A of a topological space X is said to be *bounded* if its closure $\overline{A}$ is compact. A subset $B \subseteq X$ is said to be *cobounded* if its complement $X - B$ is bounded; see Figure 14.4. A function $h\colon X \to Y$ is *proper* if it is continuous and if $h^{-1}(A)$ is bounded in X whenever A is bounded in Y.

It is immediate to check that the composition of two proper maps is proper. A proper map h between (X, A) and (Y, B) (where $A \subseteq X$ and $B \subseteq Y$) is a proper map from X to Y such that $h(A) \subseteq B$.

Definition 14.7. Let (X, A) be a pair of spaces with $A \subseteq X$. The module $A^p_c(X, A; G)$ is the submodule of $A^p(X, A; G)$ consisting of all functions $f \in A^p(X, A; G)$ such that f is locally zero on some cobounded subset B of X. If $f \in A^p(X, A; G)$ is locally zero on B, so is δf, thus the family of modules $A^p_c(X, A; G)$ with the restrictions of the δ^p is a cochain complex which is a subcomplex of $A^*(X, A; G)$. Since $A^*_0(X; G) \subseteq A^*_c(X, A; G)$, we obtain the cochain complex $A^*_{\text{A-S},c}(X, A; G)$, with

$$A^*_{\text{A-S},c}(X, A; G) = A^*_c(X, A; G)/A^*_0(X; G).$$

The *Alexander–Spanier cohomology modules of* (X, A) *with compact support* $H^p_{\text{A-S},c}(X, A; G)$ are the cohomology modules of the cochain complex $A^*_{\text{A-S},c}(X, A; G)$.

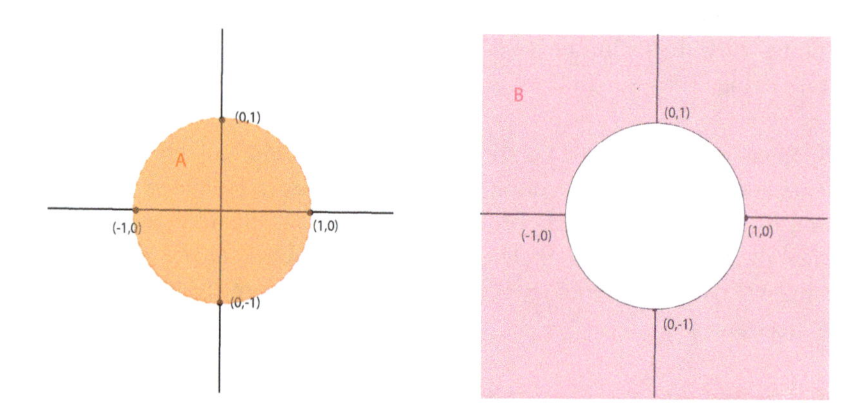

Fig. 14.4 Let $X = \mathbb{R}^2$. In the left figure, A, the open unit disk, is a bounded subset of X, while in the right figure, B, the complement of A, is a cobounded subset of X.

If $h\colon (X, A) \to (Y, B)$ is a proper map, then $h^\sharp$ maps $A^*_{\text{A-S},c}(Y, B; G)$ to $A^*_{\text{A-S},c}(X, A; G)$ and induces a homomorphism

$$h^*\colon H^p_{\text{A-S},c}(Y, B; G) \to H^p_{\text{A-S},c}(X, A; G).$$

Properties of Alexander–Spanier cohomology with compact support are investigated in Spanier [Spanier (1989)] (Chapter 6, Section 6). We just mention the following result.

Proposition 14.4. *Let (X, A) be a pair of spaces with $A \subseteq X$. If A is a cobounded subset of X, then there is an isomorphism*

$$H^*_{A\text{-}S,c}(X, A; G) \cong H^*_{A\text{-}S}(X, A; G).$$

In particular, Proposition 14.4 applies to the situation where (X, A) is a *compact pair*, which means that X is compact and A is a closed subset of X.

We conclude this section by mentioning that Alexander–Spanier cohomology enjoys a very simple definition of the cup product. Indeed, given $f_1 \in A^p(X; G)$ and $f_2 \in A^q(X; G)$ we define $f_1 \smile f_2 \in A^{p+q}(X; G)$ by

$$(f_1 \smile f_2)(x_0, \ldots, x_{p+q}) = f_1(x_0, \ldots, x_p) f_2(x_p, \ldots, x_{p+q}).$$

If f_1 is locally zero on A_1 then so is $f_1 \smile f_2$, and if f_2 is locally zero on A_1 then so is $f_1 \smile f_2$. Consequently $\smile$ induces a cup product

$$\smile\colon A^p_{\text{A-S}}(X; G) \times A^q_{\text{A-S}}(X; G) \to A^{p+q}_{\text{A-S}}(X; G).$$

One verifies that

$$\delta(f_1 \smile f_2) = \delta f_1 \smile f_2 + (-1)^p f_1 \smile \delta f_2,$$

so we obtain a cup product

$$\smile : H^p_{\text{A-S}}(X; G) \times H^q_{\text{A-S}}(X; G) \to H^{p+q}_{\text{A-S}}(X; G)$$

at the cohomology level.

It is also easy to deal with relative cohomology; see Spanier [Spanier (1989)] (Chapter 6, Section 5).

14.4 Relative Classical Čech Cohomology

In this section we deal with classical Čech cohomology, which means that given an open cover $\mathcal{U} = (U_i)_{i \in I}$ of the space X and given an R-module G, the module $C^p(\mathcal{U}, G)$ of Čech p-cochains is defined as the R-module of functions $f : I^{p+1} \to G$ such that for all $(i_0, \ldots, i_p) \in I^{p+1}$,

$$f(i_0, \ldots, i_p) = 0 \quad \text{if } U_{i_0 \cdots i_p} = \emptyset,$$

where $U_{i_0 \cdots i_p} = U_{i_0} \cap \cdots \cap U_{i_p}$. The coboundary maps are defined by

$$(\delta^p f)(i_0, \ldots, i_{p+1}) = \sum_{j=0}^{p+1} (-1)^j f(i_0, \ldots, \widehat{i_j}, \ldots, i_{p+1}),$$

for all $f \in C^p(\mathcal{U}, G)$ and all $(i_0, \ldots, i_{p+1}) \in I^{p+2}$. This is the special case of the notion of Čech cohomology with values in a presheaf discussed in Section 9.1, where the presheaf $\mathcal{F}$ is the constant presheaf G_X; see Definition 9.4.

Remark: The Čech cochain modules $C^p(\mathcal{U}, G)$ are often defined in terms of the nerve of the covering $\mathcal{U}$. The *ordered nerve* $\Delta\mathcal{N}(\mathcal{U})$ of the open covering $\mathcal{U}$ is the set of sequences $(i_0, \ldots, i_p) \in I^{p+1}$ such that $U_{i_0 \cdots i_p} \neq \emptyset$, for some $p \geq 0$. We can view $(I, \Delta\mathcal{N}(\mathcal{U}))$ as an abstract simplicial complex where the vertices are the elements of I and the ordered p-simplices are the sequences $(i_0, \ldots, i_p)$ in $\Delta\mathcal{N}(\mathcal{U})$ (recall Definition 5.19). For any given $p \geq 0$, the set of sequences $(i_0, \ldots, i_p)$ in $\Delta\mathcal{N}(\mathcal{U})$ is denoted by $\Delta\mathcal{N}_p(\mathcal{U})$. Then the cochain module $C^p(\mathcal{U}, G)$ is the set of functions $f : \Delta\mathcal{N}_p(\mathcal{U}) \to G$. Every function $f : \Delta\mathcal{N}_p(\mathcal{U}) \to G$ corresponds bijectively to the function $\widetilde{f} : I^{p+1} \to G$ obtained by extending f to I^{p+1} so that

$$\widetilde{f}(i_0, \ldots, i_p) = 0 \text{ if } U_{i_0 \cdots i_p} = \emptyset.$$

Thus it is equivalent to use functions of the form $\widetilde{f}$, and this seems simpler and more direct to us. Serre and Godement use this method.

The *nerve* $\mathcal{N}(\mathcal{U})$ of a covering $\mathcal{U}$ is defined as the set of *subsets* $\{i_0, \ldots, i_p\}$ of elements in I such that $U_{i_0 \cdots i_p} \neq \emptyset$, for some $p \geq 0$; see Spanier [Spanier (1989)], Page 109. The corresponding abstract simplicial complex is $(I, \mathcal{N}(\mathcal{U}))$. This is not the notion that we are using. The abstract simplicial complex $(I, \Delta\mathcal{N}(\mathcal{U}))$ that we are using is what Spanier calls an *ordered chain complex*; see Spanier [Spanier (1989)] (Page 170).

A last word of caution. As we explained in Section 9.1 and in the paragraph following Example 9.2, in order to deal correctly with the passage to a finer cover it is necessary to allow repetitions of indices. To eliminate repeated indices we can use alternating cochains as introduced in Definition 9.5.

Our first goal is to explain how a continuous map $h\colon X \to Y$ induces a homomorphism of Čech cohomology

$$h^{p*}\colon \check{H}^p(Y, G) \to \check{H}^p(X, G).$$

For this it necessary to take a closer look at the behavior of open covers of Y under h^{-1}.

If $\mathcal{V} = (V_i)_{i \in I}$ is an open cover of Y, then since h is continuous $h^{-1}(\mathcal{V}) = (h^{-1}(V_i))_{i \in I}$ is an open cover of X, with the same index set I. We also denote $h^{-1}(V_i)$ by $h^{-1}(V)_i$ or V'_i; see Figure 14.5.

If $\mathcal{W} = (W_j)_{j \in J}$ is a refinement of $\mathcal{V} = (V_i)_{i \in I}$ and if $\tau\colon J \to I$ is a function such that

$$W_j \subseteq V_{\tau(j)} \quad \text{for all } j \in J,$$

since

$$h^{-1}(W_j) \subseteq h^{-1}(V_{\tau(j)}),$$

if we write $W'_j = h^{-1}(W_j)$ and $V'_i = h^{-1}(V_i)$, then we have

$$W'_j \subseteq V'_{\tau(j)} \quad \text{for all } j \in J,$$

which means that $h^{-1}(\mathcal{W})$ is a refinement of $h^{-1}(\mathcal{V})$ (as open covers of X); see Figure 14.6.

Let $\mathrm{Cov}(X)$ be the preorder of open covers $\mathcal{U}$ of X under refinement and let $\mathrm{Cov}(Y)$ be the preorder of open covers $\mathcal{V}$ of Y under refinement. Observe that what we just showed implies that the map $\mathcal{V} \mapsto h^{-1}(\mathcal{V})$ between $\mathrm{Cov}(Y)$ and $\mathrm{Cov}(X)$ is an order-preserving map.

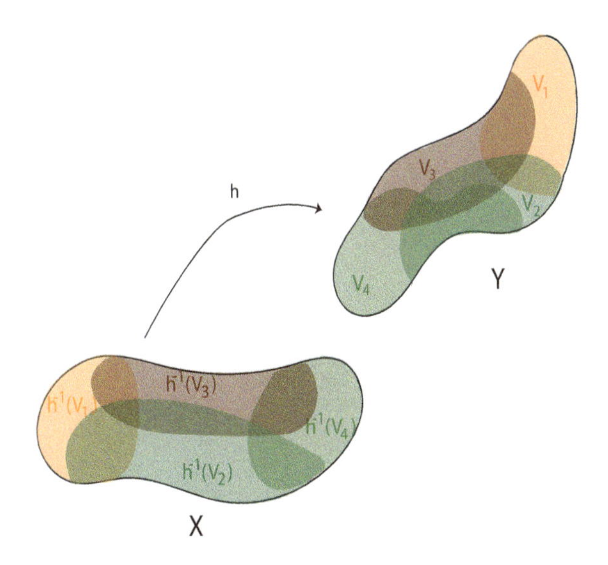

Fig. 14.5 A schematic illustration of the pullback cover $h^{-1}(\mathcal{V})$.

For any tuple $(i_0, \ldots, i_p) \in I^{p+1}$, we have

$$h^{-1}(V_{i_0 \cdots i_p}) = h^{-1}(V_{i_0} \cap \cdots \cap V_{i_p}) = h^{-1}(V_{i_0}) \cap \cdots \cap h^{-1}(V_{i_p}),$$

and if we let $h^{-1}(V)_{i_0 \cdots i_p} = h^{-1}(V_{i_0}) \cap \cdots \cap h^{-1}(V_{i_p})$, then

$$h^{-1}(V_{i_0 \cdots i_p}) = h^{-1}(V)_{i_0 \cdots i_p}.$$

Note that it is possible that $V_{i_0 \cdots i_p} \neq \emptyset$ but $h^{-1}(V_{i_0 \cdots i_p}) = h^{-1}(V)_{i_0 \cdots i_p} = \emptyset$ as evidenced by Figure 14.7.

Given a continuous map $h \colon X \to Y$ and an open cover $\mathcal{V} = (V_i)_{i \in I}$ of Y, we define a homomorphism from $C^p(\mathcal{V}, G)$ to $C^p(h^{-1}(\mathcal{V}), G)$ (where $h^{-1}(\mathcal{V})$ is an open cover of X).

Definition 14.8. Let $h \colon X \to Y$ be a continuous map between two spaces X and Y, and let $\mathcal{V} = (V_i)_{i \in I}$ be some open cover of Y. The R-module homomorphism

$$h^{p\sharp}_{\mathcal{V}} \colon C^p(\mathcal{V}, G) \to C^p(h^{-1}(\mathcal{V}), G)$$

is defined as follows: for any $f \in C^p(\mathcal{V}; G)$, for all $(i_0, \ldots, i_p) \in I^{p+1}$,

$$h^{p\sharp}_{\mathcal{V}}(f)(i_0, \ldots, i_p) = \begin{cases} f(i_0, \ldots, i_p) & \text{if } h^{-1}(V)_{i_0 \cdots i_p} \neq \emptyset \\ 0 & \text{if } h^{-1}(V)_{i_0 \cdots i_p} = \emptyset. \end{cases}$$

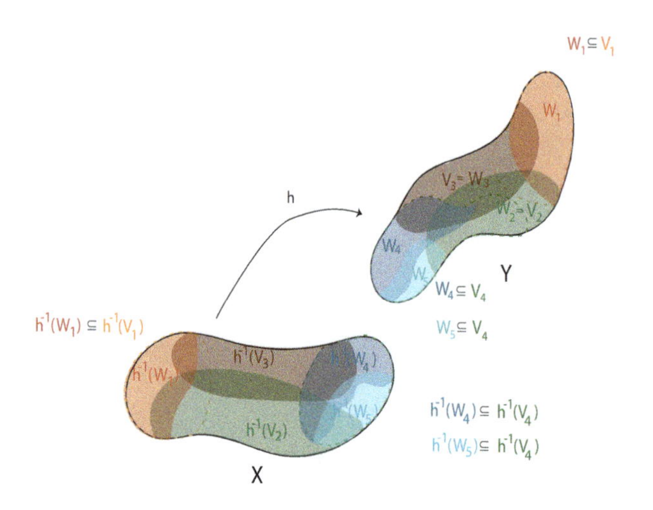

Fig. 14.6 The cover $\mathcal{W}$ is a refinement of the cover $\mathcal{V}$ illustrated in Figure 14.5. Note that $\tau(1) = 1$, $\tau(2) = 2$, $\tau(3) = 3$, $\tau(4) = 4$, and $\tau(5) = 4$. Then $h^{-1}(\mathcal{W})$ is a refinement of $h^{-1}(\mathcal{V})$.

The module homomorphism $h^{p\sharp}_{\mathcal{V}} \colon C^p(\mathcal{V}, G) \to C^p(h^{-1}(\mathcal{V}), G)$ induces a module homomorphism of Čech cohomology groups

$$h^{p*}_{\mathcal{V}} \colon \check{H}^p(\mathcal{V}; G) \to \check{H}^p(h^{-1}(\mathcal{V}); G).$$

For every refinement $\mathcal{W}$ of $\mathcal{V}$ ($\mathcal{V} \prec \mathcal{W}$), we have a commutative diagram

$$\begin{array}{ccc}
\check{H}^p(\mathcal{V}; G) & \xrightarrow{\;h^{p*}_{\mathcal{V}}\;} & \check{H}^p(h^{-1}(\mathcal{V}); G) \\
\Big\downarrow {\scriptstyle \rho^{\mathcal{V}}_{\mathcal{W}}} & & \Big\downarrow {\scriptstyle \rho^{h^{-1}(\mathcal{V})}_{h^{-1}(\mathcal{W})}} \\
\check{H}^p(\mathcal{W}; G) & \xrightarrow{\;h^{p*}_{\mathcal{W}}\;} & \check{H}^p(h^{-1}(\mathcal{W}); G),
\end{array}$$

where the restriction map $\rho^{\mathcal{V}}_{\mathcal{W}} \colon \check{H}^p(\mathcal{V}; G) \to \check{H}^p(\mathcal{W}; G)$ is defined just after 9.3 (and similarly for $\rho^{h^{-1}(\mathcal{V})}_{h^{-1}(\mathcal{W})} \colon \check{H}^p(h^{-1}(\mathcal{V}); G) \to \check{H}^p(h^{-1}(\mathcal{W}); G)$). If we define the map $\tau_h \colon \mathrm{Cov}(Y) \to \mathrm{Cov}(X)$ by $\tau_h(\mathcal{V}) = h^{-1}(\mathcal{V})$, then we see that τ_h and the family of maps

$$h^{p*}_{\mathcal{V}} \colon \check{H}^p(\mathcal{V}; G) \to \check{H}^p(h^{-1}(\mathcal{V}); G)$$

defines a map from the direct mapping family $(\check{H}^p(\mathcal{V}; G))_{\mathcal{V} \in \mathrm{Cov}(Y)}$ to the direct mapping family $(\check{H}^p(\mathcal{U}; G))_{\mathcal{U} \in \mathrm{Cov}(X)}$, and by the discussion just before Definition 8.14 we obtain a homomorphism between their direct limits,

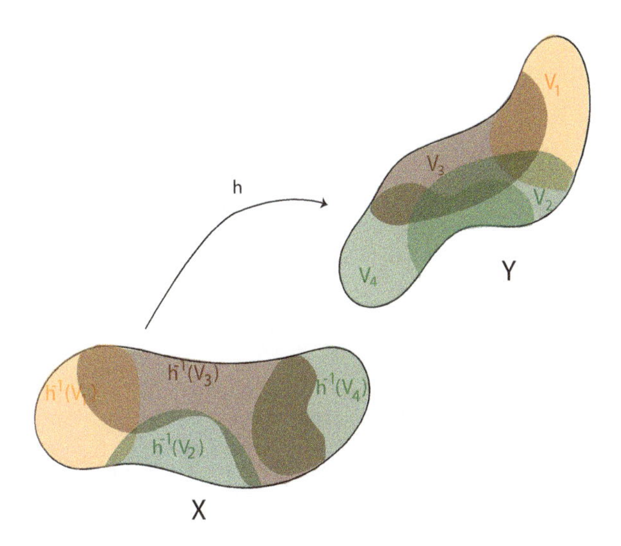

Fig. 14.7 Another schematic illustration of a pullback cover $h^{-1}(\mathcal{V})$. Note that $V_2 \cap V_4$ is a nonempty set of Y, but $h^{-1}(V_2) \cap h^{-1}(V_4) = \emptyset$.

that is, a homomorphism

$$h^{p*} \colon \check{H}^p(Y; G) \to \check{H}^p(X; G).$$

In order to define the relative Čech cohomology groups we need to consider a few more properties of the open covers of a pair (X, A). Let $h \colon (X, A) \to (Y, B)$ be a continuous map (recall that $h \colon X \to Y$ is continuous and $h(A) \subseteq B$). If $(\mathcal{V}, \mathcal{V}^B)$ is any open cover of (Y, B) (with index sets (I, I^B)) then $(h^{-1}(\mathcal{V}), h^{-1}(\mathcal{V}^B))$ is an open cover of (X, A) with the same index sets I and I^B; see Figure 14.8.

If $(\mathcal{W}, \mathcal{W}^B)$ (with index sets (J, J^B)) is a refinement of $(\mathcal{V}, \mathcal{V}^B)$ (with index set (I, I^B)) with projection function $\tau \colon J \to I$, it is immediate to check that $(h^{-1}(\mathcal{W}), h^{-1}(\mathcal{W}^B))$ is a refinement of $(h^{-1}(\mathcal{V}), h^{-1}(\mathcal{V}^B))$; see Figure 14.9.

It follows that the map

$$(\mathcal{V}, \mathcal{V}^B) \mapsto (h^{-1}(\mathcal{V}), h^{-1}(\mathcal{V}^B))$$

is an order preserving map between $\mathrm{Cov}(Y, B)$ and $\mathrm{Cov}(X, A)$. As before, for any tuple $(i_0, \ldots, i_p)$ in I^{p+1} or in $(I^A)^{p+1}$ we write

$$h^{-1}(V)_{i_0 \cdots i_p} = h^{-1}(V_{i_0 \cdots i_p}) = h^{-1}(V_{i_0}) \cap \cdots \cap h^{-1}(V_{i_p}).$$

It is possible that $V_{i_0 \cdots i_p} \neq \emptyset$ but $h^{-1}(V_{i_0 \cdots i_p}) = h^{-1}(V)_{i_0 \cdots i_p} = \emptyset$.

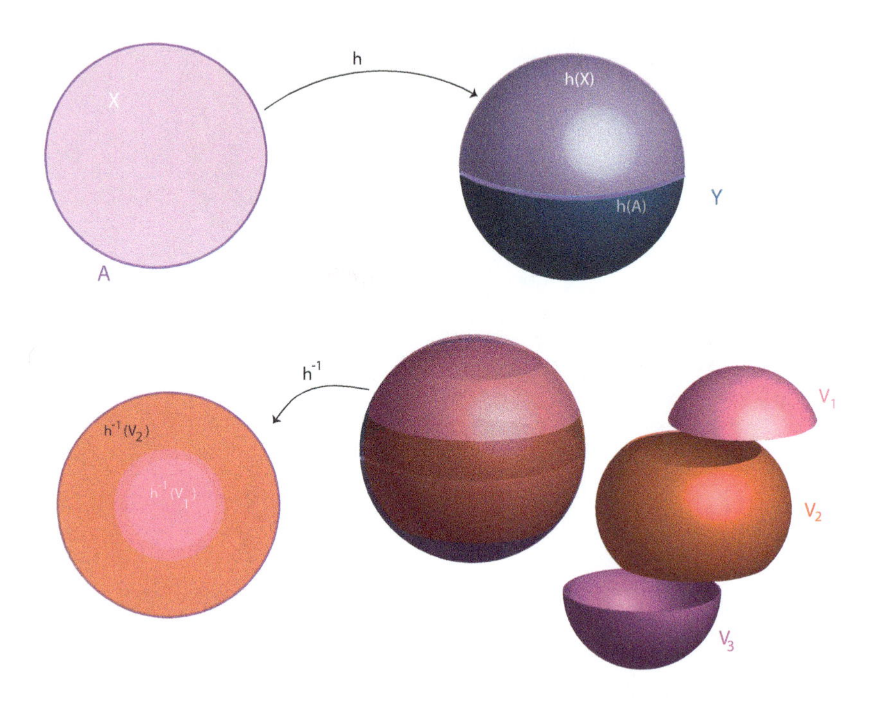

Fig. 14.8 Let X be the unit disk and A the unit circle boundary. Let Y be S^2 and B the equatorial circle. The map $h\colon (X, A) \to (Y, B)$ maps X onto the northern hemisphere of Y. For $\mathcal{V} = \{V_1, V_2, V_3\}$ with $\mathcal{V}^B = \{V_2\}$, $(h^{-1}(\mathcal{V}), h^{-1}(\mathcal{V}^B))$ is an open cover of (X, A) with $h^{-1}(V_3) = \emptyset$.

Definition 14.9. Let (X, A) be a pair of spaces with $A \subseteq X$. For every open cover $(\mathcal{U}, \mathcal{U}^A)$ of (X, A), the module $C^p(\mathcal{U}, \mathcal{U}^A; G)$ is the submodule of $C^p(\mathcal{U}; G)$ defined as follows:

$$
\begin{aligned}
C^p(\mathcal{U}, \mathcal{U}^A; G) = \{f\colon I^{p+1} \to G \mid & \text{ for all } (i_0, \ldots, i_p) \in I^{p+1}, \text{ if } U_{i_0 \cdots i_p} = \emptyset \\
& \text{or } (i_0, \ldots, i_p) \in (I^A)^{p+1} \text{ and } U_{i_0 \cdots i_p} \cap A \neq \emptyset, \text{ then} \\
& f(i_0, \ldots, i_p) = 0\}.
\end{aligned}
$$

Observe that if $A = \emptyset$, then $C^p(\mathcal{U}, \mathcal{U}^A; G) = C^p(\mathcal{U}; G)$ for any $\mathcal{U}^A$. In this case, we will restrict ourselves to covers for which $\mathcal{U}^A = \emptyset$, to ensure that direct limits are taken over $\mathrm{Cov}(X)$ in order to obtain the Čech cohomology groups of Definition 9.8.

The analogy between the above definition of $C^p(\mathcal{U}, \mathcal{U}^A; G)$ and the Alexander–Spanier modules

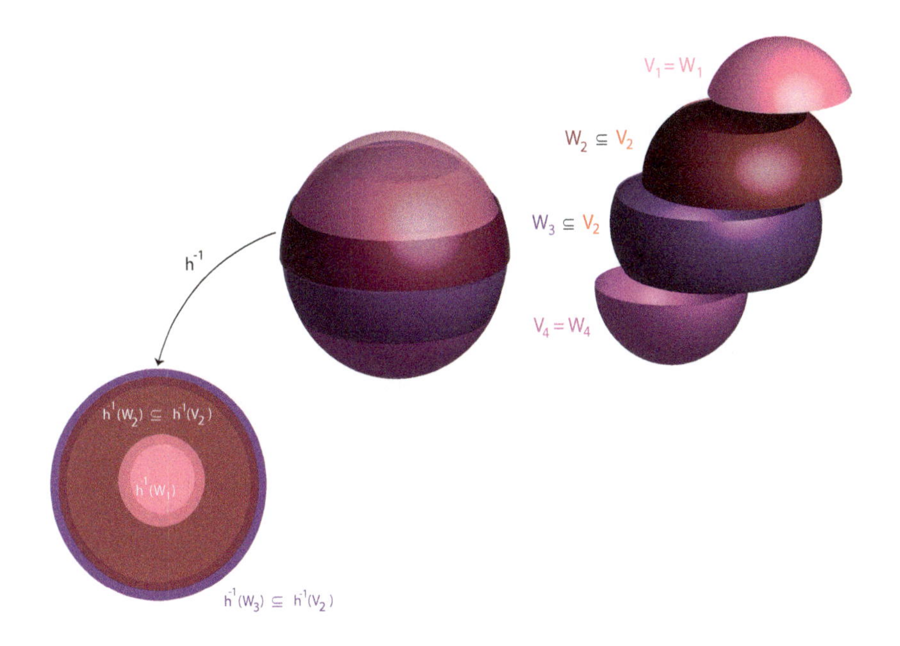

Fig. 14.9 Let X be the unit disk and A the unit circle boundary. Let Y be S^2 and B the equatorial circle. The cover $\mathcal{W}$ is a refinement of $\mathcal{V}$ from Figure 14.8 since $W_1 = V_1$, $W_2 \subseteq V_2$, $W_3 \subseteq V_2$, and $W_4 = V_3$. Observe that $\mathcal{W}^B = \{W_3\}$. Then $(h^{-1}(\mathcal{W}), h^{-1}(\mathcal{W}^B))$ is a refinement of $(h^{-1}(\mathcal{V}), h^{-1}(\mathcal{V}^B))$ with $h^{-1}(W_4) = \emptyset$.

$$\begin{aligned} A^p(\mathcal{U},\mathcal{U}^A;G) = \{f\colon \mathcal{U}^{p+1} \to G \mid f(x_0,\ldots,x_p) = 0 \\ \text{if } (x_0,\ldots,x_p) \in (\mathcal{U}^A)^{p+1} \cap A^{p+1}\} \end{aligned}$$

of Definition 14.5 is striking. Indeed, it turns out that they induce isomorphic cohomology.

It is immediately checked that the coboundary maps $\delta^p\colon C^p(U;G) \to C^{p+1}(U;G)$ restrict to the $C^p(\mathcal{U},\mathcal{U}^A;G)$ and we obtain a cochain complex $C^*(\mathcal{U},\mathcal{U}^A;G)$.

Definition 14.10. Let (X,A) be a pair of spaces with $A \subseteq X$. For every open cover $(\mathcal{U},\mathcal{U}^A)$ of (X,A), the *Čech cohomology modules* $\check{H}^p(\mathcal{U},\mathcal{U}^A;G)$ are the cohomology modules of the complex $C^*(\mathcal{U},\mathcal{U}^A;G)$.

Observe that if $A = \emptyset$, then $\check{H}^p(\mathcal{U},\mathcal{U}^A;G) = \check{H}^p(\mathcal{U};G)$ for any $\mathcal{U}^A$.

If $(\mathcal{V},\mathcal{V}^A)$ is a refinement of $(\mathcal{U},\mathcal{U}^A)$ then there is a cochain map

$$\rho^{\mathcal{U},\mathcal{U}^A}_{\mathcal{V},\mathcal{V}^A}\colon C^p(\mathcal{U},\mathcal{U}^A;G) \to C^p(\mathcal{V},\mathcal{V}^A;G).$$

One needs to prove that $\rho^{\mathcal{U},\mathcal{U}^A}_{\mathcal{V},\mathcal{V}^A}$ does not depend on the projection map $\tau \colon J \to I$, but this can be done as in Serre's FAC [Serre (1955)] or as in Eilenberg and Steenrod [Eilenberg and Steenrod (1952)] (Chapter IX, Theorem 2.13 and Corollary 2.14).

Therefore, the directed family $(C^p(\mathcal{U},\mathcal{U}^A;G))_{(\mathcal{U},\mathcal{U}^A)\in\mathrm{Cov}(X,A)}$ together with the family of maps $\rho^{\mathcal{U},\mathcal{U}^A}_{\mathcal{V},\mathcal{V}^A}$ with $(\mathcal{U},\mathcal{U}^A) \prec (\mathcal{V},\mathcal{V}^A)$ is a direct mapping family.

Remark: As usual, one has to exercise some care because the set of all covers of (X,A) is not a set. This can be dealt with as in Serre's FAC [Serre (1955)] or as in Eilenberg and Steenrod [Eilenberg and Steenrod (1952)] (Chapter IX, Page 238).

Definition 14.11. Let (X,A) be a pair of spaces with $A \subseteq X$. If $A \neq \emptyset$, then the *relative Čech cohomology modules* $\check{H}^p(X,A;G)$ are defined as the direct limits

$$\check{H}^p(X,A;G) = \varinjlim_{(\mathcal{U},\mathcal{U}^A)\in\mathrm{Cov}(X,A)} \check{H}^p(\mathcal{U},\mathcal{U}^A;G).$$

If $A = \emptyset$, then the *(absolute) Čech cohomology modules* $\check{H}^p(X;G)$ are defined as the direct limits

$$\check{H}^p(X;G) = \varinjlim_{\mathcal{U}\in\mathrm{Cov}(X)} \check{H}^p(\mathcal{U};G).$$

It is clear that the absolute Čech cohomology modules $\check{H}^p(X;G)$ are equal to the classical Čech cohomology modules $\check{H}^p(X;G_X)$ of the constant presheaf G_X as defined in Definition 9.8, since direct limits are taken over $\mathrm{Cov}(X)$.

At this stage, we could proceed with a study of the properties of the relative Čech cohomology modules as in Eilenberg and Steenrod [Eilenberg and Steenrod (1952)], but instead we will state a crucial result due to Dowker [Dowker (1952)] which proves that the relative Čech cohomology modules and the relative Alexander–Spanier cohomology modules are isomorphic; this is also true in the absolute case. This way we are reduced to a study of the properties of the Alexander–Spanier cohomology modules, which is often simpler. For example the proof of the existence of the long exact cohomology sequence in Čech cohomology is quite involved (see Eilenberg and Steenrod [Eilenberg and Steenrod (1952)] (Chapter IX), but is quite simple in Alexander–Spanier cohomology.

This does not mean that Čech cohomology is not interesting. On the contrary, it arises naturally whenever the notion of cover is involved, and it plays an important role in algebraic geometry. It also lends itself to generalizations by extending the notion of cover.

Theorem 14.5. *(Dowker) Let (X, A) be a pair of spaces with $A \subseteq X$. If $A \neq \emptyset$, then the Alexander–Spanier cohomology modules $H^p_{A\text{-}S}(X, A; G)$ and the Čech cohomology modules $\check{H}^p(X, A; G)$ are isomorphic:*

$$H^p_{A\text{-}S}(X, A; G) \cong \check{H}^p(X, A; G) \text{ for all } p \geq 0.$$

If $A = \emptyset$, then we have isomorphisms

$$H^p_{A\text{-}S}(X; G) \cong \check{H}^p(X; G) \text{ for all } p \geq 0.$$

A complete proof of Theorem 14.5 is given in Dowker [Dowker (1952)]; see Theorem 2. Dowker is careful to parametrize the Alexander–Spanier cohomology modules and the Čech cohomology modules with a directed preorder of covers Ω so that he does not run into problems when taking direct limits when $A = \emptyset$. The proof of Theorem 14.5 is also proposed as a sequence of problems in Spanier [Spanier (1989)] (Chapter 6, Problems D1, D2, D3).

14.5 Alexander–Lefschetz Duality

For any R-orientable manifold M, Alexander–Lefschetz duality is a generalization of Poincaré duality that asserts that the Alexander–Spanier cohomology group $H^p_{\text{A-S}}(K, L; G)$ and the singular homology group $H_{n-p}(M - L, M - K; G)$ are isomorphic, where $L \subseteq K \subseteq M$ and L and K are compact. Actually, the method for proving this duality yields an isomorphism between a certain direct limit $\overline{H}^p(K, L; G)$ of singular cohomology groups $H^p(U, V; G)$ where U is any open subset of M containing K and V is any open subset of M containing L, and the singular homology group $H_{n-p}(M - L, M - K; G)$.

Furthermore, it can be shown that $\overline{H}^p(K, L; G)$ and $H^p_{\text{A-S}}(K, L; G)$ are isomorphic, so Alexander–Lefschetz duality can indeed be stated as an isomorphism between $H^p_{\text{A-S}}(K, L; G)$ and $H_{n-p}(M - L, M - K; G)$. Since Alexander–Lefschetz cohomology and Čech cohomology are isomorphic, Alexander–Lefschetz duality can also be stated as an isomorphism between $\check{H}^p(K, L; G)$ and $H_{n-p}(M - L, M - K; G)$, and this is what certain authors do, including Bredon [Bredon (1993)] (Chapter 6, Section 8).

Definition 14.12. Given any topological space X, for any pair (A, B) of subsets of X, let $N(A, B)$ be the set of all pairs (U, V) of open subsets of X such that $A \subseteq U$ and $B \subseteq V$ ordered such that $(U_1, V_1) \leq (U_2, V_2)$ iff $U_2 \subseteq U_1$ and $V_2 \subseteq V_1$ (reverse inclusion); see Figure 14.10.

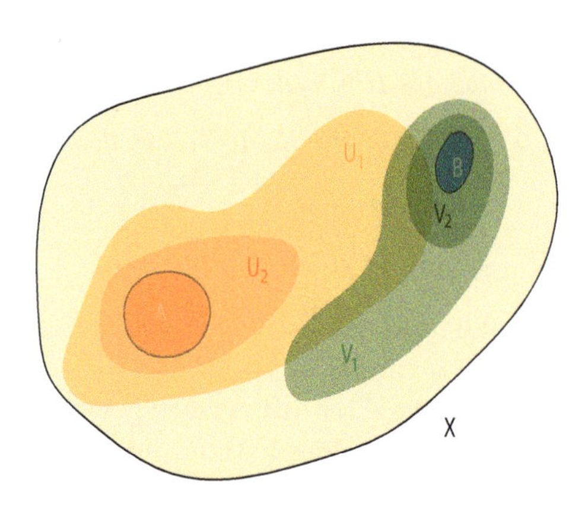

Fig. 14.10 A schematic illustration of two elements of $N(A, B)$ with $(U_1, V_1) \leq (U_2, V_2)$.

Clearly $N(A, B)$ is a directed preorder, and if $(U_1, V_1) \leq (U_2, V_2)$ then there is an induced map of singular cohomology $\rho^{U_1,V_1}_{U_2,V_2} \colon H^p(U_1, V_1; G) \to H^p(U_2, V_2; G)$, so the family $(H^p(U, V; G))_{(U,V)\in N(A,B)}$ together with the maps $\rho^{U_1,V_1}_{U_2,V_2}$ is a direct mapping family.

Definition 14.13. Given any topological space X, for any pair (A, B) of subsets of X, the modules $\overline{H}^p(A, B; G)$ are defined

$$\overline{H}^p(A, B; G) = \varinjlim_{(U,V)\in N(A,B)} H^p(U, V; G) \text{ for all } p \geq 0.$$

The restriction maps $H^p(U, V; G) \longrightarrow H^p(A, B; G)$ yield a natural homomorphism

$$i^p \colon \overline{H}^p(A, B; G) \to H^p(A, B; G)$$

between $\overline{H}^p(A, B; G)$ and the singular cohomology module $H^p(A, B; G)$. In general, i^p neither injective nor surjective. Following Spanier [Spanier (1989)] (Chapter 6, Section 1), we say that the pair (A, B) is *tautly imbedded* in X if every i^p is an isomorphism.

Remark: The notation $\overline{H}^p(A, B; G)$ is borrowed from Spanier [Spanier (1989)] (Chapter 6, Section 1). Bredon denotes the direct limit in Definition 14.13 by $\check{H}^p(A, B; G)$; see Bredon [Bredon (1993)] (Chapter 6, Section 8). He then goes on to say that if X is a manifold and A and B are closed then this group (which is really $\overline{H}^p(A, B; G)$) is naturally isomorphic to the Čech cohomology group. This is indeed true, but this is proven by showing that $\overline{H}^p(A, B; G)$ is isomorphic to the Alexander–Spanier cohomology module $H^p_{\text{A-S}}(A, B; G)$ and then using the isomorphism between the Alexander–Spanier cohomology modules and the Čech cohomology modules. Since these results are nontrivial, we find Bredon's notation somewhat confusing.

It is shown in Spanier ([Spanier (1989)], Chapter 6, Section 1, Corollary 11) that if A, B and X are compact polyhedra, then the pair (A, B) is taut in X, which means that there are isomorphisms $\overline{H}^p(A, B; G) \cong H^p(A, B; G)$, so we can simply use singular cohomology. This is the set-up in which Lefschetz duality was originally proven. We also have the following useful result about manifolds; see Spanier ([Spanier (1989)], Chapter 6, Section 9, Corollary 7).

Proposition 14.6. *If X is a manifold, then $\overline{H}^*(X; G) \cong H^*(X; G)$.*

The following result shows that when X is a manifold and (A, B) is a closed pair, the groups $\overline{H}^p(A, B; G)$ are just the Alexander–Spanier cohomology groups.

Proposition 14.7. *Let X be a manifold. For any pair (A, B) of closed subsets of X, there are isomorphisms*

$$H^p_{A\text{-}S}(A, B; G) \cong \overline{H}^p(A, B; G) \text{ for all } p \geq 0.$$

Proposition 14.7 is proven in Spanier [Spanier (1989)] (Chapter 6, Section 9, Corollary 9).

We are now ready state the main result of this chapter. Let M be an R-orientable manifold. By Theorem 7.7, for any compact subset K of M, there is a unique R-fundamental class $\mu_K \in H_n(M, M - K; R)$ of M at K. In order to state Alexander–Lefschetz duality, we need to define a relative cap product

$$\frown: H^p(U, V; G) \times H_n(M, M - K; R) \to H_{n-p}(M - L, M - K; G).$$

The derivation of this cap product is quite technical and can be skipped during a first reading.

Assume that $L \subseteq K \subseteq M$, $V \subseteq U$, $K \subseteq U$, and $L \subseteq V$, with K, L compact. Then $U - K \subseteq U - L$ and $\{V, U - L\}$ is an open cover of U. We know from Section 7.5 that there is a relative cap product

$$\frown : H^p(X, A; G) \times H_n(X, A \cup B; R) \to H_{n-p}(X; B, G),$$

so with $X = U$, $A = V$, and $B = U - K$, we have a cap product

$$\frown : S^p(U, V; G) \times S_n(U, V \cup (U - K); R) \to S_{n-p}(U, U - K; G).$$

We claim that the above cap product induces a cap product

$$\frown : S^p(U, V; G) \times S_n(U, U - K; R) \to S_{n-p}(U - L, U - K; G).$$

Since $U - K \subseteq V \cup (U - K)$, we have a homomorphism

$$i \colon S_n(U, U - K; R) \to S_n(U, V \cup (U - K); R),$$

where the equivalence class of $a \in S_n(U; R)$ mod $S_n(U - K; R)$ is mapped to the equivalence class of a mod $S_n(V \cup (U - K); R)$. Recall that a cochain $f \in S^p(U, V; G)$ is a cochain in $S^p(U; G)$ that vanishes on simplices in V. Also since $U = V \cup (U - L)$, any chain σ in $S_n(U, V \cup (U - K); R) = S_n(V \cup (U - L), V \cup (U - K); R)$ is represented by a sum of the form

$$a + b + c,$$

with $a \in S_n(V; R)$, $b \in S_n(U - L; R)$ and $c \in S_n(V \cup (U - K); R)$. Since $S_n(V; R) \subseteq S_n(V \cup (U - K); R)$, we see that $a \in S_n(V \cup (U - K); R)$ and so σ is also represented by some element $b + d$ with $b \in S_n(U - L; R)$ and $d \in S_n(V \cup (U - K); R)$. See Figure 14.11.

Then we have

$$f \frown (b + d) = f \frown b + f \frown d,$$

with $f \frown b \in S_{n-p}(U - L; G)$, and since f vanishes on V and $d \in S_n(V \cup (U - K); R)$ the term $f \frown d$ belongs to $S_{n-p}(U - K; G)$, so in the end $f \frown (b + d)$ represents a cycle in $S_{n-p}(U - L, U - K; G)$. Passing to cohomology and homology, since by excision

$$H_n(M, M - K; R) \cong H_n(U, U - K; R)$$
$$H_{n-p}(M - L, M - L; G) \cong H_{n-p}(U - L, U - K; G),$$

the cap product

$$\frown : S^p(U, V; G) \times S_n(U, U - K; R) \to S_{n-p}(U - L, U - K; G)$$

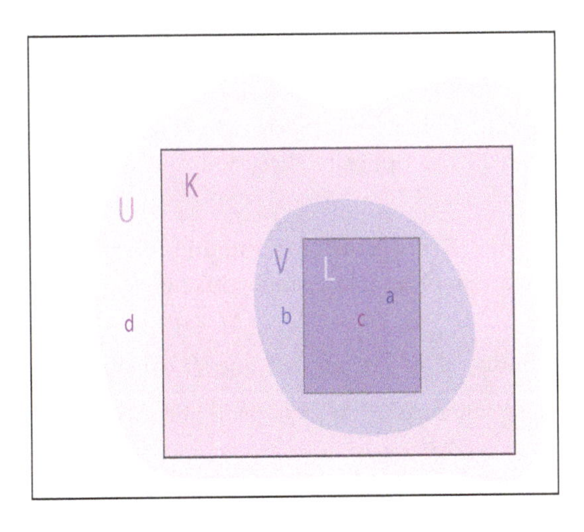

Fig. 14.11 A schematic depiction illustrating the construction of $\frown\colon S^p(U,V;G) \times S_n(U,U-K;R) \to S_{n-p}(U-L,U-K;G)$, where $a \in S_n(V;R)$, $b \in S_n(U-L;R)$, $c \in S_n(V \cup (U-K);R)$, and $d \in S_n(V \cup (U-K);R)$.

induces a cap product

$$\frown\colon H^p(U,V;G) \times H_n(M,M-K;R) \to H_{n-p}(M-L,M-K;G).$$

If M is an R-orientable manifold, for any pair (K,L) of compact subsets of M such that $L \subseteq K$ and for any pair $(U,V) \in N(K,L)$, we obtain a map

$$\frown \mu_K\colon H^p(U,V;G) \to H_{n-p}(M-L,M-K;G),$$

and by a limit argument, we obtain a map

$$\frown \mu_K\colon \overline{H}^p(K,L;G) \to H_{n-p}(M-L,M-K;G);$$

for details see Bredon [Bredon (1993)] (Chapter 6, Section 8).

Theorem 14.8. *(Alexander–Lefschetz duality) Let M be an R-orientable manifold where R is any commutative ring with an identity element. For any R-module G, for any pair (K,L) of compact subsets of M such that $L \subseteq K$, the map $\omega \mapsto \omega \frown \mu_K$ yields an isomorphism*

$$\overline{H}^p(K,L;G) \cong H_{n-p}(M-L,M-K;G) \text{ for all } p \geq 0.$$

Thus we also have isomorphisms

$$H^p_{A\text{-}S}(K,L;G) \cong \check{H}^p(K,L;G) \cong H_{n-p}(M-L,M-K;G) \text{ for all } p \geq 0.$$

Theorem 14.8 is proven in Bredon [Bredon (1993)] where it is called the *Poincaré–Alexander–Lefschetz duality* (Chapter 6, Section 8, Theorem 8.3) by using the Bootstrap Lemma (Proposition 7.6). It is also proven in Spanier [Spanier (1989)] (Chapter 6, Section 2, Theorem 17), except that the isomorphism goes in the opposite direction and does not use the fundamental class μ_K.

If we let $K = M$ and $L = \emptyset$, since for a manifold we have $\overline{H}^p(M;G) \cong H^p(M;G)$, then Theorem 14.8 yields isomorphisms

$$H^p(M;G) \cong H_{n-p}(M;G),$$

which is Poincaré duality if M is compact and R-orientable.

In the special case where $K = M$, we get a version of *Lefschetz duality* for M compact:

Theorem 14.9. *(Lefschetz Duality, Version 1) Let M be a compact R-orientable n-manifold where R is any commutative ring with an identity element. For any R-module G, for any compact subset L of M, we have isomorphisms*

$$H^p_{A\text{-}S}(M, L; G) \cong \check{H}^p(M, L; G) \cong H_{n-p}(M - L; G) \text{ for all } p \geq 0.$$

A version of Lefschetz duality where M and L are compact and triangulable, in which case singular cohomology suffices, is proven in Munkres [Munkres (1984)] (Chapter 8, Theorem 72.3).

Spanier proves a slightly more general version. A pair (X, A) is called a *relative n-manifold* if X is a Hausdorff space, A is closed in X, and $X - A$ is an n-manifold.

Theorem 14.10. *(Lefschetz Duality, Version 2) Let (X, A) be a compact relative n-manifold such that $X - A$ is R-orientable where R is any commutative ring with an identity element. For any R-module G, there are isomorphisms*

$$H^p_{A\text{-}S}(X, A; G) \cong \check{H}^p(X, A; G) \cong H_{n-p}(X - A; G) \text{ for all } p \geq 0.$$

Theorem 14.9 is proven in Spanier [Spanier (1989)] (Chapter 8, Section 2, Theorem 18).

There are also version of Poincaré and Lefschetz duality for manifolds with boundary but we will omit this topic. The interested reader is referred to Spanier [Spanier (1989)] (Chapter 8, especially Section 2).

We now turn to two versions of Alexander duality.

14.6 Alexander Duality

Alexander duality corresponds to the special case of Alexander–Lefschetz duality in which $L = \emptyset$. We begin with a version of Alexander duality in the situation where $M = \mathbb{R}^n$.

Theorem 14.11. *(Alexander–Pontrjagin duality) Let A be a compact subset of $\mathbb{R}^n$. For any commutative ring R with an identity element, for any R-module G, we have isomorphisms*

$$H^{n-p-1}_{A\text{-}S}(A;G) \cong \check{H}^{n-p-1}(A;G) \cong \widetilde{H}_p(\mathbb{R}^n - A;G) \text{ for all } p \leq n.$$

Proof. By Theorem 14.8 with $M = \mathbb{R}^n$, $K = A$ and $L = \emptyset$, there are isomorphisms

$$\check{H}^{n-p-1}(A;G) \cong H_{p+1}(\mathbb{R}^n, \mathbb{R}^n - A;G) \text{ for all } p \leq n-1.$$

We also have the long exact sequence of reduced homology of the pair $(\mathbb{R}^n, \mathbb{R}^n - A)$, which yields exact sequences

$$\widetilde{H}_{p+1}(\mathbb{R}^n;G) \longrightarrow \widetilde{H}_{p+1}(\mathbb{R}^n, \mathbb{R}^n - A;G) \longrightarrow \widetilde{H}_p(\mathbb{R}^n - A;G) \longrightarrow \widetilde{H}_p(\mathbb{R}^n;G),$$

and since $\widetilde{H}_{p+1}(\mathbb{R}^n;G) \cong \widetilde{H}_p(\mathbb{R}^n;G) \cong (0)$ (because $\mathbb{R}^n$ is contractible and by the facts stated just after Definition 4.20), we conclude that

$$H_{p+1}(\mathbb{R}^n, \mathbb{R}^n - A;G) = \widetilde{H}_{p+1}(\mathbb{R}^n, \mathbb{R}^n - A;G) \cong \widetilde{H}_p(\mathbb{R}^n - A;G),$$

which proves our result. ∎

Here is another version of Alexander duality in which $M = S^n$. Recall from Section 4.9 that the relationship between the cohomology and the reduced cohomology of a space X is

$$\begin{aligned} H^0(X;G) &\cong \widetilde{H}^0(X;G) \oplus G \\ H^p(X;G) &\cong \widetilde{H}^p(X;G), \quad p \geq 1. \end{aligned}$$

Theorem 14.12. *(Alexander duality) Let A be a proper closed nonempty subset of S^n. For any commutative ring R with an identity element, for any R-module G, we have isomorphisms*

$$\widetilde{H}_p(S^n - A;G) \cong \begin{cases} \check{H}^{n-p-1}(A;G) & \text{if } p \neq n-1 \\ \tilde{\check{H}}^0(A;G) & \text{if } p = n-1, \end{cases}$$

or equivalently

$$\tilde{\check{H}}^{n-p-1}(A;G) \cong \widetilde{H}_p(S^n - A;G) \text{ for all } p \leq n.$$

Proof. The case $n = 0$ is easily handled, so assume $n > 0$. By Theorem 14.8 with $M = S^n$, $K = A$ and $L = \emptyset$, there are isomorphisms

$$\check{H}^{n-p-1}(A; G) \cong H_{p+1}(S^n, S^n - A; G) \text{ for all } p \leq n - 1.$$

We also have the long exact sequence of reduced homology of the pair $(S^n, S^n - A)$, which yields exact sequences

$$\widetilde{H}_{p+1}(S^n; G) \longrightarrow \widetilde{H}_{p+1}(S^n, S^n - A; G) \longrightarrow \widetilde{H}_p(S^n - A; G) \longrightarrow \widetilde{H}_p(S^n; G).$$

By Proposition 4.18 the reduced homology of S^n is given by

$$\widetilde{H}_p(S^n; G) = \begin{cases} G & \text{if } p = n \\ (0) & \text{if } p \neq n. \end{cases}$$

It follows that we have isomorphisms

$$H_{p+1}(S^n, S^n - A; G) = \widetilde{H}_{p+1}(S^n, S^n - A; G) \cong \widetilde{H}_p(S^n - A; G)$$

for $p \neq n - 1$. If $p = n - 1$ we have the following commutative diagram

$$\begin{array}{ccccccccc}
0 & \longrightarrow & H^0(S^n) & \longrightarrow & \check{H}^0(A) & \longrightarrow & \widetilde{\check{H}}^0(A) & \longrightarrow & 0 \\
 & & \downarrow & & \downarrow & & \vdots & & \\
H_n(S^n - A) & \xrightarrow{0} & H_n(S^n) & \longrightarrow & H_n(S^n, S^n - A) & \longrightarrow & \widetilde{H}_{n-1}(S^n - A) & \xrightarrow{0} &
\end{array}$$

in which the left vertical solid arrow is an isomorphism by Poincaré duality, the right vertical solid arrow is an isomorphism by Theorem 14.8, the bottom row is exact by the long exact sequence of reduced homology, and the top one because

$$\check{H}^0(A) \cong \widetilde{\check{H}}^0(A) \oplus G$$

and $H^0(S^n) \cong H_n(S^n) \cong G$. We have zero maps on the bottom because the inclusion map $S^n - A \longrightarrow S^n$ factors through a contractible space $S^n - \{\text{pt}\}$. It is easy to see that the kernel of the map from $\check{H}^0(A)$ to $\widetilde{H}_{n-1}(S^n - A)$ is isomorphic to $H^0(S^n)$, so this map factors through $\widetilde{\check{H}}^0(A)$ as the dotted arrow, and using the commutative diagram and the fact that the rows are exact it is easy to show that the dotted arrow is an isomorphism. $\square$

Remark: This version involving Čech (or Alexander–Spanier) cohomology is a generalization of Alexander's original version that applies to a polyhedron in S^n, and only requires singular cohomology; see Munkres [Munkres (1984)] (Chapter 8, Theorem 72.4).

An interesting corollary of Theorem 14.9 is the following generalization of the version of the Jordan curve theorem stated in Theorem 4.21. For comparison with Theorem 14.13 below think of M as S^n and of A as C.

Theorem 14.13. *(Generalized Jordan curve theorem) Let M be a connected, orientable, compact n-manifold, and assume that $H_1(M; R) = (0)$ for some ring R (with unity). For any proper closed subset A of M, the module $\check{H}^{n-1}(A; R)$ is a free R-module such that if r is its rank, then $r+1$ is equal to the number of connected components of $M - A$.*

Proof. The number of connected components of $M - A$ is equal to the rank s of $H_0(M - A; R)$, and since $H_0(M - A; R) \cong \widetilde{H}_0(M - A; G) \oplus R$ we have $s = t + 1$ with $t = \mathrm{rank}(\widetilde{H}_0(M - A; G))$. By the long exact sequence of reduced homology of the pair $(M, M - A)$ we have the exact sequence

$$H_1(M; R) \longrightarrow H_1(M, M - A; R) \longrightarrow \widetilde{H}_0(M - A; R) \longrightarrow \widetilde{H}_0(M; R).$$

Since $H_1(M; R) = (0)$ and since M is connected $\widetilde{H}_0(M; R) = (0)$ so we get the isomorphism

$$\widetilde{H}_0(M - A; R) \cong H_1(M, M - A; R).$$

By Lefschetz duality (Theorem 14.9) we have

$$H_1(M, M - A; R) \cong \check{H}^{n-1}(A; R),$$

and thus

$$\check{H}^{n-1}(A; R) \cong \widetilde{H}_0(M - A; R),$$

which shows that $\check{H}^{n-1}(A; R)$ is a free R-module with rank $r = t = s - 1$, where s is the number of connected component of $M - A$. $\square$

Recall that given two topological spaces X and Y we say that there is an *embedding* of X into Y if there is a homeomorphism $f\colon X \to Y$ of X onto its image $f(X)$. As a corollary of Theorem 14.13 we get the following result.

Proposition 14.14. *Let M be a connected, orientable, and compact n-manifold M. If $H_1(M; \mathbb{Z}) = (0)$, then no nonorientable compact $(n-1)$-manifold N can be embedded in M.*

Proof. If the $(n-1)$-manifold N is nonorientable, then by Proposition 7.11 $H^{n-1}(N; \mathbb{Z}) \cong \mathbb{Z}/2\mathbb{Z}$, and since N is a manifold $H^{n-1}(N; \mathbb{Z}) \cong \check{H}^{n-1}(N; \mathbb{Z})$, so $\check{H}^{n-1}(N; \mathbb{Z}) \cong \mathbb{Z}/2\mathbb{Z}$, which contradicts Theorem 14.13 (since $\mathbb{Z}/2\mathbb{Z}$ is not free). $\square$

Proposition 14.14 implies that $\mathbb{RP}^{2n}$ cannot be embedded into S^{2n+1}. In particular $\mathbb{RP}^2$ cannot be embedded into S^3.

More applications of duality are presented in Bredon [Bredon (1993)] (Chapter 6, Section 10). In particular, it is shown that for all $n \geq 2$ (not just even) the real projective space $\mathbb{RP}^n$ cannot be embedded in S^{n+1}.

We conclude this chapter by stating a generalization of Alexander–Lefschetz duality for cohomology with compact support.

14.7 Alexander–Lefschetz Duality for Cohomology with Compact Support

The Alexander–Spanier cohomology modules with compact support $H_{\text{A-S},c}(X, A; G)$ were defined in Section 14.3. Alexander–Lefschetz duality (Theorem 14.8) can be generalized to arbitrary closed pairs (K, L) (not necessarily compact), using the modules $H_{\text{A-S},c}(X, A; G)$ instead of the modules $H_{\text{A -S}}(X, A; G)$, in a way which is reminiscent of the general Poincaré duality theorem (Theorem 7.16).

Theorem 14.15. *(Alexander–Lefschetz duality) Let M be an R-orientable manifold where R is any commutative ring with an identity element. For any R-module G, for any pair (K, L) of closed subsets of M such that $L \subseteq K$, there is an isomorphism*

$$H^p_{A\text{-}S,\ c}(K, L; G) \cong H_{n-p}(M - L, M - K; G) \ \textit{for all } p \geq 0.$$

Theorem 14.15 is proven in Spanier [Spanier (1989)] (Chapter 6, Section 9, Theorem 10) and in Dold [Dold (1995)] (Chapter VIII, Section 7, Proposition 7.14). It should be noted that Spanier's proof provides an isomorphism in the other direction (from homology to cohomology) and does not involve the cap product. However, Dold's version uses a version of the cap product obtained by a limit argument.

14.8 Problems

Problem 14.1. Prove that if M is an orientable connected manifold and if L is a proper compact subset of M, then $\overline{H}^n(L; G) = (0)$.

Problem 14.2. Let M be a connected, orientable, compact n-manifold and let N be a compact connected $(n-1)$-submanifold of M. Prove that if $H_1(M;\mathbb{Z}) = (0)$, then $M - N$ has exactly two components with N as the topological boundary of each.

Problem 14.3. Show that Theorem 14.13 is false if $\check{H}^{n-1}$ is replaced by H^{n-1}.

Problem 14.4. Prove that if $U \subseteq \mathbb{R}^3$ is open, then $H_1(U)$ is torsion-free.

Problem 14.5. Show that Proposition 14.14 remains true if the hypothesis that $H_1(M;\mathbb{Z}) = (0)$ is weakened to $H_1(M;\mathbb{Z}/2\mathbb{Z}) = (0)$.

Chapter 15

Spectral Sequences

A spectral sequence is a tool of homological algebra whose purpose is to approximate the cohomology (or homology) $H(M)$ of a module M endowed with a family $(F^pM)_{p\in\mathbb{Z}}$ of submodules such that $F^{p+1}M \subseteq F^pM$ for all p and

$$M = \bigcup_{p\in\mathbb{Z}} F^pM,$$

called a filtration. The module M is also equipped with a linear map $d\colon M \to M$ called differential such that $d \circ d = 0$, so that it makes sense to define

$$H(M) = \mathrm{Ker}\, d/\mathrm{Im}\, d.$$

We say that (M, d) is a differential module. To be more precise, the filtration induces cohomology submodules $H(M)^p$ of $H(M)$, the images of $H(F^pM)$ in $H(M)$, and a spectral sequence is a sequence of modules E_r^p (equipped with a differential d_r^p), for $r \geq 1$, such that E_r^p approximates the "graded piece" $H(M)^p/H(M)^{p+1}$ of $H(M)$.

Actually, to be useful, the machinery of spectral sequences must be generalized to filtered cochain complexes. Technically this implies dealing with objects $E_r^{p,q}$ involving three indices, which makes its quite challenging to follow the exposition.

Many presentations jump immediately to the general case, but it seems pedagogically advantageous to begin with the simpler case of a single filtered differential module. This the approach followed by Serre in his dissertation [Serre (2003)] (Pages 24–104, *Annals of Mathematics*, 54 (1951), 425–505), Godement [Godement (1958)], and Cartan and Eilenberg [Cartan and Eilenberg (1956)].

Spectral sequences were first introduced by Jean Leray in 1945 and 1946. Paraphrazing Jean Dieudoné [Dieudonné (1989)], Leray's definitions were cryptic and proofs were incomplete. Koszul was the first to give a clear definition of spectral sequences in 1947. This is the definition that has been used even since. Independently, in his dissertation (1946), Lyndon introduced spectral sequences in the context of group extensions.

Detailed expositions of spectral sequences do not seem to have appeared until 1951, in lecture notes by Henri Cartan and in Serre's dissertation [Serre (2003)], which we highly recommend for its clarity (Serre defines homology spectral sequences, but the translation to cohomology is immediate). A concise but very clear description of spectral sequences appears in Dieudonné [Dieudonné (1989)] (Chapter 4, Section 7, Parts D, E, F). More extensive presentations appeared in Cartan and Eilenberg [Cartan and Eilenberg (1956)] and Godement [Godement (1958)] around 1955. Every "advanced" book on algebraic topology and homological algebra published between 1960 and 1980 has a treatment of spectral sequences: Mac Lane [Mac Lane (1975)], Rotman [Rotman (1979, 2009)], Spanier [Spanier (1989)]. More recent references are Weibel [Weibel (1994)] and McCleary [McCleary (2001)]. Arm yourself with patience.

The first spectacular application of spectral sequences was made by Serre in his dissertation (1951) [Serre (2003)]. Serre used spectral sequences and other methods he invented (Serre classes of abelian groups) to prove the following results about the homotopy groups $\pi_m(S^n)$ of spheres (it was already known that $\pi_i(S^n) = (0)$ for $i < n$):

(1) The homotopy groups $\pi_m(S^n)$ are finitely generated (this was not known before Serre proved it).
(2) If n is odd, then $\pi_m(S^n)$ is finite if $m \neq n$.
(3) If n is even, then $\pi_m(S^n)$ is finite if $m \neq n$ and if $m \neq 2n - 1$.

The above results are presented in English in Spanier [Spanier (1989)] (Chapter 9).

Double complexes are a major source of spectral sequences. A double complex is a direct sum

$$C = \bigoplus_{p,q \in \mathbb{N}} C^{p,q},$$

equipped with a horizontal differential d_{I} and a vertical differential d_{II} that anticommute; see Section 15.9. A double complex yields the singly graded module $\mathrm{Tot}(C)$, called its *total space*, with

$$\mathrm{Tot}(C) = \bigoplus_{n\in\mathbb{N}} C^n, \quad \text{with} \quad C^n = \bigoplus_{p+q=n} C^{p,q},$$

and the differential $D = d_{\mathrm{I}} + d_{\mathrm{II}}$. There are two natural filtrations on C and Tot, thus two spectral sequences I and II associated with them. These spectral sequences can be used to compare the cohomology modules $H^p_{\mathrm{I}}(H^q_{\mathrm{II}}(C_{\mathrm{I}}))$ (associated with the first spectral sequence), where C_{I} is the complex C with the first filtration and the differential d_{I}, and $H^p_{\mathrm{II}}(H^q_{\mathrm{I}}(C_{\mathrm{II}}))$ (associated with the second spectral sequence), where C_{II} is the complex C with the second filtration and the differential d_{II}. Technically $H^q_{\mathrm{II}}(C_{\mathrm{I}})$ and $H^q_{\mathrm{I}}(C_{\mathrm{II}})$ are certain complexes; see Section 15.9. Under certain conditions (when the spectral sequences degenerate for $q = 0$), both $H^p_{\mathrm{I}}(H^0_{\mathrm{II}}(C_{\mathrm{I}}))$ and $H^p_{\mathrm{II}}(H^0_{\mathrm{I}}(C_{\mathrm{II}}))$ are isomorphic to the cohomology of the total space $\mathrm{Tot}(C)$ (with differential D), and we obtain an isomorphism between two kinds of cohomology because the complexes $H^0_{\mathrm{II}}(C_{\mathrm{I}})$ and $H^0_{\mathrm{I}}(C_{\mathrm{II}})$ can be recognized as known complexes. This will happen for sheaf cohomology, Čech cohomology of a cover, and Čech cohomology.

This phenomenon will be illustrated in Section 15.10 (where the row cohomology is sheaf cohomology), Section 15.11 (where the row cohomology is the Čech cohomology of a cover $\mathcal{U}$) and Section 15.12 (where the row cohomology is Čech cohomology). In all three cases, the column cohomology involves a family of sheaves $\mathcal{F}^* = (\mathcal{F}^p)$ called a differential sheaf. The column cohomology is usually harder to compute than the row cohomology precisely because it involves the whole family $\mathcal{F}^* = (\mathcal{F}^p)$. If the differential sheaf $\mathcal{F}^*$ has extra properties, for example, it is a resolution of a sheaf by special sheaves, then it can be computed.

A last comment before we launch into the presentation of spectral sequences. It seems unfortunate to us that the term "spectral" is used, since it is already used for other totally unrelated concepts: spectral theorems in linear algebra, spectral analysis in Fourier theory, *etc.* But as many other terms in mathematics ("normal"), overloading of the terminology should not stop us from moving on.

There are several methods for defining spectral sequences, including the following three:

(1) Koszul's original approach as described by Serre [Serre (2003)] and Godement [Godement (1958)]. In our opinion it is the simplest method to understand what is going on.

(2) Cartan and Eilenberg's approach [Cartan and Eilenberg (1956)]. This is a somewhat faster and slicker method than the previous method.
(3) Exact couples of Massey (1952). This somewhat faster method for defining spectral sequences is adopted by Rotman [Rotman (1979, 2009)] and Bott and Tu [Bott and Tu (1986)]. Mac Lane [Mac Lane (1975)], Weibel [Weibel (1994)], and McCleary [McCleary (2001)] also present it and show its equivalence with the first approach. It appears to be favored by algebraic topologists. This approach leads to spectral sequences in a quicker fashion and is more general because exact couples need not arise from a filtration, but our feeling is that it is even more mysterious to a novice than the first two approaches.

We will primarily follow Method (1) and present Method (2) and Method (3) in starred sections (Method (2) in Section 15.15 and Method (3) in Section 15.14). All three methods produce isomorphic sequences, and we will show their equivalence.

15.1 Case 1: Filtered Differential Modules

We begin by giving an idea of what is the spectral sequence associated with a filtered differential module. The ingredient that leads to spectral sequences is the notion of a filtration.

Definition 15.1. Given a R-module M, a *decreasing filtration* on M is a family $(F^p(M))_{p\in\mathbb{Z}}$ of R-submodules of M such that

$$F^{p+1}(M) \subseteq F^p(M) \quad \text{for all } p \in \mathbb{Z} \quad \text{and} \quad M = \bigcup_{p\in\mathbb{Z}} F^p(M).$$

An R-module M equipped with a filtration $(F^p(M))_{p\in\mathbb{Z}}$ is called a *filtered module*. For simplicity of notation, we write F^pM instead of $F^p(M)$.

In most applications we have

$$\bigcap_{p\in\mathbb{Z}} F^p(M) = (0),$$

so in an intuitive sense we can think of the elements of F^pM as being more and more negligible as p goes to $+\infty$; see Figure 15.1.

Example 15.1. Let M be a direct sum

$$M = \bigoplus_{p\in\mathbb{Z}} M^p$$

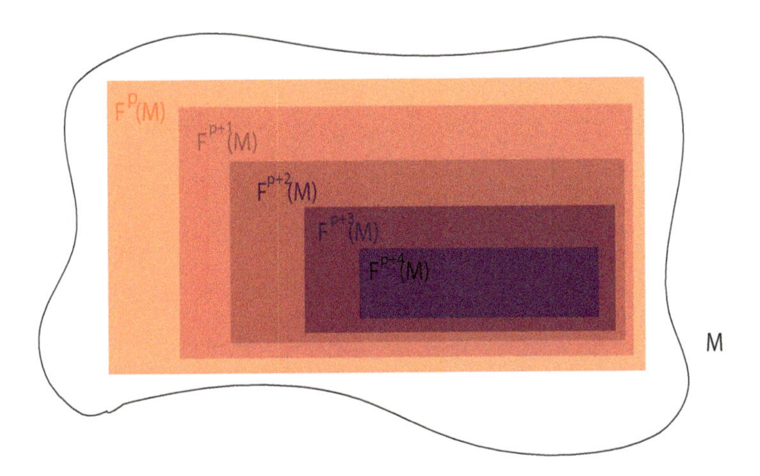

Fig. 15.1 A schematic representation of the filtered module M. As the superscripts of the filtration become more positive, the associated F^pM is smaller.

of R-modules M^p. A direct sum M as above is called a *$\mathbb{Z}$-graded module* (for short, a *graded module*). Recall that every element of M is a finite sum $\sum_{i\in I} u_i$ of elements $u_i \in M^i$ (where I is a finite subset of $\mathbb{Z}$). We have the filtration given by

$$F^pM = \bigoplus_{i\geq p} M^i.$$

Example 15.2. Let M be a direct sum

$$M = \bigoplus_{p,q\in\mathbb{Z}} M^{pq}$$

of R-modules M^{pq}. A direct sum M as above is called a *bigraded module*. We have two filtrations (F_{I}^pM) and (F_{II}^qM) given by

$$F_{\mathrm{I}}^pM = \bigoplus_{i\geq p}\bigoplus_{q\in\mathbb{Z}} M^{iq},$$

which is the direct sum of the columns of index i greater than or equal to p and

$$F_{\mathrm{II}}^qM = \bigoplus_{j\geq q}\bigoplus_{p\in\mathbb{Z}} M^{pj},$$

which is the direct sum of the rows of index greater j than or equal to q; see Figures 15.2 and 15.3.

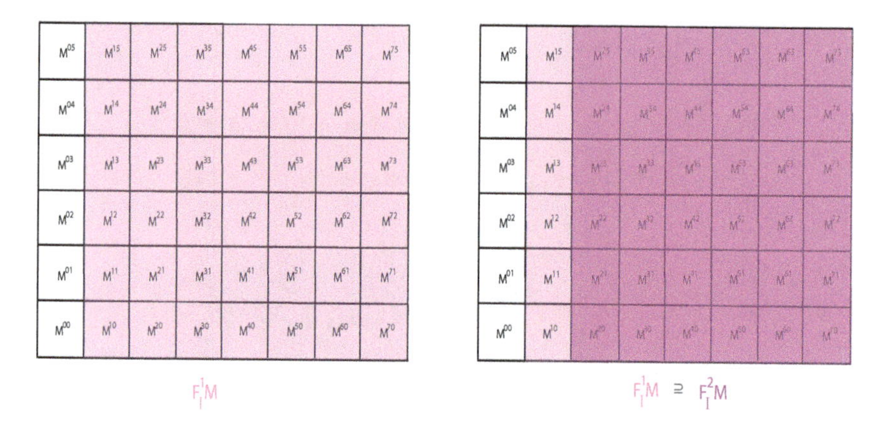

Fig. 15.2 A schematic first quadrant representation of the bigraded module M and its column filtration $(F_{\mathrm{I}}^p M)$. Each time p increases by one, the left column is excluded.

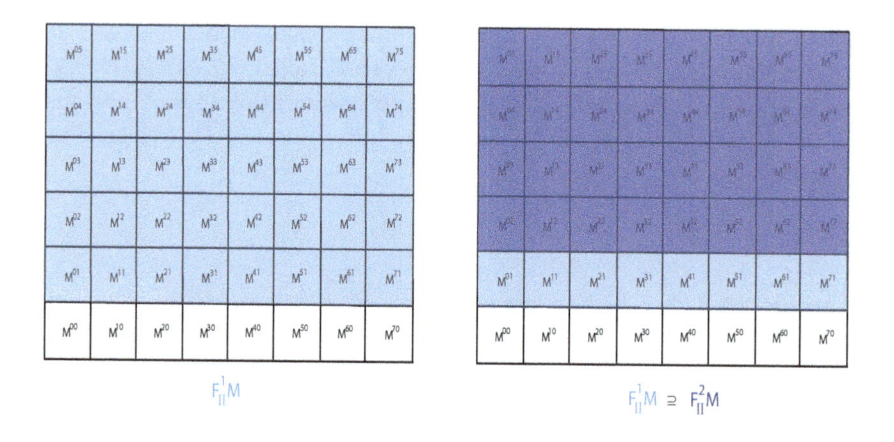

Fig. 15.3 A schematic first quadrant representation of the bigraded module M and its row filtration $(F_{\mathrm{II}}^q M)$. Each time q increases by one, the bottom row is excluded.

The key ingredient which allows the definition of cohomology (or homology) of a module M is an R-linear map such that $d \circ d = 0$.

Definition 15.2. A *differential module* is a pair (M, d), where M is an R-module and $d\colon M \to M$ is an R-linear map such that $d \circ d = 0$, called the *differential of* M. Let $Z(M) = \mathrm{Ker}\, d$, the set of *cocycles*, and $B(M) = \mathrm{Im}\, d$, the set of *coboundaries*. If no confusion is arises we write Z for $Z(M)$ and

B for $B(M)$. Since $d \circ d = 0$, we have $B \subseteq Z$, and the R-module

$$H(M) = Z/B = \operatorname{Ker} d/\operatorname{Im} d$$

is called the *derived module* of M.

If M also has a filtration (F^pM), then we require that $d(F^pM) \subseteq F^pM$ for all $p \in \mathbb{Z}$, and we call such a module a *filtered differential module*.

If (M, d) is a filtered differential module with filtration (F^pM), since $d(F^pM) \subseteq F^pM$, the restriction of the differential d to F^pM is a differential on F^pM denoted d_{F^p}.

We are interested in the cohomology $H(M)$ of M, but this may be hard to compute. This is where a filtration (F^pM) on M is useful. Indeed, a filtration (F^pM) on M induces a filtration $(H(M)^p)$ of $H(M)$. Then we can compute the quotient modules $H(M)^p/H(M)^{p+1}$.

In general it is not possible to reconstruct $H(M)$ from the modules $H(M)^p/H(M)^{p+1}$, but still this constitutes some progress. The problem is that given a module M with a filtration (F^pM), the sequence of quotient modules $F^pM/F^{p+1}M$ does not determine M uniquely, in other words, the direct sum $\bigoplus_{p\in\mathbb{Z}} F^pM/F^{p+1}M$ (denoted as $\mathrm{gr}(M)$; see Definition 15.5) does not uniquely determine M.

For example, consider the abelian groups $M_1 = \mathbb{Z}/4\mathbb{Z}$ and $M_2 = \mathbb{Z}/2\mathbb{Z} \oplus \mathbb{Z}/2\mathbb{Z}$, which are not isomorphic. Filter M_1 by

$$\mathbb{Z}/4\mathbb{Z} \supseteq \mathbb{Z}/2\mathbb{Z} \supseteq (0)$$

and M_2 by

$$\mathbb{Z}/2\mathbb{Z} \oplus \mathbb{Z}/2\mathbb{Z} \supseteq \mathbb{Z}/2\mathbb{Z} \supseteq (0).$$

Then $F^pM_1/F^{p+1}M_1 \cong F^pM_2/F^{p+1}M_2$ for $p = 0, 1$, since these groups are $\mathbb{Z}/2\mathbb{Z}, \mathbb{Z}/2\mathbb{Z}$, even though M_1 and M_2 are not isomorphic. We have what is called the *extension problem*, and such a problem lies at the heart of homological algebra (and was one of the main motivations for the development of homological algebra).

In special cases, $H(M)$ can be completely recovered from the modules $H(M)^p/H(M)^{p+1}$.

The purpose of a spectral sequence is to approximate the quotients $H(M)^p/H(M)^{p+1}$, also denoted E^p_∞, by a sequence of approximations E^p_r which gets better and better as r increases, starting with $E^p_0 = F^pM/F^{p+1}M$.

A filtration $(H(M)^p)$ of $H(M)$ can be obtained from the cohomology modules $H(F^pM)$ but an extra step is needed as we now explain.

As we said before, each submodule module F^pM of M is a differential module with d_{F^p}, the restriction of d to F^pM, as differential. Observe that $\operatorname{Ker} d_{F^p} = \operatorname{Ker} d \cap F^pM = Z \cap F^pM$ and $\operatorname{Im} d_{F^{p+1}} = dF^{p+1}M \subseteq dF^pM$.

Definition 15.3. The cohomology of the submodule F^pM is defined as

$$H(F^pM) = \operatorname{Ker} d_{F^p}/\operatorname{Im} d_{F^p}.$$

The "obvious" homomorphism from $H(F^{p+1}M)$ to $H(F^pM)$ given by $[z]_{\operatorname{Im} d_{F^{p+1}}} \mapsto [z]_{\operatorname{Im} d_{F^p}}$, with $z \in Z \cap F^{p+1}M$, is generally not injective, since if $z \in \operatorname{Im} d_{F^p} \cap F^{p+1}M \subseteq Z \cap F^{p+1}M$, then $[z]_{\operatorname{Im} d_{F^p}} = 0$; see Figure 15.4. The module $H(F^pM)$ is also not a submodule of $H(M)$. However, we can map $H(F^pM)$ in $H(M)$ (not necessarily in an injective fashion) in such a way that we obtain a filtration of $H(M)$.

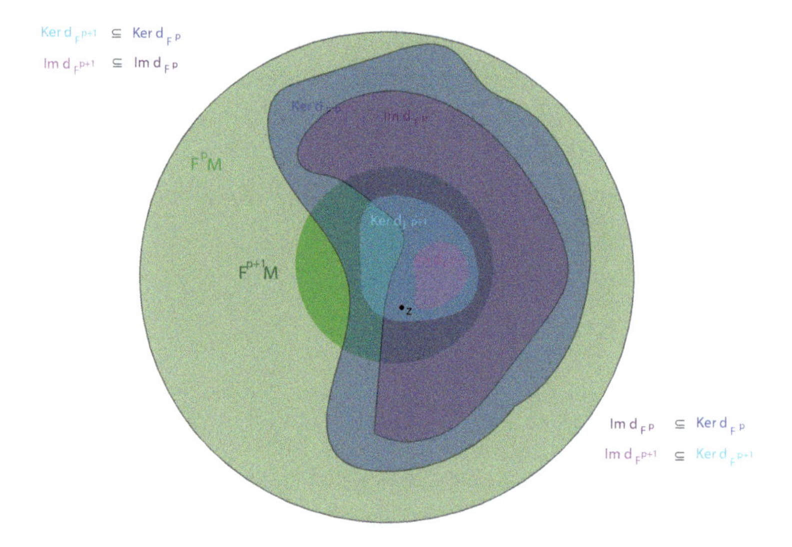

Fig. 15.4 A schematic representation illustrating the inclusion relationships between the kernels and images of d_{F^p} and $d_{F^{p+1}}$. Note there is no relationship between $\operatorname{Im} d_{F^p}$ and $\operatorname{Ker} d_{F^{p+1}}$. Furthermore since $z \in \operatorname{Im} d_{F^p} \cap F^{p+1}M \subseteq Z \cap F^{p+1}M$, then $[z]_{\operatorname{Im} d_{F^p}} = 0$.

Indeed, we have a linear map

$$\eta^p \colon \operatorname{Ker} d_{F^p}/\operatorname{Im} d_{F^p} \to \operatorname{Ker} d/\operatorname{Im} d$$

given by

$$\eta^p([z]_{\mathrm{Im}\, d_{F^p}}) = [z]_B,$$

where $[z]_{\mathrm{Im}\, d_{F^p}}$ is the equivalence class of $z \in Z \cap F^pM = \mathrm{Ker}\, d_{F^p}$ modulo $\mathrm{Im}\, d_{F^p}$ and $[z]_B$ is the equivalence class of z modulo $B = \mathrm{Im}\, d$; see Figure 15.5. Observe that if $z \in B \cap F^pM$, then $\eta^p([z]) = 0$.

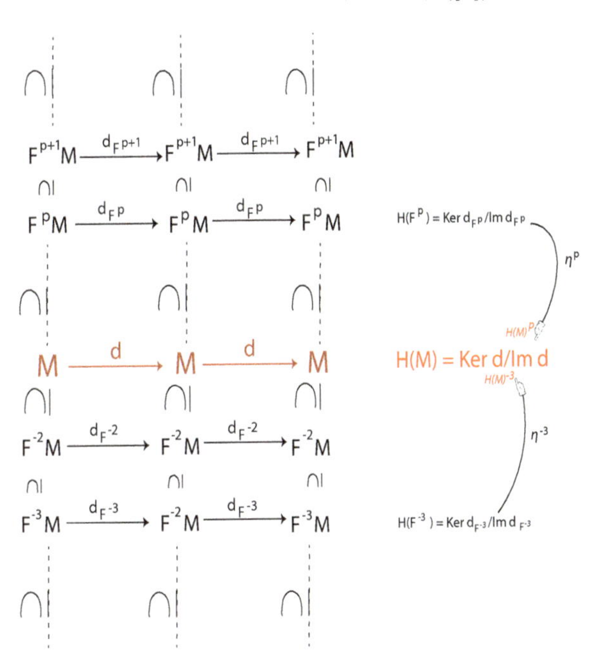

Fig. 15.5 An illustration of the η^p mapping of $H(F^pM)$ into $H(M)$.

Definition 15.4. We define the R-module $H(M)^p$ as the image of $H(F^pM)$ in $H(M)$ by η^p, namely,

$$H(M)^p = \{[z]_B \mid z \in Z \cap F^p(M)\}.$$

Since $F^{p+1}M \subseteq F^pM$, we have

$$\mathrm{Ker}\, d_{F^{p+1}} = \mathrm{Ker}\, d \cap F^{p+1}M \subseteq \mathrm{Ker}\, d \cap F^pM = \mathrm{Ker}\, d_{F^p},$$

which implies that $H(M)^{p+1} \subseteq H(M)^p$, so the modules $H(M)^p$ form a filtration of $H(M)$.

Let $Z^p_\infty = \mathrm{Ker}\, d_{F^p} = Z \cap F^pM$ and $B^p_\infty = B \cap F^pM$.

Proposition 15.1. *We have an isomorphism*

$$H(M)^p \cong Z^p_\infty / B^p_\infty.$$

Proof. The map from Z^p_∞ to $H(M)^p$ given by $z \mapsto [z]_B$ is obviously surjective. Some $z \in Z^p_\infty = Z \cap F^pM$ maps to 0 iff $z \in B \cap F^pM$. Therefore, the kernel of this map is $B \cap F^pM = B^p_\infty$ and by the first isomorphism theorem,

$$H(M)^p \cong Z^p_\infty / B^p_\infty.$$

□

15.2 Graded Modules and Their Cohomology

Having a filtration (F^pM) on an R-module M gives us the ability to form the successive quotients $F^pM/F^{p+1}M$, and this process yields a graded module defined as follows.

Definition 15.5. Let M be a filtered R-module with a filtration $(F^pM)_{p\in\mathbb{Z}}$. The *graded module* $\mathrm{gr}(M)$ induced by the filtration $(F^pM)_{p\in\mathbb{Z}}$ is defined as the direct sum

$$\mathrm{gr}(M) = \bigoplus_{p\in\mathbb{Z}} F^pM/F^{p+1}M.$$

Typically, because the quotient $F^pM/F^{p+1}M$ sets the elements in $F^{p+1}M$ to zero, the graded module $\mathrm{gr}(M)$ is simpler than the original module M.

Example 15.3. Suppose that

$$M = \bigoplus_{p\in\mathbb{Z}} M^p$$

as in Example 15.1 with the filtration given by

$$F^pM = \bigoplus_{i\geq p} M^i.$$

Obviously $F^pM/F^{p+1}M \cong M^p$, so in this case $\mathrm{gr}(M) \cong M$.

Example 15.4. Now consider the infinite dimensional vector space $M = \prod_{i\in\mathbb{N}} \mathbb{R}$, consisting of sequence $(x_i)_{i\in\mathbb{N}}$ of reals $x_i \in \mathbb{R}$ under componentwise addition and rescaling. If we define the filtration (F^pM) by $F^pM = M$ if $p \leq 0$, and

$$F^pM = \{(x_i)_{i\in\mathbb{N}} \mid x_0 = x_1 = \cdots = x_{p-1} = 0,\ p \geq 1\},$$

then

$$F^pM/F^{p+1}M \cong \mathbb{R},$$

and so

$$\mathrm{gr}(M) = \bigoplus_{p \in \mathbb{N}} \mathbb{R},$$

which is simpler that M, since the elements of $\mathrm{gr}(M)$ are sequences $(x_i)_{i \in \mathbb{N}}$ with $x_i = 0$ for all but finitely many i.

It turns out that many of the graded modules that we will encounter later, including $\bigoplus_{p \in \mathbb{Z}} F^p M$, $\mathrm{gr}(M)$, and the modules $E_r = \bigoplus_{\in \mathbb{Z}} E_r^p$ arising from spectral sequences, can be viewed as cochain complexes, except that their differential does not have degree 1. Here is the technical definition.

Definition 15.6. For any $r \in \mathbb{N}$, a *cochain complex C with differential d of degree r*, for short a *complex C with differential d of degree r*, is a direct sum $C = \bigoplus_{p \in \mathbb{Z}} C^p$ of R-modules C^p together with a map $d\colon C \to C$ such that $d \circ d = 0$ and the restriction d^p of d to C^p is a map $d^p : C^p \to C^{p+r}$. Then $d^p \circ d^{p-r} = 0$, and we define the *cohomology module* $H^p(C)$ as

$$H^p(C) = \mathrm{Ker}\, d^p / \mathrm{Im}\, d^{p-r}.$$

We write $H(C) = \bigoplus_{p \in \mathbb{Z}} H^p(C)$.

Observe that Definition 2.8 is the special case of Definition 15.6 when $r = 1$. If $r = 0$, then $d^p : C^p \to C^p$, in which case (C^p, d^p) is a differential module, so $H^p(C) = H(C^p)$, and $H(C) = \bigoplus_{p \in \mathbb{Z}} H(C^p)$. This situation arises from a filtered differential module (M, d).

Given a filtered differential module (M, d), the differential d on M induces a linear map d_0^p from $F^p M / F^{p+1} M$ to itself given by

$$d_0^p([x]_{F^{p+1}M}) = [dx]_{F^{p+1}M},$$

where $[x]_{F^{p+1}M}$ is the equivalence class of $x \in F^p M$ modulo $F^{p+1}M$ and $[dx]_{F^{p+1}M}$ is the equivalence class of dx modulo $F^{p+1}M$. This map is well-defined because $dF^pM \subseteq F^pM$ and $dF^{p+1}M \subseteq F^{p+1}M$. It is also obvious that $d_0^p \circ d_0^p = 0$, so $(F^pM/F^{p+1}M, d_0^p)$ is a differential module with cohomology $H(F^pM/F^{p+1}M)$.

The graded module $\mathrm{gr}(M) = \bigoplus_{p \in \mathbb{Z}} F^pM/F^{p+1}M$ is a complex with a differential d_0 of degree 0, with $d_0 = \bigoplus_{p \in \mathbb{Z}} d_0^p$, which means that

$$d_0 \left(\sum_{p \in I} u_p \right) = \sum_{p \in I} d_0^p(u_p), \quad u_p \in F^pM/F^{p+1}M$$

(where I is a finite index set).

Since $(\mathrm{gr}(M), d_0)$ is a complex with differential d_0 of degree 0, according to Definition 15.6 its cohomology is defined as a follows.

Definition 15.7. The cohomology of the complex $\mathrm{gr}(M)$, with $\mathrm{gr}(M) = \bigoplus_{p\in\mathbb{Z}} F^pM/F^{p+1}M$ and differential d_0 of degree 0, is defined as

$$H(\mathrm{gr}(M)) = \bigoplus_{p\in\mathbb{Z}} H(F^pM/F^{p+1}M).$$

Typically, we are more interested in the cohomology $H(M)$ of M, but this may be hard to compute. Using the filtration $H(M)^p$ on $H(M)$ we can attempt to compute the quotient modules $H(M)^p/H(M)^{p+1}$, which is usually easier to do.

Definition 15.8. The graded module $\mathrm{gr}(H(M))$ is defined by

$$\mathrm{gr}(H(M)) = \bigoplus_{p\in\mathbb{Z}} H(M)^p/H(M)^{p+1}.$$

The modules $H(M)^p/H(M)^{p+1}$ are called the *composition factors* in the filtration of $H(M)$.

The next step is to construct a sequence of families of modules $(E_r^p)_{p\in\mathbb{Z}}$ with $r \in \mathbb{N}$, where each E_r^p is a quotient of submodules of F^pM, and for each $p \in \mathbb{Z}$, an R-linear map $d_r^p\colon E_r^p \to E_r^{p+r}$ such that

$$d_r^p \circ d_r^{p-r} = 0.$$

The direct sum $E_r = \bigoplus_{p\in\mathbb{Z}} E_r^p$ is a complex with differential $d_r = \bigoplus_{p\in\mathbb{Z}} d_r^p$, but the restriction d_r^p of d_r to E_r^p is a map $d_r^p\colon E_r^p \to E_r^{p+r}$ of degree r. Therefore, (E_r, d_r) is a complex with differential d_r of degree r (as in Definition 15.6) and its pth cohomology module $H^p(E_r)$ is given by

$$H^p(E_r) = \mathrm{Ker}\, d_r^p/\mathrm{Im}\, d_r^{p-r}.$$

One of the main points of the construction of a spectral sequence is that there is an isomorphism

$$H^p(E_r) \cong E_{r+1}^p$$

for every $r \geq 0$.

There is also a sequence of "ideal" modules $(E_\infty^p)_{p\in\mathbb{Z}}$, and the E_r^p may be viewed as approximations of the E_∞^p. The significance of the E_∞^p is that we have isomorphisms

$$E_\infty^p \cong H(M)^p/H(M)^{p+1}.$$

This means that the modules E^p_∞ compute the graded pieces of the graded module $\mathrm{gr}(H(M))$, and so the E^p_r approximate the graded module $\mathrm{gr}(H(M))$. This is not as good as computing $H(M)$, but in many applications where the family of (E^p_r) *degenerates*, some E^p_r computes $H(M)^p$.

Actually, in most applications M is also a graded differential module, for example, a cochain complex, but this leads us to consider triply indexed modules $E^{p,q}_r$, and we postpone discussing this more general case.

We now give the construction of the E^p_r and E^p_∞.

15.3 Construction of the Spectral Sequence

Definition 15.9. For all $r, p \in \mathbb{Z}$, define Z^p_r, B^p_r, Z^p_∞ and B^p_∞ as follows:

$$\begin{aligned}
Z^p_r &= \{x \in F^pM \mid dx \in F^{p+r}M\},\\
B^p_r &= \{x \in F^pM \mid (\exists y \in F^{p-r}M)(x = dy)\} = (dF^{p-r}M) \cap F^pM,\\
Z^p_\infty &= \{x \in F^pM \mid dx = 0\} = Z(M) \cap F^pM\\
B^p_\infty &= \{x \in F^pM \mid (\exists y \in M)(x = dy)\} = B(M) \cap F^pM.
\end{aligned}$$

Several identities needed to justify properties of the construction can easily be derived. We classify these identities into four categories.

(1) Negative r identities. Since $F^pM \subseteq F^{p+r}M$ for $r \leq 0$, we have $Z^p_r = F^pM$ and $B^p_r \subseteq F^{p-r}M$ for all $r \leq 0$.

(2) Level p inclusions.

$$\begin{aligned}
dF^pM = B^p_0 \subseteq B^p_1 \subseteq \cdots \subseteq B^p_s \subseteq B^p_{s+1} \subseteq \cdots \subseteq B^p_\infty \subseteq Z^p_\infty \subseteq \cdots\\
\cdots \subseteq Z^p_{r+1} \subseteq Z^p_r \subseteq \cdots \subseteq Z^p_1 \subseteq Z^p_0 = F^pM
\end{aligned}$$

for all $p \in \mathbb{Z}$. See Figure 15.6.

(3) (r, p)-Boundary identity.

$$B^p_r = dZ^{p-r}_r. \qquad (*_B)$$

(4) Z-Jump identities. Since $F^{p+1}M \subseteq F^pM$, we have $Z^{p+1}_{r-1} \subseteq Z^p_r$ and $Z^{p+1}_\infty \subseteq Z^p_\infty$.

Definition 15.10. For all $r \geq 0$ and all $p \in \mathbb{Z}$, define E^p_r and E^p_∞, as follows:

$$E^p_r = Z^p_r/(B^p_{r-1} + Z^{p+1}_{r-1}) \qquad (\dagger_r)$$

and

$$E^p_\infty = Z^p_\infty/(B^p_\infty + Z^{p+1}_\infty). \qquad (\dagger_\infty)$$

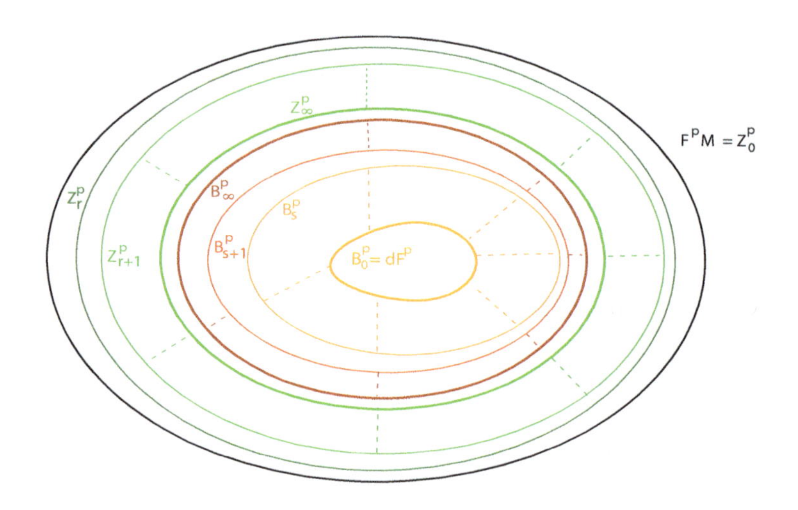

Fig. 15.6 A schematic illustration of the Level p inclusions.

Since $Z^p_{-1} = F^pM$, $Z^{p+1}_{-1} = F^{p+1}M$, and $B^p_{-1} \subseteq F^{p+1}M$, we see that $B^p_{-1} + Z^{p+1}_{-1} = F^{p+1}M$, so

$$E^p_0 = F^pM/F^{p+1}M. \tag{E_0}$$

This shows that the modules E^p_0 are the graded pieces of the graded module $\mathrm{gr}(M)$. We will see shortly that

$$E^p_1 \cong H^p(\mathrm{gr}(M)) = H(F^pM/F^{p+1}M) = H^p(E_0).$$

Now the differential d of M maps Z^p_r into Z^{p+r}_r, and $(B^p_{r-1} + Z^{p+1}_{r-1})$ into B^{p+r}_{r-1} (since $dB^p_{r-1} = (0)$ and $dZ^{p+1}_{r-1} \subseteq B^{p+r}_{r-1}$), and since

$$\begin{aligned} E^p_r &= Z^p_r/(B^p_{r-1} + Z^{p+1}_{r-1}) \\ E^{p+r}_r &= Z^{p+r}_r/(B^{p+r}_{r-1} + Z^{p+r+1}_{r-1}), \end{aligned}$$

the differential d induces a map

$$d^p_r \colon E^p_r \to E^{p+r}_r.$$

Definition 15.11. The map $d^p_r \colon E^p_r \to E^{p+r}_r$ is defined by

$$d^p_r([x]_{B^p_{r-1}+Z^{p+1}_{r-1}}) = [dx]_{B^{p+r}_{r-1}+Z^{p+r+1}_{r-1}}$$

for all $x \in Z^p_r$.

The above map is well-defined because for any $x, y \in Z_r^p$, if $x - y = z$ for some $z \in B_{r-1}^p + Z_{r-1}^{p+1}$, then $dx - dy = dz \in d(B_{r-1}^p + Z_{r-1}^{p+1}) \subseteq B_{r-1}^{p+r} \subseteq B_{r-1}^{p+r} + Z_{r-1}^{p+r+1}$.

Let us compute $\operatorname{Ker} d_r^p$ and $\operatorname{Im} d_r^{p-r}$.

Proposition 15.2. *We have*

$$\begin{aligned}\operatorname{Ker} d_r^p &= (Z_{r+1}^p + Z_{r-1}^{p+1})/(B_{r-1}^p + Z_{r-1}^{p+1}) \\ \operatorname{Im} d_r^{p-r} &= (B_r^p + Z_{r-1}^{p+1})/(B_{r-1}^p + Z_{r-1}^{p+1}).\end{aligned}$$

Proof. Since $d_r^p \colon E_r^p \to E_r^{p+r}$ and

$$E_r^p = Z_r^p/(B_{r-1}^p + Z_{r-1}^{p+1}), \qquad E_r^{p+r} = Z_r^{p+r}/(B_{r-1}^{p+r} + Z_{r-1}^{p+r+1}),$$

for any $x \in Z_r^p$, $d_r^p([x]) = 0$ iff $dx \in B_{r-1}^{p+r} + Z_{r-1}^{p+r+1} = dZ_{r-1}^{p+1} + Z_{r-1}^{p+r+1}$, which means that $dx = dy + z$, for some $y \in Z_{r-1}^{p+1}$ and some $z \in Z_{r-1}^{p+r+1}$. Since $Z_{r-1}^{p+1} \subseteq Z_r^p \subseteq F^pM$, we can write $x = y + u$ with $u = x - y$, where $u \in F^pM$, and since $dx = dy + z$ and $dx = dy + du$, we find that $du = z$. Since $du = z \in Z_{r-1}^{p+r+1}$, we have $du \in F^{p+r+1}M$, and since $u \in F^pM$, this means that $u \in Z_{r+1}^p$ as shown in Figure 15.7. Consequently, $x = u + y$, with $u \in Z_{r+1}^p$ and $y \in Z_{r-1}^{p+1}$, which shows that

$$\operatorname{Ker} d_r^p = (Z_{r+1}^p + Z_{r-1}^{p+1})/(B_{r-1}^p + Z_{r-1}^{p+1}).$$

The image of d_r^{p-r} consists of all classes modulo $(B_{r-1}^p + Z_{r-1}^{p+1})$ of elements in $B_r^p = dZ_r^{p-r}$. These are classes of the form $[x+y+z]_{(B_{r-1}^p + Z_{r-1}^{p+1})}$, where $x \in B_r^p$, $y \in B_{r-1}^p$, and $z \in Z_{r-1}^{p+1}$, but since $B_{r-1}^p \subseteq B_r^p$, these are the classes of the form $[x+z]_{(B_{r-1}^p + Z_{r-1}^{p+1})}$, where $x \in B_r^p$ and $z \in Z_{r-1}^{p+1}$, which shows that

$$\operatorname{Im} d_r^{p-r} = (B_r^p + Z_{r-1}^{p+1})/(B_{r-1}^p + Z_{r-1}^{p+1}). \qquad \square$$

Proposition 15.2 implies that

$$d_r^p \circ d_r^{p-r} = 0,$$

which allows us to make the following definition.

Definition 15.12. The graded modules E_r for $r \in \mathbb{N}$ and E_∞ are defined as

$$E_r = \bigoplus_{p \in \mathbb{Z}} E_r^p, \quad E_\infty = \bigoplus_{p \in \mathbb{Z}} E_\infty^p.$$

The graded module E_r is a complex with differential d_r of degree r defined such that the restriction of d_r to E_r^p is equal to $d_r^p \colon E_r^p \to E_r^{p+r}$.

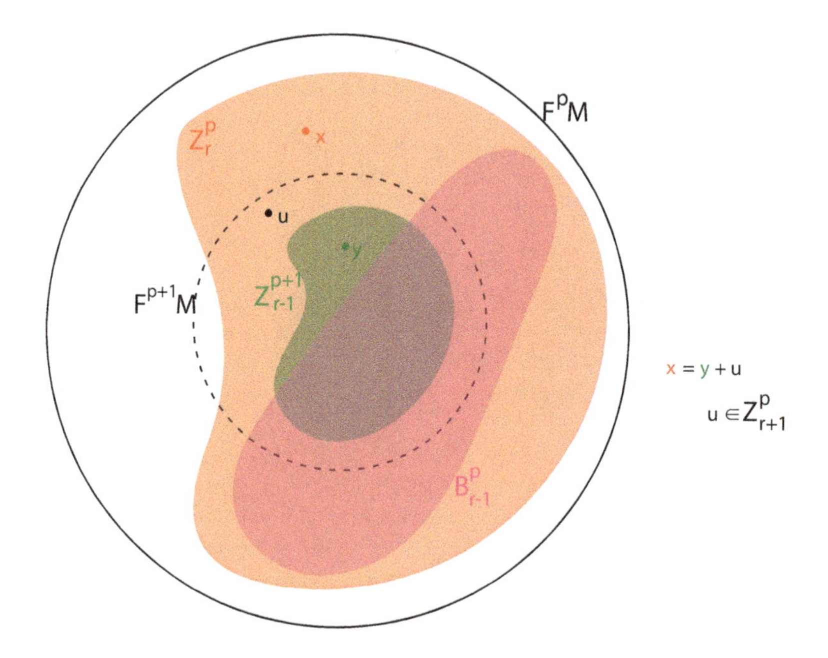

Fig. 15.7 A schematic illustration associated with $\operatorname{Ker} d_r^p = (Z_{r+1}^p + Z_{r-1}^{p+1})/(B_{r-1}^p + Z_{r-1}^{p+1})$.

Having Definition 15.12, in view of Definition 15.6, Proposition 15.2 shows that

$$H^p(E_r) = \operatorname{Ker} d_r^p / \operatorname{Im} d_r^{p-r} \cong (Z_{r+1}^p + Z_{r-1}^{p+1})/(B_r^p + Z_{r-1}^{p+1}).$$

In order to simplify the above quotient module we need the following proposition.

Proposition 15.3. *(Modular Noether isomorphism) For any R-modules A, B, Z with B a submodule of A, there is an isomorphism*

$$(A + Z)/(B + Z) \cong A/(A \cap (B + Z)).$$

Proof. To prove the above isomorphism, consider the R-linear map $\varphi\colon A \to (A + Z)/(B + Z)$ defined by

$$\varphi(a) = [a]_{B+Z}, \quad a \in A,$$

the equivalence class of $a \in A + Z$ modulo $B + Z$. This map is surjective because for every equivalence class $[a + z]_{B+Z}$, with $a \in A$ and $z \in Z$, since $a + z$ and a are equivalent modulo $B + Z$ since $a + z - a = z \in B + Z$, we have $[a+z]_{B+Z} = [a]_{B+Z} = \varphi(a)$. Since $\varphi(a) = [a]_{B+Z} = 0$ iff $a \in B + Z$, we have $\operatorname{Ker} \varphi = A \cap (B + Z)$. By the first isomorphism theorem,

$$(A+Z)/(B+Z) \cong A/(A \cap (B+Z)),$$

as claimed. $\square$

As a consequence,

$$\begin{aligned}(Z^p_{r+1} + Z^{p+1}_{r-1})/(B^p_r + Z^{p+1}_{r-1}) &\cong Z^p_{r+1}/(Z^p_{r+1} \cap (B^p_r + Z^{p+1}_{r-1})) \\ &\cong Z^p_{r+1}/(B^p_r + Z^{p+1}_r) \\ &= E^p_{r+1},\end{aligned}$$

since $B^p_r \subseteq Z^p_{r+1}$ and $Z^p_{r+1} \cap Z^{p+1}_{r-1} = Z^{p+1}_r$. Therefore, we proved that

$$H^p(E_r) \cong E^p_{r+1}, \quad r \geq 0. \tag{E_r}$$

In particular,

$$H^p(E_0) = H(F^pM/F^{p+1}M) \cong E^p_1, \tag{E_1}$$

as claimed earlier. This shows that the modules E^p_1 are the pieces of the graded module $H(\mathrm{gr}(M))$.

Recall that we showed in Proposition 15.1 that

$$H(M)^p \cong (Z(M) \cap F^pM)/(B(M) \cap F^pM) = Z^p_\infty/B^p_\infty.$$

Proposition 15.4. *We have an isomorphism*

$$E^p_\infty = Z^p_\infty/(B^p_\infty + Z^{p+1}_\infty) \cong H(M)^p/H(M)^{p+1}. \tag{E_∞}$$

Proof. Recall that

$$H(M)^p = \{[z]_B \mid z \in Z \cap F^pM\} = \{[z]_B \mid z \in Z^p_\infty\}.$$

The map from Z^p_∞ to $H(M)^p$ given by $z \mapsto [z]_B$ is obviously surjective, and by composing it with the quotient map from $H(M)^p$ to $H(M)^p/H(M)^{p+1}$ we obtain a surjection $\pi^p_\infty \colon Z^p_\infty \to H(M)^p/H(M)^{p+1}$. We have $\pi^p_\infty(z) = 0$ iff $[z]_B \in H(M)^{p+1}$ iff $z = z_1 + b$ for some $z_1 \in Z \cap F^{p+1}M = Z^{p+1}_\infty$ and some $b \in B$. Since $Z^{p+1}_\infty \subseteq Z^p_\infty$, we have $b = z - z_1 \in Z^p_\infty \subseteq F^pM$, so $b \in B \cap F^pM = B^p_\infty$. Consequently, $[z]_B \in H(M)^{p+1}$ iff $z \in B^p_\infty + Z^{p+1}_\infty$, which shows that $\operatorname{Ker} \pi^p_\infty = B^p_\infty + Z^{p+1}_\infty$. By the first isomorphism theorem,

$$E^p_\infty = Z^p_\infty/(B^p_\infty + Z^{p+1}_\infty) \cong H(M)^p/H(M)^{p+1}.$$

$\square$

Thus, the "limit modules" E_∞^p are isomorphic to the graded pieces $H(M)^p/H(M)^{p+1}$ of the graded module $\mathrm{gr}(H(M))$ and the modules E_r^p are approximations of the modules E_∞^p.

Definition 15.13. The family of graded modules $(E_r)_{r\in\mathbb{N}\cup\{\infty\}}$ is called the *spectral sequence* associated with the filtered differential module (M, d) with filtration (F^pM).

In some sense, a spectral sequence is a method for passing from the graded module $H(\mathrm{gr}(M))$ (computed by E_1) to the graded module $\mathrm{gr}(H(M))$ (computed by E_∞). We summarize the above discussion as the following theorem.

Theorem 15.5. *Let (M, d) be a filtered R-module filtered by $(F^pM)_{p\in\mathbb{Z}}$. There is a sequence of graded modules*

$$E_r = \bigoplus_{p\in\mathbb{Z}} E_r^p,\ r \geq 1, \quad E_\infty = \bigoplus_{p\in\mathbb{Z}} E_\infty^p$$

called a spectral sequence, where the graded module E_r is a complex with differential d_r of degree r, such that the following properties hold for all $r \geq 1$ and all $p \in \mathbb{Z}$:

(1) Recall that $H^p(E_r)$ is defined as

$$H^p(E_r) = \mathrm{Ker}\, d_r^p/\mathrm{Im}\, d_r^{p-r}.$$

Then there is an isomorphism

$$H^p(E_r) \cong E_{r+1}^p.$$

(2) There is an isomorphism

$$E_\infty^p \cong H(M)^p/H(M)^{p+1} = (\mathrm{gr}(H(M))^p.$$

(3) The term E_0^p is given by

$$E_0^p = F^pM/F^{p+1}M = (\mathrm{gr}(M))^p.$$

At this stage we could investigate conditions on the filtration that yield isomorphisms involving E_∞^p, but the theory of spectral sequences is most useful when applied to filtered cochain complexes. Let us mention the following case in which only finitely many E_∞^p are nonzero and the other E_∞^p are obtained from the E_r^p after finitely many steps.

Suppose that the filtration F^pM is finite, which means that there are indices $s \leq t$ such that $F^pM = M$ for all $p \leq s$ and $F^pM = (0)$ for all

$p > t$. Then the condition $F^pM = M$ for all $p \leq s$ implies that $H(M)^p = H(F^pM) = H(M)$ for all $p \leq s$, so $E^p_\infty = H(M)^p/H(M)^{p+1} = (0)$ for all $p < s$. Similarly, the condition $F^pM = (0)$ for all $p > t$ implies that $H(M)^p = H(F^pM) = (0)$ for all $p > t$, so $E^p_\infty = H(M)^p/H(M)^{p+1} = (0)$ for all $p > s$. Therefore, the only interesting E^p_∞ arise when $s \leq p \leq t$. Since

$$\begin{aligned} Z^p_r &= \{x \in F^pM \mid dx \in F^{p+r}M\}, \\ Z^p_\infty &= \{x \in F^pM \mid dx = 0\} = Z(M) \cap F^pM, \end{aligned}$$

and $F^pM = (0)$ for all $p > t$, we see that if $p+r > t$, that is, $r > t-p$, then $F^{p+r}M = (0)$, in which case $dx = 0$, and so $Z^p_r = Z^p_\infty$ and $Z^{p+1}_{r-1} = Z^p_\infty$. Since

$$\begin{aligned} B^p_r &= \{x \in F^pM \mid (\exists y \in F^{p-r}M)(x = dy)\} = (dF^{p-r}M) \cap F^pM, \\ B^p_\infty &= \{x \in F^pM \mid (\exists y \in M)(x = dy)\} = B(M) \cap F^pM, \end{aligned}$$

and $F^pM = M$ for all $p \leq s$, if $p - r \leq s$, that is, $r \geq p - s$, then $F^{p-r}M = M$, and so $B^p_r = B^p_\infty$. Since

$$\begin{aligned} E^p_r &= Z^p_r/(B^p_{r-1} + Z^{p+1}_{r-1}) \\ E^p_\infty &= Z^p_\infty/(B^p_\infty + Z^{p+1}_\infty), \end{aligned}$$

if $r > \max(p-s, t-p)$, then

$$E^p_r = E^p_\infty.$$

Since $s \leq p \leq t$, the condition $r > \max(p-s, t-p)$ is equivalent to $r > t-s$. In summary, if $r > t-s$, then

$$E^p_\infty = \begin{cases} E^p_r & \text{if } s \leq p \leq t \\ (0) & \text{if } p < s \text{ or } p > t. \end{cases}$$

This situation will be generalized in Section 15.8 dealing with degenerate spectral sequences.

Many presentations start right away with filtered and graded differential modules. The reader is hit with objects involving three indices $(E^{p,q}_r, Z^{p,q}_r, B^{p,q}_r)$ and in our opinion, it is very difficult to understand what is going on unless one already knows the subject. This is why (following Serre, Godement, and even Cartan and Eilenberg) we started with the simpler case of a single filtered module. Having done the warm up we proceed with the more general case of a filtered complex.

15.4 Case 2: Filtered Differential Complexes

Recall from Definition 2.8 that the cochain complex (C, d) is a direct sum

$$C = \bigoplus_{p \in \mathbb{Z}} C^p$$

of R-modules C^p together with an R-linear map $d\colon C \to C$ (called coboundary map) such that $dC^p \subseteq C^{p+1}$ and $d \circ d = 0$. The restriction of d to C^p is denoted by $d^p\colon C^p \to C^{p+1}$.

Some authors denote a cochain complex by $C^\bullet$ or C^*. To keep the notation as simple as possible we will omit the superscript $\bullet$ or $*$. Doing so does not appear to cause confusion.

Definition 15.14. A *filtered cochain complex* (C, d) is a cochain complex together with a family of submodules $(F^pC)_{p\in\mathbb{Z}}$ of C such that

$$\cdots \supseteq F^{-p}C \supseteq F^{-p+1}C \supseteq \cdots \supseteq F^{-1}C \supseteq F^0C \supseteq \\ F^1C \supseteq \cdots \supseteq F^pC \supseteq F^{p+1}C \supseteq \cdots$$

for all $p \in \mathbb{N} - \{0\}$. We also assume that $\bigcup_{p\in\mathbb{Z}} F^pC = C$ and $\bigcap_{p\in\mathbb{Z}} F^pC = (0)$. Moreover, if d is the coboundary map of the complex C (also called *differentiation*), we assume that

(1) The filtration (F^pC) and d are compatible, which means that $d(F^pC) \subseteq F^pC$ for all $p \in \mathbb{Z}$.
(2) The filtration (F^pC) is compatible with the grading on C, i.e.,

$$F^pC = \bigoplus_{q\in\mathbb{Z}} C^{p+q} \cap F^pC = \bigoplus_{q\in\mathbb{Z}} C^{p,q},$$

where $C^{p,q} = C^{p+q} \cap F^pC$.

An equivalent way to say that a filtration (F^pC) is compatible with the grading on C is to say that for every $p \in \mathbb{Z}$, there is a family of submodules $C^{p,q}$ of C^{p+q} $(q \in \mathbb{Z})$ such that

$$F^pC = \bigoplus_{q\in\mathbb{Z}} C^{p,q}.$$

Since $F^{p+1}C \subseteq F^pC$ and

$$F^{p+1}C = \bigoplus_{s\in\mathbb{Z}} C^{p+1,s},$$

comparing the submodules of C^{p+q}, which are $C^{p,q}$ and $C^{p+1,q-1}$, we have

$$C^{p+1,q-1} \subseteq C^{p,q} \subseteq C^{p+q}.$$

If we write $n = p+q$, we have $C^{p+1,n-p-1} \subseteq C^{p,n-p} \subseteq C^n$, so each C^n has the filtration $(C^{s,n-s})_{s\in\mathbb{Z}}$. Since

$$F^pC = \bigoplus_{q\in\mathbb{Z}} C^{p+q} \cap F^pC = \bigoplus_{q\in\mathbb{Z}} C^{p,q} = \bigoplus_{n\in\mathbb{Z}} C^{p,n-p},$$

we see that the pth module F^pC in a (compatible) filtration of C arises by picking the pth piece $C^{p,n-p}$ in the filtration of C^n and by forming the direct sum of these pieces. See Figures 15.8 and 15.9.

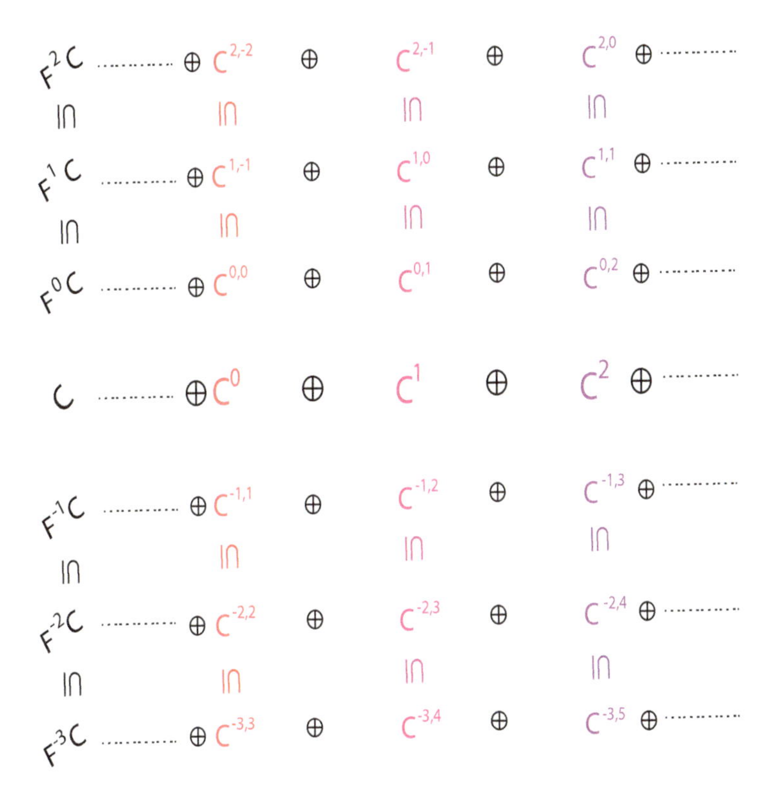

Fig. 15.8 A section of the filtered cochain complex (C, d). Each horizontal row demonstrates the graded module $F^pC = \bigoplus_{q\in\mathbb{Z}} C^{p,q}$, while each column demonstrates the $(C^{s,n-s})_{s\in\mathbb{Z}}$ filtration of C^n.

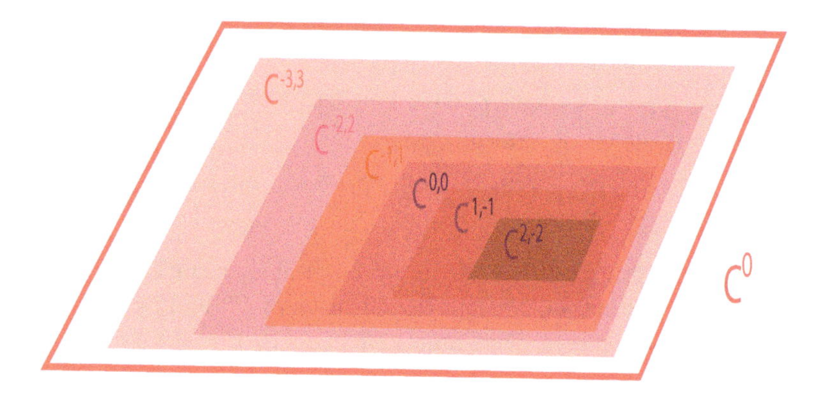

Fig. 15.9 Another view of the $(C^{s,n-s})_{s\in\mathbb{Z}}$ filtration of C^n, where $n = 0$.

The reason for using the superscript $p+q$ instead of q is technical. It has the effect that if we plot the $C^{p,q}$ as points in a grid with the index p along the x-axis and the index q along the y-axis, then the filtration of C^n appears as the descending diagonal of equation $p+q=n$ as illustrated by Figure 15.10.

Remarks:

(1) The elements in $C^{p,q} = C^{p+q} \cap F^pC$ have *degree* $n = p+q$.
(2) The $C^{p,q}$ are submodules of C^{p+q}.
(3) The $C^{p,n-p}$ filter C^n, and p is the *index of filtration*.

Example 15.5. As a naive example of a filtration compatible with the grading, we have $F^pC = \bigoplus_{n\geq p} C^n$; see also Example 15.1.

Remark: The case of a single filtered differential module (M, d) with filtration (F^pM) considered in Section 15.1 can be viewed as a special case of a filtered cochain complex by considering the cochain complex

$$C = \bigoplus_{p\in\mathbb{Z}} C^p, \quad C^p = M,$$

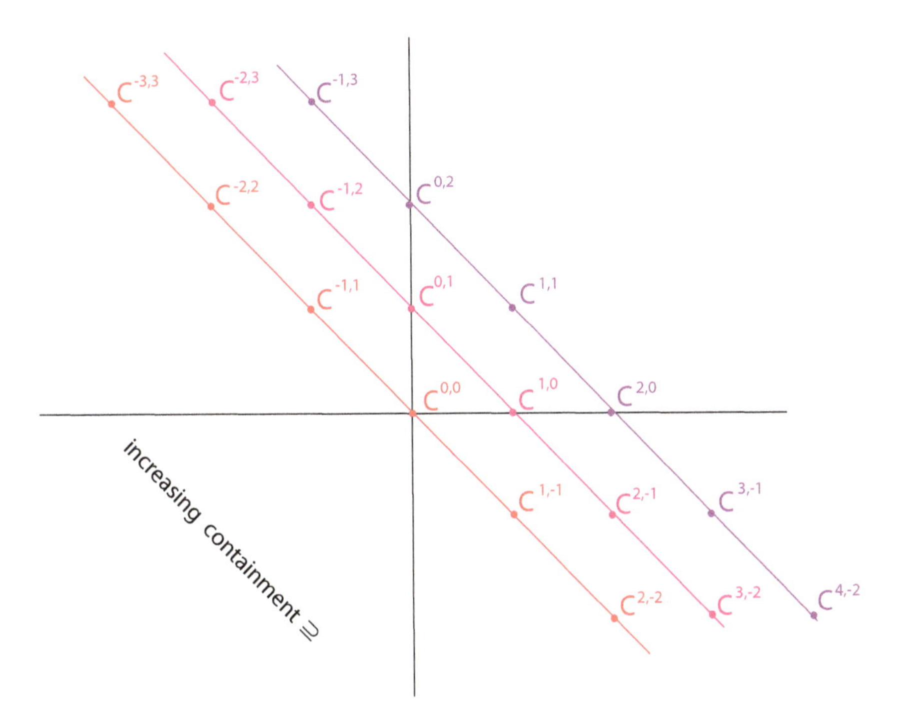

Fig. 15.10 By rotating Figure 15.8 45° counterclockwise, the vertical filtration of C^n becomes the integer coordinates on the diagonal line $p + q = n$. Also note that the graded modules $F^pC = \bigoplus_{q\in\mathbb{Z}} C^{p,q}$ are now represented as vertical columns.

with coboundary map d defined such that the restriction of d to $C^p = M$ is equal to d. The filtration (F^pC) is given by

$$F^pC = \bigoplus_{q\in\mathbb{Z}} C^{p,q}, \qquad C^{p,q} = F^pM.$$

This filtration is obviously compatible with the grading. We have $H^p(C) = H(M)$ for all p.

The reader might wonder when a filtration is not compatible with a grading. Here is an example exhibiting this behavior.

Example 15.6. Consider the module (actually, vector space) C consisting of all real polynomials of the form $P(x) + Q(y)$, where $P(x) \in \mathbb{R}[x]$ and $Q(y) \in \mathbb{R}[y]$, with the grading given by

$$C^p = \begin{cases} \{ax^{-p} \mid a \in \mathbb{R}\} & \text{if } p < 0 \\ \mathbb{R} & \text{if } p = 0 \\ \{ay^p \mid a \in \mathbb{R}\} & \text{if } p > 0. \end{cases}$$

Since every polynomial $P(x) + Q(y)$ can be expressed uniquely as the sum of monomials ax^p, a, and ay^q, with $p, q \geq 1$ and $a \in \mathbb{R}$, we have

$$C = \bigoplus_{p\in\mathbb{Z}} C^p.$$

The map d is irrelevant so we may assume that it is the zero map. Consider the filtration given by

$$F^pC = F^0C = C, \quad p < 0$$

and

$$F^pC = \left\{ \sum_{i=p}^{n} a_i(x^i + y^i) \mid a_i \in \mathbb{R},\, n \geq p \right\}, \qquad p \geq 1.$$

Since $C^{p+q} \cap F^pC = (0)$ for all p, q, the vector space F^pC is not the direct sum of the spaces $C^{p+q} \cap F^pC$. The problem is that the spaces F^pC are not "homogeneous," in the sense that the elements of F^pC do not arise from sums coming from the spaces $C^{p,q} = C^{p+q} \cap F^pC$.

As in Section 15.1, each F^pC is itself a graded complex as is $F^pC/F^{p+1}C$. To see that F^pC is a graded complex, since

$$F^pC = \bigoplus_{n\in\mathbb{Z}} C^{p,n-p},$$

with $C^{p,n-p} = C^n \cap F^pC$, since the coboundary map d of C maps C^n to C^{n+1}, we see that the restriction $d_{C^{p,n-p}}$ of d to $C^{p,n-p} = C^n \cap F^pC$ is a map $d_{C^{p,n-p}}\colon C^{p,n-p} \to C^{p,n+1-p}$ such that $d_{C^{p,n-p}} \circ d_{C^{p,n-p-1}} = 0$. Thus $F^pC = \bigoplus_{n\in\mathbb{Z}} C^{p,n-p}$ with the coboundary map $d_{F^p} = \bigoplus_{n\in\mathbb{Z}} d_{C^{p,n-p}}$ is indeed a chain complex as illustrated in Figure 15.11. More precisely, if $x = \sum_{n\in I} x_n$, with $x_n \in C^{p,n-p}$ (I is a finite index set), then

$$d_{F^p}(x) = \sum_{n\in I} d_{C^{p,n-p}}(x_n).$$

It is immediately verified that $d_{F^p} \circ d_{F^p} = 0$. Observe that when applying d_{F^p}, every component $x_n \in C^{p,n-p}$ is shifted to the right by one slot since $d_{C^{p,n-p}}(x_n)$ lands in $C^{p,n+1-p}$ as evidenced by Figure 15.12.

The following technical result will be needed to prove that the graded module associated with a filtered module which is also a direct sum can also be expressed as a direct sum of pieces.

Proposition 15.6. *Let $M = \bigoplus_{i\in I} M_i$ and $N = \bigoplus_{i\in I} N_i$ be direct sums of R-modules, with each N_i a submodule of M_i. There is an isomorphism*

$$M/N = \left(\bigoplus_{i\in I} M_i\right) \bigg/ \left(\bigoplus_{i\in I} N_i\right) \cong \bigoplus_{i\in I} M_i/N_i.$$

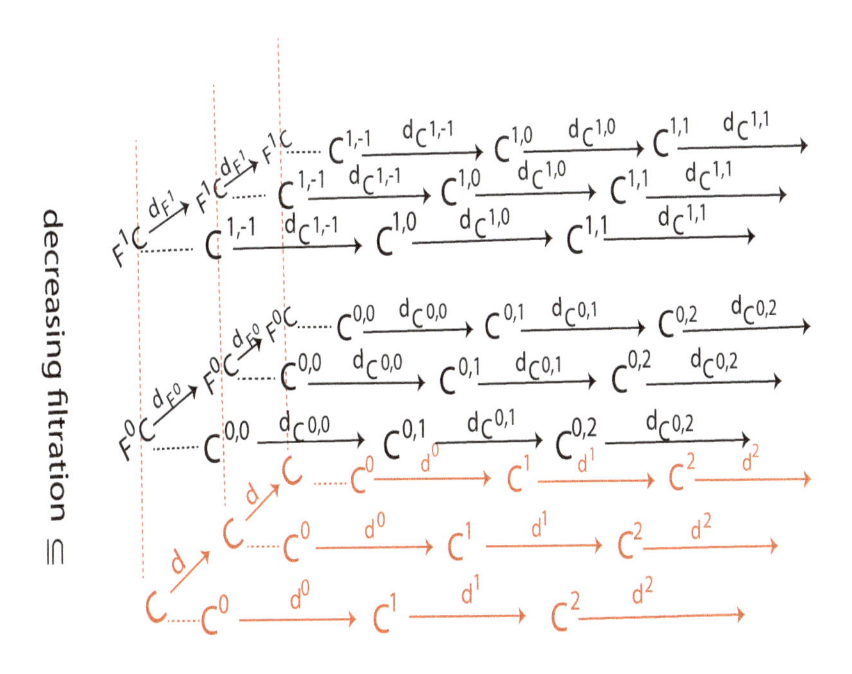

Fig. 15.11 A 3-dimensional grid which illustrates the coboundary maps d_{F^0} and d_{F^1} of the chain complexes F^0C and F^1C. If we associate the plane $z = 0$ for the chain complex (C, d) (which we denote in red), the coboundary map $d_{F^0} = \bigoplus_{n\in\mathbb{Z}} d_{C^{0,n}}$ fills the plane $z = 1$ while the coboundary map $d_{F^1} = \bigoplus_{n\in\mathbb{Z}} d_{C^{1,n-1}}$ fills the plane $z = 2$. Observe that the front face of this 3-dimensional grid is compatible with Figure 15.8.

Proof. Define the R-linear map φ from $M = \bigoplus_{i\in I} M_i$ to $\bigoplus_{i\in I} M_i/N_i$ by

$$\varphi\left(\sum_{j\in J} x_j\right) = \sum_{j\in J} [x_j]_{N_i} \quad x_j \in M_j, \quad J \text{ finite}.$$

The map φ is obviously surjective. We have $\varphi\left(\sum_{j\in J} x_j\right) = 0$ iff $x_j \in N_j$ for all $j \in J$ iff $\sum_{j\in J} x_j \in \bigoplus_{i\in I} N_i$, so $\operatorname{Ker}\varphi = \bigoplus_{i\in I} N_i$. By the first isomorphism theorem, we have the isomorphism

$$\left(\bigoplus_{i\in I} M_i\right) \Big/ \left(\bigoplus_{i\in I} N_i\right) \cong \bigoplus_{i\in I} M_i/N_i,$$

as claimed. $\square$

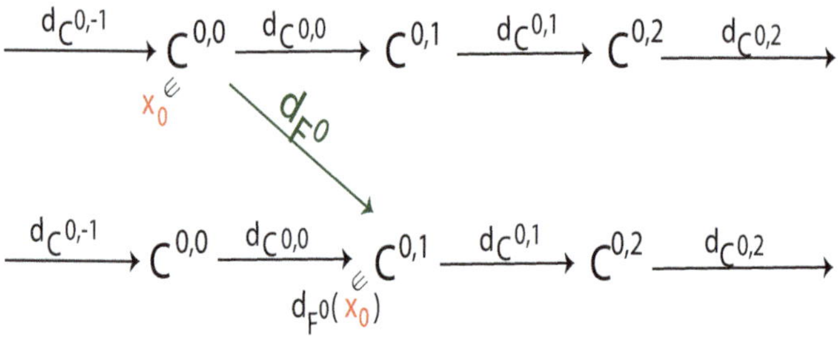

Fig. 15.12 An illustration showing the left shift phenomenon of d_{F^p} for $p = 0$ which is compatible with the first two rows in the $z = 1$ plane of Figure 15.11.

15.5 Some Graded Modules of a Filtered and Graded Complex

In preparation for the construction of a spectral sequence for a filtered cochain complex we need to generalize the definitions of the graded modules $\mathrm{gr}(M)$, $H(\mathrm{gr}(M))$ and $\mathrm{gr}(H(M))$ given in Definitions 15.5, 15.7 and 15.8. We will define the graded modules $\mathrm{gr}(C)^p = F^pC/F^{p+1}C$, $\mathrm{gr}(C) = \bigoplus_{p\in\mathbb{Z}} \mathrm{gr}(C)^p$, $H(\mathrm{gr}(C)^p)$, $H(\mathrm{gr}(C)) = \bigoplus_{p\in\mathbb{Z}} H(\mathrm{gr}(C)^p)$, $\mathrm{gr}(H(C))^p$, and $\mathrm{gr}(H(C)) = \bigoplus_{p\in\mathbb{Z}} \mathrm{gr}(H(C))^p$. Both $\mathrm{gr}(C)^p$ and $\mathrm{gr}(C)$ are complexes with a differential of degree 0.

First, mimicking Definition 15.5 we have the graded module

$$\mathrm{gr}(C) = \bigoplus_{p\in\mathbb{Z}} \mathrm{gr}(C)^p$$

induced by F on C with $\mathrm{gr}(C)^p$ defined as

$$\mathrm{gr}(C)^p = F^pC/F^{p+1}C.$$

Using Proposition 15.6, we have

$$\begin{aligned}
\mathrm{gr}(C)^p &= F^pC/F^{p+1}C \\
&= \left(\bigoplus_{q\in\mathbb{Z}} C^{p+q} \cap F^pC\right) \Big/ \left(\bigoplus_{q\in\mathbb{Z}} C^{p+q} \cap F^{p+1}C\right) \\
&\cong \bigoplus_{q\in\mathbb{Z}} (C^{p+q} \cap F^pC)/(C^{p+q} \cap F^{p+1}C) \\
&= \bigoplus_{q\in\mathbb{Z}} C^{p,q}/C^{p+1,q-1}.
\end{aligned}$$

If we let

$$\mathrm{gr}(C)^{p,q} = C^{p,q}/C^{p+1,q-1},$$

then

$$\mathrm{gr}(C)^p \cong \bigoplus_{q\in\mathbb{Z}} C^{p,q}/C^{p+1,q-1} = \bigoplus_{n\in\mathbb{Z}} \mathrm{gr}(C)^{p,n-p},$$

and $\mathrm{gr}(C)$ is a bigraded module, with

$$\mathrm{gr}(C) = \bigoplus_{p\in\mathbb{Z}} \mathrm{gr}(C)^p \cong \bigoplus_{p,q\in\mathbb{Z}} C^{p,q}/C^{p+1,q-1} = \bigoplus_{p,n\in\mathbb{Z}} \mathrm{gr}(C)^{p,n-p}.$$

Definition 15.15. If we let

$$\mathrm{gr}(C)^{p,q} = C^{p,q}/C^{p+1,q-1} \quad \text{and} \quad \mathrm{gr}(C)^p = F^pC/F^{p+1}C,$$

then $\mathrm{gr}(C)^p$ is a graded module with

$$\mathrm{gr}(C)^p \cong \bigoplus_{n\in\mathbb{Z}} \mathrm{gr}(C)^{p,n-p},$$

and $\mathrm{gr}(C)$ is a bigraded module with

$$\mathrm{gr}(C) = \bigoplus_{p\in\mathbb{Z}} \mathrm{gr}(C)^p \cong \bigoplus_{p,n\in\mathbb{Z}} \mathrm{gr}(C)^{p,n-p};$$

see Figure 15.13.

$$\mathrm{gr(C)}^{-3} = F^{-3}C/F^{-2}C = (\cdots C^{-3,3} \oplus C^{-3,4} \oplus C^{-3,5} \oplus \cdots)/(\cdots C^{-2,2} \oplus C^{-2,3} \oplus C^{-2,4} \oplus \cdots)$$

$$\mathrm{gr(C)}^{-3} \cong \cdots C^{-3,3}/C^{-2,2} \oplus C^{-3,4}/C^{-2,3} \oplus C^{-3,5}/C^{-2,4} \oplus \cdots$$

$$\mathrm{gr(C)}^{-3} \cong \cdots \oplus \mathrm{gr(C)}^{-3,3} \oplus \mathrm{gr(C)}^{-3,4} \oplus \mathrm{gr(C)}^{-3,5} \oplus \cdots$$

Fig. 15.13 The graded module $\mathrm{gr}(C)^{-3} = F^{-3}C/F^{-2}C$ obtained by taking the quotient of the last two rows of Figure 15.8.

The map $d_{C^{p,n-p}}\colon C^{p,n-p} \to C^{p,n+1-p}$ induces a quotient map $d_0^{p,n-p}\colon C^{p,n-p}/C^{p+1,n-1-p} \to C^{p,n+1-p}/C^{p+1,n-p}$ given by

$$d_0^{p,n-p}([x]_{C^{p+1,n-1-p}}) = [d_{C^{p,n-p}}(x)]_{C^{p+1,n-p}},$$

for any $x \in C^{p,n-p}$. Then by the isomorphism

$$F^pC/F^{p+1}C = \mathrm{gr}(C)^p \cong \bigoplus_{n\in\mathbb{Z}} C^{p,n-p}/C^{p+1,n-1-p},$$

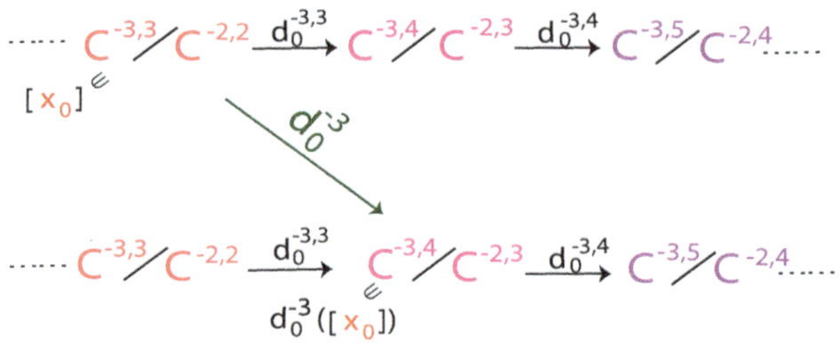

Fig. 15.14 A continuation of Figure 15.13 which illustrates the differential complex $(\mathrm{gr}(C)^{-3}, d_0^{-3} = \bigoplus_{n\in\mathbb{Z}} d_0^{-3,n+3})$.

the map $d_0^p = \bigoplus_{n\in\mathbb{Z}} d_0^{p,n-p}$ is a differential of degree 0 on the complex $\mathrm{gr}(C)^p = F^pC/F^{p+1}C$; see Figure 15.14.

Since each module C^n has the filtration $(C^{p,n-p})_{p\in\mathbb{Z}}$, we have, as demonstrated by Figure 15.15, the associated graded module $\mathrm{gr}(C^n)$, with

$$\mathrm{gr}(C^n)^p = C^{p,n-p}/C^{p+1,n-p-1} = \mathrm{gr}(C)^{p,n-p},$$

so

$$\mathrm{gr}(C^n) = \bigoplus_{p\in\mathbb{Z}} \mathrm{gr}(C)^{p,n-p}.$$

We will not need this graded module but later will consider its analog $\mathrm{gr}(H^n(C))$, so it does not hurt to consider it now. Note that the sum is performed along the diagonal $p+q=n$ (with n fixed). Since

$$\mathrm{gr}(C) = \bigoplus_{p,q\in\mathbb{Z}} \mathrm{gr}(C)^{p,q} = \bigoplus_{p,n\in\mathbb{Z}} \mathrm{gr}(C)^{p,n-q} = \bigoplus_{n\in\mathbb{Z}}\bigoplus_{p\in\mathbb{Z}} \mathrm{gr}(C)^{p,n-q},$$

we also have

$$\mathrm{gr}(C) = \bigoplus_{n\in\mathbb{Z}} \mathrm{gr}(C^n).$$

Since each $\mathrm{gr}(C)^p = F^pC/F^{p+1}C$ is a complex with differential d_0^p of degree 0, the graded module $\mathrm{gr}(C) = \bigoplus_{p\in\mathbb{Z}} \mathrm{gr}(C)^p$ is a complex with differential $d_0 = \bigoplus_{p\in\mathbb{Z}} d_0^p$ of degree 0. The cohomology $H(\mathrm{gr}(C))$ of the cochain complex $\mathrm{gr}(C)$ is given by

$$H(\mathrm{gr}(C)) = \bigoplus_{p\in\mathbb{Z}} H(\mathrm{gr}(C)^p).$$

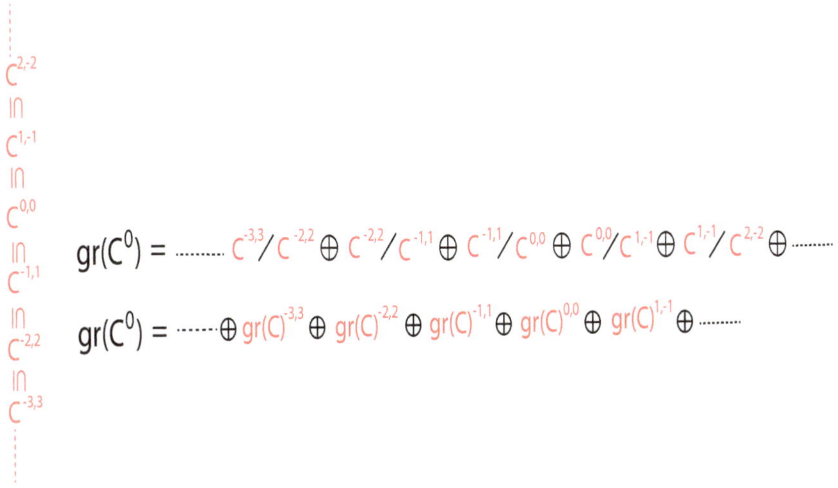

$$C^{2,-2} \subseteq C^{1,-1} \subseteq C^{0,0} \subseteq C^{-1,1} \subseteq C^{-2,2} \subseteq C^{-3,3}$$

$$\mathrm{gr}(C^0) = \cdots\cdots C^{-3,3}/C^{-2,2} \oplus C^{-2,2}/C^{-1,1} \oplus C^{-1,1}/C^{0,0} \oplus C^{0,0}/C^{1,-1} \oplus C^{1,-1}/C^{2,-2} \oplus \cdots\cdots$$

$$\mathrm{gr}(C^0) = \cdots\cdots \oplus \mathrm{gr}(C)^{-3,3} \oplus \mathrm{gr}(C)^{-2,2} \oplus \mathrm{gr}(C)^{-1,1} \oplus \mathrm{gr}(C)^{0,0} \oplus \mathrm{gr}(C)^{1,-1} \oplus \cdots\cdots$$

Fig. 15.15 An illustration of $\mathrm{gr}(C^0)$, the graded module associated with the first column of Figure 15.8.

Then we have

$$H(\mathrm{gr}(C)^p) = \bigoplus_{n\in\mathbb{Z}} H^n(\mathrm{gr}(C)^p) = \bigoplus_{n\in\mathbb{Z}} H^n(F^pC/F^{p+1}C),$$

and

$$H(\mathrm{gr}(C)) = \bigoplus_{p\in\mathbb{Z}} H(\mathrm{gr}(C)^p) = \bigoplus_{p\in\mathbb{Z}}\bigoplus_{n\in\mathbb{Z}} H^n(\mathrm{gr}(C)^p)$$
$$= \bigoplus_{p\in\mathbb{Z}}\bigoplus_{n\in\mathbb{Z}} H^n(F^pC/F^{p+1}C),$$

where

$$H^n(F^pC/F^{p+1}C) = \mathrm{Ker}\, d_0^{p,n-p}/\mathrm{Im}\, d_0^{p,n-p-1}.$$

We can make the change of variable $n = p+q$, in which case the above is written as

$$H(\mathrm{gr}(C)) = \bigoplus_{p\in\mathbb{Z}}\bigoplus_{q\in\mathbb{Z}} H^{p+q}(\mathrm{gr}(C)^p) = \bigoplus_{p\in\mathbb{Z}}\bigoplus_{q\in\mathbb{Z}} H^{p+q}(F^pC/F^{p+1}C),$$

but technically it is preferable to sum over the total index $n = p+q$, since n is the index uses in the definition of the complex

$$F^pC/F^{p+1}C \cong \bigoplus_{n\in\mathbb{Z}} C^{p,n-p}/C^{p+1,n-1-p} = \bigoplus_{n\in\mathbb{Z}} \mathrm{gr}(C)^{p,n-p}.$$

If we write

$$H(\mathrm{gr}(C))^{p,q} = H^{p+q}(F^pC/F^{p+1}C),$$

then $H(\mathrm{gr}(C)^p)$ is graded with

$$H(\mathrm{gr}(C)^p) = \bigoplus_{n\in\mathbb{Z}} H(\mathrm{gr}(C))^{p,n-p} = \bigoplus_{n\in\mathbb{Z}} H^n(F^pC/F^{p+1}C),$$

and $H(\mathrm{gr}(C))$ is bigraded with

$$H(\mathrm{gr}(C)) = \bigoplus_{p,n\in\mathbb{Z}} H(\mathrm{gr}(C))^{p,n-p}.$$

Definition 15.16. If we write

$$H(\mathrm{gr}(C))^{p,q} = H^{p+q}(F^pC/F^{p+1}C),$$

then $H(\mathrm{gr}(C))$ is bigraded and we have

$$H(\mathrm{gr}(C)) = \bigoplus_{p,n\in\mathbb{Z}} H(\mathrm{gr}(C))^{p,n-p} = \bigoplus_{p,n\in\mathbb{Z}} H^n(F^pC/F^{p+1}C).$$

Now the cochain complex $C = \bigoplus_{n\in\mathbb{Z}} C^n$ has cohomology

$$H(C) = \bigoplus_{n\in\mathbb{Z}} H^n(C).$$

The cochain complex

$$F^pC = \bigoplus_{n\in\mathbb{Z}} C^{p,n-p}$$

also possesses cohomology

$$H(F^pC) = \bigoplus_{n\in\mathbb{Z}} H^n(F^pC).$$

The inclusion map between F^pC and C is a chain map, so we have a map in cohomology $H(F^pC) \longrightarrow H(C)$ whose image is denoted $H(C)^p$. In this map, the image of $H^n(F^pC)$ in $H^n(C)$ is denoted $H(C)^{p,n-p}$. Since

$$H(C) = \bigoplus_{n\in\mathbb{Z}} H^n(C)$$

and

$$H(C)^p = \bigoplus_{n\in\mathbb{Z}} H(C)^{p,n-p}$$

(where the sum is performed along the q-axis with p fixed), observe that

$$H(C)^{p,n-p} = H^n(C) \cap H(C)^p.$$

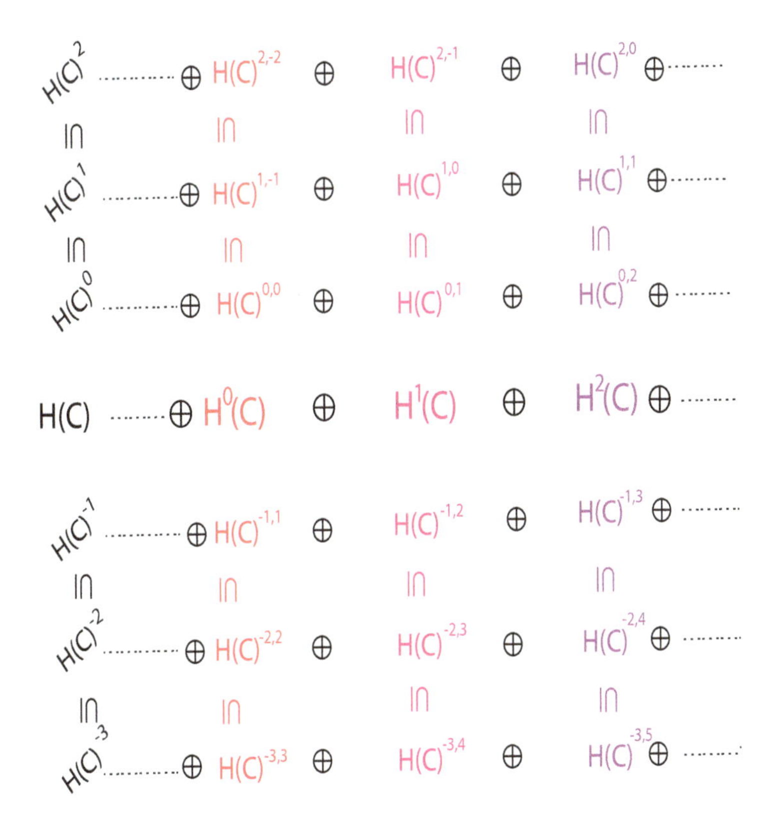

Fig. 15.16 A section of the filtered cochain complex $H(C)$, where $H(C)^{p,n-p}$ is the image $H^n(F^pC)$ in $H^n(C)$. Each horizontal row demonstrates the homology of the graded module $F^pC = \bigoplus_{q\in\mathbb{Z}} C^{p,q}$, while each column demonstrates the $(H(C)^{s,n-s})_{s\in\mathbb{Z}}$ filtration of $H^n(C)$.

The modules $H(C)^p$ filter $H(C)$ and the modules $H(C)^{p,n-p}$ filter $H^n(C)$. This is analogous to the fact that the F^pC filter C and that the $C^{p,n-p} = C^n \cap F^pC$ filter C^n; see Figures 15.8 and 15.16.

The condition $C = \bigcup_{p\in\mathbb{Z}} F^p(C)$ implies that $H^n(C) = \bigcup_{p\in\mathbb{Z}} H(C)^{p,n-p}$ because every element in $H^n(C)$ is represented by a cocycle in some $C^{p,n-p}$. However, in general $\bigcap_{p\in\mathbb{Z}} F^pC = (0)$ does not imply that $\bigcap_{p\in\mathbb{Z}} H(C)^{p,n-p} = (0)$. We will see in Section 15.7 that a sufficient condition for this property to hold is that the filtration is regular.

We now mimic the definitions of $\mathrm{gr}(C)^p$ and $\mathrm{gr}(C)$. We define the graded modules $\mathrm{gr}(H(C))^p$ and $\mathrm{gr}(H(C))$ as

$$\mathrm{gr}(H(C))^p = H(C)^p/H(C)^{p+1}$$

and

$$\mathrm{gr}(H(C)) = \bigoplus_{p\in\mathbb{Z}} \mathrm{gr}(H(C))^p.$$

Since

$$H(C)^p = \bigoplus_{n\in\mathbb{Z}} H(C)^{p,n-p} \quad \text{and} \quad H(C)^{p+1} = \bigoplus_{n\in\mathbb{Z}} H(C)^{p+1,n-p-1},$$

by Proposition 15.6, we have

$$\begin{aligned} \mathrm{gr}(H(C))^p &= H(C)^p/H(C)^{p+1} \\ &= \left(\bigoplus_{n\in\mathbb{Z}} H(C)^{p,n-p}\right) \Big/ \left(\bigoplus_{n\in\mathbb{Z}} H(C)^{p+1,n-p-1}\right) \\ &\cong \bigoplus_{n\in\mathbb{Z}} H(C)^{p,n-p}/H(C)^{p+1,n-p-1}. \end{aligned}$$

If we let

$$\mathrm{gr}(H(C))^{p,n-p} = H(C)^{p,n-p}/H(C)^{p+1,n-p-1},$$

or equivalently

$$\mathrm{gr}(H(C))^{p,q} = H(C)^{p,q}/H(C)^{p+1,q-1},$$

then

$$\mathrm{gr}(H(C))^p = H(C)^p/H(C)^{p+1} \cong \bigoplus_{n\in\mathbb{Z}} \mathrm{gr}(H(C))^{p,n-p} = \bigoplus_{q\in\mathbb{Z}} \mathrm{gr}(H(C))^{p,q};$$

see Figure 15.17.

$$\mathrm{gr}(H(C))^{-3} = H(C)^{-3}/H(C)^{-2} = \left(\cdots H(C)^{-3,3} \oplus H(C)^{-3,4} \oplus H(C)^{-3,5} \oplus \cdots\right) \Big/ \left(\cdots H(C)^{-2,2} \oplus H(C)^{-2,3} \oplus H(C)^{-2,4} \oplus \cdots\right)$$

$$\mathrm{gr}(H(C))^{-3} \cong \cdots H(C)^{-3,3}/H(C)^{-2,2} \oplus H(C)^{-3,4}/H(C)^{-2,3} \oplus H(C)^{-3,5}/H(C)^{-2,4} \oplus \cdots$$

$$\mathrm{gr}(H(C))^{-3} \cong \cdots \oplus \mathrm{gr}(H(C))^{-3,3} \oplus \mathrm{gr}(H(C))^{-3,4} \oplus \mathrm{gr}(H(C))^{-3,5} \oplus \cdots$$

Fig. 15.17 The graded module $\mathrm{gr}(H(C))^{-3} = H(C)^{-3}/H(C)^{-2}$ obtained by taking the quotient of the last two rows of Figure 15.16.

The bigraded module $\mathrm{gr}(H(C))$ is given by

$$\begin{aligned}\mathrm{gr}(H(C)) &= \bigoplus_{p\in\mathbb{Z}} \mathrm{gr}(H(C))^p \cong \bigoplus_{p\in\mathbb{Z}}\bigoplus_{n\in\mathbb{Z}} \mathrm{gr}(H(C))^{p,n-p} \\ &= \bigoplus_{p\in\mathbb{Z}}\bigoplus_{q\in\mathbb{Z}} \mathrm{gr}(H(C))^{p,q}.\end{aligned}$$

Definition 15.17. If we let

$$\mathrm{gr}(H(C))^{p,n-p} = H(C)^{p,n-p}/H(C)^{p+1,n-p-1},$$

or equivalently

$$\mathrm{gr}(H(C))^{p,q} = H(C)^{p,q}/H(C)^{p+1,q-1},$$

then the bigraded module $\mathrm{gr}(H(C))$ is given by

$$\mathrm{gr}(H(C)) \cong \bigoplus_{p\in\mathbb{Z}}\bigoplus_{n\in\mathbb{Z}} \mathrm{gr}(H(C))^{p,n-p} = \bigoplus_{p\in\mathbb{Z}}\bigoplus_{q\in\mathbb{Z}} \mathrm{gr}(H(C))^{p,q}.$$

Now $H^n(C)$ is filtered by the $H(C)^{p,n-p}$ with $p \in \mathbb{Z}$, thus the graded module $\mathrm{gr}(H^n(C))$ associated with $H^n(C)$ is given by

$$\mathrm{gr}(H^n(C)) = \bigoplus_{p\in\mathbb{Z}} H(C)^{p,n-p}/H(C)^{p+1,n-p-1} = \bigoplus_{p\in\mathbb{Z}} \mathrm{gr}(H(C))^{p,n-p},$$

so

$$\mathrm{gr}(H^n(C))^p = \mathrm{gr}(H(C))^{p,n-p};$$

see Figure 15.18.

$H(C)^{2,-2} \supseteq H(C)^{1,-1} \supseteq H(C)^{0,0} \supseteq H(C)^{-1,1} \supseteq H(C)^{-2,2} \supseteq H(C)^{-3,3}$

$$\mathrm{gr}(H^0(C)) = \cdots H(C)^{-3,3}/H(C)^{-2,2} \oplus H(C)^{-2,2}/H(C)^{-1,1} \oplus H(C)^{-1,1}/H(C)^{0,0} \oplus H(C)^{0,0}/H(C)^{1,-1} \oplus H(C)^{1,-1}/H(C)^{2,-2} \oplus \cdots$$

$$\mathrm{gr}(H^0(C)) = \cdots \oplus \mathrm{gr}(H(C))^{-3,3} \oplus \mathrm{gr}(H(C))^{-2,2} \oplus \mathrm{gr}(H(C))^{-1,1} \oplus \mathrm{gr}(H(C))^{0,0} \oplus \mathrm{gr}(H(C))^{1,-1} \oplus \cdots$$

Fig. 15.18 An illustration of $\mathrm{gr}(H^0(C))$, the graded module associated with the first column of Figure 15.16.

This is analogous to the fact that C^n is filtered by $(C^{p,n-p})_{p\in\mathbb{Z}}$ and $\mathrm{gr}(H^n(C))$ is analogous to $\mathrm{gr}(C^n)$. Observe that the sum is performed along the diagonal $p+q=n$ with n fixed. The graded module $\mathrm{gr}(H^n(C))$ consisting of a direct sum of modules appearing along the diagonal $p+q=n$ plays an important role in spectral sequences. It turns out that

$$\mathrm{gr}(H^n(C))^p = \mathrm{gr}(H(C))^{p,n-p} \cong E_\infty^{p,n-q}.$$

Note that $\mathrm{gr}(H(C))$ is also expressed as

$$\mathrm{gr}(H(C)) \cong \bigoplus_{n\in\mathbb{Z}} \mathrm{gr}(H^n(C)).$$

The rest of this chapter is replete with indices — a veritable orgy of indices. *The definitions to remember are five*: $C^{p,q}$, $\mathrm{gr}(C)^{p,q}$, $H(\mathrm{gr}(C))^{p,q}$, $H(C)^{p,q}$, *and* $\mathrm{gr}(H(C))^{p,q}$, namely:

$$\begin{aligned}
C^{p,q} &= C^{p+q} \cap F^pC \\
\mathrm{gr}(C)^{p,q} &= C^{p,q}/C^{p+1,q-1} \\
H(\mathrm{gr}(C))^{p,q} &= H^{p+q}(F^pC/F^{p+1}C) \\
H(C)^{p,q} &= H^{p+q}(C) \cap H(C)^p \\
\mathrm{gr}(H(C))^{p,q} &= \mathrm{gr}(H^{p+q}(C))^p = H(C)^{p,q}/H(C)^{p+1,q-1}.
\end{aligned}$$

15.6 Construction of a Spectral Sequence; Serre–Godement

Ideally we would like to compute the cohomology $H(C)$ of C. However, experience shows that this is usually not feasible, but instead we can begin by computing $H(\mathrm{gr}(C))$ because $\mathrm{gr}(C)$ is simpler than C. Then a spectral sequence is just the passage from $H(\mathrm{gr}(C))$ to $\mathrm{gr}(H(C))$; this is not quite $H(C)$ but is usually good enough.

Definition 15.18. A *spectral sequence* is a sequence

$$(\langle E_r, E_\infty, d_r, \alpha_r\rangle)_{r\in\mathbb{N}},$$

where

(1) Each E_r is a bigraded R-module with

$$E_r = \bigoplus_{p\in\mathbb{Z}} E_r^p \quad \text{and} \quad E_r^p = \bigoplus_{q\in\mathbb{Z}} E_r^{p,q},$$

for $r \in \mathbb{N} \cup \{\infty\}$ (the subscript r is called the *level*).

(2) For all $r \in \mathbb{N}$, the graded module $E_r = \bigoplus_{p\in\mathbb{Z}} E_r^p$ is a complex with differential d_r of degree r, where the restriction $d_r^{p,q}$ of d_r to $E_r^{p,q}$ is a map $d_r^{p,q}\colon E_r^{p,q} \to E_r^{p+r,q-r+1}$ such that $d_r^{p,q} \circ d_r^{p-r,q+r-1} = 0$, for all $p, q \in \mathbb{Z}$. The restriction d_r^p of d_r to E_r^p is the map $d_r^p = \bigoplus_{q\in\mathbb{Z}} d_r^{p,q}$, with $d_r^p\colon E_r^p \to E_r^{p+r}$.

(3) There is an isomorphism

$$\alpha_r\colon H(E_r) \to E_{r+1}$$

for all $r \in \mathbb{N}$, and more precisely,

$$H^p(E_r) = \mathrm{Ker}\, d_r^p/\mathrm{Im}\; d_r^{p-r} \cong E_{r+1}^p.$$

If we write $H^{p,q}(E_r) = \mathrm{Ker}\; d_r^{p,q}/\mathrm{Im}\; d_r^{p-r,q+r-1}$, then $H(E_r)$ is bi-graded, with

$$H(E_r) = \bigoplus_{p\in\mathbb{Z}} H^p(E_r), \quad H^p(E_r) = \bigoplus_{q\in\mathbb{Z}} H^{p,q}(E_r),$$

and we have an isomorphism

$$H^{p,q}(E_r) \cong E_{r+1}^{p,q}.$$

To understand a spectral sequence, it is useful to have in mind a pictorial representation of it in its entirety. For simplicity, assume that $E_r^{p,q} = (0)$ if $p < 0$ or $q < 0$ (a first quadrant spectral sequence). We are to imagine an infinitely tall 3-dimensional apartment house described in space by (p, q, r) coordinates where p corresponds to the x coordinate, q is the y coordinate, and r the z coordinate. The rth floor of the apartment building is on the plane $z = r$. The rooms of this rth floor are indexed by integer ordered pairs (p, q). The roof of the apartment building is the ∞-floor. In addition, there is the map $d_r^{p,q}$ on the rth floor; it goes "over r and down $r - 1$". Hence, a picture of the rth floor is shown in Figure 15.19.

The terms $E_r^{p,q}$ lie on the diagonal line (of slope -1) of equation $p+q = n$. Observe that $d_r^{p,q}$ has its range $E_r^{p+r,q-r+1}$ on the next diagonal of equation $x+y = n+1$ (where $= p+q$) and that $d^{p-r,q+r-1}$ has its domain $E_r^{p-r,q+r-1}$ on the previous diagonal of equation $x+y = n-1$. The module E_r^p corresponds to the pth column. The differential d_r maps the pth column to the $(p+r)$th column. The entire edifice is depicted in Figure 15.20.

One passes vertically directly to the floor above by forming cohomology (with respect to d_r); so one gets to the roof by repeated formings of cohomology at each higher level.

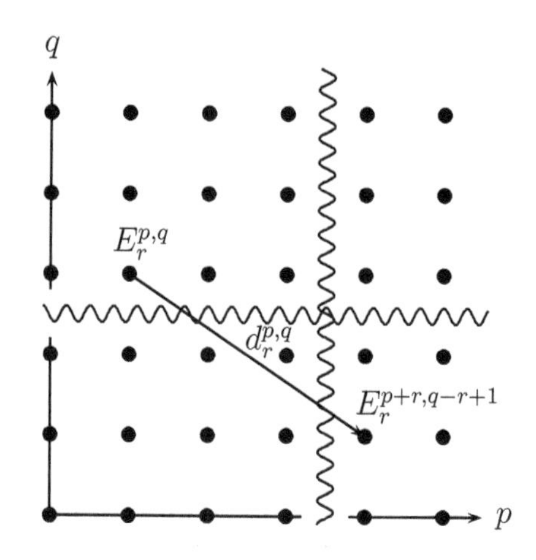

Fig. 15.19 The $E_r^{p,q}$ terms of a spectral sequence ("rth floor").

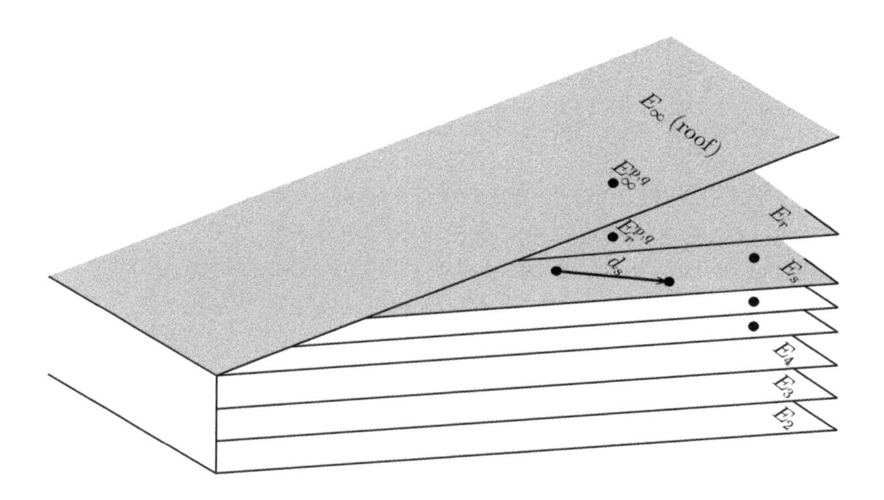

Fig. 15.20 The entire spectral sequence.

Once on the roof — at the ∞-level — the points on the line $p+q=n$, *i.e.*, the modules $E_\infty^{0,n}, E_\infty^{1,n-1}, \ldots, E_\infty^{n,0}$, are the composition factors for the filtration of $H^n(C)$:

$$H^n(C) \supseteq H(C)^{1,n-1} \supseteq H(C)^{2,n-2} \supseteq \cdots \supseteq H(C)^{n,0} \supseteq (0);$$

see Figure 15.21.

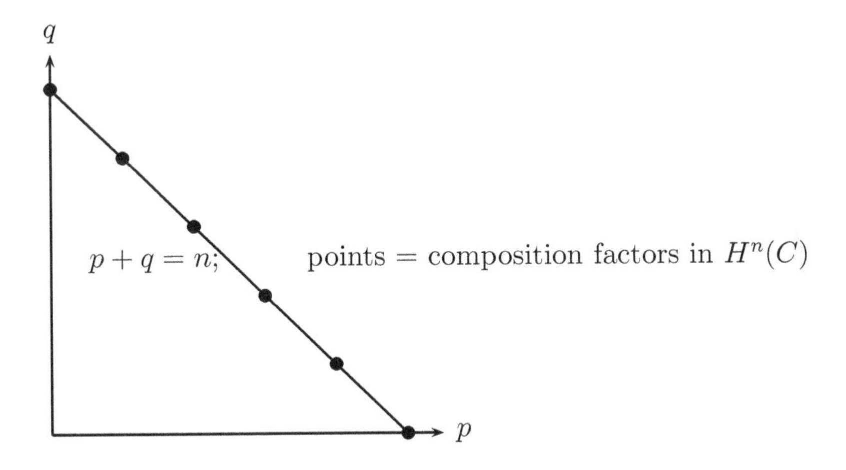

Fig. 15.21 The $E_\infty^{p,q}$ terms of a spectral sequence ("roof level").

Theorem 15.7. *Let (C,d) be a filtered and graded complex $C=\bigoplus_{n\in\mathbb{Z}}C^n$ with a differential d of degree 1 and a filtration $(F^pC)_{p\in\mathbb{Z}}$ compatible with the grading. There is a spectral sequence $(\langle E_r,E_\infty,d_r,\alpha_r\rangle)_{r\in\mathbb{N}\cup\{\infty\}}$ with the following properties:*

(1) $E_0=\mathrm{gr}(C)$ and $E_1\cong H(\mathrm{gr}(C))$. In particular,

$$E_0^{p,q}=C^{p,q}/C^{p+1,q-1}=(C^{p+q}\cap F^pC)/(C^{p+q}\cap F^{p+1}C)$$

and

$$E_1^{p,q}\cong H^{p+q}(F^pC/F^{p+1}C).$$

(2) $E_\infty\cong\mathrm{gr}(H(C))$. In particular,

$$E_\infty^{p,q}\cong\mathrm{gr}(H(C))^{p,q}=\mathrm{gr}(H^{p+q}(C))^p=H(C)^{p,q}/H(C)^{p+1,q-1}$$

and

$$\mathrm{gr}(H^n(C))\cong\bigoplus_{p\in\mathbb{Z}}E_\infty^{p,n-p}.$$

Proof. We break the proof in several steps.

Step 1: Proving the existence of a spectral sequence satisfying Conditions (1)–(3) of Definition 15.18.

We begin by constructing the $E_r^p = \bigoplus_{q\in\mathbb{Z}} E_r^{p,q}$, for $r \in \mathbb{N} \cup \{\infty\}$. The new ingredient in the construction is to apply the grading of C by the C^p to the Z_r^p and B_r^p of Definition 15.9.

Definition 15.19. For all $p, q, r \in \mathbb{Z}$, define $Z_r^{p,q}, B_r^{p,q}, Z_\infty^{p,q}$ and $B_\infty^{p,q}$ as follows:

$$\begin{aligned}
Z_r^{p,q} &= \{x \in C^{p+q} \cap F^pC \mid dx \in C^{p+q+1} \cap F^{p+r}C\}, \\
B_r^{p,q} &= \{x \in C^{p+q} \cap F^pC \mid (\exists y \in C^{p+q-1} \cap F^{p-r}C)(x = dy)\}, \\
Z_\infty^{p,q} &= \{x \in C^{p+q} \cap F^pC \mid dx = 0\}, \\
B_\infty^{p,q} &= \{x \in C^{p+q} \cap F^pC \mid (\exists y \in C^{p+q-1})(x = dy)\}.
\end{aligned}$$

We have $Z_r^{p,q} = C^{p+q} \cap F^pC = C^{p,q}$ and $B_r^{p,q} \subseteq C^{p+q} \cap F^{p-r}C = C^{p-r,q+r}$ for all $r \leq 0$. We verify easily that

$$\begin{aligned}
B_r^{p,q} &= dZ_r^{p-r,q+r-1} \\
Z_{r-1}^{p+1,q-1} &\subseteq Z_r^{p,q} \\
B_{r-1}^{p,q} &\subseteq B_r^{p,q} \subseteq Z_r^{p,q} \\
Z_\infty^{p+1,q-1} &\subseteq Z_\infty^{p,q} \\
B_\infty^{p,q} &\subseteq Z_\infty^{p,q}.
\end{aligned}$$

We define the graded modules Z_r^p and B_r^p as

$$Z_r^p = \bigoplus_{q\in\mathbb{Z}} Z_r^{p,q}, \quad B_r^p = \bigoplus_{q\in\mathbb{Z}} B_r^{p,q},$$

and the graded modules Z_∞^p and B_∞^p as

$$Z_\infty^p = \bigoplus_{q\in\mathbb{Z}} Z_\infty^{p,q}, \quad B_\infty^p = \bigoplus_{q\in\mathbb{Z}} B_\infty^{p,q}.$$

The above modules correspond precisely to the modules introduced in Definition 15.9 if we ignore the grading on C. Just as in Section 15.3 we make the following definition which is identical to Definition 15.10.

Definition 15.20. For all $r \geq 0$ and all $p \in \mathbb{Z}$, define E_r^p and E_∞^p as follows:

$$E_r^p = Z_r^p/(B_{r-1}^p + Z_{r-1}^{p+1}) \tag{$\dagger_r$}$$

and

$$E_\infty^p = Z_\infty^p/(B_\infty^p + Z_\infty^{p+1}). \tag{$\dagger_\infty$}$$

Since E_r^p and E_∞^p are direct sums, by Proposition 15.6, we get

$$\begin{aligned} E_r^p &= \left(\bigoplus_{q\in\mathbb{Z}} Z_r^{p,q}\right) \Big/ \left(\bigoplus_{s\in\mathbb{Z}} B_{r-1}^{p,s} + \bigoplus_{t\in\mathbb{Z}} Z_{r-1}^{p+1,t}\right) \\ &= \left(\bigoplus_{q\in\mathbb{Z}} Z_r^{p,q}\right) \Big/ \left(\bigoplus_{q\in\mathbb{Z}} (B_{r-1}^{p,q} + Z_{r-1}^{p+1,q-1})\right) \\ &\cong \bigoplus_{q\in\mathbb{Z}} Z_r^{p,q}/(B_{r-1}^{p,q} + Z_{r-1}^{p+1,q-1}). \end{aligned}$$

A similar computation shows that

$$E_\infty^p \cong \bigoplus_{q\in\mathbb{Z}} Z_\infty^{p,q}/(B_\infty^{p,q} + Z_\infty^{p+1,q-1}).$$

These equations suggest the following definition.

Definition 15.21. For all $r \geq 0$ and all $p, q \in \mathbb{Z}$, define $E_r^{p,q}$ and $E_\infty^{p,q}$ as follows:

$$E_r^{p,q} = Z_r^{p,q}/(B_{r-1}^{p,q} + Z_{r-1}^{p+1,q-1}) \tag{$\dagger\dagger_r$}$$

and

$$E_\infty^{p,q} = Z_\infty^{p,q}/(B_\infty^{p,q} + Z_\infty^{p+1,q-1}). \tag{$\dagger\dagger_\infty$}$$

By definition

$$\begin{aligned} E_r^p &\cong \bigoplus_{q\in\mathbb{Z}} E_r^{p,q} \\ E_\infty^p &\cong \bigoplus_{q\in\mathbb{Z}} E_\infty^{p,q}. \end{aligned}$$

Again, as in Section 15.3 (see Definition 15.11), we easily check that d induces a linear map $d_r^{p,q} \colon E_r^{p,q} \to E_r^{p+r,q-r+1}$.

We can adapt the argument used in Section 15.3 to prove Equation (E_r) to show that

$$\mathrm{Ker}\, d_r^{p,q}/\mathrm{Im}\, d_r^{p-r,q+r-1} \cong E_{r+1}^{p,q}.$$

Let use compute $\mathrm{Ker}\, d_r^{p,q}$ and $\mathrm{Im}\, d_r^{p-r,q+r-1}$. Recall that

$$\begin{aligned} B_r^{p,q} &= dZ_r^{p-r,q+r-1} \\ Z_{r-1}^{p+1,q-1} &\subseteq Z_r^{p,q} \\ B_{r-1}^{p,q} &\subseteq B_r^{p,q} \subseteq Z_r^{p,q} \\ Z_\infty^{p+1,q-1} &\subseteq Z_\infty^{p,q} \\ B_\infty^{p,q} &\subseteq Z_\infty^{p,q}, \end{aligned}$$

and

$$Z_r^{p,q} = \{x \in C^{p+q} \cap F^pC \mid dx \in C^{p+q+1} \cap F^{p+r}C\}.$$

Since $d_r^{p,q}\colon E_r^{p,q} \to E_r^{p+r,q-r+1}$ and

$$E_r^{p,q} = Z_r^{p,q}/(B_{r-1}^{p,q} + Z_{r-1}^{p+1,q-1}),$$
$$E_r^{p+r,q-r+1} = Z_r^{p+r,q-r+1}/(B_{r-1}^{p+r,q-r+1} + Z_{r-1}^{p+r+1,q-r}),$$

for any $x \in Z_r^{p,q}$, $d_r^{p,q}([x]) = 0$ iff $dx \in B_{r-1}^{p+r,q-r+1} + Z_{r-1}^{p+r+1,q-r} = dZ_{r-1}^{p+1,q-1} + Z_{r-1}^{p+r+1,q-r}$, which means that $dx = dy + z$, for some $y \in Z_{r-1}^{p+1,q-1}$ and some $z \in Z_{r-1}^{p+r+1,q-r}$. Since $Z_{r-1}^{p+1,q-1} \subseteq Z_r^{p,q} \subseteq C^{p+q} \cap F^pC$, we can write $x = y + u$ with $u = x - y$, where $u \in C^{p+q} \cap F^pC$, and since $dx = dy + z$ and $dx = dy + du$, we find that $du = z$. Since $du = z \in Z_{r-1}^{p+r+1,q-r}$, we have $du \in C^{p+q+1} \cap F^{p+r+1}C$, and since $u \in C^{p+q} \cap F^pC$, this means that $u \in Z_{r+1}^{p,q}$. Consequently, as seen by Figure 15.22, $x = u+y$, with $u \in Z_{r+1}^{p,q}$ and $y \in Z_{r-1}^{p+1,q-1}$, which shows that

$$\operatorname{Ker} d_r^{p,q} = (Z_{r+1}^{p,q} + Z_{r-1}^{p+1,q-1})/(B_{r-1}^{p,q} + Z_{r-1}^{p+1,q-1}).$$

The image of $d_r^{p-r,q+r-1}\colon E_r^{p-r,q+r-1} \to E_r^{p,q}$ consists of all classes modulo $(B_{r-1}^{p,q} + Z_{r-1}^{p+1,q-1})$ of elements in $B_r^{p,q} = dZ_r^{p-r,q+r-1}$. These are classes of the form $[x + y + z]_{(B_{r-1}^{p,q}+Z_{r-1}^{p+1,q-1})}$, where $x \in B_r^{p,q}$, $y \in B_{r-1}^{p,q}$, and $z \in Z_{r-1}^{p+1,q-1}$, but since $B_{r-1}^{p,q} \subseteq B_r^{p,q}$, these are the classes of the form $[x + z]_{(B_{r-1}^{p,q}+Z_{r-1}^{p+1,q-1})}$, where $x \in B_r^{p,q}$ and $z \in Z_{r-1}^{p+1,q-1}$, which shows that

$$\operatorname{Im} d_r^{p-r,q+r-1} = (B_r^{p,q} + Z_{r-1}^{p+1,q-1})/(B_{r-1}^{p,q} + Z_{r-1}^{p+1,q-1}).$$

In summary, we have

$$\operatorname{Ker} d_r^{p,q} = (Z_{r+1}^{p,q} + Z_{r-1}^{p+1,q-1})/(B_{r-1}^{p,q} + Z_{r-1}^{p+1,q-1})$$
$$\operatorname{Im} d_r^{p-r,q+r-1} = (B_r^{p,q} + Z_{r-1}^{p+1,q-1})/(B_{r-1}^{p,q} + Z_{r-1}^{p+1,q-1}).$$

The above formulae immediately imply that

$$d_r^{p,q} \circ d_r^{p-r,q+r-1} = 0.$$

It follows that the linear map $d_r^p = \bigoplus_{q\in\mathbb{Z}} d_r^{p,q}$ is a differential

$$d_r^p\colon E_r^p \to E_r^{p+r}$$

such that the piece $E_r^{p,q}$ is shifted left by $r-1$ slots (on the q index). As a consequence, as in Section 15.3,

$$E_r = \bigoplus_{p\in\mathbb{Z}} E_r^p$$

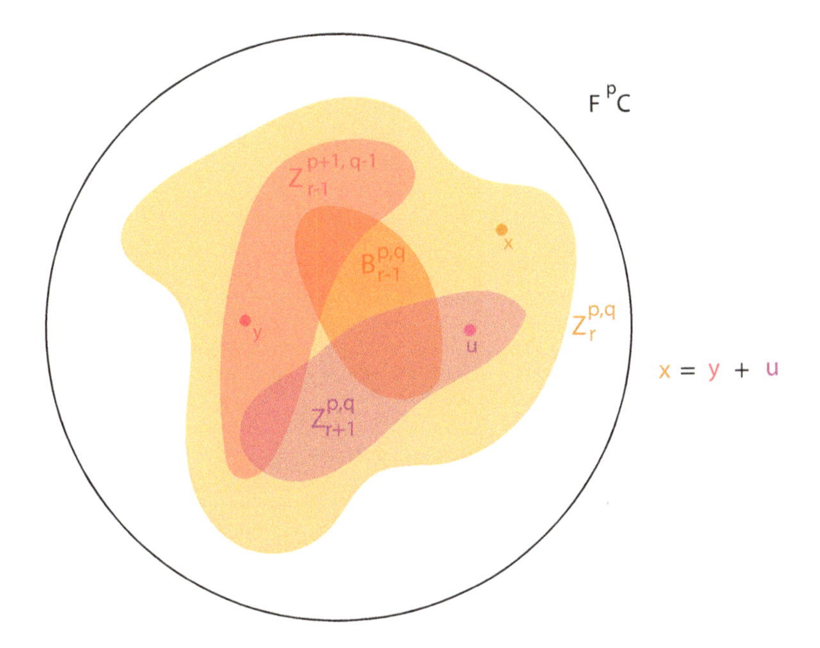

Fig. 15.22 A schematic illustration for $\mathrm{Ker}\, d_r^{p,q} = (Z_{r+1}^{p,q} + Z_{r-1}^{p+1,q-1})/(B_{r-1}^{p,q} + Z_{r-1}^{p+1,q-1})$.

is a graded complex with differential $d_r = \bigoplus_{p\in\mathbb{Z}} d_r^p$ of degree r. This proves (1) and (2) of the definition of a spectral sequence.

Using the above formulae for $\mathrm{Ker}\, d_r^{p,q}$ and $d_r^{p-r,q+r-1}$, we obtain.

$$\mathrm{Ker}\, d_r^{p,q}/\mathrm{Im}\, d_r^{p-r,q+r-1} \cong (Z_{r+1}^{p,q} + Z_{r-1}^{p+1,q-1})/(B_r^{p,q} + Z_{r-1}^{p+1,q-1}).$$

Using the modular Noether isomorphism, we have

$$\begin{aligned}(Z_{r+1}^{p,q} + Z_{r-1}^{p+1,q-1})/(B_r^{p,q} + Z_{r-1}^{p+1,q-1}) &\cong Z_{r+1}^{p,q}/(Z_{r+1}^{p,q} \cap (B_r^{p,q} + Z_{r-1}^{p+1,q-1}))\\ &\cong Z_{r+1}^{p,q}/(B_r^{p,q} + Z_r^{p+1,q-1})\\ &= E_{r+1}^{p,q},\end{aligned}$$

since $B_r^{p,q} \subseteq Z_{r+1}^{p,q}$ and $Z_{r+1}^{p,q} \cap Z_{r-1}^{p+1,q-1} = Z_r^{p+1,q-1}$. Therefore, we proved that

$$\mathrm{Ker}\, d_r^{p,q}/\mathrm{Im}\; d_r^{p-r,q+r-1} \cong E_{r+1}^{p,q} = Z_{r+1}^{p,q}/(B_r^{p,q} + Z_r^{p+1,q-1}).$$

If we define $H^{p,q}(E_r)$ as

$$H^{p,q}(E_r) = \mathrm{Ker}\, d_r^{p,q}/\mathrm{Im}\; d_r^{p-r,q+r-1},$$

then

$$H^{p,q}(E_r) \cong E_{r+1}^{p,q}. \quad (*)$$

Since

$$E_r = \bigoplus_{p,q\in\mathbb{Z}} E_r^{p,q},$$

and $d_r^p = \bigoplus_{q\in\mathbb{Z}} d_r^{p,q}$, by Proposition 15.6,

$$\begin{aligned} H^p(E_r) &= \mathrm{Ker}\, d_r^p/\mathrm{Im}\ d_r^{p-r} \cong \bigoplus_{q\in\mathbb{Z}} \mathrm{Ker}\, d_r^{p,q}/\mathrm{Im}\ d_r^{p-r,q+r-1} \\ &= \bigoplus_{q\in\mathbb{Z}} H^{p,q}(E_r) \cong \bigoplus_{q\in\mathbb{Z}} E_{r+1}^{p,q} = E_{r+1}^p. \end{aligned}$$

This proves (3) of Definition 15.18.

Step 2: Verifying Condition (1) of Theorem 15.7.

As in Section 15.3, since $Z_{-1}^{p,q} = C^{p,q}$, $Z_{-1}^{p+1,q-1} = C^{p+1,q-1}$ and $B_{-1}^{p,q} \subseteq C^{p+1,q-1} \subseteq C^{p,q}$, we find that

$$E_0^{p,q} = C^{p,q}/C^{p+1,q-1} = (C^{p+q} \cap F^pC)/(C^{p+q} \cap F^{p+1}C) = \mathrm{gr}(C)^{p,q}.$$

By definition of E_0 and $\mathrm{gr}(C)$, this proves the first part of (1).

Since

$$E_0^{p,q} = (C^{p+q} \cap F^pC)/(C^{p+q} \cap F^{p+1}C) = C^{p,q}/C^{p+1,q-1}$$

and

$$\begin{aligned} E_0^p &= \bigoplus_{q\in\mathbb{Z}} E_0^{p,q} = \bigoplus_{q\in\mathbb{Z}} C^{p,q}/C^{p+1,q-1} \\ &= \bigoplus_{n\in\mathbb{Z}} C^{p,n-p}/C^{p+1,n-p-1} = F^pC/F^{p+1}C \end{aligned}$$

is a differential module with differential $d_0^p = \bigoplus_{n\in\mathbb{Z}} d_0^{p,n-p}$ where $d_0^{p,n-p}\colon E_0^{p,n-p} \to E_0^{p,n-p+1}$ (see just after Definition 15.15), we have

$$H^n(F^pC/F^{p+1}C) = \mathrm{Ker}\, d_0^{p,n-p}/\mathrm{Im}\ d_0^{p,n-p-1},$$

or equivalently

$$H^{p+q}(F^pC/F^{p+1}C) = \mathrm{Ker}\, d_0^{p,q}/\mathrm{Im}\ d_0^{p,q-1}.$$

Now $E_0 = \bigoplus_{p\in\mathbb{Z}} E_0^p$ is a differential module with differential $d_0 = \bigoplus_{p\in\mathbb{Z}} d_0^p$ of degree 0, and by definition

$$H^{p,q}(E_0) = \mathrm{Ker}\, d_0^{p,q}/\mathrm{Im}\ d_0^{p,q-1},$$

so by $(*)$ and the above,

$$E_1^{p,q} \cong H^{p,q}(E_0) = \mathrm{Ker}\, d_0^{p,q}/\mathrm{Im}\, d_0^{p,q-1} = H^{p+q}(F^pC/F^{p+1}C),$$

proving the second part of (1).

Step 3: Verifying Condition (2) of Theorem 15.7.

The argument used in Section 15.3 can also be adapted to show that

$$E_\infty^{p,n-p} \cong \mathrm{gr}(H(C))^{p,n-p} = \mathrm{gr}(H^n(C))^p = H(C)^{p,n-p}/H(C)^{p+1,n-p-1}.$$

The key point is that since the complex F^pC is given by

$$F^pC = \bigoplus_{n\in\mathbb{Z}} C^{p,n-p}, \quad C^{p,n-p} = C^n \cap F^pC,$$

where the differential $d_{C^{p,n-p}} \colon C^{p,n-p} \to C^{p,n+1-p}$ is the restriction of d^n to $C^{p,n-p}$, and since $Z_\infty^{p,n-p} = \{z \in C^n \cap F^pC \mid dz = 0\}$, we see that the image $H(C)^{p,n-p}$ of $H^n(F^pC)$ in $H^n(C)$ is given by

$$H(C)^{p,n-p} = \{[z]_{B_n} \mid z \in Z_\infty^{p,n-p}\},$$

where $B_n = \mathrm{Im}\, d^{n-1}$. The map from $Z_\infty^{p,n-p}$ to $H(C)^{p,n-p}$ is surjective, and we obtain a surjection $\pi_\infty^{p,n-p} \colon Z_\infty^{p,n-p} \to H(C)^{p,n-p}/H(C)^{p+1,n-p-1}$. We need to figure out when $\pi_\infty^{p,n-p}(z) = 0$, which happens iff $[z]_{B_n} \in H(C)^{p+1,n-p-1}$ iff $z = z_1 + b$ for some $z_1 \in Z_\infty^{p+1,n-p-1}$ and some $b \in B_n$. Since $Z_\infty^{p+1,n-p-1} \subseteq Z_\infty^{p,n-p}$, we get $b = z - z_1 \in Z_\infty^{p,n-p} \subseteq C^n \cap F^pC$, so $b \in B_\infty^{p,n-p}$. Thus we proved that

$$\mathrm{Ker}\, \pi_\infty^{p,n-p} = B_\infty^{p,n-p} + Z_\infty^{p+1,n-p-1}.$$

By the first isomorphism theorem,

$$E_\infty^{p,n-p} = Z_\infty^{p,n-p}/(B_\infty^{p,n-p} + Z_\infty^{p+1,n-p-1}) \cong H(C)^{p,n-p}/H(C)^{p+1,n-p-1}.$$

Since

$$\mathrm{gr}(H^n(C)) = \bigoplus_{p\in\mathbb{Z}} \mathrm{gr}(H(C))^{p,n-p} = \bigoplus_{p\in\mathbb{Z}} \mathrm{gr}(H^n(C))^p,$$

$$\mathrm{gr}(H(C)) = \bigoplus_{n\in\mathbb{Z}} \mathrm{gr}(H^n(C)),$$

we have

$$\mathrm{gr}(H^n(C)) \cong \bigoplus_{p\in\mathbb{Z}} E_\infty^{p,n-p},$$

and since

$$E_\infty = \bigoplus_{p,n\in\mathbb{Z}} E_\infty^{p,n-p},$$

we obtain

$$E_\infty \cong \mathrm{gr}(H(C)),$$

establishing (2). ☐

15.7 Convergence of Spectral Sequences

Since the goal of constructing a spectral sequence is to "compute" the limit terms $E_\infty^{p,q}$, it is natural to consider what it means for a spectral sequence to converge. Various notions of convergence can be defined, weak convergence, strong convergence, *etc.*; see Cartan–Eilenberg [Cartan and Eilenberg (1956)], Weibel [Weibel (1994)] and McCleary [McCleary (2001)] for detailed expositions.

The general approach is to find conditions on filtrations that ensure some type of convergence. Ideally, we would like all $E_r^{p,q}$ (that is, for all $p, q \in \mathbb{Z}$) to stabilize to a fixed value for all $r \geq r_0$, for some fixed $r_0 > 0$, and to have $E_{r_0}^{p,q}$ isomorphic to $E_\infty^{p,q}$. This is a very strong requirement, so instead we can ask that for p, q fixed, there is some $r_0 = r(p, q)$ such that $E_{r_0}^{p,q}$ is isomorphic to $E_\infty^{p,q}$. This happens for bounded filtrations; see Theorem 15.12.

Another notion of convergence is to ask for p, q fixed whether $E_\infty^{p,q}$ can be viewed as a direct limit of the $E_r^{p,q}$. This is indeed the case when the filtration is bounded; see Proposition 15.15.

Perhaps the strongest notion of convergence is to require that for every n, there is some r such only one nonzero $E_r^{n-q,q}$ occurs on the diagonal $p + q = n$ for some $q = q(n)$. In such a case, we say that the spectral sequence degenerates at r. In this situation, if the filtration is regular, then $E_\infty^{n-q(n),q(n)}$ is isomorphic to $H^n(C)$ for every n. Under other mild conditions on $q(n)$, we actually have $E_r^{n-q(n),q(n)} \cong H^n(C)$ for every n; see Proposition 15.17. This result is quite unexpected, because in general, the $E_\infty^{p,q}$ only compute the graded pieces of $\mathrm{gr}(H(C))$. Nevertheless, degenerate spectral sequences often show up in algebraic topology.

In the rest of this chapter it is assumed that *all filtrations are compatible with the grading*. To simplify matters in stating our first "convergence" theorem, we restrict ourselves to spectral sequences induced by positive filtrations.

Definition 15.22. A filtration $(F^pC)_{p\in\mathbb{Z}}$ on a graded complex C is a *positive filtration* if $F^pC = C$ for all $p \leq 0$. It is convenient to assume that $F^{-\infty}C = C$ and that $F^{\infty}C = (0)$.

Proposition 15.8. *Let (C, d) be filtered and graded complex $C = \bigoplus_{n\in\mathbb{Z}} C^n$ with a differential d of degree 1 and a filtration $(F^pC)_{p\in\mathbb{Z}}$.*

(1) *If* $F^pC = C$ *for all* $p \leq 0$, *then* $E_r^{p,q} = (0)$ *for all* $p < 0$, *all* $q \in \mathbb{Z}$, *and all* $r \in \mathbb{N} \cup \{\infty\}$; *see Figure 15.23.*

(2) *For all* $n, p \in \mathbb{Z}$, *if* $p > n$ *implies that* $C^n \cap F^pC = (0)$, *then* $E_r^{p,q} = (0)$ *for all* $q < 0$, *all* $p \in \mathbb{Z}$, *and all* $r \in \mathbb{N} \cup \{\infty\}$; *see Figure 15.23.*

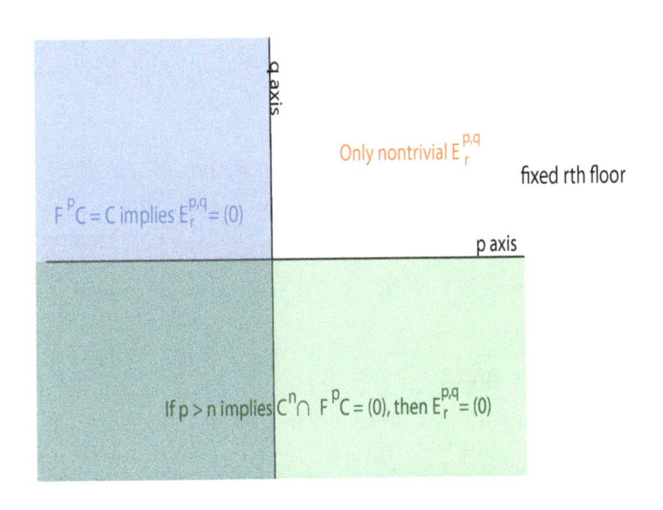

Fig. 15.23 For a fixed rth level, Proposition 15.8(1) ensures that the integer indices of blue region are associated with trivial $E_r^{p,q}$ while Proposition 15.8(2) ensures that the integer indices of the green region are with trivial $E_r^{p,q}$.

Proof. (1) Since

$$Z_r^{p,q} = \{x \in C^{p+q} \cap F^pC \mid dx \in C^{p+q+1} \cap F^{p+r}C\},$$
$$Z_{r-1}^{p+1,q-1} = \{x \in C^{p+q} \cap F^{p+1}C \mid dx \in C^{p+q+1} \cap F^{p+r}C\},$$

we have

$$Z_{r-1}^{p+1,q-1} = Z_r^{p,q} \cap F^{p+1}C,$$

so if $p < 0$, since $F^sC = C$ for $s \leq 0$, we have $F^{p+1}C = C$ and $Z_{r-1}^{p+1,q-1} = Z_r^{p,q}$. Since

$$E_r^{p,q} = Z_r^{p,q}/(B_{r-1}^{p,q} + Z_{r-1}^{p+1,q-1}),$$

we deduce that $E_r^{p,q} = (0)$ for $p < 0$.

Since

$$Z_\infty^{p,q} = \{x \in C^{p+q} \cap F^pC \mid dx = 0\},$$

we have $Z_\infty^{p+1,q-1} = Z_\infty^{p,q} \cap F^{p+1}C$, and since

$$E_\infty^{p,q} = Z_\infty^{p,q}/(B_\infty^{p,q} + Z_\infty^{p+1,q-1}),$$

we deduce that $E_\infty^{p,q} = (0)$ for $p < 0$.

(2) Since $Z_r^{p,q} \subseteq C^{p+q} \cap F^pC$, if $C^n \cap F^pC = (0)$ when $p > n$, we see that if $q < 0$, then $p > p+q$, so $C^{p+q} \cap F^pC = (0)$ and $E_r^{p,q} = Z_r^{p,q} = (0)$. We also have $Z_\infty^{p,q} \subseteq C^{p+q} \cap F^pC$, so if $q < 0$, then $E_\infty^{p,q} = Z_\infty^{p,q} = (0)$. ☐

Remark: Suppose the filtration (F^pC) is positive. Since $F^0C = C$, if the condition $C^n \cap F^pC = (0)$ holds for all $p, n \in \mathbb{Z}$ such that $p > n$, we deduce that if $n < 0$, we have $(0) = C^n \cap F^0C = C^n \cap C = C^n$, that is $C^n = (0)$.

The condition that $C^n \cap F^pC = (0)$ if $p > n$ is a special case of the following condition.

Definition 15.23. A filtration is *regular* (or *bounded below*) if for every $n \in \mathbb{Z}$, there is some integer $\mu(n)$ such that for all $p > \mu(n)$ we have $C^n \cap F^pC = (0)$. Since C^n is filtered by the $C^{p,n-p} = C^n \cap F^pC$, this means that $C^{p,n-p} = (0)$ for $p = \mu(n)+1$; that is, the filtration $(C^{p,n-p})$ of C^n stabilizes to the zero module for $p = \mu(n)+1$.

Geometrically, a sequence is regular if on every diagonal $p + q = n$ (for fixed n), as p increases, that is, as we go down along this diagonal, the modules $C^{p,n-p}$ stabilize to the zero module for $p \geq \mu(n)+1$; see Figure 15.24.

Remark: Some authors, such as Weibel [Weibel (1994)], define a spectral sequence to be regular if for each p and q, the differentials $d_r^{p,q}$ leaving $E_r^{p,q}$ are 0 for all r large enough. A spectral sequence which is bounded below is regular; see Proposition 15.10(b). This is a weaker notion of regularity that we will not use.

Proposition 15.9. *If the filtration (F^pC) on a graded complex C is regular, namely for every $n \in \mathbb{Z}$, there is some $\mu(n)$ such that for $p = \mu(n)+1$ we have $C^{p,n-p} = (0)$, then for every $n \in \mathbb{Z}$, the filtration $(H(C)^{p,n-p})$ of $H^n(C)$ is also regular and of the form*

$$H^n(C) \supseteq \cdots H(C)^{p,n-p} \supseteq H(C)^{p+1,n-p-1} \supseteq \cdots \supseteq H(C)^{\mu(n),n-\mu(n)} \supseteq (0),$$

with $H(C)^{\mu(n)+1,n-\mu(n)-1} = (0)$.

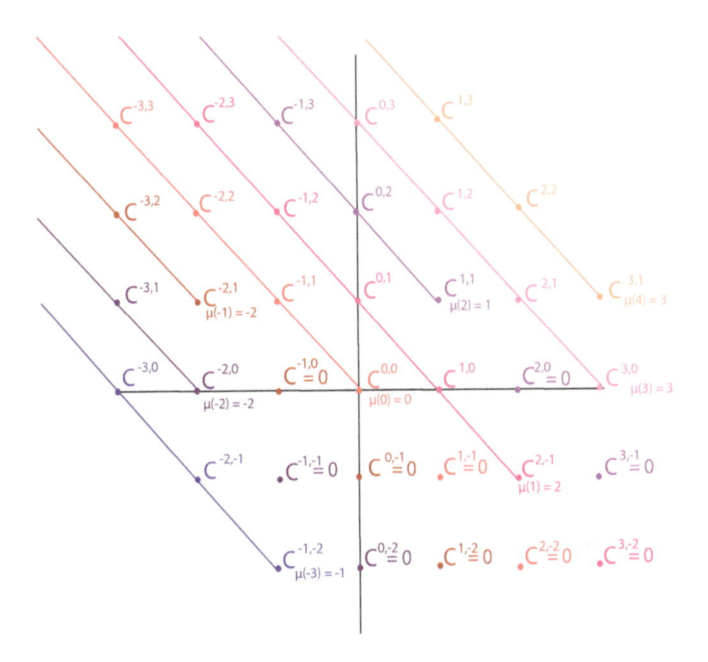

Fig. 15.24 An example of a regular filtration since along each $p+q=n$ diagonal the modules $C^{p,n-p}$ stabilize to the zero module.

Proof. Recall that $H(C)^{p,n-p}$ is the image in $H^n(C)$ of the module $H^n(F^pC)$ and that the cochain complex F^pC is given by

$$F^pC = \bigoplus_{n\in\mathbb{Z}} C^{p,n-p},$$

with the coboundary maps $d_{C^{p,n-p}}\colon C^{p,n-p} \to C^{p,n+1-p}$. Since

$$H^n(F^pC) = \mathrm{Ker}\, d_{C^{p,n-p}}/\mathrm{Im}\; d_{C^{p,n-p-1}}$$

and $C^{p,n-p} = (0)$ for all $p > \mu(n)$, we have $\mathrm{Ker}\, d_{C^{p,n-p}} = (0)$, thus $H^n(F^pC) = (0)$ for all $p > \mu(n)$, and the image $H(C)^{p,n-p}$ of $H^n(F^pC) = (0)$ is also (0) for all $p > \mu(n)$. $\square$

Proposition 15.10. *Let (C, d) be a filtered and graded complex $C = \bigoplus_{n\in\mathbb{Z}} C^n$ with a differential d of degree 1 and a filtration $(F^pC)_{p\in\mathbb{Z}}$.*

(1) If the filtration is regular, then

(a) For all $n, p \in \mathbb{Z}$, if $p > \mu(n)$, then for all $r \in \mathbb{N} \cup \{\infty\}$, we have

$$E_r^{p,n-p} = (0);$$

see Figure 15.25.

(b) *For all* $p, q \in \mathbb{Z}$, *for all* $r > \mu(p+q+1) - p$, *we have*

$$Z_r^{p,q} = Z_\infty^{p,q}$$

and $d_r^{p,q} = 0$ *on* $E_r^{p,q}$.

(2) *If the filtration is regular and positive, then for all* $p, q \in \mathbb{Z}$ *and for all* $r > max(p, \mu(p+q+1) - p)$, *we have*

$$E_r^{p,q} = E_\infty^{p,q};$$

see Figure 15.26.

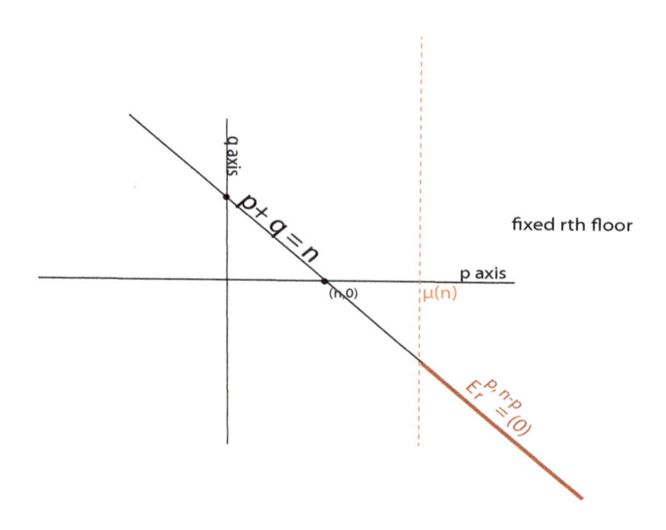

Fig. 15.25 An illustration of Proposition 15.10(1).

Proof. (1)(a) Since

$$Z_r^{p,n-p} = \{x \in C^n \cap F^pC \mid dx \in C^{n+1} \cap F^{p+r}C\}$$

and $C^n \cap F^pC = (0)$ if $p > \mu(n)$, we have $E_r^{p,n-p} = Z_r^{p,n-p} = (0)$. Since

$$Z_\infty^{p,n-p} = \{x \in C^n \cap F^pC \mid dx = 0\},$$

the same reasoning shows that $E_\infty^{p,n-p} = (0)$ if $p > \mu(n)$.

(1)(b) Since

$$Z_r^{p,q} = \{x \in C^{p+q} \cap F^pC \mid dx \in C^{p+q+1} \cap F^{p+r}C\}$$

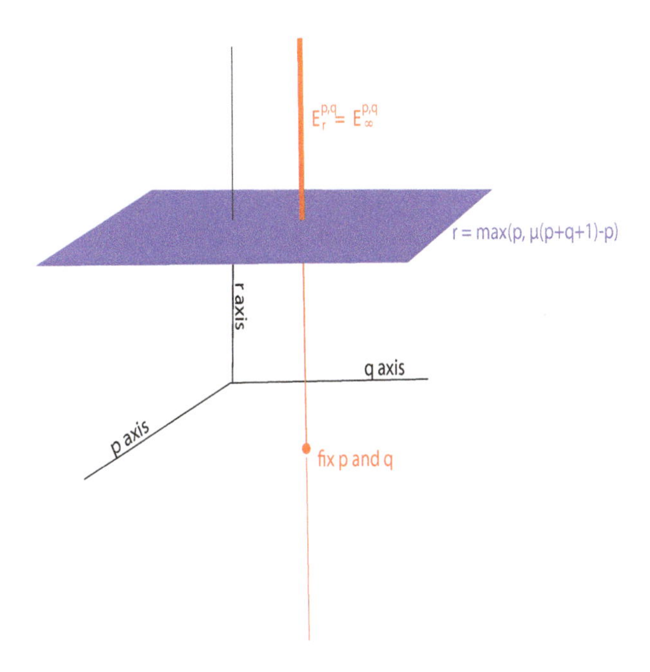

Fig. 15.26 An illustration of Proposition 15.10(2). For the integer triples (p, q, r) on the bold red line, $E_r^{p,q}$ stabilizes to $E_\infty^{p,q}$.

and since the filtration is regular, if $p + r > \mu(p + q + 1)$, that is, $r > \mu(p + q + 1) - p$, then $C^{p+q+1} \cap F^{p+r}C = (0)$, and so $Z_r^{p,q} = Z_\infty^{p,q}$.

Recall that $d_r^{p,q} \colon E_r^{p,q} \to E_r^{p+r,q-r+1}$. Since

$$Z_r^{p+r,q-r+1} = \{x \in C^{p+q+1} \cap F^{p+r}C \mid dx \in C^{p+q+2} \cap F^{p+2r}C\}$$

and the filtration is regular, if $p+r > \mu(p+q+1)$, that is, $r > \mu(p+q+1)-p$, then $C^{p+q+1} \cap F^{p+r}C = (0)$, so $Z_r^{p+r,q-r+1} = (0)$ and thus $E_r^{p+r,q-r+1} = (0)$, so $d_r^{p,q}$ is zero on $E_r^{p,q}$.

(2) Since

$$\begin{aligned} B_r^{p,q} &= \{x \in C^{p+q} \cap F^pC \mid (\exists y \in C^{p+q-1} \cap F^{p-r}C)(x = dy)\}, \\ B_\infty^{p,q} &= \{x \in C^{p+q} \cap F^pC \mid (\exists y \in C^{p+q-1})(x = dy)\} \end{aligned}$$

and the filtration is positive, if $p - r \leq 0$, that is, $r \geq p$, then $C^{p+q-1} \cap F^{p-r}C = C^{p+q-1}$, and thus

$$B_r^{p,q} = B_\infty^{p,q}.$$

Consequently, if $r > p$, then $B^{p,q}_{r-1} = B^{p,q}_{\infty}$. By (1), if $r > \mu(p+q+1) - p$, then

$$Z^{p,q}_r = Z^{p,q}_{\infty},$$

and since $r - 1 > \mu(p+q+1) - (p+1)$ because $r > \mu(p+q+1) - p$, we also have $Z^{p+1,q-1}_{r-1} = Z^{p-1,q-1}_{\infty}$. Since

$$E^{p,q}_r = Z^{p,q}_r/(B^{p,q}_{r-1} + Z^{p+1,q-1}_{r-1})$$
$$E^{p,q}_{\infty} = Z^{p,q}_{\infty}/(B^{p,q}_{\infty} + Z^{p+1,q-1}_{\infty}),$$

we see that if $r > \max(p, \mu(p+q+1) - p)$, then

$$E^{p,q}_r = E^{p,q}_{\infty},$$

as claimed. ☐

Proposition 15.10 shows that *if the filtration on C is positive and regular, then for any fixed p, q, for r large enough, the $E^{p,q}_r$ stabilize to the limit value $E^{p,q}_{\infty}$.*

Proposition 15.10(2) holds under a weaker condition than positivity of the filtration.

Definition 15.24. A filtration is *bounded* if for every $n \in \mathbb{Z}$, there are some integers $\nu(n) \leq \mu(n)$ such that

(a) For $p = \nu(n)$, we have $C^n \cap F^pC = C^n$.
(b) For $p = \mu(n) + 1$, we have $C^n \cap F^pC = (0)$.

Since C^n is filtered by the $C^{p,n-p} = C^n \cap F^pC$, this means that $C^{p,n-p} = C^n$ for $p = \nu(n)$ and $C^{n,n-p} = (0)$ for $p = \mu(n) + 1$. Thus the filtration of C^n has finite length:

$$C^n = C^{\nu(n),n-\nu(n)} \supseteq C^{\nu(n)+1,n-\nu(n)-1} \supseteq \cdots$$
$$\supseteq C^{\mu(n),n-\mu(n)} \supseteq C^{\mu(n)+1,n-\mu(n)-1} = (0).$$

Geometrically, a sequence is bounded if for every diagonal $p + q = n$ (for fixed n), there is a finite strip such that all modules $C^{p,n-p}$ above or equal to $C^{\nu(n),n-\nu(n)}$ are equal to C^n and all modules below $C^{\mu(n),n-\mu(n)}$ are zero; see Figure 15.27. Observe that a bounded filtration is bounded below (and above) and thus regular.

Proposition 15.11. *Let (C, d) be a filtered and graded complex $C = \bigoplus_{n\in\mathbb{Z}} C^n$ with a differential d of degree 1 and a filtration $(F^pC)_{p\in\mathbb{Z}}$. If the filtration is bounded, then the following facts hold:*

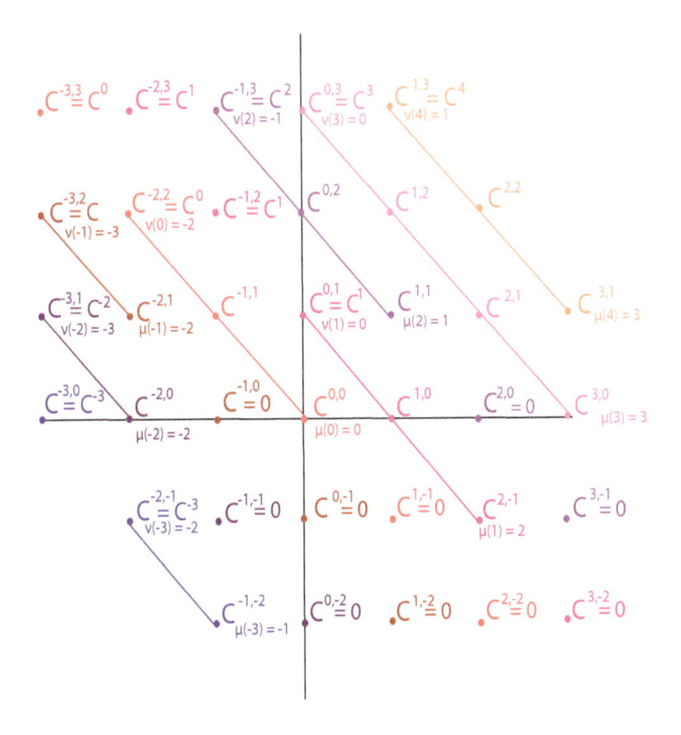

Fig. 15.27 An example of a bounded filtration. Along each diagonal $p + q = n$, if $p \leq \nu(n)$, then $C^{p,n-p} = C$, while if $p > \mu(n)$, then $C^{p,n-p} = (0)$.

(1) *For all $p, n \in \mathbb{Z}$, if $p < \nu(n)$, then for all $r \in \mathbb{N} \cup \{\infty\}$, we have*

$$E_r^{p,n-p} = (0);$$

see Figure 15.28.

(2) *For all $p, q \in \mathbb{Z}$ and for all $r > \max(p - \nu(p+q-1), \mu(p+q+1) - p)$, we have*

$$E_r^{p,q} = E_\infty^{p,q};$$

see Figure 15.26 with $r = \max(p - \nu(p+q-1), \mu(p+q+1) - p)$.

Proof. (1) Since

$$Z_r^{p,n-p} = \{x \in C^n \cap F^pC \mid dx \in C^{n+1} \cap F^{p+r}C\},$$
$$Z_{r-1}^{p+1,n-p-1} = \{x \in C^n \cap F^{p+1}C \mid dx \in C^{n+1} \cap F^{p+r}C\},$$

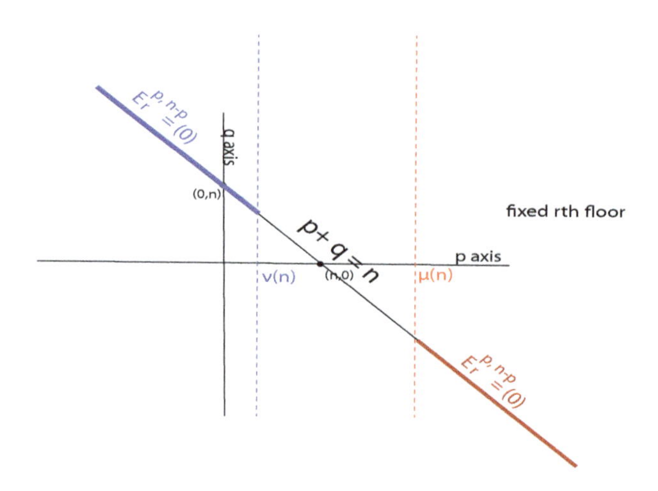

Fig. 15.28 An illustration of Proposition 15.11(1).

and since $C^n \cap F^sC = C^n$ if $s \leq \nu(n)$, if $p < \nu(n)$, then $Z_{r-1}^{p+1,n-p-1} = Z_r^{p,n-p}$. Since

$$E_r^{p,n-p} = Z_r^{p,n-p}/(B_{r-1}^{p,n-p} + Z_{r-1}^{p+1,n-p-1}),$$

we deduce that $E_r^{p,n-p} = (0)$ for $p < \nu(n)$.

Similarly,

$$Z_\infty^{p,n-p} = \{x \in C^n \cap F^pC \mid dx = 0\},$$
$$Z_\infty^{p+1,n-p-1} = \{x \in C^n \cap F^{p+1}C \mid dx = 0\},$$

and the same reasoning as above shows that $Z_\infty^{p+1,n-p-1} = Z_\infty^{p,n-p}$. Since

$$E_\infty^{p,n-p} = Z_\infty^{p,n-p}/(B_\infty^{p,n-p} + Z_\infty^{p+1,n-p-1}),$$

we deduce that $E_\infty^{p,n-p} = (0)$ for $p < \nu(n)$.

(2) Since

$$B_r^{p,q} = \{x \in C^{p+q} \cap F^pC \mid (\exists y \in C^{p+q-1} \cap F^{p-r}C)(x = dy)\},$$
$$B_\infty^{p,q} = \{x \in C^{p+q} \cap F^pC \mid (\exists y \in C^{p+q-1})(x = dy)\},$$

and the filtration is bounded, if $p-r \leq \nu(p+q-1)$, that is, $r \geq p-\nu(p+q-1)$, then $C^{p+q-1} \cap F^{p-r}C = C^{p+q-1}$, and thus

$$B_r^{p,q} = B_\infty^{p,q}.$$

Consequently, if $r > p - \nu(p+q-1)$, then $B_{r-1}^{p,q} = B_\infty^{p,q}$. The rest of the proof is identical to the proof of Proposition 15.10(2). □

Combining Proposition 15.10 and Proposition 15.11 we obtain the following result.

Theorem 15.12. *Let (C, d) be a filtered and graded complex $C = \bigoplus_{n\in\mathbb{Z}} C^n$ with a differential d of degree 1 and a filtration $(F^pC)_{p\in\mathbb{Z}}$. If the filtration is bounded, then the following facts hold:*

(1) For all $p, n \in \mathbb{Z}$, if $p < \nu(n)$ or $p > \mu(n)$, then for all $r \in \mathbb{N} \cup \{\infty\}$, we have

$$E_r^{p,n-p} = (0).$$

(2) For all $p, q \in \mathbb{Z}$ and for all $r > \max(p - \nu(p+q-1), \mu(p+q+1) - p)$, we have

$$E_r^{p,q} = E_\infty^{p,q}.$$

Geometrically, Theorem 15.12 means that for every diagonal with slope -1 given by the equation $p + q = n$, there are only finitely many nonzero terms $E^{p,n-p}$, with $\nu(n) \leq p \leq \mu(n)$, and that for n fixed, for all r large enough, namely $r > \max(p - \nu(n-1), \mu(n+1) - p)$, the $E_r^{p,n-p}$ stabilize to the limit value $E_\infty^{p,n-p}$.

The special case of a bounded filtration on a cohomological cochain complex (which means that $C^n = (0)$ for all $n < 0$) for which $\nu(n) = 0$ and $\mu(n) = n$ for all $n \in \mathbb{N}$ is of special interest. In this case, $C^{p,n-p} = C^n \cap F^pC = C^n$ for all $p \leq 0$ and all $n \in \mathbb{N}$, and since $C = \bigoplus_{n\in\mathbb{N}} C^n$, we see that $F^pC = C$ for all $p \leq 0$, which means that the filtration is positive. This case is illustrated in Figure 15.29.

Definition 15.25. A filtration (F^pC) on a cohomological cochain complex $C = \bigoplus_{n\in\mathbb{N}} C^n$ is *canonically cobounded*[1] if the filtration is positive and if $C^{p,n-p} = C^n \cap F^pC = (0)$ for all $p > n$ (equivalently, $C^{n+1,-1} = (0)$). See Figures 15.23 and 15.29.

Proposition 15.13. *If the filtration (F^pC) on a graded complex C is canonically cobounded, namely $C^n = (0)$ for all $n < 0$, and $C^{0,n} = C^n$ and $C^{n+1,-1} = (0)$ for all $n \in \mathbb{N}$, then for every $n \geq 0$, the filtration $(H(C)^{p,n-p})$ of $H^n(C)$ is also canonically cobounded and of the form*

$$H^n(C) = H(C)^{0,n} \supseteq \cdots \supseteq H(C)^{p+1,n-p-1} \supseteq \cdots \supseteq H(C)^{n,0} \supseteq (0),$$

with $H(C)^{n+1,-1} = (0)$.

[1]This notion appears to have been introduced by Mac Lane.

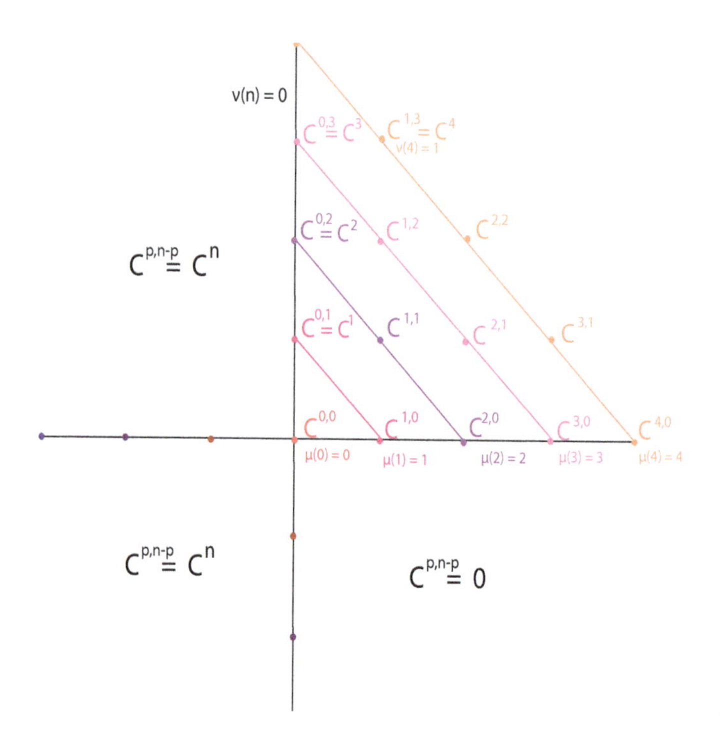

Fig. 15.29 An illustration of the canonically cobounded filtration. The pertinent information is contained within the first quadrant of the (p, q)-plane.

Proof. Recall that $H(C)^{p,n-p}$ is the image in $H^n(C)$ of the module $H^n(F^pC)$ and that the cochain complex F^pC is given by

$$F^pC = \bigoplus_{n\in\mathbb{N}} C^{p,n-p},$$

with the coboundary maps $d_{C^{p,n-p}}\colon C^{p,n-p} \to C^{p,n+1-p}$. Since $C^n = (0)$ for $n < 0$ and $C^{0,n} = C^n$ for all $n \geq 0$, the complex for F^0C is the complex C, so $H^n(F^0C) = H^n(C)$ and $H(C)^{0,n} = H^n(C)$. The fact that the filtration $H(C)^{p,n-p}$ is bounded below follows from Proposition 15.9. $\square$

Proposition 15.8 and Theorem 15.12 imply the following result.

Theorem 15.14. *Let (C, d) be filtered cohomological complex $C = \bigoplus_{n\in\mathbb{N}} C^n$ with a filtration $(F^pC)_{p\in\mathbb{N}}$. If the filtration is canonically cobounded then the following facts hold:*

(1) *For all* $p, q \in \mathbb{Z}$, *if* $p < 0$ *or* $q < 0$, *then for all* $r \in \mathbb{N} \cup \{\infty\}$, *we have*

$$E_r^{p,n-p} = (0).$$

(2) *For all* $p, q \in \mathbb{N}$ *and for all* $r > \max(p, q+1)$, *we have*

$$E_r^{p,q} = E_\infty^{p,q}.$$

Definition 15.26. A spectral sequence E for which $E_r^{p,q} = (0)$ if $p < 0$ or $q < 0$ is called a *first quadrant spectral sequence.*

15.8 Degenerate Spectral Sequences

The assumption that the bounding function ν exists was crucial for the $B_r^{p,q}$ to stabilize. If we relax this condition, the $B_r^{p,q}$ may not stabilize, but since a filtration has the property that $\bigcup_{\in\mathbb{Z}} F^pC = C$, we have

$$B_\infty^{p,q} = \bigcup_{r\in\mathbb{N}} B_r^{p,q}.$$

If we still assume that the filtration is regular, then we can prove that each $E_\infty^{p,q}$ is a direct limit of the $E_r^{p,q}$. Technically we have the following result.

Proposition 15.15. *Let* (C, d) *be a filtered and graded complex* $C = \bigoplus_{n\in\mathbb{Z}} C^n$ *with a differential* d *of degree* 1 *and a filtration* $(F^pC)_{p\in\mathbb{Z}}$. *If the filtration is regular, then for any fixed* $p, q \in \mathbb{Z}$, *for* $r_0 = \mu(p+q+1) - p + 1$, *there is a commutative diagram*

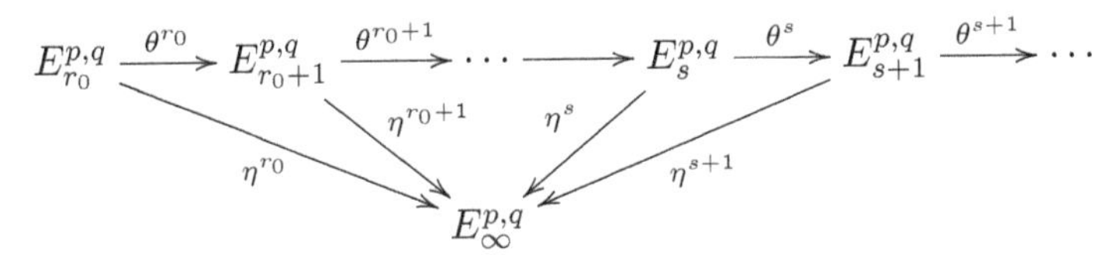

in which all the morphisms are surjective, and $E_\infty^{p,q}$ *is the direct limit of the above mapping family.*

Proof. By Proposition 15.10(1)(b), if $r > \mu(p+q+1) - p$, then we have

$$Z_r^{p,q} = Z_\infty^{p,q}$$

and $d_r^{p,q} = 0$ on $E_r^{p,q}$. Since $E_{r+1}^{p,q} \cong H^{p,q}(E_r) = \operatorname{Ker} d_r^{p,q}/\operatorname{Im} d_r^{p-r,q+r-1}$, if $d_r^{p,q} = 0$ on $E_r^{p,q}$, then $\operatorname{Ker} d_r^{p,q} = E_r^{p,q}$, so $E_{r+1}^{p,q} \cong E_r^{p,q}/\operatorname{Im} d_r^{p-r,q+r-1}$, and by composing the surjective quotient map from $E_r^{p,q}$ onto $E_r^{p,q}/\operatorname{Im} d_r^{p-r,q+r-1}$ and the isomorphism between $E_r^{p,q}/\operatorname{Im} d_r^{p-r,q+r-1}$ and $E_{r+1}^{p,q}$, we obtain a surjective map $\theta^r \colon E_r^{p,q} \to E_{r+1}^{p,q}$.

If $r > \mu(p+q+1)-p$, then $Z_r^{p,q} = Z_\infty^{p,q}$ and $Z_{r-1}^{p+1,q-1} = Z_\infty^{p+1,q-1}$, since $r-1 > \mu(p+q+1)-(p+1)$ is equivalent to $r > \mu(p+q+1)-p$. We also have $B_r^{p,q} \subseteq B_\infty^{p,q}$ for all r. Since

$$E_r^{p,q} = Z_r^{p,q}/(B_{r-1}^{p,q} + Z_{r-1}^{p+1,q-1})$$
$$E_\infty^{p,q} = Z_\infty^{p,q}/(B_\infty^{p,q} + Z_\infty^{p+1,q-1}),$$

for $r > \mu(p+q+1)-p$, we have

$$E_r^{p,q} = Z_\infty^{p,q}/(B_{r-1}^{p,q} + Z_\infty^{p+1,q-1}),$$

and since $B_r^{p,q} \subseteq B_\infty^{p,q}$, we have $B_{r-1}^{p,q} + Z_\infty^{p+1,q-1} \subseteq B_\infty^{p,q} + Z_\infty^{p+1,q-1}$, so we obtain a surjective map $\eta^r \colon E_r^{p,q} \to E_\infty^{p,q}$. It is easy to verify that $\eta^r = \eta^{r+1} \circ \theta^r$, so the diagram commutes. Since

$$B_\infty^{p,q} = \bigcup_{r\in\mathbb{N}} B_r^{p,q},$$

it can be shown that $E_\infty^{p,q}$ is a direct limit (a colimit). $\square$

Since $E_{s+1}^{p,q} \cong \operatorname{Ker} d_s^{p,q}/\operatorname{Im} d_s^{p-s,q+s-1}$, if $E_r^{p,q} = (0)$ for some r, then $\operatorname{Ker} d_r^{p,q} = (0)$, and so $E_{r+1}^{p,q} = (0)$, and by induction $E_{r+1}^{p,q} = (0)$ for all $s > r$. Then Proposition 15.15 has the following important corollary.

Proposition 15.16. *If the filtration is regular and if $E_r^{p,q} = (0)$ for some r, then $E_\infty^{p,q} = (0)$.*

This fact can be used to prove a useful fact when the spectral sequence "degenerates" for some r.

Definition 15.27. A spectral sequence *degenerates at (level) r* if for every $n \in \mathbb{Z}$ there is some $q(n) \in \mathbb{Z}$ such that

$$E_r^{n-q,q} = (0) \quad \text{for all } q \neq q(n).$$

What this means is that for every $n \in \mathbb{Z}$, there is a single nonzero $E_r^{p,q}$ on the diagonal of equation $p+q=n$. This situation is illustrated in Figure 15.30.

Remark: When the condition of Definition 15.27 holds some authors say that the spectral sequences *collapses* at level r, but other use the term "collapse" for a different condition.

Proposition 15.17. *Let (C,d) be filtered and graded complex $C = \bigoplus_{n\in\mathbb{Z}} C^n$ with a differential d of degree 1 and a filtration $(F^pC)_{p\in\mathbb{Z}}$. If*

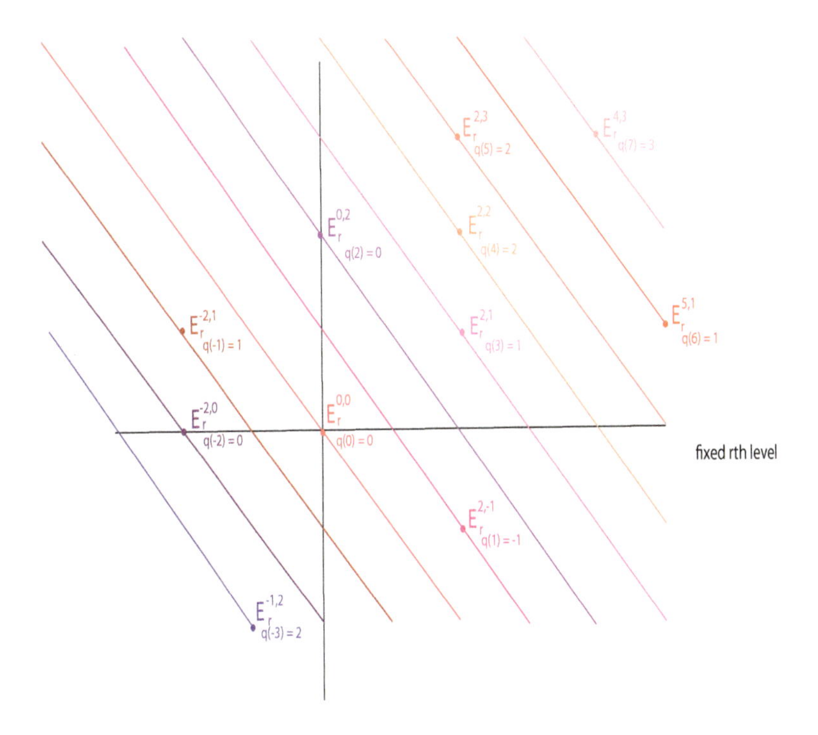

Fig. 15.30 An illustration of spectral sequence which degenerates at level r. Each diagonal line $p+q=n$ has only nonzero $E_r^{n-q,q}$ which we depicted as a colored dot.

the filtration is regular and if the spectral sequence associated with C degenerates for some $r \geq 1$, then

$$H^n(C) \cong E_\infty^{n-q(n),q(n)} \quad \textit{for all } n \in \mathbb{Z}.$$

If $q(n+1) \neq q(n) - (r-1)$ for all $n \in \mathbb{Z}$, then

$$H^n(C) \cong E_r^{n-q(n),q(n)} \quad \textit{for all } n \in \mathbb{Z}.$$

Proof. By Proposition 15.16, for every $n \in \mathbb{Z}$, if $E_r^{n-q,q} = (0)$ for all $q \neq q(n)$, then $E_\infty^{n-q,q} = (0)$ for all $q \neq q(n)$. Recall that $H^n(C)$ is filtered by the $H(C)^{p,n-p}$ with $p \in \mathbb{Z}$. Thus the graded module $\mathrm{gr}(H^n(C))$ associated with $H^n(C)$ is given by

$$\mathrm{gr}(H^n(C)) = \bigoplus_{p\in\mathbb{Z}} H(C)^{p,n-p}/H(C)^{p+1,n-p-1} = \bigoplus_{p\in\mathbb{Z}} \mathrm{gr}(H(C))^{p,n-p}.$$

Since by Theorem 15.7 we have

$$H(C)^{p,n-p}/H(C)^{p+1,n-p-1} \cong E_\infty^{p,n-p},$$

we obtain

$$\mathrm{gr}(H^n(C)) \cong \bigoplus_{p\in\mathbb{Z}} E_\infty^{p,n-p},$$

and using the change of variable $q = n - p$,

$$\mathrm{gr}(H^n(C)) \cong \bigoplus_{q\in\mathbb{Z}} E_\infty^{n-q,q}.$$

Since all terms $E_\infty^{n-q,q}$ are (0) for $q \neq q(n)$, we get

$$\mathrm{gr}(H^n(C)) \cong E_\infty^{n-q(n),q(n)}.$$

Now since $\bigcup_{p\in\mathbb{Z}} F^pC = C$, we have

$$H^n(C) = \bigcup_{q\in\mathbb{Z}} H(C)^{n-q,q},$$

and since the filtration is regular, by Proposition 15.9, the filtration $(H(C)^{p,n-p})$ of $H^n(C)$ is also regular and of the form

$$H^n(C) \supseteq \cdots H(C)^{p,n-p} \supseteq H(C)^{p+1,n-p-1} \supseteq \cdots \supseteq H(C)^{\mu(n),n-\mu(n)} \supseteq (0),$$

with $H(C)^{\mu(n)+1,n-\mu(n)-1} = (0)$.

Since $E_\infty^{p,n-p} \cong H(C)^{p,n-p}/H(C)^{p+1,n-p-1}$, we have $E_\infty^{p,n-p} = (0)$ iff $H(C)^{p,n-p} = H(C)^{p+1,n-p-1}$, that is, the two consecutive modules $H(C)^{p,n-p}$ and $H(C)^{p+1,n-p-1}$ in the filtration of $H^n(C)$ are identical. For n fixed, by the change of variable $p = n-q$, the condition $E_\infty^{n-q,q} = (0)$ for all $q \neq q(n)$ is equivalent to $E_\infty^{p,n-p} = (0)$ for all $p \neq n - q(n)$. Write $p(n) = n - q(n)$.

We may assume that $H^n(C) \neq (0)$ since otherwise we have trivially $H(C)^{p,n-p} = (0)$ for all p, so $E_\infty^{p,n-p} = (0)$ for all p, and $E_\infty^{p,n-p} = (0) = H^n(C)$ for all p. But then, since $E_\infty^{p,n-p} = (0)$ for all $p < p(n)$ and all $p > p(n)$, any two consecutive modules to the left or to the right of $H(C)^{p(n),n-p(n)}$ must be identical. Since $H(C)^{\mu(n)+1,n-\mu(n)-1} = (0)$, all the modules to the right of $H(C)^{p(n),n-p(n)}$ must be (0), so the filtration of $H^n(C)$ is of the form

$$H^n(C)\cdots \supseteq H(C)^{p(n),n-p(n)} = \cdots = H(C)^{p(n),n-p(n)} \supseteq H(C)^{p(n)+1,n-p(n)-1} = (0).$$

However, since $H^n(C) = \bigcup_{p\in\mathbb{Z}} H(C)^{p,n-p}$ we must have $H^n(C) = H(C)^{p(n),n-p(n)}$, and this is the only nonzero term in the filtration. Then

$$\begin{aligned} E_\infty^{p(n),n-p(n)} &\cong H(C)^{p(n),n-p(n)}/H(C)^{p(n)+1,n-p(n)-1} \\ &= H(C)^{p(n),n-p(n)} = H^n(C), \end{aligned}$$

and we conclude that
$$H^n(C) \cong E_\infty^{p(n),n-p(n)} = E_\infty^{n-q(n),q(n)}.$$

Since $d_r^{n-q,q}\colon E_r^{n-q,q} \to E_r^{n-q+r,q-r+1}$ and $n-q+r = n+1-(q-r+1)$, if $q-r+1 \neq q(n+1)$, we have $\mathrm{Im}\ d_r^{n-q,q} = (0)$. For $q = q(n)$ this says that if $q(n+1) \neq q(n)-(r-1)$, then $d_r^{n-q(n),q(n)} = (0)$. Similarly, since $d_r^{n-q-r,q+r-1}\colon E_r^{n-q-r,q+r-1} \to E_r^{n-q,q}$ and $n-q-r = n-1-(q+r-1)$, we have $\mathrm{Im}\ d_r^{n-q-r,q+r-1} = (0)$ if $q+r-1 \neq q(n-1)$, that is $q \neq q(n-1)-(r-1)$. For $q = q(n)$ this says that if $q(n) \neq q(n-1)-(r-1)$, then $\mathrm{Im}\ d_r^{n-q(n)-r,q(n)+r-1} = (0)$. If $d_r^{n-q(n),q(n)} = 0$ on $E_r^{n-q(n),q(n)}$, then $\mathrm{Ker}\, d_r^{n-q(n),q(n)} = E_r^{n-q(n),q(n)}$, and since
$$H^{n-q(n),q(n)}(E_r) = \mathrm{Ker}\, d_r^{n-q(n),q(n)}/\mathrm{Im}\ d_r^{n-q(n)-r,q(n)+r-1},$$
if $\mathrm{Im}\ d_r^{n-q(n)-r,q(n)+r-1} = (0)$, then
$$H^{n-q(n),q(n)}(E_r) = E_r^{n-q(n),q(n)}.$$

Since
$$E_{r+1}^{n-q(n),q(n)} \cong H^{n-q(n),q(n)}(E_r),$$
we get
$$E_{r+1}^{n-q(n),q(n)} \cong E_r^{n-q(n),q(n)}.$$
In summary, if $q(n+1) \neq q(n)-(r-1)$ and $q(n) \neq q(n-1)-(r-1)$, then
$$E_{r+1}^{n-q(n),q(n)} \cong E_r^{n-q(n),q(n)}.$$
Consequently, if $q(n+1) \neq q(n)-(r-1)$ for all $n \in \mathbb{Z}$, then
$$E_{s+1}^{n-q(n),q(n)} \cong E_s^{n-q(n),q(n)} \quad \text{for all } s \geq r,$$
that is, the sequence of $E_s^{n-q(n),q(n)}$ stabilizes for $s \geq r$, and since the filtration is regular, by Proposition 15.15, we have $E_r^{n-q(n),q(n)} \cong E_\infty^{n-q(n),q(n)}$. ☐

If $r \geq 2$ and $q(n)$ has the same value for all n, then the condition $q(n+1) \neq q(n)-(r-1)$ for all $n \in \mathbb{Z}$ is automatically satisfied, and so
$$H^n(C) \cong E_r^{n-q(n),q(n)} \quad \text{for all } n \in \mathbb{Z}.$$
In practice, this situation often arises for $r = 2$ and $q(n) = 0$, because of some acyclicity property.

The convergence of spectral sequences, including results involving weaker assumptions, is discussed in Cartan and Eilenberg [Cartan and Eilenberg (1956)], Spanier [Spanier (1989)] Mac Lane [Mac Lane (1975)],

McCleary [McCleary (2001)] and Weibel [Weibel (1994)]. Proposition 15.17 seems sufficient for all applications considered in Godement [Godement (1958)]. We warn the reader that the statement of Theorem 4.4.1 in Godement [Godement (1958)] seems incorrect. One can only claim that $H^n(C) \cong E_\infty^{n-q(n),q(n)}$, not that $H^n(C) \cong E_r^{n-q(n),q(n)}$. However, Godement only uses this result when $r \geq 2$ and $q(n) = 0$, so there is no problem.

15.9 Spectral Sequences Defined by Double Complexes

Big sources of spectral sequences are double complexes.

Definition 15.28. A *double complex* (or *bicomplex*) is a doubly-graded complex

$$C = \bigoplus_{p,q \in \mathbb{N}} C^{p,q}$$

together with two differentiations

$$\begin{aligned} d_{\mathrm{I}}^{p,q}&\colon C^{p,q} \longrightarrow C^{p+1,q}, && \text{(horizontal)} \\ d_{\mathrm{II}}^{p,q}&\colon C^{p,q} \longrightarrow C^{p,q+1}, && \text{(vertical)} \end{aligned}$$

such that

$$d_{\mathrm{I}}^{p+1,q} \circ d_{\mathrm{I}}^{p,q} = d_{\mathrm{II}}^{p,q+1} \circ d_{\mathrm{II}}^{p,q} = 0,$$

and *we also require*

$$d_{\mathrm{II}}^{p+1,q} \circ d_{\mathrm{I}}^{p,q} + d_{\mathrm{I}}^{p,q+1} \circ d_{\mathrm{II}}^{p,q} = 0, \quad \textit{for all } p, q \in \mathbb{N}. \tag{$\dagger$}$$

See Figure 15.31.

We get the (singly graded) *total complex*

$$\mathrm{Tot}(C) = \bigoplus_{n \in \mathbb{N}} C^n, \qquad \text{where} \quad C^n = \bigoplus_{p+q=n} C^{p,q},$$

with *total differential*

$$d_T = \bigoplus_{n \in \mathbb{Z}} d_T^n, \quad \text{with} \quad d_T^n = \bigoplus_{p+q=n} d_{\mathrm{I}}^{p,q} + d_{\mathrm{II}}^{p,q}.$$

Observe that $d_T^n \colon C^n \to C^{n+1}$. See Figure 15.32.

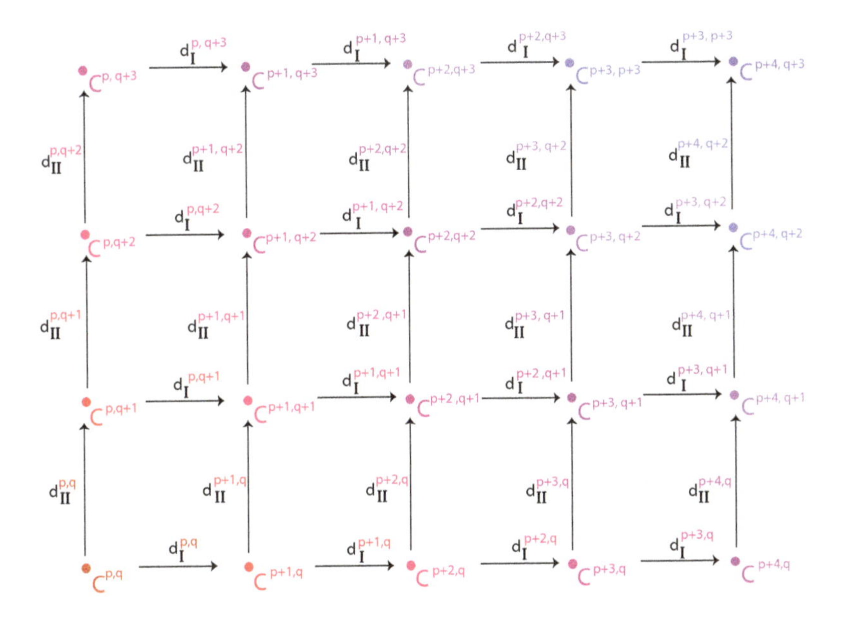

Fig. 15.31 A section of the double complex $C = \bigoplus_{p,q\in\mathbb{N}} C^{p,q}$.

Using the fact that $d_{\mathrm{II}}^{p+1,q} \circ d_{\mathrm{I}}^{p,q} + d_{\mathrm{I}}^{p,q+1} \circ d_{\mathrm{II}}^{p,q} = 0$, we immediately check that $d_T^{n+1} \circ d_T^n = 0$. For notational simplicity, we often denote $\mathrm{Tot}(C)$ as C and say that C is viewed as a total complex.

As suggested by Figure 15.31, the double complex C can be pictured as a first quadrant diagram (grid) in which the node of coordinates (p, q) is $C^{p,q}$.

Remark: It often more convenient to require that the $d_{\mathrm{I}}^{p,q}$ and $d_{\mathrm{II}}^{p,q}$ satisfy the equation

$$d_{\mathrm{II}}^{p+1,q} \circ d_{\mathrm{I}}^{p,q} = d_{\mathrm{I}}^{p,q+1} \circ d_{\mathrm{II}}^{p,q}, \quad \textit{for all } p, q \in \mathbb{N} \tag{††}$$

instead of Equation (†), namely that the following diagram commutes, rather than anti-commutes:

$$\begin{array}{ccc} C^{p,q} & \xrightarrow{d_{\mathrm{I}}^{p,q}} & C^{p+1,q} \\ {\scriptstyle d_{\mathrm{II}}^{p,q}}\big\downarrow & & \big\downarrow{\scriptstyle d_{\mathrm{II}}^{p+1,q}} \\ C^{p,q+1} & \xrightarrow[d_{\mathrm{I}}^{p,q+1}]{} & C^{p+1,q+1}. \end{array}$$

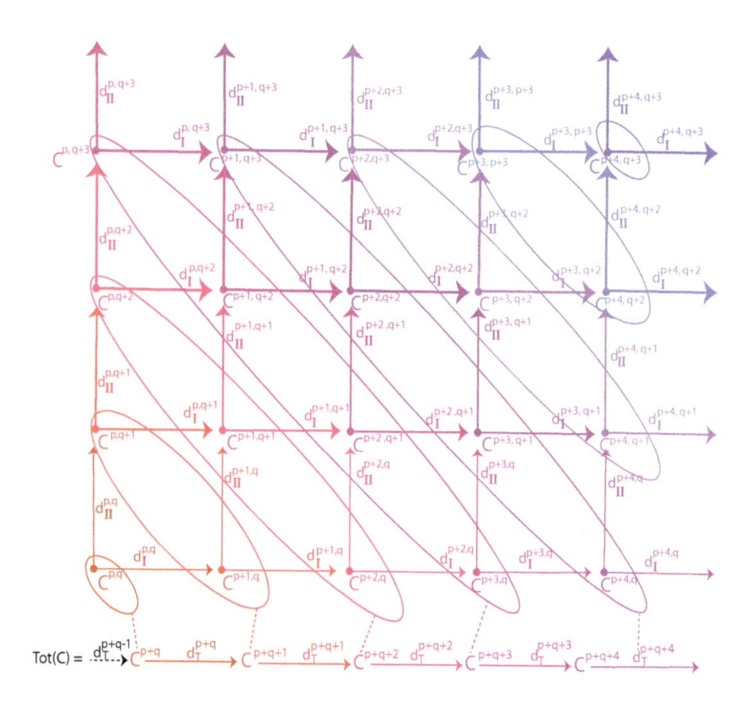

Fig. 15.32 The total complex $\mathrm{Tot}(C)$ is constructed by summing along the diagonals $p+q=n$.

In this case, we define $D_{\mathrm{I}}^{p,q}$ and $D_{\mathrm{II}}^{p,q}$ by

$$D_{\mathrm{I}}^{p,q} = d_{\mathrm{I}}^{p,q} \quad \text{and} \quad D_{\mathrm{II}}^{p,q} = (-1)^p d_{\mathrm{II}}^{p,q},$$

so that rewriting (††) in terms of $D_{\mathrm{I}}^{p,q}$ and $D_{\mathrm{II}}^{p,q}$ yields

$$D_{\mathrm{II}}^{p+1,q} \circ D_{\mathrm{I}}^{p,q} + D_{\mathrm{I}}^{p,q+1} \circ D_{\mathrm{II}}^{p,q} = 0, \quad \textit{for all } p,q \in \mathbb{N}.$$

We also set the differential of the total complex $\mathrm{Tot}(C)$ to be

$$D_T = \bigoplus_{n\in\mathbb{Z}} D_T^n, \quad \text{with} \quad D_T^n = \bigoplus_{p+q=n} D_{\mathrm{I}}^{p,q} + D_{\mathrm{II}}^{p,q}.$$

This is the approach adopted by Bott and Tu [Bott and Tu (1986)] (Chapter II), but we will stick to the first definition which seems to be the definition adopted in most books on homological algebra.

We can also view C as a cochain complex in two ways.

Definition 15.29. The complex $(C_{\mathrm{I}}, d_{\mathrm{I}})$ is the direct sum

$$C_{\mathrm{I}} = \bigoplus_{p\in\mathbb{N}} C_{\mathrm{I}}^p,$$

where C_{I}^{p} is a direct sum corresponding to the pth *column* of the diagram C, namely

$$C_{\mathrm{I}}^{p} = \bigoplus_{q\in\mathbb{N}} C^{p,q},$$

with

$$d_{\mathrm{I}}^{p,*} = \bigoplus_{q\in\mathbb{N}} d_{\mathrm{I}}^{p,q}, \quad d_{\mathrm{I}} = \bigoplus_{p\in\mathbb{N}} d_{\mathrm{I}}^{p,*};$$

see Figure 15.33.

Observe that $d_{\mathrm{I}}^{p,*}\colon C_{\mathrm{I}}^{p} \to C_{\mathrm{I}}^{p+1}$ and $d_{\mathrm{I}}^{p+1,*}\circ d_{\mathrm{I}}^{p,*} = 0$, so $(C_{\mathrm{I}}, d_{\mathrm{I}})$ is indeed cochain complex.

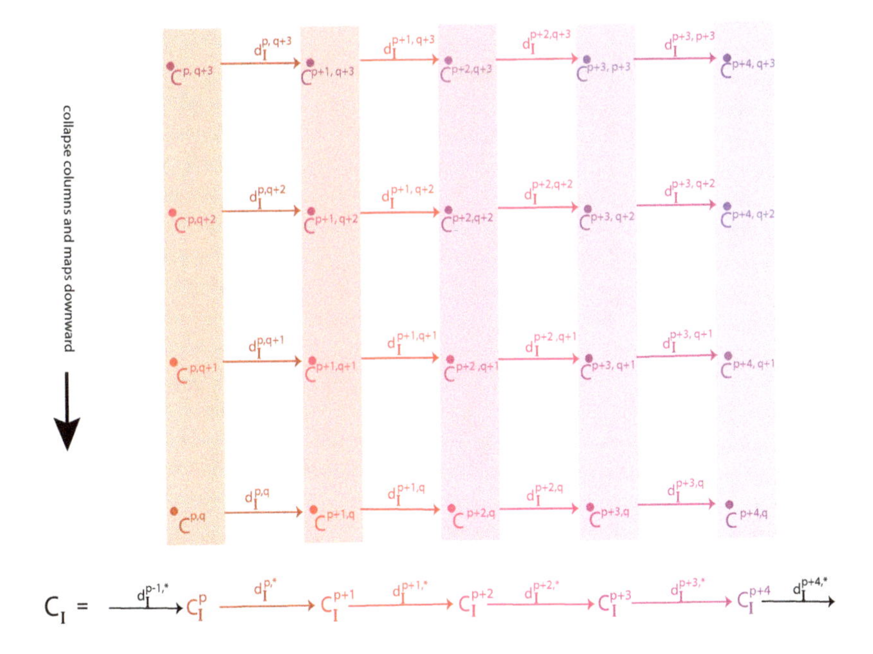

Fig. 15.33 The complex C_{I} is constructed by summing along the columns of the double complex.

Definition 15.30. The complex $(C_{\mathrm{II}}, d_{\mathrm{II}})$ is the direct sum

$$C_{\mathrm{II}} = \bigoplus_{q\in\mathbb{N}} C_{\mathrm{II}}^{q},$$

where C_{II}^q is a direct sum corresponding to the qth *row* of the diagram C, namely

$$C_{\mathrm{II}}^q = \bigoplus_{p \in \mathbb{N}} C^{p,q},$$

with

$$d_{\mathrm{II}}^{*,q} = \bigoplus_{p \in \mathbb{N}} (-1)^p d_{\mathrm{II}}^{p,q}, \quad d_{\mathrm{II}} = \bigoplus_{q \in \mathbb{N}} d_{\mathrm{II}}^{*,q};$$

see Figure 15.34.

Observe that $d_{\mathrm{II}}^{*,q} \colon C_{\mathrm{II}}^q \to C_{\mathrm{II}}^{q+1}$ and $d_{\mathrm{II}}^{*,q+1} \circ d_{\mathrm{II}}^{*,q} = 0$, so $(C_{\mathrm{II}}, d_{\mathrm{II}})$ is indeed cochain complex. The reason for inserting the sign $(-1)^p$ will become clear later. It is unnecessary if we assume that Equation (††) holds.

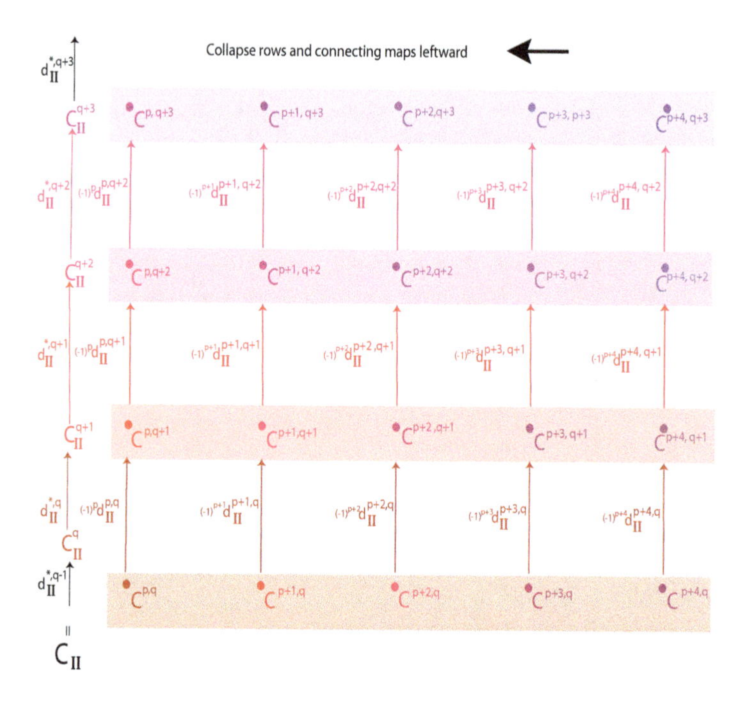

Fig. 15.34 The complex C_{II} is constructed by summing along the rows of the double complex.

In order to define spectral sequences on $\mathrm{Tot}(C)$ we use the two filtrations of Example 15.2 whose definitions are repeated for the reader's convenience.

Definition 15.31. We have the filtration

$$F_{\mathrm{I}}^{p}C = \bigoplus_{s\geq p}\bigoplus_{q\in\mathbb{N}} C^{s,q}, \quad p\in\mathbb{N}$$

on C_{I}, which amounts to the direct sum of the *columns* of index $s \geq p$, and the filtration

$$F_{\mathrm{II}}^{q}C = \bigoplus_{p\in\mathbb{N}}\bigoplus_{t\geq q} C^{p,t}, \quad q\in\mathbb{N},$$

on C_{II}, which amounts to the direct sum of the *rows* of index $t \geq q$. The filtration $F_{\mathrm{I}}^{p}C$ is illustrated by Figure 15.2 while the filtration $F_{\mathrm{II}}^{q}C$ is illustrated by Figure 15.3.

Both filtrations are also filtrations on the total complex C. In all three cases, these filtrations are compatible with the grading and regular on the total complex. In fact, they are canonically cobounded.

In order to compute the terms ${}^{\mathrm{I}}E_2^{p,q}$ and ${}^{\mathrm{II}}E_2^{p,q}$ of the spectral sequences associated with the two filtrations above, we need to define complexes whose cochain modules are the cohomology modules of the columns and of the rows of the double complex C viewed themselves as complexes.

Observe that $C_{\mathrm{I}}^{p} = \bigoplus_{q\in\mathbb{N}} C^{p,q}$ is a cochain complex with differential $d_{\mathrm{II}}^{p,*} = \bigoplus_{q\in\mathbb{N}}(-1)^p d_{\mathrm{II}}^{p,q}$. Here we view the *pth column* of the diagram C as a cochain complex.

Definition 15.32. The map $d_{\mathrm{I}}^{p,*}\colon C_{\mathrm{I}}^{p} \to C_{\mathrm{I}}^{p+1}$, where $d_{\mathrm{I}}^{p,*} = \bigoplus_{q\in\mathbb{N}} d_{\mathrm{I}}^{p,q}$, can be viewed as a chain map between C_{I}^{p} and C_{I}^{p+1} because the diagram

$$\begin{array}{ccc}
C^{p,q} & \xrightarrow{d_{\mathrm{I}}^{p,q}} & C^{p+1,q} \\
{\scriptstyle (-1)^p d_{\mathrm{II}}^{p,q}}\downarrow & & \downarrow{\scriptstyle (-1)^{p+1} d_{\mathrm{II}}^{p+1,q}} \\
C^{p,q+1} & \xrightarrow[d_{\mathrm{I}}^{p,q+1}]{} & C^{p+1,q+1}
\end{array}$$

commutes, since

$$(-1)^{p+1} d_{\mathrm{II}}^{p+1,q} \circ d_{\mathrm{I}}^{p,q} = d_{\mathrm{I}}^{p,q+1} \circ (-1)^p d_{\mathrm{II}}^{p,q}$$

is equivalent to $d_{\mathrm{II}}^{p+1,q} \circ d_{\mathrm{I}}^{p,q} + d_{\mathrm{I}}^{p,q+1} \circ d_{\mathrm{II}}^{p,q} = 0$, which is (†).

Observe that *the columns are the complexes C_{I}^{p} and the chain maps are the horizontal arrows.*

If we denote the cohomology modules of the chain complex $(C_{\mathrm{I}}^{p}, d_{\mathrm{II}}^{p,*})$ by $H_{\mathrm{II}}^{q}(C_{\mathrm{I}}^{p})$, the chain maps $d_{\mathrm{I}}^{p,*}$ from C_{I}^{p} to C_{I}^{p+1} induce homomorphisms

$$(d_{\mathrm{I}}^{p,*})^{*} \colon H_{\mathrm{II}}^{q}(C_{\mathrm{I}}^{p}) \to H_{\mathrm{II}}^{q}(C_{\mathrm{I}}^{p+1}),$$

and we easily check that $(d_{\mathrm{I}}^{p+1,*})^{*} \circ (d_{\mathrm{I}}^{p,*})^{*} = 0$. Consequently, we obtain a cochain complex.

Definition 15.33. The cochain complex $H_{\mathrm{II}}^{q}(C_{\mathrm{I}})$ is given by

$$0 \longrightarrow H_{\mathrm{II}}^{q}(C_{\mathrm{I}}^{0}) \longrightarrow H_{\mathrm{II}}^{q}(C_{\mathrm{I}}^{1}) \longrightarrow \cdots \longrightarrow H_{\mathrm{II}}^{q}(C_{\mathrm{I}}^{p}) \longrightarrow H_{\mathrm{II}}^{q}(C_{\mathrm{I}}^{p+1}) \longrightarrow \cdots ;$$

see Figure 15.35.

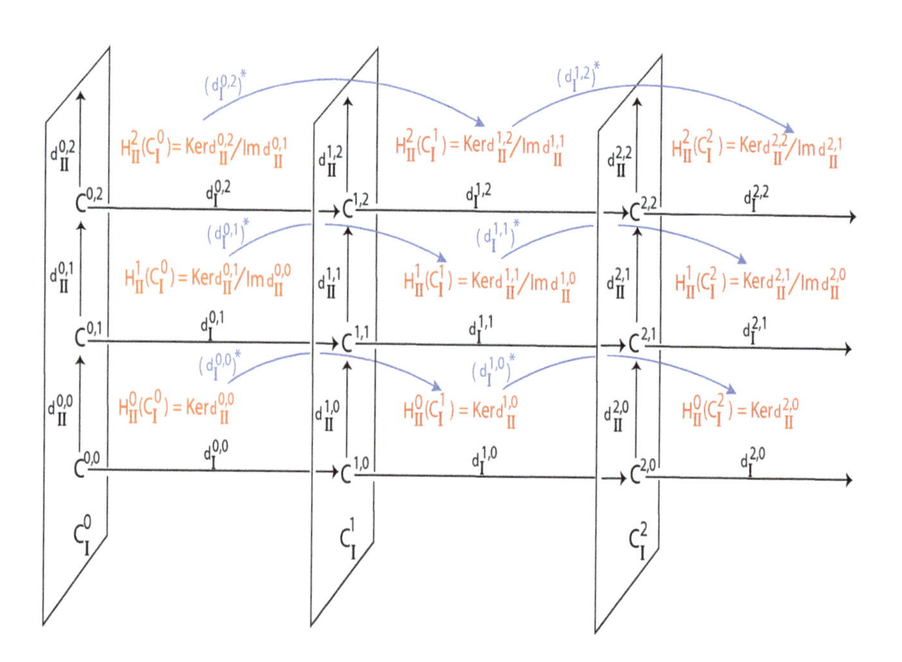

Fig. 15.35 A schematic illustration of the construction of the first three modules of $H_{\mathrm{II}}^{0}(C_{\mathrm{I}})$, $H_{\mathrm{II}}^{1}(C_{\mathrm{I}})$, and $H_{\mathrm{II}}^{1}(C_{\mathrm{I}})$.

It is important to remember that as a complex, C_{I}^{p} (the pth column) with index I, has the differential $d_{\mathrm{II}}^{p,*}$ whose index is II, and that the complex

$H^q_{\mathrm{II}}(C_{\mathrm{I}})$ whose leftmost index is II, has a differential induced by the chain map $(d_{\mathrm{I}}^{p,*})$, whose index is I. In other words, the indices I and II alternate.

As illustrated in Figure 15.36, the above complexes are the rows of a (first quadrant) diagram $H_{\mathrm{II}}(C)$, in which each node $H^q_{\mathrm{II}}(C^p_{\mathrm{I}})$ is the qth cohomology module of the *pth column* C^p_{I} of the complex C viewed as a (vertical) complex with the differential $d_{\mathrm{II}}^{p,*}$, and the complex $H^q_{\mathrm{II}}(C_{\mathrm{I}})$ corresponds to the *qth row* of this diagram viewed as a (horizontal) complex with differential induced by d_{I}.

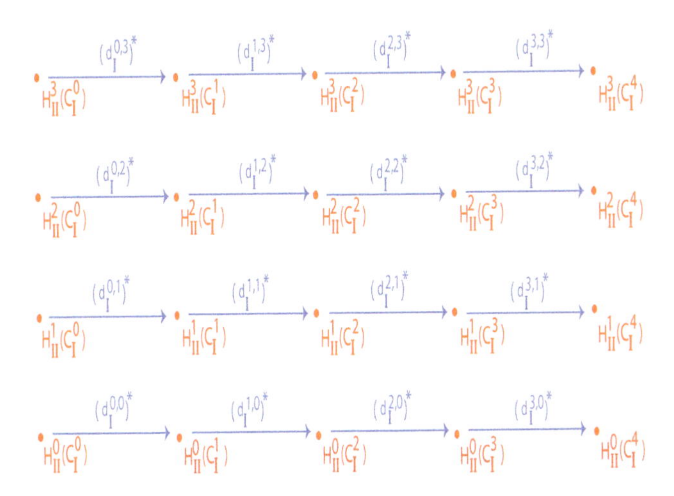

Fig. 15.36 A section first quadrant grid depiction of $H^q_{\mathrm{II}}(C_{\mathrm{I}})$, where $0 \leq q \leq 3$. Observe that the coordinates of the orange nodes have *transposed* horizontal and vertical indexing conventions.

Remark: Note that the insertion of the sign $(-1)^p$ is unnecessary if we assume that Equation (††) holds.

Similarly, observe that $C^q_{\mathrm{II}} = \bigoplus_{p\in\mathbb{N}} C^{p,q}$ is a cochain complex with differential $d_{\mathrm{I}}^{*,q} = \bigoplus_{p\in\mathbb{N}} d_{\mathrm{I}}^{p,q}$. Here we view the *qth row* of the diagram C as a cochain complex.

Definition 15.34. The map $d_{\mathrm{II}}^{*,q}\colon C^q_{\mathrm{II}} \to C^{q+1}_{\mathrm{II}}$, where $d_{\mathrm{II}}^{*,q} = \bigoplus_{p\in\mathbb{N}}(-1)^p d_{\mathrm{II}}^{p,q}$, can be viewed as a chain map between C^q_{II} and C^{q+1}_{II}

because the diagram

$$\begin{array}{ccc} C^{p,q} & \xrightarrow{d_{\mathrm{I}}^{p,q}} & C^{p+1,q} \\ {\scriptstyle (-1)^p d_{\mathrm{II}}^{p,q}} \downarrow & & \downarrow {\scriptstyle (-1)^{p+1} d_{\mathrm{II}}^{p+1,q}} \\ C^{p,q+1} & \xrightarrow[d_{\mathrm{I}}^{p,q+1}]{} & C^{p+1,q+1} \end{array}$$

commutes, as above.

Observe that *the rows are the complexes* C_{II}^q *and the chain maps are the vertical arrows.*

If we denote the cohomology modules of the chain complex $(C_{\mathrm{II}}^q, d_{\mathrm{I}}^{*,q})$ by $H_{\mathrm{I}}^p(C_{\mathrm{II}}^q)$, the chain maps $d_{\mathrm{II}}^{*,q}$ induce homomorphisms

$$(d_{\mathrm{II}}^{*,q})^* \colon H_{\mathrm{I}}^p(C_{\mathrm{II}}^q) \to H_{\mathrm{I}}^p(C_{\mathrm{II}}^{q+1}),$$

and we easily check that $(d_{\mathrm{II}}^{*,q+1})^* \circ (d_{\mathrm{II}}^{*,q})^* = 0$. Consequently, we obtain a cochain complex.

Definition 15.35. The cochain complex $H_{\mathrm{I}}^p(C_{\mathrm{II}})$ is given by

$$0 \longrightarrow H_{\mathrm{I}}^p(C_{\mathrm{II}}^0) \longrightarrow H_{\mathrm{I}}^p(C_{\mathrm{II}}^1) \longrightarrow \cdots H_{\mathrm{I}}^p(C_{\mathrm{II}}^q) \longrightarrow H_{\mathrm{I}}^p(C_{\mathrm{II}}^{q+1}) \longrightarrow \cdots;$$

see Figure 15.37.

It is important to remember that as a complex, C_{II}^q (the qth row) with index II, has the differential $d_{\mathrm{I}}^{*,q}$ whose index is I, and that the complex $H_{\mathrm{I}}^q(C_{\mathrm{II}})$ whose leftmost index is I, has a differential induced by the chain map $(d_{\mathrm{II}}^{*,q})$, whose index is II. Again, the indices I and II alternate.

As shown by Figure 15.38, the above complexes are the columns of a (first quadrant) diagram $H_{\mathrm{I}}(C)$, in which each node $H_{\mathrm{I}}^p(C_{\mathrm{II}}^q)$ is the pth cohomology module of the *qth row* C_{II}^q of the complex C viewed as a (horizontal) complex with the differential $d_{\mathrm{I}}^{*,q}$, and the complex $H_{\mathrm{I}}^p(C_{\mathrm{II}})$ corresponds to the *pth column* of this diagram viewed as a (vertical) complex with differential induced by d_{II}.

Remark: Note that the insertion of the sign $(-1)^p$ is unnecessary if we assume that Equation (††) holds.

We finally come to the spectral sequences induced by the filtrations $F_{\mathrm{I}}^p C$ and $F_{\mathrm{II}}^q C$ on C viewed as a total complex. We obtain the spectral sequences denoted ${}^{\mathrm{I}}E$ and ${}^{\mathrm{II}}E$.

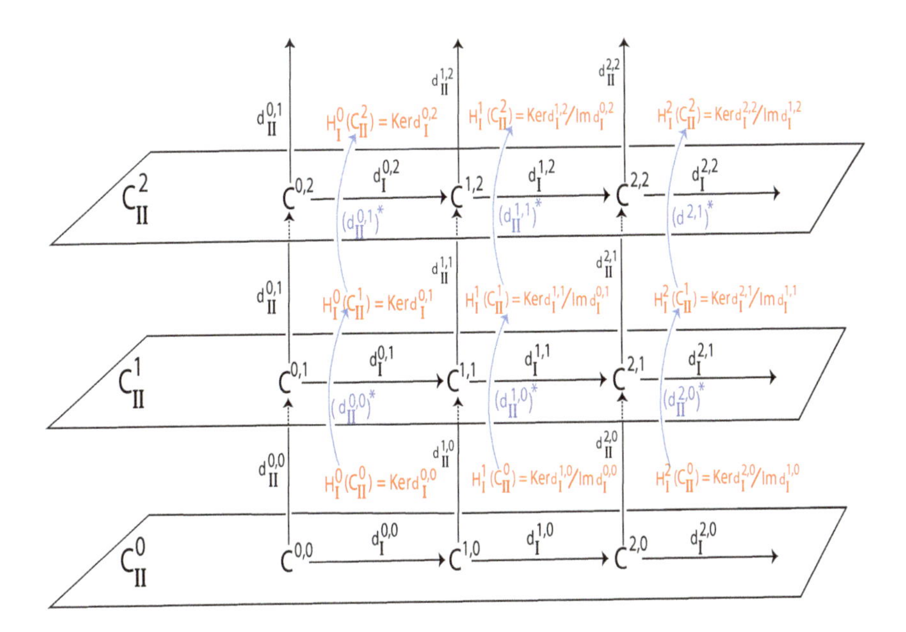

Fig. 15.37 A schematic illustration of the construction of the first three modules of $H_{\mathrm{I}}^0(C_{\mathrm{II}})$, $H_{\mathrm{I}}^1(C_{\mathrm{II}})$, and $H_{\mathrm{I}}^1(C_{\mathrm{II}})$.

It is possible to compute the terms ${}^{\mathrm{I}}E_1^{p,q}$ and ${}^{\mathrm{I}}E_2^{p,q}$ and the terms ${}^{\mathrm{II}}E_1^{p,q}$ and ${}^{\mathrm{II}}E_2^{p,q}$ with some labor.

The term ${}^{\mathrm{I}}E_0^p$ is not hard to compute since it is equal to $F_{\mathrm{I}}^pC/F_{\mathrm{I}}^{p+1}C$, which is isomorphic to

$$C_{\mathrm{I}}^p = \bigoplus_{q\in\mathbb{N}} C^{p,q},$$

the pth column, and ${}^{\mathrm{I}}E_0^{p,q} = C^{p,q}$. It follows that

$${}^{\mathrm{I}}E_1^{p,q} = H^{p+q}(F_{\mathrm{I}}^pC/F_{\mathrm{I}}^{p+1}C) = H_{\mathrm{II}}^q(C_{\mathrm{I}}^p),$$

the cohomology being computed in the complex C_{I}^p with the differential $d_{\mathrm{II}}^{p,*}$. The computation of ${}^{\mathrm{I}}E_2^p$ is more involved. By Property (3) of the definition of a spectral sequence (Definition 15.18),

$${}^{\mathrm{I}}E_2^p \cong H^p({}^{\mathrm{I}}E_1) = \mathrm{Ker}\, d_1^p/\mathrm{Im}\ d_1^{p-1},$$

so we need to compute d_1^p in spectral sequence ${}^{\mathrm{I}}E$. The key point is that for *any* graded and filtered complex C, the map d_1^p is equal to the connecting

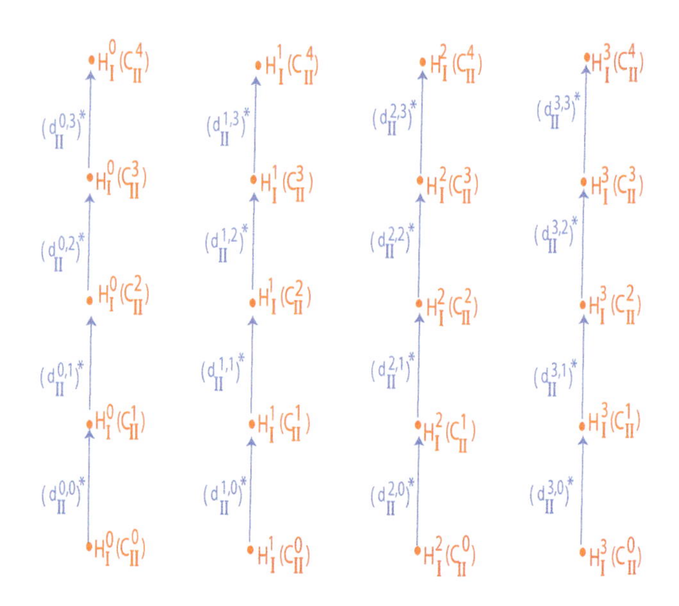

Fig. 15.38 A section first quadrant grid depiction of $H_{\mathrm{I}}^{q}(C_{\mathrm{II}})$, where $0 \leq q \leq 3$.

homomorphism arising from the short exact sequence

$$0 \longrightarrow F^{p+1}C/F^{p+2}C \longrightarrow F^{p}C/F^{p+2}C \longrightarrow F^{p}C/F^{p+1}C \longrightarrow 0.$$

Recall that

$$E_1^{p,q} = Z_1^{p,q}/(B_0^{p,q} + Z_0^{p+1,q-1}).$$

Since $Z_0^{p+1,q-1} = C^{p+1,q-1}$ and $B_0^{p,q} = dZ_0^{p,q-1} = dC^{p,q-1}$, we have

$$E_1^{p,q} = Z_1^{p,q}/(dC^{p,q-1} + C^{p+1,q-1}).$$

The qth row of the above exact sequence of complexes is

$$0 \longrightarrow C^{p+1,q-1}/C^{p+2,q-2} \longrightarrow C^{p,q}/C^{p+2,q-2} \longrightarrow C^{p,q}/C^{p+1,q-1} \longrightarrow 0.$$

For every $s \geq 1$, all differentials

$$d_{0,s}^{p,q} \colon C^{p,q}/C^{p+s,q-s} \to C^{p,q+1}/C^{p+s,q-s+1}$$

are induced by $d \colon C^{p,q} \to C^{p,q+1}$ (actually, the restriction $d_{C^{p,q}}$ of d to $C^{p,q}$), with

$$d_{0,s}^{p,q}([x]_{C^{p+s,q-s}}) = [d(x)]_{C^{p+s,q-s+1}}, \quad x \in C^{p,q}.$$

In the special case $s = 1$, $d_{0,1}^{p,q} = d_0^{p,q}$, the differential defined just after Definition 15.15. With this notation

$$H^{p+q}(F^pC/F^{p+1}C) = \mathrm{Ker}\, d_{0,1}^{p,q}/\mathrm{Im}\, d_{0,1}^{p,q-1}.$$

But $d_{0,1}^{p,q}(c) = 0$ iff $c \in C^{p,q}$ and $dc \in C^{p+1,q}$, which is equivalent to $c \in Z_1^{p,q}$. Thus we have

$$\begin{aligned}\mathrm{Ker}\, d_{0,1}^{p,q} &= (Z_1^{p,q} + C^{p+1,q-1})/C^{p+1,q-1}\\ \mathrm{Im}\; d_{0,1}^{p,q-1} &= (dC^{p,q-1} + C^{p+1,q-1})/C^{p+1,q-1}.\end{aligned}$$

We confirm (using the modular Noether isomorphism) that

$$E_1^{p,q} = Z_1^{p,q}/(dC^{p,q-1} + C^{p+1,q-1}) \cong H^{p+q}(F^pC/F^{p+1}C).$$

Similarly, we have

$$E_1^{p+1,q} = Z_1^{p+1,q}/(dC^{p+1,q-1} + C^{p+2,q-1}) \cong H^{p+q+1}(F^{p+1}C/F^{p+2}C).$$

In order to compute the connecting map

$$\delta^{p+q} \colon H^{p+q}(F^pC/F^{p+1}C) \to H^{p+q+1}(F^{p+1}C/F^{p+2}C),$$

as in the proof of the zig-zag lemma (Theorem 2.22), we use the diagram

$$\begin{array}{ccccc} & & C^{p,q}/C^{p+2,q-2} & \xrightarrow{\pi} & C^{p,q}/C^{p+1,q-1} \longrightarrow 0\\ & & \downarrow{\scriptstyle d_{0,2}^{p,q}} & & \\ 0 \longrightarrow C^{p+1,q}/C^{p+2,q-1} & \xrightarrow{i} & C^{p,q+1}/C^{p+2,q-1}. & & \end{array}$$

We start with a cocycle $\gamma = [c]_{C^{p+1,q-1}} \in C^{p,q}/C^{p+1,q-1}$, which is equivalent to $c \in Z_1^{p,q}$. Then we pull back $[c]_{C^{p+1,q-1}}$ along the projection π, obtaining $[c]_{C^{p+2,q-2}}$, push this element down along $d_{0,2}^{p,q}$, obtaining $[dc]_{C^{p+2,q-1}}$, and then pull back along i obtaining a cocycle $[dc]_{C^{p+2,q-1}}$ with $dc \in C^{p+1,q}$, which is equivalent to $dc \in Z_1^{p+1,q-1}$. In summary, we constructed a function that maps $[c] \in C^{p,q}/C^{p+1,q-1}$ with $c \in Z_1^{p,q}$ to $[dc] \in C^{p+1,q}/C^{p+2,q-1}$ with $dc \in Z_1^{p+1,q}$. The connecting map δ^{p+q} sends the equivalence class of $[c]$ modulo $\mathrm{Im}\, d_{0,1}^{p,q-1} = (dC^{p,q-1} + C^{p+1,q-1})/C^{p+1,q-1}$ (an element of $H^{p+q}(F^pC/F^{p+1}C)$) to the equivalence class of $[dc]$ modulo $\mathrm{Im}\, d_{0,1}^{p+1,q-1} = (dC^{p+1,q-1} + C^{p+2,q-1})/C^{p+2,q-1}$ (an element of $H^{p+q+1}(F^{p+1}C/F^{p+2}C)$). Since

$$\begin{aligned}&(C^{p,q}/C^{p+1,q-1})/((dC^{p,q-1} + C^{p+1,q-1})/C^{p+1,q-1})\\ &\qquad\qquad \cong C^{p,q}/(dC^{p,q-1} + C^{p+1,q-1})\end{aligned}$$

and

$$(C^{p+1,q}/C^{p+2,q-1})/((dC^{p+1,q-1}+C^{p+2,q-1})/C^{p+2,q-1}) \\ \cong C^{p+1,q}/(dC^{p+1,q-1}+C^{p+2,q-1}),$$

we deduce that we can view the map δ^{p+q} as the map $d_1^{p,q}$ from $E_1^{p,q} = Z_1^{p,q}/(dC^{p,q-1}+C^{p+1,q-1})$ to $E_1^{p+1,q} = Z_1^{p+1,q}/(dC^{p+1,q-1}+C^{p+2,q-1})$, as claimed. We leave the details as an exercise.

The rest of the proof is an adaptation of Rotman's proof which applies to homological spectral sequences.

The above exact sequence is equivalent to the exact sequence

$$0 \longrightarrow C_{\mathrm{I}}^{p+1} \longrightarrow C_{\mathrm{I}}^{p} \oplus C_{\mathrm{I}}^{p+1} \longrightarrow C_{\mathrm{I}}^{p} \longrightarrow 0.$$

The complex C_{I}^{p+1} is equipped with the differential $d_{\mathrm{II}}^{p+1,*}$ and the complex C_{I}^{p} is equipped with the differential $d_{\mathrm{II}}^{p,*}$. The complex $C_{\mathrm{I}}^{p} \oplus C_{\mathrm{I}}^{p+1}$ is equipped with the restriction of d_T to it. It is easy to show by induction that we can define the grading $(M^{p,q})_{q\in\mathbb{N}}$ of $C_{\mathrm{I}}^{p} \oplus C_{\mathrm{I}}^{p+1}$ by

$$\begin{aligned} M^{p,0} &= C^{p,0} \\ M^{p,q} &= C^{p,q} \oplus C^{p+1,q-1}, \quad q \geq 1 \end{aligned}$$

and the differential $\overline{D}^{p,q}\colon M^{p,q} \to M^{p,q+1}$ given by

$$\overline{D}^{p,q}(x_{p,q}+y_{p+1,q-1}) = d_{\mathrm{II}}^{p,q}(x_{p,q}) + d_{\mathrm{I}}^{p,q}(x_{p,q}) + d_{\mathrm{II}}^{p+1,q-1}(y_{p+1,q-1}),$$

for any $x_{p,q} \in C^{p,q}, y_{p+1,q-1} \in C^{p+1,q-1}$ (with $y_{p+1,q-1} = 0$ if $q = 0$); see Figures 15.39 and 15.40.

Since $d_{\mathrm{II}}^{p,q}(x_{p,q}) \in C^{p,q+1}$, $d_{\mathrm{I}}^{p,q}(x_{p,q}) + d_{\mathrm{II}}^{p+1,q-1}(y_{p+1,q-1}) \in C^{p+1,q}$, we have

$$\begin{aligned} \overline{D}^{p,q+1}(\overline{D}^{p,q}(x_{p,q}+y_{p+1,q-1})) &= d_{\mathrm{II}}^{p,q+1}(d_{\mathrm{II}}^{p,q}(x_{p,q})) + d_{\mathrm{I}}^{p,q+1}(d_{\mathrm{II}}^{p,q}(x_{p,q})) \\ &\quad + d_{\mathrm{II}}^{p+1,q}(d_{\mathrm{I}}^{p,q}(x_{p,q}) + d_{\mathrm{II}}^{p+1,q-1}(y_{p+1,q-1})) \\ &= d_{\mathrm{I}}^{p,q+1}(d_{\mathrm{II}}^{p,q}(x_{p,q})) + d_{\mathrm{II}}^{p+1,q}(d_{\mathrm{I}}^{p,q}(x_{p,q})) = 0, \end{aligned}$$

so $\overline{D}^{p,q+1} \circ \overline{D}^{p,q} = 0$. We also add the zero module as first module to the

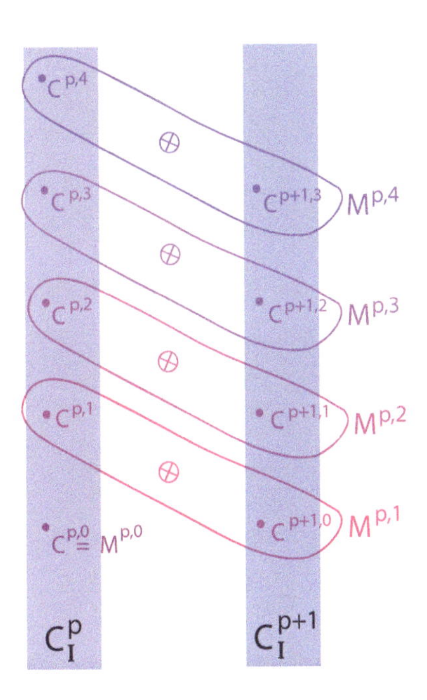

Fig. 15.39 The construction of $M^{p,q}$ for $C_{\mathrm{I}}^{p} \oplus C_{\mathrm{I}}^{p+1}$.

complex C_{I}^{p+1}, so that we have commutative diagrams with exact rows

$$
\begin{array}{ccccccccc}
0 & \longrightarrow & 0 & \longrightarrow & C^{p,0} & \longrightarrow & C^{p,0} & \longrightarrow & 0 \\
 & & \Big\downarrow{\scriptstyle 0} & & \Big\downarrow{\scriptstyle \overline{D}^{p,0}} & & \Big\downarrow{\scriptstyle d_{\mathrm{I}}^{p,0}} & & \\
0 & \longrightarrow & C^{p+1,0} & \longrightarrow & C^{p,1} \oplus C^{p+1,0} & \longrightarrow & C^{p,1} & \longrightarrow & 0
\end{array}
$$

$$
\begin{array}{ccccccccc}
0 & \longrightarrow & C^{p+1,q-1} & \longrightarrow & C^{p,q} \oplus C^{p+1,q-1} & \longrightarrow & C^{p,q} & \longrightarrow & 0 \\
 & & \Big\downarrow{\scriptstyle d_{\mathrm{I}}^{p+1,q-1}} & & \Big\downarrow{\scriptstyle \overline{D}^{p,q}} & & \Big\downarrow{\scriptstyle d_{\mathrm{I}}^{p,q}} & & \\
0 & \longrightarrow & C^{p+1,q} & \longrightarrow & C^{p,q+1} \oplus C^{p+1,q} & \longrightarrow & C^{p,q+1} & \longrightarrow & 0
\end{array}
$$

for $q \geq 1$. When we compute the connecting map

$$\delta^q \colon H_{\mathrm{II}}^q(C_{\mathrm{I}}^p) \to H_{\mathrm{II}}^q(C_{\mathrm{I}}^{p+1}),$$

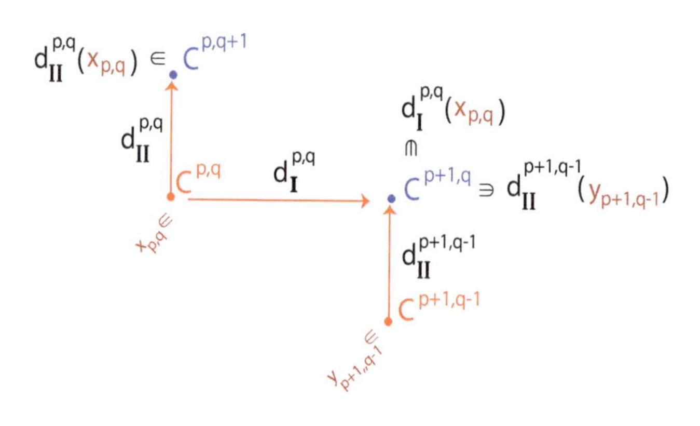

Fig. 15.40 An illustration of $\overline{D}^{p,q}\colon M^{p,q} \to M^{p,q+1}$.

as in the proof of the zig-zag lemma (Theorem 2.22), we use the diagram

$$\begin{array}{ccccccc} & & C^{p,q} \oplus C^{p+1,q-1} & \xrightarrow{\pi} & C^{p,q} & \longrightarrow & 0 \\ & & \downarrow{\scriptstyle \overline{D}^{p,q}} & & & & \\ 0 \longrightarrow C^{p+1,q} & \xrightarrow{i} & C^{p,q+1} \oplus C^{p+1,q} & & & & \end{array}$$

in which π is the projection and i is the injection, where we pick some cocycle $c \in C^{p,q}$, that is $d_{\mathrm{II}}^{p,q}(c) = 0$, pick $b = c \in C^{p,q} \oplus C^{p+1,q-1}$ such that $\pi(b) = \pi(c) = c$, push $b = c$ down using $\overline{D}^{p,q}$ obtaining $\overline{D}^{p,q}(c)$, and pull back $\overline{D}^{p,q}(c)$ using the inclusion i, which yields $a = i^{-1}(\overline{D}^{p,q}(c))$. Then

$$\delta^q([c]) = [a].$$

But by definition of $\overline{D}^{p,q}$, since $d_{\mathrm{II}}^{p,q}(c) = 0$, we see that

$$\overline{D}^{p,q}(c) = d_{\mathrm{I}}^{p,q}(c),$$

and so $\delta^q = (d_{\mathrm{I}}^{p,q})^*$, namely $\delta^q([c]) = [d_{\mathrm{I}}^{p,q}(c)]$. Consequently, the chain complex consisting of the modules $H_{\mathrm{II}}^q(C_{\mathrm{I}}^p)$ and the coboundary maps δ^q is just the chain complex $H_{\mathrm{II}}(C_{\mathrm{I}})$ of Definition 15.33, and we find that

$${}^{\mathrm{I}}E_2^{p,q} \cong H_{\mathrm{I}}^p(H_{\mathrm{II}}^q(C_{\mathrm{I}})).$$

Other (less detailed) proofs can be found in Godement [Godement (1958)] (Chapter 4, Section 4.8) and Mac Lane [Mac Lane (1975)] (Chapter XI, Section 6).

There is a subtle point which is "swept under the rug" by a number of authors, which is that if we want to keep the filtration index in the second spectral sequence to be p, we need to transpose the double complex, that is, to form the complex $C^\top$ given by

$$C^\top = \bigoplus_{p,q\in\mathbb{N}} C^{q,p},$$

as explained in Mac Lane [Mac Lane (1975)] (Chapter XI, Section 6) and Rotman [Rotman (1979, 2009)] (Chapter 11, Page 327). Indeed, to keep the index of filtration to be p, we need to consider $F^p_{\rm II}C/F^{p+1}_{\rm II}C$, which is isomorphic to

$$C^p_{\rm II} = \bigoplus_{q\in\mathbb{N}} C^{q,p},$$

the pth row, and ${}^{\rm II}E_0^{p,q} = C^{q,p}$, which explains why the transpose double complex $C^\top$ shows up. We also have to consider the differential $d_{\rm I}^\top$ and $d_{\rm II}^\top$ given by $(d_{\rm I}^\top)^{p,q} = d_{\rm II}^{q,p}$ and $(d_{\rm II}^\top)^{p,q} = d_{\rm I}^{q,p}$. Obviously $\mathrm{Tot}(C) = \mathrm{Tot}(C^\top)$, and the total differentials d_T^n and $(d^\top)_T^n$ agree. See Figures 15.41 and 15.42.

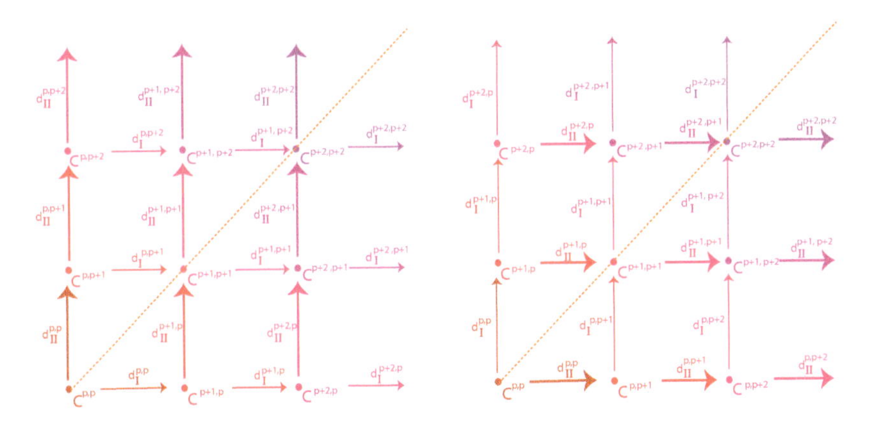

Fig. 15.41 The left diagram is the double complex C while the right diagram is its reflection over the dotted orange diagonal line. Note this reflection interchanges rows and columns and is the precursor to the double complex $C^\top$. What remains is to appropriately relabel the nodes and maps as shown in Figure 15.42.

Since the second filtration of C is equal to the first filtration of $C^\top$, we find that the ${}^{\rm II}E_2^{p,q}$ term of the spectral sequence associated with the second filtration of C is equal to the ${}^{\rm I}E_2^{'p,q}$ term of the spectral sequence associated with the first filtration of $C^\top$, so as before,

$${}^{\rm II}E_2^{p,q} = {}^{\rm I}E_2^{'p,q} \cong H_{\rm I}^{'p}(H_{\rm II}^{'q}(C_{\rm I}^\top)),$$

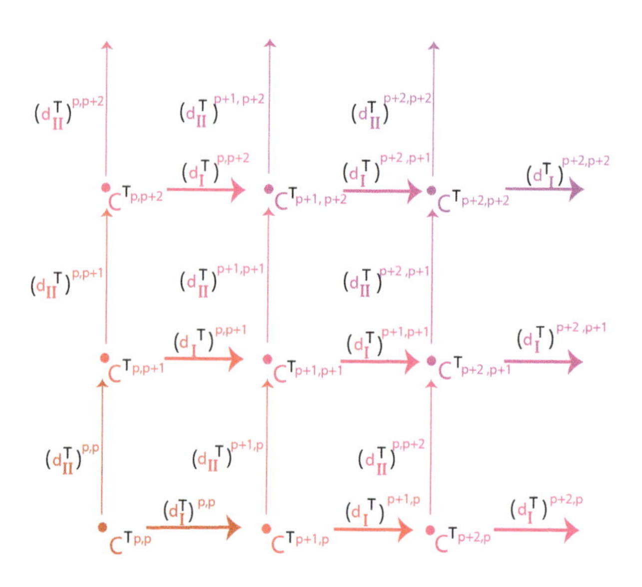

Fig. 15.42 The double complex $C^{\top}$.

where the prime indicates that we take cohomology with respect to $d_{\mathrm{I}}^{\top}$ and $d_{\mathrm{II}}^{\top}$. Let us express the above formula in terms of the complex C.

First, $C_{\mathrm{I}}^{\top} = C_{\mathrm{II}}$, and since $(d_{\mathrm{I}}^{\top})^{p,q} = d_{\mathrm{II}}^{q,p}$ and $(d_{\mathrm{II}}^{\top})^{p,q} = d_{\mathrm{I}}^{q,p}$,

$$H_{\mathrm{II}}^{'q}((C_{\mathrm{I}}^{\top})^{p}) = H_{\mathrm{I}}^{q}(C_{\mathrm{II}}^{p}),$$

so the complex $H_{\mathrm{II}}^{'q}(C_{\mathrm{I}}^{\top}))$, which is

$$0 \to H_{\mathrm{II}}^{'q}((C_{\mathrm{I}}^{\top})^{0}) \to H_{\mathrm{II}}^{'q}((C_{\mathrm{I}}^{\top})^{1}) \cdots \to H_{\mathrm{II}}^{'q}((C_{\mathrm{I}}^{\top})^{p}) \to H_{\mathrm{II}}^{'q}((C_{\mathrm{I}}^{\top})^{p+1}) \to \cdots,$$

with differential induced by the chain maps $(d_{\mathrm{I}}^{\top})^{p,*}$, is equal to

$$0 \longrightarrow H_{\mathrm{I}}^{q}(C_{\mathrm{II}}^{0}) \longrightarrow H_{\mathrm{I}}^{q}(C_{\mathrm{II}}^{1}) \cdots \longrightarrow H_{\mathrm{I}}^{q}(C_{\mathrm{II}}^{p}) \longrightarrow H_{\mathrm{I}}^{q}(C_{\mathrm{II}}^{p+1}) \longrightarrow \cdots,$$

with the differential induced by the chain maps $d_{\mathrm{II}}^{*,p}$. The above complex is just the complex $H_{\mathrm{I}}^{q}(C_{\mathrm{II}})$. Therefore, we obtain

$${}^{\mathrm{II}}E_{2}^{p,q} \cong H_{\mathrm{II}}^{p}(H_{\mathrm{I}}^{q}(C_{\mathrm{II}})),$$

where the complex $H_{\mathrm{I}}^{q}(C_{\mathrm{II}})$ is given by

$$0 \longrightarrow H_{\mathrm{I}}^{q}(C_{\mathrm{II}}^{0}) \longrightarrow H_{\mathrm{I}}^{q}(C_{\mathrm{II}}^{1}) \cdots \longrightarrow H_{\mathrm{I}}^{q}(C_{\mathrm{II}}^{p}) \longrightarrow H_{\mathrm{I}}^{q}(C_{\mathrm{II}}^{p+1}) \longrightarrow \cdots.$$

We have the following result.

Theorem 15.18. *Given a double complex $C = \bigoplus_{p,q\in\mathbb{N}} C^{p,q}$, we have two spectral sequences ${}^{\mathrm{I}}E$ and ${}^{\mathrm{II}}E$ corresponding to the filtrations $F_{\mathrm{I}}^{p}C$ and $F_{\mathrm{II}}^{q}C$ on C viewed as a total complex (in fact, these filtrations are canonically cobounded). We have*

$$^{\mathrm{I}}E_2^{p,q} \cong H_{\mathrm{I}}^p(H_{\mathrm{II}}^q(C_{\mathrm{I}}))$$

and

$$^{\mathrm{II}}E_2^{p,q} \cong H_{\mathrm{II}}^p(H_{\mathrm{I}}^q(C_{\mathrm{II}})).$$

In the next three sections we will provide several examples of double complexes.

15.10 Spectral Sequences of a Differential Sheaf

Let X be a topological space.

Definition 15.36. A *differential sheaf* $\mathcal{F}^*$ is a family $(\mathcal{F}^p)_{p\in\mathbb{N}}$ of sheaves of R-modules on X together with a family of sheaf maps $\delta^p\colon \mathcal{F}^p \to \mathcal{F}^{p+1}$ such that $\delta^{p+1}\circ\delta^p = 0$ for all $p \geq 0$. We represent a differential sheaf by the diagram

$$0 \longrightarrow \mathcal{F}^0 \xrightarrow{\delta^0} \mathcal{F}^1 \xrightarrow{\delta^1} \mathcal{F}^2 \xrightarrow{\delta^2} \cdots \longrightarrow \mathcal{F}^n \xrightarrow{\delta^n} \mathcal{F}^{n+1} \xrightarrow{\delta^{n+1}} \cdots.$$

Observe that since a resolution of a sheaf $\mathcal{F}$ is an exact sequence of sheaves

$$0 \longrightarrow \mathcal{F} \xrightarrow{j} \mathcal{F}^0 \xrightarrow{\delta^0} \mathcal{F}^1 \xrightarrow{\delta^1} \cdots \longrightarrow \mathcal{F}^n \xrightarrow{\delta^n} \cdots,$$

the sequence $\mathcal{F}^*$

$$0 \longrightarrow \mathcal{F}^0 \xrightarrow{\delta^0} \mathcal{F}^1 \xrightarrow{\delta^1} \cdots \longrightarrow \mathcal{F}^n \xrightarrow{\delta^n} \cdots$$

is also a differential sheaf. With a slight abuse of language, we also call $\mathcal{F}^*$ a resolution of $\mathcal{F}$.

Definition 15.37. Given a differential sheaf $\mathcal{F}^*$, we define the sheaves $\mathcal{Z}^n(\mathcal{F}^*)$, $\mathcal{B}^n(\mathcal{F}^*)$ and the cohomology $\mathcal{H}^n(\mathcal{F}^*)$ by

$$\begin{aligned}
\mathcal{Z}^n(\mathcal{F}^*) &= \operatorname{Ker}\delta^n \\
\mathcal{B}^n(\mathcal{F}^*) &= \operatorname{SIm}\delta^{n-1} \\
\mathcal{H}^n(\mathcal{F}^*) &= \mathcal{Z}^n\widetilde{(\mathcal{F}^*)/\mathcal{B}^n}(\mathcal{F}^*),
\end{aligned}$$

where $\mathcal{Z}^n\widetilde{(\mathcal{F}^*)/\mathcal{B}^n}(\mathcal{F}^*)$ is the sheafification of the presheaf $\mathcal{Z}^n(\mathcal{F}^*)/\mathcal{B}^n(\mathcal{F}^*)$.

Let T be an additive functor on sheaves. Given a differential sheaf $\mathcal{F}^*$, the complex $T(\mathcal{F}^*)$ is obtained by applying T to the $\mathcal{F}^n$ and to the coboundary maps δ^n, namely

$$0 \longrightarrow T(\mathcal{F}^0) \xrightarrow{T(\delta^0)} T(\mathcal{F}^1) \xrightarrow{T(\delta^1)} T(\mathcal{F}^2) \xrightarrow{T(\delta^2)} \cdots$$
$$\cdots \longrightarrow T(\mathcal{F}^n) \xrightarrow{T(\delta^n)} T(\mathcal{F}^{n+1}) \xrightarrow{T(\delta^{n+1})} \cdots .$$

The following result is proven in Godement [Godement (1958)] (Chapter 5, Page 165).

Proposition 15.19. *If the functor T is exact, for every differential sheaf $\mathcal{F}^*$, we have*

$$H^n(T(\mathcal{F}^*)) \cong T(\mathcal{H}^n(\mathcal{F}^*)), \quad n \geq 0.$$

Given a sheaf $\mathcal{F}$, recall from Definition 13.5 the canonical flasque resolution

$$0 \longrightarrow \mathcal{F} \xrightarrow{j} \mathcal{C}^0(X,\mathcal{F}) \xrightarrow{D^0} \mathcal{C}^1(X,\mathcal{F}) \xrightarrow{D^1} \mathcal{C}^2(X,\mathcal{F}) \xrightarrow{D^2} \cdots$$

of Proposition 13.4, denoted $\mathcal{C}(X,\mathcal{F})$ (or $\mathcal{C}^*(X,\mathcal{F})$). Here we have renamed the maps of the resolution as D^p instead of d^p to avoid a clash of notation below when defining d_{I}, d_{II}. Also recall from Definition 13.5 that the R-modules $C^n(X,\mathcal{F})$ are defined as

$$C^n(X,\mathcal{F}) = \Gamma(X,\mathcal{C}^n(X,\mathcal{F})) = \mathcal{C}^n(X,\mathcal{F})(X),$$

where $\Gamma(X,-)$ is the global section functor.

Definition 15.38. The cochain complex $C(X,\mathcal{F})$ (also denoted $C^*(X,\mathcal{F})$) is obtained by applying the global section functor to the canonical resolution and is given by

$$0 \longrightarrow \Gamma(X,\mathcal{F}) \xrightarrow{j^*} C^0(X,\mathcal{F}) \xrightarrow{(D^0)^*} C^1(X,\mathcal{F}) \xrightarrow{(D^1)^*} C^2(X,\mathcal{F}) \xrightarrow{(D^2)^*} \cdots .$$

We call this complex the *canonical resolution complex* of $\mathcal{F}$.

The functor $\mathcal{F} \mapsto C(X,\mathcal{F})$ from the category of sheaves to the category of cochain complexes is exact. This is proven in Godement [Godement (1958)] (Chapter 4, Page 168). This implies that the functors $\mathcal{F} \mapsto C^p(X,\mathcal{F})$ from sheaves to R-modules are also exact. As a corollary, using Proposition 15.19 with $T = C^p(X,-)$ we deduce the following result.

Proposition 15.20. *Given a differential sheaf $\mathcal{F}^*$, recall that $C^p(X,\mathcal{F}^*)$ denotes the cochain complex obtained by applying the functor $C^p(X,-)$ to the cochain complex defined by $\mathcal{F}^*$. There are isomorphisms*

$$H^q(C^p(X,\mathcal{F}^*)) \cong C^p(X,\mathcal{H}^q(\mathcal{F}^*)), \quad q \geq 0.$$

Definition 15.39. Given a differential sheaf $\mathcal{F}^*$, we define the double complex

$$C(X, \mathcal{F}^*) = \bigoplus_{p,q \in \mathbb{N}} C^p(X, \mathcal{F}^q),$$

where $C^p(X, \mathcal{F}^q)$ is the pth module in the canonical resolution $C(X, \mathcal{F}^q)$ of $\mathcal{F}^q$. We have two differentials d_{I} and d_{II}. The horizontal differential d_{I} is given by $d_{\mathrm{I}} = \bigoplus_{q \in \mathbb{N}} d_{\mathrm{I}}^q$, with $d_{\mathrm{I}}^q = \bigoplus_{p \in \mathbb{N}} d^{p,q}$, where

$$d^{p,q} = (D^{p,q})^* \colon C^p(X, \mathcal{F}^q) \to C^{p+1}(X, \mathcal{F}^q)$$

occurs in the cochain complex $C(X, \mathcal{F}^q)$ (the canonical resolution of $\mathcal{F}^q$), the qth row of the double complex $C(X, \mathcal{F}^*)$. The vertical differential d_{II} is given by $d_{\mathrm{II}} = \bigoplus_{p \in \mathbb{N}} d_{\mathrm{II}}^p$, with $d_{\mathrm{II}}^p = \bigoplus_{q \in \mathbb{N}} d_{\mathrm{II}}^{p,q}$ (a map on the pth column of $C(X, \mathcal{F}^*)$), where

$$d_{\mathrm{II}}^{p,q} \colon C^p(X, \mathcal{F}^q) \to C^p(X, \mathcal{F}^{q+1})$$

is the map induced by functoriality by the map $(-1)^p \delta^q \colon \mathcal{F}^q \to \mathcal{F}^{q+1}$. The total complex associated with $C(X, \mathcal{F}^*)$ is denoted by C. See Figure 15.43.

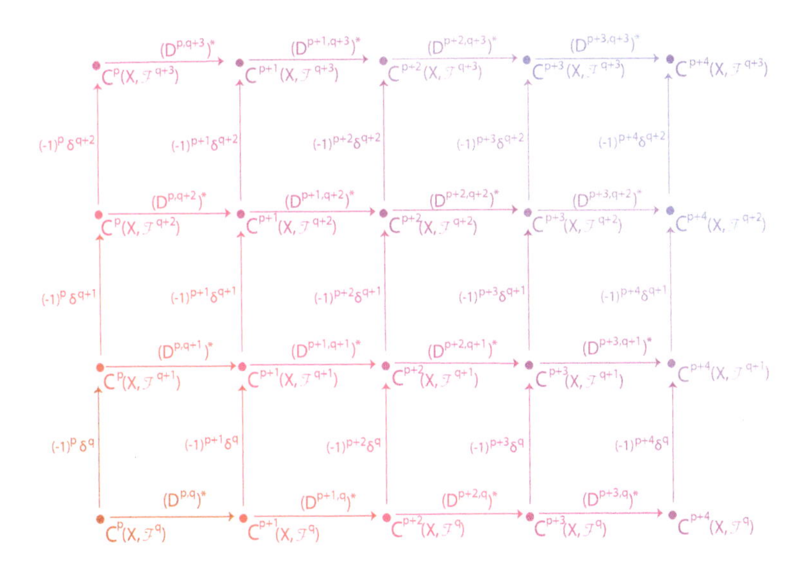

Fig. 15.43 A section of the double complex $C(X, \mathcal{F}^*) = \bigoplus_{p,q \in \mathbb{N}} C^p(X, \mathcal{F}^q)$.

Do not confuse the cochain complex $C(X, \mathcal{F})$ (the canonical resolution of Definition 15.38) and the double complex $C(X, \mathcal{F}^*)$, in which $\mathcal{F}^*$ is a differential sheaf.

The following observation applies to all the double complexes that we are considering in this section and the next two. These complexes involve a differential sheaf $\mathcal{F}^*$ in the q-coordinate. As a consequence, the row cohomology involves a single sheaf $\mathcal{F}^p$ and is usually easy to compute. It follows that computing the terms ${}^{\mathrm{II}}E_2^{p,q}$ of the second spectral sequence is also easy. On the other hand, the column cohomology involves the entire differential sheaf $\mathcal{F}^*$ and is harder to compute. The terms ${}^{\mathrm{I}}E_2^{p,q}$ of the first spectral sequence are also harder to compute.

The pth row of $C(X, \mathcal{F}^*)$ is the canonical resolution $C(X, \mathcal{F}^p)$ of $\mathcal{F}^p$, which computes the sheaf cohomology of the sheaf $\mathcal{F}^p$, and the pth column of $C(X, \mathcal{F}^*)$ computes the cohomology of the complex $C^p(X, \mathcal{F}^*)$, which is harder to handle, but can be expressed in terms of the cohomology sheaves $\mathcal{H}^q(\mathcal{F}^*)$. If $\mathcal{F}^*$ is a resolution of $\mathcal{F}$, then the first spectral sequence degenerates at $r = 2$. If this resolution consists of sheaves with special properties, then the second spectral sequence also degenerates and we obtain isomorphisms $H^n(\Gamma(\mathcal{F}^*)) \cong H^n(X, \mathcal{F})$ that allow us to compute sheaf cohomology using special resolutions $\mathcal{F}^*$.

The conditions needed for applying Theorem 15.18 are satisfied and we obtain a spectral sequence where the terms ${}^{\mathrm{I}}E_2^{p,q}$ and ${}^{\mathrm{II}}E_2^{p,q}$ are determined.

To compute ${}^{\mathrm{I}}E_2^{p,q} \cong H_{\mathrm{I}}^p(H_{\mathrm{II}}^q(C_{\mathrm{I}}))$ we need to determine the complex $H_{\mathrm{II}}^q(C_{\mathrm{I}})$, which requires computing $H_{\mathrm{II}}^q(C_{\mathrm{I}}^p)$, with C_{I}^p the complex (the pth column of $C(X, \mathcal{F}^*)$) given by

$$C_{\mathrm{I}}^p = \bigoplus_{q \in \mathbb{N}} C^p(X, \mathcal{F}^q) = C^p(X, \mathcal{F}^*).$$

By Proposition 15.20, we obtain

$$H_{\mathrm{II}}^q(C_{\mathrm{I}}^p) \cong C^p(X, \mathcal{H}^q(\mathcal{F}^*)).$$

Thus the complex $H_{\mathrm{II}}^q(C_{\mathrm{I}})$ is the complex $\bigoplus_{p \in \mathbb{N}} C^p(X, \mathcal{H}^q(\mathcal{F}^*))$, and by taking the cohomology H_{I}^p, we get

$${}^{\mathrm{I}}E_2^{p,q} \cong H^p(X; \mathcal{H}^q(\mathcal{F}^*)).$$

To compute ${}^{\mathrm{II}}E_2^{p,q} \cong H_{\mathrm{II}}^p(H_{\mathrm{I}}^q(C_{\mathrm{II}}))$, we need to determine the complex $H_{\mathrm{I}}^q(C_{\mathrm{II}})$, which requires computing $H_{\mathrm{I}}^q(C_{\mathrm{II}}^p)$, with C_{II}^p the complex (the pth row of $C(X, \mathcal{F}^*)$) given by

$$C_{\text{II}}^p = \bigoplus_{q \in \mathbb{N}} C^q(X, \mathcal{F}^p),$$

which is the canonical resolution $C(X, \mathcal{F}^p)$ of $\mathcal{F}^p$, so we have

$$H_{\text{I}}^q(C_{\text{II}}^p) \cong H^q(X, \mathcal{F}^p),$$

and the complex $H_{\text{I}}^q(C_{\text{II}})$ is the complex $H^q(X, \mathcal{F}^*)$ given by

$$0 \longrightarrow H^q(X, \mathcal{F}^0) \longrightarrow H^q(X, \mathcal{F}^1) \longrightarrow \cdots$$
$$\cdots \longrightarrow H^q(X, \mathcal{F}^p) \longrightarrow H^q(X, \mathcal{F}^{p+1}) \longrightarrow \cdots,$$

where the maps $H^q(X, \mathcal{F}^p) \longrightarrow H^q(X, \mathcal{F}^{p+1})$ are induced on cohomology by the maps $\delta^p \colon \mathcal{F}^p \to \mathcal{F}^{p+1}$. Finally, we obtain

$${}^{\text{II}}E_2^{p,q} \cong H^p(H^q(X; \mathcal{F}^*)).$$

In summary, we have shown the following result.

Proposition 15.21. *The terms ${}^{\text{I}}E_2^{p,q}$ and ${}^{\text{II}}E_2^{p,q}$ of the spectral sequences associated with the double complex $C(X, \mathcal{F}^*)$ are given by*

$$\begin{aligned} {}^{\text{I}}E_2^{p,q} &\cong H^p(X; \mathcal{H}^p(\mathcal{F}^*)) \\ {}^{\text{II}}E_2^{p,q} &\cong H^p(H^q(X; \mathcal{F}^*)). \end{aligned}$$

Observe that if the complexes $H^q(X, \mathcal{F}^*)$ are acyclic in all degrees for all $q \geq 1$, which means that $H^p(H^q(X, \mathcal{F}^*)) = (0)$ for all $p \geq 0$ and all $q \geq 1$, then we see that ${}^{\text{II}}E_2^{n-q,q} \cong H^{n-q}(H^q(X, \mathcal{F}^*)) = (0)$ for all $q \neq 0$, which is the condition for the spectral sequence ${}^{\text{II}}E$ to degenerate at $r = 2$ with $q(n) = 0$. Since the second filtration is regular, by Proposition 15.17, we have isomorphisms

$$H^n(C) \cong {}^{\text{II}}E_2^{n,0} \cong H^n(\Gamma(\mathcal{F}^*)),$$

where $\Gamma(\mathcal{F}^*)$ is the cochain complex

$$0 \longrightarrow \Gamma(X, \mathcal{F}^0) \longrightarrow \Gamma(X, \mathcal{F}^1) \longrightarrow \Gamma(X, \mathcal{F}^2) \longrightarrow \Gamma(X, \mathcal{F}^3) \longrightarrow \cdots,$$

since $H^0(X, \mathcal{F}^p) \cong \Gamma(X, \mathcal{F}^p)$. Thus we have the following theorem from Godement [Godement (1958)] (Chapter 4, Theorem 4.6.1).

Theorem 15.22. *Given any differential sheaf $\mathcal{F}^* = (\mathcal{F}^p)$, let C be the total complex associated with the complex $C(X, \mathcal{F}^*)$ of Definition 15.39. If $H^p(H^q(X, \mathcal{F}^*)) = (0)$ for all $p \geq 0$ and all $q \geq 1$, then the second* *spectral sequence* ${}^{\mathrm{II}}E$ *degenerates for* $r = 2$ *and* $q(n) = 0$ *and we have an isomorphism*

$$H^n(C) \cong H^n(\Gamma(\mathcal{F}^*)), \quad n \geq 0.$$

The first spectral sequence ${}^{\mathrm{I}}E$ *has the property that for all* $p, q \geq 0$, *for* r *large enough, we have* ${}^{\mathrm{I}}E_\infty \cong \mathrm{gr}(H(C))$, *and*

$$^{\mathrm{I}}E_2^{p,q} \cong H^p(X; \mathcal{H}^q(\mathcal{F}^*)).$$

Given a resolution $\mathcal{F}^*$

$$0 \longrightarrow \mathcal{F} \xrightarrow{j} \mathcal{F}^0 \xrightarrow{\delta^0} \mathcal{F}^1 \xrightarrow{\delta^1} \cdots \longrightarrow \mathcal{F}^n \xrightarrow{\delta^n} \cdots$$

of a sheaf $\mathcal{F}$, we have the differential sheaf $\mathcal{F}^*$ given by

$$0 \longrightarrow \mathcal{F}^0 \xrightarrow{\delta^0} \mathcal{F}^1 \xrightarrow{\delta^1} \cdots \longrightarrow \mathcal{F}^n \xrightarrow{\delta^n} \cdots,$$

and more can be said. Indeed, since the sequence

$$0 \longrightarrow \mathcal{F} \xrightarrow{j} \mathcal{F}^0 \xrightarrow{\delta^0} \mathcal{F}^1 \xrightarrow{\delta^1} \cdots \longrightarrow \mathcal{F}^n \xrightarrow{\delta^n} \cdots$$

is exact, by definition of the cohomology of a differential sheaf, we have $H^q(\mathcal{H}(\mathcal{F}^*)) = (0)$ for all $q \geq 1$, and thus

$$^{\mathrm{I}}E_2^{p,q} \cong H^p(X; \mathcal{H}^q(\mathcal{F}^*)) = (0)$$

for all $p \geq 0$ and all $q \geq 1$. We also have $\mathcal{H}^0(\mathcal{F}^*) \cong \mathcal{F}$. This means that the spectral sequence ${}^{\mathrm{I}}E$ degenerates for $r = 2$ and $q(n) = 0$. Since the first filtration is regular, by Proposition 15.17, we have isomorphisms

$$H^n(C) \cong {}^{\mathrm{I}}E_2^{n,0} \cong H^n(X, \mathcal{H}^0(\mathcal{F}^*)) \cong H^n(X, \mathcal{F}),$$

since $\mathcal{H}^0(\mathcal{F}^*) \cong \mathcal{F}$.

If we also have $H^n(C) \cong H^n(\Gamma(\mathcal{F}^*))$, as in Theorem 15.22, then if the above result holds, we obtain an isomorphism

$$H^n(\Gamma(\mathcal{F}^*)) \cong H^n(X, \mathcal{F}).$$

This means that the sheaf cohomology $H^n(X, \mathcal{F})$ can be computed using a special resolution $\mathcal{F}^*$ of $\mathcal{F}$. Indeed we have the following result (also from Godement [Godement (1958)], Chapter 4, Theorem 4.7.1).

Theorem 15.23. *Let* $\mathcal{F}^*$ *be a resolution of a sheaf* $\mathcal{F}$ *and let* C *be the total complex associated with the complex* $C(X, \mathcal{F}^*)$ *of Definition 15.39.*

The first spectral sequence degenerates for $r = 2$ and $q(n) = 0$ and we have isomorphisms

$$H^n(C) \cong H^n(X, \mathcal{F}), \quad n \geq 0.$$

If $H^p(H^q(X, \mathcal{F}^*)) = (0)$ for all $p \geq 0$ and all $q \geq 1$, then the second spectral sequence also degenerates for $r = 2$ and $q(n) = 0$ and we have an isomorphism

$$H^n(C) \cong H^n(\Gamma(\mathcal{F}^*)) \cong H^n(X, \mathcal{F}), \quad n \geq 0.$$

In particular, the above isomorphisms hold if the sheaves $\mathcal{F}^p$ are acyclic for all $p \geq 0$. The above holds in the following two cases:

(1) The sheaves $\mathcal{F}^p$ are flasque for all $p \geq 0$.
(2) The topological space X is paracompact and the sheaves $\mathcal{F}^p$ are soft for all $p \geq 0$.

Proof. Recall that a sheaf $\mathcal{F}$ is acyclic if $H^q(X, \mathcal{F}) = (0)$ for all $q \geq 1$. If all sheaves $\mathcal{F}^p$ are acyclic, then the complexes $H^q(X, \mathcal{F}^*)$ given by

$$0 \longrightarrow H^q(X, \mathcal{F}^0) \longrightarrow H^q(X, \mathcal{F}^1) \longrightarrow \cdots$$
$$\cdots \longrightarrow H^q(X, \mathcal{F}^p) \longrightarrow H^q(X, \mathcal{F}^{p+1}) \longrightarrow \cdots,$$

are trivial (all module are zero) for $q \geq 1$. It only remains to prove (1) and (2). By Proposition 13.7, flasque sheaves are acyclic. Similarly, by Proposition 13.27, a soft sheaf on a paracompact space is acyclic. $\square$

Theorem 15.23 can be applied to constant sheaves to prove that various sheaf cohomologies can be computed using special resolutions. This applies to Alexander–Spanier cochains, and de Rham cohomology. See Godement [Godement (1958)] (Chapter 4) for more details. Proposition 15.23 also implies Proposition 11.34 in the case where T is the global section functor. The proof can be generalized to any left-exact functor T.

15.11 Spectral Sequences of Čech Cohomology, I

In this section we are going to define spectral sequences induced by double complexes whose row cohomology is the Čech cohomology associated with a cover $\mathcal{U}$ on a topological space. The reader may want to review Section 9.1.

Recall from Section 13.3 that given an open cover $\mathcal{U}$ on a topological space X, for any sheaf $\mathcal{F}$ on X, we have the sheaves $\mathcal{C}^p(\mathcal{U}, \mathcal{F})$. A crucial property of the sheaves $\mathcal{C}^p(\mathcal{U}, \mathcal{F})$ is that they form a resolution

$$0 \longrightarrow \mathcal{F} \longrightarrow \mathcal{C}^0(\mathcal{U}, \mathcal{F}) \xrightarrow{\delta} \mathcal{C}^1(\mathcal{U}, \mathcal{F}) \xrightarrow{\delta} \cdots$$
$$\cdots \longrightarrow \mathcal{C}^p(\mathcal{U}, \mathcal{F}) \xrightarrow{\delta} \mathcal{C}^{p+1}(\mathcal{U}, \mathcal{F}) \xrightarrow{\delta} \cdots$$

of the sheaf $\mathcal{F}$. This is shown in Proposition 13.9, and in fact it is also a resolution if $\mathcal{U}$ is a closed cover which is locally finite. This second result is proven in Godement [Godement (1958)] (Chapter 5, Theorem 5.2.1).

Definition 15.40. The sheaves $\mathcal{C}^p(\mathcal{U}, \mathcal{F})$ constitute the differential sheaf $\mathcal{C}(\mathcal{U}, \mathcal{F})$ given by

$$0 \longrightarrow \mathcal{C}^0(\mathcal{U}, \mathcal{F}) \xrightarrow{\delta} \mathcal{C}^1(\mathcal{U}, \mathcal{F}) \xrightarrow{\delta} \cdots$$
$$\cdots \longrightarrow \mathcal{C}^p(\mathcal{U}, \mathcal{F}) \xrightarrow{\delta} \mathcal{C}^{p+1}(\mathcal{U}, \mathcal{F}) \xrightarrow{\delta} \cdots,$$

called the *Čech resolution of $\mathcal{F}$*.

Because the $\mathcal{C}^p(\mathcal{U}, \mathcal{F})$ are sheaves, applying the global section functor to the differential sheaf $\mathcal{C}(\mathcal{U}, \mathcal{F})$ we get $\Gamma(X, \mathcal{C}(\mathcal{U}, \mathcal{F})) = C(\mathcal{U}, \mathcal{F})$, which is just the Čech complex

$$0 \longrightarrow C^0(\mathcal{U}, \mathcal{F}) \longrightarrow C^1(\mathcal{U}, \mathcal{F}) \longrightarrow \cdots$$
$$\cdots \longrightarrow C^p(\mathcal{U}, \mathcal{F}) \longrightarrow C^{p+1}(\mathcal{U}, \mathcal{F}) \longrightarrow \cdots.$$

See Definitions 9.1, 9.2 and 9.3.

One should not confuse the sheaves $\mathcal{C}^p(X, \mathcal{F})$ that occur in the canonical flasque resolution of a sheaf $\mathcal{F}$, and the sheaves $\mathcal{C}^p(\mathcal{U}, \mathcal{F})$ that occur in the Čech resolution of $\mathcal{F}$. Similarly, the canonical resolution complex $C(X, \mathcal{F})$ should not be confused with the Čech complex $C(\mathcal{U}, \mathcal{F})$.

The following result is also shown in Godement [Godement (1958)] (Chapter 5, Theorem 5.2.2).

Proposition 15.24. *If the cover $\mathcal{U}$ on X is either open or closed and locally finite, then we have isomorphisms*

$$\check{H}^0(\mathcal{U}, \mathcal{F}) \cong \Gamma(X, \mathcal{F}) \cong H^0(X, \mathcal{F}).$$

The first of two spectral sequences that we will consider arise from the double complex

$$C(X, \mathcal{C}(\mathcal{U}, \mathcal{F})) = \bigoplus_{p,q \in \mathbb{N}} C^p(X, \mathcal{C}^q(\mathcal{U}, \mathcal{F})).$$

Let C be the total complex associated with $C(X, \mathcal{C}(\mathcal{U}, \mathcal{F}))$. This is an instance of Definition 15.39 with $\mathcal{F}^*$ equal to the differential sheaf $\mathcal{C}(\mathcal{U}, \mathcal{F})$ of Definition 15.40 (which involves Čech cohomology). Since the sheaves $\mathcal{C}^q(\mathcal{U}, \mathcal{F})$ form a resolution of $\mathcal{F}$, we know from Theorem 15.23 that the first spectral sequence degenerates at $r = 2$ and we obtain isomorphisms

$$H^n(C) \cong H^n(X, \mathcal{F}), \quad n \geq 0.$$

Regarding the second spectral sequence, by Theorem 15.22, we have

$${}^{\mathrm{II}}E_2^{p,q} \cong H^p(H^q(X, \mathcal{C}(\mathcal{U}, \mathcal{F}))).$$

In general this expression can't be simplified but it can be if $\mathcal{U}$ is a closed and locally finite cover. For this one needs to introduce the notion of system of coefficients, which we will do later. This leads to a result of Leray about closed and locally finite covers, but we will not pursue this matter any further and instead refer to reader to Godement [Godement (1958)] (Chapter 5, Theorem 5.2.4).

In general, $\check{H}^0(\mathcal{U}, \mathcal{F})$ and $H^0(X, \mathcal{F})$ are not isomorphic but spectral sequences can be used to prove that various conditions yield isomorphisms.

The key is that given a differential sheaf $\mathcal{F}^* = (\mathcal{F}^p)$ we can form the following double complex.

Definition 15.41. Let $\mathcal{U}$ be an open cover on a topological space X and let $\mathcal{F}^*$ be a differential sheaf on X. The double complex $C(\mathcal{U}, \mathcal{F}^*)$, illustrated by Figure 15.44, is given by

$$C(\mathcal{U}, \mathcal{F}^*) = \bigoplus_{p,q \in \mathbb{N}} C^p(\mathcal{U}, \mathcal{F}^q),$$

where $C^p(\mathcal{U}, \mathcal{F}^q)$ is the pth module in the Čech complex $C(\mathcal{U}, \mathcal{F}^q)$ of $\mathcal{F}^q$, the qth row of the double complex $C(\mathcal{U}, \mathcal{F}^*)$. The differentials d_{I} and d_{II} are defined as in Definition 15.39, except that the horizontal differential d_{I} arises from the coboundary maps of the Čech complex $C(\mathcal{U}, \mathcal{F}^q)$. The total complex associated with $C(\mathcal{U}, \mathcal{F}^*)$ is denoted by C.

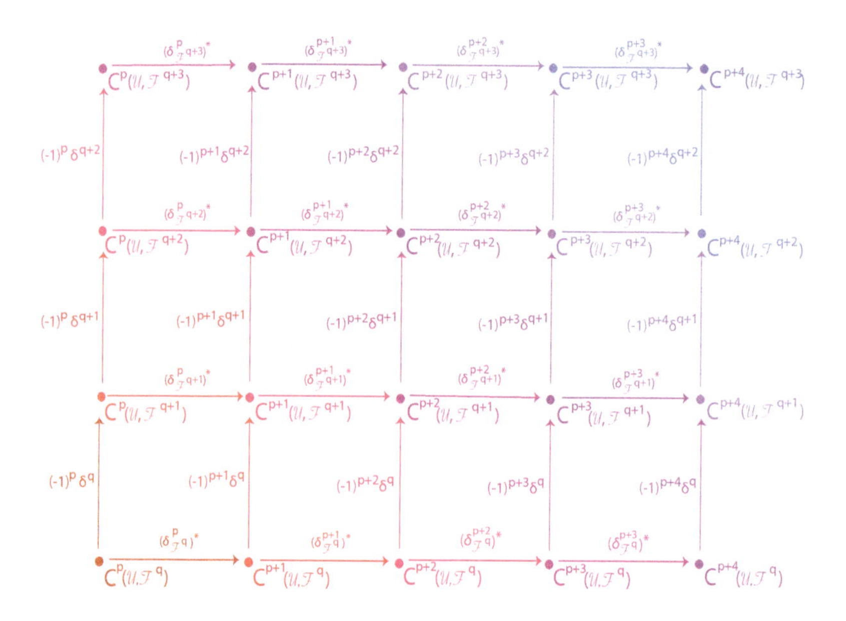

Fig. 15.44 A section of the double complex $C(\mathcal{U}, \mathcal{F}^*) = \bigoplus_{p,q\in\mathbb{N}} C^p(\mathcal{U}, \mathcal{F}^q)$, where we used the conventions of Definition 15.39 to express d_{I}.

One should not confuse the double complex $C(X, \mathcal{F}^*)$ of Definition 15.39 whose pth row involves the sheaf cohomology of $\mathcal{F}^p$, and the double complex $C(\mathcal{U}, \mathcal{F}^*)$ defined above, whose pth row involves the Čech cohomology of $\mathcal{F}^p$.

The pth row of $C(\mathcal{U}, \mathcal{F}^*)$ is the Čech complex $C(\mathcal{U}, \mathcal{F}^p)$ which computes the Čech cohomology (over $\mathcal{U}$) of the sheaf $\mathcal{F}^p$. The pth column of $C(\mathcal{U}, \mathcal{F}^*)$ computes the cohomology of the complex

$$C^p(\mathcal{U}, \mathcal{F}^*) = \bigoplus_{q\in\mathbb{N}} C^p(\mathcal{U}, \mathcal{F}^q),$$

which involves the Čech cochains $C^p(\mathcal{U}, \mathcal{F}^q)$ of the sheaves $\mathcal{F}^q$. The complex $C^p(\mathcal{U}, \mathcal{F}^*)$ is a bit complicated but it can be expressed in terms of complexes of the form $\mathcal{F}^*(U_{i_0\cdots i_s})$ obtained by applying the differential sheaf $\mathcal{F}^*$ to the open subsets $U_{i_0\cdots i_s}$. Then it is possible to express the cohomology modules of the complex $C^p(\mathcal{U}, \mathcal{F}^*)$ using what is known as systems of coefficient. If $\mathcal{F}^*$ is the canonical resolution of $\mathcal{F}$, then the second spectral sequence degenerates at $r = 2$. Under additional conditions on $\mathcal{F}^*$, the first spectral sequence also degenerates and we find isomorphisms

$$\check{H}^n(\mathcal{U}, \mathcal{F}) \cong H^n(X, \mathcal{F}), \quad n \geq 0.$$

As in the previous section, to compute ${}^{\mathrm{II}}E_2^{p,q} \cong H_{\mathrm{II}}^p(H_{\mathrm{I}}^q(C_{\mathrm{II}}))$, we need to determine the complex $H_{\mathrm{I}}^q(C_{\mathrm{II}})$, which requires computing $H_{\mathrm{I}}^q(C_{\mathrm{II}}^p)$, with C_{II}^p the complex (the pth row of $C(\mathcal{U}, \mathcal{F}^*)$) given by

$$C_{\mathrm{II}}^p = \bigoplus_{q \in \mathbb{N}} C^q(\mathcal{U}, \mathcal{F}^p),$$

which is the Čech complex $C(\mathcal{U}, \mathcal{F}^p)$ of $\mathcal{F}^p$ (over $\mathcal{U}$), so we have

$$H_{\mathrm{I}}^q(C_{\mathrm{II}}^p) \cong \check{H}^q(\mathcal{U}, \mathcal{F}^p),$$

and the complex $H_{\mathrm{I}}^q(C_{\mathrm{II}})$ is the complex $\check{H}^q(\mathcal{U}, \mathcal{F}^*)$ given by

$$0 \longrightarrow \check{H}^q(\mathcal{U}, \mathcal{F}^0) \longrightarrow \check{H}^q(\mathcal{U}, \mathcal{F}^1) \longrightarrow \cdots$$
$$\cdots \longrightarrow \check{H}^q(\mathcal{U}, \mathcal{F}^p) \longrightarrow \check{H}^q(\mathcal{U}, \mathcal{F}^{p+1}) \longrightarrow \cdots,$$

where the maps $\check{H}^q(\mathcal{U}, \mathcal{F}^p) \longrightarrow \check{H}^q(\mathcal{U}, \mathcal{F}^{p+1})$ are induced on cohomology by the maps $\delta^p \colon \mathcal{F}^p \to \mathcal{F}^{p+1}$.

Proposition 15.25. *The terms ${}^{\mathrm{II}}E_2^{p,q}$ of second spectral sequence associated with the double complex $C(\mathcal{U}, \mathcal{F}^*)$ are given by*

$${}^{\mathrm{II}}E_2^{p,q} \cong H^p(\check{H}^q(\mathcal{U}, \mathcal{F}^*)).$$

The term ${}^{\mathrm{I}}E_2^{p,q}$ is a little harder to determine. We will return to this point shortly.

The following vanishing cohomology results will be used to give sufficient conditions for the second spectral sequence to degenerate.

Proposition 15.26. *Let $\mathcal{U}$ be a cover on a topological space X, and let $\mathcal{F}$ be a sheaf on X. We have*

$$\check{H}^p(\mathcal{U}, \mathcal{F}) = (0) \quad \textit{for all } p \geq 1$$

if one of the conditions below holds:

(a) The cover $\mathcal{U}$ is open and $\mathcal{F}$ is flasque.
(b) The cover $\mathcal{U}$ is open, X is paracompact, and $\mathcal{F}$ is fine.
(c) The cover $\mathcal{U}$ is closed, locally finite, X is paracompact and $\mathcal{F}$ is soft.

Proposition 15.26 is proven in Godement [Godement (1958)] (Chapter 5, Theorem 5.2.3).

Let us now assume that $\mathcal{U}$ is open and that $\mathcal{G}^*$ is the canonical flasque resolution of the sheaf $\mathcal{F}$. By Proposition 15.26, we have

$$\check{H}^q(\mathcal{U}, \mathcal{G}^p) = (0) \quad \text{for all } p \geq 0 \text{ and all } q \geq 1,$$

so the complexes $\check{H}^q(\mathcal{U}, \mathcal{G}^*)$ are trivial for all $q \geq 1$, which implies that the spectral sequence ${}^{\mathrm{II}}E$ degenerates for $r = 2$ and $q(n) = 0$. Consequently we have isomorphisms

$$H^n(C) \cong {}^{\mathrm{II}}E_2^{n,0} \cong H^n(\check{H}^0(\mathcal{U}, \mathcal{G}^*)) \cong H^n(X, \mathcal{F}),$$

because the complex $\check{H}^0(\mathcal{U}, \mathcal{G}^*)$ is simply the complex $\Gamma(X, \mathcal{G}^*)$ since $\check{H}^0(\mathcal{U}, \mathcal{G}^p) \cong \Gamma(X, \mathcal{G}^p)$, and since $\mathcal{G}^*$ is the canonical flasque resolution of $\mathcal{F}$, the complex $\Gamma(X, \mathcal{G}^*)$ computes the sheaf cohomology of $\mathcal{F}$. In summary we proved the following result.

Proposition 15.27. *Let $\mathcal{U}$ be an open cover on a topological space X and let $\mathcal{G}^*$ be the canonical flasque resolution of a sheaf $\mathcal{F}$. The second spectral sequence associated with the double complex $C(\mathcal{U}, \mathcal{G}^*)$ degenerates at $r = 2$ and we have isomorphisms*

$$H^n(C) \cong {}^{\mathrm{II}}E_2^{n,0} \cong H^n(X, \mathcal{F}), \quad n \geq 0.$$

To proceed any further we need to determine ${}^{\mathrm{I}}E_2^{p,q}$, where we assume again that $\mathcal{F}^*$ is a differential sheaf. To compute ${}^{\mathrm{I}}E_2^{p,q} \cong H_{\mathrm{I}}^p(H_{\mathrm{II}}^q(C_{\mathrm{I}}))$ we need to determine the complex $H_{\mathrm{II}}^q(C_{\mathrm{I}})$, which requires computing $H_{\mathrm{II}}^q(C_{\mathrm{I}}^p)$, with C_{I}^p the complex (the pth column of $C(\mathcal{U}, \mathcal{F}^*)$) given by

$$C_{\mathrm{I}}^p = \bigoplus_{q \in \mathbb{N}} C^p(\mathcal{U}, \mathcal{F}^q) = C^p(\mathcal{U}, \mathcal{F}^*).$$

We now are faced with computing $H_{\mathrm{II}}^q(C^p(\mathcal{U}, \mathcal{F}^*))$. This involves cohomology groups of the form $H^q(\mathcal{F}^*(U_{i_0 \cdots i_s}))$, where as usual $U_{i_0 \cdots i_s} = U_{i_0} \cap \cdots \cap U_{i_s}$, with $v = (i_0, \ldots, i_s)$ a finite sequence of indices from the index set J of the open cover $\mathcal{U} = (U_i)_{i \in J}$. The complex $\mathcal{F}^*(U_{i_0 \cdots i_s}) = \mathcal{F}^*(U_v)$, is given by

$$0 \longrightarrow \mathcal{F}^0(U_v) \longrightarrow \mathcal{F}^1(U_v) \longrightarrow \cdots \longrightarrow \mathcal{F}^n(U_v) \longrightarrow \cdots,$$

where the coboundary maps are induced by functoriality.

The map $(i_0, \cdots, i_s) \mapsto H^q(\mathcal{F}^*(U_{i_0 \cdots i_s}))$ is not quite a presheaf. First, it is not defined on open subsets, but even it was defined on open subsets of the form $U_{i_0 \cdots i_s}$, it is not defined on *all* open subsets. Still, it shares the defining properties of a presheaf. It is called a *system of coefficients*.

Definition 15.42. Given a cover $\mathcal{U} = (U_i)_{i \in J}$ on a topological space X, a *system of coefficients* $\mathcal{K}$ *on* $J^* = \bigcup_{k \in \mathbb{N}} J^n$ consists of an assignment of an R-module $\mathcal{K}(v)$ to every finite sequence $v = (i_0, \ldots, i_s) \in J^{s+1}$, and for any two finite sequences v and w (over J^*) such that v is a subsequence of w of a restriction map $\rho^v_w \colon \mathcal{K}(v) \to \mathcal{K}(w)$. Note that since $U_w \subseteq U_v$ if v is a subsequence of w, not the other way around, the restriction map goes in the right direction. The restriction maps satisfy the following properties:

(a) If $v = w$, then $\rho^v_v = \mathrm{id}$.
(b) If t is a subsequence of v and v is a subsequence of w, then

$$\rho^t_w = \rho^v_w \circ \rho^t_v;$$

see Figure 15.45.

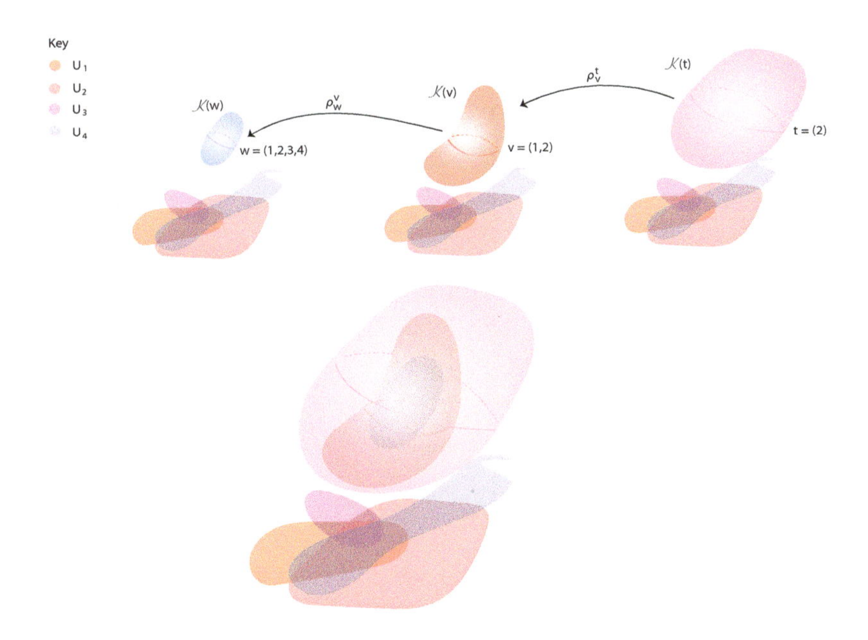

Fig. 15.45 A schematic representation of Condition (b) of Definition 15.42. The spherical objects above the plane containing the U's are the R-modules. The ρ-maps systematically collapse the larger modules onto the smaller, subcontained modules. The nesting of the modules is shown in the diagram below the maps.

Given a system of coefficients $\mathcal{K}$ on a cover $\mathcal{U}$, we define $C^p(\mathcal{U}, \mathcal{K})$ just as in Definition 9.1.

Definition 15.43. Given a topological space X, a cover $\mathcal{U} = (U_j)_{j \in J}$ of X, and a system of coefficients $\mathcal{K}$, the R-module $C^p(\mathcal{U}, \mathcal{K})$ of *Čech p-cochains with coefficients in $\mathcal{K}$* is the set of all functions f with domain J^{p+1} such that $f(i_0, \ldots, i_p) \in \mathcal{K}(i_0, \ldots, i_p)$; in other words,

$$C^p(\mathcal{U}, \mathcal{K}) = \prod_{(i_0, \ldots, i_p) \in J^{p+1}} \mathcal{K}(i_0, \ldots, i_p),$$

the set of all J^{p+1}-indexed families $(f_{i_0, \ldots, i_p})_{(i_0, \ldots, i_p) \in J^{p+1}}$ with $f_{i_0, \ldots, i_p} \in \mathcal{K}(i_0, \ldots, i_p)$.

Then as in Section 9.1 we define the Čech cohomology modules $\check{H}^p(\mathcal{U}, \mathcal{K})$ with coefficients in $\mathcal{K}$.

Returning to our differential sheaf $\mathcal{F}^*$ and to our first spectral sequence, to compute ${}^{\mathrm{I}}E_2^{p,q} \cong H_{\mathrm{I}}^p(H_{\mathrm{II}}^q(C_{\mathrm{I}}))$ we need to determine the complex $H_{\mathrm{II}}^q(C_{\mathrm{I}})$, which requires computing $H_{\mathrm{II}}^q(C_{\mathrm{I}}^p)$, where $C_{\mathrm{I}}^p = C^p(\mathcal{U}, \mathcal{F}^*)$. But we are now in the position to compute $H_{\mathrm{II}}^q(C^p(\mathcal{U}, \mathcal{F}^*))$. This is because

$$\begin{aligned} C^p(\mathcal{U}, \mathcal{F}^*) &= \bigoplus_{q \in \mathbb{N}} C^p(\mathcal{U}, \mathcal{F}^q) \\ &= \bigoplus_{q \in \mathbb{N}} \prod_{(i_0, \ldots, i_p) \in J^{p+1}} \mathcal{F}^q(U_{i_0 \cdots i_p}) \\ &\cong \prod_{(i_0, \ldots, i_p) \in J^{p+1}} \bigoplus_{q \in \mathbb{N}} \mathcal{F}^q(U_{i_0 \cdots i_p}) \\ &= \prod_{(i_0, \ldots, i_p) \in J^{p+1}} \mathcal{F}^*(U_{i_0 \cdots i_p}), \end{aligned}$$

where $\mathcal{F}^*(U_{i_0 \cdots i_p})$ is the cochain complex

$$0 \longrightarrow \mathcal{F}^0(U_{i_0 \cdots i_p}) \longrightarrow \mathcal{F}^1(U_{i_0 \cdots i_p}) \longrightarrow \cdots \longrightarrow \mathcal{F}^q(U_{i_0 \cdots i_p}) \longrightarrow \cdots \qquad (*_{\mathcal{F}^*(U_v)})$$

induced by the differential sheaf $\mathcal{F}^*$ on $U_{i_0 \cdots i_p}$ (with differential induced by functoriality by the map $(-1)^p \delta^q \colon \mathcal{F}^q \to \mathcal{F}^{q+1}$); see Figure 15.46.

Writing U_v as an abbreviation for $U_{i_0 \cdots i_p}$, we have

$$H_{\mathrm{II}}^q(C_{\mathrm{I}}^p) = \prod_{v \in J^{p+1}} H^q(\mathcal{F}^*(U_v)).$$

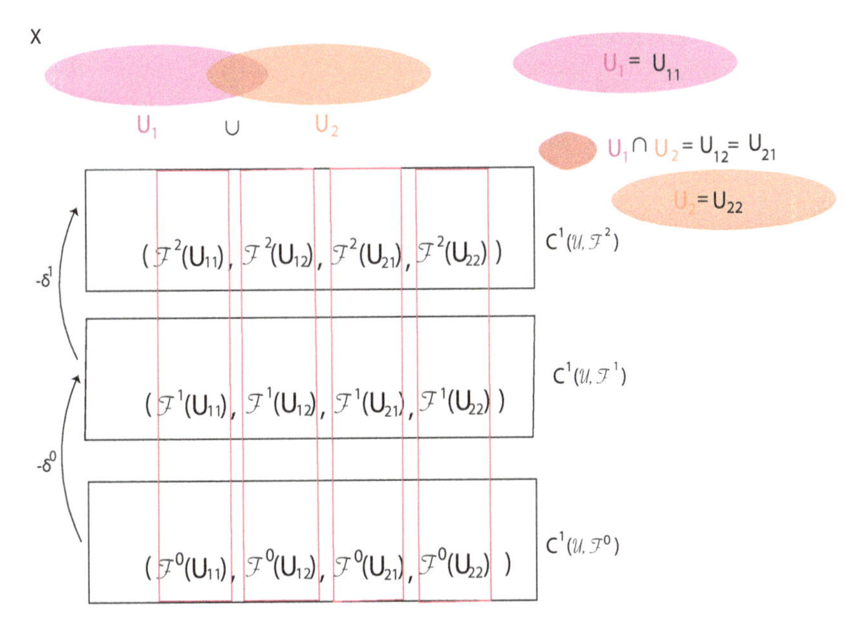

Fig. 15.46 Let X be the union of the peach and pink open disks. The three horizontal boxes are the first three elements in the second column of the double complex $C(\mathcal{U}, \mathcal{F}^*)$. A typical element of $C^1(\mathcal{U}, \mathcal{F}^q)$ is a horizontal 4-tuple. The vertical red rectangles represent the componentwise recollection used to form $\mathcal{F}^*(U_{ij})$.

This concept is illustrated by Figure 15.47. If we define the system of coefficients $\mathcal{K}^q(\mathcal{F}^*)$, by

$$\mathcal{K}^q(\mathcal{F}^*) \colon v \mapsto H^q(\mathcal{F}^*(U_v)),$$

where $\mathcal{F}^*(U_v)$ is the complex in $(*_{\mathcal{F}^*(U_v)})$, with the obvious restriction maps, then ${}^{\mathrm{I}}E_1^{p,q} = H_{\mathrm{II}}^q(C_{\mathrm{I}}^p) = C^p(\mathcal{U}, \mathcal{K}^q(\mathcal{F}^*))$, and thus we have the following result.

Proposition 15.28. *The terms ${}^{\mathrm{I}}E_2^{p,q}$ of the first spectral sequence associated with the double complex $C(\mathcal{U}, \mathcal{F}^*)$ are given by*

$${}^{\mathrm{I}}E_2^{p,q} = \check{H}^p(\mathcal{U}, \mathcal{K}^q(\mathcal{F}^*)).$$

Let us now assume that $\mathcal{U}$ is an open cover and that $\mathcal{G}^*$ is the canonical flasque resolution of $\mathcal{F}$. It can be shown that

$$H^q(\mathcal{G}^*(U_v)) \cong H^q(U_v, \mathcal{F}),$$

where $H^q(U_v, \mathcal{F})$ is the qth module of sheaf cohomology of the restriction of the sheaf $\mathcal{F}$ to U_v; see Godement [Godement (1958)] (Chapter 4,

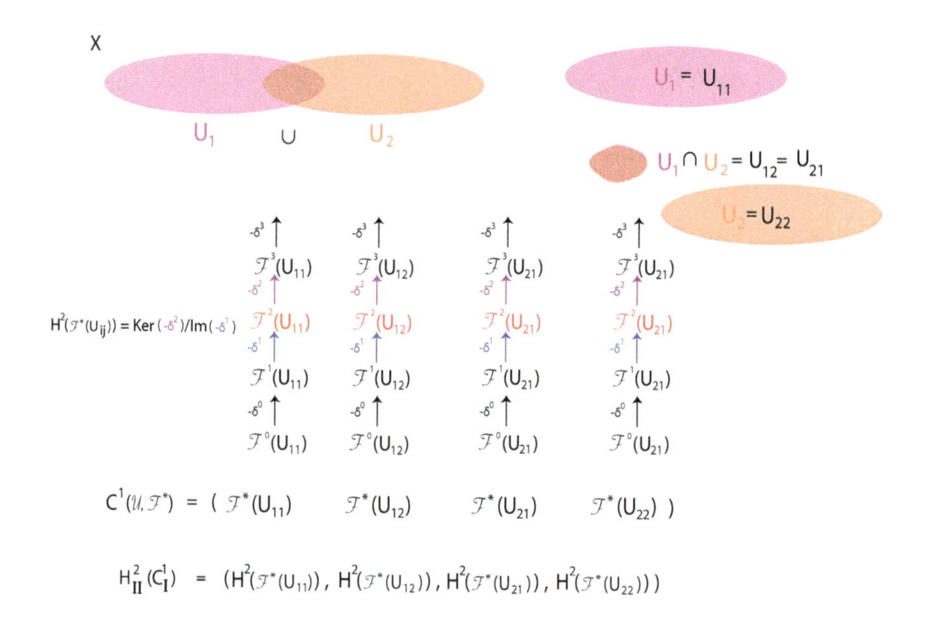

Fig. 15.47 A continuation of Figure 15.46 in which the vertical red rectangles are replaced by vertical cochains. The cohomology is via $(-1)^p\delta^q\colon \mathcal{F}^q \to \mathcal{F}^{q+1}$ as seen by the construction of $H^2_{\text{II}}(C^1_{\text{I}})$.

Lemma 4.9.1). In summary, using Proposition 15.27, we have the following theorem from Godement [Godement (1958)] (Chapter 5, Theorem 5.4.1).

Theorem 15.29. *Let $\mathcal{U}$ be an open cover on a topological space X, and let $\mathcal{F}$ be a sheaf on X. Let $\mathcal{G}^*$ be the canonical flasque resolution of a sheaf $\mathcal{F}$ and let C be the total complex associated with the double complex $C(\mathcal{U}, \mathcal{G}^*)$.*

(1) The second spectral sequence ${}^{\text{II}}E$ degenerates at $r = 2$ and there are isomorphisms

$$H^n(C) \cong H^n(X, \mathcal{F}) \quad \textit{for all } n \geq 0.$$

(2) We can define the system of coefficients $\mathcal{K}^q(\mathcal{F})$ by

$$\mathcal{K}^q(\mathcal{F})(v) = H^q(U_v, \mathcal{F}),$$

for all open subsets U_v, with $v = (i_0, \ldots, i_s) \in J^{s+1}$, and for every $q \geq 0$, and the term ${}^{\text{I}}E_2^{p,q}$ of the first spectral sequence ${}^{\text{I}}E$ is given by

$${}^{\text{I}}E_2^{p,q} = \check{H}^p(\mathcal{U}, \mathcal{K}^q(\mathcal{F})).$$

As a corollary of Theorem 15.29, we obtain a result due to Henri Cartan.

Theorem 15.30. *Let $\mathcal{U}$ be an open cover on a topological space X, and let $\mathcal{F}$ be a sheaf on X. If*

$$H^q(U_v, \mathcal{F}) = (0) \quad \textit{for all } q \geq 1$$

and all open subsets U_v, with $v = (i_0, \ldots, i_s) \in J^{s+1}$, then the first spectral sequence also degenerates at $r = 2$ and we have isomorphisms

$$\check{H}^n(\mathcal{U}, \mathcal{F}) \cong H^n(X, \mathcal{F}), \quad n \geq 0.$$

An interesting special case of the double complex $C(\mathcal{U}, \mathcal{F}^*)$ arises when $\mathcal{F}^*$ is the differential sheaf $\mathcal{A}^*$ associated with de Rham cohomology of differential forms, where $\mathcal{A}^p(U)$ consists of the p-differential forms on the open subset U. This is the *Čech–de Rham complex* double complex $\mathcal{A}C^{*,*}(\mathcal{U})$ of Section 9.3, which we now denote $C(\mathcal{U}, \mathcal{A}^*)$, with

$$C(\mathcal{U}, \mathcal{A}^*) = \bigoplus_{p,q \in \mathbb{N}} C^p(\mathcal{U}, \mathcal{A}^q)$$

and

$$C^p(\mathcal{U}, \mathcal{A}^q) = \prod_{v \in J^{p+1}} \mathcal{A}^q(U_v).$$

The pth row C^p_{II} of $C(\mathcal{U}, \mathcal{A}^*)$ is the Čech complex $C(\mathcal{U}, \mathcal{A}^p)$,

$$0 \longrightarrow C^0(\mathcal{U}, \mathcal{A}^p) \longrightarrow C^1(\mathcal{U}, \mathcal{A}^p) \longrightarrow C^2(\mathcal{U}, \mathcal{A}^p) \longrightarrow \cdots,$$

which computes the Čech cohomology modules $\check{H}^q(\mathcal{U}, \mathcal{A}^p)$ of the sheaf $\mathcal{A}^p$. However, by Proposition 8.5 of Bott and Tu [Bott and Tu (1986)], the complex

$$0 \longrightarrow \mathcal{A}^p(X) \longrightarrow C^0(\mathcal{U}, \mathcal{A}^p) \longrightarrow C^1(\mathcal{U}, \mathcal{A}^p) \longrightarrow C^2(\mathcal{U}, \mathcal{A}^p) \longrightarrow \cdots$$

is exact for all $p \geq 0$, and this implies that the complex $C(\mathcal{U}, \mathcal{A}^p)$ is acyclic, so

$$\check{H}^q(\mathcal{U}, \mathcal{A}^p) = (0) \quad \text{for all } q \geq 1 \text{ and all } p \geq 0,$$

and $\check{H}^0(\mathcal{U}, \mathcal{A}^p) \cong \mathcal{A}^p(X)$. Consequently, the complex $\check{H}^0(\mathcal{U}, \mathcal{A}^*)$ is the de Rham complex, and since by Proposition 15.25

$${}^{\mathrm{II}}E_2^{p,q} \cong H^p(\check{H}^q(\mathcal{U}, \mathcal{A}^*)),$$

the second spectral sequence degenerates for $r = 2$ and $q(n) = 0$, we have isomorphisms

$$H^n(C) \cong H^n_{\mathrm{dR}}(X).$$

Observe that we obtained a quick proof of Theorem 9.5 using spectral sequences.

The pth column $C_{\rm I}^p$ of $C(\mathcal{U}, \mathcal{A}^*)$ is the complex $C^p(\mathcal{U}, \mathcal{A}^*)$,

$$0 \longrightarrow C^p(\mathcal{U}, \mathcal{A}^0) \longrightarrow C^p(\mathcal{U}, \mathcal{A}^1) \longrightarrow C^p(\mathcal{U}, \mathcal{A}^2) \longrightarrow \cdots,$$

which computes the cohomology modules $H^q_{\rm II}(C^p(\mathcal{U}, \mathcal{A}^*))$. In general, it is not easy to compute this cohomology, but if $\mathcal{U}$ is a good cover, the open subsets U_v are contractible (with $v \in J^{p+1}, p \geq 0$), so the complex $\mathcal{A}^*(U_v)$ induced by the differential sheaf $\mathcal{A}^*$ on U_v,

$$0 \longrightarrow \mathcal{A}^0(U_v) \longrightarrow \mathcal{A}^1(U_v) \longrightarrow \cdots \longrightarrow \mathcal{A}^q(U_v) \longrightarrow \cdots,$$

is acyclic, and

$$H^q(\mathcal{A}^*(U_v)) = (0) \quad \text{for all } q \geq 1.$$

But then the system of coefficient $\mathcal{K}^q(\mathcal{A}^*)$ given by $\mathcal{K}^q(\mathcal{A}^*)\colon v \mapsto H^q(\mathcal{A}^*(U_v))$ has the property that

$$\mathcal{K}^q(\mathcal{A}^*) = (0) \quad \text{for all } q \geq 1,$$

and

$$\mathcal{K}^0(\mathcal{A}^*)(v) = H^0(\mathcal{A}^*(U_v)) = C^0(U_v, \widetilde{\mathbb{R}}_X),$$

the space of locally constant functions on U_v, where $\widetilde{R}_X$ is constant sheaf of locally constant functions with values in $\mathbb{R}$. By inspection of the definition of $\check{H}^p(\mathcal{U}, \mathcal{K}^0(\mathcal{A}^*))$, we see that

$$\check{H}^p(\mathcal{U}, \mathcal{K}^0(\mathcal{A}^*)) \cong \check{H}^p(\mathcal{U}, \widetilde{\mathbb{R}}_X),$$

the pth Čech cohomology module of $\mathcal{U}$ with coefficients in the constant sheaf $\widetilde{R}_X$ of locally constant functions with values in $\mathbb{R}$. By Proposition 15.28, the first spectral sequence degenerates for $r = 2$ and we have isomorphisms

$$H^n(C) \cong \check{H}^n(\mathcal{U}, \widetilde{\mathbb{R}}_X).$$

Therefore, spectral sequences provide a quick proof of the equivalence of de Rham cohomology and Čech cohomology for a good cover (on a manifold, Theorem 9.4), namely

$$H^n_{\rm dR}(X) \cong \check{H}^n(\mathcal{U}, \widetilde{\mathbb{R}}_X).$$

Many more applications of spectral sequences in topology are described in Bott and Tu [Bott and Tu (1986)] and McCleary [McCleary (2001)].

We now consider Čech cohomology. The reader may want to review Section 9.2.

15.12 Spectral Sequences of Čech Cohomology, II

In Definition 9.8, the Čech cohomology groups $\check{H}^p(X, \mathcal{F})$ with values in $\mathcal{F}$ are defined as the direct limits

$$\check{H}^p(X, \mathcal{F}) = \varinjlim_{\mathcal{U}} \check{H}^p(\mathcal{U}, \mathcal{F})$$

with respect to coverings $\mathcal{U} = (U_i)_{i \in I}$ whose index set I is a subset of 2^X. In order to define our double complex we need to introduce the following complexes.

Definition 15.44. The Čech cochain complexes $\check{C}(X, \mathcal{F})$ are given by the direct limits (of complexes!)

$$\check{C}(X, \mathcal{F}) = \varinjlim_{\mathcal{U}} C(\mathcal{U}, \mathcal{F}),$$

where $C(\mathcal{U}, \mathcal{F})$ is the Čech complex associated with the open cover $\mathcal{U}$ (see just after Definition 15.40 or just after Definition 9.3).

Then it is easy to see that

$$\check{H}^p(X, \mathcal{F}) = H^p(\check{C}(X, \mathcal{F})).$$

Again let us consider the canonical flasque resolution $\mathcal{G}^*$ of the sheaf $\mathcal{F}$. We define the following double complex.

Definition 15.45. Let $\mathcal{G}^*$ be the canonical flasque resolution of a sheaf $\mathcal{F}$. The double complex $\check{C}(X, \mathcal{G}^*)$ is given by

$$\check{C}(X, \mathcal{G}^*) = \bigoplus_{p,q \in \mathbb{N}} \check{C}^p(X, \mathcal{G}^q),$$

where $\check{C}^p(X, \mathcal{G}^q)$ is the pth module in the Čech complex $\check{C}(X, \mathcal{G}^q)$, the qth row of the double complex $\check{C}(X, \mathcal{G}^*)$; see Figure 15.48. We denote the total complex associated with $\check{C}(X, \mathcal{G}^*)$ as C.

The pth row of $\check{C}(X, \mathcal{G}^*)$ is the Čech complex $\check{C}(X, \mathcal{G}^p)$, which computes the Čech cohomology of the sheaf $\mathcal{G}^p$. The pth column of $\check{C}(X, \mathcal{G}^*)$ computes the cohomology of the complex $\check{C}^p(X, \mathcal{G}^*) = \bigoplus_{q \in \mathbb{N}} \check{C}^p(X, \mathcal{G}^q)$, which involves the Čech cochains $\check{C}^p(X, \mathcal{G}^q)$ of the sheaves $\mathcal{G}^q$. The modules $H^q(\check{C}^p(X, \mathcal{G}^*))$ can be expressed in terms of certain presheaves $\mathcal{H}^q(X, -)$ as

$$H^q(\check{C}^p(X, \mathcal{G}^*)) \cong \check{C}^p(X, \mathcal{H}^q(X, \mathcal{F})).$$

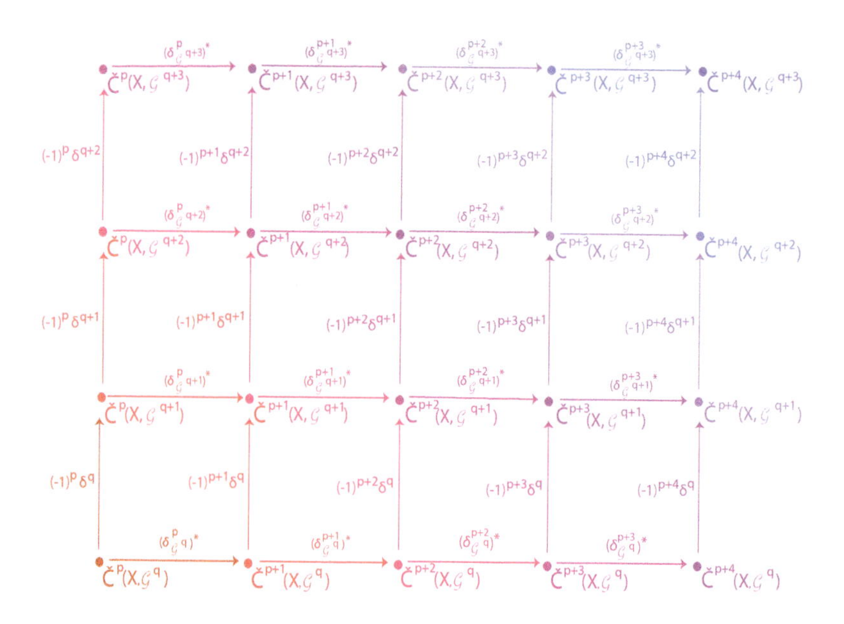

Fig. 15.48 A section of the double complex $\check{C}(X, \mathcal{G}^*) = \bigoplus_{p,q\in\mathbb{N}} \check{C}^p(X, \mathcal{G}^q)$.

Since $\mathcal{G}^*$ is the canonical resolution of $\mathcal{F}$, the second spectral sequence degenerates at $r = 2$. If X is paracompact, the first spectral sequence also degenerates and we have the celebrated isomorphisms

$$\check{H}^n(X, \mathcal{F}) \cong H^n(X, \mathcal{F}), \quad n \geq 0.$$

By mimicking the computation that we made for the second spectral sequence of the double complex $C(\mathcal{U}, \mathcal{F}^*)$, we obtain the following result.

Proposition 15.31. *The terms* ${}^{\mathrm{II}}E_2^{p,q}$ *of the second spectral sequence associated with the double complex* $\check{C}(X, \mathcal{G}^*)$ *are given by*

$${}^{\mathrm{II}}E_2^{p,q} \cong H^p(\check{H}^q(X, \mathcal{G}^*)),$$

where $\check{H}^q(X, \mathcal{G}^*)$ *is the cochain complex*

$$0 \longrightarrow \check{H}^q(X, \mathcal{G}^0) \longrightarrow \check{H}^q(X, \mathcal{G}^1) \longrightarrow \cdots$$
$$\cdots \longrightarrow \check{H}^q(X, \mathcal{G}^p) \longrightarrow \check{H}^q(X, \mathcal{G}^{p+1}) \longrightarrow \cdots.$$

Since $\mathcal{G}^*$ consists of flasque sheaves and since we are considering open covers, by passing to the limit, Proposition 15.26(a) implies that the complexes $\check{H}^q(X, \mathcal{G}^*)$ are trivial for all $q \geq 1$ (since $\check{H}^q(X, \mathcal{G}^p) = (0)$ for all

$p \geq 0$ and all $q \geq 1$). It follows that the second spectral sequence degenerates and that we have isomorphisms

$$\begin{aligned} H^n(C) &\cong {}^{\mathrm{II}}E_2^{n,0} \cong H^n(\check{H}^0(X, \mathcal{G}^*)) \\ &\cong H^n(\Gamma(\mathcal{G}^*)) = H^n(X, \mathcal{F}), \text{ for all } n \geq 0, \end{aligned}$$

since $\check{H}^0(X, \mathcal{G}^p) \cong \Gamma(X, \mathcal{G}^p)$ and since the canonical flasque resolution $\mathcal{G}^*$ compute sheaf cohomology.

Proposition 15.32. *Let $\mathcal{G}^*$ be the canonical flasque resolution of the sheaf $\mathcal{F}$. The second spectral sequence associated with the double complex $\check{C}(X, \mathcal{G}^*)$ degenerates at $r = 2$ and we have isomorphisms*

$$H^n(C) \cong {}^{\mathrm{II}}E_2^{n,0} \cong H^n(X, \mathcal{F}), \quad \textit{for all } n \geq 0.$$

It remains to determine ${}^{\mathrm{I}}E_2^{p,q}$.

By mimicking the computation that we made for the first spectral sequence of the double complex $C(\mathcal{U}, \mathcal{F}^*)$, but using the canonical resolution $\mathcal{G}^*$ of $\mathcal{F}$, we obtain

$$C_{\mathrm{I}}^p = \bigoplus_{q \in \mathbb{N}} \check{C}^p(X, \mathcal{G}^q) = \check{C}^p(X, \mathcal{G}^*),$$

and so

$${}^{\mathrm{I}}E_1^{p,q} \cong H^q(C_{\mathrm{I}}^p) = H^p(\check{C}^p(X, \mathcal{G}^*)).$$

But the functor $\mathcal{F} \mapsto \check{C}^p(X, \mathcal{F})$ is exact *on presheaves*, as shown in Godement [Godement (1958)] (Chapter 5, Theorem 5.8.1). Then if we define the *presheaves* $\mathcal{H}^q(X, \mathcal{F})$ given by

$$\mathcal{H}^q(X, \mathcal{F})(U) = H^q(\mathcal{G}^*(U)) = H^q(U, \mathcal{F}),$$

using an adaptation of Proposition 15.19, we have

$${}^{\mathrm{I}}E_1^{p,q} = H^q(\check{C}^p(X, \mathcal{G}^*)) \cong \check{C}^p(X, \mathcal{H}^q(X, \mathcal{F}))$$

and

$${}^{\mathrm{I}}E_2^{p,q} \cong \check{H}^p(X, \mathcal{H}^q(X, \mathcal{F})).$$

Beware that $\mathcal{H}^q(X, \mathcal{F})$ is not the sheaf cohomology module $\mathcal{H}^q(\mathcal{F}^*)$ of the differential sheaf $\mathcal{F}^*$ from Definition 15.37.

In summary we have the following theorem from Godement [Godement (1958)] (Chapter 5, Theorem 5.9.1).

Theorem 15.33. *Let $\mathcal{F}$ be a sheaf on a topological space X, let $\mathcal{G}^*$ be the canonical flasque resolution of $\mathcal{F}$, and let $\check{C}(X,\mathcal{G}^*)$ be the double complex $\check{C}(X,\mathcal{G}^*)$ with associated total complex C.*

(1) The second spectral sequence ${}^{\mathrm{II}}E$ degenerates at $r = 2$ and there are isomorphisms

$$H^n(C) \cong H^n(X,\mathcal{F}), \quad \text{for all } n \geq 0.$$

(2) We can define the presheaf $\mathcal{H}^q(X,\mathcal{F})$ given by

$$\mathcal{H}^q(X,\mathcal{F})(U) = H^q(U,\mathcal{F})$$

for every open subset U of X, for all $q \geq 0$, and the term ${}^{\mathrm{I}}E_2^{p,q}$ of the first spectral sequence ${}^{\mathrm{I}}E$ is given by

$${}^{\mathrm{I}}E_2^{p,q} \cong \check{H}^p(X,\mathcal{H}^q(X,\mathcal{F})).$$

Observe that since $\mathcal{H}^0(X,\mathcal{F}) \cong \mathcal{F}$, we have

$${}^{\mathrm{I}}E_2^{n,0} \cong \check{H}^n(X,\mathcal{F}).$$

Now a fundamental property of the presheaves $\mathcal{H}^q(X,\mathcal{F})$ is this.

Proposition 15.34. *The sheafification of the presheaf $\mathcal{H}^q(X,\mathcal{F})$ is the zero sheaf if $q \geq 1$.*

This fact holds because $\mathcal{H}^q(X,\mathcal{F})(U) = H^q(\mathcal{G}^*(U))$ and $\mathcal{G}^*$ is the canonical flasque resolution, so locally, every cocycle of $\mathcal{G}^*$ is a coboundary; see Godement [Godement (1958)] (Chapter 5, Page 227).

As a consequence, the first spectral sequence degenerates at $r = 2$ if we can find conditions on the space X so that

$$\check{H}^n(X,\mathcal{G}) = (0) \quad \text{for all } n \geq 0$$

for any presheaf $\mathcal{G}$ whose sheafification is the zero sheaf. This is the case if X is paracompact.

Proposition 15.35. *Let X be a topological space and let $\mathcal{G}$ be a presheaf on X. If X is paracompact and if the sheafification of $\mathcal{G}$ is the zero sheaf, then*

$$\check{H}^n(X,\mathcal{G}) = (0) \quad \text{for all } n \geq 0.$$

Proposition 15.35 is proven in Godement [Godement (1958)] (Chapter 5, Theorem 5.10,2). As a consequence, if X is paracompact, since the sheafification of the presheaves $\mathcal{H}^q(X, \mathcal{F})$ is zero for all $q \geq 1$, we see that

$$^{\mathrm{I}}E_2^{p,q} \cong \check{H}^p(X, \mathcal{H}^q(X, \mathcal{F})) = (0) \quad \text{for all } p \geq 0 \text{ and all } q \geq 1,$$

which means that the spectral sequence $^{\mathrm{I}}E$ degenerates for $r = 2$, and we have isomorphisms

$$H^n(C) \cong {}^{\mathrm{I}}E_2^{n,0} \cong \check{H}^n(X, \mathcal{H}^0(X, \mathcal{F})) = \check{H}^n(X, \mathcal{F}),$$

since by definition, $\mathcal{H}^0(X, \mathcal{F})(U) = H^0(U, \mathcal{F}) \cong \mathcal{F}(U)$. Since we also have isomorphisms

$$H^p(C) \cong H^n(X, \mathcal{F}) \quad \text{for all } n \geq 0.$$

We obtain the following important isomorphism theorem from Godement [Godement (1958)] (Chapter 5, Theorem 5.10.1).

Theorem 15.36. *Let $\mathcal{F}$ be a sheaf on a topological space X, let $\mathcal{G}^*$ be the canonical flasque resolution of $\mathcal{F}$, and let $\check{C}(X, \mathcal{G}^*)$ be the double complex $\check{C}(X, \mathcal{G}^*)$ with associated total complex C. If X is paracompact, then the first spectral sequence $^{\mathrm{I}}E$ degenerates at $r = 2$ and there are isomorphisms*

$$\check{H}^p(X, \mathcal{F}) \cong H^n(X, \mathcal{F}) \quad \textit{for all } n \geq 0.$$

Note that Theorem 15.36 is identical to Theorem 13.17(1), and we obtained another proof using spectral sequences.

Even if X is not paracompact, the spectral sequences associated with the complex $\check{C}(X, \mathcal{G}^*)$ yield interesting results. We will need the following results.

Proposition 15.37. *Let (F^pC) be a canonically cobounded filtration on a graded complex C. There is an exact sequence (called an edge sequence)*

$$0 \longrightarrow E_2^{1,0} \longrightarrow H^1(C) \longrightarrow E_2^{0,1} \longrightarrow E_2^{2,0} \longrightarrow H^2(C).$$

Proposition 15.37 is proven in Theorem 15.54 and in Godement [Godement (1958)] (Chapter 4, Theorem 4.5.1).

Proposition 15.38. *If a presheaf $\mathcal{G}$ has the zero sheaf as sheafification, then*

$$\check{C}^0(X, \mathcal{F}) = (0) \quad \textit{and} \quad \check{H}^0(X, \mathcal{F}) = (0).$$

Proposition 15.38 is proven in Godement [Godement (1958)] (Chapter 5, lemma after Theorem 5.9.1). Using these results we obtain the following theorem.

Theorem 15.39. *Let $\mathcal{F}$ be a sheaf on a topological space X. There maps*

$$\check{H}^n(X, \mathcal{F}) \cong H(X, \mathcal{F})$$

that are isomorphisms for $n = 0, 1$, and an injection for $n = 2$.

Proof. The case $n = 0$ is trivial (we have $\Gamma(X, \mathcal{F})$ in both cases). For $n = 1, 2$ we use the exact sequence

$$0 \longrightarrow E_2^{1,0} \longrightarrow H^1(C) \longrightarrow E_2^{0,1} \longrightarrow E_2^{2,0} \longrightarrow H^2(C)$$

of Proposition 15.38, where C is the total complex associated with the double complex $\check{C}(X, \mathcal{G}^*)$ and $E = {}^{\mathrm{I}}E$. Since

$${}^{\mathrm{I}}E_2^{p,q} \cong \check{H}^p(X, \mathcal{H}^q(X, \mathcal{F})),$$

and since the sheafification of the presheaf $\mathcal{H}^1(X, \mathcal{F}))$ is the zero sheaf, we have

$$\begin{aligned} {}^{\mathrm{I}}E_2^{0,1} &\cong \check{H}^0(X, \mathcal{H}^1(X, \mathcal{F})) = (0) \\ {}^{\mathrm{I}}E_2^{1,0} &\cong \check{H}^1(X, \mathcal{H}^0(X, \mathcal{F})) \cong \check{H}^1(X, \mathcal{F}) \\ {}^{\mathrm{I}}E_2^{2,0} &\cong \check{H}^2(X, \mathcal{H}^0(X, \mathcal{F})) \cong \check{H}^2(X, \mathcal{F}), \end{aligned}$$

so we have the exact sequence

$$0 \longrightarrow \check{H}^1(X, \mathcal{F}) \longrightarrow H^1(C) \longrightarrow 0 \longrightarrow \check{H}^2(X, \mathcal{F}) \longrightarrow H^2(C),$$

and since $H^n(C) \cong H^n(X, \mathcal{F})$ for all $n \geq 0$, the proposition follows. ☐

Theorem 15.33 can also be used to prove a theorem of Henri Cartan that we stated earlier (Theorem 13.19). The proof uses induction.

Theorem 15.40. *(H. Cartan) For any topological space X and any sheaf $\mathcal{F}$ on X, for any open cover $\mathcal{U}$, if $\mathcal{U}$ is a basis for the topology of X closed under finite intersections and if $\check{H}^p(U_{i_0 \cdots i_p}, \mathcal{F}) = (0)$ for all $p > 0$ and all $(i_0, \ldots, i_p)$, then*

$$\check{H}^p(X, \mathcal{F}) \cong H^p(X, \mathcal{F}), \quad \textit{for all} \ \ p \geq 0.$$

A few more interesting results can be found in Godement [Godement (1958)] (Chapter 5).

15.13 Grothendieck's Spectral Sequences of Composed Functors ⊛

This section and the next three deal with somewhat more advanced material and are optional. Before reading this section the reader may want to review Section 11.4, in particular, Definition 11.17.

Another source of double complexes appears as the answer to the following.

Problem. Given two left-exact functors $F\colon \mathcal{A} \to \mathcal{B}$ and $G\colon \mathcal{B} \to \mathcal{C}$ between abelian categories (with enough injectives, *etc.*), we have $GF\colon \mathcal{A} \to \mathcal{C}$ (left-exact); how can we compute $R^n(GF)$ if we know R^pF and R^qG?

In order to answer this question we need to introduce special kinds of injective resolutions of complexes.

Definition 15.46. A *Cartan–Eilenberg injective resolution* of a complex $C^\bullet$ (with $C^k = (0)$ if $k < 0$) is a resolution

$$0 \longrightarrow C^\bullet \longrightarrow Q^{\bullet\, 0} \longrightarrow Q^{\bullet\, 1} \longrightarrow Q^{\bullet\, 2} \longrightarrow \cdots,$$

in which each $Q^{\bullet\, j} = \bigoplus_i Q^{i,j}$ is a complex (differential d^{ij}) and every $Q^{i,j}$ is injective, so that if we write $Z^{i,j} = \mathrm{Ker}\ d^{i,j}$, $B^{i,j} = \mathrm{Im}\ d^{i-1,j}$ and $H^{i,j} = Z^{i,j}/B^{i,j}$, then we have the injective resolutions

$$(1) \qquad 0 \longrightarrow C^i \longrightarrow Q^{i,0} \longrightarrow Q^{i,1} \longrightarrow \cdots$$

$$(2) \qquad 0 \longrightarrow Z^i(C) \longrightarrow Z^{i,0} \longrightarrow Z^{i,1} \longrightarrow \cdots$$

$$(3) \qquad 0 \longrightarrow B^i(C) \longrightarrow B^{i,0} \longrightarrow B^{i,1} \longrightarrow \cdots$$

$$(4) \qquad 0 \longrightarrow H^i(C) \longrightarrow H^{i,0} \longrightarrow H^{i,1} \longrightarrow \cdots.$$

The way to remember this complicated definition is through the diagram shown in Figure 15.49.

Note that due to the exigencies of notation (we resolved our complex $C^\bullet$ horizontally) the usual conventions of horizontal and vertical were interchanged. The rows are resolutions and the columns are complexes. The same switch of rows and columns occurs in the proof of Proposition 15.41, at least as far as Cartesian coordinate notation is concerned.

Proposition 15.41. *Every complex C has a Cartan–Eilenberg resolution $0 \longrightarrow C \longrightarrow Q^\bullet$, where the $\{Q^{i,j}\}$ form a double complex. Here we have suppressed the grading indices of C and the Q^j.*

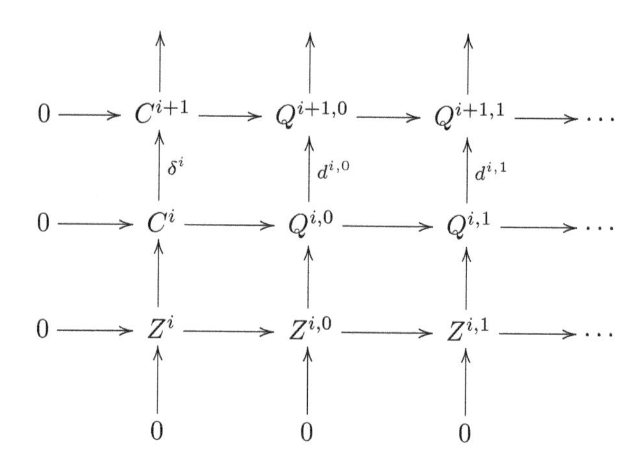

Fig. 15.49 Diagram associated with Definition 15.46.

Proof. We begin with injective resolutions $0 \longrightarrow B^0(C) \longrightarrow B^{0,\bullet}$; $0 \longrightarrow B^1(C) \longrightarrow B^{1,\bullet}$ and $0 \longrightarrow H^0(C) \longrightarrow H^{0,\bullet}$ of $B^0(C)$; $B^1(C)$; $H^0(C)$. Now we have exact sequences

$$0 \longrightarrow B^0(C) \longrightarrow Z^0(C) \longrightarrow H^0(C) \longrightarrow 0$$

and

$$0 \longrightarrow Z^0(C) \longrightarrow C^0 \xrightarrow{d^0} B^1(C) \longrightarrow 0;$$

so by Proposition 11.25, we get injective resolutions $0 \longrightarrow Z^0(C) \longrightarrow Z^{0,\bullet}$ and $0 \longrightarrow C^0 \longrightarrow Q^{0,\bullet}$, so that

$$0 \longrightarrow B^{0,\bullet} \longrightarrow Z^{0,\bullet} \longrightarrow H^{0,\bullet} \longrightarrow 0$$

and

$$0 \longrightarrow Z^{0,\bullet} \longrightarrow Q^{0,\bullet} \longrightarrow B^{1,\bullet} \longrightarrow 0$$

are exact.

For the induction step, assume that the complexes $B^{i-1,\bullet}$, $Z^{i-1,\bullet}$, $H^{i-1,\bullet}$, $Q^{i-1,\bullet}$ and $B^{i,\bullet}$ are determined and satisfy the required exactness properties ($i \geq 1$). Pick any injective resolution $H^{i,\bullet}$ of $H^i(C)$, then using the exact sequence

$$0 \longrightarrow B^i(C) \longrightarrow Z^i(C) \longrightarrow H^i(C) \longrightarrow 0$$

and Proposition 11.25, we get an injective resolution $0 \longrightarrow Z^i(C) \longrightarrow Z^{i,\bullet}$ so that

$$0 \longrightarrow B^{i,\bullet} \longrightarrow Z^{i,\bullet} \longrightarrow H^{i,\bullet} \longrightarrow 0 \qquad (*_1)$$

is exact. Next pick an injective resolution, $0 \longrightarrow B^{i+1}(C) \longrightarrow B^{i+1,\bullet}$, of $B^{i+1}(C)$ and use the exact sequence

$$0 \longrightarrow Z^i(C) \longrightarrow C^i \xrightarrow{d^i} B^{i+1}(C) \longrightarrow 0$$

and Proposition 11.25 to get an injective resolution $0 \longrightarrow C^i \longrightarrow Q^{i,\bullet}$ so that

$$0 \longrightarrow Z^{i,\bullet} \longrightarrow Q^{i,\bullet} \longrightarrow B^{i+1,\bullet} \longrightarrow 0 \qquad (*_2)$$

is exact. The differential $d_{\mathrm{II}}^{i,j}$ of the double complex $\{Q^{i,j}\}$ is the composition

$$Q^{i,j} \longrightarrow B^{i+1,j} \longrightarrow Z^{i+1,j} \longrightarrow Q^{i+1,j}$$

and the differential $d_{\mathrm{I}}^{i,j}$ is given by $d_{\mathrm{I}}^{i,j} = (-1)^i \epsilon^{i,j}$, where, $\epsilon^{i,\bullet}$ is the differential of $Q^{i,\bullet}$. The reader should check that $\{Q^{i,j}\}$ is indeed a Cartan–Eilenberg resolution and a double complex. □

In the next theorem we will resolve our Cartan–Eilenberg resolutions vertically. This has the effect of transposing the indices in Definition 15.46 and returning to the usual indexing conventions regarding the horizontal and vertical directions. This theorem provides the answer to the problem posed at the beginning of this section and has many applications throughout the field of homological algebra.

Theorem 15.42. *(Grothendieck) Let $F\colon \mathcal{A} \to \mathcal{B}$ and $G\colon \mathcal{B} \to \mathcal{C}$ be two left-exact functors between abelian categories (with enough injectives, etc.) and suppose that $F(Q)$ is G-acyclic whenever Q is injective, which means that $R^pG(FQ) = (0)$ if $p > 0$. Then for every object $A \in \mathcal{A}$, there is a double complex C of composed functors whose total complex computes the cohomology $H^n(C) \cong R^n(GF)(A)$ for all $n \geq 0$, and for the second spectral sequence, we have*

$$^{\mathrm{II}}E_2^{p,q} \cong R^pG((R^qF)(A)).$$

Proof. Pick some object $A \in \mathcal{A}$ and resolve it by injectives to obtain the resolution $0 \longrightarrow A \longrightarrow Q^\bullet(A) : 0 \longrightarrow A \longrightarrow Q^0 \longrightarrow Q^1 \longrightarrow Q^2 \longrightarrow \cdots$.

If we apply GF to $Q^\bullet(A)$ and compute cohomology, we get $R^n(GF)(A)$. If we just apply F to $Q^\bullet(A)$, we get the complex:

$$F(Q^0) \longrightarrow F(Q^1) \longrightarrow F(Q^2) \longrightarrow \cdots, \qquad (FQ^\bullet(A))$$

whose cohomology is $R^qF(A)$. Now resolve the complex $FQ^\bullet(A)$ in the vertical direction by a Cartan–Eilenberg resolution. There results a double complex of injectives (with exact columns)

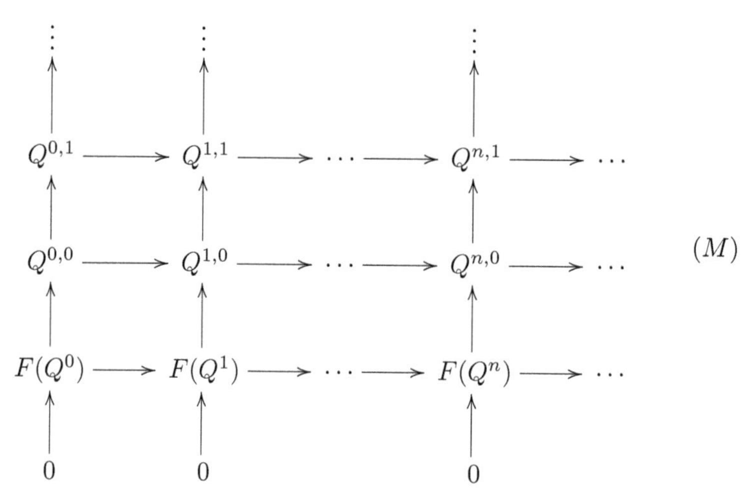

in the category $\mathcal{B}$. The rows are complexes and the columns are deleted injective resolutions. Apply the left-exact functor G to the double complex M to obtain a new double complex GM, denoted as C:

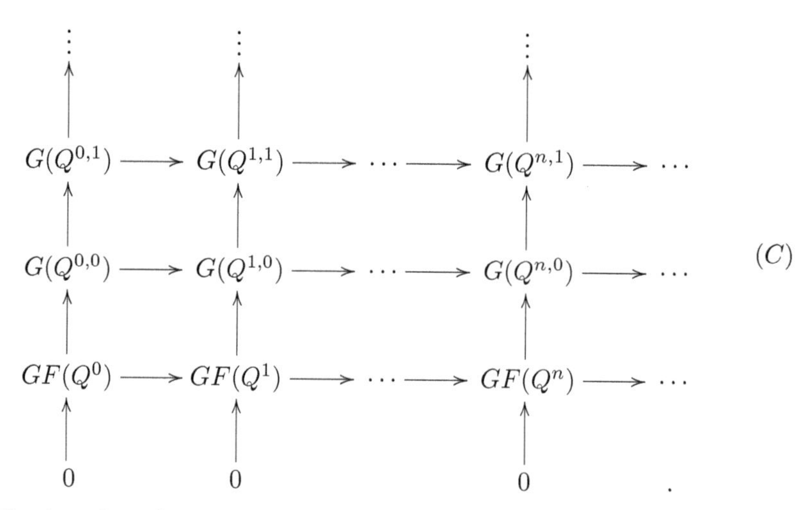

For fixed p, the pth column $M^{p,\bullet}$ of (M) is a deleted injective resolution of $F(Q^p)$ so $G(M^{p,\bullet})$ is a complex whose cohomology is R^qG, and we have

$$H_{\mathrm{II}}^{p,q}(C) = H^q(GM^{p,\bullet}) = R^qG(FQ^p).$$

But since Q^p is injective, FQ^p is G-acyclic, so $R^qG(FQ^p) = (0)$ for $q > 0$. Since G is left-exact, $R^0G = G$, and we deduce that

$$H_{\mathrm{II}}^{p,q}(C) = \begin{cases} GF(Q^p) & \text{if } q = 0 \\ (0) & \text{if } q > 0. \end{cases}$$

The only surviving terms are on the p-axis:

$$0 \longrightarrow GF(Q^0) \longrightarrow GF(Q^1) \longrightarrow GF(Q^2) \longrightarrow \cdots.$$

Applying H_{I}^p, we get

$${}^{\mathrm{I}}E_2^{p,q} = \begin{cases} R^p(GF(A)) & \text{if } q = 0 \\ (0) & \text{if } q > 0. \end{cases}$$

Therefore, the first spectral sequence degenerates for $r = 2$ and $q(n) = 0$, and we get the isomorphism

$$H^n(C) \cong R^n(GF)(A).$$

We now turn to the second spectral sequence. Since we used a Cartan–Eilenberg resolution of $FQ^\bullet(A)$, we have the following injective resolutions

$$\begin{aligned} 0 &\longrightarrow Z^p(FQ^\bullet(A)) \longrightarrow Z'^{p,0} \longrightarrow Z'^{p,1} \longrightarrow \cdots \\ 0 &\longrightarrow B^p(FQ^\bullet(A)) \longrightarrow B'^{p,0} \longrightarrow B'^{p,1} \longrightarrow \cdots \\ 0 &\longrightarrow H^p(FQ^\bullet(A)) \longrightarrow H'^{p,0} \longrightarrow H'^{p,1} \longrightarrow \cdots, \end{aligned}$$

for all $p \geq 0$. Moreover, after swapping the indices p and q in $(*_1)$ and $(*_2)$, the exact sequences

$$0 \longrightarrow Z'^{q,p} \longrightarrow Q^{q,p} \xrightarrow{d'^{q,p}} B'^{q+1,p} \longrightarrow 0$$

and

$$0 \longrightarrow B'^{q,p} \longrightarrow Z'^{q,p} \longrightarrow H'^{q,p} \longrightarrow 0$$

are split because the terms are injectives of $\mathcal{B}$. Therefore, the sequences

$$0 \longrightarrow G(Z'^{q,p}) \longrightarrow G(Q^{q,p}) \xrightarrow{Gd'^{q,p}} G(B'^{q+1,p}) \longrightarrow 0 \qquad (\dagger_1)$$

and

$$0 \longrightarrow G(B'^{q,p}) \longrightarrow G(Z'^{q,p}) \longrightarrow G(H'^{q,p}) \longrightarrow 0 \qquad (\dagger_2)$$

are still exact. From the exact sequences $(\dagger_1)$ we obtain

$$\mathrm{Ker}\,(Gd'^{q,p}) \cong G(Z'^{q,p}), \quad \mathrm{Im}\,(Gd'^{q-1,p}) \cong G(B'^{q,p}),$$

so

$$H^q_{\mathrm{I}}(C^{\bullet,p}) = \mathrm{Ker}\,(Gd'^{q,p})/\mathrm{Im}\,(Gd'^{q-1,p}) \cong G(Z'^{q,p})/G(B'^{q,p}),$$

and from the exact sequences ($\dagger_2$),

$$G(H'^{q,p}) \cong G(Z'^{q,p})/G(B'^{q,p}),$$

so we obtain

$$H^q_{\mathrm{I}}(C^{\bullet,p}) \cong G(H'^{q,p}).$$

But the $H'^{q,\bullet}$ form an injective resolution of $H^q(FQ^\bullet(A))$ and the latter is just $R^qF(A)$. So $G(H'^{q,\bullet})$ is the complex whose pth cohomology module is exactly $R^pG(R^qF(A))$. Now this cohomology module is $H^p_{\mathrm{II}}(G(H'^{q,\bullet}))$ and $H^q_{\mathrm{I}}(C^{\bullet,\bullet})$ is $G(H'^{q,\bullet})$ by the above. We obtain

$$R^pG(R^qF(A)) = H^p_{\mathrm{II}}(H^q_{\mathrm{I}}(C^{\bullet,\bullet})) = {}^{\mathrm{II}}E_2^{p,q}.$$

Since we previously found (through the calculation of the first spectral sequence) that $H^n(C) \cong R^n(GF)(A)$, we are done. $\square$

15.14 Exact Couples ⊛

Having presented spectral sequences using the Cartan–Serre method, we are now in the position to understand and appreciate Massey's approach in terms of exact couples (which are really quintuples!).

Definition 15.47. An *exact couple* is a quintuple $\mathcal{C} = (D, E, i, j, k)$ where D and E are R-modules and $i\colon D \to D$, $j\colon D \to E$ and $k\colon E \to D$ are morphisms such that the pairs (i, j), (j, k) and (k, i) are exact, that is,

$$\mathrm{Im}\, i = \mathrm{Ker}\, j, \quad \mathrm{Im}\, j = \mathrm{Ker}\, k, \quad \mathrm{Im}\, k = \mathrm{Ker}\, i.$$

A convenient way to present the above data is the triangle below:

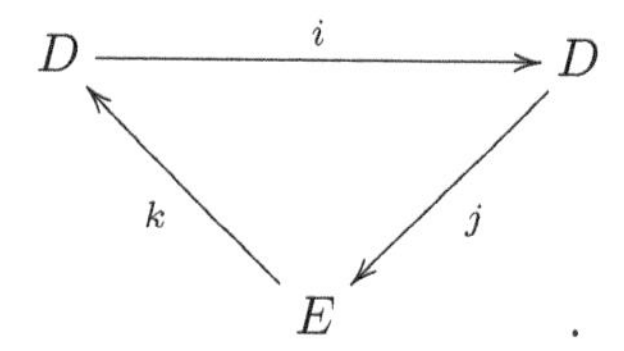

In most applications, E and D will be bigraded modules, which means that

$$D = \bigoplus_{p,q\in\mathbb{Z}} D^{p,q}, \quad E = \bigoplus_{p,q\in\mathbb{Z}} E^{p,q},$$

and the maps i, j, k are graded maps of bidegrees $(-1, 1)$, $(0, 0)$ and $(1, 0)$, which means that

$$i\colon D^{p,q} \to D^{p-1,q+1}, \quad j\colon D^{p,q} \to E^{p,q}, \quad k\colon E^{p,q} \to D^{p+1,q}.$$

More generally, j will have bidegree $(r-1, -(r-1))$ with $r \geq 1$. An exact couple of bigraded modules can be depicted as follows:

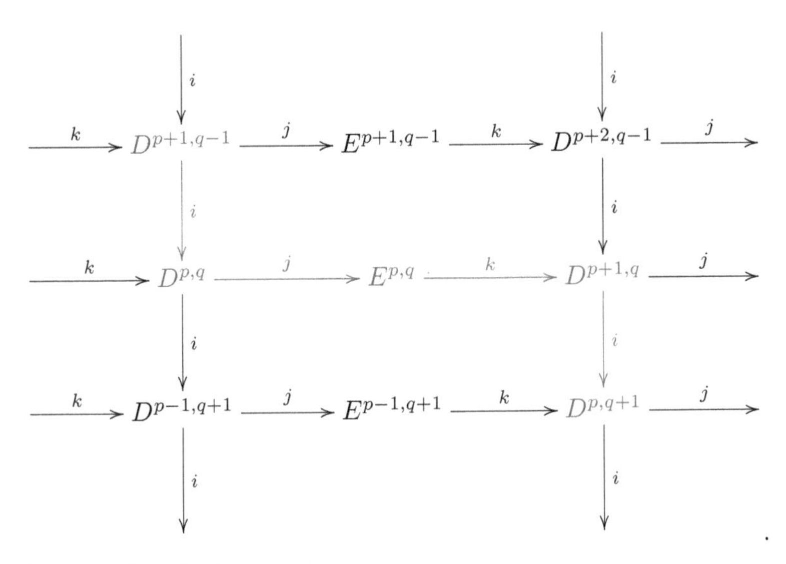

Observe that the blue path is exact.

The connection with the other definitions of a spectral sequence is that a filtration on a graded module induces an exact couple.

Example 15.7. Let C be a graded complex and let (F^pC) be a filtration of C compatible with the grading. For every p there is an exact sequence

$$0 \longrightarrow F^{p+1}C \longrightarrow F^pC \longrightarrow F^pC/F^{p+1}C \longrightarrow 0. \qquad (*_1)$$

By the zig-zag lemma, we obtain the exact sequence

$$H^{p+q}(F^{p+1}C) \xrightarrow{i^{p+1,q-1}} H^{p+q}(F^pC) \xrightarrow{j^{p,q}} H^{p+q}(F^pC/F^{p+1}C)$$
$$(*_2)$$
$$\xrightarrow{k^{p,q}} H^{p+q+1}(F^{p+1}C) \xrightarrow{i^{p+1,q}} H^{p+q+1}(F^pC).$$

If we let

$$D^{p,q} = H^{p+q}(F^pC), \quad E^{p,q} = H^{p+q}(F^pC/F^{p+1}C),$$

and

$$D = \bigoplus_{p,q\in\mathbb{Z}} D^{p,q},\ E = \bigoplus_{p,q\in\mathbb{Z}} E^{p,q},\ i = \bigoplus_{p,q\in\mathbb{Z}} i^{p,q},\ j = \bigoplus_{p,q\in\mathbb{Z}} j^{p,q},\ k = \bigoplus_{p,q\in\mathbb{Z}} k^{p,q},$$

we immediately check that we obtain an exact couple.

Observe that $E^{p,q} = E_1^{p,q} \cong H^{p+q}(F^pC/F^{p+1}C)$, the term of rank one of a spectral sequence. This should be a clue that we might be able to use exact couples to define the other terms $E_r^{p,q}$. Indeed this can be done though the introduction of the derived couple of an exact couple.

The first step is to observe that exactness of the pair (j, k) implies that $k \circ j = 0$, so if we define

$$d = j \circ k \quad \text{(note that } k \text{ comes before } j\text{)},$$

then we have a map $d\colon E \to E$ such that $d \circ d = 0$, since $d \circ d = (j \circ k) \circ (j \circ k) = j \circ (k \circ j) \circ k = 0$. Note that if D and E are bigraded modules, since j has bidegree $(r-1, -(r-1))$ and k has bidegree $(1, 0)$, $d = j \circ k$ has bidegree $(r, -(r-1))$.

The above suggests defining a new exact couple whose node E is replaced by $H(E)$. We also replace D by $i(D)$. The new maps are defined below.

Definition 15.48. Given an exact couple $\mathcal{C}$

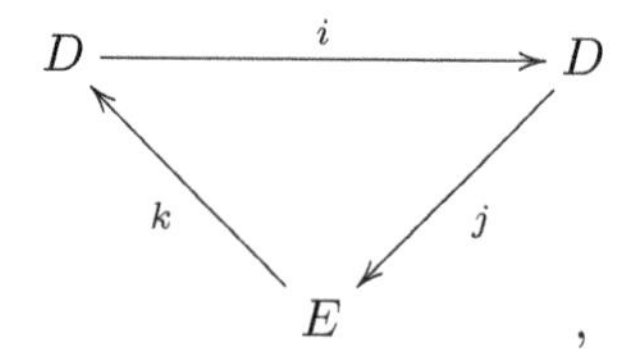

,

the *derived couple* $\mathcal{C}'$

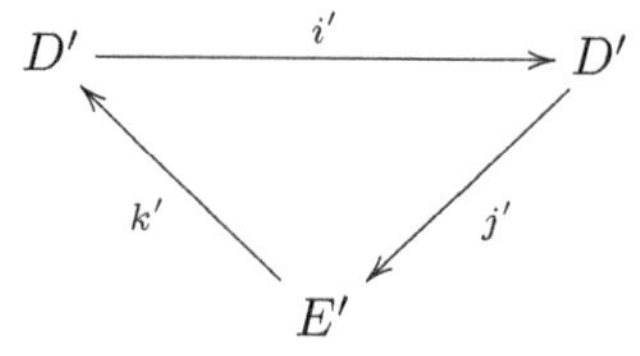

is defined as follows: $D' = i(D)$, $E' = H(E)$ (the cohomology of (E, d)), i' is the restriction of i to $i(D)$, and

$$j'(i(x)) = [j(x)], \quad x \in D$$
$$k'([e]) = k(e), \quad e \in E,$$

where $[j(x)]$ is the cohomology class of the cocycle $j(x) \in E$ and $[e]$ is the cohomology class of the cocycle $e \in E$. Note that

$$i'(i(x)) = i(i(x)), \quad x \in D.$$

We need to check that j' and k' are well-defined. Let us begin with j'. Since (j, k) is exact, $k \circ j = 0$, so we have

$$d(j(x)) = (j \circ k)(j(x)) = j((k \circ j)(x)) = 0,$$

so $j(x)$ is a cocycle.

If $i(x) = i(x')$ for some $x, x' \in D$, then $i(x - x') = 0$, so $x - x' \in \operatorname{Ker} i = \operatorname{Im} k$ (since (k, i) is exact), so there is some $y \in E$ such that $x - x' = k(y)$, and so

$$j(x - x') = j(k(y)) = d(y),$$

that is $j(x) = j(x') + d(y)$, which means that $[j(x)] = [j(x')]$. Thus j' is well-defined.

Next we check that k' is well-defined. Let e be a cocycle in E. This means that $de = 0$, that is, $(j \circ k)(e) = 0$, and since (i, j) is exact, there some $x \in D$ such that $i(x) = k(e)$, so $k'([e]) = k(e) \in i(D)$. If $[e] = [e']$ for any two cocycles $e, e' \in D$, then $e = e' + d(y)$ for some $y \in E$, so

$$k(e') = k(e + d(y)) = k(e) + k((j \circ k)(y)) = k(e) + 0 = k(e),$$

since $k \circ j = 0$ by exactness of (j, k). Consequently, k' is also well-defined.

Actually, the derived couple of an exact couple is an exact couple.

Proposition 15.43. *Given an exact couple $\mathcal{C}$*

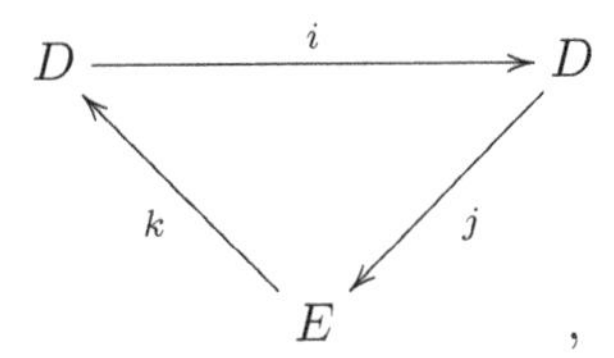

its derived couple $\mathcal{C}'$

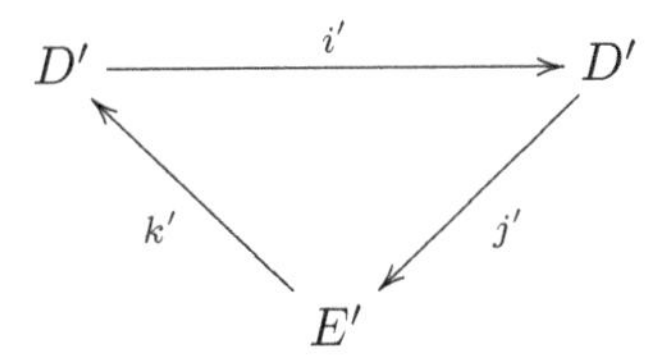

is also an exact couple.

We leave the proof that the pairs (i', j'), (j', k') and (k', i') are exact as an exercise. For help, consult McCleary [McCleary (2001)] (Chapter 2, Proposition 2.7).

The module $E' = H(E) = \mathrm{Ker}\, d/\mathrm{Im}\, d$ can be described is a way that makes it easier to understand what is the result of iterating the procedure of constructing exact couples. Indeed, since $d = j \circ k$, we have

$$\mathrm{Ker}\, d = \mathrm{Ker}\,(j \circ k) = \{x \in E \mid k(x) \in \mathrm{Ker}\, j\} = k^{-1}(\mathrm{Ker}\, j) = k^{-1}(\mathrm{Im}\, i).$$

We also have

$$\mathrm{Im}(j \circ k) = j(\mathrm{Im}\, k) = j(\mathrm{Ker}\, i).$$

Therefore, if we let

$$d^{p,q} = j^{p+1,q} \circ k^{p,q},$$

we obtain

$$\begin{aligned} E'^{p,q} &= \mathrm{Ker}\, d^{p,q}/\mathrm{Im}\, d^{p-1,q} \\ &= k^{-1}(\mathrm{Im}\, i\colon D^{p+2,q-1} \to D^{p+1,q})/j(\mathrm{Ker}\, i\colon D^{p,q} \to D^{p-1,q+1}). \quad (\dagger_1) \end{aligned}$$

The above formula provides a way of understanding graphically how $E'^{p,q}$ is constructed using the diagram

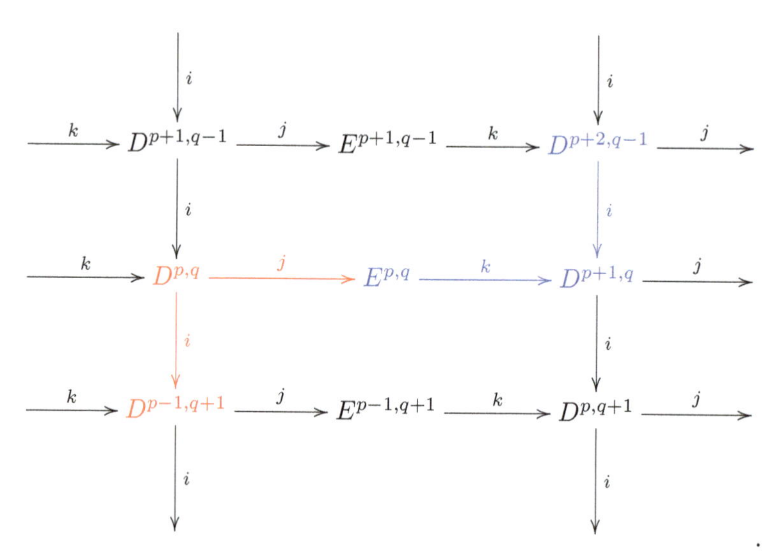

The numerator of $E'^{p,q}$ is obtained by pulling back along k (in blue) the image of the blue map i, and the denominator is obtained by pushing forward along j (in red) the kernel of the red map i.

It is now fairly obvious that a spectral sequence is obtained by iterating the construction of a derived couple.

Given an exact couple $\mathcal{C}$, we define the *nth derived couple* $\mathcal{C}^{(n)}$ inductively as follows: for $n \in \mathbb{N}$,

$$\begin{aligned}\mathcal{C}^{(0)} &= \mathcal{C}\\ \mathcal{C}^{(n+1)} &= (\mathcal{C}^{(n)})'.\end{aligned}$$

The following result can be shown.

Proposition 15.44. *Given a bigraded exact couple $\mathcal{C} = (D, E, i, j, k)$,*

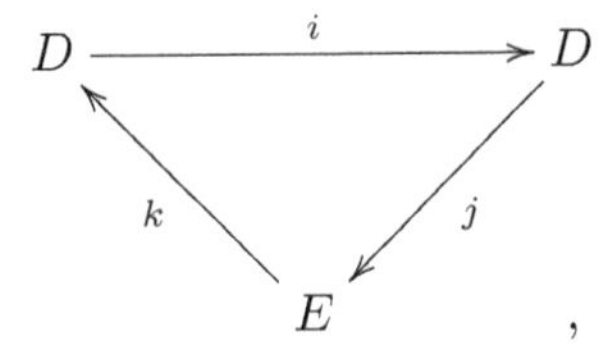

where the maps i, j, k are graded maps of bidegrees $(-1, 1)$, $(0, 0)$ and $(1, 0)$, the sequence of derived exact couples $\mathcal{E}^{(r)} = (D'_r, E'_r, i^{(r)}, j^{(r)}, k^{(r)}) = \mathcal{C}^{(r-1)}$, with $r \geq 1$, determines a spectral sequence (E'_r, d_r), with

$$d_r = j^{(r)} \circ k^{(r)},$$

and $d_r^{p,q}$ has bidegree $(r, -(r-1))$.

Proposition 15.44 is not hard to prove. By construction $E'_{r+1} \cong H(E'_r)$ so what remains to be checked is that the maps d_r have the correct bidegrees; see McCleary [McCleary (2001)] (Chapter 2, Theorem 2.8).

It is possible to describe the $E'^{p,q}_r$ in terms of some modules $Z'^{p,q}_r$ and $B'^{p,q}_r$ analogous (but not equal) to those used in the proof of Theorem 15.7. Using induction and $(\dagger_1)$, we find that for $r \geq 2$,

$$\begin{aligned}Z'^{p,q}_r &= k^{-1}(\mathrm{Im}\, i^{r-1} \colon D^{p+r,q-r+1} \to D^{p+1,q})\\ B'^{p,q}_r &= j(\mathrm{Ker}\, i^{r-1} \colon D^{p,q} \to D^{p-r+1,q+r-1})\\ E'^{p,q}_r &= Z'^{p,q}_r / B'^{p,q}_r.\end{aligned}$$

See McCleary [McCleary (2001)] (Chapter 2, Proposition 2.9). Again there is a geometric interpretation of $Z'^{p,q}_r$ and $B'^{p,q}_r$ in terms of our diagram (thanks to Mac Lane [Mac Lane (1975)], Page 338):

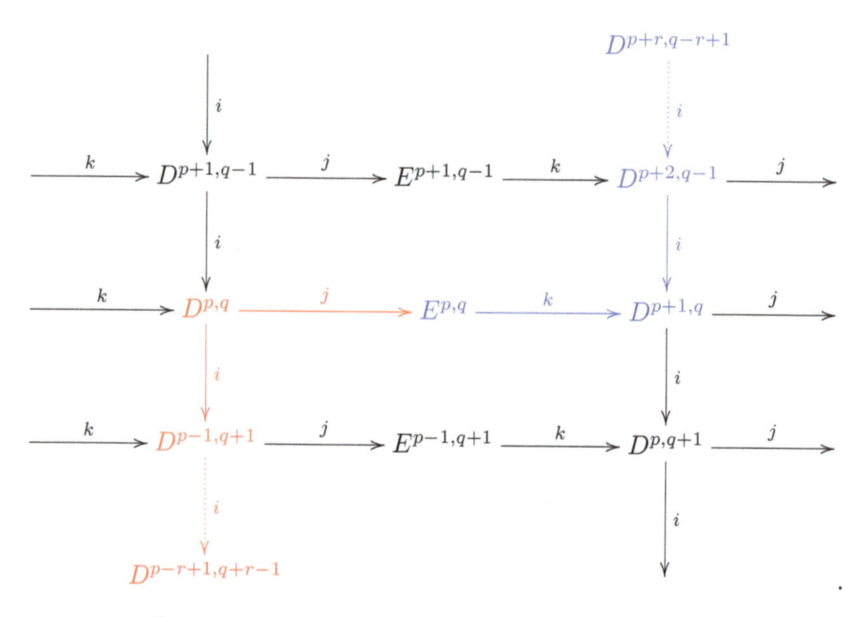

The term $Z'^{p,q}_r$ is obtained by pulling back along k (in blue) the image of the blue map i^{r-1}, and the term $B'^{p,q}_r$ is obtained by pushing forward along j (in red) the kernel of the red map i^{r-1}. It can also be shown that

$$E'^{p,q}_\infty = \left(\bigcap_{r\geq 1} Z'^{p,q}_r\right) \Big/ \left(\bigcup_{r\geq 1} B'^{p,q}_r\right).$$

Finally, it can be shown that if we start we the exact couple of Example 15.7 we obtain the *same spectral* sequence produced by Theorem 15.7.

Theorem 15.45. *Let C be a graded complex and let (F^pC) be a filtration of C compatible with the grading. If $\mathcal{C}$ is the exact couple defined in Example 15.7 by the exact sequence $(*_1)$, then the spectral sequence defined by the derived exact couples $\mathcal{E}^{(r)}$ $(r \geq 1)$ is isomorphic with the spectral sequence produced by Theorem 15.7.*

Theorem 15.45 is proven in McCleary [McCleary (2001)] (Chapter 2, Theorem 2.11) and its homological version is proven in Mac Lane [Mac Lane (1975)] (Chapter XI, Theorem 5.4). McCleary's proof contains an error but Mac Lane's proof is immediately transposed to the cohomological case. The key to the proof is that it can be shown that

$$\begin{aligned}
Z'^{p,q}_r &= k^{-1}(\operatorname{Im} i^{r-1}\colon D^{p+r,q-r+1} \to D^{p+1,q}) \\
&= (Z^{p,q}_r + (C^{p+q} \cap F^{p+1}C))/(C^{p+q} \cap F^{p+1}C) \\
B'^{p,q}_r &= j(\operatorname{Ker} i^{r-1}\colon D^{p,q} \to D^{p-r+1,q+r-1}) \\
&= (B^{p,q}_{r-1} + (C^{p+q} \cap F^{p+1}C))/(C^{p+q} \cap F^{p+1}C),
\end{aligned}$$

where $Z^{p,q}$ and $B^{p,q}_{r-1}$ are defined as in Definition 15.19. The error in McCleary is that it is claimed that $Z'^{p,q}_r = Z^{p,q}_r/(C^{p+q} \cap F^{p+1}C)$ and $B'^{p,q}_r = B^{p,q}_{r-1}/(C^{p+q} \cap F^{p+1}C)$, but the denominator is not contained in the numerator. Since

$$Z^{p+1,q-1}_{r-1} = Z^{p,q}_r \cap (C^{p+q} \cap F^{p+1}C),$$

as in Mac Lane we can use the modular Noether isomorphism (Proposition 15.3), and we have

$$\begin{aligned}
E'^{p,q}_r &= Z'^{p,q}_r / B'^{p,q}_r \\
&= k^{-1}(\operatorname{Im} i^{r-1}\colon D^{p+r,q-r+1} \to D^{p+1,q}) \\
&\quad /j(\operatorname{Ker} i^{r-1}\colon D^{p,q} \to D^{p-r+1,q+r-1}) \\
&= (Z^{p,q}_r + (C^{p+q} \cap F^{p+1}C))/(C^{p+q} \cap F^{p+1}C)) \\
&\quad /(B^{p,q}_{r-1} + (C^{p+q} \cap F^{p+1}C))/(C^{p+q} \cap F^{p+1}C)) \\
&\cong (Z^{p,q}_r + (C^{p+q} \cap F^{p+1}C))/(B^{p,q}_{r-1} + (C^{p+q} \cap F^{p+1}C)) \\
&\cong Z^{p,q}_r/(B^{p,q}_{r-1} + Z^{p,q}_r \cap (C^{p+q} \cap F^{p+1}C)) \\
&= Z^{p,q}_r/(B^{p,q}_{r-1} + Z^{p+1,q-1}_{r-1}) = E^{p,q}_r,
\end{aligned}$$

where $E^{p,q}_r$ is defined in Definition 15.21.

The definition of an exact couple and of a derived exact couple makes sense without any grading on D and E. It is also possible to define homological spectral sequences. This is achieved by changing the bidegree of the maps i, j, k so that they become $(1, -1), (0, 0)$, and $(-1, 0)$. Homological spectral sequences are discussed in Rotman [Rotman (1979, 2009)] and Mac Lane [Mac Lane (1975)].

Examples of exact couples that do not arise from filtrations are discussed in Mac Lane [Mac Lane (1975)] and McCleary [McCleary (2001)]. Among those are the *Bockstein exact couples* and another one arising from tensor products.

15.15 Construction of a Spectral Sequence; Cartan–Eilenberg ⊛

In this section we present Cartan and Eilenberg's method for defining spectral sequences [Cartan and Eilenberg (1956)] (Chapter XV). The proof due to Steve Shatz is essentially Cartan and Eilenberg's proof, except that filtered and graded complexes are considered right away rather than considering the simpler case of a filtered ungraded module first.

The theory of spectral sequences can be developed in an abelian category $\mathcal{A}$. We denote the objects of the category $\mathcal{A}$ by $\mathcal{O}b(\mathcal{A})$. In all the applications that we will consider, $\mathcal{A}$ is either the category of R-modules, the category of presheaves of R-modules, or the category of sheaves or R-modules on a topological space, so the reader may safely assume that $\mathcal{A}$ is one of these categories. We only discuss spectral sequences in the category of R-modules, leaving the adaptation to the more general case of an abelian category to the reader. One simply has to assume that we use objects, subobjects, quotient objects, and morphisms in the abelian category $\mathcal{A}$.

To simplify matters we present the construction of spectral sequences for cohomology cochain complexes and positive filtrations, which means that $C^n = (0)$ for all $n < 0$ (C is a cohomology cochain complex) and $F^pC = C$ for all $p \leq 0$ (the filtration is positive). It is convenient to assume that $F^{-\infty}C = C$ and that $F^{\infty}C = (0)$. The fact that $C^n = (0)$ for all $n < 0$ implies that $E_r^{p,q} = (0)$ for all $p + q < 0$ and the positivity of the filtration implies that $E_r^{p,q} = (0)$ for $p < 0$. In other words, we consider the first quadrant and the region of the fourth quadrant on or above the line $p + q = 0$. For the reader's convenience we give the definition of a spectral sequence in this situation.

Definition 15.49. A *spectral sequence* is a sequence

$$(\langle E_r, E_\infty, d_r, \alpha_r \rangle)_{r\in\mathbb{N}},$$

where

(1) Each E_r is a bigraded R-module with

$$E_r = \bigoplus_{p\in\mathbb{N}} E_r^p \quad \text{and} \quad E_r^p = \bigoplus_{p+q\geq 0, q\in\mathbb{Z}} E_r^{p,q}, \quad p \in \mathbb{N},$$

for $r \in \mathbb{N} \cup \{\infty\}$ (the subscript r is called the *level*).

(2) For all $r \in \mathbb{N}$, the graded module $E_r = \bigoplus_{p\in\mathbb{N}} E_r^p$ is a complex with differential d_r of degree r, where the restriction $d_r^{p,q}$ of d_r to $E_r^{p,q}$ is a map $d_r^{p,q}\colon E_r^{p,q} \to E_r^{p+r,q-r+1}$ such that $d_r^{p,q} \circ d_r^{p-r,q+r-1} = 0$, for all $p \in \mathbb{N}$ and all $q \in \mathbb{Z}$ such that $p+q \geq 0$. The restriction d_r^p of d_r to E_r^p is the map $d_r^p = \bigoplus_{p+q\geq 0} d_r^{p,q}$, with $d_r^p\colon E_r^p \to E_r^{p+r}$.

(3) There is an isomorphism
$$\alpha_r\colon H(E_r) \to E_{r+1}$$
for all $r \in \mathbb{N}$, and more precisely,
$$H^p(E_r) = \mathrm{Ker}\, d_r^p / \mathrm{Im}\, d_r^{p-r} \cong E_{r+1}^p.$$
If we write $H^{p,q}(E_r) = \mathrm{Ker}\, d_r^{p,q}/\mathrm{Im}\, d_r^{p-r,q+r-1}$ $(p \in \mathbb{N}, q \in \mathbb{Z}, p+q \geq 0)$, then $H(E_r)$ is bigraded, with
$$H(E_r) = \bigoplus_{p\in\mathbb{N}} H^p(E_r), \quad H^n(E_r) = \bigoplus_{p\in\mathbb{N}} H^{p,n-p}(E_r), \quad n \in \mathbb{N},$$
and we have an isomorphism
$$H^{p,n-p}(E_r) \cong E_{r+1}^{p,n-p}.$$

A *first quadrant spectral sequence* is a spectral sequence for which $E_r^{p,q} = (0)$ for all $q < 0$ and all $r \in \mathbb{N} \cup \{\infty\}$ (recall that in our definition, $p \in \mathbb{N}$).

Recall that the notation $r >> 0$ is an abbreviation for r is "large enough."

Remarks:

(1) In some sources the whole definition is denoted in the compact form
$$E_2^{p,q} \underset{p}{\Longrightarrow} H(C) \quad \text{or} \quad E_2^{p,q} \Longrightarrow H(C),$$
and $H(C)$ is called the *end* of the spectral sequence. In Mac Lane [Mac Lane (1975)] and Rotman [Rotman (1979, 2009)] the above notation implies that
$$E_\infty^{p,q} \cong \mathrm{gr}(H(C))^{p,q} = H(C)^{p,q}/H(C)^{p+1,q-1}.$$
The index p is called the *filtration index*, $p+q$ is called the *total* or *grading index* and q the *complementary index*.

(2) Assume that the spectral sequence is a first quadrant spectral sequence, which means that $E_r^{p,q} = (0)$ if $q < 0$. Since $d_r^{p,q}\colon E_r^{p,q} \to E_r^{p+r,q-r+1}$, if $r > q+1$, then $\mathrm{Im}\, d_r^{p,q} = (0)$, and if $r > p$, then $\mathrm{Im}\, d_r^{p-r,q+r-1} = (0)$. So if $r > \max\{p, q+1\}$, then $\mathrm{Im}\, d_r^{p,q} = (0)$ and $\mathrm{Im}\, d_r^{p-r,q+r-1} = (0)$ imply that $\mathrm{Ker}\, d_r^{p,q} = E_r^{p,q}$ and $H_r^{p,q}(E) = \mathrm{Ker}\, d_r^{p,q}/\mathrm{Im}\, d_r^{p-r,q+r-1} = E_r^{p,q}/(0) = E_r^{p,q}$. By (3), $E_{r+1}^{p,q} \cong H_r^{p,q}(E) = E_r^{p,q}$, *i.e.*, the sequence of $E_r^{p,q}$ stabilizes for $r >> 0$.

(3) In general, when $E_r^{p,q}$ stabilizes, $E_r^{p,q} \neq E_\infty^{p,q}$. Further assumptions must be made to get $E_r^{p,q} = E_\infty^{p,q}$ for $r >> 0$.

Spectral sequences can be introduced in many ways. The one chosen here leads immediately into applications involving double complexes but is weaker if one passes to triangulated and derived categories. In the existence proof given below there are many complicated diagrams and indices. We urge you to read as far as the definition of $Z_r^{p,q}$ and $B_r^{p,q}$ (one half page) and skip the rest of the proof on a first reading.

Theorem 15.46. *Let C be a filtered cohomological complex whose filtration is positive and compatible with its grading and differentiation. Then $H(C)$ possesses a filtration (and is graded) and there exists a spectral sequence*

$$E_2^{p,q} \underset{p}{\Longrightarrow} H(C)$$

with the following properties:

(1) We have $E_1^{p,q} \cong H(\mathrm{gr}(C))$ — so that

$$E_1^{p,q} = H(\mathrm{gr}(C))^{p,q} = H^{p+q}(F^pC/F^{p+1}C).$$

(2) There are isomorphisms

$$E_\infty^{p,q} \cong \mathrm{gr}(H(C))^{p,q} = \mathrm{gr}(H^{p+q}(C))^p = H(C)^{p,q}/H(C)^{p+1,q-1}.$$

(3) If the filtration is regular, the objects $\mathrm{gr}(H^{p+q}(C))^p = H(C)^{p,q}/H(C)^{p+1,q-1}$, called the composition factors in the filtration of $H^{p+q}(C)$, are exactly the $E_r^{p,q}$ when $r >> 0$.

In the course of the proof of Theorem 15.46, we shall make heavy use of the following lemma.

Lemma 15.47. *(Lemma (L)) Let*

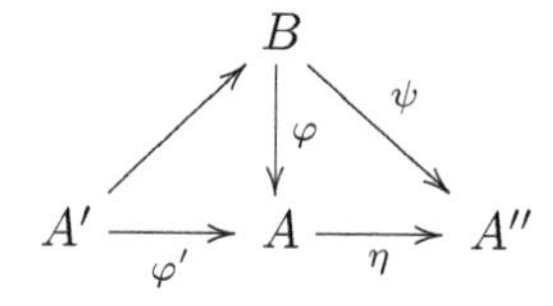

be a commutative diagram with exact bottom row, that is, $\mathrm{Im}\,\varphi' = \mathrm{Ker}\,\eta$. *Then η induces an isomorphism* $\mathrm{Im}\,\varphi/\mathrm{Im}\,\varphi' \cong \mathrm{Im}\,\psi$.

Proof. Since the left triangle commutes, $\mathrm{Im}\ \varphi' \subseteq \mathrm{Im}\ \varphi$, and since $\mathrm{Im}\,\varphi' = \mathrm{Ker}\,\eta$, we have $\mathrm{Ker}\,\eta \subseteq \mathrm{Im}\ \varphi$ and

$$\mathrm{Im}\ \varphi/\mathrm{Im}\ \varphi' = \mathrm{Im}\ \varphi/\mathrm{Ker}\,\eta.$$

By the first isomorphism theorem applied to the restriction of η to $\mathrm{Im}\ \varphi$, we have

$$\mathrm{Im}\ \varphi/\mathrm{Im}\ \varphi' = \mathrm{Im}\ \varphi/\mathrm{Ker}\,\eta \cong \eta(\mathrm{Im}\ \varphi) = \mathrm{Im}\ (\eta \circ \varphi) = \mathrm{Im}\,\psi,$$

establishing our result. ☐

Proof of Theorem 15.46. The proof makes a clever use of the zig-zag lemma (Theorem 2.22) applied to exact sequences of the form

$$0 \longrightarrow F^qC/F^sC \longrightarrow F^pC/F^sC \longrightarrow F^pC/F^qC \longrightarrow 0, \qquad (*)$$

for any p, q, s such that $p < q < s$ (including the case where $s = \infty$, in which case $F^sC = (0)$, or the case where $p < 0$, in which case $F^pC = C$), to get the exact sequence of the bottom row of the diagram in Lemma (L), and of the naturality of the zig-zag lemma (Proposition 2.23) applied to maps of exact sequences as above, to get the diagram in Lemma (L). The sequence $(*)$ is exact because by the third isomorphism theorem,

$$F^pC/F^qC \cong (F^pC/F^sC)/(F^qC/F^sC).$$

This technique is an example of a more general approach discussed in Cartan and Eilenberg [Cartan and Eilenberg (1956)] (Chapter XV, Section 7).

First we need to define $Z_r^{p,q}$ and $B_r^{p,q}$ and set $E_r^{p,q} = Z_r^{p,q}/B_r^{p,q}$.

Consider the exact sequence

$$0 \longrightarrow F^pC \longrightarrow F^{p-r+1}C \longrightarrow F^{p-r+1}C/F^pC \longrightarrow 0.$$

Upon applying cohomology (using the zig-zag lemma, Theorem 2.22), we obtain

$$\cdots \longrightarrow H^{p+q-1}(F^{p-r+1}C) \longrightarrow H^{p+q-1}(F^{p-r+1}C/F^pC) \xrightarrow{\delta^*} H^{p+q}(F^pC) \longrightarrow \cdots$$

There is also the natural map $H^{p+q}(F^pC) \longrightarrow H^{p+q}(F^pC/F^{p+1}C)$ induced by the projection $F^pC \longrightarrow F^pC/F^{p+1}C$. Moreover, we have the projection $F^pC/F^{p+r}C \longrightarrow F^pC/F^{p+1}C$, which induces a map on cohomology

$$H^{p+q}(F^pC/F^{p+r}C) \longrightarrow H^{p+q}(F^pC/F^{p+1}C).$$

Set

$$Z_r^{p,q} = \mathrm{Im}\,(H^{p+q}(F^pC/F^{p+r}C) \longrightarrow H^{p+q}(F^pC/F^{p+1}C))$$
$$B_r^{p,q} = \mathrm{Im}\,(H^{p+q-1}(F^{p-r+1}C/F^pC) \longrightarrow H^{p+q}(F^pC/F^{p+1}C)),$$

the latter map being the composition of δ^* and the map $H^{p+q}(F^pC) \longrightarrow H^{p+q}(F^pC/F^{p+1}C)$ (where $r \geq 1$). When $r = \infty$ (remember, $F^{-\infty}C = C$), we get

$$Z_\infty^{p,q} = \mathrm{Im}\,(H^{p+q}(F^pC) \longrightarrow H^{p+q}(F^pC/F^{p+1}C))$$
$$B_\infty^{p,q} = \mathrm{Im}\,(H^{p+q-1}(C/F^pC) \longrightarrow H^{p+q}(F^pC/F^{p+1}C)).$$

The inclusion $F^{p-r+1}C \subseteq F^{p-r}C$ yields a map $F^{p-r+1}C/F^pC \longrightarrow F^{p-r}C/F^pC$; hence we obtain the inclusion relation $B_r^{p,q} \subseteq B_{r+1}^{p,q}$. In a similar way, the projection $F^pC/F^{p+r+1}C \longrightarrow F^pC/F^{p+r}C$ yields the inclusion $Z_{r+1}^{p,q} \subseteq Z_r^{p,q}$. When $r = \infty$, the coboundary map yields the inclusion $B_\infty^{p,q} \subseteq Z_\infty^{p,q}$ (remember, $F^{-\infty}C = C$ and $F^\infty C = (0)$). Consequently, we can write

$$\cdots \subseteq B_r^{p,q} \subseteq B_{r+1}^{p,q} \subseteq \cdots \subseteq B_\infty^{p,q} \subseteq Z_\infty^{p,q} \subseteq \cdots \subseteq Z_{r+1}^{p,q} \subseteq Z_r^{p,q} \subseteq \cdots.$$

Set

$$E_r^{p,q} = Z_r^{p,q}/B_r^{p,q}, \quad \text{where } 1 \leq r \leq \infty, \text{ and} \quad E_n = H^n(C).$$

Then $E = \bigoplus_n E_n = H(C)$, filtered by the $H(C)^p$, as explained earlier. When $r = 1$, $B_1^{p,q} = (0)$ and

$$Z_1^{p,q} = H^{p+q}(F^pC/F^{p+1}C).$$

We obtain $E_1^{p,q} = H^{p+q}(F^pC/F^{p+1}C) = H(\mathrm{gr}(C))^{p,q}$.

Now we have the following commutative diagram with exact rows

$$\begin{array}{ccccccccc} 0 & \longrightarrow & F^pC & \longrightarrow & C & \longrightarrow & C/F^pC & \longrightarrow & 0 \\ & & \downarrow & & \downarrow & & \downarrow & & \\ 0 & \longrightarrow & F^pC/F^{p+1}C & \longrightarrow & C/F^{p+1}C & \longrightarrow & C/F^pC & \longrightarrow & 0. \end{array} \qquad (\dagger_1)$$

Applying the zig-zag lemma to the bottom row of the commutative diagram above yields the cohomology sequence

$$\cdots \longrightarrow H^{p+q-1}(C/F^pC) \xrightarrow{\delta^*}$$
$$H^{p+q}(F^pC/F^{p+1}C) \longrightarrow H^{p+q}(C/F^{p+1}C) \longrightarrow \cdots$$

and applied to the top row the connecting homomorphism $H^{p+q-1}(C/F^pC) \longrightarrow H^{p+q}(F^pC)$. Consequently, by Proposition 2.23 applied to ($\dagger_1$), we obtain the commutative diagram (with exact bottom row)

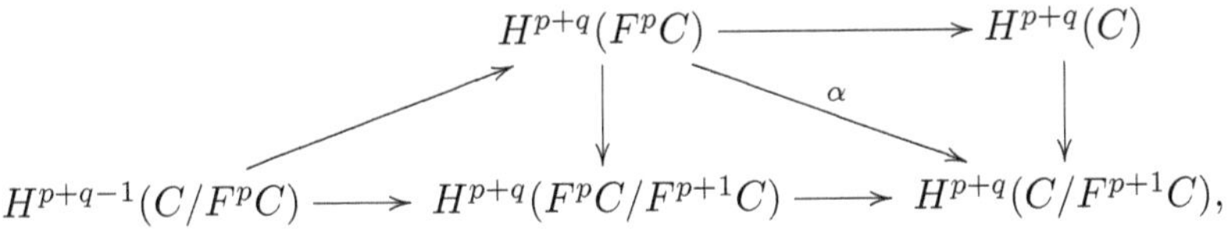

$$\begin{array}{ccccc} & & H^{p+q}(F^pC) & \longrightarrow & H^{p+q}(C) \\ & \nearrow & \downarrow & \overset{\alpha}{\searrow} & \downarrow \\ H^{p+q-1}(C/F^pC) & \longrightarrow & H^{p+q}(F^pC/F^{p+1}C) & \longrightarrow & H^{p+q}(C/F^{p+1}C), \end{array}$$

and Lemma (L) yields an isomorphism

$$\xi^{p,q} \colon E_\infty^{p,q} = Z_\infty^{p,q}/B_\infty^{p,q} \longrightarrow \operatorname{Im}(\alpha \colon H^{p+q}(F^pC) \longrightarrow H^{p+q}(C/F^{p+1}C)).$$

But another application of Lemma (L) to the diagram

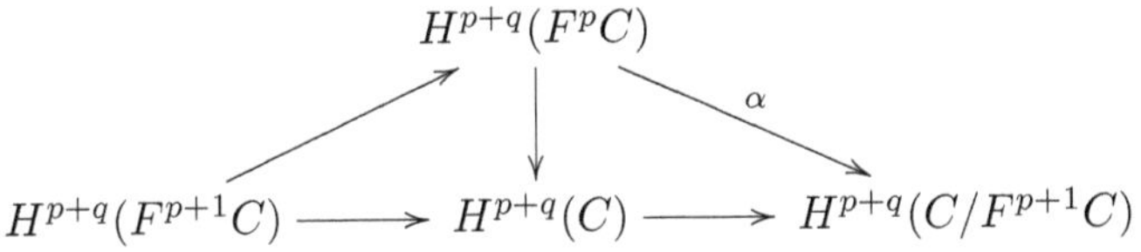

$$\begin{array}{ccccc} & & H^{p+q}(F^pC) & & \\ & \nearrow & \downarrow & \overset{\alpha}{\searrow} & \\ H^{p+q}(F^{p+1}C) & \longrightarrow & H^{p+q}(C) & \longrightarrow & H^{p+q}(C/F^{p+1}C) \end{array}$$

in which the bottom row is exact by the zig-zag lemma applied to the top row of the commutative diagram ($\dagger_1$) gives us the isomorphism

$$\eta^{p,q} \colon \operatorname{gr}(H(C))^{p,q} \longrightarrow \operatorname{Im}(\alpha \colon H^{p+q}(F^pC) \longrightarrow H^{p+q}(C/F^{p+1}C)).$$

Thus, $(\eta^{p,q})^{-1} \circ \xi^{p,q}$ is the isomorphism required by Part (2) of Theorem 15.46.

Only two things remain to be proven to complete the proof of Theorem 15.46. They are the verification of (2) and (3) of Definition 15.49, and the observation that $E_\infty^{p,q}$, as defined above, is the common value of the $E_r^{p,q}$ for $r >> 0$. The verification of (2) and (3) depends upon Lemma (L). Specifically, we have two commutative diagrams shown below. The first diagram arises from the commutative diagram with exact rows

$$\begin{array}{ccccccccc} 0 & \longrightarrow & F^{p+r}C/F^{p+r+1}C & \longrightarrow & F^pC/F^{p+r+1}C & \longrightarrow & F^pC/F^{p+r}C & \longrightarrow & 0 \\ & & \downarrow & & \downarrow & & \downarrow & & \\ 0 & \longrightarrow & F^{p+1}C/F^{p+r+1}C & \longrightarrow & F^pC/F^{p+r+1}C & \longrightarrow & F^pC/F^{p+1}C & \longrightarrow & 0, \end{array} \qquad (\dagger_2)$$

by applying Proposition 2.23 (the naturality of the zig-zag lemma),

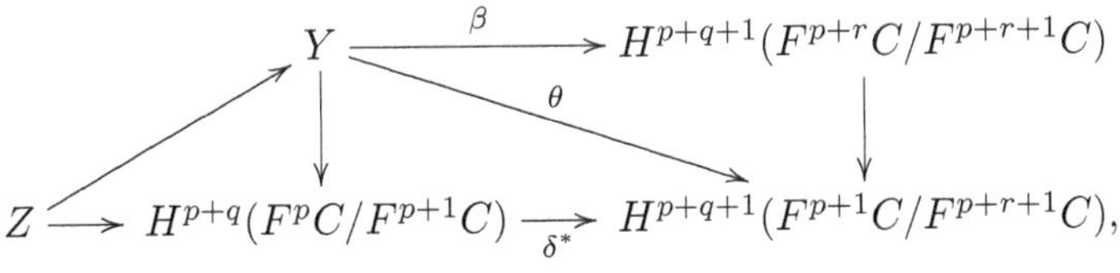

$$\begin{array}{ccccc} & & Y & \overset{\beta}{\longrightarrow} & H^{p+q+1}(F^{p+r}C/F^{p+r+1}C) \\ & \nearrow & \downarrow & \overset{\theta}{\searrow} & \downarrow \\ Z & \longrightarrow & H^{p+q}(F^pC/F^{p+1}C) & \underset{\delta^*}{\longrightarrow} & H^{p+q+1}(F^{p+1}C/F^{p+r+1}C), \end{array} \qquad (1)$$

with $Y = H^{p+q}(F^pC/F^{p+r}C)$ and $Z = H^{p+q}(F^pC/F^{p+r+1}C)$, where the left triangle obviously commutes. The second diagram arises from the commutative diagram with exact rows

$$\begin{array}{ccccccccc} 0 \to & F^{p+r}C/F^{p+r+1}C & \to & F^{p+1}C/F^{p+r+1}C & \to & F^{p+1}C/F^{p+r}C & \to 0 \\ & \downarrow & & \downarrow & & \downarrow & \\ 0 \to & F^{p+r}C/F^{p+r+1}C & \longrightarrow & F^{p}C/F^{p+r+1}C & \longrightarrow & F^{p}C/F^{p+r}C & \to 0, \end{array} \qquad (\dagger_3)$$

by applying Proposition 2.23 (the naturality of the zig-zag lemma),

$$\begin{array}{ccccc} & & Y & & \\ & \nearrow & \downarrow \beta & \searrow \theta & \\ Z & \xrightarrow[\delta^*]{} & H^{p+q+1}(F^{p+r}C/F^{p+r+1}C) & \longrightarrow & H^{p+q+1}(F^{p+1}C/F^{p+r+1}C), \end{array} \qquad (2)$$

with $Y = H^{p+q}(F^pC/F^{p+r}C)$ and $Z = H^{p+q}(F^{p+1}C/F^{p+r}C)$, where the left triangle is obtained from $(\dagger_3)$ by flipping it upside down and the right triangle comes from the upper right triangle in (1), where the map θ is the composition

$$\begin{aligned} H^{p+q}(F^pC/F^{p+r}C) &\longrightarrow H^{p+q+1}(F^{p+r}C/F^{p+r+1}C) \\ &\longrightarrow H^{p+q+1}(F^{p+1}C/F^{p+r+1}C). \end{aligned}$$

Observe that the top row of $(\dagger_2)$ is identical to the bottom row of $(\dagger_3)$, so the map β is indeed the same in (1) and (2). Now Lemma (L) yields the following facts:

$$\begin{aligned} Z_r^{p,q}/Z_{r+1}^{p,q} &\cong \operatorname{Im}\theta, && \text{by (1)} \\ B_{r+1}^{p+r,q-r+1}/B_r^{p+r,q-r+1} &\cong \operatorname{Im}\theta, && \text{by (2)}, \end{aligned}$$

that is,

$$\delta_r^{p,q}\colon Z_r^{p,q}/Z_{r+1}^{p,q} \cong B_{r+1}^{p+r,q-r+1}/B_r^{p+r,q-r+1}.$$

As $B_r^{p,q} \subseteq Z_s^{p,q}$ for every r and s, there is a surjection

$$\pi_r^{p,q}\colon E_r^{p,q} \longrightarrow Z_r^{p,q}/Z_{r+1}^{p,q}$$

with kernel $Z_{r+1}^{p,q}/B_r^{p,q}$, and there exists an injection

$$\sigma_{r+1}^{p+r,q-r+1}\colon B_{r+1}^{p+r,q-r+1}/B_r^{p+r,q-r+1} \longrightarrow E_r^{p+r,q-r+1}.$$

The composition $\sigma_{r+1}^{p+r,q-r+1} \circ \delta_r^{p,q} \circ \pi_r^{p,q}$ is the map $d_r^{p,q}$ from $E_r^{p,q}$ to $E_r^{p+r,q-r+1}$ required by (2). Observe that

$$\operatorname{Im} d_r^{p-r,q+r-1} = B_{r+1}^{p,q}/B_r^{p,q} \subseteq Z_{r+1}^{p,q}/B_r^{p,q} = \operatorname{Ker} d_r^{p,q};$$

hence

$$H(E_r^{p,q}) = \mathrm{Ker}\ d_r^{p,q}/\mathrm{Im}\ d_r^{p-r,q+r-1} \cong Z_{r+1}^{p,q}/B_{r+1}^{p,q} = E_{r+1}^{p,q},$$

as required by (3).

To prove that $E_\infty^{p,q}$ as defined above is the common value of $E_r^{p,q}$ for large enough r, we must make use of the regularity of our filtration. By Proposition 2.23 (the naturality of the zig-zag lemma), the commutative diagram with exact rows

$$\begin{array}{ccccccccc} 0 & \longrightarrow & F^{p+r}C & \longrightarrow & F^pC & \longrightarrow & F^pC/F^{p+r}C & \longrightarrow & 0 \\ & & \downarrow & & \downarrow & & \downarrow & & \\ 0 & \longrightarrow & F^{p+1}C & \longrightarrow & F^pC & \longrightarrow & F^pC/F^{p+1}C & \longrightarrow & 0 \end{array}$$

induces the commutative diagram

$$\begin{array}{ccccc} & & H^{p+q}(F^pC/F^{p+r}C) & \longrightarrow & H^{p+q+1}(F^{p+r}C) \\ & \nearrow & \downarrow & \searrow^{\lambda} & \downarrow \\ H^{p+q}(F^pC) & \longrightarrow & H^{p+q}(F^pC/F^{p+1}C) & \longrightarrow & H^{p+q+1}(F^{p+1}C), \end{array}$$

where λ is the composition

$$H^{p+q}(F^pC/F^{p+r}C) \xrightarrow{\delta^*} H^{p+q+1}(F^{p+r}C) \longrightarrow H^{p+q+1}(F^{p+1}C).$$

By Lemma (L), we have $Z_r^{p,q}/Z_\infty^{p,q} \cong \mathrm{Im}\ \lambda$. However, if $r > \mu(p+q+1)-p$, then $C^{p+q+1} \cap F^{p+r}C = (0)$, so $H^{p+q+1}(F^{p+r}C) = (0)$ (by definition of the grading on $F^{p+r}C$) and δ^* is the zero map. This shows $\mathrm{Im}\ \lambda = (0)$; hence, we have proven

$$Z_r^{p,q} = Z_\infty^{p,q} \quad \text{for } r > \mu(p+q+1)-p.$$

By our assumptions, the filtration begins with $C = F^0C$, therefore if $r > p$ we find $B_r^{p,q} = B_\infty^{p,q}$. Hence, for

$$r > \max\{p, \mu(p+q+1)-p\}$$

the term $E_r^{p,q}$ is equal to $E_\infty^{p,q}$. □

It is easy to verify that if the filtration is canonically cobounded, then we obtain a first quadrant spectral sequence.

Remarks:

(1) Even if our filtration does not start at 0, we can still understand $E_\infty^{p,q}$ from the $E_r^{p,q}$ when the filtration is regular. To see this, note that since cohomology commutes with right limits, we have
$$\varinjlim_r B_r^{p,q} = B_\infty^{p,q},$$
and this implies $\bigcup_r B_r^{p,q} = B_\infty^{p,q}$. Hence, we obtain maps
$$E_r^{p,q} = Z_r^{p,q}/B_r^{p,q} \longrightarrow Z_s^{p,q}/B_s^{p,q} = E_s^{p,q}$$
for $s \geq r > \mu(p+q+1) - p$, and these maps are surjective. (The maps are in fact induced by the $d_r^{p-r,q+r-1}$ because of the equality
$$E_r^{p,q}/\mathrm{Im}\, d_r^{p-r,q+r-1} = (Z_r^{p,q}/B_r^{p,q})/(B_{r+1}^{p,q}/B_r^{p,q}) = E_{r+1}^{p,q}$$
for $r > \mu(p+q+1) - p$.) As in Proposition 15.15 we can show that the right limit of the surjective mapping family
$$E_r^{p,q} \longrightarrow E_{r+1}^{p,q} \longrightarrow \cdots \longrightarrow E_s^{p,q} \longrightarrow \cdots$$
is the group $Z_\infty^{p,q}/(\bigcup B_r^{p,q}) = E_\infty^{p,q}$; so, each element of $E_\infty^{p,q}$ arises from $E_r^{p,q}$ if $r >> 0$ (for fixed p, q). Regularity is therefore still an important condition for spectral sequences that are first and second quadrant or first and fourth quadrant.

(2) The construction of Theorem 15.46 works just as well for an arbitrary graded and filtered complex $C = (C^n)_{n\in\mathbb{Z}}$ with a filtration $(F^pC)_{p\in\mathbb{Z}}$ compatible with the grading.

It is not true in general that $Z_\infty^{p,q} = \bigcap_r Z_r^{p,q}$ or that $\varprojlim_r Z_r^{p,q} = Z_\infty^{p,q}$. In the first case, we have a *weakly convergent* spectral sequence. In the second case, we have a *strongly convergent* spectral sequence.

Although it not obvious at all, the spectral sequences constructed by the Serre–Godement method and the Cartan–Eilenberg method are isomorphic. This is reassuring since otherwise it would be possible for some results to hold for one kind of spectral sequence but not for the other.

Let us denote the Z-terms and B-terms used to define the spectral sequence using Cartan–Eilenberg's method by $Z'^{p,q}_r$ and $B'^{p,q}_r$, and the terms of the Cartan–Eilenberg spectral sequence by $E'^{p,q}_r$. We have
$$\begin{aligned} Z'^{p,q}_r &= \mathrm{Im}\,(H^{p+q}(F^pC/F^{p+r}C) \longrightarrow H^{p+q}(F^pC/F^{p+1}C)) \\ B'^{p,q}_r &= \mathrm{Im}\,(H^{p+q-1}(F^{p-r+1}C/F^pC) \longrightarrow H^{p+q}(F^pC/F^{p+1}C)) \\ E'^{p,q}_r &= Z'^{p,q}_r/B'^{p,q}_r. \end{aligned}$$

Recall from Definition 15.19 that the terms of the Serre–Godement method are given by

$$\begin{aligned} Z_r^{p,q} &= \{x \in C^{p+q} \cap F^pC \mid dx \in C^{p+q+1} \cap F^{p+r}C\} \\ &= \{x \in C^{p,q} \mid dx \in C^{p+r,q-r+1}\} \\ B_r^{p,q} &= \{x \in C^{p+q} \cap F^pC \mid (\exists y \in C^{p+q-1} \cap F^{p-r}C)(x = dy)\} \\ &= \{x \in C^{p,q} \mid (\exists y \in C^{p-r,q+r-1})(x = dy), \end{aligned}$$

and

$$E_r^{p,q} = Z_r^{p,q}/(B_{r-1}^{p,q} + Z_{r-1}^{p+1,q-1}),$$

for all $p, q, r \in \mathbb{Z}$.

Proposition 15.48. *The spectral sequences defined by the Serre–Godement method and the Cartan–Eilenberg method are isomorphic, that is,*

$$E'^{p,q}_r \cong E_r^{p,q}$$

for all $p, q \in \mathbb{Z}$ and $r \geq 1$.

Proof. First we express $\mathrm{Im}\,(H^{p+q}(F^pC/F^{p+r}C) \longrightarrow H^{p+q}(F^pC/F^{p+1}C))$ in terms of $Z_r^{p,q}$. As in Section 15.9 we have the differentials

$$d_{0,r}^{p,q} \colon C^{p,q}/C^{p+r,q-r} \to C^{p,q+1}/C^{p+r,q-r+1}$$

induced by $d \colon C^{p,q} \to C^{p,q+1}$ (actually, the restriction $d_{C^{p,q}}$ of d to $C^{p,q}$), with

$$d_{0,r}^{p,q}([x]_{C^{p+r,q-r}}) = [d(x)]_{C^{p+r,q-r+1}}, \quad x \in C^{p,q}.$$

By definition $H^{p+q}(F^pC/F^{p+r}C) = \mathrm{Ker}\, d_{0,r}^{p,q}/\mathrm{Im}\; d_{0,r}^{p,q-1}$, so an element $\gamma \in H^{p+q}(F^pC/F^{p+r}C)$ is an equivalence class $[c]_{C^{p+r,q-r}}$ modulo $\mathrm{Im}\; d_{0,r}^{p,q-1}$ of some cocycle $[c]_{C^{p+r,q-r}} \in C^{p,q}/C^{p+r,q-r}$, which means that $d_{0,r}^{p,q}([c]) = 0$. By definition of $d_{0,r}^{p,q}$, this means that $c \in C^{p,q}$ and $dc \in C^{p+r,q-r+1}$. But these conditions are equivalent to $c \in Z_r^{p,q}$. In summary we have shown that

$$H^{p+q}(F^pC/F^{p+r}C) = \big((Z_r^{p,q} + C^{p+r,q-r})/C^{p+r,q-r}\big)/\mathrm{Im}\; d_{0,r}^{p,q-1}.$$

The inclusion

$$C^{p,q}/C^{p+r,q-r} \to C^{p,q}/C^{p+1,q-1}$$

induces the map on cohomology in which the equivalence class $[c]_{C^{p+r,q-r}}$ modulo $\mathrm{Im}\; d_{0,r}^{p,q-1}$ with $c \in Z_r^{p,q}$ (an element of $H^{p+q}(F^pC/F^{p+r}C)$) is mapped to the equivalence class $[c]_{C^{p+1,q-1}}$ modulo $\mathrm{Im}\, d_{0,1}^{p,q-1}$ (an element of $H^{p+q}(F^pC/F^{p+1}C)$). It follows that

$$\begin{aligned} Z'^{p,q}_r &= \mathrm{Im}\,(H^{p+q}(F^pC/F^{p+r}C) \longrightarrow H^{p+q}(F^pC/F^{p+1}C)) \\ &= \big((Z_r^{p,q} + C^{p+1,q-1})/C^{p+1,q-1}\big)/\mathrm{Im}\, d_{0,1}^{p,q-1}. \end{aligned} \qquad (*_1)$$

To find $\mathrm{Im}\,(H^{p+q-1}(F^{p-r+1}C/F^pC) \longrightarrow H^{p+q}(F^pC/F^{p+1}C))$ we need to compute the connecting map $\delta\colon H^{p+q-1}(F^{p-r+1}C/F^pC) \to H^{p+q}(F^pC)$ and compose it with the map $H^{p+q}(F^pC) \longrightarrow H^{p+q}(F^pC/F^{p+1}C)$ induced by the projection $F^pC \longrightarrow F^pC/F^{p+1}C$.

We have exact sequences

$$0 \longrightarrow C^{p,q-1} \longrightarrow C^{p-r+1,q+r-2} \longrightarrow C^{p-r+1,q+r-2}/C^{p,q-1} \longrightarrow 0$$

for all q and as in the proof of the zig-zag lemma (Theorem 2.22), the connecting map δ is obtained from the following diagram

$$\begin{array}{ccccccc} & & C^{p-r+1,q+r-2} & \xrightarrow{\pi} & C^{p-r+1,q+r-2}/C^{p,q-1} & \longrightarrow & 0 \\ & & \downarrow{\scriptstyle d} & & & & \\ 0 \longrightarrow C^{p,q} & \xrightarrow{i} & C^{p-r+1,q+r-1} & & & & \end{array}$$

as follows. We start with a cocycle $[c] \in C^{p-r+1,q+r-2}/C^{p,q-1}$, which means that $c \in C^{p-r+1,q+r-2}$ and $dc \in C^{p,q}$. This is equivalent to $c \in Z_{r-1}^{p-r+1,q+r-2}$. Next we pull back $[c]$ with $c \in Z_{r-1}^{p-r+1,q+r-2}$ to $b = c$, push b down to $db = dc$, and finally pull it back to $a = dc \in C^{p,q}$. Since $dZ_{r-1}^{p-r+1,q+r-2} = B_{r-1}^{p,q}$, we see that δ maps the equivalence class $[c]_{C^{p,q-1}}$ modulo $\mathrm{Im}\, d_{0,1}^{p-r+1,q+r-3}$ with $c \in Z_{r-1}^{p-r+1,q+r-2}$ (an element of $H^{p+q-1}(F^{p-r+1}C/F^pC)$) to $dc \in B_{r-1}^{p,q}$ modulo $\mathrm{Im}\, d_{C^{p,q-1}}$ (an element of $H^{p+q}(F^pC)$).

The projection map

$$C^{p,q} \to C^{p,q}/C^{p+1,q-1}$$

induces the map on cohomology defined such that the equivalence class of a cocycle $x \in C^{p,q}$ (which means that $d_{C^{p,q}}x = 0$) modulo $\mathrm{Im}\, d_{C^{p,q-1}}$ (an element of $H^{p+q}(F^pC)$) is mapped to the equivalence class of $[x]_{C^{p+1,q-1}}$ modulo $\mathrm{Im}\, d_{0,1}^{p,q-1}$ (an element of $H^{p+q}(F^pC/F^{p+1}C)$). Composing δ with the above map, we deduce that the equivalence class $[c]_{C^{p,q-1}}$ modulo $\mathrm{Im}\, d_{0,1}^{p-r+1,q+r-3}$ with $c \in Z_{r-1}^{p-r+1,q+r-2}$ (an element of $H^{p+q-1}(F^{p-r+1}C/F^pC)$) is mapped to the equivalence class

$[dc]_{C^{p+1,q-1}}$ modulo Im $d_{0,1}^{p,q-1}$, with $dc \in B_{r-1}^{p,q}$ (an element of the module $H^{p+q}(F^pC/F^{p+1}C)$). It follows that

$$\begin{aligned} B'^{p,q}_r &= \mathrm{Im}\,(H^{p+q-1}(F^{p-r+1}C/F^pC) \longrightarrow H^{p+q}(F^pC/F^{p+1}C)) \\ &= \big((B_{r-1}^{p,q} + C^{p+1,q-1})/C^{p+1,q-1}\big)/\mathrm{Im}\; d_{0,1}^{p,q-1}. \end{aligned} \qquad (*_2)$$

By $(*_1)$ and $(*_2)$, we obtain

$$E'^{p,q}_r = Z'^{p,q}_r/B'^{p,q}_r \cong (Z_r^{p,q} + C^{p+1,q-1})/(B_{r-1}^{p,q} + C^{p+1,q-1}).$$

By the modular Noether isomorphism, since $Z_r^{p,q} \cap C^{p+1,q-1} = Z_{r-1}^{p+1,q-1}$ and $B_{r-1}^{p,q} \subseteq Z_r^{p,q}$, we obtain the isomorphism

$$E'^{p,q}_r \cong Z_r^{p,q}/(B_{r-1}^{p,q} + Z_{r-1}^{p+1,q-1}) = E_r^{p,q},$$

proving the isomorphism between the Cartan–Eilenberg spectral sequence and the Serre–Godement spectral sequence. □

The way the Cartan–Eilenberg definition of the modules $Z'^{p,q}_r$ and $B'^{p,q}_r$ encodes the more explicit definition of the modules $Z_r^{p,q}$ and $B_r^{p,q}$ in the Serre–Godement's approach using maps between carefully chosen cohomology groups is definitely very clever but quite mysterious on a first encounter. This is why we chose to present the Serre–Godement approach first, but we admit that such a choice is a matter of taste.

15.16 More on the Degeneration of Spectral Sequences ⊛

First Proposition 15.17 about spectral sequences that degenerate at level r also holds, and the proof is the same.

To draw further conclusions in situations that occur in practice, we need three technical lemmas. Their proofs should be skipped on a first reading and they are only used to isolate and formalize conditions frequently met in the spectral sequences of applications. We'll label them Lemmas A, B, C as their conclusions are only used to get useful theorems on the sequences.

First observe that if for some r, there are integers n and $p_1 > p_0$ so that $E_r^{\nu,n-\nu} = (0)$ whenever $\nu \neq p_0, \nu \neq p_1$, since $H^{p,q}(E_r) = H(E_r^{p,q}) \cong E_{r+1}^{p,q}$, then $E_s^{\nu,n-\nu} = (0)$ for every s with $r \leq s \leq \infty$. If the filtration is regular, then the filtration of $H^n(C)$ is finite and since $E_\infty^{p_0,n-p_0}$ and $E_\infty^{p_1,n-p_1}$ are the only possible non-zero composition factors, the filtration of $H^n(C)$ is of the form

$$\begin{aligned} H^n(C) = \cdots = H(C)^{p_0,n-p_0} &\supseteq H(C)^{p_0+1,n-p_0-1} \\ &= \cdots = H(C)^{p_1,n-p_1} \supseteq (0). \end{aligned} \qquad (*)$$

By definition

$$\begin{aligned}E_\infty^{p_0,n-p_0} &= H(C)^{p_0,n-p_0}/H(C)^{p_0+1,n-p_0-1} = H^n(C)/H(C)^{p_1,n-p_1}\\ E_\infty^{p_1,n-p_1} &= H(C)^{p_1,n-p_1}/(0) = H(C)^{p_1,n-p_1},\end{aligned}$$

so

$$E_\infty^{p_0,n-p_0} = H^n(C)/E_\infty^{p_1,n-p_1},$$

which is equivalent to the following exact sequence:

$$0 \longrightarrow E^{p_1,n-p_1} \longrightarrow H^n(C) \longrightarrow E^{p_0,n-p_0} \longrightarrow 0. \qquad (\dagger)$$

Lemma 15.49. *(Lemma A) Let $E_2^{p,q} \Longrightarrow H(C)$ be a spectral sequence with a regular filtration. Assume there are integers $r, p_1 > p_0, n$, such that*

$$E_r^{u,v} = (0) \quad \textit{for} \quad \begin{cases} u+v = n, u \neq p_0, p_1\\ u+v = n+1, u \geq p_1 + r\\ u+v = n-1, u \leq p_0 - r;\end{cases}$$

see Figure 15.50. Then there is an exact sequence

$$E_r^{p_1,n-p_1} \longrightarrow H^n(C) \longrightarrow E_r^{p_0,n-p_0}. \qquad (A)$$

Proof. The remarks above and the first hypothesis yield sequence ($\dagger$). In the proof of Theorem 15.46, we saw that

$$\operatorname{Im} d_t^{p_0-t,n-p_0+t-1} = B_{t+1}^{p_0,n-p_0}/B_t^{p_0,n-p_0}.$$

If $r < t$, then $p_0 - r \geq p_0 - t$. Set $u = p_0 - t$ and use the third hypothesis to conclude that

$$E_r^{u,n-1-u} = E_r^{p_0-t,n-1-(p_0-t)} = (0).$$

By definition $d_t^{p_0-t,n-p_0+t-1}\colon E_t^{p_0-t,n-1-(p_0-t)} \to E_t^{p_0,n-p_0}$. But since $E_r^{p_0-t,n-1-(p_0-t)} = (0)$, this map becomes $d_t^{p_0-t,n-p_0+t-1}\colon (0) \to E_t^{p_0,n-p_0}$. So $\operatorname{Im} d_t^{p_0-t,n-p_0+t-1} = (0)$. Since we have

$$\operatorname{Im} d_t^{p_0-t,n-p_0+t-1} = B_{t+1}^{p_0,n-p_0}/B_t^{p_0,n-p_0},$$

we conclude that $B_{t+1}^{p_0,n-p_0} = B_t^{p_0,n-p_0}$. By repeating this argument for $t+1$, we deduce that $B_t^{p_0,n-p_0}$ is constant for $t \geq r$. Since the remark at the end of Section 15.15 implies that $\varinjlim_r B_r^{p,q} = B_\infty^{p,q}$, we can adapt the proof of Proposition 15.15 to deduce that $B_\infty^{p_0,n-p_0} = B_r^{p_0,n-p_0}$. Since

$$\begin{aligned}E_r^{p_0,n-p_0} &= Z_r^{p_0,n-p_0}/B_r^{p_0,n-p_0} = Z_r^{p_0,n-p_0}/B_\infty^{p_0,n-p_0}\\ E_\infty^{p_0,n-p_0} &= Z_\infty^{p_0,n-p_0}/B_\infty^{p_0,n-p_0},\end{aligned}$$

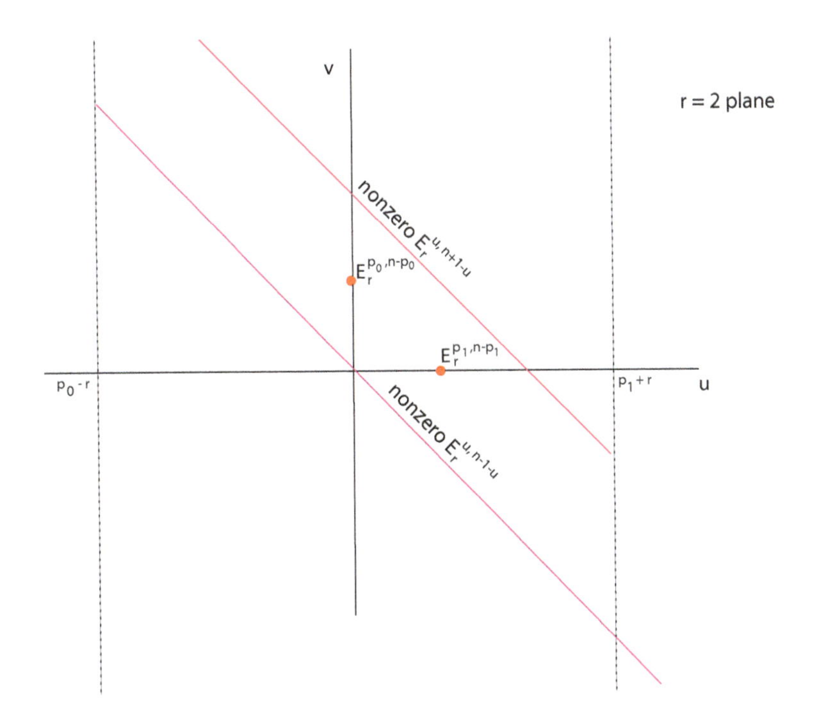

Fig. 15.50 An illustration of the $r = 2$ plane associated with Lemma A when $n = 1$, $p_0 = 0$ and $p_1 = 1$.

and since $Z_\infty^{p_0,n-p_0} \subseteq Z_r^{p_0,n-p_0}$, we obtain an injection $E_\infty^{p_0,n-p_0} \hookrightarrow E_r^{p_0,n-p_0}$.

If $r \le t$, then $p_1 + r \le p_1 + t$. Set $u = p_1 + t$ and use the second hypothesis to conclude that

$$E_r^{u,n+1-u} = E_r^{p_1+t,n+1-(p_1+t)} = (0).$$

By definition $d_t^{p_1+t,n-p_1-t+1} \colon E_t^{p_1+t,n+1-(p_1+t)} \to E_t^{p_1+2t,n-p_1-2t+2}$, which by the above equality becomes $d_t^{p_1+t,n-p_1-t+1} \colon (0) \to E_t^{p_1+2t,n-p_1-2t+2}$. Hence $\mathrm{Ker}\, d_t^{p_1+t,n-p_1-t+1} = (0)$. Since the proof of Theorem 15.46 showed that

$$\mathrm{Ker}\, d_t^{p_1+t,n-p_1-t+1} = Z_{t+1}^{p_1+t,n-p_1-t+1}/B_t^{p_1+t,n-p_1-t+1},$$

we deduce that $Z_{t+1}^{p_1+t,n-p_1-t+1} = B_t^{p_1+t,n-p_1-t+1}$. However we know that

$$\begin{aligned} B_r^{p_1+t,n-p_1-t+1} &\subseteq B_t^{p_1+t,n-p_1-t+1} \subseteq B_\infty^{p_1+t,n-p_1-t+1} \\ &\subseteq Z_\infty^{p_1+t,n-p_1-t+1} \subseteq Z_{t+1}^{p_1+t,n-p_1-t+1}, \end{aligned}$$

which we rewrite as

$$B_r^{p_1+t,n-p_1-t+1} \subseteq B_t^{p_1+t,n-p_1-t+1} \subseteq B_\infty^{p_1+t,n-p_1-t+1} \subseteq Z_\infty^{p_1+t,n-p_1-t+1} \subseteq B_t^{p_1+t,n-p_1-t+1}.$$

From this last string of containments we deduce that

$$B_{t+1}^{p_1+t,n-p_1-t+1} = B_t^{p_1+t,n-p_1-t+1}, \quad r \leq s \leq \infty.$$

Since the proof of Theorem 15.46 also provides the identity

$$Z_t^{p_1,n-p_1}/Z_{t+1}^{p_1,n-p_1} \simeq B_{t+1}^{p_1+t,n-p_1-t+1}/B_t^{p_1+t,n-p_1-t+1} = (0),$$

we conclude that $Z_t^{p_1,n-p_1}$ is constant for $r \leq t < \infty$. We can then apply the regularity argument of Theorem 15.46 to conclude that $Z_r^{p_1,n-p_1} = Z_\infty^{p_1,n-p_1}$. Since

$$\begin{aligned} E_r^{p_1,n-p_1} &= Z_r^{p_1,n-p_1}/B_r^{p_1,n-p_1} = Z_\infty^{p_1,n-p_1}/B_r^{p_1,n-p_1} \\ E_\infty^{p_1,n-p_1} &= Z_\infty^{p_1,n-p_1}/B_\infty^{p_1,n-p_1}, \end{aligned}$$

and since $B_r^{p_1,n-p_1} \subseteq B_\infty^{p_1,n-p_1}$, we obtain a surjection $E_r^{p_1,n-p_1} \longrightarrow E_\infty^{p_1,n-p_1}$, and if we combine (†), our injection for $p_0, n-p_0$, and the surjection for $p_1, n-p_1$, we get sequence (A). ☐

Lemma 15.50. *(Lemma B) Suppose that $E_2^{p,q} \Longrightarrow H(C)$ is a spectral sequence with a regular filtration. Assume that there are integers $s \geq r, p, n$, such that*

$$E_r^{u,v} = (0) \quad \textit{for} \quad \begin{cases} u+v = n-1, u \leq p-r \\ u+v = n, u \neq p \;\textit{and}\; u \leq p+s-r \\ u+v = n+1, p+r \leq u \;\textit{and}\; u \neq p+s; \end{cases}$$

see Figure 15.51. Then there is an exact sequence

$$H^n(C) \longrightarrow E_r^{p,n-p} \longrightarrow E_r^{p+s,(n+1)-(p+s)}. \tag{B}$$

Proof. We apply $d_r^{p,n-p}$ to $E_r^{p,n-p}$ and land in $E_r^{p+r,n-p-r+1}$ which is (0) by Hypothesis (3). Thus, $\operatorname{Ker} d_r^{p,n-p} = E_r^{p,n-p}$. Also, $E_r^{p-r,n-p+r-1}$ is (0) by the first hypothesis, so the image of $d_r^{p-r,n-p+r-1}$ is (0). Since

$$H^{n,n-p}(E_r^{p,n-p}) = \operatorname{Ker} d_r^{p,n-p}/\operatorname{Im} d_r^{p-r,n-p+r-1} = E_r^{p,n-p}/(0) \cong E_{r+1}^{p,n-p}$$

we conclude that $E_r^{p,n-p} = E_{r+1}^{p,n-p}$. Repeat, but with d_{r+1}; as long as $r+1 < s$ we can continue using hypotheses (1) and (3). Thus we obtain $E_r^{p,n-p} = E_s^{p,n-p}$. Now apply $d_t^{p,n-p}$ to $E_t^{p,n-p}$ where $t \geq s+1$. Hypothesis (3) shows our map is zero and similarly the map $d_t^{p-t,n-p+t-1}$ is zero by Hypothesis (1). So for all t, with $\infty > t \geq s+1$, we get

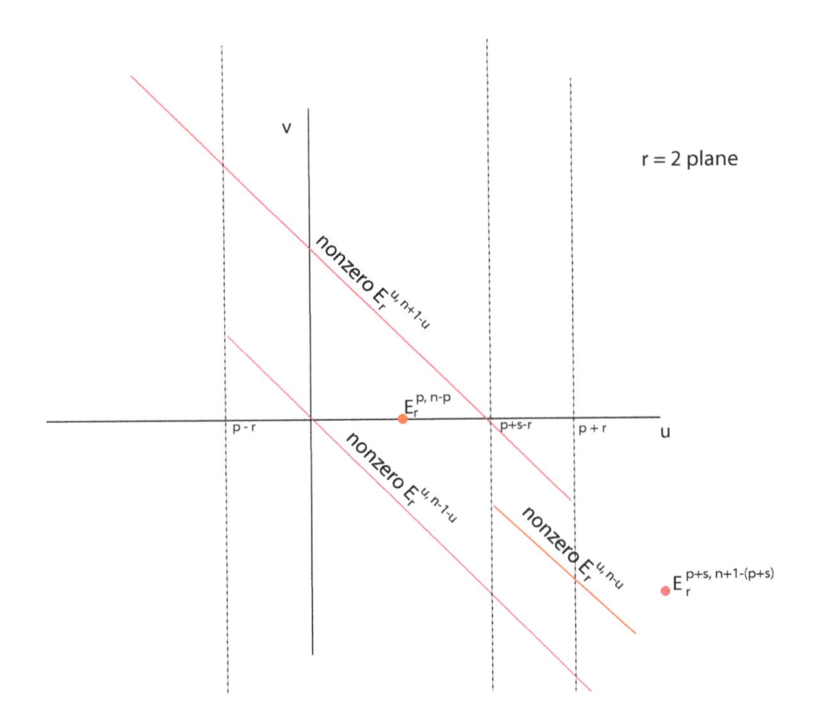

Fig. 15.51 An illustration of the $r = 2$ plane associated with Lemma B when $n = 1$, $p = 1$, and $s = 3$.

$E_t^{p,n-p} = E_{t+1}^{p,n-t}$. As the filtration is regular, we apply Proposition 15.15 to obtain $E_{s+1}^{p,n-p} = E_\infty^{p,n-p}$.

Next by Hypothesis (2) with $u = p + (s - r)$ (provided $s > r$, otherwise there is nothing to prove), we see that $\operatorname{Im} d_r^{p+s-r,n-(p+s-r)}$ is (0). Since the proof of Theorem 15.46 provides the identity

$$\operatorname{Im} d_r^{p+s-r,n-(p+s-r)} = B_{r+1}^{p+s,(n+1)-(p+s)}/B_r^{p+s,(n+1)-(p+s)},$$

we conclude that

$$B_{r+1}^{p+s,(n+1)-(p+s)} = B_r^{p+s,(n+1)-(p+s)}.$$

Should $s > r + 1$, we continue because

$$(0) = \operatorname{Im} d_{r+1}^{p+s-(r+1),n-(p+s-(r+1))}.$$

This gives

$$B_{r+2}^{p+s,(n+1)-(p+s)} = B_{r+1}^{p+s,(n+1)-(p+s)}.$$

Hence, we get

$$B_s^{p+s,(n+1)-(p+s)} = B_r^{p+s,(n+1)-(p+s)}$$

by repetition. However

$$\begin{aligned} E_s^{p+s,(n+1)-(p+s)} &= Z_s^{p+s,(n+1)-(p+s)} / B_s^{p+s,(n+1)-(p+s)} \\ &= Z_s^{p+s,(n+1)-(p+s)} / B_r^{p+s,(n+1)-(p+s)} \\ &\subseteq Z_r^{p+s,(n+1)-(p+s)} / B_r^{p+s,(n+1)-(p+s)} = E_r^{p+s,(n+1)-(p+s)}, \end{aligned}$$

since $Z_s^{p,q} \subseteq Z_r^{p,q}$ whenever $s \geq r$. Thus we obtain the *inclusion*

$$E_s^{p+s,(n+1)-(p+s)} \subseteq E_r^{p+s,(n+1)-(p+s)}.$$

Lastly, by Hypothesis (1), $E_r^{p-s,(n-1)-(p-s)} = (0)$; so Proposition 15.15 implies that $E_t^{p-s,(n-1)-(p-s)} = (0)$ for every $t \geq r$. Take $t = s$, then $d_s^{p-s,(n-1)-(p-s)}$ vanishes, and by adapting the argument in the previous paragraph we find that

$$B_{s+1}^{p,n-p} = B_s^{p,n-p}.$$

But then we obtain an inclusion

$$E_{s+1}^{p,n-p} \hookrightarrow E_s^{p,n-p}.$$

However, when doing the proof of Theorem 15.46 we found that

$$\operatorname{Ker} d_s^{p,n-p} = Z_{s+1}^{p,n-p} / B_s^{p,n-p}.$$

Since $B_{s+1}^{p,n-p} = B_s^{p,n-p}$, we may rewrite the previous identity as

$$\operatorname{Ker} d_s^{p,n-p} = Z_{s+1}^{p,n-p} / B_{s+1}^{p,n-p} = E_{s+1}^{p,n-p}.$$

Therefore we get the exact sequence

$$0 \longrightarrow E_{s+1}^{p,n-p} \longrightarrow E_s^{p,n-p} \xrightarrow{d_s^{p,n-p}} E_s^{p+s,(n+1)-(p+s)},$$

which in view of the fact that $E_{s+1}^{p,n-p} = E_\infty^{p,n-p}$ can be rewritten as

$$0 \longrightarrow E_\infty^{p,n-p} \longrightarrow E_s^{p,n-p} \xrightarrow{d_s^{p,n-p}} E_s^{p+s,(n+1)-(p+s)}. \qquad (\dagger)$$

And now we have a surjection $H^n(C) \longrightarrow E_\infty^{p,n-p}$, because $E_\infty^{u,n-u} = (0)$ when $u \leq p+s-r$ $(r \neq p)$ by Hypothesis (2). If we combine the exact sequence (†) with the inclusion $E_s^{p+s,(n+1)-(p+s)} \subseteq E_r^{p+s,(n+1)-(p+s)}$ and the surjection $H^n(C) \longrightarrow E_\infty^{p,n-p}$, we get sequence (B). ☐

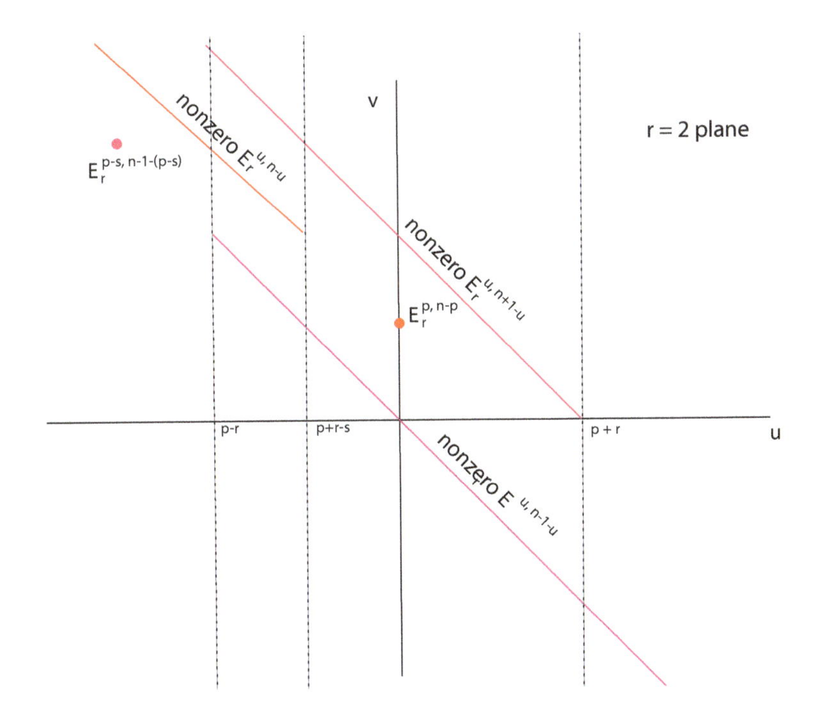

Fig. 15.52 An illustration of the $r = 2$ plane associated with Lemma C when $n = 1$, $p = 0$, and $s = 3$.

In a similar manner one proves

Lemma 15.51. *(Lemma C) If $E_2^{p,q} \Longrightarrow H(C)$ is a spectral sequence with a regular filtration and if there exist integers $s \geq r, p, n$, such that*

$$E_r^{u,v} = (0) \quad \textit{for} \quad \begin{cases} u+v = n+1, u \geq p+r \\ u+v = n, p+r-s \leq u \neq p \\ u+v = n-1, p-s \neq u \leq p-r, \end{cases}$$

(as shown in Figure 15.52), then there is an exact sequence

$$E_r^{p-s,(n-1)-(p-s)} \longrightarrow E_r^{p,n-p} \longrightarrow H^n(C). \tag{C}$$

Although Lemmas A, B, C are (dull and) technical, they do emphasize one important point: *For any level r, if $E_r^{p,q}$ lies on the line $p+q=n$, then d_r takes it to a group on the line $p+q=n+1$ and it receives a d_r from a group on the line $p+q=n-1$.* From this we obtain immediately

Corollary 15.52. *(Corollary D) Say $E_2^{p,q} \Longrightarrow H(C)$ is a regularly filtered spectral sequence and there are integers r, n such that*

$$E_r^{p,q} = (0) \quad \text{for} \quad \begin{cases} p+q = n-1 \\ p+q = n+1. \end{cases}$$

Then $E_r^{p,n-p} = E_\infty^{p,n-p}$, and the $E_r^{p,n-p}$ are the composition factors for $H^n(C)$ in its filtration.

Here are some applications of Lemmas A, B, C.

Theorem 15.53. *(Zipper Sequence) Suppose $E_2^{p,q} \Longrightarrow H(C)$ is a spectral sequence associated with a regular filtration and there exist integers p_0, p_1, r with $p_1 - p_0 \geq r \geq 1$ such that $E_r^{u,v} = (0)$ for all $u \neq p_0$ or p_1. Then we have the exact zipper sequence*

$$\cdots \longrightarrow E_r^{p_1,n-p_1} \longrightarrow H^n(C) \longrightarrow E_r^{p_0,n-p_0} \longrightarrow E_r^{p_1,n+1-p_1} \longrightarrow$$
$$H^{n+1}(C) \longrightarrow \cdots$$

Dually, if there are integers q_0, q_1, r with $q_1 - q_0 \geq r-1 \geq 1$ such that $E_r^{u,v} = (0)$ for $v \neq q_0$ or q_1, then the zipper sequence is

$$\cdots \longrightarrow E_r^{n-q_0,q_0} \longrightarrow H^n(C) \longrightarrow E_r^{n-q_1,q_1} \longrightarrow E_r^{n+1-q_0,q_0} \longrightarrow$$
$$H^{n+1}(C) \longrightarrow \cdots$$

Proof. Write $s = p_1 - p_0 \geq r$ and apply Lemmas A, B and C (check the hypotheses using $u + v = n$). By splicing the exact sequences of those lemmas, we obtain the zipper sequence. Dually, write $s = 1 + q_1 - q_0 \geq r$, set $p_0 = n - q_1$ and $p_1 = n - q_0$. Then Lemmas A, B and C again apply and their exact sequences splice to give the zipper sequence. □

The name "zipper sequence" comes from the picture in Figure 15.53. In it, the dark arrows are the maps $E_r^{p_0,n-p_0} \longrightarrow E_r^{p_1,n+1-p_1}$ and the dotted arrows are the compositions $E_r^{p_1,n+1-p_1} \longrightarrow H^{n+1} \longrightarrow E_r^{p_0,n+1-p_0}$ (one is to imagine these arrows passing through the H^{n+1} somewhere behind the plane of the page). As you see, the arrows zip together the vertical lines $p = p_0$ and $p = p_1$.

Theorem 15.54. *(Edge Sequence) Suppose that $E_2^{p,q} \Longrightarrow H(C)$ is a spectral sequence associated with a regular filtration and assume there is an integer $n \geq 1$ such that $E_2^{p,q} = (0)$ for every q with $0 < q < n$ and all p (no hypothesis if $n = 1$). Then $E_2^{r,0} \cong H^r(C)$ for $r = 0, 1, 2, \ldots, n-1$ and*

$$0 \longrightarrow E_2^{n,0} \longrightarrow H^n(C) \longrightarrow E_2^{0,n} \longrightarrow E_2^{n+1,0} \longrightarrow H^{n+1}(C)$$

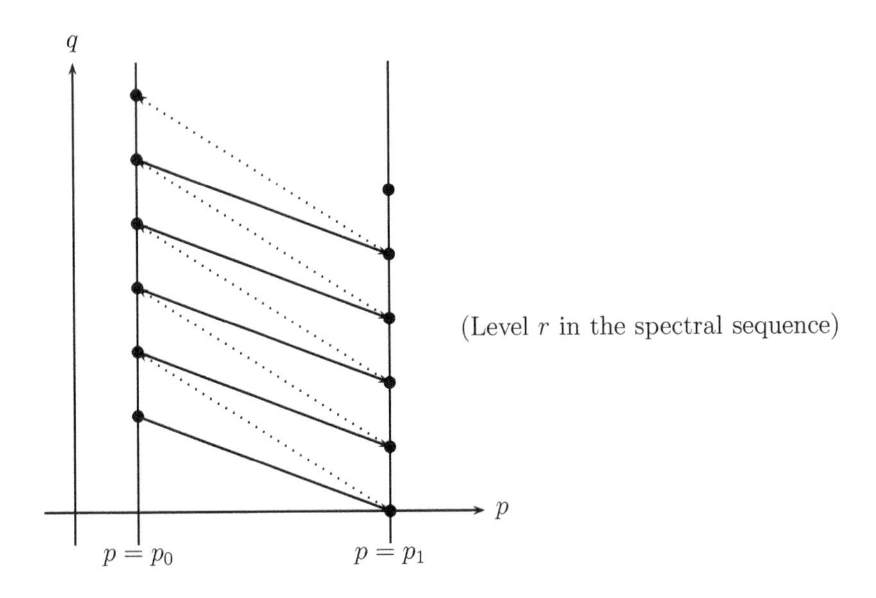

Fig. 15.53 Zipper Sequence.

is exact (edge sequence). In particular, with no hypotheses on the vanishing of $E_2^{p,q}$*, we have the exact sequence*

$$0 \longrightarrow E_2^{1,0} \longrightarrow H^1(C) \longrightarrow E_2^{0,1} \xrightarrow{d_2^{0,1}} E_2^{2,0} \longrightarrow H^2(C).$$

Proof. First, since $E_{r+1}^{p,q} \cong H^{p,q}(E_r)$, if $E_k^{p,q} = (0)$ for some k, then $E_l^{p,q} = (0)$ for all $l \geq k$. In particular, we have $E_l^{p,q} = (0)$ for every q with $0 < q < n$ and all p, for all $l \geq 2$. Since we have a cohomological (first quadrant) spectral sequence all the differentials $d_l^{r,0} \colon E_l^{r,0} \to E_l^{r+l,-l+1}$ vanish for all $l \geq 2$, so $\operatorname{Ker} d_l^{r,0} = E_l^{r,0}$. All differentials $d_l^{r-l,l-1} \colon E_l^{r-l,l-1} \to E_l^{r,0}$ vanish if $2 \leq l \leq n$, by hypothesis. Consequently, if $2 \leq l \leq n$,

$$E_{l+1}^{r,0} \cong \operatorname{Ker} d_l^{r,0} / \operatorname{Im} d_l^{r-l,l-1} = E_l^{r,0}/(0) = E_l^{r,0}.$$

If $r \leq n < l$, then all differentials $d_l^{r-l,l-1} \colon E_l^{r-l,l-1} \to E_l^{r,0}$ also vanish (since $r - l < 0$), so

$$E_{l+1}^{r,0} \cong \operatorname{Ker} d_l^{r,0} / \operatorname{Im} d_l^{r-l,l-1} = E_l^{r,0}/(0) = E_l^{r,0}.$$

Thus we proved that for any r such that $0 \leq r \leq n$ and all $l \geq 2$,

$$E_l^{r,0} \cong E_2^{r,0}.$$

Since the filtration is regular, by Proposition 15.15, we get

$$E_\infty^{r,0} \cong E_2^{r,0}, \quad 0 \le r \le n.$$

But the only non-zero term on the diagonal $r = p+q$ with $r < n$ is $E_\infty^{r,0}$ by our hypothesis on the vanishing. This is exactly the condition for the spectral sequence to degenerate at the second term for $q(n) = 0$, for that specific n, so the proof of Proposition 15.17 implies that $E_2^{r,0} \cong E_\infty^{r,0} = H^r(C)$ when $0 \le r \le n-1$.

Since $E_2^{n,0} \cong E_\infty^{n,0}$, we get an injection $E_2^{n,0} \longrightarrow H^n(C)$. Apply Lemma A with $p_0 = 0, p_1 = n, r = 2$ to find the sequence

$$0 \longrightarrow E_2^{n,0} \longrightarrow H^n(C) \longrightarrow E_2^{0,n}. \tag{$*$}$$

Next in Lemma B take $r = 2, s = n+1 \ge 2$, and $p = 0$. Sequence (B) splices to $(*)$ to yield

$$0 \longrightarrow E_2^{n,0} \longrightarrow H^n(C) \longrightarrow E_2^{0,n} \longrightarrow E_2^{n+1,0}. \tag{$**$}$$

And lastly, use Lemma C with $r = 2, s = n+1 \ge 2$, the n of Lemma C to be our $n+1 = s$ and $p = n+1$. Upon splicing Lemma C onto $(**)$ we find the edge sequence

$$0 \longrightarrow E_2^{n,0} \longrightarrow H^n(C) \longrightarrow E_2^{0,n} \longrightarrow E_2^{n+1,0} \longrightarrow H^{n+1}(C),$$

as claimed. □

Obviously, the edge sequence gets its name from the fact that the $E_2^{p,q}$ which appear in it lie on the edge of the quadrant in the picture of E_2 as points (of the first quadrant) in the pq-plane.

There is a notion of a *morphism of spectral sequences*; see Mac Lane [Mac Lane (1975)] (Chapter XI, Page 320), Weibel [Weibel (1994)] (Section 5.2, Page 123), and McCleary [McCleary (2001)] (Section 3.1). We do not need this notion for our exposition.

15.17 Problems

Problem 15.1. Finish the proof of Proposition 15.15 by proving that $E_\infty^{p,q}$ is a direct limit of the mapping family.

Problem 15.2. Using the definition of d_T^n in Definition 15.28, check that $d_T^{n+1} \circ d_T^n = 0$.

Problem 15.3. Give the details of the proof (in Section 15.9) that the connecting map $\delta^{p+q}\colon H^{p+q}(F^pC/F^{p+1}C) \to H^{p+q+1}(F^{p+1}C/F^{p+2}C)$ can be viewed as the map $d_1^{p,q}$ from $E_1^{p,q} = Z_1^{p,q}/(dC^{p,q-1} + C^{p+1,q-1})$ to $E_1^{p+1,q} = Z_1^{p+1,q}/(dC^{p+1,q-1} + C^{p+2,q-1})$.

Problem 15.4. Prove Proposition 15.19.

Hint. Use the exact sequences

$$0 \longrightarrow \mathcal{Z}^n \longrightarrow \mathcal{F}^n \longrightarrow \mathcal{B}^{n+1} \longrightarrow 0$$

$$0 \longrightarrow \mathcal{B}^n \longrightarrow \mathcal{Z}^n \longrightarrow \mathcal{H}^n \longrightarrow 0.$$

Problem 15.5. Prove Proposition 15.26.

Problem 15.6. Prove that

$$H^q(\mathcal{G}^*(U_v)) \cong H^q(U_v, \mathcal{F}),$$

where $H^q(U_v, \mathcal{F})$ is the qth module of sheaf cohomology of the restriction of the sheaf $\mathcal{F}$ to U_v (see just after Proposition 15.28).

Problem 15.7. Prove that the functor $\mathcal{F} \mapsto \check{C}^p(X, \mathcal{F})$ is exact on presheaves (see just after Proposition 15.32).

Problem 15.8. Prove Proposition 15.34.

Problem 15.9. Prove Proposition 15.38.

Problem 15.10. Prove Theorem 15.40.

Problem 15.11. Complete the proof of Proposition 15.41 by checking that $\{Q^{i,j}\}$ is indeed a Cartan–Eilenberg resolution and a double complex.

Problem 15.12. Prove Proposition 15.43.

Problem 15.13. Prove Proposition 15.44.

Problem 15.14. Prove that if we let

$$\begin{aligned} Z'^{p,q}_r &= k^{-1}(\operatorname{Im} i^{r-1}\colon D^{p+r,q-r+1} \to D^{p+1,q}) \\ B'^{p,q}_r &= j(\operatorname{Ker} i^{r-1}\colon D^{p,q} \to D^{p-r+1,q+r-1}), \end{aligned}$$

for $r \geq 2$, then

$$E'^{p,q}_r = Z'^{p,q}_r / B'^{p,q}_r.$$

Problem 15.15. Prove that

$$\begin{aligned} Z'^{p,q}_r &= k^{-1}(\operatorname{Im} i^{r-1} \colon D^{p+r,q-r+1} \to D^{p+1,q}) \\ &= (Z^{p,q}_r + (C^{p+q} \cap F^{p+1}C))/(C^{p+q} \cap F^{p+1}C) \\ B'^{p,q}_r &= j(\operatorname{Ker} i^{r-1} \colon D^{p,q} \to D^{p-r+1,q+r-1}) \\ &= (B^{p,q}_{r-1} + (C^{p+q} \cap F^{p+1}C))/(C^{p+q} \cap F^{p+1}C), \end{aligned}$$

where $Z^{p,q}$ and $B^{p,q}_{r-1}$ are defined as in Definition 15.19.

Problem 15.16. Prove Lemma C (Lemma 15.51).

Bibliography

Ahlfors, L. V. and Sario, L. (1960). *Riemann Surfaces*, Princeton Math. Series, No. 2 (Princeton University Press).

Artin, M. (1991). *Algebra*, 1st edn. (Prentice Hall).

Atiyah, M. F. and Macdonald, I. G. (1969). *Introduction to Commutative Algebra*, 3rd edn. (Addison Wesley).

Bott, R. and Tu, L. W. (1986). *Differential Forms in Algebraic Topology*, 1st edn., GTM No. 82 (Springer Verlag).

Bourbaki, N. (1980). *Algèbre, Chapitre 10*, Eléments de Mathématiques (Masson).

Bourbaki, N. (1989). *Elements of Mathematics. Commutative Algebra, Chapters 1–7* (Springer–Verlag).

Bredon, G. E. (1993). *Topology and Geometry*, 1st edn., GTM No. 139 (Springer Verlag).

Bredon, G. E. (1997). *Sheaf Theory*, 2nd edn., GTM No. 170 (Springer Verlag).

Brylinski, J.-L. (1993). *Loop Spaces, Characteristic Classes and Geometric Quantization*, 1st edn. (Birkhäuser).

Cartan, H. and Eilenberg, S. (1956). *Homological Algebra*, Princeton Math. Series, No. 19 (Princeton University Press).

Dieudonné, J. (1989). *A History of Algebraic and Differential Topology, 1900–1960*, 1st edn. (Birkhäuser).

Dold, A. (1995). *Lectures on Algebraic Topology*, 1st edn., Springer Classics in Mathematics (Springer Verlag).

Dowker, C. (1952). Homology groups of relations, *Annals of Mathematics* **56**, pp. 84–95.

Dummit, D. S. and Foote, R. M. (1999). *Abstract Algebra*, 2nd edn. (Wiley).

Eilenberg, S. and Steenrod, N. (1952). *Foundations of Algebraic Topology*, Princeton Math. Series, No. 15 (Princeton University Press).

Eisenbud, D. (1995). *Commutative Algebra With A View Toward Algebraic Geometry*, 1st edn., GTM No. 150 (Springer–Verlag).

Eisenbud, D. and Harris, J. (1992). *Schemes: The Language of Modern Algebraic Geometry*, 1st edn. (Wadsworth & Brooks/Cole).

Forster, O. (1981). *Lectures on Riemann Surfaces*, 1st edn., GTM No. 81 (Springer Verlag).

Gallier, J. H. (2011). *Geometric Methods and Applications, For Computer Science and Engineering*, 2nd edn., TAM, Vol. 38 (Springer).
Gallier, J. H. and Quaintance, J. (2020a). *Differential Geometry and Lie Groups. A Computational Perspective*, 1st edn., Geometry and Computing 12 (Springer).
Gallier, J. H. and Quaintance, J. (2020b). *Differential Geometry and Lie Groups. A Second Course*, 1st edn., Geometry and Computing 13 (Springer).
Gallier, J. H. and Xu, D. (2013). *A Guide to the Classification Theorem for Compact Surfaces*, 1st edn., Geometry and Computing, Vol. 9 (Springer).
Gelfand, S. I. and Manin, Y. I. (2003). *Methods of Homological Algebra*, 2nd edn. (Springer).
Godement, R. (1958). *Topologie Algébrique et Théorie des Faisceaux*, 1st edn. (Hermann), second Printing, 1998.
Greenberg, M. J. and Harper, J. R. (1981). *Algebraic Topology: A First Course*, 1st edn. (Addison Wesley).
Griffiths, P. and Harris, J. (1978). *Principles of Algebraic Geometry*, 1st edn. (Wiley Interscience).
Grothendieck, A. (1957). Sur quelques points d'algèbre homologique, *Tôhoku Mathematical Journal* **9**, pp. 119–221.
Grothendieck, A. and Dieudonné, J. (1961). Eléments de Géométrie Algébrique, III: Etude Cohomologique des Faisceaux Cohérents (Première Partie), *Inst. Hautes Etudes Sci. Publ. Math.* **11**, pp. 1–167.
Gunning, R. C. (1990). *Introduction to Holomorphic Functions of Several Variables, Volume III: Homological Theory*, 1st edn. (Wadsworth & Brooks/Cole).
Hartshorne, R. (1977). *Algebraic Geometry*, 1st edn., GTM No. 52 (Springer Verlag), fourth Printing.
Hatcher, A. (2002). *Algebraic Topology*, 1st edn. (Cambridge University Press).
Hilton, P. J. and Stammbach, U. (1996). *A Course in Homological Algebra*, 2nd edn., GTM No. 4 (Springer).
Hirzebruch, F. (1978). *Topological Methods in Algebraic Geometry*, 2nd edn., Springer Classics in Mathematics (Springer Verlag).
Kashiwara, M. and Schapira, P. (1994). *Sheaves on Manifolds*, 1st edn., Grundlehren der Math. Wiss., Vol. 292 (Springer–Verlag).
Lang, S. (1993). *Algebra*, 3rd edn. (Addison Wesley).
Lee, J. M. (2006). *Introduction to Smooth Manifolds*, 1st edn., GTM No. 218 (Springer Verlag).
Mac Lane, S. (1975). *Homology*, 3rd edn., Grundlehren der Math. Wiss., Vol. 114 (Springer–Verlag).
MacLane, S. and Moerdijk, I. (1992). *Sheaves in Geometry and Logic: A First Introduction to Topos Theory*, 1st edn., Universitext (Springer Verlag).
Madsen, I. and Tornehave, J. (1998). *From Calculus to Cohomology. De Rham Cohomology and Characteristic Classes*, 1st edn. (Cambridge University Press).
Massey, W. S. (1987). *Algebraic Topology: An Introduction*, 2nd edn., GTM No. 56 (Springer Verlag).

Massey, W. S. (1991). *A Basic Course in Algebraic Topology*, 1st edn., GTM No. 127 (Springer Verlag).

Matsumura, H. (1989). *Commutative Ring Theory*, 1st edn. (Cambridge University Press).

May, P. J. (1999). *A Concise Course in Algebraic Topology*, 1st edn., Chicago Lectures in Mathematics (The University of Chicago Press).

McCleary, J. (2001). *A User's Guide to Spectral Sequences*, 2nd edn., Studies in Advanced Mathematics, Vol. 58 (Cambridge University Press).

Milnor, J. W. and Stasheff, J. D. (1974). *Characteristic Classes*, 1st edn., Annals of Math. Series, No. 76 (Princeton University Press).

Morita, S. (2001). *Geometry of Differential Forms*, 1st edn., Translations of Mathematical Monographs No 201 (AMS).

Mumford, D. (1999). *The Red Book of Varieties and Schemes*, 2nd edn., LNM No. 1358 (Springer Verlag).

Munkres, J. R. (1984). *Elements of Algebraic Topology*, 1st edn. (Addison-Wesley).

Narasimham, R. (1992). *Compact Riemann Surfaces*, 1st edn., Lecture in Mathematics ETH Zürich (Birkhäuser).

Rotman, J. J. (1979). *An Introduction to Homological Algebra*, 1st edn. (Academic Press).

Rotman, J. J. (1988). *Introduction to Algebraic Topology*, 1st edn., GTM No. 119 (Springer Verlag).

Rotman, J. J. (2009). *An Introduction to Homological Algebra*, 2nd edn., Universitext (Springer).

Samuel, P. (2008). *Algebraic Theory of Numbers*, 1st edn. (Dover).

Sato, H. (1999). *Algebraic Topology: An Intuitive Approach*, 1st edn., Mathematical Monographs No 183 (AMS).

Serre, J.-P. (1955). Faisceaux algébriques cohéhents, *Annals of Mathematics* **61**, pp. 197–278.

Serre, J.-P. (2003). *Oeuvres Collected Papers. Volume I, 1949–1959*, 1st edn. (Springer).

Shafarevich, I. R. (1994). *Basic Algebraic Geometry 2*, 2nd edn. (Springer Verlag).

Shatz, S. S. and Gallier, J. H. (2016). Algebra, Tech. rep., University of Pennsylvania, Math. Department, Philadelphia, PA 19104, book in Preparation.

Spanier, E. H. (1989). *Algebraic Topology*, 1st edn. (Springer).

Tennison, B. (1975). *Sheaf Theory*, 1st edn., LMS No. 20 (Cambridge University Press).

Tu, L. W. (2008). *An Introduction to Manifolds*, 1st edn., Universitext (Springer Verlag).

Warner, F. (1983). *Foundations of Differentiable Manifolds and Lie Groups*, 1st edn., GTM No. 94 (Springer Verlag).

Weibel, C. A. (1994). *Introduction to Homological Algebra*, 1st edn., Studies in Advanced Mathematics, Vol. 38 (Cambridge University Press).

Index

www.ingramcontent.com/pod-product-compliance
Lightning Source LLC
LaVergne TN
LVHW010721170225
803784LV00010B/139